The Carpathians and Their Foreland: Geology and Hydrocarbon Resources

Edited by
Jan Golonka
and
Frank J. Picha

AAPG Memoir #84

Published by
The American Association of Petroleum Geologists
Tulsa, Oklahoma, U.S.A.

ISBN: 978-0-89181-365-1
0-89181-365-9

AAPG Editor: Ernest A. Mancini
Geoscience Director: James B. Blankenship

ON THE COVER: A view of the eastern side of the Devin tectonic klippe, near the town of Mikulov in southern Moravia (Czech Republic). The klippe is made predominantly of massive Jurassic Ernsbrunn Limestones. It is a part of the Outer klippen belt of the Waschberg sector of the Waschberg – Zdanice – Subsilesian frontal unit of the Outer West Carpathian belt in northeastern Austria and southern Moravia. The picture is taken from the hinterland (on the east) toward the foreland (on the west) of the Carpathian thrust belt. The drilling rig of the Moravian Oil Company seen in the foreground tested a subthrust prospect below the frontal zone of the thin-skinned Carpathian thrust belt. In the inset map of the Carpathians, the red-orange color denotes the older Inner Carpathian units; the yellow color denotes the younger Outer Carpathian units, the Foredeep, and the Pannonian and Transylvanian basins; and the pale green color denotes the Carpathian foreland. Photo by H. Gilikova and Z. Stranik from the Czech Geological Survey. (See related article by F. J. Picha, Z. Stranik, and O. Krejci beginning on page 49 of this Memoir.)

This publication is available from:

The AAPG Bookstore
P.O. Box 979
Tulsa, OK U.S.A. 74101-0979
Phone: 1-918-584-2555 or 1-800-364-AAPG (U.S.A. only)
Fax: 1-918-560-2652 or 1-800-898-2274 (U.S.A. only)
E-mail: bookstore@aapg.org
www.aapg.org

Geological Society Publishing House
Unit 7, Brassmill Enterprise Centre
Brassmill Lane, Bath BA13JN U.K.
Phone: +44-1225-445046
Fax: +44-1225-442836
E-mail: sales@geolsoc.org.uk
www.geolsoc.org.uk

Canadian Society of Petroleum Geologists
No. 160, 540 Fifth Avenue S.W.
Calgary, Alberta T2P OM2 Canada
Phone: 1-403-264-5610
Fax: 1-403-264-5898
E-mail: jaime.croft@cspg.org
www.cspg.org

Affiliated East-West Press Private Ltd.
G-1/16 Ansari Road, Darya Ganj
New Delhi 110 022 India
Phone: +91-11-23279113
Fax: +91-11-23260538
E-mail: affiliat@vsnl.com

Table of Contents

About the Editors

Jan Golonka is a professor of geology at the University of Science and Technology and the Jagiellonian University in Krakow, Poland. He received his M.SC. Degree (1967) from the University of Mining and Metallurgy, Krakow, Poland and Ph.D. (1978) from the Geological Institute, Warsaw, Poland. From 1967 to 1981 he was a university lecturer and research geologist in the Polish Geological Institute, Carpathian Branch in Krakow. He worked for Mobil in Dallas, Texas from 1982 to 1999. His research interests focus on paleogeography, plate tectonics, paleoclimatology, and global and regional geology, recently of the Carpathians. Jan has authored and co-authored two hundred scientific papers and also served as the editor of the SEPM book *Phanerozoic reef patterns* and Marine and Petroleum Geology *Thematic set on paleogeographic reconstruction and hydrocarbon basins: Atlantic, Caribbean, South America, Middle East, Russian Far East, Arctic*.

Frank J. Picha received his M.S. (1955) and Ph.D. (1963) degrees in geology from Charles University in Prague and a post-doctoral fellowship at the University of Wisconsin (1969–1970). From 1955 to 1979 he worked for the Geological Survey of Czechoslovakia, mainly on the Carpathian geology. In the years 1971–1976 he taught at Kuwait University in Kuwait. In 1980 he joined Gulf Oil, and in 1985 joined Chevron from which he retired in 1999. Presently he works as an international consultant. His assignments have included regional geology and exploration in Europe, the Middle East, North America, and the former Soviet Union. Since the early nineties he has again focused some of his attention on the geology and hydrocarbon resources in the Carpathian region. Picha was an AAPG Distinguished Lecturer in 1989–1990 and 1999–2000 and a recipient of the Best International Paper Award in 1991.

Acknowledgments

AAPG wishes to thank the following
for their generous contributions to

The Carpathians and their Foreland: Geology and Hydrocarbon Resources

AAPG Foundation
AGH University of Science and Technology
Chevron Corporation
Czech Geological Survey
Energy and Geoscience Institute at University of Utah
ExxonMobil Foundation
Jan Golonka
G-trend s.r.o., Bratislava
Moravské naftové doly, a.s.
OMV Exploration and Production GmbH
Frank J. Picha
Polish Geological Institute
Gabor Tari

Contributions are applied toward the production cost of publication,
thus directly reducing the book's purchase price
and making the volume available to a larger readership.

Note to the Reader

This publication is a continuation of AAPG's efforts to combine the best of both digital and print publications for its members. Memoir 84, *The Carpathians and their foreland: Geology and hydrocarbon resources*, contains 30 chapters, 17 of which are printed in full. The remaining chapters have been printed as abstracts only, with the full paper as a .pdf file on the CD-ROM in the back of this publication. In addition, the 17 printed papers can be found on the CD-ROM.

The page numbers listed in the index printed in the book refer to the page numbers of the digital version of the book on the CD-ROM. Likewise, all citations within these 30 chapters that reference other papers in this Memoir refer the reader to the page numbers of the digital version.

The dual digital and print format allows authors greater scope for both length of papers and number of color figures, and provides the reader much more versatility in using the content. For example, figures can be used directly for making slides or Microsoft PowerPoint® presentations. Individual papers can easily be printed for later reading, and key word searching through the papers is possible.

Finally, this format allows AAPG the flexibility to publish a volume that contains papers that will appeal to a wider audience with additional interests.

1

Golonka, Jan, and Frank J. Picha, 2006, Introduction, *in* J. Golonka and F. J. Picha, The Carpathians and their foreland: Geology and hydrocarbon resources: AAPG Memoir 84, p. 1–9.

Introduction

Jan Golonka
AGH University of Science and Technology, Jagiellonian University, Krakow, Republic of Poland
Frank J. Picha
International Petroleum Consultant, Walnut Creek, California, U.S.A.

INTRODUCTION

The Carpathians represent the eastern extension of the European Alps (Figure 1), but unlike the well-known classical Alps of Western Europe, the Carpathians of Central and Eastern Europe remain less known and are even somewhat mysterious to the outside world. The tumultuous political history as well as the language barriers prevented ideas and information from flowing freely through and outside the region. However, the area greatly contributed to the common knowledge in science and technology. Enormous amounts of geological work have been conducted, and thousands of papers have been published on the geology of the Carpathian region during the past 200 yr (see references in various articles of the volume). However, because of the language barriers and diverse concepts and interpretations, it is not easy for students of Carpathian geology and potential investors to cut through all the information and to get a clear picture about the geology and the hydrocarbon potential of the region. This situation has been well known to the editors of this volume, who both worked in the Carpathians and also spent a great deal of their careers in the American petroleum industry and had a chance to see the Carpathians from the view of outside world. In the early spring of 2000, they met in Krakow, where Jan Golonka, after retiring from Mobil, had just begun his new career as a professor at the Jagiellonian University and Frank Picha, after retiring from Chevron, had completed his AAPG Distinguished Lecture tour through the countries of Eastern Europe and the former Soviet Union. It was Jan Golonka's proposal to improve the worldwide visibility of the Carpathians by publishing an AAPG Memoir and Frank Picha's enthusiastic acceptance of the idea that started the hectic process of submitting proposals, forming international teams of authors, securing financial support, writing, and editing the manuscripts. It took 5 yr of intensive work to complete the task of publishing this volume, which, as we hope, would make the interesting geology of the Carpathians better known to the outside world.

The editors' intention is to overcome some of the communication problems and to provide the outside world with a comprehensive and understandable account of the complex geology and hydrocarbon potential of the entire Carpathian system, including the thrust belt, its foreland, and the late orogenic intra-Carpathian basins. Major attention is paid to petroleum systems and case studies of major oil and gas fields in order to provide an insight into past and present exploration and production activities in the region. Seventeen years have passed since the publication of the highly successful AAPG Memoir 45 on the Pannonian Basin edited by Royden and Horvath (1988). Our ambition is to provide an equally well-balanced and well-written account on the entire Carpathian system.

This publication is intended as a source of information to schools, governmental and private institutions, oil companies, and generally, to all geologists working both inside and outside the region. It should also provide qualified information about the hydrocarbon resources of the Carpathian region and give

DOI:10.1306/985605M843065

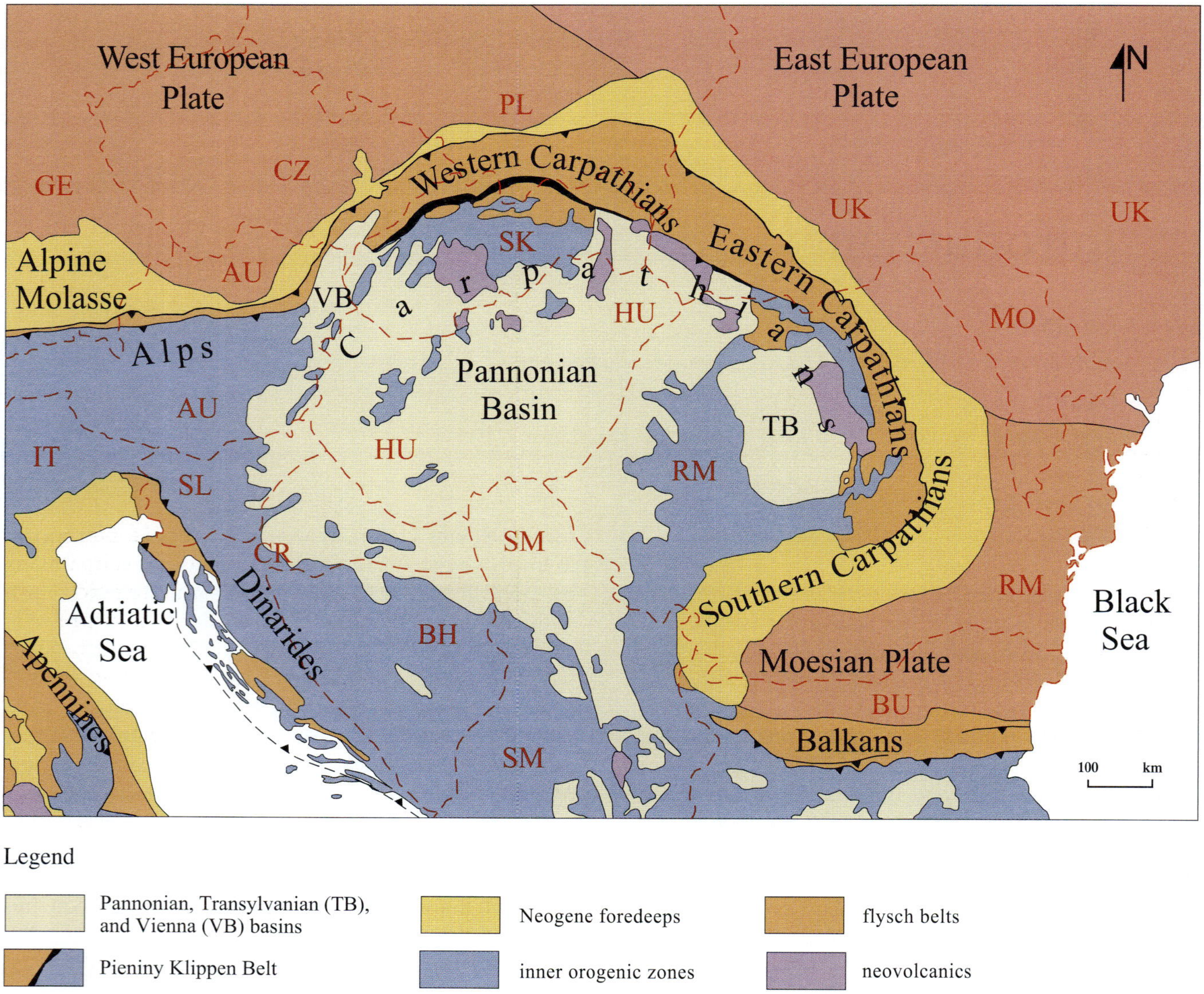

Figure 1. Index map of the Carpathians and the surrounding regions (modified after Picha, 1996). AU = Austria; BH = Bosnia and Herzegovina; BU = Bulgaria; CR = Croatia; CZ = Czech Republic; GE = Germany; HU = Hungary; IT = Italy; MO = Moldova; PL = Poland; RM = Romania; SK = Slovakia; SL = Slovenia; SM = Serbia and Montenegro; TB = Transylvanian basin; UK = Ukraine; VB = Vienna basin.

potential investors some guidance in pursuing various opportunities both in the redevelopment of the older fields and in the exploration for new reserves in the Carpathian thrust belt and its foreland.

THE PAST AND PRESENT STATUS OF THE CARPATHIAN GEOLOGY

The Carpathians extend over the territories of seven countries of Central and Eastern Europe (Figure 1), with a complex political and economic history and equally complex history of scientific and technical endeavors in the field of earth sciences and petroleum industry. The Carpathian region is one of the classical territories of the world, where some concepts of geology and petroleum geology were formulated and the petroleum industry was born. It was in the Carpathian region where, in 1853, the first oil well (Bobrka) was drilled; crude oil distillation was discovered by Lukasiewicz in Galicia, presently Poland (Karnkowski, 1999); the first industrial production of hydrocarbons was established in Romania in 1857 (e.g., Dicea, 1996); and the organic origin of oil was proposed (Zuber, 1918).

After the Second World War, for more than 40 yr, the countries of the Carpathian region were isolated from the rest of the Europe, and their geology evolved under different ideological and economic doctrines. On the one hand, enormous money was spent in search of mineral resources, including hydrocarbons. On the other hand, the ideological constraits, utmost secrecy, and limited flow of information from the western world led to theoretical and technological lagging in some areas of earth sciences. However, even during this time, significant achievements in understanding of Carpathian geology and hydrocarbon potential were accomplished. In the 1950s, the Carpathian Flysch belt served as one of the main territories in which the original concept of deep-water turbidites was formulated (Ksiazkiewicz 1954: Dzulynski et al., 1959). Thanks mainly to the Polish school (e.g., Dzulynski and Slaczka, 1958; Dzulynski, 1963, 1996; Dzulynski and Walton, 1963, 1965), the Carpathian Flysch became one of the best and thoroughly studied turbiditic facies exposed on land anywhere. It has commonly been compared with the deep-water sequences of the Apennines (e.g., Pescatore and Slaczka, 1984) and the Ouachita Mountains of Arkansas and Oklahoma (e.g., Golonka and Slaczka, 2000). In many aspects, the Carpathian Flysch, famous for its complexity and diversity, may serve as a model in exploration of hydrocarbons in deep-water deposits both along modern continental margins and in orogenic belts. Likewise, thanks to an intensive exploration for hydrocarbons and the publication of AAPG Memoir 45 (Royden and Horvath 1988), the late orogenic Pannonian Basin overlying the Alpine–Carpathian belt has become one of the best known back-arc extensional basins in the world. Numerous studies were also conducted on various aspects of the geodynamic evolution of the Carpathian region, the most comprehensive version being published by Golonka et al. (2000), and on the kinematics of the Outer Carpathian foreland fold and thrust belt (e.g., Roure et al. 1993). Since the early 1960s, growing attention was paid to the exploration for hydrocarbons underneath the thin-skinned Outer Carpathian thrust belt. Numerous deep wells have been drilled, and valuable new resources of hydrocarbons have been found, especially in the territory of the Czech Republic and Austria. The Carpathian region may thus serve as one of the best examples of exploration for hydrocarbons under the thin-skinned orogenic belts (Wessely, 1990; Picha, 1996). From numerous recent publications dealing with these accomplishments in Carpathian geology, perhaps two volumes, *Oil and gas in Alpidic thrust belts and basins of Central and Eastern Europe* (Wessely and Liebl, 1996) and *Peri-Tethys Memoir 2: Structure and prospects of Alpine basins and forelands* (Ziegler and Horvath, 1996), are the most comprehensive and informative.

The profound political changes that occurred in Eastern Europe in the early 1900s have had a significant impact, both positive and negative, on the geological activities in the Carpathian region. At the time of publishing this memoir, several nations in the Carpathian realm are still undergoing political and economic transformation, and some of them have already been admitted into the European Union. Geological institutions, universities, and extraction industries have been deeply affected by these changes. The geologists of the strong postwar generation are retiring and are not always replaced by younger colleagues. Large amounts of knowledge and experience, especially in the area of regional geology and past hydrocarbon exploration, are in a danger of being lost. At the same time, new technologies and concepts are being applied both in the theoretical and applied sciences. It is one of the intentions of this volume to bridge, at least partly, the gap between the past and present concepts and developments. For that reason, both senior traditional geologists and their younger technologically oriented colleagues were selected to contribute to this volume. Most of them have already published significant articles on geology and hydrocarbon potential of the Carpathian region and presented their contributions at various conventions and conferences. The mixture of authors from universities, national geological surveys, and industry should guarantee a balanced assembly of academic and industrial views.

THE STRUCTURE OF THE PUBLICATION

The volume is not a mere assemblage of chapters dealing with selected topics of the Carpathian geology but rather an attempt to put together a comprehensive account of the entire spectrum of regional and applied geology of the Carpathian realm (Figure 2). To accomplish this goal, we have combined main regional articles with contributions dealing with specific aspects of geology, geophysics, and hydrocarbon resources. Altogether, 30 articles authored by 107 geologists and geophysicists from Austria, Czech Republic, Hungary, Poland, Romania, Slovakia, Ukraine, U.K., and U.S.A. have been included in this volume (Figure 2). The written contributions are supplemented by numerous regional geological and geophysical maps and correlation charts, enabling better orientation throughout the entire region.

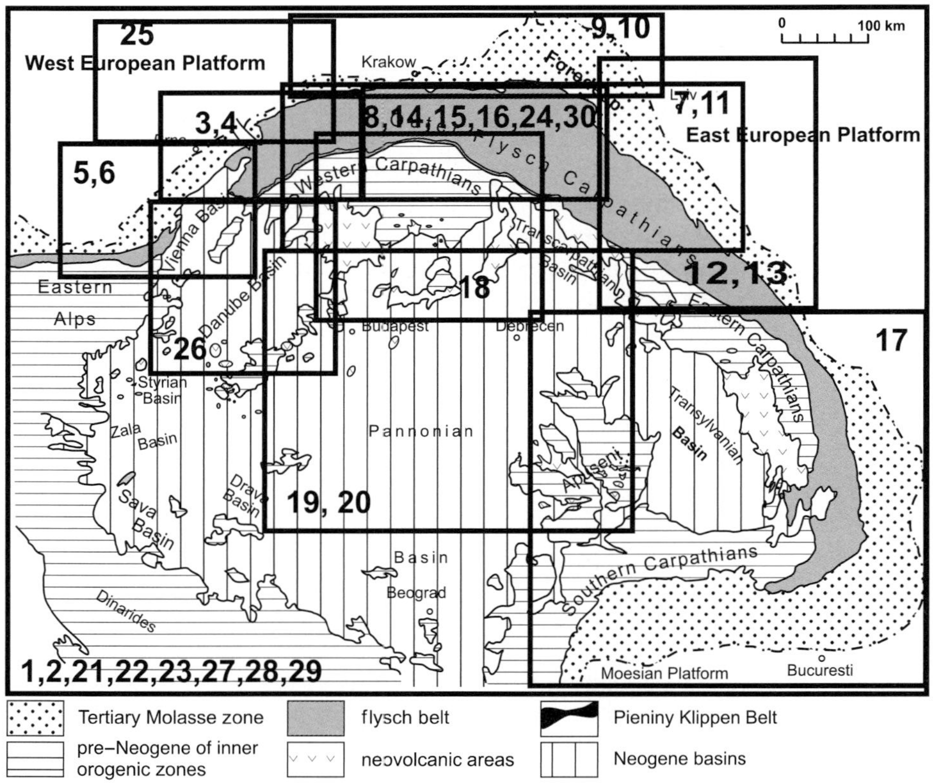

Figure 2. Index map of the Carpathians and the surrounding regions. Areas covered by various chapters of this volume are outlined and numbered. Geology simplified after Kovac et al. (1998). 1 = chapter 1 (Golonka and Picha); 2 = chapter 2 (Golonka et al.); 3 = chapter 3 (Picha et al.); 4 = chapter 4 (Kostelnicek et al.); 5 = chapter 5 (Arzmuller et al.); 6 = chapter 6 (Fuchs and Hamilton); 7 = chapter 7 (Slaczka et al.); 8 = chapter 8 (Dziadzio et al.); 9 = chapter 9 (Oszczypko et al.); 10 = chapter 10 (Mysliwiec et al.); 11 = chapter 11 (Kotarba and Koltun); 12 = chapter 12 (Popadyuk et al.); 13 = chapter 13 (Popadyuk et al.); 14 = chapter 14 (Dziadzio); 15 = chapter 15 (Dziadzio et al.); 16 = chapter 16 (Matyasik and Dziadzio); 17 = chapter 17 (Stefanescu et al.); 18 = chapter 18 (Janocko et al.); 19 = chapter 19 (Tari and Horvath); 20 = chapter 20 (Magyar et al.); 21 = chapter 21 (Pospisil); 22 = chapter 22 (Pospisil et al.); 23 = chapter 23 (Bielik and Adam); 24 = chapter 24 (Stefaniuk); 25 = chapter 25 (Hrusecky et al.); 26 = chapter 26 (Hrusecky); 27 = chapter 27 (Nemcok et al.); 28 = chapter 28 (Nemcok et al.); 29 = chapter 29 (Golonka et al.); 30 = chapter 30 (Sozanski et al.).

Considering the geological subdivision of the Carpathians as well as the historically evolved national groupings, the chapters were assembled into six major regional and thematic blocks (Figure 2).

1) *The geology and hydrocarbon resources of the Outer Western Carpathians and their foreland on the territory of northeastern Austria and Moravia in the Czech Republic*

The regional account on the westernmost sector of the Outer Carpathians adjacent to the Bohemian Massif by Picha et al. emphasizes the correlation of the Alps and Carpathians, as well as the impact of the newly defined Western Carpathian transfer zone and the Dyje–Thaya depression on the geological history of this region. Deep drilling and advanced exploration for hydrocarbons underneath the thin-skinned Outer Carpathian thrust belt enabled the formulation of critical concepts of the subthrust petroleum systems in the Carpathians and elsewhere (Wessely 1990; Picha, 1996). Examples of recently discovered subthrust fields in Moravia are presented by Kostelnicek et al. The geology and petroleum system of the highly prolific Vienna basin is dealt with in a contribution by Arzmuller et al. Fuchs and Hamilton discuss the depositional architecture of the giant Matzen field in the Vienna basin.

2) *The geology and hydrocarbon resources of the Outer Western Carpathians and their foreland on the territory of Poland and Ukraine*

This is the classical area of the flysch geology on which some of the concepts of deep-water turbiditic sedimentation were formulated (see above).

The regional geology of the Carpathian thrust belt is presented by Slaczka et al.; the hydrocarbon resources of the Polish Carpathians are described by Dziadzio et al. The Carpathian Foredeep and the European foreland are thoroughly discussed by Oszczypko et al. and their hydrocarbon resources by Mysliwiec et al. The chapters by Kotarba and Koltun and by Popadyuk et al. deal with the hydrocarbon habitat of the Polish and Ukrainian Carpathians, which are among the oldest producing provinces of the world. The oil and gas fields of the Carpathian foreland and of the Boryslav–Pokuttya zone are presented by Popadyuk et al.; the Weglowka oil field in the Polish flysch belt is described by Dziadzo. The depositional architecture of the Ciezkowice Sandstone is discussed by Dziadzio et al., and the petroleum systems of the Polish Carpathians are presented by Matyasik and Dziadzio.

3) *The geology and hydrocarbon resources of the Eastern and Southern Carpathians on the territory of Romania*

The geology and the hydrocarbon potential of the Romanian Carpathians and their foredeep are presented by Stefanescu et al. Despite the long production history, beginning in the middle of the 19th century, the Romanian territory still represents an important hydrocarbon province of Europe.

4) *The geology and hydrocarbon resources of the Inner Carpathians and the Pannonian Basin on the territory of Slovakia and Hungary*

The complex geology of the Inner Carpathians, including the late orogenic and postorogenic basins in Slovakia and some adjacent areas, is presented by Janocko et al. The geology of the Pannonian Basin, well known from the AAPG Memoir 45 (1988), is briefly overviewed and updated by Tari and Horvath. Magyar et al. describe the largest oil and gas field, Algyo, in the Pannonian Basin.

5) *Crustal and lithospheric structure of the Carpathian–Pannonian region: A geophysical perspective*

The geophysical segment begins with a historical review of the crustal and lithospheric studies in the Carpathian region compiled by Pospisil and Adam. The main results of various regional geophysical surveys and studies are assembled in an account by Pospisil et al. for the entire Carpathian region. This account, authored by numerous geophysicists from several countries, presents a vast compilation of various geophysical data, including numerous maps and diagrams. The following six interpretive contributions deal with various aspects of crustal and lithospheric studies: the structure of the lithosphere by Bielik and Adam; the magnetotelluric survey by Stefaniuk; the North European platform below the Carpathian Flysch belt by Hrusecky et al.; the bivergent lithospheric extension in the Danube basin by Hrusecky; the timing of the tectonism as related to the rollback process in the lithosphere by Nemcok et al.; and the subduction of the Carpathian remnant oceanic basin by Nemcok et al.

6) *Special thematic papers*

Three thematic chapters are present in the volume: one on the plate-tectonic evolution and paleo-geography of the Circum-Carpathian region by Golonka et al.; another on the comparison of the Carpathians with the Ouachitas of the North America by Golonka et al.; and the last one on the history of the petroleum industry in the Polish and Ukrainian Carpathians by Sozanski et al. The comparison of the Carpathians and Ouachitas is an idea considered since the mid-1960s but never truly elaborated into a consistent account. The chapter by Golonka et al. is concerned mainly with the similarities and differences in the geodynamic evolution of these two orogenic belts.

Some unavoidable overlaps among the regional chapters exist, which the authors tried to minimize by emphasizing specific aspects of each region and by referring to other chapters in the volume.

SYNOPSIS OF GEOLOGY AND HYDROCARBON RESOURCES OF THE CARPATHIAN REGION

Three primary plate-tectonic settings (divergent, convergent, and back-arc extension) can be distinguished in the evolution of the Western Carpathians. The divergent process, characterized by rifting, opening of oceanic basins, and development of passive continental margins, began in the Permian and the Triassic in the Inner Carpathians (Meliata–Hallstatt Ocean) and in the Early to Middle Jurassic in the Outer Western Carpathians (Penninic–Pieninic oceanic basin).

The convergent orogenic process marked by subduction and formation of orogenic belts began in the latest Triassic or Early Jurassic in the Inner Carpathians by the closure of the Meliata–Hallstatt Ocean and the collision of the Inner Western Carpathian with the Tisza block. During the Late Jurassic and Early Cretaceous, the deformation progressed northward by accretion of various terranes of the Inner Carpathians and the Pieniny Klippen belt. By the late Senonian, most of the Penninic–Pieninic ocean was closed, and the Inner Carpathians were wholly deformed and exhibited signs of thrust. The oceanic crust of the Penninic–Pieninic ocean was subducted, and fragments of the rifted Apulian continental lithosphere collided with the attenuated and fragmented margins of the European plate. The Late Cretaceous deformation of the Inner Carpathians marks the primary break in the evolution of the Carpathian orogenic belt, which is also reflected in the division of the Carpathians into the inner and outer domains.

In the Late Cretaceous, a major foreland basin system formed in the Outer Carpathian zone dominated by the siliciclastic shelf, and deep-water flysch sedimentation formed in the Outer Carpathian zone. It displayed a complex topography marked by an existence of intrabasinal ridges (cordilleras), such as the Silesian ridge, which became major sources of clastic material. Because of the postcompressional relaxation in the late Paleogene, a hinterland deep-water flysch basin opened in the already deformed domain of the Inner Carpathians. During the Paleogene and early Miocene, the flysch sequences of Outer Carpathian depositional system were gradually deformed and thrust over the European foreland, where the youngest molasse-type Neogene foredeep formed. As the front of deformation, and with it the depocenter of the foreland basin, gradually progressed toward the European foreland, the deformed and uplifted inner zones of the foreland basin became the main sources of the clastic material for the younger depocenters closer to the foreland. The foreland basin thus represents a dynamic system of progradational deformation, erosion, and sedimentation.

The youngest plate-tectonic setting was marked by the formation of a subduction-related magmatic arc associated with back-arc rifting, extension, and thermal subsidence, which led to the development of the Pannonian Basin in the early to late Miocene. During the late phases of the Alpine orogeny, the predominantly northward-directed thrusting, especially in the Western Carpathians, was accompanied by the orogen-parallel, strike-slip faulting and escape tectonics. This is well documented by the opening of the pull-apart Vienna basin and the existence of transpressional tectonics in the suture zone of the Pieniny Klippen belt.

The tectonic shortening of the Carpathian domain apparently occurred not only in the decoupled thin-skinned thrust belt but also at the deeper crustal level, where the various blocks of previously rifted-apart blocks of the European continental margins were accreted back to the foreland plate instead of being subducted (Picha et al., 2006).

Diverse views exist on various aspects of the Carpathian geology, particularly on the extent of the European crust and lithosphere below the Carpathian belt, on the extent of the oceanic crust and lithosphere in various basins of the Carpathian depositional system during various stages of its evolution, on the character and magnitude of the late orogenic escape tectonics, and on the amount of the compressional shortening in the Carpathian orogenic belt. Readers will find diverse interpretations and accompanying discussion on these and some other problems of stratigraphy and structure in various articles in this volume. Hopefully, the complexity of the Carpathian geology and a wide acknowledgment of the differing views in this volume will help foster further interest in Carpathian geology.

THE CORRELATION OF THE ALPS AND THE CARPATHIANS

The correlation of the Alps and the Carpathians, two segments of the European Alpine system for a long time separated by the Iron Curtain, has stood out as one of the most important tasks in the process of opening the Carpathian geology to the outside world. Many similarities exist between the Alps and the Carpathians, yet significant geological differences do exist between these two parts of the European Alpine system. Among them, the limited presence of the Penninic oceanic crust and lithosphere in the Carpathians, the absence of the Pieniny Klippen belt in the Alps, and the existence of the Inner Carpathian Paleogene basin and a major back-arc extensional Pannonian Basin system in the Carpathians seem to be the most noticeable. At least some of these differences were apparently enhanced if not fully related to the existence of the Western Carpathian transfer zone (Picha et al., 2006), which actually separated the Alps and the Carpathians during both the divergent and convergent stage of their evolution.

The correlation of strata along the strike of the Alpine–Carpathian orogenic system thus should reveal both similarities and differences, the latter being commonly related to different settings of various segments of the entire system. As an example, the evolution of the Neogene foredeeps along the entire Alpine–Carpathian belt has been controlled to a great extent by the preexisting structural history of various segments of the foreland plate. This varies greatly along the strike of the belt. Typically, prominent depocenters marked by a high rate of subsidence alternate with areas of limited subsidence and deposition. In the Alpine–Carpathian region, the Bavarian–Austrian Molasse, with 4 km (2.5 mi) of Neogene deposits, the Boryslav area in western Ukraine, with up to 7 km (4.3 mi) of deposits, and the deepest Focsany depression in Romania, with 8–10-km (5–6.2-mi) deposits, represent the prominent depocenters. However, the segments of the foredeep, which evolved on the more stable crustal blocks of the foreland, such as the Bohemian Massif, are characterized by a low rate of subsidence and deposition (Picha et al., 2006).

As noted by Trumpy (1988, p. 107), "We must abandon the cylindristic concept of parallel facies belts. Basins and swells may change their character along strike. Major paleogeographic elements, such as the Brianconnais rise, may wedge out and be replaced, in another traverse, by a similar but not homologous swell.'' The right approach to correlation of various tectonic and stratigraphic units should thus strike a balance between the general trends and local anomalies. Moreover, the various segments in the general trends should be defined by geological attributes instead of national boundaries.

THE PROBLEMS OF THE LOCAL STRATIGRAPHIC NOMENCLATURE AND THE REGIONAL CHRONOSTRATIGRAPHY OF THE NEOGENE PARATETHYS

One of the primary goals of the regional chapters is the correlation of major stratigraphic and structural units of the Carpathian belt and in its foreland across national borders. Because of closed borders and the overall limited access to information in the past, the geological activities of various countries were commonly dominated by local problems. Inevitably, such a situation led to the development of national geologies for various segments of the Carpathian belt. This is most notably seen in the stratigraphic nomenclature. A plethora of names for local stratigraphic and structural units, which are not always clearly defined, has been introduced and often published in local periodicals little known outside the particular countries. Some previously recognized units have been given new names or upgraded to higher stratigraphic ranks and further subdivided. Such a proliferation of local terms has made the regional geology increasingly difficult, if not impossible, to comprehend. In that context, even some aspects of the regional chronostratigraphic stage system of Paratethys have become increasingly debatable. The uncertainties in definition of some stages (e.g., Karpatian) led to discrepancies in the correlation of strata and tectonic events between various regions. In our opinion, an alternative comparison of the regional Carpathian stratigraphy with the standard Mediterranean stratigraphic system of the Neogene would be desirable. The Paratethys, as a system of the northern Alpine foreland and intermontane basins, was not entirely isolated from the foreland basins and marginal seas of the Mediterranean region and from the north European epicontinental depositional system. In a broader sense, it might be debatable whether a highly diversified depositional system of Paratethys, strongly affected by local depositional and structural complexities typical for foreland settings, is warranted a separate regional stratigraphic nomenclature in the first place. Furthermore, the term Tethys is generally used for oceanic settings, e.g., Paleotethys and Neotethys; its application, in the term of Paratethys, for a continental setting of the Alpine foreland basins might be misleading to those not familiar with the Alpine region.

Many discrepancies in the regional correlation became apparent even during the preparation of this volume to some authors and to the editors. However, any solution of these stratigraphic problems goes far beyond the scope of this publication. The best we could have done was to keep the stratigraphic and structural terminology in various articles simple, consistent, and compatible with conventional usage understandable to readers and, if necessary, to refer to diverse views and interpretations. Hopefully, the publication of this comprehensive regional volume may lead to serious revisions of the present practices and possibly even to the development of a simplified stratigraphic and structural nomenclature applicable to wider segments of the Carpathian region.

HYDROCARBON RESOURCES IN THE CARPATHIAN REGION

To date, 6.8 billion bbl (984 million m^3) of oil and 53.7 tcf (1522 billion m^3) of natural gas (Table 1) have been produced in various parts of the Carpathian system: the Neogene foredeep in Austria, Czech Republic, Poland, Ukraine, and Romania; the outer zones of the Carpathian thrust belt in Poland, Ukraine, and Romania; the Vienna basin in Austria, Czech Republic, and Slovakia; the Pannonian and Transylvanian basins in Hungary and Romania, respectively; and in the subthrust plays in the Czech Republic, Austria, and Ukraine. A sizable production of around 200,000 bbl of oil still continues in those areas. By applying new technologies and new scientific concepts, it may still be possible to find additional reserves of hydrocarbons in the existing fields, as well as to discover new accumulations, especially at deeper structural levels. The subthrust plays below the shallow frontal zones of the thin-skinned Carpathian thrust belt seem to be the most promising targets (Kostelnicek et al., 2006; Picha et al., 2006). Numerous deep wells have already been drilled through the thrust belt into the underlying European platform, where the presence of hydrocarbons was established in various formations, from the crystalline basement to the Neogene strata of the European platform. Commercial accumulations of oil and gas have been discovered and put on production in Austria, Czech Republic, and Ukraine. The Carpathian

Table 1. Cummulative production of oil and natural gas from the entire Carpathian system, including the Neogene foredeep and the foreland platform both in front and below the thrust belt, the Carpathian thrust belt, and the late and postorogenic intermontane basins in all countries of the Carpathian region. Based on data reported in various chapters of this volume.

	Cumulative Production			
	Oil		Gas	
	Million bbl	Million m^3	tcf	B m^3
Austria	788	113	2.7	77.4
Czech Republic	38	6	0.6	16.5
Slovakia	24	4	0.9	25.5
Hungary	634	101	7.4	208.8
Poland	167	24	6.9	196.3
Ukraine	713	102	8.3	235.9
Romania	4439	634	26.9	761.9
Total	6803	984	53.7	1522.3

region thus represents one of the best examples of successful subthrust exploration used in defining the fundamentals of hydrocarbon exploration under thrust belts elsewhere (Picha, 1996).

THE FUTURE OUTLOOK

The main objective of this volume is to provide a comprehensive overview of present knowledge on the geology and hydrocarbon resources of the Carpathians and their foreland. While pursuing this goal, we are well aware that various investigations have continued, and new data and concepts have emerged even during the publication of this volume. One of the desired goals of future investigations would be the comparison of the Carpathian Flysch depositional sequences with the deep-water turbiditic systems of modern continental margins and recently explored hydrocarbons in the Gulf of Mexico, along the margins of Africa, South America, and elsewhere. Such a comparison may help in better understanding the distribution of various facies and possibly even changing the attitude toward the ever-increasing fragmentation of the stratigraphic nomenclature of the Carpathian Flysch belt. On the practical level, the better understanding of the facies distribution may positively influence the future exploration for hydrocarbons in the Carpathians and other regions with similar structural and stratigraphic settings.

Another area of continued interest would be the investigation of the deep structure of the Carpathian belt. Significant amounts of work, including various geophysical investigations and deep exploratory drilling, have already been done (see the various contributions in this volume); however, any further progress would depend on the application of advanced geophysical, especially seismic, techniques and further deep scientific and exploratory drilling. One such deep drilling project has been recently considered in the hot spot of the Zakopane area of Poland. Its goal is to verify the existence of the late orogenic rifting in the outer zones of the Carpathian thrust belt. Of great importance, both theoretical and economic, would be the potential exploration for hydrocarbons in deeper structures (6–7 km; 3.7–4.3 mi) below the Neogene foredeep and the thin-skinned Outer Carpathian thrust belt (Dziadzio et al., 2006; Mysliwiec et al., 2006; Picha et al., 2006; and others).

We believe that geological investigation of the Carpathian belt will continue with a renewed enthusiasm; new resources of hydrocarbons will be discovered and developed; and the famed past of the Carpathians will thus be followed by an equally bright future.

ACKNOWLEDGMENTS

We express our deep thanks to all the authors for writing their contributions, to various private and governmental organizations for technical and financial support, and above all, to the AAPG for publishing this volume.

REFERENCES CITED

Dicea, O., 1996, Tectonic setting and hydrocarbon habitat of the Romanian external Carpathians, *in* P. A. Ziegler and F. Horvath, eds., Peri-Tethys Memoir 2; Structure and prospects of Alpine basins and forelands: Symposium on structure and prospects of Alpine basins and forelands: Memoires du Museum National d'Histoire Naturelle, Paris, v. 170, p. 403–425.

Dziadzio, P. S., et al., 2006, Hydrocarbon resources of the Polish Carpathians—reservoir parameters, trap types, and selected hydrocarbon fields: A stratigraphic review, *in* J. Golonka and F. J. Picha, eds., The Carpathians and their foreland: Geology and hydrocarbon resources: AAPG Memoir 84, p. 259–291.

Dzulynski, S., 1963, Directional structures in Flysch: Studia Geologica Polonica, v. 12, 136 p.

Dzulynski, S., 1996, Erosional and deformational structures in single sedimentary beds: A genetic commentary: Annales Societatis Geologorum Poloniae, v. 66, p. 101–189.

Dzulynski, S., and A. Slaczka, 1958, Directional structures and sedimentation of the Krosno beds (Carpathian Flysch): Annales Societatis Geologorum Poloniae, v. 28, p. 205–259.

Dzulynski, S., and E. K. Walton, 1963, Experimental production of sole markings: Transactions of the Edinburgh Geological Society, v. 19, p. 279–305.

Dzulynski, S., and E. K. Walton, 1965, Sedimentary features of flysch and graywackes, *in* Development in sedimentology: Amsterdam, Elsevier, v. 7, 274 p.

Dzulynski, S., M. Ksiazkiewicz, and P. H. Kuenen, 1959, Turbidites in flysch of the Polish Carpathians: Geological Society of America Bulletin, v. 70, p. 1089–1118.

Golonka, J., and A. Slaczka, 2000, Geodynamic evolution stages of two orogens: The Carpathians and Ouachita Mountains: Slovak Geological Magazine, v. 6, p. 139–142.

Golonka, J., N. Oszczypko, and A. Slaczka, 2000, Late Carboniferous–Neogene geodynamic evolution and paleogeography of the circum-Carpathian region and adjacent areas: Annales Societatis Geologorum Poloniae, v. 70, p. 107–136.

Karnkowski, P., 1999, Oil and gas deposits in Poland: Cracow, The Geosynoptics Society "Geos," University of Mining and Metallurgy, 380 p.

Kostelnicek, P., V. Ciprys, and J. Berka, 2006, Examples of recently discovered oil and gas fields in the Carpathian Foredeep and in the European foreland plate underneath the Carpathian thrust belt, Czech Republic, *in* J. Golonka and F. J. Picha, eds., The Carpathians and their foreland: Geology and hydrocarbon resources: AAPG Memoir 84, p. 177–189.

Kovac, M., A. Nagymarosy, N. Oszczypko, A. Slaczka, L. Csontos, M. Marunteanu, L. Matenco, and M. Marton, 1998, Palinspastic reconstruction of the Carpathian–Pannonian region during the Miocene, *in* M. Rakus, ed., Geodynamic development of the Western Carpathians: Bratislava, Geological Survey of Slovak Republic, p. 189–217.

Ksiazkiewicz, M., 1954, Graded and laminated bedding in the Carpathian Flysch: Annales Societatis Geologorum Poloniae, v. 22, p. 399–449.

Mysliwiec, M., et al., 2006, Hydrocarbon resources of the Polish Carpathian Foredeep: Reservoirs, traps, and selected hydrocarbon fields, *in* J. Golonka and F. J. Picha, eds., The Carpathians and their foreland: Geology and hydrocarbon resources: AAPG Memoir 84, p. 351–393.

Pescatore, T., and A. Slaczka, 1984, Evolution models of two flysch basins: The Northern Carpathians and the southern Apennines: Tectonophysics, v. 106, p. 49–70.

Picha, F. J., 1996, Exploring for hydrocarbons under thrust belts a challenging new frontier in the Carpathians and elsewhere: AAPG Bulletin, v. 89, p. 1547–1564.

Picha, F. J., Z. Stranik, and O. Krejci, 2006, Geology and hydrocarbon resources of the Outer West Carpathians and their foreland, Czech Republic, *in* J. Golonka and F. J. Picha, eds., The Carpathians and their foreland: Geology and hydrocarbon resources: AAPG Memoir 84, p. 49–175.

Roure, F., E. Roca, and W. Sassi, 1993, The Neogene evolution of the Outer Carpathian Flysch units (Poland, Ukraine and Romania): Kinematics of a foreland/fold-and-thrust belt system: Sedimentary Geology, v. 86, p. 177–201.

Royden, L., and F. Horvath, eds., 1988, The Pannonian Basin: A study in basin evolution: AAPG Memoir 45, 394 p.

Trumpy, R., 1988, A possible Jurassic–Cretaceous transform system in the Alps and the Carpathians: Geological Society of America Special Paper 218, p. 93–109.

Wessely, G., 1990, Geological results of deep exploration in the Vienna basin: Geologische Rundschau, v. 79, no. 2, p. 513–520.

Wessely, G., and W. Liebl, eds., 1996, Oil and gas in Alpidic thrustbelts and basins of Central and Eastern Europe: European Association of Geoscientists and Engineers Special Publication 5, 456 p.

Ziegler, P. A., and F. Horvath, eds., 1996, Peri-Tethys memoir 2; Structure and prospects of Alpine basins and forelands: Paris, Memoires du Museum national d'Histoire naturelle, v. 170, 547 p.

Zuber, R., 1918, Flisz i nafta. Kraków (The Flysch and the oil) (in Polish): Towarzystwo dla Popierania Nauki Poskiej we Lwowie, Drukarnia Uniwersytetu Jagiellonskiego w Krakowie, 381 p.

2

Golonka, Jan, Lisa Gahagan, Michał Krobicki, Frantisek Marko, Nestor Oszczypko, and Andrzej Ślączka, 2006, Plate-tectonic evolution and paleogeography of the circum-Carpathian region, *in* J. Golonka and F. J. Picha, eds., The Carpathians and their foreland: Geology and hydrocarbon resources: AAPG Memoir 84, p. 11–46.

Plate-tectonic Evolution and Paleogeography of the Circum-Carpathian Region

Jan Golonka
AGH University of Science and Technology, Jagiellonian University, Krakow, Republic of Poland

Lisa Gahagan
Institute for Geophysics, University of Texas at Austin, Austin, Texas, U.S.A.

Michał Krobicki
Department of Stratigraphy and Regional Geology, AGH University of Science and Technology, Krakow, Republic of Poland

Frantisek Marko
Department of Geology and Paleontology, Comenius University, Bratislava, Slovak Republic

Nestor Oszczypko
Institute of Geological Sciences, Jagiellonian University, Krakow, Republic of Poland

Andrzej Ślączka
Institute of Geological Sciences, Jagiellonian University, Krakow, Republic of Poland

ABSTRACT

Sixteen time interval maps were constructed that depict the latest Precambrian to Neogene plate-tectonic configuration, paleogeography, and lithofacies of the circum-Carpathian area. The plate-tectonic model used was based on PLATES and PALEOMAP software.

The supercontinent Pannotia was assembled during the latest Precambrian as a result of the Pan-African and Cadomian orogenies. All Precambrian terranes in the circum-Carpathian realm belonged to the supercontinent Pannotia, which, during the latest Precambrian–earliest Cambrian, was divided into Gondwana, Laurentia, and Baltica. The split of Gondwana during the Paleozoic caused the origin of the Avalonian and then Gothic terranes. The subsequent collision of these terranes with Baltica was expressed in the Caledonian and Hercynian orogenies. The terrane collision was followed by the collision between Gondwana and the amalgamation of Baltica and Laurentia known as Laurussia. The basement of most of the plates, which was an important factor in the Mesozoic–Cenozoic evolution of the circum-Carpathian area, was formed during the late Paleozoic collisional events. The older Cadomian and Caledonian basement elements experienced Hercynian tectonothermal overprint.

The Mesozoic rifting events resulted in the origin of oceanic-type basins like Meliata and Pieniny along the northern margin of the Tethys. The separation of Eurasia from Gondwana resulted in the formation of the Ligurian–Penninic–Pieniny Ocean as a continuation of the

DOI:10.1306/985606M843066

Central Atlantic Ocean and as part of the Pangean breakup tectonic system. During the Late Jurassic–Early Cretaceous, the Outer Carpathian rift developed.

The latest Cretaceous–earliest Paleocene was the time of the closure of the Pieniny Ocean. The Adria–Alcapa terranes continued their northward movement during the Eocene–early Miocene. Their oblique collision with the North European plate led to the development of the accretionary wedge of the Outer Carpathians and foreland basin. The northward movement of the Alpine segment of the Carpathian–Alpine orogen has been stopped because of the collision with the Bohemian Massif. At the same time, the extruded Carpatho-Pannonian units were pushed to the open space toward the bay of weak crust filled up by the Outer Carpathian flysch sediments. The separation of the Carpatho-Pannonian segment from the Alpine one and its propagation to the north were related to the development of the north–south dextral strike-slip faults. The formation of the Western Carpathian thrusts was completed by the Miocene. The thrust front was still progressing eastward in the Eastern Carpathians. The Carpathian loop, including the Pieniny Klippen structure, was formed. The Neogene evolution of the Carpathians resulted also in the formation of the genetically different sedimentary basins. The various basins were formed because of the lithospheric extension, flexure, and strike-slip-related processes.

INTRODUCTION

The aim of this chapter is to provide the plate-tectonic evolution and position of the major crustal elements of the area in the global framework. Sixteen time interval maps were constructed that depict the latest Precambrian to Neogene plate-tectonic configuration, paleogeography, and lithofacies of the circum-Carpathian area. We tried to apply geometric and kinematic guidelines using computer technology to model interrelations between tectonic components of the circum-Carpathian area. We restricted the number of plates and terranes modeled, trying to use the existing information and degree of certainty. This general framework will provide a basis for the future integration of the local tectonics. We skip the outline of geology of the Carpathians because the reader can easily find the necessary information in the other chapters included in this book: Slaczka et al. for the Northern Outer Carpathians; Picha et al. for the Western Carpathians; Stefanescu et al. for the Eastern and Southern Carpathians and adjacent areas in Romania; Janocko et al. for the Inner Carpathians; Oszczypko et al. for the Carpathian Foredeep; Tari and Horvath for the Pannonian Basin; and Arzmuller et al. for the Vienna basin and Carpathian–Alpine link. The details of the geodynamic evolution of the Western Carpathian are also contained in the Golonka et al. Carpathian–Ouachita comparison, whereas lithosphere plate configuration from the geophysical point of view is also contained in the Pospisil et al. chapter.

The presented maps were primarily generated as Intergraph design files and CorelDraw® files using computer software and databases. The plate-tectonic model used is based on PLATES and PALEOMAP plate models (see Golonka et al., 1994, 2000, 2003; Scotese and Lanford, 1995; Golonka and Gahagan, 1997). The plate reconstruction software programs used to create paleocontinental base maps take tectonic features in the form of digitized data files and assemble those features in accordance with user-specified rotation criteria.

The plate-tectonic model used is based on PLATES and PALEOMAP software (see Golonka et al., 1994, 2000; Scotese and Lanford, 1995; Golonka and Gahagan, 1997). The tectonic reconstruction program was used to create paleocontinental base maps. Paleomagnetic data, including the oceanic tectonic fabrics, were used to define the paleolatitudinal position of major continents, orientation of the plates with respect to the poles and equator, and rotation of plates (see, e.g., Gahagan et al., 1988; Van der Voo, 1988, 1993; Besse and Courtillot, 1991, 2002; Krs et al., 1996, Lewandowski, 2003). An attempt has also been made to use the paleomagnetic data from minor plates and allochthonous terranes (see, e.g., Balla, 1987; Channell et al., 1992; Patrascu et al., 1992, 1993; Pechersky and Safronov, 1993; Beck and Schermer, 1994; Mauritsch et al., 1995, 1996; Channell, 1996; Feinberg et al., 1996; Kondopolou et al., 1996; Krs et al., 1996; Marton and Martin, 1996; Haubold et al., 1999; Lamarche et al., 1999; Marton et al., 1999, 2000; Grabowski, 2000; Muttoni et al., 2000a, b; Lewandowski et al., 2003). However, the nature of rotation as indicated by paleomagnetism measured in sedimentary rocks in allochthonous terranes remains somewhat uncertain. It could be caused by the rotation of crustal (basement) elements, the rotation of blocks separated by dextral faults (e.g., Marton et al., 2000), or the rotation of thrust sheets (e.g., Muttoni et al., 2000b). For this reason, the authors attempt to verify the paleomagnetic data by the paleobiogeographical information (e.g., Kozur and Krahl, 1987; Paris and Robardet,

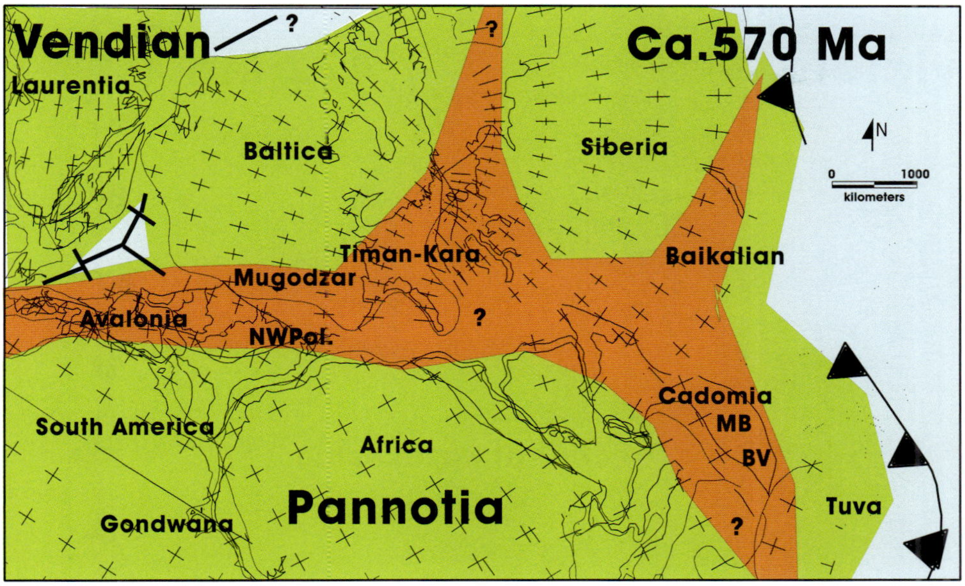

Figure 1. General paleogeography of the central Pannotia supercontinent during the latest Precambrian (Vendian)–Early Cambrian; plates position at 570 Ma. Red color = the Cadomian–Avalonian and adjacent fold belts (Timanian and Baikalian); green color = land masses; blue = oceans. Abbreviations: MB = Malopolska Block; BV = Brunovistulicum; NWPol. = northwestern Poland.

1990; Catalano et al., 1991; Krobicki, 1994; Tyszka, 1994, 2001; Vasicek et al., 1994; Pickering and Smith, 1995; Svabenicka et al., 1997; Kiessling et al., 1999; Kalvoda, 2001). Hot-spot volcanics serve as reference points for the calculation of paleolongitudes (Golonka and Bocharova, 2000).

Information from several general and regional paleogeographic papers (e.g., Dercourt et al., 1984, 1993, 2000; Ronov et al., 1984, 1989; Kazmin et al., 1986; Rakus et al., 1988; Ziegler, 1988; Zonenshain et al., 1990; Stampfli et al., 1991; Shanov et al., 1992; Kovac et al., 1993, 1998; Ellouz and Roca, 1994; Nikishin et al., 1996; Sengor and Natalin, 1996; Robertson, 1998; Von Raumer, 1998; Pharaoh, 1999; Plasienka, 1999; Golonka and Ford, 2000; Golonka et al., 2000, 2003; Wortmann et al., 2001; Stampfli and Borel, 2002; Ford and Golonka, 2003; Keppie et al., 2003; Meert, 2003; Meert and Torsvik, 2003) was filtered and used. We have also used the unpublished maps and databases from the PALEOMAP group (University of Texas at Arlington), PLATES group (University of Texas at Austin), University of Chicago, Institute of Tectonics of Lithospheric Plates in Moscow, Robertson Research in Llandudno, Wales, and the Cambridge Arctic Shelf Program. The plate and terrane separation was based on the PALEOMAP system (see Scotese and Lanford, 1995), with modifications in the Tethys area (Golonka and Gahagan, 1997; Golonka et al., 2000). The contents of the original maps were supplemented by detailed information concerning the paleogeography of the Outer Carpathian Basin (Ksiazkiewicz, 1962; Peszat, 1967; Slaczka, 1976; Birkenmajer, 1977, 1986, 1988; Elias and Eliasova, 1984; Pescatore and Slaczka, 1984; Unrug, 1984; Michalik, 1994; Michalik and Rehakova, 1995; Michalik et al., 1995, 1996; Slaczka et al., 1999; Golonka et al., 2000, 2003; Golonka and Lewandowski, 2003). The calculated paleolatitudes and paleolongitudes were used to generate computer maps in the Microstation design format using the equal-area Molweide projection. The details of the paleoreconstruction and mapping methodology were described in detail in the Golonka et al. (2003) paper. The detailed information about the database, including the paleopoles used, can be found in the PLATES homepage http://www.ig.utexas.edu/research/projects/plates/plates.htm. PLATES maintains an up-to-date oceanic magnetic and tectonic database, continuously adding new paleomagnetic, hot-spot, geological, and geophysical data to extend the span and accuracy of global plate reconstructions. PLATES' reconstructions are built around a comprehensive database of finite-difference poles of rotation, derived both from extensive plate motion research at the Institute for Geophysics, University of Texas, using the PLATES interactive plate-modeling software, and from published studies. Updated plate motion models are, in turn, applied to regional tectonic studies by PLATES investigators and collaborators and by project sponsors. The presented chapter will attempt to provide an overall review of the current state of the art on the subject of the circum-Carpathian Phanerozoic plate tectonics and paleogeography. The authors attempt to set the scene for other papers focused on petroleum geology, regional geology, and geophysics. The Paleozoic chapters and accompanying maps (Figures 1–5) provide the global and megaregional setting of the plates involved in the future Carpathian and adjacent European areas. The Mesozoic and Cenozoic part deals with the details of the circum-Carpathian plate tectonics.

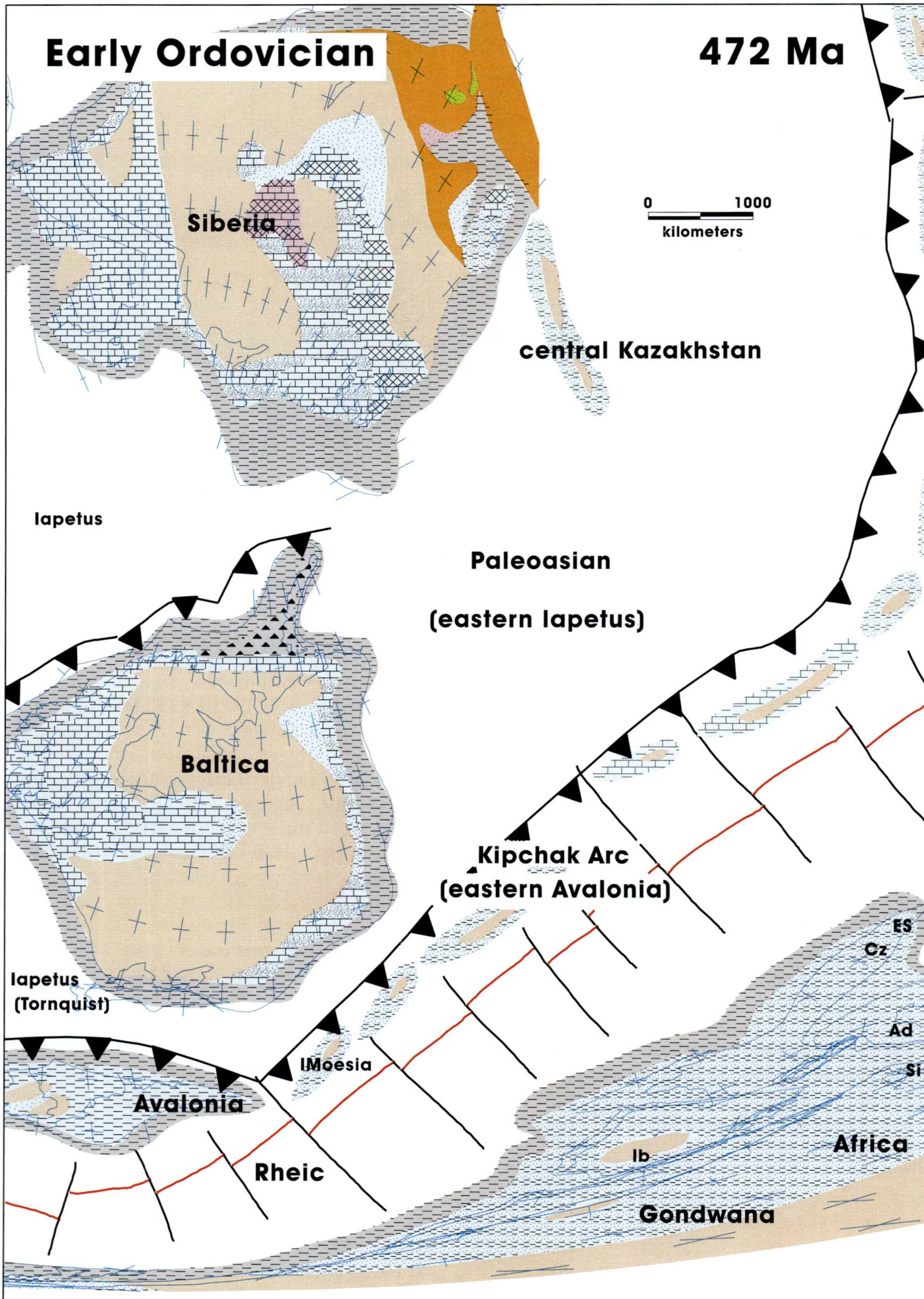

Early Ordovician
472 Ma
0
1000
kilometers
Siberia
central Kazakhstan
Iapetus
Paleoasian
(eastern Iapetus)
Baltica
Kipchak Arc
(eastern Avalonia)
ES
Cz
Iapetus
(Tornquist)
Ad
Si
Moesia
Avalonia
Africa
Ib
Rheic
Gondwana

Figure 2. Paleogeography of Gondwana–Rheic–Avalonia, Baltica, and adjacent oceans and plates during the Early Ordovician; plates position at 472 Ma. Abbreviations: Ib = Iberia; Cz = Czechia (Bohemia); ES = eastern Silesian plate; Ad = Adria; Si = Sicily. The legend shown is also the legend to Figures 3–5, 7–17.

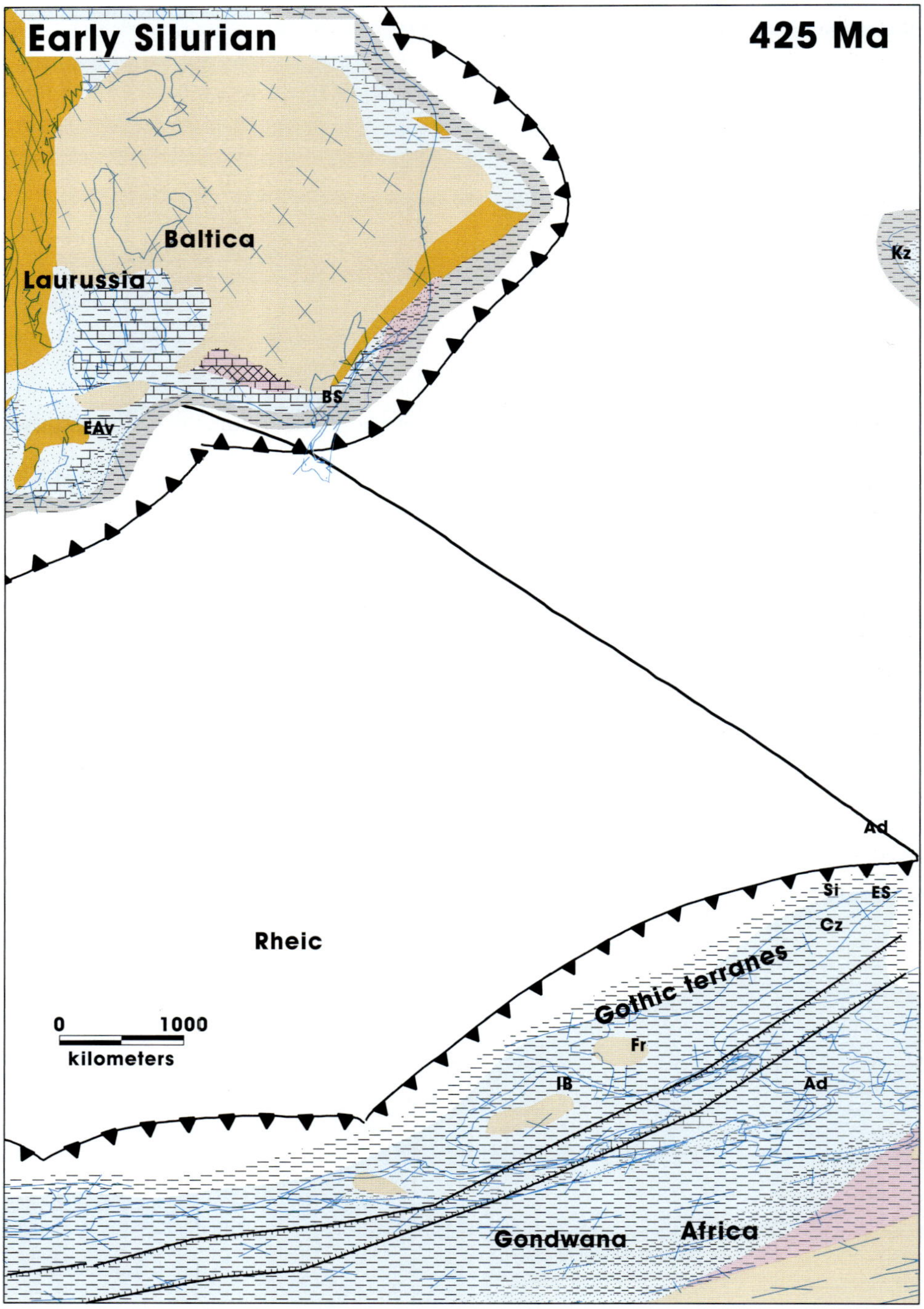

Figure 3. Paleogeography of Gondwana–Rheic–Laurussia and adjacent oceans and plates during the Early Silurian; plates position at 425 Ma. Abbreviations: IB = Iberia; Fr = France; Cz = Czechia (Bohemia); Si = western Silesia; ES = eastern Silesia; Ad = Adria; EAv = eastern Avalonia (northwestern Poland, northern Germany, and England); BS = future Black Sea site; Kz = southern Kazakhstan.

DESCRIPTION OF THE MAPS AND PLATE-TECTONIC INTERPRETATION

Latest Precambrian–Silurian

The supercontinent Pannotia (Dalziel et al., 1994; Dalziel, 1997; Golonka, 2000, 2002) was assembled during the latest Precambrian as a result of the reorganization of Rodinia (Bond et al., 1984; Torsvik et al., 1996; Poprawa et al., 1999; Kah and Bartley, 2001; Poprawa and Paczesna, 2002; Meert, 2003; Meert and Torsvik, 2003; Pesonen et al., 2003) and the Pan-African and Cadomian orogenies (Figure 1). Both Gondwana and Baltica were included in this supercontinent. The continents forming the core of Gondwana include South America, Africa, Madagascar, India, Antarctica, and Australia (Golonka, 2000, 2002). The location of numerous smaller continental blocks that bordered Gondwana is less certain. The following were adjacent to Gondwana at some time during the Paleozoic: Yucatan, Florida, Avalonia, Central European (Cadomian) terranes between the Armorica and Bohemian Massif, Moesia, Iberia, Apulia, and the smaller, southern European terranes, central Asian terranes (Karakum and others), China (several separate blocks), and the Cimmerian terranes of Turkey, Iran, Afghanistan, Tibet, and Southeast Asia. Baltica consisted of a major part of northern Europe. It was bound on the west by the Iapetus suture, on the east by the Ural suture, on the south by the Variscan–Hercynian suture, and on the southwest by a suture located close but not quite along the Teisseyre–Tornquist line (Scotese and McKerrow, 1990; Golonka et al., 1994; Soper, 1994; Pharaoh, 1999).

The Cadomian orogeny in Gondwanan Europe is a continuation of the Precambrian Pan-African event (Golonka, 2002, and references therein). This orogeny caused the deformation and magmatic events of terranes from Iberia through Armorica (northwest France), Erzgebirge (Saxoturingian zone in Germany), Bohemian Massif, Brunovistulicum and Malopolska massif (Czechia and southern Poland), and the Carpathians to the Transcaucasus area (Chalupsky, 1989; Ziegler, 1989; Compston et al., 1995; Dallmeyer et al., 1996; Moczydlowska, 1997; Unrug et al., 1999; Finger et al., 2000; Golonka, 2000; Kalvoda, 2001; Neubauer, 2002; Winchester, 2002; Kalvoda et al., 2003). The Baltica (Eastern Europe) might have collided with the Cadomian part of Gondwana during the Vendian, causing deformation in the Timan area and proto-Uralian area. The Pechora–Timan belt (Roberts and Siedlecka, 2002) and fragments of Ural, Novaya Zemlya, and Taimyr are related to the Cadomian belt (Golonka, 2000, 2002). The position of Baltica in relation to Gondwana's Cadomian terranes agrees roughly with Puchkov's (1997) analysis, suggesting Vendian rifting that occurred after the Cadomian–Timanian orogeny.

According to Torsvik et al. (1996), Baltica drifted apart from Gondwana during the late Vendian (see also Meert, 2003; Meert and Torsvik, 2003). This breakup led to the formations of new oceans. The Iapetus Ocean originated at the triple junction, somewhere between Baltica and Laurentia. The Tornquist Sea, initially opened between Baltica and Laurentia, was later situated between Baltica and Avalonia (part of Gondwana) (Figure 2). It is quite possible that this separation and the breakup of Pannotia mark the beginning of the Sauk megasequence and the general transgression of the Cambrian time (Golonka, 2000, 2002).

All Precambrian terranes in the circum-Carpathian realm belonged to the supercontinent Pannotia. After the breakup of Pannotia, most of them were located perhaps in the earliest Paleozoic Gondwana supercontinent. Shortly after their separation, they displayed characteristics of both Gondwana (or the so-called Peri-Gondwana) and Baltica. According to Finger et al. (2000), the Brunovistulicum terrane (BV, Figure 1) belonged to Cadomia or had the Amazonian provenance (Friedl et al., 2000). The Malopolska Block (MB, Figure 1) terrane could have also belonged to the Cadomia and joined Baltica during the Paleozoic (Moczydlowska, 1997; Belka et al., 2000; Winchester, 2002). The position of Moesia is somewhat uncertain. It could have belonged to the Baltica (e.g., Besutiu, 2001) or to Peri-Gondwana (e.g., Golonka, 2000). The Moesia plate could have been accreted to Baltica during the early Paleozoic. According to Tari et al. (1997), Moesian rocks younger than Devonian show no signs of metamorphism and are relatively undeformed. The accretionary Devonian complex in the North Dobrogea shows that Moesia could have joined Laurussia in the Late Devonian or post-Devonian (Golonka, 2000; V. G. Kazmin, 1999, personal communication). North Dobrogea orogen as well as the adjacent part of its foreland (Scythian platform) were strongly deformed during the Hercynian late Paleozoic orogeny (Stefanescu et al., 2006).

Advanced sea-floor spreading occurred in the Iapetus Ocean between Baltica and Gondwana during the Cambrian. Baltica moved northward and rotated. A south-dipping subduction developed along the central margin of Gondwana in the Late Cambrian–Early Ordovician. It caused the onset of rifting of the Avalonian terranes (Golonka et al., 1994; Golonka, 2000, 2002). Avalonia probably started to rift from Gondwana and move toward Baltica in the late Tremadocian, remaining in a drift stage (Figure 2) by the Llanvirnian (McKerrow et al., 1991; Torsvik et al., 1996). The continent of Avalonia consisted of northwestern and possibly southern Poland, some accreted terranes in the basement of Eastern Carpathians and their foredeep, terranes in northern Germany, the Ardennes of Belgium and northern France, England, Wales, southeastern Ireland, the Avalon Peninsula of eastern Newfoundland, much of Nova Scotia, southern New Brunswick, and some coastal parts of New England (McKerrow et al., 1991; Nance et al., 2002). The position of the Brunovistulicum and Malopolska massifs in relation to Avalonia is still a subject of debate (Moczydlowska, 1997; Belka et al., 2000; Kalvoda, 2001; Kalvoda et al., 2003). Avalonia was probably sutured to Baltica by the end of the Ordovician or in the Early Silurian (Torsvik et al., 1996). The closure of the Tornquist Sea was dominated by a strike-slip suturing of the two continents instead of a full-scale continent-continent collision. According to Pozaryski et al. (1982), the Polish part of Avalonia was

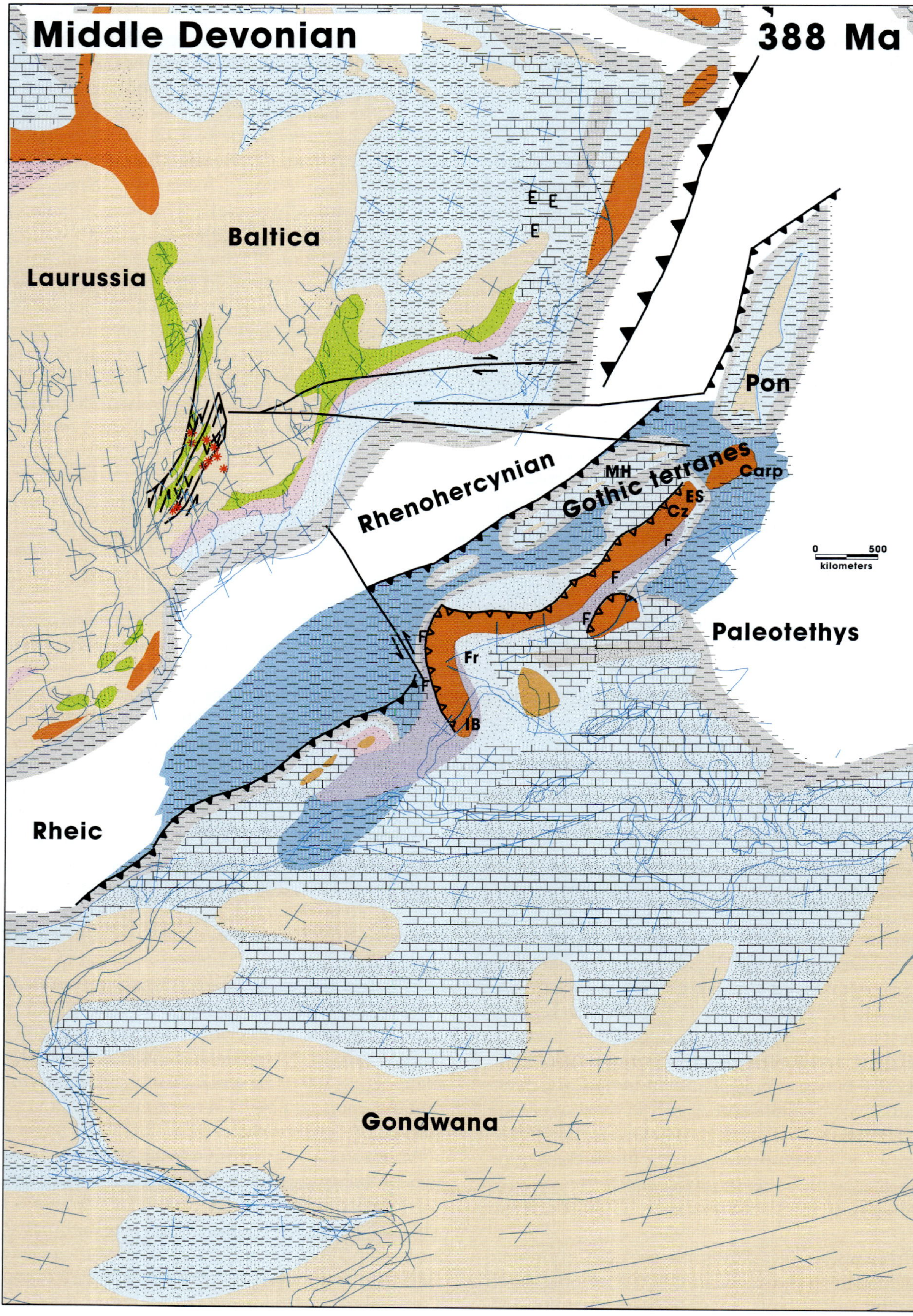
Middle Devonian
388 Ma
Baltica
Laurussia
E E
E
Pon
Rhenohercynian
MH
Gothic terranes
Carp
ES
Cz
F
F
F
F
Fr
F
IB
0 500
kilometers
Paleotethys
Rheic
Gondwana

sutured to Baltica at the end of the Ordovician, along a strike-slip fault zone known as the Tornquist–Teisseyre line (Dadlez et al., 1995; Pharaoh, 1999). The two blocks formed one larger continent Balonia (Torsvik, 1998). This represents the early Caledonian orogeny (Golonka, 2000, 2002). Granitoids of the Eastern Carpathian terranes in Romania, according to Pana et al. (2002), are of Early to Middle Ordovician age and represent the same early Caledonian event. Between Gondwana and Avalonia, a large longitudinal oceanic unit, known as the Rheic Ocean (McKerrow et al., 1991), was formed (Figures 2, 3). After the collision of Avalonia with Baltica, a southward-dipping subduction developed along the new margin of Gondwana. According to Stampfli et al. (2001), this subduction triggered detachment of a new group of terranes during the Late Silurian (see also Lewandowski, 1993, 1997, 1998, 2003; Lustrino, 2000; Stampfli and Borel, 2002). This so-called Armoric or Gothic terranes (Figure 3) included Armorica, Saxoturingia, Central German (Harz), Moldanubian (Czechia), Helvetic, Penninic, Austroalpine, Inner Carpathian, Pannonian Serbo-Macedonian, Rhodopes, and Pontides terranes. At the same time, after the complete closure of the Iapetus Ocean, the continents of Baltica, Avalonia, and Laurentia formed the continent of Laurussia (Ziegler, 1989; Paris and Robardet, 1990; Soper et al., 1992).

Devonian–Carboniferous

During the Devonian, Gondwana drifted northward and rotated clockwise (Scotese and McKerrow, 1990; Golonka, 2000, 2002). At the same time, Laurussia was rotating clockwise (Torsvik et al., 1996) at a somewhat faster rate. The Armoric (Gothic) terranes (Figure 4) began to arrive at the Laurussia margin during the Devonian. The Early to Middle Devonian convergence is detected in the mid-German crystalline high terrane and in the Bohemian Massif in Central Europe (Franke, 1989a, b, 1992). Collisional activities were also detected in the Alpine area (Dallmeyer et al., 1996; Rakus et al., 1998). The oldest deformation and metamorphic event in the Western Tatra Mountains in the Carpathians occurred about 400–380 Ma (Gaweda et al., 1998, 2003). The Western Rheic Ocean became quite narrow (Paris and Robardet, 1990). The contact between Laurussia and the Gothic terranes marks the onset of Hercynian orogeny. The Bohemian, Saxoturingian, and Malopolska high terranes (Lewandowski, 1993, 1997, 1998, 2003) moved along the strike-slip faults toward Laurussia.

The ongoing Hercynian convergence in Europe led to large-scale dextral shortening, overthrusting, and emplacement of parts of the accretionary complexes (Edel and Weber, 1995). The amount of convergence was modified by large, dextral, and sinistral transfer faults. The thrusting occurred in the Carpathians (Vozarova and Vozar, 1992).

The clockwise rotation of Gondwana during the collision with Laurussia closed the southwest remnant of the Rheic Ocean during Visean–Serpukhovian. The Hercynian orogeny in Europe was a result of the collision of several separate blocks with the Laurussian margin, followed by the Gondwana–Laurussia collision (Golonka, 2000, 2003). Widespread orogenic deformation occurred across western and Central Europe in Iberia, Ligerian, Central Massif, Sardinian–Corsican, Armorican, Harz Mountains, Saxoturingian, Bohemian, and Silesia areas (Golonka, 2000). The Rhenohercynian basin turned into foredeep, with flysch deposition. The clockwise rotation of Gondwana resulted in the involvement of this continent in the European deformation. The southwest–northeast stress direction was added to the northern one. This Gondwanan influence resulted in the convoluted shape of the Hercynian orogen, strike-slip zones, and Hercynian deformation at the eastern end in Poland (Golonka, 2000). The European foreland basin was elevated or changed its sedimentation regime from flysch to molasse.

The central Pangean mountain range was formed, which extended from Mexico to Poland. Southward, the mountain system extended to Morocco. Late Carboniferous events were also detected in the Alps and Carpathians (Vozarova and Vozar, 1992; Dallmeyer et al., 1996; Rakus et al., 1998; Gaweda et al., 1998, 2003; Sandulescu and Visarion, 2000), in the Pannonian area (Kovacs et al., 2000), the Italian realm (Lustrino, 2000), and the Rhodopes (Yanev, 1992). Most likely, the Pontides terranes were sutured to Eurasia before the Permian (Ustaomer and Robertson, 1997). According to Okay et al. (1996), the western Pontides Paleozoic sequence was folded and possibly thrust-faulted during the Late Carboniferous–Permian. As aforementioned, Moesia probably acted as terrane during the Hercynian orogeny. Paleozoic rocks were strongly deformed

Figure 4. Paleogeography of Gondwana–Rheic–Laurussia and adjacent oceans and plates during the Middle Devonian; plates position at 388 Ma. Abbreviations: IB = Iberia; Fr = France; Cz = Czechia (Bohemia); ES = eastern Silesia; MH = mid-German high; Carp = Inner Carpathian terrane; Pon = Pontide (northern Turkey) terrane; E = evaporites; F = flysch.

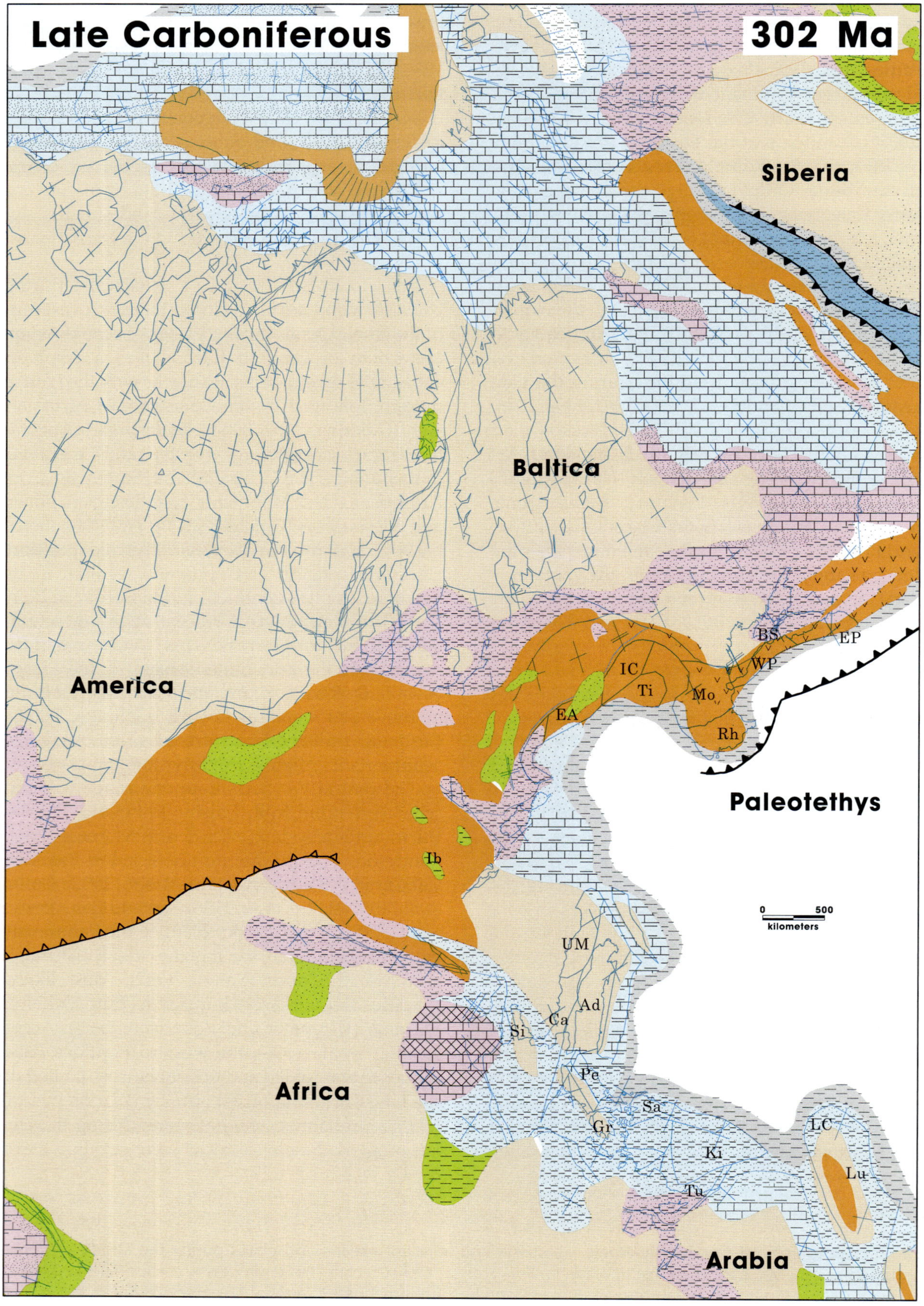
Late Carboniferous
302 Ma
Siberia
Baltica
America
IC
Ti
Mo
Rh
EA
BS
WP
EP
Paleotethys
Ib
0 500
kilometers
UM
Ad
Ca
Si
Pe
Sa
Gr
Ki
Tu
LC
Lu
Africa
Arabia

and metamorphosed in the North Dobrogea nappes (Stefanescu et al., 2006) between Moesia and Laurussia. The deformed Paleozoic rocks in the Scythian platform could represent the Hercynian foreland.

The basement of most of the plates, which became an important factor in the Mesozoic–Cenozoic evolution of the circum-Carpathian area, was formed during the late Paleozoic collisional events (Figure 5). These plates are depicted in Figure 6. The older Cadomian and Caledonian basement elements experienced Hercynian tectonothermal overprint (Vozarova and Vozar, 1992; Dallmeyer et al., 1996; Rakus et al., 1998).

The Moesia, Rhodopes, and Alcapa superterrane, which include the Eastern Alps, Inner Carpathian, Tisza, and adjacent terranes, were sutured to the Laurasian arm of Pangea, whereas Adria and the adjacent terranes were situated near the Gondwana (African arm). The Paleotethys Ocean (Figures 5, 7) was situated between the northern Laurussian (North America, Baltica, and Siberia) and southern Gondwanan branch (Africa, Arabia, and Iranian terranes) of Pangea (Golonka, 2000, 2002). The paleomagnetic studies (Krs et al., 1996) indicate the nearly equatorial position of the Bohemian Massif, adjacent to the Carpathian plates. This position agrees with the global Pangean model (Golonka et al., 1994; Golonka, 2000, 2002). Mountains that formed on the northern margin of the Paleotethys, as a result of these orogenic events, were connected with the Hercynian orogen in Europe. North-dipping subduction developed along the Paleotethys margin. This subduction was a major force driving the late Paleozoic and early Mesozoic movement of plates in this area.

Permian–Triassic

Many of the continental collisions, which began in the Carboniferous, reached maturity in the Early Permian. A major part of Pangea was assembled, and the new supercontinent, ringed by subduction zones, moved steadily northward. The Calcareous Alps and Inner Carpathians formed the marginal platform of western and Central Europe (Figure 6), whereas the Scythian–Turan platform formed the margin of Eastern Europe and central Asia.

North-dipping subduction developed along the Paleotethys margin (Robertson et al., 1991, 1996; Golonka, 2000, 2002, 2004). It caused Triassic back-arc rifting in the proto-Black Sea area and along the margins of Scythian–Turan platform (Kazmin, 1990, 1991; Zonenshain et al., 1990). The Tauric basin was formed between the Pontides and the Dobrogea–Crimea segment of the Scythian platform. A rift separated Moesia and the Ukrainian part of the Scythian platform. This rift was perhaps connected with the rift situated in Poland. According to Muttoni et al. (2000a), Moesia was probably marginally separated but not detached or rotated away from Europe by the North Dobrogea transtensional trough, which is interpreted as a back-arc basin resulting from the northward subduction of the Paleotethys Ocean.

The Carboniferous–earliest Permian rifting of the Cimmerian plates (see Dercourt et al., 1993; Golonka et al., 1994; Sengor and Natalin, 1996; Golonka and Ford, 2000) from Gondwana turned into drifting during the Permian, marking the inception of the Neotethys Ocean. Rifting and oceanic type of basin opening could also have occurred in the Mediterranean. The deep-water sediments of Sicily (Catalano et al., 1991; Kozur, 1991), Lagonegro (Marsella et al., 1993), and Crete (Kozur and Krahl, 1987) were deposited in the newly opened basins. The oceanic system was established in southern and Central Europe during the Permian–Triassic. The narrow branch of the Neotethys separated the Apulia–Taurus platform from the African continent. The Apulia platform was connected with European marginal platforms. Its northernmost part was possibly separated from the Umbria–Marche region by a rift with Verrucano type of sediments. Similar sediments were found in the Carpathian area in the Tatra Mountains and the Marmarosh Klippen zone (Mahel et al., 1968, Slaczka et al., 2006). The incipient Pindos Ocean separated the Pelagonian, Sakariya, and Kirsehir block from the Ionian–Taurus platform (Robertson et al., 1991, 1996; Stampfli et al., 1991). The proto-Transylvanian Ocean separated the Tisza (Bihor–Apuseni) block from the Moesian–Eastern European platform (Figure 8) (Sandulescu, 1988; Sandulescu and Visarion, 2000). It is possible that the northwestern embayment of this Transylvanian oceanic zone exists between the Inner Carpathian and European platform (Golonka et al., 2000, 2003). The pelagic Triassic Fleckenkalk-type pebbles in the exotic pebbles in the Pieniny Klippen (Birkenmajer, 1988; Birkenmajer et al., 1990) and Magura unit (Sotak, 1986) could have

Figure 5. Paleogeography of the circum-Carpathian area during the Late Carboniferous; plates position at 302 Ma (modified from Golonka et al., 2000). Abbreviations: Ad = Adria (Apulia); BS = Black Sea; Ca = Calabria–Campania terranes; EA = Eastern Alps; EP = eastern Pontides; Gr = Greece; Ib = Iberia plate; IC = Inner Carpathians; Ki = Kirsehir plate; LC = lesser Caucasus terrane; Lu = Lut plate; Mo = Moesia plate; Pe = Pelagonian plate; Rh = Rhodopes; Sa = Sakariya plate; Si = Sicily plate; Tu = Taurus terrane; Ti = Tisza plate; UM = Umbria–Marche; WP = western Pontides.

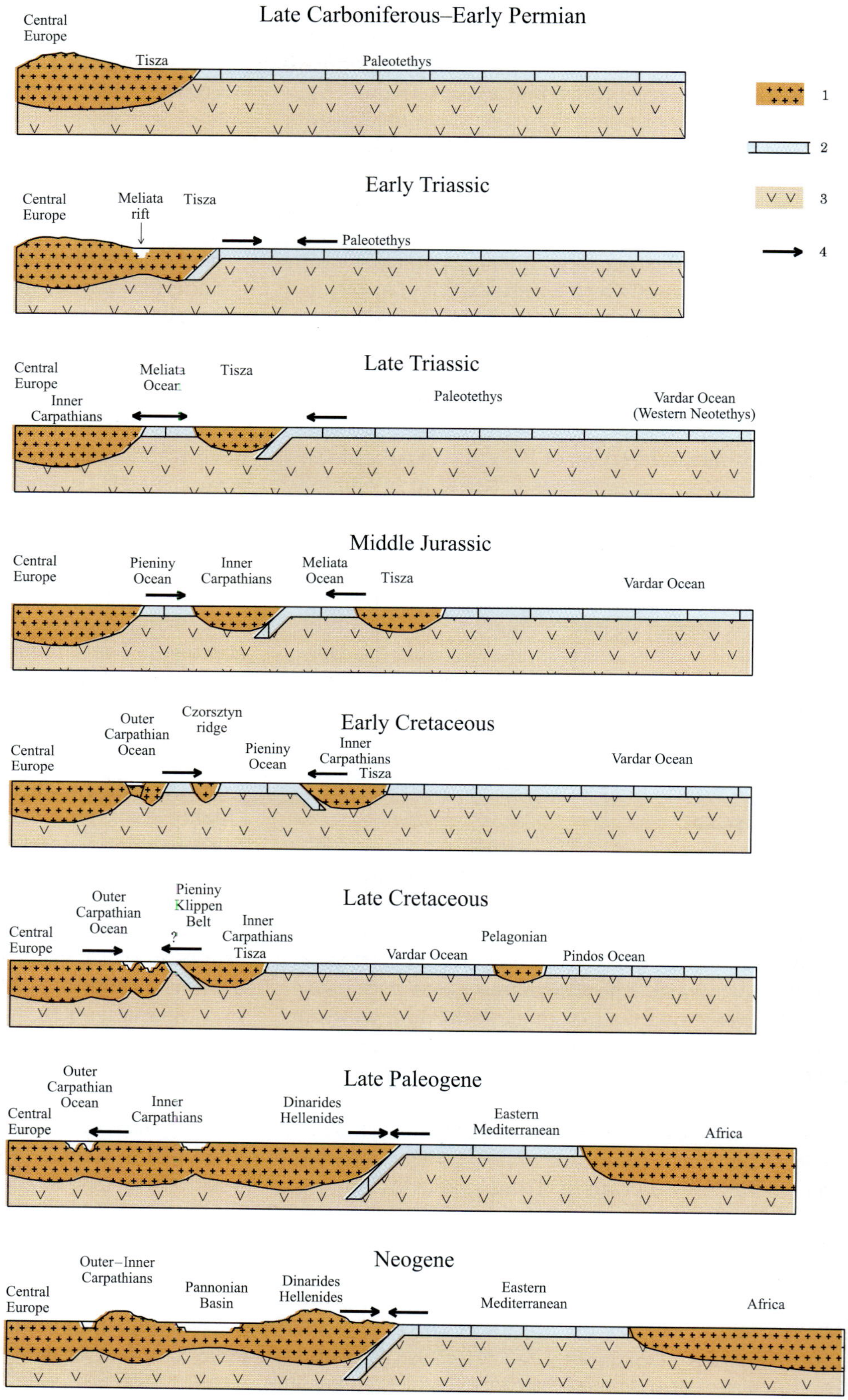

Figure 6. Highly schematic (not to scale) plate-tectonic profiles (modified from Golonka et al., 2000). Central Europe–Carpathians–Greece. 1 = Continental crust (including obducted, allochthonous rocks and sedimentary cover); 2 = oceanic crust (including deposits); 3 = upper mantle; 4 = direction of plates.

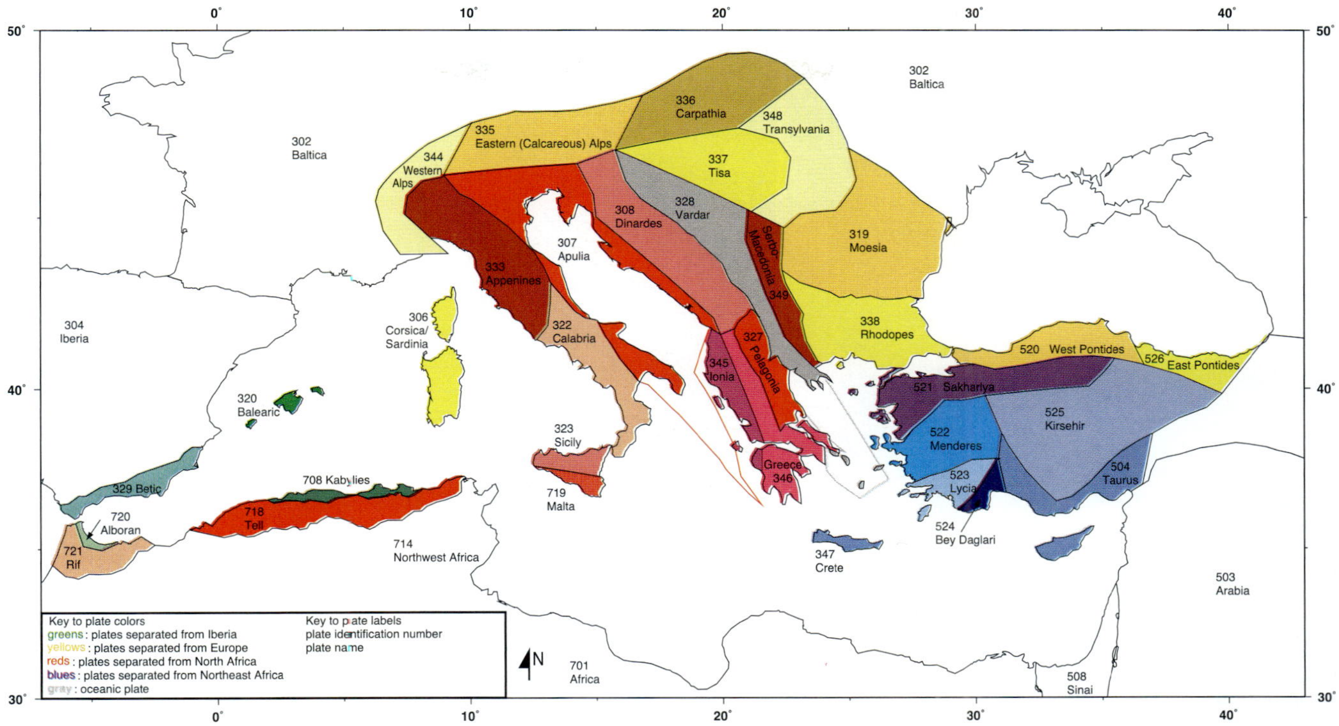

Figure 7. Microplates for the Mediterranean.

originated in this embayment. The position of this embayment of the Transylvanian Ocean and its relation to the other parts of the Tethys, Vardar Ocean, Meliata–Halstatt Ocean, Dobrogea rift, and Polish-Danish Aulacogen remain quite speculative.

According to Rakus et al. (1998), two oceanic units were located south of the Inner Carpathian plate. One was open during the Triassic and closed during the Late Triassic, as a result of the early Cimmerian collision. Another, represented by sequences at the classic profile of Meliata in Southern Slovakia, opened during the Early and Middle Jurassic as a back-arc basin and closed during the Late Jurassic. The position of the Meliata Ocean, the time of closing, and the function of the Tisza unit in the Mesozoic collisional events are still the subject of lively discussion (see Kozur, 1991; Stampfli, 1996, 2001; Channell and Kozur, 1997; Mock et al., 1998; Plasienka, 1999; Golonka et al., 2000; Wortmann et al., 2001).

Jurassic

Continued sea-floor spreading occurred during the Jurassic in the Neotethys. The Pelagonian plate, Kirsehir, and Sakariya (Robertson et al., 1991) were drifting off the Apulia–Taurus–Arabia margin. The Alpine Tethys, which is the Ligurian and Penninic oceans (Pn) and the Pieniny–Magura basin (Li, Pn, PKB, Mg, Figures 9, 10), was formed during the Jurassic. This Alpine Tethys constitutes the extension of the central Atlantic system (Golonka, 2000). The central Atlantic (Withjack et al. 1998) and Alpine Tethys went into a drifting stage during the Middle Jurassic. The oldest oceanic crust in the Ligurian–Piemont Ocean was dated as late as the Middle Jurassic in the southern Apennines and in the Western Alps (see Ricou, 1996, and literature cited therein). Bill et al. (2000) date the onset of oceanic spreading of the Alpine Tethys by isotopic methods as Bajocian. According to Winkler and Slaczka (1994), the Pieniny data fit well with the supposed opening of the Ligurian–Penninic Ocean (Li-Pn). The Pieniny–Magura basin opening is related to the closure of the Meliata Ocean (Me) and the above-mentioned enigmatic embayment of the proto-Vardar–Transylvanian Ocean (see Figure 8, Va-Tr).

The orientation of the Pieniny–Magura Ocean was southwest–northeast (see discussion in Golonka and Krobicki, 2001, 2004; Golonka et al., 2003). This ocean was divided into the northwestern and southeastern basins by the midoceanic Czorsztyn ridge (Cr, Figure 10). According to Birkenmajer (1986), the Czorsztyn ridge could be traced from the vicinity of Vienna through

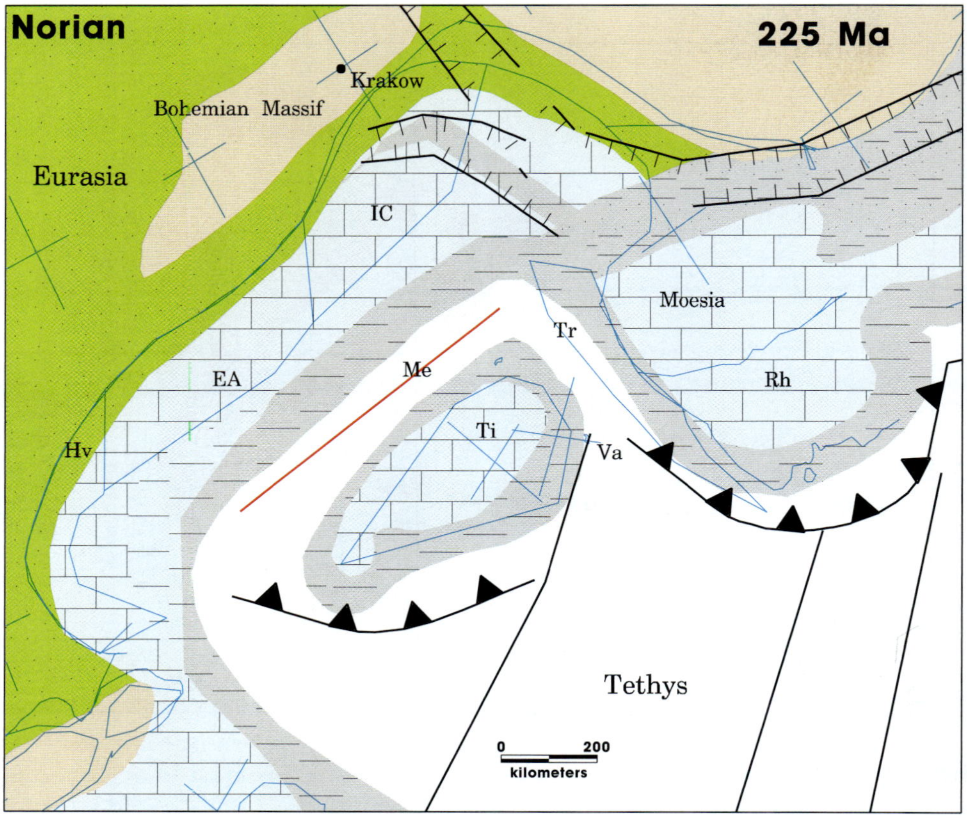

Figure 8. Paleoenvironment and lithofacies of the circum-Carpathian area during the Late Triassic; plates position at 225 Ma (modified from Golonka et al., 2000, 2003). Abbreviations: EA = Eastern Alps; IC = Inner Carpathians; Hv = Helvetic shelf; Me = Meliata–Halstatt Ocean; Rh = Rhodopes; Ti = Tisza plate; Va = proto-Vardar Ocean; Tr = Transylvanian Ocean.

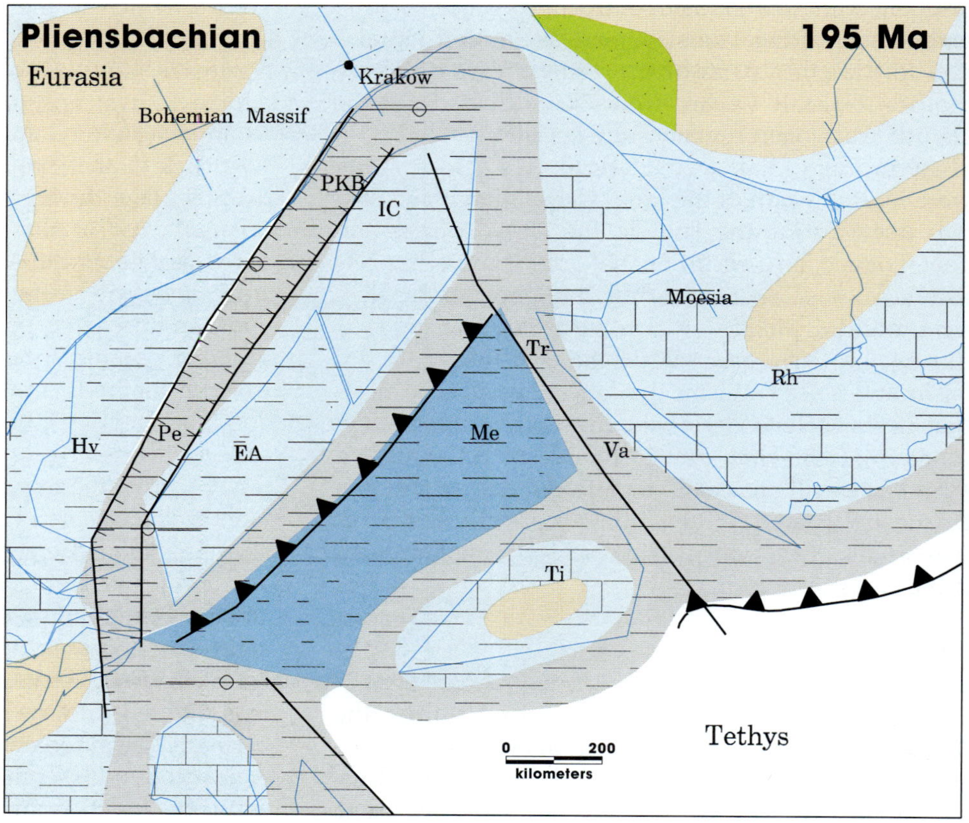

Figure 9. Paleoenvironment and lithofacies of the circum-Carpathian area during the Early Jurassic; plates position at 195 Ma (modified from Golonka et al., 2000, 2003). Abbreviations: EA = Eastern Alps; Hv = Helvetic shelf; IC = Inner Carpathians; Me = Meliata–Halstatt Ocean; PKB = Pieniny Klippen Basin; Pe = Penninic Ocean; Rh = Rhodopes; Ti = Tisza plate; Va = Vardar Ocean; Tr = Transylvanian Ocean.

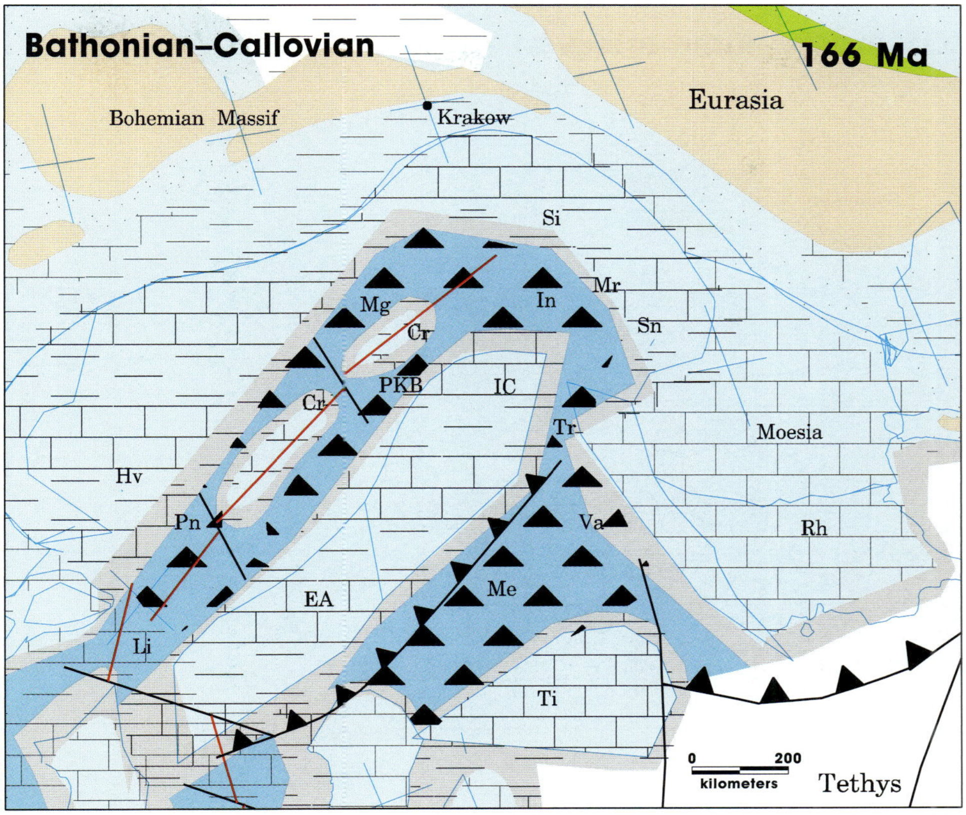

Figure 10. Paleoenvironment and lithofacies of the circum-Carpathian area during the Middle Jurassic; plates position at 166 Ma (modified from Golonka et al., 2000, 2003). Abbreviations: Cr = Czorsztyn ridge; EA = Eastern Alps; Hv = Helvetic shelf; IC = Inner Carpathians; In = Inacovce–Kricevo zone; Li = Ligurian Ocean; Me = Meliata–Halstatt Ocean; Mg = Magura basin; Mr = Marmarosh Massif; PKB = Pieniny Klippen Basin; Pn = Penninic Ocean; Si = future site of Silesian basin; Sn = future site of Sinaia basin; Rh = Rhodopes; Ti = Tisza plate; Va = Vardar Ocean; Tr = Transylvanian Ocean.

Western Slovakia, Poland, Eastern Slovakia to the Transcarpathian Ukraine, and perhaps, northernmost Romania (Bombita et al., 1992). The Jurassic–Early Cretaceous extremely reduced deep-water rocks (pelagic limestones and radiolarites) were deposited in the deepest parts of the basins. The southwestern basin is known as the Pieniny Klippen Basin (PKB in Figures 9, 10), and the northwestern basin is known as the Magura basin (Mg, Figures 10, 11) (Golonka and Sikora, 1981; Birkenmajer, 1986; Golonka et al., 2000, 2003; Golonka and Krobicki, 2004). The paleogeographic extent of the Magura basin remains somewhat enigmatic and speculative.

The shallowest ridge sequences are represented by the dark, Early Jurassic, marly (Fleckenkalk–Fleckenmergel-type) facies followed by Bajocian–Tithonian crinoidal and nodular limestones (Ammonitico rosso type) and Cretaceous, variegated marls (Scaglia rossa facies). The transitional slope sequences are known from outcrops south and north of the Czorsztyn ridge in Poland. Several successions were distinguished in the basinal and slope deposits: Zlatne, Grajcarek, Hulina, Branisko (Kysuca), Niedzica, Czertezik, and Zawiasy (Sikora, 1971; Birkenmajer, 1977; Golonka and Sikora, 1981; Krobicki et al., 2003a, b; Golonka and Krobicki, 2004; Golonka et al., 2006). The exact position of these sequences is uncertain because of the strong tectonic deformations. Ridge sequences, as well as transitional slope sequences, are also called Oravicum by Slovak geologists (Mahel et al., 1968; Plasienka, 1999).

The initial movements during Toarcian–Aalenian had to precede the appearance of the Czorsztyn ridge, which did not exist as the main paleogeographic unit before Bajocian (e.g., Aubrecht et al., 1997). This ridge emergence was connected with the postrift phase of the basin evolution (Golonka and Slaczka, 2003). The sedimentation of the younger red nodular Ammonitico rosso–type limestones was an effect of Mesozoic–Cimmerian vertical movements, which caused the Czorsztyn ridge to subside (Cr; Figures 10–12) and produced tectonically differentiated blocks, neptunian dykes, and scarp breccias (e.g., Birkenmajer, 1986; Krobicki, 1996; Aubrecht et al., 1997; Wierzbowski et al., 1999; Aubrecht, 2001; Aubrecht and Túnyi, 2001; Golonka et al., 2003; Krobicki et al., 2003a, b).

The extension of the Neotethys to the northwest into the proto-Mediterranean established a connection with the central Atlantic. A junction of the Neotethys and Atlantic–Ligurian–Penninic–Pieniny Klippen–Magura Ocean (Alpine Tethys) existed perhaps in the Eastern Slovakia–Ukrainian Carpathians and is represented by the Inacovce–Krichevo unit (Sotak et al.,

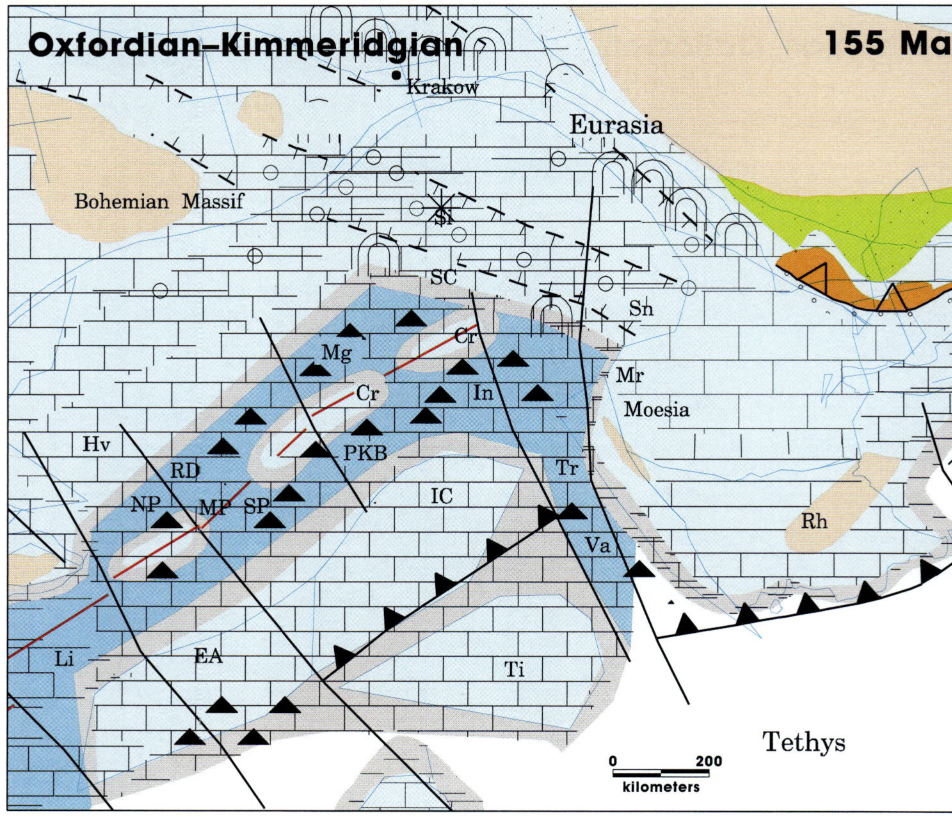

Figure 11. Paleoenvironment and lithofacies of the circum-Carpathian area during the Late Jurassic; plates position at 155 Ma (modified from Golonka et al., 2000, 2003). Abbreviations: Cr = Czorsztyn ridge; EA = Eastern Alps; Hv = Helvetic shelf; IC = Inner Carpathians; In = Inacovce–Kricevo zone; Li = Ligurian Ocean; Mg = Magura basin; MP = mid-Penninic ridge; Mr = Marmarosh Massif; NP = North Penninic Ocean; PKB = Pieniny Klippen Basin; RD = Rhenodanubian; Rh = Rhodopes; SC = Silesian ridge (Cordillera); Si = Silesian basin; Sn = Sinaia basin; SP = South Penninic Ocean; Ti = Tisza plate; Va = Vardar Ocean; Tr = Transylvanian Ocean.

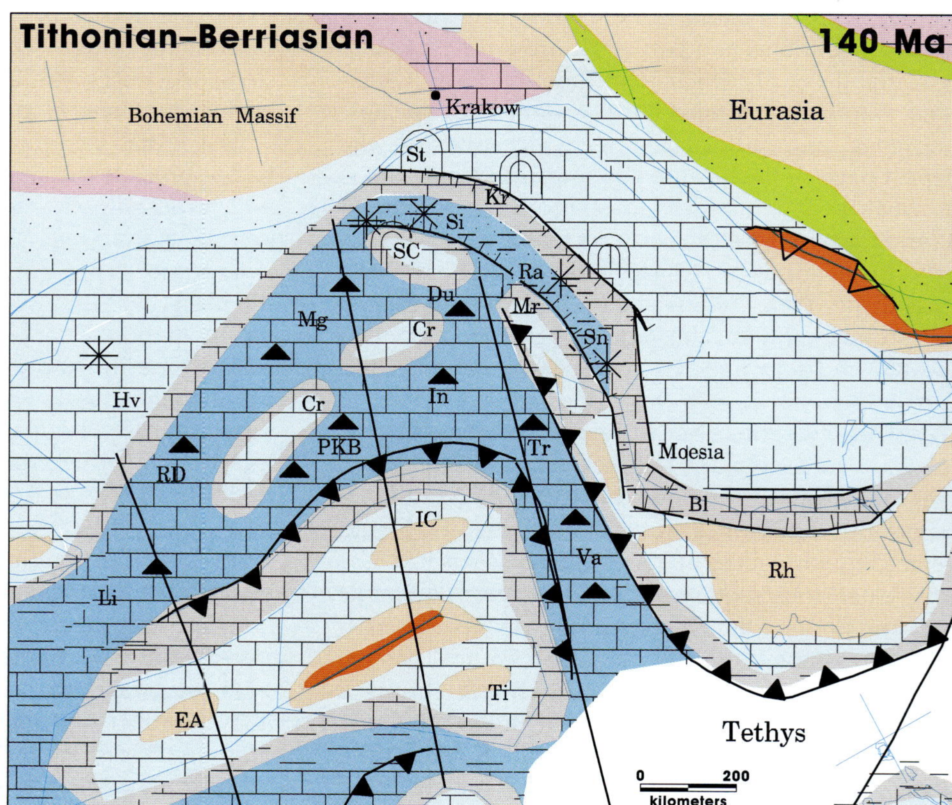

Figure 12. Paleoenvironment and lithofacies of the circum-Carpathian area during the latest Late Jurassic–earliest Early Cretaceous; plates position at 140 Ma (modified from Golonka et al., 2000, 2003). Abbreviations: Bl = Balkan rift; Cr = Czorsztyn ridge; Du = Dukla basin; EA = Eastern Alps; Hv = Helvetic shelf; IC = Inner Carpathians; In = Inacovce–Kricevo zone; Kr = Kruhel Klippe; Li = Ligurian Ocean; Mg = Magura basin; Mr = Marmarosh Massif; PKB = Pieniny Klippen Basin; Ra = Rakhiv basin; RD = Rhenodanubian; Rh = Rhodopes; SC = Silesian ridge (Cordillera); Si = Silesian basin; Sn = Sinaia basin; St = Stramberk Klippe; Ti = Tisza plate; Va = Vardar Ocean; Tr = Transylvanian Ocean.

2000). The central Atlantic was in an advanced drifting stage during the Middle and Late Jurassic (Withjack et al., 1998). Rifting continued in the North Sea and in the northern proto-Atlantic (Ziegler, 1988; Dore, 1991). The progressive breakup of the Pangea resulted in a system of spreading axes, transform faults, and rifts, which connected the ocean-floor spreading in the central Atlantic and Ligurian Sea to rifting that continued through the Polish–Danish graben to mid-Norway and the Barents Sea (Golonka et al., 1996). Tethys was connected with the Polish–Danish graben (Zytko, 1984, 1985) by a transform fault and rift system, which preceded the opening of the Outer Carpathian basins (Figure 11). The Zegocina porfiritic andesite (Slaczka, 1998; Slaczka et al., 1999) could have been a volcanic expression of the Jurassic disintegration of the southern margin of the north European plate, which later became the site of the Outer Carpathian rifting. It could also represent an early stage of the hot-spot activity in this area.

Latest Late Jurassic–Earliest Early Cretaceous

In the Alpine–Carpathian area in Europe, the subduction of the Meliata–Halstatt Ocean (Figures 11, 12) and the collision of the Pelsonian (Kovacs et al., 2000) fragment of the Tisza unit with the Inner Carpathian terranes were concluded at the end of the Jurassic (Dallmeyer et al., 1996; Froitzheim et al., 1996; Plasienka, 1999). Major plate reorganization happened during the Tithonian. The new Atlantic spreading entered the area between the Newfoundland shelf and Iberia. The Jurassic Pangean breakup system in the Alpine Tethys, Central Europe, and North Sea area was abandoned (Golonka, 2000). The closure of the Meliata–Halstatt Ocean is related to the cessation of spreading in the Alpine Tethys. In the western Alpine area in Europe, the movement of the Brianconnais terrane during the Early Cretaceous started the closing of the Ligurian Ocean (Stampfli, 1993, 1996), which then entered into its compressional phase (Marchant and Stampfli, 1997).

The Tethyan plate reorganization resulted in extensive fault movement. Several tectonic horsts and grabens were formed, rejuvenating some older, Eo- and Meso-Cimmerian faults (Krobicki, 1996; Golonka et al., 2003; Golonka and Krobicki, 2004). The Outer Carpathian rift (Silesian basin) had developed with the beginning of the Upper Jurassic–Lower Cretaceous calcareous flysch sedimentation (Slomka, 1986). The Western Carpathian Silesian basin probably extended in the Eastern Carpathian (Sinaia or black flysch) as well as in the Southern Carpathian Severin zone (Sandulescu, 1988; Krautner, 1996; Stefanescu et al., 2006). The carbonate platforms with reefs (Stramberk) developed along the margin of the Silesian basin. The Silesian ridge (Ksiazkiewicz, 1977a, b) separated the Silesian and Magura basins as a result of the fragmentation of the European platform in this area (Olszewska and Wieczorek, 2001). The subsidence in the Silesian basin was accompanied by the extrusion of basic lavas (teschenites) in the Western Carpathians (Narebski, 1990) and basic effusives in the black flysch of the Eastern Carpathians (Slaczka et al., 1999; Oszczypko et al., 2003). The Jurassic separation of Bucovino–Getic microplate, which included the Marmarosh Massif (Mr, Figure 11) from the European plate, is perhaps related to the origin of the Silesian ridge. The direct connection is obscured, however, by the remnants of the Transylvanian Ocean in the area of the eastern end of Pieniny Klippen Basin. These remnants are known from the Inacovce–Krichevo unit in Eastern Slovakia and Ukraine (Sotak et al., 2000). In this area, a junction of the different basinal units existed: the Pieniny Klippen–Magura Ocean, Transylvanian Ocean, and the Outer Carpathian basin.

The Penninic–Pieniny–Magura Ocean reached its maximum width in the latest Jurassic and then stopped spreading. The various models explaining the closing of the Pieniny Klippen Ocean were discussed by Golonka and Krobicki (2001) and Golonka et al. (2003). These authors argued for the two subduction zones, which developed at both margins of the Pieniny–Magura basin. The subduction-related metamorphism produced the latest Jurassic blueschist metamorphic rocks found as pebbles (exotics) in the Albian flysch in the Pieniny–Magura basin (Faryad, 1997). The eastward–northeastward subduction along the Silesian ridge and Bucovinian–Getic terrane was perpendicular or oblique to the southward subduction at the northern margin of the Inner Carpathian–Eastern Alps terrane. The Silesian–Sinaia basin developed as a back-arc basin. The eastward subduction of Getic terranes is connected with the northward subduction existing along the northern margin of the Neotethys. The displacement in the Transylvanian Ocean, which began in the Barremian (Stefanescu et al., 2006), is perhaps related to this subduction. The Outer Carpathian realm emerged as a combination of oceanic (part of Magura), rifted, continental, and back-arc basins separated by uplifted areas.

Early Cretaceous

During the Hauterivian–Aptian, the Ligurian Ocean entered into its compressional phase (Marchant and Stampfli, 1997). Subduction was active on the southern margin of the Pieniny Klippen Basin (Golonka et al.,

2000; Golonka and Krobicki, 2004). In the Alpine–Carpathian area, Outer Carpathian troughs, partially on the oceanic crust and partially on the attenuated continental crust, were open during this time. The mature stage of the opening of the Silesian basin was accompanied by the main phase of the formation of teschenites, which occurred in the Western Carpathians during the Hauterivian–Barremian (Lucinska-Anczkiewicz et al., 2000; Slaczka et al., 2000). These volcanic rocks display the features of oceanic islands generated by hot-spot activity. It seems that there were two hot spots in the Carpathian region. The first one, in the Western Carpathians, was connected with the Jurassic Zegocina andesites (mentioned above) and Early Cretaceous teschenites. Today, the western Turkey and Northern Aegean volcanics are located at the same latitude and longitude. The second hot spot, in the Eastern Carpathians, was connected with the Jurassic andesitic tuffites and latest Jurassic–earliest Cretaceous volcanics (Lashkevitsch et al., 1995; Lewandowski et al., 2003). Today, the Levantine (e.g., Dead Sea) hot-spot volcanics and Northern Aegean volcanics are located at the same latitude and longitude (see Golonka and Bocharova, 2000). The extension of the Silesian–Sinaia basin is marked by the beginning of the sedimentation in the Skole–Tarcau area. To the west, the Outer Carpathian troughs extended into the Valais Ocean (Golonka et al., 2000), which opened around the Aptian (Stampfli et al., 2001) and extended further into the area between Spain and France and to the Biscay Bay (Stampfli, 1993, 1996). To the east, the trough system was connected with the subsiding Balkan area.

The Outer Carpathian basin reached its greatest width during the Hauterivian–Aptian. With the widening of the basin, several subbasins (troughs) began to show their distinctive features. These subbasins, like Silesian, Subsilesian, Skole, Dukla, and Tarcau, were locally separated by uplifted areas, e.g., the Andrychow zone (Ksiazkiewicz, 1960; Olszewska and Wieczorek, 2001). The general downwarping of the Silesian basin was probably caused by the cooling effect of the underlying lithosphere (Slaczka et al., 1999). The sedimentation of calcareous sediments passed upward gradually into black shales with turbiditic sequences in the Silesian trough.

During the Aptian–Albian (Figure 13), complex tectonics began to occur in the future Alpine belt zone, between southern Europe and North Africa and Arabia. Convergent margin and north-dipping subduction along the western Pontides block caused the southward movement of this terrane and the opening of the western Black Sea as a back-arc basin (Kazmin, 1990; Banks and Robinson, 1997; V. G. Kazmin, 1997, personal communication; Golonka et al., 2000). The north-dipping subduction was consuming the main branch of the Neotethys, between the Pontides, Sakarya, and Kirsehir plates (Yilmaz et al., 1996). Opening of the western Black Sea (Figure 13) occurred by rifting and drifting of the western-central Pontides away from the Moesian and Scythian platform of Eurasia (Kazmin 1990, 1991; Banks and Robinson, 1997; V. G. Kazmin, 1997, personal communication). Continued closure of the Alpine Tethys was related to the subduction along its southern margin. Consumption of the Penninic–Pieniny led to the development of an accretionary prism in front of the moving-northward and moving-northwestward eastern Alpine and Inner Carpathian plates. In the Albian, synorogenic flysch developed in the Pieniny–Magura basin (Golonka and Sikora, 1981). The Rhenodanubian flysch could also be related to the formation of a Cretaceous accretionary prism. In the Eastern Carpathians, the compressional movement started during the Aptian and Albian, and the inner part of the Carpathians was folded and napped. In front of the moving nappes, coarse-grained sediments and olistostromes developed (Sandulescu, 1988).

Late Cretaceous: Cenomanian–Campanian

The rotation of Africa and the spreading in the eastern Mediterranean caused the Apulian plate to converge with Europe. The Cretaceous Trupchun orogeny in the Alps (Froitzheim et al., 1996) caused a subduction of the small terranes, together with the oceanic crust of the Ligurian Ocean. These terranes were subject of eclogite metamorphism. The subduction was accompanied by the decollement of ophiolites and the Ligurian–Piemont sediments and their emplacement as the earliest nappes of the Alpine evolution (Debelmas, 1989). The compressional deformation of the Inner Carpathians led to the development of nappes (Mahel et al., 1968; Ellouz and Roca, 1994; Plasienka, 1999). The emplacement of Inner and Median Dacides occurred during the Albian–Cenomanian (Stefanescu et al., 2006). The Pieniny Klippen Ocean narrowed significantly (Figure 14), whereas the Outer Carpathian basin remained widely opened and connected with the European shelf basins. The coarse clastic sediments with exotics (Burtan et al., 1984) are related to these synorogenic events. During the Cenomanian, a period of slow and uniform sedimentation embraced all Carpathian basins, and well-oxygenated conditions began to develop (Bieda et al., 1963). Slaczka et al. (1999) have distinguished the stage of the development of the Outer Carpathian basin. During the uppermost Albian–Cenomanian and Turonian stage of the development of the Outer Carpathian basin

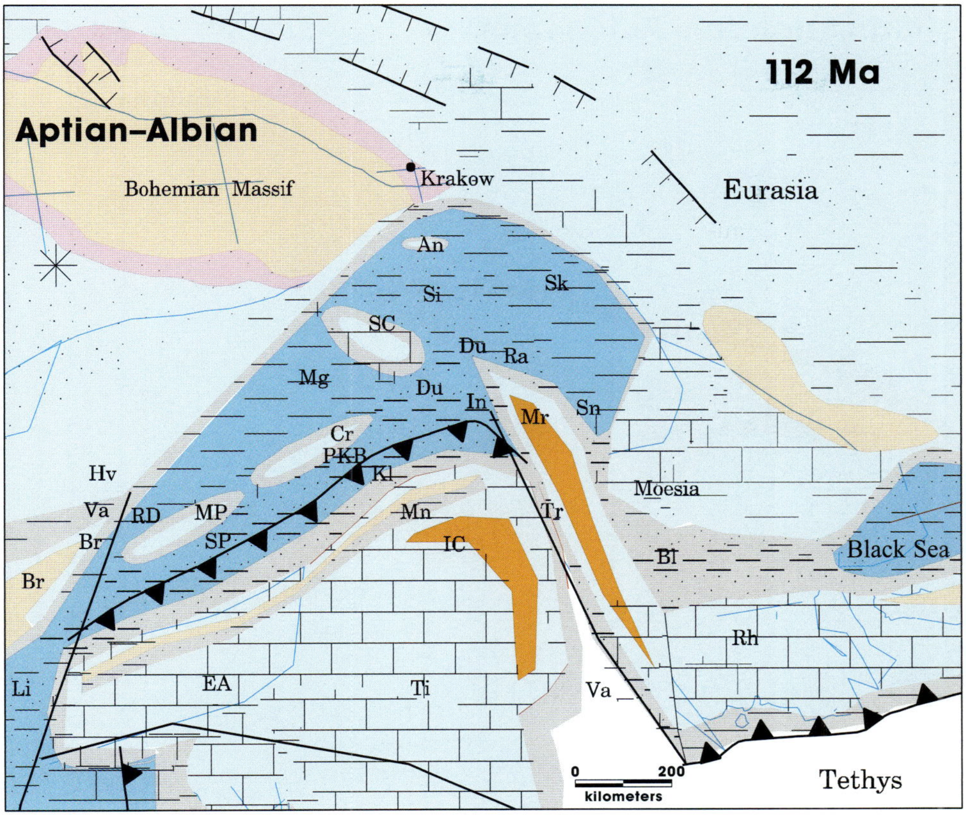

Figure 13. Paleoenvironment and lithofacies of the circum-Carpathian area during the latest late Aptian–middle Cenomanian; plates position at 112 Ma (modified from Golonka et al., 2000, 2003). Abbreviations: An = Andrychow Klippe; Bl = Balkan Basin; Br = Brianconnais Ridge; Cr = Czorsztyn ridge; Du = Dukla Basin; EA = Eastern Alps; Hv = Helvetic shelf; IC = Inner Carpathians; In = Inacovce–Kricevo zone; Kl = Klappe accretionary prism; Li = Ligurian Ocean; Mg = Magura basin; Mr = Marmarosh Massif; Mn = Manin Basin; MP = mid-Penninic ridge; PKB = Pieniny Klippen Basin; Ra = Rakhiv basin; RD = Rhenodanubian; Rh = Rhodopes; SC = Silesian ridge (Cordillera); Si = Silesian basin; Sk = Skole basin; Sn = Sinaia basin; SP = South Penninic Ocean; Ti = Tisza plate; Tr = Transylvanian Ocean; Va = Vardar Ocean.

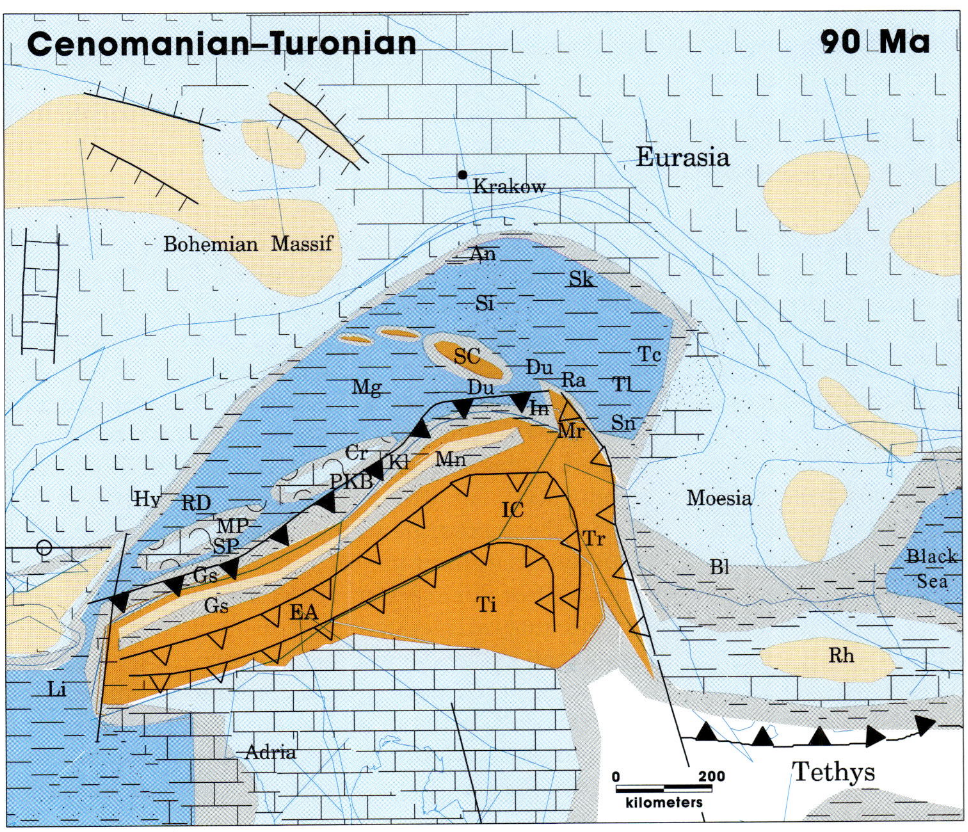

Figure 14. Paleoenvironment and lithofacies of the circum-Carpathian area during Cenomanian–Turonian; plates position at 90 Ma. Abbreviations: An = Andrychow Klippe; Bl = Balkan Basin; Cr = Czorsztyn ridge; Du = Dukla Basin; EA = Eastern Alps; Gs = Gosau Basin; Hv = Helvetic shelf; IC = Inner Carpathians; In = Inacovce–Kricevo zone; Kl = Klappe accretionary prism; Li = Ligurian Ocean; Mg = Magura basin; Mr = Marmarosh Massif; Mn = Manin Basin; PKB = Pieniny Klippen Basin; Ra = Rakhiv basin; RD = Rhenodanubian; Rh = Rhodopes; SC = Silesian ridge (Cordillera); Si = Silesian basin; Sk = Skole basin; Sn = Sinaia basin; SP = South Penninic Ocean; MP = Mid-Penninic ridge; Tc = Tarcau Basin; Ti = Tisza plate; Tl = Teleajen Basin; Tr = Transylvanian fold belt.

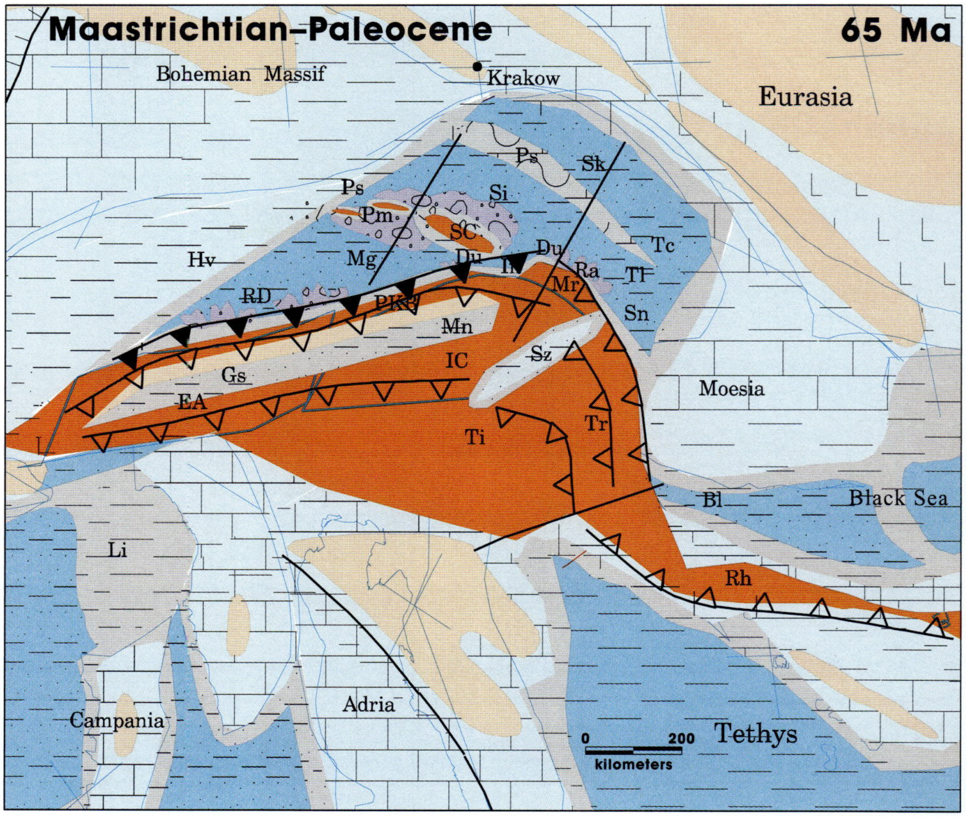

Figure 15. Paleoenvironment and lithofacies of the circum-Carpathian area during late Maastrichtian–Paleocene; plates position at 65 Ma (modified from Golonka et al., 2000). Abbreviations: Bl = Balkan Basin; Du = Dukla Basin; EA = Eastern Alps; Gs = Gosau Basin; Hv = Helvetic shelf; IC = Inner Carpathians; In = Inacovce–Kricevo zone; Li = Ligurian Ocean; Mg = Magura basin; Mr = Marmarosh Massif; Mn = Manin Basin; PKB = Pieniny Klippen Belt; Pm = Fore-Magura basin; Ps = Subsilesian ridge and slope zone; Ra = Rakhiv basin; RD = Rhenodanubian; Rh = Rhodopes; SC = Silesian ridge (Cordillera); Si = Silesian basin; Sk = Skole basin; Sn = Sinaia basin; Sz = Szolnok basin; Tc = Tarcau Basin; Ti = Tisza plate; Tl = Teleajen Basin; Tr = Transylvanian fold belt.

(Bieda et al., 1963; Slaczka et al., 1999), all the sources of siliciclastic material ceased to be active, and generally uniform pelagic sedimentation started; green, radiolarian shales with radiolarites as part of the Outer Carpathian basin were followed by red shales and marls. At the beginning of this stage, the rate of sedimentation radically decreased to 4–6 m/Ma (13–20 ft/Ma), whereas a paleobathymetry reached abyssal depth (Slaczka et al., 1999).

With the development of the Inner Carpathian nappes (Figure 14), a fore-arc basin was formed between the uplifted part of the Inner Carpathian terrane (the so-called Andrusov Ridge) and the subduction zone. The flysch of the Klappe and Zlatne successions was formed in this area. Behind the ridge, the Manin succession was deposited in the back-arc basin. The Transylvanian basin developed during the Late Cretaceous between the Apuseni Mountains and the Carpathians (Stefanescu et al., 2006).

Late Cretaceous–Earliest Paleogene

According to Froitzheim et al. (1996), the collision between the Austroalpine units and the Brianconnais terrane in the Alps started in the early Paleocene. Latest Cretaceous–earliest Paleocene was also the time of the closure of the Pieniny Klippen Ocean (Figures 6, 15) and the collision of the Inner Carpathian terranes with the Czorsztyn ridge (Birkenmajer, 1986; Winkler and Slaczka, 1994). The complex fold and thrust system has developed in the Pieniny Klippen zone. The primary shortening events in the Balkans occurred in Bulgaria (Sinclair et al., 1997). The Vardar Ocean was closed during the Paleocene (Sengor and Natalin, 1996).

The accretionary prism had overridden the Czorsztyn ridge. The subduction zone moved from the southern margin of the Pieniny Klippen Ocean (Birkenmajer, 1986) to the northern margin of the Czorsztyn ridge (Figures 6, 15) (Golonka et al., 2000). The development of the new accretionary prism in the Magura basin was related to the origin of the trench connected with this subduction zone (Oszczypko, 1998; Oszczypko et al., 2003). The submarine slumps and olistoliths along the southern margin of the Magura basin were related to the uplift of this margin. The sedimentation and subsidence rates accelerated more distinctly in the Silesian basin than in the Magura basin (Oszczypko et al., 2003) and were accompanied by a continuous uplifting of the Silesian Cordillera and of the southern margin of the European platform and the Marmarosh Massif. This uplift produced a tremendous amount of clastic material.

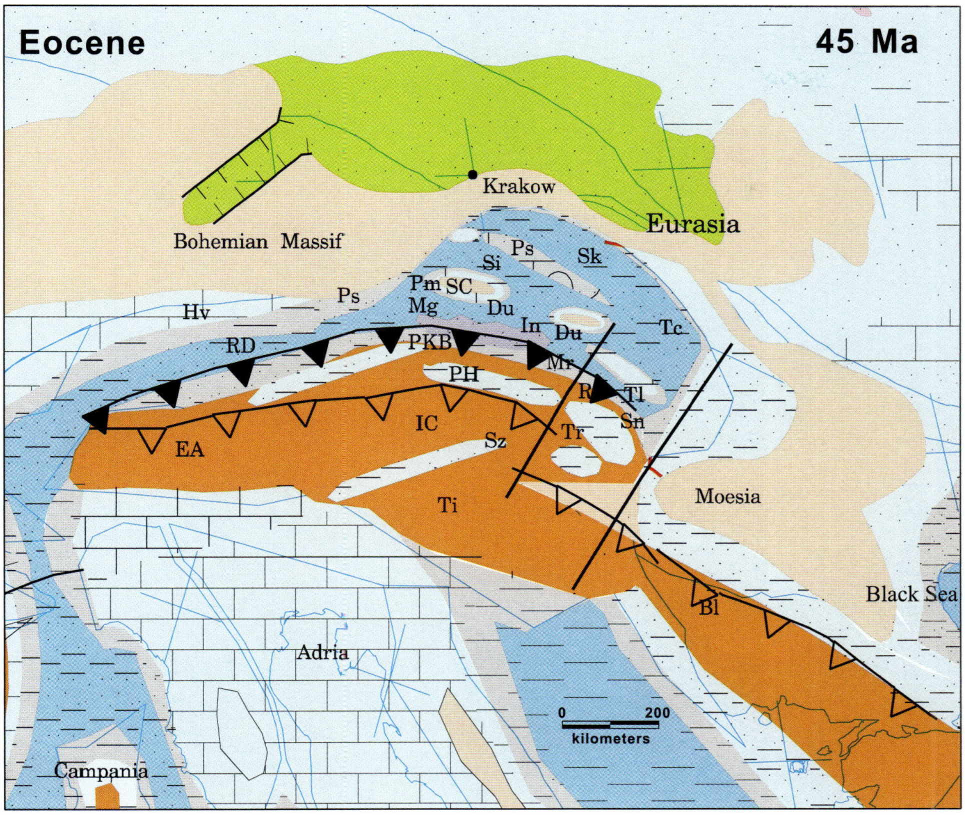

Figure 16. Paleoenvironment and lithofacies of the circum-Carpathian area during the Eocene (Lutetian–Bartonian); plates position at 45 Ma (modified from Golonka et al., 2000). Abbreviations: Bl = Balkan Basin and fold belt; Du = Dukla Basin; EA = Eastern Alps; Hv = Helvetic shelf; IC = Inner Carpathians; In = Inacovce–Kricevo zone; Mg = Magura basin; Mr = Marmarosh Massif; PH = Inner Carpathian (Podhale) Basin; PKB = Pieniny Klippen Belt; Pm = Fore-Magura basin; Ps = Subsilesian ridge and slope zone; Ra = Rakhiv basin; RD = Rhenodanubian; SC = Silesian ridge (Cordillera); Si = Silesian basin; Sk = Skole basin; Sn = Sinaia basin; Sz = Szolnok basin; Tc = Tarcau Basin; Ti = Tisza plate; Tl = Teleajen Basin; Tr = Transylvanian basin.

Paleocene–Eocene

The process of the closing of the Neotethys by the Alpine and Himalayan orogenies continued. The Adria (Apulia) plate, eastern Alpine (Austroalpine) and Inner Carpathian blocks were continuously moving northward (Figure 16). Their collision with the European plate began in the Alps at about 47 Ma (Decker and Peresson, 1996). The Valais Ocean in the Alps finally closed (Froitzheim et al., 1996; Stampfli, 1996). According to Plasienka and Kovac (1999), at that time, the Alcapa block was formed by welding together the Eastern Alps, Inner Carpathian, Tisza, as well as smaller terranes like Bukk, Transdanubian, or the Getic. The main phase of compression and formation of the thrust belt of the Balkanides in Bulgaria occurred during the Eocene (Sinclair et al., 1997; Tari et al., 1997). The closing of the Pieniny Klippen Ocean in the Carpathians was also concluded, and the Pieniny domain accreted to the Magura basin (Birkenmajer, 1986; Winkler and Slaczka, 1994). The folding of the Rhenodanubian zone occurred in the late Eocene (Figure 16).

The Magura basin narrowed significantly because of the northward Alcapian movement (Oszczypko, 1998; Oszczypko et al., 2003). The Maastrichtian–Paleocene inversion was followed by a new episode of subsidence, which began at the end of the Paleocene and accelerated during the Lutetian and Priabonian (Figure 16) (Oszczypko et al., 2003). The accretionary prism was building gradually, causing the northward migration of depocenters. Thin-bedded flysch deposits passed into a thick complex of turbidites and fluxoturbidites. The Dukla, Silesian, Fore-Silesian, and Skole–Tarcau basins remained open, with the flysch mainly in the southern part (Dukla and Silesian basins) and pelagic facies sedimentation further toward the north (Bieda et al., 1963).

Foreland basin development proceeded in southern Europe, coinciding with a general uplift of the European continent. The closure of the Pindos Ocean began (Robertson et al., 1991). Compression continued in the Balkan area in Bulgaria (Tari et al., 1997).

Oligocene

Collisions continued in the area between Africa and Eurasia. The conclusion of the compression of the Balkanides in Bulgaria occurred during the Oligocene (Sinclair et al., 1997). The collision of Apulia as well as the Alpine–Carpathian terranes (Figure 17) with the European plate continued (Decker and Peresson, 1996). The metamorphism of the undercrusted Penninic nappes

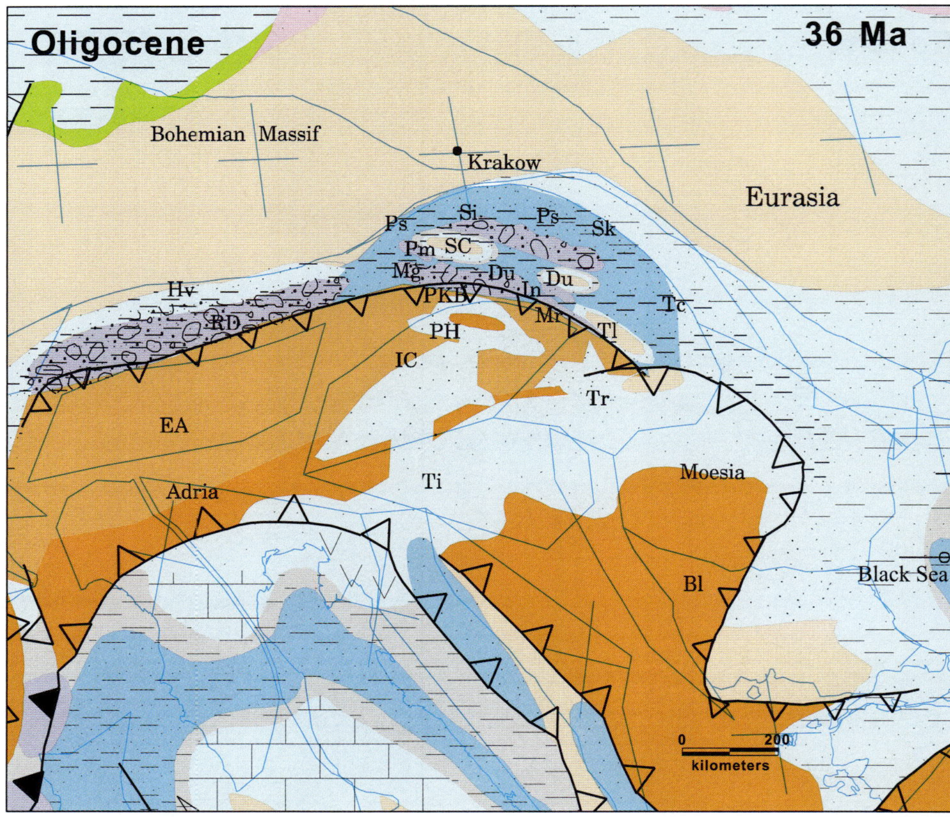

Figure 17. Paleoenvironment and lithofacies of the circum-Carpathian area during Oligocene; plates position at 36 Ma (modified from Golonka et al., 2000). Abbreviations: Bl = Balkan Basin and fold belt; Du = Dukla Basin; EA = Eastern Alps; Hv = Helvetic shelf; IC = Inner Carpathians; In = Inacovce–Kricevo zone; Mg = Magura basin; Mr = Marmarosh Massif; PH = Inner Carpathian (Podhale) Basin; PKB = Pieniny Klippen Belt; Pm = Fore-Magura basin; Ps = Subsilesian ridge and slope zone; RD = Rhenodanubian; SC = Silesian ridge (Cordillera); Si = Silesian basin; Sk = Skole basin; Tc = Tarcau Basin; Ti = Tisza plate; Tl = Teleajen Basin; Tr = Transylvanian basin.

in the Alps reached peak thermal conditions at about 30 Ma (Kurz et al., 1996). In the Carpathians, the subduction consumed part of the Magura basin (Oszczypko, 1998; Oszczypko et al., 2003). After the late Oligocene folding, the Magura nappe thrust northward and covered the remnant of the Silesian ridge and in the more outer part of the Carpathian basin (Dukla, Silesian–Subsilesian, Skole–Tarcau); flysch sedimentation continued during the Oligocene (Figure 17). Initial folding occurred in a part of the Silesian basin (Zytko, 1999a, b). Restricted basin with the sedimentation of the organic-rich Menilitic shales was formed. Sedimentation of the Podhale flysch, covering parts of the Inner Carpathian terrane, was fully developed during the Oligocene (Ksiazkiewicz, 1977b).

The Paratethys sea developed in Europe and central Asia, ahead of the progressing northward orogenic belts (Dercourt et al., 1993). Geodynamic evolution of the basins in the Alpine belt led to a transition from flysch to molasse type of sedimentation.

Rifting events were initialized during the Oligocene in several countries in Europe between France and Ukraine (Ziegler, 1988; Zytko et al., 1989; Wilson and Downes, 1991; Bois, 1993) and were associated with alkaline volcanism. According to Bois (1993), extension occurred in parts of the European plate, with the rifting of the Rhine, Limagne, and Bresse trough, contemporaneous of the climax of the Alpine compression. Part of this rift system included the Gulf of Lions, associated with the mantle plume, as expressed by volcanics in the Massif Central and Provence and on Corsica and Sardinia (Wilson and Downes, 1991). Rifting in this area was followed by oceanic sea-floor spreading and the drifting of the Corsican and Sardinian plates (Bois, 1993; Ricou, 1996). North Sea subsidence, renewed during the Tertiary (Joy, 1992), could have been related to Central European rifting.

Chattian–Burdigalian (Eggerian–Early Karpatian)

This was the time of the major Alpine orogenic phase, the formation of mountains in the Alpine–Carpathian area, the Mediterranean, central Asia, and the Himalayas (Figure 18). The morphostructural character of the Carpathians is a product of Tertiary evolution and finished up by the Neogene (Neoalpine) collisional events, which imprinted the present-day shape to the Carpathian orogenic belt. Acceleration of the ocean-floor spreading in the Atlantic Ocean was the engine of the Alpine orogenesis and led to the convergence of the Afro-Arabian lithospheric plate, including its promontory and North European one. The space in between these two megablocks where Apulia and Alcapa were located was broken up into several translating and

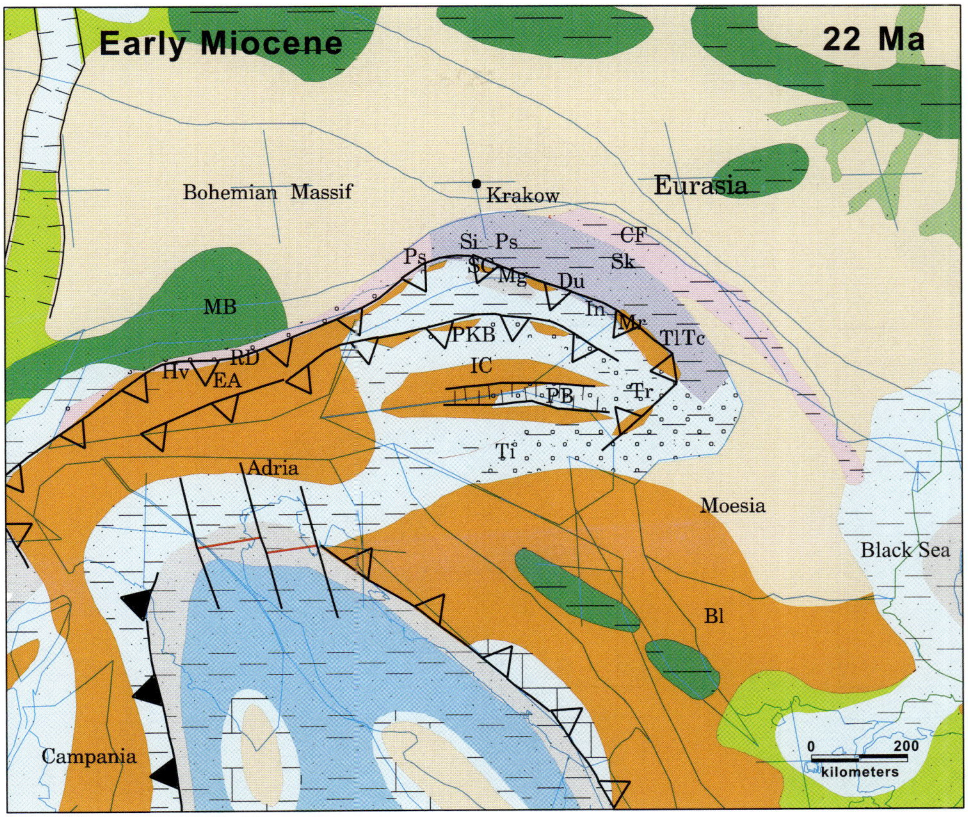

Figure 18. Paleoenvironment and lithofacies of the circum-Carpathian area during early Miocene (Chattian–Aquitanian); plates position at 22 Ma (modified from Golonka et al., 2000). Abbreviations: Bl = Balkan fold belt, CF = Carpathian Foredeep; Du = Dukla Basin; EA = Eastern Alps; Hv = Helvetic shelf; IC = Inner Carpathians; In = Inacovce–Kricevo zone; MB = Molasse basin; Mg = Magura basin; Mr = Marmarosh Massif; PB = Pannonian Basin; PKB = Pieniny Klippen Belt; Ps = Subsilesian ridge and slope zone; RD = Rhenodanubian; SC = Silesian ridge (Cordillera); Si = Silesian basin; Sk = Skole basin; Tc = Tarcau Basin; Ti = Tisza plate; Tl = Teleajen Basin; Tr = Transylvanian basin.

rotating microplates, reflecting squeezing caused by northward propagation of the Africa plate. Within the structure of the Carpathian tectogene, several typical distinct morphotectonic features can be distinguished, such as the arch-shaped orogenic belt, the core mountains alternating with the intramountain basins, the well-developed Miocene volcanic arc, the effects of intraorogen rotations, the Pieniny Klippen (Andrusov, 1938) structure, and fanlike structures. All these phenomena are believed to be products of the youngest, Neogene (Neoalpine) tectonic period. Faulting was a tremendous factor during the Neogene tectonic evolution. A very dense and regular fault net is one of the characteristic features of the Carpathians. Brittle faults, mainly strike slips, in combination with other dynamical tectonic boundaries, allowed propagation of individual detached blocks to the realm of the future Carpathian region (Figure 19).

Collisions continued in the area between Africa and Eurasia during the Chattian–Aquitanian. The movement of Corsica and Sardinia caused the plates to push eastward in the future, resulting in the deformation of the Alpine–Carpathian system. This deformation reached as far as the Romanian Carpathians (Royden, 1988; Ellouz and Roca, 1994) and continued throughout the Neogene. The Calabrian terranes in the western Mediterranean continued to progress eastward (Dewey et al., 1989; Van Dijk and Okkes, 1991). Corsica and Sardinia pushed the Umbria–Marche terrane toward collision with the Apulian block. The thrust and fold belt of the Apennines began to develop (Pialli and Alvarez, 1997).

The Apulia and the Alpine–Carpathian terranes were moving northward, colliding with the European plate, until 17 Ma (Decker and Peresson, 1996). The crust of northwardly propagated Apulia converged with the crust of the Bohemian Massif. Terranes in between Apulia and the stable Europe plate were extremely shortened and uplifted. For example, the lowermost crystalline fundament was exhumed in the Tauern window in the eastern Alpine plate. The eastward movement caused by the Sardinia–Corsica pushing force was combined with the process of lateral extrusion and gravitational spreading of the elastic Carpatho-Pannonian crustal units to the east (Ratschbacher et al., 1991). This so-called tectonic escape to the still free space in the Carpatho-Pannonian realm was controlled by strike-slip faults operating as border faults of crustal wedges moving to the east (Figure 19). The already consolidated paleomesoalpine structures were reworked during this process. Within active strike-slip shear zones in the regime of dextral transpression (Marko et al., 1991), early Miocene wrench furrows (Montenat et al., 1987) were formed at the western and eastern part of

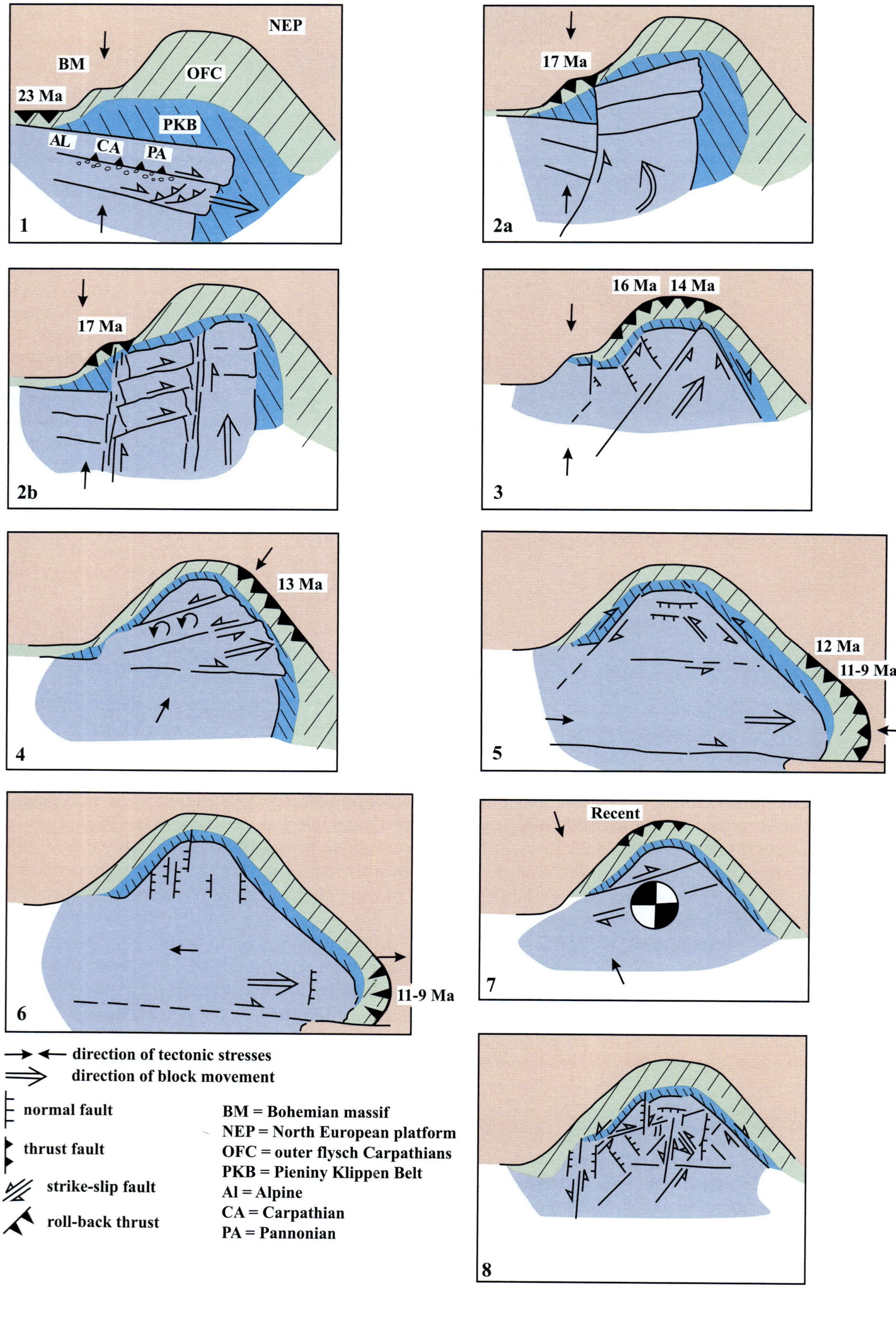
NEP
BM
OFC
23 Ma
PKB
AL
CA
PA
1
17 Ma
2a
17 Ma
2b
16 Ma
14 Ma
3
13 Ma
4
12 Ma
11-9 Ma
5
11-9 Ma
6
Recent
7
8
direction of tectonic stresses
direction of block movement
normal fault
thrust fault
strike-slip fault
roll-back thrust
BM = Bohemian massif
NEP = North European platform
OFC = outer flysch Carpathians
PKB = Pieniny Klippen Belt
Al = Alpine
CA = Carpathian
PA = Pannonian

Slovakia (Vass, 1998). These east–west-trending narrow and long furrows led to the development of basins filled up by poorly sorted and shortly transported sediments of the Eggenburgian–early Karpatian. After the formation of basins, the dextral transpression in strike-slip corridors was converted to a sinistral one, and the basins rotated in the east-northeast–west-southwest direction. This rotation did not exceed 15° in the counterclockwise direction.

The Apulia and the Alpine–Carpathian terranes' collision with the European plate also caused the foreland to propagate north. The north to north-northwest-vergent thrust system of the Eastern Alps was formed. Oblique collision between the north European plate and the overriding Western Carpathian terranes led to the development of the outer accretionary wedge, the buildup of many flysch nappes, and the formation of a foredeep (Kovac et al., 1993, 1998; Slaczka, 1996a, b). These nappes were detached from their original basement and thrust over the Paleozoic–Mesozoic deposits of the North European Platform covered partly by Tertiary deposits (Figure 18). This process was completed in the Vienna basin area and then progressed northeastward (Oszczypko, 1997). After the late Oligocene folding, the Magura nappe was thrust northward in the direction of the terminal Krosno flysch basin (Oszczypko, 1998). The synorogenic basin with the flysch sedimentation Krosno beds formed during the Oligocene as a continuation of the older Fore-Magura (with the outer part of Magura), Dukla, Silesian–Subsilesian, Skole–Tarcau, and units of the Outer Carpathians. Initial folding in this zone also occurred.

During the early Burdigalian, the front of the Magura nappe in the Outer Carpathians reached the southern part of the Silesian basin (Figure 18) (Oszczypko, 1998). This was followed by a progressive migration of axis of subsidence northward. During the course of the Burdigalian transgression, part of the Magura basin was flooded, and the seaway connection with the Vienna basin via Orava was probably established (Oszczypko et al., 2003). The thinned continental crust of the residual flysch basin was underthrust, beneath the overriding Carpathian orogen. This underthrusting was connected with the Intra-Burdigalian folding and uplifting of the Outer Carpathians. The formation of the Western Carpathian thrusts was completed (Kovac et al., 1993, 1998; Slaczka, 1996a, b; Oszczypko, 1997) (Figure 18). The Carpathians continued overriding the Eurasian platform and caused flexural depression, a peripheral foreland basin related to the moving Carpathian front (Oszczypko, 1998). The thrust front was still migrating eastward in the Eastern Carpathians.

Late Burdigalian–Langhian (Late Karpatian–Early Badenian)

The northward movement of the Alpine segment of Carpathian–Alpine orogen was stopped because of a collision with the Bohemian Massif (Figures 19, 20). At the same time, the extruded Carpatho-Pannonian units were pushed to the open space toward the bay of weak crust filled up by Outer Carpathian flysch sediments. The separation of the Carpatho-Pannonian segment from the Alpine one and its propagation to the north were related to the development of north–south dextral strike-slip faults. These faults are known from the Outer Carpathians and postulated in the basement of the Vienna basin below the Neogene sediments (Krs and Roth, 1979; Balla, 1987). The spectacular formation of the oroclinal loop began during this time (Figure 19). The loop has been traditionally regarded as a structure ductile in nature because of the bend shape of the Alpine units. The gradual formation of the Carpathian loop, which is caused by the independent propagation of detached crustal segments to the Carpathian gulf, rimmed by the Bohemian Massif, the Tornquist–Teisseyre line, and Moesia, is also possible.

The Carpathian segment of Alcapa moved to the north during the late Karpatian. The oceanic crust, underneath propagating continental Carpatho-Pannonian crustal blocks, was partially melted and produced early volcanic activity of the Western Carpathians. Distribution of sedimentation depocenters in the Vienna basin shows dramatic change of the structural plan during the late Karpatian. Structural records indicate a change of compression direction from north–south to north-northeast–south-southwest, which resulted in a change of fault-controlled sedimentation. Northeast–southwest-oriented basins overprinted older east–west-trending basins. Northeast–southwest faults were activated as dominant normal and sinistral strike-slip ones. These faults controlled sedimentation in the Vienna basin and simultaneously allowed northeastward escape of detached blocks after their collision with the platform to the north.

Figure 19. Idealized model of detached Carpathian block kinematics during the Tertiary. Succession of events is expressed by numbering of sections (from the oldest event [1] up to the youngest event [7]). Event 2 has two different possibilities (2a, 2b). The age of each event is shown in Ma in the figures; it represents the youngest movements recorded in front of the Carpathians. Section 8 shows the result of several superposed tectonics events reflected in a regular net of map-scale faults, readable from a geological map.

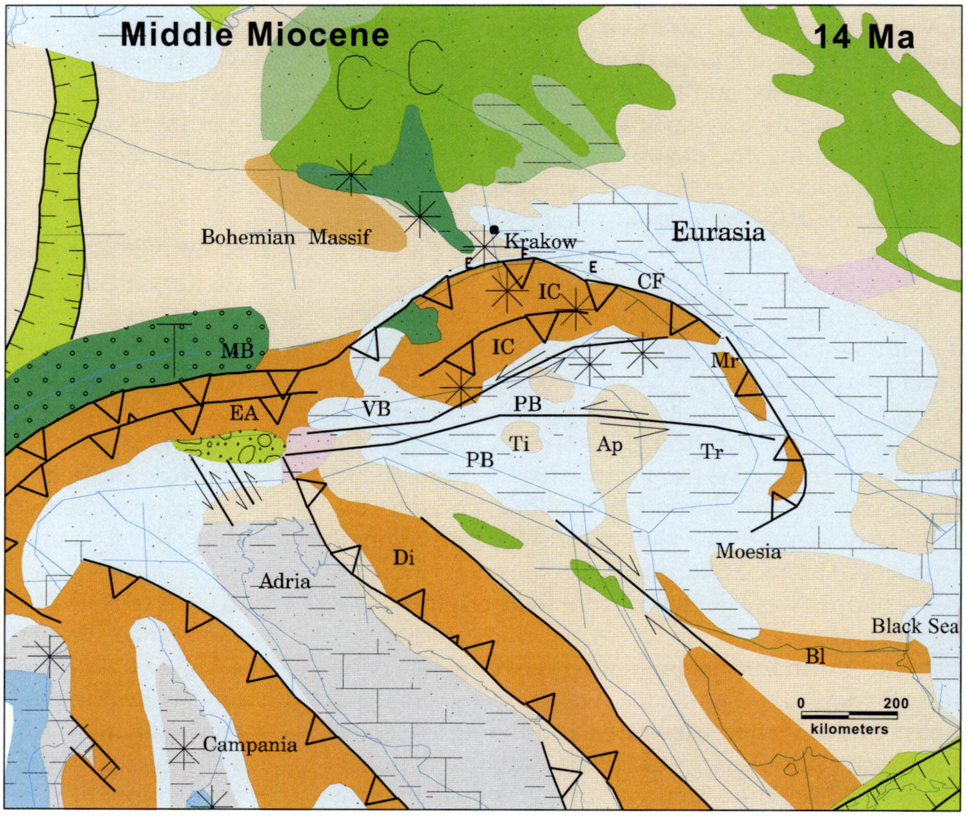

Figure 20. Paleoenvironment and lithofacies of the circum-Carpathian area during the middle Miocene (Burdigalian–Serravallian); plates position at 14 Ma (modified from Golonka et al., 2000). Abbreviations: Ap = Apuseni; Bl = Balkan fold belt; CF = Carpathian Foredeep; Di = Dinarides; EA = Eastern Alps; IC = Inner Carpathians; MB = Molasse basin; Mr = Marmarosh Massif; PB = Pannonian Basin; Ti = Tisza plate; Tr = Transylvanian basin; VB = Vienna basin.

During the late Karpatian, the first core mountains with exhumed crystalline basement emerged, which is demonstrated by fission-track ages (Kral, 1977; Kovac et al., 1994) as well as by perifault debris sedimentation along these uplifted mountains (Vass et al., 1990). The east-northeast–west-southwest-oriented wrench corridor was active as a sinistral transtensional wrench zone. Former, early Miocene wrench furrows were broken up, and separated remnants were rotated approximately 30–40° counterclockwise (Kovac and Tunyi, 1995). These rotations were coeval with rotations described in the area between the Carpathians and the Pannonian Basin (Pelso unit) (Marton and Fodor, 1995).

Serravallian (Badenian)–Holocene

During the Badenian (Serravallian) (Figures 19, 20), the Vienna basin opened because of the pull-apart activity of major northeast–southwest sinistral strike-slip faults (Royden et al., 1982; Royden, 1985). Badenian depocenters of sedimentation showed a north–south trend caused by normal faults that accommodated large magnitudes of strike-slip translations. Nevertheless, the Badenian Vienna does not have all the attributes of a pull-apart basin. The western part of the basin lies over the accretionary flysch wedge, somehow in a piggyback position, whereas the eastern part lies over the Alpine–Carpathian units deformed during the Mesozoic (internides). The Vienna basin displays some attributes of a fore-arc basin related to the Neogene volcanic arc.

The Carpathian foreland basin continued its development partly on the top of the thrust front with mainly terrestrial deposits forming the clastic wedge. This clastic wedge along the Carpathians could be comparable with the lower freshwater molasse of the Alpine foreland basin. During the Serravallian, the marine transgression flooded the foreland basin and adjacent platform.

Crustal extension of the internal zone of the Alps started in the early Miocene during the continued thrusting (Decker and Peresson, 1996). Although transtensional fore-arc basins were formed at the northern edge of the Carpathian internides, early to middle Miocene extension and back-arc-type rifting resulted in the formation of an intramountain Pannonian Basin (Royden, 1988; Decker and Peresson, 1996). The formation of Pannonian Basin is related to thermal lithospheric stretching and volcanism provoked by the heat of rising Pannonian asthenolith, which was formed in the Pannonian area.

According to paleomagnetic data, the last counterclockwise block rotations ceased in the western part of Western Carpathians in the middle Badenian

(Serravallian), whereas the last rotations in the eastern part of the Western Carpathians were recorded at the Tortonian (Sarmatian) (Orlicky, 1996). The northeast–southwest direction of compression and movement of the Carpatho-Pannonian lithospheric blocks is typical for the Serravallian. During this period, a substantial part of volcanic activity had been accomplished, and the Carpatho-Pannonian volcanic arc was formed. A huge bulk of volcanism was a result of the melting of subducted oceanic and quasioceanic crustal slab, as well as further rising of the asthenospheric asthenolith in the Pannonian region (Poka, 1988; Lexa et al., 1993). Crustal stretching caused by asthenosphere heat was accommodated by extensional normal faulting (Tari et al., 1992; Horvath and Cloething, 1996), marking the advanced stage of development of back-arc basins (Pannonian and Danube Basins), under conditions of thermal crustal extension.

After the Serravallian, the sea retreated from the Carpathian peripheral foreland basin. This was followed by the last overthrust of the Carpathians toward their present-day position (Oszczypko, 1998). The Carpathian lithospheric block propagated to the east until the collision with east European platform about 9 m.y. ago. This collision produced an east–west compressional event (Decker and Peresson, 1996), which even affected such distal terranes as the Western Carpathians (Figure 19). The east–west normal faults were reactivated in this stress period and controlled the formation of east–west-trending morphostructures horsts and depressions. During the Tortonian–Gelasian, Carpathian thrusting progressed east and southeastward, with a strong element of translation (Royden, 1988; Ellouz and Roca, 1994; Linzer, 1996). The thrusting was completed during the Pliocene–Quaternary in the Moldavides in Romania (Stefanescu et al., 2006). The eastward movement of the orogen was related to the opening of the Tyrrhenian Sea (Golonka et al., 2000). The extrusion caused by the collision of the Apulian plate with Europe could also be a factor in the eastward movement of the orogen (Decker and Peresson, 1996).

The southwest–northeast direction of the compression and the movement of the Carpatho-Pannonian lithospheric blocks is typical for the Tortonian. An active front of the orogen has been removed far from the Western Carpathian area. The northeast–southwest compression was gradually inverted to extension. The east–west extension was induced in the Carpathian realm because of the westward direction of subducting plate (Royden et al., 1982) and related rollback effect. Numerous new north–south-oriented faults developed, and many old strike-slip faults were reactivated as normal extensional faults in this youngest tensional event. Normal faults were a dominant factor as sedimentation-controlling structures. These faults are the most numerous and conspicuous brittle features in the recent architecture of the Western Carpathians.

Lacustrine modest sedimentation is typical for the Pliocene. Important regional tectonic stresses, generated by the subduction and collision of large lithospheric plates, have ceased. The Carpathian loop, including the Pieniny Klippen Belt structure, has been formed. The structural records in the Western Carpathians and the Pannonian area indicate that approximately north–south (north-northwest–south-southeast)-oriented compression affected the orogen during the last period of the Carpathian evolution (Csontos et al., 1991; Hok et al., 1995) and survived until the Holocene. The earthquakes' focal mechanism analysis of seismogenic faults supports such a recent compression.

ACKNOWLEDGMENTS

This work was partly supported by the grants from the Jagiellonian University and the University of Mining and Metallurgy (grant 11.11.140.888 and 11.11.140.159) and by grants from the Polish Committee for Scientific Research KBN (grant nos. 6 P04D 032 21 and 3 P04D 020 22). We are indebted to Lucyna Dunczyk, Elzbieta Machaniec, and Bartlomiej Rogozinski for their help in preparing the figures. We are grateful to Larry Lawver and Marek Lewandowski for their reviews.

REFERENCES CITED

Andrusov, D., 1938, Geologicky vyskum vnitrniho bradloveho pasma v Zapadnich Karpatech: III. Tektonika: Rozpravy Statniho geologickeho Ustavu CSR, v. 9, 135 p.

Aubrecht, R., 2001, New occurrences of the Krasin Breccia (Pieniny Klippen Belt, West Carpathians): Indication of Middle Jurassic synsedimentary tectonics: Acta Geologica Universitatis Comenianae, v. 56, p. 35–56.

Aubrecht, R., and I. Tunyi, 2001, Original orientation of neptunian dykes in the Pieniny Klippen Belt (Western Carpathians): The first results: Contributions to Geophysics and Geodesy, v. 31, p. 557–578.

Aubrecht, R., M. Misik, and M. Sykora, 1997, Jurassic synrift sedimentation on the Czorsztyn swell of the Pieniny Klippen Belt in Western Slovakia, *in* D. Plasienka, J. Hok, J. Vozar, and M. Ellecko, eds., Alpine evolution of the Western Carpathians: Bratislava, Geological Survey of Slovac Republic, p. 53–64.

Balla, Z., 1987, Tertiary paleomagnetic data for the Carpatho-Pannonian region in the light of Miocene rotation kinematics: Tectonophysics, v. 139, p. 67–98.

Banks, C. J., and A. G. Robinson, 1997, Mesozoic strike-slip back-arc basins of the western Black Sea region, *in* A. G. Robinson, ed., Regional and petroleum geology of the Black Sea and surrounding regions. AAPG Memoir 68, p. 53–61.

Beck, M. E., and E. R. Schermer, 1994, Aegean palaeomagnetic inclination anomalies— Is there a tectonic explanation?: Tectonophysics, v. 231, p. 281–292.

Belka Z., P. Valverde-Vaquero, H. Arendt, W. Dorr, W. Franke, and K. Wemmer, 2000, Accretion of first Gondwana derived terranes at the margin of Baltica— New data and open questions, *in* Joint Meeting of Europrobe (TESZ) and PACE Projects: Zakopane, Holy Cross Mountains, Poland, September 16–23, 2000, p. 8–9.

Besse, J., and V. Courtillot, 1991, Revised and synthetic apparent polar wander paths of the African, Eurasian, North American and Indian plates, and true polar wander since 200 Ma: Journal of Geophysical Research, v. 96, p. 4029–4050.

Besse, J., and V. Courtillot, 2002, Apparent and true polar wander and the geometry of the geomagnetic field over the last 200 Myr: Journal of Geophysical Research, v. 107, p. 6–31.

Besutiu, L., 2001, Moesia— A Baltica derived terrane?, *in* A. Adam, L. Szarka, and J. Szendroi, eds., Pancardi 2001 II Abstracts: Sopron, Hungary, Hungarian Academy of Sciences, p. CP29.

Bieda, F., S. Geroch, L. Koszarski, M. Ksiazkiewicz, and K. Zytko, 1963, Stratigraphie des Karpates externes polonaises: Biuletyn Instytutu Geologicznego, v. 182, p. 5–174.

Bill, M., L. O'Dogherty, J. Guex, P. O. Baumgartner, and H. Masson, 2000, Radiolarite ages in Alpine–Mediterranean ophiolites: Constraints on the oceanic spreading and the Tethys–Atlantic connection: Geological Society of America Bulletin, v. 113, p. 129–143.

Birkenmajer, K., 1977, Jurassic and Cretaceous lithostratigraphic units of the Pieniny Klippen Belt, Carpathians, Poland: Studia Geologica Polonica, v. 45, p. 1–158.

Birkenmajer, K., 1986, Stages of structural evolution of the Pieniny Klippen Belt, Carpathians: Studia Geologica Polonica, v. 88, p. 7–32.

Birkenmajer, K., 1988, Exotic Andrusov Ridge: Its role in plate-tectonic evolution of the West Carpathian fold belt: Studia Geologica Polonica, v. 91, p. 7–37.

Birkenmajer, K., H. Kozur, and R. Mock, 1990, Exotic Triassic pelagic limestone pebbles from the Pieniny Klippen Belt of Poland: A further evidence for early Mesozoic rifting in West Carpathians: Annales Societatis Geologorum Poloniae, v. 60, p. 3–44.

Bois, C., 1993, Initiation and evolution of the Oligo-Miocene rift basins of southwestern Europe: Contribution of deep seismic reflection profiling: Tectonophysics, v. 226, p. 227–252.

Bombita, G., E. Antonescu, E. Malata, C. Muller, and T. Neagu, 1992, Pieniny-type Mesozoic formation from Maramures, Romania (second part): Acta Geologica Hungarica, v. 35, p. 117–144.

Bond, G. C., P. A. Nickeson, and N. A. Kominz, 1984, Breakup of a supercontinent between 625 Ma and 555 Ma: New evidence and implications for continental histories: Earth and Planetary Science Letters, v. 70, p. 325–345.

Burtan, J., J. Chowaniec, and J. Golonka, 1984, Wstępne wyniki badań nad egzotycznymi skałami węglanowymi z zachodniej części polskich Karpat fliszowych; Preliminary results of studies on exotic carbonate rocks in the western part of the Polish Flysch Carpathians (in Polish with English summary): Biuletyn Instytutu Geologicznego, v. 346, p. 147–159.

Catalano, R., P. Di Stefano, and H. Kozur, 1991, Permian circum-Pacific deep-water faunas from the western Tethys (Sicily, Italy): New evidence for the position of the Permian Tethys: Palaeogeography, Palaeoclimatology, Palaeoecology, v. 87, p. 75–108.

Chalupsky, J., 1989, Major tectonostratigraphic units of the Bohemian Massif, *in* R. D. Dallmeyer, ed., Terranes in the circum-Atlantic Paleozoic orogens: Geological Society of America Special Paper 230, p. 101–114.

Channell, J. E. T., 1996, Paleomagnetism and paleogeography of Adria, *in* A. Morris and D. H. Tarling, eds., Paleomagnetism and tectonics of the Mediterranean region: Geological Society (London) Special Publication 105, p. 119–132.

Channell, J. E. T., and H. Kozur, 1997, How many oceans? Meliata, Vardar, and Pindos oceans in Mesozoic Alpine paleogeography: Geology, v. 25, p. 183–186.

Channell, J. E. T., R. Branber, A. Spieler, and J. S. Stoner, 1992, Paleogeomagnetism and paleogeography of the Northern Calcareous Alps (Austria): Tectonics, v. 11, p. 792–810.

Channell, J. E. T., O. Tuysuz, O. Bektas, and A. M. C. Sengor, 1996, Jurassic–Cretaceous paleomagnetism and paleogeography of the Pontides (Turkey): Tectonics, v. 15, p. 201–212.

Compston, W., M. S. Sambridge, R. F. Reinfrank, M. Moczydlowska, G. Vidal, and S. Claesson, 1995, Numerical ages of volcanic rocks and the earliest faunal zone within the late Precambrian of east Poland: Journal of the Geological Society (London), v. 152, p. 599–611.

Csontos, L., G. Tari, F. Bergerat, and L. Fodor, 1991, Evolution of the stress fields in the Carpatho-Pannonian area during the Neogene: Tectonophysics, v. 199, p. 73–91.

Dadlez, R., M. Narkiewicz, R. A. Stephenson, M. T. M. Visser, and J. D. van Wees, 1995, Tectonic evolution of the Mid-Polish Trough: Modelling implications and significance for Central European geology: Tectonophysics, v. 252, p. 179–195.

Dallmeyer, R. D., F. Neubauer, R. Handler, H. Fritz, W. Mueller, D. Pana, and M. Putis, 1996, Tectonothermal evolution of the internal Alps and Carpathians; evidence from $^{40}Ar/^{39}Ar$ mineral and whole-rock data, *in* S. M. Schmid, M. Frey, N. Froitzheim, R. Heilbronner, and H. Stuenitz, eds., Alpine geology: Proceedings of the Second Workshop on Alpine Geology: Eclogae Geologicae Helvetiae, v. 89, p. 203–227.

Dalziel, I. W. D., 1997, Neoproterozoic–Paleozoic geography and tectonics: Review, hypothesis, environmental speculation: Geological Society of America Bulletin, v. 109, p. 16–42.

Dalziel, I. W. D., L. H. Dalla Salda, and L. M. Gahagan, 1994,

Paleozoic Laurentia–Gondwana interaction and the origin of the Appalachian–Andean mountain system: Geological Society of America Bulletin, v. 106, p. 243–252.

Debelmas, J., 1989, On some key features of the evolution of the Western Alps, *in* A. M. C. Sengor, Y. Yilmaz, A. I. Okay, and N. Gorur, eds., Tectonic evolution of the Tethyan region: North Atlantic Treaty Organization Advanced Study Institute Series. Series C: Mathematical and Physical Sciences, v. 259, p. 23–42.

Decker, H., and H. Peresson, 1996, Tertiary kinematics in the Alpine–Carpathian–Pannonian system: Links between thrusting, transform faulting and crustal extension, *in* G. Wessely and W. Liebl, eds., Oil and gas in Alpidic thrust belts and basins of Central and Eastern Europe: European Association of Geoscientists and Engineers Special Publication 5, p. 17–21.

Dercourt, J., et al., 1984, Presentation of 9 paleogeographic maps at 20,000,000 scale from the Atlantic to the Pamir between Lias and Present: Bulletin de la Societe Geologique de France, v. 8, p. 637–652.

Dercourt, J., L. E. Ricou, and B. Vrielynck, eds., 1993, Atlas Tethys paleoenvironmental maps: Paris, Gauthier-Villars, 307 p.

Dercourt, J., M. Gaetani, B. Vrielynck, E. Barrier, B. Biju-Duval, M.-F. Brunet, J. P. Cadet, S. Crasquin, and M. Sandulescu, eds., 2000, Atlas Peri-Tethys paleoegeographical maps: Paris, Commission for the Geological Map of the World, 24 maps and explanatory note, I–XX, p. 1–269.

Dewey, J. F., M. L. Helman, E. Turco, D. H. W. Hutton, and D. Knott, 1989, Kinematics of the western Mediterranean, *in* M. P. Coward, D. Dietrich, and R. G. Park, eds., Alpine tectonics: Geological Society (London) Special Publication 45, p. 265–283.

Dore, A. G., 1991, The structural foundation and evolution of Mesozoic seaways between Europe and the Arctic: Palaeogeography, Palaeoclimatology, Palaeoecology, v. 87, p. 441–492.

Edel, J. B., and K. Weber, 1995, Cadomian terranes, wrench faulting and thrusting in the Central Europe Variscides: Geophysical and geological evidence: Geologische Rundschau, v. 84, p. 412–432.

Elias, M., and M. Eliasova, 1984, Facies and palaeogeography of the Jurassic in the western part of the Outer Flysch Carpathians in Czechoslovakia: Sbornik Geologickich Ved, v. 39, p. 105–170.

Ellouz, N., and E. Roca, 1994, Palinspastic reconstruction of the Carpathians and adjacent areas since the Cretaceous: A quantitative approach, *in* F. Roure, ed., Peri-Tethyan platform: Paris, Editiones Technip, p. 51–77.

Faryad, S. W., 1997, Petrological model for blueschist facies metamorphism in the Pieniny Klippen Belt, *in* P. Grecula, D. Hovorka, and M. Putis, eds., Geological evolution of the Western Carpathians: Bratislava, Mineralia Slovaca Monograph, p. 155–162.

Feinberg, H., B. Edel, D. Kondopoulou, and A. Michard, 1996, Implications of ophiolite palaeomagnetism for the interpretation of the geodynamics of northern Greece, *in* A. Morris and D. H. Tarling, eds., Paleomagnetism and tectonics of the Mediterranean region: Geological Society (London) Special Publication 105, p. 289–298.

Finger, F., P. Hanzl, C. Pin, A. Von Qyadt, and H. Steyer, 2000, The Brunovistulicum: Avalonian Precambrian sequence at the eastern end of the Central European Variscides?, *in* W. Franke, W. Haak, V. Oncken, and D. Tanner, eds., Orogenic processes: Quantification and modelling in the Variscan belt: Geological Society (London) Special Publication 179, p. 103–112.

Ford, D., and J. Golonka, 2003, Phanerozoic paleogeography, paleoenvironment and lithofacies maps of the circum-Atlantic margins, *in* J. Golonka, ed., Thematic set on paleogeographic reconstruction and hydrocarbon basins: Atlantic, Caribbean, South America, Middle East, Russian Far East, Arctic: Marine and Petroleum Geology, v. 20, p. 249–285.

Franke, W., 1989a, Tectonostratigraphic units in the Variscan belt of Central Europe, *in* R. D. Dallmeyer, ed., Terranes in the circum-Atlantic Paleozoic orogens: Geological Society of America Special Paper 230, p. 67–90.

Franke, W., 1989b, Variscan plate tectonics in Central Europe — Current ideas and open questions: Tectonophysics, v. 169, p. 221–228.

Franke, W., 1992, Phanerozoic structures and events in Central Europe, *in* D. Blundell, R. Freeman, and S. Mueller, eds., The European geotraverse, a continent revealed: Cambridge, University of Cambridge, p. 164–180.

Friedl, G., F. Finger, N. J. McNaughton, and I. R. Fletcher, 2000, Deducing the ancestry of terranes: SHRIMP evidence for South America-derived fragments in Central Europe: Geology, v. 28, p. 1035–1038.

Froitzheim, N., S. M. Schmid, and M. Frey, 1996, Mesozoic paleogeography and the timing of eclogite facies metamorphism in the Alps: A working hypothesis: Eclogae Geologicae Helvetiae, v. 89, p. 81–110.

Gahagan, L. M., et al. 1988, Tectonic fabric map of the ocean basins from satellite altimetry data: Tectonophysics, v. 155, p. 1–26.

Gaweda, A., K. Kozlowski, and K. Piotrowska, 1998, Tectonic development of the crystalline basement of the Polish part of the Western Tatra Mountains: Acta Universitatis Carolinae. Geologica, v. 42, p. 252–253.

Gaweda, A., J. Lefeld, M. Michalik, and A. Uchman, 2003, Inner Carpthians: Tatra Mountains, *in* J. Golonka and M. Lewandowski, eds., Geology, geophysics, geothermics and deep structure of the West Carpathians and their basement: Institute of Geophysics Polish Academy of Sciences Publication, Monographic Volume, v. M-28, no. 363, p. 57–63.

Golonka, J., 2000, Cambrian–Neogene plate tectonic maps: Krakow, Wydawnictwa Uniwersytetu Jagiellonskiego, p. 1–125.

Golonka, J., 2002, Plate-tectonic maps of the Phanerozoic, *in* W. Kiessling, E. Flugel, and J. Golonka, eds., Phanerozoic reef patterns: SEPM Special Publication 72, p. 21–75.

Golonka, J., 2004, Plate tectonic evolution of the southern margin of Eurasia in the Mesozoic and Cenozoic: Tectonophysics, v. 381, p. 235–273.

Golonka, J., and N. Y. Bocharova, 2000, Hot spot activity the break-up of Pangea: Palaeogeography, Palaeoclimatology, Palaeoecology, v. 161, p. 49–69.

Golonka, J., and D. Ford, 2000, Pangean (Late Carboniferous–Middle Jurassic) paleoenvironment and lithofacies: Palaeogeography, Palaeoclimatology, Palaeoecology, v. 161, p. 1–34.

Golonka, J., and L. Gahagan, 1997, Tectonic model of the Mediterranean terranes: AAPG Bulletin, v. 8, p. 1386.

Golonka, J., and M. Krobicki, 2001, Upwelling regime in the Carpathian Tethys: A Jurassic–Cretaceous palaeogeographic and paleoclimatic perspective: Geological Quarterly, v. 45, p. 15–32.

Golonka, J., and M. Krobicki, 2004, Jurassic paleogeography of the Pieniny and Outer Carpathian basins: Rivista Italiana di Paleontologia e Stratigrafia, v. 110, p. 5–14.

Golonka, J., and M. Lewandowski, eds., 2003, Geology, geophysics, geothermics and deep structure of the West Carpathians and their basement: Institute of Geophysics Polish Academy of Sciences Publication, Monographic Volume, v. M-28, no. 363, p. 1–184.

Golonka, J., and W. Sikora, 1981, Microfacies of the Jurassic and Lower Cretaceous sedimentarily thinned deposits of the Pieniny Klippen Belt in Poland (in Polish with English abstract): Biuletyn Instytutu Geologicznego, v. 31, p. 7–37.

Golonka, J., and A. Slaczka, 2003, Geodynamic evolution stages and climate changes in the Outer Carpathians: Mineralia Slovaca, v. 35, p. 11–14.

Golonka, J., M. I. Ross, and C. R. Scotese, 1994, Phanerozoic, paleogeographic, and paleoclimatic modeling maps, *in* A. F. Embry, B. Beauchamp, and D. J. Glass, eds., Pangea: Global environments and resources: Canadian Society of Petroleum Geologists Memoir 17, p. 1–47.

Golonka, J., M. Edrich, D. W. Ford, R. B. Pauken, N. Y. Bocharova, and C. R. Scotese, 1996, Jurassic paleogeographic maps of the world, *in* M. Morales, ed., The continental Jurassic: Museum of Northern Arizona Bulletin, v. 60, p. 1–5.

Golonka, J., N. Oszczypko, and A. Slaczka, 2000, Late Carboniferous–Neogene geodynamic evolution and paleogeography of the circum-Carpathian region and adjacent areas: Annales Societatis Geologorum Poloniae, v. 70, p. 107–136.

Golonka, J., M. Krobicki, N. Oszczypko, A. Slaczka, and T. Slomka, 2003, Geodynamic evolution and palaeogeography of the Polish Carpathians and adjacent areas during Neo-Cimmerian and preceding events (latest Triassic–earliest Cretaceous), *in* T. McCann and A. Saintot, eds., Tracing tectonic deformation using the sedimentary record: Geological Society (London) Special Publication 208, p. 138–158.

Golonka, J., A. Slaczka, and F. Picha, 2006, The West Carpathians and the Ouachitas: A comparative study of geodynamic evolution, *in* J. Golonka and F. J. Picha, eds., The Carpathians and their foreland: Geology and hydrocarbon resources: AAPG Memoir 84, p. 787–810.

Grabowski, J., 2000, Palaeo- and rock magnetism of Mesozoic carbonate rocks in the Sub-Tatric series (Central West Carpathians) palaeotectonic implications: Polish Geological Institute Special Paper 5, p. 1–88.

Haubold, H., R. Scholger, W. Frisch, H. Summesberger, and H. J. Mauritsch, 1999, Reconstruction of the geodynamic evolution of the Northern Calcareous Alps by means of paleomagnetism: Physics and Chemistry of the Earth. Part A: Solid Earth and Geodesy, v. 24, p. 697–703.

Hok, J., L. Simon, P. Kovac, M. Elecko, D. Vass, F. Verbich, and J. Halmo, 1995, Tectonic of the Hornonitrianska kotlina Basin during Tertiary: Geologica Carpathica, v. 46, p. 191–196.

Horvath, F., and S. Cloetingh, 1996, Stress-induced late-stage subsidence anomalies in the Pannonian Basin: Tectonophysics, v. 266, p. 287–300.

Joy, A. M., 1992, Right place, wrong time; anomalous post-rift subsidence in sedimentary basins around the North Atlantic Ocean, *in* B. C. Storey, T. Alabaster, and R. J. Pankhurst, eds., Magmatism and the causes of continental break-up: Geological Society (London) Special Publication 68, p. 387–393.

Kah, L. C., and J. K. Bartley, 2001, Rodinia and the Mesoproterozoic earth–ocean system: Precambrian Research, v. 111, p. 1–3.

Kalvoda, J., 2001, Upper Devonian–Lower Carboniferous foraminiferal paleobiogeography and Perigondwana terranes at the Baltica–Gondwana interface: Geologica Carpathica, v. 52, p. 205–215.

Kalvoda, J., J. Leichmann, O. Babek, and R. Melichar, 2003, Brunovistulian terrane (Central Europe) and Istanbul zone (NW Turkey): Late Proterozoic and Paleozoic tectonostratigraphic development and paleogeography: Geologica Carpathica, v. 54, p. 139–152.

Kazmin, V. G., 1990, Early Mesozoic reconstruction of the Black Sea–Caucasus region: Evolution of the northern margin of the Tethys: Memoire Societe Geologique de France, Nouvelle Series, v. 54, p. 147–158.

Kazmin, V. G., 1991, Collision and rifting in the Tethys Ocean: Geodynamic implications: Tectonophysics, v. 123, p. 371–384.

Kazmin, V. G., I. M. Sbortshikov, L.-E. Ricou, L. P. Zonenshain, J. Boulin, and A. L. Knipper, 1986, Volcanic belts as markers of the Mesozoic–Cenozoic active margin of Eurasia: Tectonophysics, v. 12, p. 123–152.

Keppie, J. D., R. D. Nance, J. B. Murphy, and J. Dostal, 2003, Tethyan, Mediterranean, and Pacific analogues for the Neoproterozoic–Paleozoic birth and development of peri-Gondwanan terranes and their transfer to Laurentia and Laurussia: Tectonophysics, v. 365, p. 195–219.

Kiessling, W., E. Flugel, and J. Golonka, 1999, Paleoreef maps: A comprehensive database of Phanerozoic reefs with graphic presentations: AAPG Bulletin, v. 83, p. 1552–1587.

Kondopolou, D., A. Atzemoglou, and S. Pavlides, 1996, Palaeomagnetism as a tool for testing geodynamic models in the North Aegean: Convergences, controversies and a further hypothesis, *in* A. Morris and D. H. Tarling, eds., Palaeomagnetism and tectonics of the Mediterranean region:

Geological Society (London) Special Publication 105, p. 277–288.

Kovac, M., and I. Tunyi, 1995, Interpretácia paleomagnetických údajov zo západnej časti Centrálnych Západných Karpát: Mineralia Slovaca, v. 27, p. 213–220.

Kovac, M., A. Nagymarosy, J. Sotak, and K. Sutovska, 1993, Late Tertiary paleogeographic evolution of the Western Carpathians: Tectonophysics, v. 226, p. 401–415.

Kovac, M., J. Kral, E. Marton, D. Plasienka, and P. Uher, 1994, Alpine uplift history of the Central Western Carpathians: Geochronological, paleomagnetic, sedimentary and structural data: Geologica Carpathica, v. 45, p. 83–96.

Kovac, M., A. Nagymarosy, N. Oszczypko, A. Slaczka, L. Csontos, M. Marunteanu, L. Matenco, and M. Marton, 1998, Palinspastic reconstruction of the Carpathian–Pannonian region during the Miocene, *in* M. Rakus, ed., Geodynamic development of the Western Carpathians: Bratislava, Geological Survey of Slovac Republic, p. 189–217.

Kovacs, S., J. Haas, G. Csaszar, T. Szederkenyi, G. Buda, and A. Nagymarosy, 2000, Tectonostratigarphic terranes in the pre-Neogene basement of the Hungarian part of the Pannonian area: Acta Geologica Hungarica, v. 43, p. 225–328.

Kozur, H., 1991, The evolution of the Meliata–Halstatt Ocean and its significance for the early evolution of the Eastern Alps and Western Carpathians: Palaeogeography, Palaeoclimatology, Palaeoecology, v. 87, p. 109–130.

Kozur, H., and W. Krahl, 1987, Erster Nachweis von Radiolarien in tethyalen Perm Europas: Neues Jahrbuch fur Mineralogie, Geologie und Paleontologie, Abhandlungen, v. 174, p. 357–372.

Kral, J., 1977, Fission track ages of apatites from some granitoid rocks in West Carpathians: Geologicky Zbornik, Geologica Carpathica, v. 28, p. 269–276.

Krautner, H. G., 1996, Alpine rifting, subduction and collision in the Romanian Carpathians, *in* G. Amman, K. W. Handler, and H. P. Steyrer, eds., Abstracts VI Symposium fur Tektonik Struktur- und Kristallingeologie: University of Salzburg, University Press, p. 230–234.

Krobicki, M., 1994, Stratigraphic significance and palaeoecology of the Tithonian–Berriasian brachiopods in the Pieniny Klippen Belt, Carpathians, Poland: Studia Geologica Polonica, v. 106, p. 89–156.

Krobicki, M., 1996, Neo-Cimmerian uplift of intraoceanic Czorsztyn pelagic swell (Pieniny Klippen Belt, Polish Carpathians) indicated by the change of brachiopod assemblages, *in* A. C. Riccardi, ed., Advances in Jurassic Research: GeoResearch Forum, v. 1–2, p. 255–264.

Krobicki M., R. Aubrecht, and J. Golonka, 2003a, Pieniny Klippen Belt, *in* J. Golonka and M. Lewandowski, eds., Geology, geophysics, geothermics and deep structure of the West Carpathians and their basement: Institute of Geophysics Polish Academy of Sciences Publication, Monographic Volume, v. M-28, no. 363, p. 25–33.

Krobicki, M., S. S. Kruglov, B. A. Matyja, A. Wierzbowski, R. Aubrecht, A. Bubniak, and I. Bubniak, 2003b, Relation between Jurassic klippen successions in the Polish and Ukrainian parts of the Pieniny Klippen Belt: Mineralia Slovaca, v. 35, p. 56–58.

Krs, M., and Z. Roth, 1979, Insubrický karpatský tret'ohorný blokový systém, jeho vznik a rozloženie: Geologicky Zbornik, Geologica Carpathica, v. 30, p. 3–17.

Krs, M., M. Krsova, and P. Pruner, 1996, Palaeomagnetism and palaeogeography of the Western Carpathians from the Permian to the Neogene, *in* A. Morris and D. H. Tarling, eds., Palaeomagnetism and tectonics of the Mediterranean region: Geological Society (London) Special Publication 105, p. 175–184.

Ksiazkiewicz, M., 1960, Outline of the paleogeography in the Polish Carpathians: Prace Instytutu Geologicznego, v. 30, p. 209–231.

Ksiazkiewicz, M., ed., 1962, Geological atlas of Poland: Stratigraphic and facial problems: Warsaw, Poland, Instytut Geologiczny, Wydawnictwa Geologiczne, 24 p.

Ksiazkiewicz, M., 1977a, Hypothesis of plate tectonics and the origin of the Carpathians: Annales Societatis Geologorum Poloniae, v. 47, p. 329–353.

Ksiazkiewicz M., 1977b, Tectonics of the Carpathians, *in* W. Pozaryski, ed., Geology of Poland, v. IV: Tectonics, Warsaw, Poland, Wydawnictwa Geologiczne, p. 476–604.

Kurz, W., F. Neubauer, and J. Genser, 1996, Kinematics of Penninic nappes (Glockner nappe and basement-cover nappes) in the Tauern window (Eastern Alps, Austria) during subduction and Penninic–Austroalpine collision: Eclogae Geologicae Helvetiae, v. 89, p. 573–605.

Lamarche, J., J. L. Mansy, F. Bergerat, O. Averbuch, M. Hakenberg, M. Lewandowski, E. Stupnicka, J. Swidrowska, B. Wajsprych, and J. Wieczorek, 1999, Variscan tectonics in the Holy Cross Mountains (Poland) and the role of structural inheritance during Alpine tectonics: Tectonophysics, v. 313, p. 171–186.

Lashkevitsch, Z. M., A. P. Medvedev, and Y. Z. Krupskiy, 1995, Tectonomagmatic evolution of Carpathians (in Russian): Kiev, Naukova Dumka, p. 1–113.

Lewandowski, M., 1993, Paleomagnetism of the Paleozoic rocks of the Holy Cross Mountains (Central Poland) and the origin of the Variscan orogen: Institute of Geophysics Polish Academy of Sciences Publication, v. A-23, p. 3–84.

Lewandowski, M., 1997, Paleomagnetism of early Paleozoic terranes between Baltica and Avalonia: Terra Nostra, v. 97, p. 69–72.

Lewandowski, M., 1998, Assembly of Pangea: Combined paleomagnetic and paleoclimatic approach, *in* M. Ginter and M. H. Wilson, eds., Circum-Arctic Paleozoic faunas and facies: Ichtyolith Issues Special Publication 4, p. 29–32.

Lewandowski, M., 2003, Assembly of Pangea: Combined paleomagnetic and paleoclimatic approach: Advances in Geophysics, v. 46, p. 199–236.

Lewandowski, M., M. Krobicki, B. A. Matyja, and A. Wierzbowski, 2003, Palaeogeography of the Pieniny Klippen Basin: Middle Jurassic to Early Cretaceous evolution on the basis of palaeomagnetic data from the Kamenets section (Transcarpathian Ukraine) (in Polish only): Tomy Jurajskie, v. 1, p. 29–34.

Lexa, J., V. Konecny, M. Kaliciak, and V. Hojstricova, 1993, Distribúcia vulkanitov karpato-panónskeho regiónu v priestore a čase, *in* M. Rakus, ed., Geodynamický model a hlbinná stavba Zapadných Karpát, Konference-symposium-seminar: Bratislava, Geologicky Ustav Dionyza Stura, p. 57–69.

Linzer, H.-G., 1996, Kinematics of retreating subduction along the Carpathian arc, Romania: Geology, v. 24, p. 167–170.

Lucinska-Anczkiewicz, A., A. Slaczka, I. M. Villa, and R. Anczkiewicz, 2000, $^{39}Ar/^{40}Ar$ dating of the teschenite association rocks from the Polish Outer Carpathians: Mineralogical Society of Poland Special Paper 17, p. 241–242.

Lustrino, M., 2000, Phanerozoic geodynamic evolution of the circum-Italian realm: International Geology Review, v. 42, p. 724–757.

Mahel, M., et al., 1968, Regional geology of Czechoslovakie: Part II. The West Carpathians: Praha, Geological Survey of Czechoslovakia, p. 1–723.

Marchant, R. H., and G. M. Stampfli, 1997, Subduction of continental crust in the Western Alps: Tectonophysics, v. 269, p. 217–235.

Marko, F., L. Fodor, and M. Kovac, 1991, Miocene strike-slip faulting and block rotation in Brezovske Karpaty Mountains (Western Carpathians): Mineralia Slovaca, v. 23, p. 189–200.

Marsella, E., H. Kozur, and B. D'Argenio, 1993, Monte Facito Formation (Scythian–middle Carnian). A deposit of the ancestral Lagonegro Basin in southern Apennines: Bolletino de Servisio Geologico Italia, v. 119, p. 225–248.

Marton, E., and L. Fodor, 1995, Combination of paleomagnetic and stress data— A case studies from north Hungary: Tectonophysics, v. 242, p. 99–114.

Marton, E., and P. Martin, 1996, Large-scale rotations in north Hungary during the Neogene as indicated by palaeomagnetic data, *in* A. Morris and D. H. Tarling, eds., Paleomagnetism and tectonics of the Mediterranean region: Geological Society (London) Special Publication 105, p. 153–173.

Marton, E., L. Mastella, and A. K. Tokarski, 1999, Large counterclockwise rotation of the Inner West Carpathian Paleogene flysch— Evidence from paleomagnetic investigations of the Podhale flysch (Poland): Physics and Chemistry of the Earth, Part A: Solid Earth and Geodesy, v. 24, p. 645–649.

Marton, E., J. Kuhlemann, W. Frisch, and I. Dunkl, 2000, Miocene rotations in the Eastern Alps— Palaeomagnetic results from intramontane basin sediments: Tectonophysics, v. 323, p. 163–182.

Mauritsch, H. J., R. Scholger, S. L. Bushati, and H. Ramiz, 1995, Palaeomagnetic results from southern Albania and their significance for the geodynamic evolution of the Dinarides, Albanides and Hellenides: Tectonophysics, v. 242, p. 5–18.

Mauritsch, H. J., R. Scholger, S. L. Bushati, and A. Xhomo, 1996, Palaeomagnetic investigations in northern Albania and their significance for the geodynamic evolution of the Adriatic Aegean realm, *in* A. Morris and D. H. Tarling, eds., Paleomagnetism and tectonics of the Mediterranean region: Geological Society (London) Special Publication 105, p. 265–275.

McKerrow, W. S., J. F. Dewey, C. R. Scotese, 1991, The Ordovician and Silurian development of the Iapetus Ocean, *in* M. G. Bassett, P. D. Lane, and D. Edwards, eds., The Murchison Symposium; Proceedings of an International Conference on the Silurian System: Special Papers in Palaeontology, v. 44, p. 165–178.

Meert, J. G., 2003, A synopsis of events related to the assembly of eastern Gondwana: Tectonophysics, v. 362, p. 1–40.

Meert, J. G., and T. H. Torsvik, 2003, The making and unmaking of a supercontinent: Rodinia revisited: Tectonophysics, v. 375, p. 261–288.

Michalik, J., 1994, Notes on the paleogeography and paleotectonics of the Western Carpathian area during the Mesozoic: Mitteilungen der Osterreichischen Geologischen Gesellschaft, v. 86, p. 101–110.

Michalik, J., and D. Rehakova, 1995, Sedimentary records of Early Cretaceous tectonic activity in the Alpine–Carpathian region: Slovak Geological Magazine, v. 2, p. 159–164.

Michalik, J., D. Rehakova, and Z. Vasicek, 1995, Early Cretaceous sedimentary changes in West Carpathian area: Geologica Carpathica, v. 46, p. 285–296.

Michalik, J., D. Rehakova, and J. Jablonsky, 1996, Geodynamic setting of fluxoturbidites in West Carpathian Upper Jurassic and Lower Cretaceous sedimentary basins: Slovak Geological Magazine, v. 3–4, p. 325–329.

Mock, R., M. Sykora, R. Aubrecht, L. Ozvoldova, B. Kronome, P. Reichwalder, and J. Jablonsky, 1998, Petrology and stratigraphy of the Meliaticum near the Meliata and Jaklovce villages, Slovakia: Slovak Geological Magazine, v. 4, p. 223–260.

Moczydlowska, M., 1997, Proterozoic and Cambrian successions in upper Silesia: An Avalonian terrane in southern Poland: Geological Magazine, v. 134, p. 679–689.

Montenat, C., O. D'Estvou, and P. Masse, 1987, Tectonic-sedimentary characters of the Betic Neogene basins evolving in a crustal transcurrent shear zone (SE Spain): Pau, France, Societe Nationale Elf Aquitaine, p. 1–22.

Muttoni, G., M. Gaetani, K. Budurov, I. Zagorchev, E. Trifonova, D. Ivanova, L. Petrounova, W. Lowrie, 2000a, Middle Triassic paleomagnetic data from northern Bulgaria: Constraints on Tethyan magnetostratigraphy and paleogeography: Palaeogeography, Palaeoclimatology, Palaeoecology, v. 160, p. 223–237.

Muttoni, G., L. Lanci, A. Argnani, A. M. Hirt, U. Cibin, W. Abrahamsen, and W. Lowrie, 2000b, Paleomagnetic evidence for a Neogene two-phase counterclockwise tectonic rotation in the northern Apennines (Italy): Tectonophysics, v. 326, p. 241–253.

Nance, R. D., J. B. Murphy, and J. D. Keppie, 2002, A Cordilleran model for the evolution of Avalonia: Tectonophysics, v. 352, p. 11–31.

Narebski, W., 1990, Early rifts in the evolution of western part of the Carpathians; geochemical evidence from limburgite and teschenite rock series: Geologicky Zbornik, Geologica Carpathica, v. 41, p. 521–528.

Neubauer, F., 2002, Evolution of late Neoproterozoic to early Paleozoic tectonic elements in central and southeast

European Alpine mountain belts: Review and synthesis: Tectonophysics, v. 352, p. 87–103.

Nikishin, A. M., et al., 1996, Late Precambrian to Triassic history of the east European craton: Dynamics of sedimentary basin evolution: Tectonophysics, v. 268, p. 23–63.

Okay, A. I., M. Satir, H. Maluski, M. Siyako, P. Monie, R. Metzger, and S. Akyuz, 1996, Paleo- and Neo-Tethyan events in northwestern Turkey: Geologic and geochronologic constraints, *in* A. Yin and T. M. Harrison, eds., The tectonic evolution of Asia: Cambridge, Cambridge University Press, p. 420–441.

Olszewska, B., and J. Wieczorek, 2001, Jurassic sediments and microfossils of the Andrychow Klippes (Outer Carpathians): Geologica Carpathica, v. 52, p. 217–228.

Orlicky, O., 1996, Paleomagnetism of neovolcanics of the east Slovak lowlands and Zemplinske vrchy Mountains: A study of the tectonics applying the paleomagnetic data (W. Carpathians): Geologica Carpathica, v. 47, p. 13–20.

Oszczypko, N., 1997, The early–middle Miocene Carpathian peripheral foreland basin (Western Carpathians, Poland), *in* M. Krobicki and W. Zuchiewicz, eds., Dynamics of the Pannonian–Carpathian–Dinaride system; Panonian–Carpathian–Dinaride 97: Przeglad Geologiczny, v. 45, p. 1054–1063.

Oszczypko, N., 1998, The Western Carpathian Foredeep—Development of the foreland basin in front of the accretionary wedge and its burial history (Poland): Geologica Carpathica, v. 49, p. 415–431.

Oszczypko, N., J. Golonka, T. Malata, P. Poprawa, T. Slomka, and A. Uchman, 2003, Tectonostratigraphic evolution of the Outer Carpathian basins (Western Carpathians, Poland): Mineralia Slovaca, v. 35, p. 17–20.

Pana, D., I. Balintoni, L. Heaman, and R. Creaser, 2002, The U-Pb and Sm-Nd dating of the main lithotectonic assemblages of the Outer Eastern Carpathians, Romania: Geologica Carpathica, v. 53, special issue, p. 177–180.

Paris, F., and M. Robardet, 1990, Early Paleozoic palaeobiogeography of the Variscan regions: Tectonophysics, v. 177, p. 193–213.

Patrascu, S., M. Bleahu, C. Panaiotu, and C. E. Panaiotu, 1992, The paleomagnetism of the Upper Cretaceous magmatic rocks in the Banat area of South Carpathians—Tectonic implications: Tectonophysics, v. 213, p. 341–352.

Patrascu, S., M. Seclaman, and C. Panaiotu, 1993, Tectonic implications of paleomagnetism in Upper Cretaceous deposits in the Hateg and Rusca Montana basins (South Carpathians, Romania): Cretaceous Research, v. 14, p. 255–264.

Pechersky, D. M., and V. A. Safronov, 1993, Palinspastic reconstruction of the mountainous Crimea position in the Middle Jurassic–Early Cretaceous from paleomagnetic data: Geotectonika, v. 1, p. 96–105.

Pescatore, T., and A. Slaczka, 1984, Evolution models of two flysch basins: The Northern Carpathians and the southern Apennines: Tectonophysics, v. 106, p. 49–70.

Pesonen, L. J., S.-A. Elming, S. Mertanen, S. Pisarevsky, M. S. D'Agrella-Filho, J. G. Meert, P. W. Schmidt, N. Abrahamsen, and G. Bylund, 2003, Palaeomagnetic configuration of continents during the Proterozoic: Tectonophysics, v. 375, p. 289–324.

Peszat, C., 1967, The lithological development and conditions of sedimentation of the Cieszyn limestones: Prace Geologiczne (Poska Akademia Nauk Oddzial w Krakowie), v. 44, p. 1–111.

Pharaoh, T. C., 1999, Paleozoic terranes and their lithospheric boundaries within the trans-European suture zone (TESZ): A review: Tectonophysics, v. 314, p. 17–41.

Pialli, G., and W. Alvarez, 1997, Tectonic setting of the Miocene northern Apennines; the problem of contemporaneous compression and extension, *in* A. Montanari, G. S. Odin, and R. Coccioni, eds., Miocene stratigraphy; an integrated approach: Developments in Palaeontology and Stratigraphy, v. 15, p. 167–185.

Pickering, K. T., and A. G. Smith, 1995, Arcs and backarc basins in the early Paleozoic Iapetus Ocean: The Island Arc, v. 4, p. 1–67.

Plasienka, D., 1999, Tektonochronológia a paleotektonický model jursko-kriedového vývoja centralnych Západných Karpat: Tectonochrology and Paleotectonic model of the Jurassic–Cretaceous evolution of the Central Western Carpathians: Bratislava, Veda, 127 p.

Plasienka, D., and M. Kovac, 1999, How to loop Carpathians—An attempt to reconstruct Meso-Cenozoic palinspastic history of the Carpathian orocline: Geologica Carpathica, v. 50, p. 163–165.

Poka, T., 1988, Neogene and Quaternary volcanism of the Carpathian–Pannonian region: Changes in chemical composition and its relationship to basin formation, *in* L. H. Royden and F. Horvath, eds., The Pannonian Basin—A study in basin evolution: AAPG and Hungarian Geological Society, p. 257–278.

Poprawa, P., and J. Paczesna, 2002, Late Neoproterozoic to early Paleozoic development of a rift and the Lublin-Podlasie slope of the east European craton—Analysis of subsidence and facies record (eastern Poland): Przeglad Geologiczny, v. 50, p. 49–63.

Poprawa, P., S. Sliaupa, R. Stephenson, and J. Lazauskiene, 1999, Late Vendian–Paleozoic tectonic evolution of the Baltic basin: Regional tectonic implications from subsidence analysis: Tectonophysics, v. 314, p. 219–239.

Pozaryski, W., W. Brochwicz-Lewinski, and H. Tomczyk, 1982, Sur le caractere heterochronique de la Ligne Teisseyre-Tornquist, entre Europe centrale et orientale: Time-transgressive character of the Teisseyre–Tornquist lineament between Central and Eastern Europe: Comptes Rendus des Seances de l'Académie des Sciences. Serie 2, Mecanique, Physique, Chimie, Sciences de la Terre, Sciences de l'Univers, v. 295, p. 691–696.

Puchkov, N., 1997, Structure and geodynamics of the Uralian orogen, *in* J.-P. Burg and M. Ford, eds., Orogeny through time: Geological Society (London) Special Publication 121, p. 201–236.

Rakus, M., M. Misik, J. Michalik, R. Mock, T. Durkovic, T. Korab, R. Marschalko, J. Mello, M. Polak, and J. Jablonsky, 1988, Paleogeographic development of the West Carpathians: Anisian to Oligocene: Memoires de la Societe

Geologique de France, Nouvelle Series, v. 154, p. 39–62.

Rakus, M., M. Potfaj, and A. Vozarova, 1998, Basic paleogeographic and paleotectonic units of the Western Carpathians, *in* M. Rakus, ed., Geodynamic development of the Western Carpathians: Bratislava, Geological Survey of Slovak Republic, p. 15–26.

Ratschbacher, L., G. Merle, P. Davy, and P. Cobbold, 1991, Lateral extrusion in the Eastern Alps: Part I. Boundary conditions and experiments scaled for gravity: Tectonics, v. 10, p. 245–266.

Ricou, L.-E., 1996, The plate tectonic history of the past Tethys Ocean, *in* A. E. M. Nairn, L.-E. Ricou, B. Vrielynck, and J. Dercourt, eds., The ocean basins and margin. The Tethys Ocean: New York, Plenum Press, v. 8, p. 3–70.

Roberts, D., and A. Siedlecka, 2002, Timanian orogenic deformation along the northeastern margin of Baltic, northwest Russia and northeast Norway, and Avalonian–Cadomian connections: Tectonophysics, v. 352, p. 169–184.

Robertson, A. H. F., 1998, Mesozoic–Tertiary tectonic evolution of the easternmost Mediterranean area: Integration of the marine and land evidence, *in* A. H. F. Robertson, C. Richter, and C. A. Camerlenghi, eds., Proceedings of the Ocean Drilling Program. Scientific Results, v. 60, p. 723–782.

Robertson, A. H. F., P. D. Clift, P. Degnan, and G. Jones, 1991, Paleogeographic and paleotectonic evolution of eastern Mediterranean Neotethys: Palaeogeography, Palaeoclimatology, Palaeoecology, v. 87, p. 289–344.

Robertson, A. H. F., J. E. Dixon, S. Brown, A. Collins, A. Morris, E. A. Pickett, I. Sharp, and T. Ustaomer, 1996, Alternative tectonic models for the late Paleozoic–early Tertiary development of Tethys in the eastern Mediterranean region, *in* A. Morris and D. H. Tarling, eds., Palaeomagnetism and tectonics of the Mediterranean region: Geological Society (London) Special Publication 105, p. 239–263.

Ronov, A., V. E. Khain, and A. Seslavinski, 1984, Atlas of lithological paleogeographical maps of the world: Late Precambrian and Paleozoic of the continents: Leningrad, U.S.S.R. Academy of Sciences, 70 p.

Ronov, A., V. E. Khain, and A. N. Balukhovski, 1989, Atlas of lithological paleogeographical maps of the world: Mesozoic and Cenozoic of the continents: Leningrad, U.S.S.R. Academy of Sciences, p. 1–70.

Royden, L. H., 1985, Vienna basin, a thin skinned pull-apart basin, *in* K. Biddle and N. Christie Blick, eds., SEPM Special Publication 37, p. 319–338.

Royden, L. H., 1988, Late Cenozoic tectonics of the Pannonian Basin system, *in* L. H. Royden and F. Horvath, eds., The Pannonian Basin: A study in basin evolution: AAPG Memoir 45, p. 27–48.

Royden, L. H., F. Horvath, and B. C. Burchfiel, 1982, Transform faulting, extension and subduction in the Carpathian–Pannonian region: Geological Society of America Bulletin, v. 93, p. 715–725.

Sandulescu, M., 1988, Cenozoic tectonic history of the Carpathians, *in* L. H. Royden and F. Horvath, eds., The Pannonian Basin: A study in basin evolution: AAPG Memoir 45. p. 17–25.

Sandulescu, M., and M. Visarion, 2000, Crustal structure and evolution of the Carpathian–western Black Sea area: Romanian Geophysics, v. 7, suppl. 1, p. 228–231.

Scotese, C. R., and R. P. Lanford, 1995, Pangea and the paleogeography of the Permian, *in* P. A. Scholle, T. M. Peryt, and D. S. Ulmer-Scholle, eds., The Permian of northern Pangea, v. 1. Paleogeography, paleoclimate, stratigraphy: Berlin, Springer Verlag, p. 3–19.

Scotese, C. R., and W. S. McKerrow, 1990, Revised world maps and introduction, *in* W. S. McKerrow and C. R. Scotese, eds., Paleozoic paleogeography and biogeography: Geological Society Memoir 12, p. 1–21.

Sengor, A. M. C., and B. A. Natalin, 1996, Paleotectonics of Asia: Fragment of a synthesis, *in* A. Yin and T. M. Harrison, eds., The tectonic evolution of Asia: Cambridge, Cambridge University Press, p. 486–640.

Shanov, S., E. Spassov, and T. Georgiev, 1992, Evidence for existence of a paleosubduction zone beneath the Rhodopean massif (Central Balkans): Tectonophysics, v. 206, p. 307–314.

Sikora, W., 1971, Ocherk tektogeneza Peninskoy Utesovoy Zony v Polshe v svetle novykh geologicheskikh dannych: Outline of the tectonogenesis of the Pieniny Klippen zone in Poland in the light of the new geological data (in Russian): Rocznik Polskiego Towarzystwa Geologicznego, v. 6, p. 221–238.

Sinclair, H. D., S. G. Juranov, G. Georgiev, P. Byrne, and N. P. Mountney, 1997, The Balkan thrust wedge and foreland basin of eastern Bulgaria: Structural and stratigraphic development, *in* A. G. Robinson, ed., Regional and petroleum geology of the Black Sea and surrounding regions: AAPG Memoir 68, p. 91–114.

Slaczka, A., ed., 1976, Atlas of paleotransport of detrital sediments in the Carpathian–Balkan Mountain system: Warsaw, Poland, Instytut Geologiczny, Wydawnictwa Geologiczne, 12 p.

Slaczka, A., 1996a, Oil and gas in the Northern Carpathians, *in* G. Wessely and W. Liebl, eds., Oil and gas in Alpidic thrust belts and basins of Central and Eastern Europe: European Association of Geoscientists and Engineers Special Publication 5, p. 187–195.

Slaczka, A., 1996b, Oil and gas in the Ukrainian part of the Carpathians and their foredeep, *in* G. Wessely and W. Liebl, eds., Oil and gas in Alpidic thrust belts and basins of Central and Eastern Europe: European Association of Geoscientists and Engineers Special Publication 5, p. 17–21.

Slaczka, A., 1998, Age of the andesitic rocks in the Subsilesian unit (Outer Carpathians) (abs.): Carpathian–Balkan Geological Association XVI Congress Abstracts, Vienna, p. 559.

Slaczka, A., N. Oszczypko, E. Malata, and M. Cieszkowski, 1999, An early history of the Outer Carpathian basin: Geologica Carpathica, v. 50, p. 170–172.

Slaczka, A., M. Cieszkowski, and N. Oszczypko, 2000, Development of the Early Cretaceous deposits in the

Outer Carpathians and their westward prolongation: Terra Nostra, v. 1, p. 110.

Slaczka, A., S. Kruglow, J. Golonka, N. Oszczypko, and I. Popadyuk, 2006, Geology and hydrocarbon resources of the Outer Carpathians, Poland, Slovakia, and Ukraine: The general geology of the Outer Carpathians, Poland, Slovakia, and Ukraine, *in* J. Golonka and F. J. Picha, eds., The Carpathians and their foreland: Geology and hydrocarbon resources: AAPG Memoir 84, p. 221–258.

Slomka, T., 1986, Statistical approach to study of flysch sedimentation— Kimmeridgian–Hauterivian Cieszyn beds, Polish Outer Carpathians: Annales Societatis Geologorum Poloniae, v. 56, p. 227–336.

Soper, N. J., 1994, Neoproterozoic sedimentation on the northeast margin of Laurentia and the opening of Iapetus: Geological Magazine, v. 131, p. 291–299.

Soper, N. J., R. A. Strachan, R. E. Holdsworth, R. A. Gayer, and R. O. Greiling, 1992, Sinistral transpression and the Silurian closure of Iapetus: Journal of the Geological Society (London), v. 14, p. 871–880.

Sotak, J., 1986, Stratigraphy and typology of the Upper Triassic development in outer units in the West Carpathians (reconstructions from redeposits localised in the Silesian Cordillera area): Knihovnicka Zemniho Plynu a Nafty, v. 31, p. 1–53.

Sotak, J., A. Biron, R. Prokesova, and J. Spisiak, 2000, Detachment control of core complex exhumation and back-arc extension in the east Slovakian basin: Slovak Geological Magazine, v. 6, p. 130–132.

Stampfli, G. M., 1993, Le Briançonnais, terrain exotique dans les Alpes?; The Briançonnais, exotic terrane in the Alps?: Eclogae Geologicae Helvetiae, v. 86, p. 1–45.

Stampfli, G. M., 1996, The intra-Alpine terrain; a Paleotethyan remnant in the Alpine Variscides, *in* S. M. Schmid, M. Frey, N. Froitzheim, R. Heilbronner, and H. Stuenitz, eds., Alpine geology: Proceedings of the Second Workshop on Alpine Geology: Eclogae Geologicae Helvetiae, v. 89, p. 13–42.

Stampfli, G. M., ed., 2001, Geology of the western Swiss Alps— A guide book: Memoires de Geologie (Lausanne), v. 361, p. 1–195.

Stampfli, G. M., and G. D. Borel, 2002, A plate tectonic model for the Paleozoic and Mesozoic constrained by dynamic plate boundaries and restored synthetic oceanic isochrons: Earth and Planetary Science Letters, v. 196, p. 17–33.

Stampfli, G. M., J. Marcoux, and A. Baud, 1991, Tethyan margins in space and time, *in* J. E. T. Channell, E. L. Winterer, and L. F. Jansa, eds., Palaeogeography, Palaeoclimatology, Palaeoecology, v. 87, p. 373–409.

Stampfli, G. M., J. von Raumer, G. D. Borel, and F. Bussy, 2001, The Variscan and pre-Variscan evolution, *in* G. M. Stampfli, ed., Geology of the western Swiss Alps— A guide book: Memoires de Geologie (Lausanne), v. 361, p. 28–41.

Stefanescu, M., O. Dicea, A. Butac, and D. Ciulavu, 2006, Hydrocarbon geology of the Romanian Carpathians, their foreland, and the Transylvanian basin, *in* J. Golonka and F. J. Picha, eds., The Carpathians and their foreland: Geology and hydrocarbon resources: AAPG Memoir 84, p. 521–567.

Svabenicka, L., M. Bubik, O. Krejci, and Z. Stranik, 1997, Stratigraphy of Cretaceous sediments of the Magura group of nappes in Moravia (Czech Republic): Geologica Carpathica, v. 48, p. 179–191.

Tari, G., F. Horvath, and J. Rumpler, 1992, Styles of extension in the Pannonian Basin: Tectonophysics, v. 208, p. 203–216.

Tari, G., O. Dicea, J. Faulkerson, G. Georgiev, S. Popov, M. Stefanescu, and G. Weir, 1997, Cimmerian and Alpine stratigraphy and structural evolution of the Moesian Platform (Romania/Bulgaria), *in* A. G. Robinson, ed., Regional and petroleum geology of the Black Sea and surrounding regions: AAPG Memoir 68, p. 63–90.

Torsvik, T. H., 1998, Palaeozoic palaeogeography: A North Atlantic viewpoint: Geologiska Foreningens i Stockholm Forhandlingar, v. 120, p. 107–116.

Torsvik, T. H., M. A. Smethurst, J. G. Meert, R. Van der Voo, W. S. McKerrow, M. D. Brasier, B. A. Sturt, and H. J. Walderhaug, 1996, Continental break-up and collision in the Neoproterozoic and Palaeozoic; a tale of Baltica and Laurentia: Earth-Science Reviews, v. 40, p. 229–258.

Tyszka, J., 1994, Response of Middle Jurassic benthic foraminiferal morphogroups to disoxic/oxic conditions in the Pieniny Klippen Basin, Polish Carpathians: Palaeogeography, Palaeoclimatology, Palaeoecology, v. 110, p. 55–81.

Tyszka, J., 2001, Microfossil assemblages as bathymetric indicators of the Toarcian/Aalenian "Fleckenmergel"— Facies in the Carpathian Pieniny Klippen Belt: Geologica Carpathica, v. 52, p. 147–158.

Unrug, R., 1984, Geodynamic evolution of the Carpathians: Annales Societatis Geologorum Poloniae, v. 52, p. 39–66.

Unrug, R., C. Haranczyk, and M. Chocyk-Jaminska, 1999, Easternmost Avalonian and Armorican–Cadomian terranes of Central Europe and Caledonian–Variscan evolution of the polydeformed Kraków mobile belt: Geological constraints: Tectonophysics, v. 302, p. 133–157.

Ustaomer, T., and A. Robertson, 1997, Tectonic-sedimentary evolution of the north Tethyan margin in the central Pontides of northern Turkey, *in* A. G. Robinson, ed., Regional and petroleum geology of the Black Sea and surrounding region: AAPG Memoir 68, p. 255–290.

Van der Voo, R., 1988, Paleozoic paleogeography of North America, Gondwana, and intervening displaced terranes: Comparison of paleomagnetism with paleoclimatology and biogeographical pattern: Geological Society America Bulletin, v. 100, p. 311–324.

Van der Voo, R., 1993, Paleomagnetism of the Atlantic, Tethys and Iapetus: Cambridge, Cambridge University Press, 411 p.

Van Dijk, J., and M. Okkes, 1991, Neogene tectonostratigraphy and kinematics of Calabrian basins; implications for the geodynamics of the central Mediterranean: Tectonophysics, v. 196, p. 23–60.

Vasicek, Z., J. Michalik, and D. Rehakova, 1994, Early Cretaceous stratigraphy, paleogeography and life in Western Carpathians: Beringeria, v. 10, p. 1–170.

Vass, D., 1998, Neogene geodynamic development of the Carpathian arc and associated basins, *in* M. Rakus, ed., Geodynamic development of the Western Carpathians: Geological Survey of Slovak Republic: Bratislava, Dionyz Stur Publishers, p. 155–158.

Vass, D., A. Nagy, M. Kohut, and I. Kraus, 1990, Granitoid clastics on the southeast margin of the Vienna basin and basin genesis, *in* D. Minarikova and H. Lobitzer, eds., Thirty years of the geological cooperation between Austria and Czechoslovakia: Praha, Ustredni Ustav Geologicky, p. 179–184.

Von Raumer, J. F., 1998, The Palaeozoic evolution in the Alps: From Gondwana to Pangea: Geologische Rundschau, v. 87, p. 407–435.

Vozarova, A., and J. Vozar, 1992, Variscan terranes and evolution of the late Paleozoic sedimentary basins (Western Carpathians): International Geological Correlation Programme, no. 276, Special Volume: Bratislava, Geologicky Ustav Dionyza Stura, p. 147–154.

Wierzbowski, A., M. Jaworska, and M. Krobicki, 1999, Jurassic (upper Bajocian–lowest Oxfordian) ammonitico rosso facies in the Pieniny Klippen Belt, Carpathians, Poland: Its fauna, age, microfacies and sedimentary environment: Studia Geologica Polonica, v. 115, p. 7–74.

Wilson, M., and H. Downes, 1991, Tertiary–Quaternary extension related alkaline magmatism in western and Central Europe: Journal of Petrology, v. 32, p. 811–849.

Winchester, J. A., 2002, Palaeozoic amalgamation of Central Europe: New results from recent geological and geophysical investigations: Tectonophysics, v. 360, p. 5–21.

Winkler, W., and A. Slaczka, 1994, A Late Cretaceous to Paleogene geodynamic model for the Western Carpathians in Poland: Geologica Carpathica, v. 45, p. 71–82.

Withjack, M. O., R. W. Schlische, and P. O. Olsen, 1998, Diachronous rifting, drifting, and inversion on the passive margin of central eastern North America: An analog for other passive margins: AAPG Bulletin, v. 82, p. 817–835.

Wortmann, U. G., H. Weissert, H. Funk, and J. Hauck, 2001, Alpine plate kinematics revisited: The Adria problem: Tectonics, v. 20, p. 134–147.

Yanev, S. N., 1992, Contribution to the elucidation of pre-Alpine evolution in Bulgaria (based on sedimentological data from the marine Paleozoic): Geologica Balcanica, v. 22, p. 3–31.

Yilmaz, P. O., I. O. Norton, D. Leary, and R. J. Chuchla, 1996, Tectonic evolution and paleogeography of Europe, *in* P. A. Ziegler and F. Horvath, eds., Peri-Tethys Memoir 2: Structure and prospects of Alpine basins and forelands: Symposium on Structure and Prospects of Alpine Basins and Forelands: Memoires du Museum National d'Histoire Naturelle 170, p. 47–60.

Ziegler, P. A., 1988, Evolution of the Arctic–North Atlantic and the western Tethys: AAPG Memoir 43, p. 1–198.

Ziegler, P. A., 1989, Evolution of Laurussia: Dordrecht, The Netherlands, Kluwer Academic Publishers, 102 pp.

Zonenshain, L. P., M. L. Kuzmin, and L. N. Natapov, 1990, Geology of the U.S.S.R.: A plate-tectonic synthesis, *in* B. M. Page, ed., Geodynamics Series: Washington, D.C., American Geophysical Union, v. 21, p. 1–242.

Zytko, K., 1984, The Atlantic, the Indian Ocean and main linear fracture zones of the post-Variscan Europe: Annales Societatis Geologorum Poloniae, v. 52, p. 3–38.

Zytko, K., 1985, Some problems of a geodynamic model of the Northern Carpathians: Kwartalnik Geologiczny, v. 29, p. 85–108.

Zytko, K., 1999a, Correlation of the main structural units of the Western and Eastern Carpathians (English summary): Prace Panstwowego Instytutu Geologicznego, v. 168, p. 135–164.

Zytko, K., 1999b, Symmetrical pattern of the late Alpine features of the Northern Carpathian basement, their foreland and hinterland; orogen and craton suture: Prace Panstwowego Instytutu Geologicznego, v. 58, p. 165–194.

Zytko, K., et al., 1989, Geological map of the Western Outer Carpathians and their foreland without Quaternary formations, *in* D. Poprawa and J. Nemcok, eds., Geological atlas of the Western Carpathians and their foreland: Warsaw, Poland, Panstwowy Instytut Geologiczny, p. 1.

THE GEOLOGY AND HYDROCARBON RESOURCES OF THE OUTER WESTERN CARPATHIANS AND THEIR FORELAND ON THE TERRITORY OF NORTHEASTERN AUSTRIA AND MORAVIA IN THE CZECH REPUBLIC

3

Picha, Frank J., Zdeněk Stráník, and Oldřich Krejčí, 2006, Geology and hydrocarbon resources of the Outer Western Carpathians and their foreland, Czech Republic, *in* J. Golonka and F. J. Picha, eds., The Carpathians and their foreland: Geology and hydrocarbon resources: AAPG Memoir 84, p. 49–175.

Geology and Hydrocarbon Resources of the Outer Western Carpathians and Their Foreland, Czech Republic

Frank J. Picha
International Petroleum Consultant, Walnut Creek, California, U.S.A.

Zdeněk Stráník
Czech Geological Survey, Brno, Czech Republic

Oldřich Krejčí
Czech Geological Survey, Brno, Czech Republic

ABSTRACT

The Western Carpathians in the territory of Moravia (the eastern part of the Czech Republic) and northeastern (Lower) Austria represent the westernmost segment of the entire Carpathian orogenic system linked to the Eastern Alps. Based on differences in their depositional and structural history, the Carpathians are divided into two primary domains: the Inner Carpathians deformed and thrusted in the Late Jurassic to Early Cretaceous, and the Outer Carpathians deformed and thrusted over the European foreland during the Paleogene and Neogene. These two domains are separated by the Pieniny Klippen Belt, which bears signatures of both these domains and stands out as a primary suture in the Western Carpathians. Only the Outer Carpathians, including the thin-skinned thrust belt partly overlain by the Vienna basin and the undeformed Neogene foredeep, are present in the territory of Moravia and, as such, are subjects of our deliberation.

The foreland of the Carpathians in Moravia is represented by the Bohemian Massif, which is a part of the West European plate. It consists of the Hercynian orogenic belt and the late Precambrian (Cadomian) foreland terrane of the Brunovistulicum. The unmetamorphosed sedimentary cover of the cratonic basement of the Bohemian Massif in Moravia extends through two plate-tectonic cycles, the Paleozoic Hercynian and the Mesozoic to Cenozoic Tethyan–Alpine. The Bohemian Massif continues far below the Carpathian foredeep and the thin-skinned Outer Carpathian thrust belt. Various deep antiformal structures have been identified in the subthrust plate by seismic methods and drilling. Some of these structures apparently formed during the Hercynian orogeny, whereas others are related either to the Jurassic rifting or to the compressional Alpine tectonics extending from the Late Cretaceous to Miocene. During the Laramide uplifting of the European foreland, in the Late Cretaceous

DOI:10.1306/985607M843067

to early Paleogene, two large paleovalleys and submarine canyons were cut into the foreland plate and filled with deep-water Paleogene strata.

The Carpathian orogenic system, as we know it today, evolved during the late Paleozoic, Mesozoic, and Cenozoic through the divergent and convergent processes of the plate-tectonic cycle. In the Outer Western Carpathians of Moravia, the divergent stage began in the Middle to Late Jurassic by rifting, opening of Tethyan basins, and development of the passive margins dominated by the carbonate platforms and basins. Further rifting and extension occurred in the Early Cretaceous. The convergent orogenic process in the Outer Carpathians began in the Late Cretaceous by the subduction of the Penninic–Pieninic oceanic basin and collision of the Inner Carpathians with the fragmented margins of the European plate. Since the Late Cretaceous, a major foreland basin dominated by the siliciclastic shelf and deep-water flysch sedimentation has formed in the Outer Carpathian domain. The Carpathian foreland basin, especially during the Late Cretaceous to the early Eocene, displayed a complex topography marked by an existence of intrabasinal ridges (cordilleras) such as the Silesian cordillera. We interpret them as preexisting rift-related crustal blocks activated during the Late Cretaceous–early Paleocene uplifting as foreland-type compressional structures.

During the Paleogene and early Miocene, the Upper Jurassic to lower Miocene sequences of the Outer Carpathian depositional system were gradually deformed and thrusted over the European foreland. The tectonic shortening occurred not only in the decoupled thin-skinned thrust belt but also at the deeper crustal level, where various blocks of the previously rifted margins were apparently at least partly accreted back to the foreland plate instead of being subducted.

Since the early Miocene, the synorogenic, predominantly deep-water flysch sedimentation was replaced by the shallow-marine and continental molasse-type sedimentation of the Neogene foredeep, which remained mostly undeformed. Also during the Miocene, the Vienna basin formed in the Carpathian belt of southern Moravia and northeastern Austria as a result of subsidence, back-arc extension, and the orogen-parallel pull-apart strike-slip faulting.

During its entire history, the evolution of Outer Western Carpathians in Moravia was significantly affected by the existence of two main structural elements, the Western Carpathian transfer zone and the Dyje–Thaya depression. The southwest–northeast-trending Western Carpathian transfer zone actually separated the Alps from the Carpathians. During the divergent stage, in the Early Cretaceous, the dextral motion in this zone accommodated a significant extension in the Outer Carpathian domain. Conversely, during the convergent stage in the Paleogene and Neogene, the sinistral transpressional motion in this zone facilitated the northeastern translation (escape) of the Carpathian belt and the opening of the pull-apart depocenter in the Vienna basin.

The northwest–southeast-trending Dyje–Thaya depression, in southern Moravia and northeastern Austria, formed, or at least was activated, during the Jurassic rifting. Within the fault-bounded limits of this depression, thick, organic-rich marls were deposited in the Late Jurassic, shallow-marine clastic strata were laid down and preserved in the Late Cretaceous, two paleovalleys were excavated in the Late Cretaceous–early Paleogene, and finally, the Vienna basin formed in the Miocene. The complex structural and depositional history of the depression and its surroundings created one of the most prolific petroleum systems in the entire Carpathian region, from which more than 850 million bbl of oil has been produced to date. Historically, the Vienna basin has been the dominant producer in Austria and Moravia. More recently, however, the subthrust European platform with multiple hydrocarbon plays has become the main producing province in Moravia. Some of the identified deep subthrust structures represent significant exploration prospects, which yet have to be tested.

INTRODUCTION

The main objective of this chapter is to provide a comprehensive and intelligible account on the geology and hydrocarbon resources of the Carpathian belt and its foreland on the territory of Moravia (eastern part of the Czech Republic) and northeastern Austria. To achieve this goal, we not only summarize the present knowledge on stratigraphy, structure, and petroleum systems in that particular area but also integrate it into a broader context of the Alpine–Carpathian geology and make it compatible with modern geological concepts. We also take this opportunity to present new original ideas, particularly on the structure and the depositional systems of the Outer Carpathians and their foreland.

The Western Carpathians in the territory of southern Moravia and northeastern Austria represent the westernmost segment of the entire Carpathian orogenic belt linked to the Eastern Alps. The location of this sector in the contact zone between the Alps and the Carpathians and in the proximity of the well-exposed Cadomian and Hercynian foreland terranes of the Bohemian Massif provides a unique opportunity to address some general problems of the Alpine–Carpathian geology. Among them, the correlation of various units of the Eastern Alps and Western Carpathians has been given a high priority. Southern Moravia is the only place in the entire Carpathian belt where the Paleogene strata have been found in their original unquestionable autochthonous position below the Neogene foredeep and the frontal zones of the thrust belt. Their existence not only enables better correlation with the Alpine Molasse but also provides a critical starting point for a more realistic palinspastic reconstruction of the Carpathian depositional system.

During both the divergent and convergent stages of the development of the Western Carpathians, the Austrian–Moravian sector functioned as a southwest–northeast-trending transfer corridor; we call it the Western Carpathian transfer zone, which separated the Alpine and Carpathian domains. The strike-slip movements in this zone were instrumental in the opening and closing parts of the Western Carpathian depositional system and in the late orogenic northeastward translation (escape) of the Carpathians and formation of the Vienna basin.

Special attention is paid to the existence of a southeast–northwest-trending Dyje–Thaya depression, which apparently formed during the Jurassic rifting at the territory of southern Moravia and northeastern Austria. Repeatedly subsiding during its geological history, this depression became one of the most important generative kitchens for hydrocarbons in the entire Carpathian system.

Thanks to a relatively shallow depth and the presence of important resources (coal and hydrocarbons), the subthrust European plate in Moravia has been explored intensively by drilling. It is, in fact, the most heavily drilled sector of the subthrust plate in the Alpine–Carpathian belt and possibly in the entire world and, as such, may serve as a model for other similar settings elsewhere.

In addition to the above-mentioned aspects specific to the Austrian–Moravian area, attention is paid to some general issues of the Outer Carpathian belt. Among them, the characterization of the Outer Carpathian foreland basin, which has evolved since the Late Cretaceous, seems to be the most important. Although bearing many primary characteristics of foreland basins elsewhere, the complex architecture of the Carpathian foreland basin, marked by the existence of internal ridges (cordilleras) and intervening troughs, as well as the prominent presence of both the deep-water flysch and the shallow-marine to continental molasse stages, sets it apart as a specific model among the foreland depositional systems.

In our discussions of the tectonic history of the Carpathians, we avoid using the traditional terminology of specific orogenic phases, such as Austrian, Pyrenean, Savian, etc. (e.g., Stille, 1936). Although pulses of higher and lower tectonic activity are well documented in various regions of Alps and Carpathians, it does not seem possible to correlate and generalize these events over the complex plate-tectonic setting of the entire Alpine–Carpathian system. The only phase we still use under its original name is the so-called Laramide orogeny marked by an extensive crustal uplifting and inversion of the Carpathian foreland plate at the Cretaceous–Paleogene transition. It is the predominantly vertical character of this far-reaching and deeply rooted compressional tectonism, which, in our opinion, singles out the Laramide orogeny from the progradational thin-skinned tectonism that shaped the Outer Carpathian thrust belt.

In addition, the term "sequence" is used for distinct depositional units without any reference to the specific terminology of the sequence stratigraphy.

The constructions of maps and cross sections are supported by seismic and well data. Stratigraphic records of 37 critical deep wells are listed in Appendix 1.

Finally, the local geographical and geological terms, as well as the names of authors in our article, are printed without diacritics. For better communication, lists of geographical and geological terms and authors' names with and without diacritics are provided in Appendix 2.

REGIONAL GEOLOGICAL SETTING

Part of the Tethyan–Alpine orogenic system of Europe, the Carpathian fold and thrust belt extends from the Eastern Alps in northeastern (Lower) Austria to southern Romania (Figure 1). Considering some regional differences in the evolution of the Carpathians as well as the diverse character of their foreland, the Carpathian belt is traditionally divided into three parts: the Western, Eastern, and Southern Carpathians. The Western Carpathians extend from the Danube

Figure 1. Generalized geologic map of the Alpine–Carpathian orogenic system of Europe. Study area is located in the box. The inner zones of the Alpine–Carpathian orogen are shown in blue. Modified from Picha (1996). Cross section of AA′ is shown in Figure 2.

Valley in Austria through Moravia and Slovakia to eastern Poland and the Ukrainian border (Uh River valley) and are bounded by the Hercynian West European platform. The Eastern Carpathians in Ukraine and eastern Romania are attached to the Late Proterozoic (Cadomian) East European platform. The boundary between the West and East European crustal domains is traditionally attributed to the Teisseyre–Tornquist line as marked by the course of the Polish–Danish trough. The Southern Carpathians constitute the southward-verging segment of the Carpathian belt bounded by the Moesian Platform. They extend westward to the Iron Gate on the Danube River in southwestern Romania.

The Carpathians, based on main differences in their depositional and structural history, have been divided into two main domains: the Inner Carpathians deformed and thrusted in the Late Jurassic to the Late Cretaceous and the Outer Carpathians deformed and thrusted during the Paleogene and Neogene (Figures 1, 2). In Romania, the Inner and Outer Carpathians are known as the Dacides and Moldavides, respectively (Sandulescu, 1988). Some authors, e.g., Plasienka (1995), further divided the Inner Carpathians into the Inner Carpathians sensu stricto, related to the subduction of the Meliata–Hallstatt ocean in the Late Jurassic, and the Central Carpathians consolidated in the Late Cretaceous. Because our article is not concerned with the geology of either of these systems, we prefer to use the simpler version, the Inner Carpathians, for both of them. Included into the Inner Carpathian domain are also the unfolded deposits of the Paleogene flysch basin and the Neogene molasse deposits of the Pannonian Basin that rest unconformably on the Mesozoic nappes of the Inner Carpathians.

The Inner and Outer Carpathians are separated by the Pieniny Klippen Belt (Figure 3), which has been traditionally considered to be a part of the Outer Carpathians (Ksiazkiewicz, 1977; Picha, 1996), although it bears signatures of both these domains. Structurally, the Pieniny Klippen Belt represents a complex suture, along which some elements of the oceanic and continental

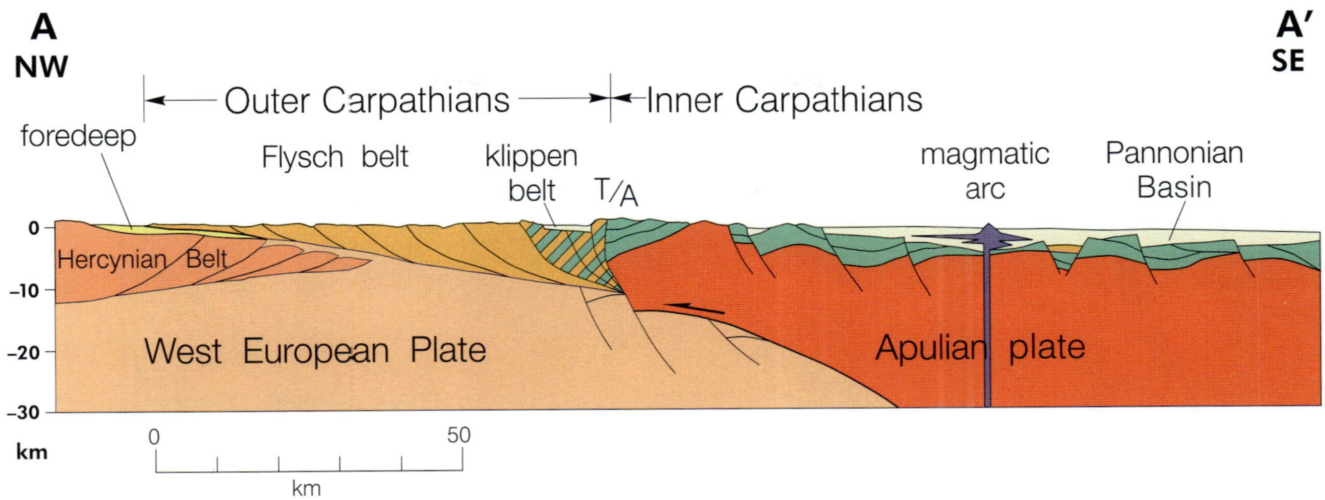

Figure 2. Generalized cross section through the Western Carpathians and their foreland. The term "Apulian plate" (Adria) is used here as a general name for microplates of the Carpathian region separated from North Africa. Location in Figure 1. Modified from Picha (1996).

lithosphere were subducted during the Late Cretaceous and early Miocene convergency (e.g., Grecula and Roth, 1978; Misik, 1979; Birkenmajer, 1986).

The Outer Western Carpathians thus consist of the allochthonous Pieniny Klippen Belt, the Flysch belt, the autochthonous Neogene foredeep, and the late orogenic and postorogenic Vienna basin, which rests transgressively on both the Outer and Inner Carpathians (Figure 3). Included into the Outer Carpathian domain are also the autochthonous Tethyan Jurassic, Cretaceous, and Paleogene strata preserved in their original depositional site on the Carpathian foreland. All these Outer Carpathian units, with the sole exception of the Pieniny Klippen Belt, are widely present on the territory of Moravia and, as such, remain the main subjects of our deliberations. The Pieniny Klippen Belt does not extend into this area; however, its complex depositional and structural history is critical for understanding the Outer Carpathian system in its entirety and, therefore, will be discussed as necessary.

The Flysch belt consists of a stack of several rootless tectonostratigraphic units, which, as a whole, is thrusted over the Neogene foredeep and the underlying European foreland. This wedge-shaped, southward-thickening thrust edifice comprises predominantly clastic sequences, whose age extends from the Jurassic to the early Miocene. It includes deposits of both the rifted continental margins and the synorogenic foreland basins, which formed after the Late Cretaceous collision of the Inner Carpathians with the fragmented margins of the West European plate.

The foreland of the Western Carpathians, including its part buried below the thrust belt, belongs to the West European plate (platform). In the territory of Moravia, the West European plate is represented by the Bohemian Massif, which consists of both Hercynian and Precambrian (Cadomian) terranes. Within the Outer Carpathian depositional system, other crustal blocks were present, such as the Silesian and Czorsztyn cordilleras, whose existence and character have been acknowledged only indirectly from the distribution of facies and composition of conglomerates in the Pieniny Klippen Belt and the Flysch belt. Tentatively, we interpret them as rifted away highstanding blocks of the European plate, further inverted during the Laramide uplifting of the foreland. Their adherence to either Hercynian or Cadomian systems may be debated.

THE HISTORY OF INVESTIGATION OF THE CARPATHIAN REGION IN MORAVIA

The geology of the Western Carpathians in Moravia has been studied for more than one-and-a-half centuries. The oldest contributions had a character of rather general observations by naturalists and travelers; more systematic geological investigation began in the second part of the 19th century. Most of the geological mapping and regional work was originally done by the Austrian Geological Survey (K.K. Geologische Reichsanstalt), which was founded in 1849. Foetterle (1866) published the geological map of Moravia and Silesia, and Hauer (1867–1874, 1869 [Western Carpathians]) published the first geological map of the Austro-Hungarian monarchy. Hohenegger (1852, 1855, 1861), based on his studies of macrofauna, laid down the foundation of stratigraphy of the Moravo-Silesian Beskydy Mountains, which, in principle, is valid to the present day.

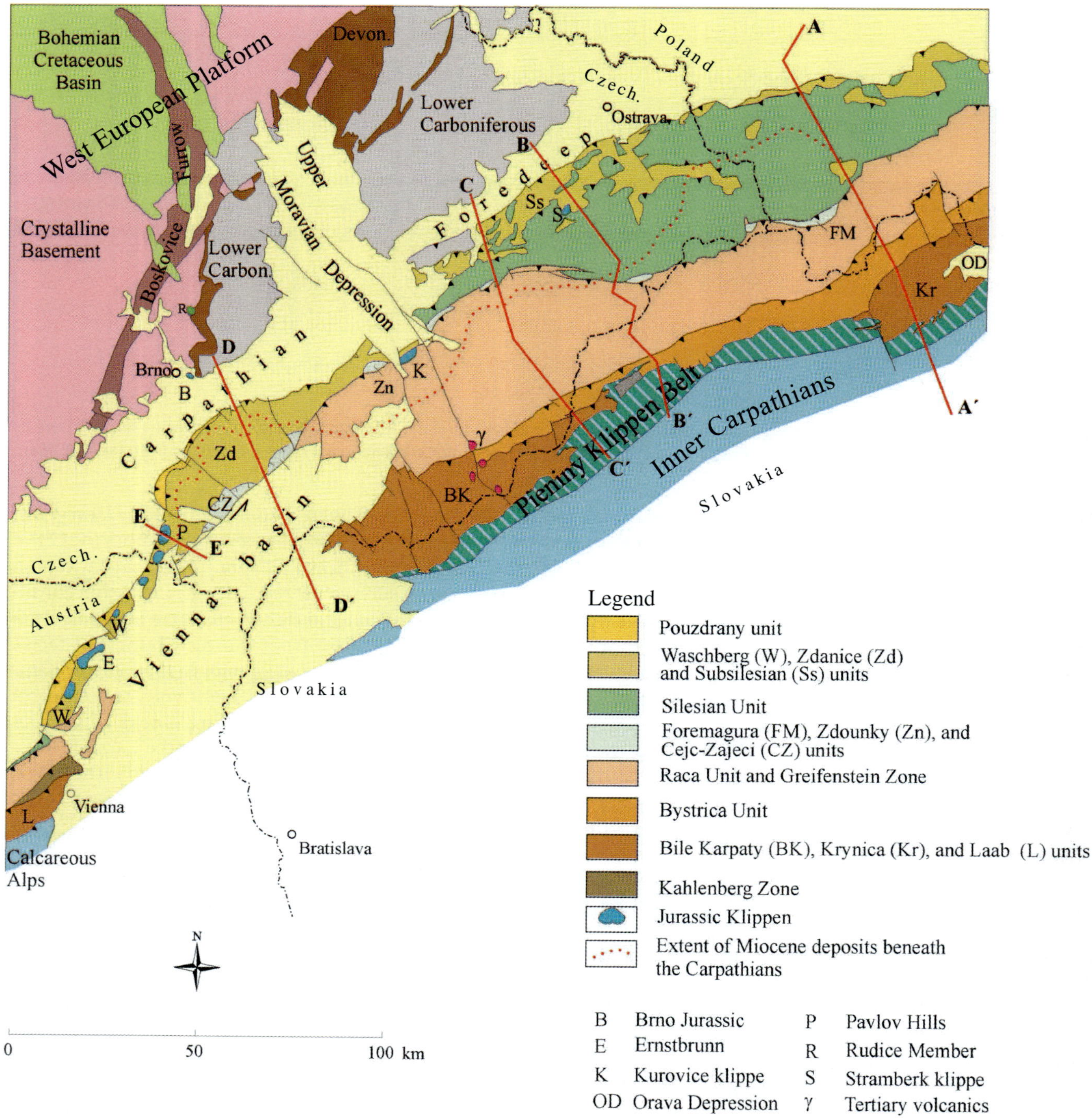

Figure 3. Surface geological map of the Western Carpathians and their foreland (West European platform) in northeastern Austria, Moravia (the eastern part of the Czech Republic), and western Poland. Cross sections of AA′, BB′, CC′, and DD′ are shown in Figure 20 (located on page 118).

In the second part of the 19th century, following the discovery of oil (e.g., Borislav oil field) in Galicia (presently southeastern Poland and southwestern Ukraine), the investigations of the Carpathian flysch belt intensified. Paul and Tietze (1877), Walter and Dunikowski (1883), and Paul (1890, 1893) established the stratigraphy of the flysch belt in Moravia, Slovakia, Poland, and Ukraine (Galicia). At the turn of the century, the Carpathian geology was significantly influenced by the nappe theory established in the Alps by Bertrand (1898), Argand (1911, 1916, 1924), and others. Based on Uhlig's (1897, 1903) observations, Lugeon (1903) published a nappe synthesis of the Carpathians, which, after much dispute, was accepted by Uhlig (1907). The existence of overthrusting was then proven by wells in the Beskydy foothills (Petraschek and Fuchs, 1912).

A new phase of investigation of the Outer Carpathians in Moravia and Silesia began after World War I. The Czechoslovak Geological Survey, as one of the successors of the Austrian Geological Survey, was established in 1919 in the newly founded Czechoslovak Republic. The geological mapping and studies in the Carpathians were motivated mainly by the search for oil and other mineral resources and the construction of railways. A significant factor in further development of the Carpathian geology was the Carpatho-Balkanian Association established in 1922. The Third Congress held in 1931 in Czechoslovakia was one of the most memorable. Beck and Gotzinger (1932), in their explanations to the geological map of the Ostrava–Karvina coal basin, interpreted the nappe structure of the Beskydy Mountains. Andrusov (1938) published the results of his investigation of the Pieniny Klippen Belt, which he considered to be the substratum of the Magura flysch.

A large expansion of geological activities both in the Carpathian belt and its foreland began after the Second World War. The geological activities were driven mainly by the search for raw materials and fuels, including coal, oil, and natural gas. The geological investigations were aided by the application of new techniques and scientific methods, including microbiostratigraphy, sedimentology, geochemistry, and geophysics. The application of seismic, gravimetric, and magnetic geophysical methods represented a significant step in interpreting the architecture of the deeper structural levels of the Carpathian thrust belt and its foreland. Of major importance was the publication of a comprehensive volume on geology of the Czechoslovak Carpathians by Andrusov (1959).

The completion of the geological maps and explanatory texts of all of Czechoslovakia on a scale of 1:200,000 during the 1950s and early 1960s was a major accomplishment. It was followed by a publication of several volumes on the regional geology of Czechoslovakia in Czech and English versions (Svoboda et al., 1966; Mahel and Buday, eds., et al., 1968). In recognition of these achievements, the 23rd International Geological Congress was held in Prague in 1968.

After the publication of general maps, the geological activity concentrated on the detailed investigation and mapping on scales of 1:50,000 and 1:25,000. An intensive exploration for coal and hydrocarbons under the Carpathian foredeep and the thrust belt, which began in the 1950s, invigorated the geological and geophysical activities in Moravia. Deep drilling (see the overview by Suk et al., 1991) and geophysical investigations (Ibrmajer and Suk, 1989) provided critical information about the stratigraphy and structure of both the Carpathian thrust belt and the subthrust plate. Large reserves of coal and several oil and gas fields have been found in various formations below the Carpathian thin-skinned belt. Numerous maps and papers, including the geological atlas of the Western Carpathians (Poprawa and Nemcok, 1988–1989), the palinspastic maps of the Western Carpathian Neogene (Kovac et al., 1998), and the accounts on the geodynamic evolution of the circum-Carpathian region (Golonka et al., 2000) and on the subthrust exploration for hydrocarbons (Picha, 1996), have been published. A great deal of work has also been done in the range of environmental geology, namely, engineering geology (construction of reservoirs, roads, and railroads) and hydrogeology. Traditionally, attention has been paid to geological hazards, especially landslides, which affect large areas of the Carpathian Flysch belt.

After the division of Czechoslovakia into two countries, the Czech Republic and Slovakia, in 1992, the geological mapping and regional investigations in the Moravian Carpathians and their foreland have been conducted by the Czech Geological Survey, which is also one of the primary sponsors of this publication.

GEODYNAMIC EVOLUTION OF THE OUTER WESTERN CARPATHIANS AND THEIR FORELAND IN MORAVIA (CZECH REPUBLIC)

The Tethyan–Alpine orogenic system, as we know it today, evolved during the late Paleozoic, Mesozoic, and Cenozoic through the processes of rifting of the former Pangea, opening of Tethyan oceanic basins, which separated Eurasia from Africa–Apulia, and the development of passive continental margins, followed by the subduction of the oceanic basins, continental collision, and formation of the Carpathian orogenic belt.

Plate kinematic reconstructions of the Alpine–Carpathian system have been published by Dewey et al. (1973), Biju-Duval et al. (1977), Tollmann (1978, 1980), Roth (1980b, 1986), Ziegler (1982, 1988), Dercourt et al. (1986), Le Pichon et al. (1988), Winkler and Slaczka (1994), Plasienka (1995), Yilmaz et al. (1996), Golonka (2000), and Golonka et al. (2000, 2003, 2006a), among others. The architecture of the Alpine orogen was newly interpreted, e.g., by Ziegler and Roure (1996). Still, some differences in the interpretation of the geodynamic history and structure of the region do exist. In this chapter, we concentrate on problems relevant to the Outer Western Carpathians and their foreland in Moravia and only briefly mention some general aspects of the geodynamic evolution of the Alpine–Carpathian region as a whole.

The Hercynian Setting of the Western Carpathian Region

The Mesozoic and Cenozoic evolution of the Carpathian region resulted from the interplay of numerous lithospheric plates, such as the West and East European platforms, Adria (Apulia), Tisza, and Moesia. All of these plates were parts of the Pangea assembly, which formed during the Hercynian orogeny in the Late Carboniferous. Unlike the linear fold belts, such as the Uralides and the Appalachians, the Hercynian fold belt of Europe displays a complex arcuate architecture. This apparently resulted from the irregular geometry of the preexisting Ordovician to Early Carboniferous rift systems, from draping of the belt around several internal microcratons, as well as from the widespread wrench faulting and rotation of individual blocks (Ziegler, 1988). The curvature of the Hercynian belt is well evident in the Silesian–Moravian sector of the Carpathian foreland, where the generally northwest–southeast-trending front of the Hercynian belt suddenly turns into an anomalous north-northeast–south-southwest direction (orocline bend) apparently being deflected by the Cadomian block of the Brunovistulicum (Figure 4).

In the Moravian–Silesian area, the Hercynian convergent tectonism ended in the Late Carboniferous (Stephanian). The subsequent wrench faulting, rifting, and opening of grabens accompanied by magmatic activity began the process of the fragmentation of the previously assembled orogenic belt. This process is documented by the opening of the arcuate, generally north–south-trending Boskovice Furrow (graben) (Figure 3), filled with the Late Carboniferous (Stephanian C) to Early Permian (Autunian) clastic deposits (Svoboda, ed., et al., 1966; Maly, 1993; Jaros and Maly in Pesek et al., 2001). According to Malkovsky (1976) and Ziegler (1988), this late Hercynian fault system became reactivated time and again. It became a major factor during the Mesozoic rifting stage preceding the opening of the Tethys and the North Atlantic Ocean.

The Late Permian and Mesozoic Rifting and Formation of Passive Tethyan Margins

The evolution of the Eastern Alps–Western Carpathians began in the Late Permian to Early Triassic by rifting of the Tisza plate from Eurasia and the opening of the Meliata–Hallstatt ocean (Channell and Kozur, 1997; Golonka and Gahagan, 1997). The process of rifting and fragmentation of the Pangea assembly gradually spread through the entire Inner Carpathian zone and, in the Early to Middle Jurassic, reached into the Outer Carpathian zone. It led to the opening of the Penninic–Pieninic oceanic basin, the formation of the passive continental margins, and the establishment of a connection with the central Atlantic rift system. The newly formed continental margins were characterized by the development of carbonate platforms and intervening basins (Figures 5, 6).

According to Funk et al. (1987), the crustal stretching in the Tethyan realm was discontinuous, resulting in a system of structural highs with almost normal crustal thickness and structural lows with a thin continental or oceanic crust. This complexity might have

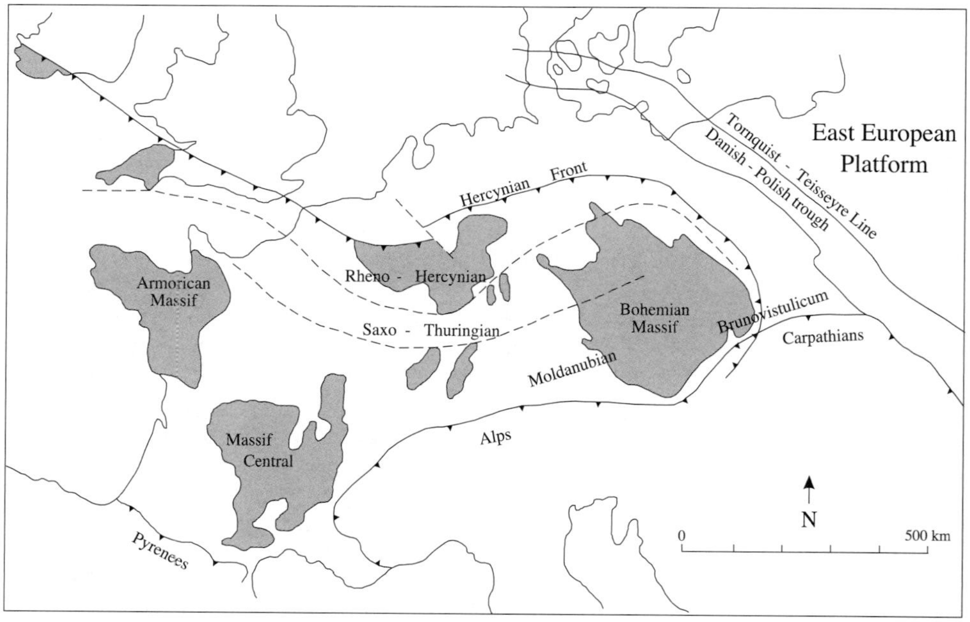

Figure 4. Hercynian system of Western and Central Europe. The Hercynian massifs are in gray pattern. The Bohemian Massif consists of the Rhenohercynian, Saxo-Thuringian, and Moldanubian zones of the Hercynian belt and of the late Precambrian terrane of the Brunovistulicum.

been, at least partly, related to the sinistral strike-slip movements associated with the opening of Tethys between the African and European continents (Boillot et al., 1984). In such a complex and relatively narrow system of Tethyan margins commonly dominated by transcurrent faulting, little space was present for the opening of broad oceanic domains (Plasienka, 1995). The structural and depositional pattern of the complex Tethyan continental margins thus differed markedly from the classical models elaborated for the relatively simple Atlantic-type continental margins.

The Pieninic oceanic basin fully opened by the Middle–Late Jurassic (Birkenmajer, 1986). From the Magura basin and the rest of the Outer Carpathian realm, the Pieniny ocean was separated by the Czorsztyn ridge (e.g., Birkenmajer, 1988). This positive structural feature was marked by a relatively shallow-water carbonate sedimentation, which contrasted with the deep-water basinal environment both north and south of the ridge (Golonka and Sikora, 1981). Golonka et al. (2006a) interpreted the Czorsztyn ridge as a midoceanic ridge. We tend to believe that the Czorsztyn ridge was rather a rifted-apart fragment of the European crust. Its position is similar to that of the Brianconnais zone of the Western Alps (Stampfli, 1993; Tomek, 1993), which is interpreted as a distal element of the European margin (Trumpy, 1988). The Magura flysch basin would then be an equivalent of the Valais trough, which separated the Brianconnais zone from the main part of the European margin. Our further postulations are based on an assumption that the substratum of the Magura basin was made mostly by an attenuated continental crust, and only the deepest zones might have been underlain by a basaltic basement (Figures 5, 6). However, such an interpretation is not accepted by everybody, and different views on this matter do exist even among the authors of this article.

In the Outer Carpathians, the process of divergency and subsidence peaked in the Hauterivian–Aptian. This period was marked by the deposition of black shales and silicites elsewhere in the Outer Carpathian realm. Teschenite intrusive and extrusive magmatism occurred in the highly attenuated Silesian basin.

The Dyje–Thaya Rift-related Depression of Southern Moravia and Northeastern Austria

In southern Moravia and northeastern Austria, the Jurassic (Liassic–Doggerian) rifting and extension led to the opening of a deep depression filled with as much as 2000 m (6600 ft) of Jurassic and Upper Cretaceous strata, including as much as 1500 m (4950 ft) of organic-rich Malmian Marls (Picha et al. 1978; Picha, 1979a). This depression seems to trend in the northwest–southeast direction (Picha, 1979a) (Figure 5), parallel with the Danish–Polish trough (e.g., Pozaryski and Zytko, 1980) (Figure 4), the wrench fault system on the southwestern side of the Bohemian Massif (Nachtmann and Wagner, 1987; Schroder, 1987), and some other northwest–southeast-trending Jurassic and Cretaceous rift-related grabens of the Central and Western Europe (e.g., Ziegler, 1988). We call this major depression of the Carpathian foreland the Dyje–Thaya depression according to the river, which flows through it and is called Dyje in Czech and Thaya in German.

The Dyje–Thaya depression of southern Moravia and northeastern Austria is one of the most persistent structural features in the entire Carpathian system. Formed, or at least reactivated, during the Jurassic rifting, this depression and its northwest–southeast-trending bounding faults became a significant factor in the evolution of the Western Carpathian region. During most of the Early Cretaceous, the western part of the depression was uplifted and exposed, whereas sedimentation continued in the adjacent parts of the Silesian, Magura, and Pieniny realms. The subsidence and sedimentation in the depression resumed again toward the end of the Early Cretaceous (Albian) to be interrupted in the early Cenomanian and again at the Cretaceous–Paleogene transition during the Laramide uplifting and inversion of the foreland. Renewed subsidence occurred during the Paleogene to early Miocene. Finally, the Vienna basin formed in the realm of this depression during the Miocene. This repeated history of subsidence indicates that the continental crust of the depression had remained weakened and fragmented since the time of Jurassic rifting.

The complex structural and depositional evolution of the depression also created one of the most important petroleum systems in the entire Carpathian region, from which more than 850 million bbl of oil, generated mainly from the Jurassic source rocks, has been produced to date in Austria, Moravia, and Slovakia.

The Jurassic to Late Cretaceous Convergency in the Inner Carpathians and the Formation of the Outer Carpathian Foreland Basins in the Late Cretaceous

In the Inner Carpathians, the convergence began in the latest Triassic or the earliest Jurassic (Hettangian) by gradual subduction of the Meliata–Hallstatt ocean below the Tisza block (Bukk) (Kozur, 1991; Plasienka, 1995). During the Late Jurassic to the Early Cretaceous, the Meliata–Hallstatt ocean-derived units were deformed, thrusted, and accreted, and the subduction and deformation gradually progressed into the Penninic–Pieninic domain of the Inner Carpathians (Plasienka, 1995).

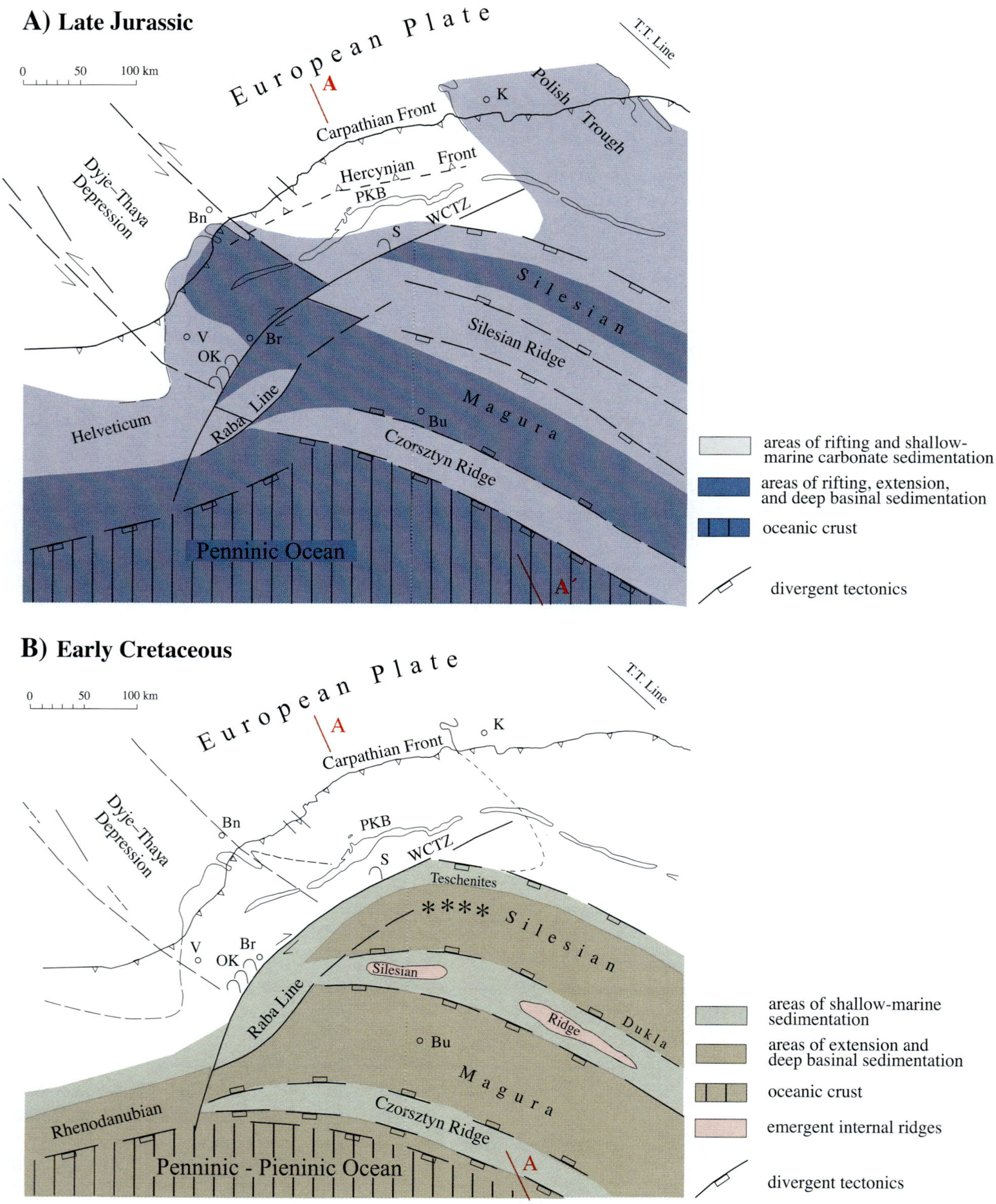

Figure 5. Palinspastic reconstruction of depositional systems of the Western Outer Carpathians during the Late Jurassic (A), the Early Cretaceous (B), the Late Cretaceous (C), and the late Oligocene (D). The distribution of the Middle to the Late Jurassic strata (A) indicates the existence of a northwest–southeast-trending structural pattern of rift-related grabens, such as the Dyje–Thaya depression in southern Moravia and northeastern Austria or the Polish trough and intervening ridges, e.g., the incipient Silesian ridge. In the Early Cretaceous (B), the depositional system was affected by the continuing rifting and extension as well as by the dextral motion and counterclockwise rotation along the southwest–northeast-trending Western Carpathian transfer zone (WCTZ), which separated the Carpathians from the Alps. During this period, the divergent depositional system of the Outer Carpathians reached its maximum extent.

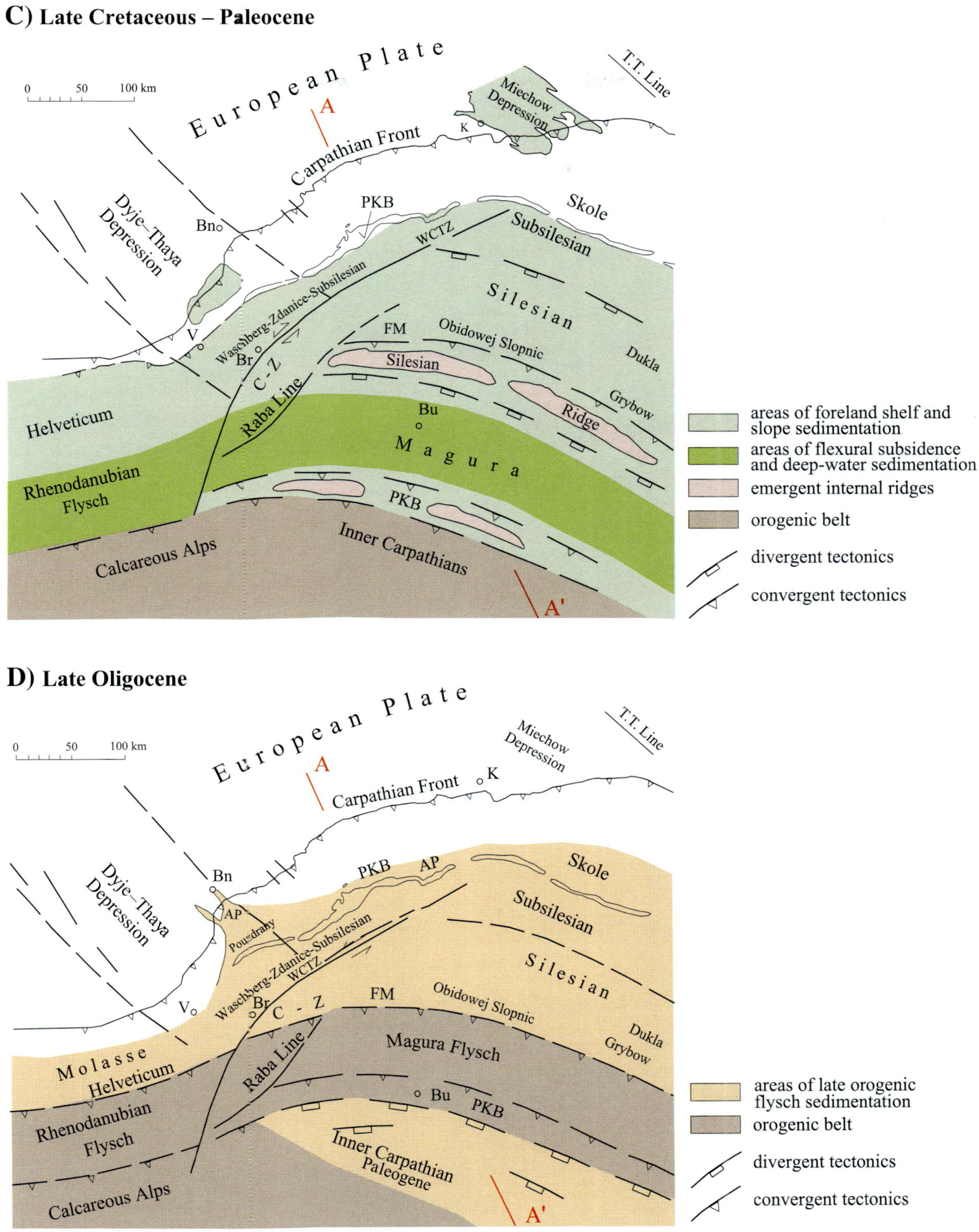

Figure 5. (cont.). In the Late Cretaceous (C), the divergent regime of the Tethyan margins changed into a convergent regime; and the motion along the Western Carpathian transfer zone reversed from dextral to sinistral. The sedimentary system spread farther northwest over the foreland. In the late Oligocene (D), the inner Magura unit was deformed and uplifted, whereas the Krosno-type flysch synorogenic sedimentation continued in the external zones of the Outer Carpathian system. The Inner Carpathian Paleogene basin formed on the top of the Inner Carpathian nappes. Line AA′ marks the section used for the geotectonic reconstructions in Figure 6. AP = autochthonous Paleogene; Bn = Brno; Br = Bratislava; Bu = Budapest; C-Z = Cejc–Zajeci unit; FM = Fore-Magura unit; K = Krakow; OK = Outer Klippen Belt carbonate buildups; PKB = Pieniny Klippen Belt; S = Stramberk carbonate buildup; V = Vienna; WCTZ = Western Carpathian transfer zone.

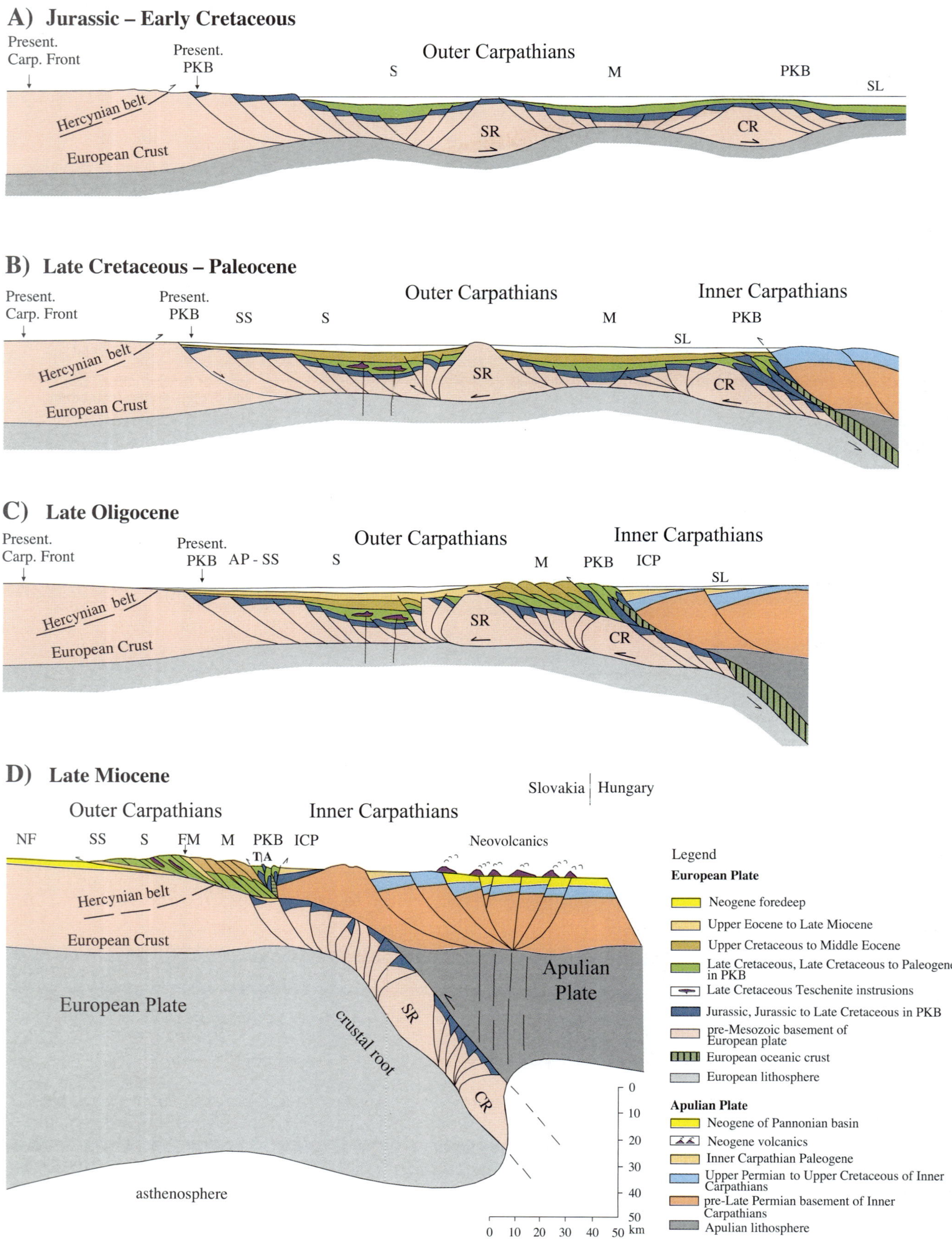
A) Jurassic – Early Cretaceous
Present. Carp. Front
Present. PKB
Outer Carpathians
S
M
PKB
SL
Hercynian belt
European Crust
SR
CR
B) Late Cretaceous – Paleocene
Present. Carp. Front
Present. PKB
SS
S
Outer Carpathians
M
Inner Carpathians
PKB
SL
Hercynian belt
European Crust
SR
CR
C) Late Oligocene
Present. Carp. Front
Present. PKB
AP - SS
S
Outer Carpathians
M
PKB
ICP
Inner Carpathians
SL
Hercynian belt
European Crust
SR
CR
D) Late Miocene
Slovakia
Hungary
Outer Carpathians
Inner Carpathians
NF
SS
S
FM
M
PKB
ICP
TA
Neovolcanics
Hercynian belt
European Crust
European Plate
Apulian Plate
SR
crustal root
CR
asthenosphere
0
10
20
30
40
50
0 10 20 30 40 50 km
Legend
European Plate
Neogene foredeep
Upper Eocene to Late Miocene
Upper Cretaceous to Middle Eocene
Late Cretaceous, Late Cretaceous to Paleogene in PKB
Late Cretaceous Teschenite instrusions
Jurassic, Jurassic to Late Cretaceous in PKB
pre-Mesozoic basement of European plate
European oceanic crust
European lithosphere
Apulian Plate
Neogene of Pannonian basin
Neogene volcanics
Inner Carpathian Paleogene
Upper Permian to Upper Cretaceous of Inner Carpathians
pre-Late Permian basement of Inner Carpathians
Apulian lithosphere

By the Senonian, all the Inner Carpathian basement-cored units, Gemeric, Veporic, and Tatric were thrusted and overridden by cover nappes of Silicic, Hronic, and Fatric units. Likewise, in the Alps, the basement-cored Austroalpine and Penninic nappes were emplaced on the Helvetic shelf of northern Europe (e.g., Tollman, 1978; Trumpy, 1980).

In the Pieniny Klippen Belt zone, the subduction of the oceanic and attenuated continental crust began in the Senonian (Figure 6). The detached Jurassic and Early Cretaceous strata were accreted toward the front of the Inner Carpathian nappe stack, whereas the sedimentation continued in a piggyback fashion, and synorogenic flysch was deposited at the front of the progressing nappes.

The deformation of the Inner Carpathians and their collision with fragmented margins of the European foreland in the Late Cretaceous marked the beginning of a major rearrangement of the Outer Carpathian depositional realm, which gradually passed from a divergent stage of rifts and passive margins to a convergent stage of prograding foreland basins and foredeeps. These two stages differ by the direction of the depositional progradation and supply of detrital material: in the passive margins, from the continent into the ocean, and in the foreland basin in the opposite direction, from the migrating thrust belt onto the continental foreland.

The depositional system of the foreland basin, which gradually spread over the Outer Carpathian domain and the adjacent part of the European foreland, was dominated by siliciclastic sedimentation of flysch and molasse. Although the synorogenic deep-water flysch sedimentation prevailed in the Late Cretaceous to early Miocene, the late orogenic and postorogenic shallow-marine and continental molasse-type sedimentation continued into the late Miocene and Pliocene.

The Late Cretaceous to Early Paleogene Laramide Uplifting of the Carpathian Foreland

The compressional stresses associated with the Late Cretaceous to early Paleocene deformation and thrusting of the Inner Carpathians and the Pieniny Klippen Belt were transmitted far into the European foreland, which was uplifted and deeply eroded. This Late Cretaceous–early Paleogene deformation of the Alpine–Carpathian foreland is commonly referred to as the Laramide orogenic phase or Laramide orogeny. During this event, some preexisting Jurassic to Lower Cretaceous extensional structures in the foreland as well as the Late Permian to Cretaceous Polish trough were inverted (Roure, et al., 1993; Kusmierek, 1994; Krzywiec, 2002). The Bohemian Massif, including the Carpathian foreland in Moravia, was dissected and, in part, uplifted along a set of wrench and reverse faults (e.g., Roth, 1978; Malkovsky, 1987). The magnitude of uplifting and erosion is well documented by the incision of two paleovalleys and submarine canyons, more than 1500 m (5000 ft) deep, in the Dyje–Thaya Jurassic rift-related depression of southern Moravia and northeastern Austria. The paleovalleys were filled with Paleogene marine deposits, which remained in the autochthonous position below the Carpathian thrust belt and the Neogene foredeep (see the section on the Autochthonous Paleogene and Paleovalleys). The Laramide uplifting and cutting of the deep paleovalleys further increased the structural complexity of the Jurassic Dyje–Thaya depression.

Within the Outer Carpathian depositional system, the Laramide phase led to a further uplifting of the inner ridges (cordilleras), as documented by a sudden influx of the coarse clastic material from the ridges into the Carpathian basins, most prominently into the Silesian and Magura basins.

The Laramide uplifting and inversion of the Carpathian foreland and the rise of internal ridges in the Outer Carpathian depositional domain during the Late Cretaceous to early Paleogene tectonism may be compared with the Late Cretaceous–early Paleogene (Laramide) foreland-type deformation in the North American Rocky Mountains in Colorado and Wyoming. The easternmost of these thick-skinned crystalline basement-involving structures is located more than 500 km (300 mi) east of the front of the thin-skinned Sevier orogenic belt in Utah and Wyoming. By analogy, the foreland-type deformation in the Carpathian

Figure 6. Geotectonic reconstruction of the depositional and structural history of the Western Outer Carpathians from the time of maximum divergence in the Aptian–Albian (A) to the time of the late orogenic convergence in the late Miocene (D). The present interpretation is based on assumptions that the Magura flysch was laid down on a continental and transitional crust, and that the Silesian and Czorsztyn ridges were rifted apart from the European plate and during the convergence accreted back to the European plate instead of being subducted. The term "Apulian plate" (Adria) is used here as a general name for microplates (terranes) of the Carpathian region separated from North Africa. The thickness of sedimentary formations is vertically exaggerated, the European crust is not internally differentiated, and the Inner Carpathians are shown schematically in three colors only. AP = autochthonous Paleogene, CR = Czorsztyn ridge, M = Magura flysch, NF = Neogene foredeep, PKB = Pieniny Klippen Belt, S = Silesian unit, SR = Silesian ridge, SS = Subsilesian unit, SL = sea level. The reconstructions are made along the section AA′ located in Figure 5.

region apparently also extended for hundreds of kilometers from the thin-skinned front of the Inner Carpathians into the Outer Carpathian depositional system and its foreland.

The Internal (Intrabasinal) Ridges (Cordilleras)

When formed in the Late Cretaceous, the depositional realm of the Outer Carpathian foreland basin displayed a complex rugged relief of deep basins and intervening ridges (cordilleras), such as the Silesian and Andrychow ridges of the Western Carpathians and the Marmarosh Massif of the Eastern Carpathians. The Silesian ridge (cordillera) (Ksiazkiewicz, 1960) separated the Magura flysch basin from the more external Silesian and Subsilesian depositional realm and shed clastics into both of these depositional systems (Figures 5, 6). The less prominent Andrychow ridge, recognized in the Polish sector of the Western Carpathians (Ksiazkiewicz, 1960, 1977), emerged between the Silesian and Subsilesian-Skole basins.

The sudden rise of the Silesian ridge in the Cenomanian–Turonian (Poprawa et al., 2002) is documented by the deposition of thick coarse clastics of the Godula and Istebna formations (Cenomanian–Turonian to Paleocene) in the Silesian basin. Only later, since the late Campanian, were the coarse clastics from the Silesian cordillera supplied into the Solan Formation (Maastrichtian to Paleocene) of the most proximal Raca subunit of the Magura flysch basin. The existence of a significant time gap between the beginning of the deposition of coarse cordillera-derived clastics into the Silesian and Magura units remains unexplained. In southern Moravia, where the Silesian unit is not present, the coarse clastics from the Silesian ridge were shed directly into the adjacent Cejc–Zajeci unit.

Attempts were made to reconstruct the geological history of these internal ridges from the composition of sandstones and conglomerates supposedly sourced from these ridges. Based on studies of Paleocene to Eocene conglomerates in the Cejc Zajeci unit in southern Moravia, Picha et al. (1966) suggested that the part of the Silesian ridge that sourced these conglomerates consisted of epizonal metamorphic rocks with stressed granites and mylonitized zones. The sedimentary cover was represented by Jurassic clastic and carbonate rocks, similar to those found in the Jurassic platforms and basins elsewhere in the Carpathian system. Fragments of crystalline rocks commonly found in some Jurassic limestones indicate that the Jurassic deposits, at least in some areas, were laid down directly on the crystalline basement. Sporadically found pebbles of basic effusive rocks might be assigned to the Permian or Triassic volcanism. In northern Moravia, the conglomerates of the Subsilesian unit contain abundant pebbles and cobbles of Paleozoic rocks (Boucek, 1952; Boucek and Pribyl, 1954, and others). The overall architecture of the inner ridges, their drainage system, and the way the quantities of coarse clastic were transported into the surrounding basins are still poorly understood. The formation of thick but spatially limited subsea fans, such as the Godula and Istebna formations in the Silesian unit, adjacent to the hypothetical ridges, would require an existence of major submarine canyon and channel systems connected with an active drainage network. Further studies of the architecture of these subsea fans combined with the application of modern analogs may provide a better understanding of the character of these ridges and the ways of transportation of the ridge-derived clastics into the flysch basins.

The origin and geological history of the internal ridges have been discussed in various publications. Sandulescu (1988) interpreted the Silesian ridge (cordillera) and corresponding Middle Dacides in the Eastern Carpathians as a thrust belt, which formed during the middle Cretaceous. Picha and Stranik (1999) explained the Silesian cordillera as a foreland-type compressional structure whose rise in the Late Cretaceous was caused by compressional stresses associated with the early collision of the Inner Carpathians with fragmented margins of Europe in the Late Cretaceous. These crustal deformations penetrated into the foreland along the deep decollements, possibly at the base of the continental crust, and led to compression and shortening of the previously attenuated European crust (Figure 6). At least some rise of the Silesian cordillera and other internal ridges might be related to the Laramide uplifting of the European foreland in the Late Cretaceous and early Paleogene (see previous section).

Based on our new paleogeographical reconstructions, we tend to believe that the internal ridges of the Outer Carpathian depositional domain may have initially formed during the Jurassic to Early Cretaceous rifting and extension of the European margins. Thus, already during the divergent stage, these highstanding crustal blocks (horsts) of the rift system separated various depocenters, e.g., the Silesian and Magura basins, in the passive continental margins (Figure 6). At least sporadically, these ridges sourced some flysch-type depositional sequences, e.g., the Hradiste Formation in the Silesian subbasin. The elevated ridges also became sites of the development of carbonate buildups and platforms, such as the Stramberk and Ernsbrunn–Pavlov Outer Klippen (Figure 5). Following the early Late Cretaceous to early Paleogene collision of the Inner Carpathians with the fragmented European margins, these ridges were uplifted and apparently deeply eroded, shedding quantities of coarse clastics into the adjacent flysch basins. Toward the end of the Eocene, the ridges submerged and were buried below the progressing Carpathian thrust belt (Figure 6). Whether they were subducted or accreted to the European crust, the explanation we prefer, remains a matter of discussion.

The Paleogene and Early Neogene Deformation and Thrusting of the Outer Carpathians

The structural and depositional evolution of the Outer Carpathian foreland basin and its equivalents in the Alpine region was primarily controlled by the subduction of the Penninic–Pieninic oceanic and attenuated continental lithosphere underneath the Eastern Alps–Inner Carpathian system (jointly referred to as ALCAPA). This process resulted in a northward shortening and stacking of thrust sheets and a gradual transfer of the depocenters of the foreland basins toward the European foreland. By the end of the Late Cretaceous, the Penninic–Pieninic oceanic lithosphere was consumed, and in the Paleocene, the subduction progressed into the Pieniny–Magura zone of the Outer Carpathians (Birkenmajer, 1986). The deformation and uplifting of the Magura flysch basin, as documented by deposition of Krosno-type strata in the external units, continued into the early Miocene. In the more external Silesian and Waschberg–Zdanice–Subsilesian zones, the deformation and thrusting began in the late Oligocene and lasted into the early Miocene (Karpatian) to the early middle Miocene (Badenian). By then, the Outer Carpathian Flysch belt in Moravia was thrusted over the undeformed Neogene foredeep and the European foreland. However, in the Vrancea Mountains of the Eastern Carpathians, the compressional Alpine tectonics continued into the Pliocene–Quaternary. According to Nemcok et al. (1998a), the Carpathian subduction was ended by a weak collision during the Pannonian (11.5–6.2 Ma).

The advancement of the compressional deformation and thrusting into the Outer Carpathian realm was accompanied by the extension and subsidence in the Inner Carpathian hinterland. Following the marine transgression in the middle Eocene (Lutetian), a deep-water flysch basin formed in the northern zones of the Inner Carpathian nappe stack (Figures 5D, 6). The sedimentation in this Inner Carpathian Paleogene basin continued into the early Miocene. With respect to its previously consolidated Inner Carpathian substratum, this was a postorogenic basin only marginally deformed by the transpressional tectonics along the contact zone with the Pieniny Klippen Belt. No comparable equivalents of the Inner Carpathian Paleogene basin are present in the Alpine region, where the more advanced continental collision apparently did not allow for any significant subsidence and extension during the middle to late Paleogene and early Miocene.

The Late Orogenic Extension and Subsidence in the Carpathian Orogenic System

The youngest phase of the evolution of the Western Carpathians is marked by the melting of the subducting lithosphere and the formation of magmatic arc accompanied by the asthenospheric upwelling (e.g., Lexa and Konecny, 1974; Stegena et al., 1975) and the opening of the back-arc Pannonian Basin (Figure 6D). The development of the Pannonian Basin began with the transtensional rifting and opening of deep pull-apart grabens filled with the middle Miocene (Badenian) strata. After the late Miocene (Pannonian), this fault-dominated extensional stage was followed by the thermally induced regional subsidence, which gave the Pannonian Basin its present form. According to Burchfiel and Royden (1982), the back-arc extension in the Pannonian Basin area was compensated for by a long-range thrusting of the Outer Carpathians over the European foreland and further bending of the Carpathian arc.

The northward thrusting in the Carpathian belt and the back-arc extension in the Pannonian Basin were further modified by a lateral, southwest–northeast-directed material movement (escape) from the Eastern Alps into the Carpathian–Pannonian realm (e.g., Le Pichon et al., 1988; Nemcok et al., 1998b; Ratschbacher et al., 1991a, b). The Pannonian Basin system thus developed in a combined back-arc and escape tectonic setting (Burchfiel and Royden, 1988).

In the early to late Miocene, the Vienna basin formed in the Outer Carpathian and partly Inner Carpathian nappe stack. Its formation and development is related to the combining effect of the continuous subsidence in the Dyje–Thaya depression and the transtensional wrench faulting associated with the northeastern translation (escape) of the Western Carpathians along the Western Carpathian transfer zone (see the next section). In addition, a small Pliocene Orava–Nowy Targ basin formed in a downwarped zone of the Pieniny Klippen Belt and adjacent zones of the Inner Carpathian Paleogene and the Magura flysch in the Orava–Podhale region.

In the late Miocene to Pliocene, the Upper Moravian Depression formed in central Moravia. Bounded by northwest–southeast-trending faults, it extends from the Hercynian foreland across the Neogene foredeep into the Carpathian thrust belt (Figure 3).

Although undeformed, these various late orogenic and postorogenic basins formed primarily in response to the tectonic movements in the orogenic belt and its foreland and, as such, have to be considered as components of the overall Carpathian orogenic system. Of those, only the Vienna basin and the Upper Moravian Depression are present in the territory of Moravia and, hence, the subject of a more substantial treatment in the following sections. The Pannonian Basin is more thoroughly discussed by Tari and Horvath (2006).

The Western Carpathian Transfer Zone

Most structural reconstructions and geophysical interpretations of the Western Carpathians in Moravia and Western Slovakia indicate an existence of a southwest–northeast-trending fault zone (wrench corridor) that roughly parallels the Pieniny Klippen Belt. Various lineaments, and fault zones referred in the literature as the Peripieniny lineament of Maska (Buday et al., 1961; Roth, 1980b), the Vah line, the Zahori fault (O. Fusan, J. Ibrmajer, and J. Plancar, 1979, personal communication), the Verona–Semmering–Vah fault system (Schenk et al., 1994), the Mur–Murtz–Zilina line of Bada et al. (2001), or even the Raba line and some other southwest–northeast-trending faults of western Hungary (Szafian et al., 1999) may, in fact, be elements of this broad structural corridor, which we propose to call the Western Carpathian transfer zone (Figure 5). We understand it as a wide structural zone in which southwest–northeast motion occurred along widely spaced (en echelon) individual wrench faults alternatively activated and newly created during the changing geodynamic settings of the main lithospheric units.

The lateral motion along the strike-slip faults in the transfer zone, predominantly dextral during the divergent stage and sinistral during the convergent stage, was an important factor in the structural and depositional evolution of Western Carpathians and their differentiation from the Eastern Alps. Among the many differences between the external zones of the Eastern Alps and those of the Western Carpathians, the absence of the Pieniny Klippen Belt in the Alpine sector seems to be the most obvious. The typical Pieniny Klippen Belt, as one of the most characteristic features of the Western Carpathians, does not continue into the Alpine domain beyond the projected trend of the transfer zone near the city of Vienna. In addition, the Early Cretaceous rifting and extension, which led to a wide opening of Outer Carpathian basins, the Silesian basin in particular, and possibly to an incipient formation of the internal ridges, was more prominent in the Carpathian domain. The continental collision began in the Western Alps in the late Eocene and led to a significant inversion of the European foreland plate and development of large foreland molasse basins. In the Carpathian realm, east of the transfer zone, the deep-water flysch sedimentation continued into the early Miocene, and the weak continental collision did not lead to a significant activation of the foreland plate and to the uplifting of the external massifs. The sinistral motion along the transfer zone also accommodated the late orogenic extrusion of the Western Carpathians toward the northeast in a process referred to as escape tectonics (e.g., Ratschbacher et al., 1991a, b; Nemcok, 1993; Sperner et al., 2002). The northeastern escape of the Western Carpathians, which began in the early Badenian (middle Miocene), reflects the final adjustment of various crustal blocks to orogenic stresses, which occurred during the late stages of the Alpine orogeny. Associated with this sinistral motion in the transfer zone is also the opening of the pull-apart Vienna basin in the Western Carpathian thrust belt.

Accommodated by the lateral motion in the transfer zone, thus, were all significant differences in the rate of divergency and convergency between the west–east-trending external Alps and the southwest–northeast-running Carpathians. Because of many uncertainties in the palinspastic reconstruction of the Western Carpathians and their foreland, the location of various elements of the Western Carpathian transfer zone can only be guessed at. Tentatively, based on our reconstruction, we would locate this broad southwest–northeast-trending transfer zone in a broad corridor extending from the deep strike-slip faults of the Vienna basin on the west to the faults on the western and eastern side of the present Male Karpaty range and the Raba line on the east.

STRATIGRAPHY AND STRUCTURE OF THE EUROPEAN FORELAND PLATE

The foreland of the Outer Western Carpathians is represented by the West European platform (plate), which is, in numerous publications, referred to as the Epivariscan platform (e.g., Stranik et al., 1993). It is a collage of Precambrian, Caledonian, and Hercynian cratonic terranes assembled in the late Paleozoic and partly covered by the Paleozoic, Mesozoic, and Cenozoic strata (Figures 7, 8). In the Moravian part of the Carpathian foreland, the unmetamorphosed sedimentary cover of the cratonic basement extends over two plate-tectonic cycles, the Paleozoic Hercynian and the Mesozoic to Cenozoic Tethyan–Alpine cycle (Figure 9).

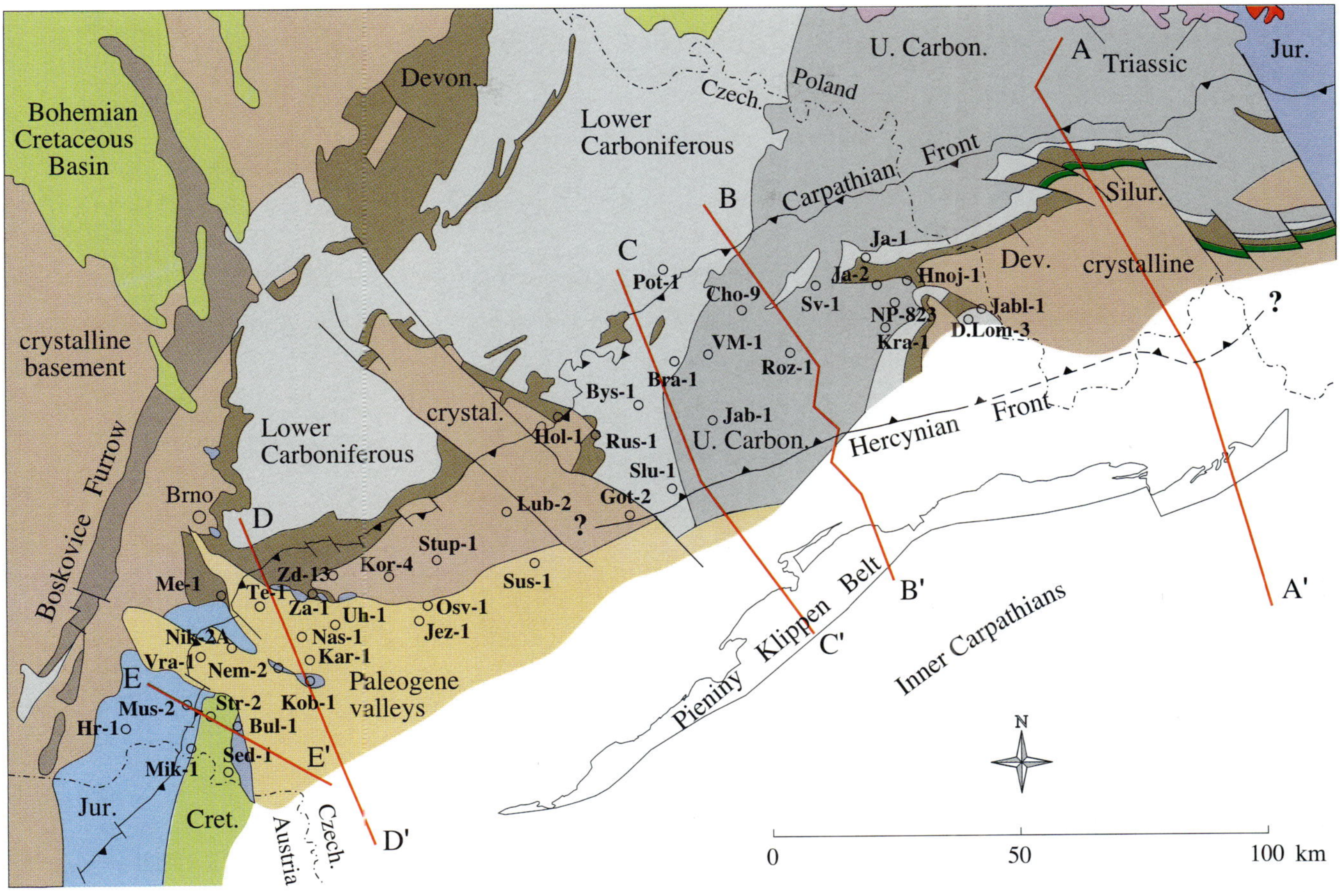

Figure 7. Pre-Neogene subcrop map showing the continuation of Paleozoic, Mesozoic, and Paleogene strata of the West European foreland plate (platform) below the Carpathian thrust belt. Modified from Poprawa and Nemcok (1988) and Picha (1996). Also shown is the anticipated front of the Hercynian orogenic belt below the Carpathian thrust belt. Stratigraphic records of all wells shown on the map are reported in Appendix 1. Cross sections of AA′, BB′, CC′, and DD′ are shown in Figure 20 (located on page 118).

The Cratonic Basement

Bohemian Massif, Brunovistulicum

At the territory of northeastern Austria, Moravia, and Silesia, the cratonic basement of the Carpathian foreland, both in front and under the Carpathian thrust belt, is represented by a complex terrane traditionally called the Bohemian Massif (Figure 4). It actually consists of two primary units: the Hercynian (Variscan) orogenic belt and the late Precambrian (Cadomian) terrane. In the Czech literature, this Precambrian terrane is most commonly known as the Brunovistulicum (Dudek, 1980), a term coined with an emphasis on the present similarity of various segments juxtaposed from southern Moravia to Vistula River in Poland. It is also called Brunnia, a broad paleogeographic term for a continental wedge between the typical Baltica and the Gondwana-related segments (Zapletal, 1931), Brno–Malopolska Block (Suk et al., 1984), having similar meaning as the voluminous assemblage of terrane segments of Brunnia by Zapletal or the East Silesian Massif of Ziegler (1989). This Cadomian assemblage of segments consists of the Brunovistulian, Malopolska, and Lysogory suspect terranes (or groups) (e.g., Valverde-Vaquero et al., 2000). The Hercynian orogenic system is further divided into the internal Moldanubian and Saxo-Thuringian and the external Rheno-Hercynian (Moravo-Silesian) zones (Figure 4). In most of the publications, however, the loosely defined term Bohemian Massif is used for both the Hercynian and late Precambrian terranes. Because, following the Hercynian orogeny and prior to the development of the Tethyan–Alpine system, the Cadomian and Hercynian terranes were accreted into a single West European plate, the usage of the common term Bohemian Massif for both of these terranes in the Alpine–Carpathian literature is acceptable.

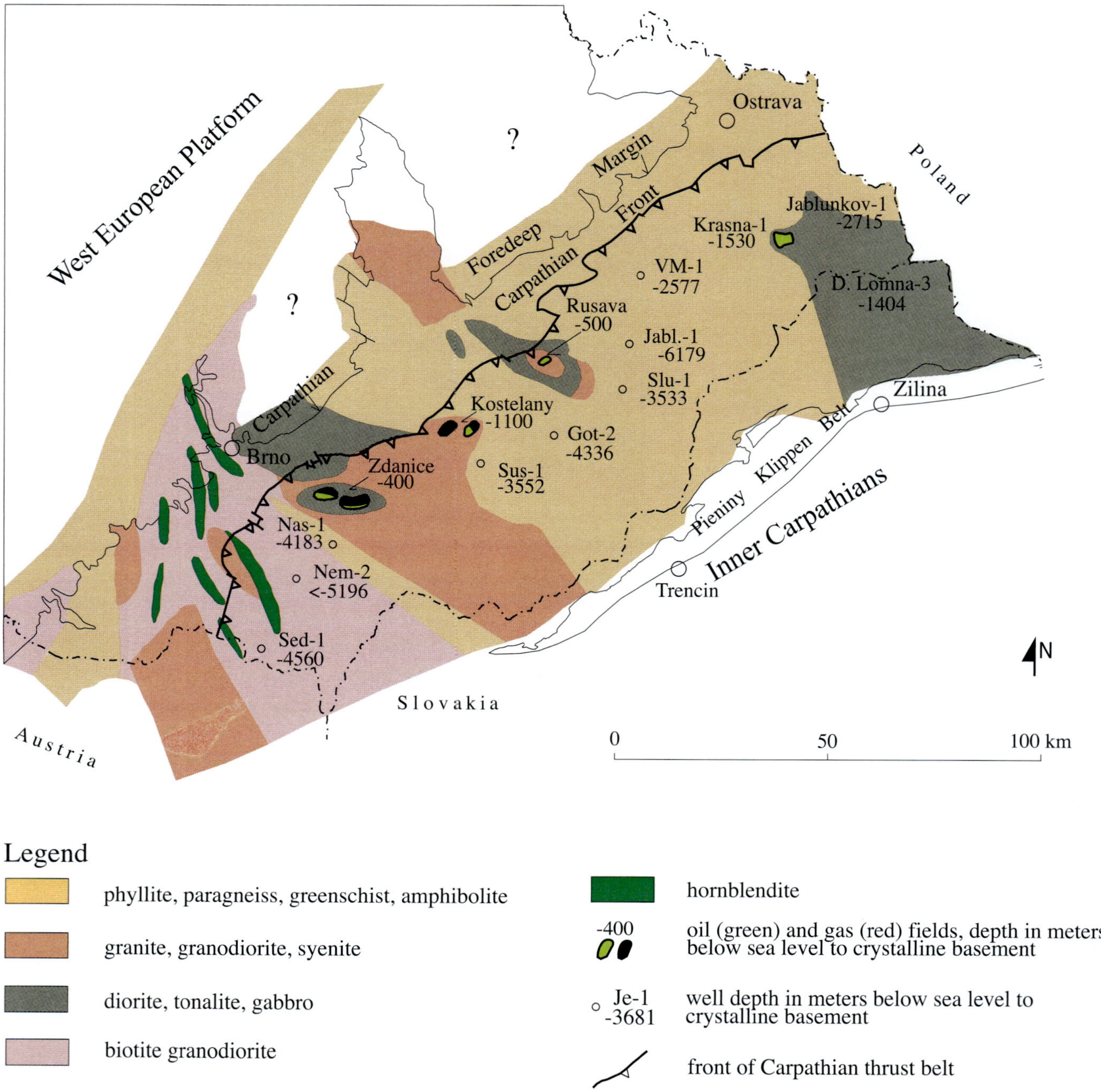

Figure 8. Geological map of the crystalline basement below the Neogene foredeep and the Carpathian thrust belt in Moravia. Modified from Dudek (1980). Stratigraphic records of wells shown on the map are reported in Appendix 1.

Because of the limited amount of seismic data and the overprint by younger tectonic events, the deep contact between the Hercynian system of the Bohemian Massif and the Brunovistulicum in its foreland is little known. Tentatively, the Sternberk–Benesov suture zone of northern Moravia and Silesia and the southern extension of the Boskovice graben in southern Moravia are considered as superficial expressions of such a main geological boundary (e.g., Dudek, 1980). Eastward, a series of Cadomian basement blocks continues into the Upper Silesian block and the Malopolska Massif and underneath the Carpathian Foredeep and thrust belt at least to the suture of the Pieniny Klippen Belt. The relation of the Brunovistulicum to the Carpathian–Pannonian block remains a subject of speculations. Zapletal (1954) suggested that the Brunovistulicum is a promontory of the Fenosarmatian platform inserted between the Bohemian Massif and the Western Carpathians. Others

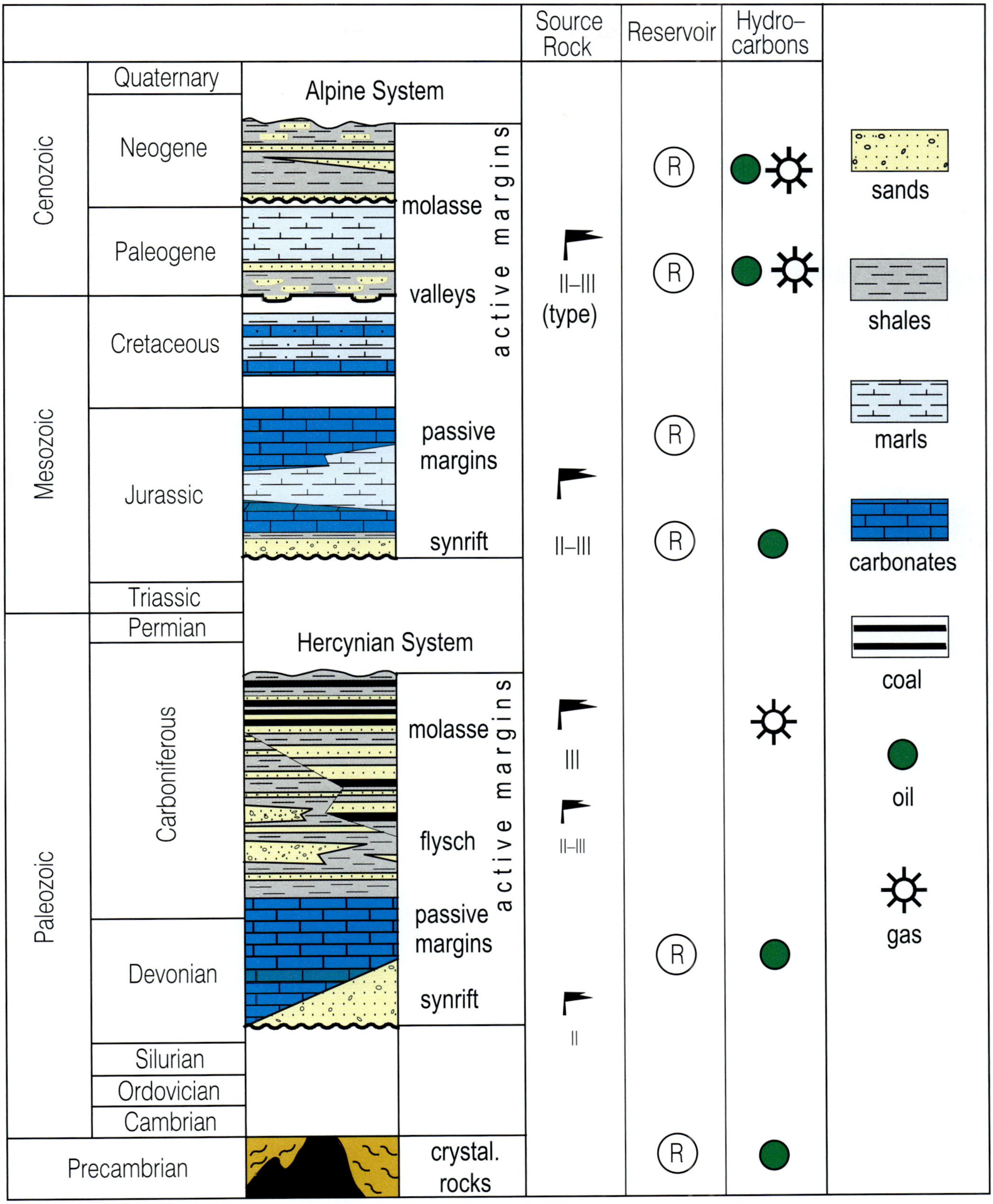

Figure 9. Stratigraphy and hydrocarbon habitat of the foreland zone of the European platform concealed underneath the Neogene foredeep and the Western Carpathian thrust belt in Moravia. Modified after Picha (1996).

believe that the Brunovistulicum is a microcontinent separated from the Gondwana at the beginning of the Paleozoic and positioned at the eastern side of the hypothetical PeriGondwana terrane of Avalonia (e.g., Kalvoda, 1995). The regional extent of the Brunovistulicum has been discussed by Buday and Suk (1989) and Suk (1993). The latest account on the structure of the Brunovistulicum was published by Gnojek and Hubatka (2001).

The Brunovistulicum and other Cadomian terranes thus actually make the essential part of the Western Carpathian foreland basement, both in front and below the thrust belt, from the Danube River in the southwest to the Teisseyre–Tornquist line separating the Western and Eastern European platforms in the northeast. In Moravia, the Brunovistulicum attains a very special position as a cratonic foreland for both the eastward-verging Hercynian and westward-verging Carpathian orogenic belts and as such was tectonically affected by both these orogenies. This is obvious especially on the western side of the Brunovistulian block, where the Early Devonian to Late Carboniferous deposits of the former passive margins of the Brunovistulicum, together with their crystalline basement, were detached and incorporated into the Hercynian thrust belt.

The crystalline basement of the Brunovistulicum consists of both magmatic and metamorphic rocks, the former prevailing in southern Moravia and the latter in central and northern Moravia. Several Cadomian plutons, such as the Brno, Zdanice, and Lubna (Kostelany) massifs, have been recognized and partly mapped below the Carpathian foredeep and thrust belt (Figure 8). The largest of them, the Brno massif, partly exposed in the foreland, is composed predominantly of granites and granodiorites, with subordinate quartz diorites, tonalites, diorites, and gabbros. Sandwiched in the middle of the Brno massif is a north–south-trending unit of tholeitic metabasalts of an oceanic character. Radiometric measurements (Dudek and Melkova, 1975; Finger et al., 2000) indicate that the granitic massifs originated during the final stages of the Cadomian orogeny (580–590 Ma). The metabasalts, including the quartz keratophyre suite and the secondary quartzites, are more than 725 Ma (Finger et al., 2000). Their contact with granites is tectonic, possibly accretionary (Hanzl and Melichar, 1997). The metamorphic rocks, prevalent in northern Moravia, are represented by various types of slightly migmatized paragneisses, greenschists, phylites, and amphibolites.

The relief map of the crystalline basement of the Brunovistulicum below the Carpathian foredeep and thrust belt (Gnojek and Hubatka, 2001) displays a rugged relief marked by southeastward-running erosional-like features, most commonly present in the southern and central Moravia and less visible in the northern Beskydy region.

Commercial accumulations of oil and gas have been found in the weathered and fractured surface of the granitic rocks of the Zdanice, Lubna, and Krasna elevations (Krejci, 1993; Blizkovsky et al., 1994).

The Hercynian Cycle

The predominantly unmetamorphosed Paleozoic rocks of the external zones of the Hercynian belt thrusted eastward over the bulged Brunovistulicum are exposed in the Drahany and Nizky Jesenik highlands of Moravia and Silesia. The extension of Paleozoic strata, both allochthonous and autochthonous, further east is buried below the Neogene foredeep and the Carpathian thrust belt. However, numerous deep wells drilled in search of coal and hydrocarbons in this area provide information about the distribution of Paleozoic rocks below the Carpathian system (Figure 7). The stratigraphy and structure of the Paleozoic strata adjacent and below the Carpathian thrust belt have been discussed in numerous papers, e.g., Kettner (1950, 1970), Dvorak and Ptak (1963), Dvorak (1973, 1978, 1993, 1995), Zukalova (1976), Adamek et al. (1980), Zukalova and Chlupac (1982), Hladil (1986, 1988), Chlupac (1989, 1994), and Hladil et al. (1994, 1999).

Considering their geodynamic evolution, the Paleozoic strata of the Carpathian foreland could be divided into three major sequences: (1) the Lower Devonian to Lower Carboniferous, synrift and passive margins sequence; (2) the Lower Carboniferous synorogenic flysch (Culm) sequence; and (3) the Upper Carboniferous late orogenic and postorogenic sequence of the Upper Silesian coal basin, including the Namurian relicts in southern Moravia.

The Lower Devonian to Lower Carboniferous Synrift and Passive Margins Sequence

The Paleozoic Hercynian plate-tectonic cycle in eastern Moravia begins in late Early Devonian (Emsian) with continental rifting and deposition of terrestrial and shallow-marine synrift clastics, commonly especially in the older literature, called the "Old Red Sandstone" (Figure 10). The thickness of these deposits ranges from a few meters to as much as more than 1000 m (3300 ft), e.g., in the deep well Menin-1. Based on finds of acritarch fauna, some of the synrift deposits have been assigned to the Lower Cambrian (Jachowicz and Prichystal, 1997). However, the existence of presumably Cambrian clastics below the lithologically indistinguishable Devonian clastics in the same wells (e.g.,

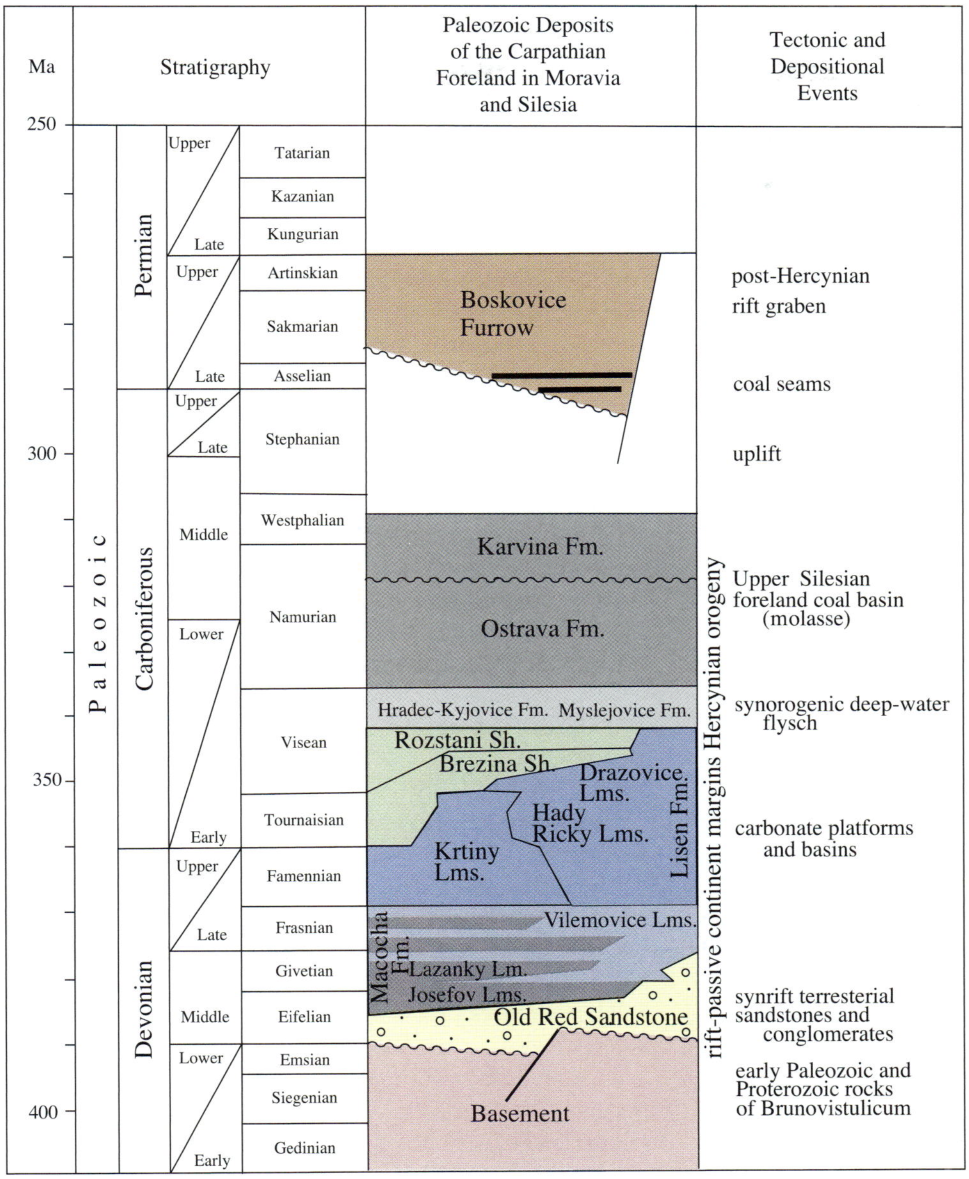

Figure 10. Stratigraphy of the Paleozoic strata of the European foreland plate in Moravia and Silesia, Czech Republic.

Menin-1) and the possibility of redeposition of the acritarch fauna make the presence of Cambrian strata disputable.

The synrift clastic sedimentation of terrestrial conglomerates, arcosic, and quartzitic sandstones was followed by a widespread marine transgression and development of carbonate platforms and ramps on the passive continental margins of the Brunovistulicum. The predominantly carbonate sedimentation continued through the Late Devonian into the Early Carboniferous (Tournaisian–Visean). The lower part of the carbonate section is represented by the Macocha Formation (Eifelian to Frasnian) deposited in an environment of large carbonate platforms with lagoons and reefs. Hladil (1983, 1986) recognized in the Macocha Formation four cycles (depositional sequences) separated by hiatuses marked by karstified surfaces, fossil soils, or layers of sandstones. These sequences commonly begin with dark-gray dolomitic limestones with brachiopods (Josefov Limestones), followed by dark-layered Amphipora limestones (Lazanky Limestones), and terminate with light-gray massive limestones with abundant coral and stromatoporoid fauna (Vilemovice Limestones). The overall thickness of the Macocha Formation is about 400 m (1300 ft); locally, it may exceed 1000 m (3300 ft). The highly diagenetically altered and karstified Vilemovice Limestones, especially their thinner equivalents that developed on paleogeographic highs, represent one of the most promising Paleozoic reservoirs for hydrocarbons, especially in the area underneath the Carpathian foredeep and thrust belt (Hladil et al., 1994).

The overlying Lisen Formation of the Famennian to Visean age is marked by the disappearance of coral and stromatoporoids apparently caused by the global Kellwasser cooling event at the Frasnian–Famennian transition. In the absence of carbonate buildups, the sedimentation tended to smooth the relief of the carbonate platforms. Unlike the underlying Macocha Formation, the Lisen Formation is characterized by a large diversity of facies. Several types of limestones, such as the nodular, micritic, and argillaceous Krtiny limestones, the turbiditic and argillaceous bituminous cherty Hady–Ricky Limestones, and the shallow-marine partly oolitic Drazovice Limestones, have been distinguished. The thickness of the Lisen Formation ranges from less than 50 to about 300 m (160 to about 1000 ft). Because of a fine-grained structure and a higher content of argillaceous material, the reservoir properties of the Lisen Formation are poor; the formation may be considered as a sealing horizon rather than a reservoir.

The overlying, mostly argillaceous deposits marked by occasional tuffaceous horizons and numerous hiatuses are known under the local names of Ostrov shale, Brezina shale, and Rozstani Shale. They represent a condensed transitional series between the predominantly carbonate sedimentation of the divergent continental margins and the clastic sedimentation of the convergent stage of the Hercynian cycle (Figure 10). This transition, however, did not occur simultaneously in the entire area, and consequently, the stratigraphic relationship of these transitional strata to both the underlying carbonates and the overlying younger clastic deposits might be diachronous. The most elevated blocks of the depositional environment have been emergent since the late Frasnian to the late Visean or even to the earliest Namurian.

The Lower Carboniferous Synorogenic Flysch (Culm)

At the onset of the Hercynian orogeny, in the Visean, the synorogenic clastic deposits supplied from the inner zones of the orogenic belt gradually spread over the carbonate platforms of the Hercynian foreland. They are represented by the deep-water turbiditic flysch facies, of alternating conglomerates, sandstones, and shales, called Culm.

The flysch deposits were laid down in a predominantly deep-water system of prograding synorogenic foredeeps. Their overall thickness decreases from several thousand meters at the western orogenic side of the depositional system, e.g., in the Nizky Jesenik Mountains, to only a few hundred meters on the distal eastern and northeastern side of the Hercynian foreland. In addition, the onset of the Culm sedimentation tended to be progressively younger toward the more stable zone of the Brunovistulican platform. In the zone adjacent to the present front of the Carpathians, the carbonate sedimentation may have lasted into the very late Visean (Dvorak, 1978), and the overlying Culm deposits are represented by a distal predominantly argillaceous synorogenic deep-water flysch facies of the Hradec–Kyjovice Formation and the Myslejovice Formation (Figure 10). The distribution of facies in the Myslejovice Formation indicates an existence of a large subsea fan system in this part of the Hercynian foredeep. The proximal part of this fan, located in the Drahany Highland, is made by the Racice and Lulec conglomerates of the channelized inner fan, which, toward the north-northwest, pass into a predominantly sandy turbiditic facies and even further into a distal, predominantly shaly facies of the foreland.

In the Carpathian region, the predominantly distal facies of the Lower Carboniferous strata of the Culm do not possess any source rock or reservoir properties; however, they may function as a regional seal for the potential underlying carbonate reservoirs.

The Upper Carboniferous, Upper Silesian Coal Basin

In the early Namurian, the deep-water flysch-type Culm sedimentation was replaced by the late and postorogenic shallow-marine and nonmarine molasse of the Upper Silesian coal basin, which formed in the distal zone of the Brunovistulican foreland in northern Moravia and Silesia. The transition from the marine Culm facies into the paralic molasse sedimentation of the Upper Silesian basin was gradual, reflecting on the gradual marine regression during the Namurian. As documented by remnants of the Upper Carboniferous strata in the Nemcicky-1 well in southern Moravia and occurrences of pebbles of Carboniferous rocks in the Cretaceous and Paleogene conglomerates of the Outer Carpathians, the original extent of the Upper Carboniferous molasse basin was apparently much larger. Situated on the outer side of the Hercynian orogen, the Upper Silesian basin has a similar position as some other coal basins of Western Europe, e.g., in Wales, Belgium, and Ruhrland.

A systematic investigation of the Upper Silesian coal basin began in the 1860s. Stur (1875, 1885) described marine fauna and flora of the Carboniferous strata. Jicinsky (1885) and Folprecht and Pateisky (1928) published monographic studies on the basin. After the Second World War, systematic work was done by Havlena (1964, 1982), Rehor and Rehorova (1972), Dopita and Havlena (1977), and Dopita and Kumpera (1993), among others. The present knowledge of the basin has been compiled in a comprehensive monography by Dopita et al. (1997).

Stratigraphically, the Upper Carboniferous strata in Moravia extend from Namurian A to Westphalian A. They are divided into the Ostrava and Karvina formations, each of them further subdivided into several members (Figure 10).

The Ostrava Formation, as much as 3200 m (10,500 ft) thick, belongs to the lower Namurian. It is a coal-bearing paralic molasse consisting of numerous cyclothems, formed through the transition from the fluviatile to lacustrine, swampy, and lagoonal environment. A typical cyclothem begins with the deposition of basal fluviatile sandstones and/or conglomerates, followed by lacustrine siltstones, a coal seam formed in a swampy environment and completed with deposition of lagoonal clays with fresh water, brackish, and marine fauna. Four major marine horizons in the Ostrava Formation are used as primary correlation horizons. The correlation of strata is further helped by the presence of several horizons of acidic tuffs and tuffites (Tonsteins). Swarms of dikes and sills of acidic and intermediate magmatic rocks are found locally.

As many as 500 coal seams are present in the Ostrava Formation, of which about one quarter is mineable. The overall thickness of the Ostrava Formation and, with that, also the number and thickness of coal beds decrease eastward toward the foreland.

In southern Moravia, the Ostrava Formation is known from deep wells. In the Damborice-1 well, the Ostrava Formation, 523 m (1715 ft) thick, evolves gradually from the underlying Myslejovice Formation of the Culm without any apparent interruption. However, in the Nemcicky-1 and Nemcicky-2 wells, the 150-m (500-ft)-thick basal conglomerates of the 1300-m (4300-ft)-thick Ostrava Formation rest on an erosional surface of the thin shaly Myslejovice Formation. The upper part of the formation consists of depositional cycles with coal seams similar to those found in the northern part of the Upper Silesian basin. The existing evidence, however, would exclude the presence of marine horizons in this part of the basin.

The overlying Karvina Formation, several hundred meters thick, also consists of numerous cycles, but unlike the Ostrava Formation, it does not have any marine horizons and therefore represents a typical continental coal-bearing molasse. Twenty-three freshwater horizons important for correlation have been distinguished in the formation. The coal seams, of which about 90 are mineable, are typically thicker but less numerous than in the Ostrava Formation. The Karvina Formation is more widely present in the Polish part of the Upper Silesian basin.

In the Czech part of the Upper Silesian basin, the sedimentation ended in the lower Westphalian, whereas in the Polish territory, sedimentation continued into the Stephanian. The top of the Karvina Formation in the Czech sector of the basin is deeply eroded and covered by Miocene deposits. Two large paleovalleys, Detmarovice and Bludovice, were cut into the Upper Carboniferous strata, apparently during the Late Cretaceous–Paleocene Laramide uplifting of the Carpathian foreland, and filled with Karpatian to Badenian detrital deposits.

The Upper Silesian basin was partly deformed during the last stages of the Hercynian orogeny, which, in the Moravo-Silesian zone, came to an end during the Westphalian. The Westphalian strata are only lightly folded, and the Stephanian strata remained undeformed. Locally, they rest transgressively on the older, moderately folded Westphalian strata. The frontal compressional structures, such as the Orlova fold, are bounded by reverse faults rather than the low-angle thrust faults.

The Upper Silesian basin is a significant black coal-producing province. Evidence exists that coal was used there during the last ice age by people of the Gravettian culture, about 23,000 yr ago. The present length of the subsurface tunnels is in a range of several thousands of kilometers. The first deep well was drilled in 1867; the present length of all wells drilled in the basin exceeds 1800 km (1100 mi).

The Upper Carboniferous strata of the Upper Silesian coal basin, with the exception of some weathered surfaces, are not considered to be reservoir rocks for hydrocarbons. However, they represent a significant source of gas, some of which accumulated in the overlying Tertiary deposits. So far, any attempts to produce commercial quantities of the coalbed methane, both in the Czech and Polish parts of the basin, have not been successful.

The Extent of Hercynian Deformations underneath the Outer Western Carpathians

Compressional Hercynian structures underneath the Carpathian foredeep and the Flysch belt in eastern Moravia have been reported by Jurkova (1979), Mencik (in Poprawa and Nemcok, 1988–1989), Cizek and Tomek (1991), Stranik et al. (1993), Picha (1996), Grygar and Jelinek (1999), and Hladil et al. (2000). Using well data and regional seismic lines, Picha (1996) tentatively mapped the front of the Hercynian belt below the Carpathian belt in northeastern Moravia. The frontal edge of the Hercynian belt, as mapped, roughly parallels the front of the overlying Carpathian belt but verges in an opposite southwest direction (Figure 7). The continuation of the Hercynian structures to southern Moravia and Austria, however, remains uncertain. Outside the Carpathian realm, the Hercynian thrusting has been recognized in the Drahany Highland west of Brno, where the high-grade metamorphic rocks of Moldanubicum are thrusted over the low-grade metamorphic rocks of Moravicum. Similarly, in the Moravian karst north of Brno, the Cadomian granitic rocks of the Brno massif are thrusted over Devonian carbonates (Kettner, 1950). Below the Carpathian thrust belt, the existence of a steep Hercynian thrust was proven by the Nemcicky wells (Krejci et al., 2002). South of the tectonically induced Nesvacilka paleovalley, the recognition of the Hercynian structures in the Carpathian foreland is further complicated by the absence of Paleozoic strata (Figure 7), which, in that area, were either eroded or initially not deposited. Tentatively, we interpret some of the structural features seen on the seismic lines in southern Moravia as Hercynian, thus extending the frontal zone of the Hercynian belt to the western margins of the Vienna basin (Figure 20 DD′, shown on page 118).

An existence of moderate Hercynian deformation in a zone adjacent to the Carpathian belt in southern Poland has been reported, among others, by Dadlez et al. (1994). Kotanski (1997) proposed to call this zone, where the Hercynian folds overprint the older Caledonian and Cadomian deformations, the Peri-Variscicum (Peri-Hercynicum).

The frontal anticlinal structures of the Hercynian belt are bounded by reverse faults and high-angle thrust faults rather than by the low-angle thrust faults. Typically, the thrusting involves both the Paleozoic (Devonian to Late Carboniferous) strata and the Cadomian crystalline basement. We interpret these thick-skinned folds as foreland-type structures detached along a deeper decollement in the crystalline basement. They differ markedly from the thin-skinned structures known from the external zone of the Hercynian belt in the Odra Hills and the Nizky Jesenik Mountains, as reported by Cizek and Tomek (1991). This implies that in the external Hercynides in northern Moravia and adjacent zones of Poland, two sets of compressional structures are present: the thin-skinned structures, comprising only the sedimentary rocks, and the thick-skinned foreland-type structures involving the crystalline basement. In that sense, the frontal Hercynides resemble some other orogenic belts, e.g., the Rocky Mountains of North America or the Ural Mountains of Russia, where both types of structures have been recognized.

We caution, however, that not all deep antiformal structures bounded by northwestward-dipping reverse faults visible on seismic lines are necessarily of Hercynian age. The origin of some of the antithetic reverse faults (with respect to the Carpathians) and associated anticlinal structures in the subthrust basement may also be related to the Laramide (Late Cretaceous–Paleogene) uplifting of the foreland or to the late orogenic transpressional strike-slip faulting, commonly identified in the Pieniny Klippen Belt. Depending on the distribution of compressional stresses, the sinistral motion of crustal blocks along the Western Carpathian transfer fault zone might have generated both synthetic and antithetic compressional structures. One such questionable structure (well documented on the seismic sections 8HR, 2T) is the Orava structure. The antithetic faults bounding this structure on its eastern side may be either Hercynian, as we interpret it on the AA′ cross section (Figure 20, shown on page 118), or much younger, related to the Alpine convergent orogeny. Krzywiec (2001) explained similar south-verging structures, seen on seismic lines in the eastern part of the Polish Carpathian foreland, by rotation of basement blocks along inherited synthetic southward-dipping normal faults. Such a rotation will induce extension on the northern side of the rotating blocks and compressional reverse faulting at their southern edges.

The potential Hercynian structures below the thin-skinned Carpathian belt may represent an interesting target for exploration. These structures and potential reservoirs in them were in place prior to the generation and migration of hydrocarbons in the Carpathian system. Unless further disrupted by younger tectonism, they may still hold hydrocarbons (Picha, 1996).

Similarities of the Hercynian and the Carpathian Orogenic Belts in Moravia

The external zone of the Hercynian belt in Moravia is a mirror image of the Outer Carpathian belt. Despite the different age, the late Paleozoic for the Hercynian belt and the Mesozoic–Cenozoic for the Carpathian belt, the opposite verging frontal zones of these two orogenic belts display many similarities in their depositional and structural setting (Figure 7). Both the Hercynian and the Carpathian external zones are represented by the synorogenic and late to postorogenic flysch and molasse sequences. These are separated from the older, more internal zones by narrow structurally complex sutures, the Pieniny Klippen Belt in the Carpathians and the Sternberk–Benesov suture in the Hercynian belt. The territory of central and northern Moravia thus provides a unique opportunity to study the frontal zones of these two major orogenic belts of Europe in the same relatively small area.

The Tethyan–Alpine Cycle

In the Carpathian foreland of Moravia and northeastern Austria, the Mesozoic to Cenozoic Tethyan–Alpine cycle began in the Early to Middle Jurassic with continental rifting and extension, followed by the marine transgression and development of passive continental margins dominated by carbonate sedimentation in the uppermost Middle and Late Jurassic (Figure 9).

No direct evidence of the existence of a Triassic strata in the Outer Carpathians and their foreland in Moravia is available, although clasts of Triassic detrital, carbonate, and possibly volcanic rocks have been reported from conglomerates in the Carpathian Flysch belt (Sotak, 1985). Considering the long-distance thrusting and orogen-parallel lateral translation of various units of the Flysch belt, the original source of these Triassic rocks should be sought in areas quite removed from the present location of the conglomeratic bodies in the Flysch belt. The most likely source would be the Silesian cordillera, a tectonically active ridge that separated the external units of the Outer Carpathians from the Magura flysch or the collisional zone between the Inner and Outer Carpathians, which comprised elements of the Czorsztyn ridge, Pieniny Klippen Belt, and the frontal units of the Inner Carpathians.

During most of the Early Cretaceous, the area of southern Moravia and northeastern Austria was uplifted. The marine carbonate sedimentation marginally resumed in the Aptian–Albian, followed by a major Cenomanian transgression and deposition of siliciclastic rocks. During the Laramide uplifting and erosion, at the Cretaceous–Paleogene transition, large paleovalleys were excavated in the Carpathian foreland of Moravia and later filled with transgressive Paleogene strata of the Carpathian foreland basin, which has gradually evolved since the latest Cretaceous. The geological history of the foreland region ended with the formation of the Carpathian foredeep in the early to middle Miocene.

During the various phases of the Alpine–Carpathian orogeny, most of the Mesozoic and Cenozoic strata were deformed and thrusted to form the Carpathian thrust belt. Only on the foreland periphery, some of the Jurassic, Upper Cretaceous, and Paleogene strata, as well as most of the Neogene foredeep, remained in their autochthonous position. Although integral parts of the Carpathian depositional assembly, these autochthonous strata are dealt with in this section on the European foreland, whereas all the deformed and thrusted Jurassic to Neogene sequences, including the undeformed Vienna basin, are discussed in the section on the Outer Carpathian thrust belt.

The Jurassic Rifting and Passive Margins

With the exception of a few smaller exposures near Brno, the autochthonous Jurassic strata on the territory of southern Moravia and northeastern Austria are buried below the Neogene foredeep and the Carpathian thrust belt (Figures 3, 7). Numerous wells, both in Moravia and Austria, indicate that the autochthonous Jurassic strata are confined to a relatively narrow zone of the Carpathian foreland between Brno and the Danube Valley. Picha (1979a) interpreted this zone, roughly identical with the extent of the Vienna basin, as a remnant of a larger northwest–southeast-trending rift-related depression, which originally extended farther into the Bohemian Massif. We call it the Dyje–Thaya depression (Figure 5A). On its northern side, this depression is bounded by a fault that parallels the axis of the Nesvacilka graben and the superimposed Nesvacilka Paleogene valley and submarine canyon and, on its southern side, by faults defined by, e.g., Zimmer and Wessely (1996) on the northeastern side of the Bohemian Massif spur south of the Danube Valley (Figure 5). These northwest–southeast-trending faults, which apparently formed or were activated during the Jurassic extensional phase, seem to continue further into the Bohemian Massif or at least are on a trend with some other significant northwest–southeast faults in the massif. As indicated by the erosional remnants of Jurassic rocks in the vicinity of Brno (Figure 7) and along the Luzice (Lusatian) fault in northern Bohemia, the Jurassic strata may have originally extended across the Bohemian Massif into northern Europe (Picha, 1979a). The continuation of the Jurassic depression toward the southeast underneath the deeper zones of the Carpathian thrust belt is not known, but the palinspastic reconstruction of the Outer Carpathian depositional system (Figure 5A) indicates that the Jurassic strata of the Dyje–Thaya depression most likely represent an extension of the Silesian–Magura–Pieniny depositional system into the foreland.

The distribution of carbonate platforms and basins along the northwestern margin of the depression (Figure 5A, B) and orientation of rift-related fault blocks (e.g., Wessely, 1988) indicate an existence of southwest–northeast-oriented structural trends in the presumably northwest–southeast-running Dyje–Thaya depression. Such a discrepancy may be explained by an existence of a transtensional rather than an orthogonal tectonic regime during the rifting phase and opening of the depression. The southwest–northeast lateral motion in the Western Carpathian transfer zone and the Late Cretaceous–early Paleocene uplifting and inversion of the Carpathian foreland might have further modified the original structural grain. The resulting distribution and orientation of Jurassic faults and rotated fault blocks in the Dyje–Thaya depression might then be complex. The detailed mapping of the structural pattern of the Jurassic strata based on three-dimensional (3-D) seismic data may be of great importance for any future exploration of deeper subthrust structures within the depression.

The Dyje–Thaya depression and the Polish Trough (Dadlez, et al., 1995) are the two areas in the foreland of the Western Carpathians where significant Jurassic rift-related faults have been positively identified and mapped. In the Alpine Molasse foredeep southwest of the Dyje–Thaya depression and west of the Bohemian Massif spur, most of the faults affecting Jurassic strata in the subthrust plate (e.g., Wagner, 1996; Zimmer and Wessely, 1996) apparently formed by the flexural downbending of the European foreland during the formation of the Neogene Molasse Basin rather than by the Mesozoic rifting and extension. North of the Nesvacilka paleovalley, in central and northern Moravia, and the adjacent part of Poland, the Jurassic strata are not known. They were either not deposited in this area or later removed by erosion. Only in the deep zone adjacent to the Pieniny Klippen Belt, some normal faults visible on seismic lines might be considered as potential rift-related structures. In addition, the Miocene faults of the Carpathian foreland in western Poland, according to Krzywiec (2001), only seldom show any evidence of being reactivated from preexisting Jurassic rift-related structures. However, in the area adjacent to the Polish trough, large normal faults, dissecting both Paleozoic basement and the Miocene foredeep infill, have been interpreted as Mesozoic rift-related structures reactivated during the Miocene (Krzywiec, 2001). In the Polish trough, the Oxfordian–Kimmeridgian event represents the second phase of rifting, which followed the previous Late Permian–Triassic phase of crustal and lithospheric thinning associated possibly with extensive regional wrenching (Dadlez et al., 1995). The existence of Jurassic rifting in the Dyje–Thaya depression and the Polish trough and its apparent absence in the area between these two features and in the eastern Alpine foreland strengthen the assumption that during the Jurassic, the northwest–southeast-trending rift-related faulting and sedimentation might have prevailed in the Western Carpathian region (Figure 5A).

The autochthonous Jurassic strata of the Carpathian foreland in southern Moravia were studied by Elias (1981), Adamek (1986, 2002), and Elias and Wessely (1990), among others. The Jurassic sequence begins with synrift terrestrial and deltaic deposits more traditionally known as the Gresten Formation but also named the Divaky Formation by M. Elias (1971, personal communication) and assigned to the Doggerian age (Figure 11). Their thickness ranges from tens of meters on the uplifted edges of rotated fault blocks to more than 1000 m (3300 ft) on the subsided sides of the fault blocks and in the local grabens. So far, the very thick synrift clastic deposits are known only from the territory of Austria, where they are subdivided into four members: the lower and upper quartzarenite series and lower and upper shaly horizons, representing the alternating continental and deltaic sequences, respectively (Elias and Wessely, 1990). The basal transgressive member contains intercalations of coaly shales and coals. The sandy facies of the Gresten Formation proved to be a good reservoir in the Damborice oil field and the Uhrice gas field.

In the Callovian, the terrestrial and deltaic synrift sedimentation of the Gresten Formation was followed by a further marine incursion and gradual development of the predominantly carbonate depositional environment of the passive continental margins. The transition was marked by deposition of 70–250-m (230–820-ft)-thick dolomitic sandstones known as the Nikolcice Member in Moravia and the Hoflein beds in Austria. In the early Oxfordian, two main facies, (1) the marginal, high-energy carbonate platform facies, and (2) the deeper basinal anoxic facies, evolved at the territory of southern Moravia and northeastern Austria (Stranik, et al., 1968; Brix, et al., 1977; Ladwein, 1988; Elias and Wessely, 1990) (Figure 11).

The marginal carbonate facies, including the erosional remnants of Jurassic strata on the exposed margins of the Bohemian basin, formed in a shallow sea along the western side of the Jurassic basin. At the surface localities near Brno, e.g., Stranska skala, Hady, Nova hora, and Svedske sance, the Jurassic limestones rest directly on the granitic rocks of the Brno massif or the Devonian limestones. These basal sandy and brecciated limestones pass into the bedded micritic, cherty, and bioclastic (crinoidal) limestones with glauconite and phosphates. The rich, shallow-marine fauna, studied by Oppenheimer (1907, 1926, 1932), among others, would assign these strata to the Callovian and Oxfordian. The known thickness of these Jurassic deposits near Brno is about 50 m (160 ft). In the area below the Carpathian foredeep and the frontal zone of the thrust belt, the carbonate facies is represented by the Altenmarkt Group, in Moravia formerly known as the Hrusovany Limestones, Novosedly Limestones and Dolomites, and Pasohlavky Limestones. This platform sequence, about 450-m (1476-ft)-thick, begins with bedded limestones and dolomites, an equivalent of the Vranovice Limestones of the basinal facies, which grade upward into bioclastic limestones and finally into algal and coral patch reefs alternating laterally with oolitic and bioclastic limestones. Basinward, the carbonate facies is fringed by a transitional facies of marly limestones distinguished in Austria as the Falkenstein Formation.

In the basinal facies, the full carbonate sedimentation began with the deposition of the fine-grained, cavernous, and partly silicified calcareous dolomites and limestones called the Vranovice Limestones (formerly the lower carbonate series in Austria). The thickness of

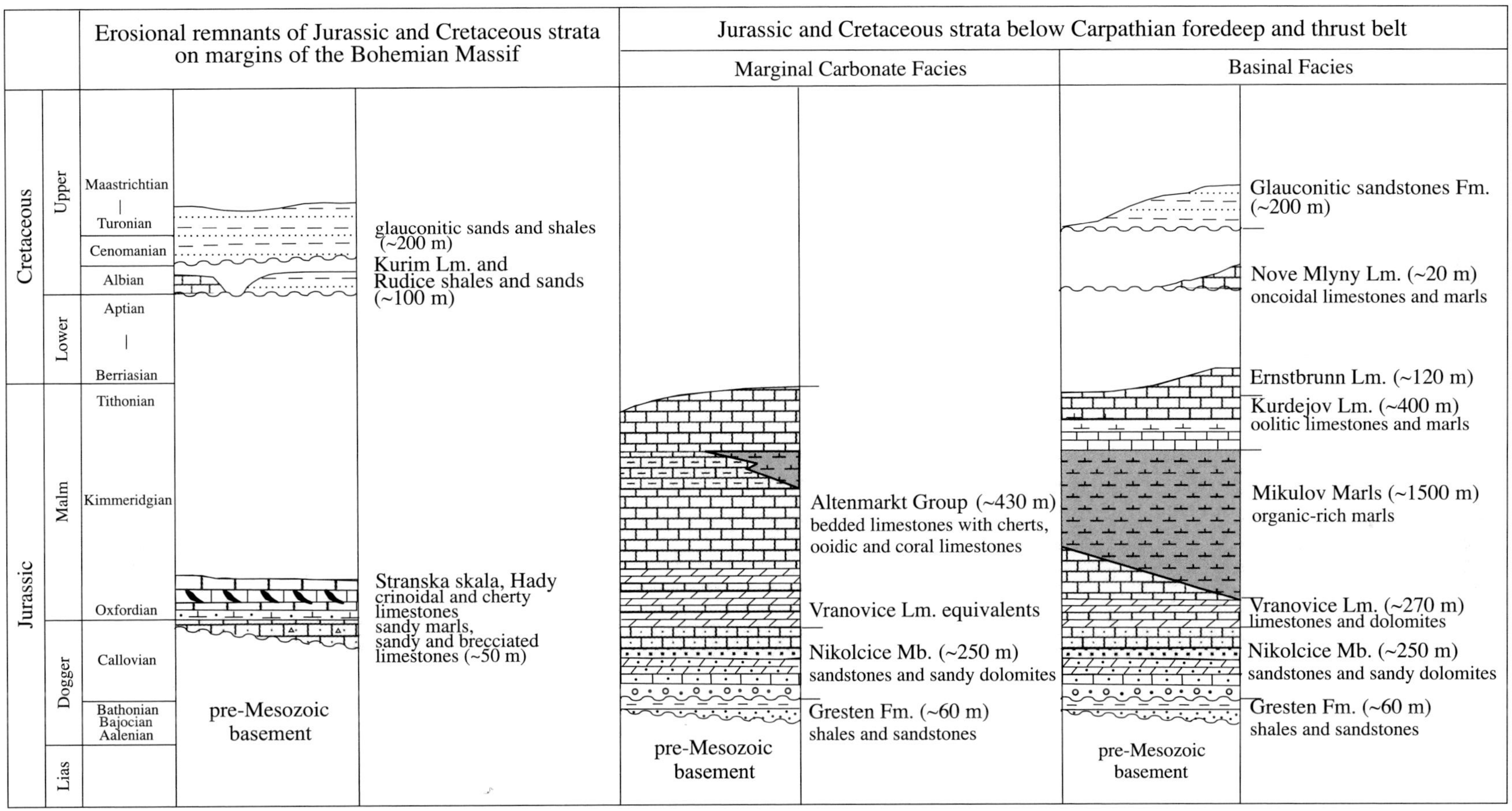

Figure 11. Stratigraphy of the autochthonous Mesozoic strata of the European foreland plate predominantly concealed below the Neogene foredeep and the Western Carpathian thrust belt in Moravia.

these strata, assigned to Oxfordian, decreases eastward toward the deeper parts of the basin. As documented by the newly discovered Zarosice oil field in central Moravia, the dolomitic rocks of the Vranovice Limestones have good reservoir properties.

The Vranovice Limestones pass gradually into the full basinal facies represented by the monotonous sequence of dark, organic-rich marls with subordinate lenses and intercalations of organodetritic limestones. They are called Mikulov Marls and correlated with the Klentnice Formation of the Outer Klippen. As established in numerous wells, the thickness of these strata increases basinward to as much as 1500 m (4950 ft). The Malmian Mikulov Marls represent a world-class source rock, which sourced most of the oils found elsewhere in northeastern Austria and southern Moravia (Ladwein, 1988; Francu et al., 1996, Picha and Peters, 1998).

Upward, the Mikulov Marls gradually pass into an approximately 400-m (1300-ft)-thick sequence of dark organodetritic limestones, dolomites, and minor marls called the Kurdejov Limestones. They are followed by the upper Tithonian, organodetritic, partly dolomitized limestones with rich mollusk, algal, and coral fauna. In wells, Wessely (in Brix et al., 1977) and Adamek (1986) named these strata the Ernstbrunn Limestones, as an equivalent of the well-known allochthonous Ernstbrunn Limestones of the Outer Klippen in northeastern Austria and southern Moravia (Figure 3).

Similarly, the autochthonous Mikulov Marls show many similarities with the coeval allochthonous Klentnice Formation of the Outer Klippen. Whether these two sets of strata, autochthonous and allochthonous, originated in a single depositional system, from which the allochthonous Klentnice Formation and the Ernstbrunn Limestones were tectonically detached, or whether the allochthonous strata were tectonically transported from a more distal carbonate environment remains uncertain. Apparently, other more internal, carbonate platforms and basins formed on the rifted and fragmented continental margins of Europe (Picha, 1996), from which, during the Alpine orogeny, the tectonic klippen might have been detached, integrated into the Outer Carpathian belt, and tectonically transported for a considerable distance.

The Cretaceous Period

The deposition of the Ernstbrunn Limestones in the Tithonian was followed by a major regression and hiatus, which, in the area of southern Moravia and northeastern Austria, lasted through most of the Lower Cretaceous. A new limited marine incursion, documented by rare finds of Aptian to Albian limestones at Kurim and sands and shales at Rudice both north of Brno, occurred toward the end of the Lower Cretaceous (Figure 11). The relicts of the Aptian to Albian limestone breccias near Kurim, described by Krystek and Samuel (1979), are found in open fractures of granitic rocks of the Brno massif. The locality is situated in a northwest–southeast-trending fault zone, which delineates the northeastern margin of the Dyje–Thaya depression. The upper Albian bioclastic and micritic limestones with algae and oysters have been also reported by Rehanek (1984) from the deep well Nove Mlyny-1 (Nove Mlyny Limestones by Adamek, 1986). No equivalents of these Lower Cretaceous rocks have been found in the tectonic klippen of the Pavlov and Ernsbrunn Hills in southern Moravia and northeastern Austria. Most of the European foreland in Moravia apparently remained exposed during the Early Cretaceous.

The uplifting of the foreland during the Early Cretaceous might have been associated with the fundamental reorganization of the Outer Carpathian realm from one dominated by the northwest–southeast-trending rift pattern of the Middle to Late Jurassic to another associated with the activation of the northeast–southwest-trending Western Carpathian transformation zone. On the Carpathian side of this zone, the dextral motion along the sheer zone led to further rifting, extension, and formation of a system of grabens and ridges (Figure 5), whereas the area west of the transfer zone in southern Moravia and northeastern Austria was uplifted and remained relatively stable throughout the rest of the Mesozoic history.

In the Cenomanian, a major global transgression engulfed most of the European platform, including its marginal parts adjacent to the Tethyan realm. The Upper Cretaceous strata are known from the Dyje–Thaya depression in southern Moravia and northeastern Austria and from the Miechow depression in Poland (Figures 5C, 7). In Moravia and Austria, they have been encountered in numerous wells below the Waschberg–Zdanice nappe. There, the Upper Cretaceous strata rest transgressively on the karstified Jurassic carbonates. The basal glauconitic sandstones pass upward into the calcareous claystones and siltstones with beds of sandy limestones and sandstones. Typically, these strata are more than 200 m (660 ft) thick; the highest known thickness of 517 m (1696 ft) was recorded in the well Ameis-1 in Austria. According to Stranik et al. (1996), the age of these deposits in Moravia extends from the Turonian to the early Campanian. In Austria, the Upper Cretaceous strata are known as the Klement Supergroup, which is subdivided into the Ameis (Cenomanian–Santonian) and the Poysdorf (Campanian–Maastrichtian) formations (Fuchs and Wessely, 1977, 1996). Both lithologically and stratigraphically, these autochthonous strata may be compared with the Klement and Palava formations (Stranik et al., 1996) found on the top of the Tithonian Ernstbrunn Limestones in the tectonic klippen of the Pavlov Hills. As indicated by faunas, these Upper Cretaceous strata of southern Moravia and northeastern Austria represent a transitional facies between the epicontinental deposits of the boreal sea of northern Europe, such as the Brezno Formation of the Bohemian Massif, and the coeval deposits of the Tethyan continental margins. Their deposition seems to be related more to the global Cenomanian transgression rather than to a progression of the Tethyan Sea onto the European foreland. Nevertheless, the distribution of these strata in the Dyje–Thaya depression in southern Moravia and northeastern Austria and in the Miechow trough in Poland indicates that the deposition and/or preservation of these strata was at least partly controlled by the renewed subsidence in these depressions.

The Autochthonous Paleogene and Paleovalleys

The autochthonous Paleogene in southern Moravia is represented predominantly by the depositional fill of two large paleovalleys buried below the Neogene foredeep and the Carpathian thrust belt (Figures 7, 12). These features, also called Nesvacilka and Vranovice depressions or canyons, were originally interpreted as tectonic grabens (Homola et al., 1961; Adam et al., 1968; Nemec, 1973), with some erosional component in their formation (Dlabac and Mencik, 1964). Picha (1974, 1979a) and Picha et al. (1978) fully recognized their erosional origin and interpreted them as paleovalleys and submarine canyons. In addition to these two main valleys, several smaller side tributaries, mainly located on the northeastern side of the Nesvacilka canyon, have been mapped (Jiricek, 1987, 1994). Indications exist that a similar paleovalley, named the Tulln canyon by Jiricek (1994), may exist at the Austrian territory west of Vienna (Figure 22, shown on page 128). The recognition of the Paleogene depressions as erosional features has been obscured by their location in a broader northwest–southeast-trending Dyje–Thaya Jurassic tectonic depression (Picha et al., 1978) (see previous section). In particular, the Nesvacilka paleovalley is located in the confines of an asymmetrical northwest–southeast-trending Jurassic rift-related Nesvacilka half graben (Picha 1996), bounded on its southwestern side by a steep fault, which apparently marks the northeastern margin of the Dyje–Thaya depression (Figure 13). In the literature, the relationships of these two genetically and morphologically different features are commonly confused, and the term Nesvacilka graben has been alternatively used for both the rift-related Jurassic graben and the Paleogene valley. So far, no indication is found that a similar rift-related graben does exist below the Vranovice paleovalley.

The tectonically controlled valleys are interpreted (Picha, 1979a, 1996) as being excavated by rivers during the Late Cretaceous to early Paleogene Laramide uplifting of the Carpathian foreland and later inundated by the sea and converted into estuaries and submarine canyons through which clastic material was transported into the foreland basin. Like some of the modern river-submarine channel systems, e.g., the Amazon and Niger, the Paleogene valleys in Moravia evolved preferentially in the preexisting depression, and their course was affected by an active fault system.

The Paleogene valleys are as much as 12 km (7 mi) wide and more than 1500 m (4900 ft) deep. By drilling and seismic data, these paleovalleys have been traced for more than 40 km (24 mi). In their size, the paleovalleys in southern Moravia are comparable with the Grand Canyon of Arizona or the Hudson submarine canyon of the East Coast of North America. They are cut into Mesozoic and Paleozoic carbonate and clastic strata and the Precambrian crystalline basement rocks and are filled with Paleogene clastic deposits (Figures 12, 13). The excavation of these gigantic valleys testifies to a major uplifting of the Carpathian foreland and sea level changes during the Late Cretaceous and early Paleogene.

The sedimentary fill of the canyons, a few hundred meters to more than 1500 m (4900 ft) thick, is made predominantly of organic-rich (1–9% total organic carbon [TOC]), laminated mudstones and siltstones with subordinate channelized sands, proximal and distal turbidites, and debris flows. The coarser clastics, sandstones, and conglomerates are distributed mainly in the lower axial parts of the Nesvacilka valley, where they form a discontinuous turbiditic sequence as much as 300 m (1000 ft) thick. They were deposited during the early active phase, dominated by erosion and transport of clastics into the foreland basin (Figure 14). The overlying mudstones then represent the later phase of abandonment and hemipelagic drape sedimentation. So far, no significant coarser deposits have been found in the less-drilled Vranovice valley. The common presence of slump folds, pebbly mudstones, and chaotic slump bodies indicates that the mass movement was a significant factor in sediment transport inside the canyons. The uppermost parts of the canyon fill are commonly deformed and possibly detached and tectonically transported along the base of the allochthonous thrust sheets (Picha, 1979a).

Because of a complex and discontinuous depositional pattern, Picha (1979a, b) assigned the whole fill, including its tectonized upper part, to one lithostratigraphic unit and called it the Nesvacilka Formation after the Nesvacilka-1, 2, and 3 wells, in which the

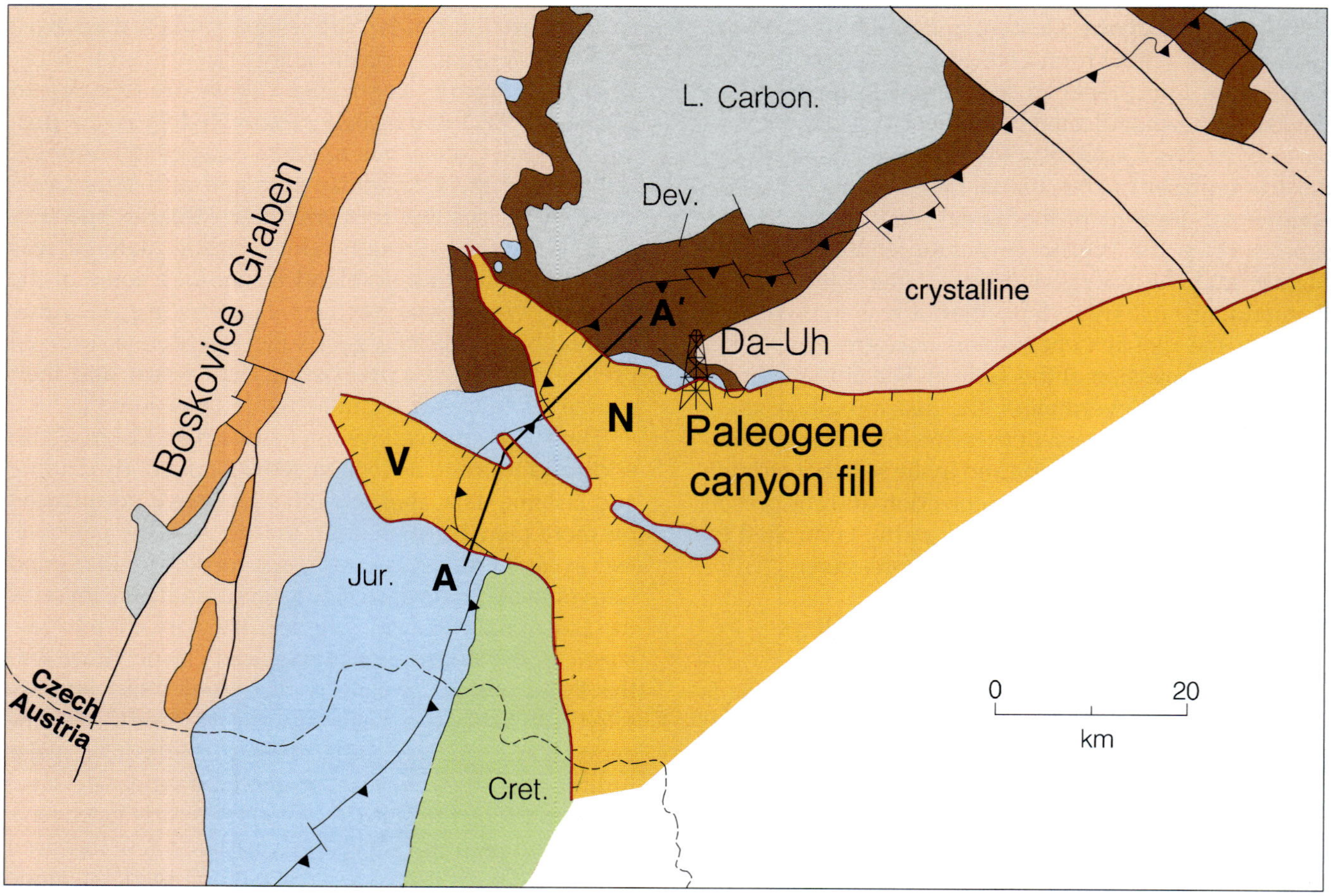

Figure 12. Pre-Neogene subcrop map showing the Nesvacilka (N) and Vranovice (V) paleovalleys cut into the European foreland plate, filled with Paleogene deposits, and later buried below the edges of the Western Carpathian thrust belt and the Neogene foredeep. Da-Uh marks the location of the Damborice and Uhrice oil and gas fields, respectively (Picha, 1996). Cross section of AA′ shown in Figure 13.

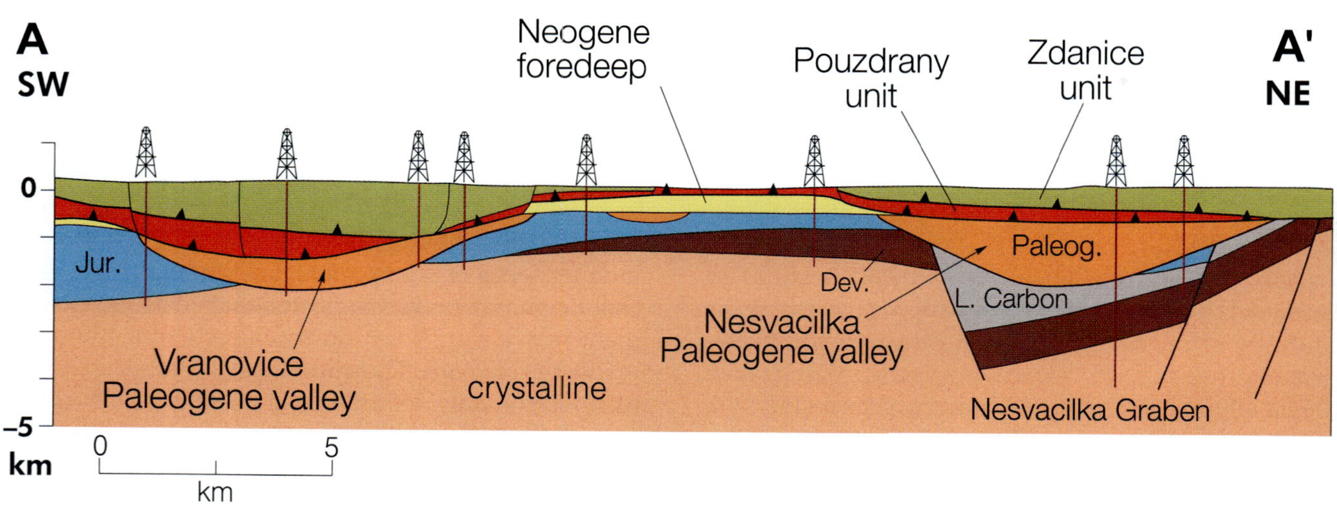

Figure 13. Cross section through the Vranovice and Nesvacilka Paleogene valleys-submarine canyons along the line AA′ (location in Figure 12). The erosional features are cut into the Mesozoic and Paleozoic strata and the Precambrian crystalline basement rocks of the Brunovistulicum. Note the position of the Nesvacilka paleovalley in the Jurassic rift-related Nesvacilka graben.

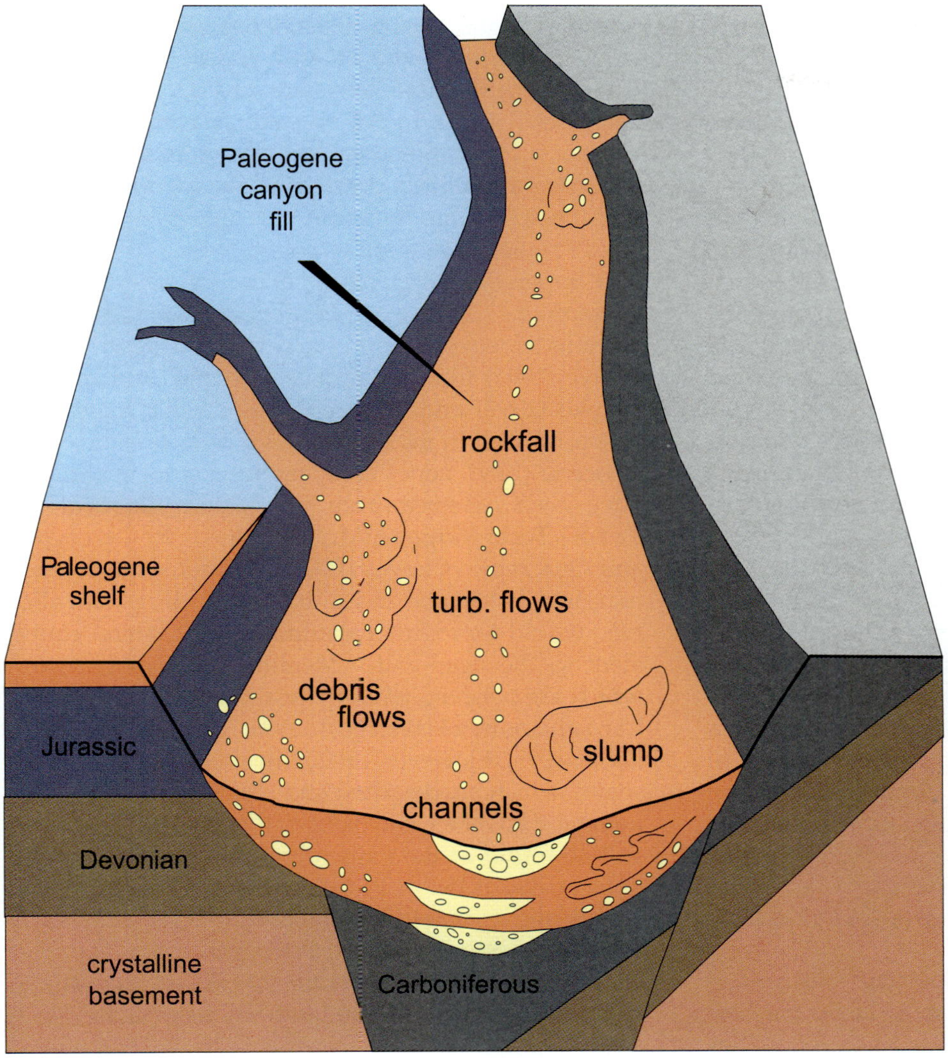

Figure 14. Schematic reconstruction of the deep-water depositional system of the Nesvacilka paleovalley and submarine canyon located in the confines of the Nesvacilka graben. The dynamic depositional regime of the early stages of the development of the submarine canyon was characterized by the activity of gravity-driven currents (turbidites and debris flows) and widespread slumping. The best reservoirs are found among the channelized sands. Modified after Picha (1979a).

autochthonous Paleogene was originally defined (Holzknecht in Homola et al., 1961). Later, Jiricek (1994) and Rehanek (1994) introduced a new name, the Damborice Group, for the whole autochthonous valley fill and subdivided it into two formations, keeping the term Nesvacilka Formation for the hemipelagic upper part of the sequence and using the name Tesany Formation for the turbidity-dominated lower part of the sequence.

The age determination of the paleovalleys fill has become a subject of discussions as well. Based on paleontological evidence, Pokorny (in Dlabac, 1946), Grill (1947), Holzknecht (in Homola et al., 1961), and Cicha and Hozknecht (1964) dated the autochthonous Paleogene deposits as late Eocene to early Oligocene. Hanzlikova (in Picha et al., 1973) confirmed the late Eocene to early Oligocene age but also pointed to a complex character of microfaunal assemblages that, in addition to highly diversified indigenous species, include faunas redeposited from the older, Jurassic, Cretaceous, and early Paleogene strata. Hamrsmid and Krhovsky (1987), Holzknecht and Krhovsky (1987), Hamrsmid et al. (1990), and Krhovsky et al. (1990) disregarded the possibility of redeposition of all the Late Cretaceous and early Paleogene microfaunas and extended the stratigraphic range of the Nesvacilka Valley fill from the Late Cretaceous (Campanian to Maastrichtian) to early Paleogene (Paleocene to early Eocene). Later, Jiricek (1994) reassigned the age of the paleovalley fill to the early Paleocene to early Oligocene.

From the regional viewpoint, the late Paleogene age of the paleovalley fill seems to be more logical. The marine incursion and depositional filling of paleovalleys could then be associated with the late Eocene transgression, which initiated the development of the Molasse basin in the Alpine foreland. Such an interpretation also enables better correlation of the autochthonous Paleogene with the allochthonous Pouzdrany unit, as well as with the lower marine Molasse of the Alps in Austria and Bavaria (Picha and Stranik, 1999). Picha (1979b) loosely correlated the autochthonous

Paleogene with the allochthonous late Eocene to early Oligocene Pouzdrany Formation of the Pouzdrany unit and with the coeval part of the Nemcice (Submenilitic) Formation and the early Oligocene Menilitic Formation of the Zdanice unit (Figure 17A, shown on page 89). All these formations belong to the late Eocene to early Oligocene depositional sequence of the Carpathian foreland basin, the autochthonous Nesvacilka Formation with the highest proportion of coarser clastics being associated with the estuarine to submarine canyons environment and the coeval deposits in the Pouzdrany and Zdanice units being deposited in the open shelf and slope to basin environments, respectively (Figure 5D).

Questionable with respect to the known stratigraphy of the valley fill is the position of the Malesovice Formation, recognized in one well at the northwestern end of the Vranovice paleovalley (Ctyroky, 1987). This formation, made by only a few meters of brownish gray marine shales with Egerian fauna, most likely represents the youngest members of the paleovalley fill, which may be correlated with the upper part of the Uhercice Formation of the allochthonous Pouzdrany unit (Figure 17A, shown on page 89). Another less likely solution would be to interpret these strata as the oldest members of the Neogene foredeep. However, because the evolution of the Neogene foredeep in that particular region began later during the Eggenburgian (Burdigalian) transgression, such an explanation would not be consistent with the overall regional stratigraphic pattern.

The rivers, which cut these paleovalleys as well as the successor submarine canyons, must have funneled large amounts of clastic material into the foreland basin. According to Picha (1969), the Late Cretaceous to the late Eocene kaolinite-rich pelitic deposits of the external Zdanice–Subsilesian and Silesian units were, to a great extent, supplied from the deeply weathered foreland of the Bohemian Massif.

Large paleovalleys buried below the Neogene foredeeps and the Carpathian thrust belt are not unique to southern Moravia. They have been identified in northern Moravia (Detmarovice and Bludovice valleys) and elsewhere in eastern Poland. However, unlike the Nesvacilka and Vranovice canyons, these valleys are filled with the middle Miocene (Badenian) deposits. Furthermore, the large Hystria and Kamchia depressions in the Black Sea shelf of Romania and Bulgaria, respectively, were apparently at least partly shaped by erosion. They are filled with 2.5–3.0 km (1.5–1.8 mi) of Eocene and Oligocene deposits.

The sedimentary fill of the paleovalleys and adjacent shelf deposits represent the northernmost marginal facies of the Carpathian foreland basin in the Paleogene. As indicated by the presence of coeval organic-rich marginal facies in the Alpine Molasse and in the Black Sea region, these deposits may have extended along the entire length of the Alpine–Carpathian belt (Picha, 1996). In the Carpathian region, with the exception of southern Moravia, these marginal Paleogene strata would be buried below the thin-skinned thrust belt. Toward the south, this autochthonous marginal Paleogene facies may extend to the present Pieniny Klippen Belt (e.g.,Tomek, 1976, 1979). In the Polish Carpathians, limited occurrences of assumed autochthonous Paleogene deposits have been described by Moryc (1995) from wells near Rzeszow and by Oszczypko (1997) and Oszczypko et al. (2000) from the Zawoja-1 well in the Orava region. Oligocene to early Miocene (NN1 zone) conglomerates and mudstones have been reported from the eastern slope of the Pribor–Tesin Ridge (Andrychow area) by Oszczypko and Oszczypko-Clowes (2003). At least some of these marginal Paleogene facies, including the paleovalley fills, would represent a shallow-marine lateral equivalent to the deep-water Menilitic (Dysodile) shales, which are believed to be the main source rocks in the Carpathians (ten Haven et al., 1993; Lafargue et al., 1994; Dicea, 1995). However, unlike the more distal Menilitic shales, which are detached and integrated into the Carpathian thrust belt, the marginal Paleogene facies may still be found in the autochthonous or parautochthonous positions below the allochthonous belt.

After being buried below the thrust belt in the early Miocene, these organic-rich Paleogene strata may have generated significant quantities of oil and natural gas, which would have migrated into any structural or stratigraphic trap, which, at the time of the migration, existed both in the Carpathian thrust belt and the underlying European foreland. The assumed presence of paleovalleys and organic-rich Paleogene deposits underneath the Carpathian thrust belt may provide a critical ingredient to a variety of potential hydrocarbon plays and, thus, greatly widen the prospectivity of the entire Carpathian system.

The Carpathian Neogene Foredeep

The Carpathian Neogene foredeep in Moravia is a part of Paratethys as originally defined by Laskerev (1924) and later further elaborated by Cicha et al. (1975a, b), Rogl and Steininger (1983), and Rogl (1998), among others, as an autonomous, depositional system separated from the Mediterranean realm by the rising Alpine–Carpathian mountain chain. However, episodic communication between these two areas did exist, especially in the middle Miocene. At this time, through the Rhine Graben, the Paratethys was also connected with the North Sea. Geodynamics and paleoclimatic changes of the Central Paratethys were discussed by Cicha and Kovac (1990), among others.

With respect to its somewhat isolated position, an autonomous regional stratigraphic nomenclature, distinct from the general Mediterranean classification, has been established for the Central Paratethyan realm (e.g., Cicha et al., 1975a, b; Steininger et al., 1988).

The evolution of foreland basins and foredeeps has been, to a great extent, controlled by the preexisting structural pattern of the foreland, which varied significantly along the length of the Alpine–Carpathian thrust belt. Typically, prominent depocenters marked by a high rate of subsidence alternate with areas of limited subsidence and deposition. In the Alpine–Carpathian region, the Bavarian–Austrian Molasse with more than 4 km (2.4 mi) of Neogene deposits, the Boryslav area in Western Ukraine with up to 7 km (4.2 mi) of Neogene deposits, and, the deepest of all, the Focsany depression in Romania with 8–10 km (5–6 mi) of Neogene deposits (Sandulescu, 1984) represent prominent depocenters. However, the segments of the foredeep, which evolved on more stable crustal blocks of the foreland, such as the Bohemian Massif on the territory of Moravia, are characterized by a low rate of subsidence and sedimentation.

As, in time, the collision along the Carpathian arc moved progressively eastward, so did the rate of sedimentation in the foredeep (Meulenkamp et al., 1996; Oszczypko, 1998). In Austria and southern Moravia, the highest rate of subsidence occurred in the Eggenburgian to Karpatian, in northern Moravia and western Poland in the Badenian, and in Romania in the Dacian and Romanian.

In the Carpathian terminology, the term foredeep is used for all principally undeformed late orogenic and postorogenic, shallow-marine and nonmarine Neogene deposits found in front, underneath, and marginally on the top of the frontal zone of the Alpine–Carpathian thrust belt and also in tectonic slices incorporated into the thrust belt. The Neogene foredeep in Moravia displays a broad range of depositional environments, including the typical depocenters adjacent to the front of the thrust belt as well as the distal facies deposited on the European foreland. For such an assembly of depositional settings, the term foreland basin would be more proper, but for the sake of nomenclatorial unity, we continue to use the traditional term Neogene foredeep for all the undeformed late orogenic and postorogenic sequences laid down on the Carpathian foreland. The northwestern landward margin of the Neogene foredeep in Moravia is delineated by outcrops of the Bohemian Massif (Figure 3). The southeastern margin is outlined on the surface by the front of the Carpathian thrust belt, but the undeformed strata of the Neogene foredeep continue farther below the thrust belt. At least 30 km (18 mi) of overthrusting over the undeformed strata of the Neogene foredeep has been verified by deep drilling (Stranik et al., 1993; Elias and Palensky, 1997; Oszczypko et al., 2006).

As established by numerous wells, the overall thickness of the Neogene foredeep on the territory of Moravia ranges in several hundred meters. Only locally, e.g., at the edge of the thrust belt near the Austrian border, in the Bludovice and Detmarovice depressions and paleovalleys in the Ostrava region, and possibly in the Roznov area underneath the Carpathian nappes (Palensky et al., 1995), the thickness of the foredeep strata exceeds 1000 m (3300 ft). The deepest zone of the foredeep is typically situated along the present erosional edge of the thrust belt, and from there, the thickness of Neogene strata generally decreases toward both the foreland and the hinterland underneath the Carpathian nappes.

The stratigraphic range of the undeformed Carpathian Foredeep in Moravia extends from the Eggenburgian (Burdigalian) to Badenian (Langhian and early Serravalian) (Jiricek and Seifert, 1990). In addition, the latest Miocene and Pliocene strata deposited in the Upper Moravian Depression are considered to be a part of the foredeep. In a broad sense, all these deposits can be correlated with the upper marine and upper nonmarine Alpine Molasse of Switzerland, Bavaria, and Austria (Wagner, 1996, 1998; Burkhard and Sommaruga, 1998). They also can be correlated with similar deposits of the Neogene foredeep on the territory of Poland (Oszczypko, 1998; Oszczypko et al., 2006).

The stratigraphic subdivision and correlation of the Neogene strata is based primarily on biostratigraphy. In terms of the Paratethyan stratigraphy, three major depositional units(?), (1) the Eggenburgian–Ottnangian, (2) the Karpatian, and (3) the Badenian, might be distinguished (Figure 15). Each of these units is characterized by a specific depositional regime, apparently dependent on the progression of the Carpathian thrusting and global sea level changes, as well as on the preexisting structural pattern.

The lithostratigraphy of the Carpathian foredeep in Moravia has not been fully developed. The fragmentary nomenclature we present in our chapter, including the chart in Figure 15, may undergo significant changes in the future.

The Eggenburgian–Ottnangian (Burdigalian)

The Eggenburgian to Ottnangian strata in Moravia are distributed mainly in the southern part of the Carpathian Foredeep between the Austrian border and Brno. They are also found on top of the Zdanice nappe

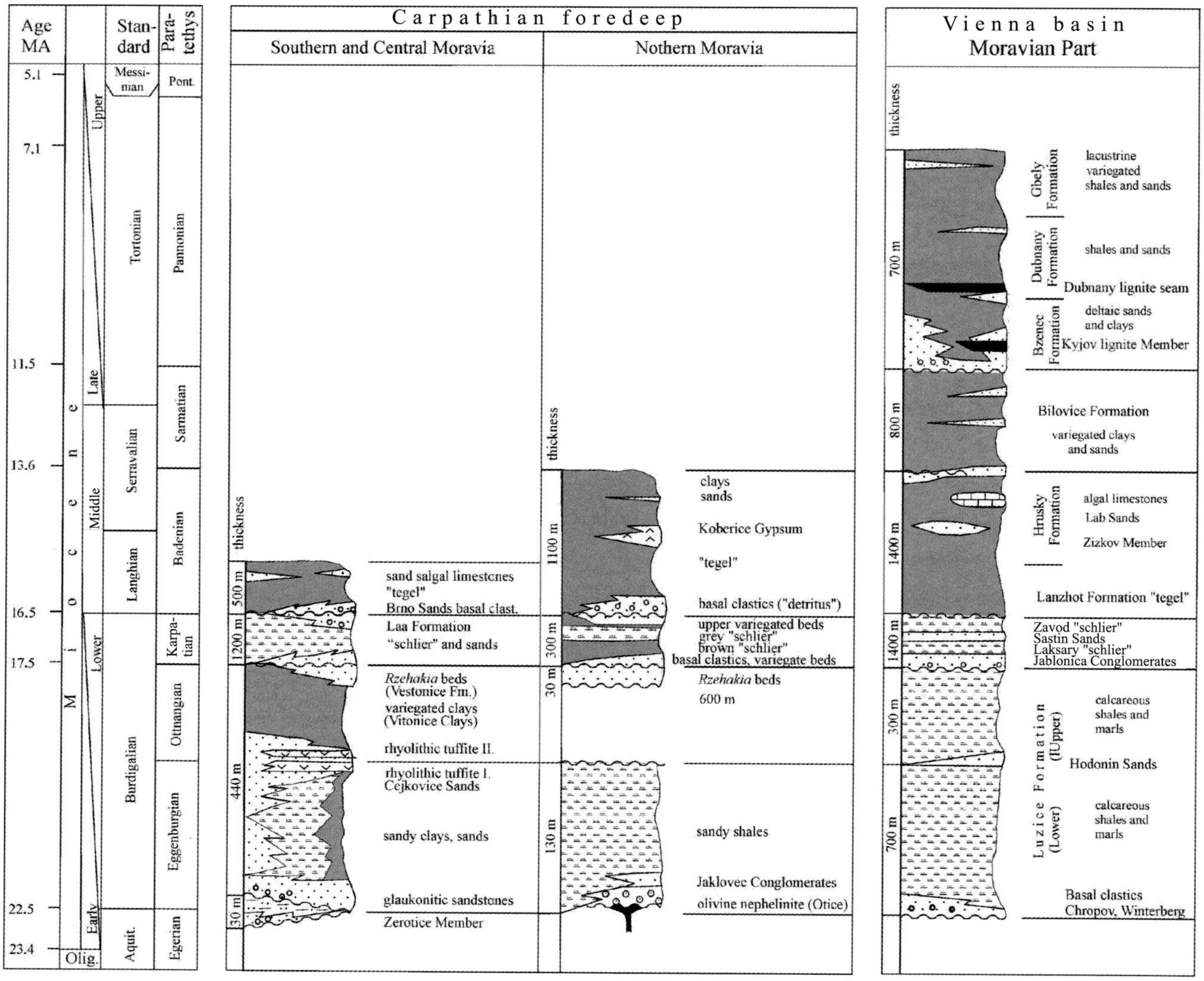

Figure 15. Stratigraphy of the Neogene foredeep and the Vienna basin in Moravia.

(Sakvice Marls and Pavlovice Formation) and in the Vienna basin (Luzice Formation). The distribution of the Eggenburgian and Ottnangian strata in southern Moravia was, to a great degree, controlled by the preexisting northwest–southeast-trending fault system of the Jurassic Dyje–Thaya depression, which underlies the Carpathian foredeep and the thrust belt. In northern Moravia, the Eggenburgian fluvial sands and overlying marine shales are known only from the Detmarovice depression and paleovalley and from a few small erosional remnants (Jurkova et al., 1983).

The Eggenburgian–Ottnangian strata are exposed along the western margin of the foredeep. They rest transgressively on kaolinized crystalline rocks of the Bohemian Massif marked by a rugged erosional relief with ridges and canyons. In the eastern zone adjacent to the thrust belt, the Eggenburgian–Ottnangian strata are covered by younger Miocene deposits. Their extent and stratigraphy were established from well data, e.g., during the exploration of the Dolni Dunajovice gas field (Adamek, 1979).

Two main facies, a marginal, shallow-marine and a deeper basinal facies, have been distinguished in the Eggenburgian deposits of southern Moravia (Ctyroky 1982, 1991, 1993). A great lithological variability characterizes the shallow marginal facies; the basal coarse conglomerates and kaolinitic sands are followed by silts, sands, shales, and occasional coaly shales. At the western margin of the basin near Znojmo, the sporadically variegated proluvial clastic deposits were described and named the Zerotice Member by L. Prachar (1970, personal communication). P. Ctyroky and J. Adamek (1988, personal communication) assigned these strata to the Egerian(?) to Eggenburgian (Figure 15). In the basinal facies, the basal transgressive conglomerates and glauconitic sands are overlain by alternating sands and shales. The Ottnangian deposits, otherwise similar to those of the underlying Eggenburgian strata, were laid down in a shallower, more restricted depositional environment (Jiricek, 1983) marked by the presence of dark shales with remnants of fish. Characteristic for the Ottnangian strata is the presence of the *Rzehakia* beds represented by sandstones and gravels with pebbles of Jurassic limestones and endemic mollusks (*Rzehakia socialis*). The *Rzehakia* beds, recently renamed by Adamek (2003) the Vestonice Formation, mark a transgressional-regressional event (Brzobohaty and Cicha, 1993), whose stratigraphic adherence to either Ottnangian or Karpatian, at least in the outer zone of the foredeep in Moravia, is not clear. In northern Moravia, the Ottnangian strata were recognized recently below the Carpathian nappes near Frydek–Mistek (P. Ctyroky, 1996, personal communication). The maximum known thickness of the Eggenburgian–Ottnangian strata of about 440 m (1400 ft) has been recorded in a zone adjacent to the front of the Carpathian thrust belt near the Austrian border.

Occasional layers of rhyolitic tuffs associated with the arc volcanism of the Carpatho-Pannonian province are found in the uppermost Eggenburgian and the lowermost Ottnangian strata in Moravia (Nehyba, 1997; Nehyba and Roetzel, 1999).

The Karpatian (Upper Burdigalian)

The chronostratigraphic stage Karpatian is used in terms of the original definition (Cicha et al., 1967), although the new lithostratigraphic division of strata deposited in the time span of the Karpatian stage in the Moravian sector of the foredeep is also considered (Brzobohaty et al., 2003).

In the Karpatian, the paleogeography of the foredeep changed significantly. The front of the deformation progressed far onto the foreland, and the most external units of the Outer Carpathian belt were thrusted over the inner zones of the Neogene foredeep. Related with the flexural downbending of the crust in front of the thrust belt, the foredeep took the shape of a continuous depositional wedge, which rimmed the front of the Western Carpathian thrust belt along its entire length (Cicha and Kovac, 1990). The highest rate of subsidence occurred in southern and central Moravia, where the thickness of the Karpatian deposits may exceed 1200 m (3900 ft). Whether all this thickness is depositional or partly enhanced by tectonic stacking remains uncertain.

The predominantly marine Karpatian deposits of southern Moravia are attributed to the Laa Formation defined by Kapounek et al. (1960) in the Molasse Basin of northeastern Austria. The lower part of the formation, represented by laminated shales and silts, known as "schlier" and recently named the Musov Member (Adamek et al., 2003), was deposited in a deeper neritic environment of the transgressive stage. The upper part of the Laa Formation, named the Novy Prerov Member (Adamek, 2003), is characterized by a higher proportion of sands and silts. It was laid down in a deeper neritic to bathyal environment during the maximum extent of the Karpatian transgression. With a distinct discordance, the Novy Prerov Member is overlain by a transgressive sequence, the Ivan Member (Adamek, 2003), believed to be the highest part of the Karpatian.

In northern Moravia, the Karpatian strata transgressed over a more diverse relief, marked by the existence of the large Bludovice and Detmarovice paleovalleys and intervening ridges. The basal sands of different age and composition pass upward into the variegated and brown silty shales deposited in lagoons and coastal lakes with variable salinity. They are followed by marine gray laminated silty shales (schlier) (e.g., Jurkova and Novotna, 1974; Palensky, 1988; Palensky et al., 1995), which correspond to the Stryszawa Formation (Slaczka, 1977) in Poland. For the latest information on the stratigraphy of the Karpatian stage of the Central Paratethys, we refer to a book by Brzobohaty et al., 2003.

Toward the end of the Karpatian, the foredeep shrank to a narrow and shallow basin stretched along the front of the thrust belt. Variegated shales with anhydrite and unsorted sands were deposited in northern Moravia, whereas shales, sands, and gravels with clasts of rocks brought in from the Carpathian belt were laid down in central Moravia.

In the uplifted southeastern margin of the Bohemian Massif, an extensive erosion of Miocene deposits led to the development of a rugged relief with deep northwest–southeast-trending valleys. An uninterrupted sedimentation between the Karpatian and Badenian continued only at the territory of Austria (Grund Member) and locally in southern Moravia (Cicha and Ctyroka, 1995).

The Badenian (Langhian–Serravalian)

The Badenian began with a local transgression originally limited to the deepest zone of the foredeep along the front of the Carpathian front. Only gradually, the lower Badenian transgression spread over the more distal zones of the foredeep and adjacent parts of the Bohemian Massif. After a brief retreat of the sea, the whole foreland submerged, and the Badenian sea spread over vast areas of the Bohemian Massif.

The Badenian sedimentation began with continental breccias followed by marine sands and gravels known under local names, e.g., Brno Sands in the surroundings of Brno or "detritus" in the Ostrava region

(Jurkova, 1971). Calcareous clays called "tegel" were deposited in deeper and farther offshore-located parts of the basin. Following the second transgression, mainly calcareous clays (tegels) were laid down in the deeper parts of the basin, whereas thin biostromes and bioherms of algal (Lithothamnion) and bryozoan limestones formed on elevated parts of the relief. Volcanic ashes from the Carpathian–Pannonian volcanic province fell repeatedly into the basin (Nehyba, 1997).

The rate of subsidence varied greatly in various parts of the foredeep and the foreland. In contrast with the Karpatian, the subsidence of the foredeep in the Badenian was more intense in northern Moravia, where as much as 1100 m (3600 ft) of the lower Badenian strata accumulated in the Bludovice and Detmarovice deep depressions and paleovalleys. At the same time, a chain of southwest–northeast-trending elevations, previously known as the Slavkov–Tesin Ridge (Dlabac and Mencik, 1964), kept rising. According to Krejci et al. (2002), these elevations have diverse origin and do not represent a forebulge.

In the Opava subbasin of northern Moravia, the lower Badenian is represented by basal clastics, followed by variegated clays with intercalations of lignite. Subsequent deposition of gray calcareous shales with rich foraminiferal fauna marks the maximum extent of the lower Badenian sea. At this time, basaltic volcanism was active in the area.

In southern Moravia, the sedimentation ended in the lower Badenian, whereas in the northern part of Moravia, the sedimentation continued until the end of the Badenian (Figure 15). There, the middle Badenian gray clays are overlain by evaporites (Koberice Gypsum) as much as 65 m (213 ft) thick (Cicha et al., 1985). The precipitation of evaporites marks the middle Badenian shallowing of the sea and the formation of restricted evaporitic conditions (saline crisis) that occurred elsewhere in the outer basins of the Central Paratethys (Oszczypko et al., 2006). The upper Badenian clays with abundant remnants of plants and with thin beds of limestones at the base represent the terminal marine sediments in the northern part of the Carpathian Foredeep in Moravia. Since then, only continental fluviatile and lacustrine deposits of different ages accumulated in local depressions, both on the Bohemian Massif (e.g., in the Boskovice Furrow) and in the Carpathian Foredeep (e.g., in the Upper Moravian Depression). Among them are also the gravels with tectites, e.g., at Dukovany.

The Late Miocene to Pliocene Upper Moravian Depression

In the late Miocene and in the Pliocene, a nonmarine basin, the Upper Moravian Depression, formed in the Carpathian foreland and the marginal zone of the thrust belt (Figure 3). The bounding faults of this depression trend in the northwest–southeast direction, perpendicular to the direction of the Carpathian belt. These faults, extending from the foreland of the Bohemian Massif across the Neogene foredeep into the marginal parts of the thrust belt, apparently follow the preexisting fault system reactivated during the postorogenic rebound of the foreland plate. The sedimentary fill of the Upper Moravian Depression consists of fluviatile and lacustrine deposits, which may be divided into the lower and upper parts (Ruzicka, 1989). The lower part of the fill is represented by as much as 100-m (330-ft)-thick reddish deposits, supplied from the deeply weathered surface. The equally thick upper part is composed of monotonous gray and green clays and sands with numerous layers of coaly clays and lignite. The subsidence in the Upper Moravian Depression continued into the Pleistocene.

The Paleovalleys with Neogene Fill

Numerous paleovalleys filled with Miocene strata have been reported from the Carpathian foreland. On the territory of the Czech Republic, large, tectonically controlled Bludovice and Detmarovice paleovalleys are cut into the upper Paleozoic strata of the Upper Silesian coal basin and filled with Karpatian to Badenian strata of the Carpathian foredeep (Jurkova, 1961, 1976). These features, as well as the intervening Ostrava–Karvina and Pribor–Tesin ridges, trend generally in the west–east direction.

Paleovalleys with Neogene fill have been identified in the Carpathian foreland in eastern Poland (Czernicki and Karnkowski, 1987), in Western Ukraine (Sovchik and Vul, 1996), and in the Moesian Platform of Romania (Paraschiv, 1979). Although no obvious Paleogene sediments have been found in these paleovalleys in the territory of Poland, Krzywiec (2001) believes that these features in eastern Poland and the adjacent part of Ukraine were incised in the Paleogene in conjunction with the inversion and erosion of the Polish trough. This, however, would not determine the time of excavation of paleovalleys in other parts of the Carpathian foreland. Whether both the paleovalleys with the Paleogene fill, like the Nesvacilka and Vranovice canyons of southern Moravia, and those filled with Neogene deposits were cut at the same time during the Late Cretaceous to early Paleogene uplifting of the Carpathian foreland and filled with either Paleogene or Neogene strata or whether there were successive stages of paleovalleys cutting and filling remains to be discovered. Paleogeographic reconstruction of the Paleogene strata in the Alpine Molasse and in southern Moravia would indicate that the Paleogene transgression might have been limited to the southern zone of the foreland, including the Nesvacilka and Vranovice paleovalleys and the Hystria and Kamchia depression in the Black Sea shelf. The younger, Miocene transgressions then progressed farther north, into the paleovalleys of northern Moravia, eastern Poland, and Western Ukraine. The different regional extent of Paleogene and Neogene transgressions may thus explain why various paleovalleys, possibly incised at the same time, during the Late Cretaceous–Paleocene uplifting of the Carpathian foreland, were filled with either Paleogene or Neogene strata.

The Structure of the Neogene Foredeep

The Neogene strata of the foredeep remain generally undeformed. Just along the very front of the Waschberg–Zdanice–Subsilesian nappe, e.g., in central Moravia and in the Pavlov Hills of southern Moravia, the Neogene strata of the foredeep are detached and integrated into the imbricated frontal zone of the thrust belt. The structural pattern of the Neogene foredeep is dominated by a system of synthetic and some antithetic faults formed during the flexural downbending of the foreland in front of the progressing thrust belt. The presence of antithetic faults seems to be critical for the formation of structural traps (Adamek, 1979). However, the character of some of these faults is not clear and remains a subject of disputes. One of them is the antiformal feature, which underlies the Dunajovice gas field located in the Neogene foredeep west of Mikulov (Figure 26, shown on page 137). On its western side, the structure is bounded by an eastward steeply dipping Vestonice fault, which penetrates through the Jurassic strata into the crystalline basement but does not continue directly into the overlying Neogene foredeep deposits, which drape over this structure. The thinning of the lower Miocene (Eggenburgian) strata over the structure would suggest that the fault was active during the early Miocene. Adamek (1979) and Kostelnicek et al. (2006) interpret this structure as an antithetic westward-dipping normal fault. However, the overall geometry of the structure, as seen on the seismic data, suggests that this might be rather a steep reverse and/or strike-slip fault, whose origin may be related to the compressive or transpressive stresses, which propagated into the foreland from the progressing thrust belt at a deeper structural level.

The structural pattern of the Neogene foredeep in Moravia has been only partly mapped and interpreted (e.g., Jurkova, 1961, 1971; Jurkova and Novotna, 1974; Jurkova, 1976; Elias and Palensky, 1997). The full evaluation of the structural architecture of the foredeep may yet provide new opportunities for the exploration for hydrocarbons.

THE STRATIGRAPHY AND DEPOSITIONAL SYSTEM OF THE OUTER WESTERN CARPATHIAN THRUST BELT

The Outer Western Carpathian thrust belt may be defined as a stack of rootless, thin-skinned nappes thrusted over the Neogene foredeep and the underlying West European platform (Figures 2, 3). In a broad sense, it consists of the Pieniny Klippen Belt, the Flysch belt, and the successor Vienna basin, which, mostly undeformed, rests on the top of the Carpathian belt. The extent of tectonic shortening and the character of the original depositional sites of various thrust sheets are little known and remain subjects of alternate geodynamic reconstructions. Different views also exist on the character of the crust underlying the depositional sites, which apparently varied from continental to transitional and, according to some views, possibly to oceanic, and on the amount of its shortening, accretion, and subduction.

In terms of the Carpathian terminology, the Flysch belt includes not only the typical deep-water turbiditic flysch deposits of alternating shales and sandstones found mainly in the Magura flysch and the Silesian units but also the hemipelagic marginal deposits characteristic for the external Pouzdrany, Waschberg–Zdanice–Subsilesian units, and parts of the Silesian unit. The definition of Carpathian Flysch belt is thus based more on structural rather than lithological criteria. In that respect, the Carpathian terminology differs from that used in the Alpine region. There, the term flysch is used exclusively for typical synorogenic deep-water turbiditic deposits of the inner zones of the depositional system, whereas the hemipelagic and shallow-marine strata of the more external zones of the Outer Alpine depositional system are referred to as Helveticum or even included into the Alpine Molasse (Figure 16).

In the Alpine nomenclature, the term Molasse is traditionally used for undeformed and marginally deformed (imbricated Molasse) late Eocene to Oligocene marginal marine deposits (lower marine molasse) of the external zones of the depositional system, as well as for the shallow-marine and terrestrial Miocene deposits of the Neogene foredeep (lower nonmarine, upper marine, and upper nonmarine molasse). In the Carpathians, the deformed Paleogene to early Miocene (Egerian) strata are included into the Flysch belt, whereas the undeformed Neogene deposits are attributed to the Carpathian foredeep.

The depositional assemblies of the Outer Carpathians comprise sequences formed during both the divergent and convergent stages of the evolution of the Tethyan–Alpine system. The preorogenic depositional assembly of divergent margins encompasses the Jurassic to the Lower Cretaceous carbonate and siliciclastic

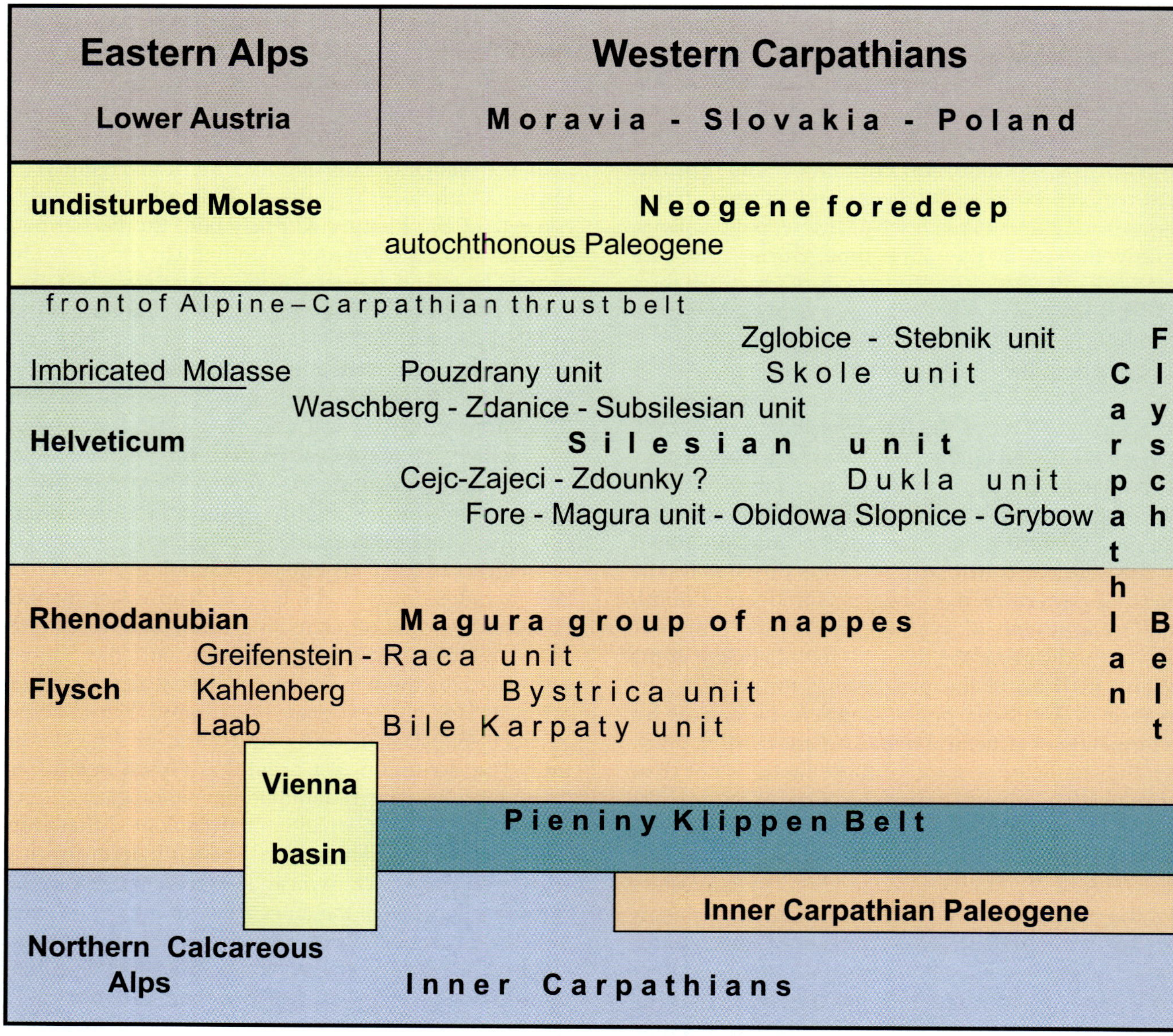

Figure 16. Regional distribution and correlation of principal tectonostratigraphic units of the external Eastern Alps in northeastern Austria and the Outer Western Carpathians in Moravia (Czech Republic), Slovakia, and Poland. Arranged from the foreland (top) to the hinterland (bottom).

strata, including some deep-water turbidites. However, with the exception of the Silesian unit, these strata are seldom exposed on the surface. Only scattered information about their character and distribution are available from wells and tectonic klippen and olistoliths incorporated into the thrust belt. The Late Cretaceous to early Miocene synorogenic flysch and molasse deposits of the convergent tectonic setting represent the bulk of the Flysch belt depositional assembly. They were deposited and redeposited in a series of troughs and foredeeps and sourced from internal ridges (cordilleras), the prograding thrust belt, as well as from the European foreland.

The conversion from the divergent to the convergent tectonic setting in the Outer Carpathian realm, which resulted from the collision of the Inner Carpathians with the rifted and attenuated margins of Northern Europe in the Late Cretaceous, was gradual and not always clearly defined. In the outermost zones, e.g., in the autochthonous Paleogene, the Pouzdrany, and the Waschberg–Zdanice–Subsilesian domains, the Late Cretaceous and Paleogene synorogenic sequences are separated from the Jurassic to the Lower Cretaceous strata of the passive margins by a major hiatus. In the more internal Silesian and Magura units, however, the sedimentation continued without major interruption from one regime to another. The transition from the divergent to the active convergent setting, especially in the external zones of the Outer Carpathian depositional system, was also partly obscured by the worldwide transgression and high-water stand in the Cenomanian–Turonian. Possibly, the best indicator of the inception of the synorogenic regime is the sudden appearance of thick packages of coarse turbiditic sands, such as the Godula sandstones in the Silesian unit or Solan sandstones in the Magura unit (Figure 17A, B).

Reasoning on the fact that the convergent depositional system of the Outer Carpathians gradually progressed toward the foreland and was supplied with detrital material mainly from the tectonically activated inner ridges and the progressing orogenic belt, Picha and Stranik (1999) defined the entire convergent depositional system since the Upper Cretaceous to Neogene as a foreland basin. Despite the possibility that during the Late Cretaceous to early Paleogene, the innermost part of this foreland basin, particularly the Magura deep-water depositional site, might have been still at least partly underlain by an attenuated transitional crust or even, as some believe, an oceanic crust, the term foreland basin seems to be justified. Under such a broad definition, the Carpathian foreland basin would include both the Late Cretaceous to early Miocene synorogenic, predominantly deep-water flysch sequences constituting the Flysch belt and the early to late Miocene late orogenic and postorogenic shallow-marine and continental molasse deposits of the Neogene foredeep and of the Vienna basin. In that respect, the Alpine–Carpathian foreland basin, by having two stages, namely, the deep-water flysch and the shallow-marine and continental molasse, compares well with other foreland basins of the Tethyan system, such as the Apennines and the Dinarides–Hellenides, or with the Ouachitas of North America (Golonka et al., 2006b). However, it differs from some other classical foreland basins, such as the well-studied Cretaceous to Paleogene Rocky Mountain foreland basin of North America, which does not have the typical deep-water flysch phase and is filled only with molasse-type deltaic and other shallow-marine and continental deposits sourced from the Sevier orogenic belt as well as the foreland structures of the Rocky Mountains.

The terms foreland basin and foredeep are commonly used interchangeably for various foreland depositional settings. In this chapter, following the previous usage by Picha and Stranik (1999), we keep the term foredeep for a fast subsiding wedge-shaped depocenter bordering the structural front of the thrust belt and supplied primarily from the emerging thrust belt. The foredeeps are formed typically by flexural downbending of the foreland plate caused by the tectonic load of the progressing accretionary prism (Allen and Allen, 1990). The foreland basin, as defined, for example, by DeCelles and Giles (1996), is a more complex depositional setting, which, in addition to the foredeep proper, may also include the flexural forebulge and the broad landward part of the basin supplied predominantly from the foreland. That means that the foredeep is merely the innermost, deepest, and most active part of a foreland depositional setting of the foreland basin. As the front of deformation moved toward the foreland, so did the foredeep, the forebulge, and the rest of the foreland basin.

The typical foreland basin succession records a marine transgression, followed by a deepening phase, and later by regression to alluvial and lacustrine conditions (Lickorish and Ford, 1998). The foreland basin thus passes from an underfilled to an overfilled stage. The Carpathian foreland basin, in general terms, fits this classical model with the hemipelagic and deep-water turbiditic flysch deposits representing the underfilled synorogenic stage and the molasse shallow-marine and continental deposits exemplifying the overfilled stage. However, only the outer zones of the Carpathian foreland basin began with transgression over the foreland, whereas in the inner zones, the sedimentation passed from the divergent to the convergent stage continuously without any major interruption.

As indicated by the distribution of various lithostratigraphic facies, the structural and depositional pattern in the Carpathian basin was complex. The basin architecture was characterized by a rugged topography, partly inherited from the previous divergent stage and further enhanced by the foreland-type crustal uplifting, caused by compressional and transpressional orogenic stresses. The depositional systems in the foreland basin were thus determined not only by the progression of the orogenic belt, global fluctuations of sea level, and climatic changes, but also by the orogenic activation of preexisting basement structures.

THE WESTERN CARPATHIAN FLYSCH BELT

The thin-skinned Outer Western Carpathian Flysch belt consists of numerous tectonostratigraphic units and subunits. In the Polish Carpathians, Nowak (1927) separated them into three groups: the marginal, central (Menilite–Krosno), and inner (Magura) groups. However, the delimitation of the marginal group, which, in Moravia, will be represented only by the Pouzdrany unit, proved to be questionable. The Pouzdrany unit, as well as the autochthonous Paleogene, may be better classified as a continuation of the Alpine Molasse into the territory of southern Moravia. To simplify the matter, Swidzinski (1934) distinguished only two groups of units: the external and internal. The external or Menilite–Krosno group of units is characterized by the Late Cretaceous to late Eocene succession of variegated shales, early Oligocene menilitic silicites, and the late Oligocene to early Miocene Krosno-type flysch. In a broad sense, based mainly on their structural position, these units are comparable with the Helvetic units of the Alps. In Moravia, this external group includes the Pouzdrany, Waschberg–Zdanice–Subsilesian, Silesian, Cejc–Zajeci, Zdounky, and Fore-Magura units (Figures 3; 16; 17A, B). Thick Late Cretaceous to Eocene deep-water flysch deposits characterize the internal Magura unit, which, during the sedimentation, was separated from the external units by the Silesian ridge. The Magura unit, also referred to as the Magura group of nappes, is correlated with the Rhenodanubian Flysch of the Alps (Elias et al., 1990).

The stratigraphic extent of these tectonic units of the Carpathian Flysch belt varies but, with the exception of the Silesian nappe, only seldom encompasses the complete stratigraphic range of the original depositional assembly from the Jurassic to the early Miocene. The various units of the Carpathian flysch belt are composed predominantly of the Late Cretaceous to Miocene clastic deposits, both shelf and deep-water flysch facies, with only a token presence of Jurassic and Lower Cretaceous carbonates and clastics. These older synrift and continental margin strata laid down in a structurally diversified setting of rift-related basins and platforms might have been cut off by the sole detachment and left behind in their autochthonous position. Occasionally, tectonic slivers of these Jurassic and Cretaceous carbonate and clastic strata were detached and incorporated into the Carpathian Flysch belt as the so-called "tectonic klippen." Typically well exposed on the surface, these klippen provide valuable information about the character and distribution of Jurassic and Cretaceous strata along the margins of the European plate now deeply buried below the thrust belt.

Some incompetent depositional sequences were reduced tectonically, smeared, or piled in duplexes at the base of major thrust sheets. The stratigraphic extent of sequences was also reduced by erosion, which occurred either prior or during and after the deformation. The reconstruction of the full original depositional record thus requires integration of fragments of evidence gathered from numerous sites of tectonically separated units. Without strong support from biostratigraphy, such a reconstruction would be very difficult if not impossible.

The Pouzdrany Unit

The marginal Pouzdrany unit, as defined by Cicha et al. (1964, 1965), is the outermost tectonostratigraphic unit of the Carpathian thrust belt in Southern Moravia. It appears as a narrow imbricated thrust sheet trending along the frontal edge of the Waschberg–Zdanice nappe in southern Moravia and northeastern Austria (Stranik, 1996) (Figure 3). However, it has not been officially recognized as a separate unit in Austria, where its equivalents are traditionally included into the Waschberg zone.

The Pouzdrany unit comprises strata ranging from the late Eocene to the early Miocene. They have been divided into five lithostratigraphic units: (1) the Pouzdrany Marls, (2) the Uhercice Formation, including the diatomites, (3) the Boudky Marls, (4) the Krepice Formation, and (5) the Sakvice Marls (Figure 17A, B).

The oldest known strata in the Pouzdrany unit are the brown fossiliferous shales and *Globigerina* Marls described as the Pouzdrany Marls by Rzehak (1880, 1895a) and dated as the late Eocene to early Oligocene by Pokorny (1954), Cicha (1975), Krhovsky (1981), Krhovsky et al. (1992), and others. Locally, these predominantly pelitic strata contain intercalations and concretional bodies of limestones with rich late Eocene faunas (Oppenheim, 1922; Ctyroky, 1966). These limestones have been named the Moutnice Limestones and correlated with the Hollingstein Limestones of the Waschberg zone in Austria by Ctyroky (1966). The Pouzdrany Marls are overlain by diatomites with cherty concretion and pelagic micritic limestones and dolomites, followed by brown and gray shales with limonite and jarosite coatings and gypsum crystals (weathering products of pyrite) and scattered lenses of glauconitic sands with shark teeth and brachiopods. The whole sequence, about 200 m (660 ft) thick, was studied by Pokorny (1960), Cicha et al. (1965), and Stranik (1981) and named the Uhercice Formation by Krhovsky (1981). The diatomites have been correlated with the Menilitic Formation of the Zdanice unit and assigned to the lower Oligocene by Picha and Stranik (1999). In the autochthonous Paleogene series, the early Egerian Malesovice Member might be considered as an equivalent of the Uhercice Formation (Figure 17A, B).

The Uhercice Formation is followed by an approximately 30-m (100-ft)-thick sequence of gray pelagic marls with dolomitic concretions. Cicha et al. (1964, 1965) named these strata the Boudky Marls and correlated them with the Egerian (Aquitanian) Michelstetten beds (Grill, 1962) of the Waschberg zone in Austria. Cicha (1975) and Cicha et al. (1975b) assigned them to the late Egerian. Krhovsky (1981) and Stranik et al. (1981a) dated the Boudky Marls as the late Egerian to early Eggenburgian. Picha and Stranik (1999) tentatively suggested the late Oligocene to early Miocene (Egerian) age for these open-marine marls as the most likely. The Boudky Marls may represent a distal landward facies of the Outer Carpathian foreland basin. It might have been deposited on or close to a foreland

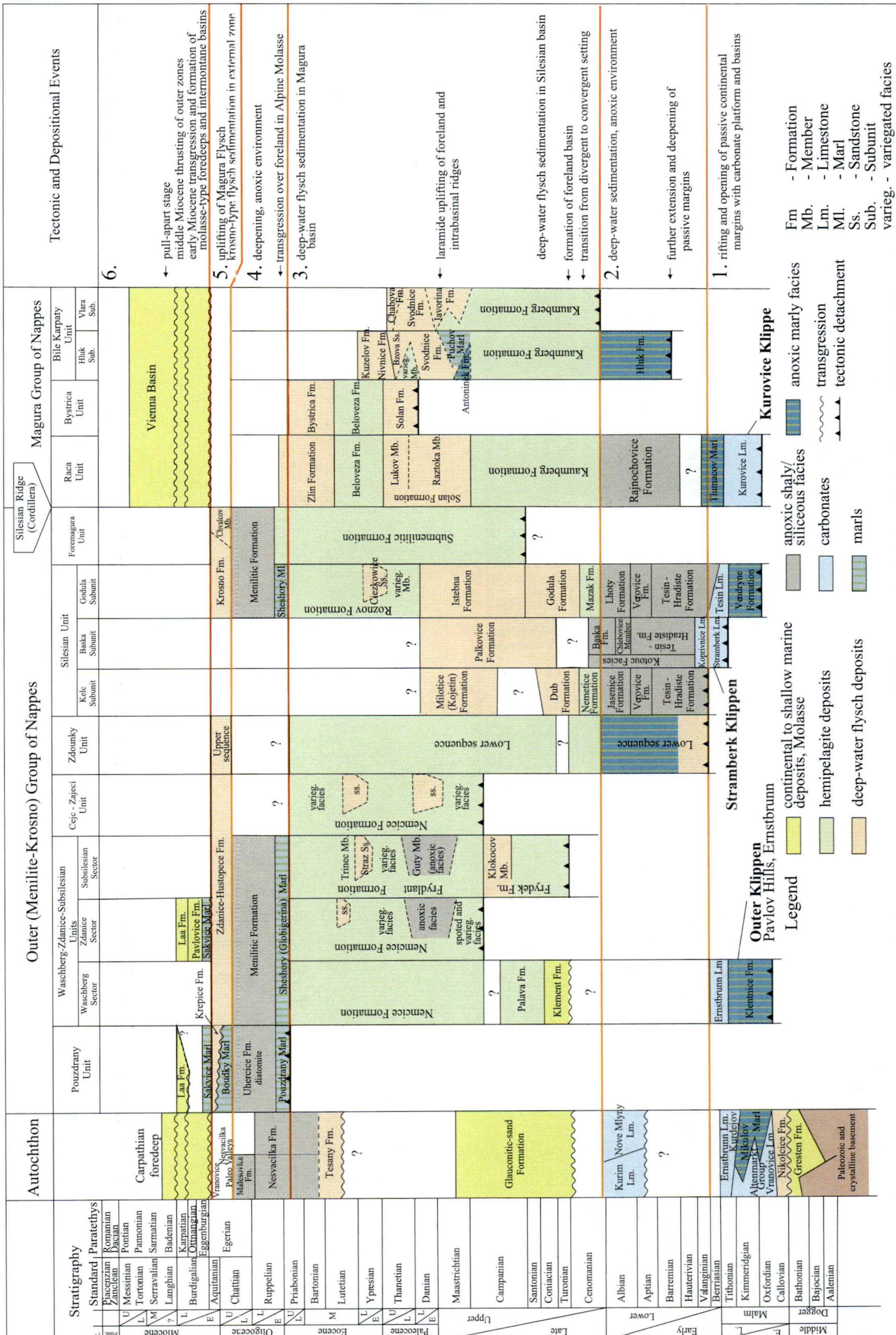

Figure 17. (A) Stratigraphic correlation of principal tectonostratigraphic units of the Outer Western Carpathians in Moravia. Major depositional sequences distinguished and numbered 1–6.

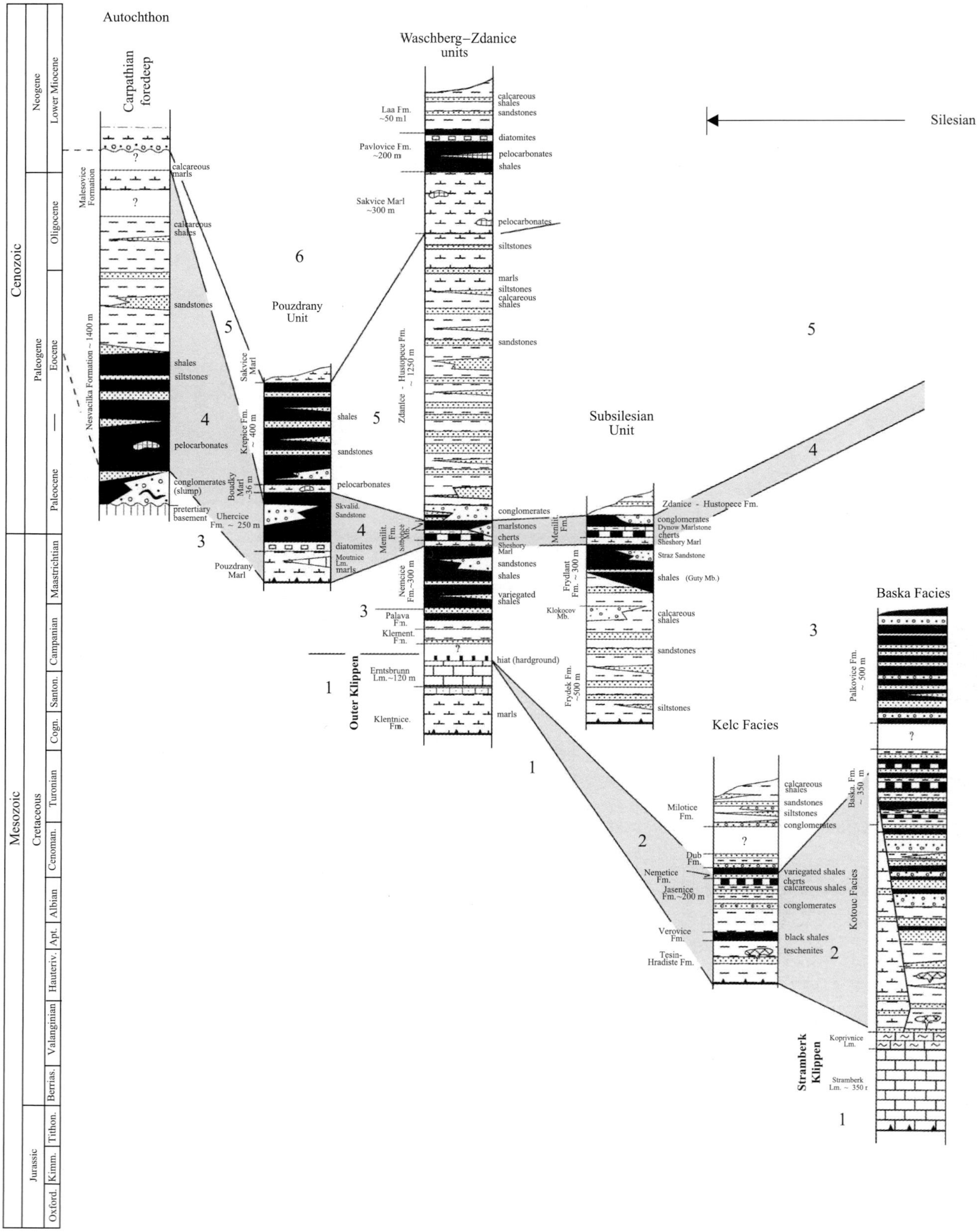

Figure 17. (cont.) (B) Lithologic columns of various tectonostratigraphic units of the Outer Western Carpathian Flysch belt in Moravia. For time (chronostratigraphic) correlation of various stratigraphic units and major depositional sequences 1–6, see Figure 17A. The lithologic columns for the Godula facies of the Silesian unit and all units of the Magura group of nappes are reduced by 50% with respect to other columns.

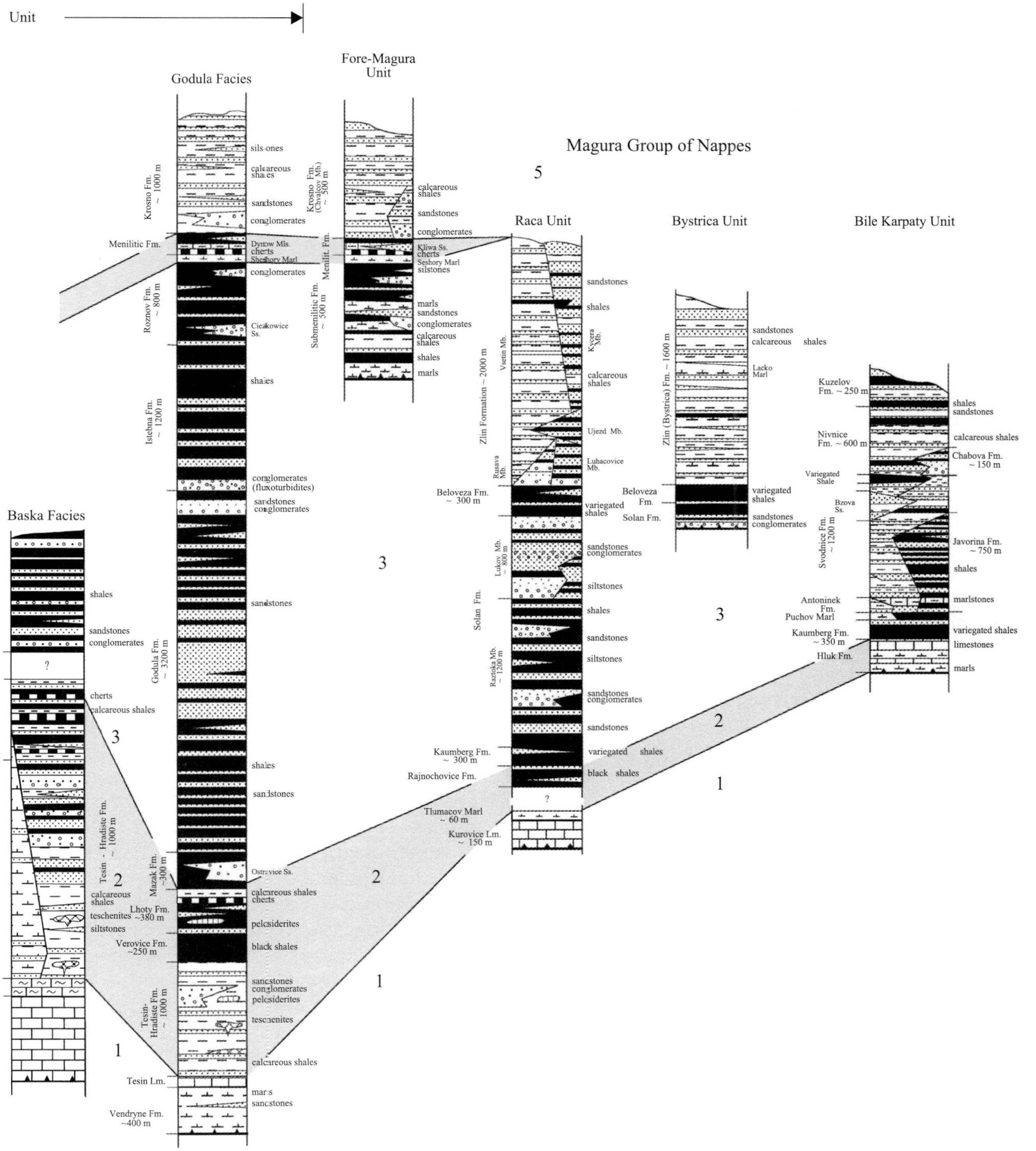
Unit
Godula Facies
Fore-Magura Unit
Magura Group of Nappes
Raca Unit
Bystrica Unit
Bile Karpaty Unit
Baska Facies
Krosno Fm. ~ 1000 m
siltstones
calcareous shales
sandstones
conglomerates
Menilitic Fm.
Dynow Mls.
cherts
Sheshory Marl
conglomerates
Roznov Fm. ~ 800 m
Cieszkowice Ss.
Istebna Fm. ~ 1200 m
shales
conglomerates (fluxoturbidites)
sandstones conglomerates
Godula Fm. ~ 3200 m
sandstones
shales
sandstones
Mazak Fm. ~300 m
Ostravice Ss.
calcareous shales
cherts
Lhoty Fm. ~380 m
pelosiderites
Verovice Fm. ~250 m
black shales
Tesin-Hradiste Fm. ~ 1000 m
sandstones
conglomerates
pelosiderites
teschenites
calcareous shales
Tesin Lm.
marls
sandstones
Vendryne Fm. ~400 m
Krosno Fm. (Chvalcov Mb.) ~ 500 m
Menilit. Fm.
Submenilitic Fm. ~ 500 m
calcareous shales
sandstones
conglomerates
Kliwa Ss.
cherts
Seshory Marl
silstones
marls
sandstones
conglomerates
calcareous shales
shales
marls
5
3
2
1
Zlin Formation ~ 2000 m
Vsetin Mb.
Kyczera Mb.
Ujezd Mb.
Luhacovice Mb.
Rusava Mb.
sandstones
shales
calcareous shales
Beloveza Fm. ~ 300 m
variegated shales
Solan Fm.
Lukov Mb. ~ 800 m
Raztoka Mb. ~ 1200 m
sandstones conglomerates
siltstones
shales
sandstones
siltstones
sandstones conglomerates
sandstones
Kaumberg Fm. ~ 300 m
Rajnochovice Fm.
variegated shales
black shales
Thmacov Marl ~ 60 m
Kurovice Lm. ~ 150 m
Zlin (Bystrica) Fm. ~ 1600 m
sandstones
calcareous shales
Lacko Marl
Beloveza Fm.
Solan Fm.
variegated shales
sandstones conglomerates
Kuzelov Fm. ~ 250 m
Nivnice Fm. ~ 600 m
Variegated Shale
Bzova Ss.
Svodnice Fm. ~ 1200 m
Antoninek Fm.
Puchov Marl
Kaumberg Fm. ~ 350 m
Hluk Fm.
shales sandstones
calcareous shales
Chabova Fm. ~ 150 m
Javorina Fm. ~ 750 m
shales
marlstones
variegated shales
limestones
marls
Tesin - Hradiste Fm. ~ 1000 m
shales
sandstones conglomerates
cherts
calcareous shales
calcareous shales
teschenites
siltstones

bulge, which formed in response to the tectonic loading and downbending of the foreland crust in front of the progressing Carpathian thrust belt in the late Oligocene and early Miocene.

The Boudky Marls are overlain by the flyschlike Krepice Formation (Cicha et al., 1964, 1965), a sequence as much as 400 m (1300 ft) thick of alternating shales and fine micaceous sandstones and siltstones. Based on similarities in lithology, the Krepice Formation has been interpreted as a distal facies of the Oligocene to early Miocene (Egerian) Zdanice–Hustopece Formation of the Zdanice unit (Cicha et al., 1964, 1965; Picha, 1979b). This is a typical synorogenic turbiditic flysch facies, whose base may become progressively younger toward the foreland, thus reflecting on the gradual progression of deformation and foredeep deposition. An existence of a local erosional event at the base of the Krepice Formation (Stranik et al., 1981a) would support such an interpretation. The distal Krepice Formation of the Pouzdrany unit may, in fact, correlate only with the upper part of the Zdanice–Hustopece Formation of the Zdanice unit (Figure 17A, B). The underlying Boudky Marls, possibly deposited on the foreland bulge, may then be coeval with the lower part of the Zdanice–Hustopece Formation elsewhere (Picha, 1979b; Stranik, 1983). Because of the lack of diagnostic fossils and the structural complexity of the Pouzdrany unit, the stratigraphic position of the Krepice Formation has been disputed. The gray marls and calcareous shales overlying the Krepice Formation have been compared with the Eggenburgian Sakvice Marls of the Zdanice unit (Stranik and Molcikova, 1980) (Figure 17A, B). Such a superposition would support the Egerian age of the underlying Krepice Formation. However, based on the presence of rusty-colored limonitic siltstones and claystones with sparse laminae of diatoms in its upper part, Krhovsky et al. (1995) compared the Krepice Formation with the Ottnangian to Karpatian Pavlovice Formation of the Zdanice unit (Stranik, 1983). Considering the overall lithologic similarity with the Zdanice–Hustopece Formation and the stratigraphic position below the Eggenburgian strata, the interpretation of the Krepice Formation as a distal facies of the Zdanice–Hustopece Formation of the Egerian age seems to be more likely.

Cicha et al. (1965) defined the Pouzdrany unit as a tectonic sliver detached from the more internal zones of the autochthonous Paleogene series and tectonically transported at the front of the thrust belt during the youngest early Miocene phase of thrusting. The obvious spatial proximity of the depositional sites of these two units (Figure 5D) and similarities, both lithological and faunal, of their upper Eocene to lower Oligocene strata would advocate such an interpretation. Because no equivalents of the younger, lower Miocene members of the Pouzdrany unit, the Boudky Marls and the Krepice Formation, have been found in the autochthonous series, it is speculated that these younger members of the sequence were detached and tectonically transported at the front of the thrust belt with the rest of the Pouzdrany unit, whereas the older members, mainly the Eocene fill of the paleovalleys, were mostly preserved in their autochthonous position.

Stratigraphically, the strata of the Pouzdrany unit would correlate with the Lower Marine Molasse of the Alps. Structurally, the Pouzdrany unit may be compared with the imbricated (Gefaltette) Molasse of Austria, as described, e.g., by Wagner (1996).

The Waschberg–Zdanice–Subsilesian Units

This major frontal thrust system of the Outer Western Carpathian belt consists of three sectors: the Waschberg sector (unit) of northeastern Austria and southernmost Moravia (south of the Dyje River), the Zdanice sector (unit) of southern and central Moravia, and the Subsilesian sector (unit) of northern Moravia (north of the Upper Moravian Depression) and western Poland (Figure 3). On the Polish territory, the Subsilesian unit may be traced to the Bielsko–Biala area; its continuation further to the east is not clear. Included in the Subsilesian unit in Poland are also some Lower Cretaceous strata (Slaczka et al., 2006), which Elias (1998) compares with the coeval strata of the Kelc subunit of the Silesian unit of central Moravia. Together with some elements of the Silesian unit, the Waschberg–Zdanice–Subsilesian thrust system is comparable with the Helvetic nappes of the Alps.

The inner structure of the rootless, tectonically transported Waschberg–Zdanice–Subsilesian units is complex. This lowermost unit of the Outer Western Carpathian nappe stack consists of numerous imbricates, duplexes, and partial thrust sheets of the Upper Cretaceous to lower Miocene strata. In the Washberg sector of northeastern Austria and southernmost Moravia, tectonic slivers (klippen) of Jurassic marls and carbonates and Upper Cretaceous (Turonian to Campanian) clastic rocks are tectonically incorporated into the thrust sheets of younger, Upper Cretaceous to lower Miocene sequences (e.g., Juttner, 1922, 1928, 1933, 1942; Stranik, 1963; Stranik et al., 1999, and others). These so-called "Outer Klippen" of the Palava Hills in Moravia and of the Ernstbrunn area in Austria were detached from the underlying European foreland and tectonically integrated into the frontal zones of the Carpathian thrust belt during the last stages of the thrusting.

For better clarity and in compliance with some traditional views, the Outer Klippen of the Waschberg sector in Moravia, the Waschberg–Zdanice sector, and the Subsilesian sector of the Waschberg–Zdanice–Subsilesian thrust system in Moravia are discussed in three separate sections. The Waschberg sector of northeastern Austria is not covered in our chapter.

The Outer Klippen of the Waschberg Sector

The Outer Klippen of the Waschberg unit varies in size from small tectonic slivers to large bodies several kilometers long and several hundred meters thick. They consist of several formations of the Late Jurassic and Late Cretaceous age (Figure 17A, B). The oldest Klentnice Formation of Oxfordian to early Tithonian age (Hanzlikova, 1965b) is composed predominantly of dark-gray marls and calcareous shales with an upward-increasing proportion of fine-grained limestones. The poorly preserved fauna of ammonites, bellemnites, brachiopods, pelecypods, crinoids, sponges, bryozoans, and ostracods was described by K. Matzka (1934, personal communication), Bachmayer (1957), Pokorny (1959, 1973), and Vasicek (1971a), among others. Both biostratigraphically and lithologically (Elias, 1991), the Klentnice Formation is comparable with the Mikulov Marls of the autochthonous foreland plate. The known thickness of the Klentnice Formation does not exceed 200 m (660 ft). The marls of the Klentnice Formation are organic rich; however, because of their limited extent and the insufficient burial history, they have practically no potential as source rocks for hydrocarbons.

The Klentnice Formation passes gradually into the overlying Ernsbrunn Limestones of the Tithonian to Berriasian and, according to Elias and Eliasova (1984), up to Hauterivian(?) age. The lower part of the formation is dominated by brecciated organodetritic limestones with matrix of calcareous shales and occasional large clasts of limestones up to several meters in diameter. This facies may represent a detrital talus of a carbonate platform dominated by gravitational transport, including slides, debris flows, and turbidites. Thick-bedded, partly dolomitized calcarenites (locally oolitic) and micritic limestones, which apparently originated in the shallow-water environment of the carbonate platform, make up the upper part of the formation. Occasional hardgrounds and karstifications testify to sporadic emergencies of parts of the platform. The rich fauna, in addition to forms found in the Klentnice Formation, also includes fragments of corals, stromatoliths, and calcareous algae and fish. The known thickness of the Ernsbrunn Limestones is about 120 m (400 ft).

Traditionally, the Ernstbrunn Limestones and the underlying Klentnice Formation of the Outer Klippen have been interpreted as a tectonically detached part of a carbonate succession, which evolved on the rifted passive margins in the Oxfordian and Tithonian. An alternative interpretation by Elias and Eliasova (1984) assumes that the Ernstbrunn Limestones represent a pile of carbonate debris derived from a preexisting hypothetical Tithonian Pavlov platform and redeposited into the Zdanice basin in time of a eustatic drop of the sea level.

Equivalents of Ernsbrunn Limestones, if preserved in an autochthonous position underneath the Carpathian thrust belt, would represent a potential reservoir conveniently charged by hydrocarbons from the underlying organic-rich basinal deposits of the Klentnice Formation.

The Tithonian to Berriasian Ernsbrunn Limestones in the Pavlov Hills and the Waschberg zone are transgressively overlain by the Turonian–Coniacian Klement Formation (Klement Supergroup in Austria) composed of shales, glauconitic sandstones, and sandy limestones. Both lithologically and biostratigraphically, the Klement Formation resembles the Upper Cretaceous epicontinental boreal deposits of northern Europe, specifically the Brezno Formation of the Bohemian Massif (Stranik et al., 1996). The maximum known thickness of the Klement Formation in southern Moravia is 32 m (105 ft). The juxtaposition of the Ernstbrunn and the Klement formations documents the existence of a stratigraphic gap, which most likely lasted from the Valanginian to the Cenomanian and is marked by distinct hardgrounds (Stranik et al., 1996). In that sense, the marginal depositional zone of the Outer Klippen differed from the Silesian basin (Godula subunit), where the sedimentation continued from the Late Jurassic to the early Neogene without any significant interruption (Figure 17A, B).

The overlying Palava Formation (Stranik et al., 1996), previously known as the Mucronata Marls (Abel, 1899), is composed of gray calcareous shales assigned to the late Coniacian to Campanian (Stranik et al., 1996). Fuchs and Wessely (1996) dated the Mucronata Marls of the Outer Klippen in northeastern Austria as the late Maastrichtian. Despite the differences in dating, the Palava Formation and the Mucronata Marls in both these areas are believed to have evolved without an interruption from the underlying Klement Formation (Klement Supergroup in Austria). An existence of a hiatus in the Santonian (Glaessner, 1931) has thus not been proved (Stranik et al., 1999).

The Waschberg–Zdanice Units (Sectors) in Moravia

Excluding the Jurassic and Upper Cretaceous strata of the tectonic klippen, the deposits of the Waschberg–Zdanice units in southern Moravia (the Austrian part of the Waschberg sector is not discussed) range in age from the Late Cretaceous (Campanian–Maastrichtian) to the early Miocene (Egerian to Karpatian). These strata have been divided into six lithostratigraphic formations: (1) the Nemcice Formation, including the Sheshory Marls; (2) the Menilitic Formation, including the Subchert Marls, the Menilitic Cherts, the Dynow Marls, and the Sitborice Member; (3) the Zdanice–Hustopece Formation; (4) the Sakvice Marls; (5) the Pavlovice Formation; and (6) the Laa Formation (Figure 17A, B). The first three formations represent the essentially continuous Late Cretaceous to early Miocene (Egerian) sequence of the Western Carpathian Flysch belt; the other three belong to the late orogenic transgressional Miocene molasse-type deposits comparable to the coeval deposits of the Vienna basin.

The oldest Nemcice Formation of southern Moravia, in recent publications more commonly known as the Submenilitic Formation (Swidzinski, 1948; Roth, 1962), was defined by Rzehak (1880), and its fauna was described by Rzehak (1895b) and Oppenheim (1922). It is composed predominantly of hemipelagic, gray, green, brown, black, and red calcareous shales with subordinate laminae and lenticular bodies of siltstones and sandstones. The proportion of sandstones increases from the northwest to the southeast and markedly steps up in the Cejc–Zajeci zone, interpreted by present authors as a separate unit of the Outer Carpathian Flysch belt. The age of the Nemcice Formation extends from the Late Cretaceous (Campanian–Maastrichtian) to the early Oligocene. However, the oldest Upper Cretaceous strata have been reported mostly from the innermost Cejc–Zajeci zone (unit) (Picha et al., 1968).

In the late Eocene–early Oligocene transition, the deepening and submergence of sources of clastic material, combined with global cooling, created an environment favorable for accumulation of biogenic deposits. In the external Waschberg–Zdanice–Subsilesian and Silesian basins, this new regime led to the deposition of the *Globigerina* Marls (Grzybowski, 1897), which was renamed to Sheshory Marls by Vialov (1951). In the Zdanice sector (unit), the thickness of these predominantly brown laminated marls and shales does not exceed 30 m (100 ft). They correlate well with the coeval Pouzdrany Marls of the Pouzdrany unit and also with the upper part of the autochthonous Nesvacilka Formation (Figure 17A, B). The overall depositional character of Nemcice Formation, including the Sheshory Marls, refers to an underfilled basin, in which the subsidence outpaced the sediment supply.

The Sheshory Marls pass gradually into a variety of biogenic deposits summarily assigned to the Menilitic Formation (Glocker, 1843). Stranik (1981), following the subdivision of the Menilitic Formation in Poland (Jucha and Kotlarczyk, 1961), distinguished in the Menilitic Formation of Moravia the Subchert Member, the Chert Member (Menilitic Cherts), and the Dynow Marls. He also named the uppermost predominantly shaly part of the Menilitic Formation the Sitborice Member (Figure 18).

The Subchert Marls, about 10 m (33 ft) thick, is represented by brown stratified marls and shales with fish scales. Vertically, they pass into laminated cherts and organic-rich siliceous shales of the Chert Member (Menilitic Cherts), which represents the most characteristic member of the Menilitic Formation. In the outermost zones of the Zdanice unit, the cherts are partly substituted by diatomites and in the marginal Pouzdrany unit fully replaced by diatomites (Figure 19). This would suggest that the typical Menilitic Cherts of the Zdanice–Subsilesian units were formed by the diagenetic alteration of laminated diatomites. The Menilitic Cherts reach their maximum thickness of approximately 4 m (13 ft) in the northwestern part of the Zdanice unit. Their thickness decreases both toward the south into the Waschberg zone and toward the east. The Menilitic Cherts have not been found in the Cejc–Zajeci unit.

According to Picha and Stranik (1999), the Menilitic Cherts were deposited in a zone of upwelling of nutrient-rich deep waters and proliferation of marine life (diatoms), combined with anoxic conditions on the bottom of the sea and a very limited influx of detrital material both from the foreland and the orogenic belt (Figure 19). The sedimentation was essentially a function of siliceous phyto- and zooplankton manufacture. The presence of remnants of fish with light-emitting organs (Kalabis, 1949) indicates that the depositional environment of the Menilitic Formation was relatively deep. Based on studies of the ichthyofauna, Brzobohaty (1981) and Gregorova (1988) suggested a mesopelagic depositional environment with a water depth between 200 and 1000 m (660 and 3300 ft). Picha and Stranik (1999) compared the depositional environment of the Menilitic Cherts to that which prevailed along the active margins of coastal California, where conditions for deposition of organic-rich diatomites have repeatedly occurred since the Late Cretaceous to the Miocene. The most prominent Monterey Formation of Miocene age was deposited in a neritic to bathyal environment (water depth 50–1200 m [160–4000 ft]) of silled basins formed along the active continental margins of North America (Smith, 1968).

The Menilitic Cherts are overlain by pelagic limestones and marls of the Dynow Marls (10 m; 33 ft), which are followed by a predominantly argillaceous noncalcareous sequence of the Sitborice Member

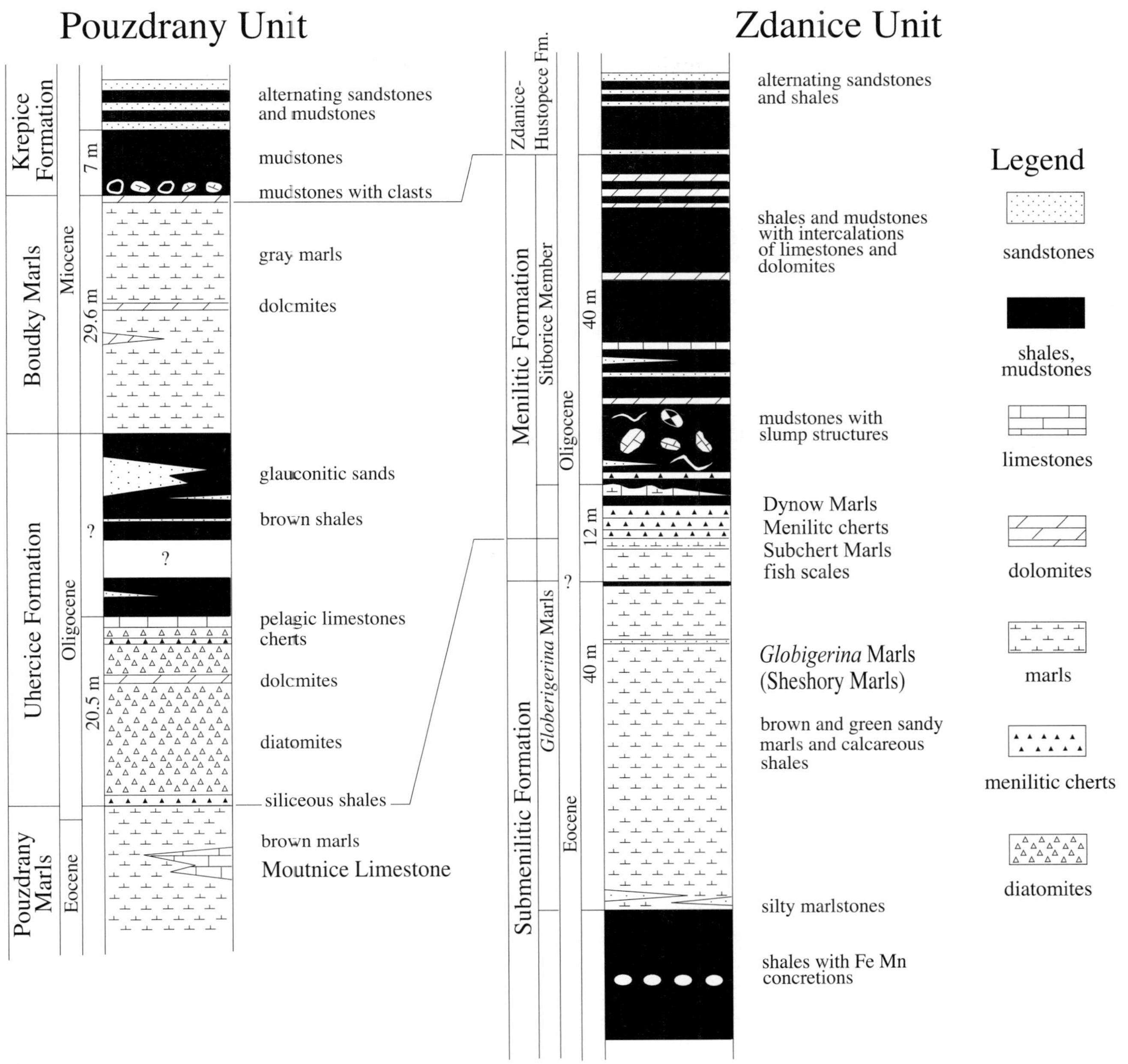

Figure 18. Correlation of the upper Eocene to the lower Miocene strata in two measured sections, one in the Pouzdrany unit at the Wine cellars in the village Pouzdrany and the second in the Zdanice unit in the road cut near the village Velke Nemcice. Modified after Picha and Stranik (1999).

(Stranik, 1981). The base of this sequence is commonly marked by an erosional surface, most likely associated with the activity of turbidity currents and debris flows. In its lower part, the Sitborice Member is made mostly of brown noncalcareous shales with occasional slump bodies of mudstones and debris flows with intraclasts of Menilitic cherts. Locally, large blocks of carbonates and crystalline rocks, apparently derived from the adjacent slopes or from the slump-prone deposits of the submarine canyons, are found in this sequence. The upper part of the Sitborice Member is represented by alternating brown and green shales with manganese and jarosite coatings, scattered crystals of anhydrite and gypsum, and intercalations and concretions of limestones and dolomites (Stranik et al., 1981b). Picha and Stranik (1999) correlated this upper part of the Menilitic Formation with the lithologically similar Uhercice Formation of the Pouzdrany unit

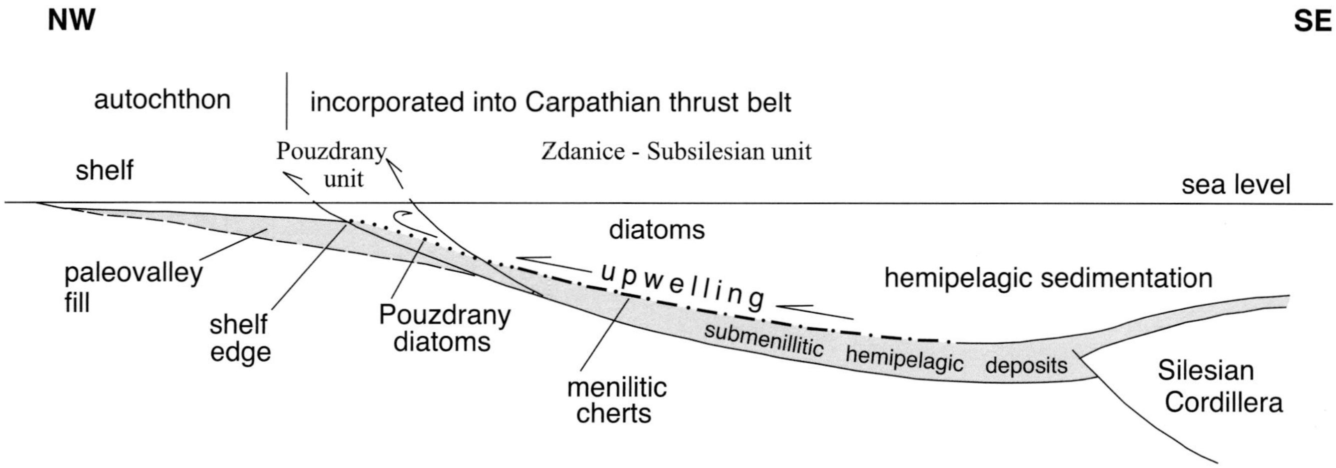

Figure 19. Depositional setting of the lower Oligocene organic-rich diatomites and Menilitic Cherts of the Pouzdrany and Zdanice units, respectively. The slope-to-basin environment was characterized by upwelling of nutrient-rich waters, proliferation of diatoms, and prevalence of anoxic conditions on the bottom (Picha and Stranik, 1999, reprinted with permission).

(Figure 17A, B). The overall thickness of the Menilitic Formation in the Zdanice unit ranges from several tens to 200 m (660 ft).

The presence of submarine slump conglomerates and debris flows in the Sitborice Member of the Zdanice unit and in the Menilitic Formation of the Subsilesian unit marks the end of the quiet period and the beginning of a new phase of tectonism. In the Outer Western Carpathians, this tectonic phase resulted in the deformation and uplifting of the Magura flysch and the formation of the Krosno-type synorogenic foredeeps.

Because of the lack of diagnostic fossils both in the Menilitic Formation and the overlying Krosno-type flysch, the exact timing of these events and even the age of the Menilitic Formation have become a subject of discussion. The Menilitic Formation of the Zdanice unit was assigned variably to the late Eocene (Pokorny, 1947), to the Eocene–Oligocene boundary (Hanzlikova in Mahel and Buday, eds., et al., 1968), and to the early Oligocene (Rzehak, 1922; Pokorny, 1960; Cicha et al., 1971; Hanzlikova, 1981; Krhovsky, 1981; Bubik, 1987; Jurasova, 1987; Krhovsky et al., 1992). The deposition of the Menilitic Formation in the Outer Carpathians falls into a critical period in Earth history, when the greenhouse climates of the Eocene were replaced by the icehouse conditions associated with a major extinction in the early Oligocene (Prothero et al., 2000). The onset of deposition of the Menilitic Cherts marks the beginning of the cooling period on the northern hemisphere (Krhovsky et al., 1992).

The organic-rich Menilitic Formation represents one of the most important source rocks for hydrocarbons in the entire Carpathian realm. In the Zdanice–Subsilesian unit, however, with the exception of the inner zones buried deeper below the Silesian and Magura nappes, the organic-rich strata of the Menilitic Formation remain immature.

The Menilitic Formation of the Zdanice unit is overlain by the as much as 1200-m (4000-ft)-thick Zdanice–Hustopece Formation (Chmelik and Matejka in Kalasek et al., 1963). It is a typical late orogenic Krosno-type turbiditic sequence of gray shales with laminae of siltstones and beds of sandstones and local conglomerates. The proportion of sands and shales varies both vertically and horizontally from predominantly shaly (Hustopece Marls; Rzehak, 1881) to predominantly sandy facies (Zdanice Sandstones; Paul, 1893) with a widespread transitional facies between these two end members.

The overall northwestward fining and thinning of turbiditic sequences, as well as the orientation of the paleocurrent markings, indicate that the Zdanice–Hustopece Formation was deposited in a system of northwestward-prograding and overlapping subsea fans. Occasional bodies of conglomerates, interpreted as debris flows and channel fills, are more commonly found in the proximal southeastern parts of these fans. In addition to abundant larger clasts of sandstones reworked from the Magura flysch, these conglomerates contain pebbles of crystalline rocks and Mesozoic shales and carbonates, most likely supplied from the tectonically activated (inverted) crustal blocks located at the front of the Magura thrust system. In composition, the conglomerates of the Zdanice–Hustopece Formation differ markedly from the Paleocene–Eocene conglomerates found in the pericordilleran Cejc–Zajeci unit. These older conglomerates were apparently sourced from the Silesian cordillera (Picha et al., 1966; Sotak, 1992).

Similarly, the monotonous garnet-dominated assemblages of heavy minerals of the upper Oligocene to lower Miocene Zdanice–Hustopece Formation, as well as of the Krepice Formation of the Pouzdrany unit, differ markedly from the complex garnet-zircon-tourmaline-staurolite assemblage of heavy minerals found in the older, Upper Cretaceous to lower Oligocene strata of the Zdanice and Pouzdrany units and in the autochthonous Paleogene (Krystek in Homola et al., 1961; Picha, 1963, 1965; Stranik et al., 1968) (Table 1).

As documented by differences in composition of heavy minerals and conglomerates (Picha et al., 1966), the primary restructuring of the Outer Carpathian depositional system in the late Oligocene also led to primary changes of the provenance of the clastic material. The older pre-late Oligocene strata were supplied from the foreland and the uplifted Silesian ridge, whereas the younger, late Oligocene flysch deposits were recycled mainly from the uplifted inner zones of the Flysch belt, particularly the Magura flysch.

In addition, most of the microfauna found in the Zdanice–Hustopece Formation was redeposited from the older Cretaceous and lower Paleogene flysch deposits, whereas the diagnostic indigenous species are scarce. The age determination of the formation, thus, to a great extent, depends on dating of the underlying and overlying strata and the overall tectonostratigraphic position of the formation. Considering all these aspects, the age of the Zdanice–Hustopece Formation was assigned to the late Oligocene to early Miocene (Egerian) (Cicha et al., 1964, 1975b).

The Zdanice–Hustopece Formation and its equivalent, the Krepice Formation of the Pouzdrany unit, are typical synorogenic flysch sequences deposited in a system of foredeeps, which formed by flexural downbending of the foreland plate in front of the northwesterly progressing nappe stack of the Magura flysch. In a broader sense, these formations belong to the late Oligocene to early Miocene Krosno-type synorogenic depositional system characteristic for the external units of the Outer Carpathian belt.

The deposition of the Zdanice–Hustopece formation in Chattian–Aquitanian (Egerian) marks the end of the continuous sedimentation in most of the Waschberg–Zdanice–Subsilesian zone, although in the Dyje–Thaya depression of southern Moravia and the Waschberg zone, the sedimentation continued into the Burdigalian (Eggenburgian, Ottnangian, and Karpatian). These strata, however, belong to a new molasse-type phase associated with a major marine transgression and formation of the Neogene foredeep in the foreland and the Vienna basin in the hinterland of the Outer Carpathian thrust belt in Moravia.

Table 1. Assemblages of heavy minerals in the autochthonous unit and in various formations of the Pouzdrany and Zdanice allochthonous units in southern Moravia (Krystek in Homola et al., 1961; Picha, 1963). Note the principal differences in composition of heavy-mineral assemblages between the pre-late Oligocene (Nesvacilka, Pouzdrany, and Submenilitic formations and Sitborice Member) and the late Oligocene and early Miocene (Krepice and Zdanice–Hustopece formations) (Picha and Stranik, 1999). All numbers are percentages.

	Autochthonous Unit (%)	Pouzdrany Unit (%)		Zdanice Unit (%)		
	Nesvacilka Formation Eocene–Lower Oligocene	**Pouzdrany Formation** Eocene–Lower Oligocene	**Krepice Formation** Upper Oligocene–Lower Miocene	**Submenilitic Formation** Upper Cretaceous–Upper Eocene	**Sitborice Member** Lower Oligocene	**Zdanice–Hustopece Formation** Upper Oligocene–Lower Miocene
Garnet	40.6	17.3	80.5	31.3	26.4	79.1
Zircon	30.7	23.1	3.1	33.3	25.1	4.0
Staurolite	4.9	25.2	3.8	4.1	13.3	6.7
Tourmaline	7.9	12.1	1.7	7.6	16.0	2.2
Rutile	6.5	10.7	4.9	8.9	8.9	2.7
Kyanite	2.3	7.3		1.7	4.7	1.2
Anatase	0.7			1.8		0.4
Apatite	5.5			10.1	3.9	1.4
Monazite	0.3			0.4		
Amphibole				0.2		
Other	0.9		1.5		1.7	

In the Zdanice unit, this new phase began with deposition of as much as 200 m (660 ft) of shallow-marine calcareous shales and marls, which were assigned to lower Burdigalian (Eggenburgian) and correlated with the Luzice Formation of the Vienna basin by Benesova et al. (1963) and Cicha and Picha (1964) and called Sakvice Marls by Cicha et al. (1975b) (Figure 17A, B). These strata are found mainly in an east–west-trending depression (Figure 3), which is generally identical with the extent of the Vranovice paleovalley buried underneath the Zdanice nappe. This coincidence may be explained by the preservation of the Sakvice Marls in a zone of a higher subsidence apparently related to a higher compaction rate of the underlying paleovalley fill. Another much smaller occurrence of the Sakvice Marls is located near the village of Kobyli, just in front of the Cejce–Zajeci thrust unit (Cicha et al., 1964). The Sakvice Marls, in some areas, e.g., at the type locality near Sakvice, seem to evolve from the underlying Zdanice–Hustopece Formation without interruption (Benesova et al., 1963); in other areas, however, the Sakvice Marls rest transgressively on the underlying older strata without any evident angular discontinuity. Elsewhere in Moravia, the Eggenburgian strata represent a transgressive sequence, which was laid down on the folded and thrusted Magura flysch in the Vienna basin area or on the newly submerged zone of the foreland as a basal sequence of the late orogenic and postorogenic molasse-type foredeep.

Locally, e.g., in the vicinity of the Zajeci railway station, the Eggenburgian Sakvice Marls pass vertically into a 300-m (1000-ft)-thick sequence of greenish and brownish gray claystones with lenses of silty Fe and Mn pelocarbonates followed by diatomites (Pokorny, 1961; Benesova et al., 1963; Cicha and Picha, 1964). Stranik (1983) named this sequence the Pavlovice Formation and assigned it to the Eggenburgian to Karpatian (Figure 17A, B). The Pavlovice Formation is overlain by gray calcareous silty shales with rich microfauna, which Pokorny (1961), Benesova et al. (1963), and Cicha and Picha (1964) attributed to the Karpatian. Stranik (1983) assigned these about 100-m (330-ft)-thick strata to the Laa Formation (including the Korneuburg Member) (Figure 17A, B), as known from the Neogene foredeep and the Vienna basin in Austria (Grill, 1953, 1962; Kapounek et al., 1960; Wessely, 1998).

During the last stages of the Carpathian thrusting in Moravia, at the end of the early Miocene, the lower Miocene (Eggenburgian to Karpatian) strata, together with the underlying strata of the Zdanice and Pouzdrany units, were gently folded and tectonically transported toward the foreland. The level of deformation of these lower Miocene deposits seems to be similar to that of the underlying relatively competent Zdanice–Hustopece Formation. The older, less competent strata of the Nemcice and Menilitic formations were deformed more intensely (Hrouda and Stranik, 1985; Stranik, 1999).

The Subsilesian Unit (Sector)

The Subsilesian unit of northern Moravia comprises strata ranging in age from the Late Cretaceous (Turonian–Maastrichtian) to the early Miocene (Egerian). They have been divided into four lithostratigraphic units: (1) the Frydek Formation, (2) the Frydlant Formation, (3) the Menilitic Formation, and (4) the Zdanice–Hustopece Formation (Figure 17A, B).

The oldest, Turonian to Paleocene Frydek Formation, about 500 m (1600 ft) thick, is made by a monotonous sequence of gray laminated silty shales and siltstones with a rich foraminiferal microfauna (Hanzlikova, 1969). Turbiditic sandstones and slump bodies of conglomerates with fragments of corals are found locally, especially in the upper part of the formation in the vicinity of Pribor, where they are known as the Klokocov Member. Elias (1998) interpreted these coarse deposits as a proximal facies of the lower slope environment, whereas he assigned the finer predominantly pelitic strata of the Frydek Formation to a deep-water basinal facies. According to the senior author of this article, the Frydek Formation may be interpreted as a distal facies of a major submarine fan system, whose more proximal facies are represented by the Istebna Formation of the Silesian unit. The Klokocov Member would then be a local facies related to a proximity of the locally uplifted ridge holding the Jurassic Stramberk reef complex (fragments of corals).

The overlying Frydlant Formation (Elias, 1998), also known as the Submenilitic Formation, is facially more diversified. Hanzlikova et al. (1963) distinguished three facies in this formation: (1) the anoxic facies (Guty Member of E. Mencik, M. Elias, I. Zurkova, F. Jurasova, and L. Rybarova, 1973, personal communication) of dark-gray shales, which prevailed in the late Paleocene; (2) the variegated facies (Roth, 1962) of gray, green, and red shales, which dominated in the late Eocene; and (3) the transitional facies of spotted gray, green, and red shales (spotted facies of Roth, 1962). Elias (1998) added a facies of thick sandstones and conglomerates (Straz Sandstones of Hanzlikova et al., 1955), which may be interpreted as channel and overbank deposits, similar to those found in the Cejc–Zajeci unit.

The Menilitic Formation of the Subsilesian unit, otherwise similar to that of the Zdanice unit, locally includes about 100-m (330-ft)-thick facies of slump conglomerates and debris flows. They contain clasts of crystalline basement rocks, Devonian and Carboniferous carbonates and clastics, as well as intraclasts of Upper Cretaceous to upper Eocene strata derived from the margins of the Subsilesian basin proper. The nannoplankton from the locality Dolni Tesice indicates the early Oligocene age of the Menilitic Formation (Jurasova, 1974). Thin intercalations of laminated shaly limestones (Jaslo Limestones, Uhlig, 1882, 1883b; Jucha, 1958) occur in the upper part of the Menilitic Formation near Bystrice nad Olsi (Nowak, 1965). Bubik (1987) compared them with the Oligocene Zagorza horizon (Haczewski, 1989) of the Polish Carpathians. The overall thickness of the Menilitic Formation in the Subsilesian unit may reach as much as 200 m (660 ft).

The Zdanice–Hustopece Formation (Elias, 1998), previously called the Krosno Formation, is coeval and lithologically similar to the Zdanice–Hustopece Formation of the Zdanice sector, but its extent and thickness are limited. The known thickness of this formation in the Subsilesian sector does not exceed several tens of meters.

Additional information about the stratigraphy and structure of the Subsilesian unit may be found in publications by Hanzlikova et al. (1953), Mencik and Pesl (1955), and Mencik and Hanzlikova (1983).

The Facies Variations along the Strike of the Waschberg–Zdanice–Subsilesian Units

The character of the Waschberg–Zdanice–Subsilesian nappe system significantly changes along the strike of the Carpathian thrust belt. The Upper Cretaceous strata are prominently present everywhere in the Subsilesian sector, whereas their extent in the Waschberg–Zdanice sector is limited. However, the late Oligocene to early Miocene Krosno-type flysch deposits are dominant in the Zdanice and partly in the Waschberg sectors and only marginally present in the Subsilesian sector of northern Moravia. The lower Miocene Sakvice, Pavlovice, and Laa formations, typical for the Waschberg–Zdanice sectors, have not been found in northern Moravia. Their presence seems to be limited to the confines of the Dyje–Thaya depression. The Pouzdrany marginal unit has been recognized only in the Zdanice sector, but it is apparently also present in the Waschberg zone of Austria (Stranik, 1996) but evidently absent in northern Moravia. Finally, the Jurassic and Cretaceous tectonic klippen is found only in the Waschberg zone.

These stratigraphic and facial differences obscured the correlation and led to the development of separate terminologies for the various sectors. The most apparent terminalogical differences exist between the Austrian and Czech parts of the Outer Carpathians (Figure 16). Whereas in Austria, the Waschberg zone is generally considered to be a part of Helveticum, the Zdanice unit, thanks to the dominant presence of the flysch-type Zdanice–Hustopece Formation, is referred to as a part of the Carpathian Flysch belt.

The three sectors also differ in their structural architecture. The southernmost Waschberg sector is characterized by a rather steep internal structure (Matejka in Buday et al., 1961). The Zdanice sector displays a typical wedge-shaped geometry, and the northern Subsilesian sector, as documented by numerous wells, has an extremely low-angle flat structural pattern (Figure 20, shown on page 118). This is apparently related to the distribution of compressional and transpressional stresses along the strike of the Western Carpathian belt (Nemcok et al., 1998a). The amount of shortening accommodated by the sinistral strike-slip motion is the highest in the north–south-trending relatively steep Waschberg sector. Toward the north, the amount of transpressional strike-slip motion decreases, and the shortening is accommodated mainly by the normal forward thrusting, resulting in a flat structural grain of the thrust system.

The Silesian Unit

The Silesian thrust unit, positioned between the more external Subsilesian and the more internal Magura units, is the stratigraphically most complete unit of the Outer Western Carpathians (Figures 3; 17A, B). Its more or less continuous stratigraphic section extends from the Late Jurassic to the late Oligocene and possibly even to the early Miocene. As a distinct depositional site, the Silesian basin formed during the Late Jurassic and the Early Cretaceous rifting and extension of the European plate. Jurassic carbonate platforms marked the northwestern margin of the basin, whereas the Silesian ridge bordered the Silesian basin on the south, thus separating it from the more internal Magura depositional domain (Figure 5). Carbonates, deep-water shales, silicites, and turbidites were deposited during the divergent stage of the basin, whereas thick turbiditic flysch sequences were laid down during the convergent stage. The Silesian ridge functioned as the main source of clastics especially during the convergent synorogenic phase of the evolution of the Silesian basin. Both organic-rich source rocks and good reservoirs are present in the stratigraphic records of the Silesian unit, which, especially on the territory of Poland, proved to be the most prolific unit of the Western Carpathian Flysch belt.

The Silesian unit is most prominently developed in the Beskydy region of northern Moravia and western

Poland, where it forms the highest mountain peaks (more than 1300 m [4300 ft]). The continuation of the Silesian unit to the southwest into southern Moravia and the Alpine domain is not clear. Lower Cretaceous strata, similar to those of the Silesian unit, have been described from the Zdounky unit in central Moravia (Matejka in Buday et al., 1963) and from the Hauptzone of the Wienerwald (Vienna Forest) in Austria. However, the typical Silesian unit, as it is known from northern Moravia and western Poland, apparently originally ended at the southwest–northeast-trending Western Carpathian transfer zone, which separated the Alpine region from the more attenuated and differentiated Carpathian realm (Figure 5).

Three stratigraphically and structurally distinct subunits (facies), Godula, Baska (Matejka and Roth, 1949a, 1955), and Kelc (Elias, 1970, 1979), have been recognized in the Silesian unit of Moravia. They differ in the stratigraphic extent and overall thickness of their depositional sequences as well as in the proportion of detrital material derived from the marginal carbonate platforms. Although the dominant Godula facies makes up the bulk of the Silesian unit in Moravia and continues on in the territory of Poland, the areal extent of the other two subunits is limited. The Baska subunit represents the local marginal facies of the Silesian unit adjacent to the Stramberk carbonate platform, and the relatively thin Kelc subunit occupies the southwestern end of the Silesian basin. Only the Godula facies extends over the entire stratigraphic range from the Upper Jurassic to Oligocene, whereas the stratigraphic sequences of the Baska and the Kelc facies lack most of the Paleogene section, which most likely was tectonically detached during the thrusting. The structurally complex frontal zone of the Silesian thrust system, in addition to elements of the Baska and Kelc subunits, also comprises tectonic slivers of the incompetent Lower Cretaceous strata of the Godula subunit as well as the tectonically incorporated slivers of the Subsilesian unit. The recognition of various elements of these units and subunits is not always easy.

The Godula Subunit

The Godula subunit has been divided into numerous formations and members (Figure 17A, B). The oldest known strata of the Godula subunit, originally known as the Lower Tesin (Teschen in German; Cieszyn in Polish) Member (Hohenegger, 1861) and recently renamed to the Vendryne Formation (Elias et al., 2003), are represented by dark-brown calcareous shales with occasional thin beds of siltstones and limestones. Slump conglomerates (debris flows) with clasts of limestones (Ropice horizon of Mencik et al., 1983) occur in the upper part of the formation. Vasicek (1972a, b) assigned the Vendryne Formation, about 350–600 m (1100–2000 ft) thick, to the Oxfordian to Tithonian and possibly to the Berriasian(?) age. The Vendryne Formation is thus coeval with the Klentnice Formation of the Jurassic klippen in southern Moravia. Like the Klentnice Formation, it also contains an increased content of organic matter and, as such, may be considered as a source rock for hydrocarbons.

The Vendryne Formation passes upward into the Tesin (Teschen) Limestone, which, mainly in its lower part, is composed of pelagic thinly bedded micritic limestones alternating with calcareous shales. The presence of the *Calpionella alpina* allocates these strata to the uppermost Tithonian to the early Berriasian (Hanzlikova and Roth, 1964). Laterally, these no more than 20–30-m (66–100-ft)-thick pelagic limestones pass into a facies of detrital limestones with occasional cherts, debris flows, grain flows, and intercalations of calcareous shales. The grains and larger clasts of these detrital limestones are made of carbonate rocks derived from the Stramberk carbonate platform. In a broad sense, the Tesin Limestones might be an equivalent of the Ernsbrunn Limestones of the Outer Klippen Belt in southern Moravia (Figure 17A, B).

The following Valanginian to early Aptian Tesin (Teschen)–Hradiste Formation (sensu Matejka and Roth, 1954) consists of three distinct lithologies: (1) the calcareous shales with subordinate turbiditic sandstones, limestones, and pelosiderites in the lower part [the Hohenegger's (1861) Upper Tesin (Teschen) Member]; (2) the thick packets (tens of meters) of coarse turbiditic sandstones and conglomerates separated by thin layers of claystones in the middle part (the Hradiste Sandstone of Hohenegger, 1861); and (3) the calcareous shales with abundant lenses and concretions of pelosiderites in the upper part (the lower part of the Verovice shales, sensu Hohenegger, 1861). The coarse sandstones and conglomerates of the Hradiste Sandstone are composed predominantly of quartz grains and fragments of the Stramberk Limestones, apparently supplied mainly from the uplifted Stramberk platform in the northwest. To simplify the complex historically evolved nomenclature, Elias et al. (2003) suggested the use of the term Hradiste Formation for the entire complex of the Tesin (Teschen)–Hradiste Formation. For the sake of consistency with the stratigraphy of the Baska and Kelc subunits and the compatibility with numerous previous publications, in our article, we adhere to the traditional term Tesin–Hradiste Formation, sensu Matejka and Roth (1954) (Figure 17A, B).

Characteristic for the Tesin (Teschen)–Hradiste Formation is the presence of dikes, veins, lavas, pillow lavas, and pyroclastic rocks of the teschenite rift-related submarine alkalic, calc-alkalic, and basic volcanism. Smid (1962) and Smid and Mencik in Mencik et al. (1983) distinguished three groups of volcanic rocks: picrites, teschenites, and monchiquites. Hovorka and Spisiak (1988) associated the teschenite volcanism with a short-term rifting of the continental crust. Dostal and Owen (1998) pointed to similarities of these rocks with basalts, basanites, and nephelinites derived from the upper mantle. The volcanic activity peaked during the deposition of the lower part of the Tesin (Teschen)–Hradiste Formation in the early Berriasian to Hauterivian time, although teschenite volcanic rocks are sporadically found also in the underlying Tesin Limestone and the Vendryne Formation.

The Tesin (Teschen)–Hradiste Formation was deposited during the phase of rifting, extension, and further diversification of the Outer Carpathian depositional system. From the Magura basin, the Silesian depositional site was separated by the Silesian ridge (cordillera), which emerged as a horst between these two basins. As documented by the increase of the thickness of the Tesin–Hradiste Formation, from several hundred meters in the northwest to more than 1000 m (3300 ft) in the southeast. Most of the clastic material into the Tesin–Hradiste Formation was apparently supplied from the Silesian ridge. Despite the intensive rifting and extension, the Silesian basin remained underlain by a thinned continental crust.

The stratigraphy and fauna of the Lower Cretaceous sequences of the Silesian unit were studied by Uhlig (1883a, 1902) and recently by Eliasova (1962a), Hanzlikova (1965a), Vasicek (1971b, 1973, 1979, 1999), Housa (1975, 1978), Vasicek et al. (1994), Skupien and Vasicek (2002), and Elias et al. (2003).

In the 19th century, these iron-rich pelosiderites of the upper part of the Tesin–Hradiste Formation provided a raw material for the nascent iron and steel industry of Silesia (Roth and Matejka, 1953).

The overlying Verovice Formation (sensu Matejka and Roth, 1949a), several tens of meters to as much as 250 m (820 ft) thick, is composed predominantly of black organic-rich cherty shales with rusty iron oxide coatings, locally interbedded with thin beds of quartzitic sandstones and concretions of pelosiderites. Given the total absence of diagnostic fossils, the assignment of the Verovice Formation to the Aptian age is based solely on the determination of the ages of the underlying and overlying strata. The depositional environment of the Verovice Formation was extremely anoxic, favorable for preservation of organic matter. The high content of organic matter makes the Verovice Formation a potential source rock for hydrocarbons; however, its role in the petroleum systems of the Carpathians has yet to be better understood.

The deposition of the Verovice Formation in the Silesian basin and the coeval Rajnochovice Formation (Gault Flysch) in the Magura flysch marks a period of maximum deepening of the Outer Carpathian basins and the submergence of all sources of detrital material, both in the European platform and in the nascent Silesian ridge. According to Slaczka et al. (1999), the downwarping of the Silesian basin was most likely caused by the cooling effect of the underlying lithosphere, previously dominated by the high heat-flow regime of the rift stage.

The Verovice Formation passes upward into the 100–380-m (330–1250-ft)-thick Lhoty Formation assigned to the Albian (e.g., Hanzlikova, 1966). In its lower part, it consists of bioturbated shales with subordinate thinly bedded distal turbiditic sandstones, whose proportion increases upward. The upper part of the Lhoty Formation is characterized by the predominance of turbiditic sandstones and the occurrence of cherts (Mikuszowice Cherts of Szajnocha, 1884). The Lhoty Formation was still deposited in a deep-water uncompensated environment although, as indicated by the presence of bioturbations, less anoxic than that of the underlying Verovice Formation. Even further, a less restricted environment prevailed during the deposition of the overlying Cenomanian pelagic and hemipelagic red and green shales of the Mazak Formation (sensu Roth, 1980a), also known as the Variegated Godula Member (Zahalka and Koutek, 1927; Matejka and Roth, 1949c). These shaly deposits are interbedded with thick coarse turbiditic sandstones and conglomerates of the Ostravice Sandstone (Andrusov, 1933). The deposition of the variegated shales of the Mazak Formation assigned to the Cenomanian age (Hanzlikova, 1973) thus marks the change from the deep anoxic conditions of the Verovice and Lhoty formations into the more dynamic environment of the deep-water turbiditic flysch facies. The Ostravice Sandstone, interpreted by Elias (1995) as a prograding turbiditic subsea fan, represents the oldest clearly defined synorogenic sequence of the Outer Carpathians in Moravia and actually marks the transition from the passive-margin environment into an active-margin depositional setting (Figure 17A, B).

The overlying Godula Formation of the Cenomanian(?)–Turonian–Santonian age is a typical flysch sequence of alternating sandstones and shales with a variable proportion of these two main lithological components. The lithological subdivision of the Godula Formation has been elaborated by Burtan et al. (1937), Matejka (1949, 1952), Mencik et al. (1983), Elias (2000), and others. The lower part of the formation is typically represented by a turbiditic facies of thinly bedded glauconitic sandstones and shales followed in the middle part of the formation by a facies dominated by coarse glauconitic sandstones and conglomerates. The upper part of the Godula Formation is again represented by a facies of thinly interbedded glauconitic sandstones and shales, although coarse sandstones and conglomerates, such as those of the Malinowska skala Sandstone (Burtan et al., 1937; Elias, 2000) and of the Pustevny Sandstone (Mencik et al., 1983), occur locally. The vertical and horizontal variability in proportion of sandstones and shales reflects on the dynamic depositional environment of the Godula Formation dominated by prograding and shifting subsea fans. The overall thickness of the Godula Formation ranges from more than 3000 m (10,000 ft) at the southern proximal side of the Silesian basin adjacent to Silesian ridge to only a few hundred meters in the distal northern side of the basin facing the platform. The Godula Formation thus appears as a depositional wedge, whose thick part formed in a zone of maximum subsidence at the front of the emerging Silesian ridge, which supplied most of the clastic material into the Godula Formation. The sudden rise of the Silesian ridge was apparently caused by compressional stresses associated with the accelerated subduction of the Penninic–Pieninic ocean and the early collision of the Inner Carpathians with the fragmented margins of Europe in the early Late Cretaceous. The deposition of the Godula Formation in the Silesian basin thus marks the beginning of the convergency and the concomitant formation of the foreland depositional regime in the Outer Carpathian realm.

In the Campanian–Maastrichtian, the foreland depositional system progressed further onto the European foreland, where the Waschberg–Zdanice–Subsilesian basin came into existence. In the Silesian basin, the Istebna Formation of the Campanian to Maastrichtian to Danian(?) age (Liebus and Uhlig, 1902; Hanzlikova, 1972b) evolved from the underlying Godula Formation. The presence of conglomerates and erosional surfaces at the boundary with the underlying Godula Formation was, by some geologists (Zahalka and Koutek, 1927), mistakenly taken as evidence of a transgressive character of the Istebna Formation. The 1000–1200-m (3300–4000-ft)-thick Istebna Formation is represented by a typical flysch facies of alternating sequences (70–200 m; 230–660 ft) of arkosic sandstones, slump conglomerates, and sand flows (fluxoturbidites) with equally thick sequences of dark shales. The common presence of pelocarbonate concretions, authigenic siderite, pyrite, and organic matter indicates that the depositional environment of the Istebna Formation was anoxic. An abundance of gravity-driven slide conglomerates and sandstones testifies to an increased tectonic activity along the slopes of the rising Silesian ridge. Further north away from the Silesian ridge, the proximal facies of the Istebna Formation passes into more distal turbiditic facies. Moreover, according to the senior author of this account, the Frydek Formation of the Subsilesian basin may represent the most distal facies of the Istebna Formation fan system. The highly dynamic depositional environment of the Istebna Formation was apparently related to the further rise of the Silesian ridge during the Laramide orogeny in the Late Cretaceous to the early Paleocene. As documented by the depositional record both in the Silesian and Magura basins (Figure 17A, B), the Laramide uplifting and inversion of crustal blocks in the Outer Carpathian basin and in the European foreland had a great impact on the evolution of the entire Carpathian region.

Sporadically found rhyolitic and andesitic tuffitic rocks in the Istebna Formation and also in the underlying Godula Formation indicate that the volcanic activity in the Silesian basin continued into the Turonian–Maastrichtian.

The overlying Paleocene to Eocene Roznov Formation (Elias, 2002), formerly known as the Submenilitic Formation, about 800 m (2600 ft) thick, is composed predominantly of hemipelagic variegated, red, green, and gray shales with subordinate thin sandstone beds. Encased into this monotonous sequence, thick packets (as much as 150 m [500 ft]) of coarse sands (Ciezkowice Sandstone) are present, which may be interpreted as subsea channel fills and levee deposits of the basin floor. Numerous oil fields in the Polish sector of the Silesian unit are reservoired in these channelized sands (Karnkowski and Ozimkowski, 1998; Dziadzio et al., 2006).

The upper part of the Roznov Formation is marked by the presence of the Sheshory Marls (previously known as *Globigerina* Marls), composed of laminated brown siltic marls alternating with greenish pelagic clays. They represent a transition into the lower Oligocene Menilitic Formation, composed of dark-gray and brown silicified shales, cherts, marls, and micritic limestones. Lithologically, the Menilitic Formation of the Silesian unit is comparable with its equivalents in the Waschberg–Zdanice–Subsilesian unit, including its subdivision into the Subchert and Chert members, the Dynow Marls, and the Sitborice Member. Locally, in the zone adjacent to the front of the Magura nappe, thick beds of greenish glauconitic sandstones are found in the Menilitic Formation. They are comparable with the Kliwa Sandstones of the Polish Carpathians, where they represent a significant oil-bearing reservoir. However, the association of these sandstones with either the Silesian or the Fore-Magura units in the structurally complex zone at the front of the Magura nappe remains uncertain. The organic-rich Menilitic Formation of the

Silesian unit, if properly buried, may represent an important source rock for hydrocarbons.

The highest strata of the Godula subunit are represented by the late Oligocene to early Miocene Krosno Formation (Tietze, 1889). It is a typical flysch facies of alternating turbiditic sandstones, shales, and occasional conglomerates. As in the Zdanice–Hustopece Formation of the Waschberg–Zdanice–Subsilesian unit, with which the Krosno Formation is comparable, the proportion of sandstones and shales varies both laterally and vertically from the predominantly shaly to the predominantly sandy facies. On the territory of Moravia, the Krosno Formation is more than 1000 m (3300 ft) thick.

The overall maximum thickness of the Godula subunit may reach 4000 m (13,000 ft) on its southern proximal side but would decrease significantly toward the more distal northern side of the Silesian unit. These variations in thickness are caused mainly by the wedge-shaped Godula and Istebna formations, whose fast subsiding depocenters were adjacent to the emerging Silesian cordillera in the south.

The Baska Subunit

Unlike the dominant and widespread Godula subunit, the Baska subunit is restricted to a relatively smaller area of the Palkovice Hills and the Stramberk klippen (Figure 3). It consists of several tectonic imbricates comprising tectonic klippen detached from the Tithonian to Valanginian carbonate platform and the younger Hauterivian to Paleocene strata characterized by the abundant presence of clastic material derived from this carbonate platform as well. The overall stratigraphic thickness of the Baska subunit does not exceed 1600–1800 m (5200–5900 ft) (Figure 17A, B). The carbonate platform formed on an elevated block of the rifted margins of the European plate in the Tithonian and possibly lasted into the Berriasian–Valanginian (Figure 5). Two main types of limestones, the whitish gray Stramberk Limestones of Tithonian age and the red-brown and green Koprivnice Limestones with the uppermost Tithonian to Valanginian fossils, have been recognized both in the tectonic klippen and the carbonate debris found in the Valanginian–Cenomanian Kotouc facies and the Chlebovice Member of the Tesin–Hradiste Formation (Figure 17A, B). The Stramberk Limestones, as much as 350 m (1100 ft) thick, represent the typical platform assemblage of grainstones, reefal framestones, boundstones, and detrital slope deposits. They display many similarities with the Ernsbrunn Limestones of the Outer Klippen in southern Moravia. The younger Koprivnice Limestones, made of olistoliths, brecciated limestones, and marls, are thought to be associated with the last stages of the platform growth, its exposure, and partial destruction. The large klippen of the Stramberk and Koprivnice limestones exposed in several large active and inactive quarries are well-known paleontological sites. The rich fauna was studied by Zittel (1868, 1873), Blaschke (1911), Spath (1933), Housa (1961, 1975, 1976, 1978), Eliasova (1962b), Zitt (1974), Nekvasilova (1977), and others. As summarized by Mencik et al. (1983), more than 600 species, among them more than 50 species of ammonites, have been described from these localities.

The origin of some large carbonate blocks, e.g., the Kotouc Hill in Stramberk (more than 1000 m [3300 ft] in diameter), became a subject of controversy. Matejka and Roth (1955), Eliasova (1962b), Housa (1976), and Mencik et al. (1983) interpreted the large carbonate blocks as tectonic klippen detached from the carbonate platform during thrusting. According to Elias and Stranik (1963), Elias (1979), and Elias and Eliasova (1984, 1986), the large bodies (klippen) in Stramberk formed by accretion of bigger and smaller blocks of Stramberk and Koprivnice limestones derived from the disintegrated platform and redeposited into the younger strata of the Kotouc facies and the Chlebovice Member. Neither of these interpretations can fully explain the character of these chaotic deposits. On one hand, it is difficult to accept that all the large bodies of the very clean limestones, up to several hundred meters in diameter, formed by the accretion of the carbonate detritus eroded from an older platform and redeposited in the younger strata, as proposed by Elias and Stranik (1963), Elias (1979), and others. However, it is also unlikely that all the carbonate blocks represent tectonic slivers with preserved stratigraphic succession, as advocated by Housa (1976). In our opinion, a more plausible interpretation may be found somewhere between these extreme views. Like any other carbonate platform, the Jurassic Stramberk platform apparently had a flat plateau dominated by the sedimentation of carbonate grainstones, elevated rims with coral reefs, and slopes composed predominantly of debris derived from the edge of the carbonate platform (e.g., Eliasova, 1962b). Gravity slides and turbidity currents transported smaller and bigger debris detached from the edges of the platform farther into the adjacent basin. The tectonic slivers of the Baska facies may thus comprise both tectonically detached pieces of the original carbonate platform, including the coeval slope and the more distal talus deposits, as well as debris derived from the exposed platform into the younger strata. During the tectonic transport, the rigid pieces of the carbonate platform may have been further separated from the softer, less competent rocks present in the slopes of the platform. This may have resulted in the formation of a melange in which the larger blocks are true tectonic fragments (klippen) of the original platform, whereas the smaller blocks and debris are part of the coeval platform talus or even clastic material redeposited from the emergent and disintegrated platform into some younger strata.

The carbonate debris from the Stramberk and Koprivnice limestones are prominently distributed in the Valanginian to Albian strata of the Tesin–Hradiste Formation, i.e., in the lower part of the Kotouc facies and the Chlebovice Member, which represents the upper (Albian to Cenomanian) part of the Tesin–Hradiste Formation. In the Baska subunit, the Tesin–Hradiste Formation is only 150–500 m (500–1600 ft) thick as compared to the 500–1200-m (1600–4000-ft)-thick equivalent of this formation in the Godula subunit. To a lesser degree, the material derived from the Jurassic to Lower Cretaceous carbonate platform is also present in the Baska (Albian–Cenomanian) and the Palkovice (Coniacian–Danian) flysch formations. The Baska Formation, 250–350 m (800–1100 ft) thick, consists of sandstones, allodapic limestones with spongilitic cherts, and greenish gray shales. The Palkovice Formation, about 500 m (1600 ft) thick, is the youngest known stratigraphic member of the Baska unit. It is composed of black shales alternating with thick-bedded sandstones and conglomerates containing pebbles, cobbles, and blocks of the Stramberk Limestones. The presence of clasts of Stramberk Limestones in the Baska and Palkovice formations indicates that the Stramberk carbonate platform located at the western margin of the Silesian basin remained at least partly exposed until the Late Cretaceous. At that time, the remnants of the platform were gradually inundated by the sea and possibly became a part of the Subsilesian depositional system.

According to Michalik and Sotak (1990), an erosional event at the Jurassic–Cretaceous transition apparently partly destroyed the Jurassic carbonate buildups, including the platforms in the Waschberg sector and the Stramberk platform, which formed on the elevated blocks of the rifted Outer Carpathian depositional realm (Figure 5A, B). Abundant clasts from the Jurassic platforms are found elsewhere in the Cretaceous and Paleogene conglomerates of the Waschberg–Zdanice–Subsilesian unit and the Magura flysch (e.g., Picha et al., 1966; Elias, 1998).

Elias (1998) situated the original Tithonian to Valanginian Stramberk platform on a hypothetical Baska ridge, allegedly situated between the Silesian and Subsilesian basins. Because the Subsilesian basin in Moravia came into existence much later, in the Turonian, it seems to be more prudent to locate the carbonate platform more generally on the northern rifted margin of the Silesian basin without implying an existence of the Baska ridge. The similarly positioned Andrychow ridge between the Silesian and Subsilesian–Skole units in the Polish sector of the Western Carpathians (Ksiazkiewicz, 1960) apparently emerged in the Aptian (Golonka et al., 2000; Slaczka et al., 2006).

In the Stramberk area, the Baska subunit is thrusted over the Subsilesian unit and again over the autochthonous strata of the Neogene foredeep and the Hercynian basement. In the vicinity of Stramberk, more than 100 wells penetrated the Baska subunit and the Subsilesian unit of the Carpathian belt and drilled into the subthrust Neogene foredeep, in which some clastic reservoirs were developed into a gas storage.

The Kelc Subunit

The Kelc subunit, as defined by Elias (1970, 1979), represents the southwestern marginal part of the Silesian basin. It comprises strata of the Valanginian to Paleocene age subdivided into several formations (Figure 17A, B). The Valanginian to Aptian (Tesin–Hradiste and Verovice formations) are identical with their stratigraphic equivalents in the Godula subunit. The Albian Jasenice Formation, corresponding to the Lhoty formations in the Godula subunit, is characterized by the predominance of gray and green spotted shales with subordinate turbiditic sandstones and micritic limestones. Its thickness ranges between 100 and 200 m (330 and 660 ft). The overlying Nemetice Formation of Albian to Cenomanian age is composed of hemipelagic variegated (gray, green, and red) shales with occasional thin beds of sandstones. It is followed by the Cenomanian to Santonian(?) Dub Formation (Hanzlikova and Matejka, 1958), represented by a proximal facies of coarse- to fine-grained sandstones, calcareous sandy shales, and slump bodies of conglomerates with blocks of the Stramberk Limestones. Lithologically similar to the Dub Formation are the Turonian to lower Senonian deposits at Stary Jicin (Stranik et al., 1997) and also the younger Campanian to Paleocene Milotice Formation (Elias, 1979) and the Kojetin Formation mentioned by Mencik et al. (1983). They consist of gray and greenish calcareous sandy shales, calcareous sandstones, and slump conglomerates. The thickness of these various lithologic units is about 300 m (1000 ft).

The Kelc subunit, typically only less than 1000 m (3300 ft) thick, may be interpreted as marginal facies of the Silesian unit formed in a zone of limited subsidence and limited impact of the Stramberk carbonate platform. Its lower members bear similarities with the Lower Cretaceous strata of the Zdounky unit. The areal extent and the structural position of the Kelc subunit remain uncertain. Some doubts even exist about the recognition of the Kelc subunit as a separate entity in the Silesian unit.

The External Units Occurring in Front and Below the Magura Nappe

This is a widely defined group of tectonostratigraphic units, into which we include various minor tectonic thrust sheets and tectonic slivers, such as the Cejc–Zajeci unit, the Zdounky unit, and the Fore-Magura unit, typically distributed along the frontal edges of the Magura nappe in Moravia, as well as the units exposed in tectonic slivers and windows (window units) in the Magura nappe in Moravia, Slovakia, and Poland. Stratigraphically, these units bear many similarities with the adjacent external units of the Flysch belt, e.g., the Waschberg–Zdanice–Subsilesian and Silesian units, but differ facially. Some of these units contain a higher proportion of coarser clastics, including slump conglomerates and olistoliths, which indicate a depositional proximity to a tectonically active source. They are interpreted as being deposited on the northern and western side of the Silesian cordillera (Figure 5). During thrusting, the various strata of these units, including some older Jurassic and Early Cretaceous carbonate members, were detached and tectonically transported at the edge of the Magura frontal nappe or tectonically piled below the Magura nappe as duplexes and occasionally exposed in tectonic windows. The surface appearances of some of these units in Moravia are associated with the late orogenic transpressional faults. Although small in the areal extent, these units provide important information about the character of the Outer Carpathian depositional system as well as about the range of overthrusting of the out-of-sequence Magura nappe over the external domains of the Flysch belt.

The Cejc–Zajeci Unit

The Cejc–Zajeci unit is exposed along the eastern margin of the Zdanice unit and western edge of the Magura nappe in southern Moravia (Figure 3). From the Zdanice unit, into which it was originally included, the Cejc–Zajeci unit differs by a wider presence of the Upper Cretaceous strata (Campanian–Maastrichtian), by the absence of typical Menilitic cherts, and by the abundant occurrence of thick, discontinuous bodies (several meters to tens of meters) of coarse sandstone and conglomerates in the Nemcice (Submenilitic) Formation (Picha et al., 1968). The sandstones contain a high proportion of the organodetritic material, mainly fragments of lithothamnia. The discontinuous bodies of coarse clastics are interpreted as either submarine slumps or submarine channel fills and overbank deposits distributed at the mouth of submarine canyons. Their distribution, as well as the northwest-directed paleocurrent markings, suggests that these coarser clastics were supplied from the Silesian ridge, which separated the outer Waschberg–Zdanice–Subsilesian and Silesian basins from the inner Magura depositional system (Figure 5C, D). The conglomerates thus provide information about the composition and the geological history of the Silesian ridge (Picha et al., 1966). As a source of clastics for both the Magura unit and the external units, the Silesian ridge has been active since the Late Cretaceous. The influx of the coarse clastic material into the Cejc–Zajeci unit peaked in the Paleocene to middle Eocene and then gradually decreased and, in the late Eocene, when the Silesian cordillera submerged, ceased entirely. The channelized sandstones of the Cejc–Zajeci unit bear similarities with the Paleogene Ciezkowice Sandstones of the Silesian unit of northern Moravia and western Poland.

The structural position of the Cejc–Zajeci unit was established by the deep well Kobyli-1 (Appendix 1; Figure 20, shown on page 118, section DD′). At the depth of 702 m (2303 ft), the Cejc–Zajeci unit is thrusted over two stacked thrust sheets of the Zdanice unit, which, at the depth of 3135 m (10,285 ft), are thrusted over the autochthonous Jurassic strata of the foreland plate. Incorporated into the base of the lowest thrust sheet is a tectonic sliver (18 m; 59 ft) of the lower Miocene strata tentatively compared with the Krepice Formation of the Pouzdrany unit (Picha et al., 1971). On the southeastern side, the Cejc–Zajeci unit is covered by the Miocene strata of the Vienna basin (Figure 3). Small erosional remnants of the Magura flysch sandstones found elsewhere on the top of the Cejc–Zajeci unit indicate that the Magura nappe was originally thrusted over the Cejc–Zajeci unit. The contact between these two units was further modified by the strike-slip faulting related to the opening of the pull-apart Vienna basin in the middle Miocene.

Picha and Hanzlikova (1965) described loose blocks of Jurassic radiolaritic limestones and marls at two localities near Pritluky and Zajeci and interpreted them as weathered tectonic klippen incorporated into the frontal zone of the Cejc–Zajeci unit. Z. Stranik et al. (1982, personal communication) reinterpreted these occurrences of Jurassic rocks as olistoliths. They are comparable with other small Jurassic klippen, such as Cetechovice and Lukovecek, found in the Solan

Formation at the edges of major thrust units of the Magura nappe and recently interpreted as olistoliths.

We understand the Cejc–Zajeci unit as a marginal pericordilleran facies of the Waschberg–Zdanice–Subsilesian depositional system, adjacent to the southern end of the Silesian ridge, which separated this external zone from the Magura depositional realm and shed coarse clastics to both sides (Figure 6C). Cicha et al. (1964) tentatively correlated the Cejc–Zajeci unit with the Zdounky unit in central Moravia.

The Zdounky Unit

Tectonic slivers of the Zdounky unit are distributed along the northwestern edge of the Magura nappe in the Chriby Mountains of central Moravia (Figure 3). The largest surface exposure of the unit has been mapped near the village Zdounky; other smaller appearances have been found at Rostin, Cetechovice, and Bohuslavice near Kyjov. Small slivers of the Zdounky unit are also found incorporated into the frontal zone of the Magura nappe at Korycany and Stare Hute in the Chriby Mountains (Mencik and Pesl, 1958). The distinct lithological character of the Zdounky unit was recognized by Paul and Tausch (1899) and Dreger (1899), and further established by Uhlig (1903), Petraschek (1907), Pesl and Mencik (1956), Hanzlikova and Matejka (1962), Cicha et al. (1964), and Chmelik (1971).

The known strata of the Zdounky unit have been divided into the lower sequence (Lower Cretaceous–lower Oligocene) and the upper sequence (upper Oligocene) (Figure 17A, B). The lower sequence has sandstones and conglomerates [Berriasian–Barremian(?)] at the base, overlain by grayish calcareous shales and limestones. This is further followed by variegated shales with beds of organodetritic sandstones, in which Hanzlikova (in Chmelik, 1971) distinguished Aptian and Albian benthonic foraminifera. Greenish gray calcareous shales with foraminifera and radiolaria represent the Cenomanian. They are overlain by gray, green, and red shales with intercalations of organodetritic sandstones extending into the lower Oligocene. The strata of the lower sequence are strongly deformed; their depositional thickness would not exceed a few hundred meters. The upper sequence, 500–600 m (1600–2000 ft) thick, is represented by the Krosno facies of alternating proximal turbiditic sandstones with shales and occasional slump conglomerates that contain redeposited Eocene corals and echinoderms (Oppenheim, 1913).

The Berriasian to Cenomanian strata of the Zdounky unit display many similarities with the coeval strata of the Baska and Kelc facies of the Silesian unit; the Senonian to lower Oligocene strata, marked by the absence of the Menilitic Formation, are comparable with the Cejc–Zajeci unit of southern Moravia. Like in the Cejc–Zajeci unit, the presence of coarse organodetritic and siliciclastic sandstones and slump conglomerates indicates that the depositional site of the Zdounky unit was adjacent to the Silesian cordillera, which supplied the coarse clastics.

The tectonic slivers of the Zdounky unit in the Chriby Mountains are distributed along, and apparently structurally associated with, the late orogenic strike-slip fault, which outlines the frontal zone of the Magura nappe in southern Moravia and in the Chriby Mountains (Figure 3). This southwest–northeast-trending orogen-parallel fault is apparently a component of the Western Carpathian transfer zone. In the middle to late Miocene, the sinistral strike-slip motion in the transfer zone led to the opening of the pull-apart Vienna basin in northeastern Austria and southern Moravia and to the northeastern escape of the Western Carpathians.

The Fore-Magura Unit

The Fore-Magura unit appears as a narrow discontinuous band of tectonic slivers sandwiched between the Magura and the Silesian units in northern Moravia and western Poland (Figure 3). Its existence in the territory of Moravia was suggested by Burtan et al. (1937) and further confirmed by Hanzlikova et al. (1962), Pesl et al. (1964), Pesl (1967), Mencik (1973), Plicka (1978), Mencik et al. (1983), and Pesl and Hanzlikova (1983). The presence of the Fore-Magura unit below the Magura nappe was confirmed by the deep wells Jablunka-1 and Gottwaldov-2 (Appendix 1; Figure 20, section CC′ shown on page 118).

The known stratigraphic sequence of the Fore-Magura unit extends from the Campanian to the Oligocene (Figure 17A, B). The oldest Submenilitic Formation (Campanian to lower Oligocene), several hundred meters thick, is represented by variegated shales with sparse thin beds of sandstones. Thick bodies of sandstones and conglomerates in this formation are known from the vicinity of Bystrice pod Hostynem. The overlying Oligocene Menilitic Formation, several tens of meters to as much as 150 m (500 ft) thick, in its lower part consists of siliceous shales with fish scales, Dynow Marls with laminae of cherts, and discontinuous bodies of sandstones comparable with the Kliwa Sandstones of the Flysch belt in Poland, Ukraine, and Romania. The upper part of the formation, made by cyclical green and brown shales and subordinate sandstones, resembles the Sitborice Member of the Zdanice unit. The youngest Krosno Formation (500 m; 1600 ft) of the

Fore-Magura unit is characterized by an alternation of facies with a variable proportion of sandstones and shales. The predominantly sandstone facies with slump bodies of pebbly mudstones was named the Chvalcov beds by Pesl and Hanzlikova (1983). It may be correlated with the Pochodzita Sandstones (Burtan et al., 1937) of the Polish Carpathians.

The deposits of the Fore-Magura unit show many similarities with the coeval strata of the Waschberg–Zdanice–Subsilesian and Silesian units. However, the presence of discontinuous, apparently channelized, sandstones and conglomerates in the Submenilitic Formation indicates that the depositional site of the Fore-Magura unit was situated closer to the Silesian ridge, which remained the source of coarse clastics until the middle to the late Eocene (Figure 5).

Units of the Tectonic Windows

Eleven tectonic windows in the Magura flysch nappe have been recognized in the territory of Poland and Eastern Slovakia. In Moravia, only one tectonic window is present in the Magura nappe near Rajnochovice in the Hostyn Hills, but equivalents of the window units are known as minor thrust sheets and tectonic slivers in the Hostyn Hills and the Chriby and Beskydy mountains of Moravia.

In those windows, the Paleogene strata of the underlying window units (Sikora, 1970), such as the Obidowa Slopnice and Grybow (inclusive Smilno) units, were brought to the surface during the compressional thrusting in the early Miocene. For more information, see Slaczka et al. (2006). The known Eocene to Oligocene strata of these window units resemble the coeval strata of the external Silesian or Fore-Magura units. The existence of the window units below the Magura nappes indicates that in the Eocene and early Oligocene, the depositional realm of the external units extended far south, possibly beyond the present site of the Pieniny Klippen Belt (Figure 5C, D).

During compressional thrusting, the rigid Magura flysch nappes moving northward greatly distorted the incompetent, mostly shaly strata of these external domains by detaching and transporting some of them at their fronts and piling the others in tectonic duplexes at their bottom. The simultaneous movement of the crustal blocks underneath the thin-skinned structure apparently further complicated this tectonic process. The soft, external units thus served as a lubricant between the upper thin-skinned and the lower basement, involving structural levels of the Outer Carpathian thrust belt.

The existence of the low-gravity zone at the front of the Pieniny Klippen Belt might then be explained by the accumulation of light rocks of these external units below the Magura nappes in this zone (Figure 20, shown on page 118). Such an interpretation is supported by the results of the Oravska Polhora-FPJ-1 well in northwestern Slovakia (M. Zakovic, D. Bodis, M. Fendek, M. Potfaj, G. Gebauer, and J. Balint, 1989, personal communication; Potfaj, 2003), which, below the Magura nappe at the depth of 1323 m (4340 ft) to the final depth of 2417 m (7929 ft), penetrated the Obidowa–Slopnice or Grybow unit (Potfaj, 2003).

The Magura Group of Nappes

The Magura group of nappes is the dominant tectonostratigraphic unit of the Outer Western Carpathians. It forms a continuous belt along the Western Carpathian arc from the Vienna Forest in Austria to the Western Ukraine. Its southwestern part in northeastern Austria and southern Moravia is partly buried below the Neogene strata of the Vienna basin (Figure 3). The Rhenodanubian Flysch of the Eastern Alps is considered to be an equivalent of the Magura flysch in the Alpine region (Elias et al., 1990) (Figure 16).

Most of the known depositional sequences of the Magura flysch evolved during the Late Cretaceous and Paleogene convergent stage of the Tethyan–Alpine cycle. The Jurassic and Lower Cretaceous deposits of the passive continental margins are known only from the tectonic klippen Kurovice and Hluk and olistoliths (originally described as klippen) Cetechovice (Neumayer, 1870; Neumann, 1907; Chmelik, 1957) and Lukovecek (Uhlig, 1903; Rzehak, 1904; Oppenheimer, 1913; Rakus, 1987). They are represented by deep-water carbonates and marls, as well as by some flyschlike turbiditic deposits. The sparse occurrence of the older Jurassic and Lower Cretaceous deposits in the Magura nappe, however, may not be fully indicative of the actual distribution of these strata in their original depositional setting. The lower part of the Magura depositional sequence, deposited on an uneven substratum of a rifted basin, might have been decoupled from the upper part by the sole thrust (decollement) and left behind as an autochthon and possibly partly subducted with its substratum (Figure 6). Thus, only the upper part of the Magura depositional sequence, made predominantly by the Late Cretaceous to the early Oligocene synorogenic flysch deposits was integrated into the present Magura nappes. These synorogenic deposits are represented by deep-water turbiditic facies consisting predominantly of hemipelagic muds and a variety of gravity-driven deposits, including proximal and distal turbidites, debris flows, and occasional olistoliths. The proportion of shales and sandstones varies both vertically and laterally, reflecting on the dynamic and complex depositional environment of the subsea fans. As such, the deposits of the Magura unit may be considered as one of the most typical flysch facies in the entire Western Carpathian belt. The overall architecture of the Magura basin was constrained by the existence of the Silesian ridge on its northwestern side and the Czorsztyn ridge on its southeastern side (Figure 5). The Magura sedimentary basin thus consisted of a variety of depositional environments from the steep slopes of the Silesian and Czorsztyn ridges to the deep-water environment in the axial part of the basin (Figure 6). The Silesian ridge and, to a lesser extent, the Czorsztyn ridge supplied most of the coarse clastic material into the Magura basin.

Structurally, the Magura unit consists of numerous thrust sheets and imbricates. Because of a relatively competent stratigraphic section with a high proportion of sandstones, the internal structure of the individual thrust sheets is relatively simpler than that of the most external units of the Flysch belt.

Based on significant lithostratigraphic and structural differences, Matejka and Roth (1950) subdivided the complex Magura flysch in Moravia into three major units: the Raca, Bystrica, and Bile Karpaty–Krynica (Oravska Magura) units (Figures 3; 17A, B). The Bile Karpaty unit of southeastern Moravia is, on the surface, separated from the Krynica unit of northwestern Slovakia by a wide gap. However, an existence of some connection between these two segments underneath the Pieniny Klippen Belt cannot be excluded. According to Mencik (1969), the transverse segmentation of the Magura flysch units might be related to deep northwest–southeast-trending faults in the underlying subthrust platform.

The Raca Unit

The Raca unit is the most external and the most widespread unit of the Magura nappe system. It appears as a continuous belt along the entire length of the Western Carpathian Flysch belt (Figure 3). The original depositional site of this unit was on the southeastern side of the Silesian ridge, which separated the Magura flysch basin from the depositional realm of the external units (Figure 5A, B). The Raca unit comprises strata ranging in age from the Late Jurassic to the Oligocene (Figure 17A, B). Based on the lithological differences, Pesl and Krystek (1965) distinguished in the Raca unit six lithofacial zones: Hostyn, Tri kameny, Trnava–Staskov, Vsetin, Luhacovice, and Kycera.

The oldest known strata of the Raca subunit are found in the Kurovice tectonic klippe located at the frontal edge of the Magura nappe (Figure 3). They are represented by the deep-water Kurovice Limestones and the overlying Tlumacov Marls discordantly overlapped by conglomerates and breccias (Elias et al., 1996) (Figure 17A, B). The Kurovice Limestones (Glocker, 1840), as much as 150 m (500 ft) thick, are composed of medium- to thick-bedded and fine-grained limestones with sparse lenses of cherts. The common presence of graded bedding (in some beds, the fine-grained limestones even pass into marls) would indicate a deposition from the gravity-driven turbidity currents. Based on studies of foraminifera (Benesova et al., 1962, 1968) and the aptychi and calpionellids (Vasicek and Rehakova, 1994), the age of the Kurovice Limestones was established in a wide range from the Oxfordian to Tithonian [Berriasian(?)]. The Kurovice Limestones pass upward into the about 60-m (200-ft)-thick Tlumacov Marls (Elias and Eliasova, 1985), represented by flysch-type alternation of thinly to medium-bedded biomicritic limestones and marls. Based on the evaluation of aptychi and calpionellids (Vasicek and Rehakova, 1994) and nannofossils (Svabenicka et al., 1997), the Tlumacov Marls were dated as Berriasian to Valanginian. The conglomerates and breccias with intercalations of variegated shales, which overlap the Kurovice Limestones and Tlumacov Marls, are interpreted as debris flows and assigned to the Senonian (Svabenicka et al., 1997).

The oldest known flysch strata of the Raca unit, so far identified only in tectonic slivers at the fault plane along which the Tri kameny lithofacies zone is thrusted over the frontal Hostyn zone, are the black shales alternating with siliceous sandstones. These strata have been assigned to a wide stratigraphic range from the Hauterivian to the Cenomanian (Svabenicka et al., 1997) and called the Gault flysch, referring thus to the lithological and stratigraphical resemblance of these strata with the Lower Cretaceous Gault flysch of the Rhenodanubian Flysch belt in Eastern Alps. Stranik et al. (2004, personal communication) renamed these strata the Rajnochovice Formation after the village Rajnochovice in the Hostyn Hills, where the type locality of this formation is located in the banks of the Juhyna River.

The continuous succession of the Raca unit begins with the Kaumberg Formation, named after the type locality in the Vienna Forest (Wiennerwald) in Austria (Gotzinger, 1954). Formerly, these strata were denoted as the lower variegated beds (Matejka and Roth, 1949b). In the Polish Carpathians, the Kaumberg Formation corresponds to the Upper Cretaceous variegated beds, known also as the Malinowa Formation. The Kaumberg Formation is composed predominantly of red and green hemipelagic to pelagic shales with intercalations of siltstones and fine-grained sandstones. Typically, the

Kaumberg Formation, about 300 m (1000 ft) thick, is found at the base of individual thrust sheets in the Raca unit (Svabenicka et al., 1997). Its age ranges from the Cenomanian to the Campanian–Maastrichtian transition. The lower part of the formation is comparable with the variegated beds of the Mazak Formation of the Silesian unit (Figure 17A, B). The stratigraphic position of the Kaumberg Formation was well established in the deep well Jarosov-1 (Figure 7; Appendix 1), which penetrated 5540 m (18,175 ft) of the Raca unit and was thus instrumental in defining the stratigraphic succession of the Magura flysch especially at the deeper Cretaceous level (Hanzlikova, 1976).

The Solan Formation (Matejka and Roth, 1949b) of the Maastrichtian to Paleocene age represents a typical flysch sequence with a variable proportion of sandstones and shales. Based primarily on lithology, the Solan Formation has been subdivided into the Raztoka, Hostyn, and the Lukov members (V. Pesl, V. Cekan, J. Kolejka, M. Ruzicka, L. Rybarova, and V. Volsan, 1984, personal communication). The Raztoka Member of alternating shales and sandstones with abundant biotite, about 1200 m (4000 ft) thick, occurs typically in the Tri kameny lithofacial zone of the Raca unit. Within this member at the locality Ustgrun, the Cretaceous–Paleogene boundary was identified for the first time in the Carpathian Flysch belt (Svabenicka et al., 1997). The Hostyn Member, about 800 m (2600 ft) thick, is the stratigraphic equivalent of the Raztoka Member in the Hostyn lithofacial zone. It is made predominantly by thick-bedded calcareous sandstones with fragments of lithothamnia and nummulites. The uppermost Lukov Member, 200–800 m (660–2600 ft) thick, is composed of thick beds of sandstones and conglomerates interpreted as sand flows and debris flows of the proximal part of a subsea fan (Elias, 1963). The weathering-resistant coarse sandstones and conglomerates of the Lukov Member typically stand out as morphological ridges, which mark the trend of individual thrust sheets.

The general upward coarsening of the Solan Formation possibly reflects on a gradual progression of proximal turbiditic facies over the more distal parts of the deep-water turbidite fan. The overall thinning and fining of the Solan Formation toward the southeast, as well as the orientation of the paleocurrent markings and the distribution of slump conglomerates, indicate that the Solan Formation was sourced mainly from the Silesian cordillera located on the northern side of the Magura basin (Pesl and Krystek, 1965; Pesl, 1968) (Figure 6). The sandstones and conglomerates of the Solan Formation contain a high proportion of grains and pebbles of Jurassic limestones and fragments of lithothamnia apparently derived from the cordillera and its shallow shelf. In composition, the sandstones and conglomerates of the Solan Formation show many similarities with the channelized sandstones and conglomerates of the Istebna Formation of the Silesian unit deposited on the other (northern) side of the Silesian ridge.

The overlying (about 300-m [1000-ft]-thick) Beloveza Formation (Paul, 1869) was assigned to the Paleocene to middle Eocene (Hanzlikova in Matejka and Roth, 1956). It is composed predominantly of green and red shales with thin beds of sandstones. Occasional packets of arcosic sandstones, up to several tens of meters thick, may be interpreted as submarine channel fills similar to those found in the Paleocene to Eocene strata of the Silesian unit (Ciezkowice Sandstones) and in the Cejc–Zajeci unit both deposited on the opposite northwestern side of the Silesian ridge.

The youngest Zlin Formation (Zapletal, 1937) of the middle to late Eocene and possibly to the early Oligocene age is as much as 2500 m (8200 ft) thick in the Raca unit. Its lower part, overlying the Beloveza Formation, has been subdivided into several lithologically defined members, such as Krive, Rusava, Ujezd, Luhacovice, Vsetin, and Kycera (Pesl, 1968). The Krive Member, exposed at the front of the Magura nappe in the Valaske Mezirici area, is characterized by the predominance of sandstones with a significant biodetritic component. The Rusava Member, about 500 m (1600 ft) thick, is limited to the front of the Raca unit in the Hostyn Hills; the Ujezd and Luhacovice members, 500–700 m (1600–2300 ft) thick, are typically present in the more internal zones of the Raca unit. All these members, characterized by the presence of coarse sandstones and conglomerates, were laid down in the proximal part of the turbiditic fans close to the mouth of submarine canyons. The upper part of the Zlin Formation is represented by the Vsetin Member of the middle Eocene to the early Oligocene age. It is composed predominantly of calcareous shales and marls alternating with subordinate beds of fine- to medium-grained glauconitic sandstones, locally as much as 10 m (33 ft) thick. Typically, the Vsetin Member overlies the previously mentioned lower members of the Zlin Formation dominated by sandstones and conglomerates, but locally, it rests directly on the Beloveza Formation. The Vsetin Member represents a typical foredeep facies supplied predominantly from the uplifted, more internal zones of the Magura flysch basin rather than from the Silesian cordillera, which, in the late Eocene, was already submerged. The paleocurrent measurements in the Vsetin Member indicate a longitudinal transport (Pesl and Krystek, 1965). The Vsetin Member, as much as 2000 m (6600 ft) thick, is the most typical stratigraphic unit of the Zlin Formation and of the entire Magura unit. Its equivalent in the innermost part of the Raca unit is the Kycera Member, characterized by thick-bedded, muscovite-rich sandstones alternating with mostly calcareous shales.

The Bystrica Unit

The distinct character of the Bystrica unit was recognized by Kodym (1923), who called it the Bylnice nappe. The name Bystrica unit was introduced by Matejka and Roth (1949b). The Bystrica unit forms a relatively narrow belt between the Raca and the Bile Karpaty–Krynica (Oravska Magura) units along most of the Western Carpathian arc. Only in the Kysuca area, where the Bile Karpaty unit is missing, is the Bystrica unit in direct contact with the Pieniny Klippen Belt (Figure 3). It consists of several thrust sheets, which comprise the Solan, Beloveza, and Bystrica formations only (Figure 17A, B). It cannot be excluded, however, that some older members of the Bystrica unit not exposed on the surface might be present at a greater depth. The lowermost Solan Formation, whose known rudimentary thickness does not exceed 400 m (1300 ft), consists of coarse-grained arcosic sandstones (Matejka and Roth, 1949b). Vujta et al. (1991), however, included these presumably Solan sandstones into the overlying Beloveza Formation. The Beloveza Formation, more than 300 m (1000 ft) thick, of the late Paleocene to early Eocene age (Hanzlikova in Matejka and Roth, 1956) is composed of green and red shales with subordinate thin-bedded sandstones. The highest Bystrica Formation, as much as 1600 m (5200 ft) thick, is comparable with the Vsetin Member of the Zlin Formation in the Raca unit. It is represented by turbiditic sequences of calcareous shales, marls, and subordinate sandstones with nummulites. The individual turbiditic sequences commonly begin with glauconitic sandstones, which pass into laminated siltstones, calcareous shales, marls, and thin layers of pelagic shales. Characteristic for the Bystrica Formation and the Bystrica unit, in general, are the thick intervals of marls (up to several meters) known as the Lacko Marls in Poland. Based on foraminifera and nannofossils, the Bystrica Formation has been assigned to the Eocene (Hanzlikova, 1955; Bubik and Svabenicka in Vujta et al., 1991).

The Bile Karpaty–Krynica (Oravska Magura) Units

The Bile Karpaty–Krynica (Oravska Magura) units represent the innermost zone of the Magura flysch nappe system adjacent to the Pieniny Klippen Belt. It has two segments: the Bile Karpaty unit in the south and the Krynica unit, also known as the Oravska Magura unit, in the north (Figure 3). Only the southern Bile Karpaty segment is present on the territory of Moravia and, as such, a subject of our deliberation. Matejka and Roth (1956) divided the Bile Karpaty unit into the Hluk and Vlara subunits.

The Hluk Subunit. The Hluk subunit is found mainly in the southwestern part of the Bile Karpaty unit; toward the north, it ends, at least at the surface, at the Nezdenice fault. It comprises strata from the Barremian to the Eocene, which, based on lithological and micropaleontological criteria, are divided into several lithostratigraphic units: the Hluk and Kaumberg formations, Puchov Marls, and the Antoninek, Svodnice, Nivnice, and Kuzelov formations (Figure 17A, B).

The oldest, the Hluk Formation of the Barremian–Albian age (Vasicek, 1947; Hanzlikova in Matejka and Roth, 1956; Stranik et al., 1995; Svabenicka et al., 1997), is made by pelagic and hemipelagic black shales, marls, and deep-water turbiditic limestones with cherts. The shales prevail in the lower part of the Hluk Formation, and the limestones prevail in its upper part. Both lithologically and biostratigraphically, the Hluk Formation, more than 120 m (400 ft) thick, may be correlated with the Wolfpassing Member and the Bartberg Member of the Nord zone (Grun et al., 1972) and their stratigraphic equivalents in the Schottenhof zone in Wienerwald (Vienna Forest) in Austria.

The overlying Kaumberg Formation, which, in the Hluk facies, is about 100 m (330 ft) thick, is composed of green and red noncalcareous shales with intercalations of fine-grained laminated sandstones. The rich assemblages of agglutinated foraminifera indicate the Cenomanian to Maastrichtian age of the Kaumberg Formation (Svabenicka et al., 1997).

The Puchov Marls are composed of red calcareous shales and marls with rich calcareous and agglutinated benthic foraminiferal and nannofossil fauna, which indicate the Maastrichtian age (Hanzlikova, 1972a; Bubik and Svabenicka in Stranik et al., 1995; Svabenicka et al., 1997). On the surface and in the wells drilled in the Hluk gas field, the Puchov Marls appear as tectonic slivers, about 100 m (330 ft) thick. Their lithology and biofacies seem to be identical with the Puchov Marls of the Pieniny Klippen Belt and some deposits of the Hauptklippen zone of the Wienerwald (Vienna Forest) in Austria. The depositional and structural adherence of the Puchov Marls to the Bile Karpaty unit thus remains uncertain. These strata may be interpreted as a part of the Pieniny Klippen Belt or as an extension of the Hauptklippen zone of the Wienerwald tectonically incorporated into the Bile Karpaty flysch unit or even as an extension of the klippen belt depositional facies into the Magura realm (Bubik, 1995; Svabenicka et al., 1997).

Equally questionable is the structural position of the Antoninek Formation (Vujta et al., 1989) made by sandy limestones, detrital limestones, marls, and calcareous shales and, based on planktonic foraminifera and nannofossils, assigned to the Campanian–Maastrichtian age. Its known thickness is only a few tens of meters. The Antoninek Formation shows many similarities with the Sneznica Formation of the Pieniny Klippen Belt. Vujta et al. (1989) thus considered the Antoninek Formation to be a tectonic sliver of the klippen belt incorporated into the Bile Karpaty flysch unit.

The Svodnice Formation, as much as 1000 m (3300 ft) thick, was defined by Pesl (1968) and attributed to the Maastrichtian–Paleocene by Svabenicka and Bubik (1992). It is characterized by a facies of rhythmically alternating calcareous shales, subordinate sandstones, and fucoidal limestones, apparently deposited in a distal environment of turbiditic fans. Thicker beds of sandstones named Bzova Sandstones (Potfaj, 1993) occur in the upper part of the formation. Stranik in Elias et al. (1990) compared the Svodnice Formation with the Laab Formation of the Wienerwald (Vienna Forest) in Austria.

The overlying Nivnice Formation (Stranik et al., 1989), about 600 m (2000 ft) thick, is composed of brown-gray and green-gray calcareous shales alternating with subordinate, mostly thin-bedded sandstones and occasional debris flows. Well-preserved foraminiferal and nannoplankton fauna attributes the Nivnice Formation to the late Paleocene–early Eocene (Svabenicka and Bubik in Stranik et al., 1995).

The Kuzelov Formation defined by Stranik et al. (1989) is characterized by a predominance of greenish, brownish, and reddish shales with mostly thin-bedded laminated sandstones. Diagnostic for the Kuzelov Formation are the commonly found beds of pelocarbonates. The Kuzelov Formation is about 250 m (820 ft) thick, and its age ranges from the late Paleocene to the early Eocene (Svabenicka and Bubik in Stranik et al., 1995).

The structural position of the Eocene, the so-called "variegated strata," which occur as narrow bands in some outcrops and wells and are commonly tectonically combined with the Kaumberg Formation, is questionable. Lithologically and stratigraphically, they resemble the Beloveza Formation of the Raca and Bystrica units or the variegated shales of the Proc Formation in the Kopanice facies of the Pieniny Klippen Belt (Stranik et al., 1995).

The Vlara Subunit. The known stratigraphic extent of the Vlara subunit of the Bile Karpaty unit is limited to the Kaumberg, Javorina, Svodnice, and the Chabova formations (Figure 17A, B). The Kaumberg Formation (350 m; 1100 ft) does not differ from its equivalent in the Hluk subunit. Potfaj (1993) called the upper part of the formation, made by alternating thin-bedded sandstones and variegated shales, the Ondrasovec Member.

The Javorina Formation (750 m; 2500 ft) is composed predominantly of sandstones alternating with thin layers of silty shales. The sandstones are characterized by a high proportion of carbonate, mostly dolomitic detritus, apparently supplied from the klippen belt and the nappes of the Inner Carpathians. Because of the poorly preserved microfauna, the age determination of this formation is uncertain and varies in a wide range from the Campanian to Paleocene (Svabenicka, 1990). The Javorina Formation, apparently deposited in a proximal part of a subsea fan system, has no coeval lithologic equivalent in the more distal Hluk facies.

The Svodnice Formation, about 1200 m (4000 ft) thick, is comparable with its equivalent in the Hluk subunit. Its Maastrichtian(?) to Paleocene age has been confirmed by Hanzlikova in Stranik et al. (1989), Svabenicka (1990), and Potfaj (1993). Thick-bedded coarse sandstones with carbonate clasts alternating with thin layers of shales represent the youngest Chabova Formation. Like the Javorina Formation, the Chabova Formation was possibly deposited in a channelized proximal part of the subsea turbiditic fan system. The calcareous nannoplankton indicates the uppermost Paleocene to early Eocene age (Potfaj, 1993). The preserved part of the Chabova Formation is 150 m (500 ft) thick.

The Pieniny Klippen Belt

The Pieniny Klippen Belt is a narrow (only a few kilometers wide) zone of complex steeply dipping fanlike structure, which parallels the Carpathian arc from northeastern Austria to Romania and separates the Inner Carpathians from the Outer Carpathians (Figure 3). It is built of smaller and bigger blocks of weathering-resistant Jurassic and Lower Cretaceous, predominantly carbonate rocks (klippen) that protrude from the less resistant Upper Cretaceous and Paleogene shales, marls, and flyschlike strata. The contact between the klippen and surrounding rocks is mostly tectonic. Excluding the clasts in conglomerates, no pre-Mesozoic sedimentary strata or rocks have been found in the klippen belt. The complex stratigraphy and structure of the Pieniny Klippen Belt were studied by many authors, most prominently by Andrusov (1938) and Birkenmajer (1960).

Until the Late Cretaceous, the depositional history of the Pieniny Klippen Belt was similar to that of the Inner Carpathians. The Jurassic to Lower Cretaceous strata of the klippen belt were apparently laid down on both the attenuated continental crust of the European continental margins and the oceanic crust of the Penninic–Pieninic ocean, which formed during the Triassic and Jurassic rifting north of the Tatric realm of the Inner Carpathians. From the Magura depositional site, the Pieninic oceanic basin was at least partly separated by the Czorsztyn ridge (Figures 5, 6), whose character and function became a subject of discussion. Some authors, e.g., Sotak (1992), combined these two depositional realms into one, the Pieniny–Magura depositional basin, and assigned it to the northern Penninic zone (Valais domain).

In the Late Cretaceous and early Paleogene, the Penninic–Pieninic oceanic crust and lithosphere were gradually subducted, and the Inner Carpathians collided with the Czorsztyn ridge. In this process, the Jurassic to Lower Cretaceous strata of the klippen belt were decoupled from their subducting substratum and piled in front of the progressing Inner Carpathian thrust system, whereas the sedimentation of the Late Cretaceous and Paleogene clastic sequences continued over the top of the klippen belt structures. The appearance of flysch facies with exotic conglomerates would indicate that during the progression of deformation, internal ridges emerged occasionally in the klippen belt depositional realm.

Little direct evidence, at least at the surface, is present for the existence of the Penninic–Pieninic-derived oceanic units in the Inner Western Carpathians. The Upper Cretaceous to Paleogene flysch complexes exposed along the northern Tatric edge in the peri-klippen belt may represent the Penninic ocean in the Carpathians; however, their original basement is not known (Plasienka, 1995). According to several tectonic models, this basement, called Vahicum by Mahel (1981), was underthrusted beneath the Inner Carpathians. Plasienka (1995) suggested that some small tectonic units on the northern side of the Tatric zone of the Inner Carpathians, such as Borinka and Belice, comprise oceanic rock complexes and thus might be considered as parts of the Penninic–Pieninic oceanic realm.

Since the Campanian, the Pieniny Klippen Belt became part of the Outer Carpathian foreland depositional system and, as such, was further affected by the Paleogene and Neogene tectonic phases of the Alpine orogeny. It was the scraping and intense deformation of the Jurassic to Lower Cretaceous strata and simultaneous deposition of younger, Upper Cretaceous and Paleogene strata that resulted in the complex contacts between the older, Jurassic to Lower Cretaceous rigid klippen and their softer Upper Cretaceous and Paleogene siliciclastic depositional cover. The structural complexity of the klippen belt was further enhanced by the strike-slip faulting and backthrusting, which occurred during the latest stages of the Alpine orogeny (e.g., Birkenmajer, 1985).

The Pieniny Klippen Belt is subdivided into the shallow-marine Czorsztyn and the deep-water Kysuca (Pieniny) units and a multitude of transient units, such as the Drietoma, Bosaca, Haligovce, Klape, and Manin units of the so-called peri-klippen belt (Andrusov, 1938). The Czorsztyn unit is characterized by the presence of Jurassic crinoidal and nodular limestones, the existence of an Early Cretaceous hiatus, and the presence of Albian to Maastrichtian variegated marly facies (couches rouges). It was deposited on an elevated site, interpreted either as a fragment of the continental crust (our view) or as a midoceanic ridge (Golonka et al., 2003). The Kysuca unit is distinguished by the presence of the Doggerian spotted marls and radiolarites, the Late Jurassic to Early Cretaceous pelagic cherty limestones, and the Late Cretaceous to Paleogene conglomeratic flysch sequences. The deep-water depositional environment of the Kysuca unit is commonly believed to be oceanic and a part of the Penninic–Pieninic ocean, which separated Apulia from the West European platform (Stampfli, 2001). The transient units of the peri-klippen belt are composed mainly of Cretaceous and, in part, Paleogene couches rouge marls and flysch sequences with frequent conglomerates.

The contact of the Pieniny Klippen Belt with the Inner Carpathians is tectonic. The klippen belt is steeply thrusted back over the Inner Carpathian units, including the Inner Carpathian Paleogene in the area north of Zilina. The extent of backthrusting of the klippen belt over the Inner Carpathian Paleogene typically ranges within a few kilometers but may reach as much as 8–10 km (4.8–6 mi) in the northern part of the Orava region (Chmelik in Mahel and Buday, eds., et al., 1968). Our interpretation of seismic line 2AT84 in the northern Orava region (Figure 20, section AA′, shown on page 118), suggests at least 12 km (7.2 mi) of tectonic transport of the klippen belt and the inner zone of the Magura unit over the projected edge of the Inner Carpathians. As evident from seismic data, the angle of the leading backthrust faults ranges between 30 and 45°. South of Zilina, the direct contact between the Pieniny Klippen Belt and the most external Tatric units of the Inner Carpathians is obscured by the presence of the peri-klippen belt.

The relationship of the Pieniny Klippen Belt to the inner units of the Magura flysch belt is complex. As demonstrated by the detailed fieldwork and some wells, e.g., Klanecnica-1 (Potfaj, 1993, 1998) and Lubina-1

(Lesko et al., 1978) near Stara Tura in Western Slovakia, the depositional relationship between various units of the Magura flysch and the klippen belt is obscured by enormous structural complexities of the contact zone. In the Bile Karpaty area, Stranik et al. (1989) included the Upper Cretaceous and Paleogene strata of the contact zone into the Kopanice facies. The older Cenomanian to Coniacian strata of this facies, represented by the variegated shales with subordinate sandstones, are comparable with the Kaumberg Formation of the Bile Karpaty unit of the Magura flysch. They are overlain by a turbiditic flysch sequence of alternating sandstones and calcareous shales with sparse intercalations of variegated shales and thin beds of limestones. The sandstones are characterized by a high proportion of clasts from carbonate rocks and by a high amount of calcitic cement. Foraminifers (E. Hanzlikova, 1989, personal communication) and nannofossils (Svabenicka in Stranik et al., 1989) would suggest the Campanian to Eocene age for these several-hundred-meter-thick flysch deposits, which show similarities with the Proc Formation of Eastern Slovakia (Lesko, 1960), and the Upper Cretaceous to Paleogene Jarmuta Formation of the Polish Carpathians (Horwitz and Rabowski, 1929). The whole Kopanice facies is comparable with the Kyjov facies (Stranik and Roth, 1959), the Lackovce facies (Pesl and Mencik, 1959), or the Inovce facies (Lesko, 1960) of Eastern Slovakia, where all these facies are attributed to the cover units of the Pieniny Klippen Belt. In our opinion, the Kopanice facies, as well as other facies of the contact zone, represent the Cretaceous to Paleogene cover of the Jurassic to Lower Cretaceous klippen of the Pieniny Klippen Belt. Considering the complex structure of the contact zone between the Pieniny Klippen Belt and the Magura unit, it is not inconceivable that at least the younger Upper Cretaceous strata known from the klippen belt might also be present in the deeper parts of the complex, imbricated structure of the Bile Karpaty and Krynica (Oravska Magura) units of the Magura flysch. The existence of tectonic slivers of Puchov Marls and Antoninek Limestones, the apparent members of the klippen belt, within the Bile Karpaty unit, would support such an interpretation.

The Inner Carpathian Paleogene

In the middle Eocene (Lutetian), the hinterland spreading and subsidence resulted in a marine transgression over the northern parts of the already deformed and consolidated Inner Carpathians and the formation of the Inner Carpathian Paleogene basin (Podhale basin in Poland) in which thick flysch sequences were laid down until the early Miocene (Figure 5D). The transgressive sequence began with conglomerates, followed by nummulitic limestones, shallow-marine sands, and minor coals. In the late Eocene and Oligocene, a rapid subsidence led to the deepening of the Inner Carpathian Paleogene basin and the deposition of deep-water shales with occasional submarine debris flows and slumps. Locally, e.g., in the Orava, Zilina, and Handlova regions, thin menilitic shales and cherts coeval with the lower Oligocene Menilitic Formation of the Outer Carpathians were laid down. Layers of sedimentary manganese ores are found in the Spis (Kisovce–Svabovce), Handlova, Rajec, and Orava regions (Picha, 1964a). These predominantly shaly strata were followed by an accumulation of turbiditic flysch deposits as much as 2000 m (6600 ft) thick characterized by an upward-increasing proportion of sandstones.

Because the Inner Carpathian Paleogene is not a subject of our study, we restrict our deliberation to a few notes and instead refer to some more comprehensive papers dealing with this subject, e.g., Radomski (1958), Picha (1964b), Chmelik (in Mahel and Buday, eds., et al., 1968), Marschalko (1978), Gross et al. (1984), Nemcok et al. (1996), and Janocko et al. (2006).

The Inner Carpathian Paleogene basin is asymmetrical with its deepest part adjacent to the Pieniny Klippen Belt and shallowing toward the south. The primary source of the clastic material was the actively moving frontal zone of deformation, which, at the time of deposition of the Inner Carpathian Paleogene in the late Eocene to Oligocene, possibly included the northern edge of the Inner Carpathians, the Pieniny Klippen Belt, and the inner zones of the Magura flysch. The overall architecture of the Inner Carpathian Paleogene thus resembled that of a foreland basin; only, it formed in the hinterland of the orogenic belt and did not prograde toward the foreland. Despite the different tectonic setting, the Inner Carpathian Paleogene is a mirror image of the Outer Carpathian late orogenic foreland basin, from which it was separated by the tectonically active Pieniny Klippen Belt and the already uplifted inner zone of the Magura flysch. In the Inner Carpathian Paleogene, only the zones adjacent to the Pieniny Klippen Belt are marginally folded and locally (e.g., in the Orava region) overthrusted by the Mesozoic rocks of the klippen belt. These deformations are at least partly related to the late orogenic transpressional strike-slip motion in the Pieniny Klippen Belt.

The Vienna Basin

The late orogenic to postorogenic Vienna basin is located in the transitional zone between the Eastern Alps and the Western Carpathians on the territory of southern Moravia, Slovakia, and northeastern Austria (Figures 1, 3). Its extent along the strike of the orogenic belt is confined to the limits of the northwest–southeast-trending Dyje–Thaya depression, whose tectonic history goes back at least to the Jurassic rifting and extension. The Vienna basin is superimposed on various nappe units of the Northern Calcareous Alps and the Inner Carpathians, the Pieniny Klippen Belt, and various Outer Carpathian units, namely, the Rhenodanubian and Magura flysch and the Waschberg–Zdanice external units. Geological maps of the basement of the Vienna basin were published by Kroll et al. (1993), Zimmer and Wessely (1996), and Arzmuller et al. (2006).

The depositional fill of the Vienna basin, as much as 5500 m (18,000 ft) thick, is composed mainly of molasse-type shallow-marine and nonmarine detrital deposits of the Eggenburgian (early Burdigalian) to the Pliocene age (Figure 15). The composition and distribution of various facies in the basin reveal the complex geotectonic history of the Vienna basin, which may be divided into three main stages: (1) the Eggenburgian to Ottnangian stage of synorogenic subsidence of the Carpathian thrust belt in the realm of the Dyje–Thaya depression; (2) the Karpatian to early Badenian tectonic transport and pull-apart extension along the orogen-parallel strike-slip faults, and (3) the late Badenian to Pliocene postorogenic subsidence. During its entire history, the Vienna basin was connected with the Pannonian Basin and, until the early Badenian, also with the Neogene foredeep (Jiricek, 1988). The structural and depositional history of the Vienna basin was discussed by Royden (1985), Tomek and Thon (1988), Wessely (1988), Jiricek and Seifert (1990), Fodor (1995), Seifert (1996), and Kovac (2000), among others. Our deliberations are limited only to some general aspects of the stratigraphy and structure of the Moravian part of the Vienna basin. For more information, we refer to the article by Arzmuller et al. (2006).

The Eggenburgian to Ottnangian (Burdigalian) Stage

The depositional history of the Vienna basin began with a marine transgression in the Eggenburgian. The basal conglomerates and sandstones (e.g., Chropov and Winterberg members) are followed by a monotonous sequence of calcareous shales and marls of the lower Luzice Formation. These 300–700-m (1000–2300-ft)-thick strata can be correlated with the coeval Sakvice Marls of the Zdanice unit preserved in the Sakvice depression above the subthrust Vranovice paleovalley and in the Kobyli lake and also with some deposits in the Carpathian foredeep. Two deltas have been recognized in the Eggenburgian depositional system. They prograde from the south-southwest into the deep basinal part of the Vienna basin located along the present Austrian–Slovak border. The boundary between the Eggenburgian and Ottnangian is locally, e.g., in the northern part of the basin, marked by an erosion and deposition of calcareous sands (Hodonin and Luzice sands), which locally rest directly on the Magura flysch basement. The Ottnangian calcareous shales and marls of the upper Luzice Formation, about 300 m (1000 ft) thick, show an upward trend of gradual shallowing and regression of the sea, apparently caused by the tectonic motion in the progressing belt. As a result of this tectonic activity, the piggyback transported Eggenburgian and Ottnangian strata of the Vienna basin were partly deformed and eroded.

The Karpatian to Early Badenian (Burdigalian–Langhian) Stage

The new marine transgression in the Karpatian surpassed the previous extent of the Vienna basin. Through the Sakvice depression in the Zdanice unit, the Vienna basin was connected with the foredeep. The southern part of the basin was dominated by the lacustrine and fluviatile deposits, which, farther north, passed into the deltaic and basinal deposits of the central depression, where the Karpatian strata reached their maximum thickness of about 1400 m (4600 ft). Generally, the Karpatian section of the Vienna basin is characterized by a large lithological variability. According to Jiricek (1988), the lower part of the Karpatian section is represented by the Jablonica Conglomerates, the Tynec Sands, and/or the Laksary schliers; the middle part is made of the Sastin Sands; and the upper part is made of the Zavod schliers and/or the Lab *ostracoda* beds (Figure 15). Kovac (2000) generalized the stratigraphic record and divided the Karpatian deposits of the Vienna basin in Slovakia into the lower part, the Laksary Formation, and the upper part, the Zavod Formation. The regression and deposition of brackish sediments with anhydrite in the Zavod Formation mark the end of the Karpatian. In southern Moravia, this was the time of the final thrusting of the outermost units of the Western Carpathian Flysch belt, including the overlying (piggyback) Eggenburgian to Karpatian strata of the Vienna basin, over the eastern part of the Neogene foredeep. However, because the rigid foreland plate of the Bohemian Massif increasingly inhibited the forelandward thrusting, the compressional stresses tended to be released by the northeast-directed strike-slip motion (escape tectonics). At the Karpatian–Badenian transition, the Vienna basin thus began to attain the character of a pull-apart basin dominated by the orogen-parallel strike-slip faults (Burchfiel and Royden, 1982; Royden, 1985).

The Badenian stage began with a new marine transgression and deposition of thick calcareous shales (tegels) of the Lanzhot Formation. The existence of analogous deposits in the foredeep would affirm to a renewed communication between these two depositional systems. The continuing pull-apart mechanism led to a further deepening of the Moravian central depression, which formed between the Steinberg and the Lanzhot–Hrusky fault systems, and which was filled with more than 4000 m (13,100 ft) of Neogene deposits. The subsidence of the depression peaked during the middle Badenian, when thick, variegated, predominantly shaly deposits of the Zizkov Member of the Hrusky Formation were laid down (Figure 15). A large, deltaic system with lobes extending into the central depression formed at the western side of the Vienna basin. Two smaller deltas have been recognized in the northeastern and southwestern ends of the basin. In the upper part of the Zizkov Member, shallow-marine Lab Sands were deposited in coastal areas, whereas bioherms of lithothamnion limestones formed on the elevations. The Lab Sands represent the most important hydrocarbon-bearing reservoir in the Vienna basin.

The Late Badenian (Serravalian) to Pannonian (Tortonian) Stage

In the late Badenian, because of the gradual decrease of orogenic stresses, the strike-slip fault system of the Vienna basin became gradually inactive. The subsidence, rather than the pull-apart extension, increasingly dominated the depositional regime after the late Badenian. The Vienna basin shallowed and was converted into a brackish embayment, which only sporadically communicated with the open sea. Following the deposition of the variegated fresh-water clays and sands at the Badenian–Sarmatian transition, the basin spread into the Hradiste graben. There, the Sarmatian deposits rested directly on the Magura flysch. The brackish sediments of the Bilovice Formation, as much as 800 m (2600 ft) thick, are represented by coarser clastics, lumachellas, and oolitic limestones in the coastal areas and calcareous clays and sands in the basinal facies.

During the Pannonian, the connection with the Pannonian Basin was further restricted, and the Vienna basin became a slightly brackish lake with marginal lagoons and swamps. The lowest Bzenec Formation consists of deltaic gravels and sands at its base and calcareous clays in its upper part. The Kyjov lignite seams (Figure 15) overlain by thick sands and variegated calcareous clays formed in the marginal marshes. The existence of a thick and widespread lignite bed in the lower part of the overlying Dubnany Formation indicates that the marshes and swamps expanded over most of the remnant Vienna basin. The lignites have been mined near Ratiskovice, Dubnany, Hodonin, and Mikulcice. The deposition of the lacustrine coaly clays, sands, and variegated shales of the freshwater Gbely Formation concluded the depositional history in the Vienna basin at the territory of Moravia.

During the Pliocene–Holocene, fluviatile sediments accumulated in the valleys of the Morava and Dyje rivers, and lacustrine deposits several hundred meters thick were laid down in the Zohor–Plavecky Mikulas graben at the western side of the Male Karpaty Mountains.

Structure of the Vienna Basin

The internal architecture of the Vienna basin is complex. It is dominated by a system of grabens and horsts that formed and modified in response to the compressional and tensional stresses in the underlying Carpathian thrust belt and the subthrust foreland plate. During its dynamic structural history, the Vienna basin passed from the early stage of subsidence on the top of the progressing orogenic belt to the stage of the pull-apart extension and finally to the stage of postextensional subsidence.

The kinematic character of the strike-slip faults in the Vienna basin became a subject of discussions. According to Royden (1985) and Royden and Dovenyi (1988), the pull-apart extension of the Vienna basin occurred in the underlying thin-skinned thrust belt above a relatively shallow sole detachment of the thrust belt. Such an interpretation would be consistent with the low heat flow (Cermak, 1979) and only a limited thermal subsidence following the pull-apart extension. Seismic data, however, indicate that at least some of the major strike-slip faults, e.g., the Steinberg and Leopoldsdorf faults, do not sole out in the basal detachment of the thin-skinned thrust belt but rather continue deeper into the subthrust basement (Wessely, 1988). Lankreijer et al. (1995), based on their structural analysis, suggested that the structural pattern of the Vienna basin changes from the thin-skinned extension in the northwestern part to the whole lithospheric extension in the central and southeastern part of the basin. Picha (1996) suggested that the deep-rooted strike-slip faults in the Vienna basin may copy the fault planes of the preexisting normal faults formed at the time of the Jurassic and Early Cretaceous rifting and extension of the European plate (Figure 20, section DD′, shown on page 118). During the convergency, these and some other preexisting faults might have been reactivated as reverse faults and/or as strike-slip faults in the lower level of the orogenic system, which apparently existed below the thin-skinned thrust belt. Locally, these deep faults penetrated into the overlying thrust belt and created a new structural regime in the thin-skinned belt. Apparently, it was the late orogenic northeastward translation of the entire Carpathian system along the Western Carpathian transfer zone, during which the actively moving deep faults propagated into the overlying thin-skinned belt and opened the pull-apart structure of the Vienna basin. Picha (2002) compared the structural setting of the Vienna basin with that of the Albanian foredeep in the Adriatic region. Like the Vienna basin, the Albanian foredeep formed in a major northwest–southeast-trending strike-slip transfer zone, which existed between the Dinarides and the Apulian plate and provided a means of tectonic transport from the collision zone of Apulia with Europe toward the subduction zone of the Hellenic trench. Both the Vienna basin and the Albanian foredeep thus represent a specific type of pull-apart basin, associated with the escape strike-slip tectonics. They evolved on a thick continental crust devoid of magmatic centers and thermal anomalies.

Neovolcanics in the Outer Carpathian Thrust Belt

The only neovolcanic rocks known from the Outer Carpathian thrust belt in Moravia are the trachyandesites and alkalic basalts at Banov, Bojkovice, and Hrozenkov in the vicinity of Uhersky Brod (Figure 3). They have been investigated by Schmidt (1858), Tschermak (1858), and Klvana (1891), among others, and more recently by Krystek (1955), Shrbeny (1974), Prichystal (1993), Adamova et al. (1995), and Prichystal et al. (1998).

These neovolcanics appear as dikes and sills, several meters to several tens of meters thick in all partial units of the Magura flysch and locally in the Pieniny Klippen Belt. They cut through the thrust structures without being affected by any tectonic movement, thus clearly documenting their post-thrust origin. Radiometric analyses indicate the early Badenian (16.8 ± 0.6 Ma) age of these volcanic rocks. Their origin thus falls in the time of the late orogenic strike-slip faulting and opening of the pull-apart stage of the Vienna basin. The country rocks of the dikes and sills are affected by contact metamorphism. So far, no remnants of surface volcanism have been found, although at several localities near Banov and Nezdenice, volcanic breccias that apparently formed in volcanic vents are present. Associated with the volcanic activity are mineralogical occurrences of polymetallic ores and asphaltic organic materials.

The chemical composition, close to that of the typical alkalic volcanic rocks, indicates that these Outer Carpathian volcanics attain a transitional position between the alkalic neovolcanic rocks of the Bohemian Massif and of the calc-alkalic neovolcanic rocks of the Inner Carpathians in Slovakia. The strontium (Sr) isotopic ratio points to the mantle as a source of magma for these Outer Carpathian neovolcanics.

Rhyolite and rhyodacite tuffites are known from the Badenian strata of the Neogene foredeep and from the Eggenburgian strata of the Vienna basin in Moravia. These tuffitic rocks were apparently supplied from volcanoes located in the Inner Carpathians.

THE STRUCTURE OF THE OUTER WEST CARPATHIAN THRUST BELT

The Outer Carpathian thrust belt on the territory of Moravia is a thin-skinned structure composed entirely of sedimentary sequences. However, below this thin-skinned belt and the Neogene foredeep, deeper compressional structures exist involving both the crystalline basement and its sedimentary cover (Picha, 1996; Krejci et al., 2002) (Figure 20). As indicated by numerous wells and regional seismic lines, the thin-skinned Outer Carpathian thrust belt consists of numerous imbricates, duplexes, and thrust sheets (nappes). The internal architecture of these various thrust units is strongly controlled by the stratigraphy. Whereas the incompetent, predominantly shaly successions display a complex internal deformation marked by tight folds, imbricates, and duplexes, the thick competent sequences with a high proportion of sandstones or carbonates are less internally deformed and form larger scale folds and thrust sheets (nappes). Architectural differences also exist between the internal and external zones of the Carpathian thrust belt. Various thrust sheets of the external zones typically dip rather gently toward the hinterland; in the inner zones, as evident in the relatively consistent Magura nappe, the inclination of thrust units gradually increases. Some strata are even overturned and locally thrusted backward. The fanlike complex structure of the Pieniny Klippen Belt and of the adjacent zone of the Flysch belt is partly thrusted back over the edges of the Inner Carpathians (Figure 20, section AA′).

The major thrust faults bounding major tectonostratigraphic units might have initiated along the preexisting normal faults, which, during the divergent phase, separated various depocenters and intervening ridges in the Outer Carpathian depositional system. However, not every juxtaposition of lithologically distinct units should be interpreted as a major, tectonically induced separation of those units in their original depositional setting. Significant lithological changes between tectonically separated units, such as the different proportion of sands and shales, may as well be explained by lithological variations in the channelized depositional fan system.

Like in other thrust belts, the deformation of various units of the Outer Western Carpathians progressed from the hinterland toward the foreland. The regular sequential succession of the forelandward-heading thrust sheets is locally disjoined by an appearance of the out-of-sequence thrust imbricates and duplexes, such as the window units in the Magura complex nappe system. The Magura unit itself is a major out-of-sequence thrust sheet, which glides over the structures of the external units (Figure 20).

The timing of deformation and thrusting of various major units of the Outer Carpathians is typically established from the age of the strata in the footwall and the termination of the continuous sedimentation in the hanging wall. In the Pieniny Klippen Belt zone and the innermost Bile Karpaty unit of the Magura flysch in Moravia, the sedimentation ended in the Eocene; in the more external Raca subunit of the Magura flysch, the sedimentation continued until the early Oligocene; and in the most external Zdanice and Pouzdrany units, the uninterrupted sedimentation lasted into the early Miocene. In time, the deformation of the Outer Carpathians not only progressed from the internal to the external zones but also along the strike of the arcuate Carpathian belt from the west to the east (Buday et al., 1961). At its western end in northeastern Austria and southern Moravia, the frontal Pouzdrany and Waschberg–Zdanice units are thrusted over the lower Miocene (Eggenburgian to Karpatian) deposits of the Neogene foredeep; in northern Moravia, the lowermost middle Miocene (lower Badenian) strata are found below the edges of the thrust belt (Jurkova, 1976). In the Vrancea zone of the Eastern Carpathians, the thrusting has continued into the Pliocene–Quaternary(?).

Strike-slip Faulting and Backthrusting

The thrusting of the Outer Western Carpathian units over the European foreland in Moravia, especially during the last stages of the orogenesis in the late early to middle Miocene, was associated with the southwest–northeast-trending left-lateral strike-slip motion related to the northeastward extrusion (escape) of the Carpathian–Pannonian block (Roth, 1980b; Tomek et al., 1987; Ratschbacher et al., 1991a). As a consequence of this lateral slip, the various units of the thrust belt progressed toward the foreland increasingly in an oblique direction, generating transtensional and transpressional stresses elsewhere in the thrust system. The presence of the transtensional stresses is well documented by the opening of the pull-apart stage of the Vienna basin in southern Moravia and northeastern Austria, whereas the existence of the transpressional folding, faulting, and backthrusting is best evident in the Pieniny Klippen Belt and the adjacent zone of the Magura nappe in Poland (Swidzinki, 1953) and Eastern Slovakia (Stranik, 1965). The stress measurements by Nemcok et al. (1998a, b) indicate that the proportion of shortening accommodated by the sinistral strike slip is the highest in the southwestern part of the Western Carpathian arc and decreases toward the north, where most of the shortening is accommodated by the frontal compression. Most of the strike-slip motion apparently occurred in the southwest–northeast-trending Western Carpathian transfer zone especially along the deep faults of the Pieniny Klippen Belt. The Western Carpathian transfer zone, which was a major factor in opening the Outer Carpathian depositional system during the divergent stage, in a reverse sense of the motion, also facilitated a great deal of the tectonic shortening and escape tectonics during the convergent stage (Figure 5).

The combination of normal forward thrusting and strike-slip motion resulted into an extremely complex structural pattern of the Pieniny Klippen Belt, which is partly thrusted back over the Inner Carpathians (Figure 20). The backthrusting of the Outer Carpathian units over the Inner Carpathian Paleogene most likely occurred during the last stages of the Alpine orogeny in the late early to middle Miocene and was apparently coeval with the opening of the pull-apart stage of the Vienna basin. The extension in one area was thus compensated by the compressional shortening in other parts of the interconnected orogenic system.

Deep Structures Below the Foredeep and the Thin-skinned Belt

Below the Neogene foredeep and the thin-skinned Outer Carpathian belt in Moravia, deeper compressional and extensional structures exist whose geometries are not related to the dominant structural trends of the overlying belt. The normal and reverse faults bounding these features, with a few exceptions, such as the strike-slip faults in the Vienna basin, do not continue into the thrust belt. The sole decollement of these deep

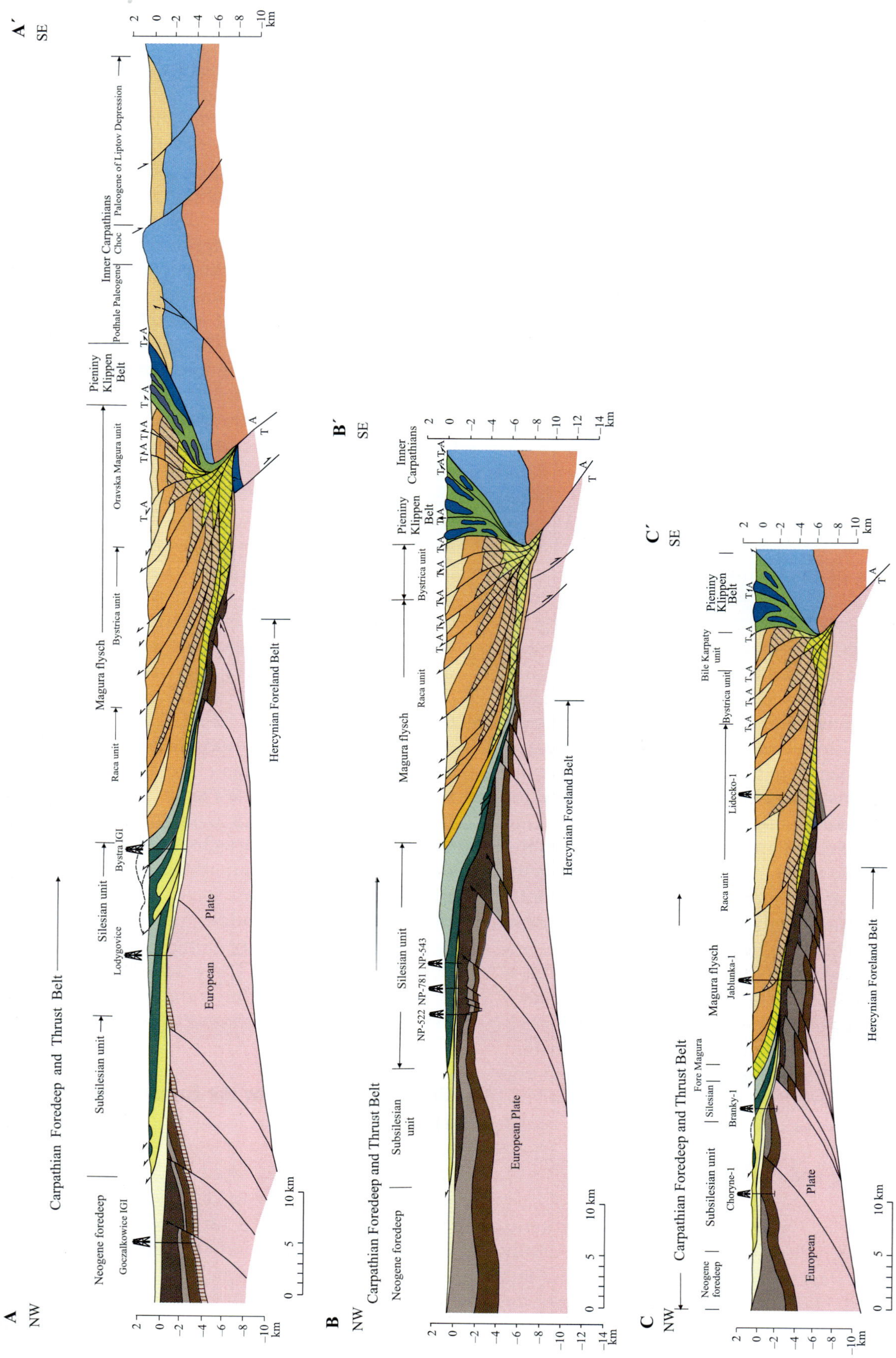
A
NW
A′
SE
Carpathian Foredeep and Thrust Belt
Neogene foredeep
Subsilesian unit
Silesian unit
Magura flysch
Raca unit
Bystrica unit
Oravska Magura unit
Pieniny Klippen Belt
Inner Carpathians
Podhale Paleogene
Choc
Paleogene of Liptov Depression
Goczalkowice IGI
Lodygovice
Bystra IGI
European
Plate
Hercynian Foredland Belt
B
NW
B′
SE
European Plate
NP-522 NP-781 NP-543
C
NW
C′
SE
Neogene foredeep
Silesian
Fore Magura
Bile Karpaty unit
Chorync-1
Branky-1
Jablunka-1
Lidecko-1
km
10 km

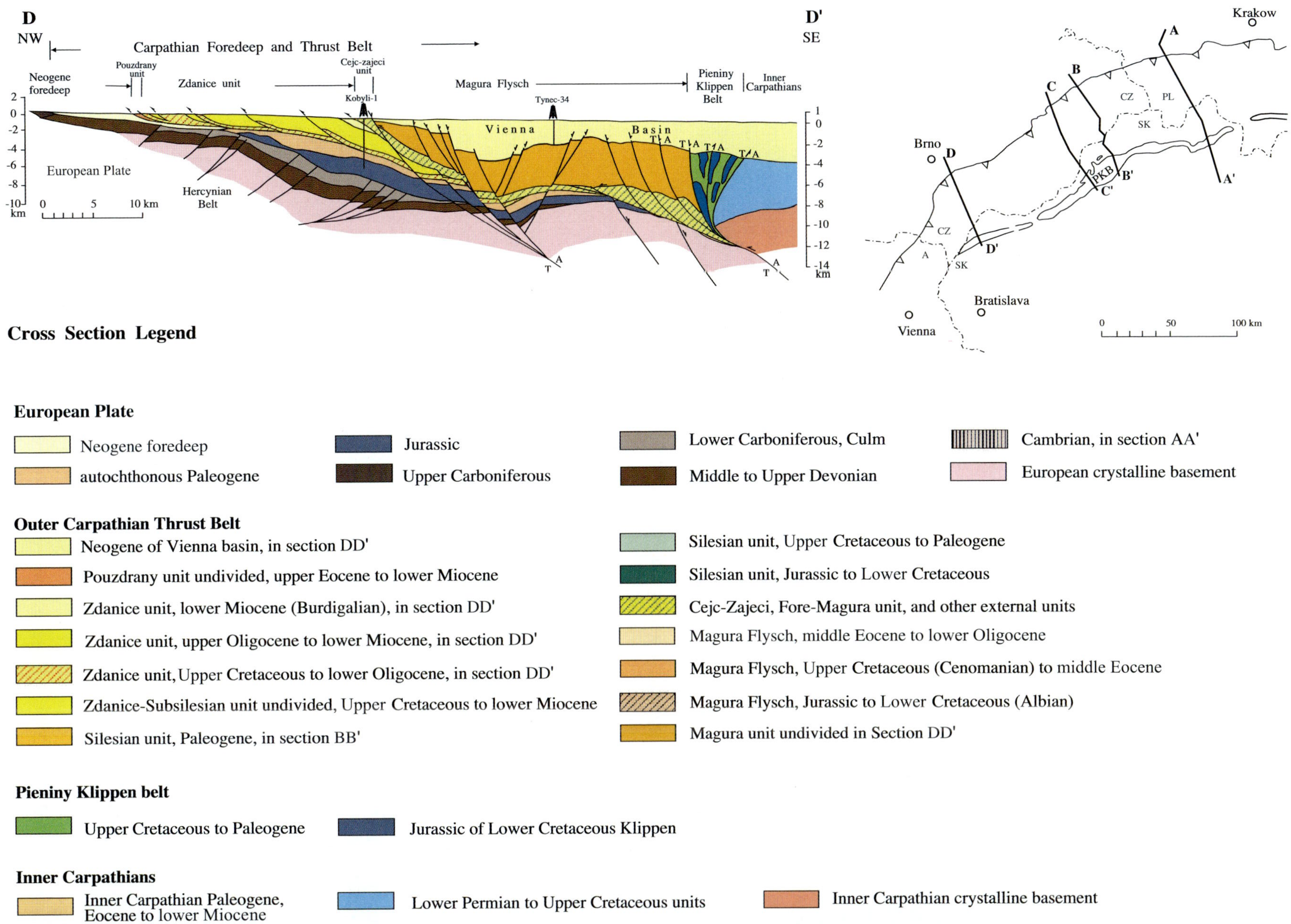

Figure 20. Four regional partly balanced cross sections through the Western Carpathian thrust belt and its foreland along the lines: AA′ Bielsko Biala–Low Tatra Mountains, BB′ Odra Hills–Manin, CC′ Hranice–Ilava, DD′ Brno–Kuty. Location of the lines is shown in Figures 3 and 7. The same vertical and horizontal scales apply. Stratigraphic records of some of the wells shown on cross sections CC′ and DD′ are reported in Appendix 1.

structures may exist at greater depths in the brittle to ductile transition zone or even at the base of the crust. Structural analyses indicate that these structures belong to several genetically diverse types. The oldest recognized, deep antiformal features are the Hercynian (late Paleozoic) anticlinal structures typically bounded by the southeastward-verging high-angle reverse faults, which typically terminate in the youngest Carboniferous (Stephanian) strata and do not continue into the overlying Mesozoic and Cenozoic strata of the European foreland plate. However, in areas without a significant autochthonous cover, which would document their age, the origin of these structures remains uncertain. They may as well be interpreted as Alpine structures associated with antithetic back-arc faulting in the foreland plate. An example of such a structure of questionable origin is the Orava structure evident on the seismic line 2AT/84, whose interpretation is presented at the cross section AA′ in Figure 20. The association of this structure with either the Hercynian or Alpine orogeny is difficult to ascertain from the existing data. In the cross section AA′, we tentatively interpret the structure as Hercynian but do not exclude its younger Alpine origin.

Within the Jurassic Dyje–Thaya depression of southern Moravia and northeastern Austria, deep antiformal structures exist, which are apparently generated by the rotation of crustal blocks during the Jurassic rifting and extension of the European plate. At least one of them, the large antiformal Tynec structure (Figure 20, section DD′), was reactivated during the late orogenic strike-slip faulting. The original Jurassic fault bounding the western side of this structure might have been a precursor of the major orogen-parallel Steinberg fault, along which the pull-apart depocenter of the Vienna basin opened in the early to middle Miocene. In the territory of Austria, some of these rift-related structures have been explored by deep drilling (Wessely, 1990). They represent the third depth level of exploration in the Vienna basin area, the first being the Vienna basin and the second the Carpathian thrust belt.

Some of the southeastward-dipping normal and reverse faults identified on the seismic lines in the European platform, along the western side of the Pienniny suture zone (Figure 20, sections AA′, BB′, and DD′), may also be genetically related to the Jurassic rifting. Typically, these faults do not propagate into the overlying nappes, although some of them might have been activated by transpressional stresses during the last stages of the Alpine orogeny.

Lastly, below the thrust belt and the Neogene foredeep, deeper compressional structures exist, whose origin is clearly related to the Alpine orogeny. Some of them may be associated with the Late Cretaceous–early Paleocene Laramide uplifting of the Carpathian foreland; the others may have been active during the late Paleogene to the middle Miocene convergency and lateral escape. Because of the compressional stresses, the previously attenuated European crust was compressed and thrusted back toward the foreland, thus forming a lower structural level of the Carpathian orogenic belt, as suggested among others by Roure et al. (1993), Roca et al. (1995), and Roure and Sassi (1995). One such recognized structure is the overthrust of plutonic rocks of the Brno massif over Cenomanian sediments near the town of Blansko (Krejci et al., 2002). The chain of antiformal structures, which parallels the frontal zone of the thrust belt and the foredeep in Moravia known as the Slavkov–Tesin ridge (Dlabac and Mencik, 1964), and the positive structure that underlies the Dolni Dunajovice gas field (Figure 26, shown on page 137) are possibly other examples of the young Alpine compressional tectonics in the foreland plate. The orientation of some of the foreland-type structures is not necessarily parallel with the Alpine–Carpathian structural trends; their directions might have been inherited from the preexisting structural trends in the European foreland. The Alpine foreland-type deformations apparently propagated far into the foreland. F. Chmelik (1977, personal communication) even anticipated that the Alpine deformations may continue westward as far as the eastern side of the Boskovice Furrow (graben) in central Moravia (Figure 3).

The compressional deformation of the underlying European plate in the Carpathians resembles the structural pattern of southern Apennines, where, below the thin-skinned Apenninic thrust belt, deeper parautochthonous structures exist in the Apulian plate at a depth of 4–6 km (2.4–3.6 mi). Some of these deep structures have been drilled, and significant accumulations of hydrocarbons, such as the Tempa Rosa, Monte Alpi, and Volturno fields, have been found in these deep prospects.

The origin and age of the various deep structures in the sub-Carpathian foreland in Moravia remain uncertain and deserve further attention. Some of these structures represent potential targets of exploration for hydrocarbons and, therefore, the timing of their formation remains possibly one of the most critical factors in any future evaluation of their hydrocarbon potential (Picha, 1996).

Tectonic Shortening

Restoration of orogenic belts to their original predeformation stage represents a complex task that is commonly addressed by construction of balanced cross sections and calculation of the rate of shortening by dividing the length of shortening by the time during which shortening occurred. Attempts also have been made to establish the shortening using other indicators such as the plate-tectonic reconstructions or the distance of the flexural wave migration in the foreland and the width of the orogenic wedge (DeCelles and DeCelles, 2001). The balanced cross sections, possibly the most useful tools in restoring the thrust belts, are normally valid only in the unmetamorphosed external zones of orogenic belts, e.g., the Outer Carpathians, and even in these relatively simple zones, small-scale compressional structures may produce significant shortening that hardly can be integrated into the balanced restorations.

A significant factor that has to be considered in restoring the Carpathian depositional and structural settings is the amount of erosion that occurred both prior to and during the orogeny. As the deformation and uplifting progressed gradually from the hinterland into the foreland of the thrust belt, the more internal, already uplifted and deformed units, such as the Magura flysch, shed clastic material into the more external zones of the thrust belt and eventually into the autochthonous foredeep. The present frontal edges of various thrust sheets thus do not fully represent the original extent of these units. In the Beskydy region of northern Moravia, for example, the frontal zones of the Magure nappe were removed by erosion, and the underlying Silesian nappe was exposed on the surface in a large half window (Figure 3).

To calculate the amount of shortening of the Outer Carpathians in Moravia, four partly balanced cross sections have been constructed (Figure 20). Their geometry is constrained by the combination of surface geology, well data, and interpretation of seismic lines. The amount of shortening can be more reliably established in the less deformed more competent units with a high proportion of sandstones, such as the Magura unit, whereas the shortening of the highly deformed, incompetent, predominantly shaly units, e.g., Subsilesian sector of the Waschberg–Zdanice–Subsilesian unit, can only be guessed. Numerous deep and shallow wells drilled in the Subsilesian unit of central and northern Moravia indicate that its internal structure is extremely complex. Detailed folding and faulting and the existence of duplexes known, for example, from the tectonic windows in the Beskydy region would suggest a very high rate of shortening, whereas an absence of a full stratigraphic section in other areas would indicate an existence of tectonic stretching. As the lowermost unit of the Outer Carpathian nappe stack in Moravia, the incompetent Subsilesian unit (sector) was squeezed and smeared along the base of the more competent Silesian and Magura nappes. In the process, even some Neogene foredeep strata were detached and integrated into the frontal zone of the thrust belt. The overall amount of shortening of the Subsilesian unit thus cannot be reliably established from the existing evidence.

The internal structure of the Silesian nappe, which, as indicated by seismic data, does not seem to continue far below the Magura nappe, is also complex. Locally, tectonic slices of crystalline basement rocks, Devonian and Carboniferous strata, as well as Neogene deposits of the foredeep are incorporated into the basal part of the Silesian unit. The Silesian unit comprises both massive, several-kilometers-thick, competent strata of the Upper Cretaceous Godula and Istebna flysch formations and the predominantly incompetent Upper Jurassic and Lower Cretaceous strata laid down on the rifted passive margins. During the deformation and tectonic transport, these two lithologically different sets of strata were locally decoupled, deformed, and thrusted disharmonically. The structural discrepancies between the competent upper and incompetent lower parts of the Silesian unit were originally explained by an existence of two separate nappes, the lower highly deformed and imbricated lower Tesin (Teschen) nappe and the upper Godula nappe with a relatively simple structural pattern (Mencik, 1966). According to some authors, e.g., Jurkova (1971), the complex disharmonic structure of the Silesian nappe resulted from two phases of deformation of the Silesian and Subsilesian units in terms of Stille's (e.g., 1936) concept of orogenic phases. We believe that the complexities of the internal structure of the Silesian unit can better be explained by the disharmonic folding, faulting, and decoupling of competent and incompetent strata during a continuous process of shortening and overthrusting of the Outer Carpathians onto the European foreland. During this process, the competent younger strata of the Silesian unit were locally thrusted over the older incompetent members of the unit, thus invoking the idea of an existence of separate nappes formed during the subsequent stages of the deformation.

The relatively competent Magura nappe displays a more regular internal structure, consisting of numerous imbricates and thrust sheets. Its relatively simpler internal structure enables a more reliable restoration and calculation of the range of the tectonic shortening. As a whole, however, the Magura unit is an out-of-sequence unit (e.g., Nemcok at al., 1998a) superimposed on the external units of the Outer Carpathian thrust stack. This further increases the amount of tectonic shortening of the whole Outer Carpathian thrust system.

Based on the interpretation of the well data and the regional seismic profiles, we tend to believe that the structural units of the Outer Carpathian belt (Subsilesian, Silesian, Magura, and the Pieniny Klippen Belt) do not continue far underneath the Inner Carpathians. The progressing stack of the Inner Carpathian nappes apparently scraped off most of the Late Cretaceous and Paleogene Outer Carpathian flysch strata and piled them at its front. Some of the older Upper Jurassic and Lower Cretaceous deposits of the Outer Carpathian basin might have been left behind on the pre-Jurassic substratum and, with it, partly subducted. In that sense, the Carpathian sector seems to differ from the Eastern Alps in Austria, where the Helvetic and Flysch units continue further below the nappes of the Calcareous Alps (e.g., Zimmer and Wessely, 1996). With respect to these differences, it is necessary to stress that the contact between the Inner and Outer Carpathians was greatly affected by the strike-slip motion in the Pieniny Klippen Belt.

The tectonic shortening in the Outer Carpathians is apparently not limited to the thin-skinned sedimentary cover but encompasses the decoupled basement and possibly the whole crust as well. The crustal-scale convergency of the previously attenuated and rifted crust of the passive margins and underthrusting of the oceanic crust, as proposed for the Inner Carpathians by Plasienka (1995), seems to explain better the geodynamic evolution of the Carpathian thrust belt without resorting to multiple subductions, whose existence is not evidenced by subduction-related Mesozoic and Paleogene volcanism.

The process of rifting and extension, which formed the Outer Carpathian passive continental margins, was reversed during the convergent phase. The various rift-apart crustal blocks, including the Silesian and Czorsztyn internal ridges, were compressed and accreted back to the European platform (Figure 6). In such a process, the original extensional normal faults might have been reactivated as thrust and reverse faults. Both during the extensional and compressional phase, the various crustal blocks were apparently decoupled from the mantle, and during the conversion, the depositional sequences of the thin-skinned belt were also decoupled from their substratum. Such an interpretation largely reduces the amount of subduction of the lower crust needed to compensate for the shortening in the thin-skinned thrust belt. This is compatible with a similar situation in Eastern Alps, where, according Helwig (1976), a significant subduction of the continental crust did not occur either.

However, in comparison with the Western Alps, where the advanced collision led to the inversion of the external massifs, the inversion of the European foreland below the Carpathian belt is minor, and major basement structures of the Aar Massif type are lacking (Roure et al., 1993). According to Ziegler and Roure (1996), the absence of major post-Paleocene compressional foreland structures beneath the Carpathians suggests only a low level of coupling between the orogenic wedge and its foreland.

Based on the restoration of balanced cross sections (Figures 6, 20), our calculations suggest that the minimum width of the Outer Western Carpathian depositional system, excluding the Pieniny Klippen Belt, in the time of its maximum extent in the late Cretaceous was at least 230 km (143 mi). Because the present width of the thin-skinned Outer Carpathian belt in Moravia, excluding the impact of erosion, is about 70 km (43 mi) (Figure 3), the minimum cumulative shortening would be about 160 km (99 mi). Thus, during 17 m.y. of convergency from the late Oligocene to early Badenian, the extent of the original depositional environment of the Outer Carpathians was reduced by about 70%. Such a shortening of the sedimentary sequences must then be compensated by an equal reduction of the basement. Our interpretations (Figure 6) indicate that the original depositional basement of the Outer Carpathians, including both rifted and attenuated continental and possibly some oceanic crust, was shortened 35% by compression and accretion to the foreland plate and an additional 35% by underthrusting the Carpathian plate and partial subduction into the asthenosphere (see discussion in the following section). The total shortening of the basement by 70% would thus compensate for about 70% of the shortening of its sedimentary cover accreted into the thin-skinned Outer Carpathian thrust belt. Considering about 17 m.y. of the duration of convergency from the middle Oligocene to the middle Miocene (Badenian) and the minimum amount of shortening (160 km; 99 mi), the rate of shortening of the Outer Carpathians in Moravia was about 9.4 mm/yr (0.37 in./yr). A similar rate of shortening of about 10.6 mm/yr (0.42 in./yr) may be calculated from the amount of shortening (from 235 km [146 mi] in the middle Oligocene to 55 km [34 mi] at the present time) anticipated by Roca et al. (1995) for the Polish sector of the Outer Western Carpathians. The rate of shortening of 7–10 mm/yr (0.27–0.39 in./yr) during the Eocene–Miocene determined from restored balanced cross sections in the North Alpine foreland basin by Pfiffner (1986) is also quite comparable.

The minimum anticipated width of the Outer Carpathian depositional system of about 230 to 250 km (about 142 to 155 mi) is thus comparable with the present width of the North Sea basin between Norway and the Shetland Islands. If inverted and thrusted, that part of the North Sea would possibly form a thrust belt of a width similar to that of the present Outer Western Carpathian belt.

The minimum ranges of shortening of the Outer Carpathians based on construction of balanced cross sections, however, differ substantially from much larger amounts of tectonic shortening anticipated by other authors, e.g., Roth (1987). Based on the mutual position of the Cretaceous paleomagnetic poles, he calculated that since the Cretaceous, the total north-south approach of the present-day North European and African plates amounted to 2200–2300 km (1375–1438 mi). The existing width of the Outer Carpathian belt (about 70 km [43 mi]) and the volume of sediments comprised in it would not allow for such an enormous shortening.

The Extent of the European Plate Underneath the Carpathian Belt

More than 100 yr ago, Suess (1875) proposed that the northwestern Europe was partly buried below the Alps and Carpathians, and somewhat later, Ampferer (1906) introduced the concept of crustal subduction. Zoubek (1948) and Stille (1953), among others, recognized that the crystalline basement of the European plate (Bohemian Massif) differs from the basement of the Inner Carpathians and postulated that during the Alpine convergency, the European plate was underthrusted below the Carpathian orogenic belt. More recently, the character of the deep contact of the Carpathian belt with the underlying platform has been examined by seismic (Tomek and Hall, 1993), seismotectonic (Schenk et al., 1994), magnetotelluric (Cerv et al., 1994), magnetic (Gnojek and Heinz, 1993), electromagnetic (Jankowski et al., 1985), and other methods. For more information on the subject, see chapters by Hrusecky et al. (2006), Nemcock et al. (2006), Pospisil and Adam (2006), and Pospisil et al. (2006) in this publication. Despite these intensive studies, the extent of the subthrust European plate below the Carpathians and the amount of its subduction have remained subjects of discussion. Some authors (e.g., Slaczka, 1975; Roth, 1977, 1980b; Dudek, 1980; Stranik et al., 1993) have suggested that the European plate extends to the axis of a distinct zone of gravity lows, which, in the Western Carpathians, roughly parallels the northwestern side of the Pienniny Klippen Belt from Hodonin to Namestovo and then continues northeastward toward the Nowy Sacz. The existence of such a low-gravity zone has been alternatively explained by a deep gravity minimum, deep faulting and fracturing in the Moho, and even by inhomogeneity in the top of the upper mantle. Interpretation of reflection seismic data, however, indicates that the gravity minimum is most likely caused by shallow factors such as the maximum thickness (10–12 km; 6–7.2 mi) of the accretionary wedge of the low-gravity flysch and molasse rocks (e.g., Ibrmajer, 1971; M. Krs and A. Sutora, 1974, personal communication; Pospisil and Filo, 1980; Tomek, 1982), and that the European platform continues underneath this zone farther east below the Pieniny Klippen Belt (Figure 6D). The gravity data indicate that the lower part of the accretionary wedge has lower gravity than its upper part. This can be explained by the piling of lighter, predominantly shaly duplexes of the external units of the Flysch belt, such as the Fore-Magura, Grybow, and Obidowa Slopnice, below the heavier thrust sheets of the Magura flysch in the low-gravity zone (Figure 20).

Another widely considered line of termination of the European plate is the peri-Pieniny lineament of Maska (in Buday et al., 1961; Svoboda, ed., et al., 1966; Pecova et al., 1979). Originally located in the Pieniny Klippen Belt, the peri-Pieniny lineament was later relocated into the western side of the Male Karpaty Mountains and called the Zahorie fault (O. Fusan, J. Ibrmajer, and J. Plancar, 1979, personal communication).

Other authors, e.g., Uhlig (1907), Roth (1978), and Lesko et al. (1980), suggested that the European platform extends farther east and south beyond the Pieniny Klippen Belt and below the Inner Carpathians. According to Roth (1978), the North European platform below the Inner Carpathians continues as far as the centers of the volcanic activity in central Slovakia. The continuation of the European platform below the Calcareous Alps–Inner Carpathians is at least partly documented by results of the deep wells Berndorf-1 and Aderklaa-UT1 in Austria (Wessely, 1990), which, below the edges of the Calcareous Alps, drilled into the Bohemian Massif. In addition, the reflection profiles from the Danube basin (western part of the Pannonian Basin) would indicate that the European plate may extend far below the Austroalpine and Penninic units of Alps to the Mihalyi high and the Raba line, about 130 km (80 mi) southeast from the frontal edge of the thrust belt (Szafian et al., 1999).

Our geological interpretation of regional seismic lines across the Outer Carpathians in Moravia and Slovakia shows that the European platform continues uninterrupted to the vicinity of the Pieniny Klippen Belt, where it is intersected by normal and reverse faults, which do not continue into the thrust belt (Figure 20). At least some of these faults might have originated during the Jurassic to Early Cretaceous rifting and extension of the European plate. Later, during the Alpine orogeny, some of these normal faults might have been inverted by compressional and transpressional stresses. We believe that these faults mark the break between the thick platform-type crust of the European plate and the rifted attenuated crust of the European continental margins. During the compressional orogeny, the edges of the thick European platform acted as a buttress, causing the piling of the rootless slices of the Flysch belt and the Pieniny Klippen Belt in the low-gravity zone. Given the significant component of strike-slip motion in the Pieniny Klippen Belt, it is likely that the pieces of the European crust juxtaposed on both sides of the klippen belt had moved laterally. Their present fit may thus differ from the original one especially prior to the late orogenic northeastward translation of the Western Carpathians.

The attenuated crust of the European continental margins, unless completely subducted, must continue beyond the Pieniny Klippen Belt underneath the Inner Carpathian nappes, possibly to the volcanic centers of central Slovakia, as proposed by Roth (1978). We anticipate that the European plate bends under the Inner Carpathian (Apulian) plate and descends at an angle of about 45° into the asthenosphere. At a depth of 80–100 km (50–60 mi), it partly melts and generates the subduction-type volcanic activity (Figure 6D). The European plate thus underplates the northwestern part of the Inner Carpathians in Slovakia, which may explain the relatively high mountainous relief of that region. Such an interpretation would suggest the existence of a significant crustal root down to depth of about 100 km (60 mi) below the Inner Carpathians, similar to the crustal root predicted for the Western Alps. According to Marchant and Stampfli (1997), the eclogized crustal root of the European plate overthrusted by the Adriatic crust and lithosphere extends below the Po plain.

Nemcok et al. (1998b, 2006) excluded the existence of a crustal root along the western part of the Carpathian arc. Instead, they proposed that the end of the subduction of the remnant oceanic flysch basin and the beginning of the collision were accompanied by the detachment of the subducting plate and by the occurrence of the break-off-related volcanism. The detachment began in the early Miocene in the western part of the Carpathian arc and propagated eastward to the Vrancea area of a presently active tear. Such an interpretation assumes the existence of a large oceanic domain in the Outer Carpathian Flysch basin. As previously discussed, the extent of the oceanic crust and lithosphere in the Outer Western Carpathian basin was apparently limited and confined mainly to the margins of the Penninic–Pieninic ocean south of the Czorsztyn ridge (Figure 6). The subduction of the limited oceanic realm would thus not compensate for the large shortening in the Outer Carpathian belt. Neither would the interpretation proposed by Nemcok et al. (1998b) satisfactorily explain the fate of the large portions of the continental lithosphere, which originally underlined the Outer Carpathian depositional system. These discussions indicate that the geodynamic reconstruction of the Carpathian region remains a challenging task, whose satisfactory solution will require additional geological and geophysical studies addressing both local and wider regional problems of the Alpine–Carpathian system of Europe.

Neotectonics and Topography of the Carpathian Region

In comparison with the Alps, the topographic elevation of the Carpathians above the sea is relatively low. According to Roure et al. (1993), the subdued relief of the Carpathians might be caused by a low level of collision (coupling) between the orogenic wedge and the foreland. Unlike the Alps, the Carpathians underwent two periods of subsidence and extension. The first one, marked by normal faulting and formation of the Inner Carpathian Paleogene basin on the top of the Mesozoic nappes, occurred in the middle Eocene to early Miocene in the northern part of the Inner Carpathian belt. The second, associated with the formation of the volcanic arc and the back-arc Pannonian Basin in the Miocene, affected southern zones of the Inner Carpathians.

The Outer Carpathian belt superimposed on the relatively stable European platform remained mostly unaffected by the postorogenic extension and subsidence. Only the southwestern part of the Carpathian belt, mainly confined to the extent of the Dyje–Thaya depression on the territory of northeastern Austria and southern Moravia (Figure 3), underwent a significant extension and subsidence most prominently marked by the opening of the Vienna basin. The formation of the Vienna basin during the Miocene was apparently related both to the transtensional tectonics in the Western Carpathian transfer zone and to the inherited weakness of the crust in the Dyje–Thaya Jurassic depression. The continuous subsidence of this area is documented by the accumulation of the Pliocene to Holocene alluvial deposits of the Dyje and Morava Rivers. The measurements of the horizontal dynamics in the contact zone between the Bohemian Massif and the Western Carpathians indicate that extension prevails in the southern part of the Carpathian foredeep, whereas compression dominates its northern part. The former is apparently associated with a wide-range subsidence of southern Moravia (Dyje–Thaya depression), the latter with the uplifting of the Beskydy and Jesenik Mountains of northern Moravia (Bucha and Blizkovsky, 1994).

Elsewhere outside the Vienna basin, the Flysch belt in southern Moravia is characterized by a low relief of

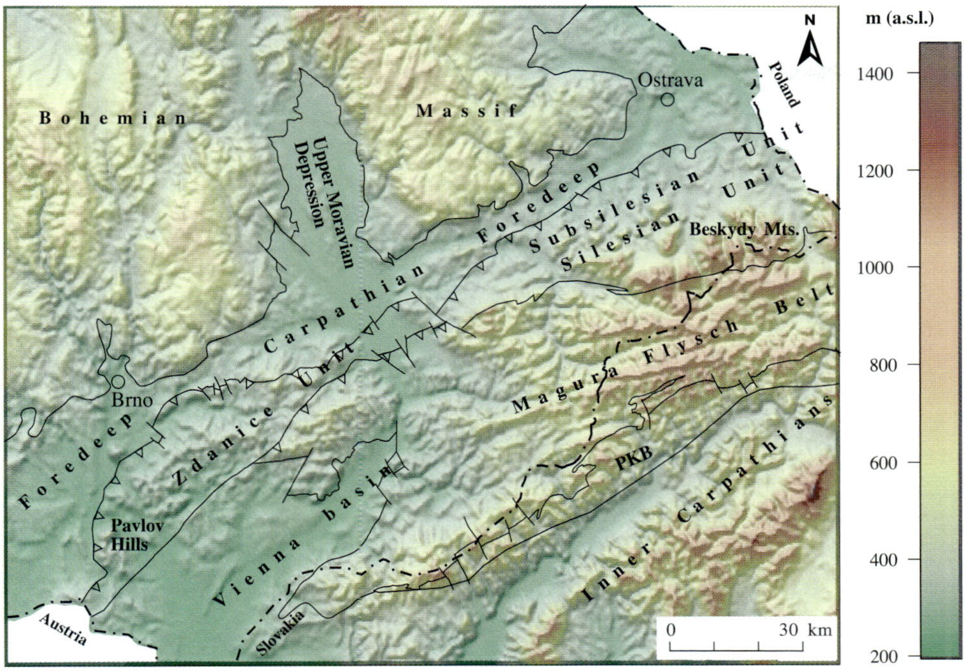

Figure 21. Topographic map of the Western Carpathians and their European foreland in Moravia.

rolling hills, from which only the carbonate klippen of Ernstbrunn and Pavlov Hills stand out morphologically. In central Moravia and especially in the Beskydy Mountains of northern Moravia, the relief of the Outer Carpathian Flysch belt is much more mountainous with the highest peak, the Lysa hora (Bold Mountain), reaching 1328 m (4356 ft). This relatively high relief might be explained by the combination of the crustal uplifting of the Beskydy Mountains (Figure 21) and the presence of thick, weathering-resistant Upper Cretaceous sandstone beds of the Godula and Istebna formations. The mountain ranges of the Flysch belt, in general, and the Magura flysch, in particular, trend mostly in the southwest–northeast direction, parallel with the thrust-related structures. Their topography is further enhanced by the alternation of packages of weathering more resistant sandstones and less resistant shales. The Pliocene to Holocene fluvial erosion dissected the uplifted ranges by deep northwest–southeast-running valleys.

The earthquake foci in the Western Carpathians, according to Bucha and Blizkovsky (1994), are confined to three main structural systems: (1) southwest–northeast-trending sliding surfaces between the crustal blocks; (2) northwest–southeast-trending faults in the underlying European platform; and (3) thrust planes and shears in the Carpathian thrust belt. The first two systems are apparently related to two primary structural trends, known from the Bohemian Massif, and expressed also by the southwest–northeast-directed Western Carpathian transfer zone and the northwest–southeast-trending Jurassic Dyje–Thaya depression, respectively. Most of the recorded earthquake activity is associated with the peri-Pieniny lineament of Maska (in Buday et al., 1961). The links between neotectonics and the continental topography in the Pannonian Basin and the Carpathian orogenic arc have become a subject of intensive studies (Bada et al., 2001; Cloetingh et al., 2002).

The Pleistocene continental glaciation reached into the northernmost part of the Carpathian foredeep in Moravia. The permafrost, to the depth of several tens of meters, formed elsewhere in the periglacial regions. The unstable slopes of the flysch mountain ranges are predisposed to extensive mass wasting and landsliding.

MAJOR DEPOSITIONAL SEQUENCES OF THE OUTER WEST CARPATHIANS

To better comprehend the geological history of the Outer Carpathians and their foreland and to integrate it into the evolution of the broader Tethyan–Carpathian system, attempts have been made to define the main depositional sequences and to relate them to major tectonostratigraphic events, such as rifting, major transgressions, episodic progradation of the Carpathian orogenic belt, and tectonic mobilization of the cratonic foreland (e.g., Picha and Stranik, 1999; Slaczka et al., 2006).

Based on such criteria, we have divided the Jurassic to Neogene strata of the Outer Western Carpathians in Moravia into six main sequences related to the critical tectonic and depositional events in the evolution of the Outer Carpathians (Figure 17A, B). The term sequence is used here in a broad sense, without any reference to the terminology of the sequence stratigraphy.

1) The oldest, Middle Jurassic (Dogger) to the Early Cretaceous [Berriasian–Valanginian(?)] sequence encompasses the oldest strata of the Tethyan–Alpine tectonic cycle found in the European foreland and in the Outer Carpathians in Moravia and northeastern Austria. Their depositional system was controlled by the Middle Jurassic rifting of the European platform, followed by the encroachment of the Tethyan Sea and development of the passive continental margins in the outer zone of the Carpathian system. The rifting of the European foreland in Moravia and northeastern Austria was predominantly confined to the northwest–southeast-trending Dyje–Thaya depression (Figure 5). The synrift phase was marked by deposition of the clastic Gresten Formation, followed by the development of carbonate platforms and basins on the rifted continental margins. A chain of carbonate buildups, known from the tectonic klippen such as Ernsbrunn, Pavlov Hills, and Stramberk, evolved along the shallow northwestern side of the rift basin (Figure 5). Included into this oldest sequence are the autochthonous Jurassic strata of the foreland in Moravia and northeastern Austria, the Outer Klippen of the Waschberg–Zdanice unit, the Jurassic strata of the Silesian unit including the Stramberk klippe, the Kurovice klippe of the Magura unit, and the Jurassic to Lower Cretaceous klippen of the Pieniny Klippen Belt. Reworked rocks of this sequence are commonly found elsewhere in the conglomerates of the Flysch belt.

2) The Early Cretaceous (Neocomian to Albian) sequence began with a primary change of the structural pattern of the Outer Western Carpathian region, which became dominated by the dextral motion along the Western Carpathian transform zone (Figure 5). Whereas the area west of this zone was uplifted and partly eroded, the sedimentation continued uninterrupted in the rest of the Tethyan continental margins east of this zone. The continuing rifting and extension resulted in a further differentiation of the Outer Carpathian depositional realm into a system of basins and ridges. In the Silesian basin, the further rifting and extension of the attenuated crust was accompanied by the occurrence of the teschenite volcanism. The Silesian ridge separated the Silesian basin from the Magura basin, which evolved on a highly attenuated continental and, according to some views, even oceanic crust (Figure 6). The Magura basin was then separated from the oceanic realm of the Pieniny Klippen Belt by the Czorsztyn ridge. The extension and crustal thinning in the Outer Carpathian realm peaked during the Early Cretaceous. In the Aptian–Albian, a widespread anoxic event, marked by the deposition of deep-water black shales, silicites, and turbidites, occurred in the entire Tethyan system. In the Western Carpathians, this anoxic event is evidenced by the deposition of the Verovice and Lhoty formations in the Silesian basin and the Rajnochovice Formation and Gault flysch in the Magura and the Rhenodanubian Flysch, respectively.

3) The Late Cretaceous (Cenomanian) to the late Eocene (Bartonian) sequence reflects the continuous deformation of the Inner Carpathians and their collision with the Czorsztyn ridge and other fragments of the European plate. In the Outer Carpathian realm, this was marked by a gradual transition from a divergent to a convergent phase and development of a progradational foreland basin. The compressional stresses at the Albian–Cenomanian transition (formerly the Austrian orogeny) caused the sudden rise of the intrabasinal ridges, which shed big amounts of clastics into the adjacent deep-water flysch basins. In the Cenomanian, this process was combined with the global transgression (high water stand), during which the epicontinental shallow seas spread over vast territories of the continents, including the Carpathian foreland in Moravia. In the Campanian–Mastrichtian, the Alpine–Carpathian seaway progressed over the European foreland to form the Waschberg–Zdanice–Subsilesian basin (Figure 5). At the Cretaceous–Paleogene transition, during the so-called Laramide orogeny, the European foreland adjacent to the Outer Carpathian basin was uplifted, deeply weathered, and eroded. The magnitude of the uplifting and erosion is well documented by the incision of two large paleovalleys in southern Moravia. The hemipelagic variegated (commonly red) muds, derived mainly from the weathered platform, were deposited elsewhere in the external units of the Outer Carpathians. At the same time, deep-water turbiditic sedimentation prevailed in the Silesian and Magura flysch basins supplied with clastics from the Silesian cordillera. Characteristic for the Paleocene to middle Eocene variegated muddy deposits of the external units is the occasional presence of thick, channelized sands, which are significant reservoirs, especially in the Silesian unit of Poland.

The Late Cretaceous to late Eocene sequence represents a large part of the depositional stack of the Magura flysch (Kaumberg, Solan, Beloveza, and Zlin formations), of the Silesian unit (Godula, Istebna, and

Roznov formations of the Godula subunit and their equivalents in the Kelc and Baska subunits), and of the Waschberg–Zdanice–Subsilesian units (Nemcice and Frydlant formations) (Figure 17A, B). In addition, the lower part of the autochthonous fill of the Nesvacilka paleovalley containing the questionable Late Cretaceous to Paleocene fauna might belong to this sequence. The Late Cretaceous to late Eocene sequence of the Outer Carpathians also encompasses most of the Helveticum and the Rhenodanubian Flysch of the Austrian, Bavarian, and Swiss Alps.

4) The late Eocene (Priabonian) to early Oligocene (Rupelian) sequence is associated with the extensive late Eocene marine transgression over the European foreland, deepening of the external Carpathian basins, submergence of inner ridges, and the termination of sedimentation in the Magura flysch. Euxinic facies developed in the Carpathians and throughout the entire Tethyan region from the Western Mediterranean to the Black and Caspian Seas region and beyond.

In the Alpine sector, the late Eocene transgressive event marks the formation of the Molasse basin with the deposition of a variety of lagoonal and marine facies, including the Lattorfian Fish shales and Rupelian bituminous marls of the lower marine Molasse in Austria and Bavaria (Bachmann and Muller, 1991; Wagner, 1996). The Menilitic cherts, typical for the Carpathians, are not found in the Alpine Molasse. The more advanced collision and thickening of the crust in the Alpine region apparently prevented more substantial deepening of the Molasse basin needed for upwelling and accumulation of silica-forming organisms, such as diatomites (Figure 22A).

In the Moravian sector of the Western Carpathians, the late Eocene marine transgression reached into the most external zones of the Pouzdrany and the Waschberg–Zdanice–Subsilesian basins, and the paleovalleys and submarine canyons were filled with hemipelagic deposits, which, according to Picha and Stranik (1999), exemplify the continuation of the lower marine Molasse of the Alps into the Carpathian realm (Figure 22A). The late Eocene to early Oligocene sequence in Moravia extends from the autochthonous fill of the Vranovice and Nesvacilka paleovalleys into the Pouzdrany and Uhercice formations, and the Boudky Marls of the Pouzdrany unit, the *Globigerina* (Sheshory) Marls, and the Menilitic Formation of the Waschberg–Zdanice–Subsilesian and Silesian units and their equivalents in the Fore-Magura and Cejc Zajeci units to the youngest strata of the Zlin Formation in the outer zone of the Magura flysch (Figure 17A, B). In the territory of Slovakia and Poland, equivalents of these strata are also found in the Grybow, Obidowa Slopnice, and other window units. In the Inner Carpathian Paleogene, the Lutetian transgressive shallow-marine facies was followed by the deep-water shaly deposits, including some Menilitic cherts and manganese ores (Picha, 1964a). Equivalents of these strata are also present in the Hungarian Paleogene basin, known as the euxinic Tard Clay (Tari et al., 1993).

The early Oligocene, Menilitic cherts, and other organic-rich shales and marls and their equivalents elsewhere in the Tethyan region represent a significant correlation horizon. They were deposited in a stagnant environment with a very limited influx of coarser clastics both from the submerged orogenic belt and from the foreland. Typically, these deposits contain a high content of organic matter, which makes them one of the most important sources of hydrocarbons in the entire Tethyan–Alpine system from Western Europe to the Caucasus, the Caspian region (Maykop Formation), the Himalayas, and Southeast Asia.

5) The late Oligocene to early Miocene (Chattian–Aquitanian, Egerian) sequence was deposited in a system of foredeeps, which formed in the late Oligocene to early Miocene throughout the Carpathian orogenic system. Their formation was linked to the uplifting and deformation of the inner zones of the Outer Carpathians, e.g., the Pieniny Klippen Belt and the Magura flysch, and to the flexural downbending of the foreland crust (Figures 6C, 22B). The sequence is dominated by the deep-water turbiditic Krosno-type flysch facies, supplied with clastic material predominantly from the folded and uplifted inner zones of the Magura flysch. In Moravia, these deposits are represented by the Krepice Formation of the Pouzdrany unit and its equivalents, the Eisenschussige clays and sands (Grill, 1962) of the Waschberg sector in Austria, the Zdanice–Hustopece Formation of the Waschberg–Zdanice–Subsilesian unit (the Auspitzer Mergel in Austria; Grill, 1962), and the Krosno Formation of the Silesian unit. Similarities also exist between the Zdanice–Hustopece Formation of southern Moravia and the deep-water turbidite facies of the Puchkirchen beds of the Bavarian Molasse (Bachmann and Muller, 1991). Further west in Switzerland, this turbiditic facies gradually passes into nonmarine deposits of the lower fresh-water Molasse (Schlunegger et al., 1997) (Figure 22B). In the Inner Carpathians, a major deep-water flysch basin, characterized by the turbiditic sedimentation and transport of the clastic material from the uplifted zones of the Magura flysch and the Pieniny Klippen Belt, evolved in the late Oligocene to early Miocene. Its deepest and tectonically most active part was adjacent to the Pieniny Klippen Belt zone. The Krosno-type flysch deposits of the late Oligocene to early Miocene

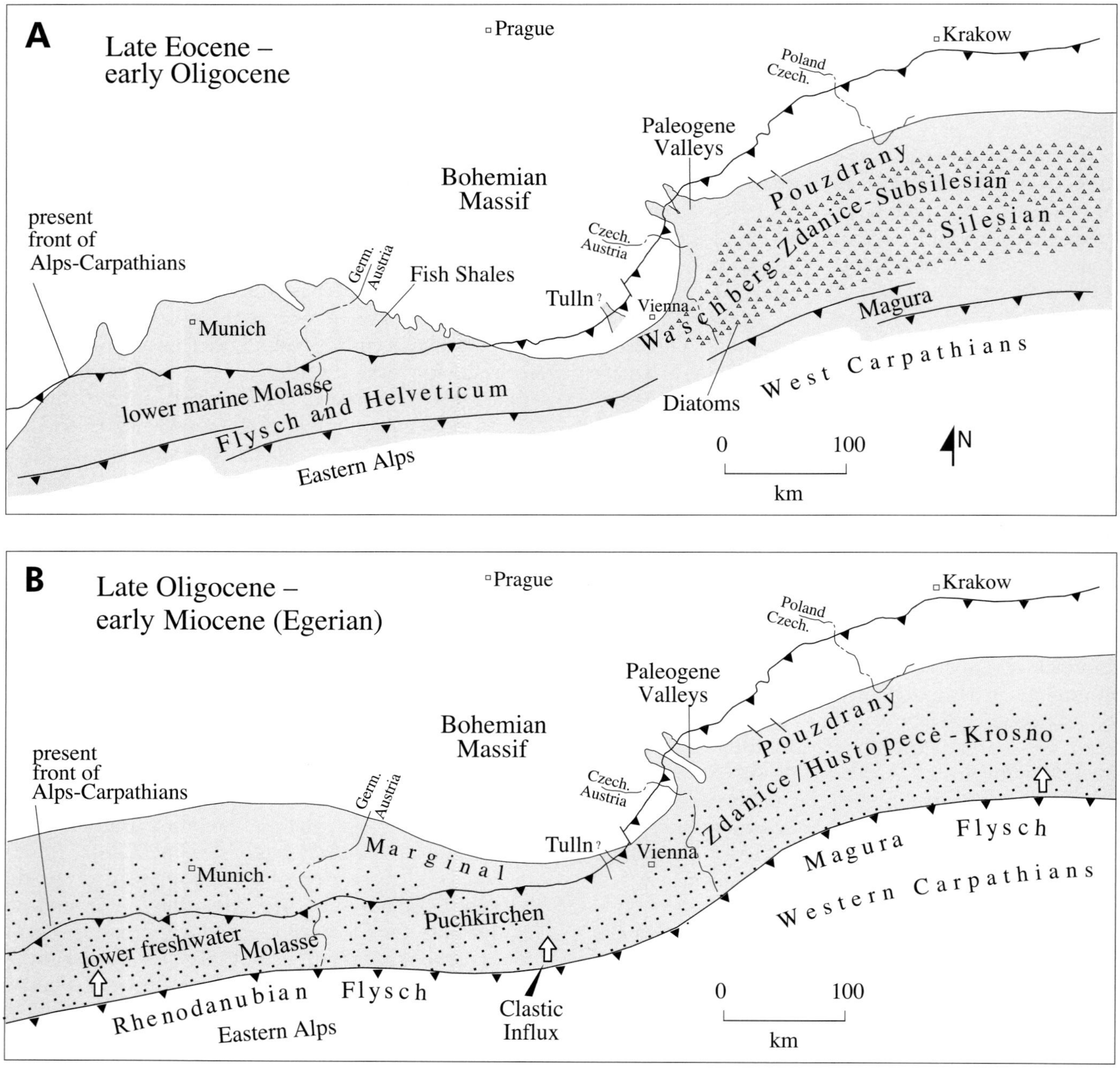

Figure 22. Correlation of (A) late Eocene to early Oligocene and (B) late Oligocene to early Miocene strata in the Eastern Alps and Western Carpathians (Picha and Stranik, 1999, reprinted with permission).

age, typically present in the external zones of the Outer Carpathians and the Inner Carpathian Paleogene basin, represent the youngest deep-water flysch deposits found anywhere in the Western Carpathians.

6) The early Miocene (Eggenburgian) to Pliocene sequence represents the youngest stage of development of the Tethyan–Alpine system. It is related to final phases of folding and thrusting of the Outer Carpathians and the development of a system of the molasse-type foredeeps on the European foreland, referred to as the Paratethys (Laskerev, 1924). Within the Carpathian thrust belt, intramontane basins, such as the pull-apart Vienna basin and the back-arc Pannonian Basin, formed as an outcome of the late orogenic strike-slip (escape) tectonics and back-arc extension and subsidence. The molasse-type, shallow-marine, and continental deposits of this sequence reflect both on the tectonic activity of the Carpathian orogenic belt and the global sea level oscillations. In Moravia, this sequence is represented by the whole depositional fill of the Neogene foredeep and the Vienna basin and also by the youngest members

of the Waschberg–Zdanice–Subsilesian unit, such as the Eggenburgian Sakvice Marls, the Ottnangian and Karpatian Pavlovice and Laa formations, and their equivalents in the Pouzdrany unit (Figure 17A, B). All these strata may be correlated with the deposits of the upper marine and upper nonmarine Molasse of Bavaria and Austria and with other coeval strata of Paratethys elsewhere.

THE HYDROCARBON RESOURCES OF THE CARPATHIAN THRUST BELT AND ITS FORELAND IN MORAVIA

The first attempts to find hydrocarbons on the territory of Moravia go back to the turn of the 19th century. In 1899–1900, J. May drilled the first well to 450.7 m (1478.6 ft) and the second well in 1902 to 645.2 m (2116.7 ft) in the Flysch belt near Bohuslavice nad Vlarou without any apparent economic success, although the presence of oil and natural gas was confirmed. In 1900, the first well drilled in the Vienna basin near Hodonin to a depth of 217 m (711 ft) encountered noncommercial shows of oil and gas. Natural gas produced from a well was, for the first time, used as a fuel in a sugar mill in Slavkov (Austerlitz) in 1908. The first commercial oil in the Vienna basin was found in 1914 at Gbely (now in the territory of Slovakia) at the depth of 164 m (538 ft). In 1919, oil was discovered in two horizons, Sarmatian and upper Badenian, near the town Hodonin. This was followed by discoveries of the Vacenovice field in the Vienna basin and the Zatcany and Sokolnice fields in the shallow part of the Carpathian foredeep. The production of oil in the former Czechoslovak Republic reached 10,110 t in 1920 and 22,796 t in 1930. The first official report about the production of natural gas (242,300 m^3; 8,556,742 ft^3) was issued in 1930. During World War II, the intensified exploration resulted in discoveries of the oil and gas fields Bilovice, Zizkov, Luzice, Tynec, Kostice, and Breclav in the Vienna basin.

After the Second World War, a major exploration effort by the newly established state oil company Moravske naftove doly led to discoveries of several small and midsized fields, e.g., Brodske, Poddvorov, and Mutenice, and, in 1959, of the largest oil and gas reservoir in the Moravian part of the Vienna basin, the Hrusky field. The Hrusky field and its extension, the Josefov field, held initial reserves of 15.5 million bbl of oil and 90 bcf of gas. The Vienna basin remained the main producer of hydrocarbons in Czechoslovakia until the end of 1970. In the 1960s, a pioneering deep drilling project, conducted initially by the Czechoslovak Geological Survey, revealed the presence of hydrocarbons in various stratigraphic horizons in the foreland plate of the Bohemian Massif underneath the thin-skinned Carpathian thrust belt in Moravia. The exploration activity gradually moved from the Vienna basin to these deeper subthrust plays, where several fields, such as Kostelany, Damborice, Uhrice, Zdanice, and Zarosice, have been found to date. For more information on some of these fields, see the account by Kostelnicek et al. (2006). Following the development of some of these fields, the production of oil in the Czech Republic increased to more than 6000 bbl/day (more than 315,000 t/yr) in 2002. The remaining undrilled potential of the subthrust play in Moravia is still substantial, especially at greater depth. Most recently, the application of the new seismic techniques, such as 3-D survey, enabled the reassessment of the previously explored plays in the Vienna basin and led to additional discoveries of oil and gas even in this mature area.

To date, about 5.5 million t (38.4 million bbl) of oil and 9 billion m^3 (320 bcf) of natural gas have been produced from all the fields in the territory of Moravia. An additional 7.5 billion m^3 (265 bcf) of coal gas from the degasification of the coalfields was produced in the Moravian part of the Upper Silesian basin by 1998. However, several attempts to produce economic quantities of coalbed methane from the coal seams of the Upper Silesian basin, in the territory of both the Czech Republic and Poland, so far have not been successful. Some of the depleted oil and gas fields, e.g., Dunajovice, Hrusky, Bojanovice, and Uhrice, have been converted into gas storages.

Practically all the oil and gas fields discovered at the territory of Moravia are listed in Tables 2–4 (shown on pages 136, 139, and 143, respectively), and their locations are shown in Figure 25, shown on page 135.

Numerous papers and internal reports have been written about various aspects of the hydrocarbon prospectivity of Moravia; however, only a fraction of them are available in the public domain and published in English. Among them, the recent publications by Durica et al. (1986), Krejci (1993), Krejci et al. (1994, 1996), Ciprys et al. (1995), Dorman (1995), Brzobohaty et al. (1996), Francu et al. (1996), Picha (1996), and Picha and Peters (1998) seem to be the most informative and relatively easily accessible to readers.

THE HYDROCARBON SYSTEMS OF THE CARPATHIAN THRUST BELT AND ITS FORELAND IN MORAVIA

In Moravia, hydrocarbons have been found in the Neogene foredeep, the Vienna basin, the Flysch belt, and the subthrust European platform. Their genesis is related to a variety of petroleum systems, which exist in this relatively small but geologically complex area. As defined by Demaison and Huizinga (1994), among others, the concept of hydrocarbon systems integrates all components involved in the generation, migration, accumulation, and preservation of hydrocarbons. In principle, that includes source rocks, reservoirs, seals, and traps.

The Source Rocks

In the territory of Moravia and lower Austria, potential source rocks are found both in the European foreland plate and in the Carpathian thrust belt.

The Source Rocks in the European Foreland Plate

In the European foreland plate, organic-rich deposits occur in the Middle Devonian, Upper Carboniferous, Jurassic, and Paleogene (Figures 23, 24).

The generative potential of the Paleozoic source rocks, such as the Devonian carbonates (e.g., the Lazanky Limestones) and the Carboniferous shales, is still little understood. Some Carboniferous rocks, especially those associated with the coal-bearing sequences of the Upper Silesian basin, have a fair source rock potential. On average, they contain 1.2 wt.% of the predominantly gas-prone type III kerogen (Krejci et al., 1994). The Carboniferous coal beds evidently sourced some accumulations of gases in northern Moravia. However, the generative potential of these mostly postmature Paleozoic rocks seems to be limited to areas where the source rocks did not pass into the generative window prior to the emplacement of the Carpathian nappes and the formation of the Neogene foredeep (Figure 23). According to Lafargue et al. (1994), the possibility that the organic-rich Paleozoic rocks may have sourced some potential deep subthrust plays in the Carpathians should not be dismissed entirely.

The Upper Jurassic (Malmian) organic-rich Mikulov marls found in the Jurassic Dyje–Thaya depression of southern Moravia and northeastern Austria (Figures 23, 24) represent a world-class source rock whose generative potential has been well established (Ladwein, 1988; Krejci et al., 1994; Francu et al., 1996; Picha and Peters, 1998). These marls, as much as 1500 m (4900 ft) thick, were deposited in a marine basinal restricted environment with a high rate of accumulation of the predominantly algal-derived type II organic matter. The total organic carbon (TOC) of these rocks ranges from 0.2 to more than 10.0 wt.% and averages around 1.9 wt.% (Ladwein et al., 1991; Krejci et al., 1994). These Jurassic organic-rich rocks sourced most of the oils in the Vienna basin (Ladwein, 1988) and in the subthrust foreland plate of southern Moravia and northeastern Austria (Ciprys et al., 1995; Francu et al., 1996; Zimmer and Wessely 1996; Picha and Peters, 1998).

The Jurassic black shales and marls are also locally found in the Polish and Ukrainian part of the Carpathian foreland. High organic contents (2.0–3.0 wt.% of TOC) and hydrogen index values between 100 and 400 mg hydrocarbons (HC)/g TOC in these potential source rocks have been reported, e.g., by Lafargue et al. (1994).

The organic-rich shales and marls of the autochthonous Paleogene represent the second most important source rock in Moravia (Figures 23, 24). They are found predominantly in the Paleogene fill, as much as 1500 m (4900 ft) thick, of the two paleovalleys and submarine canyons (Picha 1979a, 1996). The TOC content in these rocks ranges between 1 and 9 wt.% and averages around 3 wt.% (Brzobohaty, 1993; Krejci et al., 1994). The organic matter is composed mostly of the terrigenous gas-prone type III kerogen with a varying amount of the oil-prone type II kerogen of algal and bacterial origin, whose proportion, however, may increase further down the slope away from the river estuaries. The Paleogene source rocks apparently sourced the Krumvir and Karlin gas fields in southern Moravia and possibly the oil fields Kostelany, Krasna, and Dolni Lomna in central and northern Moravia (Figure 25, shown on page 135). The organic-rich autochthonous Paleogene deposits of southern Moravia are partly coeval with the organic-rich Menilitic shales of the Carpathian flysch belt and the Lattorfian (Priabonian) Fish shales (Fischschiefer) of the Alpine Molasse in Bavaria and Austria. The Fish shales, as much as 100 m (330 ft) thick, contain 1–8 wt.% TOC dominated by the type II kerogen. They pass into a generative window at a depth of 4000 m (13,000 ft), mainly in the zone buried below the Alpine thrust belt (Schmidt and Erdogan, 1996; Wagner, 1996; Zimmer and Wessely, 1996). The Fish shales are considered to be the main source rocks for the oil and gas fields in the Alpine Molasse of Austria. Until the end of 1992, the cumulative production from those fields amounted to 56 million bbl of oil and 12.5 billion m^3 (462 bcf) of gas (Wagner, 1996).

The organic-rich autochthonous and parautochthonous marginal Paleogene strata extend from the Alpine Molasse of Bavaria and Austria to southern Moravia, where they fill two paleovalleys, and then possibly below the Flysch belt to the Black Sea and further east through the Caucasus to the Caspian Sea. The potential existence of these organic-rich deposits below the Outer Carpathian belt may enhance the prospectivity of the entire Carpathian region. The hydrocarbons generated from these rocks could have

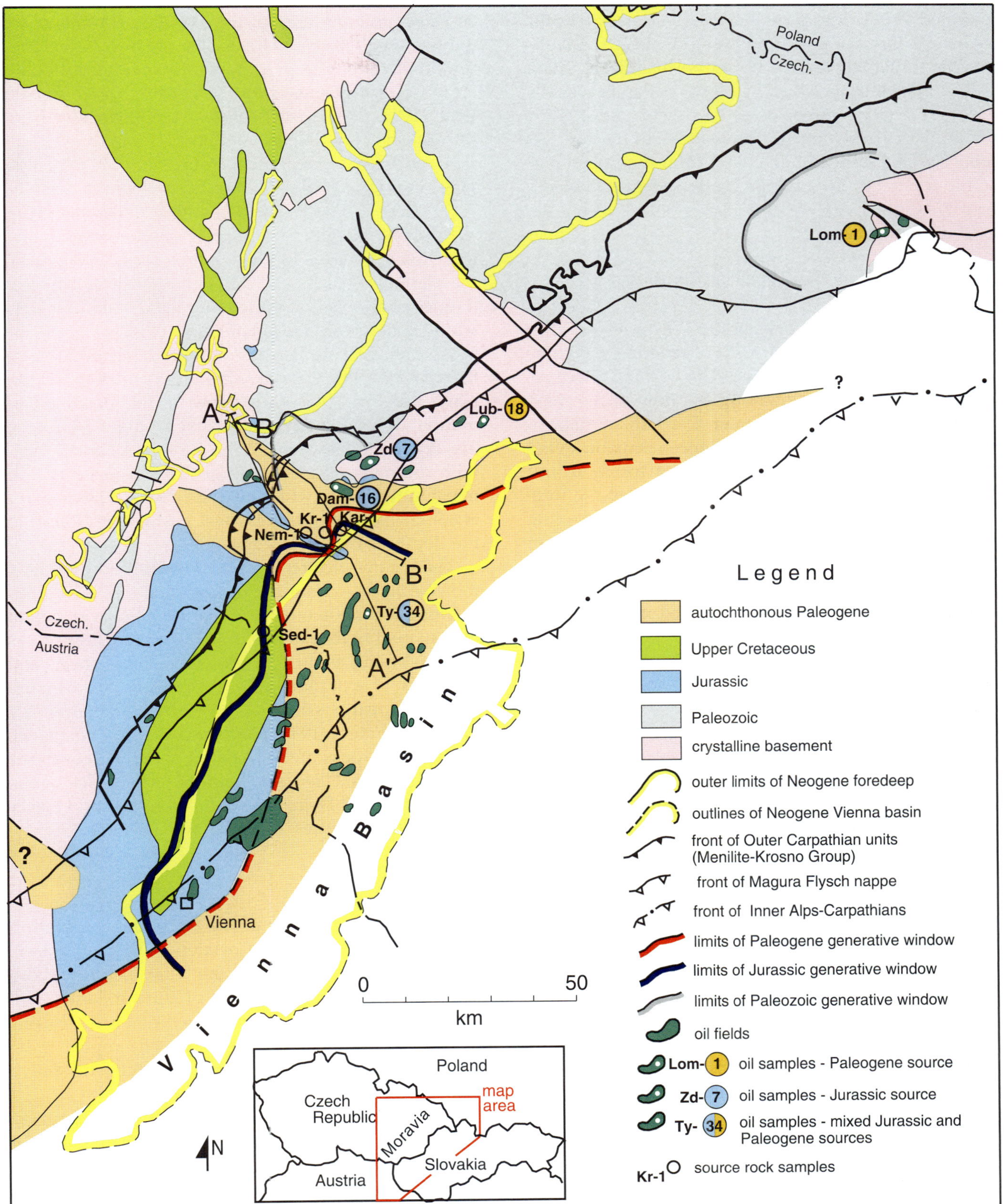

Figure 23. Pre-Neogene subcrop geological map of the European foreland plate in Moravia and northeastern Austria showing the distribution of Paleozoic, Jurassic, and Paleogene strata with potential source rocks. Oil samples (e.g., Lom-1) analyzed for biomarkers (modified from Pícha and Peters, 1998).

migrated updip into potential reservoirs both in the foreland subthrust plate and possibly along the faults into the thrust belt and successor basins (Picha, 1996).

The Neogene foredeep in Moravia has practically no organic-rich source rocks; moreover, the organic material in the Neogene deposits is immature. As reported by Kotarba et al. (1987) from the Polish part of the foredeep, the organic matter is mainly of terrestrial origin (type III) and shows only a feeble degree of maturation.

The Source Rocks in the Carpathian Thrust Belt

Two potential source rocks in the Western Carpathian thrust belt are related to two prominent euxinic events in the Tethyan region: (1) the Lower Cretaceous (Aptian–Albian) event marking the time of the maximum divergency in the Outer Carpathian depositional realm and (2) the lower Oligocene euxinic event related to the transgression over the foreland and the Inner Carpathians and the deepening of the foreland basins combined with a period of global cooling and proliferation of siliceous phyto- and zooplankton (Figure 17A, B).

The Lower Cretaceous (Aptian–Albian) black shales and cherts are prominently developed in the Silesian basin, but their equivalents are also found in the Magura and Rhenodanubian Flysch (Rajnochovice Formation and Gault Flysch, respectively) and in the external Carpathian units of eastern Poland and Ukraine (Koltun et al., 1998). All these rocks are organic rich, but because of the low hydrogen index, their generative potential seems to be limited. According to Lafargue et al. (1994), the average TOC value of the Lower Cretaceous black shales is 2.5 wt.%, and their average hydrogen index is 150 mg HC/g TOC. Likewise, ten Haven et al. (1993), in a sample from the Polish Carpathians, report a high TOC value of 3.47 wt.% but a low hydrogen index of only 53 mg HC/g TOC. According to Slaczka (1996) and Koltun et al. (1998), at least in some areas, the Lower Cretaceous shales passed into the generative window prior to the thrusting and may thus be considered as potential source rocks.

The early Oligocene Menilitic shales and cherts of the Waschberg–Zdanice–Subsilesian, Silesian, and other external units have long been recognized as the most important source rock in the entire Carpathian belt. They are partly coeval with the autochthonous marginal Paleogene deposits of the subthrust plate, but unlike the autochthonous Paleogene, the Menilitic shales were detached, incorporated into complex structures of the thrust belt, and tectonically transported over the foreland (Figure 19). In Moravia, the organic-rich laminated siliceous shales, cherts, marls, and micritic limestones of the Menilitic Formation are only about 20 m (66 ft) thick. The organic matter present in these rocks is predominantly marine, with only a limited terrigenous input. According to ten Haven et al. (1993), five extracts of the Menilitic shales from the Polish Carpathians contain 2.15–22.15 wt.% of TOC, with hydrogen indices from 300 to 650 mg HC/g TOC. Lafargue et al. (1994) published similar results, 2–20 wt.% TOC (average 6%) with 300–600 mg HC/g TOC, from Menilitic shales in Poland, Ukraine, and Romania. Of all the studied source rocks in Moravia, the Menilitic shales have the highest liptinite contents (Francu et al., 1996). If buried into the oil window, they would generate significant amounts of oil.

Geochemical studies by Koltun et al. (1998) indicate that in the prolific Boryslav–Pokuttya unit of the Ukrainian Carpathians, the oil window for the Menilitic Formation (assuming the geothermal gradient 24°C/km) lies at a depth between 4.2 and 5.8 km (2.6 and 3.6 mi) with the peak oil generation at 5.3 km (3.3 mi). The maximum depth at which oil (with <50° API) may occur is 6.9 km (4.2 mi), whereas wet gas may be found to the depth of 7.1 km (4.4 mi).

The organic-rich rocks of the Menilitic Formation in the Waschberg–Zdanice–Subsilesian and Silesian units in Moravia remain mostly immature. However, lateral migration of hydrocarbons from the more mature rocks of the Menilitic Formation present in thrust sheets and duplexes of external units underneath the Magura nappe is a real possibility. The biomarker analyses by Picha and Peters (1998) confirmed that the Paleogene organic-rich rocks of the Menilitic Formation or of the autochthonous Paleogene sourced oils in the Lubna (Kostelany), and Dolni Lomna fields central and northern Moravia (Figures 23, 25 [shown on page 135]).

Migration

The petroleum systems of the Carpathian region in Moravia are primarily associated with the Jurassic and Paleogene source rocks, the latter including both the autochthonous Paleogene of the foreland plate and the Menilitic Formation of the thrust belt. The oils from the Vienna basin (Tynec-34 well) appear to be sourced from both the Jurassic and Paleogene source rocks (Picha and Peters, 1998).

Oils and gases generated from the deeply buried source rocks in the subthrust European plate migrated both laterally into the foreland, where they supplied several oil and gas fields, and vertically through the thrust belt into the Miocene reservoirs in the Vienna basin (Figure 24). The lateral migration was probably helped by the existence of impermeable shaly deposits at the base of the Carpathian thrust belt, which

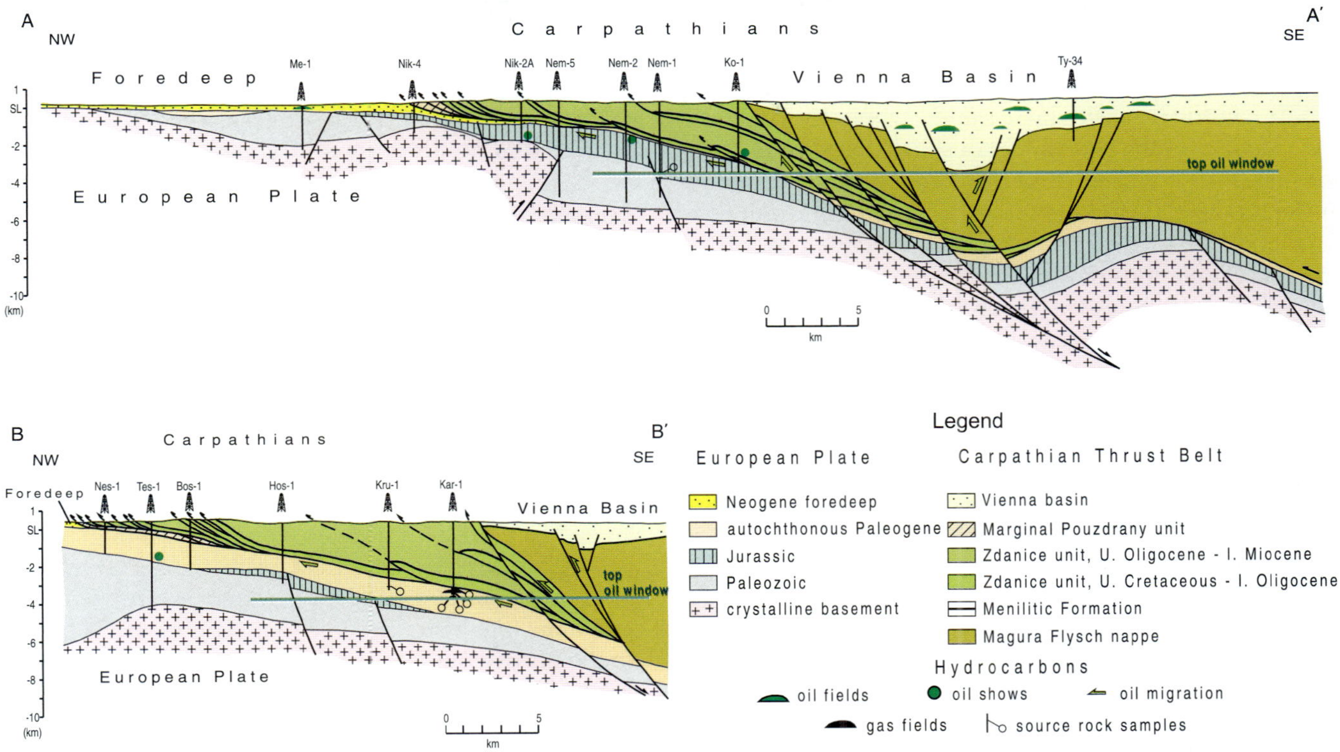

Figure 24. Cross sections AA′ and BB′ (location in Figure 23) through the West European plate, the overlying Outer Western Carpathian thrust belt, and the successor Vienna basin showing the top of the generative window for the organic-rich source rocks in the Jurassic and Paleogene strata of the subthrust plate and the Menilitic Formation of the thrust belt. With the exception of deep zones below the thick Magura nappe, the Menilitic organic-rich shales and cherts of the Zdanice unit remain mostly immature in the frontal zones of the Flysch belt in Moravia. The cross section BB′ passes through the Paleogene fill of the Nesvacilka paleovalley and submarine canyon. In the Vienna basin area, some of the Jurassic rift-related faults were apparently reactivated as transtensional faults during the pull-apart stage of the evolution of the basin in the middle Miocene and served as conduits for migration of hydrocarbons from the Jurassic and Paleogene source rocks of the subthrust plate through the thick flysch sequences of the thrust belt into the Neogene reservoirs of the Vienna basin. Modified after Picha and Peters (1998). Stratigraphic records of some of the wells shown on cross sections are reported in Appendix 1.

acted as a regional seal and prevented vertical escape and dispersal of hydrocarbons. The distribution of hydrocarbon occurrences in the foreland plate and the Neogene foredeep indicates that the tectonically controlled Paleogene valleys may have served as conduits for migration of hydrocarbons from the deeper zones of the Carpathian system into the foreland (Picha, 1979a). The vertical migration from the deep Jurassic and Paleogene source rocks in the subthrust plate through the thick flysch belt into reservoirs in the Neogene Vienna basin was most likely enabled by the continuing activity of the late orogenic wrench faults, such as the Steinberg fault. The strike-slip motion along these orogen-parallel deep faults started the pull-apart stage of the Vienna basin. In a transtensional regime, these faults may have served as conduits for the vertical hydrocarbon migration, whereas in the transpressional regime, these strike-slip faults may have actually sealed some of the reservoirs. The proper timing of generation and migration of hydrocarbons with respect to the tectonic regime on these faults seems to be critical for the accumulation and preservation of hydrocarbons in the reservoirs (Picha and Peters, 1998).

Some short-range migration from mature source rocks, e.g., the organic-rich Menilitic shales, into nearby reservoirs, e.g., the Kliwa Sandstones or the Paleocene–Eocene channelized sandstones, apparently also occurred in the Carpathian thrust belt. According to Lafargue et al. (1994), this type of migration was a significant factor in the deeper parts of the Eastern Carpathian thrust belt in Ukraine and Romania.

The Reservoirs

In Moravia, proven reservoirs have been found in the European foreland plate, the Neogene foredeep, the Carpathian thrust belt, and the Vienna basin. In the

European foreland plate, oils are reservoired in the fractured and weathered surface of the Precambrian granitic rocks, in the Devonian to Carboniferous carbonates, in the Jurassic clastics and carbonates, and in the clastics of the autochthonous Paleogene and the Neogene foredeep (Figure 9). Numerous sandstones are present in the Carpathian Flysch belt, which might have good reservoir properties, but so far, no significant accumulations of hydrocarbons have been found in these rocks in Moravia. Potentially, the Paleocene to Eocene channelized sandstones enclosed in the impermeable claystones in the external units of the Carpathian thrust belt seem to be the most promising targets. Similar sandstones (Ciezkowice Sandstones) are proven producers in the Polish Carpathians. Good, predominantly clastic reservoirs exist in the Vienna basin, of which the middle Miocene (Badenian and, to a lesser degree, Sarmatian) sands and algal (lithothamnian) limestones are the most prolific.

The character of these various reservoirs is further discussed in the context of the next section on the various types of hydrocarbon plays.

THE HYDROCARBON PROVINCES, PLAYS, AND FIELDS

During the long history of exploration, several distinct hydrocarbon plays have been identified in the Neogene foredeep, the autochthonous European plate, the thin-skinned Carpathian thrust belt, and the Vienna basin. Numerous oil and gas fields have been discovered in those various settings in the territory of Moravia. They are listed in Tables 2, 3, and 4 (shown on pages 136, 139, and 143, respectively) and their locations are shown in Figure 25.

The Neogene Foredeep

In comparison with the Molasse Basin of Bavaria and Austria, the Neogene foredeep in Moravia is relatively shallow. The average thickness of the Neogene strata ranges within a few hundred meters. Besides the biogenic gas, the foredeep does not have any generative potential of its own. The small- and middle-sized oil and gas fields (Figure 25; Table 2), found in the clastic reservoirs of the Neogene foredeep, were sourced predominantly from the older, organic-rich, and deeply buried formations. In northern Moravia, the gas fields were supplied from the underlying coal-bearing Upper Carboniferous strata; in central and southern Moravia, the hydrocarbons migrated into the reservoirs in the foredeep from the Jurassic and Paleogene source rocks buried deeply below the Carpathian thrust belt.

The largest field found in the Neogene foredeep in Moravia is the Dolni Dunajovice gas field, discovered in 1973. It is an antiformal structure associated with the Vestonice fault, which cuts through the crystalline basement and Jurassic strata into the lowermost Miocene deposits of the foredeep (Figure 26). The fault is interpreted as an antithetic Neogene fault related to flexural downwarping of the foredeep (Adamek, 1979) or as a foreland-type compressional and possibly transpressional structure associated with the propagation of compressional orogenic stresses in the foreland plate, the interpretation we prefer. The gas is reservoired in Eggenburgian basal sandstones draping over the basement structure and sealed by the overlying Eggenburgian Marls. More than 810 million m^3 (28.5 bcf) of gas had been produced from the field prior to its conversion into a gas storage in 1989. For more information on the Dolni Dunajovice field, see Kostelnicek et al. (2006). On the border with Austria, another gas field, the Alt Prerau–Novy Prerov, was found in clastic rocks of the lower Miocene and produced from the Austrian side.

The very shallow (50 m; 160 ft) oil fields Menin and Zatcany (Figure 25) contain heavy biodegraded oils in the Karpatian and Badenian sands. The fields are situated at the margin of the Nesvacilka paleovalley, which apparently served as a migration path from the generative centers below the Carpathian belt into the foreland. The heavy oils from the Menin–Zatcany fields were partly exploited by the surface mining operations.

Hydrocarbons have been also found in the coastal and lagoonal Karpatian and possibly Ottnangian (Salaj, 1996) sands deposited at the slopes of the crystalline basement elevations in central Moravia. These reservoirs are hydrodynamically connected with the fractured reservoirs in the crystalline basement rocks. Commercial reserves of hydrocarbons in these Karpatian sands have been found in the Zdanice, Mourinov, Korycany, Kostelany (Lubna) and Rusava fields. Small, noncommercial occurrences of hydrocarbons in the Karpatian strata not associated with the reservoirs in the crystalline basement rocks have been encountered in numerous wells, e.g., Kozusice-4, Mikulov-2, Jezov-3, and Rostin-2.

In northern Moravia, the gas fields in the Neogene foredeep are linked to the migration of gases from the underlying Upper Carboniferous coal-bearing strata of the Upper Silesian coal basin. The largest of them, the Horni Zukov field, has produced more than 1.1 billion m^3 (39 bcf) of gas. It is followed by the Pribor–Klokocov field (1.0 billion m^3; 35 bcf), the Choryne field (640 million m^3; 21 bcf), and the Stramberk field (600 million m^3; 21 bcf)

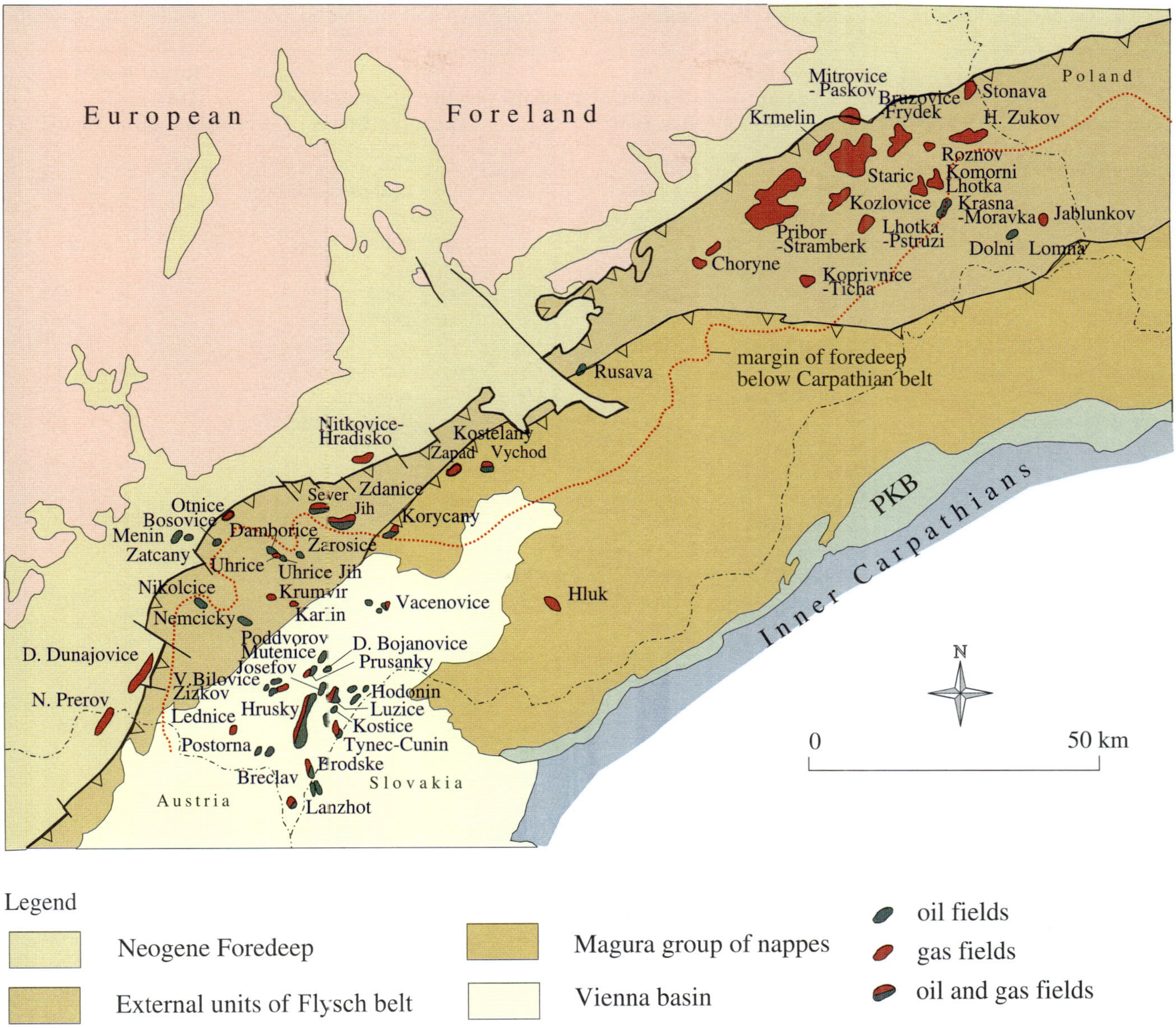

Figure 25. Map of oil and gas fields in the Neogene foredeep, the Western Carpathian thrust belt, the Vienna basin, and the European subthrust plate in Moravia. PKB = Pieniny Klippen Belt.

(Table 2). With the exception of the Horni Zukov field, which is reservoired in the lower Badenian, all the other fields mentioned above are located in the Karpatian reservoirs. The Horni Zukov and Stramberk gas fields have been converted into the gas storages, with potential capacity of 400 and 420 million m^3/yr (14.1 and 14.8 bcf/yr), respectively. Natural gas was also found in the Karpatian strata resting marginally on the Devonian to Lower Carboniferous carbonates in the Krasna oil field.

The Neogene foredeep, although heavily drilled, may still have some additional hydrocarbon potential, especially in the subthrust zone around the basement elevations and in the fault-related traps. In addition, the depositional fill and the slopes of the Bludovice and Detmarovice paleovalleys in the Upper Silesian basin have not been fully explored for gas.

The Carpathian Flysch Belt

The hydrocarbon potential of the Carpathian Flysch belt in the territory of Moravia is very limited (Table 2). However, the imposition of the wedge-shaped flysch belt over the European platform was critical for the

Table 2. Oil and gas fields in the Carpathian Neogene foredeep and the Flysch belt in Moravia. Location of fields in Figure 25.

Field Name	Year of Discovery	Age of Reservoirs	Type of Reservoir	Porosity (%)	Field Type	Production Status	Initial Reserves Oil (million bbl)	Gas (bcf)
Carpathian foredeep								
Bruzovice–Frydek	1952	Miocene	clastic	15–20	gas	coal degasation		5
		Carboniferous	clastic	7–15				
Choryne	1908	Karpatian	clastic	15–20	gas	conservation		21
		Carboniferous	clastic	7–15				
Dolni Dunajovice	1973	Eggenburgian	clastic	23–37	gas	gas storage		35.5
Horni Zukov	1915	lower Badenian	clastic	15–20	gas	gas storage		39.1
		Carboniferous	clastic	7–15				
Komorni Lhotka	2001	Miocene	clastic	15–20	gas	development ongoing		3.7
		Carboniferous	clastic	7–15				
Kozlovice–Lhotka	1975	Silesian unit	clastic	10–15	gas	development ongoing		0.9
		Miocene	clastic	15–20				
		Carboniferous	clastic	7–15				
Krmelin	1958	Miocene	clastic	15–20	gas	coal degasation		1.5
		Carboniferous	clastic	7–15				
Menin–Zatcany	1944	Badenian	clastic	30	oil	producing	1	
	1930	Karpatian	clastic	25–30				
Mitrovice–Paskov	1909	Miocene	clastic	15–20	gas	coal degasation		10
		Carboniferous	clastic	7–15				
Novy Prerov	1986	lower Miocene	clastic	18–25	gas	producing		0.76
Pribor Jih (Stramberk)	1965	Karpatian	clastic	15–20	gas	gas storage		21
		Carboniferous	clastic	7–15				
Pribor–Klokocov	1908	Karpatian	clastic	15–20	gas	conservation		35
	1912	Carboniferous	clastic	7–15				
Stonava	1952	Miocene	clastic	15–20	gas	coal degasation		1.5
		Carboniferous	clastic	7–15				
Staric–Liskovec–Sviadnov	1913	Miocene	clastic	15–20	gas	coal degasation		7
		Carboniferous	clastic	7–15				
Carpathian Flysch belt								
Hluk	1943	Bile Karpaty Flysch	clastic	14–24	gas	abandoned		0.4
Lhotka–Pstruzi	1975	Silesian unit	clastic	10–15	gas	producing		6
Roznov	1983	Subsilesian unit	clastic	10–15	gas	development ongoing		0.3
Koprivnice–Ticha	1982	Silesian unit	clastic	10–15	gas	producing		17.6
		Miocene	clastic	15–20				
		Carboniferous	clastic	7–15				

generation of hydrocarbons from source rocks located in the underlying platform. The predominantly impermeable rocks of the flysch belt also provide an important regional seal.

The marginal Pouzdrany unit, as presently known, does not have any economic hydrocarbon potential. The Oligocene organic-rich rocks of this unit are immature; however, the tectonic slivers of this unit buried deeper below the Outer Carpathian thrust belt, together with the remnants of the autochthonous Paleogene, may represent hydrocarbon source rocks.

Likewise, the Waschberg–Zdanice–Subsilesian unit has practically no hydrocarbon potential. Numerous wells, drilled both in the southern Zdanice sector and the northern Subsilesian sector, encountered practically no noticeable signs that oil and/or gas might be present in commercial quantities in these units. Some rocks of the Menilitic Formation are organic rich but remain mostly immature. Only limited parts of the Zdanice–Subsilesian unit buried deeper below the Silesian and Magura nappes may have generated some hydrocarbons. However, the impermeable strata of the lower members of this

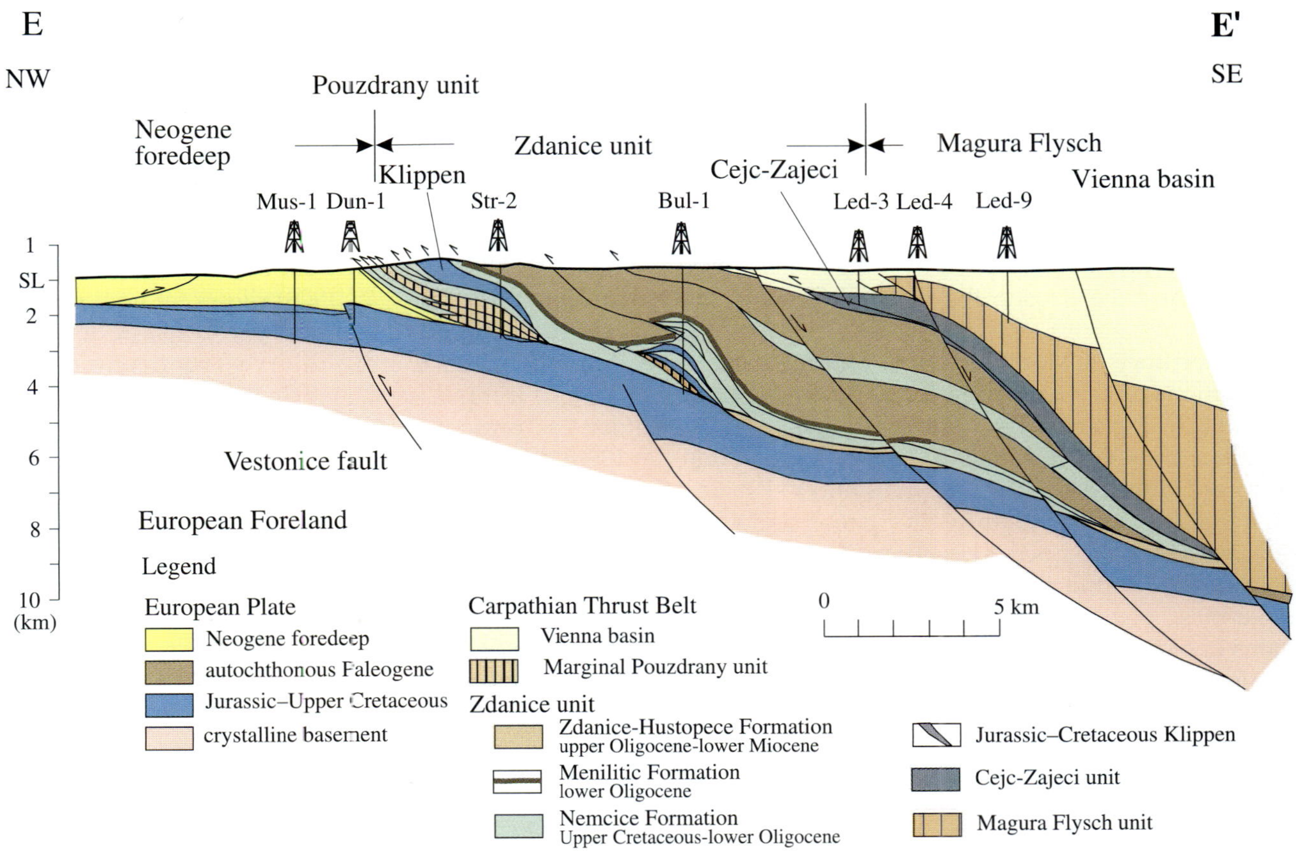

Figure 26. Cross section EE′ (location in Figures 3, 7) through the Neogene foredeep, the frontal units of the Western Carpathian thrust belt, the Vienna basin, and the underlying European plate in southern Moravia. Note the structural position of the Jurassic–Cretaceous klippen in the Waschberg–Zdanice unit and the relationship of the allochthonous Pouzdrany unit to the autochthonous Paleogene. The Vestonice fault is interpreted as a compressional or transpressional structure. It may have originated as a normal fault during the Jurassic rifting and was activated again as a reverse fault during the Alpine convergency in the early Miocene (Eggenburgian). The Dolni–Dunajovice gas field is reservoired in the lower Miocene sands draping over the antiformal structure associated with the hanging wall of the Vestonice fault. Modified from Picha and Stranik (1999). Stratigraphic records of some wells shown on the cross section are reported in Appendix 1.

unit represent an excellent regional seal, critical for the lateral migration of hydrocarbons from the deep generative kitchens into the reservoir in the foreland and their preservation in the otherwise open reservoirs, such as the fragmented and weathered crystalline basement rocks.

Some Paleocene to Eocene sandstones and conglomerates of the Nemcice (Submenilitic) Formation in the Cejc–Zajeci unit possess good porosities and may be considered as reservoir rocks. The thickness of the individual sand bodies ranges from several meters up to several tens of meters; however, their lateral extent seems to be limited. These coarse clastics enclosed in the impermeable shaly deposits may actually represent the deep-water channels similar to those explored in the North Sea (e.g., Alba and Frigg fields) and elsewhere along the present continental margins. Recognition of these bodies at depth and outlining of closures in the structurally complex (duplex-type) Cejc–Zajeci thrust unit, however, would require 3-D seismic survey. A shallow well drilled into the Eocene sands west of the village of Kobyli encountered several gas shows (88% methane and 9% nitrogen) (Picha et al., 1968). The deep Kobyli-1 well drilled nearby, however, did not register any hydrocarbon shows in the entire Cejc–Zajeci unit (Picha et al., 1971). The Paleocene to Eocene sandstones in the Cejc–Zajeci unit bear many similarities with the Ciezkowice Sandstones of the Silesian unit in Poland.

The Silesian unit has been the most significant oil producer in the Polish Western Carpathians. It contains both significant source rocks in the Lower Cretaceous strata and the lower Oligocene Menilitic Formation and numerous reservoir sandstones. Among them, the Paleogene Ciezkowice Sandstones are the most prolific (Karnkowski and Ozimkowski, 1998; Dziadzio et al., 2006). In numerous smaller fields, they produced almost half of the total oil recovered in the entire Polish Carpathians (Wdowiarz, 1985; Slaczka, 1996). Additional reservoirs are present in the Lower Cretaceous Tesin–Hradiste, Verovice, and Lhoty formations, the Upper Cretaceous Istebna Formation (locally called Czarnorzeki beds), and in the Oligocene Kliwa Sandstones and the Krosno Formation (Karnkowski, 1993).

In the territory of Moravia, so far, only two small accumulations of gas have been found in the Silesian unit (Table 2). The Roznov-1 well encountered 45 gas-bearing horizons, 0.8–11.4 m (2.6–37.4 ft) thick, in the Silesian and Subsilesian units. In the Lhotka–Pstruzi field, in addition to the subthrust Karpatian and Carboniferous reservoirs, accumulations of natural gas have also been found in the overlying Silesian unit. Despite these poor results, the Silesian unit possibly remains one of the more promising subjects of exploration in the Flysch belt of Moravia.

As documented by numerous wells, the hydrocarbon potential of the Magura flysch in Moravia seems to be very limited. It has only minor source rocks in the Zlin Formation of the Raca unit and the Hluk facies of the Bile Karpaty unit, and poor reservoirs elsewhere. The porosity of sandstones ranges from 7 to 12% in the unweathered and more than 12% in the weathered rocks (Ondra and Hanak, 1992). Besides seepages and oil and gas shows in wells, only two small fields have been found so far in the Magura flysch (Table 2; Figure 25). A small gas field, Hluk, located in the Bile Karpaty unit at a depth of 200–600 m (660–2000 ft), has been exploited since 1943. In the abandoned Vacenovice field (Table 4, shown on page 143), oil and gas were produced from the weathered surface of the Paleogene flysch deposits underneath the depositional fill of the Vienna basin. On the positive side, the Magura nappe represents a significant regional seal.

In our opinion, the duplexes of Paleogene strata of the external units buried below the major Silesian and Magura nappes remain the most promising targets of exploration in the entire Western Carpathian Flysch belt.

The Subthrust Plays

The hydrocarbon potential of the European plate was greatly enhanced by its burial below the accretionary wedge of the Carpathian thrust belt. The overthrusting by the thrust belt not only enhanced the maturation of source rocks and generation of hydrocarbons but also improved the trapping mechanism by providing a good regional seal (Picha, 1996). The increasing thickness of the thrust belt toward southeast, however, limits the extent of the subthrust plate accessible to exploration. Considering the economic and technical feasibility of drilling to a depth of 6–7 km (3.7–4.3 mi), the width of the drillable subthrust zone in Moravia typically would not exceed 50 km (31 mi) (Figure 27). On the positive side, the relatively low heat flow in the Carpathian foreland (Cermak, 1975, 1979) allows for liquid hydrocarbons to be generated and preserved at depths of more than 5 km (3 mi) and dry gas at depths of 9 km (5.6 mi) (Stranik et al., 1993). Thus, both oil and gas may be present in the deep subthrust plays (Picha, 1996).

Hydrocarbons have been found in the Paleogene, Jurassic, and Paleozoic strata and the crystalline basement of the subthrust European plate (Table 3; Figure 25).

The Nesvacilka and Vranovice Paleovalleys and Their Paleogene Fill

The large Nesvacilka and Vranovice paleovalleys represent a complex hydrocarbon province with multiple potential hydrocarbon plays. So far, only the more thoroughly explored Nesvacilka paleovalley yielded commercial discoveries. The Nesvacilka paleovalley is located in a broader structure of the northwest–southeast-trending Nesvacilka graben, whose origin is most likely related to the Jurassic rifting (Figure 13). The Paleogene fill not only holds source rock and reservoirs but also functions as a regional seal for hydrocarbon accumulations located in the fractured crystalline rocks and the Paleozoic and Jurassic carbonate and clastic reservoirs situated along the northeastern more gently dipping side of the Nesvacilka graben and the northern side of the Nesvacilka paleovalley (Picha, 1996) (Figure 12) (see also the account by Kostelnicek et al., 2006). Last but not least, the paleovalleys apparently also served as conduits for migration of hydrocarbons from the generative kitchen located deeper below the thrust belt into shallower reservoirs in the foreland.

The autochthonous Paleogene fill of the large Nesvacilka and Vranovice paleovalleys possesses all the main components of a petroleum system: organic-rich source rocks, deep-water channelized reservoirs enclosed in impermeable rocks, and proper depth of burial below the thickening wedge of the thrust belt (Figure 28). The best reservoirs are the meandering channels of subsea fans similar to those explored for hydrocarbons in deep-water turbiditic deposits elsewhere along the passive continental margins of South America, Africa and Gulf of Mexico; only the

Table 3. Oil and gas fields in the European platform underneath the Outer Carpathian belt in Moravia. Location of fields in Figure 25.

Field Name	Year of Discovery	Age of Reservoir	Type of Reservoir	Porosity (%)	Field Type	Production Status	Initial Reserves Oil (million bbl)	Initial Reserves Gas (bcf)
Bosovice	1990	Paleogene	clastic	15–25	oil	development ongoing		
Damborice	1986	Paleogene	clastic	15–23	oil and gas	producing	12	5.5
		Jurassic	carbonate	10				
		Jurassic	clastic	15–25				
Dolni Lomna	1985	Silesian unit	clastic	10–15	oil	development ongoing	0.15	
		crystalline	paragneiss	3–5				
Karlin	1991	Paleogene	clastic	15–20	gas	producing	0.15	3
Korycany	1978	lower Miocene	clastic	15–18	oil and gas	producing	0.2	3
		crystalline	granodiorite	3–5	oil and gas	producing		
Kostelany Vychod (East)	1968	lower Miocene	clastic	15	oil and gas	producing	1	30
		crystalline	granite	3–5				
Kostelany Zapad (West)	1971	lower Miocene	clastic	15	gas	producing		24.9
		crystalline	granite	3–5				
Krumvir	1997	Paleogene	clastic	15–25	gas and oil	producing	0.3	2.7
Krasna–Moravka	1980	lower Miocene	clastic	15–20	oil and gas	development ongoing	0.3	0.73
		crystalline	carbonate	4				
Nemcicky	1978	Devonian	carbonate	3–5	oil and gas	producing(?)	0.01	0.03
Nikolcice	1969	Paleogene	clastic	15–23	oil and gas	producing		0.01
Nitkovice–Hradisko	1971	Devonian	carbonate	3–5	oil and gas	producing		
Otnice	1994	Paleogene	clastic	15–25	gas	development ongoing		1.5
Rusava	1976	lower Miocene	clastic	12	gas	abandoned		0.1
		crystalline	gabbro	2–5				
Uhrice	1982	Paleogene	clastic	15–23	oil and gas	gas storage	1.5	7.9
		Jurassic	carbonate	10				
		Jurassic	clastic	15–25				
	1978	Devonian	carbonate	3–5	oil and gas			
Uhrice Jih (South)	2001	Jurassic	clastic	21	oil and gas	producing	2.5	2.3
Zarosice	2001	Jurassic	carbonate	10	oil and gas	producing	3.8	3.5
Zdanice Jih–Zdanice	1984	lower Miocene	clastic	25	oil and gas	producing	3.5	4.5
Sever–Kloboucky	1973	crystalline	diorite–tonalite	3–6				

dimensions of those features in Moravia seem to be smaller.

The hydrocarbon potential of the autochthonous Paleogene fill in the Nesvacilka paleovalley has been only partly explored (Brzobohaty, 1993; Brzobohaty et al., 1996). Numerous gas shows and some oil shows have been encountered in several wells, e.g., Bosovice-1, Hosteradky-1, Milesovice-1, Tesany-1, Nasedlovice-1, and Susice-1, but so far, only two smaller gas-condensate-oil fields, Krumvir and Karlin, have been found and put into production. The Krumvir-2 well, drilled to the total depth of 3600 m (11,811 ft), encountered a 30-m (100-ft) section of good-quality gas- and oil-bearing sand with 15–25% porosity. An initial drillstem test flowed 750 bbl of oil/day of 30° API, low-sulfur crude oil against a restricted 0.22-in. (5.5-mm) choke. High-pressure, gas-bearing sand horizons have also been encountered in the Karlin-1 well located further east of the Krumvir field (Benada and Blazej, 1991). The deeper zones of the Nesvacilka valley fill with a higher proportion of sandstones remain unexplored. Some measurements, however, indicate that the quality of these reservoirs may deteriorate with the increasing depth. The Vranovice paleovalley has been explored only marginally,

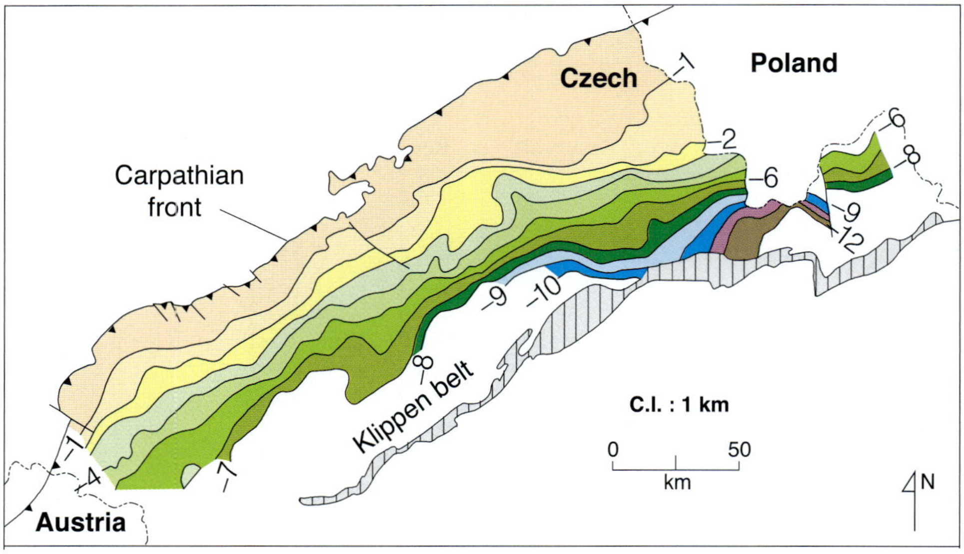

Figure 27. Depth (in km) to the base of the Outer Carpathian thrust belt in Moravia. Modified after Picha (1996).

without any apparent success. Application of new technologies, such as the 3-D seismics, may provide a tool for better assessment of the geometry and quality of reservoirs and a lowering of the exploration risk. We believe that small- and middle-sized oil and gas-condensate fields can be found in the predominantly stratigraphic traps, especially in the deeper parts of the paleovalley fill. Moreover, existence of channelized subsea fans may be anticipated at the mouths of these large submarine canyons. Presently, these Paleogene fan deposits are either deeply buried below the Carpathian thin-skinned nappes or detached and integrated into the allochthonous belt itself. If accessible by drilling, these potential fans may yet become targets of a future exploration.

As such, the autochthonous Paleogene remains an interesting object of the subthrust exploration in southern Moravia. In many aspects, including the reservoirs and generation and migration of hydrocarbons, the Paleogene play of southern Moravia resembles the hydrocarbon habitat of the Molasse Basin in Austria and Bavaria. The structural setting of these two hydrocarbon provinces, however, is different; it is predominantly a subthrust play in Moravia, whereas in

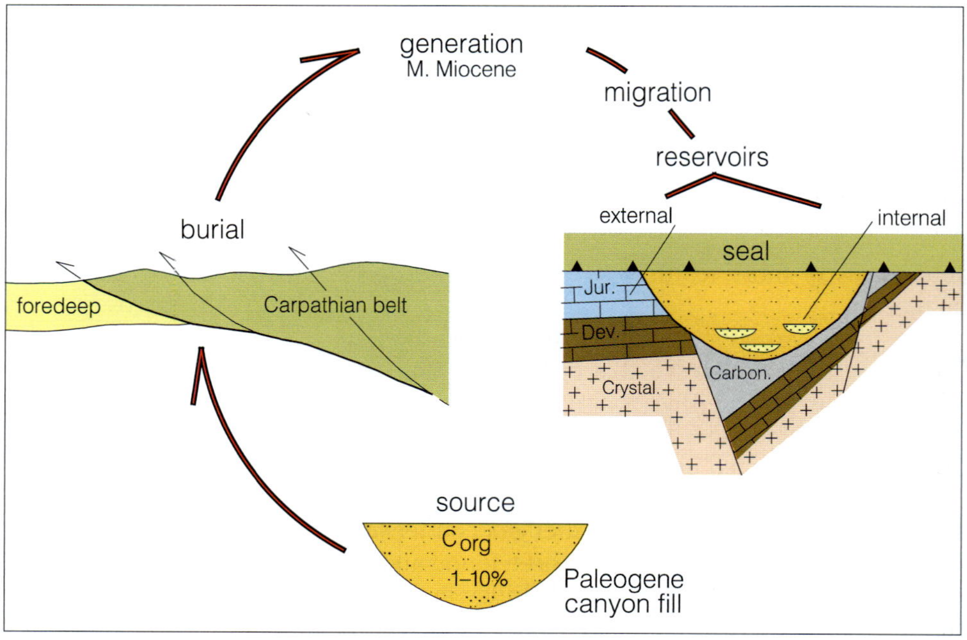

Figure 28. Hydrocarbon habitat of the Nesvacilka graben and the Paleogene canyon system (Picha, 1996).

Austria and Bavaria, the productive Paleogene to Neogene reservoirs found in the Molasse Basin are located in front of the Alpine thrust belt.

The Jurassic Play

By far, the largest subthrust oil and gas accumulations in Moravia have been found in the erosionally isolated blocks and pinnacles of Jurassic strata in the northeastern slopes of the Nesvacilka graben and the Nesvacilka paleovalley, buried below the autochthonous Paleogene deposits and the Zdanice nappe (Figures 7, 12, 13; Table 3). The hydrocarbons are reservoired in the clastic rocks of the basal Gresten Formation (Damborice oil field, Uhrice gas field, and Uhrice South oil and gas field) or in the Vranovice Limestones and dolomites (Zarosice oil field). The porosities of the Gresten sandstones may reach as much as 25%. Additional hydrocarbons have been found in adjacent Paleozoic carbonates and clastics. The Damborice field, holding about 30 million bbl of oil in place, represents the largest subthrust discovery in the Czech Republic; more than 70% of the total present production in the country comes from this field. The Uhrice gas field with initial reserves of 224 million m^3 (7.9 bcf) of natural gas has been exploited and converted into a gas storage. All these fields were sourced from the Jurassic organic-rich Mikulov Marls. Because no generative kitchen is present in the proximity of these fields, the hydrocarbons must have migrated laterally updip from the zones of the subthrust plate buried deeper below the Carpathian thrust belt and the Vienna basin. Similar prospects in erosional blocks and pinnacles may also exist in the Vranovice paleovalley, which has not been explored for this type of play. Shows of hydrocarbons in Jurassic strata have been also encountered in the deep wells Dolni Dunajovice-1, Hrusovany-1, Nikolcice-1, Kobyli-1, and Nove Mlyny-2.

Within the Dyje–Thaya depression, large subthrust antiformal structures, such as Tynec, Holic, and Lednice, have been identified on seismic data. Estimated depths to the tops of these structures range from 4000 m (13,000 ft) for the Lednice structure to 6000–7000 m (19,700–23,000 ft) for the Tynec and Holic structures. These structures apparently formed as tilted blocks and horsts during the Jurassic rifting, and at least some of them, e.g., Tynec structure, were further activated during the late orogenic strike-slip faulting. In southern Moravia, these structures are situated in the generative zone of the Jurassic and Paleogene source rocks. The Tynec deep structure (Figure 20, section DD′) underlies the shallow oil and gas fields in the Vienna basin. Biomarker studies (Picha and Peters, 1998) indicate that the oil from the Tynec-34 well has a mixed Jurassic and Paleogene signature. Based on the interpretation of limited seismic data, the size of the Tynec structure has been estimated to be about 70 km^2 (27 mi^2). The seismically better defined Holic structure, located further northeast mostly on the territory of Slovakia, may cover as much as 90 km^2 (35 mi^2). The deep Lednice (16 km^2; 6.2 mi^2) structure interpreted from seismic, geomagnetic, and gravity data (Ciprys and Thon, 1990) is located below the western margin of the Vienna basin in southern Moravia. Potential reservoirs in these structures may be found in the Paleogene clastics (channels and fans), Jurassic clastics and carbonates, Devonian dolomites, and the Precambrian crystalline rocks. All these rocks are proven reservoirs in the shallower zones of the subthrust platform. The thrust belt with predominantly shaly sequences on its bottom provides an excellent seal. With respect to a great depth of these structures, most likely, gas and condensate rather than oil may be found in these prospects.

In the territory of Austria, the gas fields Klement and Hoflein and the oil field Grunau have been found in Jurassic and Cretaceous clastic and carbonate reservoirs in the shallower zone of the foreland plate below the Waschberg and Flysch belt nappes (Grun, 1984; Zimmer and Wessely, 1996). Very deep, large, presumably rift-related structures, such as Maustreng–Zistersdorf and Aderklaa, have been partly explored by deep drilling without any commercial success (Wessely, 1990; Milan and Sauer, 1996).

The presence of potential reservoirs in all these deep structures is considered to be the most critical risk factor. However, because of their large size and location in the generative kitchen for hydrocarbons, these structures remain tempting exploration targets.

Fractured Malmian Marls, as much as 1500 m (4900 ft) thick, if overpressured, could also be considered as unconventional producers of natural gas (tight-gas play). Gas generation apparently still continues in these organic-rich source rocks.

The Jurassic play, especially in combination with the Nesvacilka and Vranovice paleovalleys and their impermeable Paleogene fill, is undoubtedly the most promising subthrust exploration play in Moravia. The chances of finding additional fields in this play remain quite high.

The Paleozoic Play

Only small accumulations of hydrocarbons have been found in the Paleozoic sequences: the Middle Devonian clastics, the Upper Devonian to the Lower Carboniferous karstified and fractured limestones and dolomites, and the Lower Carboniferous Culm deposits (Thonova and Benada, 1990). In northern Moravia, a small Krasna field (Figure 25), reservoired in the Devonian carbonates and crystalline basement rocks, has been developed and put on production. Other noncommercial Paleozoic subthrust accumulations of hydrocarbons have been encountered in the wells at Nitkovice (small gas field), Slavkov, Nemcicky, and Drazuvky, all located in central Moravia. Some of these occurrences are just minor extensions of larger fields, hydrodynamically connected with other more productive main reservoirs, e.g., the crystalline basement rocks (Zdanice field) or Jurassic clastics and carbonates (Damborice field).

In the Upper Silesian basin, coal gases accumulated in elevations of the Carboniferous paleorelief. They are reservoired in the weathered surface of the Upper Carboniferous rocks, as well as in the overlying Miocene deposits of the Carpathian foredeep. Additional coal gases are found dissolved in waters of the lower Badenian detritic fill of the Bludovice and Detmarovice erosional depressions.

The Paleozoic strata of the Carpathian foreland were deformed during the Hercynian orogeny. Foreland-type compressional structures of the frontal zone of the southwest–northeast-trending Hercynian belt in Moravia extend far below the Carpathian thrust belt (Figure 7). These Hercynian structures with potential reservoirs in the Devonian and Carboniferous clastics and carbonates did exist prior to the development of the Mesozoic Tethyan rifted margins and the Carpathian thrust belt. Unless disrupted by erosion and further tectonism, these structures, buried and possibly sealed by the impermeable rocks of the Neogene foredeep and the Carpathian thrust belt, may have become receptacles for hydrocarbons generated during the Carpathian tectogenesis (Picha, 1996). The reservoir properties of the Paleozoic rocks may actually improve eastward toward the stable foreland of the Hercynian belt, which, during the Devonian to Lower Carboniferous, was the site of a shallow-marine carbonate sedimentation favorable for generation and preservation of porosities. At least some of the Hercynian compressional structures might have been reactivated during the Late Cretaceous–Paleogene (Laramide) and/or the Miocene late orogenic mobilization of the European foreland plate.

Despite the lack of commercial discoveries to date, it is still possible that accumulation of hydrocarbons may still be found in Paleozoic structures deeply buried below the Carpathian thrust belt. Some interpretations of seismic data indicate a potential existence of Devonian and Carboniferous reefal limestones and dolomites in the subthrust plate at depths greater than 5000 m (16,400 ft). However, any further exploration of the Paleozoic potential will require application of more advanced seismic techniques combined with a thorough structural and stratigraphic analysis of the potential prospects.

Elevations of the Crystalline Basement

Several oil and gas fields, such as Zdanice, Kloboucky, Lubna, Kostelany, Krasna, and Dolni Lomna, have been found in elevations of the crystalline basement (buried hills) below the Neogene foredeep and the frontal zones of the thin-skinned Carpathian thrust belt in Moravia (Figure 8; Table 3). These accumulations are reservoired in the weathered and fractured surface zones of granitic rocks and in the overlying lower Miocene (Karpatian) sands of the Neogene foredeep sealed by the impermeable deposits of the Zdanice–Subsilesian nappe (Krejci, 1993; Blizkovsky et al., 1994). The Precambrian crystalline rocks have low porosities, on average 2–4%; however, the thickness of the oil-saturated zone may reach several tens up to several hundred meters. The overlying lower Miocene sands are only a few tens of meters thick; their porosities range from 15 to 25%. These relatively shallow, less than 1500-m (4900 ft)-deep oil accumulations are charged by a long-range migration from the Jurassic and Paleogene organic-rich source rocks in the deeper zones of the European platform and the Carpathian thrust belt.

The largest of these crystalline basement-reservoired accumulations, the Zdanice field located in central Moravia (Figure 25), holds more than 35 million bbl of 19–33° API oil in place, which, however, is difficult to extract (Krejci, 1993; Buchta and Dohnal, 1996). In 1998, the entire production from the Zdanice elevation was only about 300 bbl/day. Chances are that the rate of production may improve by application of more advanced technologies, such as horizontal drilling or steam flooding.

The Vienna Basin

The Vienna basin, superimposed on the Carpathian thrust belt in southern Moravia and northeastern Austria (Figures 1, 3), has been one of the most prolific producers of hydrocarbons in the entire Carpathian region. More than 800 million bbl of oil has been produced from the numerous fields of the Vienna basin, mainly in the territory of Austria. The majority of this production has come from the giant Matzen field (Fuchs and Hamilton, 2006).

The oils of the Vienna basin were sourced predominantly from the Jurassic organic-rich marls and shales and, to a lesser degree, from the Paleogene strata, both found in the underlying European platform (Ladwein, 1988; Ladwein et al., 1991; Picha and Peters, 1998). The vertical migration through the Flysch belt into the reservoirs in the Vienna basin was apparently enabled by the continuing tectonic activity along the major wrench faults (Figure 24, section AA′). The potential source rocks in the Vienna basin, although locally buried to a depth of 4 km (2.4 mi), would remain mostly immature. Numerous structural and stratigraphic traps were created during the complex structural and depositional history of the Vienna basin. The middle Miocene (Badenian) sands represent the best reservoirs, but oil and gas accumulations have been also found in the lower Miocene and Sarmatian deposits.

Numerous smaller fields have been found in the Moravian part of the Vienna basin (Table 4; Figure 25). The Hrusky–Josefov field, with initial reserves of 15.5 million bbl of oil and 90 bcf of gas, is the largest. It has been partly converted into a gas storage. The geology and hydrocarbon potential of the Vienna basin is more thoroughly discussed by Seifert (1996), Arzmuller et al. (2006), and Fuchs and Hamilton (2006).

The Key Role of the Dyje–Thaya Depression in the Formation of Petroleum Systems

The existence of the Dyje–Thaya depression and its evolution during the Mesozoic and Cenozoic times apparently was a key factor in the development of several petroleum systems in southern Moravia and

Table 4. Oil and gas fields in the Moravian part of the Vienna basin. Location of fields in Figure 25.

Field Name	Year of Discovery	Age of Reservoir	Type of Reservoir	Porosity (%)	Field Type	Production Status	Initial Reserves Oil (million bbl)	Initial Reserves Gas (bcf)
Breclav	1946	Badenian	clastic	15	oil and gas	producing	0.25	1.9
Brodske	1951	Badenian	clastic	20	oil	producing	2.5	1.9
Dolni	1989	Sarmatian	carbonates	15–25		producing		
Bojanovice		Badenian	clastics	20–25	oil and gas	gas storage		
Hodonin	1920	Sarmatian	clastic	15–20	oil and gas	producing	2.5	0.7
		Badenian	clastic	20				
		lower Miocene	clastic	18–20				
Hrusky–Josefov	1959	Sarmatian	clastic	15–25	oil and gas	producing	15.5	90
		Badenian	clastic	20				
		Karpatian	clastic	18–20				
Kostice	1953	Sarmatian	clastic	15–25	oil and gas		2.5	2
		Badenian	clastic	20–25				
Lanzhot	1957	Sarmatian	clastic	20	oil and gas	conservation	1	4.5
		Badenian	clastic	15				
Lednice–Valtice	1979	Badenian	clastic	20	gas	producing		3
		Badenian	carbonates	5–10				
Luzice	1921	Sarmatian	clastic	15–20	oil and gas	producing	2.75	10
	1944	Badenian	clastic	20				
		lower Miocene	clastic	18–20				
Poddvorov–Mutenice	1951	Sarmatian	clastic	20–25	oil and gas	producing	3.5	30
		Badenian	clastic	20–25				
Postorna	2001	Badenian	clastic	15	oil	exploration	0.45	
Prusanky	2001	Sarmatian	clastic	20–25	gas	producing		3.5
Tynec–Cunin	1959	Raca unit	clastic	15–20	oil and gas	producing	1.4	3.6
	1945	lower Miocene	clastic	15–18				
Vacenovice	1930	Sarmatian	clastic	20	oil and gas	abandoned	0.4	0.6
		Raca unit	clastic	15–20				
Velke Bilovice–Zizkov–Podivin	1944	Sarmatian	clastic	20	oil and gas	producing	0.4	30.8
		Badenian	clastic	20–25				

northeastern Austria. The existing evidence indicates that this depression was formed or at least activated during the Jurassic rifting and repeatedly uplifted in the Lower Cretaceous and in the Late Cretaceous–early Paleogene. Prior to the Jurassic rifting, the area of the depression was situated in the frontal zone of the late Paleozoic Hercynian orogenic belt. The potential compressional Hercynian structures, still possibly evident on some regional sections (Figure 20, section DD′), might have been reactivated during the subsequent tectonic events. During the Laramide uplifting of the European foreland at the Cretaceous–Paleogene transition, two large paleovalleys were cut inside the depression and later filled with the deep-water Paleogene deposits. In the early Miocene, the depression was mostly overthrust by the thin-skinned Carpathian belt. The Jurassic and Late Cretaceous strata were partly detached from the autochthonous cover of the European platform and integrated into the Waschberg sector of the thrust belt as the Outer klippen of Ernsbrunn and Pavlov Hills. Finally, in the middle Miocene, the pull-apart phase of the Vienna basin formed along the orogen-parallel strike-slip faults within the confines of the depression. Any interpretation of structural history and resulting structural architecture of the area, especially at deeper levels below the Carpathian thrust belt, thus remains a daunting task requiring a four-dimensional analysis (including timing).

Within the structural limits of the depression, thick, organic-rich marls accumulated in the Late Jurassic, and the Paleogene organic-rich deposits filled two paleovalleys and submarine canyons. After being buried below the thrust belt and the Vienna basin, these source rocks matured, and the Dyje–Thaya depression became a world-class generative center, from which the hydrocarbons migrated both vertically and horizontally into various reservoirs in the Vienna basin and the subthrust foreland, respectively. More than 850 million bbl of oil and 4 trillion cubic feet of gas have been produced from various stratigraphic and structural levels in the realm of this depression. With respect to its limited size, the Jurassic Dyje–Thaya depression may be regarded as one of the most productive provinces in the world. Although heavily drilled to the depth of about 3000 m (10,000 ft), the recent discoveries, especially in the subthrust settings, indicate that the Dyje–Thaya depression still remains one of the most promising places for exploration in the entire Carpathian region.

CONCLUDING REMARKS

We have attempted to present a comprehensive and up-to-date account on the geology and the hydrocarbon potential of the Carpathian thrust belt and its foreland in the territory of the Czech Republic. Some readers may find it too detailed, others too short and superficial. In our account, we emphasized those aspects of geology that are specific for the region, critical for integration of local geology into a wider framework of the Alpine–Carpathian region, and important for understanding the hydrocarbon potential. However, we have also included more detailed description of stratigraphy, including logs of numerous deep wells, which may be useful to future students of Carpathian geology. Several new concepts and original ideas, e.g., about the functioning of the Western Carpathian transfer zone, the existence of the Dyje–Thaya depression, and the foreland depositional setting, have been broadly formulated in this chapter for the first time.

In our deliberation, we tried to be objective and present a balanced picture of interpretations and ideas. Given the enormous amount of published and unpublished data, we did not have a choice but to be selective. Unintentionally, we may have omitted some important papers or interpretation, and we apologize for it. Despite of these challenges, we hope that the article will provide useful information to geologists interested in Carpathian geology and enhance further discussions and additional work on some of the controversial issues. Last but not least, we anticipate that the comprehensive account on the hydrocarbon systems may encourage further exploration for oil and natural gas in this prolific and still promising region.

ACKNOWLEDGMENTS

We thank the Czech Geological Survey for technical and financial support of the publication and the Moravian Oil Company for the financial contribution for printing the colored figures. We thank the AAPG, Springer Verlag, and the Geological Society (London) for allowing us to reprint illustrations from previous publications. We also thank R. Brzobohaty and P. Palensky for the constructive reviews of the Neogene stratigraphy, J. Hladil for contributing useful remarks on the geology of the crystalline basement and its Paleozoic cover, J. Adamek for updating the stratigraphy of the Jurassic platform and Neogene foredeep, N. Oszczypko and A. Slaczka for reviewing the Carpathian foredeep and the Carpathian Flysch belt from the Polish perspective, G. Wessely and W. Hamilton for reviewing the manuscript from the Austrian perspective, and H. Repkova and F. Tamo for drafting the illustrations.

APPENDIX 1: SELECTED DEEP WELLS IN THE MORAVIAN CARPATHIANS (CZECH REPUBLIC)

Branky - 1 (Bra-1)

0-378 m	Silesian and Subsilesian units
--------------	basal overthrust of the Flysch nappes
-748 m	lower Miocene (Karpatian) of the Carpathian foredeep
-1625 m	Lower Carboniferous (Visean) - Culm facies
-2372 m	Devonian (Frasnian) to Lower Carboniferous (Visean) - Carbonate facies
-2427 m	Devonian (Givetian) - basal clastics
-2540 m	crystalline basement (metamorphites)

Bulhary - 1 (Bul-1)

0-32 m	Badenian Vienna basin (transgression)
-1192 m	Zdanice-Hustopece formations (Waschberg sector)
-2295 m	Menilitic and Nemcice formations (imbricated) (Waschberg sector)
-2391 m	Mikulov Marls, Jurassic
--------------	overthrust
-2688 m	Nemcice Formation (Waschberg sector)
-2706 m	Mikulov Marls, Jurassic with incorporated Boudky Marls of the Pouzdrany unit
--------------	overthrust
-3119 m	Nemcice Formation (Waschberg sector)
--------------	overthrust
-3126 m	Pouzdrany Marls of the Pouzdrany unit
--------------	overthrust
-3231 m	Nemcice Formation (Waschberg sector)
--------------	basal overthrust of the Flysch nappes
-3500 m	Mikulov Marls (autochthonous Jurassic)

Bystrice Pod Hostynem - 1 (Bys-1)

0-400 m	Fore-Magura unit	
	-47 m	Krosno Formation
	-400 m	Submenilitic Formation
--------------	overthrust	
-691 m	Silesian unit	
	-639 m	Krosno Formation
	-691 m	Roznov Formation
----------------	overthrust	
-1002 m	Silesian(?) or Waschberg - Zdanice - Subsilesian units(?) (Zdanice sector)	
	-915 m	Zdanice-Hustopece Formation
	-1002 m	Nemcice Formation
-----------	basal overthrust of the Flysch nappes	
-1242 m	lower Miocene (Karpatian) of the Carpathian foredeep	
-1897 m	Lower Carboniferous (Culm facies)	
-2600 m	Devonian (Frasnian) - Lower Carboniferous (carbonate facies)	

Choryne - 9 (Cho-9)

-914 m	Lower Carboniferous	
-1595.9 m	Devonian	
	-961.7 m	Lisen Formation (Krtiny Limestones)
	-1471 m	Macocha Formation (Lazanky and Vilemovice limestones)
	-1595.9 m	Basal clastics (Devonian)
-1711.6 m	crystalline basement (metamorphite)	

D. Lomna - 3 (D.Lom-3)

0-1639 m	Silesian unit (Godula subunit)
-1919 m	Subsilesian unit (Submenilitic Formation)
-1962 m	lower Miocene (Karpatian)
-2151 m	crystalline basement (metamorphite)

Gottwaldov - 2 (Got-1)

0-3893 m	Raca unit of the Magura Group of nappes	
	-1340 m	Vsetin Member of Zlin Formation
	-1735 m	Ujezd Member of Zlin Formation
	-1980 m	Beloveza Formation
	-3835 m	Solan Formation
	-3893 m	Kaumberg Formation
--------------	overthrust of the Magura Group of nappes	
-4215 m	Fore-Magura unit	
--------------	basal overthrust of the Flysch nappes	
-4725 m	lower Miocene (Karpatian) of the Carpathian foredeep	
-4850 m	crystalline basement (metamorphite)	

Hnojnik - 1 (Hnoj-1)

0-284 m	Silesian unit

-890 m Subsilesian unit
basal overthrust of the Flysch nappes
-942 m lower Miocene (Karpatian) of the Carpathian foredeep
-1100.5 m Upper Carboniferous (Namurian A), Ostrava Formation
-1100.5 m Lower Carboniferous (Toumaisian) to Upper Carboniferous (Namurian), Culm facies
-1214 m Devonian (Eifelian) to Lower Carboniferous (Visean), Carbonate facies

Holesov - 1 (Hol-1)

0-272 m Silesian unit (Roznov Formation)
-------------- basal overthrust of the Flysch nappes
-728 m lower Miocene (Karpatian) of the Carpathian foredeep
-907 m Macocha Formation (Devonian)
-924 m crystalline basement (Diorite)
overthrust
-940 m Macocha Formation (Devonian)
-995 m basal clastics (Devonian)
-1050 m crystalline basement (Diorite)

Hrusovany - 1 (Hr-1)

-561 m Carpathian foredeep
    -169 m Badenian
    - 490 m Karpatian
    - 534 m Eggenburgian to Ottnangian
- 742.5 m - 561 m basal clastics, Eggenburgian
- 816 m autochthonous Mesozoic, Altenmarkt, and Nikolcice formations
crystalline basement (Granodiorite)

Jablunka - 1 (Jab-1)

0-2505 m Raca unit of the Magura Group of nappes
    -100 m Beloveza Formation
    -1276 m Solan Formation
    -2505 m Kaumberg Formation
------------ overthrust of the Magura Group of nappes
-2895 m Fore-Magura unit
------------ basal overthrust of the Flysch nappes
-3950 m Upper Carboniferous (Namurian A) - Ostrava Formation
-5405 m Lower to Upper Carboniferous (Visean - Namurian A) - Culm facies
    -5283 m Hradec-Kyjovice Formation
    -5405 m Moravice Formation
-6278 m Devonian to Lower Carboniferous (Givetian - Visean) - carbonate facies
    -5836 m Lisen Formation (Krtiny and Hady - Ricky Limestones)
    -6278 m Macocha Formation (Lazanky Limestones), Givetian
-6318 m basal clastics
-6506 m crystalline basement (metamorphites with weins of Granite)

Jablunkov - 1 (Jabl-1)

0-1190 m Silesian unit
    -96 m Krosno Formation
    -515 m Menilitic Formation
    -294 m Roznov Formation
------------ overthrust
    -619 m Krosno Formation
    -725 m Roznov Formation
------------ overthrust
    -100 m Lhoty Formation
    -1190 m Mazak Formation
------------ -2091 m basal overthrust of Flysch nappes (melange [109 m] from Silesian and Subsilesian units)
-2278 m lower Miocene (Karpatian) of the Carpathian foredeep
-2345 m Lower Carboniferous - Culm facies (Hradec - Kyjovice Formation)
-3002 m Devonian to Lower Carboniferous
    -2793 m Lisen Formation (Krtiny Limestones and Drazovice Limestones)
    -3002 m Macocha Formation (Lazanky and Vilemovice limestones.)
-3156 m basal clastics (Old Red) - Devonian
-3200 m crystalline basement (Diorite)

Janovice - 2 (Ja-2)

0-964.7 m Silesian and Subsilesian units
------------ basal overthrust of Flysch nappes
-969.2 m lower Miocene (Karpatian) of the Carpathian foredeep
-1485 m Upper Cretaceous (Namurian A), Ostrava Formation
-1694.5 m Lower Carboniferous (Visean) to Upper Carboniferous (Namurian), Hradec-Kyjovice Formation (Culm)

Karlin - 1 (Kar-1)

-3133 m Zdanice-Hustopece Formation
-3196 m Submenilitic Formation

-----------	overthrust (Zdanice unit)		
-3985 m	Upper Cretaceous and autochthonous Paleogene		

Kobyli - 1 (Kob-1)

0-702 m	Cejc - Zajeci unit, Nemcice Formation (Upper Cretaceous to lower Eocene)		
-----------	overthrust		
-3113 m	Waschberg-Zdanice-Subsilesian units (Zdanice sector)		
		-1444 m	Zdanice-Hustopece Formation (Egerian)
		-1504 m	Menilitic Formation (Oligocene)
		-1607 m	Nemcice Formation (Paleocene - Eocene)
-----------	overthrust		
		-2897 m	Zdanice-Hustopece Formation (Egerian)
		-3113 m	Nemcice Formation (Paleocene - Eocene)
-----------	overthrust		
-3131 m	Pouzdrany unit, Krepice Formation(?) (Oligocene-Miocene),		
-----------	overthrust		
-3135 m	Waschberg-Zdanice-Subsilesian units (Zdanice sector)		
-----------	basal overthrust of the Flysch nappes		
-4351 m	autochthonous Mesozoic (Mikulov Marls)		

Korycany - 4 (Kor-4)

0-1638 m	Raca unit of the Magura Group of nappes		
		-256 m	Zlin and Beloveza formations
		-312 m	Lukov Member of the Solan Formation
		-1638 m	Raztoky Member of the Solan Formation
------------	overthrust of the Magura Group of nappes		
-1812 m	Zdounky unit (Upper Cretaceous to Eocene)		
------------	basal overthrust of the Flysch nappes		
-1834 m	lower Miocene (Karpatian) of the Carpathian foredeep		
-2250 m	crystalline basement (Granodiorite)		

Kozlovice - 1 (SV-1)

-411 m	Silesian unit
-837 m	autochthonous Karpatian
-1076 m	Ostrava Formation (Upper Carboniferous)
-1421 m	Petrikovice Member (Upper Carboniferous)
-1422 m	Stur Member (Upper Carboniferous)
-1880 m	Culm facies
-2274 m	carbonate facies (Devonian)
-2283 m	crystalline basement

Krasna - 1 (Kra-1)

0-1707.5 m	Silesian unit (Godula subunit)		
		-1613 m	Godula Formation (Turonian - Santonian)
		-1707.5 m	Godula Lhoty Formation (Albian)
--------------	overthrust		
-1712.5 m	basal clastics (Devonian)		
-1812.5 m	crystalline basement (metamorphite)		
-------------	overthrust (up to 1847 m tectonic breccia zone with Devonian to Upper Carboniferous and Cenomanian rocks		
-1969 m	Upper Carboniferous (Namurian) - Ostrava Formation		
-2074 m	Lower Carboniferous (Visean) - Culm facies		
		-2001 m	Hradec - Kyjovice Formation
		-2050 m	Moravice Formation
		-2074 m	Lisen Formation
-2190 m	Devonian to Lower Carboniferous - carbonate facies		
		-2079 m	Hady - Ricky Limestone (Visean)
		-2199 m	Macocha Formation (Givetian)
-2211 m	Devionian (Givetian) - basal clastics		
-2561 m	crystalline basement (metamorphites with diorite intrusions)		

Lubna - 2 (Lub-2)

0-350 m	Raca unit of the Magura Group of nappes		
		-80 m	Solan Formation (Lukov Member)
		-350 m	Solan Formation (Raztoky Member)
-----------	overthrust of the Magura Group of nappes		
-426 m	Upper Cretaceous (Maastrichtian)		
-----------	overthrust		
-1140 m	Waschberg - Zdanice - Subsilesian units (Zdanice sector)		
		-972 m	Zdanice - Hustopece Formation
		-1140 m	Nemcice Formation
-----------	overthrust of the Zdanice unit		
-1870 m	Pouzdrany unit (Oligocene to Miocene strata)		
-----------	basal overthrust of the Flysch nappes		
-1928 m	crystalline basement (Granite)		

Menin - 1 (Me-1)

- 0-49 m middle Miocene (lower Badenian) of the Carpathian foredeep
- -410 m Devonian (Givetian - Frasnian) - Macocha Formation
- -2100 m basal clastics (Cambrian(?), Devonian)

Mikulov - 1 (Mik-1)

- 0-1700 m lower Miocene of the Carpathian foredeep
  - -1100 m Karpatian, Laa Formation
  - -1700 m Eggenburgian - Ottnangian, (basal clastics,146 m)
- -2520 m autochthonous Mesozoic
  - -1990 m Kurdejov Limestones (Kimmeridgian-Tithonian)
  - -2410 m Mikulov Marls (Oxfordian-Kimmeridgian)
  - -2478 m Vranovice Member (Callovian-Oxfordian)
  - -2520 m Nikolcice Member (Callovian)
- -2589 m crystalline basement (Granite)

Musov - 2 (Mus-2)

- 0-973 m lower Miocene of the Carpathian foredeep
  - -890.3 m Karpatian (Laa Formation)
  - -973 m Eggenburgian (Vestonice Member)
- -1895 m autochthonous Mesozoic
  - -1185 m Kurdejov Limestones (Kimmeridgian-Tithonian)
  - -1647 m Mikulov Marls (Oxfordian-Kimmeridgian)
  - -1814 m Vranovice Member (Callovian-Oxfordian)
  - -1895 m Gresten Formation (bazal clastics)
- -1910.25 m crystalline basement (Quarzdiorite)

Nasedlovice - 1

- 0-2200 m Zdanice unit of the Magura Group of nappes
  - -2200 m Zdanice-Hustopece Formation
  - -2245 m Menilitic Formation
  - -2465 m Submenilitic Formation
  - -3190 m autochthonous Paleogene
  - -3560 m Carboniferous (Culm facies)
  - -4099 m Carbonate facies (Devonian, Carboniferous)
  - -3792 m Lisen Formation
  - -4367 m Macocha Formation
  - -4367 m basal clastics (Old Red)
  - -4450 m crystalline basement

Nikolcice - 2A (Nik-2A)

- 0-1025 m Waschberg - Zdanice - Subsilesian units (Zdanice sector)
  - -775 m Zdanice - Hustopece Formation (Egerian)
  - -853 m Menilitic Formation (Oligocene)
  - -1025 m Nemcice Formation (Paleocene - Oligocene)
- ------------ basal overthrust of the Flysch nappes
- -1265 m autochthonous Paleogene, Nesvacilka Formation with clastics at the base (1201–1265 m)
- -2225 m autochthonous Mesozoic
  - -1560 m Kurdejov Limestones (Kimmeridgian)
  - -1936 m Mikulov Marl (Oxfordian to Kimmeridgian)
  - -2137 m Vranovice Member (Callovian to Oxfordian)
  - -2225 m Nikolcice Member (Callovian)
- -2308 m crystalline basement (Granodiorite)

Osvetimany - 1 (Osv-1)

- 0-2278 m Raca unit of the Magura Group of nappes
  - -130 m Luhacovice Member of the Zlin Formation (Eocene)
- ------------ overthrust
  - -310 m Lukov Member of the Solan Formation
  - -520 m Raztoky Member of the Solan Formation
- ------------ overthrust
  - -1230 m Zlin Formation (Eocene)
  - -1450 m Beloveza Formation (Paleocene to Eocene)
  - -1660 m Lukov Member of the Solan Formation (Paleocene)
  - -2278 m Raztoky Member of the Solan Formation (Upper Cretaceous to Paleocene)
- -2480 m Zdounky unit
- ------------ basal overthrust of the Flysch nappes
- -2518 m autochthonous Paleogene
- -2820 m crystalline basement (Granitoid)

Potstat - 1 (Pot-1)

- 0-2866.5 m Lower Carboniferous (Visean), Culm facies (Moravice Formation)
- -3741 m Devonian (Givetian) to Lower Carboniferous (Visean), Carbonate facies (Macocha Formation)

-3221.5 m Vilemovice Limestones
------------ overthrust
-3741 m Devonian to Lower Carboniferous (Visean), Carbonate facies (Lisen Formation)
-3658 m Ricky Limestones
-3741 m Krtiny Limestones
-4100 m Devonian (Frasnian to Famenian) Limestones

Roznov - 1 (Roz-1)

0-437.5 m Verovice Formation (Aptian), Silesian unit
-2062 m Hradiste Formation (Aptian-Valanginian), Silesian unit
------------ overthrust of the Silesian unit
-2240 m Frydek Formation (Campanian), Subsilesian unit
-2544 m Frydland Formation (Eocene-Paleocene), Subsilesian unit
------------ overthrust
-2809 m Frydek Formation (Maastrichtian-Campanian), Subsilesian unit
-2872 m Frydland Formation (Paleocene), Subsilesian unit
------------ overthrust
-2948 m Frydek Formation (Upper Cretaceous), Subsilesian unit
-2989 m Frydland Formation (Oligocene–Eocene), Subsilesian unit
------------ basal overthrust of the Flysch nappes
-3015.5 m lower Miocene (Karpatian), Neogene foredeep

Sedlec - 1 (Sed-1)

0-3320 m Waschberg - Zdanice - Subsilesian units (Waschberg sector):
-1530 m Zdanice-Hustopece Formation
-1680 m Nemcice Formation
------------ overthrust
-3320 m Waschberg - Zdanice - Subsilesian units (Waschberg sector)
-3025 m Zdanice-Hustopece Formation
-3149 m Menilitic Formation
-3320 m Nemcice Formation
overthrust
-3489 m Pouzdrany unit
-4765 m autochthonous Mesozoic
-3600 m Glauconitic sand Formation Upper Cretaceous
-3860 m Kurdejov Limestone
-4720 m Mikulov Marls
-4765 m Gresten Formation
-4850 m crystalline basement (Granitoids)

Slusovice - 1 (Slu-1)

0-3080 m Raca unit of the Magura Group of nappes
-785 m Zlin Formation
-1717 m Beloveza Formation
-2900 m Lukov Member of the Solan Formation
-3080 m Raztoky Member of the Solan Formation
----------- basal overthrust of the Magura Group of nappes
-3785 m lower Miocene (Karpatian?)
-3850 m carbonate rocks (Devonian - Lower Carboniferous)
-3877 m basal clastics (Old Red) - Devonian
-4000 m crystalline basement (Metamorphites)

Strachotin - 2 (Str-2)

0-805 m Waschberg - Zdanice - Subsilesian units (Waschberg sector)
-100 m Zdanice-Hustopece Formation
-400 m Nemcice Formation
-485 m Upper Cretacious (Palava and Klement formations)
-660 m Ernstbrunn Limestone
-805 m Klentnice Formation
------------ overthrust
-1675 m Pouzdrany unit (Pouzdrany and Boudky Marls with slivers of Jurassic rocks)
------------ basal overthrust of the Flysch nappes
-3075 m autochthonous Mesozoic
-1880 m Glauconitic sand Formation (Upper Cretaceous)
-2208 m Kurdejov Limestones
-2750 m Mikulov Marls
-2910 m Vranovice Carbonates
-2997 m Nikolcice Formation
-3075 m Gresten Formation
-3147 m crystalline basement (Granitoids)

Stupava - 1 (Stup-1)

0-1969 m Raca unit of the Magura Group of nappes
-205 m Beloveza Formation

	-365 m	Lukov Member
	-1245 m	Raztoky Member
------------	overthrust	
	-182.05 m	Lukov Member of the Solan Formation
	1965 m	Roztoky Member of the Solan Formation
------------	overthrust of the Magura Group of nappes	
-2156 m	Zdounky unit	
	-2065 m	Lower sekvence (Albian)
	-2156 m	Upper sekvence (Eocene)
------------	basal overthrust of the Flysch nappes	
-2431 m	lower Miocene (Karpatian) of the Carpathian foredeep	
-2500 m	crystalline basement (Granodiorite)	

Susice - 1 (SU-1)

0-215 m	Neogene of the Vienna Basin	
-3525 m	Raca unit of the Magura group of nappes	
	-2540 m	Zlin Formation
	-2590 m	Beloveza Formation
	-3320 m	Solan Formation
	-3525 m	Kaumberg Formation
	overthrust of the Magura Group of nappes	
-3670 m	Zdounky unit	
------------	basal overthrust of the Flysch nappes	
-3760 m	autochthonous Paleogene	
-3810 m	crystalline basement	

Tesany - 1 (Te-1)

0-465 m	Waschberg - Zdanice - Subsilesian units (Zdanice sector)	
------------	overthrust	
-610 m	Pouzdrany unit	
------------	basal overthrust of the Flysch nappes	
-1905 m	autochthonous Paleogene (Nesvacilka Formation) with basal clastics (505 m)	
-2980 m	Lower Carboniferous (Culm facies)	
-4090 m	Devonian to Early Carboniferous (Carbonate facies)	
-4290 m	basal clastics (Old Red) - Devonian	
-4500 m	crystalline basement (Granodiorite)	

Uhrice - 1 (Uh-1)

- 2140 m	Waschberg-Zdanice-Subsilesian units, Zdanice sector	
	-1978 m	Zdanice-Hustopece Formation
	-2002 m	Menilitic Formation
	-2140 m	Nemcice Formation
------------	basal overthrust of the Flysch nappes	
-2763 m	autochthonous Paleogene, Nesvacilka Formation with clastics at the base (2697–2763 m)	
-3043 m	Upper Carboniferous	
-3574 m	Devonian	
-3831 m	Devonian, basal clastics	
-3960 m	crystalline basement, metamophites	

Valasske Mezirici - 1 (VM-1)

	0-1169 m	Silesian unit
	basal overthrust of the Flysch nappes	
	-2948 m	carbonate facies (Devonian, Carboniferous)
	-2980 m	basal clastic (Old Red)
	-3036 m	crystalline basement

Vranovice - 1 (VRA-1)

0–841 m	Waschberg - Zdanice - Subsilesian units (Zdanice sector)	
	-600 m	Zdanice - Hustopece Formation (Egerian)
	-632 m	Menilitic Formation (Oligocene)
	-841 m	Nemcice Formation (Paleocene - lower Oligocene)
------------	basal overthrust of the Flysch nappes	
-1172 m	autochthonous Paleogene - Nesvacilka Formation	
-1565 m	autochthonous Mesozoic	
	-1355 m	Mikulov Marls (Oxfordian - Kimmeridgian)
	-1450 m	Vranovice Member (Callovian - Oxfordian)
	-1565 m	Nikolcice Member a Gresten Formation
-1750 m	crystalline basement (Granodiorite)	

Zdanice - 13 (Zd-13)

0-929 m	Waschberg - Zdanice - Subsilesian units (Zdanice sector)	
	-650 m	Zdanice-Hustopece Formation (Egerian)
	-730 m	Menilitic Formation (Oligocene)
	-929 m	Nemcice Formation (Paleocene - lower Oligocene)

----------- basal overthrust of the Flysch nappes
-975 m lower Miocene of the Carpathian foredeep
-1054 m crystalline basement (Plutonic rocks)

Zarosice - 1 (Za-1)

0-1509 m Waschberg - Zdanice - Subsilesian units (Zdanice sector)
-1257 m Zdanice - Hustopece Formation (Oligocene - lower Miocene)
-1509 m Nemcice Formation (Paleocene - Eocene)
basal overthrust of the Flysch nappes
-1873 m autochthonous Paleogene (Nesvacilka Formation)
-2274 m Carboniferous (Visean - Namurian) - Ostrava Formation
-2330 m Lower Carboniferous (Tournaisian - Visean), Culm facies (Hradec-Kyjovice Formation)
-2760 m Devonian (Givetian) to Lower Carboniferous (Visean), carbonate facies (Macocha Formation)
-2867 m Devonian - basal clastics (Old Red).

APPENDIX 2: AUTHOR'S NAMES IN ASCII AND WITH DIACRITICS

Author's names not containing diacritics are not listed.

ASCII	With Diacritics
Adamek, J.	Adámek, J.
Adamova, M.	Adamová, M.
Arzmuller, G.	Arzmüller, G.
Baldi, T.	Báldi, T.
Baldi-Beke, M.	Báldi-Beke, M.
Benesova, E.	Benešová, E.
Beranek, B.	Beránek, B.
Blazej, J.	Blažej, J.
Blizkovsky, M.	Blížkovský, M.
Bodis, D.	Bodiš, D.
Bohacek, Z.	Boháček, Z.
Boucek, B.	Bouček, B.
Brzobohaty, J.	Brzobohatý, J.
Brzobohaty, R.	Brzobohatý, R.
Bubik, M.	Bubík, M.
Cahelova, J.	Cahelová, J.
Caslavsky, J.	Čáslavský, J.
Cech, S.	Čech, S.
Cejchan, P.	Čejchan, P.
Cekan, J.	Čekan, J.
Cermak, V.	Čermák, V.
Cerv, V.	Červ, V.
Chlupac, I.	Chlupáč, I.
Chmelik, F.	Chmelík, F.
Cizek, P.	Čížek, P.
Ctyroka, J.	Čtyroká, J.
Ctyroky, P.	Čtyroký, P.
Dlabac, M.	Dlabač, M.
Dostal, J.	Dostál, J.
Durica, D.	Ďurica, D.
Durkovic, T.	Ďurkovič, T.
Dvorak, J.	Dvořák, J.
Dvorakova, L.	Dvořáková, L.
Elias, M.	Eliáš, M.
Eliasova, H.	Eliášová, H.
Fusan, O.	Fusán, O.
Francu, J.	Franců, J.
Gasparikova, V.	Gašparíková, V.
Gotzinger, G.	Götzinger, G.
Gregorova, R.	Gregorová, R.
Hamrsmid, B.	Hamršmíd,B.
Grun, W.	Grün, W.
Hanak, J.	Hanák, J.
Hanzlikova, E.	Hanzlíková, E.
Helesicova, K.	Helešicová, K.
Hladikova, M.	Hladíková, M.
Hloska, M.	Hloška, M.
Horvath, F.	Horváth, F.
Housa, V.	Houša, V.
Hrubanova, J.	Hrubanová, J.
Jablonsky, J.	Jablonský, J.
Janocko, J.	Janočko, J.
Jaros J.	Jaroš J.
Jelinek, J.	Jelínek, J.
Jicinsky, W.	Jičínský, W.
Jiricek, R.	Jiříček, R.
Jurasova, F.	Jurášová, F.
Jurkova, A.	Jurková, A.
Juttner, K.	Jüttner, K.
Kalasek, J.	Kalášek, J.
Klvana, J.	Klvaňa, J.
Kohler, E.	Köhler, E.
Konecny, V.	Konečný, V.
Korab, T.	Koráb, T.
Kovac, M.	Kováč, M.
Krejci, J. J.	Krejčí, J. J.
Krejci, O.	Krejčí, O.
Krhovsky, J.	Krhovský, J.
Kroll, A.	Kröll, A.
Ksiazkiewicz, M.	Książkiewicz, M.
Kulmannova, A.	Kulmanová, A.
Lesko, B.	Leško, B.
Mahel, M.	Mahel', M.
Malkovsky, M.	Malkovský, M.
Maly, L.	Malý, L.
Martinek, K.	Martínek, K.
Maska, M.	Maška, M.
Matejka, A.	Matějka, A.
Mencik, E.	Menčík, E.
Michalik, J.	Michalík, J.
Minarilova, D.	Minaříková, D.
Molcikova, V.	Molčíková, V.
Muller, C.	Müller, C.

Muller, P.	Müller, P.
Mysik, M.	Mišík, M.
Nekvasilova, O.	Nekvasilová, O.
Nemcok, M.	Němčok, M.
Nemec, F.	Němec, F.
Novotna, E.	Novotná, E.
Obstova, V.	Obstová, V.
Pagac, I.	Pagáč, I.
Palensky, P.	Pálenský, P.
Pecova, J.	Pečová, J.
Peslova, H.	Peslová, H.
Picha, F.	Pícha, F.
Pitonak, P.	Pitoňák, P.
Plancar, J.	Plančár, J.
Plasienka, D.	Plašienka, D.
Plicka, M.	Plička, M.
Pokorny, V.	Pokorný, V.
Policky, J.	Polický, J.
Pospisil, L.	Pospíšil, L.
Pozaryski, W.	Požaryski, W.
Pribyl, A.	Přibyl, A.
Prichystal, A.	Přichystal, A.
Ptak, J.	Pták, J.
Rakus, M.	Rakús, M.
Rehakova, D.	Reháková, D.
Rehakova, Z.	Řeháková, Z.
Rehanek, J.	Řehánek, J.
Rehor, F.	Řehoř, F.
Rehorova, M.	Řehořová, M.
Repcok, I.	Repčok, I.
Rogl, F.	Rögl, F.
Ruzicka, M.	Růžička, M.
Rybarova, L.	Rybářová, L.
Schroder, B.	Schröder, B.
Sedlak, J.	Sedlák, J.
Senes, J.	Seneš, J.
Shrbeny, O.	Shrbený, O.
Sikula, J.	Šikula, J.
Slaczka, A.	Ślączka, A.
Smid, B.	Šmíd, B.
Sotak, J.	Soták, J.
Spisiak, J.	Spišiak, J.
Stankova, E.	Staňková, E.
Stastny, M.	Šťastný, M.
Stranik, Z.	Stráník, Z.
Svabenicka, L.	Švábenická, L.
Svancara, J.	Švancara, J.
Swidzinski, H.	Świdziński, H.
Tazler, R.	Tázler, R.
Tezky, A.	Těžký, A.
Thonova, H.	Thonová, H.
Trumpy, R.	Trümpy, R.
Vasicek, M.	Vašíček, M.
Vasicek, Z.	Vašíček, Z.
Volsan, V.	Volšan, V.
Vujta, M.	Vůjta, M.
Zahalka, B.	Zahálka, M.
Zakovic, M.	Zakovič, M.
Zukalova, V.	Zukalová, V.
Zurkova, I.	Žůrková, I.

APPENDIX 3: GEOGRAPHICAL AND GEOLOGICAL TERMS IN ASCII AND WITH CZECH DIACRITICS

Terms not containing Czech diacritics are not listed.

ASCII	With Diacritics
Antoninek Formation	Antonínek Formation
Banov	Bánov
Baska Formation, facies, unit	Baška Formation, facies, unit
Baska ridge	Baška ridge
Beloveza Formation	Beloveža Formation
Bile (Biele in Slovak) Karpaty Mountains	Bílé (Biele in Slovak) Karpaty Mountains
Bilovice Formation, field	Bílovice Formation, field
Bohuslavice nad Vlarou	Bohuslavice nad Vlárou
Bosaca unit	Bošáca unit
Bosovice-1 well, field	Bošovice-1 well, field
Breclav field	Břeclav field
Brezina shale	Březina shale
Brezno Formation	Březno Formation
Brodske field	Brodské field
Bruzovice–Frydek field	Bruzovice–Frýdek field
Bystrice nad Olsi	Bystřice nad Olší
Bystrice pod Hostynem	Bystřice pod Hostýnem
Bystrice pod Hostynem-1 well	Bystřice pod Hostýnem-1 well
Bzova Sandstone	Bzová Sandstone
Cejc–Zajeci unit	Čejč–Zaječí unit
Chabova Formation	Chabová Formation

Choryne-9 well, field	Choryně-9 well, field
Chriby Mountains	Chřiby Mountains
Chvalcov Member	Chvalčov Member
Damborice field	Dambořice field
Damborice Group	Dambořice Group
Damborice-1 well	Dambořice-1 well
Detmarovice paleovalley	Dětmarovice paleovalley
Divaky Formation	Diváky Formation
Dolni Bojanovice	Dolní Bojanovice
Dolni Dunajovice-1 well, field	Dolní Bojanovice-1 well, field
Dolni Lomna-3 well, field	Dolní Lomná-3 well, field
Dolni Tesice	Dolní Těšice
Drazovice Limestone	Dražovice Limestone
Drazuvky	Dražůvky
Dubnany Formation	Dubňany Formation
Frydek Formation	Frýdek Formation
Frydlant Formation	Frýdlant Formation
Hady	Hády
Hady–Ricky Limestones	Hády–Říčky Limestones
Handlova	Handlová
Hnojnik-1 well	Hnojník-1 well
Hodonin	Hodonín
Hodonin field	Hodonín field
Holesov fault	Holešov fault
Holesov-1 well	Holešov-1 well
Holic structure	Holíč structure
Horni Zukov field	Horní Žukov field
Hosteradky-1 well	Hostěrádky-1 well
Hostyn Hills	Hostýn Hills
Hostyn Member, zone	Hostýn Member, zone
Hradiste Formation, graben	Hradiště Formation, graben
Hrusky Formation, field	Hrušky Formation, field
Hrusky–Josefov field	Hrušky–Josefov field
Hrusovany Limestones	Hrušovany Limestones
Hrusovany-1 well	Hrušovany-1 well
Hustopece Marls	Hustopeče Marls
Istebna Formation	Istebná Formation
Ivan Member	Iváň Member
Jablunka	Jablůnka
Jablunka-1 well	Jablůnka-1 well
Jarosov	Jarošov
Jarosov-1 well	Jarošov-1 well
Javorina Formation	Javořina Formation
Jesenik	Jeseník
Jezov-3 well	Ježov-3 well
Juhyne River	Juhyně River
Karlin-1 well, field	Karlín-1 well, field
Karvina Formation	Karviná Formation
Kelc subunit, facies	Kelč subunit, facies
Kisovce–Svabovce	Kišovce–Švábovce
Klanecnica	Klanečnica
Kloboucky field	Kloboučky field
Klokocov Member	Klokočov Member
Koberice Gypsum	Kobeřice Gypsum
Kobyli	Kobylí
Kobyli-1 well	Kobylí-1 well
Kojetin Formation	Kojetín Formation
Komorni Lhotka field	Komorní Lhotka field
Koprivnice Limestones	Kopřivnice Limestones

Koprivnice–Ticha field	Kopřivnice–Tichá field
Korycany-4 well, field	Koryčany-4 well, field
Kostelany Vychod field	Kostelany Východ field
Kostelany Zapad field	Kostelany Západ field
Kotouc facies	Kotouč facies
Kotouc Hills	Kotouč Hills
Kozusice-4 well	Kožušice-4 well
Krasna-1 well, field, elevation	Krásná-1 well, field, elevation
Krasna–Moravka field	Krásná–Morávka field
Krepice Formation	Křepice Formation
Krive Member	Křivé Member
Krmelin field	Krmelín field
Krtiny limestones	Křtiny limestones
Krumvir field	Krumvíř field
Kurdejov Limestones	Kurdějov Limestones
Kurim	Kuřim
Kuty	Kúty
Kuzelov Formation	Kuželov Formation
Kycera Member, zone	Kyčera Member, zone
Lab sands, ostracoda beds	Láb sands, ostracoda beds
Laksary Formation, schlier	Lakšáry Formation, schlier
Lanzhot Formation, field	Lanžhot Formation, field
Lanzhot–Hrusky fault	Hrušky–Lanžhot fault
Lazanky Limestones	Lažánky Limestones
Lhotka–Pstruzi field	Lhotka–Pstruží field
Lisen Formation	Líšeň Formation
Lower Tesin Member	Lower Těšín Member
Lubna-2 well, field, massif, elevation	Lubná-2 well, field, massif, elevation
Luhacovice Member, zone	Luhačovice Member, zone
Lukovecek klippe	Lukoveček klippe
Lulec Conglomerates	Luleč Conglomerates
Luzice Formation, sands, field	Lužice Formation, sands, field
Luzice (Lusatian) fault	Lužice (Lusatian) fault
Lysa hora Mountain	Lysá hora Mountain
Malesovice Formation	Malešovice Formation
Manin unit	Manín unit
Mazak Formation	Mazák Formation
Menin-1 well	Měnín-1 well
Menin–Zatcany field	Měnín–Žatčany field
Mikulcice	Mikulčice
Milesovice-1 well	Milešovice-1 well
Moravske naftove doly, oil company	Moravské naftové doly, oil company
Mourinov field	Mouřínov field
Mur–Murtz–Zilina line	Mur–Murtz–Žilina line
Musov Member	Mušov Member
Musov-2 well	Mušov-2 well
Mutenice field	Mutěnice field
Nasedlovice-1 well	Násedlovice-1 well
Nemcice Formation	Němčice Formation
Nemcicky	Němčičky
Nemcicky-1, 2 wells, field	Němčičky-1, 2 wells, field
Nemetice Formation	Němetice Formation
Nesvacilka Formation, paleovalley, graben	Nesvačilka Formation, paleovalley, graben
Nesvacilka-1, 2, 3 wells	Nesvačilka-1, 2, 3 wells
Nikolcice Member	Nikolčice Member
Nikolcice-1, 2A wells, field	Nikolčice-1, 2A wells, field
Nitkovice	Nítkovice
Nitkovice–Hradisko field	Nítkovice–Hradisko field
Nizky Jesenik Mountains	Nízký Jeseník Mountains

Nova Hora	Nová Hora
Nove Mlyny limestones	Nové Mlýny limestones
Nove Mlyny-1 well	Nové Mlýny-1 well
Novy Prerov field	Nový Přerov field
Novy Prerov Member	Nový Přerov Member
Ondrasovec Member	Ondrášovec Member
Oravska Magura	Oravská Magura
Oravska Polhora	Oravská Polhora
Orlova fold	Orlová fold
Ostrava–Karvina ridge	Ostrava–Karviná ridge
Osvetimany-1 well	Osvětimany-1 well
Palava Formation	Pálava Formation
Palkovice Formation, Hills	Pálkovice Formation, Hills
Pasohlavky Limestones	Pasohlávky Limestones
Podvorov–Mutenice field	Podvorov–Mutěnice field
Postorna field	Poštorná field
Potstat-1 well	Potštát-1 well
Pouzdrany Formation, Marls, unit	Pouzdřany Formation, Marls, unit
Pribor Jih (Stramberk) field	Příbor Jih (Štramberk) field
Pribor–Klokocov field	Příbor–Klokočov field
Pribor–Tesin ridge	Příbor–Těšín ridge
Pritluky	Přítluky
Proc Formation	Proč Formation
Prusanky field	Prušánky field
Puchov Marls	Púchov Marls
Raca unit	Rača unit
Racice Conglomerates	Račice Conglomerates
Rajec	Rájec
Ratiskovice	Ratíškovice
Raztoka Member	Ráztoka Member
Rostin	Roštín
Rostin-2 well	Roštín-2 well
Roznov Formation, area	Rožnov Formation, area
Roznov-1 well, field	Rožnov-1 well, field
Rozstani Shales	Rozstání Shales
Sakvice Marls, depression	Šakvice Marls, depression
Sastin sands	Šaštín sands
Sitborice Member	Šitbořice Member
Slavkov–Tesin ridge	Slavkov–Těšín ridge
Slusovice-1 well	Slušovice-1 well
Sneznica Formation	Sněžnica Formation
Solan Formation	Soláň Formation
Spis	Spiš
Stara Tura	Stará Turá
Stare Hute	Staré Hutě
Staric–Liskovec–Sviadnov field	Staříč–Lískovec–Sviadnov field
Stary Jicin	Starý Jičín
Sternberk–Benesov suture	Šternberk–Benešov suture
Strachotin-2 well	Strachotín-2 well
Stramberk klipppe, reef, platform, field	Štramberk klippe, reef, platform, field
Stramberk Limestones	Štramberk Limestones
Stranska skala	Stránská skála
Straz Sandstones	Stráž Sandstones
Susice-1 well	Sušice-1 well
Svedske Sance	Švédské Šance
Tesany Formation	Těšany Formation
Tesany-1 well	Těšany-1 well
Tesin–Hradiste Formation	Těšín–Hradiště Formation
Tesin nappe	Těšín nappe

Tesin (Teschen) Limestones	Těšín (Teschen) Limestones
Tlumacov Marls	Tlumačov Marls
Tri Kameny zone	Tři Kameny zone
Trnava–Staskov zone	Trnava–Staškov zone
Tynec–Cunin field	Týnec–Cunín field
Tynec structure, field	Týnec structure, field
Tynec-34 well	Týnec-34 well
Uhercice Formation	Uherčice Formation
Uhersky Brod	Uherský Brod
Uhrice Jih (South) field	Uhřice Jih (South) field
Uhrice-1 well, field	Uhřice-1 well, field
Ujezd Member	Újezd Member
Ustgrun	Ustgruň
Vah line	Váh line
Valasske Mezirici	Valašské Meziříčí
Valasske Mezirici-1 well	Valašské Meziříčí-1 well
Velke Bilovice–Zizkov–Podivin fields	Velké Bílovice–Žižkov–Podivín fields
Vendryne Formation	Vendryně Formation
Verona–Semmering–Vah fault	Verona–Semmering–Váh fault
Verovice Formation, Shales	Veřovice Formation, Shales
Vestonice Formation, fault	Věstonice Formation, fault
Vilemovice Limestones	Vilémovice Limestones
Vlara subunit	Vlára subunit
Vsetin Member, zone	Vsetín Member, zone
Waschberg–Zdanice–Subsilesian units, nappes	Waschberg–Ždánice–Subsilesian units, nappes
Zahori fault	Záhoří fault
Zajeci	Zaječí
Zarosice-1 well, field	Žarošice-1 well, field
Zatcany field	Žatčany field
Zavod Formation, schlier	Závod Formation, schlier
Zdanice–Hustopece Formation	Ždánice–Hustopeče Formation
Zdanice Jih–Zdanice Sever–Kloboucky field	Ždánice Jih–Ždánice Sever–Klobouky field
Zdanice Sandstone, unit, nappe, massif, elevation, basin	Ždánice Sandstone, unit, nappe, elevation, basin
Zdanice-13 well, field	Ždánice-13 well, field
Zerotice Member	Žerotice Member
Zilina	Žilina
Zizkov Member, field	Žižkov Member, field
Zlin Formation	Zlín Formation
Zohor–Plavec graben	Zohor–Plaveč graben

REFERENCES CITED

Abel, O., 1899, Die Beziehung des Klippengebietes zwischen Donau und Thaya zum alpin-karpatischen Gebirgssysteme: Verhandlungen der Geologischen Reichsanstalt, Wien, v. 15–16, p. 374–381.

Adam, Z., M. Dlabac, F. Chmelik, and Z. Stranik, 1968, Geologische Interpretation der seismischen Profilen im sudlichen Teil der Zdanice Einheit: Sbornik geologickych ved. Uzita Geofyzika, Praha, v. 14, no. 1, p. 69–76.

Adamek, J., 1979, Plynove lozisko Dolni Dunajovice a geologicka stavba jizni casti karpatske predhlubne (in Czech): Zemni Plyn a Nafta, Hodonin, v. 24, no. 1, p. 1–22.

Adamek, J., 1986, Geologicke poznatky o stavbe mesozoika v useku Jih jv. svahu Ceskeho masivu (in Czech): Zemni Plyn a Nafta, Hodonin, v. 31, no. 4, p. 453–484.

Adamek, J., 2002, Regional-geological evaluation of Jurassic sediments on the south-east margin of the Bohemian Massif (in Czech with English summary): Zpravy o geologickych vyzkumech v roce 2001, Praha, p. 9–11.

Adamek, J., 2003, The Miocene of the Carpathian foredeep in southern Moravia, geological development and lithostratigraphic classification (in Czech with English summary): Zpravy o geologickych vyzkumech v roce 2002, Praha, p. 9–11.

Adamek, J., J. Dvorak, and J. Kalvoda, 1980, Prispevek ke geologicke stavbe a naftove-geologickemu hodnoceni nikolcicko-kurdejovskeho hrbetu (in Czech): Zemni Plyn a Nafta, Hodonin, v. 25, no. 4, p. 441–474.

Adamek, J., R. Brzobohaty, P. Palensky, and J. Sikula, 2003, The Karpatian in the Carpathian foredeep (Moravia), *in* R. Brzobohaty, I. Cicha, M. Kovac, and F. Rogl, eds., The Karpatian, a lower Miocene stage of the central Paratethys: Brno, Masaryk University, p. 75–88.

Adamova, M., O. Krejci, and A. Prichystal, 1995, Neovolcanite east of the town Uhersky Brod: Geologicke vyzkumy na Morave a ve Slezsku v roce 1994: Moravske Zemske Muzeum a Sekce geologickych ved Prirodovedecke fakulty Masarykovy university, Brno, v. 2, p. 12–15.

Allen, P. A., and J. R. Allen, 1990, Basin analysis: Oxford, Blackwell, 451 p.

Ampferer, O., 1906, Uber das Bewegungsbild von Faltengebirgen: Jahrbuch der geologischen Reichsanstalt, Wien, v. 56, p. 539–622.

Andrusov, D., 1933, Communications sommaires sur la geologie des Carpathes moravo-silesiennes (in Czech with French summary): Vestnik Statniho geologickeho ustavu CSR, Praha, v. 9, p. 144–149.

Andrusov, D., 1938, Etude geologique de la zone des Klippes internes des Karpates occidentales: III partie: Tectonique (in Czech with French summary): Rozpravy Statniho geologickeho ustavu CSR, Praha, v. 9, p. 1–135.

Andrusov, D., 1959, Geologia ceskoslovenskych Karpat (in Slovak): Bratislava, Slovenska Akad. Vied, II, 375 p.

Argand, E., 1911, Les nappes de recouvrements des Alpes Pennines et leurs prolongement structureaux: Materiaux Carte Geologique Beitrage Suisse (n.s.), v. 31, 26 p.

Argand, E., 1916, Sur l' arc des Alpes Occidentales: Eclogae Geologicae Helvetiae, v. 14, p. 145–191.

Argand, E., 1924, La tectonique de l' Asie: Congres Geological International, Compte Rendu, Bruxelles, 1922, p. 171–372.

Arzmuller, G., S. Buchta, E. Ralbovsky, and G. Wessely, 2006, The Vienna basin, *in* J. Golonka and F. J. Picha, eds., The Carpathians and their foreland: Geology and hydrocarbon resources: AAPG Memoir 84, p. 191–204.

Bachmann, G. H., and M. Muller, 1991, The Molasse Basin, Germany: Evolution of a classic petroliferous foreland basin, *in* A. M. Spencer, ed., Generation, accumulation, and production of Europe's hydrocarbons: European Association of Petroleum Geologists Special Publication 1, p. 263–276.

Bachmayer, F., 1957, Das Mesozoikum der niederösterreichischen Klippen: Zeitschrift Deutschen geologischen Gesellschaft, Hannover, v. 109, no. 2, p. 659–660.

Bada, G., F. Horvath, S. Cloetingh, D. D. Coblentz, and T. Toth, 2001, Role of topography-induced gravitational stresses in basin inversion: The case study of the Pannonian Basin: Tectonophysics, v. 20, no. 3, p. 343–363.

Beck, H., and G. Gotzinger, 1932, Erlauterungen zur geologischen Karte des Ostrau-Karwiner Steinkohlenbeckens, der West Beskiden und des sudetischen Randgebietes: Wien, Geologische Bundesanstalt, 84 p.

Benada, S., and J. Blazej, 1991, Results from the Karlin-1 well in the south Moravia (in Czech): Geologicky pruzkum, Praha, v. 33, p. 257–260.

Benesova, E., E. Hanzlikova, and A. Matejka, 1962, Prispevek ke geologii kurovickeho bradla (in Czech): Zpravy o geologickych vyzkumech v roce 1961, Praha, p. 185–186.

Benesova, E., I. Cicha, F. Picha, Z. Rehakova, and Z. Stranik, 1963, Profil durch die Zdanice-Einheit zwischen Velke Pavlovice und Nosislav (in Czech with German summary): Sbornik geologickych ved, Geologie, Praha, v. 1, p. 139–156.

Benesova, E., M. Elias, and A. Matejka, 1968, Geology of the Kurovice Klippe (in Czech with English summary): Sbornik geologickych ved, Geologie, Praha, v. 13, p. 7–36.

Bertrand, M., 1898, La nappe de recouvrement des environs de Marseille: Lame de chariage et raprochement avec le bassin houllier de Silesie: Bulletin de la Societe Geologique de France 3me Serie, Paris, v. 26, p. 232–252.

Biju-Duval, B., J. Dercourt, and X. LePichon, 1977, From the Tethys Ocean to the Mediterranean Seas: A plate tectonic model of the evolution of the Western Alpine system, *in* B. Biju-Duval and L. Montadert, eds., Structural history of the Mediterranean basins: Paris, Editions Technip, p. 143–164.

Birkenmajer, K., 1960, Geology of the Pieniny Klippen Belt of Poland: Jahrbuch der Geologischen Bundesanstalt, Wien, v. 103, p. 1–36.

Birkenmajer, K., 1985, Major strike-slip faults of the Pieniny Klippen Belt and the Tertiary rotation of the Carpathians: Publications of the Institute of Geophysics, Polish Academy of Sciences, Warsaw, Ser. A-16, p. 101–115.

Birkenmajer, K., 1986, Stages of structural evolution of the Pieniny Klippen Belt, Carpathians: Studia Geologica Polonica, v. 88, p. 7–32.

Birkenmajer, K., 1988, Exotic Andrusov ridge: Its role in plate-tectonic evolution of the West Carpathian fold belt: Studia Geologica Polonica, v. 91, p. 7–37.

Blaschke, F., 1911, Zur Tithonfauna von Stramberg in Mahren: Annalen Naturhistorischen Hofmuseum, Wien, v. 25, no. 1–2, p. 143–221.

Blizkovsky, M., et al., 1994, Exploration history, geology and hydrocarbon potential in the Czech Republic and Slovakia, *in* B. M. Popescu, ed., Hydrocarbons of eastern Central Europe: Berlin, Springer-Verlag, p. 71–117.

Boillot, G., L. Montadert, M. Lemoine, and B. Biju-Duval, 1984, Les marges continentales actuelles et fossils autourde de la France: Paris, Masson, 342 p.

Boucek, B., 1952, Vyzkum exotickych bloku na Tesinsku (in Czech): Prirodovedecky Sbornik Ostravskeho kraje, Opava, v. 13, no. 3–4, p. 587.

Boucek, B., and A. Pribyl, 1954, Uber dem subsilesischen Paleogene in der Umgebung von Bystrice no. Olsi und seiner exotitschen Blocken, hauptsachlich des Kohlenkalkesmit der Fauna (Vise) (in Czech with Russian and German summary): Prirodovedecky Sbornik Ostravskeho kraje, Opava, v. 15, no. 2–3, p. 220–235.

Brix, F., A. Kroll, and G. Wessely, 1977, Die Molassezone und deren Untergrund in Niederosterreich: Erdol und Erdgas Zeitschrift, Wien, v. 93, p. 12–36.

Brzobohaty, J., 1993, Autochthonous Paleogene fill of the Nesvacilka depression and its importance for hydrocarbon exploration (in Czech with English summary): Zemni Plyn a Nafta, Hodonin, v. 38, no. 3, p. 153–184.

Brzobohaty, J., S. Benada, J. Berka, and J. Rehanek, 1996, Hydrocarbon habitat of the Paleogene Nesvacilka trough, Carpathian foreland basin, Czech Republic, *in* P. A. Ziegler and F. Horvath, eds., Peri-Tethys Memoir 2: Structure and prospects of Alpine basins and forelands: Memoires du Museum National d'Histoire Naturelle, Paris, v. 170, p. 305–319.

Brzobohaty, R., 1981, Fischreste aus den Menilit-Schichten der Zdanice-Einheit in Mahren (Oligozan, Aussere Westkarpaten) (in Czech with German summary): Zemni plyn a nafta, Hodonin, v. 26, p. 75–87.

Brzobohaty, R., and I. Cicha, 1993, Carpathian foredeep: Geologie Moravy a Slezska, *in* A. Prichystal, V. Obstova, and M. Suk, eds., Geologie Moravy a Slezska (in Czech with English summary): Moravske zemske muzeum a Sekce geologickych ved Prirodovedecke fakulty Masarykovy university, Brno, p. 123–128.

Brzobohaty, R, I. Cicha, M. Kovac, and F. Rogl, eds., 2003, The Karpatian, a lower Miocene stage of the central Paratethys: Brno, Masaryk University, 360 p.

Bubik, M., 1987, Oligocene calcareous nannoplankton of the Menilitic Formation with the Jaslo limestones horizon from Bystrice nad Olsi (Subsilesian unit, West Carpathians) (in Czech with English summary): Miscellanea Micropaleontologica, II/2, Knihovnicka zemniho plynu a nafty, Hodonin, v. 6b, p. 45–55.

Bubik, M., 1995, Cretaceous to Paleogene agglutinated foraminifera of the Bile Karpaty unit (West Carpathians, Czech Republic), *in* M. A. Kaminski, S. Geroch, and M. A. Gasinski, eds., Proceedings of the Fourth International Workshop of Agglutinated Foraminifera: Grzybowski Foundation Special Publication 3, p. 71–116.

Bucha, V., and M. Blizkovsky, eds., 1994, Crustal Structure of the Bohemian Massif and West Carpathians: Praha, Academia-Springer Verlag, 355 p.

Buchta, S., and K. Dohnal, 1996, Loziska tezke ropy ve Zdanicke oblasti (in Czech with English abstract): Zemni plyn a nafta, Hodonin, v. 41, nos. 1–2, p. 73–83.

Buday, T., and M. Suk, 1989, Problems of the contact area of the Bohemian Massif and West Carpathians: *in* J. Ibrmajer and M. Suk, eds., Geophysical pattern of the Czechoslovakia (in Czech with English summary): Praha, Ustredni ustav geologicky, p. 296–300.

Buday, T., O. Kodym st., M. Mahel, M. Maska, A. Matejka, J. Svoboda, and V. Zoubek, 1961, Tektonicky vyvoj Ceskoslovenska: Praha, Ustredni ustav geologicky, 256 p.

Buday, T., et al., 1963, Vysvetlivky k prehedne geologicke mape CSSR, scale 1:200,000 M-33 XXX-Gottwaldov (in Czech): Praha, Ustredni ustav geologicky, 238 p.

Burchfiel, B. C., and L. Royden, 1982, Carpathian foreland thrust belt and its relation to Pannonian and other basins: AAPG Bulletin, v. 66, p. 1179–1195.

Burchfiel, B. C., and L. Royden, 1988, Afterword: A general approach to basin analysis, *in* L. H. Royden and F. Horvath, eds., The Pannonian Basin: AAPG Memoir 45, p. 373–375.

Burkhard, M., and A. Sommaruga, 1998, Evolution of the western Swiss Molasse basin: Structural relations with the Alps and the Jura belt: *in* A. Mascle, C. Puigdefabregas, H. P. Luterbacher, and M. Fernandez, eds., Cenozoic foreland basins of Western Europe: Geological Society (London) Special Publication 134, p. 279–298.

Burtan, J., K. Konior, and M. Ksiazkiewicz, 1937, Mapa geologiczna Karpat Slaskich (in Polish): Krakow, Wydawnictwo slaskie PAU, 104 p.

Cermak, V., 1975, Temperature-depth profiles in Czechoslovakia and some adjacent areas derived from heat-flow measurements, deep seismic sounding and other geophysical data: Tectonophysics, v. 26, p. 103–119.

Cermak, V., 1979, Review of heat flow in Czechoslovakia, *in* V. Cermak and L. Ryback, eds., Terrestrial heat flow in Europe: Berlin, Springer-Verlag, p. 152–160.

Cerv, V., J. Pek, B. Picha, O. Praus, and M. Tobiasova, 1994, Magnetotelluric models of inhomogeneity zones, *in* V. Bucha and M. Blizkovsky, eds., Crustal structure of the Bohemian Massif and West Carpathians: Praha, Academia-Springer-Verlag, p. 147–157.

Channell, J. E. T., and H. W. Kozur, 1997, How many oceans? Meliata, Vardar, and Pindos oceans in Mesozoic Alpine paleogeography: Geology, v. 25, p. 183–186.

Chlupac, I., 1989, The Devonian of Czechoslovakia and its stratigraphical significance, *in* N. J. McMillan, A. F. Embry, and D. J. Glass, eds., Devonian of the world: Canadian Society of Petroleum Geologists Memoir 14, no. 1, p. 481–497.

Chlupac, I., 1994, Facies and biostratigraphic relationship in Devonian of the Bohemian Massif: Courier Forschungsinstitut Senckenberg, v. 169, p. 299–317.

Chmelik, F., 1957, The geology of the klippen of Cetechovice and their surroundings (in Czech with English and Russian summary): Sbornik Ustredniho ustavu geologickeho, v. 27, Geologie 2, p. 533–568.

Chmelik, F., 1971, Die Geologie der Zdounky Einheit und ihre Stellung im alpin-karpatischen system (in Czech with Russian and German summary): Sbornik geologickych ved, G., v. 19, p. 123–149.

Cicha, I., 1975, Die Entwicklung der Pouzdrany Schichtengruppe und des Egerien in der Zdanice-Podslezska (Steinitz-Subsilesischer) Einheit in der Tschechoslowakei, *in* T. Baldi and J. Senes, eds., Chronostratigraphie und Neostratotypen, Miozan der Zentralen Paratethys: Bratislava, Bd. V, Veda, Slovakische Akademie der Wissenschaten, p. 65–70.

Cicha, I., and J. Ctyroka, 1995, Problemy hranicnich vrstev karpatu a badenu v jizni casti karpatske predhlubne (33–24 Hnanice) (in Czech): Zpravy o geologickych vyzkumech v roce 1994, Praha, p. 19.

Cicha, I., and M. Holzknecht, 1964, Einige neuen Erkenntnisse uber die Stratigraphie des Oligozan in der Pouzdrany (Pausramer) Einheit (in Czech with German summary): Geologicke prace, Bratislava, v. 36, p. 87–94.

Cicha, I., and M. Kovac, 1990, Neogene climatic changes and geodynamics of the Central Paratethys, *in* D. Minarikova and H. Lobitzer eds., Thirty years of geological cooperation between Austria and Czechoslovakia: Vienna, Federal Geological Survey; Prague, Geological Survey, p. 70–78.

Cicha, I., and F. Picha, 1964, Beitrag zur Kenntnis der stratigraphischen und lithologischen Entwicklung des SO-Teiles der Zdanice Einheit (in Czech with German summary):

Sbornik geologickych ved, Praha, Geologie, v. 4, p. 137–158.

Cicha, I., F. Chmelik, F. Picha, and Z. Stranik, 1964, Ubersicht uber der heutigen Stand der Forschungen in der Molassezone Zdanice und Pouzdraner Einheit Sud-Mahren: Mitteilungen der Osterreichischen Geologischen Gesellschaft, Wien, v. 56, no. 2, p. 445–468.

Cicha, I., F. Chmelik, F. Picha, and Z. Stranik, 1965, Eine neue tektonische Einheit der ausseren Karpaten in Sudmahren: Geologicke prace, Zpravy, Bratislava, v. 36, p. 85–104.

Cicha, I., J. Senes, and J. Tejkal, eds., 1967, M3 (Karpatien)-Die Karpatische Serie und ihr Stratotypus: Chronostratigraphie und Neostratotypen, 1. Veda, Bratislava, p. 1–312.

Cicha, I., H. Hagn, and E. Martini, 1971, Das Oligozan und Miozan der Alpen und Karpaten. Ein Vergleich mit Hilfe planktonisher Organismen: Mitteilungen der Bayerischen Staatssammlung fur Palaontologie und Histor. Geologie, Munchen, v. 11, p. 279–293.

Cicha, I., F. Marinescu, and J. Senes, 1975a, Correlation du Neogene de la Paratethys centrale— Stratigraphic correlation Tethys–Paratethys Neogene: International Geoscience Programme–International Union of Geological Sciences–United Nations Educational, Scientific and Cultural Organization Project No. 25: Geological Survey, Prague, p. 1–33.

Cicha I, A. Papp, and J. Senes, 1975b. Beziehungen des Egerien zu den ubrigen stratigraphischen Einheiten des Mediterranean und Borealen Raumes, *in* T. Baldi and J. Senes eds., Chronostratigraphie und Neostratotypen: Miozan der Zentralen Paratethys, Bd. V, OM Egerien, Die Egerer, Pouzdraner, Purchkirchnerer Schichtengruppe und die Bretkaer Formation: Bratislava, Veda, Slovak Academy of Sciences, p. 43–48.

Cicha, I., et al., 1985, Neogen in OS-1 Kravare and OS-2 Hat boreholes in the Opava area (in Czech with English, and Russian summary): Praha, Sbornik geologickych ved, Geologie, v. 40, p. 183–229.

Ciprys, V., and A. Thon, 1990, Deep-seated structures of the Bohemian Massif in the region between the Vranovice graben and the Czechoslovak–Austrian frontier, *in* D. Minarikova and H. Lobitzer eds., Thirty years of geological cooperation between Austria and Czechoslovakia: Vienna, Federal Geological Survey; Prague, Geological Survey, p. 18–23.

Ciprys, V., J. Adamek, and S. Benada, 1995, Petroleum geology of the Carpathian foredeep and overthrust zones in the Czech Republic: Petroleum Geoscience, v. 1, p. 89–94.

Cizek, P., and C. Tomek, 1991, Large-scale thin-skinned tectonics in the eastern boundary of the Bohemian Massif: Tectonics, American Geophysical Union, v. 10, no. 2, p. 273–286.

Cloetingh, S., F. Horvath, G. Bada, and A. Lankreijer, eds., 2002, Neotectonics and surface processes: The Pannonian Basin and Alpine/Carpathian System: American Geophysical Union Special Publication 3, 295 p.

Ctyroky, P., 1966, Die eozane Fauna der Moutnice Kalke (Sudmahren): Sbornik geologickych ved, Paleontologie, Praha, v. 8, p. 71–107.

Ctyroky, P., 1982, Das Untermiozan (Egenburg–Ottnang) im SW teil der Karpatischen Vortiefe in Mahren (in Czech with German summary): Zemni plyn a nafta, Hodonin, v. 27, no. 4, p. 379–394.

Ctyroky, P, 1987, Eggenburgian, Ottnangian and Karpatian (early Miocene) along the Bohemian Massif in Moravia (Czechoslovakia), *in* Proceedings of the VIIIth Congress of the Regional Committee on Mediterranean Neogene Stratigraphy: Annales Instituti Geologici Publici Hungarici, Budapest, v. 70, p. 119–123.

Ctyroky, P., 1991, Division and correlation of the Eggenburgian and Ottnangian in the southern Carpathian foredeep in southern Moravia (in Czech with English summary): Zapadne Karpaty, Geologia, v. 15, p. 67–109.

Ctyroky, P., 1993, Das Tertiar der Bohmische Masse in Sudmahren: Jahrbuch der Geologischen Bundesanstalt, v. 136, no. 4, p. 707–713.

Czernicki, J., and P. Karnkowski, 1987, The possibility of hydrocarbon prospecting in paleo-highs Bachorzec and Dynow in Przemysl woiewodiship (in Polish with English summary): Nafta, Krakow, v. 43, no. 9, p. 221–228.

Dadlez, R., Z. Kowalczewski, and Z. Znosko, 1994, Some key problems of the pre-Permian tectonics of Poland: Kwartalnik geologiczny, Warszawa, v. 38, no. 2, p. 169–189.

Dadlez, R., M. Narkiewicz, R. A. Stephenson, M. T. M. Visser, and J. D. van Wees, 1995, Tectonic evolution of the Mid-Polish Trough: Modelling implications and significance for Central European geology: Tectonophysics, v. 252, p. 179–195.

DeCelles, P. G., and P. C. DeCelles, 2001, Rates of shortening, propagation, underthrusting, and flexural wave migration in continental orogenic systems: Geology, v. 29, no. 2, p. 135–138.

DeCelles, P. G., and A. K. Giles, 1996, Foreland basin system: Basin Research, v. 8, p. 105–123.

Demaison, G., and B. J. Huizinga, 1994, Genetic classification of petroleum systems using three factors; Charge, migration, and entrapment, *in* L. B. Magoon and W. G. Dow, eds., The petroleum system— From source to trap: AAPG Memoir 60, p. 73–89.

Dercourt, J., et al., 1986, Geological evolution of the Tethys belt from the Atlantic to the Pamirs since the Lias: Tectonophysics, v. 123, p. 241–315.

Dewey, J. F., W. C. Pitman, W. B. F. Ryan, and J. Bonin, 1973, Plate tectonics and the evolution of the Alpine system: Geological Society of America Bulletin, v. 63, p. 3137–3180.

Dicea, O., 1995, The structure and hydrocarbon geology of the Romanian East Carpathian border from seismic data: Petroleum Geoscience, v. 1, p. 135–143.

Dlabac, M., 1946, Geological results of drilling work in the Vyskov basin, between Vyskov and Slavkov during the year 1943 (in Czech with English and Russian summary): Sbornik Statniho geologickeho ustavu CSR, Praha, v. 13, p. 23–39.

Dlabac, M., and E. Mencik, 1964, Geological structure of the autochthonous basement of the western part of the Outer

Carpathians on the territory of the Czechoslovak Socialistic Republic (in Czech with English summary): Rozpravy Ceskoslovenske akademie ved. Praha, v. 74, no. 1, p. 1–58.

Dopita, M., and V. Havlena, 1977, Geology of the Ostrava–Karvina coal field, *in* Symposium on Carboniferous Stratigraphy: Praha, Ustredni ustav geologicky, p. 183–192.

Dopita, M., and O. Kumpera, 1993, Geology of the Ostrava–Karvina coal-field, Upper Silesian basin, Czech republic, and its influence on mining: International Journal of the Coal Geology, Amsterdam, v. 23, p. 291–321.

Dopita, M., et al., 1997, Geology of the Czech part of the Upper Silesian basin (in Czech with English summary): Ministry of the Environment of the CR (Czech Republic), Praha, 278 p.

Dorman, J. T., 1995, The Moravian foredeep and its hydrocarbon potential: ESRI Occasional Publication 11A, Slovakian Geology: Korab Memorial Volume, p. 89–116.

Dostal, J., and J. V. Owen, 1998, Cretaceous alkaline lamprophyres from northeastern Czech Republic, geochemistry and petrogenesis: Geologische Rundschau, v. 87, no. 1, p. 67–77.

Dreger, I., 1899, Erlauterungen zu geologischen Karten Austerlitz: Wien, Verlag der Geologischen Reichsanstalt, 26 p.

Dudek, A., 1980, The crystalline basement block of the Outer Carpathians in Moravia: Bruno–Vistulicum: Rozpravy Ceskoslovenske akademie ved, Rada matematicko-prirodnich ved, Praha, v. 8, p. 1–85.

Dudek, A., and J. Melkova, 1975, Radiometric age determination in the crystalline basement of the Carpathian foredeep and of the Moravian Flysch: Bulletin of the Czech Geological Survey, Praha, v. 50, p. 257–264.

Durica, D., J. G. Namestnikov, I. Pagac, and Z. Roth, 1986, Oil and gas fields in the Central Europe (in Slovak with English summary): Bratislava, Alfa, 284 p.

Dvorak, J., 1973, Synsedimentary tectonics of the Paleozoic of the Drahany Upland (Sudeticum, Moravia, Czechoslovakia): Tectonophysics, v. 17, p. 359–391.

Dvorak, J., 1978, Geologie paleozoika v podlozi Karpat jv. od Drahanske vrchoviny: Zemni plyn a nafta, Hodonin, v. 23, no. 2, p. 185–203.

Dvorak, J., 1993, Moravske paleozoikum, *in* A. Prichystal, V. Obstova, and M. Suk, eds., Geologie Moravy a Slezska (in Czech): Moravske Zemske Muzeum a Sekce Geologickych Ved Prirodovedecke fakulty Masarykovy university, Brno, p. 41–58.

Dvorak, J., 1995, Tectonic and morphological development of the southeast margin of the Bohemian Massif during its subduction under Carpathians (in Czech with English summary): Knihovnicka Zemniho plynu a nafty, Hodonin, v. 16, p. 15–24.

Dvorak J., and J. Ptak, 1963, Geologische Entwicklung und Tektonic des Devons und Unteren Karbons im Mahrischen Karst (in Czech with German and Russian resume): Sbornik Geologickych ved, Geologie, Praha. v. 3, p. 49–84.

Dziadzio, P. S., M. A. Enfield, M. P. Watkinson, and S. J. Porebski, 2006, The Ciezkowice Sandstone: Examples of basin-floor fan-stacking patterns from the main (upper Paleocene to Eocene) reservoir in the Polish Carpathians, *in* J. Golonka and F. J. Picha, eds., The Carpathians and their foreland: Geology and hydrocarbon resources: AAPG Memoir 84, p. 477–496.

Elias, M., 1963, The problem of the origin of the Solan beds in the western area of the Magura flysch in Czechoslovakia (in Czech with English summary): Vestnik Ustredniho ustavu geologickeho, Praha, v. 34, no. 4, p. 253–260.

Elias, M., 1970, Lithology and sedimentology of the Silesian unit in the Moravo-Silesian Beskydy Mts. (in Czech with English summary): Sbornik geologickych ved, Geologie, Praha, v. 8, p. 7–99.

Elias, M., 1979, Facies and paleogeography of the Silesian unit in the western part of the Czechoslovak Flysch Carpathians: Vestnik Ustredniho ustavu geologickeho, Praha, v. 54, no. 6, p. 327–339.

Elias, M., 1981, Facies and paleogeography of the Jurassic of the Bohemian Massif: Sbornik geologickych ved, Geologie, Praha, v. 35, p. 75–144.

Elias, M., 1991, Sedimentology of the Klentnice Formation and the Ernstbrunn Limestones (Zdanice–Subsilesian unit) of the of the Outer West Carpathians: Bulletin of the Czech Geological Survey, Praha, v. 67, p. 179–193.

Elias, M., 1995, Sedimentologie Mazackych vrstev a Ostravickeho piskovce (Godske souvrstvi) (in Czech): Zpravy o geologickych vyzkumech v roce 1994, Cesky geologicky ustav, Praha, p. 44–45.

Elias, M., 1998, Sedimentology of the Subsilesian unit (in Czech with English summary): Czech Geological Survey Special Papers 8, 48 p.

Elias, M., 2000, Relation between Pustevny Sandstone and Malinowska skala Sandstone (Godula Formation) in the Beskydy Mts. (in Czech with English abstract): Geologicke vyzkumy na Morave a ve Slezsku v roce 1999, Moravske zemske muzeum a Sekce geologickych ved Prirodovedecke Fakulty Masarykovy University, Brno, v. 19, p. 64–66.

Elias, M., 2002, The Roznov Formation— New formal name for the Sumenilitic formation of the Silesian Unit (Moravskoslezske Beskydy Mts., Kelcska pahorkatina Upland, Outer West Carpathians) (in Czech with English abstract): Geologicke vyzkumy na Morave a ve Slezsku v roce 2001, Moravske zemske muzeum a Sekce geologickych ved Prirodovedecke Fakulty Masarykovy University, Brno, v. 9, p. 27–28.

Elias, M., and H. Eliasova, 1984, Facies and Paleogeography of the Jurassic in Western part of the Outer Flysch Carpathians in Czechoslovakia: Sbornik geologickych ved, Geologie, Praha, v. 35, p. 75–144.

Elias, M., and H. Eliasova, 1985, Tlumacovske slienovce, *in* D. Andrusov and O. Samuel, eds. Stratigraficky slovnik Zapadnych Karpat (in Slovak): v. 2, p. 272.

Elias, M., and H. Eliasova, 1986, Elevation facies of the Malm in Moravia: Geologicky Zbornik, Geologica Carpathica, v. 37, no. 4, p. 533–550.

Elias, M., and P. Palensky, 1997, Model vzniku miocennich predhlubni na Ostravsku (in Czech with English title): Zpravy o geologickych vyzkumech v roce 1997, Cesky geologicky ustav, Praha, p. 65–66.

Elias, M., and Z. Stranik, 1963, A contribution to the origin of the Stramberk Limestone (in Czech with English summary): Vestnik Ustredniho ustavu geologickeho, Praha, v. 38, no. 2, p. 133–135.

Elias, M., and G. Wessely, 1990, The autochthonous Mesozoic on the eastern flank of the Bohemian Massif— An object of mutual geological efforts between Austria and CSSR, *in* D. Minarikova and H. Lobitzer, eds., Thirty years of geological cooperation between Austria and Czechoslovakia: Vienna, Federal Geological Survey; Prague, Geological Survey, p. 78–83.

Elias, M., W. Schnabel, and Z. Stranik, 1990, Comparison of the flysch zone of the Eastern Alps and the Western Carpathians based on recent observations, *in* D. Minarikova and H. Lobitzer eds., Thirty years of geological cooperation between Austria and Czechoslovakia: Vienna, Federal Geological Survey; Prague, Geological Survey, p. 37–46.

Elias, M., P. Martinec, D. Rehakova, and Z. Vasicek, 1996, Geology and stratigraphy of Kurovice Limestone and Tlumacov Marl Formation at the Kurovice quarry (Upper Jurassic–Lower Cretaceous, Outer Western Carpathians, Czech Republic): Bulletin of the Czech Geological Survey, Praha, v. 71, no. 3, p. 259–275.

Elias, M., P. Skupien, and Z. Vasicek, 2003, Proposal of the arrangement of the lithostratigraphy in the lower part of the Silesian unit on territory of Czech Republic (Outer Western Carpathians) (in Czech with English summary): Sbornik vedeckych praci Vysoke skoly banske-TU, R. hornicko-geologicka, Monografie 8, Ostrava, p. 7–13.

Eliasova, H., 1962a, Anmerkungen zu den Funden Ammoniten-Fauna der Teschener-Hradiste Schichten der Mahrisch-Schlesischen Beskiden (in Czech with German summary): Casopis pro mineralogii a geologii, Praha, v. 7, no. 2, p. 130–131.

Eliasova, H., 1962b, Remarks on the character of the Stramberk Limestone (in Czech with English summary): Vestnik Ustredniho ustavu geologickeho, Praha, v. 37, no. 1, p. 11–19.

Finger, F., M. Tichomirowa, C. Pin, and P. Hanzl, 2000, Relics of an early-Panafrican metabasite-metarhyolite formation in the Brno massif, Czech Republic: International Journal of Earth Sciences, v. 89, p. 328–335.

Fodor, L., 1995, From transpression to transtension: Oligocene–Miocene structural evolution of the Vienna basin and the East Alpine–Western Carpathian junction: Tectonophysics, v. 242, p. 151–182.

Foetterle, F., 1866, Geologische Karte der Markgraftschaft Mahren und des Herzogthumes Schlesien: scale 1:280,000.

Folprecht, J., and K. Pateisky, 1928, Geologie ostravsko-karvinskeho kamenouhelnehc reviru, *in* Kamenouhelne doly ostravsko-karvinskeho reviru (in Czech): Moravska Ostrava, v. 1, p. 27–340.

Francu, J., M. Radke, R. G. Schaefer, H. S. Poelchau, J. Caslavsky, and Z. Bohacek, 1996, Oil-oil and oil-source rock correlations in the northern Vienna basin and adjacent Carpathian Flysch zone (Czech and Slovak area), *in* G. Wessely and W. Liebl, eds., Oil and gas in Alpidic thrustbelts and basins of Central and Eastern Europe: Geological Society (London) Special Publication 5, p. 343–353.

Fuchs, R., and W. Hamilton, 2006, New depositional architecture for an old giant: The Matzen field, Austria, *in* J. Golonka and F. J. Picha, eds., The Carpathians and their foreland: Geology and hydrocarbon resources: AAPG Memoir 84, p. 205–219.

Fuchs, R., and G. Wessely, 1977, Die Oberkreide des Molasseuntergrundes im nördlichen Niederosterreich: Jahrbuch der Geologischen Bundesanstalt, Wien, v. 120, p. 426–435.

Fuchs, R., and G. Wessely, 1996, The autochthonous Cretaceous at the southern edge of the Bohemian Massif (Austria), *in* G. Wessely and W. Liebl, eds., Oil and gas in Alpidic thrustbelts and basins of Central and Eastern Europe: Geological Society (London) Special Publication 5, p. 249–253.

Funk, H., R. Oberhansli, A. Pfiffner, S. Schmid, and W. Wildi, 1987, The evolution of the northern margin of Tethys in eastern Switzerland: Episodes, v. 19, no. 2, p. 102–106.

Glaessner, M., 1931, Geologische Studien in der ausseren Klippenzone: Jahrbuch der Geologischen Bundesanstalt, Wien, v. 81, p. 84–97.

Glocker, E. F., 1840, Ueber den Jurakalk von Kurowitz in Mahren und ueber darin vorkommenden Aptychus imbricatus: Acta Acad. Leop.-Carol., Breslau, v. 19, Suppl. 2, p. 275–307.

Glocker, E. F., 1843, Die Menilitformation in Mahren: Amtlicher Bericht uber die einundzwanzigste Versammlung deutschen Naturforschern und Aerzte in Gratz in September 1843, Gratz, p. 139–141.

Gnojek, I., and H. Heinz, 1993, Central European (Alpine–Carpathian) belt of magnetic anomalies and its geological interpretation: Geologica Carpathica, v. 45, no. 1, p. 135–142.

Gnojek, I., and F. Hubatka, 2001, Structure of the Brunovistulian unit (E part of the Czech Republic) on the basis of seismic and magnetic data: Bulletin of the Czech Geological Survey, Praha, v. 76, no. 3, p. 157–168.

Golonka, J., 2000, Cambrian–Neogene plate tectonic maps: Wydawnictwo Universytetu Jagiellonskiego, Krakow, 125 p.

Golonka, J., and L. Gahagan, 1997, Tectonic model of the Mediterranean terranes: AAPG Bulletin, v. 65, no. 8, p. 1378.

Golonka, J., and W. Sikora, 1981, Microfacies of the Jurassic and Lower Cretaceous sedimentarily thinned deposits of the Pieniny Klippen Belt in Poland (in Polish with English abstract): Biuletyn Instytutu Geologicznego, Warszawa, v. 31, p. 7–37.

Golonka, J., N. Oszczypko, and A. Slaczka, 2000, Late Carboniferous–Neogene geodynamic evolution and paleogeography of the circum-Carpathian region and adjacent areas: Annales Societatis Geologorum Poloniae, v. 70, p. 107–136.

Golonka, J., M. Krobicki, N. Oszczypko, A. Slaczka, and T. Slomka, 2003, Geodynamic evolution and paleogeography of the Polish Carpathians and adjacent areas during

Neo-Cimmerian and preceding events (latest Triassic–earliest Cretaceous), *in* T. McCann and A. Saintot, eds., Tracing tectonic deformation using the sedimentary record: Geological Society (London) Special Publication 208, p. 138–158.

Golonka, J., L. Gahagan, M. Krobicki, F. Marko, N. Oszczypko, and A. Slaczka, 2006a, Plate-tectonic evolution and paleogeography of the circum-Carpathian region, *in* J. Golonka and F. J. Picha, eds., The Carpathians and their foreland: Geology and hydrocarbon resources: AAPG Memoir 84, p. 11–46.

Golonka, J., A. Slaczka, and F. J. Picha, 2006b, The West Carpathians and Ouachitas: A comparative study of geodynamic evolution, *in* J. Golonka and F. J. Picha, eds., The Carpathians and their foreland: Geology and hydrocarbon resources: AAPG Memoir 84, p. 787–810.

Gotzinger, G., 1954, Die Flyschzone, *in* R. Grill and H. Kupper, eds., Erlauterungen zur geologischen Karte der Umgebung von Wien: Geologisches Bundesanstalt, Wien, p. 43–93.

Grecula, P., and Z. Roth, 1978, Kinematic model of the West Carpathians (in Czech with English summary): Sbornik geologickych ved, Geologie, Praha, v. 82, p. 96–102.

Gregorova, R., 1988, The ichthyofauna of the Menilitic Formation in the locality Litencice and its stratigraphic position (in Czech with English summary): Casopis moravskeho muzea v Brne, Vedy prirodni, v. 73, p. 83–88.

Grill, R., 1947, Uber erdolgeologische Arbeiten in der Molassezone von Osterreich: Verhandlungen der Geologischen Bundesanstalt, Wien, v. 1945, p. 4–28.

Grill, R., 1953, Der Flysch, die Waschbergzone und Jungtertiar um Ernstbrunn: Jahrbuch der Geologischen Bundesanstalt, Wien, v. 96, p. 65–116.

Grill, R., 1962, Erlauterungen zur Geologischen Karte der Umgebung von Korneuburg und Stockerau: Geologisches Bundesanstalt, Wien, 52 p.

Gross, P., E. Kohler, and O. Samuel, 1984, A new lithostratigraphical division of the Inner-Carpathian Paleogene (in Slovak with English summary): Geologicke prace, Spravy, Bratislava, v. 81, p. 103–117.

Grun, W., 1984, Die Erschliessung von Lagerstatten im Untergrundder alpin-karpatischen Stirnzone Niederosterreichs: Erdol-Errdgas Zeitschrift, v. 100, p. 292–295.

Grun, W., G. Kittler, G. Lauer, A. Papp, and W. Schnabel, 1972, Studien in der Unterkreide des Wienerwaldes: Jahrbuch der geologischen Reichsanstalt, Wien, v. 115, p. 103–185.

Grygar, R., and J. Jelinek, 1999, Morphostructural analysis of the mutual relationships and tectonic influence of the Variscan and Alpine orogeny in the northern part of the Moravosilesian Zone: Geolines, v. 8, p. 25–26.

Grzybowski, J., 1897, Mikroskropowe badania namulow wiertnicznych z kopaln naftowych I, Pas potocki i okolice Krosna, II. Uwagi ogolne (in Polish): Kosmos, v. 22, p. 293–439.

Haczewski, G., 1989, Coccolith limestone horizons in the Menilite–Krosno series (Oligocene, Carpathians)—Identification, correlation, and origin (in Polish with English summary): Annales Societatis Geologicorum Poloniae, v. 59, p. 435–523.

Hamrsmid, B., and J. Krhovsky, 1987, Calcareous nannoplankton of autochthonous Paleogene sediments of Nesvacilka-1 borehole (West Carpathian Flysch belt, south Moravia, Czechoslovakia) (in Czech, English summary): Knihovnicka Zemniho plynu a nafty, Miscellanea Micropaleontologica, v. 6b, no. 2, p. 217–238.

Hamrsmid, B., J. Krhovsky, and L. Svabenicka, 1990, Biostratigraphic evaluation of calcareous nannoplankton from the autochthonous Upper Cretaceous and Paleogene of the Nesvacilka and Vranovice Grabens, SE margin of the Bohemian Massif: Vestnik Ustredniho ustavu geologickeho, Praha, v. 65, no. 3, p. 129–141.

Hanzl, P., and R. Melichar, 1997, The Brno massif: a section through the active continental margin or compressed terrane: Krystalinikum, Brno, v. 23, p. 33–58.

Hanzlikova, E., 1955, Prispevek k mikrobiostratigrafii magurskeho flyse v povodi Oravy (in Czech): Zpravy o geologickych vyzkumech v roce 1954, p. 22–25.

Hanzlikova, E., 1965a, Stratigraphie der Kreide und des Paleogens der Flyschzone der Westkarpathen: Geologicky Sbornik Slovenskej akademie vied, Geologica Carpathica, v. 16, no. 1, p. 33–64.

Hanzlikova, E., 1965b, The foraminifera of the Klentnice beds (Malm): Sbornik geologickych ved, Paleontologie, Praha, v. 5, p. 39–106.

Hanzlikova, E., 1966, Die Foraminiferen der Lhoty-Schichten: Casopis Moravskeho muzea, v. 51, p. 95–132.

Hanzlikova, E., 1969, The foraminifera of the Frydek Formation (Senonian): Sbornik geologickych ved, Paleontologie, Praha, v. 11, p. 7–84.

Hanzlikova, E., 1972a, Carpathian Upper Cretaceous foraminiferida of Moravia (Turonian–Maastrichtian): Rozpravy Ustredniho ustavu geologickeho, Praha, v. 39, p. 1–160.

Hanzlikova, E., 1972b, Mikropalaontologische Zoneneinteilung und Stratigraphie der Istebna-Schichten und ihres unmittelbaren Hangenden: Vestnik Ustredniho ustavu geologickeho, Praha, v. 47, no. 2, p. 69–77.

Hanzlikova, E., 1973, Foraminifera of the Variegated Godula Member in Moravia (Cenomanian–Turonian): Sbornik geologickych ved, Paleontologie, Praha, v. 15, p. 119–184.

Hanzlikova, E., 1976, Biostratigraphy of the Cretaceous and Paleogene borehole Jarosov-1: Vestnik Ustredniho ustavu geologickeho, Praha, v. 51, no. 3, p. 153–162.

Hanzlikova, E., 1981, Biostratigraphy and ecology of the Menilitic Formation from Moravia (in Czech with English summary): Zemni plyn a nafta, Hodonin, v. 26, no. 1, p. 29–44.

Hanzlikova, E., and A. Matejka, 1958, Das Cenoman von Dub in den Vorbergen des mahrisch-schlesischen Beskydy-Gebirges (in Czech with German summary): Vestnik Ustredniho ustavu geologickeho, Praha, v. 33, p. 170–177.

Hanzlikova, E., and A. Matejka, 1962, Prispevek ke geologii zdouneckeho bradla (in Czech): Zpravy o geologickych vyzkumech v roce 1961, p. 183–185.

Hanzlikova, E., and Z. Roth, 1964, Ein Fund der Art Lorenz in den Tesin Kalksteinen (in Czech with German summary): Vestnik Ustredniho ustavu geologickeho, Praha, v. 39, no. 5, p. 387–389.

Hanzlikova, E., E. Mencik, and V. Pesl, 1953, Poznamky ke stratigrafii a tektonice subslezskeho a slezskeho prikrovu na listu Novy Jicin (in Czech): Zpravy o geologickych vyzkumech v roce 1952, Praha, p. 15–18.

Hanzlikova, E., A. Matejka, Z. Roth, and F. Chmelik, 1955, Zpravy o vyzkumech v udoli Becvy mezi Valasskym Meziricim a Cernotinem (in Czech): Zpravy o geologickych vyzkumech v roce 1954, p. 30–40.

Hanzlikova, E., E. Mencik, and V. Pesl, 1962, Contribution to the knowledge of the Fore-Magura unit in the Moravskoslezske Beskydy Mts. (in Czech with German summary): Geologicky Sbornik Slovenskej akademie vied, Bratislava, v. 13, no. 1, p. 81–90.

Hanzlikova, E., F. Picha, and I. Cicha, 1963, Geologie der karpatischen Deckformationen auf dem Bohrfeld von Chlebovice-Rychaltice (in Czech with German and Russian summary): Sbornik Ustredniho ustavu geologickeho, Praha, v. 28, p. 435–455.

Hauer, F. v. R., et al., 1867–1874, Geologische Uebersichts-Karte der Oesterreichisch-ungarischen Monarchie: 1:576,000, K. k. geologische Reichsanstalt: Wien, zweite unveranderte Auflage: Verlag der Beck'schen Universitats—Buchhandlung (Alfred Hoelder).

Hauer, F. v. R., 1869, Geologische Ubersichtskarte der osterreichisch-ungarischen Monarchie, Blatt III, Westkarpathen: Jahrbuch der geologischen Reichsanstalt, Wien, v. 19, no. 4, p. 485–566.

Havlena, V., 1964, Geologie uhelnych lozisek, dil II (in Czech): Ceskoslovenska akademie ved, Praha, p. 1–439.

Havlena, V., 1982, The Namurian deposits of the Upper Silesian coal basin: Rozpravy Ceskoslovenske akademie ved: Rada matematicko-prirocnich Ved, Praha, v. 92, no. 7, p. 1–88.

Helwig, J., 1976, Shortening of continental crust in orogenic belts and plate tectonics: Nature, v. 260, p. 768–770.

Hladil, J., 1983, Cyclic sedimentation in the Devonian carbonates in the Macocha (in Czech with English summary): Zemni plyn a nafta, Hodonin, v. 28, no. 1, p. 1–15.

Hladil, J., 1986, Trends in development and cyclic patterns of Middle and Upper Devonian buildups: Facies, v. 15, p. 1–34.

Hladil, J., 1988, Structure and microfacies of the Middle and Upper Devonian carbonate buildups in Moravia, Czechoslovakia, *in* N. J. McMillan, A. F. Embry, and D. J. Glass, eds., Devonian of the world: Canadian Society of Petroleum Geologists Memoir 14, v. 2, p. 607–618.

Hladil, J., K. Helesicova, J. Hrubanova, P. Muller, and M. Ures, 1994, Devonian island elevations under the scope; Central Europe, basement of the Carpathian Mountains in Moravia: Jahrbuch der Geologischen Bundesanstalt, Wien, v. 136, no. 4, p. 741–750.

Hladil, J., R. Melichar, J. Otava, A. Galle, M. Krs, O. Man, P. Pruner, P. Cejchan, and P. Orel, 1999, The Devonian in the easternmost Variscides, Moravia: A holistic analysis directed towards comprehension of the original context, North Gondwana: Mid-Paleozoic terranes, stratigraphy and biota: Abhandlungen der Geologischen Bundesanstalt, Wien, v. 54, p. 27–47.

Hladil, J., L. F. Jansa, A. Tezky, K. Helesicova, and J. Hrubanova, 2000, Stratigraficke doklady o nasunovem zlomu ve vrtu Raskovice Ja-7 (in Czech): Geologicke vyzkumy na Morave a ve Slezsku v roce 1999, Moravske zemske muzeum a Sekce geologickych ved Prirodovedecke fakulty Masarykovy university, Brno, v. 7, p. 87–90.

Hohenegger, L., 1852, Geognostische Skizzen der Nordkarpaten von Schlesien und den nachsten Angrenzungen (nach den geggenwartigen Standpunkte meiner Erfahrungen): Jahrbuch der geologischen Reichsanstalt, Wien, v. 3, no. 3, p. 135–148.

Hohenegger, L., 1855, Neuere Erfahrungen aus Nordkarpaten: Jahrbuch der geologischen Reichsanstalt, Wien, v. 6, p. 304–312.

Hohenegger, L., 1861, Die geognostischen Verhaltnisse der Nordkarpathen in Schlesien und den angrenzenden Teilen von Mahren uns Galizien als Erlauterung zu der geognostischen Karte der Nordkarpathen: Gotha, p. 1–50.

Holzknecht, M., and J. Krhovsky, 1987, Paleocene to early Eocene foraminifera of the Nesvacilka Formation from Nesvacilka-1 borehole (the autochthon of the external Western Carpathians, southern Moravia, Czechoslovakia) (in Czech with English summary): Knihovnicka Zemniho plynu a nafty, Miscellanea Micropaleontologica, v. 6b, no. 2, p. 127–215.

Homola, V., M. Holzknecht, J. Juranek, I. Krystek, M. Michalicek, V. Simanek, J. Smeral, J. Uhmann, and I. Zapletalova, 1961, Die Stutzbohrung Nesvacilka-1 im SW-Teile des aussenkarpatischen Beckens in Mahren (in Czech with German summary): Prace Vyzkumneho ustavu Ceskoslovenskych naftovych dolu, Brno, v. 17, p. 4–132.

Horwitz, L., and F. Rabowski, 1929, Przewodnik do vycieczki Polskiego Tow. Geologicznego w Pieniny (in Polish): Rocznik Polskiego Towarzystwa Geologicznego, v. 6, p. 109–195.

Housa, V., 1961, Das Alter des Stramberger und Nesselsdorfers Kalksteines (in Czech with German resume): Casopis pro mineralogii a geologii, Praha, v. 6, no. 4, p. 406–418.

Housa, V., 1975, Geology and paleontology of the Stramberk Limestone (upper Tithonian) and the associated Lower Cretaceous beds: Memoires du Bureau de Recherches Geologiques et Minieres, Paris, v. 86, p. 342–349.

Housa, V., 1976, The Lower Cretaceous formations accompanying the Tithonian limestone bodies near Stramberk (in Czech with English summary): Casopis Slezskeho muzea, Vedy prirodni, Serie A, v. 25, no. 2, p. 119–131.

Housa, V., 1978, Calpionellids of the Tithonian and the Lower Cretaceous from Strambersk (Czechoslovakia) (in Czech with English summary): Zemni plyn a nafta, Hodonin, v. 23, no. 4a, p. 525–531.

Hovorka, D., and J. Spisiak, 1988, Mesozoic volcanism of the

Western Carpathians (in Slovak, English summary): Bratislava, Veda, Slovenska akademia vied, 263 p.

Hrouda, F., and Z. Stranik, 1985, The magnetic fabric of the Zdanice thrust sheet of the Flysch belt of the West Carpathians: Sedimentological and tectonic implication: Sedimentary Geology, v. 45, p. 125–145.

Hrusecky, I., D. Plasienka, and L. Pospisil, 2006, Identification of the North European platform below the eastern part of the Western Carpathian Flysch belt, *in* J. Golonka and F. J. Picha, eds., The Carpathians and their foreland: Geology and hydrocarbon resources: AAPG Memoir 84, p. 717–727.

Ibrmajer, J., 1971, Gravity anomalies and the structures of the Earth's crust on the territory of Czechoslovakia: Upper Mantle Project programme in Czechoslovakia, 1963–1970 Geophysics, Final Report: Praha, Academia, p. 47–50.

Ibrmajer, J., and M. Suk, eds., 1989, Geophysical pattern of Czechoslovakia (in Czech with English summary): Praha, Ustredni ustav geologicky, 354 p.

Jachowicz, M., and A. Prichystal, 1997, Lower Cambrian sediments in deep boreholes in South Moravia: Bulletin of the Czech Geological Survey, Praha, v. 72, no. 4, p. 329–332.

Jankowski, J., Z. Tarlowski, O. Praus, J. Pecova, and V. Petr, 1985, The results of deep electromagnetic sounding in the West Carpathians: Geophysical Journal of the Royal Astronomical Society, London, v. 80, p. 561–574.

Janocko, J., M. Pereszlenyi, D. Vass, V. Bezak, S. Jacko Jr., S. Jacko, M. Kohut, M. Polak, and J. Mello, 2006, Geology and hydrocarbon resources of the Inner Western Carpathians, Slovakia, and Poland, *in* J. Golonka and F. J. Picha, eds., The Carpathians and their foreland: Geology and hydrocarbon resources: AAPG Memoir 84, p. 569–603.

Jicinsky, W., 1885, Geognostischer Theil in: Monographie des Ostrau-Karwiner Steinkohlen-Reviers: Teschen, Berg und Huttenmannischen Verhandlungen, p. 7–42.

Jiricek, R., 1983, Geologicka stavba spodniho miocenu celni hlubiny v useku JIH (in Czech): Zemni plyn a nafta, Hodonin, v. 28, no. 2, p. 197–212.

Jiricek, R., 1987, Die stratigraphische und fazielle Unterteilung der autochtonen Sedimenten des Palaogens and SO Abhangen der Bohmische Masse (in Czech with German summary): Knihovnicka Zemniho plynu a nafty, Miscellanea Micropaleontologica, v. 6b, no. 2, p. 247–314.

Jiricek, R., 1988, Stratigrafie, paleogeografie a mocnosti sedimentu v neogenu videnske panve (in Czech): Zemni plyn a nafta, Hodonin, v. 33, no. 4, p. 583–622.

Jiricek, R., 1994, Progress in geological studies of autochthonous Paleogene deposits in south Moravia: Zemni plyn a nafta, Hodonin, v. 38, no. 3, p. 185–246.

Jiricek, R., and P. H. Seifert, 1990, Paleogeography of the Neogene in the Vienna basin and the adjacent part of the foredeep, *in* D. Minarikova and H. Lobitzer, eds., Thirty years of geological cooperation between Austria and Czechoslovakia: Vienna, Federal Geological Survey; Prague, Geological Survey, p. 89–105.

Jucha, S., 1958, Contribution on Jaslo shaly limestones in Polish Carpathians: Bulletin de l'Academie Polonaise des Sciences, Serie des Sciences Chimiques, Geologiques and Geographiques, v. 5, p. 681–688.

Jucha, S., and J. Kotlarczyk, 1961, La serie des couches a menilite et des couches de Krosno dans le flysch des Karpates (in Polish with French resume): Prace Geologiczne, Poska Akademia Nauk, oddzial v Krakowie, v. 4, 115 p.

Jurasova, F., 1974, Nannoplankton from the Menilitic Formation (lower Oligocene) at Dolni Tesice: Vestnik Ustredniho ustavu geologickeho, Praha, v. 49 no. 2, p. 91–96.

Jurasova, F., 1987, The biostratigraphy and ecology of the Sitborice beds encountered in Krepice-no-5 borehole (in Czech with English summary): Knihovnicka Zemniho plynu a nafty, Hodonin, v. 6b, p. 27–44.

Jurkova, A., 1961, Das Relief der Karbonoberflache und die klastischen Basalsedimente des Miozans im Ostrava-Karvina Revier (in Czech with German summary): Prirodovedecky casopis slezsky, Opava, v. 22, no. 3, p. 311–327.

Jurkova, A., 1971, Vyvoj badenske celni hlubiny v Moravske brane a na Ostravsku (in Czech): Geologicke prace, Spravy, Bratislava, v. 57, p. 155–160.

Jurkova, A., 1976, The structure of the Carpathian foredeep and Flysch nappes in north-eastern Moravia (in Czech with English summary): Casopis pro mineralogii a geologii, Praha, v. 21, no. 4, p. 349–362.

Jurkova, A., 1979, Confrontation of geological structure of Neoid and Variscan structural levels in the Moravian–Silesian Beskides and their foothills, *in* M. Mahel, ed., Tectonic profiles through the West Carpathians (in Czech with English summary): Bratislava, Geologicky ustav D. Stura, p. 31–36.

Jurkova, A., and E. Novotna, 1974, Facie a stavba karpatu na sv. Morave (in Czech): Sbornik Geologickeho pruzkumu (Ostrava), v. 12, p. 73–88.

Jurkova, A., V. Molcikova, P. Ctyroky, and J. Policky, 1983, New Eggenburgian finds in northeastern Moravia (in Czech with English summary): Geologicke prace, Spravy, Bratislava, v. 79, p. 153–168.

Juttner, K., 1922, Enstehung und Bau der Pollauer Berge: Nikolsburg, H. Bartosch, 67 p.

Juttner, K., 1928, Tektonik und geologische Geschichte der Pollauer Berge: Verhandlungen des Naturforschenden Vereines in Brunn, Brno, v. 61, p. 1–19.

Juttner, K., 1933, Zur Stratigraphie und Tektonik des Mesozoikums der Pollauer Berge: Verhandlungen des Naturforschenden Vereines in Brunn, Brno, v. 64, p. 15–31.

Juttner, K., 1942, Beitrage zur Geologie des Gebietes Pollauer Berge: Niederdonau Natur und Kultur, K. Kuhne, Wien, v. 16, p. 1–34.

Kalabis, V., 1949, Les poissons avec les organs lumiers du paleogene de Moravie (couches menilitiques) (in Czech with French resume): Casopis Zemskeho musea v Brne, v. 32, p. 131–174.

Kalasek, J., et al., Vysvetlivky k prehledne geologicke mape

CSSR, scale 1:200,000, M-33-XXIX-Brno, Praha, Ustredni ustav geologicky, 256 p.

Kalvoda, J., 1995, Devonske panve pri okraji vychodni Avalonie na Morave: Geologicke vyzkumy na Morave a ve Slezsku v roce 1994 (in Czech): Moravske zemske muzeum a Sekce geologickych ved prirodovedecke fakulty Masarykovy University, Brno, v. 2, p. 48–50.

Kapounek, J., A. Papp, and K. Turnowsky, 1960, Grundzuge der Gliederung von Oligozan und alteren Miozan in Niederosterreich: Verhandlungen der Geologischen Bundesanstalt, Wien, p. 217–226.

Karnkowski, P., 1993, Carpathians and Carpathian foredeep, *in* Oil and gas fields in Poland (in Polish with English resume): Krakow, Towarzystwo Geosynoptykow "Geos," v. 2, 256 p.

Karnkowski, P. H., and W. Ozimkowski, 1998, The distribution of oil- and gasfields in relation to satellite image interpretation: An example from the Polish East Carpathians and the adjacent foredeep: Journal of Petroleum Geology, v. 21, no. 2, p. 213–231.

Kettner, R., 1950, Geologicka stavba severni casti Moravskeho krasu a oblasti prilehlych (in Czech): Rozpravy II. Tridy Ceskoslovenske Akademie ved, Praha, v. 59, no. 11, p. 1–29.

Kettner, R., 1970, Geologicky a geomorfologicky vyvoj Moravskeho krasu a jeho okoli, *in* K. Absolon, ed., Moravsky kras (in Czech): Praha, v. 2, p. 261–284.

Klvana, J., 1891, Das sudmahrische Eruptivgebiet: Verhandlungen des Naturforschenden Vereines in Brunn, Brno, v. 29, p. 1–3.

Kodym, O., 1923, Geologie du Flysch dans les environs de Valasske Klobouky en Moravie (in Czech with French resume): Sbornik Statniho geologickeho ustavu Praha, v. 2, p. 43–54.

Koltun, Y. U., J. Espitalie, M. Kotarba, F. Roure, N. Ellouz, and P. Kosakowski, 1998, Petroleum generation in the Ukrainian external Carpathians and the adjacent foreland: Journal of Petroleum Geology, v. 21, no. 3, p. 265–288.

Kostelnicek, P., V, Ciprys, and J. Berka, 2006, Examples of recently discovered oil and gas fields in the Carpathian foredeep and in the European foreland plate underneath the Carpathian thrust belt, Czech Republic, *in* J. Golonka and F. J. Picha, eds., The Carpathians and their foreland: Geology and hydrocarbon resources: AAPG Memoir 84, p. 177–189.

Kotanski Z., 1997, Geology of Poland on the map of horizontal cutting (in Polish with English summary): Przeglad Geologiczny, Warszawa, v. 45, no. 6, p. 605–618.

Kotarba, M., S. Szafran, and J. Espitalie, 1987, A study of organic matter and natural gases of the Miocene sediments in the Polish part of the Carpathian foredeep: Chemical Geology, v. 64, p. 197–207.

Kovac, M., 2000, Geodynamicky, paleogeograficky, a strukturny vyvoj Karpato-pannonskeho regionu v miocene: Novy pohled na neogenne panvy Slovenska (in Slovak): Bratislava, Veda, Vydavatelstvo Slovenskej Akademie vied, 202 p.

Kovac, M., A. Nagymarosy, N. Osczypko, L. Czontos, A. Slaczka, M. Marunteanu, L. Matenco, and E. Marton, 1998, Palinspastic reconstruction of the Carpathian–Pannonian region during the Miocene, *in* M. Rakus, ed., Geodynamic development of the Western Carpathians: Special Publication, Geological Survey of the Dionyz Stur Institute, Bratislava, p. 189–217.

Kozur, H., 1991, The evolution of the Meliata–Hallstatt Ocean and its significance for the early evolution of the Eastern Alps and Western Carpathians: Palaeogeography, Palaeoclimatology, Palaeoecology, v. 87, p. 109–135.

Krejci, J. J., 1993, Zdanice–Krystalinikum field–Czechoslovakia, Carpathian foredeep, Moravia, *in* N. N. Foster and E. A. Beaumont, eds., Structural Traps VIII: Treatise of Petroleum Geology Atlas of Oil and Gas Fields: AAPG, p. 153–175.

Krejci, O., J. Francu, P. Muller, M. Pereszlenyi, and Z. Stranik, 1994, Geologic structure and hydrocarbon generation in the Carpathian flysch belt of southern Moravia: Bulletin of the Czech Geological Survey, v. 69, p. 13–26.

Krejci, O., J. Francu, H. S. Poelchau, P. Muller, and Z. Stranik, 1996, Tectonic evolution and oil and gas generation at the border of the North European platform with the West Carpathians (Czech Republic), *in* G. Wessely and W. Liebl, eds., Oil and gas in Alpidic thrustbelts and basins of Central and Eastern Europe: European Association of Geoscientists and Engineers Special Publication 5, p. 177–186.

Krejci, O., P. Hanzl, F. Hubatka, J. Sedlak, and J. Svancara, 2002, Hercynian and Alpine brittle deformation of the Bruno–Vistulicum and its sedimentary cover units in the footwall of the Outer Western Carpathians (Eastern part of the Czech Republic): Krystalinikum, Moravian Museum, Brno, v. 28, p. 145–167.

Krhovsky, J., 1981, Microbiostratigraphic correlations in the Outer Flysch units of the southern Moravia and influence of the eustacy on their paleogeographical development (in Czech with English summary): Zemni plyn a nafta, Hodonin, v. 26, p. 665–688.

Krhovsky, J., B. Hamrsmid, and L. Svabenicka, 1990, Micropaleontological evidence of the autochthonous Upper Cretaceous, Paleocene, Eocene and Oligocene in the sediments of Vranovice and Nesvacilka trenches (in Czech with English resume): Zbornik paleontologickej konferencie "Sucasne problemy a trendy v ceskoslovenskej paleontologii:" Bratislava, Geologicky Ustav D. Stura, v. 1, p. 45–51.

Krhovsky, J., M. Adamova, J. Hladikova, and H. Maslowska, 1992, Paleoenviromental changes across the Eocene/Oligocene boundary in the Zdanice and Pouzdrany units (Western Carpathians, Czechoslovakia): The long-term trend and orbitally forced changes in calcareous nannofossil assemblages, *in* B. Hamrsmid and J. Young, eds., Proceedings of the Fourth International Nannoplankton Association Conference, Prague 1991, Nannoplankton Research, v. 2: Tertiary Biostratigraphy and Paleoecology; Quaternary coccoliths: Knihovnicka zemniho plynu a nafty, Hodonin, v. 14b, p. 105–187.

Krhovsky, J., M. Bubik, B. Hamrsmid, and M. Stastny, 1995, Lower Miocene of the Pouzdrany unit, the West Carpathian Flysch belt, Southern Moravia: New results in Tertiary of West Carpathians II: Knihovnicka Zemniho plynu a nafty, Hodonin, v. 16, p. 73–83.

Kroll, A., H. Heinz, R. Jiricek, B. Meurers, and W. Seibere, 1993, Erlauterungen zu den Karten uber den Untergrund des Wiener Beckens und der angrenzenden Gebiete: Karten 1:200,000, Geologische Bundesanstalt, Wien.

Krystek, I., 1955, Alkalische Eruptivgesteine im sudostlichen Mahren (in Czech with German and Russian resume): Geologicke prace Geologickeho ustavu Dionyza Stura, Bratislava, v. 41, p. 103–130.

Krystek, I., and A. Samuel, 1979, Stratifical and tectonical-paleogeographical position of limestones at the north of Brno (Kurim) (in Czech with English resume): Geologicke prace, Spravy, Bratislava, v. 71, p. 93–110.

Krzywiec, P., 2001, Contrasting tectonic and sedimentary history of the central and eastern parts of the Polish Carpathian foredeep basin— Results of seismic data interpretation: Marine and Petroleum Geology, v. 18, p. 13–38.

Krzywiec, P., 2002, Mid-Polish Trough inversion-seismic examples, main mechanism and its relationship to the Alpine–Carpathian collision: European Geophysical Society Special Publication, Stephan Mueller Series 1, p. 233–258.

Ksiazkiewicz, M., 1960, Outline of the paleogeography in the Polish Carpathians (in Polish with English summary): Prace Instytutu Geologicznego, Krakow, v. 30, no. 2, p. 209–231.

Ksiazkiewicz, M., 1977, Tectonics in the Carpathians, *in* W. Pozaryski, ed., Geology of Poland, v. IV: Tectonics: Warsawa, Poland, Wydawnictwa Geologiczne, p. 476–604.

Kusmierek, J., 1994, Evolution of the central Carpathian oil basin: A quantitative interpretation, *in* F. Roure, F. Shein, and A. Sidorov, eds., Geodynamic evolution of sedimentary basins: Paris, Editions Technip, p. 281–304.

Ladwein, H. W., 1988, Organic geochemistry of Vienna basin: Model for hydrocarbon generation in overthrust belts: AAPG Bulletin, v. 72, no. 5, p. 586–599.

Ladwein, H. W., F. Schmidt, P. Seifert, and G. Wessely, 1991, Geodynamics and generation of hydrocarbons in the region of the Vienna basin, Austria, *in* A. M. Spencer, ed., Generation, accumulation, and production of Europe's hydrocarbons: European Association of Petroleum Geoscientists Special Publication 4, p. 289–305.

Lafargue, E., N. Ellouz, and F. Roure, 1994, Thrust-controlled exploration plays in the Outer Carpathians and their foreland (Poland, Ukraine and Romania): First Break, v. 12, no. 2, p. 69–79.

Lankreijer, A., M. Kovac, S. Cloetingh, P. Pitonak, M. Hloska, and C. Biermann, 1995, Quantitative subsidence analysis and forward modelling of the Vienna and Danube basins: Thin-skinned versus thick-skinned extension: Tectonophysics, v. 252, p. 433–451.

Laskerev, V. N., 1924, Sur les equivalents du Sarmatian superieur en Serbie, *in* Recueil de travaux offert a M. Jovan Cvijic par ses amis et collaborateurs, Drzavna Stamparia Kralovine Srba, Chrvata i Slovenata, Beograd, p. 73–75.

Le Pichon, X., F. Bergerat, and M. J. Roulet, 1988, Plate kinematics and tectonics leading to the Alpine belt formation; a new analysis, *in* S. Clark Jr., B. C. Burchfiel, and J. Suppe, eds., Processes in continental lithospheric deformation: Geological Society of America Special Paper 218, p. 111–131.

Lesko, B., 1960, Palaogen der Klippenzone in der Ostslowakei (in Slovak with German resume): Geologicky Sbornik Slovenskej akademie vied, v. 11, no. 1, p. 95–103.

Lesko, B., T. Durkovic, V. Gasparikova, A. Kulmannova, and O. Samuel, 1978, New data on geology of the Myjavska pahorkatina Upland based on the results of the drill hole Lubina-1) (in Slovak with English summary): Geologicke prace, Spravy, v. 70, p. 35–55.

Lesko, B., B. Beranek, and Z. Varga, 1980, Cisaillements horizontaux profonds sous les Carpates occidentales a la lumiere des connaisances geophysiques: Geophysiques: Revue de Geologie Dynamique et de Geographie Physique, Paris, v. 22, nos. 4–5, p. 225–266.

Lexa, J., and V. Konecny, 1974, The Carpathian volcanic arc: A discussion: Acta Geologica Academiae Scientiarium Hungaricae, Budapest, v. 18, nos. 3–4, p. 279–293.

Lickorish, W. H., and M. Ford, 1998, Sequential restoration of the external Alpine Digne thrust system, SE France, constrained by kinematic data and synorogenic sediments, *in* A. Mascle, C. Puigdefabregas, H. P. Luterbacher, and M. Fernandez, eds., Cenozoic foreland basins of Western Europe: Geological Society (London) Special Publication 134, p. 189–211.

Liebus, A., and V. Uhlig, 1902, Uber einige Fossilien aus der karpatische Kreide: Beitrage zur Palaontologie und Geologie Ostereich-Ungarns und des Orients, v. 14, p. 113–130.

Lugeon, M., 1903, Les nappes de recouvrement de la Tatra et l'origine des Klippes des Carpathes: Bulletin des Laboritoires Geologie Geographie, Physique, Mineralogie et Pertrographie de l'Universite Lausanne, v. 4, p. 1–51.

Mahel, M., 1981, Island character of Klippen Belt; Vahicum— Continuation of southern Penninicum in West Carpathians: Geologicky Zbornik, Geologica Carpathica, v. 32, p. 293–305.

Mahel, M., and T. Buday, eds., et al., 1968, Regional geology of Czechoslovakia: Part II— The West Carpathians: Praha, Geological Survey of Czechoslovakia in Academia, 723 p.

Malkovsky, M., 1976, Saxonische Tektonik der Bohmischen Masse: Geologische Rundschau, Stuttgart, v. 65, p. 127–143.

Malkovsky, M., 1987, The Mesozoic and Tertiary basins of the Bohemian Massif and their evolution: Tectonophysics, v. 137, p. 31–42.

Maly, L., 1993, Generation of the Boskovice Furrow Permo-Carboniferous sedimentary basin and development of upper Stephanian sedimentation in the Rosice-Oslavany coal basin, *in* A. Prichystal, V. Obstova, and M. Suk, eds., Geologie Moravy a Slezska (in Czech, English summary): Moravske zemske muzeum a Sekce geologickych ved prirodovedecke fakulty Masarykovy university, Brno, p. 187–199.

Marchant, R. H., and G. M. Stampfli, 1997, Subduction of continental crust in the Western Alps: Tectonophysics, v. 269, p. 217–235.

Marschalko, R., 1978, Vyvoj sedimentarnych bazenov a paleotektonicke rekonstrukcie Zapadnych Karpat, *in* J. Vozar, R. Marschalko, M. Misik, and J. Nemcok, eds., Paleogeograficky vyvoj Zapadnych Karpat: Konferencie, Sympozia, Seminare, Geologicky ustav Dionyza Stura, Bratislava, p. 49–80.

Matejka, A., 1949, Geologicka studie z okoli Valasskeho Mezirici: Sbornik. Statniho geologickeho ustavu, Praha, v. 16, p. 643–693.

Matejka, A., 1952, Contribution to the geology of the basin of the Horni Ostravice in the Moravskoslezske Beskydy (in Czech with English and Russian summary): Sbornik Statniho geologickeho ustavu, Praha, v. 19, p. 621–646.

Matejka, A., and Z. Roth, 1949a, Predbezne poznatky ke geologii Moravskoslezskych Beskyd (in Czech with English and Russian summary): Sbornik Statniho geologickeho ustavu, Praha, v. 16, p. 293–328.

Matejka, A., and Z. Roth, 1949b, The geology of the Magura Flysch Group in the drainage basin of the Kysuca (in Czech with Russian and English summary): Sbornik Statniho geologickeho ustavu, Praha, v. 16, p. 521–619.

Matejka, A., and Z. Roth, 1949c, Zprava o geologickem mapovani v okoli Roznova pod Radhostem (in Czech): Vestnik Statniho geologickeho ustavu, Praha, v. 24, p. 45–56.

Matejka, A., and Z. Roth, 1950, Remark on the main tectonic units of the Magura Flysch in Czechoslovakia (in Czech, English summary): Vestnik Statniho geologickeho ustavu, Praha, v. 25, p. 301–308.

Matejka, A., and Z. Roth, 1954, Mapovaci zprava z terenu mezi Ostravici a Stonavkou (Moravskoslezske Beskydy) (in Czech): Zpravy o geologickych vyzkumech v roce 1953, p. 97–106.

Matejka, A., and Z. Roth, 1955, Predbezna zprava o geologickem mapovani okoli Stramberka (in Czech): Zpravy o geologickych vyzkumech v roce 1954, p. 110–113.

Matejka, A., and Z. Roth, 1956, The geology of the Magura Flysch Group in the northern river basin of the Vah between Bytca and Trencin (in Czech with English and Russian summary): Rozpravy Ustredniho ustavu geologickeho, Praha, v. 22, 332 p.

Mencik, E., 1966, Tektonische Gliederung der silesischen Einheit in der Mahrisch-Schlesischen Beskiden: Vestnik Ustredniho ustavu geologickeho, Praha, v. 41, no. 5, p. 379–382.

Mencik, E., 1969, Tectonics of the Magura Flysch Group of Czechoslovakia (in Czech with English summary): Geologicke prace, Spravy, Bratislava, v. 47, p. 91–128.

Mencik, E., 1973, Problems of the Fore-Magura unit in the Moravskoslezske Beskydy Mts. (in Czech with English and Russian resumes): Vestnik Ustredniho ustavu geologickeho, Praha, v. 48, no. 2, p. 73–77.

Mencik, E., and E. Hanzlikova, 1983, Geology of the Outer Flysch belt in Moravsko-slezske Beskydy Mts., *in* O. Samuel and V. Gasparikova, eds., European Colloquy on Micropaleontology (excursion guide: Conferencie, Sympozia, Seminare): Bratislava, Geologicky Ustav D. Stura, p. 131–140.

Mencik, E., and V. Pesl, 1955, Stratigraphy and tectonic of the Subsilesian nappe on the map of Novy Jicin (in Czech with English and Russian summary): Sbornik Ustredniho ustavu geologickeho, Praha, v. 21, Geologie, no. 1, p. 149–180.

Mencik, E., and V. Pesl, 1958, Die Oberkreideschichten in dem Magura-Flysch in Mahren (in Czech with German summary): Prace Brnenske zakladny Ceskoslovenske Akademie Ved, 30/1958, 370, v. 5, p. 203–217.

Mencik, E., et al., 1983, Geology of the Moravskoslezske Beskydy Mountains and Podbeskydska pahorkatina Upland (in Czech with English summary): Praha, Oblastni regionalni geologie CSR, Ustredni ustav geologicky, 304 p.

Meulenkamp, J. E., M. Kovac, and I. Cicha, 1996, On late Oligocene to Pliocene depocenter migrations and the evolution of Carpathian–Pannonian system: Tectonophysics, v. 266, p. 301–317.

Michalik, J., and J. Sotak, 1990, Lower Cretaceous marine buildups in the Western Carpathians and their relationship to pelagic facies: Cretaceous Research, v. 11, p. 211–227.

Milan, G., and R. Sauer, 1996, Ultra-deep drilling in the Vienna basin— A review of geological results, *in* G. Wessely and W. Liebl, eds., Oil and gas in Alpidic thrustbelts and basins of Central and Eastern Europe: European Association of Geoscientists and Engineers Special Publication 5, p. 109–117.

Misik, M., 1979, Pieniny Klippen Belt and the global tectonics model, *in* M. Mahel and P. Reichwalder, eds., Czechoslovak geology and global tectonics G: Bratislava, Veda, p. 89–101.

Moryc, W., 1995, Terrestrial Paleogene formations in the area of Carpathian foredeep (in Polish with English summary): Nafta-Gaz, Krakow, v. 61, no. 5, p. 181–195.

Nachtmann, W., and L. Wagner, 1987, Mesozoic and early Tertiary evolution of the Alpine foreland in Upper Austria and Salzburg, Austria: Tectonophysics, v. 137, p. 61–76.

Nehyba, S., 1997, Miocene volcanoclastics of the Carpathian foredeep in Czech Republic: Bulletin of the Czech Geological Survey, Praha, v. 72, no. 4, p. 311–324.

Nehyba, S., and R. Roetzel, 1999, Lower Miocene volcanoclastics in South Moravia and Lower Austria: Jahrbuch der geologischen Bundesanstalt, Wien, v. 141, p. 473–490.

Nekvasilova, O., 1977, Rhynchonellida (Brachiopoda) from the Lower Cretaceous of Stramberk (Czechoslovakia): Sbornik geologickych ved, Paleontologie, v. 19, p. 45–78.

Nemcok, K., J. F. Keith Jr., and D. G. Neese, 1996, Development and hydrocarbon potential of the Central

Carpathian Paleogene basin, West Carpathians, Slovak Republic, *in* P. A. Ziegler and F. Horvath, eds., Peri-Tethys Memoir 2: Structure and prospects of Alpine basins and forelands: Memoires du Museum National d'Histoire Naturelle, N.S. Serie C, Paris, 170, p. 321–342.

Nemcok, M., 1993, Transition from convergence to escape: field evidence from the West Carpathians: Tectonophysics, v. 217, p. 117–142.

Nemcok, M., J. J. Houghton, and M. P. Coward, 1998a, Strain partitioning along the western margin of the Carpathians: Tectonophysics, v. 292, p. 119–143.

Nemcok, M., L. Pospisil, J. Lexa, and R. A. Donelick, 1998b, Tertiary subduction and slab break-off model of the Carpathian–Pannonian region: Tectonophysics, v. 295, nos. 3–4, p. 307–340.

Nemcok, M., L. Pospisil, I. Hrusecky, and T. Zsiros, 2006, Subduction in the remnant Carpathian Flysch Basin, *in* J. Golonka and F. J. Picha, eds., The Carpathians and their foreland: Geology and hydrocarbon resources: AAPG Memoir 84, p. 767–785.

Nemec, F., 1973, Geologie des autochthonen Palaogene an den sudostlichen Hangen der Bohmischen Masse in Mahren (in Czech with German summary): Sbornik Geologickych ved, Praha, Geologie, v. 24, p. 125–174.

Neumann, J., 1907, Die Oxfordfauna von Cetechowitz: Beitrage zur Palaontologie und Geologie Osterreich-Ungarns und des Orients, Wien, v. 20, no. 1, p. 1–67.

Neumayer, M., 1870, Jurastudien I. Die Klippe von Cetechowitz in Mahren: Jahrbuch der geologischen Reichsanstalt, Wien, v. 20, no. 4, p. 549–556.

Nowak, J., 1927, Zarys tektoniki Polski (in Polish): Krakow, Zjazd Geografov i Etnografov Slowianskich v Polsce, 160 p.

Nowak, W., 1965, Sur l' origine organique des calcaires de Jaslo des couches menilitiques et de Krosno dans les Karpates flyscheuses (Oligocene), *in* Carpatho-Balcan Association, VII Congress, Sofia, September 1965, Reports, pt. 2, v. 1, p. 287–290.

Ondra, P., and J. Hanak, 1992, Correlation of some flysch sequences of eastern Moravia based on petrophysical parameters (in Czech with English summary): Bulletin of the Czech Geological Survey, Praha, v. 67, no. 2, p. 117–130.

Oppenheim, P., 1913, Fauna und Alter der Konglomerate von Zdaunek bei Kremsier: Jahrbuch der geologischen Reichsanstalt, Wien, v. 63, no. 4, p. 695–710.

Oppenheim, P., 1922, Uber Alter und Fauna des Tertiarhorizontes der Niemtschitzer Schichten in Mahren: Selbstverlag des Verfassers, Berlin, 106 p.

Oppenheimer, J., 1907, Der Malm der Schwedenschanze bei Brunn: Beitrage zur Palaontologie und Geologie Osterreich-Ungarns und des Orients, Wien, v. 20, p. 221–271.

Oppenheimer, J., 1913, Der Malm von Freistadtl in Mahren: Verhandlungen des Naturforschenden Vereines in Brunn, Brno, v. 52, p. 277–288.

Oppenheimer, J., 1926, Der malm der Stranska skala bei Brunn: Casopis Moravskeho zemskeho musea, Brno, v. 24, p. 145–176.

Oppenheimer, J., 1932, Der Malm des Hady Berges bei Brunn: Verhandlungen des Naturforschenden Vereines in Brunn, Brno, v. 63, p. 1–35.

Oszczypko, N., 1997, The early–middle Miocene Carpathian peripheral foreland basin (Western Carpathians), Poland: Przeglad Geologiczny, Warszawa, v. 45, no. 10, p. 1054–1063.

Oszczypko, N., 1998, The Western Carpathian foredeep—Development of the foreland basin in front of the accretionary wedge and its burial history (Poland): Geologica Carpathica, v. 49, no. 6, p. 415–431.

Oszczypko, N., and M. Oszczypko-Clowes, 2003, The Aquitanian arine deposits in the basement of Polish Western Carpathians and its paleogeographical and paleotectonic implications: Acta Geologica Polonica, v. 53, no. 2, p. 101–122.

Oszczypko, N., A. Luczinska-Anckiewicz, P. Gedl, and E. Malata, 2000, Paleogene autochthonous deposits at the basement of the Polish Outer Carpathians and their paleogeographical implications: Slovak Geological Magazine, v. 6, nos. 2–3, p. 143–145.

Oszczypko, N., P. Krzywiec, I. Popadyuk, and T. Peryt, 2006, Carpathian foredeep basin (Poland and Ukraine): Its sedimentary, structural, and geodynamic evolution, *in* J. Golonka and F. J. Picha, eds., The Carpathians and their foreland: Geology and hydrocarbon resources: AAPG Memoir 84, p. 293–350.

Palensky, P., 1988, Prispevek k poznani litofacialniho vyvoje autochtonniho karpatu severni casti karpatske predhlubne na Morave (in Czech): Zemni plyn a nafta, Hodonin, v. 33, no. 1, p. 90–116.

Palensky, P., J. Sikula, and E. Novotna, 1995, Paleografie a litologie autochtonniho miocenu na severni Morave (in Czech): Zemni plyn a nafta, Hodonin, v. 16, p. 119–128.

Paraschiv, D., 1979, Romanian oil and gas fields: Institute of Geology and Geophysics, Bucharest, A Series, Geological Prospecting and Exploration, v. 13, 382 p.

Paul, C. M., 1869, Die geologischen Verhaltnisse des nordlichen Saroser-und Zempliner Comitates: Jahrbuch der geologischen Reichsanstalt, Wien, v. 19, p. 265–279.

Paul, C. M., 1890, Die Karpathensandsteine des mahrisch-ungarischen Grenzgebirges: Jahrbuch der geologischen Reichsanstalt, Wien, v. 40, p. 447–513.

Paul, C. M., 1893, Das Sudwestende der Karpathensandsteinzone: Jahrbuch der geologischen Reichsanstalt, Wien, v. 43, p. 199–256.

Paul, C. M., and L. Tausch, 1899, Geologische Spezialkarte Austerlitz (4358): Scale 1:75,000, Neu Aufgenommen und Bearbeit in den Jahren 1890–1892: Wien, Verlag der geologischen Reichsanstalt, 1 sheet.

Paul, C. M., and E. Tietze, 1877, Studien in der Sandsteinzone der Karpathen: Jahrbuch der geologischen Reichsanstalt, Wien, v. 27, p. 33–130.

Pecova, J., W. Petr, and O. Praus, 1979, Internal electrical conductivity distribution on Czechoslovak territory: Geodynamic investigation in Czechoslovakia: Bratislava, Veda, p. 119–127.

Pesek, J., V. Holub, J. Jaros, L. Maly, K. Martinek, V. Prouza, J. Spudil, and R. Tazler, 2001, Geologie a loziska svrchnopaleozoickych limnickych panvi Ceske republiky (in Czech): Praha, Cesky geologicky ustav, 243 p.

Pesl, V., 1967, Die Vormagura-Einheit auf dem Gebiet der CSSR: Vestnik Ustredniho ustavu geologickeho, Praha, v. 42, no. 1, p. 45–48.

Pesl, V., 1968, Lithofacies of Paleogene in the Magura unit of the Outer Carpathian Flysch on the territory of Czechoslovakia and Poland (in Czech with English summary): Zbornik geologickych vied, Zapadne Karpaty, Bratislava, v. 3, p. 179–212.

Pesl, V., and E. Hanzlikova, 1983, Chvalcov Schichten-ein neues Formationglied der Vor-Magura-Einheit: Vestnik Ustredniho ustavu geologickeho, Praha, v. 58, no. 5, p. 263–274.

Pesl, V., and I. Krystek, 1965, Ergebnisse der Fazialanalyse und sedimentologischen Studien im Paleogen des westlichen Teils der Magura-Flysch-Gruppe (Westkarpaten) (English, French, and Russian resume): Geologische Rundschau, v. 55, no. 3, p. 643–665.

Pesl, V., and E. Mencik, 1956, Zprava o vyzkumech v sirsim okoli zdouneckeho bradla (in Czech): Zpravy o geologickych vyzkumech v roce 1955, p. 170–172.

Pesl, V., and E. Mencik, 1959, Prehledne geologicke mapovani magurskeho flyse na vychodnim Slovensku (in Czech): Zpravy o geologickych vyzkumech v roce 1957, p. 181–186.

Pesl, V., E. Mencik, and E. Hanzlikova, 1964, Die Vormagura-Serie sudlich von Jablunkov (Moravskoslezske Beskydy) (in Czech with German resume): Vestnik Ustredniho Ustavu geologickeho, Praha, v. 39, no. 3, p. 189–199.

Petraschek, W., 1907, Die Kreideklippe von Zdaunek bei Kremsier: Verhandlungen der k. k. geologischen Reichsanstalt, Wien, no. 13, p. 307–312.

Petraschek, W., and T. Fuchs, 1912, Die tertiaren Schichten im Liegenden der Kreide des Teschener Hugellandes: Verhandlungen der k. k. geologischen Reichsanstalt, Wien, no. 2, p. 75–95.

Pfiffner, O. A., 1986, Evolution of the north Alpine foreland basin in the Central Alps, *in* P. A. Allen and P. Homewood, eds., Foreland basins: International Association of Sedimentologists Special Publication 8, p. 219–228.

Picha, F., 1963, Results of the sedimentological investigation of the Pouzdrany unit on the map M-33-106-C Zidlochovice (in Czech with English resume): Zpravy o geologickych vyzkumech v roce 1963, p. 234–236.

Picha, F., 1964a, Manganese ores in the Paleogene of the Central Carpathians and their genesis (in Czech with English resume): Vestnik Ustredniho ustavu geologickeho, Praha, v. 34, p. 251–259.

Picha, F., 1964b, Ergebnisse der sedimentologischen Untersuchungen im Paleogen der Zentralkarpaten (in Czech with German summary): Sbornik geologickych vied, Zapadne Karpaty, Bratislava, v. 2, p. 77–125.

Picha, F., 1965, Sedimentologic methods applied for the reconstruction of the Carpathian Outer Flysch zone sedimentation basins in southern Moravia: Carpatho-Balkan Geological Association, VII Congress, Sofia, September 1965, Reports, Part II/1, p. 315–320.

Picha, F., 1969, Clay minerals in the Carpathian Flysch: Acta Universitatis Carolinae, Geologica, Praha, v. 1, p. 31–36.

Picha, F., 1974, Ancient submarine canyons of the Carpathian miogeosyncline, *in* R. H. Dott Jr. and R. H. Shaver, eds., Modern and ancient geosynclinal sedimentation: SEPM Special Publication 19, p. 126–127.

Picha, F., 1979a, Ancient submarine canyons of Tethyan continental margins, Czechoslovakia: AAPG Bulletin, v. 63, no. 1, p. 67–86.

Picha, F., 1979b, Nesvacilka Formation— A new name for the strata of the autochthonous Paleogene in south Moravia (in Czech with English summary): Casopis pro mineralogii a geologii, Praha, v. 24, no. 3, p. 305–311.

Picha, F., 1996, Exploring for hydrocarbons under thrust belts— A challenging new frontier in the Carpathians and elsewhere: AAPG Bulletin, v. 80, p. 1547–1564.

Picha, F. J., 2002, Late orogenic strike-slip faulting and escape tectonics in frontal Dinarides–Hellenides, Croatia, Yugoslavia, Albania, and Greece: AAPG Bulletin, v. 86, p. 1659–1671.

Picha, F., and E. Hanzlikova, 1965, Die Juraklippen in der Zdanice-Einheit bei Zajeci: Vestnik Ustredniho ustavu geologickeho, Praha, v. 40, no. 3, p. 175–179.

Picha, F., J. Weiss, and E. Hanzlikova, 1966, Petrographical investigation of several conglomerate localities in Zdanice unit (in Czech with English resume): Geologicke prace, Zpravy, Bratislava, v. 38, p. 31–45.

Picha, F., E. Hanzlikova, and I. Cicha, 1968, Geology of the Cejc–Zajeci zone (in Czech with English resume): Sbornik geologickych ved, Praha, Geologie, v. 13, p. 37–74.

Picha, F., Z. Adam, and J. Krejci, 1971, Geologic evaluation of borehole Kobyli-1 and the seismic profile Nikolcice–Hodonin (in Czech with English summary): Sbornik geologickych ved, Praha, Series G, v. 19, p. 151–166.

Picha, F., E. Hanzlikova, and J. Cahelova, 1978, Fossil submarine canyons of the Tethyan margins of the Bohemian Massif in southern Moravia: Vestnik Ustredniho ustavu geologickeho, Praha, v. 53, p. 257–272.

Picha, F. J., and K. E. Peters, 1998, Biomarker oil-to-source rock correlation in the Western Carpathians and their foreland, Czech Republic: Petroleum Geoscience, v. 4, p. 289–302.

Picha, F. J., and Z. Stranik, 1999, Late Cretaceous to early Miocene deposits of the Carpathian foreland basin in southern Moravia: International Journal of Earth Sciences, v. 88, p. 475–495.

Plasienka, D., 1995, Passive and active margin history of the northern Tatricum (Western Carpathians, Slovakia): Geologische Rundschau, v. 84, p. 748–760.

Plicka, M., 1978, On the problem of the Fore-Magura unit near Valasske Mezirici (in Czech with English abstract): Geologicke prace, Spravy, v. 70, p. 177–182.

Pokorny, V., 1947, Contribution a l'etude de la geologie des environs de Hustopece (feuille 4457/2) (in Czech with French resume): Sbornik Statniho Geologickeho Ustavu CSR, Praha, v. 14, p. 325–360.

Pokorny, V., 1954, Microstratigraphic division of Eocene strata with planktonic foraminifera in the area of Zdanice Forest: Vestnik Ustredniho ustavu geologickeho, Praha, v. 29, p. 266–270.

Pokorny, V., 1959, Beitrag der Mikropalantologie zur Tektonik der Pallauer Berge (in Czech with German resume): Casopis pro mineralogii a geologii, Praha, v. 4, no. 1, p. 40–45.

Pokorny, V., 1960, Microbiostratigraphie et biofacies du flysch Carpatique de la Moravie meridionale (Tchecoslovaquie): Revue de l'Institut Francais du Petrole, Paris, v. 5, nos. 7–8, p. 1099–1141.

Pokorny, V., 1961, Contribution to the microstratigraphical division of the "Hustopece Marls" in the vicinity of Trkmanec and Zajeci (Zdanice unit, southern Moravia, Czechoslovakia) (in Czech with English resume): Casopis pro mineralogii a geologii, Praha, v. 4, no. 3, p. 305–315.

Pokorny, V., 1973, The ostracoda of the Klentnice Formation (Tithonian?), Pavlovske vrchy Mountains, Czechoslovakia: Rozpravy Ustredniho ustavu geologickeho, Praha, v. 40, p. 1–107.

Poprawa, D., and J. Nemcok, eds., 1988–1989, Geological atlas of the Western Outer Carpathians and their foreland: Warszawa, Panstwowy Institut Geologiczny, 9 p.

Poprawa, P., T. Malata, and N. Oszczypko, 2002, Tectonic evolution of the Polish part of Outer Carpathian's sedimentary basins— Constraints from subsidence analysis: Przeglad Geologiczny, v. 50, no. 11, p. 1092–1108.

Pospisil, L., and M. Filo, 1980, Centralne tiazove minimum Zapadnych Karpat a jeho interpretacia (in Slovak): Mineralia Slovaca, Spisska Nova Ves, v. 12, no. 1, p. 325–335.

Pospisil, L., and A. Adam, 2006, Review of the crust-lithosphere research in the Carpathians, *in* J. Golonka and F. J. Picha, eds., The Carpathians and their foreland: Geology and hydrocarbon resources: AAPG Memoir 84, p. 635–649.

Pospisil, L., et al., 2006, Crustal and lithospheric structure of the Carpathian-Pannonian region— a geophysical perspective: Regional geophysical data on the Carpathian–Pannonian lithosphere, *in* J. Golonka and F. J. Picha, eds., The Carpathians and their foreland: Geology and hydrocarbon resources: AAPG Memoir 84, p. 651–697.

Potfaj, M., 1993, Position and role of the Biele Karpaty unit in the Flysch zone of the West Carpathians (in Slovak with English resume): Geologicke prace, Spravy, Bratislava, v. 98, p. 55–78.

Potfaj, M., 1998, Geodynamics of the klippen belt and Flysch belt of the Western Carpathians, *in* M. Rakus, ed., Geodynamic development of the Western Carpathians: Special Publication of the Geological Survey D. Stur, Bratislava, p. 143–154.

Potfaj, M., 2003, Geology of the Slovakian part of the Orava— An overview, *in* J. Golonka and M. Lewandowski, eds., Geology, geophysics, geothermics and deep structure of the West Carpathians and their basement: Institute of Geophysics, Polish Academy of Sciences Publication, Monographic Volume M-28, no. 363, p. 51–63.

Pozaryski, W., and K. Zytko, 1980, On the Mid-Polish aulacogen and the Carpathian geosyncline: Bulletin de l'Academie Polonaise des Sciences, Serie des Sciences de la Terre, Varsovie, v. 28, no. 4, p. 303–316.

Prichystal, A., 1993, Paleozoic to Quaternary volcanism in geological history of Moravia and Silesia, *in* A. Prichystal, V. Obstova, and M. Suk, eds., Geologie Moravy a Slezska (in Czech with English summary): Moravske zemske muzeum a Sekce geologickych ved prirodovedecke fakulty Masarykovy University, Brno, p. 59–70.

Prichystal, A., I. Repcok, and O. Krejci, 1998, Radiometric dating of trachyandesite near the town of Uhersky Brod (Magura Group of the Carpathian Flysch belt): Geologicke vyzkumy na Morave a ve Slezsku v roce 1997 (in Czech with English abstract): Moravske Zemske Muzeum a Sekce Geologickych ved Prirodovedecke Fakulty Masarykovy University, Brno, v. 5, p. 33–34.

Prothero, D., L. Ivany, and E. Nesbitt, 2000, The marine Eocene–Oligocene transition: Penrose Conference Report: Geological Society of America Today, v. 10, July 2000, p. 10–11.

Radomski, A., 1958, Charakterystyka sedimentologiczna fliszu podhalanskiego: Acta Geologica Polonica, v. 8, p. 335–409.

Rakus, M., 1987, Cephalopod Funa of the Lias and lower Dogger from olistoliths of the Raca unit of the Magura Flysch (locality Lukovecek): Zapadne Karpaty, Paleontologia, v. 12, p. 730.

Ratschbacher, L., W. Frisch, and H. G. Linzer, 1991a, Lateral extrusion in the Eastern Alps: Part 2, Structural Analysis: Tectonics, v. 10, no. 2, p. 257–271.

Ratschbacher, L., O. Merle, P. Davy, and P. Cobbold, 1991b, Lateral extrusion in the Eastern Alps: Part 1, Boundary conditions and experiments scaled for gravity: Tectonics, v. 10, no. 2, p. 245–256.

Rehanek, J., 1984, The marine upper Albian of the Bohemian Massif in South Moravia (in Czech with English summary): Geologicke prace, Spravy, Bratislava, v. 81, p. 87–101.

Rehanek, J., 1994, Lithostratigraphic units, depositional pattern and facies developments of the autochthonous Paleogene fill within the Nesvacilka trough (in Czech with English summary): Zemni plyn a nafta, Hodonin, v. 38, no. 3, p. 105–151.

Rehor, F., and M. Rehorova, 1972, Die Makrofauna des Kohlenfuhrenden Karbons im tschechoslovakischen Teil des Oberschlesischen Beckens (in Czech with German resume): Nakladatelstvi Profil Ostrava, 140 p.

Roca, E., G. Bessereau, E. Jawor, M. Kotarba, and F. Roure, 1995, Pre-Neogene evolution of the Western Carpathians: Constraints from the Bochnia–Tatra Mountains section (Polish Western Carpathians): Tectonics, v. 14, no. 4, p. 855–873.

Rogl, F., 1998, Paleogeographic considerations for Mediterranean and Paratethys seaways (Oligocene to Miocene): Annales des Naturhistorischen Museums, Wien, v. 99A, p. 279–310.

Rogl, F., and F. F. Steininger, 1983, Vom Zerfall der Tethys zu Mediterran und Paratethys. Die Neogene Palaeogeographie und Palinspastik des zirkum-mediterranen Raumes: Annalen des Naturhistorischen Museum, Wien, v. 85A, p. 135–163.

Roth, Z., 1962, Contribution to the stratigraphy and paleogeography of the Subsilesian unit in eastern Moravia (in Czech with English and Russian resumes): Sbornik Ustredniho ustavu geologickeho, Praha, v. 27, p. 447–473.

Roth, Z., 1977, Structure of the North European platform below the Carpathian foredeep and the Carpathians in the CSSR: Vestnik Ustredniho ustavu geologickeho, Praha, v. 52, no. 3, p. 129–136.

Roth, Z., 1978, Geology of the Moravian margin of the platforms and its relation to the structure of the Carpathians: Casopis pro mineralogii a geologii, Praha, v. 23, no. 4, p. 349–356.

Roth, Z., 1980a, The stratigraphy of the Godula Group of the Moravskoslezske Beskydy Mountains in the light of deep boreholes (in Czech with English summary): Vestnik Ustredniho ustavu geologickeho, Praha, v. 55, no. 2, p. 75–83.

Roth, Z., 1980b, Zapadni Karpaty-tercierni struktura stredni Evropy (in Czech): Knihovna Ustredniho ustavu geologickeho, v. 55, 128 p.

Roth, Z., 1986, Kinematic model of the tectonic development of the Carpathians and Alps in Cenozoic times: Casopis pro mineralogii a geologii, Praha, v. 31, no. 1, p. 1–15.

Roth, Z., 1987, Kinematic analysis of the of the mutual position of the Cretaceous paleomagnetic poles of the European epi-Variscan and African platforms with respect to the Alpine movements in the Mediterranean Alpides: Mineralia Slovaca, v. 19, no. 3, p. 193–202.

Roth, Z., and A. Matejka, 1953, The Pelosiderites of the Moravosilesian Beskydy (in Czech with English and Russian summary): Geotechnica, v. 16, 111 p.

Roure, F., and W. Sassi, 1995, Kinematics of deformation and petroleum system appraisal in Neogene foreland fold-and-thrust belts: Petroleum Geoscience, v. 1, p. 253–269.

Roure, F., E. Roca, and W. Sassi, 1993, The Neogene evolution of the Outer Carpathian flysch units (Poland, Ukraine and Romania): Kinematics of a foreland/fold-and-thrust belt system: Sedimentary Geology, v. 86, p. 177–201.

Royden, L. H., 1985, The Vienna basin: A thin-skinned pull-apart basin, *in* K. T. Biddle and M. Christie-Blick, eds., Strike-slip deformation, basin formation and sedimentation: SEPM Special Publication 37, p. 319–338.

Royden, L. H., and P. Dovenyi, 1988, Variations in extensional styles at depth across the Pannonian Basin system, *in* L. H. Royden and F. Horvath, eds., The Pannonian Basin, a study in basin evolution: AAPG Memoir 45, p. 235–255.

Ruzicka, M., 1989, Pliozan der Obermarchsenke und der Mohelnice-Furche (in Czech with German resume): Sbornik geologickych ved, Antropozoikum, Praha, v. 19, p. 129–151.

Rzehak, A., 1880, Uber die Gliederung und Verbreitung der alteren Mediterranstufe in der Umgebung von Gr. Seelowitz in Mahren: Verhandlungen der geologischen Reichsanstalt, Wien, v. 16, p. 300–303.

Rzehak, A., 1881, Uber die Gliederung und Verbreitung des Oligozans in der Gegend sudostlich von Gr. Seelowitz in Mahren: Verhandlungen der geologischen Reichsanstalt, Wien, v. 11, p. 211–216.

Rzehak, A., 1895a, Das Alter des Pausraumer Mergels: Verhandlungen der k. k. geologischen Reichsanstalt, Wien, v. 14, p. 363–365.

Rzehak, A., 1895b, Die "Niemtschitzer Schichten:" Verhandlungen des naturforschenden Vereines in Brunn, Brno, v. 34, p. 207–254.

Rzehak, A., 1904, Das Liasvorkommen von Freistadtl in Mahren: Zeitschrift des mahrisches Landesmuseum, Brno, v. 4, p. 89–152.

Rzehak, A., 1922, Das mahrische Tertiar: Knihovna Statniho geologickeho ustavu CSR, Praha, v. 3, p. 3–39.

Salaj, J., 1996, Vysledky biostratigraficke studie miocenu z vrtu v rasovicke depresi (oblast Zdanickeho lesa) (in Czech): Zemni plyn a nafta, Hodonin, v. 41, no. 3, p. 125–135.

Sandulescu, M., 1984, Geotectonica Romaniei (in Romanian): Bucharest, Editura Tehnica, 336 p.

Sandulescu, M., 1988, Cenozoic tectonic history of the Carpathians, *in* L. H. Royden and F. Horvath, eds., The Pannonian Basin: AAPG Memoir 45, p. 17–25.

Schenk, V., D. Prochazkova, and Z. Schenkova, 1994, Seismotectonic studies of the Bohemian Massif and the West Carpathians. *in* V. Bucha and M. Blizkovsky, eds., Crustal structure of the Bohemian Massif and Western Carpathians: Praha, Academia-Springer Verlag, p. 76–84.

Schlunegger, F., A. Matter, D. W. Burbank, W. Leu, M. Manges, and J. Matyas, 1997, Sedimentary sequences, seismofacies and evolution of depositional systems of the Oligo/Miocene Lower Fresh-water Molasse group, Switzerland: Basin Research, v. 9, p. 1–26.

Schmidt, F., and T. Erdogan, 1996, Paleohydrodynamics in exploration, *in* G. Wessely and W. Liebl, eds., Oil and gas in Alpidic thrustbelts and basins of Central and Eastern Europe: European Association of Geoscientists and Engineers Special Publication 5, p. 255–265.

Schmidt, J., 1858, Uber erloschene Vulcane Mahrens: Jahrbuch der geologischen Reichsanstalt, Wien, v. 9, p. 1.

Schroder, B., 1987, Inversion tectonics along the western margin of the Bohemian Massif: Tectonophysics, v. 137, p. 93–100.

Seifert, P., 1996, Sedimentary-tectonic development and Austrian hydrocarbon potential of the Vienna basin, *in* G. Wessely and W. Liebl, eds., Oil and gas in Alpidic thrustbelts and basins of Central and Eastern Europe: European Association of Geoscientists and Engineers Special Publication 5, p. 331–341.

Shrbeny, O., 1974, The petrochemical relation of the south-Moravian neovolcanic rocks to the neighbouring volcanic areas: Vestnik Ustredniho ustavu geologickeho, Praha, v. 49, no. 5, p. 75–279.

Sikora, W., 1970, The structure of the Magura nappe between Szymbark Ruski and Nawojowa (in Polish with

English summary): Biuletyn Instytut Geologiczny, Warszawa, v. 235, p. 5–12.

Skupien, P., and Z. Vasicek, 2002, Lower Cretaceous ammonite and dinocyst biostratigraphy and paleoenvironment of the Silesian basin (Outer Western Carpathians): Geologica Carpathica, v. 53, no. 3, p. 179–189.

Slaczka, A., 1975, Remarks on morphology of the substratum of the Polish Carpathians, *in* Proceedings of the 10th Jubilee Congress, Carpathian–Balkan Geological Association: Proceedings, Section III, Geologicky ustav Dionyza Stura, Bratislava, p. 281–290.

Slaczka, A., 1977, The Miocene deposits in the sucha IG-1 borehole (in Polish): Kwartalnik Geologiczny, v. 21, p. 405–406.

Slaczka, A., 1996, Oil and gas in the Northern Carpathians, *in* G. Wessely and W. Liebl, eds., Oil and gas in Alpidic thrustbelts and basins of Central and Eastern Europe: European Association of Geoscientists and Engineers Special Publication 5, p. 187–195.

Slaczka, A., N. Oszczypko, E. Malata, and M. Cieszkowski, 1999, An early history of the Outer Carpathian basin: Geologica Carpathica, v. 50, p. 170–172.

Slaczka, A., S. Kruglow, J. Golonka, N. Oszczypko, and I. Popadyuk, 2006, Geology and hydrocarbon resources of the Outer Carpathians, Poland, Slovakia, and Ukraine: The general geology of the Outer Carpathians, Poland, Slovakia, and Ukraine, *in* J. Golonka and F. J. Picha, eds., The Carpathians and their foreland: Geology and hydrocarbon resources: AAPG Memoir 84, p. 221–258.

Smid, B., 1962, Prehled geologie a petrografie hornin tesinitove asociace na severnim upati Beskyd (in Czech): Geologicke prace Geologickeho ustavu D. Stura, Bratislava, Zosit, v. 63, p. 53–60.

Smid, B., and E. Mencik, 1983, The Teschenite volcanism of the Silesian unit, *in* E. Mencik et al., eds., Geology of the Moravskoslezske Beskydy Mountains and the Podbeskydska pahorkatina Upland (in Czech with English summary): Academie, Praha, Ustredni ustav geologicky, p. 69–74.

Smith, P. B., 1968, Paleoenvironment of phosphate-bearing Montery Shale in Salinas Valley, California: AAPG Bulletin, v. 52, no. 9, p. 1785–1791.

Sotak, J., 1985, Basinal developmental types of the Triassic (Reifling and "Pseudohallstatt" limestones) in outer units of the West Carpathians: Scripta Facultatis Scientiarum Naturalium Universitatis Purkynianae Brunensis, Brno, v. 15, no. 2 (Geologia), p. 89–114.

Sotak, J., 1990, Study of Mesozoic rocks in flysch zones, and its contribution to information about microfacies, paleogeography and paleotectonics of the West Carpathians, *in* M. Sykora, J. Jablonsky, and O. Samuel, eds., Sedimentological problems of the West Carpathians (in Slovak with English summary): Bratislava, Geologicky ustav D. Stura, p. 45–68.

Sotak, J., 1992, Evolution of the Western Carpathian suture zone— Principal geotectonic events: Geologicky Zbornik Geologica Carpathica, v. 43, no. 6, p. 355–362.

Sovchik, Y. V., and M. A. Vul, 1996, New data on the structure and hydrocarbon prospects of the Ukrainian Carpathians and their foreland, *in* P. A. Ziegler and F. Horvath, eds., Peri-Tethys Memoir 2: Structure and prospects of Alpine basins and forelands: Memoires du Museum National d'Histoire Naturelle, Paris, v. 170, p. 391–401.

Spath, L. F., 1933, Revision of the Jurassic cephalopod faunas of Kachh (Cutch): Paleontologia Indica, Calcutta, v. 9, no. 6, p. 659–945.

Sperner, B., L. Ratschbacher, and M. Nemcok, 2002, Interplay between subduction retreat and lateral extrusion: Tectonics of the Western Carpathians: Tectonics, v. 21, no. 6, p. 1–24.

Stampfli, G. M., 1993, "Le Brianconnais, terrain exotique dans les Alpes?": Eclogae Geologicae Helvetiae, v. 86, p. 1–45.

Stampfli, G. M., ed., 2001, Geology of the western Swiss Alps— A guide book: Memoirs de Geologie (Lausanne), v. 361, p. 1–195.

Stegena, L., B. Geczy, and F. Horvath, 1975, Late Cenozoic evolution of the Pannonian Basin: Tectonophysics, v. 26, p. 71–91.

Steininger, F. F., C. Muller, and F. Rogl, 1988, Correlation of Central Paratethys, Eastern Paratethys, and Mediterranean Neogene Stages, *in* L. H. Royden and F. Horvath, eds., The Pannonian Basin, a study in basin evolution: AAPG Memoir 45, p. 79–87.

Stille, H., 1936, Present tectonic state of the Earth: AAPG Bulletin, v. 20, p. 849–880.

Stille, H., 1953, Der geotektonische Werdegang der Karpaten: Geologische Jahrbuch, v. 8, p. 1–139.

Stranik, Z., 1963, Tectonic structure of the southern part of the Zdanice unit: Geologicke prace, Spravy, Bratislava, v. 28, p. 155–160.

Stranik, Z., 1965, Geology of the Magura flysch in Cerhovske pohori Mountains and western part of the Ondava vrchovina Hills (in Czech with English summary): Sbornik geologickych ved, Zapadni Karpaty v. 3, p. 125–178.

Stranik, Z., 1981, Lithofacies and correlation of the Menilitic Formation in the Carpathian Flysch belt of Moravia (in Czech with English summary): Zemni plyn a nafta, Hodonin, v. 26, no. 1, p. 9–18.

Stranik, Z., 1983, Outer Carpathians in South Moravia, *in* O. Samuel and V. Gasparikova, eds., European colloquy on micropaleontology (excursion guide): Konferencie, Sympozia, Seminare: Bratislava, Geologicky Ustav D. Stura, p. 167–175.

Stranik, Z., 1996, Bericht uber geologische Aufnahmen im Tertiar der Neogenen Vortiefe und Waschbergzone auf Blatt 23 Hadres: Jahrbuch der geologischen Bundesanstalt, Wien, v. 139, p. 301.

Stranik, Z., 1999, Tectonic events of the Neoalpine orogeny in the Carpathian Flysch belt (South Moravia): Geolines, v. 8, p. 65–66.

Stranik, Z., and V. Molcikova, 1980, Untermiozan nordostlich von Pouzdrany: Vestnik Ustredniho ustavu geologickeho, Praha, v. 55, no. 2, p. 93–100.

Stranik, Z., and Z. Roth, 1959, Predbezna zprava o

geologickych vyzkumech v Cechoskem pohori a zapadni casti Ondavske vrchoviny (in Czech): Zpravy o geologickych vyzkumech v roce 1957, p. 228–231.

Stranik, Z., E. Benesova, and F. Ficha, 1968 Geology of the deep boring Bulhary-1 (in Czech with English summary): Sbornik geologickych ved, Geologie, Praha, v. 13, p. 75–131.

Stranik, Z., E. Hanzlikova, and F. Jurasova, 1981a, Stratigraphic position of the Boudky Marls within the framework of the Oligocene–Miocene stage (in Czech with English summary): Zemni plyn a nafta, Hodonin, v. 26, no. 4, p. 689–699.

Stranik, Z., F. Jurasova, and H. Peslova, 1981b, Sitborice Member in the borehole Krepice-5 (in Czech with English summary): Zemni plyn a nafta, Hodonin, v. 26, p. 701–710.

Stranik, Z., O. Krejci, and E. Mencik, 1989, Prispevek ke geologii Bilych Karpat (in Czech): Zpravy o geologickych vyzkumech v roce 1986, p. 145–147.

Stranik, Z., J. Dvorak, O. Krejci, P. Muller, A. Prichystal, M. Suk, and C. Tomek, 1993, The contact of the North European Epivariscan platform with the West Carpathians: Journal of the Czech Geological Society, Praha, v. 38, nos. 1–2, p. 21–29.

Stranik, Z., M. Bubik, R. Marschalko, L. Svabenicka, and M. Vujta, 1995, New lithostratigraphy of the Hluk development of the Bile Karpaty unit: Geologicke prace, Spravy, Bratislava, v. 100, p. 57–69.

Stranik, Z., M. Bubik, S. Cech, and L. Svabenicka, 1996, The Upper Cretaceous in south Moravia: Bulletin of the Czech Geological Survey, Praha, v. 71, no. 1, p. 1–30.

Stranik, Z., M. Bubik, and L. Svabenicka, 1997, Svrchnokridove stari konglomeratu od Stareho Jicina: Geologicke vyzkumy na Morave a ve Slezsku v roce 1997 (in Czech): Moravske zemske muzeum a Sekce geologickych ved Prirodovedecke Fakulty Masarykovy University, Brno, v. 4, p. 43–44.

Stranik, Z., P. Ctyroky, and P. Havlicek, 1999, Die geologische Vergangenheit der Pollauer Berge (in Czech with German summary): Journal of Geological Sciences (Sbornik geologickych ved), Geology, v. 49, p. 5–32.

Stur, D., 1875, Vorkommnisse mariner Petrefacten in den Ostrauer schichten in der Umgebung vom Mahrischen Ostrau: Verhandlungen der k.k. geologischen Reichsanstalt, Wien, v. 8, p. 153–155.

Stur, D., 1885, Die Carbon-Flora der Schatzlaren Schichten: Abhandlungen der k. k. geologischen Reichsanstalt, Wien, B, 11 p.

Suess, E., 1875, Die Entstehung der Alpen: Wien, Verlag Braumuller, 168 p.

Suk, M., 1993, Geological regionalisation of the Moravian and Silesian Region, *in* A. Prichystal, V. Obstova, and M. Suk, eds., Geology of the Moravia and Silesia region (in Czech with English summary): Brno, Moravske zemske museum-Masarykova Universita, p. 157–161.

Suk, M., et al., 1984, Geological history of the territory of the Czech Socialist Republic: Academia, Geological Survey, Prague, 387 p.

Suk, M., D. Durica, V. Obstova, and E. Stankova, eds., 1991, Deep boreholes in Bohemia and Moravia and their geological applications (in Czech with English summary): Ministerstvo pro Hospodarskou politiku a rozvoj, Nakladatelstvi Gabriel, Praha, 171 p.

Svabenicka, L., 1990, Cretaceous nannoplankton in flysch sediments of the Bile Karpat unit (the West Carpathians): Journal of Geological Sciences, Paleontology, Praha, v. 31, p. 9–65.

Svabenicka, L., and M. Bubik, 1992, The fossil record in single intervals of the flysch sequence (Upper Cretaceous, Czechoslovakia): Cretaceous Research, v. 13, p. 583–590.

Svabenicka, L., M. Bubik, O. Krejci, and Z. Stranik, 1997, Stratigraphy of Cretaceous sediments of the Magura Flysch in Moravia (Czech Republic): Geologica Carpathica, v. 48, no. 3, p. 179–191.

Svoboda, J., ed., et al., 1966, Regional geology of Czechoslovakia: Part I, The Bohemian Massif: Praha, Geological Survey of Czechoslovakia, 668 p.

Swidzinski, H., 1934, Remarques sur la structure des Carpathes flyscheuses: Sprawozdania Polskiego Instytutu Geologicnego, Warszawa, v. 8, no. 1, p. 75–199.

Swidzinski, H., 1948, Stratigraphical index of the northern Flysch Carpathians: Biuletyn Panstwowy Instytut Geologiczny, Warszawa, v. 37, p. 1–128.

Swidzinski, H., 1953, Karpaty fliszowe miedzy Dunajcem i Sanem (in Polish): Regionalna geologia Polski, I: Tektonika, v. 2, p. 362–422.

Szafian, P., G. Tari, F. Horvath, and S. Cloetingh, 1999, Crustal structure of the Alpine–Pannonian transition zone: A combined seismic and gravity study: International Journal of Earth Sciences, v. 88, p. 98–110.

Szajnocha, W., 1884, Studia geologiczne v Karpatach Galicyi zachodnej (in Polish): Kosmos, v. 9.

Tari, G., T. Baldi, and M. Baldi-Beke, 1993, Paleogene retroarc flexural basin beneath the Neogene Pannonian Basin: a geodynamic model: Tectonophysics, v. 226, p. 433–455.

Tari, G. C., and F. Horvath, 2006, Alpine evolution and hydrocarbon geology of the Pannonian Basin: An overview, *in* J. Golonka and F. J. Picha, eds., The Carpathians and their foreland: Geology and hydrocarbon resources: AAPG Memoir 84, p. 605–618.

ten Haven, H. L., E. Lafargue, and M. Kotarba, 1993, Oil/oil and oil/source rock correlations in the Carpathian foredeep and overthrust, south-east Poland: Organic Geochemistry, v. 20, no. 7, p. 935–959.

Thonova, H., and S. Benada, 1990, Occurrence of natural hydrocarbons at the Variscan level of the central and adjacent southern parts of southeastern slopes of the Bohemian Massif, *in* D. Minarikova and H. Lobitzer, eds., Thirty years of geological cooperation between Austria and Czechoslovakia: Vienna, Federal Geological Survey; Prague, Geological Survey, p. 239–244.

Tietze, E., 1889, Beitrage zur Geologie von Galizien: Jahrbuch der geologischen Reichsanstalt, Wien, v. 39, p. 289–404.

Tollmann, A., 1978, Plattentektonische Fragen in den

Ostalpen und der plattentektonische Mechanismus des Mediterranen Orogen: Mitteilungen der osterreischen geologischen Gesellschaft, Wien, v. 69, p. 291–351.

Tollmann, A., 1980, Grosstectonische Ergebnisse aus den Ostalpen im Sinne der Plattentektonik: Mitteilungen der osterreischen geologischen Gesellschaft Wien, v. 71–72, p. 37–44.

Tomek, C., 1976, Tektonika litosferickych desek a tercierni vyvoj Zapadnich Karpat, *in* Zborniik referatov z vedeckej konference Ceskoslovenska geologia a globalna tektonika: Smolenice 1976: Bratislava, Geologicky ustav Dionyza Stura, p. 118–123.

Tomek, C., 1979, Plate tectonics and Neogene development of the Carpathians— A review, *in* Czechoslovak geology and global tectonics: Bratislava, Veda, p. 41–55.

Tomek, C., 1982, Zdroj zapadokarpatskeho tihoveho minima a nektere nove moznosti pro pruzkum zivic (in Czech): Geologicky pruzkum, Praha, v. 24, no. 4, p. 105–108.

Tomek, C., 1993, Deep crustal structure beneath the Central and Inner West Carpathians: Tectonophysics, v. 226, p. 417–431.

Tomek, C., and J. Hall, 1993, Subducted continental margin imaged in the Carpathians of Czechoslovakia: Geology, v. 21, p. 535–538.

Tomek, C., and A. Thon, 1988, Interpretation of seismic reflection profiles from the Vienna basin, the Danube basin and the Transcarpathian depression in Czechoslovakia, *in* L. H. Royden and F. Horvath, eds., The Pannonian Basin, a study in basin evolution: AAPG Memoir 45, p. 171–182.

Tomek, C., L. Dvorakova, I. Ibrmajer, R. Jiricek, and T. Korab, 1987, Crustal profiles of active continental collisional belt: Czechoslovak deep seismic reflection profiling in the West Carpathians: Geophysical Journal of the Royal Astronomical Society of London, v. 89, p. 383–388.

Trumpy, R., 1980, Geology of Switzerland— A guide book: Part A, Outline of the geology of Switzerland: Basel, Wepf. & Co. Publ., 102 p.

Trumpy, R., 1988, A possible Jurassic–Cretaceous transform system in the Alps and the Carpathians: Geological Society of America Special Paper 218, p. 93–109.

Tschermak, G., 1858, Die Trachyte bei Banov in Mahren: Jahrbuch der geologischen Reichsanstalt, Wien, v. 9, p. 63.

Uhlig, V. 1882, Reisebericht aus Westgalizien. Funde cretazischer und alttertiarer Versteinerungen: Verhandlungen der geologischen Reichsanstalt, v. 15–16, p. 306–307.

Uhlig, V., 1883a, Die Cephalopodenfauna der Wernsdorfer Schichten: Denkenschriften osterreichische Akademie der Wissenschaften, Mathematisch-Naturwissenschaftliche Klasse, v. 46, p. 127–290.

Uhlig, V., 1883b, Beitrage zur Geologie der Westgalizischen Karpathen: Jahrbuch der geologischen Reichsanstalt, Wien, v. 33, p. 443–563.

Uhlig, V., 1897, Die Geologie des Tatragebirges I: Denkenschriften osterreichische Akademie der Wissenschaften, Mathematisch–Naturwissenschaftliche Klasse, v. 64, p. 643–684.

Uhlig, V., 1902, Uber die Cephalopodenfauna der Teschener und Grodister Schichten: Denkenschriften osterreichische Akademie der Wissenschaften, Mathematisch–Naturwissenschaftliche Klasse, v. 72, p. 1–88.

Uhlig, V., 1903, Bau und Bild der Karpathen, *in* C. Diener, R. Hoernes, F. E. Suess, and V. Uhlig, eds., Bau und Bild Osterreichs: Wien–Leipzig, Verlag Tempsky and Frytag, p. 651–911.

Uhlig, V., 1907, Uber die Tektonik der Karpathen: Sitzungsberichte der Kaiserlichen Akademie der Wissenschaften, Mathematisch–Naturwissenschaftliche Klasse, Wien, v. 116, p. 871–982.

Valverde-Vaquero, P., W. Dorr, Y. Belka, W. Franke, J. Wiszniewska, and J. Schastok, 2000, U-Pb single-grain dating of detrital zircon in the Cambrian of central Poland: Implications for Gondwana versus Baltica provenance studies: Earth and Planetary Science Letters, v. 184, p. 225–240.

Vasicek, M., 1947, Remarks on the microstratigraphy of the Magura Flysch in Moravia (in Czech with English summary): Vestnik Statniho geologickeho ustavu, Praha, v. 22, no. 4, p. 235–256.

Vasicek, Z., 1971a, Die makropaleontologische Untersuchung von Bohrkernen aus dem Mesozoikum des Untergrunds in Sudmahren: Vestnik Ustredniho ustavu geologickeho, Praha, v. 46, no. 2, p. 83–91.

Vasicek, Z., 1971b, Report on the results of paleontological studies of the Lower Cretaceous of the western part of the Silesian unit in the Moravskoslezske Beskydy Mountains (in Czech with English, German, and Russian resume): Sbornik vedeckych praci Vysoke skoly banske v Ostrave, Rada hornicko-geologicka, v. 15, p. 121–143.

Vasicek, Z., 1972a, The Ammonoidea of the Tesin–Hradiste Formation (Lower Cretaceous) in the Moravskoslezske Beskydy Mountains: Rozpravy Ustredniho ustavu geologickeho, Praha, v. 38, p. 1–103.

Vasicek, Z., 1972b, Contribution to the fauna of the lower Tesin Formation (Upper Jurassic) in the Trinec district: Sbornik vedeckych praci Vysoke skoly banske v Ostrave, Rada hornicko-geologicka, v. 17, no. 2, p. 97–107.

Vasicek, Z., 1973, Zur Barreme-Apt-Grenze in der Schlesischen Einheit: Sbornik vedeckych praci Vysoke skoly banske v Ostrave, Rada hornicko-geologicka, v. 18, p. 101–107.

Vasicek, Z., 1979, Die Cephalopoden der schlesischen Unterkreide und ihre palaogeographische Bedeutung, *in* J. Wiedmann, ed., Aspekte der Kreide Europas: International Union of Geological Sciences, Series A 6, Schweizerbartsche Verlags, Stuttgart, p. 323–334.

Vasicek, Z., 1999, Early Cretaceous ammonoid paleobiogeography of the West Carpathian part of Paleoeuropean shelf margin: Neues Jahrbuch fur Geologie und Palaontologie, Abhandlungen, Stuttgart, v. 212, p. 241–262.

Vasicek, Z., and D. Rehakova, 1994, Biostratigraficke vyzkumy v lomu u Kurovic v roce 1993 (Vnejsi Karpaty, Tithonspodni Valangin); Geologicke vyzkumy na Morave a ve Slezsku v roce 1993 (in Czech): Moravske zemske muzeum a Sekce Geologickych Ved Prirodovedecke Fakulty Masarykovy University, Brno, v. 1, p. 28.

Vasicek, Z., J. Michalik, and D. Rehakova, 1994, Early Cretaceous stratigraphy, paleogeography and life in the Western Carpathians: Beringeria, v. 10, p. 1–169.

Vialov, O. S., 1951, Zametchania o paleogenovom flishe Borislava (in Russian): Doklady Akademii Nauk SSSR (Soviet Union), Novaia seria, v. 77, no. 3, Moskva, p. 465–468.

Vujta, M., Z. Stranik, M. Bubik, and L. Svabenicka, 1989, Novy vyskyt sedimentu svrchni kridy v magurskem flysi (in Czech): Zpravy o geologickych vyzkumech v roce 1987, p. 135–136.

Vujta, M., M. Bubik, and L. Svabenicka, 1991, Zprava o geologickem mapovani v racanske a bytricke jednotce magurskeho flyse na listu Puchov (moravska cast) (25-43 Puchov) (in Czech): Zpravy o geologickych vyzkumech v roce 1990, p. 115–158.

Wagner, L., 1996, Stratigraphy and hydrocarbons in the upper Austrian Molasse foredeep (active margin), *in* G. Wessely and W. Liebl, eds., Oil and gas in Alpidic thrustbelts and basins of Central and Eastern Europe: European Association of Geoscientists and Engineers Special Publication 5, p. 217–235.

Wagner, L. R., 1998, Tectono-stratigraphy and hydrocarbons in the Molasse Foredeep of Salzburg, upper and lower Austria, *in* A. Mascle, C. Puigdefabregas, H. P. Luterbacher, and M. Fernandez, eds., Cenozoic foreland basins of Western Europe: Geological Society (London) Special Publication 134, p. 339–369.

Walter, H., and E. Dunikowski, 1883, Geologiczna budowa naftonosnego obszaru zachodnio-galicyjiskich Karpat: Czast I (in Polish): Kosmos, v. 7, p. 300–334.

Wdowiarz, S., 1985, On some aspects of geological structure and oil and gas accumulation of the Central Carpathian Synclinorium in Poland: Instytut Geologiczny Biultyn, Warszawa, v. 350, p. 5–52.

Wessely, G., 1988, Structure and development of the Vienna basin in Austria, *in* L. H. Royden and F. Horvath, eds., The Pannonian Basin, a study in basin evolution: AAPG Memoir 45, p. 333–346.

Wessely, G., 1990, Geological results of deep exploration in the Vienna basin: Geologische Rundschau, v. 79, no. 2, p. 513–520.

Wessely, G., 1998, Geologie des Korneuburger Beckens: Beitage zur Paleontologie und Geologie Osterreich-Ungarns und des Orients, v. 23, p. 9–23.

Winkler, W., and A. Slaczka, 1994, A Late Cretaceous to Paleogene geodynamic model for the Western Carpathians in Poland: Geologica Carpathica, v. 45, p. 71–82.

Yilmaz, P. O., I. O. Norton, D. Leary, and R. J. Chuchla, 1996, Tectonic evolution and paleogeography of Europe, *in* P. A. Ziegler and F. Horvath, eds., Peri-Tethys Memoir 2: Structure and prospects of Alpine basins and forelands: Memoires du Museum National d'Histoire Naturelle, Paris, 170, p. 47–69, enclosures 1–13.

Zahalka, B., and J. Koutek, 1927, Geologie Moravskych Beskyd v okoli Roznova pod Radhostem (in Czech with French resume): Sbornik Statniho geologickeho ustavu CSR, Praha, v. 7, p. 1–50.

Zapletal, K., 1931, Postaveni zeme moravskoslezske ve Variscidach a Alpidach (in Czech): Sbornik Prirodovedecke spolecnosti v Moravske Ostrave, v. 6, p. 257–292.

Zapletal, K., 1937, Vyzkum zeme provadeny Batovymi zavody (in Czech): Priroda, Brno, v. 30, no. 5, p. 148–155.

Zapletal, K., 1954, Ke geologii krystalinika a odersko-viselske panve (v ramci prvohornich pohori), karpatskeho flyse, molasy a mladych vulkanitu (in Czech): Spisy prirodovedecke fakulty Masarykovy university v Brne, R. L7 G4, no. 1 (352), p. 37–60.

Ziegler, P. A., 1982, Geological atlas of Western and Central Europe: The Hague, Shell Internationale Petroleum Maatschappij B. V., 130 p.

Ziegler, P. A., 1988, Evolution of the Arctic–North Atlantic and the western Tethys: AAPG Memoir 43, 198 p.

Ziegler, P. A., 1989, Evolution of Laurasia, a study in late Paleozoic plate tectonics: Dordrecht, Kluwer Academic Publishers, 102 p.

Ziegler, P. A., and F. Roure, 1996, Architecture and petroleum systems of the Alpine orogen and associated basins, *in* P. A. Ziegler and F. Horvath, eds., Peri-Tethys Memoir 2: Structure and prospects of Alpine basins and forelands: Memoires du Museum national d'Histoire naturelle, Paris, v. 170, p. 15–45.

Zimmer, W., and G. Wessely, 1996, Exploration results in thrust and subthrust complexes in the Alps and below the Vienna basin in Austria, *in* G. Wessely and W. Liebl, eds., Oil and gas in Alpidic thrustbelts and basins of Central and Eastern Europe: European Association of Geoscientists and Engineers Special Publication 5, p. 81–107.

Zitt, J., 1974, Sclerocrinus Jekel, 1891 and Proholopus Jaekel, 1907 (Crinoidea, Cyrtocrinida) from the Lower Cretaceous of Stramberk (Czechoslovakia): Sbornik geologickych ved, Paleontologie, Praha, v. 16, p. 7–34.

Zittel, K. A., 1868, Die Cephalopoden der Stramberger Schichten: Paleontologische Mitteilungen Museum des Bayerischen Staates, v. 2, no. 1, p. 1–118, Stuttgart.

Zittel, K. A., 1873, Die Gastropoden der Stramberger Schichten: Palaeontographica, Abteilung B, Suppl. 2, Kassel, p. 193–373.

Zoubek, V., 1948, Poznamky ke geologii krystalinika Ceskeho masivu (in Czech): Sbornik Statniho geologickeho ustavu Ceskoslovenske Republiky, Praha, v. 15, p. 339–398.

Zukalova, V., 1976, Biostratigraphy of the Palaeozoic in the basement and foreland of the Carpathians east of Brno: Casopis pro mineralogii a geologii, Praha, v. 21, no. 4, p. 369–385.

Zukalova, V., and I. Chlupac, 1982, Stratigraficka klasifikace nemetamorfovaneho devonu moravskoslezske oblasti (in Czech): Casopis pro mineralogii a geologii, Praha, v. 27, no. 3, p. 225–241.

4

Kostelníček, Petr, Vladimír Ciprys, and Jiří Berka, 2006, Examples of recently discovered oil and gas fields in the Carpathian foredeep and in the European foreland plate underneath the Carpathian thrust belt, Czech Republic, *in* J. Golonka and F. J. Picha, eds., The Carpathians and their foreland: Geology and hydrocarbon resources: AAPG Memoir 84, p. 177–189.

Examples of Recently Discovered Oil and Gas Fields in the Carpathian Foredeep and in the European Foreland Plate underneath the Carpathian Thrust Belt, Czech Republic

Petr Kostelníček
Moravske Naftove Doly, A.S., Hodonin, Czech Republic

Vladimír Ciprys
Moravske Naftove Doly, A.S., Hodonin, Czech Republic

Jiří Berka
Moravske Naftove Doly, A.S., Hodonin, Czech Republic

ABSTRACT

Several oil and gas fields have been discovered recently in the Neogene foredeep, in the European foreland plate, which underlies the foredeep and the thin-skinned Carpathian thrust belt. The fields are reservoired in the Neogene strata of the foredeep plate and in the crystalline basement rocks and the Jurassic, Paleogene, and Neogene strata of the European platform. The most significant accumulations of hydrocarbons in the subthrust plate have been found in the erosional relicts (buried hills) of the Jurassic rocks on the northern side of the Nesvacilka graben and paleovalley and in the fractured and weathered surface of the Precambrian granite massifs of the Chriby and Zdanice elevations. The thickness of the saturated parts of these reservoirs typically ranges from several tenths of a meter to as much as 200 m (660 ft). The geological reserves of the discovered fields are in a range of hundreds of thousands to a few million cubic meters (less than 1 million bbl to several million barrels) of crude oil and hundreds of millions to billions of cubic meters (several to tens of billions cubic feet) of gas. They represent a significant economical potential for petroleum companies operating in the territory of the Czech Republic.

INTRODUCTION

The main objective of this chapter is to present examples of oil and gas fields found recently in the Neogene foredeep and in various reservoirs of the European foreland plate underneath the thin-skinned Carpathian thrust belt. A systematic exploration of the sub-Carpathian potential began only in the early 1960s, although several shallow gas fields had been found previously in the Neogene foredeep below the thrust

DOI:10.1306/985731M843140

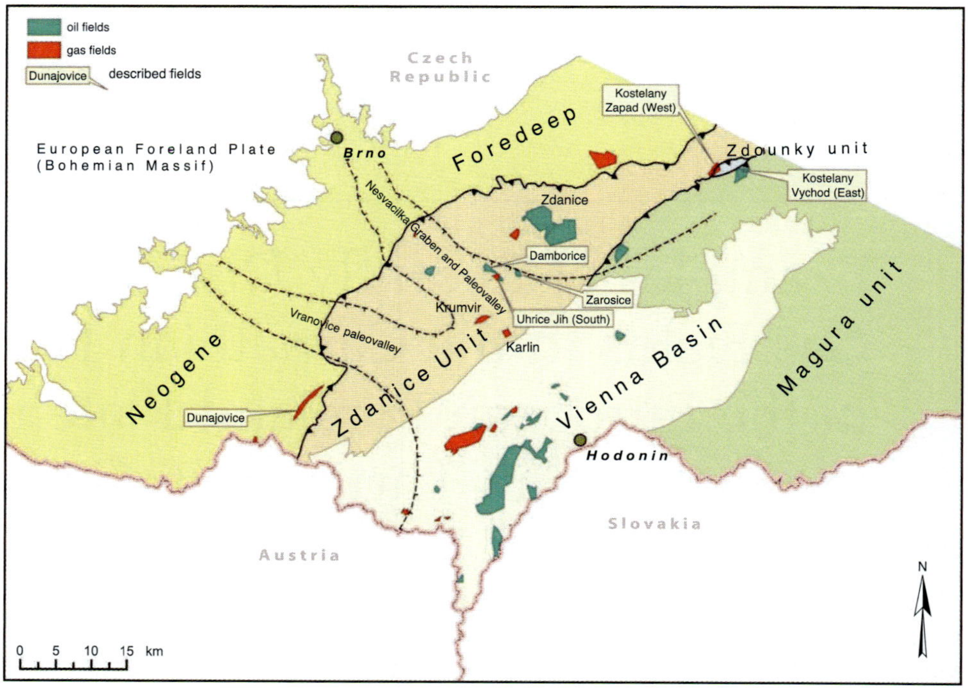

Figure 1. Index map showing the main tectonic units and major oil and gas fields in the territory of southeastern Moravia (Czech Republic).

belt in northern Moravia. The early studies and deep drilling, done mainly by the Czechoslovak Geological Survey, led to an initial stratigraphic reconstruction of the subthrust plate and to a discovery of the first commercial oil field Kostelany Vychod (East) (Lubna-1 well) in 1966. Further detailed exploration work was conducted by the Moravske Naftove Doly (Moravian Oil Company; MND) state oil company. The systematic exploration of the hydrocarbon potential of the subthrust plate began in the late 1970s. Several oil and gas fields in various subthrust stratigraphic and structural settings have been found: the Zdanice oil fields in the granitic basements rocks; the Uhrice gas and oil field in the Devonian and Jurassic carbonates and Jurassic and Paleogene clastics; the Korycany oil and gas field in the lower Miocene clastics; the Uhrice gas field and the largest oil field Damborice in the Jurassic clastics; and the Karlin and Krumvir gas fields in the autochthonous Paleogene. In addition, the largest gas field Dolni Dunajovice was found in the Neogene foredeep just in front of the Carpathian thrust belt.

In 1992–1998, the former state company was gradually privatized under the original name MND. The string of successes has continued with recent discoveries of the Uhrice Jih (South) and Zarosice fields in the erosional relicts of the Jurassic strata with reserves in the range of 1–3 million m^3 crude oil in place. The exploration work has been greatly improved by the application of three-dimensional (3-D) seismic acquisition, which considerably decreased the risk of exploratory drilling. Exploration works have concentrated on reservoirs in the Jurassic clastic and carbonate sequences and the Paleogene sandstones in the Nesvacilka graben and paleovalley in depths ranging from 1000 to 3500 m (3300 to 11,500 ft). Several typical examples of the subthrust fields and one example of a field found in the foredeep in front of the thrust belt are discussed in the following sections.

GEOLOGICAL SETTING

The area considered by this chapter covers the contact zone between the west European foreland plate and the Western Carpathian thrust belt in the territory of southern Moravia (southeastern part of the Czech Republic) (Figure 1). The European foreland plate in southern Moravia is represented by the Bohemian Massif, which consists of the Hercynian orogenic system of Western Europe and its late Precambrian (Cadomian) foreland terrain called the Brunovistulicum that extends far below the Carpathian belt. The Brunovistulicum underlies the entire exploration area of southern Moravia. Since the Middle Devonian, the Brunovistulicum passed through two full plate-tectonic cycles, the Hercynian and the Tethyan–Alpine cycles (Picha, 1996). The Paleozoic Hercynian cycle in the area began with rifting and deposition of the late Early to Middle Devonian clastics. In the Late Devonian to Early

Carboniferous, carbonate platforms and basins evolved on passive continental margins. At the onset of the Hercynian orogeny in the Early Carboniferous, the carbonate sedimentation was replaced by synorogenic flysch (Culm) facies, which, in the Late Carboniferous, was followed by the late orogenic and postorogenic molasse of the upper Silesian coal basin. During the Hercynian orogeny, these Paleozoic deposits were folded and thrust over the foreland of the Brunovistulicum.

The Mesozoic to Cenozoic Alpine cycle in southern Moravia began with continental rifting in the Early Jurassic and deposition of synrift detrital deposits, followed by marine transgression and deposition of mixed carbonate and siliciclastic rocks of the passive margins. The carbonate platforms occupied the shallow marginal zone and uplifted rift blocks, whereas the organic-rich marly facies (Mikulov marls) formed in a deeper basinal environment.

During the Late Cretaceous to early Paleogene, the European foreland in Moravia was uplifted and deeply eroded. Rivers cut the deep Nesvacilka and Vranovice valleys that turned into submarine canyons more than 1500 m (5000 ft) deep, which were gradually filled with the Paleogene detrital organic-rich deposits (Picha, 1979, 1996; Jiricek, 1987) (Figure 1).

In the early Late Cretaceous, the Carpathian passive margins were converted into a foreland-type synorogenic basin marked by deep-water turbiditic flysch sedimentation, followed by a late orogenic and post-orogenic shallow-marine and continental molasse sedimentation. During the late phases of the Alpine orogeny, in the late Paleogene and early Miocene, the flysch sequences were deformed and thrust over the European foreland, including the inner zone of the Neogene foredeep. The Carpathian thrust belt in Moravia is composed of Late Jurassic to early Miocene synorogenic sequences of the flysch belt, with several rootless tectonostratigraphical units, including the marginal Pouzdrany, the Zdanice–Subsilesian, the Silesian, and the Magura units. The Neogene foredeep (Figure 1), with the exception of the innermost zones adjacent to the front of the thrust belt, remained undeformed. The thickness of the sedimentary fill of the Neogene foredeep in Moravia typically does not exceed a few hundred meters. Superimposed on the thrust belt is the Vienna basin. For more information on the geology of the region, we refer to the account by Picha et al. (2006).

During this complex geological history, conditions for deposition of organic-rich source rocks occurred in the Middle Devonian, Late Carboniferous, Late Jurassic, and early Oligocene (Francu et al., 1996; Krejci et al., 1996). After rock burial below the thin-skinned Carpathian thrust belt, hydrocarbons were generated from some of these source rocks. They migrated into potential reservoirs in the foreland plate and in the Neogene foredeep. Oil and gas accumulations have been found in the weathered and fractured surface of basement granitic rocks, in the Devonian carbonates, in the Jurassic clastics and carbonates, and in the Paleogene and Neogene clastics. Early accounts of petroleum systems and characteristics of some of the fields were published by Krejci (1993) and Ciprys et al. (1995).

In the following sections, we present examples of three typical subthrust fields: Kostelany Vychod (East) and Kostelany Zapad (West), which were reservoired in the crystalline basement, and the newly discovered Uhrice Jih (South) and Zarosice fields, with reservoirs in Jurassic clastics and carbonates, respectively. We also describe here Dolni Dunajovice, which is, by far, the largest gas field found in the Neogene foredeep.

Kostelany Oil and Gas Fields

Kostelany Vychod (East) oil and gas field and Kostelany Zapad (West) gas field are situated on the northeastern part of the Chriby Height in central Moravia (Figures 1–3). The fields are reservoired mainly in the weathered and fractured surface of the basement granitic rocks, which form two elevations (buried hills) with relicts of metamorphosed Proterozoic shales and Paleozoic carbonate rocks on the slopes of the elevations and in the surrounding depressions. The strongly weathered surface zone of granites commonly has a character of sandstones with very high permeability. The basement rocks are covered by transgressive deposits of the Karpatian (lower Miocene), which are composed predominantly of shales with some sandstones that occur on the eastern side of the elevation. The Miocene strata make a good cap rock. Superimposed on these autochthonous deposits are the Carpathian nappes represented by four distinctive tectonic units: the Magura flysch unit, the Zdounky unit, the Zdanice–Subsilesian unit, and the Pouzdrany unit, which may actually be a tectonically transported slice of the autochthonous Paleogene (Figure 3). The Pouzdrany unit is composed predominantly of shales and mudstones with a few enclosed discontinuous sandstone beds that represent additional reservoirs. The shaly deposits of the Zdanice and Pouzdrany units are considered to be an excellent regional seal.

Biomarker studies (Picha and Peters, 1998) indicate that the hydrocarbons in the Kostelany fields were sourced from organic-rich Paleogene rocks. The autochthonous Paleogene and detached slices of it, such as the Pouzdrany unit and/or the Menilitic shales of the Zdanice unit, are the most likely sources of these hydrocarbons.

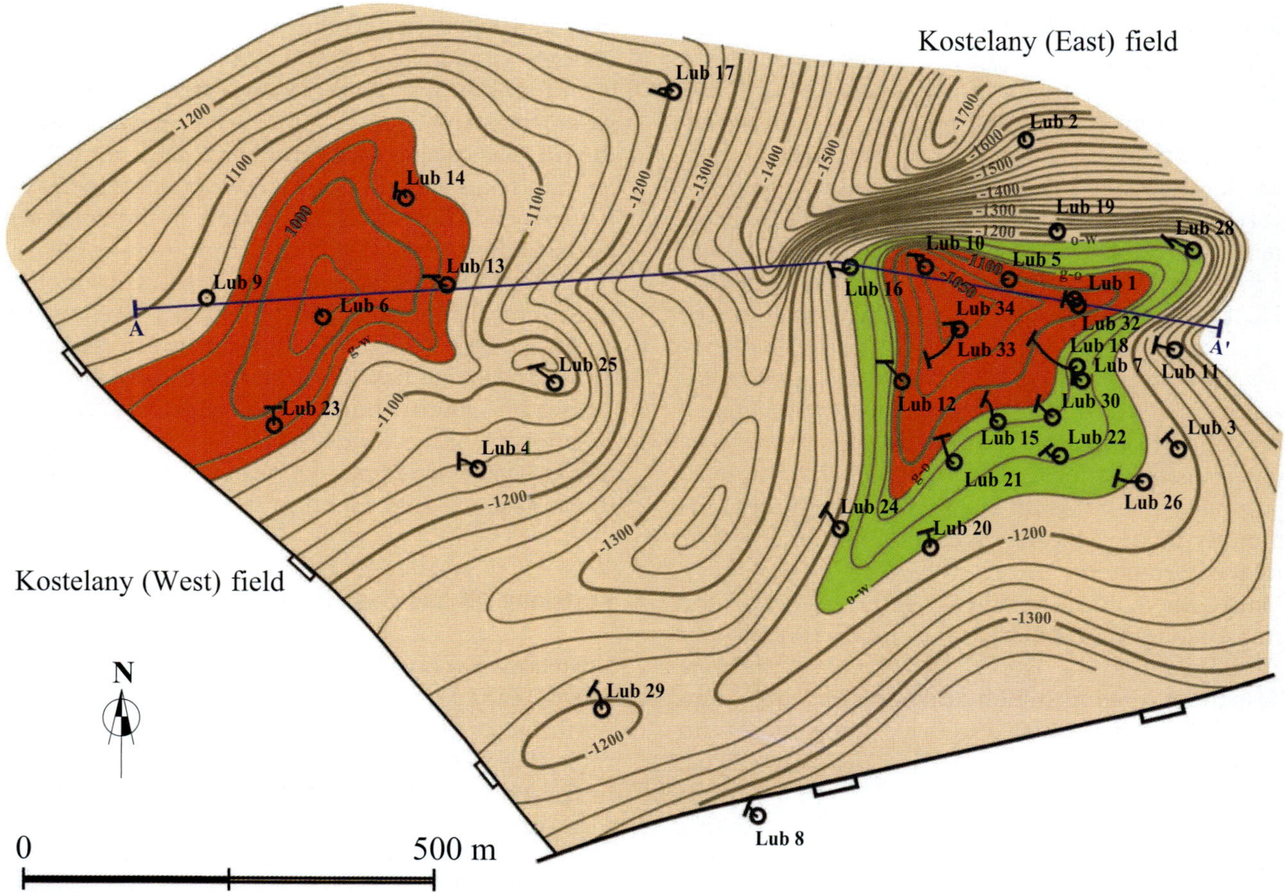

Figure 2. Kostelany oil and gas fields. Structure map is on the top of crystalline basement. Hydrocarbons are reservoired in the weathered and fractured surface of granitic rocks. The oil and gas columns reach as much as 75 m (250 ft) in the Kostelany Zapad (West) field and as much as 175 m (570 ft) in the Kostelany Vychod (East) field.

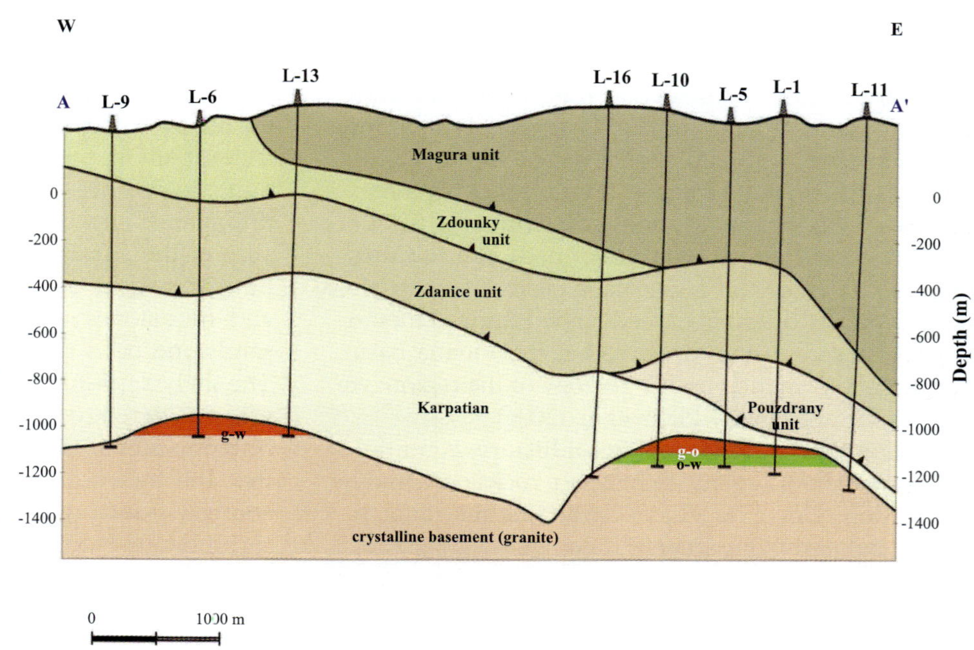

Figure 3. Kostelany oil and gas fields. Cross section AA′ is located in Figure 2. Oil and gas accumulations in the surface zone of the crystalline basement are sealed by Karpatian strata of the Neogene foredeep and further overthrust by four tectonic units of the Carpathian thrust belt. g-w = gas-water contact; g-o = gas-oil contact; o-w = oil-water contact.

Table 1. The characteristics of significant deposits.

Parameters \ Fields		Kostelany Zapad (West)	Kostelany Vychod (East)	Dunajovice	Zarosice	Uhrice Jih (South)
Discovery		1971	1966	1973	2001	2001
Exploration Period		1971–1975	1966–1975	1973–1977	2001–present	2001–2003
Production Period		1976–2000	1968–present	1977–1984*	2001–present	2001–present
Oil in place		---	1,4 MM m^3	---	2,9 MM m^3	1,1 MM m^3
Total oil exploited		---	185 M m^3	---	71,4 M m^3	100,3 M m^3
Gas in place		840 MM m^3		1600 MM m^3	140 MM m^3	98 MM m^3
Total gas exploited		579 MM m^3		811 MM m^3	cca 5 MM m^3	5,5 MM m^3
Field area		2,3 km^2	3,1 km^2	5,4 km^2	0,8 km^2	0,6 km^2
Geological unit		crystalline basement	crystalline basement	Eggenburgian	Jurassic-Vranovice carbonates formation	Jurassic-Gresten formation
Reservoir rocks		granite	granite,sandstones	transgressive basal sandstones	dolomites	beach sandstones
Sealing		Karpatian shales	Karpatian shales	Eggenburgian shales and the Vestonice fault	Mikulov marls and Paleogene shales	Paleogene shales
Source rocks		Paleogene	Paleogene	assumed Jurassic marlstones	Paleogene, Jurassic	Paleogene, Jurassic
Depth of the reservoir		1270–1350 m	1450–1550 m	1050–1120 m	1600–1850 m	1850–1950 m
Effective thickness of the reservoir		max. 83 m, average 55 m	max. 107 m, average 75 m	11–37 m	max. 250 m, average 150 m	max. 90 m, average 50 m
Porosity		3–10 %	3–15 %	23–35 %	average 10%	21%
Permeability		5–15 mD	5–20 mD	15–1300 mD	min. 500 mD	300–400 mD
Gas saturation		65%	65%	67%	65–70 %	60–70 %
Reservoir pressure		13,2 MPa	13,5 MPa	10,966 MPa	18,038 MPa	17,336 MPa
Reservoir temperature		54°C	60°C	42°C	51,4°C	55,5°C
Oil API gravity		---	26.2	---	23.5	24.4
Gas composition	methane	65.9%	87.1%	98%	85%	94%
	higher hydrocarbones	2.2%	6.2%	0.8%	8%	4.9%
	$CO_2 + N_2$	31.9%	6.7%	1.20%	7%	1.1%
	H_2S	0	0	0	0	0

*Converted to the gas storage.

The Kostelany Vychod (East) field, which was discovered in 1966, was the first commercial field found in the subthrust foreland plate below the Carpathian thrust belt in Moravia. Although in production for more than 30 yr and despite the high stage of reservoir flooding, production is still commercial.

Basic parameters of these fields as well as further fields are shown in Table 1.

The Zarosice and Uhrice Jih (South) Subthrust Oil and Gas Fields

The Zarosice and Uhrice Jih (South) fields are located in the Nesvacilka graben (Figures 1, 4), an asymmetrical northwest–southeast-trending depression whose southwestern side is bounded by a steep fault, whereas the northeastern side is stepping down gradually. The Cadomian crystalline basement of the graben area is covered by depositional sequences of Paleozoic to Miocene age. They include the late Lower Devonian basal clastics, the Middle Devonian to Lower Carboniferous carbonates, the Lower Carboniferous flysch, the Upper Carboniferous coal-bearing molasse, and the Middle–Upper Jurassic synrift clastics and carbonates. In the Late Cretaceous to early Paleogene, the foreland area was uplifted and deeply eroded. A large, as much as 1500-m (5000-ft)-deep, paleovalley and submarine canyon was cut within the confines of the Nesvacilka graben and filled with Paleogene organic-rich deep-water deposits. Finally, the lower Miocene clastic deposits of the Carpathian foredeep were laid down in the northwestern part of the graben. During the latest phases of Alpine orogeny, the Nesvacilka graben, including the paleovalley fill, was buried below the Zdanice–Subsilesian nappe, which thus formed a regional seal. The most promising exploration targets proved to be the basal sandstones and the carbonates of Jurassic, occurring as isolated erosional relicts on the northern slope of the graben (Damborice, Uhrice, and Zarosice fields) (Figure 4). These relicts (buried hills) are sealed by impermeable shales and mudstones of the paleovalley fill. Thanks to differences in velocities between the Jurassic and the overlying Paleogene rocks, these erosional relicts can be recognized and outlined on the 3-D seismic survey. Also small in their areal extent, the saturated reservoirs of these relicts are relatively thick (several tens of meters to 200 m [660 ft]) and typically have very good reservoir properties. The largest discovery so far is the Damborice field with more than

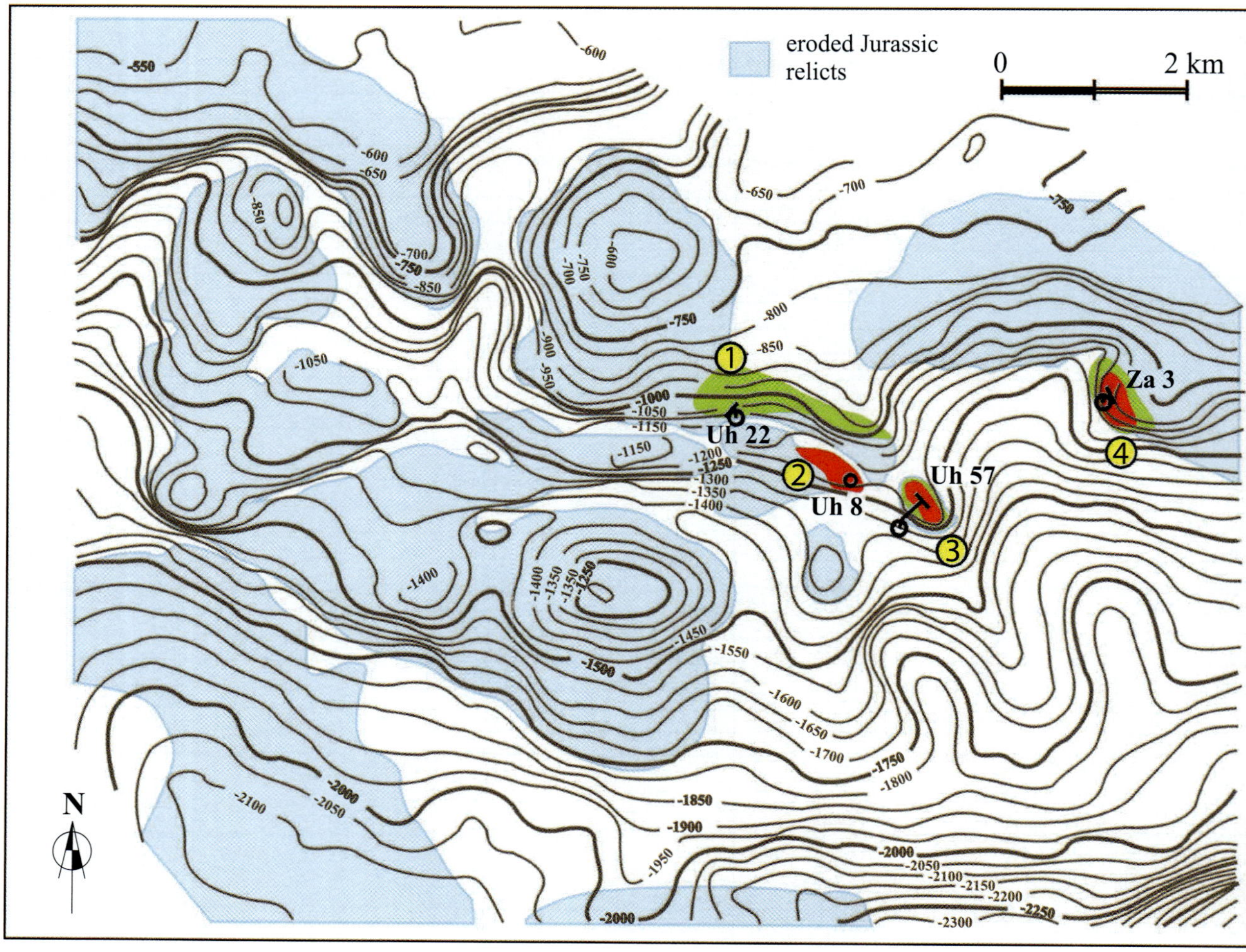

Figure 4. Time map of the pre-Tertiary relief of the northeastern slope of the Nesvacilka graben with delimited erosional relicts of Jurassic strata and proved oil and gas fields: (1) Damborice oil field; (2) Uhrice gas field; (3) Uhrice Jih (South) oil and gas field; and (4) Zarosice oil and gas field.

6 million m^3 (42 million bbl) of oil in place reservoired in the as much as 100-m (330-ft)-thick basal Jurassic clastics (Gresten Formation).

The isolated bodies of turbiditic and channelized sandstones enclosed within the Paleogene fill of the Nesvacilka paleovalley represent the secondary exploration play (Ciprys and Kostelnicek, 1990). The 3-D seismic survey enables us to see these isolated bodies of sandstones and, thus, lower the exploration risk. Interesting accumulations of gas and condensate have been found in these sandstones in the deeper parts of the Nesvacilka paleovalley (Krumvir and Karlin gas fields) (Figure 1). The reservoir pressure in the Karlin field is extremely high; it doubles the hydrostatic pressure.

Thus far, only small uneconomic accumulations of hydrocarbons have been found in the deeper Paleozoic sequences, namely, in the Devonian carbonates.

Geochemical studies (Francu et al., 1996; Krejci et al., 1996; Picha and Peters, 1998) suggest that the hydrocarbons in the Nesvacilka graben area were predominantly sourced from Jurassic, organic-rich Mikulov marls. The Paleogene shales and mudstones may be considered as additional source rocks, especially for gas.

ZAROSICE FIELD

Zarosice field (Figures 1, 5–8) is located in an erosional relict of Jurassic rocks (buried hill) on the northern slope of the Nesvacilka graben. It is reservoired in the Jurassic Vranovice carbonates (mostly dolomites with some limestones and sandstones) and sealed by the mudstones of the Paleogene valley fill and the overlying Zdanice nappe.

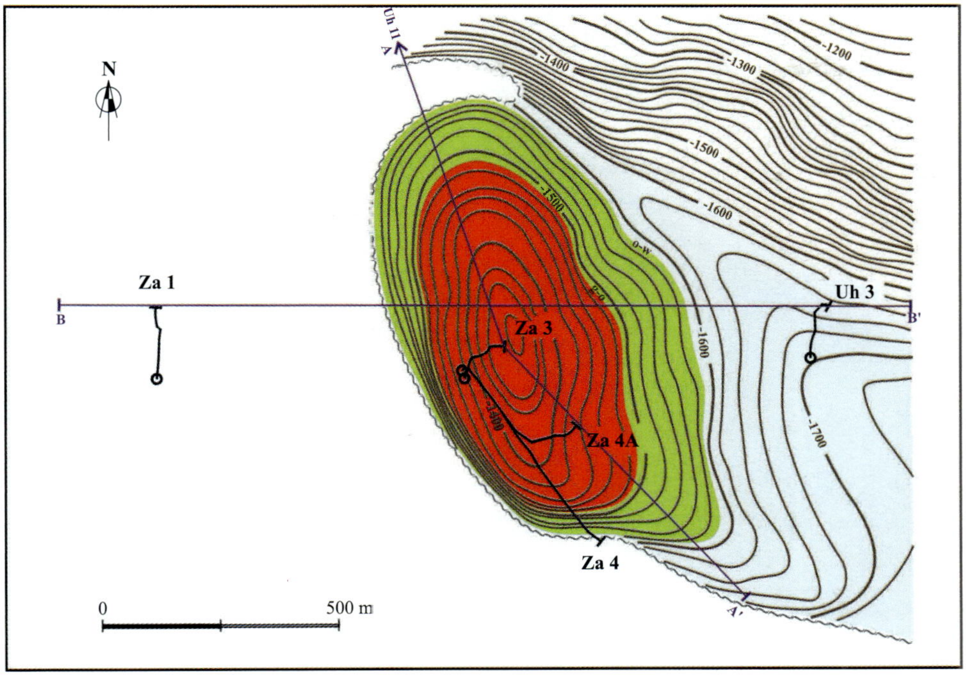

Figure 5. Zarosice oil and gas field. Structure map is on the top of the Vranovice carbonates in an erosional relict of the Jurassic rocks sealed by the Paleogene valley fill and overthrust by the Zdanice nappe. Combined oil and gas column reaches as much as 250 m (820 ft). o-w = oil-water contact.

UHRICE JIH (SOUTH) FIELD

The Uhrice Jih (South) field is located in an erosional relict of Jurassic rocks (buried hill) on the northern slope of the Nesvacilka graben, southeast of the major Damborice field (Figure 4). It is reservoired in the basal Jurassic clastics (quartzose sandstones) of the Gresten Formation and sealed by the mudstones of the Paleogene valley fill and the overlying Zdanice nappe (Figures 9–12).

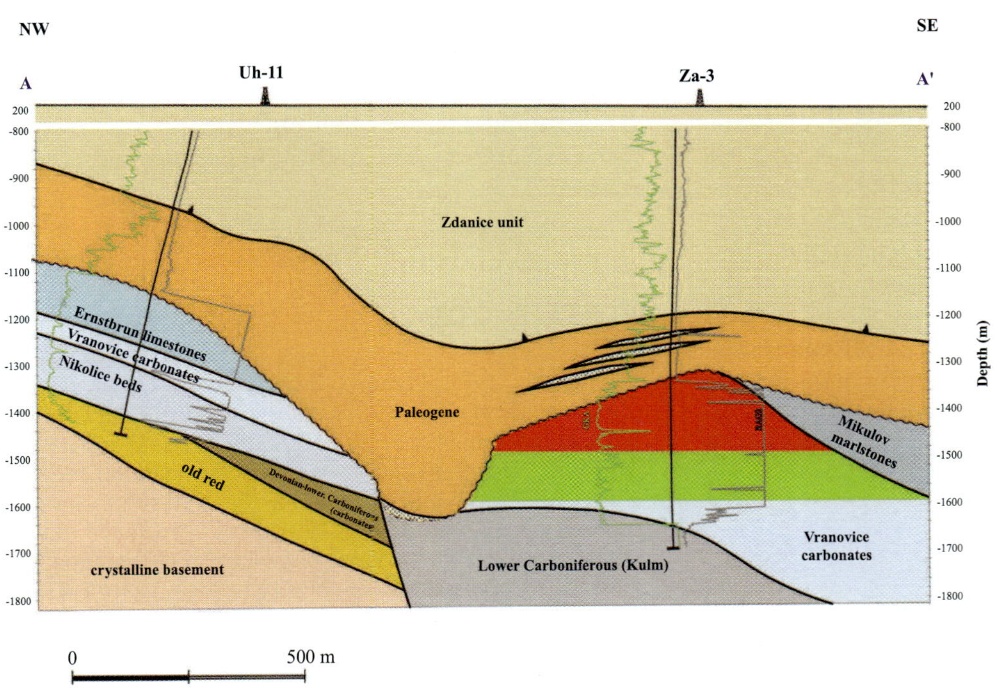

Figure 6. Zarosice oil and gas field. Cross section AA′ is located in Figure 5.

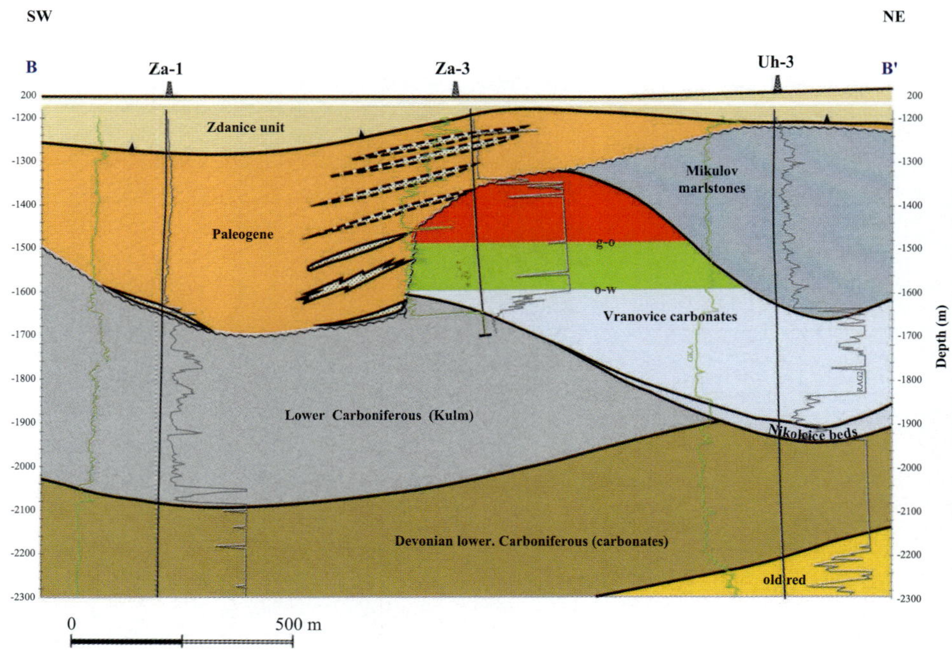

Figure 7. Zarosice oil and gas field. Cross section BB′ is located in Figure 5. g-o = gas-oil contact; o-w = oil-water contact.

DOLNI DUNAJOVICE GAS FIELD

The gas field Dolni Dunajovice, which was discovered in 1973 in the Neogene foredeep in front of the Carpathian thrust belt in southern Moravia (Figure 1), is one of the largest gas fields found in the territory of the Czech Republic. The field is reservoired in the Eggenburgian sandstones draping over a southwest–northeast-trending upthrown block of the foreland plate below the Neogene foredeep (Figures 13–15). On its western side, the block is bounded by the Vestonice fault interpreted in this account as an antithetic normal fault formed during the flexural downbending of the foreland plate during the progression of the Carpathian thrust belt onto the European platform (Adamek, 1979). Some other geologists (Picha et al., 2006) rather interpret this fault as a foreland-type compressional and possibly transpressional structure generated by compressional stresses in the foreland plate during the orogeny. The thinning of Eggenburgian strata over the uplifted block indicates that the block was actively rising during that time. The pre-Miocene basement is represented by the crystalline rocks of the Brunovistulicum, overlain by Jurassic strata; the Paleozoic is not present in this southernmost part of Moravia.

CONCLUSIONS

Discoveries of several oil and gas fields in a complex depositional and structural setting of the subthrust foreland plate underneath the thin-skinned Carpathian thrust belt at the territory of Moravia (Czech Republic) represent a major accomplishment. They resulted from long-term systematic geological and exploration work conducted by various state organizations and more recently by a private oil company (MND). Although small in size, the territory of Moravia has become one of the most successful exploration provinces in the entire Carpathian region in recent years.

ACKNOWLEDGMENTS

We gratefully acknowledge the management of the Moravske Naftove Doly, A.S. (MND) for permission to release the data for publication. MND is an owner of the exploration and production permits and is sole investor.

REFERENCES CITED

Adamek, J., 1979, Plynove lozisko Dolni Dunajovice a geologicka stavba jizni casti karpatske predhlubne (in Czech): Zemni plyn a nafta, Hodonin, v. 24, no. 1, p. 1–22.

Ciprys, V., and P. Kostelnicek, 1990, New data on the extent, structures and deposits of the autochthonous Paleogene in the Nesvacilka graben, *in* D. Minarikova and H. Lobitzer, eds., Thirty years of geological cooperation between Austria and Czechoslovakia: Federal Geological Survey, Vienna, and the Geological Survey, Prague, p. 198–202.

Ciprys, V., J. Adamek, and S. Benada, 1995, Petroleum geology of the Carpathian foredeep and overthrust zones

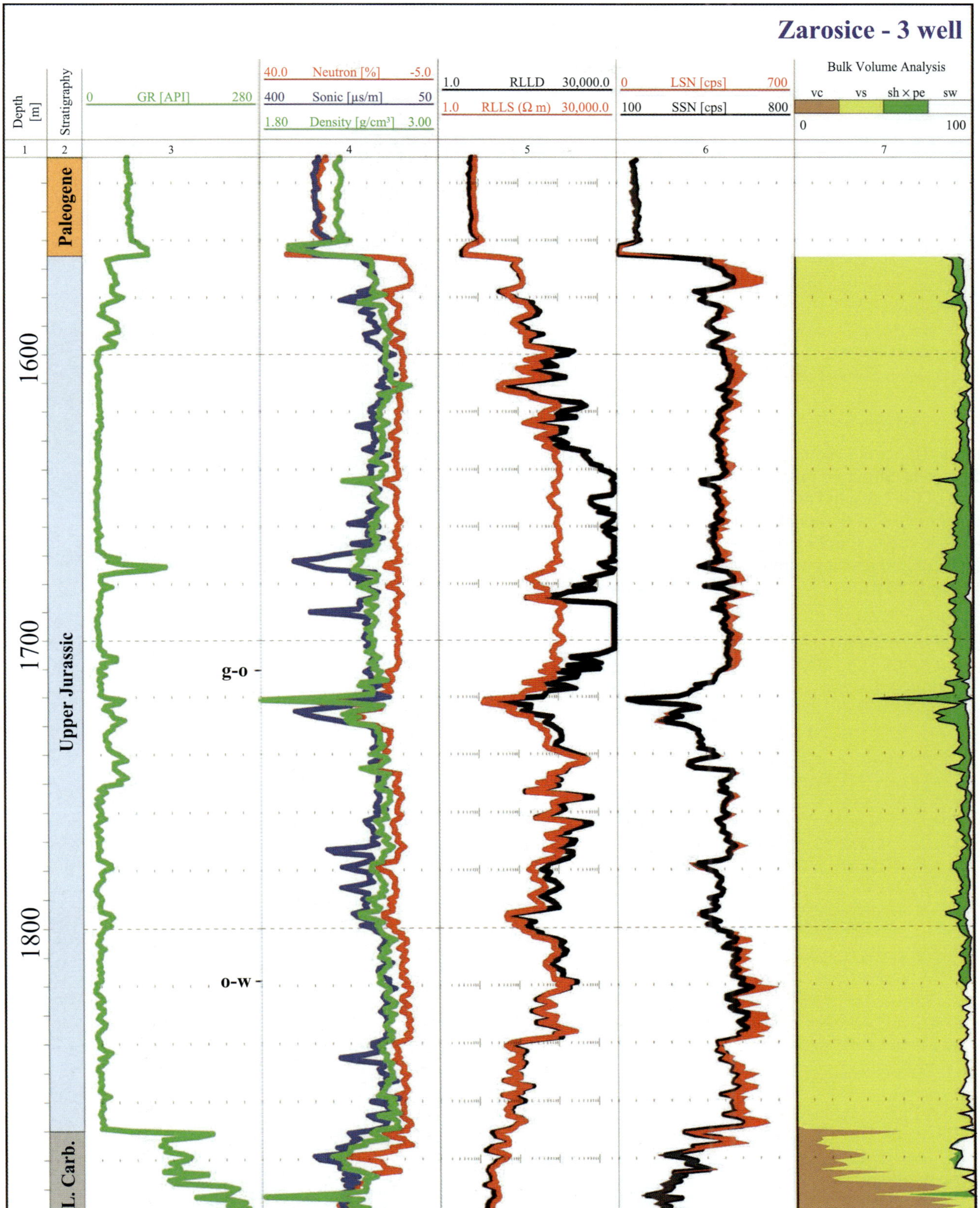

Figure 8. Log diagrams and saturation estimation from the Zarosice-3 well. vc = clay volume; vs = sand (matrix) volume; sh × pe = hydrocarbon saturation multiplied by the porosity; sw = water saturation; o-w = oil-water contact; g-o = gas-oil contact; RLLD = resistivity laterolog–deep; RLLS = resistivity laterolog–short; LSN = long-spaced count neutron; SSN = short-spaced count neutron; GR = gamma ray.

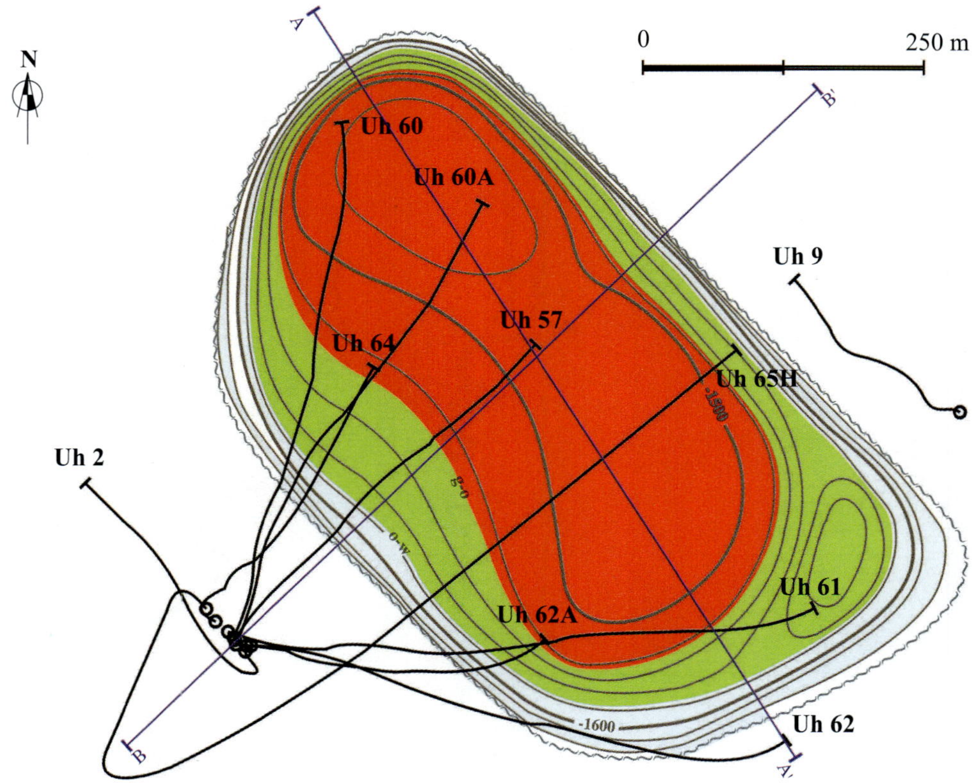

Figure 9. Uhrice Jih (South) oil and gas field. Structure map is on the top of the Gresten Formation (sandstones) in an erosional relict of Jurassic rocks. Combined oil and gas column reaches as much as 100 m (330 ft).

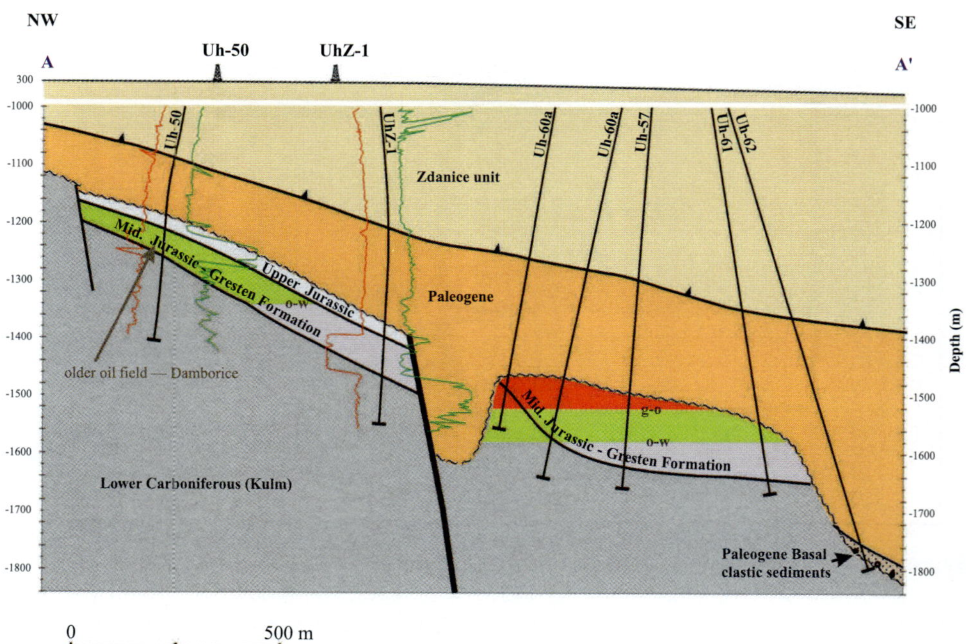

Figure 10. Uhrice Jih (South) oil and gas field. Cross section AA′ is located in Figure 9. o-w = oil-water contact; g-o = gas-oil contact.

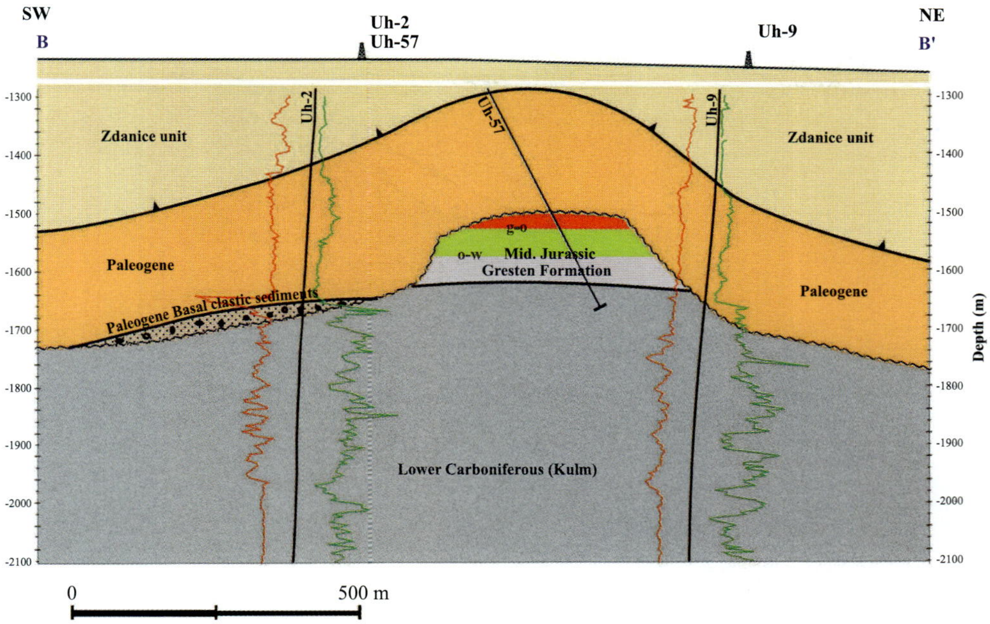

Figure 11. Uhrice Jih (South) oil and gas field. Cross section BB′ is located in Figure 9. g-o = gas-oil contact; o-w = oil-water contact.

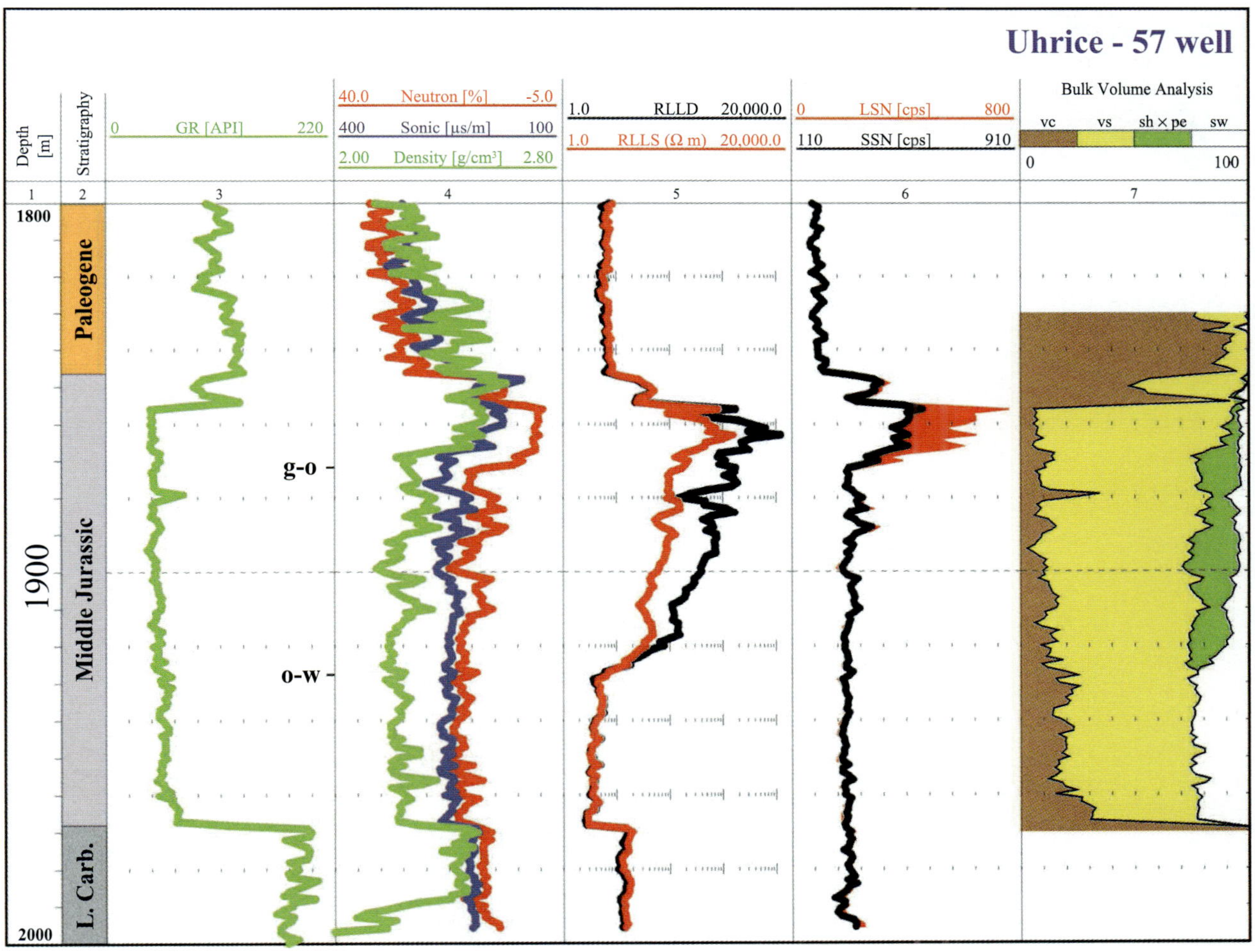

Figure 12. Uhrice Jih (South) oil and gas field. Log diagrams and saturation estimation from the Uhrice-57 well. vc = clay volume; vs = sand (matrix) volume; sh × pe = hydrocarbon saturation multiplied by the porosity; sw = water saturation; o-w = oil-water contact; g-o = gas-oil contact; RLLD = resistivity laterolog – deep; RLLS = resistivity laterolog – short; LSN = long-spaced count neutron; SSN = short-spaced count neutron; GR = gamma ray.

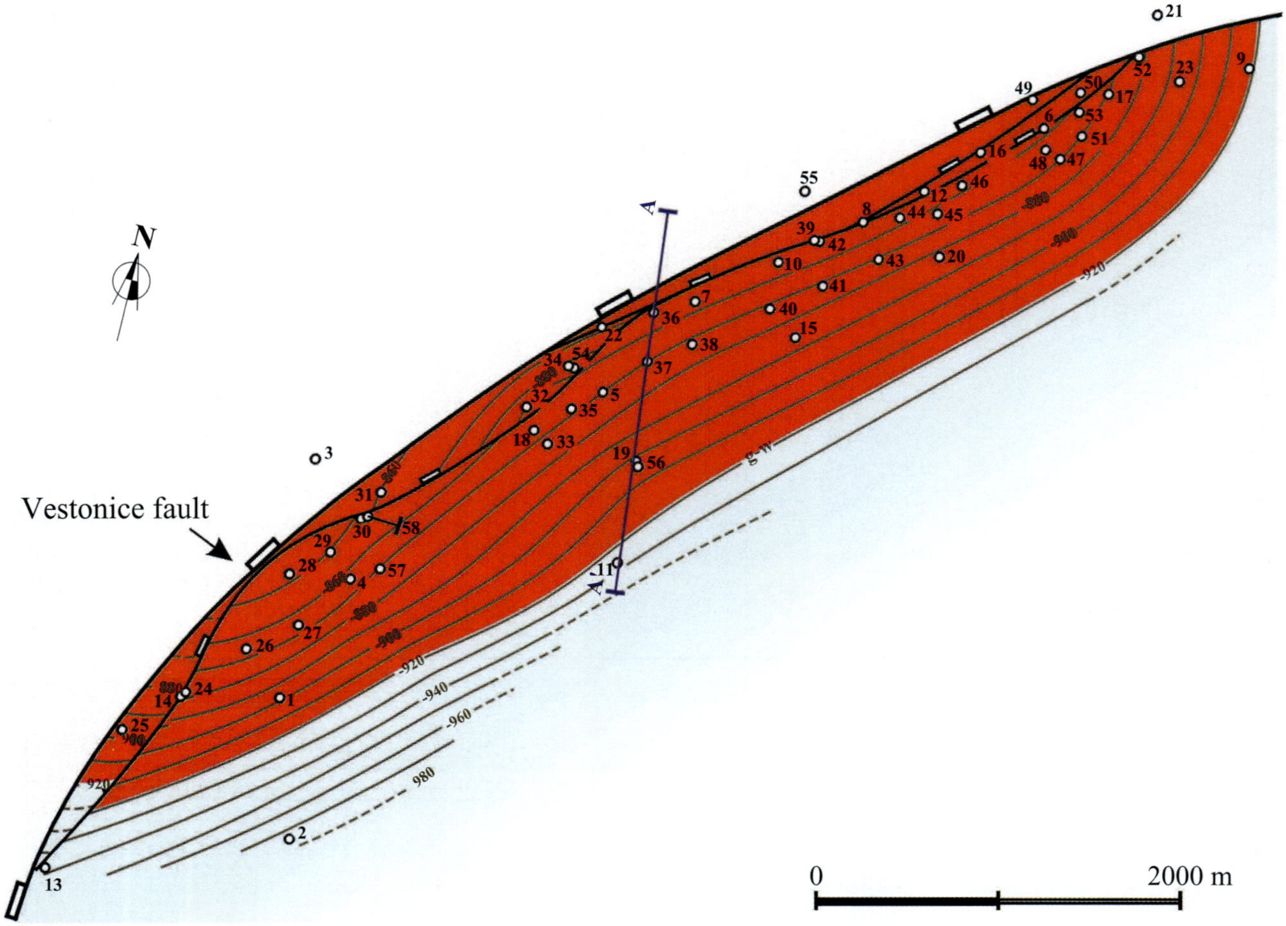

Figure 13. Dolni Dunajovice gas field. Structure map is on the top of the Eggenburgian sandstones.

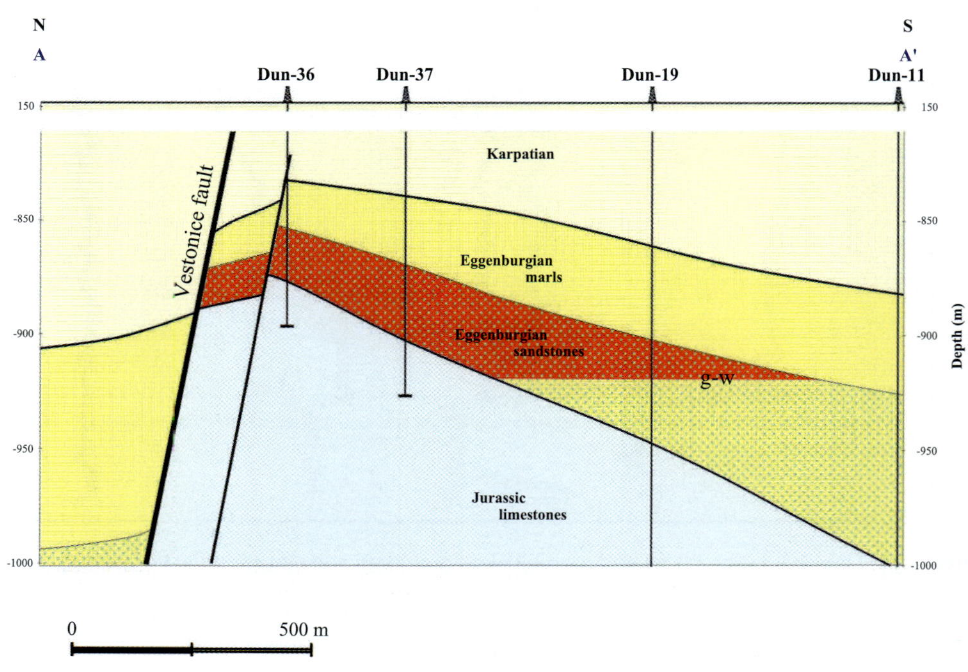

Figure 14. Dolni Dunajovice gas field. Cross section AA′ is located in Figure 13. g-w = gas-water contact.

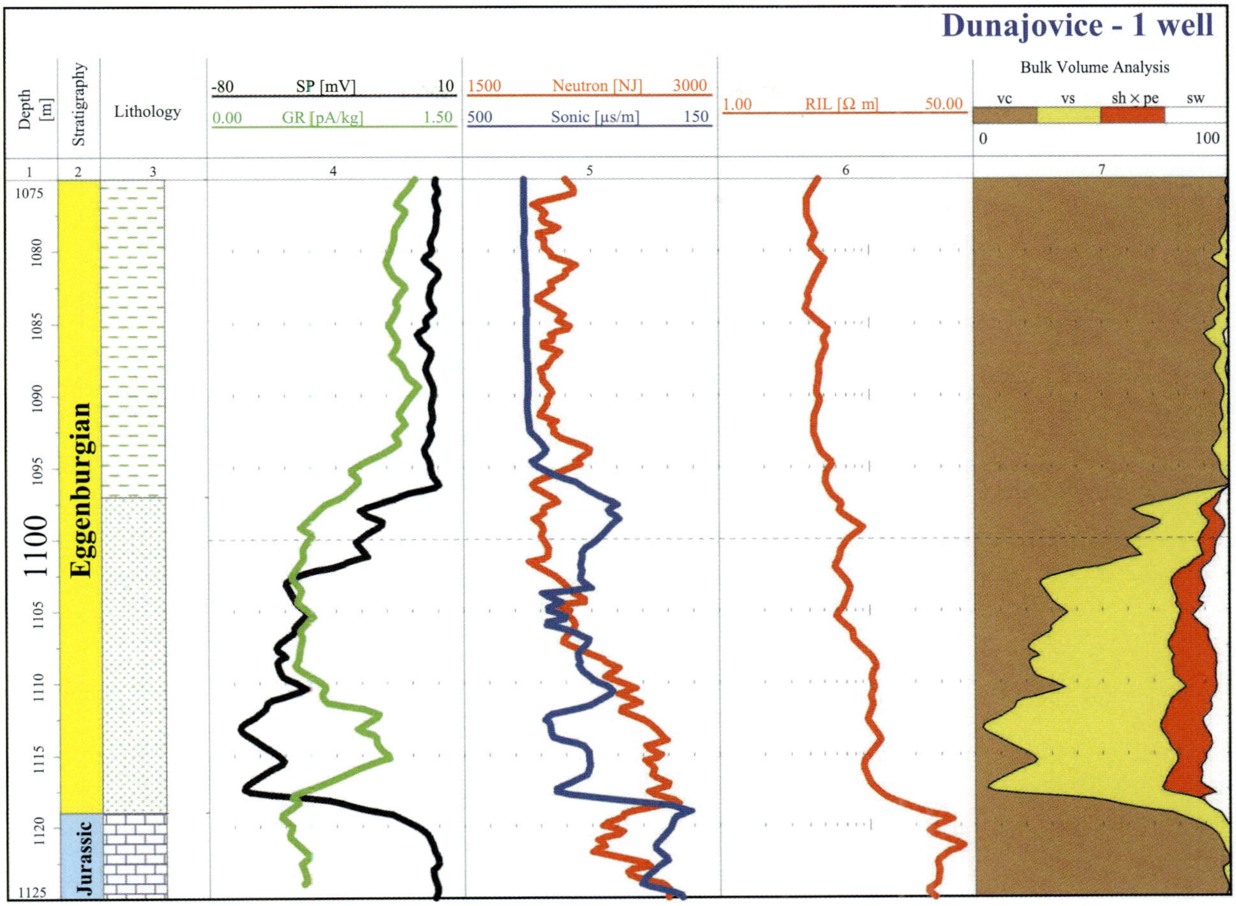

Figure 15. Dolni Dunajovice gas field. Log diagrams and saturation estimation from the Dunajovice-1 well. vc = clay volume; vs = sand (matrix) volume; sh × pe = hydrocarbon saturation multiplied by the porosity; sw = water saturation; SP = spontaneous potential; RIL = resistivity induction log; GR = gamma ray.

in the Czech Republic: Petroleum Geoscience, v. 1, p. 89–94.

Francu, J., M. Radke, R. G. Schaefer, H. S. Poelchau, J. Caslavsky, and Z. Bohacek, 1996, Oil-oil and oil-source rock correlations in the northern Vienna basin and adjacent Carpathian Flysch Zone (Czech and Slovak area), *in* G. Wessely and W. Liebl, eds., Oil and gas in Alpidic thrustbelts and basins of Central and Eastern Europe: European Association of Geoscientists and Engineers Special Publication 5, p. 343–353.

Jiricek, R., 1987, Die stratigraphische und fazielle Unterteilung der autochtonen Sedimenten des Palaogens and SO Abhangen der Bohmische Masse (in Czech with German summary): Knihovnicka Zemniho plynu a nafty 6b, Miscellanea Micropaleontologica, Hodonin, v. II/2, p. 247–314.

Krejci, J., 1993, Zdanice-Krystalinikum field—Czechoslovakia, Carpathian foredeep, Moravia, *in* N. N. Foster and E. A. Beaumont, eds., Treatise of petroleum geology atlas of oil and gas fields, structural traps VIII: AAPG, p. 153–175.

Krejci, O., J. Francu, H. S. Poelchau, P. Muller, and Z. Stranik, 1996, Tectonic evolution and oil and gas generation at the border of the north European platform with the West Carpathians (Czech Republic), *in* G. Wessely and W. Liebl, eds., Oil and gas in Alpidic thrustbelts and basins of Central and Eastern Europe: European Association of Geoscientists and Engineers Special Publication 5, p. 177–186.

Picha, F., 1979, Ancient submarine canyons of Tethyan continental margins, Czechoslovakia: AAPG Bulletin, v. 63, no. 1, p. 67–86.

Picha, F., 1996, Exploring for hydrocarbons under thrust belts—A challenging new frontier in the Carpathians and elsewhere: AAPG Bulletin, v. 80, p. 1547–1564.

Picha, F., and K. E. Peters, 1998, Biomarker oil-to-source rock correlation in the Western Carpathians and their foreland, Czech Republic: Petroleum Geoscience, v. 4, p. 289–302.

Picha, F., Z. Stranik, and O. Krejci, 2006, Geology and hydrocarbon resources of the Outer West Carpathians and their foreland, Czech Republic, *in* J. Golonka and F. J. Picha, eds., The Carpathians and their foreland: Geology and hydrocarbon resources: AAPG Memoir 84, p. 49–175.

5

Arzmüller, Gerhard, Štěpán Buchta, Eduard Ralbovský, and Godfrid Wessely, 2006, The Vienna basin, *in* J. Golonka and F. J. Picha, eds., The Carpathians and their foreland: Geology and hydrocarbon resources: AAPG Memoir 84, p. 191–204.

The Vienna Basin

Gerhard Arzmüller
Oesterreichische Mineraloel Verwaltung Exploration, Vienna, Austria

Štěpán Buchta
Moravske Naftove Doly, Hodonin, Czech Republic

Eduard Ralbovský
Slovak Gas Industry, VVNP Research Oil Company, Gbely, Slovakia

Godfrid Wessely
Siebenbrunnengasse 29. Vienna, Austria

ABSTRACT

Through conventional geoscientific research, a fundamental knowledge of the Vienna basin was acquired with the limited data available at the time. Since then, vast exploration programs of the Austrian, Czech, and Slovakian oil industries have contributed significantly to a more detailed understanding of the geological evolution of the basin.

The pull-apart or piggyback nature of the basin at present is well understood and commonly accepted. Basically, it resulted from an easterly directed extrusion of the Central Alpine block alternating with compressional events during the final stages of the Alpine convergent phase.

The basin evolved in several stages that finally resulted in an intricate arrangement of prominent highs and partly deeper subsided depocenters. The basin was filled by Miocene to Pleistocene sediments that can be subdivided into sequences separated by unconformities; the most pronounced are between the early and the middle Miocene.

The basin-floor section below the Neogene fill consists of the Alpine–Carpathian imbricated system. From north to south, these individual thrust piles are the Waschberg–Zdanice zone, the Flysch zone, the Calcareous Alps (including its Paleozoic base, the Grauwacken zone), the Central Alps, and the Tatrides. All these units lie on top of the Miocene Molasse, a Mesozoic series, and the crystalline basement. This section is well known from wells drilled in the molasse zone sensu stricto but were also drilled in the Vienna basin. Ultradeep wells targeting the autochthonous basement reached total depths of between 6.3 and 8.5 km (3.9 and 5.3 mi).

Oil and gas are trapped in all units of the basin, from the Neogene fill down to the autochthonous sedimentary cover. Generally, the traps are structural, but recently, stratigraphic traps have also been drilled successfully.

The main source rocks are autochthonous Jurassic marls, which seem to form a substantially thickening package in an easterly direction, according to the sporadic well information available. Although the ultradeep well program was believed to be uneconomic at that time, indications point to an unconventional tight-gas play type in the autochthonous Jurassic marls.

DOI:10.1306/985608M843068

The contribution of all Vienna basin member countries made it possible to present a comprehensive surface and subsurface compilation. A cross section through the deepest known part of the basin without border limitations was an additional result of this cooperative effort.

INTRODUCTION

For more than 150 yr, intense research has been done on the Vienna basin. Recent overviews have been written by Sauer et al. (1992), Brix and Schultz (1993), Hlavaty (1996), and Hamilton et al. (2000).

Previous generations of geoscientists have gained fairly good knowledge of the Vienna basin, but it was the oil industry that contributed significantly to understanding the basin evolution. This was achieved by drilling more than 6000 wells (the deepest to a total depth of 8553 m [28,061 ft]) and acquiring a vast coverage of two- and three-dimensional (2-D and 3-D) seismic data.

Early scientific geological investigations in the Vienna basin were done by Hornes, Prevost, Boue, Suss, Schaffer, Friedl, Janoschek, Grill, and Papp in the first half of the 20th century. More recent structural and sedimentological studies were done by Austrian, Slovak, and Czech geoscientists in the second half of the 20th century. This improved geological wisdom, as well as the implementation of 3-D seismic technology, resulted in new discoveries and more effective exploitation of existing reservoirs.

Further success is expected with more extensive use of other modern geoscience methods. These include the application of modern sedimentology, neural network technology, prestack depth migration, seismic inversion, etc.

BASIN FORMATION

Tectonic-Paleogeographic Preconditions

The Vienna basin evolved when the Alpine–Carpathian thrust front reached the European forelands during the early Miocene. At the Alpine–Carpathian junction, the Bohemian Promontory forms a recess in which the thrust units could advance progressively, whereas to the west of the reentry, thrust movement stagnated earlier.

Consequently, thrust ages become younger from west to east along the Alpine–Carpathian front. Final thrusting of the easternmost Alpine nappes terminated by the end of Karpatian, whereas the Romanian Carpathians protracted thrusting beyond the Pannonian (Jiricek, 1979) (Figure 1). This resulted in a complex succession of thrusting, wrenching, and normal faulting along the Alpine–Carpathian junction. It was Royden (1985) who first summarized the tectonic features as belonging to a pull-apart basin. Accordingly, the Vienna basin was described as having developed between a system of sinistral transfer faults. More detailed interpretations were done by Wessely (1988), Decker and Perreson (1996), Seifert (1996), and Kovac (2000).

The southern edge of the basin is delimited by a northeast-trending strike-slip fault system with typical transtensional patterns, such as negative flower structures and small pull-apart basins. This lineament still represents a zone of increased seismicity. It encloses a graben-in-graben structure with a considerably thicker Quaternary fill, which is another clear indication for its current activity. However, no evidence exists for this strike-slip zone ever acting as a major fault system throughout the basin evolution.

A structural counterpart north of the basin has not been located so far.

Some faults sole out into individual Alpine thrust sheets or into the main decollement of the whole Alpine thrust complex. The majority of these fault systems follow the strike of the Alpine–Carpathian nappes. Minor lateral offsets occur where the divergent stepover faults deviate from these thrust elements and transect through them.

In summary, the criteria for the pull-apart mechanism are met and are quite obvious. These criteria include the rhombohedral shape of the basin, pronounced depocenters, and en echelon faults. The right-stepping arrangement clearly indicates the left lateral movement. A further controlling element, which has influenced the basin development, is the pre-Alpine morphotectonic features of the subthrust.

Geodynamics

Three development stages are recognized in the evolution of the Vienna basin: the pre-, the proto-, and the neo-Vienna basin (Figure 2).

Pre-Vienna Basin

The area of today's Vienna and Alpine–Carpathian Molasse Basin was a rifting zone during the Middle Jurassic, a passive-margin basin between the Malmian

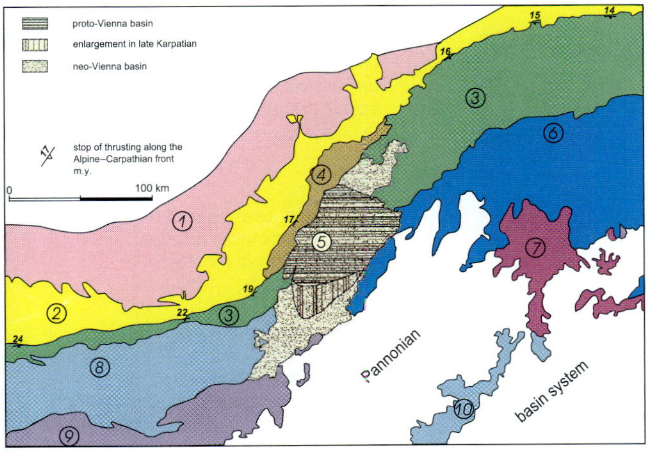

Figure 1. The Vienna basin in the Alpine–Carpathian thrust belt.

and the Cretaceous and finally a molasse foredeep basin for the gradually evolving thrust belt in the south up to the Egerian. This foredeep was partly overthrusted by Alpine-Carpathian nappes stratigraphically ranging from the Triassic to the Cretaceous.

Post-collisional thrust activity continued to the Oligocene, as evidenced by the exploration well OMV Berndorf 1, 40 km (25 mi) south of the Alpine thrust front, which encountered Oligocene Molasse sediments below the Alpine nappes at a depth of approximately 6000 m (20,000 ft).

Proto-Vienna Basin

The first tensional forces initiated during the thrust movement and led to the development of the Proto-Vienna piggyback basin.

Between the Eggenburgian and the Karpatian, the thrust belt advanced into the molasse foredeep. Corresponding series were therefore overthrusted as well as deposited onto the Alpine thrust front. In the Proto-Vienna basin, restricted to the northern part of the present basin, horizontal extension has been illustrated by synsedimentary normal faulting since the early Miocene. Czech as well as Austrian seismic data show synsedimentary faults with considerable offsets in the early Miocene deposits (Ladwein et al., 1991) east of the subsequently evolving Steinbergsystem. To the west, the Korneuburg basin is also genetically related to that period.

Toward the end of the Karpatian, the southwest–northeast-orientated Matzen Flysch ridge separated the marine environments in the north from nonmarine in the south.

The facies boundary was related to an uplift culminating in subaerial exposure of the ridge. Early Miocene sediments were tilted and, along the crest, entirely removed by erosional processes that also substantially affected the flysch itself. Structurally associated subsidence south of the high created an additional accommodation space, which was filled by the limnic Aderklaa Formation. Off the tilted areas, the limnic package is conformably overlain by the Badenian Aderklaa Conglomerate, indicating that the late Karpatian readjustments essentially predefined the geometry of the modern Vienna basin. Only a small rim of pre-Neogene basement, close to the southern margin, was later transgressed by the basal Badenian complex (Weissenback, 1996).

Neo-Vienna Basin

The thrust movements ceased at the end of the Karpatian period and the basin subsequently subsided into a consolidated substrate. The distribution of the Aderklaa Conglomerate or time-equivalent sediments mirror the established Neo-Vienna basin that retained its northeast–southwest-oriented trough axes throughout the deposition of the Sarmatian and Pannonian series that blanketed the Badenian sequences without major depositional breaks.

TECTONIC ELEMENTS

The main tectonic elements that occurred as a result of these geodynamic processes are the Steinberg fault, with a vertical displacement of about 6000 m (20,000 ft); the Schrattenberg fault; the Bisamberg fault; the Leopoldsdorf fault (~4000-m [~13,100-ft] displacement); the Markgrafneusiedl fault; the Laksary fault; the Farsky fault; the Lanzhot–Hrusky fault; and other younger elements with offset of several hundred meters (Kroll et al., 1993; Cekan et al., 1990). Remarkably, the regional highs are always up against prominent lows. Examples of this are the Mistelbach (upthrown)

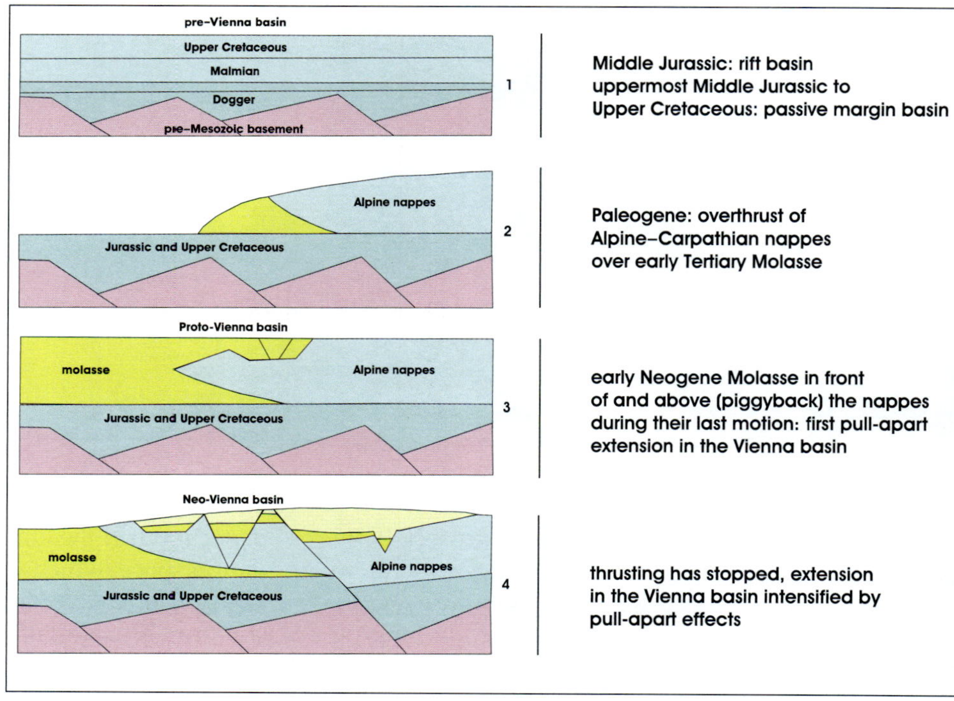

Figure 2. Stages of basin development.

block with the Steinberg high vs. the Zistersdorf depression, or the Oberlaa–Laxenburg high on the Modling block against the Schwechat deep (Figure 3).

Further tectonic features similar to small pull-apart basins and grabens, which are related to sinistral transtensional deformation, are preferentially aligned with the southern basin margin and supplement the structural inventory of the Vienna basin. The surface expressions of this active zone are the Wiener Neustadt subbasin, the Mitterndorf depression, the Lassee graben, and its continuation toward Slovakia, the Zohor–Plavecky graben. These are flanked by the northwest-dipping Pottendorf, Kopfstetten, and Engelhardstetten faults. The Hradiste graben is a structural equivalent in the northernmost (Czech) part of the Vienna basin.

The most prominent fault in the northwestern part of the basin is the Steinberg fault, which has been active through almost the entire basin evolution. The syntectonic deposition led to a significant contrast in thicknesses of the Badenian to Pannonian sections on both sides of the fault. Several thousand meters on the downthrown block correspond to an extremely condensed, time-equivalent section of several hundred meters only in the upthrown block. The fault complex has a listric geometry and most likely roots into a detachment plane in the autochthonous, sedimentary cover below the Alpine–Carpathian thrust assembly.

An increased accumulation of sediments could have also been observed in stationary depocenters subsiding through time without contribution of normal faults. In the Schwechat low, for example, isopachs of the lower Badenian Aderklaa Conglomerate indicate an eightfold increase of the interval thickness from the depocenter to the adjoining areas.

The fault block arrangement, as well as tectonic events along the basinal boundary, created a variety of different, areally restricted, facies zones bordering each other. Thus, in-situ banks of Lithothamnia, lumachelles (Badenian), or even oolithic developments (Sarmatian) of elevated areas correlate with silty, shale-dominated layers in the drowning regions.

BASIN FILL

An overview of the sediments of the Vienna basin has been presented by Tollmann (1985), Jiricek (1988), Kreutzer and Hlavaty (1990), Kreutzer (1993a), and Wessely (2000). Biostratigraphic investigations have been done by people like Grill, Papp, Turnovsky, Senes, Cicha, Steininger, Rogl, Fuchs, Rupp, and others.

The basin fill of the early Miocene is interpreted as foredeep deposits in front and on top of the gradually emerging thrust belt. From Eggenburgian to Ottnangian (Luzice formation), the trough was sourced mainly from the south. In terms of facies, these sediments correspond to the so-called "Schlier" of the molasse zone sensu stricto, with which the proto-Vienna basin region still had connection (Steininger et al., 1986). Shales, sands, and conglomerates of open-marine or restricted marine origin in the north were deposited

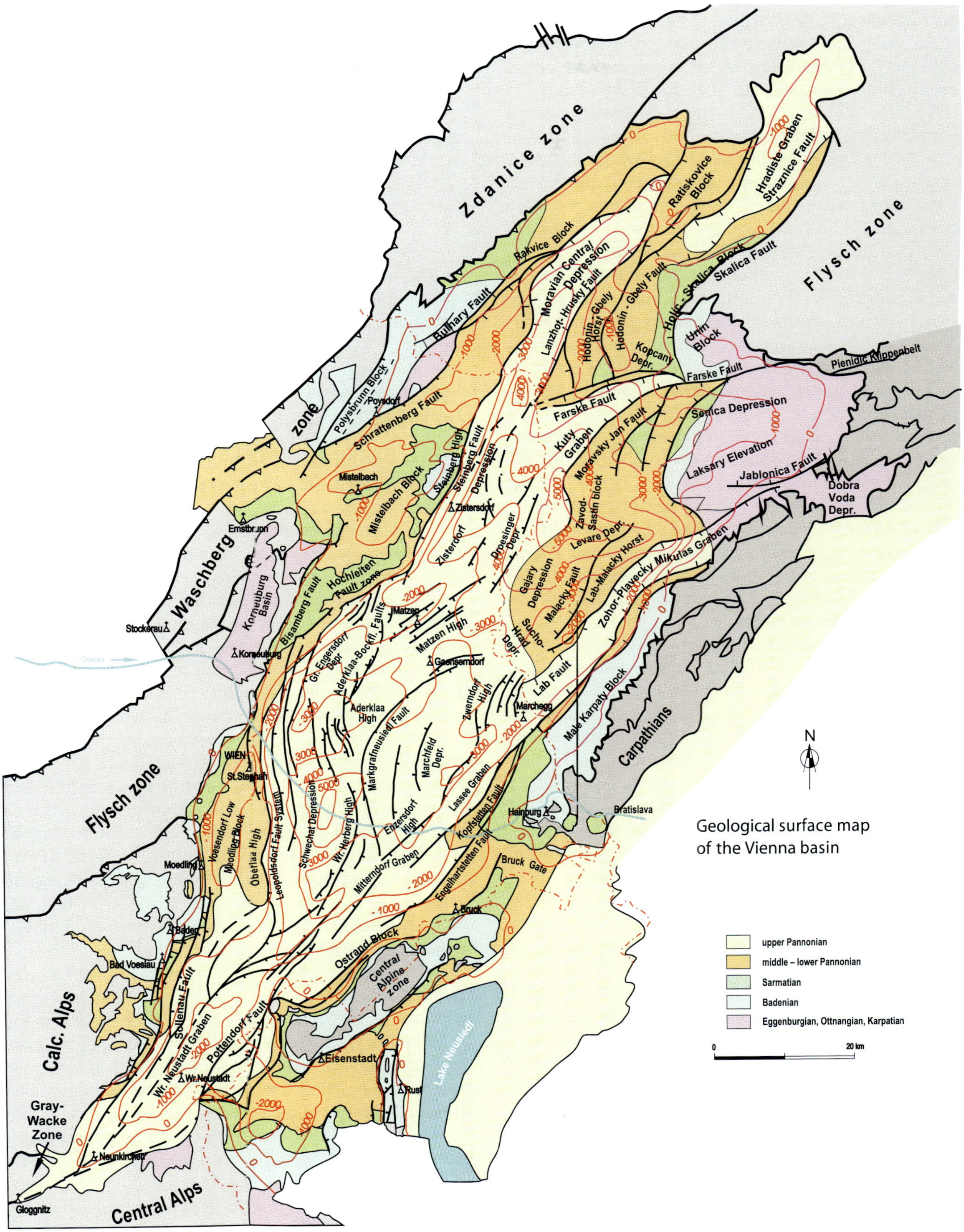

Figure 3. Geological surface map of the Vienna basin with contours of the pre-Neogene basin fill.

concurrently with shallow-marine or terrigenous sediments along the southern margin of the trough. Toward the late Eggenburgian, an impoverished faunal variety could be observed.

The areal extent of the Karpatian is still covered by open-marine deposits (Laa Formation) similar to those of the Eggenburgian. At the lower Karpatian, a brachyhyaline, sandy-marly development (Bockfliess Formation) predominated the deposition in the area south of the Matzen ridge and within the Korneuburg (Sub) basin. The Matzen ridge represented a facies boundary, which was effective until the late Karpatian. The Bockfliess Formation was followed by a sequence of coarse-grained layers, sands, and marls containing a limnic, Ostracoda fauna. This group comprises the Gansendorf conglomerate at the base, variegated marls of the Gansendorf Formation, and the Aderklaa Formation on top.

The whole Karpatian group underwent an erosional truncation and was subsequently covered by the fluvial Aderklaa Conglomerate (see their figure 2, Fuchs and Hamilton, 2006). Older stratigraphic tables assigned this lithostratigraphic unit to the top Karpatian period. Presently, it is more likely representing a lowstand equivalent at the onset of the open-marine Badenian cycle that progressively affected all parts of the evolving neo-Vienna basin.

During the Badenian, sediments were derived from the northwest and, to a lesser degree, also from the east and south. Isopachs indicate a readjustment of trough axes through time to a northeast–southwest orientation (Jiricek and Seifert, 1990). Simultaneously, the development of the depocenters occurred.

Lithothamnium reefs grew on shoals or along the coast. Partially, they were replaced by other typical shore deposits. The distal realm of the basin, dominated by thick, fine-grained intervals throughout the transgressional phase, was covered afterward by prograding fans fed by river systems emptying into the basin mainly at the western fringe.

Hamilton and Johnson (1999) and Fuchs et al. (2001) reported on a reevaluation of the highly mature Matzen oil field. The comprehensive revision included traditional (core and log analysis) as well as advanced methods like 3-D seismic and areal petrophysical evaluation (based on neural networks). The study came up with a new, detailed sequence-stratigraphic frame for the Badenian interval. Thus, two cycles could be distinguished: a transgressive wedge, followed by a pronounced highstand systems tract.

The 16th horizon in the Badenian, because of its thickness and distribution (still the basin's main producing interval), is seen as a diachronous remnant of the transgressive lag.

Sourced from a large delta located in the northwest, a subsequent series of subfans were deposited in the area of Matzen, Spannberg, Ebenthal, and Gajary. These fans spread out toward the southwest into central parts of the basin, but lobes pinch out rapidly as seen in corresponding deposits of the adjacent areas.

Substantial post-Pannonian uplift of the Alpine–Carpathian foredeep, from the molasse zone to the Western Carpathians, tilted the prograding sequence and created prolific stratigraphic traps (Figure 4).

These formations contain specific faunal associations, and diverse biostratigraphic concepts were developed; a microfaunal subdivision provided the most detailed insight. Although microfaunas are highly dependent on prevailing regional conditions, they provide excellent results wherever a basinwide homogeneous pelitic facies is dominating. The microfaunal evolution of the entire Paratethys bioprovince is preserved, mirroring the continuously decreasing salinity from marine Badenian up to the freshwater facies of the upper Pannonian.

The sand-shale ratio of the Sarmatian is quite similar to the previous period, whereas the contrast of diverse facies zones bordering each other is not that strongly developed. The basin margins still offered good conditions for limestone developments. They were predominantly built up by algae and bryozoans and even locally by ooliths. Ooliths and lumachelles are, furthermore, indicative for deposition on internal paleohighs.

The character of sediment influx did not change essentially throughout the Pannonian stage. The basin, meanwhile, was completely cut off from the open-marine Paratethys, which led to a rapid decrease in salinity and to a stop of carbonate production, as well as a significant change in faunal associations. For the basal part of the lower Pannonian, Kreutzer (1990) recognized mouthbar sand bodies (partly reaching substantial thicknesses in excess of 100 m [330 ft]) that overlap and/or replace each other in a relief-compensating way.

A forerunner of the Danube (Urdonau) flowed across the molasse zone, which was mainly land tied since the end of early Miocene, and left its marks in the form of the giant Hollabrunn–Mistelbach alluvial fan (Roetzel, 1999). The apron gradually prograded toward the east, depositing the majority of the coarse-grained material west of the Steinberg fault. The fluvial systems then built up huge deltaic complexes and distributed sands across the basin in a south-southeasterly direction.

The correlation of the individual units is possible because of the continuous shale intervals, which separate the genetically related packages.

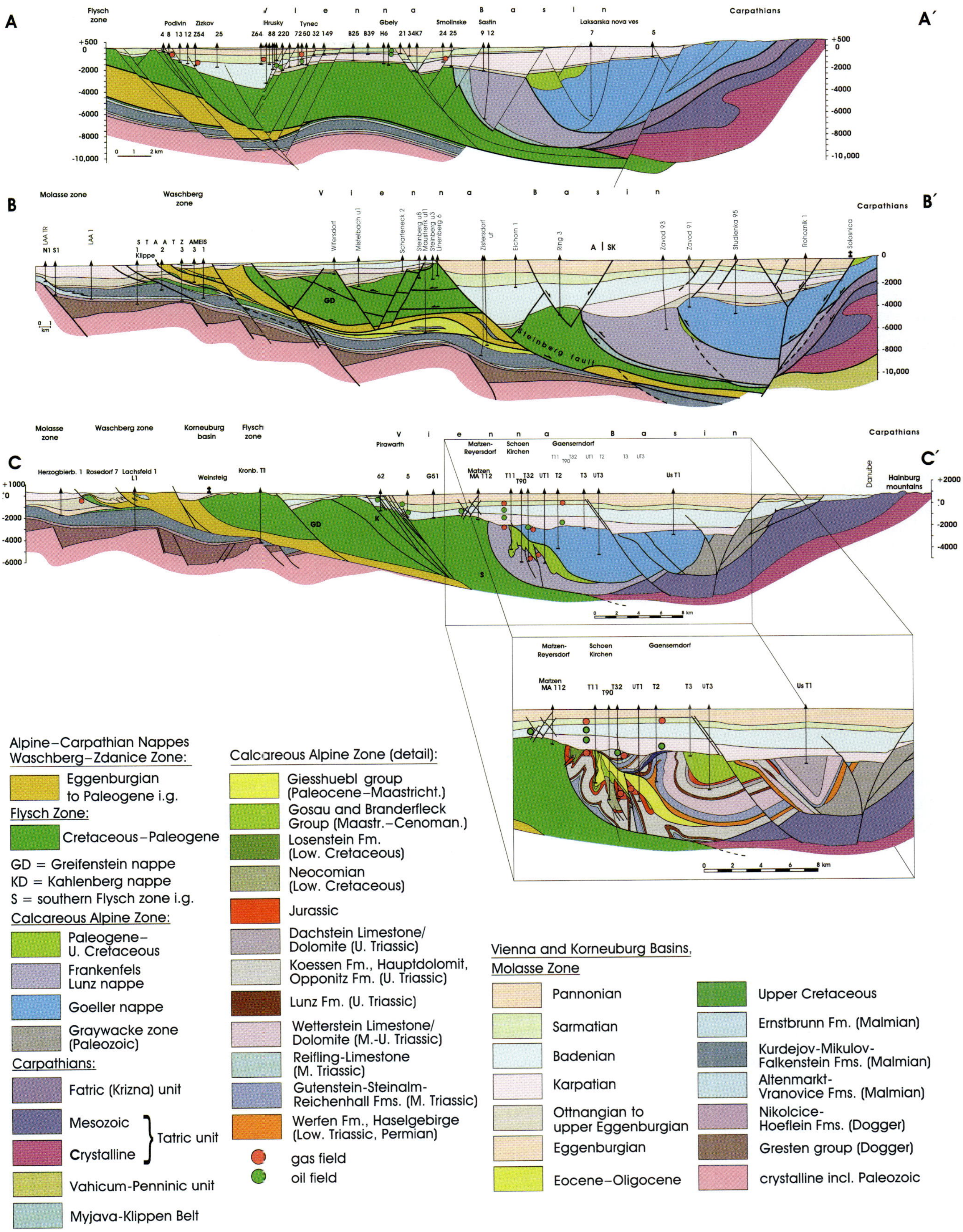

Figure 4. Geological sections through the Vienna basin and its pre-Neogene floor.

Within the middle to upper Pannonian section, a broad variety of terrestrial environments, ranging from lacustrine to fluvial, is preserved. Therefore, only a local correlation is possible.

Thick Pliocene and Quaternary alluvial deposits could be found in depressions like the Mitterndorf Graben, which are associated with the active fault zone at the southeastern basin margin. Presently, the area is undergoing normal faulting, currently superimposed by sinistral strike-slip movement.

As a consequence of the glacial era during the Pleistocene, the Danube (like all Alpine rivers) cut deeply into the underlying sediments and left several generations of terraces. Thick eolic loess deposits were accumulated adjacent to the river-dominated landscapes.

PRE-NEOGENE FLOOR

Allochthonous Alpine–Carpathian Floor

From southwest to northeast, the following allochthonous Alpine–Carpathian thrust elements form the floor below the Neogene basin fill (Figure 5): the Central Alpine–Tatride zone, the Grauwacken zone, the Calcareous Alps, the Flysch zone, and the Waschberg zone (Jiricek, 1979; Wessely, 1992, 1993; Kroll et al., 1993). The individual thrust sheets of the Waschberg zone comprise intervals from early Miocene Molasse deposits down to Paleogene and Jurassic sediments, which were originally deposited on the post-Hercynic, consolidated Bohemian Massif. The thrust belt developed during the final stages of the Alpine orogeny, when the main decollement switched down to the Tertiary Molasse, the autochthonous Mesozoic sedimentary cover, or even the crystalline basement.

The Calcareous Alps and a northward-thinning wedge of their former Paleozoic substrate, the Grauwacken zone, represent the highest tectonic unit overlying the Flysch in the north and the Central Alpine–Tatride zone in the south. They were derived from a realm south of the Central Alpine system and were sheared off the Adriatic microplate during the continental collision by the end of the Eocene.

The Flysch zone is made up of the Raca, Greifenstein, Kahlenberg, and the Laab nappe, stratigraphically ranging from upper Lower Cretaceous up to the Eocene. It belongs to the northern, youngest part of the Penninic system, an oceanic province that separated the European continent and the Afro-Adriatic terrane since the Jurassic. The sedimentological composition, as well as the structural imprint of the Flysch zone, reflects the most important stages of the Alpine tectonism. Throughout the convergent phase, the trough sediments were uprooted from their basement and formed an accretionary wedge that was finally thrusted onto the European foreland. Because of the strong internal deformation, the tectonic position of individual units, such as the Kahlenberg–St. Veit system, is currently unknown and under investigation. The oil-bearing thrust sheets (Paleocene–Eocene) of the Steinberg high, which belong to the Greifenstein nappe, are of economic significance.

Among all of the basin-floor units, the Calcareous Alps still offers the greatest potential for prolific reservoirs. They range stratigraphically from Permoskythian to Paleocene; the whole succession is dominated by Middle to Upper Triassic limestones and dolomites, which could reach a primary thickness of several thousand meters. Within the whole suite, the Norian Hauptdolomit proved to be the most important reservoir rock for gas and oil.

The tectonic style and the facies distribution do not differ significantly from the outcrops at the western basin margin.

The intensively deformed Frankenfels–Lunz nappe (Figure 5) represents the frontal and tectonically deepest element along the northwest border and belongs to the Bajuvarikum master unit. Folds and internal secondary thrusts are unconformably overlain by the Giesshubl Gosau, covering a time span from Upper Cretaceous to Paleocene. Whereas the Upper Cretaceous varies from river-dominated facies with conglomerates (>1000 m; >3300 ft) to marine facies with sands and marly limestones, the Paleocene was deposited in a deep-water, flyschoid environment. The latter is of particular relevance for the sealing of internal reservoirs.

The Giesshubl Gosau is overthrusted by the Goller nappe, a member of the Tirolikum assemblage. In addition, this nappe carries as a postdeformational layer group, the Glinzendorf Gosau. In contrast to the Giesshubl series, the stratigraphic range is limited to the Upper Cretaceous and shows a variety of different facial settings from predominantly limnic to shallow-marine and even minor, open-marine deposits.

The tectonically highest structures are a deep-reaching syncline and an isolated body of Lower to Middle Triassic. The syncline consists mainly of Wetterstein (Ladinian) and Dachstein limestone (Norian). The Triassic rocks are piled up against the Glinzendorf Gosau in the area of Tallesbrunn. Accordingly, these structures are equated to the Tirolikum system outside the basin.

In the south (southwest), on the rear side, the Calcareous Alps show a sedimentary contact to their former Paleozoic basement. The wedge-shaped sliver mainly consists of low metamorphic, terrigenous layers and is therefore of minor interest in an exploratory

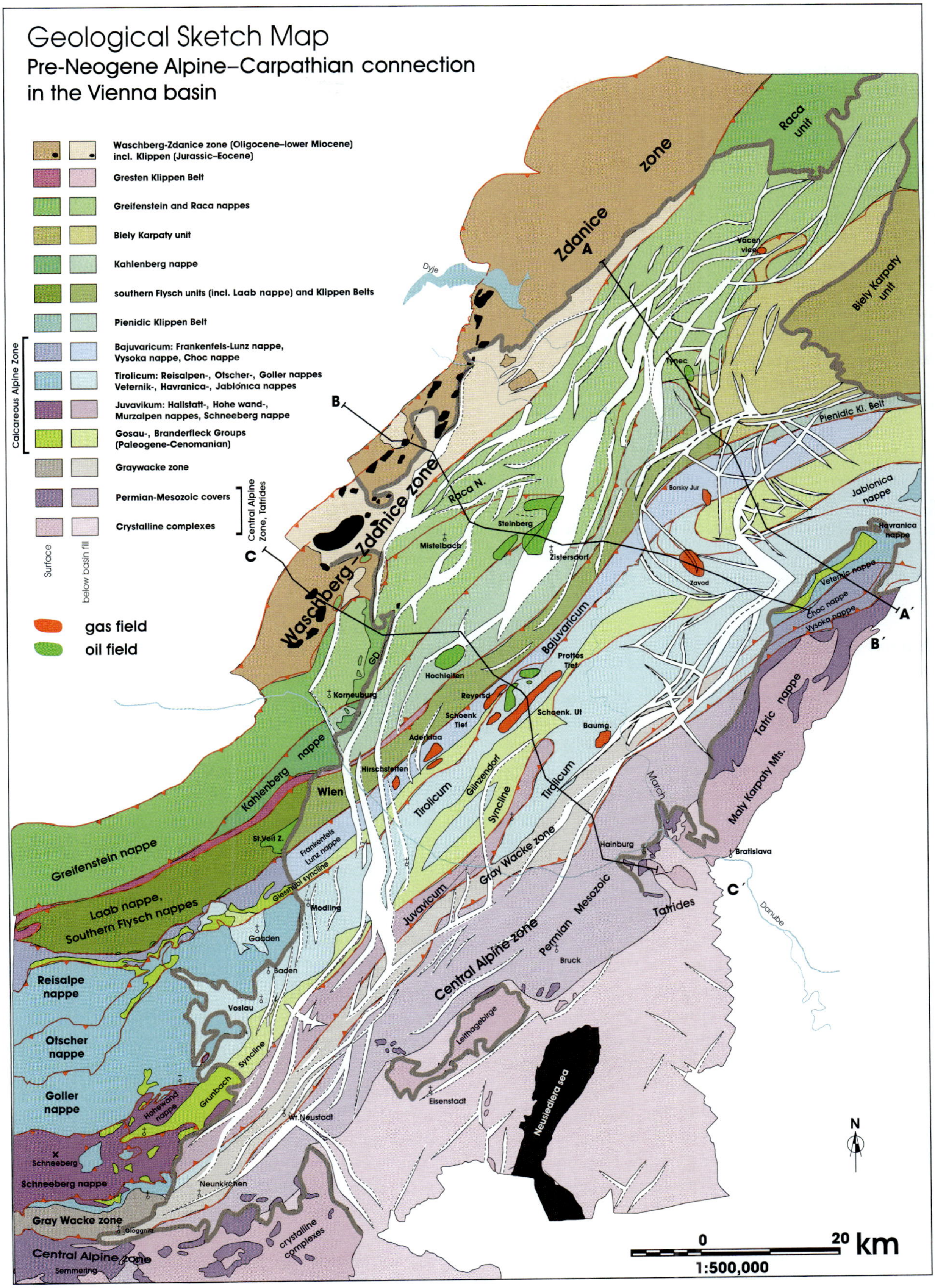

Figure 5. Geological sketch map showing the Alpine–Carpathian connection under the Neogene fill of the Vienna basin.

sense. The Grauwacken zone ends at the Slovak part of the basin without any comparable continuation further to the northeast. The relationship between the Calcareous Alps and their equivalents in the Carpathian mountain belt has been discussed several times by previous authors (e.g., Wessely, 1992).

The Central Alpine–Tatride system in the Vienna basin (represented by the Lower Austroalpine complex) was penetrated by only a few wells, which encountered Permian–Triassic strata of the Semmering Mesozoic type. Because of the complex internal structure and a low well density, the individual tectonic features could not be confidently traced across the basin.

The Alpine–Carpathian thrust belt evolved as a consequence of postcollisional north–south shortening between the Afro-Adriatic upper plate and the European lower plate. The climax of the collision is dated by the end of the deposition, and subsequent overthrust of the Helvetic unit, representing the shelf area of the European continent, is dated by the end of the Eocene. Ongoing convergence caused the Penninic–Austroalpine bulge to advance further onto the European foreland. As mentioned, these processes terminated at the end of early Miocene, about 17 m.y. (Figure 1).

Autochthonous Floor

The autochthonous floor widely extends toward east-southeast below the foreland molasse and the Alpine–Carpathian thrust bulge. It consists of the crystalline basement, an unconformably overlying Mesozoic cover, and the Tertiary autochthonous Molasse (Figure 4).

The Mesozoic succession starts with deltaic sequences of the early Dogger, deposited in half grabens, formed during the Jurassic rifting stage.

The observed strata growth from east to west is a result of active synsedimentary faulting.

Sandy dolomites with lenses of chert were deposited in the uppermost Dogger. Fault activity must have been terminated or at least significantly reduced at that time, as the interval is regionally present with a low variation in thickness. This formation represents the reservoir rock of the gas-condensate field Hoflein below the Flysch zone near Vienna.

A north–south-trending carbonate platform (Altenmarkt Formation) developed in the Malmian. Toward the east, the platform environment interfingers with the basinal marls (the Mikulov Formation), which is, by far, the most important source rock of the Vienna basin hydrocarbon province. The transition between both realms is represented by the slope development of the Falkenstein formation, marly limestones with carbonatic debris. The whole section was covered by a prograding platform complex, the Ernstbrunn Formation, and reefoidal and lagoonal carbonates, which developed during the Upper Jurassic (Tithonian) regression.

Within the basin, only four wells reached the autochthonous Mesozoic: OMV Zistersdorf UT1a (total depth 7544 m [24,750 ft]), OMV Zistersdorf UT2a (total depth 8553 m [28,061 ft]), OMV Maustrenk UT1 (total depth 6563 m [21,532 ft]), and OMV Aderklaa UT1a (total depth 6630 m [21,752 ft]): Wessely (1990), Milan and Sauer (1996).

A gas kick with a considerable flow of overpressured gas in OMV Zistersdorf UeT1a occurred at 7544 m (24,750 ft) and was related to the Ernstbrunn Formation and its cover of basal breccia of Paleogene Molasse. The gas could not be produced because the borehole collapsed, and the replacement well UeT2a did not repenetrate the gas reservoir.

Nearby Zistersdorf, the well OMV Maustrenk UeT1 was drilled into the Jurassic Mikulov marls. Moderate amounts of oil and gas could be produced from a fractured zone close to an internal detachment plane, just above the undisturbed autochthonous interval. It turned out that the Malmian source rock could serve as an example of an unconventional, tight-gas play type.

Next to the border of Vienna, another ultradeep well (OMV Aderklaa UT1) was drilled through the Mesozoic section into the crystalline basement but did not encounter any reservoirs. The reservoir rocks were removed either by a tectonically or an erosional event.

Oil and Gas Occurrence

The oil and gas provinces of the Vienna basin (Friedl, 1957; Kroll, 1980; Betnarikova and Thon, 1984; Ladwein et al., 1991; Kreutzer, 1993b; Ciprys et al., 1996) coincide with distinct structural features. A concentration occurs along the depocenter of deeply buried autochthonous Malmian marls below the northern Vienna basin. Additional source rocks have been referenced by Francu et al. (1996).

The structural trapping features are the large fault systems of the northern Vienna basin in Austria; in particular, the Steinberg fault system, the median highs of Matzen–Aderklaa–Enzersdorf, including the pre-Neogene Calcareous Alpine floor, and the southern and southeastern fault systems (Figure 6).

The northern and central provinces contain oil and gas of thermocatalytic origin, whereas predominantly biogenic gas was found in the south and southeastern regions.

Lower and middle Miocene transgressive and regressive sandstone cycles resulted in a stack of multiple productive zones (Kreutzer, 1993b).

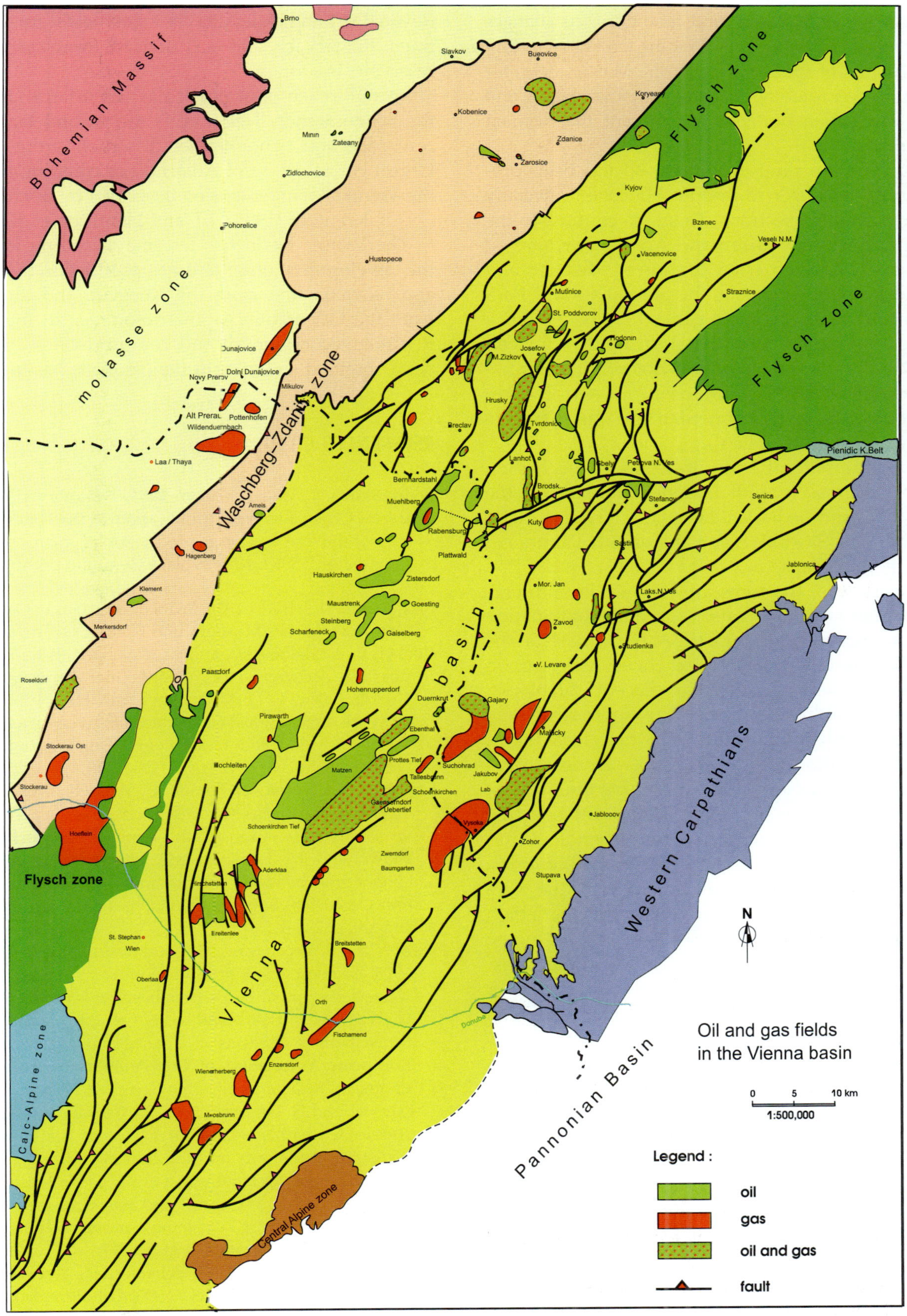

Figure 6. Oil and gas fields in the Vienna basin.

The number of Miocene sandstone horizons changes from province to province, as does the thickness and internal structure of the horizons. In the Matzen field, for example, 9 lower Miocene, 16 Badenian, 9 Sarmatian, and 4 lower Pannonian horizons are present that contain hydrocarbons.

In the northern fault provinces, oil has been produced from the lower to middle Miocene, especially from the Sarmatian sections. This is in contrast to the central provinces of Matzen and Aderklaa, where oil production is derived from the lower Miocene and Badenian horizons, whereas the Sarmatian and Pannonian horizons only contain gas.

In the southern and southeastern fault provinces, gas only occurs in the Badenian (Zwerndorf), Sarmatian (Fischamend, Orth, and Moosbrunn), and also, to a minor degree, the lower Pannonian horizons (Obersiebenbrunn, Markgrafneusiedl).

The most important zones in terms of production are the transgressive 16th Badenian horizon in the Matzen field, middle Badenian (called the Lab horizon in Czechia and Slovakia), and the eighth Pannonian horizon in Slovakia.

The majority of the traps are tectonically induced. Tilting of huge fault blocks also caused the combination of stratigraphic and structural traps, particularly in the lower Miocene. Along the faulted zones, hydrocarbon accumulations were discovered in adjacent fault blocks. The Steinberg fault creates a classic rollover structure.

Extended anticlines form traps in the central highs, such as in the Matzen, Aderklaa, and Zwerndorf fields. Faults in these areas are partly controlling the shape of the structures but do not act as significant trapping components. The use of 3-D seismic technology has increased the number of stratigraphic trap discoveries.

The Flysch zone below and juxtaposed to the Neogene is only productive in the area of the Steinberg high and related structures. It contains reservoirs in the Paleocene as well as the Eocene turbiditic sandstones.

The oil and gas fields of the Calcareous–Alpine floor are mainly located along the central highs. The main reservoirs are Upper Triassic dolomites (Hauptdolomit) and, in only one case, dolomitic limestones (Dachstein Limestone). The dolomites could reach thicknesses of as much as 500 m (1600 ft). Two types of traps can be distinguished: relief or buried hill traps and thrust internal traps that developed in the imbricated system. For the first type, Neogene marls act as a cap rock, whereas for the second, tight sediments of the Calcareous Alpine complex (Cretaceous to Paleocene shales and sandstones) seal the reservoirs.

The Schonkirchen Tief and Prottes Tief oil and the Aderklaa, Baumgarten, Zavod, and Borsky Jur gas fields are relief pools, whereas the Schonkirchen Ubertief, Reyersdorf, and Aderklaa Tief gas fields are internal traps. Remarkably, the gas derived from the internal reservoirs is sour gas.

Schonkirchen is of particular importance, because the Schonkirchen Tief pool is the second largest oil reservoir, and Schonkirchen Ubertief is the second largest gas reservoir of Austria. The trapping types in the Vienna basin have been described by Buchta et al. (1990), Kreutzer (1993a, b), and Hamilton et al. (2000).

The Vienna basin has become a leading factor in the development of oil and gas production in Austria and Slovakia first from the Neogene and then later from the Calcareous Alpine dolomites. In the Czech Republic, economic oil and gas production has only been obtained so far from the Neogene basin fill.

REFERENCES CITED

Bednarikova, J., and A. Thon, 1984, Naftovy prumysl na uzemi Ceskoslovenska (publication at the 70th anniversary of the first economic production of hydrocarbons in the area of CSSR): Knihovnicka Zemniho plynu a nafty, v. 5, 368 p.

Brix, F., and O. Schultz, 1993, Erdöl und Erdgas in Österreich: Naturhistorisches Museum und F. Berger Horn, 688 p.

Buchta, S., E. Ralbovsky, and M. Kolarova, 1990, Zones of possible occurrence of nonanticlinal deposits in the Vienna basin, *in* D. Minarikova and H. Lobitzer, eds., 30 years of geological cooperation between Austria and Czechoslovakia: Federal Geological Survey, Vienna, and Geological Survey, Prague, p. 203-206.

Cekan, V., A. Kocak, C. Tomek, G. Wessely, and D. Zych, 1990, Czechoslovak–Austrian cooperation in geophysical structural exploration in the Vienna basin, *in* D. Minarikova and H. Lobitzer, eds., 30 years of geological cooperation between Austria and Czechoslovakia: Federal Geological Survey, Vienna, and Geological Survey, Prague, p. 23–31.

Ciprys, V., P. Prochazka, M. Morkovsky, and V. Kolejka, 1996, Tectonics and hydrocarbon accumulations of the western margin of the Vienna basin in south Moravia, *in* G. Wessely and W. Liebl, eds., Oil and gas in Alpidic thrust belts and basins of Central and Eastern Europe: European Association of Petroleum Geoscientists and Engineers Special Publication 5, p. 391.

Decker, K., and H. Peresson, 1996, Tertiary kinematics in the Alpine–Carpathian–Pannonian system: Links between thrusting, transform faulting and crustal extension, *in* G. Wessely and W. Liebl, eds., Oil and gas in Alpidic thrust belts and basins of Central and Eastern Europe: European Association of Petroleum Geoscientists and Engineers Special Publication 5, p. 69–77.

Francu, J., M. Radke, R. G. Schaefer, H. S. Poelchau, J. Caslavsky, and Z. Bohacek, 1996, Oil-oil and oil-source rock correlations in the northern Vienna basin and adjacent Carpathian Flysch zone (Czech and Slovak area), *in* G. Wessely and W. Liebl, eds., Oil and gas in Alpidic thrust belts and basins of Central and Eastern Europe: European Association of Petroleum Geoscientists and Engineers Special Publication 5, p. 343–353.

Friedl, K., 1957, Das Wiener Becken, *in* F. Bachmayer, ed., Erdöl in Österreich: Wien, Natur und Technik, 92 p.

Fuchs, R., and W. Hamilton, 2006, New depositional architecture for an old giant: The Matzen field, Austria, *in* J. Golonka and F. J. Picha, eds., The Carpathians and their foreland: Geology and hydrocarbon resources: AAPG Memoir 84, p. 205–219.

Fuchs, R., R. Ramberger, and C. H. Veit, 2001, Das Matzen Projekt-Renaissance des größten Öl- und Gasfeldes in Österreich (Wiener Becken): Erdol, Erdgas, Kohle, v. 117, no. 11, p. 528–540.

Hamilton, W., and N. E. G. Johnson, 1999, The Matzen Projekt— Rejuvenation of a mature field: Petroleum Geoscience, v. 5, no. 2, p. 119–125.

Hamilton, W., L. Wagner, and G. Wessely, 2000, Oil and gas in Austria: Aspects of geology in Austria— Mitteilungen der Osterreichischen Geologischen Gesellschaft, v. 92, p. 235–262.

Hlavaty, V., 1996, The Slovakian part of the Vienna basin— Exploration results, *in* G. Wessely and W. Liebl, eds., Oil and gas in Alpidic thrust belts and basins of Central and Eastern Europe: European Association of Petroleum Geoscientists and Engineers Special Publication 5, p. 41–42.

Jiricek, R., 1979, Tectogenetic development of the Carpathian arc in the Oligocene and Neogene, *in* M. Mahel, ed., Tectonic profiles through the West Carpathians: Bratislava, Geologicky ustav D. Stura, p. 203-212.

Jiricek R., 1988, Stratigrafie, paleogeografie a mocnost sedimentu h Neogenu Videnske panve: Zemny plyn a nafta, v. 33, no. 4, p. 583–622.

Jiricek, R., and P. Seifert, 1990, Paleogeography of the Neogene in the Vienna basin and the adjacent part of the foredeep, *in* D. Minarikova and H. Lobitzer, eds., 30 years of geological cooperation between Austria and Czechoslovakia: Federal Geological Survey, Vienna, and Geological Survey, Prague, p. 89–105.

Kovac, M., 2000, Geodynamický, paleografický a štruktúrný vývoj karpatsko-panonského regionu v miocéne. Nový pohlad na neogénne panvy, Slovenska: Bratislava, Veda 5, p. 202.

Kreutzer, N., 1990, The lower Pannonian sands and the Pannonian–Sarmatian boundary in the Matzen area of the Vienna basin, *in* D. Minarikova and H. Lobitzer, eds., 30 years of geological cooperation between Austria and Czechoslovakia: Federal Geological Survey, Vienna, and Geological Survey, Prague, p. 105–111.

Kreutzer, N., 1993a, Das Neogen des Wiener Beckens, *in* F. Brix and O. Schultz, eds., Erdöl und Erdgas in Österreich: Naturhistorisches Museum Wien Und F. Berger Horn, p. 232–248.

Kreutzer, N., 1993b, Lagerstätten im Neogen des Wiener Beckens und dessen Untergrund, *in* F. Brix and O. Schultz, eds., Erdöl und Erdgas in Österreich: Naturhistorisches Museum Wien und F. Berger Horn, p. 403–434.

Kreutzer, N., and V. Hlavaty, 1990, Sediments of the Miocene (mainly Badenian) in the Matzen area in Austria and in the southern part of the Vienna basin in Czechoslovakia, *in* D. Minarikova and H. Lobitzer, eds., 30 years of geological cooperation between Austria and Czechoslovakia, Federal Geological Survey, Vienna, and Geological Survey, Prague, p. 112–124.

Kroll, A., 1980, Die österreichischen Erdöl- und Erdgasprovinzen: Das Wiener Becken, *in* F. Bachmayer, ed., Erdöl und Erdgas in Österreich: Naturhistorisches Museum Wien und F. Berger Horn, p. 147–179.

Kroll, A., H. Heinz, R. Jiricek, B. Meurers, W. Seiberl, P. Steinhauser, G. Wessely, and D. Zych, 1993, Karten 1:200.000 über den Untergrund des Wiener Beckens und der angrenzenden Gebiete (mit Erläuterungen 1–22): Wien, Geologische Bundesanstalt.

Ladwein, H. W., F. Schmidt, P. Seifert, and G. Wessely, 1991, Geodynamics and generation of hydrocarbons in the region of the Vienna basin, Austria, *in* A. M. Spencer, ed., Generation, accumulation and production of Europe's hydrocarbons: European Association of Petroleum Geoscientists and Engineers Special Publication 1, p. 289–305.

Milan, G., and R. Sauer, 1996, Ultra-deep drilling in the Vienna basin— A review of geological results, *in* G. Wessely and W. Liebl, eds., Oil and gas in Alpidic thrust belts and basins of Central and Eastern Europe: European Association of Petroleum Geoscientists and Engineers Special Publication 5, p. 109-117.

Roetzel, R., ed., 1999, Arbeitstagung Geologische Bundesanstalt 1999— Retz-Hollabrunn-Geol.: BA, Wien, 366 p.

Royden, L. H., 1985, The Vienna basin. A thin-skinned pull-apart basin, *in* K. T. Biddle and N. Christie-Blick, eds., Strike slip deformation, basin formation and sediments: SEPM Special Publication No. 37, p. 319-338.

Sauer, R., P. Seifert, and G. Wessely, 1992, Guidebook to excursions in the Vienna basin and the adjacent Alpine–Carpathian thrust belt in Austria: Mitteilungen der Osterreichischen Geologischen Gesellschaft, v. 85, p. 5–39.

Seifert, P., 1996, Sedimentary-tectonic development and Austrian hydrocarbon potential of the Vienna basin, *in* G. Wessely and W. Liebl, eds., Oil and gas in Alpidic thrust belts and basins of Central and Eastern Europe: European Association of Petroleum Geoscientists and Engineers Special Publication 5, p. 331–342.

Steininger, F. F., G. Wessely, F. Rogl, and L. Wagner, 1986, Tertiary sedimentary history and tectonic evolution of the eastern Alpine foredeep: Giornale Geologico, ser. 3, Bologna, v. 48, no. 1–2, p. 285–297.

Tollmann, A., 1985, Geologie von Österreich, Bd. 2: Wien, F. Deuticke, p. 482–532.

Weissenback, M., 1996, Miocene sedimentation in the Vienna basin, *in* G. Wessely and W. Liebl, eds., Oil and gas in Alpidic thrust belts and basins of Central and Eastern Europe: European Association of Petroleum Geoscientists and Engineers Special Publication 5, p. 355–363.

Wessely, G., 1988, Structure and development of the Vienna basin in Austria, *in* L. H. Royden and F. Horvath, eds.,

The Pannonian system. A study in basin evolution: AAPG Memoir 45, p. 333–346.

Wessely, G., 1990, Geological results of deep exploration in the Vienna basin: Geologische Rundschau, v. 79, no. 2, p. 513–520.

Wessely, G., 1992, The Calcareous Alps below the Vienna basin in Austria and their structural and facial development in the Alpine–Carpathian borderzone: Geologia Carpathica, v. 43, no. 6, p. 347–353.

Wessely, G., 1993, Der Untergrund des Wiener Beckens, *in* F. Brix and O. Schultz, eds., Erdöl und Erdgas in Österreich: Naturhistorisches Museum und F. Berger Horn, p. 249–280.

Wessely, G., 2000, Sedimente des Wiener Beckens und seiner alpinen und subalpinen Unterlagerung-Exkursionsfühurer Sediment 2000; Mitteilungen der Gesellschaft der Geologie- und Bergbaustudenten in Wien, v. 44, p. 191–214.

6

Fuchs, Reinhard, and Walter Hamilton, 2006, New depositional architecture for an old giant: The Matzen field, Austria, *in* J. Golonka and F. J. Picha, eds., The Carpathians and their foreland: Geology and hydrocarbon resources: AAPG Memoir 84, p. 205–219.

New Depositional Architecture for an Old Giant: The Matzen Field, Austria

Reinhard Fuchs
OMV-Austria Exploration and Production GmbH, Vienna, Austria
Walter Hamilton
OMV, Exploration and Production GmbH, Vienna, Austria

ABSTRACT

Geophysical, sequence-stratigraphic, and petrophysical evaluation of the main oil and gas reservoirs of the Matzen field in the central Vienna basin resulted in a revised field architecture, and provided new insights into middle Miocene (Badenian, Sarmatian) depositional settings. The latest analysis suggests that the reevaluation of this mature field not only allows a better explanation of production history, but has also led to further successful redevelopment activities.

The structural setting of the Vienna basin is the result of subsidence, extension, and a final phase of subsidence. Compressional, transgressional, and extensional events had a strong influence on the basin configuration and sediment distribution. The Matzen field illustrates these pull-apart and piggyback mechanisms clearly. It is an elongated anticline (Matzen anticline), bounded by a pull-apart graben in the north (Matzen fault system), a fault complex in the west (Bockfliess fault system), and another fault zone in the south (Markgraneusiedl fault zone).

Three depositional cycles (Matzen cycles) have been subdivided into some 30 sequences.

The first Matzen cycle culminates with the transgressive Matzen sand, the main producing horizon in the field.

Produced during the Matzen main cycle (second Matzen cycle), marine prograding delta systems of Badenian age contain the main oil reservoirs. The 9th Tortonian horizon is a textbook example of a shelf-slope basin sucession. Within six sublayers, the shelf edge advances some 2 km (1.2 mi) from north to south. Transgressive meandering channel systems of Sarmatian age hold the main gas reservoirs that are partly used as gas storage. The 5th Sarmatian horizon was deposited in a shallow-marine deltaic setting. During a sea level drop, the depocenter shifted southward with simultaneous incision of a widespread channel system into the underlying 6th Sarmatian horizon. The base of the system represents an important sequence boundary. Final isolation of the Vienna basin created brackish conditions.

The third Matzen cycle developed as a thick series of lacustrine prograding delta sequences of the Pannonian (upper Miocene).

DOI:10.1306/985609M843069

INTRODUCTION

The giant Matzen oil field is located in the Vienna basin about 20 km (12 mi) to the northeast of the Austrian capital of Vienna (Figure 1). The multipool oil and gas field was discovered in 1949 (Table 1). The field comprises 400 producing intervals generating more than 70 horizons. To date, it has produced 516 million bbl oil and 1.1 tcf gas. The oil and gas are located in shallow-marine to fluvial clastic sediments of middle Miocene age (Badenian, Sarmatian, and Pannonian). Gas is predominantly found in the shallower Pannonian and upper Sarmatian horizons. The lower Sarmatian and the Badenian are oil prone. The Miocene depositional environments can be subdivided into three main stratigraphic sequences. These sequences govern the complex relationships of reservoir and cap rock distribution. Different pools are described by characteristic depositional environments. These environments define the porosity and permeability distribution.

New studies reveal a complex relationship between depositional environment, fault pattern, and observed production heterogeneities. These studies also show new production opportunities and promise a longer field tail production than originally anticipated.

This chapter combines the structural and depositional setting of the Matzen field with modern geological and geophysical tools. It provides an example of how modern research techniques, coupled with

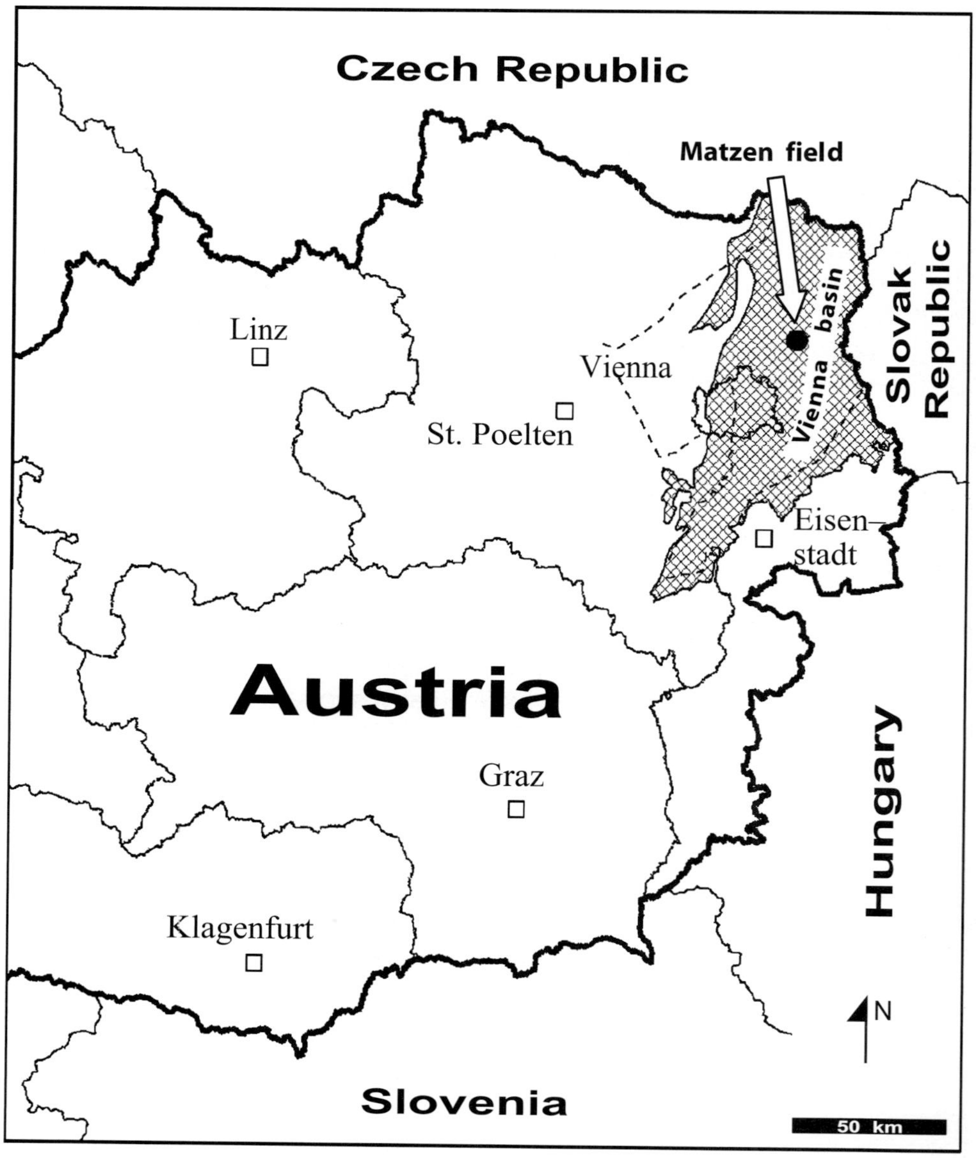

Figure 1. Matzen field location map.

Table 1. Matzen field: Summary sequence of events.

1948	62 shallow stratigraphic wells (200–300 m [660–1000 ft] deep)
1949	discovery well Matzen 3 (total depth: 1742.4 m [5716.5 ft])
	drillstem test in Matzen sand: 142 tons/day (1000 bbl/day)
1950	first production
1949–1955	425 exploration, appraisal, and production wells drilled
1955	peak production: 2,873,143 tons/year (20.2 million bbl/year)
1955–1992	approximately 1000 wells drilled, including injection and gas-storage wells
1994–1996	3-D seismic data acquisition throughout the entire field
1995	Matzen Project kickoff
1996	first new oil from Matzen Project
2001	approximately 90% of the established initial reserves have been produced: 69.1 million tons (487.4 million bbl)

new interpretations of old and long-available data, lead to revised field architecture, thereby creating new exploitation opportunities and prolonging the economic life of the field (i.e., Ebenthal satellite).

The detailed geologic and stratigraphic evaluation of the field's producing horizons and the integration of their production behavior provide new insights to the overall depositional setting of the Badenian and Sarmatian. This giant field in the heart of Europe got a new economic boost (sidetracks, recompletions, and new wells) on the basis of the geological and geophysical reevaluation, as set out in this chapter.

HISTORY AND PRODUCTION DATA

The Matzen field is a giant field by all standards. Initial recoverable reserves are estimated at 550 million bbl oil and 1.5 tcf gas. The field shows a typical production profile of accelerated early production followed by a long decline and an extended tail production. To date, it is has produced 516 million bbl oil and 1.1 tcf gas from more than 400 producing units (a production unit is defined by a characteristic oil-water contact, and pressure and production behavior) in approximately 70 stratigraphic horizons. Some 1500 wells have been drilled, the majority of them during the initial phase of production between 1949 and 1954. Current production amounts to 12,000 bbl oil/day and 66 mmcf gas/day. This production is achieved by currently 500 oil- and 70 gas-producing wells. The average water cut is 93%. The production of 160,000 bbl water/day constitutes one of the technical challenges of today's production. Matzen can be considered a mature field.

REGIONAL GEOLOGICAL SETTING

Stratigraphy

The Neogene sediments of the Vienna basin are part of the central Paratethys. Separated by the emerging Alps and the Dinarides, the basin shows an independent depositional development. The Vienna basin stratigraphy exhibits a complex, shallow-marine to fluviatile and limnic depositional setting, and it is subdivided into a pre-, proto-, and neo-Vienna basin development (Wessely, 2000).

The first stage of the development in the area of the Vienna basin comprises all sediments older than Eggenburgian and is called the pre-Vienna basin.

The oldest sediments of today's Vienna basin (proto-Vienna basin) date from Eggenburgian to Karpatian, with subsidence along synsedimentary faults and the beginning of the pull-apart effect (Hamilton et al., 2000). These marine sediments are part of the parautochthonous molasse and are located in the northern part of the basin. During the late Karpatian, lacustrine sediments were deposited south of the Matzen area. The boundary between marine and lacustrine and fluviatile sediments is suggested to be the Matzen–Spannberg ridge.

The next stage, the neo-Vienna basin, started during the Badenian with a widespread transgression over tilted and partly eroded sediments of the proto-Vienna basin. It lasted up to the Pliocene. The Badenian transgression marked the onset of a regional subsidence. As much as 5500 m (18,000 ft) of Miocene clastic sediments was deposited in two depocenters, one in the northern part and one in the southern part of the basin. Between the depocenters and on their flanks, horsts and highs were developed. The most pronounced is the Matzen high containing the Matzen field.

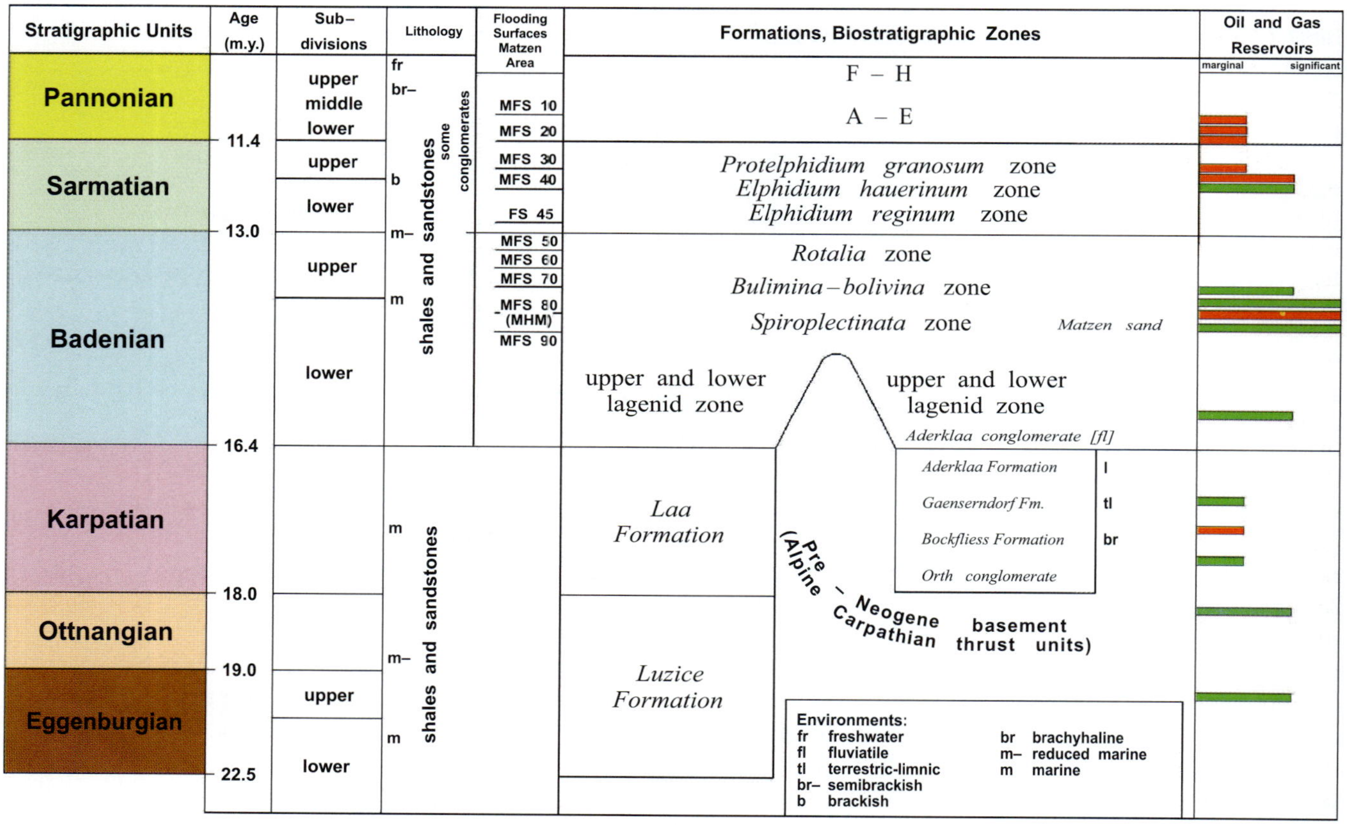

Figure 2. Neogene stratigraphy of the Vienna basin (Austria). MFS = Maximum Flooding Surface; FS = Flooding Surface; MHM = Matzener Hauptmarker.

The transgression starts with fluvial lowstand conglomerates (Aderklaa conglomerate) and is followed by a sequence of marine shales, marls, and sandstones (Figure 2). The first marine transgression occurred in lows of the central Vienna basin during the Lower Lagenid zone of the Badenian (Papp et al., 1973). The highly marine ingression came through a seaway connecting the Vienna basin with the Mediterranean area in the south. The depositional environments during Badenian vary from highly marine shaly basin-floor fans to shallow-marine, delta-plain, sand-rich deposits. Delta-front sediments create lobes, which are extremely important for hydrocarbon trapping. At the end of the Badenian, the depositional environment not only got shallower but also changed from high to less salinity. Marine conditions ended during the late Badenian.

The Sarmatian is predominantly characterized by meandering-channel, splay, and other delta-plain deposits. This formation correlates with the final isolation of the Paratethys from the Mediterranean Sea. During the late Sarmatian, the Vienna basin became a remote bay of the Paratethys. The isolation from the open sea can be seen in the fossil record, specifically in the evolution of endemic faunas. The ancient Danube River transported sediments from the Bohemian Massif to the east trough of the Zaya furrow into the area of Mistelbach, from where it extended into the Matzen area.

The Pannonian, the youngest regional depositional phase in the Vienna basin, shows restricted salinity (brackish) in the early Pannonian with coarse clastic and pelitic depositions and mostly fluvial and lacustrine freshwater sediments in the late Pannonian. Lacustrine sedimentation ceased during the late Pannonian. Pontian age sedimentation has not been proved by the fossil record so far (Rogl and Daxner-Hock, 1996).

During the Pliocene and Pleistocene, local gravel and eolian loess were deposited.

Structural Evolution

Today's structural setting of the Vienna basin is the result of a series of compressional, transgressional, and extensional events. The proto-Vienna basin (Eggenburgian, Karpatian) was dominated by pull-apart forces and defined the early development of the basin.

Syndepositional faults, partly with strike-slip components, segment this early basin. Late Alpine thrusting transported parts of this basin to the northeast (piggyback structures).

A subsequent compressional phase pushed the sediments up, tilted, and largely eroded the Eggenburgian and Karpatian cover. The early Badenian transgression marked the beginning of the neo-Vienna basin and terminated this compressional phase. Additional synsedimentary faults developed. Their en echelon arrangement and the overall rhombic fault pattern strongly support the idea of a further widening of the basin. Regionally, the Alps got stuck south of the Bohemian Massif, whereas the Carpathians moved on northward, thereby creating a void and extensional movement at the hinge between these two orogenic systems. The most prominent tectonic feature of the Vienna basin is the Steinberg fault system with a throw as much as 6000 m (20,000 ft).

REDEVELOPMENT ACTIVITIES

A complete reevaluation of the geological architecture of the Badenian of the Matzen field was initiated in 1996. Three-dimensional (3-D) seismic data and modern interpretation of up to 50-yr-old wire-line logs (neural network), in combination with new core descriptions, were used as the basis for determining the depositional and structural framework of the field. Depositional units and facies distribution were defined by log shapes and seismic data, and a modern sequence-stratigraphic framework was established. Petrophysical data were incorporated into the reservoir units, creating new opportunities of exploiting bypassed oil and gas (Veit et al., 1999; Fuchs et al., 2001). The natural production decline could be stopped through a variety of activities (e.g., work-over procedures, sidetracks for attic oil, and infill wells for acceleration of production).

Several Sarmatian horizons were also evaluated. The newly developed depositional architecture of the reservoirs provides the basis for operating optimization of some gas-storage horizons.

FIELD GEOLOGY

Stratigraphy

Neogene sediments (Badenian, Sarmatian, and Pannonian) have been subdivided into three major depositional cycles (Fouad and Knox, 2001). These Matzen cycles are composed of more than 30 sequences divided by flooding surfaces.

The first Matzen cycle culminated with the transgression of the 16th Tortonian horizon (16th Tortonian horizon; Matzen sand), the main producing horizon in the Matzen field. The maximum flooding surface 80 (MFS 80 = MHM [Matzener Hauptmarker]) marked the peak of the transgression and supports the theory of a further opening of the Vienna basin (Figures 2, 3).

The second cycle (Matzen Main Cycle [Matzener Hauptzyklus]) started with MFS 80 (Figure 4). A regressive prograding delta system developed, providing sediment input from the northwest. It culminated at the Badenian–Sarmatian boundary, including the deposition of high-energy conglomerates. This prograding system houses the oil and gas reservoirs of the 15th, 14th, 13th, 12th, 11th, 9th, 8th, 7th, 6th, and 5th Tortonian horizons. At the Badenian–Sarmatian boundary, the depositional trend reversed again into a retrograde transgressive phase, developing widespread meandering-channel systems, providing traps for gas reservoirs (i.e., 3rd/4th, and 5th Sarmatian). The Sarmatian–Pannonian boundary is characterized by a pronounced highstand (MFS 20), which deposited mudstones.

The third Matzen cycle (and closing cycle) started with MFS 20 at the beginning of the Pannonian and developed into thick series of lacustrine prograding delta sequences. These sequences produced shallow gas reservoir traps (3rd, 4th, and 5th Unter Pannon [UP]) (early Pannonian).

Tectonic Evolution

The Matzen field in the center of the Vienna basin illustrates the pull-apart and piggyback mechanisms very clearly. It shows evidence of a complex series of extensional deformational events during Badenian to Pannonian (Decker and Peresson, 1998).

Three structure features are dominant in the Matzen field area: an elongated anticline in the central part of the field (Matzen anticline); a pull-apart graben in the north (Matzen fault system); the fault complex in the Bockfliess area in the west (Bockfliess fault system); and the Markgrafneusiedl fault zone in the south (Figure 5).

The Matzen elevation is an asymmetric elongated anticline of about 10 km (6.2 mi) in length and 5 km (3.1 mi) in breadth. The northeast axis of the anticline migrates from the deeper to shallower horizons toward the southeast. The northwestern limb of the anticline is dissected by intersecting faults dipping to the northwest and to the southeast, which form the pull-apart graben. The southeastern limb dips toward the southeast-dipping Markgrafneusiedl fault.

The main graben of the Matzen field in the north is formed at the northwest limb of the anticline and is rotated to the north. It plunges in the northeast.

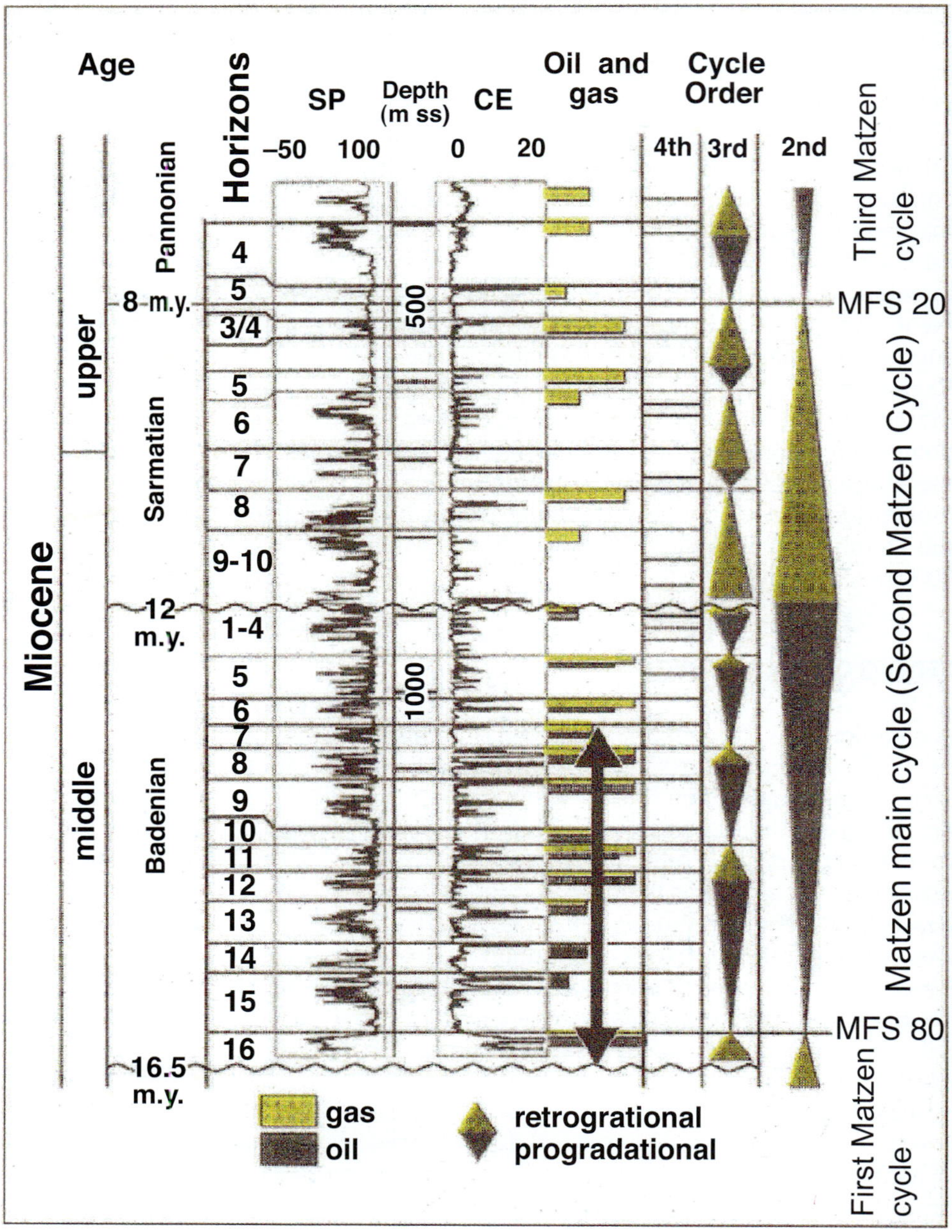

Figure 3. Type log of Matzen field. Matzener Hauptzyklus = main Matzen cycle; double arrow = main oil and gas horizons (modified after K. Fouad and P. Knox, 2001).

As with the main pull-apart graben, the Bockfliess structure seems to have been active from before the Matzen sand (16th Tortonian horizon) deposition through the Sarmatian to partly Pannonian.

Detailed tectonics of the Matzen field were evaluated resulting in three sets of fractures developed in the Matzen area: early east-northeast-trending sinistral R1 Riedel shears; late north- to north-northeast-trending dextral R2 Riedel shears (controlling the locations where canyons have incised the shelf and slope); and S-shaped duplex structures (M. Light, BEG Texas, 1998, personal communication).

MAIN PRODUCING HORIZONS

Badenian

16th Tortonian Horizon (Matzen Sand)

The 16th Tortonian horizon is the most important reservoir horizon in the Matzen field. It is a blocky transgressive sand as much as 50 m (160 ft) with excellent porosity (as much as 33%) and permeability (darcy range). It was deposited on top of a regional unconformity separating the Karpatian and older formations

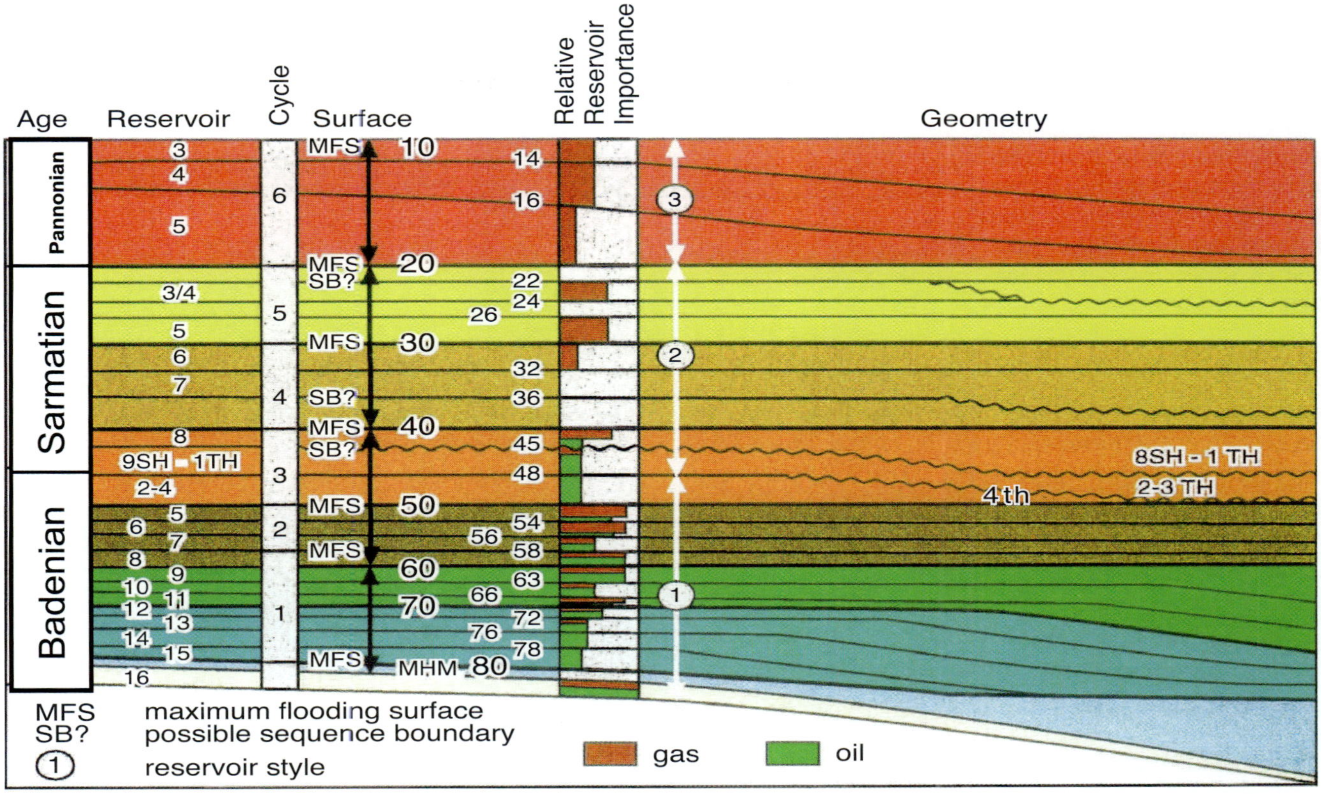

Figure 4. Stratal surface nomenclature and stratigraphy of Matzen field (P. Knox, 2000, personal communication). MFS = Maximum Flooding Surface; SB = sequence boundary; TH = Tortonian horizon; SH = Sarmatian horizon.

from the younger Badenian transgression. The sand can be subdivided into at least seven onlapping layers separated by minor flooding events. Each layer is composed of prograding, sand-rich, delta-front deposits and transgressive shelfal sediments. The older units fill the erosional relief, and each younger unit transgresses a little further to the north and buries older rocks. Main sediment input is believed to have derived from the northwest. The fill of the relief explains the high thickness variations over short distances. Basinward to the south,

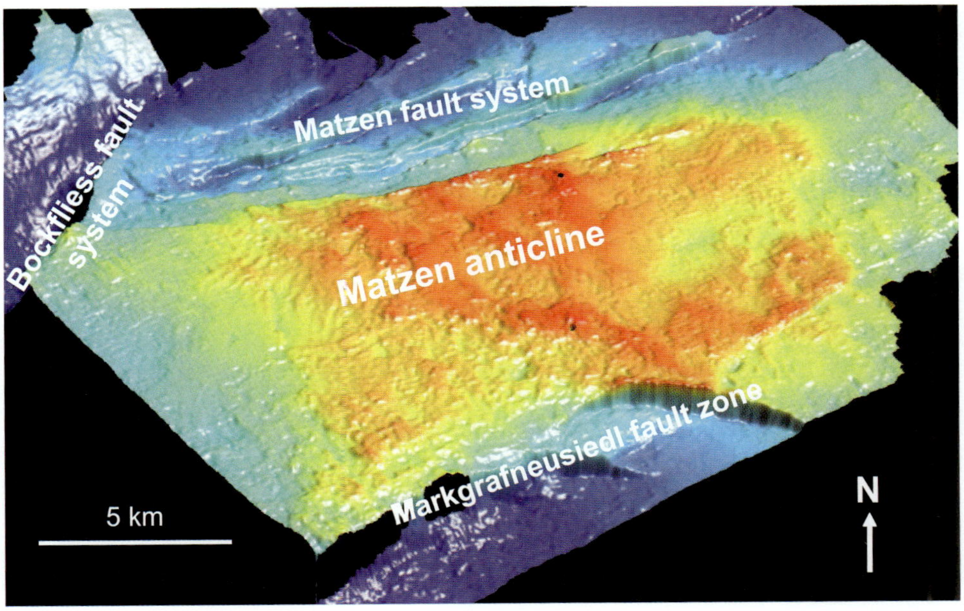

Figure 5. Main structural pattern of the Matzen field area.

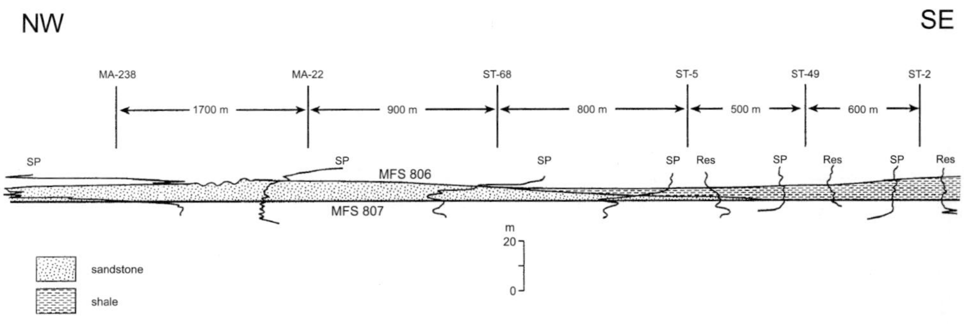

Figure 6. Change of log shape in Matzen Sand (16th Tortonian horizon) because of gradual transition of facies development in dip direction from northwest to southeast (P. Knox, 2000, personal communication).

the layers get thinner and shale out in minor flooding surfaces. Hence, the top of the 16th Tortonian horizon clearly is not a chronostratigraphic surface but a lithostratigraphic top of varying age (Figure 6). The mentioned minor flooding surfaces in the sandstone show bioturbation and higher shale content than the sands above and below and can constitute flow and permeability heterogeneities.

At least six facies types can be recognized from core and log analysis (P. Knox, BEG Texas, 2000, personal communication):

- coarsening-upward sandstones (thickness <10 m [<33 ft]), deposited during progradational periods in frontal parts of the delta; these sandstones can contain thin conglomerates (some 20 cm [8 in.]) at the base
- fining-upward sandstones, deposited during storm floods; these sands are bioturbated and only as much as 1.5 m (5 ft) thick
- 2–8-m (6.6–26-ft) amalgamated fining-upward sands, representing the retreat of shoreface sediments
- reworked shoreface, barrier bars, and submerged offshore bar sediments, cross-stratified coarsening-upward sands in a high-energy setting, 3–5 m (10–16 ft) in thickness
- laminated, partly bioturbated siltstone (5–6 m; 16–20 ft) deposited in a low-energy, shallow, open-marine setting
- shallow-marine limestones represented by algal to bivalve wackestones and packstones with shoaling-upward evidence deposited during a sea level highstand event. The patchy distribution of these limestones, also referred to as hard layers in reports and the literature (S. Schaub, Proseis, Zurich, 2003, personal communication), reduces reservoir quality in certain areas (Figure 7).

15th and 14th Tortonian Horizon

The 15th and the 14th Tortonian horizons are part of a highstand systems tract in the lower part of the Matzen main cycle (second cycle). The reservoirs belong to a prograding delta complex, developing from the north to the south. At the northeastern flank of the Matzen high, these delta lobes were tilted and form updip stratigraphic pinch-out traps, the only commercial stratigraphic reservoirs found so far in the Vienna basin. The reservoirs (Ebenthal) are satellite fields to Matzen, have been discovered during the recent redevelopment phase, and are still under development (Fuchs et al., 1999).

13th, 12th, and 11th Tortonian Horizon

These horizons are also part of the widespread prograding Matzen main cycle delta system. The 13th Tortonian horizon is divided into eight layers. Predominantly prograding sands and shales, very similar to the 14th and the 15th Tortonian horizons, were deposited at the 13th Tortonian horizon level. The main facies types are distributary channels, mouth bars, levees, crevasse splays, and bay fill sediments. Tidal sands and barrier bars were reworked, and the sediments were deposited down the delta slope via canyons as slope fans and basin-floor fans.

More and more aggradational sediment patterns can be observed during the deposition of the 12th and the 11th Tortonian horizons. The 12th Tortonian horizon is divided into five sublayers. The lower two layers (layers 5 and 4) are mainly composed of shelf-edge, high-energy sediments. The third, middle layer is characterized by a transgressional event. The shelf and the shelf edge move toward the north. The uppermost two layers (layers 2 and 1) reverse this trend again, and the shelfal sediments advance to the south.

Layer 2 carves canyons into the shelf slope and supports the idea of a regressional lowstand event. Syndepositional extensional duplex faults govern the deposition of the channel and mouth bar sediments parallel to the faults.

The 11th Tortonian horizon marks a pronounced transgression and can be subdivided into three layers separated by minor flooding surfaces.

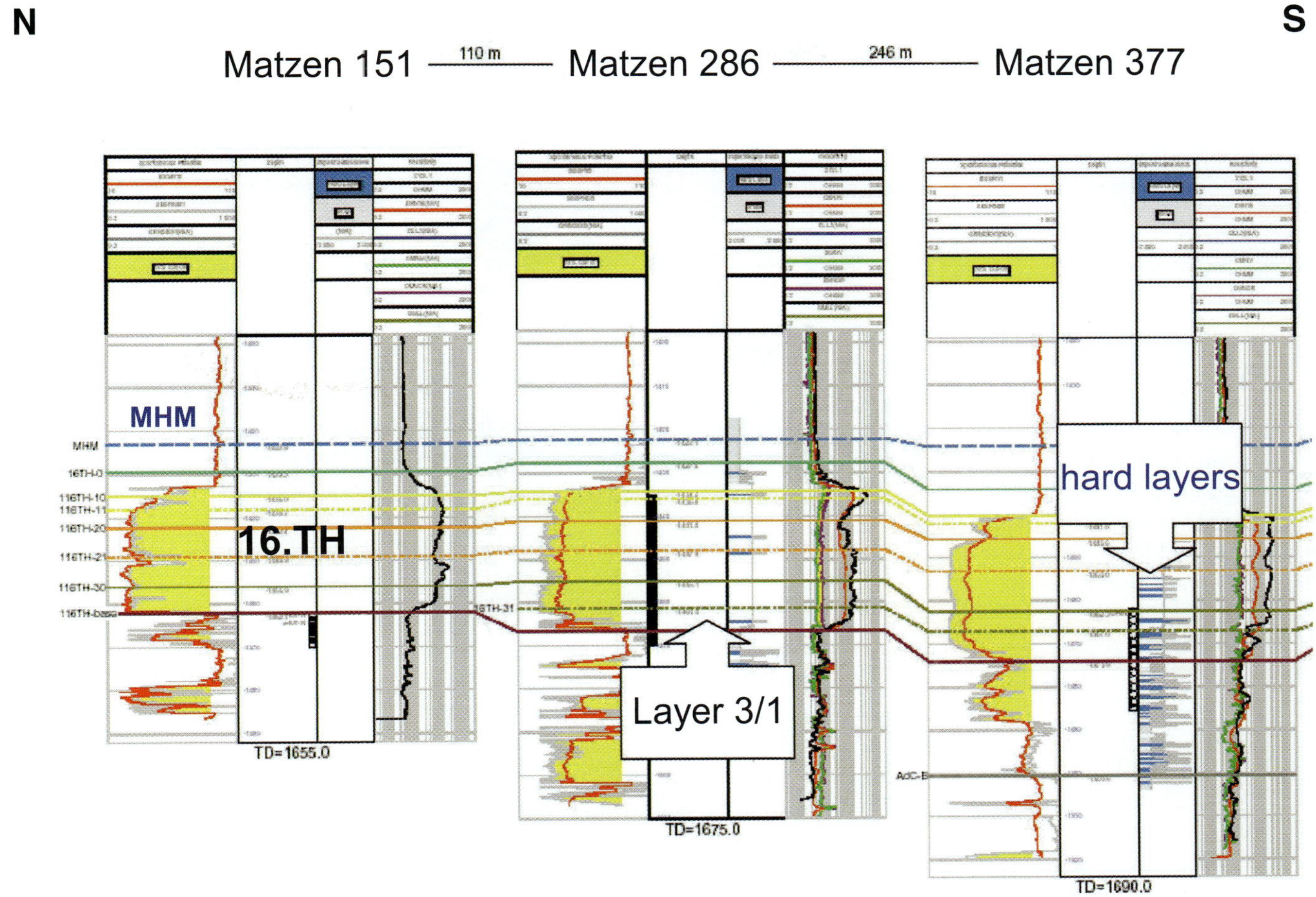

Figure 7. Cross section Matzen 151–Matzen 286–Matzen 377 in north–south dip direction representing a typical pattern of 16th Tortonian horizon. Blue dotted line = MHM (Matzener Hauptmarker); top of 16th Tortonian horizon = green line (layer 0); base of 16th Tortonian horizon = red line. Onlapping character can be seen in base Ma 286 to base Ma 151, where the lowermost layer 30/1 disappears (S. Schaub, Proseis, Zurich, 2003, personal communication).

10th Tortonian Horizon

Two main sand layers, A and B, separated by a shale intercalation (<10 m; <33 ft) several meters thick build up the 10th Tortonian horizon. These sands create a small low-relief platform on top of the 11th Tortonian horizon shelf platform. The limited accommodation space of this inner platform results in a quick downdip sand pinch-out of the 10th Tortonian horizon. The thickest sand occurs close to the faults in the north, suggesting that ongoing movement of the faults may have created controls on depositional patterns. The top of the 10th Tortonian horizon marks the highstand of the sea level.

9th Tortonian Horizon

The prograding sequence of the 9th Tortonian horizon was deposited during a sea level drop period (falling stage systems tract and lowstand systems tract). At the top of the horizon, a transgression created aggradational and retrogradational sedimentation. Several pronounced high-frequency marine flooding surfaces subdivided the horizon into six layers. The 9th Tortonian horizon is a textbook example of a shelf-slope-basin succession, laterally as well as vertically (fourth-order cycle) (Knox et al., 1999). The shelf edge advances from north (layer 6) to south (layer 1) some 2 km (1.2 mi) (Figures 8, 9).

The 9th Tortonian horizon is one of the most important oil producers of the Matzen field. Revisiting this important reservoir is one of the major technical tasks for the future. The complex depositional architecture of the reservoir provides various opportunities ranging from reperforating reservoir units to water shutoff measures and even drilling new wells. The detailed knowledge of the reservoir depositional environment (see below) is a prerequisite for future successful exploitation activities.

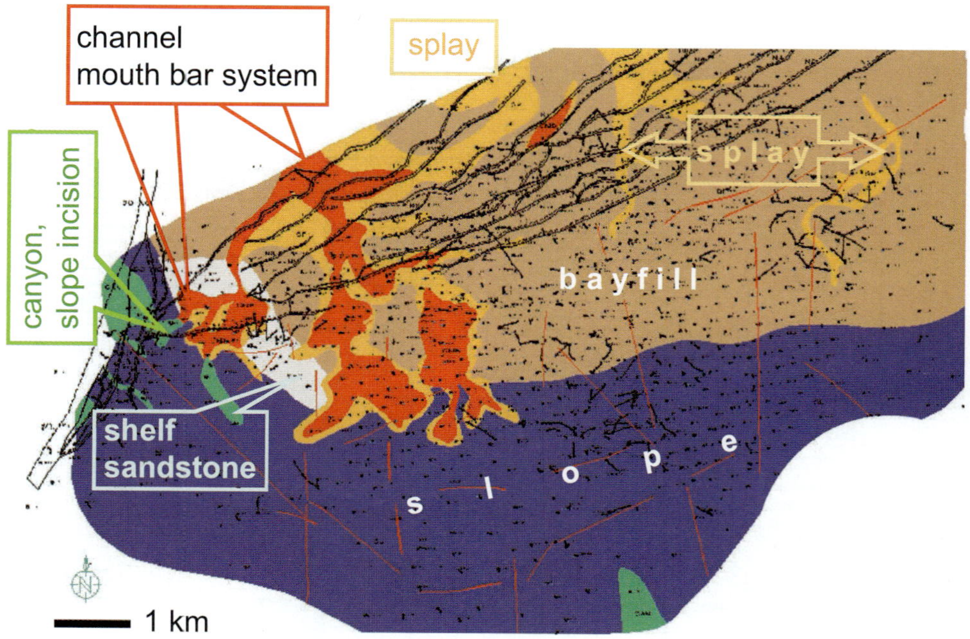

Figure 8. *9th Tortonian Horizon, Layer 6.* Paleogeographic setting: channel mouthbar system in the west and bayfill and splays in the east. Shale (blue) marking slope striking from west to east.

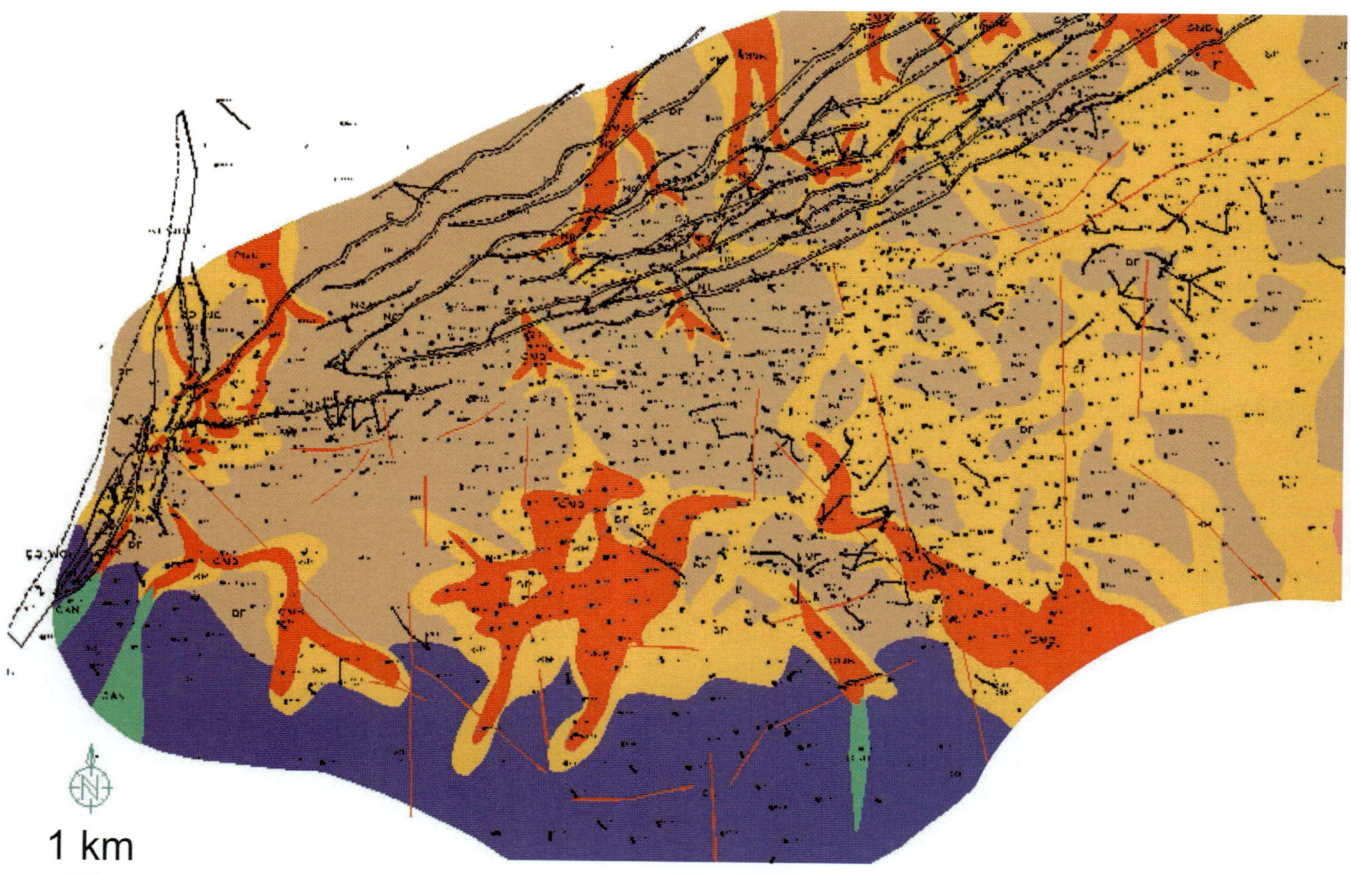

Figure 9. *9th Tortonian Horizon, Layer 1.* Widespread channel mouthbar systems extended far to the south. Some canyons (green) scour into the slope. Compared with the extension of Layer 6, the slope moved some 2 km (1.2 mi) southward. (For an explanation of colors see Figure 8).

Within each reservoir, thin sand layers can be identified. Examples exist of watered-out sands between oil-bearing sands, indicating vertical isolation, and suggest possible further subdivision on a chronostratigraphic basis. These thin, shale-prone intervals represent high-frequency flooding surfaces. These surfaces form a clinoform shape as they follow the contour of the depositional surface from the flat-lying shelf crossing the dipping slope to the flat basin floor (Knox et al., 1998).

Shales along these high-frequency flooding surfaces are thin and silty; they are more likely to have allowed fluid or pressure communication over geologic time. There may be areas of sand-to-sand contacts. Thus, these layers in a reservoir have similar original fluid contacts, at least in the same structural feature.

In other cases (like the shale below layer 1), the seal is good enough to separate two reservoirs, the upper reservoir contains gas in layer 1 and oil in layer 2.

Layer 6, the bottom layer, was mainly deposited in the northwestern part of the Matzen field, with an obvious difference between the western and the eastern part. The western area is characterized by southward-directed channel and channel mouth bar sediments, and the eastern part is characterized by thin splay and bay fill sediments (Figure 8).

Layer 5 shows an opposite character. Now, the main deposition is located in the east, and the western part shows only thin splay shales and siltstones. Channel and mouth bar systems were deposited along the Matzen fault system, filled lows, and eventually formed a pronounced sand-rich delta shelf edge in the main parts of the Matzen field. Even further south, the sands deposited along the shelf front as canyon deposits, providing a very strong aquifer for the reservoir.

Layer 4 is composed of three areas, each showing a different depositional setting in a sea level drop environment. The main (northeastern) part, the delta plain, is characterized by intense reworking and erosion, leaving behind only 2-m (6.6-ft)-thick sandstone. The second zone, the shelf edge, shifted southward and partly buried the underlying basin sediments, thereby gaining a thickness of more than 100 m (330 ft) in some places. The sediment input came from the north through channels along the Matzen fault system. The third area, the westernmost field area, remained in deeper water and shalier delta slope sediments.

Layer 3 shows an almost uniform channel mouth bar facies distribution. Only a few canyons and shelf edge deltas developed during this period. In some areas, the sandstones of layer 3 rest on the sandstones of layer 4, thereby creating fluid and pressure communication.

Layer 2 is also composed of a thick complex of channel mouth bar facies. Channel axes are perpendicular to the Matzen faults. Sometimes, channels cut into layer 3, providing pressure communication with the underlying reservoirs.

In layer 1, called Nullipora limestone, low sediment input from the north and a very shallow-marine environment allowed for patchy algal limestone development in the highest parts of layer 1. Layer 1 is characterized by aggradational and slightly transgressional sedimentation (Figure 9).

8th Tortonian Horizon

The 8th Tortonian horizon is divided into four sublayers, separated by thin shaly marl layers representing flooding events. The sublayers show aggradational character. Only in the southwestern (Bockfliess) and southeasternmost parts (Tallesbrunn) they show the progradational setting of clinoforms. Sediment input was from the north and east. The sands were governed by a fluvial-dominated shelf environment moving to the south. The 8th Tortonian horizon is a very important oil reservoir, and further exploitation is essential.

Layer 4, the lowermost layer, is a continuation of the 9th Tortonian horizon with a widespread channel mouth bar setting. Crevasse splays and bay fill siltstones were deposited between the channels. The channel direction is (sub)parallel to the Matzen fault system. Deltaic shelf-edge sediments developed southward, sometimes cut by submarine canyons.

Layer 3 is largely comparable to layer 4, only the shelf edge is located further southeast.

Layer 2 documents the shelf edge even further south. The same overall depositional environment as in layers 4 and 3 can be observed.

Layer 1, the uppermost 8th Tortonian horizon layer, shows a series of northwestern-oriented channels and channel mouth bar systems, mostly not controlled by faults. An interdistributary bay setting developed south of the channel facies, and southward is again replaced by shelf-edge deltas and fans.

7th Tortonian Horizon

The 8th Tortonian horizon builds a stable delta-plain platform, and the 7th Tortonian horizon rests regionally on this platform. Two layers are separated by an impermeable marl layer (maximum flooding surface). The bottom layer is characterized by channel mouth bar sediments with sandstone reservoirs as much as 11 m (36 ft) thick and accompanying dirty splay sandstones. Sediment input came from the east. Delta-front sediments are interpreted as products of long-shore drift currents or as storm flood sediments. Low-energy bay fill sediments characterize the southern part of the field. The top layer shows two dominant sediment input directions: east and west. The channel mouth bar environment advanced southward and buried the bay fill sediments of the bottom layer. Another front of delta sands developed south of this setting.

6th and 5th Tortonian Horizon

The upper Badenian (Bulmina–Rotalia zone) represents an overall regressive cycle (highstand systems tract). The 5th and 6th Tortonian horizons appear to be of rather uniform thickness (parallel bedding of mostly thin layers with no obvious large-scale erosional events) in the Matzen area.

The 6th and 5th Tortonian horizons can be subdivided into 15 layers (S. Schaub, Proseis, 2002, personal communication). These intervals represent short-term regressive cycles with layer-cake geometry. The sequence has been deposited in a shallow-marine, deltaic top set setting of a reduced marine (brachyhaline) environment. The main sediment transport direction is from northeast to southwest and from north to south. A complex arrangement of channels and channel-mouth bar systems represents the transport mechanism to the south toward a distinct shelf break line. No evidence for major slope channels exists in that area.

Sarmatian

6th Sarmatian Horizon

In the main Matzen area, the 6th Sarmatian horizon is deposited as intercalations of sands and shales containing some gas reservoirs with different contacts. Gas production was some 70 million m^3 (2.4 bcf). In the area of Prottes and Tallesbrunn, the top of the 6th Sarmatian horizon is scoured by basal channels with layers of gravel of the 5th Sarmatian horizon.

5th Sarmatian Horizon

The 5th Sarmatian horizon contains two main gas-storage areas, one in the central Matzen area and another in the southeast in the Tallesbrunn area. Gas production during the 1950s to the 1970s was 1.1 billion m^3 (38.83 bcf) in Matzen and, from 1962 to 1974, 290 million m^3 (10.24 bcf) in Tallesbrunn. Working gas volume is about 210 million m^3 (7.41 bcf) in the main Matzen storage and 300 million m^3 (10.59 bcf) in the Tallesbrunn storage.

The 5th Sarmatian horizon in the Matzen area was deposited in a subaerial to shallow-marine deltaic top set setting. Two distinct sediment transport directions can be observed by mapping two prominent channel systems: one in a northeast–southwest direction from the Matzen fault area and a second one in a northwest–southeast direction along the axis Matzen–Tallesbrunn. Four layers can be correlated through the entire area (Figure 10). An important sequence boundary between layers 2 and 3 represents the base of a prominent channel system. It separates layers 3 and 4 of a lower part of the unit from an upper part (layers 1 and 2) (S. Schaub, Proseis, Zurich, 2003, personal communication).

The base of the 5th Sarmation horizon sand lies below a maximum flooding surface, recognized in a low-resistivity shale as MFS 30. From this surface to the sequence boundary, the sea level was stable or slightly rising (highstand systems tract). During the sea level drop at the sequence boundary, the depocenter shifted southward with the simultaneous incision of a thick channel system into the underlying 6th Sarmatian horizon.

Scouring the 6th Sarmatian horizon causes a cross-flow of water between the two horizons. A water-drive effect in the 5th Sarmatian horizon is amplified by the fact that coarse-grained 5th Sarmatian horizon sands directly communicate with the underlying porous 6th Sarmatian horizon sands. The scouring of the 5th Sarmatian horizon can be seen across large distances in 3-D seismic cross sections (Figures 11, 12).

The sand-rich channel system represents a lowstand systems tract. The change from lowstand systems tract to a transgressive systems tract is seen in the upper part of the 5th Sarmatian horizon. The deposition of the uppermost sand interval is an indication of the onset of that transgressive systems tract.

Facies. Layer 4 developed from the north with a thin sand accumulation fed by small channels creating splays and flood plains. These sands either shale out rapidly toward the south or are missing because of erosion in the areas of the 5th Sarmatian horizon channels (layers 1 and 2).

Layer 3 again occurs only in small areas (near Tallesbrunn) as levee, point bar, and flood plain. Huge areas in the east of the main field, specifically north of Tallesbrunn, were largely eroded by the channel system of layers 1 and 2. The main interval consists of shale originated from marshes, lakes, and flood plains and, further to the south, also from shales of marine origin. The top of the layer is defined by the erosive base of the channel systems of layers 1 and 2 (sequence boundary).

In layer 2, a large channel system developed along the fault system in the north. One branch was shedding to the Bockfliess area into the southwest direction, and another main system developed into the

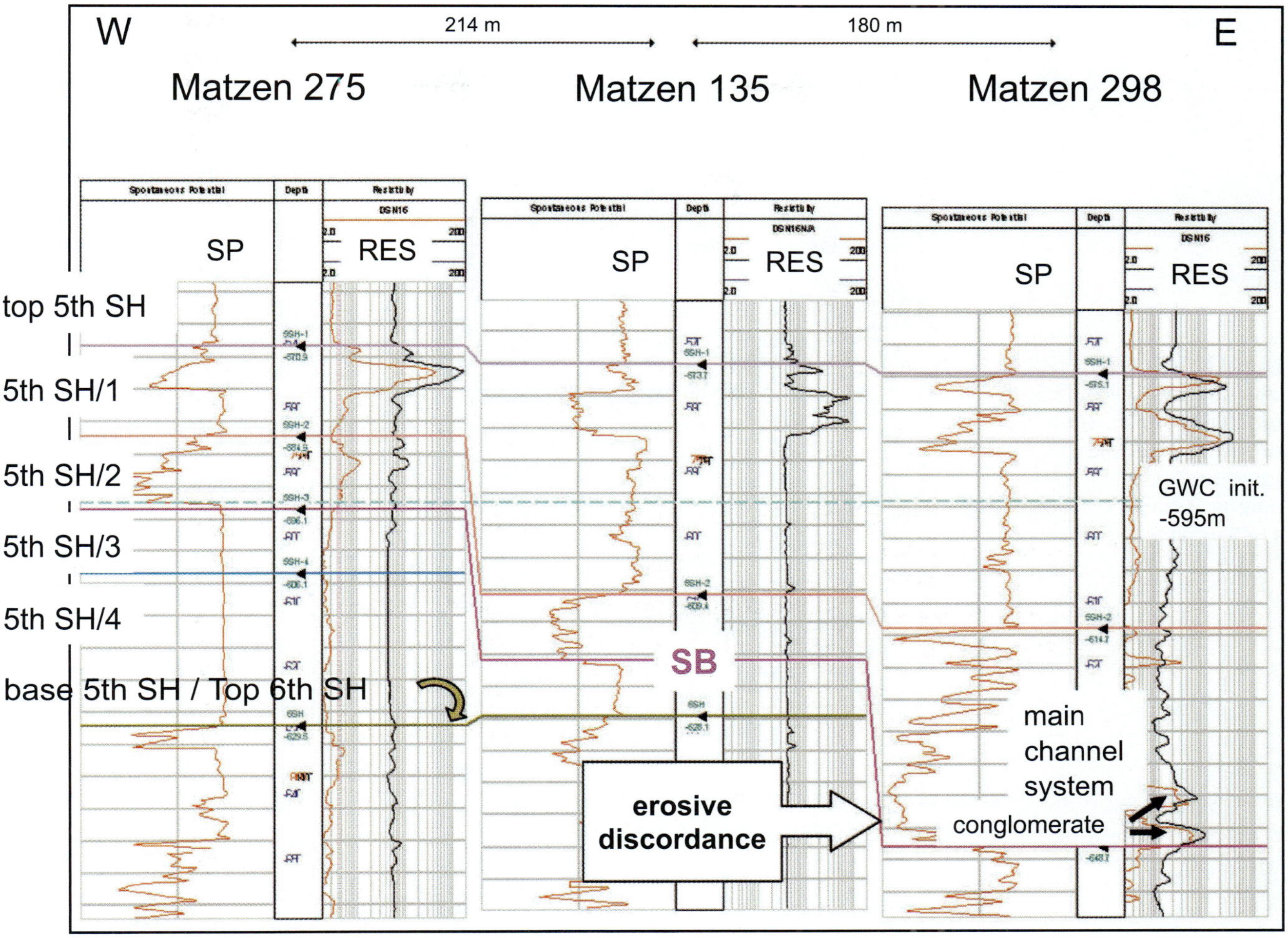

Figure 10. West–east cross section of the 5th Sarmatian horizon and underlying 6th Sarmatian horizon. The main channel system of the 5th Sarmatian horizon in the east scours into the 6th Sarmatian horizon, marking a prominent erosive discordance. SP = self potential; RES = resistivity; SB = sequence boundary; GWC = gas-water contact (S. Schaub, Proseis, Zurich, 2003, personal communication).

Tallesbrunn area to the southeast. In the Tallesbrunn area, the channel system widened considerably, developing into a broad channelized system, consisting of many laterally amalgamating channels. The erosive base is the sequence boundary. The channels are accompanied by point bars, levees, splays, and flood plains. Shaly intervals sometimes separate sand accumulations.

The top of layer 1 coincides with the top of the 5th Sarmatian horizon. The two-channel systems of layer 2 persist through layer 1 in a similar shape but less extended. In the central area of the Matzen field, a major splay system develops to the southwest into a large flood-plain area.

3rd/4th Sarmatian Horizon

The deposition of the overlying 3rd/4th Sarmatian horizons is part of the same transgressive systems tract. The 3rd/4th Sarmatian horizons consist of an intercalation of shales and sands and, in some areas, of basal coarse-grained sands and pebbles. Within the central area of the Matzen field, a thick shale separates the horizon, giving rise to the expression 3rd/4th Sarmatian horizons.

The base layer with gravels is a meandering-channel system extending from the area of Spannberg in the north to the area of Gansendorf in the south. The base of the 3rd/4th Sarmatian horizons coincides with flooding surface 24. Gross thickness of the 3rd/4th Sarmatian horizons is 20 m (66 ft), and the main gas reservoir comprises an initial gas in place of 610 million m^3 (21.54 bcf).

Pannonian

Early Pannonian deltaic systems created thick sand accumulations with gas reservoirs in the 3rd and 4th upper Pannonian horizons, which currently act as

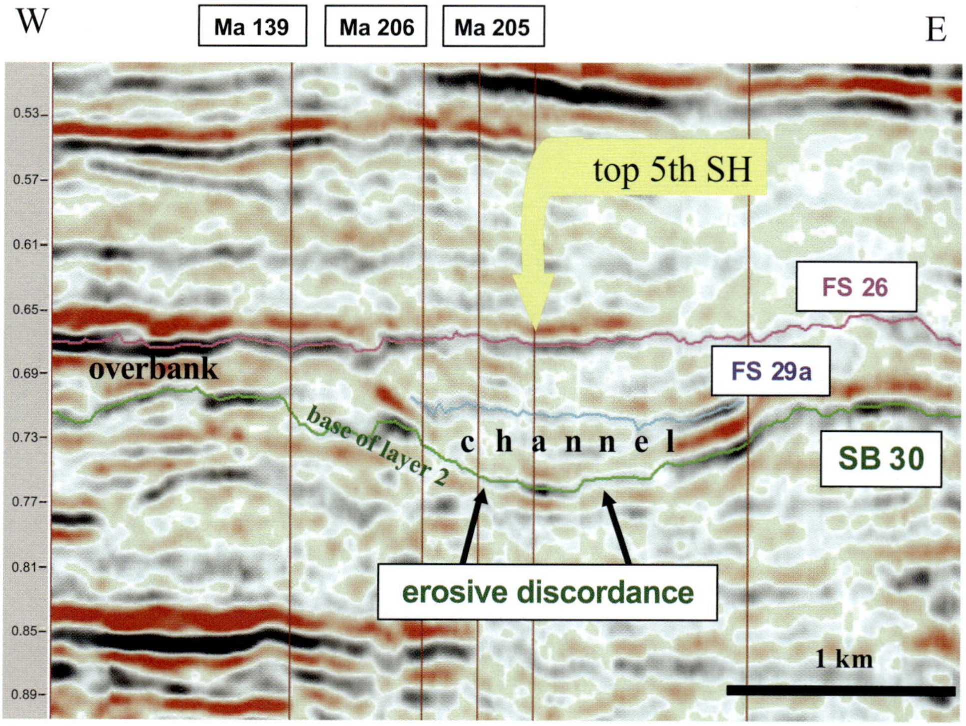

Figure 11. Scouring of 5th Sarmatian horizon into the 6th Sarmatian horizon; seismic west–east cross section. SB = Sequence Boundary; FS = Flooding Surface; Ma = Matzen; SH = Sarmatian horizon.

gas storages. Prominent channel systems developed from north to south, creating extended lacustrine delta systems.

CONCLUSION

A new detailed geological and petrophysical model based on 3-D seismic data and sequence stratigraphy of the main Matzen field hydrocarbon horizons was established resulting in better understanding of flow units, permeability barriers, and water drive mechanisms. The new model, as a starting point for simulation studies, not only gave input for a better explanation of production behavior but also led to successful work-over projects, sidetracks, and new wells. Since the implementation of the results of these studies, oil production could be stabilized. For a reservoir in the 9th Tortonian horizon, a two-compartment material balance model could be built, which caused a successful redesign of production. New oil and gas from this old field support slowing down the decline.

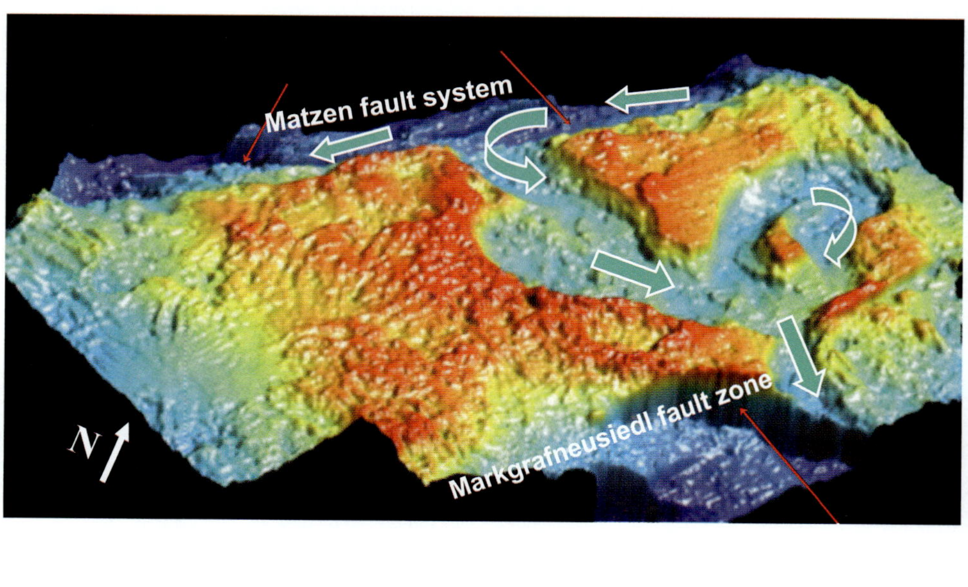

Figure 12. 5th Sarmatian horizon, sequence boundary 30, 3-D view from south–southwest, time map; arrows indicate flow direction in prominent channel system.

FURTHER READING

A detailed overview of the literature from more than 150 yr of intense research of the Vienna basin is given in Arzmuller et al. (2006).

REFERENCES CITED

Arzmuller, G., S. Buchta, E. Ralbovsky, and G. Wessely, 2006, The Vienna Basin, *in* J. Golonka and F. J. Picha, eds., The Carpathians and their foreland: Geology and hydrocarbon resources: AAPG Memoir 84, p. 191–204.

Decker, K., and H. Peresson, 1998, Miocene to present-day tectonics of the Vienna basin transform fault: Links between the Alps and the Carpathians: Carpathian-Balkan Geological Association, XVI Congress, Geologische Bundesanstalt, Vienna, Austria, p. 33–36.

Fouad, K., and P. R. Knox, 2001, Neural network analysis of 3-D seismic data delineates slope- to basin-floor fan bodies in a high-frequency falling-stage systems tract, Matzen Field, Austria (abs.), *in* Offshore Technology Conference, Houston OTC 2001 Proceedings CD, Paper 13160, 10 p.

Fuchs, R., F. Schoffmann, and W. E. Sackmaier, 1999, Feldesentwicklung mit Richtbohrungen in Weiner Becken (Österreich): Hamburg/Wien, Erdöl Erdgas Kohle, v. 115, no. 12, p. 571–574.

Fuchs, R., R. Ramberger, and C. Veit, 2001, Das Matzen Projekt: Renaissance des größten Öl-und Gasfeldes in Österreich (Wiener Becken): Hamburg/Wien, Erdöl Erdgas Kohle, v. 117, no. 11, p. 528–540.

Hamilton, W., L. Wagner, and G. Wessely, 2000, Oil and gas in Austria: Mitteilungen Österreichische Geologische Gesellschaft, v. 92, p. 235–262.

Knox, P. R., M. H. Holtz, M. Mercer, and R. J. Finley, 1998, Identifying facies-controlled preferential pathways of water encroachment: Integrated characterization of Miocene deltaic reservoirs, Matzen field, Vienna Basin, Austria (abs.), *in* Bridges to discovery: Extended abstracts, AAPG Annual Meeting Program, v. 2, p. A370.

Knox, P. R., E. Von Lunen, and R. J. Finley, 1999, Shelf-to-basin transition: Architecture of Miocene falling-stage deltaic to basinal reservoirs, Matzen field, Vienna basin, Austria (abs.), *in* AAPG Annual Meeting Program, p. A74.

Papp, A., W. Krobot, and W. Hladecek, 1973, Zur Gliederung des Neogens im Zentralen Wiener Becken: Mitteilungen Gesellschaft Geologie-Bergbaustudenten, v. 22, p. 191–199.

Rogl, F., and G. Daxner-Hock, 1996, Late Miocene Paratethys correlations, *in* R. L. Bernor, V. Fahlbusch, and H.-W. Mittmann, eds., The evolution of Western Eurasian Neogene mammal faunas: Columbia Press, New York, v. 3, p. 47–55.

Veit C., R. Fuchs, and R. Finley, 1999, "Matzen Project" a field re-development in Central Europe: Proceedings Society of Petroleum Engineers–Pakistan Association of Petroleum Geologists, Annual Technical Conference, Islamabad, Pakistan, p. 223–234.

Wessely, G., 2000, Sedimente des Wiener Beckens und seiner alpinen und subalpinen Unterlagerung-Exkursionsführer Sediment 2000: Mitteilungen Gesellschaft Geologie-und Bergbaustudenten in Österreich, v. 44, p. 191–214.

THE GEOLOGY AND HYDROCARBON RESOURCES OF THE OUTER WESTERN CARPATHIANS AND THEIR FORELAND ON THE TERRITORY OF POLAND AND UKRAINE

Ślączka, Andrzej, Stanisław Kругłov, Jan Golonka, Nestor Oszczypko, and Igor Popadyuk, 2006, Geology and Hydrocarbon Resources of the Outer Carpathians, Poland, Slovakia, and Ukraine: General Geology, *in* J. Golonka and F. J. Picha, eds., The Carpathians and their foreland: Geology and hydrocarbon resources: AAPG Memoir 84, p. 221–258.

7

Geology and Hydrocarbon Resources of the Outer Carpathians, Poland, Slovakia, and Ukraine: General Geology

Andrzej Ślączka
Institute of Geological Sciences, Jagiellonian University, Krakow, Republic of Poland

Stanisław Krugłov*
Ukrainian State Geological Research Institute, Lviv, Ukraine

Jan Golonka
AGH University of Science and Technology, Jagiellonian University, Krakow, Republic of Poland

Nestor Oszczypko
Institute of Geological Sciences, Jagiellonian University, Krakow, Republic of Poland

Igor Popadyuk
Ukrainian State Geological Research Institute, Lviv, Ukraine

ABSTRACT

The purpose of this chapter is to provide the general overview of the stratigraphy and tectonics of the Polish, Ukrainian, and adjacent parts of the Slovakian Outer Carpathians. The Polish and Ukrainian Outer Carpathians form the north and northeastern part of the Carpathians that expand from the Olza River on the Polish–Czech border to the Ukrainian–Romanian border. Traditionally, the Northern Carpathians are subdivided into an older range, known as the Inner Carpathians, and the younger ones, known as the Outer Carpathians. These ranges are separated by a narrow, strongly tectonized belt, the Pieniny Klippen Belt. The Outer Carpathians are made up of a stack of nappes and thrust sheets showing a different lithostratigraphy and tectonic structures. Generally, each Outer Carpathian nappe represented separate or partly separate sedimentary subbasin. In these subbasins, enormous continuous sequence of flysch-type sediments was deposited; their thickness locally exceeds 6 km (3.7 mi). The sedimentation spanned between the Late Jurassic and early Miocene. During the folding and overthrusting, sedimentary sequences were uprooted, and generally, only sediments from the central parts of basins are preserved.

The Outer Carpathian nappes are overthrust on each other and on the North European platform and its Miocene–Paleocene cover. In the western part, overthrust plane is relatively flat and becomes more and more steep eastward. Boreholes and seismic data indicate a minimal distance of the overthrust of 60–80 km (37–50 mi).

*Deceased.

DOI:10.1306/985610M843070

The evolution of the Northern Outer Carpathian Flysch basins shows several tectonostratigraphic stages. The first period (Early Jurassic–Kimmeridgian) began from the incipient stage of rifting and formation of local basins. The next stage (Tithonian–Early Cretaceous) is characterized by rapid subsidence of local basins where calcareous flysch sedimentation started. The third period (Late Cretaceous–early Miocene) is characterized by compression movements, appearance of intensive turbiditic sedimentation, and increased rate of subsidence in the basins.

INTRODUCTION

The purpose of this chapter is to provide the general overview of the stratigraphy and tectonics of the Outer Carpathians in Poland, Ukraine, and adjacent Slovakia. The details concerning petroleum geology will be provided by other chapters in this Memoir by employees of the Polish Oil and Gas Company and the Ukrainian State Geological Research teams. These chapters include as follows:

- *Hydrocarbon resources of the Polish Outer Carpathians—reservoir parameters, trap types, and selected hydrocarbon fields: A stratigraphic review* by Dziadzio et al. (2006b, chapter 8)
- *Petroleum geology of the Boryslav–Pokuttya zone, the Ukrainian Carpathians* by Popadyuk et al. (2006, chapter 13)
- *The geology of the Weglowka oil field, Subsilesian unit, Polish Outer Carpathians* by Dziadzio (2006, chapter 14)
- *The Ciezkowice sandstone: Examples of basin-floor fan-stacking patterns from the main (upper Paleocene to Eocene) reservoir in the Polish Carpathians* by Dziadzio et al. (2006a, chapter 15)
- *Reconstruction of petroleum systems based on integrated geochemical and geological investigation: Selected examples from Middle, Outer Polish Carpathians* by Matyasik and Dziadzio, (2006, chapter 16)

A special chapter also deals with the geochemistry of the Carpathian source rocks: *The origin and habitat of hydrocarbons of the Polish and Ukrainian parts of the Carpathian province* by Kotarba and Koltun (2006, chapter 11).

THE POSITION OF THE OUTER CARPATHIANS IN THE ALPINE–CARPATHIAN FOLD AND THRUST BELT

The Polish and Ukrainian Outer Carpathians form the northeastern part of the great arc of mountains, which stretch more than 1300 km (807 mi) from the Vienna Forest to the Iron Gate on the Danube (Figure 1). Traditionally, the Carpathians are subdivided into an older range, known as the Inner Carpathians, and the younger ones, known as the Outer Carpathians. From the point of view of the plate-tectonic evolution of the basins, the following major elements could be distinguished in the Outer Carpathians and the adjacent part of the Inner Carpathians.

Inner Carpathian terrane: A continental plate built of the continental crust of Hercynian (Variscan) age and Mesozoic–Cenozoic sedimentary cover. The Inner Carpathians form a prolongation of the Northern Calcareous Alps and are related to the Apulia plate (in a regional sense; Picha, 1996). The uppermost Paleozoic–Mesozoic continental and shallow-marine sedimentary sequences of this plate are folded and thrust into a series of nappes. They are divided into the Tatric, Veporic, and Gemeric nappes (Figure 2) that are the prolongation of the lower, middle, and upper Austroalpine nappes, respectively. The nappes and the Hercynian basement are unconformably covered by middle Eocene–Oligocene flysch and early–middle Miocene marine and terrestrial (continental) molasses. Another terrane with the Hercynian basement, known as the Tisza–Dacia, is amalgamated with the Inner Carpathian terrane. According to Golonka et al. (2000), the Inner Carpathians, Eastern Alps, and Tisza–Dacia form Alcapa superterrane; according to, e.g., Kovac et al. (1998), Alcapa and Tisza constitute different plates.

North European platform: Large continental plate amalgamated during the Precambrian–Paleozoic. Proterozoic, Vendian (Cadomian), early Paleozoic (Caledonian), and late Paleozoic (Hercynian) fragments could be distinguished in the folded and metamorphosed basement of this plate. Beneath the Outer Carpathians, the sedimentary cover consists of the autochthonous upper Paleozoic, Mesozoic, and Cenozoic sequences covered by the allochthonous Jurassic–Neogene rocks (see Oszczypko et al., 2006).

These allochthonous rocks are uprooted and overthrust onto the southern part of the North European platform at a distance of at least 60–100 km (37–62 mi) (Ksiazkiewicz, 1977; Oszczypko and Slaczka, 1985). They form stacks of nappes and thrust sheets arranged in several tectonic units. In Poland, these allochthonous, mainly

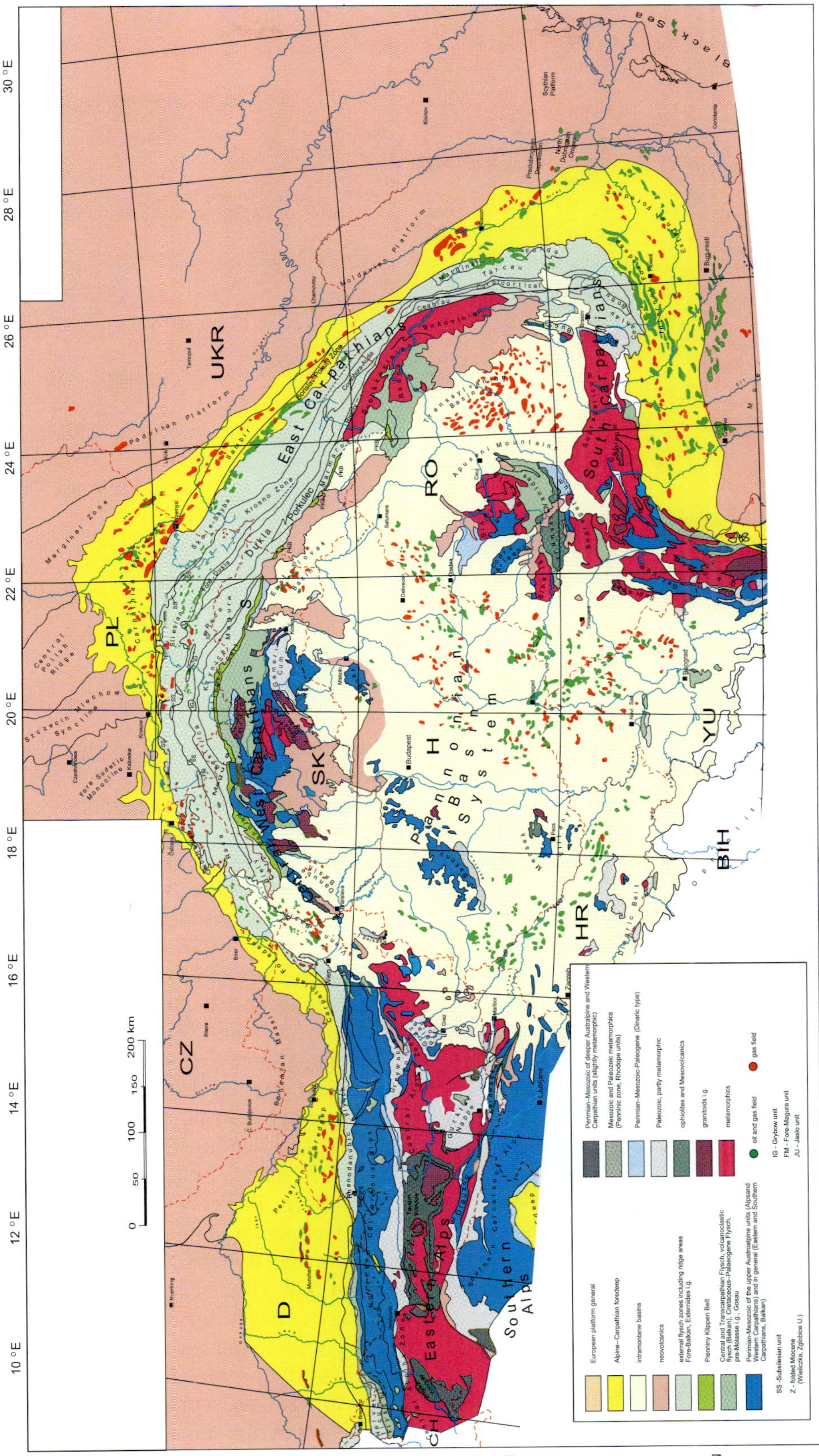

Figure 1. General overview and distribution of oil and gas in the Circum Carpathian Region of Central Europe.

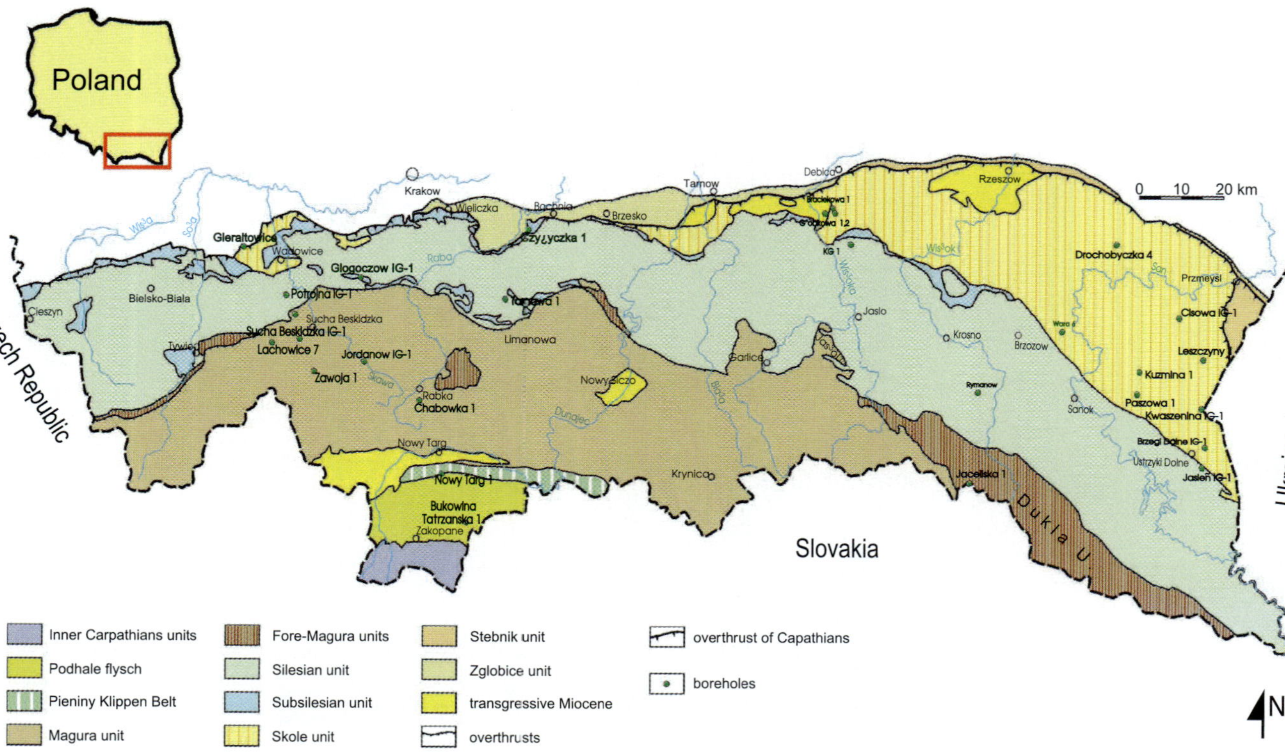

Figure 2. Tectonic scheme of the Polish Carpathians.

flysch units are being regarded as the Carpathian Flysch. Along the frontal Carpathian thrust, a narrow zone of folded Miocene deposits was developed (Figures 2, 3): the Zglobice unit in the Polish Carpathians, the Sambir nappe of the Ukrainian part of the Carpathians, and the Subcarpathian unit in Romania.

The Penninic realm is a part of the Alpine Tethys (e.g., Birkenmajer, 1986; Sandulescu, 1988; Oszczypko, 1992; Plasienka, 1999, 2000; Stampfli, 2001; Golonka, 2004; Golonka et al., 2006) that developed as a basin during the Jurassic between the Inner Carpathian–Eastern Alpine terrane and the North European platform. In the western part, it contains the ophiolitic sequences indicating the truly oceanic crust. In the eastern part, the ophiolitic sequences are known only as pebbles in the flysch; the basement of the Penninic realm was partly formed by the attenuated crust. In Poland, Slovakia, and Ukraine, the Penninic realm is represented by the sedimentary sequences of Jurassic, Cretaceous, Paleogene, and Miocene age belonging to the Pieniny Klippen Belt and the Magura unit (Golonka et al., 2003). Some of these sequences have been recently located in the suture zone between the Inner Carpathian terrane forming the Pieniny Klippen Belt; other sequences are involved in the allochthonous units covering the North European platform (Magura nappe) or accreted to the Inner Carpathian terrane. Because of the evolutionary connotations of the Penninic realm, the Pieniny Klippen Belt could also be regarded as belonging to the Outer Carpathians (e.g., Ksiazkiewicz, 1977; Picha, 1996). The Czorsztyn submerged ridge was a part of the Penninic realm dividing the oceanic basin into two subbasins. The southern subbasin and the ridge traditionally constitute the Pieniny domain. Its sequences are involved in the Pieniny Klippen Belt, whose strongly tectonized structure is about 800 km (500 mi) long and 1–20 km (0.6–12 mi) wide, which stretches from Vienna in the west to the Poiana Botizei (Maramures, northeast Romania) in the east (Figure 1). The largest part of the northern subbasin forms the Magura unit, traditionally belonging to the Outer Carpathians. The Pieniny Klippen Belt is separated from the Magura nappe by the Miocene subvertical strike-slip fault (e.g., Birkenmajer, 1986, 1988).

The Transylvanian domain formed perhaps the extension of the eastern Tethys between the European platform and Tisza–Dacia terrane (Sandulescu, 1988; Sandulescu and Visarion, 2000). In Romania, it was truly an oceanic realm as indicated by the existence of ophiolites. It developed during the Triassic and was closed during the Cretaceous. In Poland, the presumed Transylvanian domain sedimentary sequences are represented only by pebbles in the flysch of the Pieniny Klippen Belt and the Magura unit.

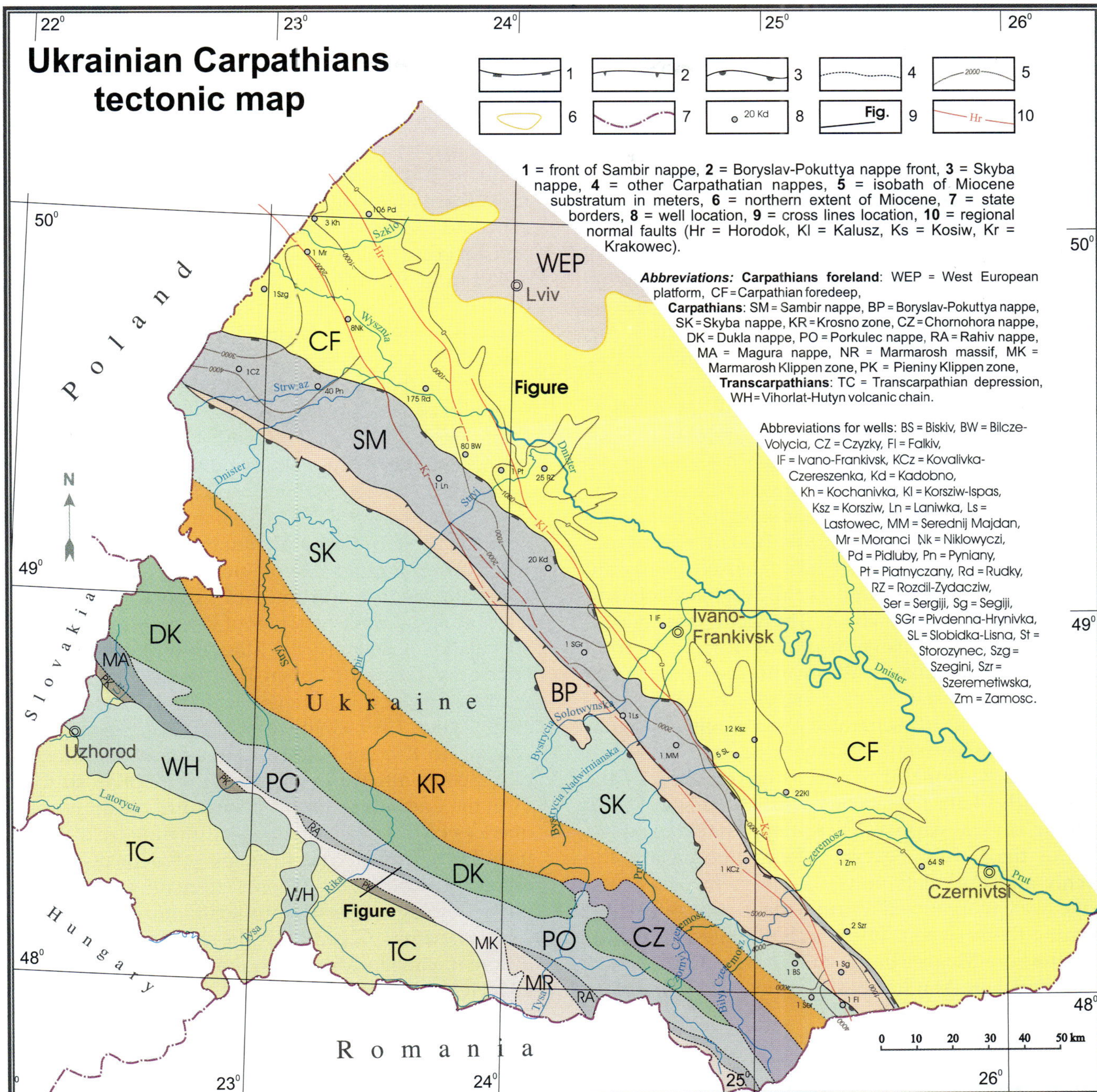

Figure 3. Ukrainian Carpathians tectonic map. Modified after Glushko and Kruglov (1986).

The Severin–Moldavidic realm (Balintoni, 1998), also known in Romania as the Outer Dacides and the Moldavides (Sandulescu, 1988; Stefanescu et al., 2006), developed in the North European platform as rift and/or back-arc basin. The Severin–Moldavidic basement is represented by the attenuated crust of the North European plate with perhaps incipient oceanic fragments. The sedimentary cover is represented by several sequences of Late Jurassic–early Miocene age belonging recently to Dukla, Silesian, Subsilesian, Zdanice, and Skole tectonic units in Poland (Figure 2) and Czech Republic (Pescatore and Slaczka, 1984; Zytko et al., 1989; Stranik et al., 1993; Slaczka and Kaminski, 1998) and to Marmarosh, Rachiv, Porkulets, Chornohora, Skyba, and Boryslav–Pokuttya (Figure 3) nappes in Ukraine (Kruglov, 1989; Slaczka, 1996b). The subbasins in the Severin–Moldavidic are divided by ridges and uplifted zones. The most prominent one is the Subsilesian bulge between Silesian and Skole subbasins. Another ridge was located between Dukla and Silesian subbasins. The Subsilesian bulge is connected in the west with the slope sequences between the epicontinental part of the North European platform and the Outer Carpathian basins (Silesian and Magura basins). The Severin–Moldavidic basinal realm ends in Moravia, whereas slope sequences extend further westward. The Severin part (Severinides, see, e.g., Balintoni, 1998, 2001; or Outer Dacides, Sandulescu, 1988; Stefanescu et al., 2006) of the basin is represented in Ukraine by the Kaminnyi Potik and Rachiv units. This basin was closed during the Cretaceous. The Moldavic part (Moldavides, see, e.g., Sandulescu, 1988; Balintoni, 1998; Stefanescu et al., 2006) was closed during the Neogene.

The Getic–Marmarosh ridge (Golonka et al. 2003), also known as the Median Dacides (Sandulescu, 1988; Stefanescu et al., 2006), constitutes a fragment of the North European platform that rifted away during the opening of the Severin–Moldavidic basins. It includes Precambrian and early Paleozoic (Caledonian) granites and metamorphic rocks, late Paleozoic (Variscan) metamorphic rocks, as well as the late Paleozoic and Mesozoic sedimentary cover. The Getic–Marmarosh ridge separated the Severin–Moldavidic basin from the Transylvanian basin. Westward, the similar position has the Silesian ridge (Sandulescu, 1988; Oszczypko, 1992), which separated the Penninic–Magura realm and the Severin–Moldavidic realm. The eastern part of the Getic–Marmarosh ridge collided in the latest Early–Late Cretaceous with the Tisza–Dacia, forming several nappes. These nappes also included part of the Civcin–Severin–Moldavidic basins, the Rachiv and Porkulets units. The western part of the ridge was reorganized during the Late Cretaceous, forming the basement of the uplifted Silesian Cordillera, the Fore-Magura basin; a ridge separated Magura basin and Fore-Magura basin, as well as a marginal part of the Magura basin.

OUTER CARPATHIAN STRATIGRAPHY

The Outer Carpathians are made up of a stack of nappes and thrust sheets showing a different lithostratigraphy and tectonic structures (Figures 2, 3). Part of them can be traced along the whole of the Outer Carpathians, and a part can be traced only in the Eastern Carpathians. These tectonic units are overthrust one on the other from the south and built of sediments representing the time span between the Late Jurassic and the early Miocene (Ksiazkiewicz, 1962, 1977; Bieda et al., 1963; Mahel et al., 1968; Koszarski and Slaczka, 1976). They correspond to more or less separated sedimentary basins, and every basin generally displays a different lithostratigraphic development. During the overthrusting movements, the tectonic units became uprooted, and generally, only central parts of the basins are preserved.

The western and northern parts of the Outer Carpathians (Figure 4) consist of four main longitudinal units (basins): the Magura, Dukla–Fore-Magura, Silesian, and Skole belonging to two realms, the Penninic and Severin–Moldavidic, originally divided by the Silesian ridge (Cordillera). The other ridges and swells existed in the realms; the most distinctive were the Subsilesian swell and Andrychow ridge (Ksiazkiewicz, 1962). The Magura, Dukla, Silesian, and Skole sedimentary units (basins) extend eastward to the Ukrainian part of the Outer Carpathians. Several units, e.g., Marmarosh, Rachiv, Porkulets, Chornohora, Skyba (Skole), and Boryslav–Pokuttya units (Subbasin) in the Severin–Moldavidic realm, extend from Ukraine southeastward and southward to Romania; their correlation with the Western Carpathian units still remains somewhat uncertain and disputable. The relationship between the Magura unit and the Pieniny Klippen Belt in Ukraine and the adjacent part of Romania also remains speculative. Generally, the Magura unit is located between the Pieniny Klippen Belt (zone) and the Marmarosh zone (e.g., Sandulescu et al., 1981; Sandulescu, 1988; Zytko, 1999; Golonka et al., 2003); however, according to Kruglov (2001), fragments of the Magura nappe could also exist north from the Marmarosh zone.

In the Carpathian basins, enormous, continuous sequences of flysch were deposited mainly by different turbidity currents. The thickness of deposits locally exceeds 10 km (6.2 mi). These sediments were divided on a lithological basis into many formations with separate names. Commonly, the same formation has a different name in adjacent countries (Bieda et al., 1963; Geroch et al., 1967; Mahel et al., 1968; Slaczka, 1980; Kruglov, 1989, 2001; Pasternak, 1989; Golonka and Lewandowski, 2003). The source areas for the clastic material were situated on the outer and inner margin as well as on the intrabasinal ridges. The sedimentary sequences display lateral and vertical variations in the thickness and lithofacies, reflecting changes in subsidence and sedimentation rates and shifting of source areas.

Within the sedimentary sequences, generally, three main megastages of development exist, connected with the global events: an early stage from Jurassic to Albian, characterized by development of black shales; the second stage (Cenomanian–Eocene) characterized by the occurrence of red and variegated shales; and the final stage (Oligocene–early Miocene), when brown bituminous shales appeared and variegated shales disappeared.

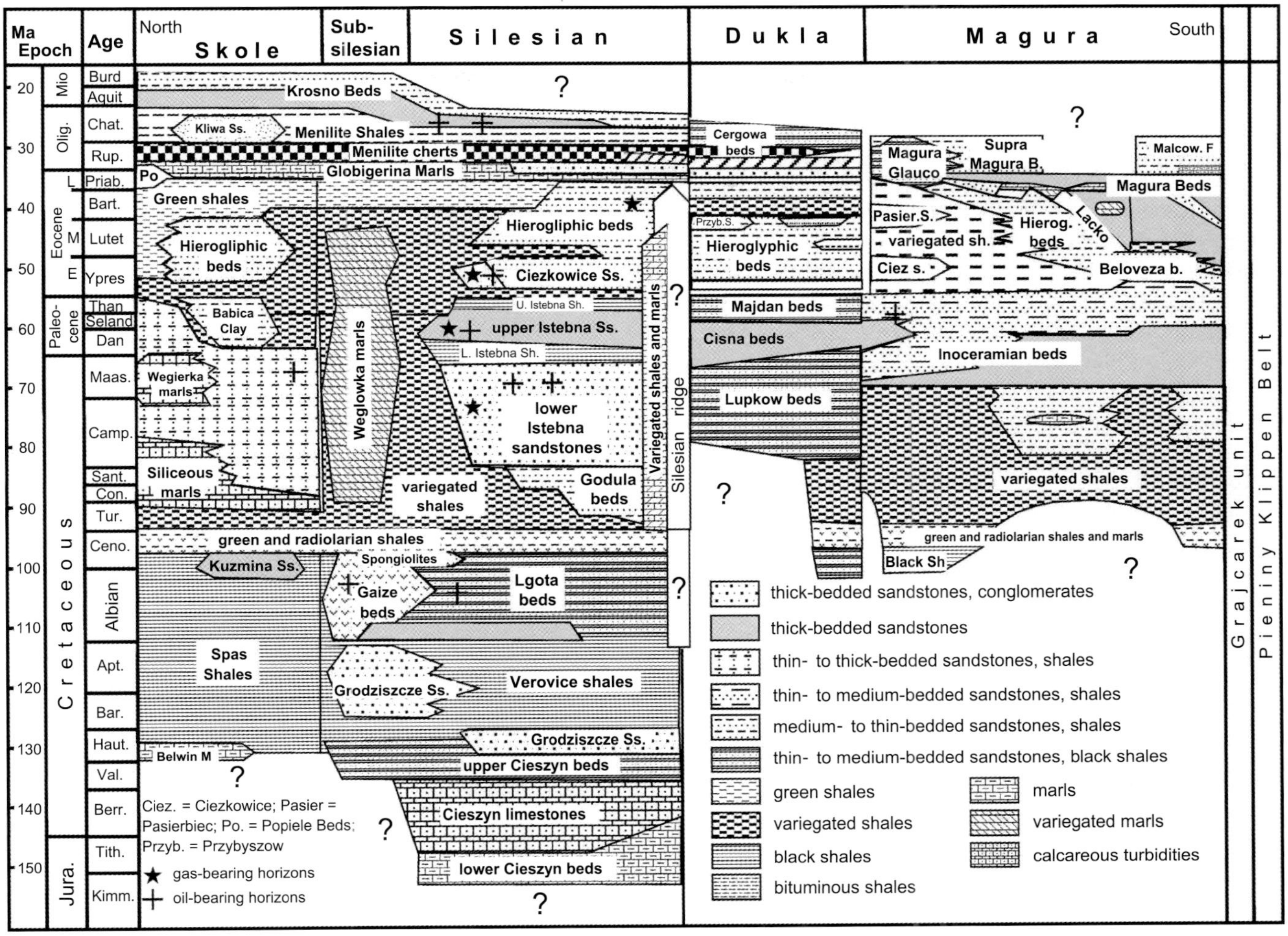

Figure 4. Lithology and chronostratigraphy of the Polish Outer Carpathians.

PENNINIC REALM

The Magura Nappe

The Magura nappe is the largest tectonic unit of the Western Carpathians and is linked with the Rhenodanubian flysch of the Eastern Alps. During the overthrust movements, the Magura nappe has been completely uprooted from its substratum along the ductile Upper Cretaceous rocks. The Tithonian–Barremian limestones and marls from the Kurovice Klippe are regarded as the basal portion of the marginal part of the Magura nappe in southern Moravia. More or less complete sections of this unit are known only from that part of the basin, which was incorporated into the Pieniny Klippen Belt (i.e., the Grajcarek, Birkenmajer, 1977; or Hulina unit, Burtan et al., 1981; Golonka and Sikora, 1981; Golonka and Raczkowski, 1984; see also Birkenmajer and Oszczypko, 1989). This succession is represented by Toarcian–Aalenian black shales, Bathonian–Oxfordian radiolarites, Kimmeridgian limestones (Ammonitico rosso), Tithonian–Barremian cherty limestones, Aptian spotty shales, Albian–Cenomanian black flysch, and Cenomanian green radiolarian shales (Birkenmajer, 1977). A single, Jurassic–Early Cretaceous Klippe of Hulina–Grajcarek type is also known from the Krynica unit (Golonka and Sikora, 1981).

The Albian–Cenomanian spotty shales remain in the southern margin of the Mszana Dolna tectonic window (Birkenmajer and Oszczypko, 1989). More recently, Hauterivian–Albian deposits have been recognized in a few localities in southern Moravia (Svabenicka et al., 1997). On the basis of facial differentiation in the Paleogene deposits, the Magura nappe in Poland, Slovakia, and Czech Republic (Moravia) has been subdivided into four facies-tectonic subunits: the Krynica, Bystrica (Nowy Sacz), Raca, and Siary (Figure 1, see also Koszarski et al., 1974). Additionally, in the west Slovakia segments of the Magura nappe, the Biele Karpaty unit has been distinguished.

The Cretaceous–Paleogene deposits of the Magura nappe may be subdivided into three turbidites complexes (see Oszczypko, 1992): Upper Cretaceous–Paleocene, lower–upper Eocene, and upper Eocene–Oligocene. Each of them begins with pelitic basinal deposits (variegated shales) that pass into thin- and medium-bedded turbidites with intercalations of allodapic limestones and marls, thick-bedded deposits, and finally, into thin-bedded turbidites. The Upper Cretaceous deposits begin with Cenomanian–Turonian, variegated, hemipelagic mudstones, with intercalations of thin-bedded turbidites (Malinowa Shale Formation). The thickness of the formation ranges from 70 m (230 ft) in the Raca subunit (Sikora, 1970) to as much as 200 m (660 ft) in the Krynica subunit (Birkenmajer and Oszczypko, 1989; Oszczypko et al., 1990). In the Raca and Bystrica subunits, sedimentation of variegated clays was finished at the Santonian–Campanian boundary, whereas in the Krynica subunit, it still existed in the Maastrichtian. The Malinowa Formation passes upward into thin-bedded turbidites with sporadic intercalations of thick-bedded sandstones. In the Grajcarek–Hulina unit, variegated shales pass upward into coarse-grained sandstones and exotic conglomerates (Burtan et al., 1984) of the Jarmuta Formation (Maastrichtian–Paleocene; see Birkenmajer and Oszczypko, 1989). This formation, as much as 400 m (1300 ft) thick, passes upward into medium- to thick-bedded calcareous turbidites of the Szczawnica Formation (Paleocene–lower Eocene), whose thickness also reaches 400 m (1300 ft). In the Krynica subunit, the thickness of the Jarmuta Formation reaches at least a few dozen meters (Oszczypko et al., 1990), whereas the thickness of the Szczawnica Formation oscillates about 300–400 m (1000–1300 ft) (Oszczypko et al., 1990, 1999a). In the more northern subunits of the Magura nappe, the Malinowa Formation is followed up in the section by thin- to medium-bedded turbidites of as much as 50 m (160 ft) in thickness, containing sometimes numerous 5–7- to 30-cm (2–3- to 12-in.)-thick intercalations of turbiditic limestones (Kanina beds; see Cieszkowski et al., 1989). These deposits pass upward into thick-bedded sandstones and conglomerates, which are as much as 100–400 m (330–1300 ft) thick (Szczawina sandstones) of the Maastrichtian–Paleocene age. The youngest unit of the complex, 100–200 m (330–660 ft) thick, is composed of medium- to thick-bedded turbidites of Paleocene age (Ropianka beds; see Malata et al., 1996), which are an equivalent of the Szczawnica Formation of the Krynica subunit.

The Ropianka beds are overlain by the lower to middle Eocene variegated, noncalcareous shales of Labowa Formation (Oszczypko, 1991). The thickness of this formation ranges from a few meters in the northern part of the Krynica subunit to as much as 130 m (430 ft) in the Siary subunit (see Sikora, 1970; Oszczypko, 1973, 1991). This formation represents the basinal plane facies, deposited below calcite compensation depth (CCD) (Oszczypko, 1991; Malata, 2001). The variegated shales pass upward into thin-bedded turbidites of the Zarzecze, Beloveza, and Hieroglyphic formations (Oszczypko, 1991; Oszczypko et al., 1999a). The thickness of these deposits varies from 500 to 600 m (1640 to 2000 ft) in the Krynica subunit (Zarzecze Formation), 350 m (1100 ft) in the Bystrica subunit (Beloveza Formation), and 200 m (660 ft) in the Raca subunit (Hieroglyphic beds).

In the Bystrica Subunit, the Beloveza Formation is overlain by thin- to medium-bedded turbidites with intercalations of the Lacko-type marls. In the Krynica, Bystrica, and Raca subunits, the youngest deposits of the Eocene complex belong to the Magura Formation, which is of lower to upper Eocene–Oligocene age. Thickness of this formation reached 1200–1400 m (4000–4600 ft) in the Krynica subunit, 500–2000 m (1600–6600 ft) in the Bystrica, and 1000 m (3300 ft) in the Raca subunit (Birkenmajer and Oszczypko, 1989; Oszczypko, 1999). The Magura Formation is represented by the thick-bedded turbidites and fluxoturbidites. The Magura Formation is locally overlain by the *Globigerina* marls (upper Eocene–lower Oligocene), Menilite shales, and the Malcov Formation (late Oligocene). In the northernmost part of the Magura nappe (Siary subunit), middle–upper Eocene variegated shales pass upward into the marls and thin-bedded flysch of the Zembrzyce beds, which contain a foraminiferal horizon of *Globigerina* marls (upper Eocene–lower Oligocene), thick-bedded glauconitic sandstones (Watkowa Sandstones) of the upper Eocene–lower Oligocene age, and finally, marls with intercalations of glauconitic sandstones (Budzow beds, Oligocene; see Oszczypko-Clowes, 2001). Traditionally, the Oligocene Malcov Formation was regarded as the youngest strata of the Magura nappe (Oszczypko-Clowes, 1998). However, in the vicinity of Nowy Sacz, the early Miocene Zawada Formation, belonging to the Magura nappe, has recently been discovered (Oszczypko et al., 1999b; Oszczypko-Clowes, 2001; Oszczypko and Oszczypko-Clowes, 2002). This formation reaches at least 450 m (1470 ft).

The Biele Karpaty unit was traditionally regarded as an equivalent of the Oravska Magura–Krynica subunit, which is in a similar tectonic position. Recently, Potfaj (1993) arranged the Biele Karpaty unit into Hluk and Vlara successions (groups). The Hluk succession, extended mostly in the southwest part of the Biele Karpaty Mountains, consists of several lithostratigraphic units. The Hluk Formation [Barremian(?)–Aptian], as much as 120 m (400 ft) thick, is represented by dark-gray, black-greenish shales and turbiditic limestones at the top. This formation passes upward into the Gbely (Kaumberg) Formation [Albian(?)–Cenomanian–Turonian], composed of variegated claystones and marls about 200 m (660 ft) thick. Higher up in sequence occurs the Puchov Formation [Coniacian(?)–Campanian–Maastrichtian], represented by 100 m (330 ft) of red and green shaly marls. These marls are interfingered with a 10-m (33-ft)-thick packet of gray-brown limy claystones, marls, and limestones (Antoninek Member, Campanian). The siciliclastic flysch begins with the Svodnice Formation (Paleocene–early Eocene), a 700-m (2300-ft)-thick complex of medium-bedded turbidites with thick intercalations of gray and brown-gray calcareous claystones. The upper part of the formation (Bzova Member, Paleocene–Eocene) contains a 600-m (2000-ft)-thick complex of thick-bedded, fine- to medium-grained sandstones, which resemble the Magura sandstones of the Krynica (Orava) subunit. Toward the north, the Svodnice Formation is replaced by thin-bedded, calcareous turbidites, about 600 m (2000 ft) thick (Nivnice Formation, late Paleocene), with two horizons of variegated shales at the base. This formation is overlain by the Kuzetov Formation (latest Paleocene to early Eocene), which is 250 m (800 ft) thick, as represented by thick horizons of the variegated shales and calcareous claystones and sandstone intercalations.

The Kaumberg (Lopenik) Formation (Campanian–Maastrichtian) belongs to the Vlara Group. It is built of thick sandy flysch with variegated claystones and marls at the base. In the upper part of the formation, two members were distinguished by Potfaj (1993). The Javorina Member (late Campanian–Maastrichtian) consists of thin-bedded turbidites with lenses of microconglomerates. The muscovite sandstones are composed of quartz, metamorphic rocks, and a large amount of carbonate clasts. The thickness of this member is about 600 m (2000 ft). The Drietomica Member (Maastrichtian) is characterized by the domination of thick-bedded, fine- to coarse-grained sandstones and microconglomerates of mixed carbonate-siliciclastic composition. The estimated thickness of this member is 200 m (660 ft). The Rajkovec beds [Paleocene–early Eocene(?)] are predominantly a sandstone-flysch sequence with thin intercalations of claystones and limy mudstones. The sandstones are fine to medium grained and of mixed carbonate-siliciclastic composition. The maximum thickness of the Rajkovec Member is as much as 500 m (1600 ft). The youngest deposits of the Vlara succession belong to the Chabova beds (late Paleocene–early Eocene), developed as thick beds of medium- to coarse-grained siliciclastic sandstones with carbonate clasts.

THE GETIC–MARMAROSH UNITS

The Marmarosh Unit

The Marmarosh unit is located in the Transcarpathian Ukraine next to the Romanian border (Figures 3, 5). The development of this unit differs from the other Outer Carpathian flysch units, and as was mentioned above, the exact position of this unit is disputable; it is placed adjacent to Pieniny Klippen Belt or north to the Magura unit. It consists of two parts: the Marmarosh Klippen zone (internal) and the Marmarosh crystalline massif (external).

The Marmarosh Klippen zone (Figures 3, 5) is situated southwest and west from the Marmarosh Massif. It is divided into two subzones: the external Vezhany subzone, containing olistolithes (Soimul olistostrome), and the internal Monastyrets subzone, built mainly by flysch. The giant blocks (as much as 1 km [0.62 mi]), olistolithes, and olistoplakes (Kruglov, 1965) occur in the marginal part of the Vezhany subzone (Soimul olistostrome). They are built of Paleozoic and Proterozoic schists and gneisses, Paleozoic, Triassic, Jurassic, and Barremian–Aptian dolomites and limestones of age, and Permian–Triassic quarzitic sandstones and conglomerates. The blocks of serpentinites, diabases, gabbro-diabases, pebbles and boulders of granites, quartz porphyries, and granitoids are also numerous there. The olistostromes pass upward into polimiktic conglomerates with a thickness of about 1000 m (3300 ft) covered by dark gray mudstones intercalated by thin sandstones (Soimul beds; Aptian–Albian). Complexes of thick sandstones are developed only locally. The Soimul beds pass upward into red marls (Puchov beds; Turonian–Campanian) and thin-bedded sandstones with green-gray and sparse red shales, correlated by Kruglov (1965, 1969) with Jarmuta beds (Maastrichtian) known from the Pieniny Klippen Belt in Poland and Slovakia. Paleogene sediments start with conglomerates and thick-bedded sandstones (Metovo beds) covered by thick-bedded marls of the Eocene. They pass into Oligocene black marls and shale (Dusina Beds) and thick-bedded gray sandstones. In the inner Monastyrets subzone, the lithostratigraphic profile starts with Paleogene variegated shales intercalated with thin-bedded and sporadically bedded sandstones (Shopurka beds) as much as 1000 m (3300 ft) thick. They are covered by thick-bedded sandstones (Dragov beds). According to Sandulescu et al. (1981), this sequence represents the southeast prolongation of the Magura unit.

The Marmarosh Massif is built by complex, dislocated, mesozonally metamorphosed Riphean–Vendian rocks and by sedimentary, volcanic, and epizonally

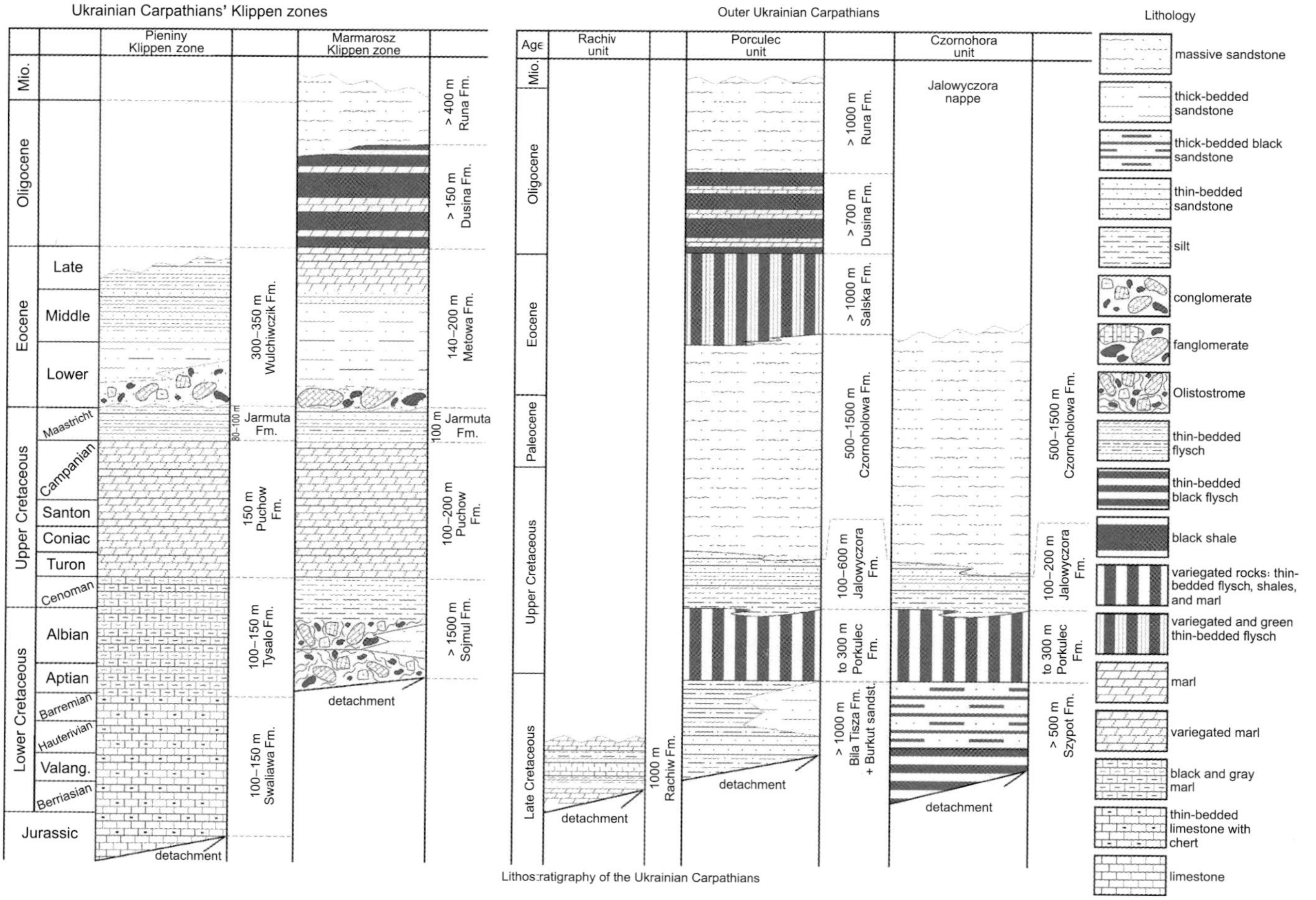

Figure 5. Lithology and chronostratigraphy of the internal part of the Ukrainian Carpathians.

metamorphosed Carboniferous, Triassic, and Jurassic formations. The Cretaceous conglomerates, organogenic limestones, and marls of age discordantly overlie older rocks. According to Romanian geologists, the Marmarosh Massif belongs to the large Getic realm (e.g., Sandulescu, 1988; Balintoni, 1998, 2001; Golonka et al., 2003) described above as the Getic–Marmarosh ridge. The metamorphic basement sequences of this realm are thrust over the flysch sequences of the Outer Dacides (Severinides). The nappe character of the Ukrainian part of the Getic–Marmarosh ridge was described in detail by Kropotkin (1991).

OUTER DACIDES–SEVERINIDES

The Kaminnyi Potik Unit

In the frontal part of the Marmarosh Massif, a tectonic thrust fold (Glushko and Kruglov, 1986) is present that is composed of dark, thin-bedded limestones, black shales, sandstones, and conglomerates (Kaminnyi Potik beds; Tithonian–Valanginian) containing the Upper Jurassic effusives of the basic type. It belongs probably to the basins of the Outer Dacides–Severinides and is the prolongation of the Black Flysch unit of the Romanian Carpathians.

Rachiv Unit

The Rachiv unit is represented mainly by the Valanginian–Hauterivian deposits (Rachiv beds) 1000 m (3300 ft) thick. The lower part is represented by black shales, calcareous thin- and medium-bedded turbiditic sandstones, and limestones passing upward into the complex that contains an intercalation of thick-bedded sandstones and conglomerates. Exotic blocks of Mesozoic limestones and diabases are also present. The whole sequence is terminated by black shales and thin-bedded calcareous sandstones. The uppermost part can also represent the Barremian. The Rachiv unit continues southward into the Cuk digitation of the Ceahlau nappe in the Romanian Carpathians containing basic volcanites.

Porkulets Unit

The oldest, Early Cretaceous deposits of the Porkulets unit (Figures 3, 5) are represented by gray shales, marly shales, thin- to thick-bedded laminated sandstones (Bila Tisza beds; Barremian–Albian, more than 1000 m [3300 ft] thick), and lenses of conglomerates that contain blocks of metamorphic and volcanic rocks (Bogdan and Bronka conglomerates). The Bila Tisza beds are locally replaced by massive, thick-bedded sandstones, more than 500 m (1600 ft) thick (Burkut conglomerates). The Porkulets beds, about 50–300 m (160–1000 ft) thick, are represented by uppermost Albian, Cenomanian, and Turonian red, green, and brown shales and marls intercalated by thin- and medium-bedded sandstones. They pass upward into a series of gray, medium-bedded sandstones, shales, and marls [Jalowychora beds; Coniacian, Santonian, Campanian(?)]. Their thickness increases to as much as 600 m (2000 ft) toward the northwest. In this area, they are known as lower Berezna beds. Higher up, a thick (as much as 1500 m [5000 ft]) complex of thick-bedded sandstones is present, representing the time span from late Campanian up to middle Eocene (Chornogolova beds). The higher part of the middle Eocene and upper Eocene is developed as green-gray and red shales (Skalska beds) with intercalations of Sol beds; as much as 1000 m (3300 ft), whereas the Oligocene is developed as black and yellowish marls and black shales (Dusina beds; as much as 700 m [2300 ft]) locally with intercalations of cherts and sandstones (Turica beds). The higher part of the Oligocene rocks is represented by a complex of thick-bedded sandstones as much as 1000 m (3300 ft) thick (Malavyzna beds). Toward the southeast, the Porkulets unit continues into the Bodoc digitation of the Ceahlau unit in Romania.

Dukla Unit

The oldest, Early Cretaceous sediments are known only from the eastern part of the Dukla unit (Figures 4–6). They are represented by black shales, siliceous mudstones, and sandstones with intercalations of cherts and siderites (lower Shypot beds; Barremian–Aptian) passing upward into a series of siliceous sandstones and black shales (upper Shypot beds; Albian). In the Cenomanian, green and red shales with subordinate thin turbidites started to develop (lower Yalovetz beds). The Senonian sequence commences (Figure 6) with a 900-m (2900-ft) complex of dark shales and medium-bedded, fine-grained, calcareous sandstones (lower Berezna beds in Ukraine). This lithofacies, known in the southern part of the Dukla unit as Lupkow beds (Campanian–Maastrichtian and Campanian–Paleocene), represents the oldest sediment of the Dukla unit in Poland and Slovakia (Lesko and Samuel, 1968; Slaczka, 1971; Korab and Durkovic, 1978). Uppermost Cretaceous and Paleocene deposits in almost the entire unit are represented by thick-bedded calcareous, coarse-grained, feldspathic sandstones and/or fine-grained, micaceous sandstones intercalated with dark clay shales (Cisna beds or upper Berezna beds in Ukraine, mainly Paleocene). This complex disappears toward the west and toward the north (data from Jasliska 2 borehole, southeast from Krosno). In the western part of the unit, between villages of Cisna and Jasliska, black shales and dark siliceous sandstones (Majdan

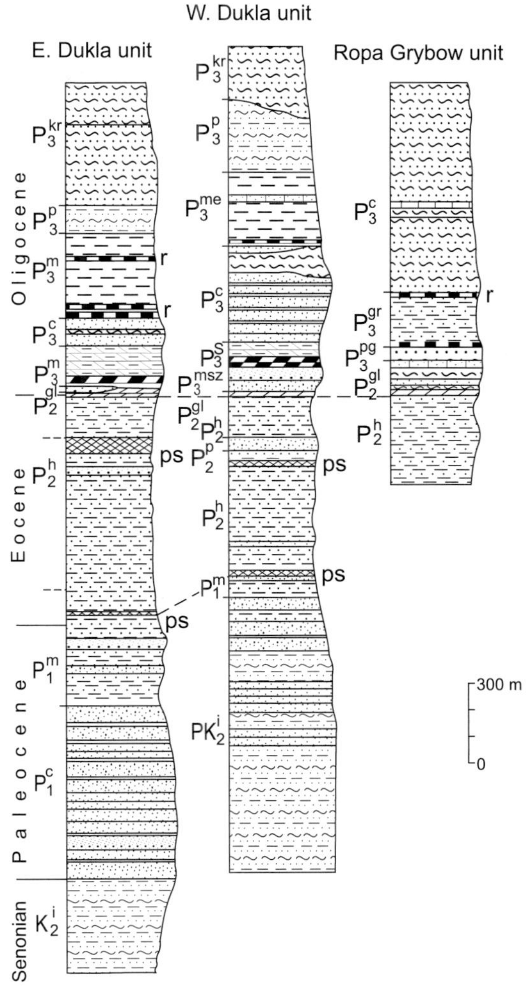

Figure 6. Comparative lithostratigraphic profiles of the Dukla and Grybow units. P_3^{kr} = Krosno beds; P_3^{p} = transition zone between Krosno and Menilite beds; P_3^{me} = Menilite shales; r = cherts; P_3^{c} = Cergowa sandstones; P_3^{s} = Jawornik marls; P_3^{msz} = Mszanka sandstones; P_2^{gl} = *Globigerina* marls; P_2^{h} = Hieroglyphic beds; ps = red and green shales; P_1^{m} = Majdan beds; PK_2^{c} = Inoceramian beds; P_1^{c} = Cisna beds; K_2^{I} = Cisna beds.

beds; Paleocene) cover the Cisna beds. Locally, they contain the intercalation of thick-bedded sandstone complexes (Welka Berezna in Slovakia, Ljuta beds in Ukraine). Sometimes, these thick-bedded sandstones dominate. Starting from the youngest Paleocene, a striking change of sedimentation appears. The Eocene deposits are developed as a complex of green shales, medium to thin bedded, fine and medium grained, as much as 1000 m (3300 ft) thick (Hieroglyphic beds in Polish, sub-Menilite beds in Slovakian, and Strichava beds in the Ukraine parts of the Dukla unit) with intercalations of red shales (Sol and Vyshka beds in Ukraine). Locally, intercalations of thick-bedded sandstone lenses exist (as much as 200 m [660 ft]) (Przybyszow sandstones in Poland and Stavne beds in Ukraine). The upper Eocene Hieroglyphic type of sediments passes upward into the green shales and higher up into the *Globigerina* marls. At the beginning of the Oligocene, dark bituminous sediments started to develop in the Dukla unit as well as in the northern units. These sediments are represented by dark brown shales (Menilite shales) with horizons of cherts. Within these shales, several lithological members have developed. In the lowermost part, a lenticular complex (as much as 250 m [820 ft]) of thick-bedded, coarse-grained, or conglomeratic sandstones (Mszanka sandstones) is present; higher up, the complex-brown siliceous marls (Jawornik marls) and a thick complex (as much as 300 m [1000 ft]) of thick-bedded, medium-grained, calcareous sandstones intercalated by gray calcareous shales are present (Cergowa sandstones; Slaczka and Unrug, 1976). Toward the south, the sandstone complex is replaced by gray marly shales and thin-bedded, calcareous, laminated sandstones, calcareous shales, and subsequently replaced by a complex composed predominantly of shales. The dark Menilite shales pass gradually upward into a 1000-m (3300-ft) series of calcareous, thin-bedded, fine-grained sandstones and gray calcareous shales (Krosno beds; Oligocene), which terminate the sequence of the Dukla unit.

Fore-Magura Group of Units

In the Western Carpathians, north from the Magura unit, several units are characterized by the occurrence of the Upper Cretaceous–Paleocene sediments similar to those of the Magura unit and the Oligocene deposits similar to deposits from the Silesian unit. From the west, these are the Fore-Magura sensu stricto, Obidowa–Slopnice, Jaslo, and Grybow units. The relation between these units is not clear, but it is supposed that the Grybow unit was located in the more internal position than the Dukla unit or represents a prolongation of the southern part of the Dukla unit.

Fore-Magura Unit Sensu Stricto

This unit is known only from the western part of the Polish Carpathians in front of the Magura unit (Burtan, 1968; Golonka, 1981). The sequence commences with the Upper Cretaceous thick-bedded sandstones, which contain intercalations of green and gray shales. The lower and middle Eocene is developed as argillaceous variegated shales, locally with variegated marls, and the upper Eocene is developed as green shales. The Oligocene deposits are characterized by the occurrence of thick-bedded, gray, micaceous, and calcareous sandstones passing upward into gray, marly shales with subordinated thin, laminated micaceous sandstones (Krosno beds).

Obidowa–Slopnice Unit

This unit (Figure 7) is situated below the Magura nappe and is recognized only from the boreholes (Slopnice and Obidowa areas) (Cieszkowski et al., 1985). It shows a resemblance to the Dukla and Zboj units and probably represents its western equivalent. The oldest sediments of the Obidowa–Slopnice unit represent the Upper Cretaceous and are developed as thick-bedded, coarse-grained sandstones (1000 m [3300 ft] thick) showing similarity to a part of the Inoceramian beds in the Dukla unit. They pass upward into a complex with pebbly mudstones (Obidowa beds), covered again by thick-bedded sandstones (70 m [230 ft] thick) that belong to the Paleocene. The Eocene is developed mainly as green-gray shales and thin sandstones similar to those in the other Fore-Magura units. Intercalations of black shales, thick-bedded conglomerates, and sandstones reaching 700 m (2300 ft) (Rdzawka beds) occur in the upper Eocene deposits. The Eocene sequence is terminated by dark, black, calcareous mudstones and micaceous sandstones that show similarity to some

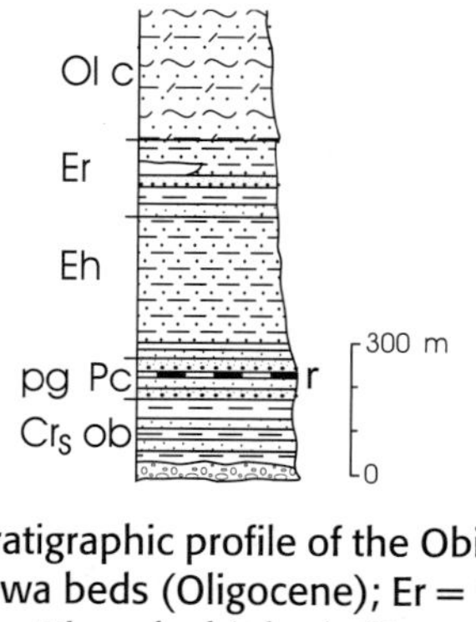

Figure 7. Lithostratigraphic profile of the Obidowa–Slopnice unit. Olc = Cergowa beds (Oligocene); Er = Rdzawka beds (late Eocene); Eh = Hieroglyphic beds (Eocene); pg Pc = thick-bedded sandstones (Paleocene); Cr$_s$ob = Obidowa beds (Late Cretaceous); r = cherts.

Oligocene deposits from the Dukla unit. This unit is correlated with the sediments found in Zboj 1 borehole near the Slovakian–Ukrainian border. That suggests a rather great extension of the Obidowa–Slopnice lithofacies and its position in front of the Dukla–Grybow units.

Grybow Unit

The Grybow unit (Figure 6) represents a western prolongation of the Dukla unit and is only known from the tectonic windows in the Magura nappe (Kozikowski, 1956; Sikora, 1970) and also from the boreholes. The sequence commences with the Eocene sediments that are developed as green, gray, and black shales, intercalated by thin- and medium-bedded sandstones (Hieroglyphic beds). The Eocene sequence terminates the *Globigerina* marls in a similar way as in other external units of the Outer Carpathians. The Oligocene is developed as gray marls with sporadic intercalations of sandstones (sub-Grybow marls, about 150 m [500 ft] thick). They are covered by black, hard shales with horizons of black hornstones at the top (Grybow shales, about 200 m [660 ft] thick). The youngest sediments are represented by gray, calcareous, thin- and medium-bedded sandstones with gray calcareous shales that prevail toward the top (Cergowa beds; as much as 400 m [1300 ft] thick). Toward the east, the Grybow unit passes probably into the inner part of the Dukla unit, and its prolongation can be the Porkulets unit. Both of them show similarity in the development of the Oligocene sediments.

Jaslo Unit

This unit, lately delimited in the Harklowa peninsula of the Magura nappe (Koszarski and Koszarski, 1985), is represented by Eocene variegated and green-brown shales covered by greenish gray marly shales with intercalations of sandstones (Dulabka beds; Eocene–Oligocene). The higher part of the Oligocene is developed as brownish, commonly siliceous shales followed by grayish marly shales and sandstones (Krosno beds). The occurrence of olistostromes is very typical, and some geologists (L. Jankowski, 2003, personal communication) state that all deposits of the Jaslo unit represent olistostromes in front of the Magura nappe.

Zboj Unit

The Zboj unit was found in borehole Zboj 1 at the depth of 3800–5002 m (12,500–16,410 ft) in Eastern Slovakia near the Ukrainian border below the Dukla unit (Korab and Durkovic, 1978). Mainly massive, homogenous, arkosic sandstones that are strongly siliceous represent this unit. Subordinately, complexes of greenish shales with thin-bedded sandstones and intercalations of mudstones with slump structures are present. These deposits, called Zboj beds, represent the upper Eocene and probably also the Oligocene. The possibility cannot be excluded that they represent the innermost part of the Silesian unit.

Chornohora Unit

The Chornohora unit (Figures 3, 5) is situated in the southeast part of the Ukrainian Carpathians and disappears toward the northwest. It can be correlated with the Audia zone in the Romanian Carpathians. Its situation outside the Dukla unit and the development of the Lower Cretaceous imply that the Chornohora unit can represent the prolongation of the inner part of the Silesian unit (Slaczka, 1959). However, according to one of the authors of this chapter (Kruglov), it can also form a prolongation of the Dukla unit. The Chornohora unit is divided in two subunits: the Skupiv subzone in the north and the Yalovychor, or the Hoverla, subzone in the south. This division is based on some differences in the development of the Late Cretaceous sediments. The oldest sediments (Barremian–Aptian) are represented by black, calcareous shales with siderites, fucoid marls, and thin-bedded sandstones (lower Shipot beds; as much as 300 m [1000 ft]) passing upward into series to as much as 200 m (660 ft) of dark, quartzitic sandstones with black shales (upper Shipot beds of Albian age). The Cenomanian–Turonian is developed, similarly to other Carpathian areas, as green and red shales, with thin sandstones and red marls (Porkulets beds; as much as 200 m [660 ft]). In the lower part of this sequence, intercalations of black shales, radiolarites, and siderites are present. The variegated sediments pass upward into gray shales, marls, and sandstones of Coniacian–Santonian age (Yalovychor beds; as much as 400 m [1300 ft]). The late Senonian is characterized in the Skupiv subzone by medium-bedded, gray sandstones and shales (Skupiv beds; as much as 500 m [1600 ft]) and in the Yalovychor–Hoverla subzone by a complex as much as 1000 m (3300 ft) of very thick- to thin-bedded sandstones and gray shales (Chornohora beds) that terminate the whole sequence of this subzone. The younger sediments are known only from the Skupiv subzone. Paleocene–early Eocene thin- to thick-bedded sandstones, limestones, and gray-green and red shales (Hniletz beds, about 400 m [1300 ft] thick) are present, covered by middle Eocene massive micaceous sandstones (Topilchany beds; as much as 250 m [800 ft] thick) and late Eocene thin- to medium-bedded sandstones and variegated shales (Probijeny beds; as much as 350 m [1150 ft]). The sedimentary sequence terminates by the Oligocene dark brown shales (Menilite beds). Recently, olistostrome formations of insignificant thickness were discovered in front of the Chornohora nappe. They are being compared with the Miocene facies of Slon in the Vinecisu strata in the Romanian Carpathians. This formation is characterized by the wide development of variegated, red rocks.

Silesian Unit

The oldest sediments of the Silesian (Figures 4, 8) are known only in the Moravia and Silesia areas in the Western Carpathians (see Picha et al., 2006). They are represented by the lower Tithonian dark gray, calcareous mudstones (lower Cieszyn shales) that began the euxinic cycle that lasted without major interruption until the Albian. Some of these mudstones represent slump deposits. The mudstones pass upward into turbiditic limestones and marls (Cieszyn limestones; Tithonian–Berriasian; as much as 200 m [660 ft] thick). The material of the detritic limestones was derived from the adjacent shallow-water, calcareous platforms, mainly from the north. The younger, Valanginian–Hauterivian dark gray, black calcareous shales with intercalations of dark, thin- and medium-bedded calcareous sandstones (upper Cieszyn beds; as much as 300 m [1000 ft] thick) are known from the whole Silesian unit. They generally pass into Barremian–Aptian black shales (Verovice shales; as much as 300 m [1000 ft]). During the Hauterivian and Barremian, several complexes (as much as 200 m [660 ft] thick) of sandstones and conglomerates have developed (Grodziszcze sandstones). During the late Tithonian and Early Cretaceous, the opening of the western part of the Silesian basin alkaline magma (teschenites association rocks) intruded the flysch deposits (Lucinska-Anczkiewicz et al., 2000). The black shale turbiditic sedimentation (Lgota beds; as much as 450 m [1500 ft]) developed in the Silesian basin in the beginning of the Albian and lasted during the whole Albian. It started with thick-bedded, coarse-grained sandstones and conglomerates that passed upward into thin- and medium-bedded, quarzitic sandstones intercalated by black, greenish shales. Locally, especially in the inner part of the Silesian basin, the Lgota beds are entirely developed as thin-bedded facies. North from Sanok, Lgota beds are replaced by fine-grained, light gray sandstones with sponge needles (Gaize beds, 150 m [500 ft] thick). At the beginning of the Cenomanian, slow and uniform sedimentation occurred, and green shales, with an intercalation of black shales and radiolarites, developed. Similar facies also developed in other parts of the Outer Carpathian flysch basin. During the Cenomanian–Turonian, the green shales and radiolarites passed upward to red and variegated shales. During the early part of the Late Cretaceous, a supply of clastic material commenced. The uppermost Turonian and the lower Senonian are represented by a very thick flysch series of Godula beds (approximately 2000 m [6600 ft] thick) developed as thick- and thin-bedded sandstones intercalated by green shales. The Godula beds occur mainly in the inner zone of the western and central parts of the Silesian unit. Toward the north and east, they are replaced laterally by variegated shales (Godula shales). The Godula beds are covered by another sandy complex, the Istebna beds (upper Senonian–Paleocene), which reaches 1700 m (5600 ft) locally. It is represented in the lower part by noncalcareous, coarse-grained, thick-bedded fluxoturbidites and conglomerates with subordinate intercalations of gray shales. The upper part is developed as a complex of slightly calcareous, thick-bedded, and coarse-grained conglomerates and sandstones covered by dark gray shales reaching 200 m (660 ft). These lithofacies are known from Moravia up to the Bieszczady Mountains in the east, where they were found in deep boreholes. The Paleocene–Eocene deposits of the Silesian unit are characterized by the occurrence, in the lower part, of a thick complex that attains 500 m (1600 ft) of thick-bedded fluxoturbidites and conglomerates in the form of elongated lenses more than 200 m (660 ft) thick (Ciezkowice sandstones) with a thin intercalation of red shales. The middle and upper Eocene consist principally of alternating green gray shales and thin-bedded sandstones (Hieroglyphic beds) that pass upward into green shales and *Globigerina* marls. Only in the southeast part of the Silesian unit is the upper Eocene developed as thick-bedded sandstones (Mszanka beds). Near the Roznow Lake, the Eocene deposits are more sandy and could be regarded as a separate unit, the Michalczowa unit (Cieszkowski, 1992).

During the early Oligocene, the sedimentation of brown, generally bituminous shales started that display horizons of cherts (Menilite shales, 200–400 m [660–1300 ft]). Locally in the southern area, the shales are replaced by thick-bedded sandstones (Magdalena sandstones) or partly replaced by thick-bedded calcareous sandstones (Cergowa sandstones). The Menilitic shales pass upward into the Krosno beds of the Oligocene–early Miocene age. They are developed in the lower part as thick- and medium-bedded, calcareous sandstones that pass upward into a complex of medium- to thin-bedded sandstones alternating with gray marly shales and covered by a complex of gray marly shales with subordinate thin-bedded sandstones. These lithofacies are diachronous across the Silesian unit (Jucha and Kotlarczyk, 1961; Koszarski and Zytko, 1961); in the northeastern part of this unit,

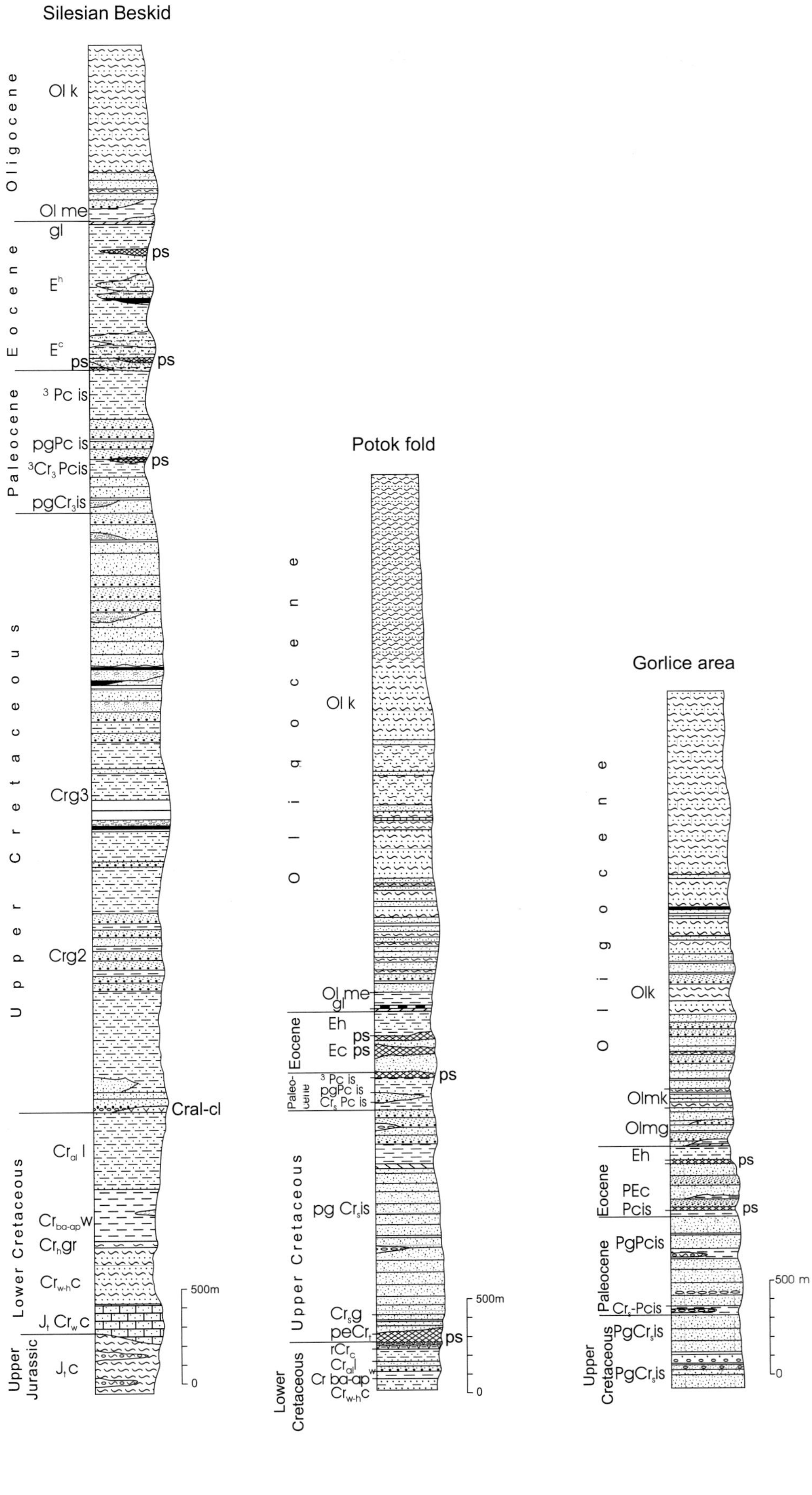

Figure 8. Comparative lithostratigraphic profiles of the Silesian unit from west to east. Olk = Krosno beds; Olme = Menilite beds; Olmg = Magdalena sandstones; gl = *Globigerina* marls; Eh = Hieroglyphic beds; Ec = Ciezkowice sandstones; ps = variegated shales; lPc is = upper Istebna shales; pgPcis = upper Istebna sandstones; lCr_3Pcis = lower Istebna shales; $pgCr_2is$ = lower Istebna sandstones; Cr_2g = Godula beds; Crg3 = upper Godula beds; Crg2 = middle Godula beds; Crg1-cl = lower Godula beds and white marls; $peCr_2$ = red shales (Godula shales); rCr = cherts; $Cr_{al.}l$ = Lgota beds; $Crba\text{-}ap^{w}$ = Verovice shales; $Cr_{h}gr$ = Grodziszcze sandstones; $Cr_{wh}c$ = upper Cieszyn shales; $J_3Cr_{w}c$ = Cieszyn limestones; J_3c = lower Cieszyn shales.

they also represent the upper Miocene. However, in the inner part of the Silesian unit, in front of the Magura nappe, the Krosno beds also pass into the upper Miocene deposits that contain olistolithes derived from the frontal part of the Magura nappe (Slaczka and Oszczypko, 1987). In addition, in the southeastern part of Poland, in the Bieszczady Mountains in the lower part of the Krosno beds, horizons of olistostromes occur with blocks of metamorphic rocks and limestones, and near Sanok town, horizons of olistostromes occur with blocks of granites.

Toward the east, near the Polish–Ukrainian border, as an effect of plunging of the Silesian unit, the deposits that are older than the Oligocene almost completely disappear from the surface, and the unit passed in the Ukrainian Carpathians into the internal part of the Krosno zone (the Turka subzone that wedges out near the Black Tisza). In addition, a change in lithofacies exists. The Late Cretaceous (Coniacian–Maastrichtian) sediments are represented by a 200-m (660-ft) sequence of dark, medium-bedded quartzitic sandstones and dark shales that pass into the complex of dark medium- to thick-bedded sandstones and dark greenish shales (Soimy beds; Paleocene–middle Eocene; 400–500 m [1300–1600 ft]). The latter lithofacies of the Krosno zone are similar to those known from the inner part of the Silesian unit from the wells Wetlina Geological Institute (IG 1) and Wetlina IG 2 near the village of Wetlina in the southeastern Polish Carpathians. The late Eocene and Oligocene–early Miocene deposits (as much as 2000 m [6600 ft]) are similar to those of the Silesian and Skole–Skyba units.

Andrychow Ridge

This unit is represented by several huge blocks on the boundary between the Silesian and Subsilesian units, near the town of Andrychow. Probably, they are remnants of a calcareous platform that was situated between Silesian and Subsilesian sedimentary areas or represented a part of Subsilesian substratum. The composition of the klippes differs from the adjacent units, although the Upper Cretaceous sediments show a similarity to the sequences of the Subsilesian unit. The nonflysch, calcareous facies are very characteristic for the Andrychow ridge sequences (Ksiazkiewicz, 1951; Gasinski, 1998). The basement of the ridge was made up of granite-gneiss or mylonitized rocks. The sedimentary sequences are represented by crinoidal and shallow-water limestones of Late Jurassic age covered by transgressive early Campanian conglomerates and marls, limestones, and shaly marls of Campanian and Maastrichtian age. Paleocene and early Eocene are represented by organogenic limestones and shales. The more basinal or slope facies are represented by Maiolica-type Late Jurassic–Early Cretaceous cherty limestones (Olszewska and Wieczorek, 2001).

Subsilesian Unit

During the Early Cretaceous and the lower part of the Late Cretaceous, the sedimentation in the Subsilesian realm (Figures 3, 9) had a basinal character, very similar to the Silesian basin, and both basins were connected. The Subsilesian realm became generally uplifted later on, and on the slope of this uplift, marly lithofacies developed. The oldest known deposits (Barremian–Aptian) of this unit are euxinic black shales (Verovice shales), with intercalations of complexes of thick-bedded, coarse-grained, and conglomeratic sandstones (Grodziszcze sandstones; as much as 200 m [660 ft]). These sediments show similarity to the deposits of the same age in Silesian subbasin. During the Albian, turbiditic sedimentation developed which displayed a thinning-upward sedimentary sequence. The lower part of the sequence is represented by a complex of thick-bedded, fine- to coarse-grained sandstones more than 150 m (500 ft) thick, with intercalations of black shales (lower Lgota beds). These sandstones are overlain by a 400-m (1300-ft)-thick complex of fine-grained sandstones varying in thickness. A part of these beds represent typical gaizes, with numerous spicules (Gaize beds), in places with lenses of spongiolites. At the beginning of the Cenomanian, green radiolarian shales and radiolarites developed, similar to the adjacent basins. They are covered by variegated shales of the Cenomanian–Turonian age that pass upward into a thick complex (about 700 m [2300 ft]) of red and green marls (Weglowka marls) that are Senonian to middle Eocene (Mitura and Birecki, 1966). In the western part of the Subsilesian unit, intercalations of sandy and conglomeratic complexes of the upper Senonian and/or Paleocene are present. During the late Senonian, gray marls (Frydek marls), commonly with exotic rocks (Burtan et al., 1984), developed in this area. The marly complex passed upward into variegated shales and/or a series of shales and thin-bedded sandstones terminated by green shales and *Globigerina* marls representing the uppermost part of the Eocene. The Oligocene begins, as in adjacent areas, with brown, bituminous shales (Menilite beds) that grade upward into a complex of thick- and medium-bedded, calcareous sandstones and marly shales (Krosno beds). Locally, near the Zegocina village, gray mudstones similar to the Krosno beds of the early Miocene have been found (W. Gorczyk, 2003, personal communication).

Toward the east, the Subsilesian, Late Cretaceous variegated marls extend probably to the Ukrainian Carpathians. A narrow strip of marly deposits at the

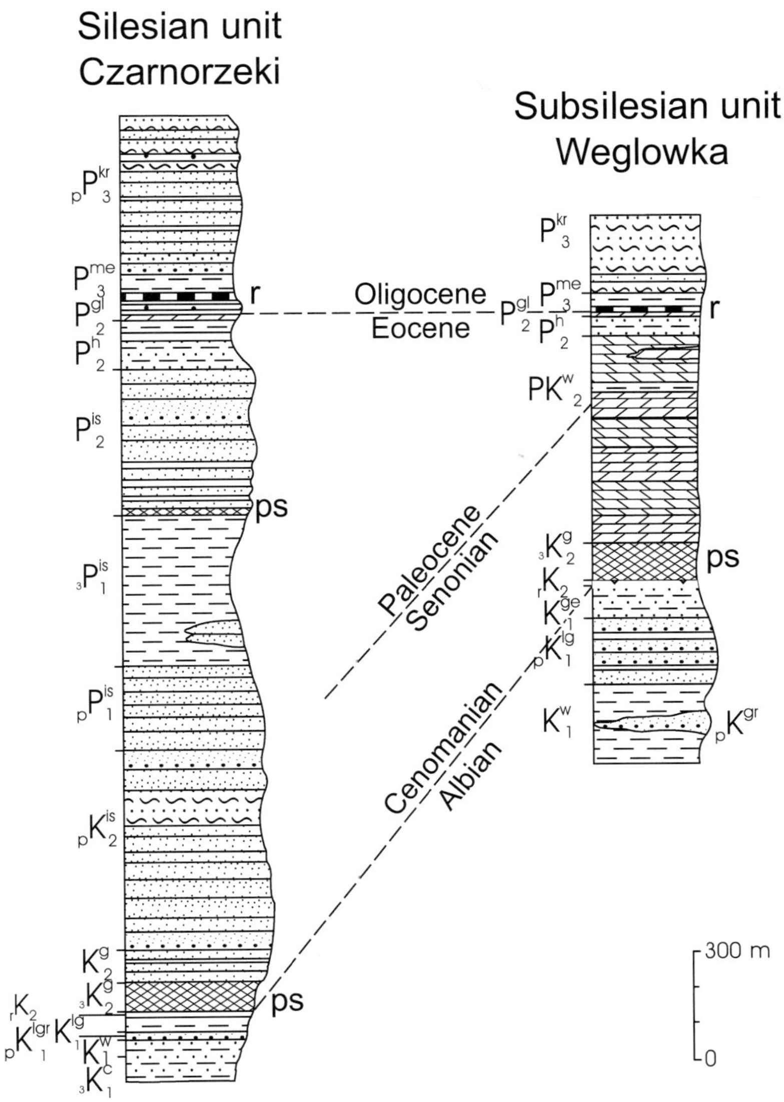

Figure 9. Comparative lithostratigraphic profiles of the Silesian and Subsilesian units, north from the Krosno town. $_pP^{kr}{}_3$ = Krosno beds, sandstones; $P^{kr}{}_3$ = Krosno beds; P_3^{me} = Menilite shales; r = cherts; P_2^{gl} = *Globigerina* marls; P_2^{h} = Hieroglyphic beds; PK_2^{w} = Weglowka marls; ps = red and green shales; P_2^{c} = Ciezkowice sandstones; $_lP_1^{is}$ = upper Istebna shales; $_pP_1^{is}$ = upper Istebna sandstones; $_pK_2^{is}$ = lower Istebna beds; K_2^{g} = Godula beds; $_rK_2$ = radiolarian cherts; K_1^{lg} = Lgota beds; K_1^{ge} = Gaize beds; $_pK_1^{gr}$ = Grodziszcze sandstones; K_1^{w} = Verovice shales; K_1^{c} = Cieszyn beds.

Rozluch scale can represent a prolongation of the Weglowka marls.

To the west, the Subsilesian sediments and especially red marls continue to the Helvetic zone of the Alpine foreland. In the Moravian Carpathians, the Subsilesian lithofacies are partly replaced by sediments of the Zdanice unit (Picha et al., 2006).

Zdanice Unit

For a description of this unit, see Picha et al. (2006).

SKOLE-SKYBA UNIT AND BORYSLAV-POKUTTYA UNITS

The Skole–Skyba units and more external Boryslav–Pokuttya unit (Figures 4, 5, 10) are described together because they developed in the same basin and show generally similar sequences of sediments. The differentiation of both units is based mainly on their tectonic position.

The oldest sediments of the Skole unit (Figures 4, 5) are represented by anoxic black shales, thin, laminated siltstones with layers and lenses of siderites (Spas shales; Barremian–Albian) (Kotlarczyk, 1985; Kruglov, 2001). The subordinate intercalations of fine-grained, thin-bedded, quartzitic, and sparse calcareous sandstones also occur here. The Kuzmina borehole, located southwest from Przemysl, encountered a complex of thick-bedded sandstones several tens of meters thick in the upper part of the Spas shales (Kuzmina sandstones, Borys and Cisek, 1989). At the boundary between the Albian and Cenomanian, a thin layer of green radiolaritic shales developed, which passed upward to the red shales and whitish, siliceous turbiditic marls (Holownia marls). Above them, a thick complex of siliciclastic

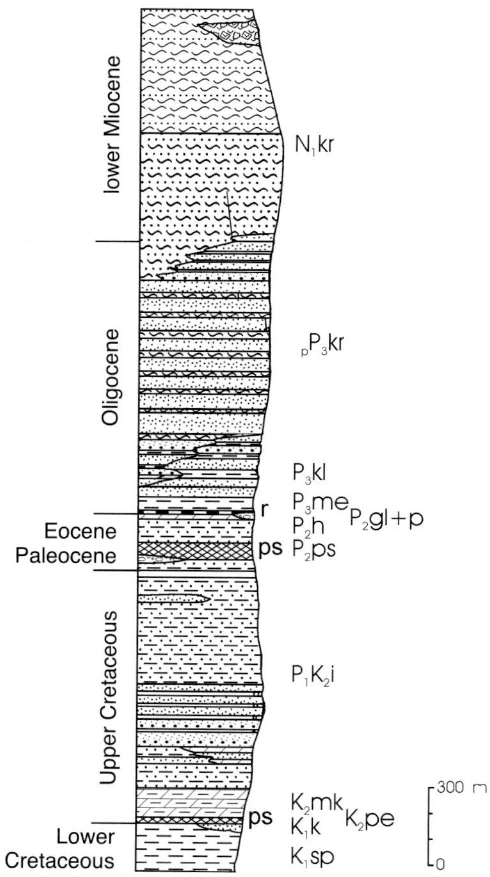

Figure 10. Lithostratigraphic profile of the Skole unit. N_1kr = Krosno beds; $_pP_3kr$ = sandstones, Krosno beds; P_3kl = Kliwa sandstones; P_3me = Menilite shales; r = cherts; P_2gl + p = *Globigerina* marls and Popiele beds; P_2h = Hieroglyphic beds; P_2ps = variegated shales; P_1K_2 i = Inoceramian beds; K_2mk = Holownia marls; ps = red and green shales, radiolarian cherts; K_1k = Kuzmina sandstones; K_1sp = Spas shales.

turbidites is developed (Inoceramian beds = Ropianka Formation [Kotlarczyk, 1978] in Poland, Stryj beds in Ukraine; Turonian–Paleocene; more than 1500 m [5000 ft] thick). It is represented by thin-bedded, calcareous sandstones alternating with gray shales in the lower part. The middle part consists of thick-bedded sandstones, and the upper part consists of thin- and medium-bedded, calcareous sandstones and gray shales. Locally, in the western part of the Skole–Skyba unit, in the uppermost part (Paleocene) of the sequence, lenses of black pebbly mudstones (Babica Clays) are present. Above the turbiditic sequence, a horizon of red and green shales appears (Jaremche horizon; Paleocene). The higher part of the Paleogene displays distinct lateral changes of lithofacies. In the Polish and inner part of the Ukrainian area above the red horizon, developed green and gray clayey shales with intercalations of thin layers of sandstones are present (Hieroglyphic beds in Poland, Witwica beds in Ukraine; late Paleocene–middle Eocene about 150 m [500 ft]). In the outer part of the Ukrainian part of Skyba and in the Boryslav–Pokuttya units, the red horizon is covered by thick-bedded fluxoturbidites and conglomerates (Jamna beds; late Paleocene; as much as 350 m [1150 ft]). These sediments pass upward into thin-bedded turbidites and shales (Maniava beds; early Eocene; as much as 350 m [1150 ft]) covered by a complex of thick-bedded turbidites (Wyhoda beds; middle Eocene; as much as 300 m [1000 ft]) and locally by a series of light limestones and calcareous turbidites (Pasieczna beds; middle Eocene; as much as 200 m [660 ft]). The sedimentation of green shales with subordinate thin sandstones (Bystryci Formation) with local intercalations of gray marls and slump deposits with exotic material (Popiele beds) prevailed during the late Eocene. Marls with *Globigerina* sp. (*Globigerina* marls; late Eocene) terminate these sequences.

The Oligocene sequence begins with dark brown, bituminous shales (Menilite beds) with cherts in its lower part and lenses of thick, coarse-grained, fluxoturbiditic sandstones (Kliwa sandstones) as much as 250 m (800 ft) thick. The Menilite beds pass gradually upward through a transitional zone into a complex of thick-bedded, medium-grained, calcareous sandstones (lower Krosno beds, as much as 800 m [2600 ft] thick). They are overlain by medium- and thin-bedded, calcareous sandstones and gray marly shales. Toward the top of the sequence, sandstones gradually disappear, and the upper part of the Krosno beds is represented mainly by marly shales (upper Krosno Beds; early Miocene). The lithological boundary is diachronous. It migrates upward from the inner to the outer part of the Skole–Skyba unit. The thickness of the Krosno beds exceeds 2400 m (7900 ft). In the outer area of the Ukrainian part of the Skyba unit and in the Boryslav-Pokuttya unit, the dark brown shales are divided by gray marls and thin sandstones (Lopianetz beds; Rupelian–Eggerian; as much as 200 m [660 ft]) into lower and upper Menilite beds. They are covered by gray marly shales and sandstones (Polianycia Formation; Eggenburgian, more than 600 m [2000 ft]). In the Boryslav-Pokuttya unit above the Polanytsia beds, marls and shales (Vorotyshcha beds) are present with evaporates (gypsum and salt) and with lenses of conglomerates (Sloboda conglomerate).

OUTER CARPATHIAN TECTONICS

During the late Oligocene and Miocene orogenesis, several nappes corresponding to the lithostratigraphic units were formed with prevalent northern direction of thrusting in the Western Carpathians, northeastern and eastern in the Eastern Carpathians (e.g., Koszarski et al., 1974; Ksiazkiewicz, 1977) (Figures 1–3, 11–13).

In the Western Carpathians, from the south, they are

1) Magura nappe
2) Fore-Magura Group of nappes that consist of Fore-Magura unit, namely, the Obidowa–Slopnice unit, Grybow unit, and Jaslo unit
3) Dukla nappe
4) Silesian nappe
5) Subsilesian nappe
6) Skole–Skyba nappe

In the Ukraine Carpathians, they are

1) Magura nappe
2) Marmarosh unit
3) Rachiv nappe
4) Porkulets nappe
5) Dukla nappe
6) Chornohora nappe
7) Silesian–Krosno zone
8) Skyba unit
9) Boryslav–Pokuttya nappe

The Magura Nappe

The Magura nappe is the largest tectonic unit of the Outer Western Carpathians (Figure 1) linked with Rhenodanubian flysch of the Eastern Alps (Schnabel, 1992). This unit runs as an arc from the Wienerwald in Austria through Moravia, Slovakian and Polish Beskids, and Eastern Slovakia. It narrows and disappears east of Uzhorod (Transcarpathian Ukraine and adjacent Romania) beneath the Miocene volcanic rocks (Ksiazkiewicz, 1977).

At the east of Latorica River, the prolongation of the Magura nappe is uncertain and speculative. The Monastyrets–Petrova unit of the Marmarosh Klippen zone, situated between the Shopurka and Latorica Rivers, could be regarded as the prolongation of the Magura nappe in the Eastern Carpathians (Sandulescu et al., 1981; Sandulescu, 1988; Bombita and Pop, 1991; Bombita et al., 1992; Oszczypko, 1992; Zytko, 1999; Airoldi, 2001). The eastern termination of this unit is known from the Poiana Botizei area in the Romanian Carpathians.

During the overthrust movements, the Magura nappe has been completely uprooted from its substratum mainly along ductile Upper Cretaceous rocks (Birkenmajer 1986; Oszczypko, 1992). On the base of the facial differentiation of the Paleogene deposits, the Magura nappe has been subdivided into four facies-tectonic subunits: the Krynica, Bystrica, Raca, and Siary (see Koszarski et al., 1974). The Krynica and Bystrica subunits form regional thrust sheets. These separate thrust sheets are especially visible in the western part of the Magura nappe in Slovakia. Slovak geologists use the term "Magura Group of Nappes" (e.g., Kovac and Plasienka, 2002). These thrust sheets are also well visible in the adjacent part of Poland in Beskidy west of Babia Gora (Golonka, 1981; see also Zytko et al., 1989). Eastward in the very well-recognized Polish sector of the Magura nappe, the independent subsidiary nappes are hard to distinguish (Ksiazkiewicz, 1977). In the western prolongation of the Magura nappe (Slovak and Czech Republics), the position of the Krynica zone is occupied by the Bile Karpaty zone (Potfaj, 1993). The Magura nappe is separated from the Pieniny Klippen Belt in Poland and the adjacent part of Slovakia by a steeply dipping southward strike-slip boundary (Birkenmajer, 1986; Plasienka, 1999b). Westward, this boundary has a mixture of strike-slip and thrust character (Picha, 1996; Kovac and Plasienka, 2002). The Magura nappe is flatly thrust over its foreland, and made up of the Fore-Magura group of units and partly by the Silesian unit. The amplitude of the overthrust is at least 50 km (30 mi), and the post-middle Badenian thrust displacement is more than 12 km (7 mi) (Oszczypko and Zuchiewicz, 2000; Oszczypko, 2001). The northern limit of the nappe has an erosional character. The subthrust morphology of the Magura foreland is very distinctive. The shape of the northern limit of the Magura nappe and the distribution of the tectonic windows inside the nappe are connected with the denivellation of the Magura basement. As a rule, the embayments of the marginal thrust are related to transverse bulges in the Magura basement, whereas the peninsulas are located in the depressions of the basement (Oszczypko, 2001). The zone of the tectonic windows, which are connected with the uplifted Fore-Magura basement, is located at a distance of 10–15 km (6–9 mi) south from the northern limit of the unit (e.g., Mszana Dolna, Szczawa, Kleczany, Ropa, Ujscie Gorlickie, and Swiatkowa tectonic windows). The Obidowa unit occurs in the tectonic windows, which is regarded as the western prolongation of the Dukla unit (Cieszkowski et al., 1985) and the Grybow unit, known also as the Kleczany–Pisarzowa unit (Kozikowski, 1958). These units are composed predominantly of the late Eocene–Oligocene deposits and sometimes of the Upper Cretaceous–Paleocene deposits. The Mszana Dolna tectonic window is the biggest. It is situated in the central part of the Polish Carpathians. In the Mszana Dolna tectonic window, the Grybow unit is made up of several, south-dipping thrust sheets, composed of hinterland-dipping duplexes (see also Mastella, 1988). This duplex structure had developed during the middle Miocene thrusting, between the Magura and Dukla units, which

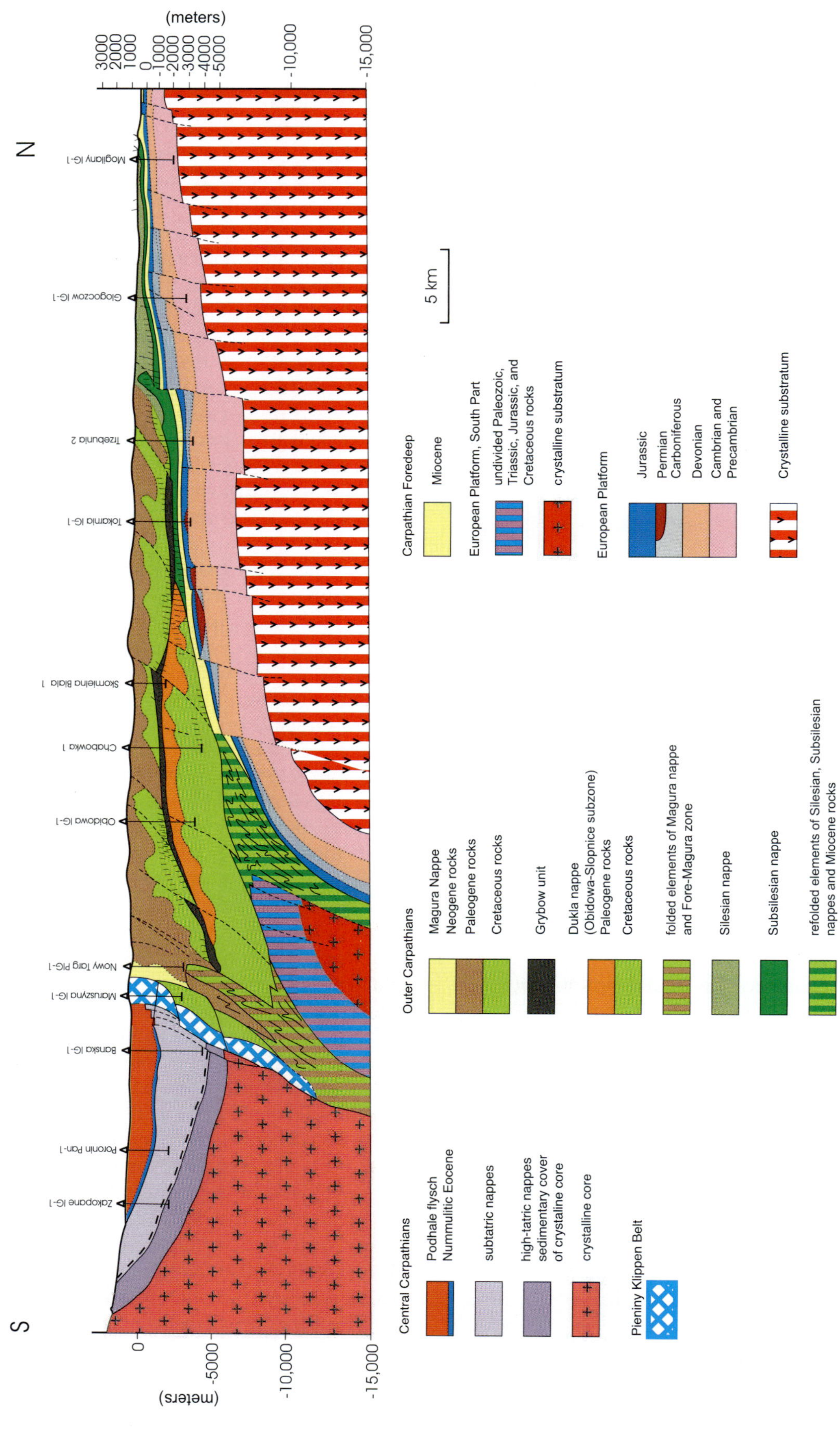

Figure 11. Cross section through the western part of the Polish Carpathians between Krakow–Zakopane (M. Cieszkowski, 2002, personal communication).

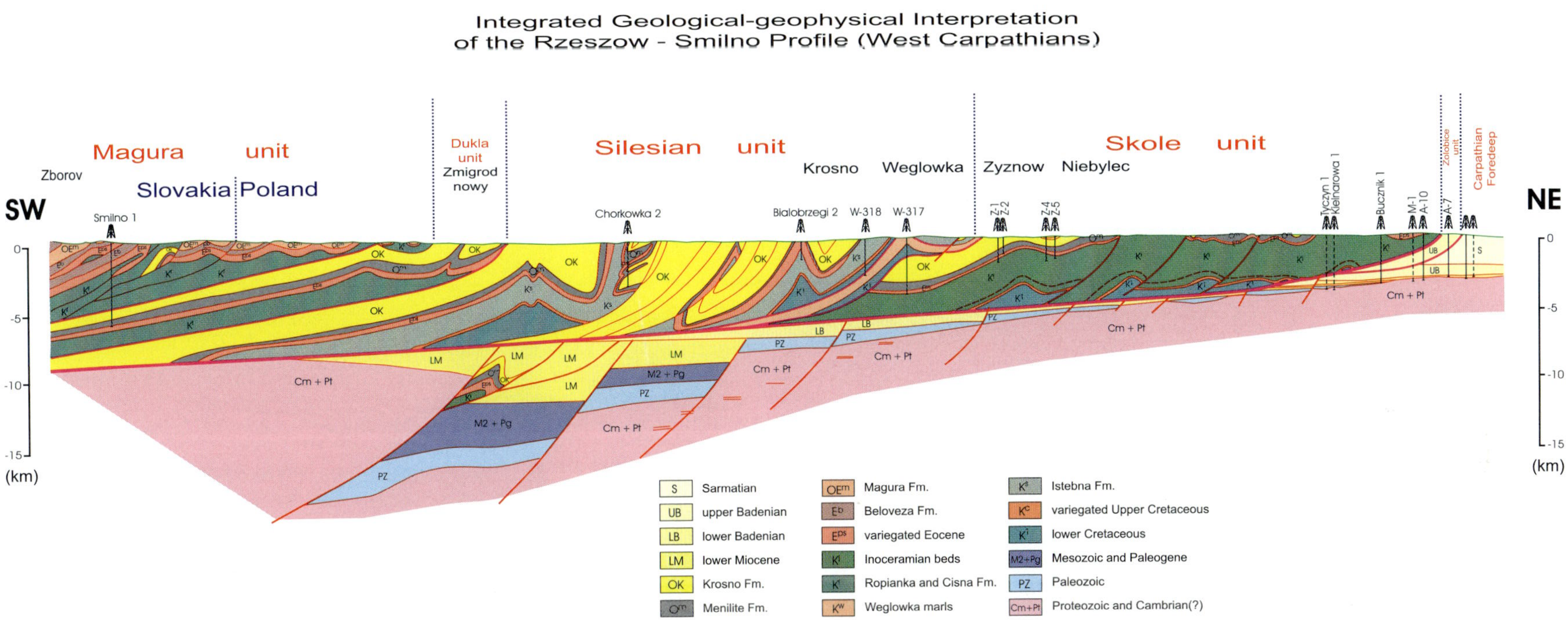

Figure 12. Cross section through the eastern part of the Polish and Slovak Carpathians between Rzeszow and Smilno (modified after Oszczypko et al., 1998).

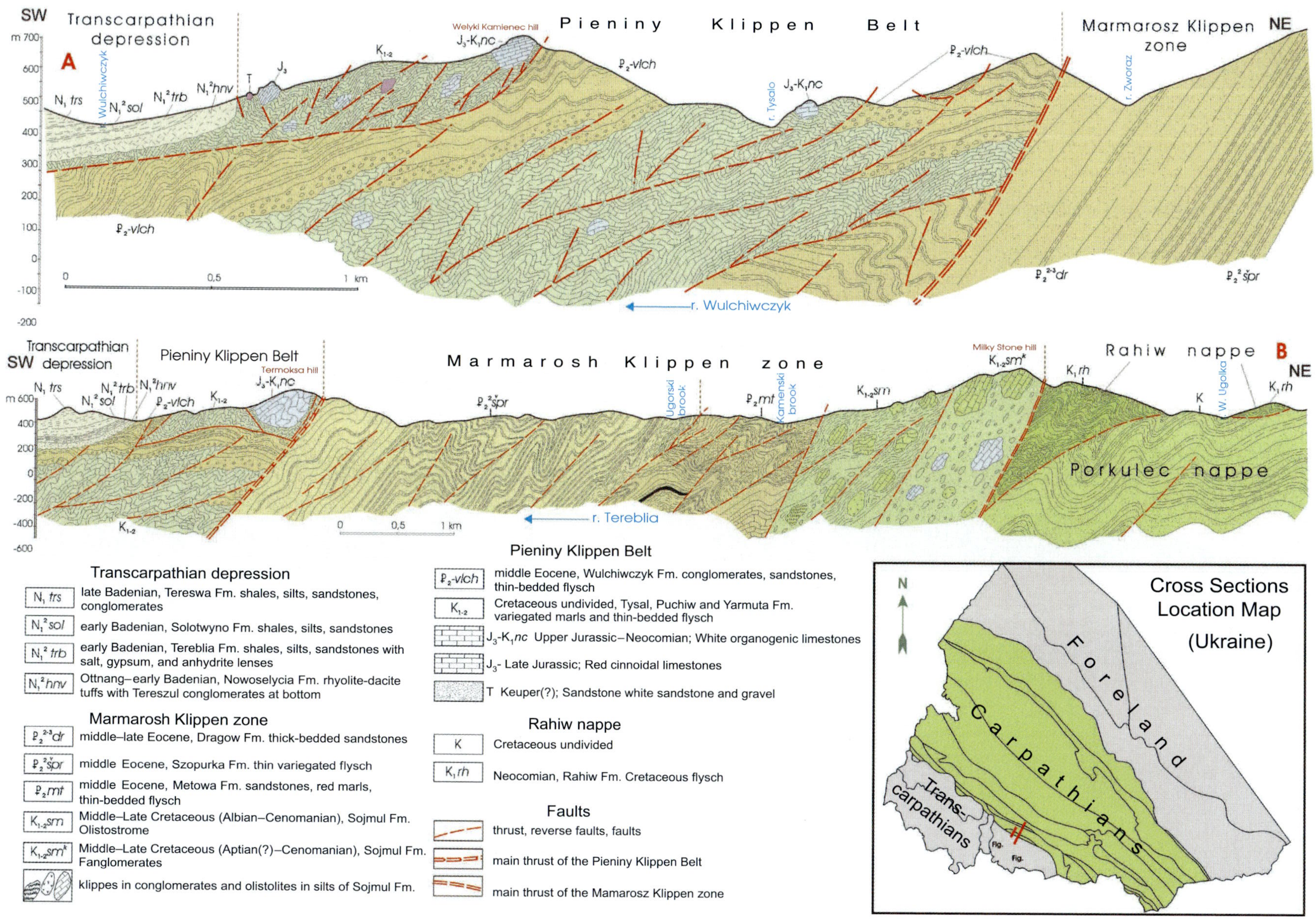

Figure 13. Cross sections through the inner part of the Ukrainian Carpathians.

formed the roof and floor thrust, respectively (Oszczypko, 2001). South of the zone of tectonic windows, the inclination of the Magura thrust surface increases, and at the northern boundary of the Pieniny Klippen Belt, the thickness of the Magura nappe is more than 5 km (3 mi). The Krynica, Bystrica (Nowy Sacz), Raca, and Siary subunits coincide, to a large extent, with the corresponding facies zone. In the Magura unit, three structural complexes can be distinguished. The first was formed during the Late Cretaceous–Paleocene, the second was formed during the early to late Eocene, and the last was formed during the Oligocene to early Miocene (Oszczypko and Thomas, 1985; Oszczypko, 1992, 1999; Malata et al., 1996). These complexes revealed a decreasing degree of tectonic deformation from the base to the top of the nappe. In the area surrounding the Mszana Dolna and Szczawa tectonic windows, the basal part of the nappe, made up of Upper Cretaceous–Paleocene flysch rocks, is strongly deformed. In the lower to upper Eocene flysch of the Raca and Krynica subunits, the broad west–east-trending synclines and narrow anticlines dominate. The southern limbs of synclines are commonly reduced and overturned. In the Bystrica (Nowy Sacz) subunit, subvertical thrust sheets are common. Both the northern limbs of the anticlines and the southern limbs of the synclines are tectonically reduced and commonly overturned. In the Krynica and Raca subunits, the youngest (Malcov Formation, late Eocene–early Oligocene), weakly deformed deposits of the Magura nappe unconformably overlie the older Eocene flysch deposits.

The Biele Karpaty (White Carpathians) were traditionally regarded as an integral part of the Magura nappe and correlated with the Oravska Magura (Krynica subunit), which is in a similar tectonic position with respect to the Pieniny Klippen Belt. According to Potfaj (1993), the Biele Karpaty forms a separate subunit. This subunit is overthrust onto the Raca and Bystrica subunits of the Magura (at least to 25 km [15 mi]). The inner structure of the Biele Karpaty unit has a fold and thrust character in its western part. The biggest thrust, known as the Javorina nappe, forms a flat-lying body approximately 50 km (30 mi) long, about 900 m (2900 ft) thick, and 20–25 km (12–16 mi) wide.

The Marmarosh Unit

The Marmarosh unit stretching along the Pieniny Klippen Belt is divided into two parts:

the Marmarosh Klippen zone and Marmarosh crystalline massif
the Marmarosh Klippen zone

The Marmarosh Klippen zone is situated northeast from the Pieniny Klippen zone; its northwest part is overthrust onto the Rachiv and Porkulets nappes, and its southeast part is overthrust onto the Marmarosh crystalline massif. The surface of the overthrust, traced by several boreholes, is generally steep, and the whole zone dips steeply toward the southwest; only in the area of Chornyj Cheremosh does the overthrust become locally flat, and its visible distance is more than 3.5–4 km (2.1–2.5 mi). The Marmarosh Klippen zone is divided into two subzones on the base of tectonic structures and age of folding: the external subzone of Vezhany, which contains huge olistolithes (Soimul olistostrome), and the internal subzone of Monastyrets, which is built mainly by flysch. The first subzone is made up of a few thrust folds, whereas the second subzone is a large monocline with small secondary folds. The internal subzone is the prolongation of the Petrova nappe and is regarded by some geologists as the southeast prolongation of the Magura unit (Sandulescu, 1988; Oszczypko, 2001).

The Marmarosh crystalline massif is the northwest prolongation of units known from the Romanian Carpathians (Kropotkin, 1991; Golonka et al., 2003) as Bucovinian–Getic nappes (Sandulescu et al., 1981) or Median Dacides (e.g., Stefanescu et al., 2006). It is overthrust on the Black Flysch unit or Rachiv unit and dipping under the Marmarosh Klippen zone. The massif displays a nappe structure, and two subnappes were distinguished (Kruglov, 2001): the Bilyi Potik (the lower subnappe) and the Dilove (the upper subnappe). Minimal displacement magnitude of the Dilove nappe is more than 10 km (6 mi), whereas in the Romanian Carpathians, its displacement of more than 25 km (16 mi) is observed. In front of the massif, the narrow Kaminnyi Potik unit is present, composed of calcareous-terrigenous deposits containing the Upper Jurassic effusives of the basic type. Glushko and Kruglov (1986) consider this thrust sheet as a part of the massif.

The Rachiv Nappe

The Rachiv nappe is represented by a narrow zone made up of secondary folded Lower Cretaceous flysch dipping generally toward the southwest under the Kaminnyi Potok–Marmarosh unit and is overthrust on the Porkulets nappe. This overthrust is very gentle in the Chyvchyny area and becomes very steep to the northwest near the village of Dovhe. Magnitude of the overthrust is apparently great, and its minimal value, according to the existing data, is of several kilometers. The Rachiv nappe forms a direct continuation of the upper internal thrust sheet or the Ciuc digitation of the Ceahlau nappe in the Romanian Carpathians. Toward the northwest, it disappears under Vezany subzone north from the village of Dovhe, and its west continuation is not known.

Porkulets Nappe

The Porkulets nappe is the largest among the internal thrust sheets of the Ukrainian Carpathian Flysch. It overthrusts the Dukla nappe and also the Chornohora nappe farther to the southeast. The magnitude of horizontal displacement of the Porkulets nappe reaches at least 12 km (7 mi) locally. This displacement is evidenced by the existence of a tonguelike tectonic semi-outlier in the Mount Petros area and by borehole data. According to the data from the Chornoholova parametric borehole, the thrust surface is steep and can be estimated as 80° in the inner part of the Porkulets nappe. The Porkulets nappe plunges to the southwest, below the fragmented strip of the Rachiv nappe and at the westernmost part below the Magura nappe. The thrust surface is steep, and its dip can be estimated as 80°. The inner part of the Porkulets nappe is characterized by a general submergence of the axis of folds from the southeast, where only Early Cretaceous deposits are exposed to the northwest where younger deposits are also visible. It abruptly terminates near the Slovak border, and its northwest prolongation is not clear.

The four subzones separated by cross-faults are distinguished based on tectonic structures and stratigraphic profiles. They are, from the northwest to the southeast, Chornoholova, Turya Polyana, Lysychiv, and Bila Tisza. A distinctive feature of the Chornoholova subzone is its general elevation and planar structures. Toward the southeast, this subzone terminates on a transverse wrench fault that is manifested as the Turya sigmoid (flexure) in the more external Dukla nappe. The thrust surface of the Turya Polyana subzone is steep and can be estimated as 80°. This zone consists of two thrust folds: the internal Turycia and the external Zvir thrust folds. A wrench fault separates this subzone from the next, more uplifted Lysychiv subzone, the main subzone of the Porkulets nappe. This subzone is made up of several thrust folds secondarily folded and cut by several cross-faults and is terminated by a cross-fault along the Luzanka River.

The Bila Tisza subzone, built mainly of the Early Cretaceous deposits, is located further to the southeast. The distinctive feature of this subzone is the presence of the greenstone basic igneous and tufogenic rocks in its frontal part. The Bila Tisza subzone is strongly tectonized, and its internal structures are still not satisfactorily recognized. It prolongs southeastward in the Romanian territory into the Bodok digitation of the Ceahlau nappe.

Dukla Nappe

The Dukla nappe stretches from the Polish to Ukrainian Carpathians. In its southeastern part, this nappe consists of several imbricated, thrust-faulted folds (Figure 12) with a northwest–southeast strike. The Dukla nappe is maximally elevated in the east. The axis of folds plunges gradually toward the northwest, and eventually, the whole nappe disappears below the Magura nappe. Data from boreholes (Zboj 1, Smilno 1) show that the Dukla nappe extends under the Magura nappe far to the south. Southeast from the Slovak–Ukraine border, where the Magura nappe disappears, the inner part of the Dukla nappe is hidden below the Porkulets nappe. From the more external Silesian nappe and/or Zboj unit, the Dukla nappe is separated by a thrust plane more distinct in the eastern than in the western part. Data from the deep boreholes Jasliska 2 and Wetlina 3 show that the thrust plane in the Polish part of the nappe is very steep. However, data from the deep borehole Zboj 1 indicate that thrust exceeds 15 km (9 mi), and the thrust plane beneath the more internal part of the Dukla nappe becomes more flat.

In the Polish and Slovakian part of the Dukla nappe, two subunits can be distinguished: the internal and the external subunits. The folds in the internal subunit are generally gently dipping toward the southwest and the unit's overthrust is low dipping, whereas in the external subunit, folds are steep and commonly with a reversed (southwestern) vergence. The internal subunit disappears on the border between Slovakia and Ukraine; however, the fact that this subunit prolongs into the Porkulets nappe cannot be excluded.

In the southeastern Ukrainian part, the Dukla nappe has a general direction of submergence from the southeast to the northwest. It is made up of several thrust folds commonly with a southwest vergence. The presence of the gentle brachysynclinal folds is the most distinctive structural feature in the Ukrainian part of the Dukla unit. The best known brachysyncline of the Polonyna Runa was interpreted as a late Oligocene superimposed trough or as an isometric fold formed by an uplift of a crystalline basement at its basis or as a rootless tectonic outlier. Kruglov and Tarasenko (1984) interpret it as a normal brachysyncline with disharmonic folds at its base because of diverse physical properties of the substratum. Three structural-facial en echelon subzones are delimited in the Dukla nappe: the Stuzhycia in the northwest, the median Luzhanka, and the Blyznyca in the southeast. The Dukla nappe disappears from the surface in the valley of Bily Cheremosh, and its continuation toward the southeast is unknown. It could be connected with the Romanian Teleajen nappe, which is assigned to the Moldavides (Sandulescu et al., 1981).

Beneath the Dukla nappe, a separate tectonic unit, the Zboj unit, was described from the borehole Zboj 1, situated in east Slovakia. Only a fragment of a limb of anticline represents this unit. Its internal structures and relation to the more outer tectonic units, especially the Silesian nappe, are unknown.

Fore-Magura Group of Units

The small tectonic units encountered in the Western Carpathians in front or beneath the Magura nappe belong to this group. The Fore-Magura nappe, located near the town of Zywiec, occupies the most western position in the Fore-Magura group of units. It consists of two narrow, asymmetrical anticlines with thrust faults, the inner anticline being strongly deformed and disharmonic. Toward the east, this unit disappears completely. Further toward the east, another unit, the Grybow unit, is exposed in several tectonic windows in the Magura nappe, from the Mszana Dolna tectonic window to the Smilno tectonic window in Slovakia. It was also encountered in several boreholes below the Magura nappe. This unit is strongly folded, with several disharmonic thrust-faulted folds. The Grybow unit is itself thrust over the Obidowa–Slopnice unit or on the Silesian unit. The Obidowa–Slopnice unit is present in several boreholes between the Obidowa and Slopnice (Obidowa IG 1 and Chabowka 1) and in several boreholes in the Slopnice area. The strata of this unit are fairly gently dipping toward the south, commonly without any intense tectonic deformation, except in the higher part. The Jaslo nappe (Koszarski, 1985) forms a rather poorly defined unit existing in front of the Magura nappe to the west of the town of Jaslo. It was distinguished in the Harklova and Luzna peninsulas of the Magura nappe. The Jaslo nappe is flatly thrust over the Silesian nappe with complicated internal structures. However, L. Jankowski's (2003, personal communication) opinion states that rocks regarded as the Jaslo nappe represent olistostromes in the youngest deposits of the Silesian unit. Some of these units (Grybow and Slopnice-Obidowa) display a general affinity and probably belonged to a bigger nappe, which was divided into separate units during the Neogene folding.

Chornohora Nappe

The Chornohora nappe stretches from the Romanian border to the Teresva–Chorna Tisza area, where it disappears under the Dukla nappe. It is connected to the Audia nappe in Romania. Two smaller elements are distinguished here, according to the character of tectonic structures and differences in their lithological development: these are the Skupiv subzone in the north and the Yalovychor, or the Hoverla, subzone in the south. The first one is represented by a complicated secondary folded and faulted monoclinal structure dipping toward the northwest. The more internal subzone is made up of numerous narrow, long, commonly steep thrust folds. It is also called the subzone of small thrust folds. Existing data imply that the Chornohora nappe was displaced more than 10 km (6 mi) to the northeast along a gentle overthrust on the Krosno zone. The Chornohora nappe was completely penetrated at a depth of 3800 m (12,500 ft) by the Hryniava parametric borehole, which was started in the Skupiv zone, 6 km (4 mi) from the nappe front. After penetrating through the Chornohora nappe, the borehole reached into the Oligocene–Lower Cretaceous deposits of the Verhovyna depression belonging to the Krosno zone. The Chornohora nappe thrust surface is dipping at an angle of 30°.

Silesian Nappe

The Silesian nappe stretches from the Moravia area (Czech Republic) to Ukraine, where it loses its individuality. In the western segment of the Polish Carpathians, the Silesian nappe is flatly overthrust onto the substratum (Figure 11). Within the Silesian nappe, several tectonic windows are present where the Subsilesian nappe is exposed. Toward the east, the thrust plane gradually plunges, and at the same time, the character of the tectonic structures changes in the Silesian nappe. In the western part, the structures are generally shallow and gently folded, whereas toward the east, they pass into long, narrow, steeply dipping, imbricated folds. The southern part of the Silesian nappe is hidden beneath the Magura nappe and the Dukla–Fore-Magura nappes (Figures 11, 12).

To the west of the Sola River, near the western border of Poland, the Silesian nappe is composed of two subunits: the Cieszyn subunit is built of a strongly folded Lower Cretaceous strata, whereas the Godula subunit is composed of Upper Cretaceous and Paleogene deposits that dip monoclinally southward. This part of the Silesian nappe is cut by several transverse faults. Farther to the east, the Cieszyn subunit and the Godula subunit join, and the Silesian nappe is built of several gently folded structures. East of the Dunajec River, these structures pass into imbricated folds. The more important structures are the Stroze, Jankowa, and Ciezkowice folds. Small oil fields are associated with these folds. Toward the east, the Stroze fold passes into the broad Gorlice fold (Figure 12), where one of the oldest oil fields in the Carpathians exists. Within the culmination of the next fold to the north, the Ciezkowice–Biecz fold, small oil fields were found. The eastern part of the Silesian nappe, east of the Wislok river, plunges toward the southeast and is represented by a synclinorium (Central Carpathian Synclinorium), which is made up mainly of Oligocene deposits (Wdowiarz, 1985). The Central Carpathian Synclinorium is made of several long, narrow, imbricated, thrust-faulted folds that are commonly disharmonic (Figure 12). These folds are cut by several transverse faults that divide them into separate blocks. The folds display, along the strike, several axial culminations, where, along the northern and southern margins of the synclinorium, the Cretaceous and Eocene strata are exposed. Several folds and thrust folds were distinguished in the Central Carpathian Synclinorium: the Folusz–Bukowica–Fore-Dukla zone, Zboiska, the Lubatowka–Iwonicz Spa–Tokarnia fold, the Osobnica–Bobrka Rogi–Suche Rzeki fold, the Lubienka–Mokre–Zatwarnica fold, the Roztoki–Potok–Turaszowka–Kroscienko–Tarnawa–Wielopole–Czarna fold, the Sanok–Zmiennica–Strachocina–Czarnorzeki fold, and the Ustianowa–Miedzybrodzie–Grabownica fold. Several oil fields are connected with these folds (Slaczka, 1996a).

The Central Carpathian Synclinorium passes into the Krosno zone on the territory of Ukraine. Southeast from the town of Sanok, the amplitudes of marginal thrusts of the Silesian nappe and of the Subsilesian nappe diminish and eventually terminate, and this zone passes into a normal fold. Therefore, the prolongation of the boundary between the Silesian and Skole units toward the southeast is not so clear as it is in the west. The Silesian nappe prolongs on the Ukrainian territory into the Krosno zone.

The boundary between the Krosno zone and the Skyba nappe in Ukraine is generally placed along the large rupture dislocation that stretches along the internal boundary of the Skyba nappe and ends somewhere in the median part of the Ukrainian Carpathians. However, the dislocation does not limit but merely cuts the large synclinorium buildup of thick Oligocene–lower Miocene flysch strata and the intramountain molasse of the Krosno beds.

Within the Krosno zone, three subzones are distinguished according to structural and lithological-facial peculiarities. The most inner subzone (the Bytliana subzone) is characterized by the development of the Oligocene olistostrome formations. It forms a narrow strip along the front of the Dukla nappe and corresponds generally to the Fore-Dukla subunit of the Polish Carpathians. Pre-Oligocene formations are exposed only at the Stryi headwaters in the core of the Smole structure.

The Turka subzone that wedges out near the Black Tisza occupies the intermediate location. It is a nappe thrust over the most external Slavsko–Verhovyna subzone. This subzone is elevated at its southeastern part (at the interflow of the Black Tisza and the Rika) and is submerged at its northwestern part. It was also proposed to relate the area located to the northeast from this nappe (Slavsko–Verhovina subzone) already to the Skyba nappe.

A distinctive feature of the Slavsko–Verhovyna subzone is its general submergence. The subzone is divided into the Slavsko depression in the northwest and the Verhovyna depression in the southeast. These elements are separated by a gentle transversal elevation located between the headwaters of the Tereblia and the Bystrytsia Nadvirnianska.

Andrychow Zone

In the western part of the Outer Carpathians near the town of Andrychow, along the Silesian nappe, several huge blocks exist built mainly by Jurassic limestones. They were regarded as tectonic klippen that were sheared off during the movements of the Silesian nappe (Ksiazkiewicz, 1977); however, new data suggest that they are olistolithes in the uppermost part of the Krosno beds of the Subsilesian nappe (Slaczka, 1998).

The Subsilesian Nappe

The Subsilesian nappe is exposed in a narrow, strongly tectonized zone in front of the Silesian nappe and also in the Western Polish Carpathians in several tectonic windows in that nappe. The presence of the Subsilesian nappe was also established in numerous boreholes beneath the Silesian and the Magura nappes (Figures 11, 12).

In the frontal part of the Silesian nappe, north of the town Krosno, the Subsilesian nappe is exposed in the Weglowka tectonic half window. Deep wells connected with the Weglowka oil field show that the tectonic window is built of a refolded thrust-faulted anticline. The Subsilesian nappe is steeply overthrust onto the Skole nappe. Further to the east, the Subsilesian nappe forms once more a narrow zone in front of the Silesian nappe (Figure 12). Near the town of Ustrzyki Dolne, the Subsilesian nappe disappears from the surface, and the frontal part of the Silesian nappe becomes a thrust-faulted fold and eventually joins with the Skole nappe. A possibility also exists that tectonic prolongation of the Subsilesian nappe is the Rosluch thrust fold in Ukraine.

THE SKOLE–SKYBA NAPPE

The Skole–Skyba nappe forms a large portion of the eastern part of the Northern Carpathians and is overthrust onto the Miocene sediments that cover the North European platform (Figure 12). This nappe is the largest unit in the Outer Carpathians (as much as 40 km [25 mi] wide) and is the only structural unit that stretches from Poland, where it is called the Skole nappe, through the whole of the Ukrainian Carpathians, where it is named the Skyba nappe, and passes into Romania as the Tarcau nappe. The most distinctive structural feature of the zone is the occurrence of large thrust-folds skybas thrust over each other in the northeast direction and traced for several hundreds of kilometers along the stretch of the Carpathian arc. The width of such skybas ranges from a few kilometers to as much as 12 km (7 mi).

Toward the west, the Skole–Skyba nappe becomes more and more narrow, and eventually, the Skole nappe plunges under the Subsilesian and Silesian nappes near the town of Brzesko, and in this area, these nappes reach the Carpathian margin. Its prolongation further toward the west is not clear. However, sometimes, the marginal flysch unit, near the town of Wadowice, is considered as a western prolongation of the Skole nappe.

The outer part of the Skole nappe is elevated and consists of numerous, narrow, steep, thrust-faulted anticlines where the Lower Cretaceous Spas shales appear at the surface (Figure 12). The synclines are narrow and made up of the Menilite or Krosno beds. Southwest from the town of Przemysl, the folds create a sigmoidal arc that reflects a similar sigmoidal bend of the Carpathian margin. Near the town of Rzeszow, the marginal part of the Skole nappe is covered by Miocene molasses, which form a piggyback basin. Farther toward the west, near the town of Pilzno, the northern part of the Skole nappe is probably folded together with the Miocene cover.

The inner part of the Skole nappe is represented by a synclinal area made of several folds (Tyrawa–Lodyna, Paszowa–Wankowa, Witrylow, and Wara) with broad synclines composed of the Oligocene–lower Miocene Krosno beds (Figure 12). In this part of the Skole nappe, the deepest borehole in the Polish Carpathians, the Kuzmina 1 (7541 m; 24,740 ft), was situated. Near the town of Strzyzow, the axis of folds plunges, and synclines join together, creating a vast Strzyzow synclinorium.

The structures observed in the western part of the nappe continue toward the southeast. Six thrust folds (skybas), secondary folded, were distinguished in the Skole–Skyba nappe in Ukraine (external thrust folds: Marginal, Oriv, and Skole; and internal thrust folds: Parashka, Zelemjanka, and Rozhanka). The frontal parts of thrust folds in Skole–Skyba nappe are commonly made up of the Upper Cretaceous deposits (Holovnia and Stryi beds) and sometimes of Paleogene formations. The inner parts of three external thrust folds are made of the Oligocene and the lower Miocene deposits (Menilite shales), whereas the inner parts of the internal thrust folds are represented by the lower Miocene Krosno beds. Two inner thrust folds are distinctly traced only to the Lomnytsia River. Farther to the east, they are hidden below the Oligocene lower Krosno beds. The Marginal (Berehova) thrust fold disappears in the area of the Pokuttya Carpathians, where it was completely destroyed by erosion. Hence, only two thrust folds are distinctly traced up to the Romanian border: the Oriv and the Skole thrust folds. The continuation of the more inner thrust folds is not very clear.

The Skole–Skyba nappe is thrust over the Boryslav–Pokuttya nappe in the east and over the folded and/or autochthonous Miocene covering the North European platform in the west. The thrust plane of the Skyba nappe is different in diverse regions and changes from very gentle (where the thickness of the nappe is small) to very steep. Especially, a steep thrust plane exists at the most internal part of the nappe, where it has been discovered in recent years by superdeep boreholes. In the Bytkov region, the magnitude of a vertical displacement of the Skyba nappe exceeds 17 km (11 mi). However, in the west, the thrust plane is more flat. The Kuzmina 1 borehole (Figure 2), situated 30 km (18 mi) from the northern boundary of the Skole–Skyba nappe, reached the thrust plane at a depth of 6399 m (20,994 ft). Further to the west, in borehole Szufnarowa 1 (Figure 2) situated 15 km (9 mi) from the northern boundary, the thrust plane was encountered at the depth 4000 m (13,000 ft).

The Skole nappe is locally covered by Miocene molasses, which form piggyback basins. Near the town of Pilzno, the northern part of the Skole nappe is probably folded together with the Miocene cover.

Boryslav–Pokuttya Nappe

This nappe developed only in the Eastern Carpathians. Within the Ukrainian segment, it is commonly designated as the Boryslav–Pokuttya nappe, and on Romanian territory, it is designated as the Marginal fold unit. This nappe is composed of Cretaceous–lower Miocene flysch, as well as early Miocene molasse. Therefore, that unit is commonly regarded as the inner foredeep zone.

The Boryslav–Pokuttya nappe consists of four main thrust folds: the Boryslav, the Maydan, the Pokuttya, and the Runhur (Glushko and Kruglov, 1986). They have a very complicated, refolded, internal structure typical for duplexes. The majority of the folds display a northeast vergence as well as an absence of tucked-in limbs. Very commonly, two or three folds are overthrust one on top of the other and display stacked structures. The presence of thick clay and salt-containing deposits substantially complicates the local structures of those folds. Sometimes, two or three digitations of such folds (partial thrust sheets) could be delimited in the profile. Sometimes, a diagonal strike of the main folds in relation to the main structure of the nappe could be observed, and five to seven separate groups of such folds are distinguished. The largest groups of folds are the Blazhiv, the Boryslav, the Dolyna, the Bytkiv, and the Pokuttya–Dzviniany.

The Boryslav–Pokuttya nappe is overthrust on the folded Miocene [Sambir (Stebnik) nappe] laying on the autochthonous Miocene of the Ukrainian platform (Kruglov, 1999). The thrust plane is very steep or even overturned near the surface, but at depths of 3–4 km (2–2.5 mi), it became almost flat. The overthrust, documented by boreholes, is of a magnitude of at least 20 km (12 mi). The continuation of the Boryslav–Pokuttya nappe below the Polish part of the Skole–Skyba nappe is still controversial. No real traces of this structure have been present. If it exists, it is hidden far to the south below the Carpathians.

THE DEEP STRUCTURE OF THE OUTER CARPATHIANS

The deep structure of the Polish Outer Carpathians and its basement has been recognized by deep borehole results as well as by magnetotelluric, gravimetric, magnetic, geomagnetic, and deep seismic sounding profiles (Woznicki and Sucha, 1989; Guterch et al., 2001; Stefaniuk, 2002; Stefaniuk and Slaczka, 2002).

Deep boreholes (as much as 9000 m [29,500 ft]), which reached the Carpathian substratum, were drilled along the Outer Carpathians (Figures 2, 11, 12) (see also Oszczypko et al., 2006; Picha et al., 2006). They allowed the recognition of the deep structures of the Carpathians, the depth of Carpathian thrust plane, its minimal range, as well as the character of the substratum. First of all, they proofed the thin-skinned character of the Outer Carpathians orogen, which is thrust over the autochthonous Miocene deposits covering the eastern and western parts of the North European platform. They also documented the occurrence of several uprooted nappes thrust upon each other and the existence of new tectonic and lithostratigraphic units that were not known from the surface data.

Generally, the thrust plane of the Carpathians dipped slowly to the south in their western part (see Oszczypko and Tomas, 1985). The borehole Bystre IG 1 located 30 km (18 mi) southward from the northern margin of the Carpathians crossed this plane at the depth of −3131 m (−10,272 ft) below sea level. Toward the east, this plane plunges gradually. West of Krakow, borehole Zawoja 1, also situated 30 km (18 mi) from the Carpathian margin, crossed the plane on the depth of −3225 m (−10,580 ft) below sea level. In the eastern part of the Polish Carpathians (Oszczypko and Tomas, 1985), borehole Brzozowa 1, situated 10 km (6 mi) from the margin, reached substratum at the depth of −2575 m (−8448 ft) below sea level. Borehole Szufnarowa 1, 15 km (9 mi) from the margin, passed the plane at depth of −3455 m (−11,335 ft) below sea level, and borehole Kuzmina 1, 25 km (16 mi) from the margin, reached the plane at the depth of −6885 m (−22,588 ft) below sea level. The southeast plunging of the thrust plane of the Carpathians continues also in the Ukrainian Carpathians, where borehole Schevthenkovo 1, situated 16 km (10 mi) from the Carpathian margin, did not reach the Carpathian flysch up to the depth of 7520 m (24,672 ft).

Seismic data provide similar values of the depth of the thrust plane. These seismic data were obtained from hundreds of reflection and refraction profiles crossing the Outer Carpathians, especially along their outer part.

Because of data from deep boreholes, new tectonostratigraphic units were found, which were hidden completely below the nappes and not known previously from the surface data. Part of them are an important factor in the hydrocarbon exploration. In the middle part of the Outer Carpathians (Figure 2, 11), several boreholes, e.g., Obidowa IG 1, Chabowka 1, and Slopnice 20, encountered a new unit (Obidowa–Slopnice) completely hidden below the Magura nappe. In addition, in the eastern part of Slovakia, the borehole Zboj 1 came across an unknown unit (Zboj) below the Dukla nappe.

The magnetotelluric soundings in the Polish Carpathians revealed a high resistivity horizon connected to a consolidated crystalline basement (Kucharski et al., 1990; Rylko and Tomas, 1995; Stefaniuk, 2002). According to these authors, the depth of the crystalline basement ranges from 4 to 8 km (2.5 to 5 mi) in the northern part of the Carpathians, dips to approximately 15–20 km (9–12 mi) at its deepest point, then rises to about 8–10 km (5–5 mi) in the southernmost part of the orogen. The axis of the basement depression is located along the line Orava–Nowy Targ–Krynica–Dukla–Jablonka and more or less coincides with the axis of the gravimetric minimum. The obtained results are generally in agreement with the seismic and borehole data.

The results of gravimetric studies show distinct gravity minimum (as much as 60 mGal) along the Carpathian Flysch. The axis of this anomaly runs approximately parallel to the marginal Carpathian thrust in the west, along the northern boundary of the Pieniny Klippen Belt. East from the town of Nowy Targ, it shifts toward the northwest along a perpendicular dislocation and diagonally crosses the structures of the Outer Carpathians. In the Krosno area, the en echelon relocation of the axis toward the north is observed, probably along a strike-slip fault. Further eastward in the Ukrainian Carpathians, the axis runs toward Staryj Sambir and almost approaches the outer margin of the Carpathians. That gravity minimum is related either to the presence of low-density molasse deposits below the Carpathians or to the structures in Moho (Woznicki and Sucha, 1989). The integrated geophysical modeling along the Rzeszow–Bardejov geotraverse (Figure 12) (Oszczypko et al., 1998) documented that regional gravimetric low south of Krosno is a result of a combined effect of the thick Carpathian nappes, which cover thick early Miocene molasses, and the occurrence of the Mesozoic and Paleogene deposits related to the passive margin of the European platform.

The geomagnetic soundings provided the interesting data that helped to recognize the Carpathian substratum. They revealed the presence of a zone of zero values of the Wieses vectors (see Woznicki and Sucha, 1989) that runs south from the Pieniny Klippen Belt near Kroscienko upon Dunajec; westward, it crosses the klippen zone and runs beneath the inner part of the Outer Carpathians. This zone of zero values is connected to a high conductivity body 2.5–6 km (1.5–3.7 mi) thick, at a depth of 15–30 km (9–18 mi). It probably indicates the position of the southern extent of the North European platform and its contact with the Inner Carpathian basement (see Stranik et al., 1993) and probably the southern extent of the Outer Carpathian overthrust. The observed data in the Eastern Carpathians imply a change of the depth of the asthenosphere that sinks from 80 km (50 mi) in the west to more than 150 km (93 mi) in the Eastern Carpathians.

The deep seismic profile located in the Slovakian Carpathians (see Tomek, 1993) gives an approximate thickness of the lithosphere and depth of the Moho below the Carpathians. It shows that two groups of south-dipping reflectors north of Pieniny Klippen Belt exist that probably are related to the middle Miocene subduction of the Moldavidic part of the Severin–Moldavidic realm (see also Tomek and Hall, 1993). The upper reflection between 1 and 3 s (about 4.5–8 km [2.8–5 mi]) probably belongs to a plate boundary between the upper plate (the thrust Penninic realm, mainly the Magura wedge) and the accretionary wedge complex (Moldavides or Dukla–Silesian–Subsilesian group of units). The lower reflectors represent the crystalline basement of the lower plate (North European plate) and its sedimentary cover.

The position of the crust–mantle boundary (Moho) has been recognized along several seismic profiles (Woznicki and Sucha, 1989; Guterch et al., 2001). The depth to the Moho discontinuity generally ranged from 30 to 40 km (19 to 25 mi) at the front of the central part of the Carpathians and increases to 50 km (31 mi) south of the Nowy Sacz. South of the Pieniny Klippen Belt, this value decreases to 36–37 km (22–23 mi) and to only 24–30 km (15–18 mi) beneath the Pannonian Basin. The consolidated basement with velocity $V_p > 6.0$ km/s (3.7 mi/s) is situated at a depth of 10–18 km (6–11 mi) beneath Carpathians and 5–8 km (2.8–5 mi) beneath the Pannonian Basin. The Moho discontinuity in Ukrainian part of the Outer Carpathians is determined at a depth of about 60 km (37 mi), with the P-wave velocity at 8.25 km/s (5 mi/s). It implies the existence of a cross-fault between the Western and Eastern Carpathians. The seismic data show the existence of a low-velocity zone beneath the Tatra that can be connected with the prolongation of the Outer Carpathians flysch below the Tatras. According to Lillie et al. (1994), the thickness of the lithosphere in the Polish Carpathians ranges from 160 km (99 mi) near Krakow to 100 km (62 mi) in the Pieniny Klippen Belt, with a temperature at the depth of the Moho discontinuity ranging from 400°C beneath the Pieniny Klippen Belt to 660–700°C beneath the Outer Carpathians.

Jurassic to Early Miocene Tectonostratigraphic Evolution of the Outer Carpathian Flysch Basins

The Outer Carpathian basin (Figure 14) was created during the course of disintegration of the North European plate. It was connected with rifting and a formation of thinned or oceanic crust in the Penninic realm–Magura basin (Early–Middle Jurassic) and also in the Severin Moldavides realm–Outer Dacides and Silesian basin (Jurassic–Cretaceous boundary) (see Birkenmajer, 1986; Rakus, et al., 1988; Sandulescu, 1988; Tolman, 1990; Dercourt et al., 1993; Golonka et al., 2000, 2003, 2004, 2006). During the sedimentary-tectonic evolution of the Outer Carpathian basins, several periods can be distinguished.

The first period began from the incipient stage of rifting of the southern part of the European plate and the formation of local basins (Early Jurassic–Kimmeridgian). The new configuration of plates replaced the older one, represented by the Transylvanian domain. The Transylvanian oceanic basin developed during the Triassic

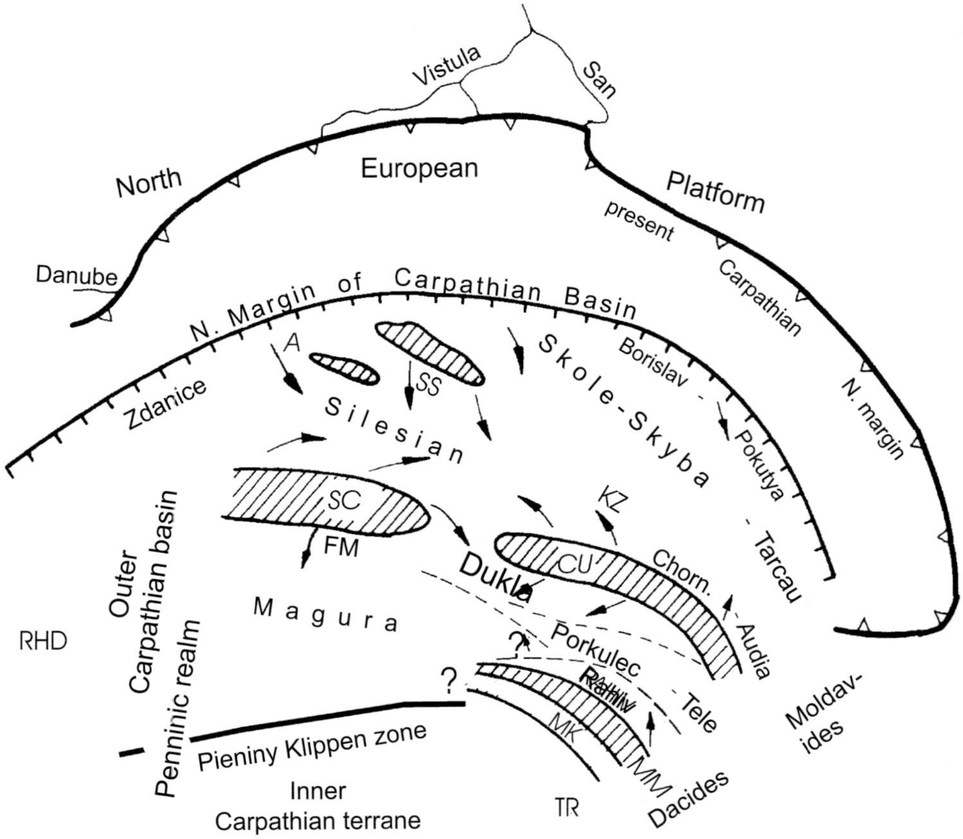

Figure 14. Sketch of probable distribution of ridges and subbasins in Outer Carpathian basin, partly palinspastic (based on Slaczka, 2000). A = Andrychow ridge; CU = Kuman ridge; FM = Fore-Magura basin; MK = Marmarosh Klippen; KZ = Krosno zone; MM = Marmarosh massive; RHD = Rhenodanubian flysch; SC = Silesian Cordillera; SS = Subsilesian realm; Chorn. = Chornohora subbasin; Tele. = Teleajen subbasin.

and was closed during the Cretaceous. The exotic material of the Late Triassic pelagic spotty limestones, which occur as pebbles in Cretaceous–Paleogene gravelstones in the Pieniny Klippen Belt (Birkenmajer et al., 1990) and Magura unit (Sotak, 1986), could have originated in this basin. The Transylvanian basin position and its relation to the other parts of the Tethys and Vardar Ocean, Meliata–Halstatt Ocean, Dobrogea rift, and Polish–Danish Aulacogen remain quite speculative. The Penninic domain, which belongs to the Alpine Tethys, was opening during the Early–Middle Jurassic. The Alpine Tethys, that is, the Ligurian, the Penninic, and the Pieniny Klippen Belt and Magura oceans, constitutes the extension of the Central Atlantic system. Bill et al. (2001) date the onset of oceanic spreading of the Alpine Tethys by isotopic methods as Bajocian. According to Winkler and Slaczka (1994), the Pieniny data fit well with the supposed opening of the Ligurian–Penninic Ocean. During the Early Jurassic rift stage, the single basin existed in the future Pieniny–Magura realm. During the Bajocian–Bathonian, the Czorsztyn ridge originated. Plasienka (2002) postulated the thermal uplift above the distal, subcrustal part of detachment fault. The origin of the Czorsztyn ridge is coeval with the spreading phase of the Pieniny–Magura Ocean. However, it should be stressed that the existence of the oceanic crust beneath the Magura basin could be disputable (Winkler and Slaczka, 1994). The occurrence of the mafic (basalt) intrusions in the eastern termination of the Czorsztyn ridge in Noselica Klippen (Lashkevitsch et al., 1995) seems to support the thermal origin of the ridge related to the oceanic spreading. The orientation of the Pieniny Ocean was southwest–northeast (see discussion in Aubrecht and Tunyi, 2001; Golonka and Krobicki, 2001). The Pieniny–Magura Ocean was divided into the northwestern and southeastern branches. The deepest parts of both basins are documented by deep-water, extremely condensed, Jurassic–Early Cretaceous pelagic limestones and radiolarites. The shallowest ridge sequences are known as the Czorsztyn succession. In this succession, the Early Jurassic *Posidonia* marls are followed by Middle Jurassic–earliest Cretaceous crinoidal and nodular limestones and Late Cretaceous Ammonitico rosso and marly facies. The transitional slope sequences between the deepest basinal units and ridge units consist of mixed cherty, limestone, and marly facies. The northeastern part of the basin developed later into the Magura basin. The detailed study of the basinal facies (Golonka and Sikora, 1981) revealed an enormous condensation of *Nannoconus* limestones and radiolarites. The Upper Jurassic–Lower Cretaceous profiles do not exceed more than a dozen or so meters. In the external case, 2 m (6.6 ft) was deposited during a time span of 50 m.y. This Jurassic Ocean was connected with the older, Triassic Transylvanian Ocean. A junction of the eastern and Atlantic Tethys existed perhaps in the Eastern Slovakian–Ukrainian Carpathians and is represented by the Inacovce–Krichevo unit (Sotak et al., 2000). The opening of the Severin–Moldavidic basins was related to the gradual closing of the Transylvanian Ocean, which disappeared entirely during the Cretaceous. Shallow-water, calcareous sediments prevailed in the Severin–Moldavidic realm. Riftogenic magmatic rocks were connected with the Jurassic period: calc-alkaline rocks (andesites of Zegocina; Slaczka, 1998) and basic, ultrabasic rocks, e.g., Biala Woda, Kaminnyi, Potok, and Poiana Botizei, along the southern margin of the platform.

The next stage (Tithonian–Valanginian) is characterized by a rapid subsidence of local basins (Slaczka, et al., 1999), where calcareous flysch sedimentation started (Cieszyn area; Cieszyn beds and Eastern Carpathians Kamiennyi Potik–Black Flysch and Rachiv units; Slaczka, 1976). This calcareous detritus was probably derived from the erosion of the reef carbonate platform and deposited by turbidity currents. Pelagic, deep-water sedimentation occurred mainly in the Magura basin at the same time. The Penninic–Pieniny–Magura Ocean reached its maximum width in the latest Jurassic and then stopped spreading. Subduction jumped to the northern margin of the Inner Carpathian terranes and began to consume the Pieniny Klippen Belt Ocean (Birkenmajer, 1986; Golonka et al., 2000). The closing process of the Pieniny Klippen Belt basin is a matter of discussion, as summarized by Golonka and Krobicki (2001) according to earlier elaborations (e.g., Birkenmajer, 1986; Plasienka, 1999; Golonka et al., 2000). The latest Jurassic blueschist metamorphic rocks found as pebbles (exotics) in the Albian flysch in the Pieniny Basin indicate the existence of a subduction below the northern margin of the Inner Carpathian plate (Fayrad, 1997).

The following stage (Valanginian–Aptian) discerned the continuous spreading of the basin and regional subsidence. Sedimentation of black shales embraced almost the whole Outer Carpathian basin with local submarine clastic fans. Clastic material was derived or from the outer margin of the basin or from the uplifted remnants of the North European platform in the basin. Before the Albian, the Magura basin was separated from the Silesian basin by the Silesian submerged ridge (latest Jurassic–earliest Cretaceous). Basalt rocks started to intrude in the Eastern Carpathians (Kamiennyi Potik–Black Flysch; see Sandulescu, 1975), and diabase, teschenites, and syenites started to intrude in the Western Carpathians (Ksiazkiewicz, 1977). In the latter area, it lasted up to the Aptian (Lucinska-Anczkiewicz et al., 2002).

The next stage (Aptian–Albian) is characterized in the Eastern Carpathians by compressional movements that caused folding and overthrusting in the Marmarosh area during the Aptian and development of huge olistolithes in Marmarosh Klippen zone. The existence of Late Jurassic metamorphosed limestones also suggests periods of metamorphism during that time. It is still not certain whether the folding processes also embraced more outer areas (the Cordillera in front of the Dukla subbasin and Silesian Cordillera). On the western and more outer subbasins during the Albian, acceleration of the turbiditic sedimentation occurred. Only in the Skole–Skyba zone was continuation of black siliceous shales observed. At the beginning of the Aptian, a ridge between Silesian–Subsilesian subbasins and Skole subbasin developed, which gave a material to gaize sandstones in the Subsilesian area. South-dipping subduction was active on the southern margin of the Magura basin. Consumption of the Penninic–Pieniny Ocean led to the development of accretionary prism in front of the moving plates. In the Albian, synorogenic flysch developed in the Magura basin (Golonka and Sikora, 1981).

The following period lasted during the Cenomanian–Turonian. It was characterized by subsidence of all the subbasins, cessation of clastic sedimentation, slow and uniform sedimentation, and well-oxygenated conditions. It marked the beginning of a period of intensified, well-oxygenated, deep-water circulation in Tethyan and Proto-Atlantic Oceans (Slaczka and Kaminski, 1998). This period was characterized by very low rates of sedimentation (0.5–5 m/m.y.; 1.6–16 ft/m.y.; Poprawa et al., 2002). At the turn of the Cenomanian, radiolarian shales followed by red clays with intercalations of basinal turbidites were deposited below the CCD (Oszczypko, 1999). The oxic conditions prevailed, and the appearance of the red and green shales like the Malinowa Formation in the Magura basin (Birkenmajer and Oszczypko, 1989) is characteristic. In the northern and middle part of the Magura basin, this type of sedimentation persisted up to the Campanian, whereas in the Krynica zone, this type of sedimentation persisted up to the Maastrichtian (Oszczypko, 2001). The rate of sedimentation of variegated shales oscillated between 15 and 25 m/m.y. (49 and 82 ft/m.y.). In the northern part of the Outer Carpathian, in the Skole–Skyba subbasin, this type of sedimentation terminated during the Turonian (Kotlarczyk, 1985).

The Late Cretaceous–early Miocene period is characterized by compressional movements, appearance of intensive turbiditic sedimentation, and increased rate of subsidence in the basins. Late Cretaceous Laramian tectonic movements, very intensive in the Inner Carpathians, resulted in the enlivening of intrabasinal source areas and in the creation of several sedimentary subbasins, each with specific deep-water clastic sedimentation. The uplift of the southern margin of the North European platform resulted in a tectonic inversion, a general regression of the sea, and erosion. The eroded material supplied the Skole subbasin, and the first appearance of intensive clastic sedimentation during the Turonian occurred in this outermost subbasin.

In the more southern subbasin, the Silesian subbasin, clastic sedimentation appeared on the Turonian–Coniacian boundary in its western part. The clastic material was delivered from the Silesian ridge (Cordillera) and rejuvenated along a system of longitudinal faults, and several clastic submarine fans or faulted submarine slope aprons developed, each with a specific heavy-mineral composition (Wieser, 1970). Probably, the Silesian Cordillera was, in reality, a system of ridges divided by more local basins, some of which later became the Fore-Magura units. The most coarse-grained sediments were deposited along the southern deepest part of the Silesian subbasin. Toward the north and northeast, these deposits were gradually replaced, at first by variegated shales, and eventually by variegated marls that deposited on a slope of the Subsilesian ridge. This slope, also known as Subsilesian basin, was occupied by a pelagic marly sedimentation from the Coniacian–Santonian to Paleocene. Sedimentation of variegated marls also occurred on the southern slope of the Silesian subbasin (Slaczka and Gasinski, 1985) that belonged to the Silesian ridge system, which separated the Silesian subbasin from more inner Fore-Magura subbasins. A diachronous migration of the first appearance of clastic material in the Silesian subbasin from the west toward the east occurred, near the Polish–Ukrainian border; it started during the Santonian–Maastrichtian, and it was probably connected with the migration of the uplifting of the Silesian Cordillera from the west toward the east. In the Dukla and Porkulets subbasins, clastic sedimentation started during the Coniacian from the source areas that were created along the northeastern margin of the Dukla subbasin on the prolongation of the Silesian Cordillera and continued toward the southeast.

During the Maastrichtian–Paleocene, a considerable reorganization of the Magura basin occurred. This was connected both with the compression at the southern margin of the basin and an inversion of the northern margin (Silesian uplifted ridge system). It was accompanied by a deposition of the upper Senonian–Paleocene flysch (the so-called Inoceramian beds). The rate of sedimentation oscillated between 50 and 75 m/m.y. (160 and 250 ft/m.y.). This clastic sedimentation started in the Magura basin during the Campanian along its northern margin and during the Maastrichtian along the southern margin, where clastic material was delivered generally from the south (Oszczypko, 1975). At the convergent, southern margin of the Magura basin, subduction of the oceanic and thinned continental crust beneath the Czorsztyn ridge occurred during the Maastrichtian and Paleocene, and the Pieniny Klippen Belt was formed. As a result, the Magura basin became trenchlike in its character.

The intensive sedimentation from uplifting source areas lasted generally up to the end of the Paleocene, except in the Silesian subbasin, where it lasted until the late Eocene. Such distribution of the first period of clastic sedimentation implies that, in response to the tectonic movement in the Inner Carpathians, several bulges were created in their foreland, earlier in more distal parts and consequently later and later toward the inner margin of the Outer Carpathian basin. Since the Paleocene–early Eocene, the accretionary prism began to develop in the southern margin of the Magura basin (Oszczypko, 1992, 1999, 2001), close to the fold and thrust Pieniny Klippen Belt. The moving load in front of this accretionary prism has caused subsidence and a progressive northward shift of depocenters. The early Eocene axis of deposition was located in the Krynica zone and then, during the middle and late Eocene, it migrated northward, toward the Bystrica and Raca zones, respectively. In this part of the basin, a narrow and very long submarine fan (a few hundred kilometers) was formed. The clastic immature material of the fan was supplied from a southeast direction and was derived from the erosion of the exposed part of the accretionary prism. During the early to middle Eocene, the deepest part of the basin, commonly beneath the CCD, was located in the northern part of the basin. The rate of sedimentation varied from 10–15 m/m.y. (32–50 ft/m.y.) on the abyssal plain to 75–100 m/m.y. (246–330 ft/m.y.) in the outer fan and between 200 and 300 m/m.y. (660 and 1000 ft/m.y.) in the area affected by the middle fan-lobe system (Oszczypko, 2001). Simultaneously, along the northern margin of the basin (Siary zone), small fans developed. In the Moldavidic realm during the early and middle Eocene, the intense tectonic movement ceased, and the remaining Outer Carpathian subbasins were dominated by a hemipelagic deposition (green and variegated shales) with an intercalation of thin-bedded turbidites. Only locally within the southeast part of the Silesian and in Dukla subbasins did sporadically clastic fans develop. Sedimentation in the Magura basin was strongly influenced by the development of an accretionary prism along its inner margin (see also Misik et al., 1991).

In the late Eocene, the southern part of the Magura basin was affected by tectonic movements (see Ksiazkiewicz and Lesko, 1959; Oszczypko and Zytko, 1987; Royden and Baldi, 1988; Seifert, 1992) and connected with the continuous subduction of the southern margin of the basin beneath the Pieniny Klippen Belt (Oszczypko, 1992). At the turn of Eocene, in the whole Outer Carpathian basin, unification of the sedimentary condition occurred, and green shales and, later on, *Globigerina* marls were deposited.

As a result of the late Eocene tectonic movements, the Carpathian basins were partly isolated from the Mediterranean area. This was the beginning of the Early Parathethys formation (Kovac et al., 1993). This isolation caused the deposition of the black shales that are rich in organic matter (Menilite shales). This type of anoxic sedimentation persisted in the Skole basin until the Oligocene–Miocene boundary. The Magura accretionary wedge was completed before the late Oligocene. The late Oligocene folding and northward thrusting in the Magura nappe was almost contemporaneous with that of the Northern Calcareous Alps and Rhenodanubian flysch. Simultaneously, the southern part of the Magura nappe was transformed into the piggyback basin flooded during the Aquitanian–early Burdigalian (Oszczypko et al., 1999b; Oszczypko and Oszczypko-Clowes, 2002). At this point in time, the front of the overriding Magura nappe reached the Silesian ridge, which collapsed as a result of the lithosphere flexure, and in the southern part of the Silesian subbasin, in front of the advancing nappe, huge olistolithes were deposited (Slaczka and Oszczypko, 1987). The Outer Carpathian flysch basin attained features of a relatively deep-water foreland basin (Golonka et al., 2000). During the Oligocene, the main area of subsidence and deposition of the Menilite and Krosno beds were shifted from the Dukla to the Silesian subbasin (Slaczka, 1969). In the next stage (early Burdigalian), the axis of deposition was transferred to the Skole; the frontal part of the Silesian–Subsilesian nappes became uplifted, and the inner part subsided, creating probably locally piggyback basins. Blocks of Weglowka marls slipped into the Skole subbasin from the uplifted Subsilesian nappe. In the Boryslav–Pokuttya residual flysch basin, the turbiditic deposition was replaced by salt-bearing molasses deposits (Worotysche Formation). Sedimentation in this basin was probably completed during the Ottnangian, whereas the whole former flysch basin was uplifted. According to Veto (1987), the anoxia in the Carpathian flysch basin was caused by very high plankton productivity, whereas nutrients necessary for it were most likely supplied by upwelling.

ACKNOWLEDGMENTS

This work was partly supported by the grant from the Jagiellonian University and by grants from the Polish Committee for Scientific Research Komitet Badan Naukowych grants no. 6 P04D 02022 19 and AGH grant 11.11.140.159. We are indebted to Lucyna Dunczyk, Elzbieta Machaniec, and Bartlomiej Rogozinski for their help in preparing the figures. We are grateful to the referees for their reviews.

REFERENCES CITED

Airoldi, C., 2001, The Pienides in Maramures—Sedimentation, tectonics and paleogeography: Cluj-Napoca, Romania, Cluj University Press, 156 p.

Aubrecht, R., and I. Tunyi, 2001, Original orientation of neptunian dykes in the Pieniny Klippen Belt (Western Carpathians): The first results: Contributions to Geophysics and Geodesy, v. 31, p. 557–578.

Balintoni, I., 1998, An evolutionary model for the rift of external Carpathian flysch basin: Studia Universitatis Babes-Bolyai: Geologia, v. 42, p. 119–131.

Balintoni, I. C., 2001, The structure of the Romanian territory, *in* I. I. Bucur, S. Filipescu, and E. Sasaran, eds., Algae and carbonate platforms in the western part of Romania: 4th Regional Meeting of International Fossil Algae Association, August 29–September 5, 2001: Field Trip Guide Book: Cluj-Napoca, Romania, Cluj University Press, p. 1–8.

Bieda, F., S. Geroch, L. Koszarski, M. Ksiazkiewicz, and K. Zytko, 1963, Stratigraphie des Karpates externes polonaises: Biuletyn Instytutu Geologicznego, v. 182, p. 5–174.

Bill, M., L. O'Dogherty, J. Guex, P. O. Baumgartner, and H. Masson, 2001, Radiolarite ages in Alpine–Mediterranean ophiolites: Constraints on the oceanic spreading and the Tethys–Atlantic connection: Geological Society of America Bulletin, v. 113, p. 129–143.

Birkenmajer, K., 1977, Jurassic and Cretaceous lithostratigraphic units of the Pieniny Klippen Belt, Carpathians, Poland: Studia Geologica Polonica, v. 45, p. 1–158.

Birkenmajer, K., 1986, Stages of structural evolution of the Pieniny Klippen Belt, Carpathians: Studia Geologica Polonica, v. 88, p. 7–32.

Birkenmajer, K., 1988, Exotic Andrusov Ridge: Its role in plate-tectonic evolution of the West Carpathian fold belt: Studia Geologica Polonica, v. 91, p. 7–37.

Birkenmajer, K., and N. Oszczypko, 1989, Cretaceous and Palaeogene lithostratigraphic units of the Magura nappe, Krynica subunit, Carpathians: Annales Societatis Geologorum Poloniae, v. 59, p. 145–181.

Birkenmajer, K., H. Kozur, and R. Mock, 1990, Exotic Triassic pelagic limestone pebbles from the Pieniny Klippen Belt of Poland: A further evidence for early Mesozoic rifting in West Carpathians: Annales Societatis Geologorum Poloniae, v. 60, p. 3–44.

Bombita, G. E., and G. Pop, 1991, Mesozoic formations from Poiana Botizii, Pieniny Klippen Belt of Romania: Geologica Carpathica, v. 42, p. 139–146.

Bombita, G., E. Antonescu, E. Malata, C. Muller, and T. Neagu, 1992, Pieniny-type Mesozoic formation from Maramures, Romania (second part): Acta Geologica Hungarica, v. 35, p. 117–144.

Borys, Z., and B. Cisek, 1989, New perspectives on finding hydrocarbon deposits in the Lower Cretaceous sandstones in the Skole unit: Nafta (Katowice), v. 10–12, p. 142–147.

Burtan, J., 1968, The pre-Magura scale: XXIII International Geological Congress (Prague, 1968): Guide to Excursion C44—Geology of the Polish Flysch Carpathians: Warsaw, Poland, Wydawnictwa Geologiczne, p. 14–16.

Burtan, J., J. Golonka, N. Oszczypko, and Z. Paul, 1981, Mapa Geologiczna Polski, Arkusz Nowy Sacz: Instytut Geologiczny geological map of Poland (in Polish): Warszawa, Geological Institute Publishing House, scale 1:200,000, 1 sheet.

Burtan, J., J. Chowaniec, and J. Golonka, 1984, Wstepne wyniki badan nad egzotycznymi skalami weglanowymi z zachodniej czesci polskich Karpat fliszowych (Preliminary results of studies on exotic carbonate rocks in the western part of the Polish Flysch Carpathians): Biuletyn Instytutu Geologicznego, v. 346, p. 147–159.

Cieszkowski, M., 1992, Michalczowa zone: A new unit of the Fore-Magura zone, Outer West Carpathians, south Poland (in Polish with English summary): Zeszyty Naukowe Akademia Gorniczo-Hutnicza, v. 1463, p. 1–125.

Cieszkowski, M., A. Slaczka, and S. Wdowiarz, 1985, New data on structure of the Flysch Carpathians: Przeglad Geologiczny, v. 6, p. 313–333.

Cieszkowski, M., N. Oszczypko, and W. Zuchiewicz, 1989, Upper Cretaceous siliciclastic-carbonate turbidites at Szczawa, Magura nappe, West Carpathians, Poland: Bulletin of the Polish Academy of Sciences. Earth Sciences, v. 37, p. 231–245.

Dercourt, J., L. E. Ricou, and B. Vrielynck, eds., 1993, Atlas Tethys paleoenvironmental maps: Paris, Gauthier-Villars, 307 p.

Dziadzio, P. S., 2006, The Geology of the Weglowka oil field, Subsilesian unit, Polish Outer Carpathians, *in* J. Golonka and F. J. Picha, eds., The Carpathians and their foreland: Geology and hydrocarbon resources: AAPG Memoir 84, p. 467–476.

Dziadzio, P. S., M. A. Enfield, M. P. Watkinson, and S. J. Porebski, 2006a, The Ciezkowice Sandstones: Examples of basin-floor fan-stacking patterns from the main (upper Paleocene to Eocene) reservoir in the Polish Carpathians, *in* J. Golonka and F. J. Picha, eds., The Carpathians and their foreland: Geology and hydrocarbon resources: AAPG Memoir 84, p. 477–496.

Dziadzio, P. S., et al., 2006b, Hydrocarbon resources of the Polish outer Carpathians—reservoir parameters, trap types, and selected hydrocarbon fields: A stratigraphic review, *in* J. Golonka and F. J. Picha, eds., The Carpathians and their foreland: Geology and hydrocarbon resources: AAPG Memoir 84, p. 259–291.

Fayrad, S. W., 1997, Petrological model for blueschist facies metamorphism in the Pieniny Klippen Belt, *in* P. Grecula, D. Hovorka, and M. Putis, eds., Geological evolution of the Western Carpathians: Mineralia Slovaca Monograph, Bratislava, p. 155–162.

Gasinski, M. A., 1998, Campanian–Maastrichtian palaeoecology and palaeobiogeography of the Andrychow Klippes, Outer Carpathians, Poland: Wydaw: Krakow, Uniwersytetu Jagiellonskiego, 90 p.

Geroch, S., A. Jednorowska, M. Ksiazkiewicz, and J. Liszkowa, 1967, Stratigraphy based upon foraminifera of the Western Polish Carpathians: Biuletyn Instytutu Geologicznego, v. 211, p. 185–267.

Glushko, W. W., and S. S. Kruglov, 1986, Tectonic map of the Ukrainian Carpathians: Kiev, Ministerstvo Geologii, scale 1:200,000, 6 sheets.

Golonka, J., 1981, Arkusz Bielsko-Biala; Objasnienia do mapy geologicznej Polski (Geological map of Poland, explanations; Bielsko-Biala sheet): Warsaw, Geological Institute Publishing House, p. 1–63.

Golonka, J., 2004, Plate tectonic evolution of the southern margin of Eurasia in the Mesozoic and Cenozoic: Tectonophysics, v. 381, p. 235–273.

Golonka, J., and M. Krobicki, 2001, Upwelling regime in the Carpathian Tethys: A Jurassic–Cretaceous palaeogeographic and paleoclimatic perspective: Geological Quarterly, v. 45, p. 15–32.

Golonka, J., and M. Lewandowski, eds., 2003, Geology, geophysics, geothermics and deep structure of the West Carpathians and their basement: Institute of Geophysics Polish Academy of Sciences Monographic Volume, v. M-28, no. 363, p. 1–184.

Golonka, J., and W. Raczkowski, 1984, Arkusz Piwniczna; Objasnienia do szczegolowej mapy geologicznej Polski (Detailed geological map of Poland) (in Polish): Warsaw, Geological Institute Publishing House Wydawnictwa Geologiczne, p. 85, scale 1:50,000, 1 sheet.

Golonka, J., and W. Sikora, 1981, Microfacies of the Jurassic and Lower Cretaceous sedimentarily thinned deposits of the Pieniny Klippen Belt in Poland (in Polish with English abstract): Biuletyn Instytutu Geologicznego, v. 31, p. 7–37.

Golonka, J., N. Oszczypko, and A. Slaczka, 2000, Late Carboniferous–Neogene geodynamic evolution and paleogeography of the circum-Carpathian region and adjacent areas: Annales Societatis Geologorum Poloniae, v. 70, p. 107–136.

Golonka, J., M. Krobicki, N. Oszczypko, A. Slaczka, and T. Slomka, 2003, Geodynamic evolution and palaeogeography of the Polish Carpathians and adjacent areas during neo-Cimmerian and preceding events (latest Triassic–earliest Cretaceous), *in* T. McCann and A. Saintot, eds., Tracing tectonic deformation using the sedimentary record: Geological Society (London) Special Publication 208, p. 138–158.

Golonka, J., L. Gahagan, M. Krobicki, F. Marko, N. Oszczypko, and A. Slaczka, 2006, Plate-tectonic evolution and paleogeography of the circum-Carpathian region, *in* J. Golonka and F. J. Picha, eds., The Carpathians and their foreland:

Geology and hydrocarbon resources: AAPG Memoir 84, p. 11–46.

Guterch, A., M. Grad (reporters) and Polonaise '97 and Celebration 2000 Working Group, 2001, New deep seismic studies of the lithosphere in Central Europe: Polonaise '97 and Celebration 2000 seismic experiments: Biuletyn Panstwowego Instytutu Geologicznego, v. 396, p. 61.

Jucha, S., and J. Kotlarczyk, 1961, Le series des couches a Menilite et des couches de Krosno dans le flysch des Karpates: Pace Geologiczne Polska Academia Nauk, Oddzial a Krakowie, v. 4, p. 1–115.

Korab, T., and T. Durkovic, 1978, Geology of the Dukla unit (east Slovakian flysch) (in Slovakian with English summary): Bratislava, Geologicky Ustav D. Stura, 194 p.

Koszarski, L., ed., 1985, Geology of the middle Carpathians and the Carpathian foredeep: Guide to Excursion 3, Carpatho-Balkan Geological Association XIII Congress, Krakow, Poland, 254 p.

Koszarski L., and A. Koszarski, 1985, Dukla unit and its relation to Silesian nappe: Marginal zone of the Magura nappe: Tectonic windows with Grybow unit series: Jaslo nappe, *in* L. Koszarski, ed., Geology of the Middle Carpathian Foredeep, Guide to excursion 3, Carpatho-Balkan Association XIII Congress, Cracow, Poland, Geological Institute, Poland, p. 185–225.

Koszarski L., and A. Slaczka, 1976, The Outer (Flysch) Carpathians: The Cretaceous, *in* S. Cieslinski, ed., Geology of Poland, v. I, Stratigraphy part 2: Instytut Geologiczny, Warszawa, p. 495–498, 657–679, 740–748.

Koszarski, L., and K. Zytko, 1961, Jaslo shales within the Menilite-Krosno series in the Middle Carpathians: Biuletyn Instytutu Geologicznego, v. 166, p. 87–213.

Koszarski, L., W. Sikora, and S. Wdowiarz, 1974, The Flysch Carpathians: Polish Carpathians, *in* M. Mahel, ed., Tectonics of the Carpathian–Balkan regions: Bratislava, Geologicky Ustav Donuza Stura, p. 180–197.

Kotarba, M., and Y. V. Koltun, 2006, The origin and habitat of hydrocarbons of the Polish and Ukrainian parts of the Carpathian province, *in* J. Golonka and F. J. Picha, eds., The Carpathians and their foreland: Geology and hydrocarbon resources: AAPG Memoir 84, p. 395–442.

Kotlarczyk, J., 1978, Stratigraphy of the Ropianka Formation in the Skole unit of the Flysch Carpathians (in Polish with the English summary): Polska Academia Nauk Oddzial w Krakowie. Komisja Nauk Geologicznych Prace Geologiczne, v. 108, p. 1–82.

Kotlarczyk, J., 1985, An outline of the stratigraphy of marginal tectonic units of the Carpathian orogen in the Krakow–Przemysl area: Guide to Excursion 4. Carpatho-Balkan Geological Association—XIII Congress, Krakow, Poland, 171 p.

Kovac, M., and D. Plasienka, eds., 2002, Geological structure of the Alpine–Carpathian–Pannonian junction and neighbouring slopes of the Bohemian Massif: Bratislava, Comenius University, 84 p.

Kovac, M., A. Nagymarosy, J. Sotak, and K. Sutovska, 1993, Late Tertiary paleogeographic evolution of the Western Carpathians: Tectonophysics, v. 226, p. 401–415.

Kovac, M., A. Nagymarosy, N. Oszczypko, A. Slaczka, L. Csontos, M. Marunteanu, L. Matenco, and M. Marton, 1998, Palinspastic reconstruction of the Carpathian–Pannonian region during the Miocene, *in* M. Rakus, ed., Geodynamic development of the Western Carpathians: Bratislava, Geological Survey of Slovac Republic, p. 189–217.

Kozikowski, H., 1956, Pisarzowa unit, a new tectonic unit of the Polish Flysch Carpathians: Biuletyn Instytutu Geologicznego, v. 110, p. 93–121.

Kozikowski, H., 1958, Geologia Centralnej Depresji Karpackiej między Żmigrodem a Sanokiem: Acta Geologica Polonica, v. 8, p. 471–500.

Kropotkin, P. N., 1991, Nappes of the Maramures zone of the Eastern Carpathians: Geotectonics, v. 25, p. 46–52.

Kruglov, S. S., 1965, O prirode marmaroshskikh utiosov Sovetskikh Karpat (About the nature of the Marmarosh Klippens of the Soviet Carpathians) (in Russian): Geologicheskiy Sbornik Lvovskogo Geologiceskogo Obsestva, v. 9, p. 41–54.

Kruglov, S. S., 1969, Pro stratyhrafichne polozhennya pudynhovych piskovykiv i hravelitiv v zoni peninskyh skel (Zakarpattia) (About stratigraphical position of pudding sandstones and gritstones within the Pieniny Klippen zone (Zakarpatia) (in Ukrainian): Proceedings of the Academy of Sciences of the Ukrainian SSR, v. 10, p. 874–877.

Kruglov, S. S., 1989, Geodynamics of the Ukrainian Carpathians: Geologica Carpathica, v. 40, p. 101–123.

Kruglov, S. S., 1999, V zashchitu Predkarpatskoho prohiba (In advocacy of the pre-Carpathian Foredeep) (in Russian): Geodynamics (L'viv), v. 1, p. 88–91.

Kruglov, S. S., 2001, The problems of tectonics and paleogeodynamics of western Ukraine (a critical survey of the new publications) (in Ukrainian): L'viv, Ukrainian Interdisciplinary Committee for Tectonics, 83 p.

Kruglov, S. S., and W. I. Tarasenko, 1984, Brachysynclinal folds in the Sovjet Carpathians (in Russian): Gielogiceskij Zurnal, v. 2, p. 28–37.

Ksiazkiewicz, M., 1951, Kreda Karpat Zewnętrznych, *in* M. Ksiazkiewicz, ed., Regionalna Geologia Polski, Karpaty, Stratygrafia: Rocznik Polskiego Towarzystwa Geologicznego, v. 1, p. 86–110.

Ksiazkiewicz, M., ed., 1962, Geological atlas of Poland: Stratigraphic and facial problems: Instytut Geologiczny: Warsaw, Poland, Wydawnictwa Geologiczne, 16 maps.

Ksiazkiewicz, M., 1977, Tectonics of the Carpathians, *in* W. Pozaryski, ed., Geology of Poland, v. IV, Tectonics: Warsaw, Poland, Wydawnictwa Geologiczne, p. 476–604.

Ksiazkiewicz, M., and B. Lesko, 1959, On relation between the Krosno and Magura Flysch: Bulletin of the Polish Academy of Sciences Series, Chemistry, Geology, and Geography, v. 7, p. 773–780.

Kucharski, R., S. Doktor, M. Graniczny, M. Molek, and B. Dabrowska, 1990, Wgłębna budowa geologiczna Karpat w świetle kompleksowej analizy teledetekcyjno-geofizycznej: Przegląd Geologczny, v. 11, p. 469–476.

Lashkevitsch, Z. M., A. P. Medvedev, and Y. Z. Krupskiy, 1995, Tectonomagmatic evolution of Carpathians (in Russian): Kiev, Naukova Dumka, p. 1–113.

Lesko, B., and O. Samuel, 1968, Geology of the Eastern Slovakian

Flysch (in Slovakian with English summary): Bratislava, Slovenska Akademia ved, 256 p.

Lillie, R. J., M. Bielik, V. Babuska, and J. Plomerova, 1994, Gravity modelling of the lithospere in the eastern Alpine–Western Carpathian–Pannonian Basin region: Tectonophysics, v. 231, p. 215–235.

Lucinska-Anczkiewicz, A., A. Slaczka, I. M. Villa, and R. Anczkiewicz, 2000, $^{39}Ar/^{40}Ar$ dating of the teschenite association rocks from the Polish Outer Carpathians: Mineralogical Society of Poland Special Papers 17, p. 241–242.

Lucinska-Anczkiewicz, A., I. M. Villa, R. Anczkiewicz, and A. Slaczka, 2002, $^{40}Ar/^{39}Ar$ dating of alkaline lamprophyres from the Polish Western Carpathians: Geologica Carpathica, v. 53, p. 45–52.

Mahel, M., et al., 1968, Regional geology of Czechoslovakie: Part II, The West Carpathians: Praha, Geological Survey of Czechoslovakia, 723 p.

Malata, E., 2001, Interpretacja biostratygraficzna zespołów malych otwornic w środkowej części płaszczowiny magurskiej (polskie Karpaty zewnętrzne), Rozprawa doktorska w Instytucie Geologicznym Uniwersytetu Jagiellońskiego: Ph.D. dissertation, Jagiellonian University, Krakow, 273 p.

Malata E., T. Malata, and N. Oszczypko, 1996, Litho- and biostratigraphy of the Magura nappe in the eastern part of the Beskid Wyspowy Range (Polish Western Carpathians): Annales Societatis Geologorum Poloniae, v. 66, p. 269–283.

Mastella, L., 1988, Structure and evolution of Mszana Dolna tectonic window, Outer Carpathians, Poland (in Polish with English summary): Annales Societatis Geologorum Poloniae, v. 58, p. 53–173.

Matyasik, I., and P. S. Dziadzio, 2006, Reconstruction of petroleum systems based on integrated geochemical and geological investigations: Selected examples from the Middle, Outer Carpathians in Poland, *in* J. Golonka and F. J. Picha, eds., The Carpathians and their foreland: Geology and hydrocarbon resources: AAPG Memoir 84, p. 497–518.

Misik M., M. Sykora, and J. Jablonsky, 1991, Strihovske zlepence a juhomagurska kordilera: Zapadne Karpaty: Seria Geologicka, v. 14, p. 7–72.

Mitura, F., and T. Birecki, 1966, Geological structure of the Carpathians between Korczyna and Domaradz (in Polish with English summary): Prace Instytutu Naftowego, 3, Instytut Naftowy, Krakow, Poland, 52 p.

Olszewska, B., and J. Wieczorek, 2001, Jurasic sediments and microfossils of the Andrychow Klippes (Outer Western Carpathians): Geologica Carpathica, v. 52, p. 217–228.

Oszczypko, N., 1973, Budowa geologiczna Kotliny Sądeckiej: Biuletyn Instytutu Geologicznego, v. 271, p. 101–197.

Oszczypko, N., 1975, Exotic rocks in the Palaeogene of the Magura nappe between Dunajec and Poprad rivers, Carpathians, Poland: Rocznik Polskiego Towarzystwa Geologicznego, v. 45, p. 403–431.

Oszczypko, N., 1991, Stratigraphy of the Palaeogene deposits of the Bystrica subunit (Magura nappe, Polish Outer Carpathians): Bulletin of the Polish Academy of Sciences. Earth Sciences, v. 39, p. 415–431.

Oszczypko, N., 1992, Late Cretaceous through Paleogene evolution of Magura basin: Geologica Carpathica, v. 43, p. 333–338.

Oszczypko, N., 1999, From remnant oceanic basin to collision-related foreland basin—A tentative history of the Outer Western Carpathians: Geologica Carpathica, v. 40, special issue, p. 161–163.

Oszczypko, N., 2001, Magura unit—General geology, *in* K. Birkenmajer and M. Krobicki, eds., Carpathian palaeogeography and geodynamics: A multidisciplinary approach: Field Trip C, 2nd Meeting of the Association of European Geological Societies, Krakow 2001, Panstwowy Instytut Geologiczny, p. 173–177.

Oszczypko, N., and M. Oszczypko-Clowes, 2002, The new findings of lower Miocene deposits in the Nowy Sacz area (Magura nappe, Polish Outer Carpathians): Geological Quarterly, v. 46, p. 15–30.

Oszczypko, N., and A. Slaczka, 1985, An attempt at palinspastic reconstruction of Neogene basin in the Carpathian foredeep: Annales Societatis Geologorum Poloniae, v. 55, p. 55–75.

Oszczypko, N., and A. Thomas, 1985, Tectonic evolution of marginal part of the Polish Flysch Carpathians in the middle Miocene: Kwartalnik Geologiczny, v. 29, p. 109–128.

Oszczypko, N., and W. Zuchiewicz, 2000, Jointing in Eocene flysch strata of the mid-eastern Magura nappe, Polish Outer Carpathians: Implications for the timing deformation: Slovak Geological Magazine, v. 6, p. 441–456.

Oszczypko, N., and K. Zytko, 1987, Main stages of evolution of the Polish Carpathians during the late Palaeogene and Neogene times, *in* Y. G. Leonov and V. E. Khain, eds., Global correlation of tectonic movements: Chichester, J. Wiley and Sons Ltd., p. 187–198.

Oszczypko, N., J. Dudziak, and E. Malata, 1990, Stratygrafia osadów płaszczowiny magurskiej (kreda-paleogen) w Beskidzie Sądeckim, Karpaty: Studia Geologica Polonica, v. 97, p. 109–181.

Oszczypko, N., P. Krzywiec, M. Lemberger, M. Stefaniuk, K. Pietsch, and H. Trygar, 1998, Integrated geological-geophysical interpretation of the Rzeszow-Smilno profile (Western Carpathians) (abs.): Carpathian–Balkan Geological Association XVI Congress, Vienna, 1998, Abstracts, p. 446.

Oszczypko, N., E. Malata, and M. Oszczypko-Clowes, 1999a, Revised position and age of the Eocene deposits on the northern slope of the Gorce Range (Bystrica subunit, Magura nappe): Slovak Geological Magazine, v. 5, p. 235–254.

Oszczypko N., A. S. Andreyeva-Grigorovich, E. Malata, and M. Oszczypko-Clowes, 1999b, The lower Miocene deposits of the Raca subunit near Nowy Sacz (Magura nappe, Polish Outer Carpathians): Geologica Carpathica, v. 50, p. 419–433.

Oszczypko, N., P. Krzywiec, I. Popadyuk, and T. Peryt, 2006, Carpathian foredeep basin (Poland and Ukraine): Its sedimentary, structural, and geodynamic evolution, *in* J. Golonka and F. J. Picha, eds., The Carpathians and their foreland: Geology and hydrocarbon resources: AAPG Memoir 84, p. 293–350.

Oszczypko-Clowes, M., 1998, Late Eocene–Oligocene calcareous nannoplankton and stable isotopes (^{13}C, ^{18}O) of the

Globigerina marls in the Magura nappe (West Carpathians): Slovak Geological Magazine, v. 4, p. 107–120.

Oszczypko-Clowes, M., 2001, The nannofossils biostratigraphy of the youngest deposits of the Magura nappe (east of the Skawa river, Polish Flysch Carpathians) and their palaeoenvironmental conditions: Annales Societatis Geologorum Poloniae, v. 71, p. 139–188.

Pasternak, S. I., 1989, Stratotypes of the Cretaceous and Palaeogene deposit of the Ukrainian Carpathians: Kijev, Naukova Dumka Publishers, 204 p.

Pescatore, T., and A. Slaczka, 1984, Evolution models of two flysch basins: The Northern Carpathians and the southern Apennines, Tectonophysics, v. 106, p. 49–70.

Picha, F. J., 1996, Exploring for hydrocarbons under thrust belts: A challenging new frontier in the Carpathians and elsewhere: AAPG Bulletin, v. 80, no. 10, p. 1547–1564.

Picha, F. J., Z. Stranik, and O. Krejci, 2006, Geology and hydrocarbon resources of the Outer Western Carpathians and their foreland, Czech Republic, *in* J. Golonka and F. J. Picha, eds., The Carpathians and their foreland: Geology and hydrocarbon resources: AAPG Memoir 84, p. 49–175.

Plasienka, D., 1999, Tektonochronológia a paleotektonický model jursko-kriedového vývoja centralnych Západných Karpat (Tectonochrology and paleotectonic model of the Jurassic–Cretaceous evolution of the Central Western Carpathians): Bratislava, Veda, 127 p.

Plasienka, D., 2000, Paleotectonic controls and tentative palinspastic restoration of the Carpathians realm during the Mesozoic: Slovak Geological Magazine, v. 6, p. 200–204.

Plasienka, A., 2002, Early stages of tectonic evolution of the Pieniny Klippen Belt: Geolines, v. 14, p. 75–77.

Popadyuk, I., M. Vul, Y. Ladyzhensky, and P. Shpak, 2006, Petroleum geology of the Boryslav–Pokuttya zone, the Ukrainian Carpathians, *in* J. Golonka and F. J. Picha, eds., The Carpathians and their foreland: Geology and hydrocarbon resources: AAPG Memoir 84, p. 455–466.

Poprawa, P., T. Malata, and N. Oszczypko, 2002, Ewolucja tektoniczna basenów sedymentacyjnych polskiej części Karpat zewnętrznych w świetle analizy subsydencji: Przegląd Geologiczny, v. 50, p. 1092–1105.

Potfaj, M., 1993, Position and role of the Biele Karpaty unit in the Flysch Zone of the Western Carpathians (in Slovakian with English summary): Geologicke Prace, Spravy, v. 98, p. 55–78.

Rakus, M., M. Misik, J. Michalik, R. Mock, T. Durkovic, T. Korab, R. Marschalko, J. Mello, M. Polak, and J. Jablonsky, 1988, Paleogeographic development of the West Carpathians: Anisian to Oligocene: Memoires de la Societe Geologique de France, Nouvelle Series, v. 154, p. 39–62.

Royden, L. H., and T. Baldi, 1988, Early Cenozoic tectonics in paleogeography of the Pannonian and surrounding regions, *in* L. H. Royden and F. Horvath, eds., The Pannonian Basin: A study in basin evolution: AAPG Memoir 45, p. 1–16.

Rylko, W., and A. Tomas, 1995, Morfologia powierzchni stropowej podłoża krystalicznego Karpat w świetle badañ magneto-telurycznych: Kwartalnik Geologiczny, v. 39, p. 1–16.

Sandulescu, M., 1975, Essai de synthese structurale des Carpathes: Bulletin de la Societe Geologique de France, v. 17, p. 299–358.

Sandulescu, M., 1988, Cenozoic tectonic history of the Carpathians, *in* L. Royden and F. Horvath, eds., The Pannonian Basin: A study in basin evolution: AAPG Memoir 84, v. 45, p. 17–25.

Sandulescu, M., and M. Visarion, 2000, Crustal structure and evolution of the Carpathian–western Black Sea area: Romanian Geophysics, v. 7, suppl. 1, p. 228–231.

Sandulescu, M., H. Krautner, I. Balintoni, D. Russo-Sandulescu, and M. Micu, 1981, The structure of the East Carpathian (Moldavia–Muramures area): Carpathian–Balkan Geological Association, XII Congress Bucharest–Romania 1981, Guide to Excursion B1, p. 1–92.

Schnabel, W. G., 1992, New data on the flysch zone of the Eastern Alps in the Austrian sector and new aspects concerning the transition to the flysch zone of the Carpathians: Cretaceous Research, v. 13, p. 405–419.

Seifert, P., 1992, Palinspastic reconstruction of the Easternmost Alps between upper Eocene and Miocene: Geologica Carpathica, v. 43, p. 327–331.

Sikora, W., 1970, Geology of the Magura nappe between Szymbark Ruski and Nawojowa: Biuletyn Instytutu Geologicznego, v. 235, p. 5–121.

Slaczka, A., 1959, Stratigraphy of the Bystra scale, Middle Carpathians (in Polish with English summary): Biuletyn Instytutu Geologicznego, v. 131, p. 203–286.

Slaczka, A., 1969 Final stages of geosynclinal development in the southeast-part of the Polish Carpathians: Acta Geologica Academiae Scientiarum Hungaricae, v. 13, p. 331–335.

Slaczka, A., 1971, Geology of the Dukla unit (in Polish with English summary): Prace Instytutu Geologicznego, v. 1, p. 1–63.

Slaczka, A., ed., 1976, Atlas of paleotransport of detrital sediments in the Carpathian–Balkan Mountains system: Warszawa, Geological Institute, scale 1:2,000,000, 6 sheets.

Slaczka, A., 1980, Les Carpathes flysheuses, *in* R. Osika, ed., La geologie et les gites mineraux en Pologne: Biuletyn Instytutu Geologicznego, v. 251, p. 491–497, 604–615.

Slaczka, A., 1996a, Oil and gas in the Northern Carpathians, *in* G. Wessely and W. Liebl, eds., Oil and gas in Alpidic thrust belts and basins of Central and Eastern Europe: European Association of Geoscientists and Engineers Special Publication 5, p. 187–195.

Slaczka, A., 1996b, Oil and gas in the Ukrainian part of the Carpathians and their foredeep, *in* G. Wessely and W. Liebl, eds., Oil and gas in Alpidic thrust belts and basins of Central and Eastern Europe: European Association of Geoscientists and Engineers Special Publication 5, p. 17–21.

Slaczka, A., 1998, Age of the andesitic rocks in the Subsilesian unit (Outer Carpathians) (abs.): XVI Congress, Carpatho-Balkan Geological Association, Vienna, August 30–September 2, 1998, Abstracts, p. 559.

Slaczka, A., 2000, Dukla unit: A key between Western and Eastern Carpathians: Terra Nostra, Schriften der Alfred-Wegener-Stiftung 2000/1, Geology 2000, Vienna, April 14–17, 2000, p. 109.

Slaczka, A., and M. A. Gasinski, 1985, Paleobathymetric model of the upper Senonian flysch sediments of a part of the Silesian basin (Polish Carpathians): Cretaceous Research, v. 6, p. 207–218.

Slaczka, A., and M. A. Kaminski, 1998, Guidebook to excursions in the Polish Flysch Carpathians: Grzybowski Foundation Special Publication 6, p. 1–171.

Slaczka, A., and N. Oszczypko, 1987, Olistostromes and overthrusting in the Polish Carpathians: Annales Instituti Geologici Hungarici, v. 70, p. 287–292.

Slaczka, A., and R. Unrug, 1976, Trends of textural and structural variations in turbidite sandstone (Oligocene, Outer Carpathians): Rocznik Polskiego Towarzystwa Geologicznego, v. 46, p. 55–75.

Slaczka, A., N. Oszczypko, E. Malata, and M. Cieszkowski, 1999, An early history of the Outer Carpathians basin: Geologica Carpathica, v. 50, p. 170–172.

Sotak, J., 1986, Stratigraphy and typology of the Upper Triassic development in outer units in the West Carpathians (reconstructions from redeposits localised in the Silesian Cordillera area): Knihovnicka Zemniho Plynu a Nafty, v. 31, p. 1–53.

Sotak, J., A. Biron, R. Prokesova, and J. Spisiak, 2000, Detachment control of core complex exhumation and back-arc extension in the east Slovakian basin: Slovak Geological Magazine, v. 6, p. 130–132.

Stampfli, G. M., ed., 2001, Geology of the western Swiss alps—A guide book: Memoires de Geologie, Lausanne, v. 361, p. 1–195.

Stefanescu, M., O. Dicea, A. Butac, and D. Ciulavu, 2006, Hydrocarbon geology of the Romanian Carpathians, their foreland, and the Transylvanian basin, *in* J. Golonka and F. J. Picha, eds., The Carpathians and their foreland: Geology and hydrocarbon resources: AAPG Memoir 84, p. 521–567.

Stefaniuk, M., 2002, Basic structures of the Carpathian basement as a result of magnetotelluric sounding interpretation: Geologia Carpathica, v. 53, special issue (CD-ROM).

Stefaniuk, M., and A. Slaczka, 2002, Interpretation of the flysch cover structures based on magnetotelluric data (Polish Outer Carpathians): Geologica Carpathica, v. 53, special issue (CD-ROM).

Stranik, Z., J. Dvorak, O. Krejci, P. Muller, A. Prichystal, M. Suk, and C. Tomek, 1993, The contact of the North European platform with the West Carpathians: Journal of the Czech Geological Society, v. 38, p. 21–29.

Svabenicka, L., M. Bubik, O. Krejci, and Z. Stranik, 1997, Stratigraphy of Cretaceous sediments of the Magura group of nappes in Moravia (Czech Republic): Geologica Carpathica, v. 48, p. 179–191.

Tolman, A., 1990, Paleogeographic maps and profiles in the Eastern Alps and relationship of the Eastern Alps and neighboring terrains, *in* A. E. M. Nairn, ed., Evolution of the northern margin of Tethys: The results of International Geological Correlation Project 198, v. 3: Memoires de la Societe Geologique de France, Nouvelle Series, v. 154, p. 23–38.

Tomek, C., 1993, Deep crustal structure beneath the Central and Inner West Carpathians: Tectonophysics, v. 226, p. 417–431.

Tomek, C., and J. Hall, 1993, Subducted continental margin in the Carpathians of Czechoslovakia: Geology, v. 21, p. 535–538.

Veto, I., 1987, An Oligocene sink for organic carbon: Upwelling in the Paratethys? Palaeogeography, Palaeoclimatology, Palaeoecology, v. 60, p. 143–153.

Wdowiarz, S., 1985, On same aspects of the geological structure and oil and gas accumulation of the Carpathian synclinorium: Biuletyn Instytutu Geologicznego, v. 350, p. 5–52.

Wieser, T., 1970, Exotic rocks from the deposits of the Magura nappe: Biuletyn Instytutu Geologicznego, v. 235, p. 123–161.

Winkler, W., and A. Slaczka, 1994, A Late Cretaceous to Paleogene geodynamic model for the Western Carpathians in Poland: Geologica Carpathica, v. 45, p. 71–82.

Woznicki, J., and P. Sucha, 1989, Geophysical map of the Western Outer Carpathians and their foreland, *in* D. Porawa and J. Nemcok, eds., Geological atlas of the Western Outer Carpathians and their foreland: Warszawa, Panstwowy Instytut Geologiczny, Table 6.

Zytko, K., 1999, Correlation of the main structural units of the Western and Eastern Carpathians (in Polish with English summary): Prace Panstwowego Instytutu Geologicznego, v. 168, p. 135–164.

Zytko, K., et al., 1989, Geological map of the Western Outer Carpathians and their foreland without Quaternary formations, *in* D. Poprawa and J. Nemcok, eds., Geological atlas of the Western Carpathians and their foreland: Warszawa, Poland, Panstwowy Instytut Geologiczny, Table 1.

8

Dziadzio, Piotr S., Zenon Borys, Stanisław Kuk, Emil Masłowski, Jaromir Probulski, Małgorzata Pietrusiak, Adam Górka, Jan Moryc, Anna Baszkiewicz, Piotr Karnkowski, Paweł H. Karnkowski, Marek Pietrusiak, 2006, Hydrocarbon resources of the Polish Outer Carpathians – Reservoir parameters, trap types, and selected hydrocarbon fields: A stratigraphic review, *in* J. Golonka and F. J. Picha, eds., The Carpathians and their foreland: Geology and hydrocarbon resources: AAPG Memoir 84, p. 259–260.

Hydrocarbon Resources of the Polish Outer Carpathians—Reservoir Parameters, Trap Types, and Selected Hydrocarbon Fields: A Stratigraphic Review

Piotr S. Dziadzio, Zenon Borys, Stanisław Kuk, Emil Masłowski, Jaromir Probulski, Małgorzata Pietrusiak, Adam Górka, Jan Moryc
Polish Oil and Gas Company, Exploration Department, South Regional Division, Jaslo, Republic of Poland

Anna Baszkiewicz
Polish Oil and Gas Company, Exploration Department, Warsaw, Republic of Poland

Piotr Karnkowski
Polish Oil and Gas Company, Exploration Department, Warsaw, Republic of Poland

Paweł H. Karnkowski
Polish Oil and Gas Company, Exploration Department, Warsaw, Republic of Poland

Marek Pietrusiak
Polish Oil and Gas Company, Exploration Department, Branch Sanok, Republic of Poland

ABSTRACT

This chapter presents a stratigraphic review of reservoirs and their parameters, trap types, and important fields in all of the tectonic units (nappes) in the Polish Outer Carpathians, where hydrocarbon deposits have been discovered and exploited for more than 150 yr.

The first part of this chapter is an introduction to the information about the occurrence of reservoir rocks in the Carpathians; however, the variability of these parameters is commonly surprising. Well-known examples are present where reservoir parameters vary greatly even in the same field, but this is a separate problem, and this is only mentioned here. Despite the very large number of wells drilled in the Polish Carpathians, the quantity of detailed petrophysical data is not so large. Good-quality data have been obtained only during the last 30 yr. The most recent and best quality data originate from wells drilled by Polish Petroleum Industry in areas of known fields during research of deeper prospects and from recognized wells.

DOI:10.1306/985611M843071

The best reservoir data are from the Skole and Silesian nappes, and these are presented in this chapter in great detail. Within the region of the Outer Polish Carpathians, reservoir rocks are found to have good potential as in shallow as well as deep structures in deposits of Lower Cretaceous to lower Oligocene in age.

The majority of hydrocarbon accumulations in the thrusted and folded Carpathians are within structural style traps. Exploration for them throughout the past 150 yr has enabled geologists to recognize their many different types, such as those related to thrust anticlines and folds, but which before were only interpreted as related to folds. Most of the oldest exploited oil accumulations in the Carpathians are of the contractional anticline type, commonly associated with thrusting. Most of these fields can only be illustrated by line-drawn sections based on drilling information because no seismic data are available. Some of the more spectacular traps in the Outer Carpathians are connected to disharmonic thrusted folds, tilted thrust faults, overturned frontal parts of thrust sheets, imbricate fan types, sandstone pinch-out, and traps sealed by asphalt.

In this chapter, selected and more important oil and gas fields that can be examples of characteristic hydrocarbon accumulations are also described. The most southern nappe is the Dukla unit, which lies beneath the Magura nappe, where six hydrocarbon accumulations have been discovered in the Oligocene Cergowa sandstone to date. A good example of the hydrocarbon accumulations and tectonic styles of the fields in the Dukla unit is the Slopnice–Limanowa oil and gas field. Here, the hydrocarbons have accumulated in recumbent thrust folds.

The Silesian nappe is represented by two important fields: the Bobrka oil field, which is located in the Bobrka anticline, and the Potok oil and gas field in the Potok anticline. Both of these fields have hydrocarbon accumulations in the Ciezkowice and Istebna reservoirs, which are trapped by thrust-related anticlines. The minor tectonic elements, such as thrust-related anticlines and synclines that separate the two fields, however, do not yield hydrocarbons.

The Bobrka oil field lies in the world's oldest area of petroleum exploration and production. This field is taken as the symbol of the Polish and international oil industry and is presented here from a historical point of view.

The Potok oil and gas field is located approximately 10 km (6 mi) to the north of the Bobrka oil field. It is one of the six most productive oil fields in the region and produces from an anticlinal structure more than 40 km (25 mi) long.

The Skole nappe is a similarly important unit for hydrocarbon exploration. Four oil fields have produced 1.7 million t of oil and more than 180 million m^3 (3.8 bcf) of gas from accumulations in the Menilite sandstone. One example of such an accumulation is seen in the Lodyna oil field. Here, the hydrocarbons are accumulated in a series of almost vertical beds of menilites within pinching out of Kliwa Sandstone.

9

Oszczypko, N., P. Krzywiec, I. Popadyuk, and T. Peryt, 2006, Carpathian Foredeep Basin (Poland and Ukraine): Its Sedimentary, Structural, and Geodynamic Evolution, *in* J. Golonka and F. J. Picha, eds., The Carpathians and their foreland: Geology and hydrocarbon resources: AAPG Memoir 84, p. 261 – 318.

Carpathian Foredeep Basin (Poland and Ukraine): Its Sedimentary, Structural, and Geodynamic Evolution

N. Oszczypko
Institute of Geological Sciences, Jagiellonian University, Krakow, Republic of Poland

P. Krzywiec
Polish Geological Institute, Warsaw, Republic of Poland

I. Popadyuk
Ukrainian State Geological Research Institute, L'viv, Ukraine

T. Peryt
Polish Geological Institute, Warsaw, Republic of Poland

ABSTRACT

The Polish and Ukrainian Carpathian Foredeep, about 600 km (376 mi) long and as much as 100 km (62 mi) wide, is part of the large sedimentary basin that stretches for more than 1300 km (816 mi) from the Danube in Vienna (Austria) to the Iron Gate on the Danube (Romania). To the west, the Carpathian Foredeep is linked with the Alpine Molasse Basin, and to the east, it passes into the Balkan foreland basin. Like other foreland basins, the Carpathian Foredeep is asymmetric and filled with predominantly clastic sediments of the Miocene age as much as 3 and 6 km (1.8 and 3.6 mi) thick at the Carpathian front in Poland and Ukraine, respectively. The molasse deposits of the Carpathian Foredeep are underlain by the basement of the European Platform, covered mainly by Permian–Mesozoic terrestrial and shelf sediments and locally by the Paleogene deposits. According to seismic, magneto-telluric, and well data, the platform basement with Miocene molasse cover dips southward underneath the Outer Carpathian nappes to a distance of at least 50 km (31 mi). The early to middle Miocene Carpathian Foredeep developed as a peripheral foreland basin related to the moving Carpathian front. The Paleozoic–Mesozoic and Tertiary strata of the Carpathian Foredeep are oil and gas productive.

DOI:10.1306/985612M843072

INTRODUCTION

The Polish and Ukrainian Carpathian Foredeep basins (PCFB and UCFB, respectively), about 600 km (380 mi) long and as much as 100 km (60 mi) wide, are part of a large sedimentary basin that stretches for more than 1300 km (810 mi) from the Danube in Vienna (Austria) to the Iron Gate on the Danube (Romania). In the west, the Carpathian Foredeep is linked with the Alpine Molasse Basin, and on the east, it passes into the Balkan foreland basin (Figure 1). Like other foreland basins, the Carpathian Foredeep is asymmetric and is filled with predominantly clastic sediments of Miocene age as much as 3 and 6 km (1.8 and 3.7 mi) thick at the Carpathian front in Poland and Ukraine, respectively. The molasse deposits of the Carpathian Foredeep are underlain by the basement of the West and East European platform. The platform basement is covered mainly by Permian–Mesozoic terrestrial and shelf sediments and locally by the Paleogene deposits. According to seismic, magnetotelluric, and well data, the platform basement with Miocene molasse cover dips southward underneath the Outer Carpathian nappes to a distance of at least 50 km (31 mi) (Oszczypko and Slaczka, 1985; Oszczypko, 1998). The Paleozoic–Mesozoic and Tertiary strata of the Carpathian Foredeep are oil and gas productive.

PRE-TERTIARY SUBSTRATUM: OUTLINE OF STRATIGRAPHY AND GEOTECTONIC EVOLUTION

The substratum of the PCFB is highly diversified (Karnkowski, 1974; Oszczypko et al., 1989) (Figure 2), which is related to the pre-Neogene history of this area. Carboniferous and Permian disintegration of the Variscan Europe followed by Mesozoic extension led to the development of the Tethyan rift system with its important element, the Danish–Polish basin system (Pozaryski and Brochwicz-Lewinski, 1978; Ziegler, 1990; Dadlez et al., 1995). It extended at least as far as the present-day Carpathian thrust front in eastern Poland and western Ukraine (Pozaryski and Zytko, 1980; Kutek, 1994; Hakenberg and Swidrowska, 1997, 1998). During the Late Cretaceous to Paleogene, the Mid-Polish Trough was inverted and uplifted. Inversion of the mid-Polish Trough can be correlated with collision

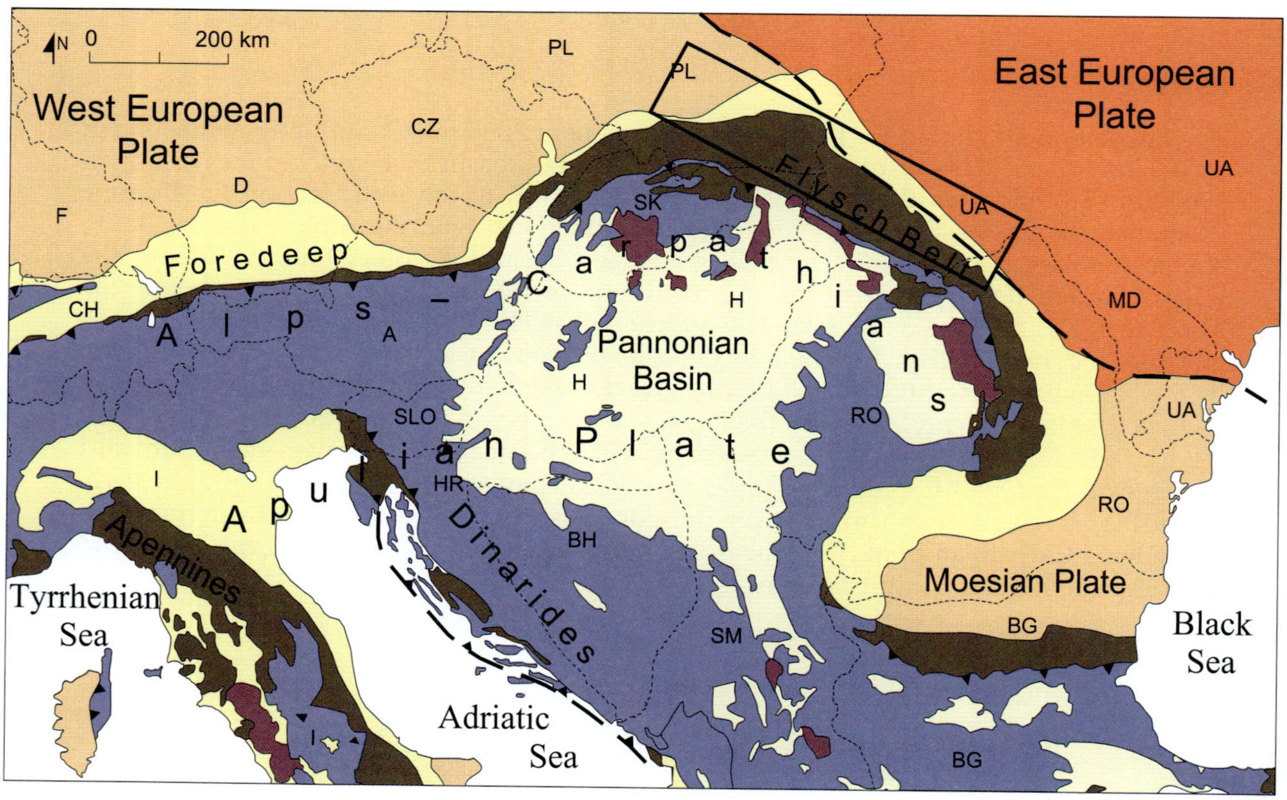

Figure 1. Position of the Polish and Ukrainian Carpathian Foredeep in the Alpine–Carpathian system (modified after Picha, 1996).

events in the Alpine–Carpathian collision zone (Kutek and Glazek, 1972; Krzywiec, 2002). It led to significant uplift and partial or total erosion of the Mesozoic sedimentary cover. The tectonic grain of the Carpathian foreland plate connected to this rifting, and inversion tectonic activity is related to the so-called Teisseyre–Tornquist zone, generally follows a northwest–southeast direction, and is oblique to the Polish Carpathians. It can therefore be hypothesized that during the Miocene development of the PCFB, some form of reactivation of older inherited structures of the foreland plate had influenced the basin formation.

Precambrian

The pre-Tertiary substratum of the Carpathian Foredeep and the marginal part of the Outer Carpathians are composed of the Precambrian consolidated basement and the Paleozoic–Mesozoic sedimentary cover (Figure 2) (see also Oszczypko et al., 1989). The Precambrian metamorphic and plutonic rocks form the basement of the Paleozoic sediments and, in a broad sense, the so-called "consolidated basement" recognized by numerous wells, seismic data, and the magnetotelluric soundings as a high-resistivity horizon (Figure 3) (see also Rylko and Tomas, 1995; Zytko, 1997). The depth of the magnetotelluric basement varies from a few hundred meters in the northeast part of the Polish Carpathian Foredeep to more than a few thousand meters in the middle part of the foredeep. Beneath the Carpathians, this depth ranges from 1–1.5 km (0.6–1 mi) in the area of Zywiec–Cieszyn (Cieszyn–Slavkov ridge) to 3–5 km (1.8–3 mi) in the Rzeszow and as much as 15–20 km (9–12 mi) at its deepest point, which coincides approximately with the axis of gravimetric minimum. The Precambrian crystalline rocks, together with the Cambrian–Carboniferous sedimentary rocks that deformed during the Caledonian and Variscan orogenies, are also known as the epi-Variscan platform.

The oldest structural elements of the crystalline basement belong to the Brunovistulicum and Malopolska terranes. The Brunovistulicum terrane is known from the basement of the upper Silesian basin, and its southern margin is located beneath the Outer Carpathians (Slavkov–Cieszyn ridge), as well as from the Rzeszotary–Dobczyce zone. The southern boundary of Brunovistulicum is roughly consistent with the Alcapa–North European suture zone or peri-Pieninian fault zone (Figure 3).

In the upper Silesia, the Precambrian foredeep basement is composed of Proterozoic greenschists, amphibolite-epidote facies, mica-chlorite-quartz schists, and hornfelse gneisses accompanied by granitoid intrusions and basic effusives (diallague-olivine gabbro) (Heflik and Konior, 1974; Slaczka, 1976; Moryc and Heflik, 1998). On the sub-Neogene surface, similar rocks appear in the Cieszyn–Slavkov ridge (Figure 2). Proterozoic (Riphean) greenschists are also known from core data of Chidnovychi, Mostyska, Pyniany, and Chyzhki wells drilled in the southwestern part of Ukrainian Carpathian Foredeep, where they subcrop the Miocene sediments (Glushko and Kruglov, 1977).

The consolidation of Precambrian rocks of the Brunovistulicum probably occurred during the final orogenic phase of the Cadomian movements (latest Proterozoic). This is supported in the Polish part of the Brunovistulicum by the K/Ar determination (542–566 m.y.) of muscovite from the Lower Cambrian sandstones in the Goczalkowice IG 1 borehole (Belka et al., 2000) (IG is the acronym of wells that have been drilled by the Polish Geological Institute). The material of these sandstones was derived from the batholite of Cadomian age.

The Vendian–Lower Cambrian (sub-Holmian) folded metaargillites and metaaleurites, which occur on the southeastern prolongation of the mid-Polish Swell, are known as the Malopolska terrane. In the PCFB, they occur in a zone 50–70 km (31–43 mi) wide immediately beneath the Neogene sediments (Figures 2, 3). To the northeast of the upper Silesian Massif in the basement of the Miechow (Nida) trough (MNT) and of the Carpathian Foredeep occur strongly folded metaargillites and metaaleurites of the latest Proterozoic (Vendian). They outcrop locally in the southern margin of the Holy Cross (Swietokrzyskie) Mountains and were documented also in numerous boreholes in the basement of the Miechow trough. The thickness of these rocks is estimated to be 3–4 km (1.8–2.4 mi) (Znosko, 1983). According to Belka et al. (2000), the Malopolska terrane was also consolidated during the Cadomian orogeny.

The Malopolska terrane is separated from the Brunovistulicum by the tectonic unit known as the Krakow–Lubliniec fault zone (Figure 2) (see also Zaba, 1999; Kalvoda, 2001). Northwestward, it continues in the direction of Hamburg and is known as the Hamburg–Krakow fault zone. The continuation of this zone southeast of Krakow below the Carpathian thrust is speculative. Assuming its straight prolongation, it should reach the Slovak border in the Bieszczady area. The discussion about the Baltican or Gondwanan affinities of the Brunovistulicum and Malopolska terranes (e.g., Kalvoda, 2001) is perhaps irrelevant. According to Golonka (2000), Gondwana and Baltica formed one supercontinent, Pannotia, during the Cadomian orogeny, and their separation occurred around the Proterozoic–Cambrian boundary. The Cambrian could display characteristics of both Baltica and Peri-Gondwana. The

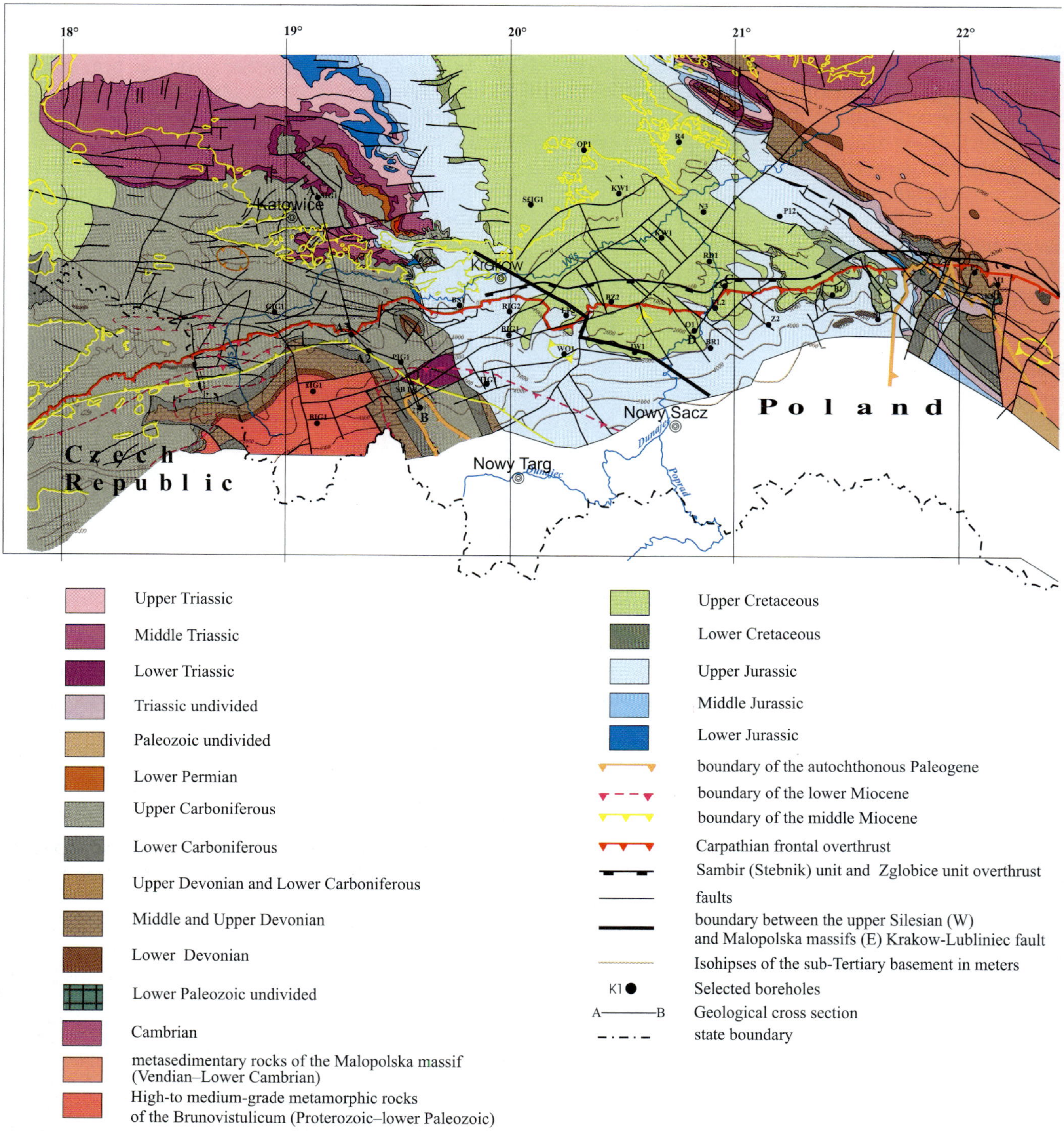

Figure 2. Geological map of the sub-Tertiary basement of the Polish and Ukrainian Carpathian Foredeep (compiled by Oszczypko and Popadyuk, modified after Oszczypko et al., 1989; Maksym et al., 2001).

Malopolska and Brunovistulicum terranes could have different locations in the Peri-Gondwanan or Avalonian realm (Finger et al., 2000), as well as different early Paleozoic history (Belka et al., 2000; Kalvoda, 2001). Together, they joined the supercontinent of Laurussia prior to the Variscan orogeny.

Lower Paleozoic

On the Brunovistulicum (upper Silesian Massif), deposition occurred since the Early Cambrian (Kotas, 1973). Eastward of the upper Silesian Massif, to the Teisseyre–Tornquist zone, folded Lower Paleozoic rocks occurred,

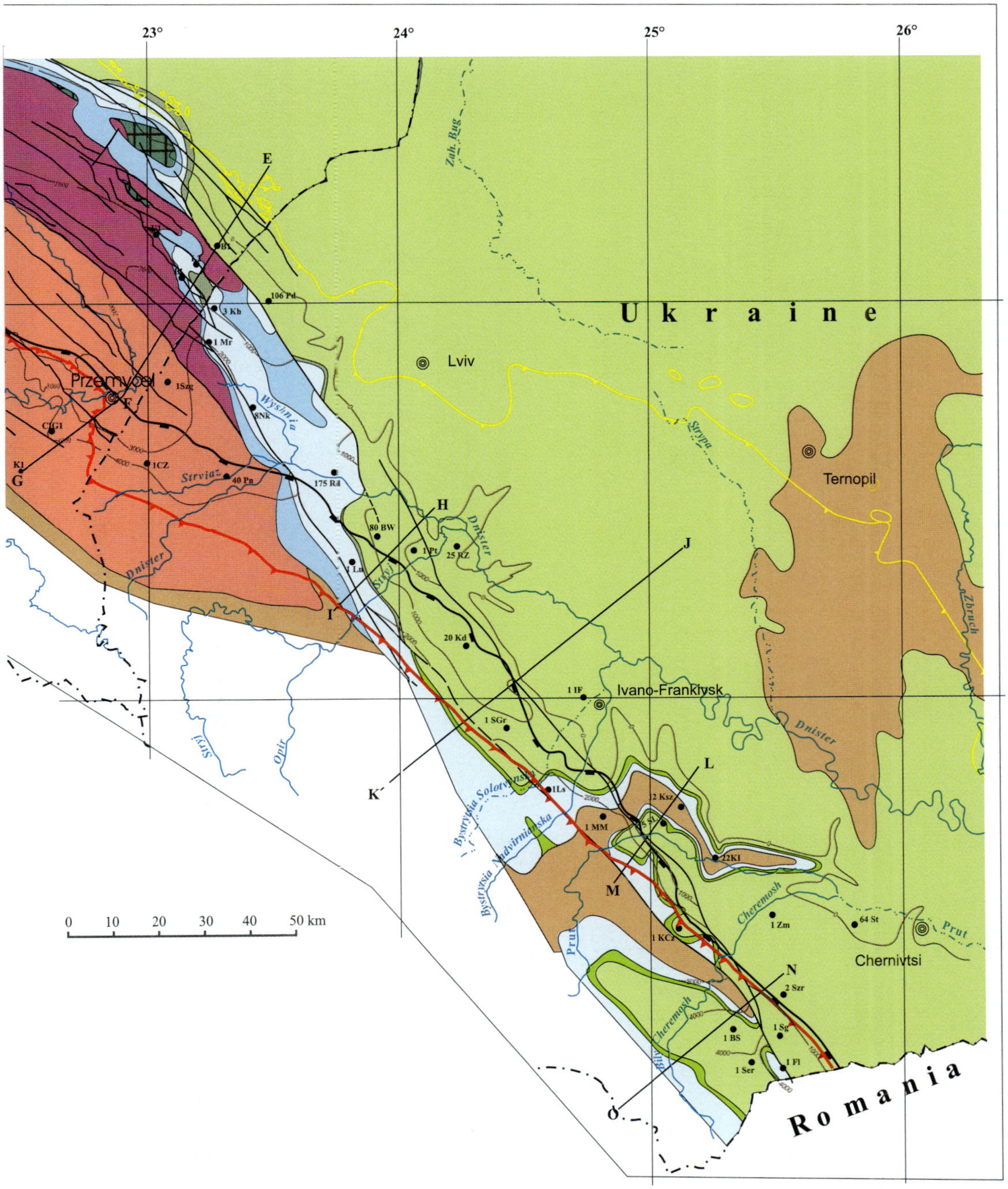

Figure 2. (cont.).

known from boreholes. They are cropping out to the sub-Neogene surface only in the northeast part of Lower San High (Figures 2, 4, 5). Eastward, in the Ukrainian foreland, northeast from the Proterozoic core of Lower San High, Cambrian black, noncalcareous shales with subordinate quartz sandstones were documented in wells where they underlie Mesozoic strata, whereas further northeast under Mesozoic cover, Silurian and Devonian rocks appear (Glushko and Kruglov, 1977).

In the Lubliniec–Zawiercie–Krakow zone (Bula et al., 1999), the Precambrian basement is unconformably covered with Cambrian rocks about 1000 m (3300 ft) thick, represented by phyllites and metaargillites with quartzite intercalations (Bula, 2000). These rocks are, in turn, unconformably covered with Ordovician biotite schists, quartz feldspar schists, and sericite-chlorite schists. They pass upward to similar Landoverian and Wenlockian sediments and to dark shales and

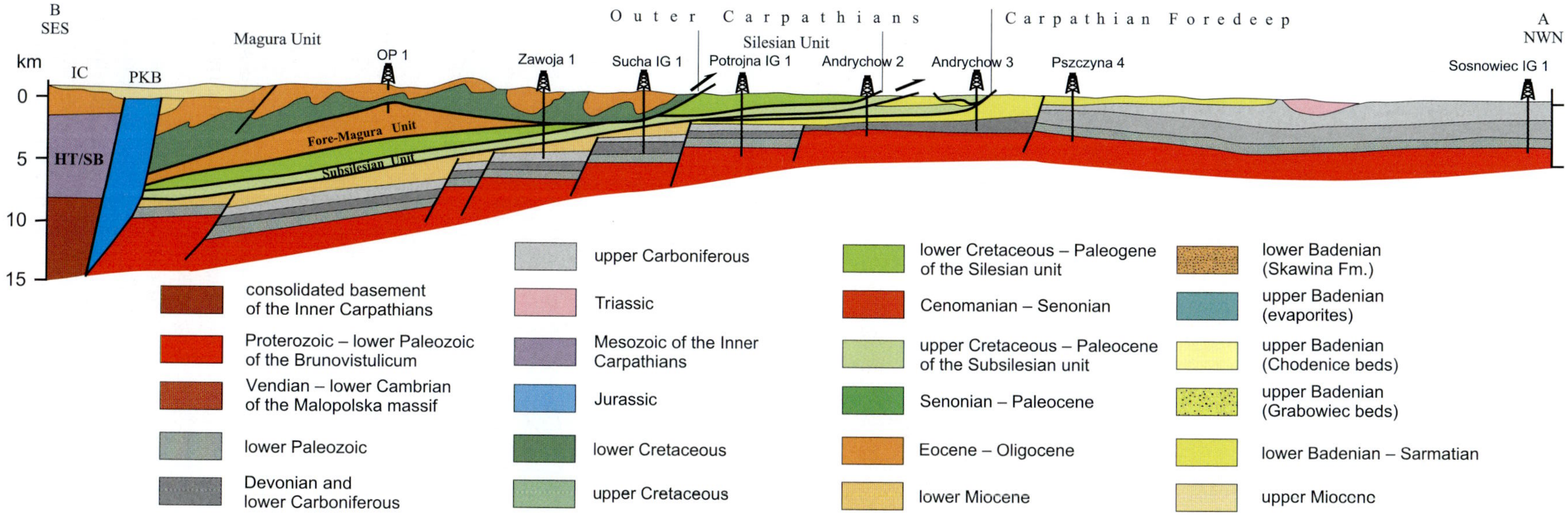

Figure 3. Geological cross section AB (shown in Figure 2) through the western part of the Polish Outer Carpathians and their foreland (modified after Oszczypko, 1997). IC = Inner Carpathians; PKB = Pieniny Klippen Belt; HT/SB = High Tatra and Sub-Tatra units of the Inner Carpathians.

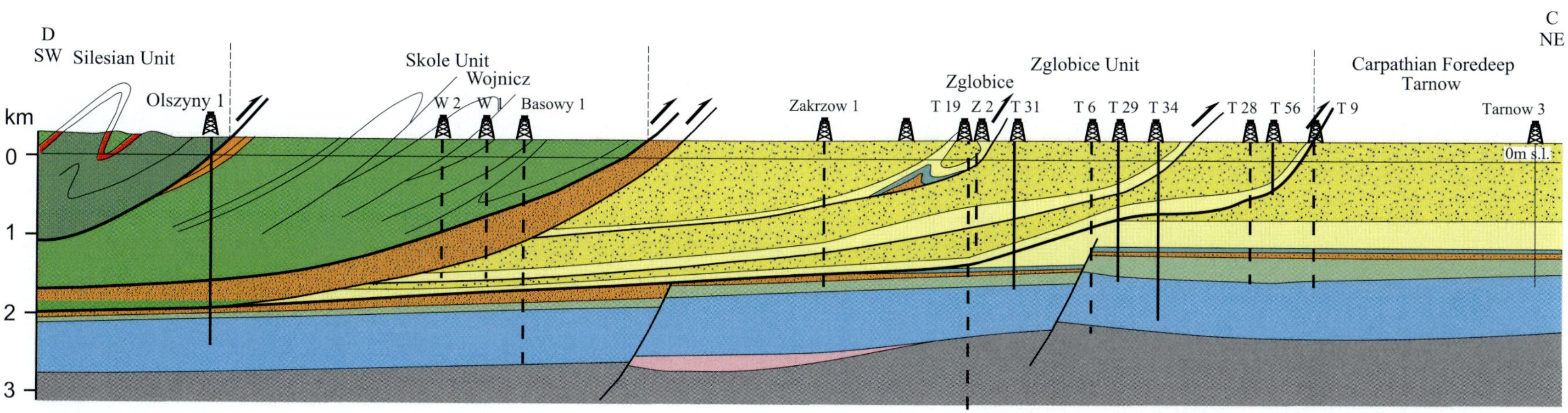

Figure 4. Geological cross section CD (shown in Figure 2) through the marginal part of the Polish Carpathians west of Tarnow (modified after Oszczypko, 1998). For explanations, see Figure 3.

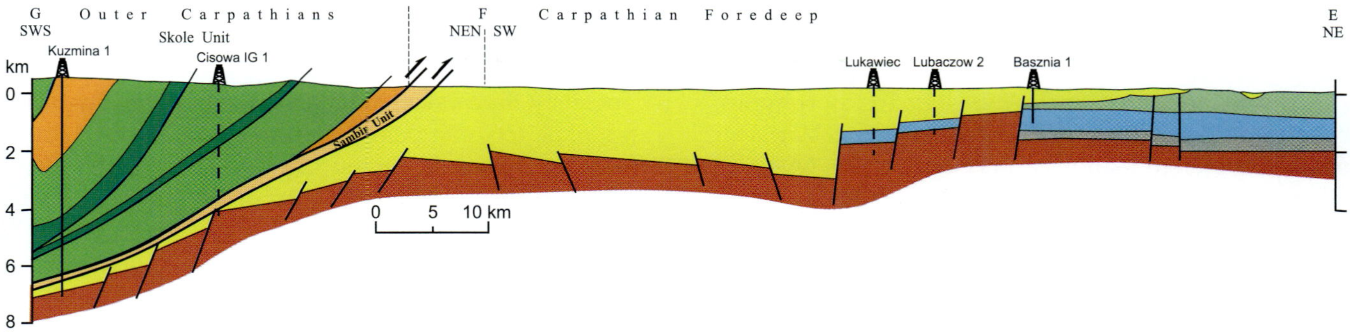

Figure 5. Geological cross section EF (shown in Figure 2) through the eastern part of the Polish Outer Carpathians and their foreland (modified after Oszczypko et al., 1989). For explanations, see Figure 3.

mudstones of the lower Ludlovian. These strata are unconformably covered with the upper Ludlovian graywackes, variegated claystones, and pebbly claystones. The summary thickness of the Ordovician–Silurian strata exceeds 1000 m (3300 ft). The pre-Permian magmatism represented by gabbrodiabases, quartz monzonites, granodiorites, and microgranites, as well as by subsequent volcanism (diabases and quartz porphyres), is clearly manifested in this zone. In the basement of the MNT and of the Carpathian Foredeep, the Vendian–Lower Cambrian slates are overlain with large unconformity by the Ordovician sediments, preserved in a few erosional outliers. The locally occurring Silurian strata are represented by clayey mudstone facies, with intercalations of limestones and cherts.

In the southern margin of Holy Cross Mountains, the uppermost Proterozoic–Vendian rocks are represented by claystones and mudstones, as well as graywacke sandstones and polymictic conglomerates with traces of anchimetamorphism. A similar development is also typical for the Cambrian strata. At the Tremadocian–Arenigian boundary, distinct angular unconformity is manifested in this area. In the basement of the MNT and Carpathian Foredeep, the Arenigian overlies directly the Vendian strata (Znosko, 1983). During the Ordovician in Holy Cross Mountains area, in the shallower Kielce facies zone, the carbonate sedimentation dominated, whereas in the Lysogory facies zone, the deep-water clastic sedimentation occurred. During the Silurian, the unification of sedimentary conditions occurred in both facies zones. This resulted in the deposition of the pelagic graptolite shales followed by the deposition of upper Ludlovian synorogenic turbidites varying in thickness from 500 m (1600 ft) in the Kielce facies zone to 2000 m (6600 ft) in the Lysogory facies zone. At the same time, diabase sills were formed.

Upper Paleozoic

After the late Ludlovian (Ardennian and Erian phases) and before the Emsian, the Kielce zone and the adjacent basement of the MNT and the Carpathian Foredeep were uplifted, whereas in the Lysogory zone, brackish Gedinnian and Siegenian deposition persisted. During the Early Devonian, the area of Caledonian folding and thrusting was subject to intense denudation. A new sedimentary cycle began in the Emsian or locally in the Middle Devonian. The Emsian clastic deposition was followed by the Eifelian carbonate sedimentation. In the Carpathian foreland, between the Holy Cross Mountains and the upper Silesian Massif, this type of deposition persisted to the Visean. In the MNT, the thickness of the Devonian and Carboniferous exceeds 1500 m (5000 ft) (Jawor, 1970; Zajac, 1981). During the late Visean, the sea gradually retreated from the area situated southeast of the Holy Cross Mountains.

In the north Moravia and the upper Silesian basin, the Culm flysch deposits pass continually to paralic and freshwater (deltaic and alluvial, see Gradzinski, 1982), coal-bearing sediments (Namurian A–Westphalian D).

As a result of the Variscan folding, the Silesian–Moravian fold belt, as much as 100 km (60 mi) wide, developed during the Namurian A. Traditionally, the Orlova thrust is regarded as the eastern boundary of this belt. Simultaneously with this folding, the upper Silesian basin was formed in the final stages of the Variscan cycle, filled with paralic and freshwater molasses of the Namurian and Westphalian. This basin was formed both in the marginal part of the Moravian–Silesian Variscides and on the platform deposits of the upper Silesian Massif (Gradzinski, 1982; Gradzinski and Porebski, 1995; Kotas, 1995). The coal-bearing series of upper Silesian basin are as much as 4000 m (13,100 ft) thick. At the northeast margin of the upper

Silesian basin, the final stages of the Variscan folding were accompanied by the Late Carboniferous to Early Permian volcanism (granitoides, porphyries, rhyodacites, diabases, melaphyres, quartz porphyries, rhyolites, lamprophyres, and tuffs). The Variscan movements were manifested not only in the Moravo-Silesian externides and in the upper Silesian basin but also in the remaining area of the Carpathian foreland and in the Holy Cross (Swietokrzyskie) Mountains. During that time, the Lysogory overthrust and several west-northwest–east-southeast-trending folds were formed.

During the Late Permian to the Early Triassic, the area of both the PCFB and UCFB was affected by intense denudation of the Variscan relief. Only locally are the Upper Carboniferous faulted and partly folded sediments unconformably overlain by the Lower Permian conglomerates. The late Variscan northwest–southeast-trending Slawkow Graben is filled with Lower Permian conglomerates, travertines, and volcanic tuffs (Kiersnowski, 1991) as much as 600 m (2000 ft) thick. The Liplas–Tarnawa pull-apart basin (Poprawa et al., 2001) developed during the Late Permian–Early Triassic on the southeast prolongation of the Krakow–Lubliniec suture zone (Zaba, 1999; Bula, 2000, 2001). This basin was filled with fine clastic sediments with anhydrite intercalations, as much as 1400 m (4600 ft) thick (Moryc and Senkowiczowa, 1968). In the UCFB, the post-Variscan denudation persisted until the Early Jurassic.

Mesozoic

In the Krakow region, the Mesozoic epicontinental deposits as much as 400 m (1300 ft) thick unconformably overlie the epi-Variscan platform. In the MNT, the thickness of the Mesozoic deposits reached 2000 m (6600 ft). This sequence, recognized by drillings, reflects the above-mentioned rifting. This sequence begins with the Lower Triassic variegated conglomerates, sandstones, and claystones, locally up 1000 m (3300 ft) thick (Moryc, 1971). This was followed by the late Scythian shallow-marine ingression, which covered the Cracovian–Silesian monocline and the Carpathian foreland. The shallow-water carbonate sedimentation of the so-called Germanic facies continued until the late Norian. The PCFB was a strongly eroded land area during the Early Jurassic; only locally, the freshwater sediments were deposited. The late Bathonian marine transgression entered the area of the Carpathian foreland gradually from the north along the axis of the MNT, and in the Callovian, it covered the area east of Krakow.

In the MNT and the adjacent part of the Carpathian foreland, the Middle Jurassic sandstone gaize series, mudstones, and dark clays (Golonka, 1972, 1978; Morycowa and Moryc, 1976) passed upward into the Callovian nodular limestones covered by the lower Oxfordian deposits (Olszewska, 2001). Paleogeographic analysis and plate-tectonic computer modeling indicate that the Upper Jurassic in the Carpathian foreland was deposited in the failed rift arm of the Tethys Ocean (Golonka et al., 2006).

In southern Poland and the adjacent part of Ukraine, marine Upper Jurassic–Early Cretaceous carbonate series exceeding 1000 m (3300 ft) in thickness are partially covered by Tertiary molasse series or by thrust units of the Outer Carpathians (Moryc, 1992). The maximum thickness of 1500 m (5000 ft) of the Jurassic sediments was encountered in the Nawsie 1 well under the Carpathian thrust south of Rzeszow (Maksym et al., 2001; Zdanowski et al., 2001). A present-day thickness map reflects original sedimentation with a depocenter along the rift axis and, later, erosional sediment removal. The rift was originated during the Triassic and later renewed during the Late Jurassic opening of the Outer Carpathians.

From a lithostratigraphic point of view, six facies units can be recognized in these Oxfordian to Valanginian carbonates, namely, calcareous plankton, calcareous sponge, algal-oolitic, calcareous-marly coquina, dolomitic-calcareous, and upper algal series. During the earlier time of the Late Jurassic, deeper neritic deposits prevailed (calcareous plankton series). These gave way to upward shoaling conditions (calcareous sponge series), a deeper basin (calcareous-marly series), and a final shallow-water environment (algal-oolitic series and younger series) (Zdanowski et al., 2001). The barrier reefs and associated fore-reef and back-reef facies occur in an algal-oolitic series. They trend northwest–southeast and reach 50 m (160 ft) of thickness. The reefs were built mainly by algae. Algal reefs under the molasse or overthrust Carpathians are potential hydrocarbon exploration targets. The paleogeographic position of the Jurassic of the Carpathian foreland in the Tethyan realm, which contains a large part of the world's producible oil reserves, supports this potential.

During the late Early Cretaceous, southern Poland was an area of denudation. This area was locally invaded by marine ingressions from the Outer Carpathian basin along narrow embayments (Moryc and Wisniowska, 1965).

A marine transgression from the north began in the Albian. During the Cenomanian, the sea covered the area situated east of Krakow and entered the area of the Sudetes and the Fore-Sudetian monocline (Opole trough). These sediments attain the maximum thickness of 700 m (2300 ft) in the axial part of the MNT. After the early Maastrichtian, the sea retreated from the Carpathian foreland. During the Late Cretaceous and Tertiary inversion, the Mesozoic deposits were removed from the Lower San High, which separated the Lublin and Miechow troughs.

The main pre-Neogene tectonic units of the platform area of southern Poland were formed as a consequence of the Late Cretaceous–Paleogene Laramian inversion movements. These units comprise the Opole trough, Silesian–Cracovian monocline, MNT, lower San uplift, and Lublin trough (see Oszczypko et al., 1989).

In UCFB, the Mesozoic sequence starts with Early to Middle Jurassic deltaic to marine siliciclastics as thick as 2500 m (8200 ft) (Table 1). These sediments rest on Proterozoic or Paleozoic rocks in the northwestern part of the Ukrainian foreland and do not extend eastward to the Podolian margin of the east European platform (Figures 6, 7). Late Jurassic carbonates overstep the Middle Jurassic clastic sediments and Paleozoic rocks in the adjacent east and west European platforms (Dulub et al., 2000). Several lithofacies have been recognized and grouped into three facies zones: fore-reef, barrier-reef, and back-reef facies.

The barrier reef that fringed the southwestern margin of the carbonate platform consists of bioconstructed limestones of Oxfordian, Kimmeridgian, and Tithonian ages of total thickness of as much as 1000 m (3300 ft). These sediments pass into lagoon dolomites, anhydrites, and shallow-marine bioclastic limestones of the back-reef zone northeastward and laminated lime mudstone, marls, and bioclastic limestones of fore-reef zone southwestward, respectively (Izotova et al., 1991; Izotova and Popadyuk, 1996).

The Lower Cretaceous (Neocomian) rocks cover local areas of UCFB. They consist of carbonate or marly lithologies and are sparsely encountered in the subsurface. The lack of Barremian deposits over the UCFB may be caused by post-Neocomian regional uplift (Popadyuk and Bondarenko, 2001).

Upper Jurassic to Hauterivian shallow-water limestones are overlain by late Aptian(?)–Albian shales with subordinated sand lithologies. These strata are generally thin (Figures 8, 9).

Upper Cretaceous rocks form mostly shallow-marine sedimentary succession (Table 1). Cenomanian glauconite-rich sandstones 10–50 m (33–160 ft) thick are followed by Turonian chalk, which, in turn, is overlain by shallow-marine mixed siliciclastic-carbonate facies assigned to the Senonian (Pasternak et al., 1987). The Upper Cretaceous succession attains a thickness of 800 m (2600 ft).

Tertiary inversion of the Lower San High and the adjacent area uplift led to subaerial exposure and erosion of the Mesozoic and, locally, the Paleozoic rocks. These were important controls on the development of erosional topography in the UCFB and PCFB alike.

Paleogene

The Paleogene was traditionally regarded as a period of intensive inversion and erosion of the area of the Carpathian foreland. According to Kutek and Glazek (1972), during that time, the Mesozoic and Paleozoic cover was removed from the Lower San High, whereas the material resulting from erosion was transported by rivers southward to the flysch basin. This concept has been modified after finding the Paleogene autochthonous deposits at the base of the Miocene molasses.

The autochthonous Paleogene deposits have been known from morphotectonic structures of the Vranovice and Nesvacilka troughs in southern Moravia (Picha, 1979, 1996) since the early 1960s.

In the Polish Carpathian Foredeep basin, the autochthonous middle–upper Eocene terrestrial deposits were documented by Moryc (1995) in a few boreholes located in the frontal part of the Skole nappe near Rzeszow (Figure 2). These deposits, as much as 302 m (991 ft) thick, filled the southward-dipping paleodepression as much as 20 km (12 mi) wide. The Paleogene sequence consists of 96–203-m (315–666-ft)-thick basal, variegated coarse conglomerates, composed of clasts derived from the basement, followed by the organic-rich mudstones with sandstone intercalations.

Recently, the Oligocene marine deposits have been described by Gedl (2000) from the northeast part of the PCFB. These are several-meters-thick brownish, grayish, and whitish loose sands discordantly overlying the Cambrian substratum.

In the western part of the foredeep, the remnants of the Oligocene–early Miocene(?) deposits are also known from the vicinity of Zory in upper Silesia (Figure 2), where the menilitelike bituminous shales with fish scales were found above the Upper Carboniferous. These shales are transgressively overlapped by lower Badenian (Langhian) deposits (Oszczypko et al., 1989).

In 1989, in the deep borehole Zawoja 1, at a depth of 4689.0–4825.5 m (15,383.9–15,831.6 ft) beneath the Polish Western Flysch Carpathians (Figures 2, 3) above Lower Triassic(?) and Upper Carboniferous, coarse clastic deposits have been found (all of these wells have been drilled by Polish Oil Industry, whose name, at present, is Polish Oil and Gas Company). These deposits were described by Moryc (1989) as the Zawoja Formation of the early Miocene age (see also Poltowicz, 1995). Recently, the Zawoja Formation has been redefined and assigned to Paleogene (Oszczypko, 1997; Oszczypko et al., 2000). This formation begins with a 10.5-m (34.4-ft)-thick layer of gray-greenish to black conglomerates composed of the fine- to medium-grained, well-rounded clasts. Conglomerates are composed of rocks derived from the Paleozoic and Proterozoic

Table 1. Regional stratigraphic scheme of the Mesozoic deposits of the Polish and Ukrainian Carpathian Foredeep basin (compiled by Maksym, Oszczypko, Popadyuk).

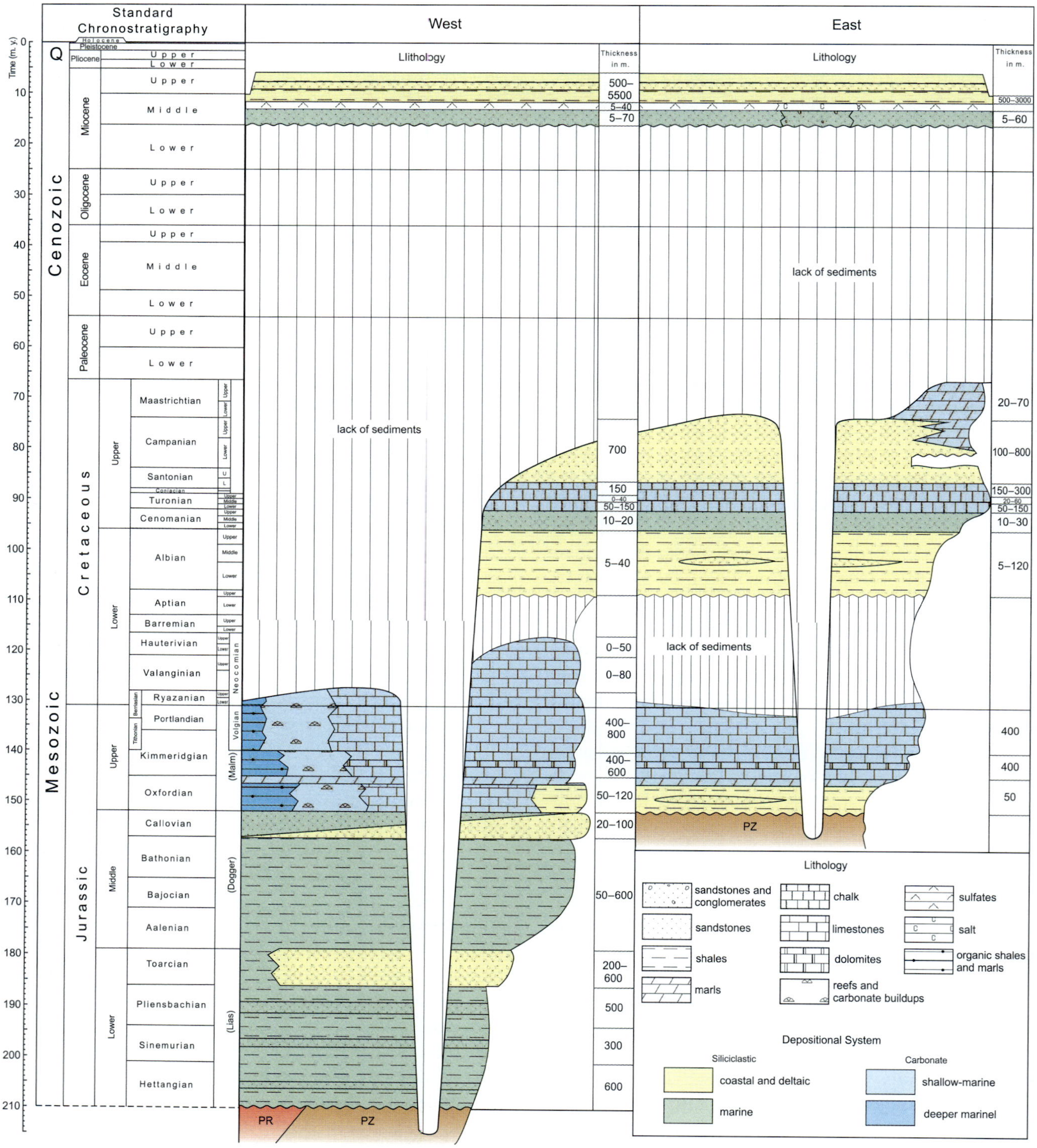

basement rocks (Moryc, 1989). The sandy-silty matrix is rich in coalified flakes. These conglomerates are followed by a 115-m (377-ft)-thick sequence of dark-colored deposits, which display a few fining- and thinning-upward sequences, 25–40 m (82–131 ft) thick. Conglomerates made of pebbles and boulders with sandy-silty matrix containing considerable amount of organic matter and FeS_2 dominate the lower part of these

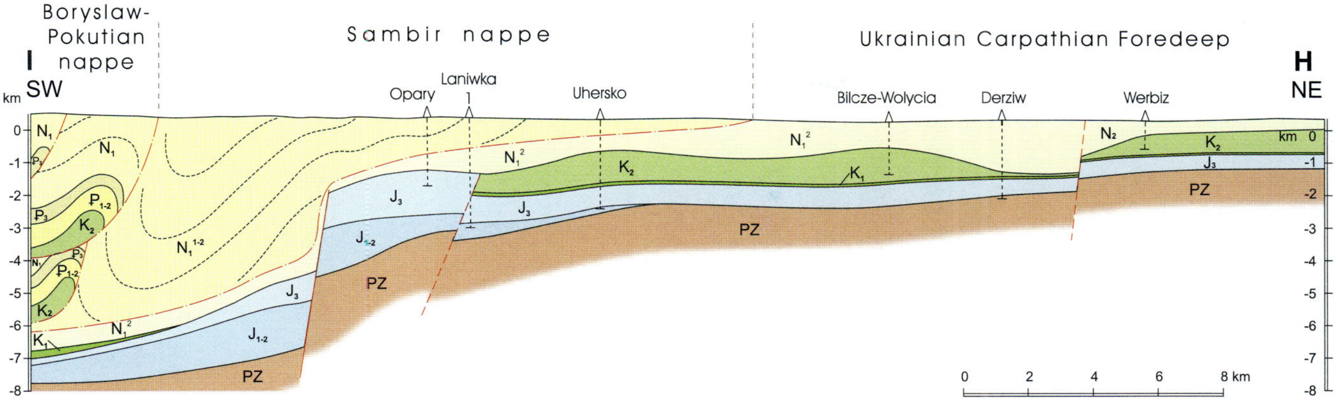

Figure 6. Geological cross section IH (shown in Figure 2) through the northwest marginal part of the Ukrainian Carpathians and their foreland (modified after Glushko and Kruglov, 1986; supplemented by Kruglov and Popadyuk).

sequences. According to dinoflagellate determinations, these strata belong to the Eocene–Oligocene.

These conglomerates are covered by parautochthonous Paleocene–middle Eocene deposits, represented by laminated mudstone, and very fine, thin-bedded sandstones with intercalations of dark gray pebbly mudstones. The mudstones are rich in organic matter (cement) and clay minerals (matrix). The presence of pyrite (framboides on micas) indicates the reductive conditions of the sedimentary environment. The subvertically dipping and folded dark mudstones (interval 4751.0–4759.0 m [15,587–15,613 ft]) yielded only a few specimens of poorly preserved Upper Cretaceous to Paleocene planktonic and agglutinated foraminifera. The core material from a depth of 4689.0–4697.0 m (15,383.8–15,410.1 ft) consists of green calcareous-free siltstones with red, irregular lamination or hematite coating. The green and variegated siltstones contained mixed, two-colored specimens of the Late Cretaceous *Globotruncana* and agglutinated early and middle Eocene species.

The deposits that penetrated in the boreholes in the Rzeszow and Zawoja areas probably represent sedimentary infill of the southeast-trending paleovalleys, which supplied the northern part of the Carpathian Flysch Basin, and could be correlated with the autochthonous Paleogene deposits from southern Moravia (see Picha and Stranik, 1999).

In UCFB, the Paleogene siliciclastics occur, which are as much as 40 m (130 ft) thick, mainly sandstones, resting between the Mesozoic strata and the middle Miocene Tyras gypsum (Kudrin, 1964). Later strata of similar position and lithologies in the wells Sheremetivska 4, Sviatoslavska 1, 2, and 5, and Lopushna 9 in the southeastern part of UCFB were also interpreted as being of Paleogene (Eocene) age (Figures 2, 9) (This group of wells has been drilled by the Ukrainian Oil Industry, whose name, at present, is the Ukrainian Oil and Gas Company [NaftoGas Ukraine]). However, fossiliferous content of these facies included mixed Cretaceous, Paleogene, and Miocene microfauna supporting the Miocene age of sediments.

STRATIGRAPHY AND DEPOSITONAL ENVIRONMENTS

Introduction

The early to middle Miocene PCFB and UCFB developed as a peripheral foreland basin related to the advancing Carpathian front (Figures 2, 10). Similar to other orogenic belts, the Outer Carpathians were progressively folded toward the continental margin (see Jurkova, 1979; Oszczypko and Slaczka, 1985, 1989; Oszczypko and Tomas, 1985; Kovac et al., 1989, 1998; Oszczypko, 1982, 1996, 1997, 1998; Tari et al., 1997). The final flysch deposition (Polyanytsa Formation, see also Koszarski et al., 1995) occurred in the late early Burdigalian (Ottnangian) in the northernmost part of the Outer Carpathian basin (Zdanice unit, Czech Republic; Boryslav–Pokuttya unit, Ukraine; and marginal folds unit, Romania; Micu, 1982; Krhovsky et al., 1995). This basin finally underwent desiccation [evaporite of the upper Vorotyscha Formation in Ukraine (Andreyeva-Grigorovich et al., 1997) and the Salt Formation in Romania]. This salinary crisis was followed by the intra-Burdigalian phase (Ottnangian compressive tectonic event), when the Outer Carpathian basin was folded and uplifted. During this event, the oceanic or thinned continental crust of the Outer Carpathian residual flysch basin was subducted below the overriding Carpathian orogen (Alcapa and Tisza-Dacia microplates). This was accompanied by the outward overthrusting and formation of the flexural foreland basin at the moving orogenic front and partly on top of the orogenic wedge (Tomek and Hall, 1993). The active

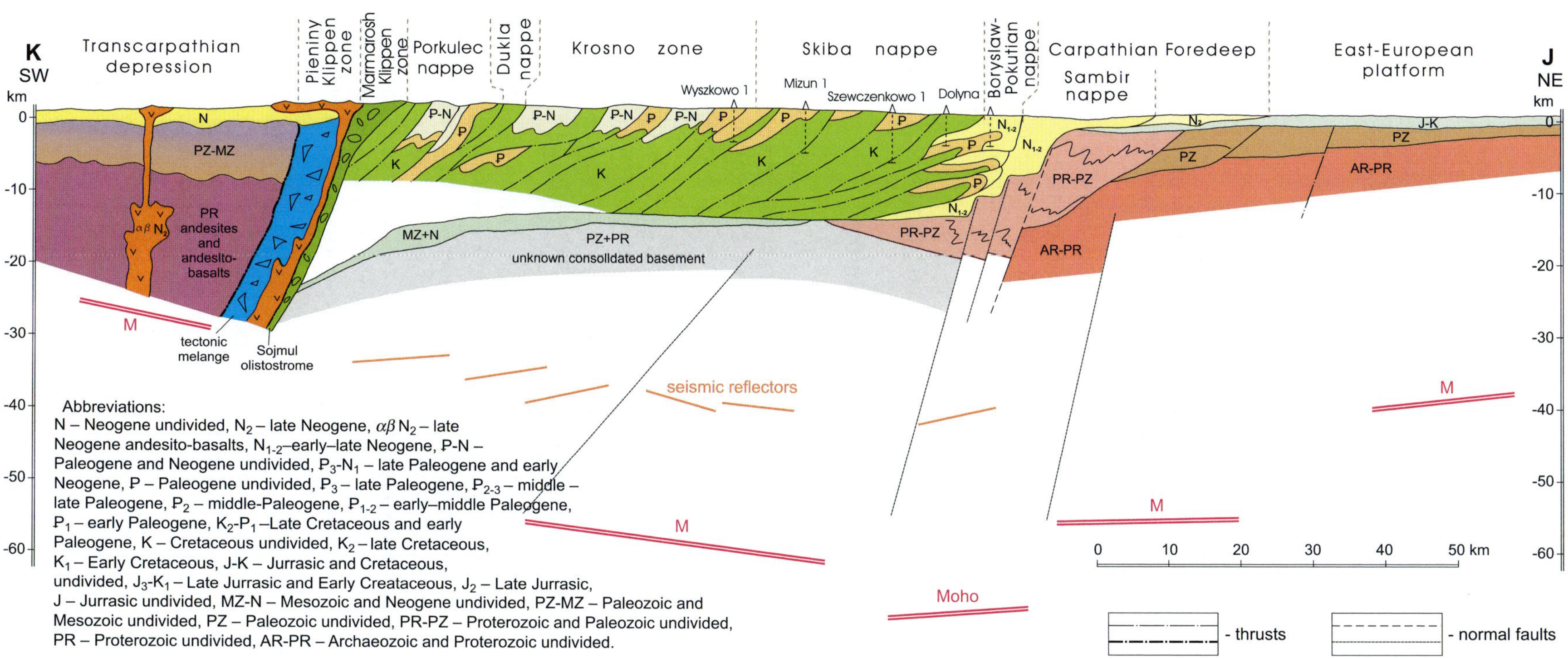

Figure 7. Geological cross section JK (shown in Figure 2) through the middle part of the Ukrainian Carpathians and their foreland (modified after Glushko and Kruglov, 1986; supplemented by Kruglov and Popadyuk).

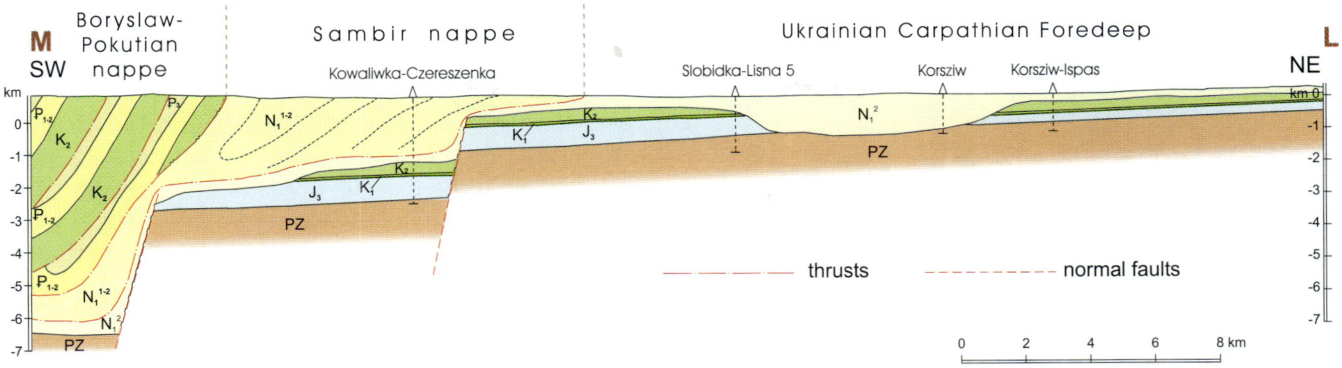

Abbreviations:
N_1^2 – middle Miocene, $N_1^{1\text{-}2}$ – early–middle Miocene, N_1 – Miocene, P_3 – late Paleogene, $P_{1\text{-}2}$ – early–middle Paleogene, K_2 – Late Cretaceous, K_1 – Early Cretaceous, J_3 – Late Jurrasic, $J_{1\text{-}2}$ – Early–middle Jurrasic, PZ – Paleozoic undivided.

Figure 8. Geological cross section ML (shown in Figure 2) through the southeast marginal part of the Ukrainian Carpathians and their foreland (modified after Glushko and Kruglov, 1986; supplemented by Kruglov and Popadyuk).

thrust front of the Outer Carpathians moved to the Silesian–Subsilesian and Skole units front (see Kovac et al., 1998). The Carpathians overrode the platform and caused flexural depression of the foreland. The flysch olistoplaque, recognized in the Sucha IG 1 and Zawoja 1 boreholes, probably records that period of overthrusting (Table 2; Figure 3).

The Carpathian Foredeep can be subdivided into two parts, the outer and inner parts (Oszczypko, 1982, 1997, 1998). The boundary between the outer and inner foredeep runs along the Carpathian marginal thrust in Poland and Sambir thrust in Ukraine (Figures 3–6, 8, 10). The width of the Polish outer foredeep, which covers about 18,000 km^2 (6900 mi^2), ranges between 30 and 40 km (18 and 25 mi) in the western segment and as much as 90 km (56 mi) on the Rzeszow meridian and around 50 km (31 mi) along the Polish–Ukrainian state boundary, whereas the width in the central part of the Ukrainian segment oscillates around 20–30 km (12–19 mi). The outer foredeep is filled with the middle Miocene (Badenian and Sarmatian; see also Alexandrowicz, 1965) marine deposits, which range from a few hundred meters in thickness in the northern marginal part to as much as 3500 and 5000 m (11,500 and 16,400 ft) in the southeastern part of the Polish and Ukraine segments, respectively. The inner foredeep, located beneath the Carpathian nappes, is more than 50 km (31 mi) wide (Figures 3, 5) (see also Oszczypko and Slaczka, 1989), and their axes coincide approximately with the Carpathian gravimetric low (see Oszczypko, 1997, 1998). The Polish inner foredeep is composed of lower to middle Miocene autochthonous deposits. Although these deposits were tectonically eroded by the Carpathian nappes, their preserved thickness is as much as 1500 m (5000 ft). The lower Miocene strata are mainly terrestrial in origin, whereas the Badenian and Sarmatian strata are marine. The allochthonous deposits of the Ukrainian inner foredeep were incorporated into the Boryslav–Pokuttya and Sambir units (Figures 8, 10). Traditionally, in the Central Paratethys stages (see Rogl, 1996, 1999), the boundary between Badenian and Sarmatian is dated as 13.5 m.y. and is located at the top of Neogene nannoplankton (NN) 5 zone. In this chapter, we accepted the eastern Paratethys stages where this boundary is situated at the top of NN6 zone and is dated as 11.8 m.y. (Marunteanu et al., 1999). This stratigraphic division fits the generally accepted lithostratigraphy based on forams.

The Miocene deposits of the Carpathian Foredeep are very weakly exposed. The best, but not numerous, exposures are located along the northern edge of the foredeep (the Krakow–Miechow uplands, the southern slope of the Holy Cross [Swietokrzyskie] Mountains, and the Roztocze range in Poland and Ukraine), whereas in the central and southern part of the foredeep, exposures are very scarce. The geological structure and stratigraphy of the PCFB (Figure 11) and UCFB are based on the borehole explorations for hydrocarbons (especially gas), numerous drillings of the coal mining industry (upper Silesian coal basin), and sulfur (Tarnobrzeg–Grzybow area), as well as seismic surveys. In the PCFB, between the years 1946–1990, about 2000 wells were completed (Karnkowski, 1999).

Lower Miocene

In PCFB, the oldest Miocene deposits belong to the Zebrzydowice Formation (Table 2), which has been found in a few boreholes in the western outer and inner foredeep. The formation is marine and is composed of pelitic gray-greenish deposits of the early

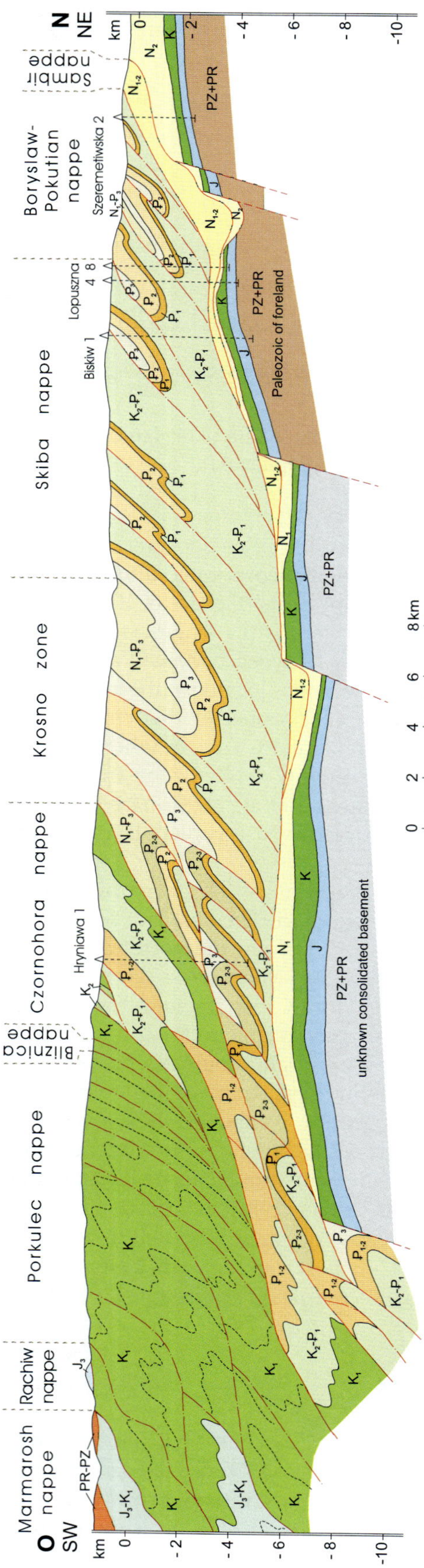

Figure 9. Geological cross section NO (shown in Figure 2) through the southeast part of the Ukrainian Carpathians and their foreland (based on Vul, Kruglov, and Shlapinski interpretation and supplemented by Kruglov and Popadyuk).

Burdigalian (N5–N6 Foraminifera zone) age (Garecka et al., 1996). The same early Burdigalian (NN2–NN3 zone) marine deposits were also discovered in northeast Moravia (Jurkova et al., 1983). The Zebrzydowice Formation probably corresponds to Aquitanian–early Burdigalian dark marine mudstones pierced in the Andrychow-6 borehole (Oszczypko and Oszczypko-Clowes, 2003). In the Cieszyn area, this formation is overlapped by the flysch olistoplaque of the Zamarski Formation.

In the southwest part of the PCFB (Figure 3; Table 2), at the base of the upper Burdigalian sequence, the flysch-derived olistoplaque has been documented in a few deep boreholes (Slaczka, 1977; Bula and Jura, 1983; Moryc, 1989; Slaczka and Oszczypko, 1987; Oszczypko, 1998). This olistoplaque overlaps both the Paleozoic basement, Paleogene Zawoja Formation, as well as the early Burdigalian marine deposits of the Zebrzydowice Formation (see Oszczypko et al., 2000).

In the Zawoja–Sucha Beskidzka area, a 260–370-m (853–1214-ft)-thick flysch-derived olistoplaque is referred to the Sucha Formation (Slaczka, 1977; Moryc, 1989) and known from the Sucha IG 1, Zawoja 1, and Lachowice 1 and 2 wells (Figure 3). In the Sucha IG 1 borehole, this formation is composed of a few separate flysch olistoliths of various age (Early Cretaceous to Paleocene), whereas in the Zawoja 1 borehole, olistostrome formation is represented by a uniform sequence of black shales with sparse intercalations of thin-bedded sandstones (Aptian–late Albian; Gedl, 1997). These deposits correspond to the Verovice shales and Lgota Formation of the Subsilesian–Silesian units and the Spas Shales of the Skole unit. In the Sucha–Zawoja area, the age of the flysch olistoplaque development could be estimated as Ottnangian–Karpatian(?) (Garecka et al., 1996).

In the Cieszyn area, the flysch olistoplaque (Zamarski Formation) has been preserved in the central part of the paleovalley beneath the frontal part of the Carpathian overthrust (Bula and Jura, 1983). The Zamarski olistoplaque, which overlapped both the Upper Carboniferous basement as well as the lower Miocene Zebrzydowice Formation, is 25–150 m (80–500 ft) thick, reaches at least 50 km^2 (19 mi^2), and is composed of elements of the Subsilesian unit. An isolated fragment of the Subsilesian unit, 112 m (43 mi^2) thick, has also been found in a subthrust position in borehole K-3 near Bielsko–Biala (Figure 2).

The flysch olistoplaque (tectonosome) (Sucha and Zamarski formations) from the western part of PCFB could be correlated with the old Styrian overthrust in the marginal part of the Moravo-Silesian Carpathians (Jurkova, 1971). In the Pribor area, the front of the Old Styrian nappe, composed of the Subsilesian sequence,

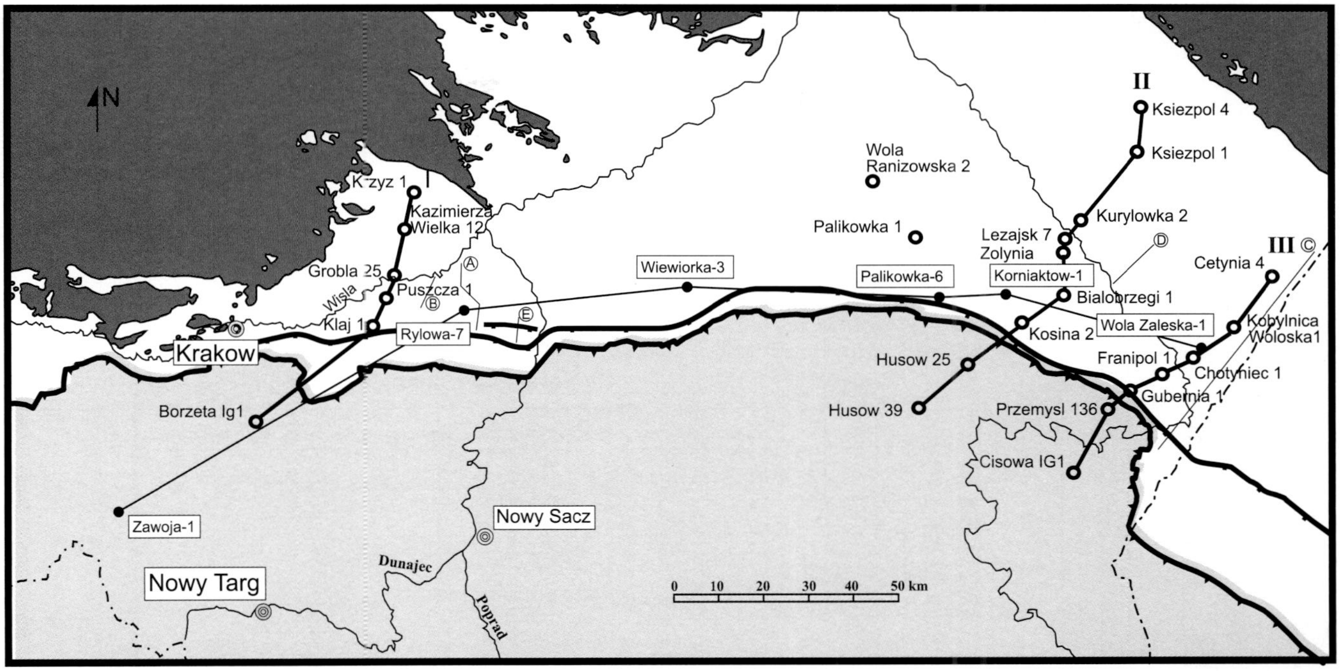

Figure 10. Sketch map of the Polish Carpathian Foredeep basin with location of selected boreholes and seismic profiles. Zawoja-1 — Wola Zaleska 1 gamma-ray cross section shown in Figure 11; A = seismic profile shown in Figure 23; B = seismic profile shown in Figure 24; C = seismic profile shown in Figure 25; D = seismic profile shown in Figure 26; E = seismic profile shown in Figure 27; I, II, III = subsidence cross sections and boreholes shown in Figures 29–32.

is located 1–3 km (0.6–1.8 mi) north of the present-day margin of the Carpathians. Close to the Polish–Czech border, the Old Styrian nappe is presently overlapped by a margin of the Carpathians. Meanwhile, east of Cieszyn, this nappe has been described as the Zamarski and Sucha formations.

The Sucha olistoplaque is overlapped by the Stryszawa Formation that reaches the thickness of 650 m (2130 mi) (Table 2) (see also Slaczka, 1977; Strzepka, 1981; Moryc, 1989; Oszczypko, 1998). These deposits are composed of coarse- to medium-grained, polimictic conglomerates with carbonate and, locally, gypsum-anhydrite cement. The thickness of these conglomerates ranges from 140 m (460 ft) (Sucha IG 1) to 229 m (751 ft) (Lachowice 2), rising as much as 650 m (2130 ft) in borehole Slemien 1 (Baran et al., 1997; Laskowicz, 1997).

The lower part of the formation is represented by polimictic conglomerates of the Stachorowka Member. In the Zawoja 1 borehole, the Stachorowka Member, 180 m (590 ft) thick, is composed of conglomerates and coarse- to fine-grained arenites (Figure 12A). The conglomerate clasts are dominated by sandstones, mudstones, carbonates, and crystalline rocks derived from both the Carpathian flysch as well as from the Paleozoic basement. These conglomerates pass into the upper part of the Stryszawa Formation (210–240 m [690–790 ft] thick). This part of the formation is composed of variegated, conglomerate-sandy-mudstone strata (Figure 12B). The Stryszawa Formation contains a relatively frequent microfauna of the Early Cretaceous–early Miocene age recycled from the Carpathian flysch and showing connection with the sediments of the Subsilesian development (Oszczypko, 1998). From this formation, Karpatian calcareous nannoplankton (NN4) has also been reported (Garecka et al., 1996). The upper part of the Stryszawa Formation resembles features of the alluvial-fan deposits. In the Sucha IG-1 borehole (Table 2), this formation passes upward into the lower Badenian Skawina Formation.

In the Cieszyn area, the Zamarski olistoplaque is covered by a 10–110-m (33–360-ft)-thick layer of variegated conglomerates composed of flysch-derived clasts. These conglomerates resemble the Stryszawa Formation from the Sucha–Zawoja area (Oszczypko et al., 2000).

In the Zawoja–Sucha Beskidzka and Cieszyn areas, the upper part of the subthrust molasse sequence is represented by a 40–90-m (130–300-ft)-thick complex of transgressive Debowiec conglomerates, which are composed of Upper Carboniferous clasts. In the Cieszyn area, these conglomerates pass upward into the Skawina Formation. According to Garecka et al. (1996), both the Debowiec conglomerate and the lowermost part of the Skawina Formation belong to the late Karpatian. The inner PCFB filled with upper Burdigalian

Table 2. Regional stratigraphic scheme of the Miocene deposits from the Polish and Ukrainian Carpathian Foredeep basin (modified after Oszczypko, 1998; Andreyeva-Grigorovich et al. 1997). Time scale after Berggren et al. (1995).

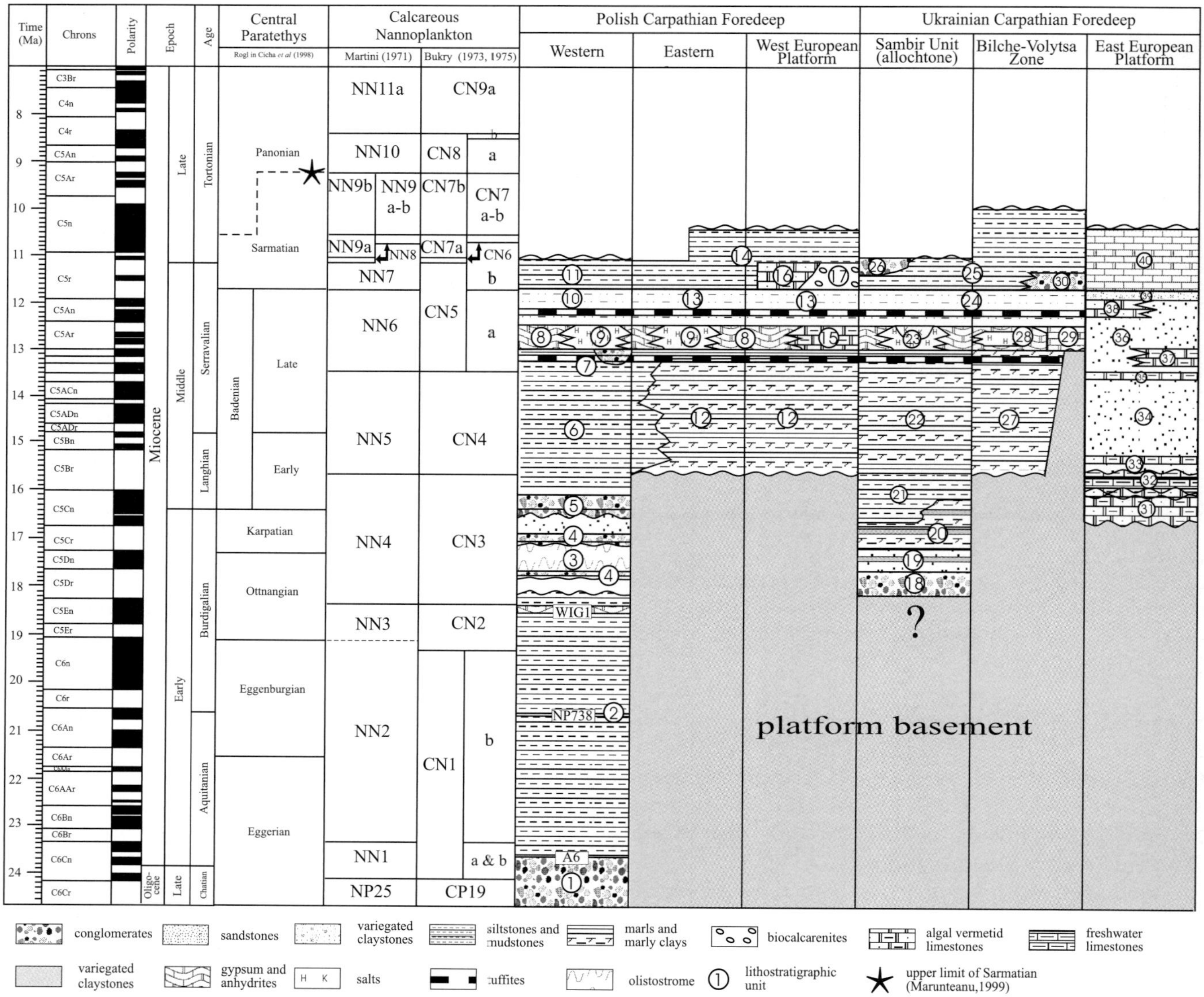

Lithostratigraphic unit: 1 = Andrychow Formation; 2 = Zebrzydowice Formation; 3 = Sucha Formation-flysch olistoplaque; 4 = Stryszawa Formation; 5 = Debowiec Conglomerates; 6 = Skawina Formation; 7 = Sypka Gora Conglomerates; 8 = Krzyzanowice Formation; 9 = Wieliczka Formation; 10 = Gliwice beds; 11 = Kedzierzyn beds; 12 = Baranow beds; 13 = Chodenice and Grabowiec beds; 14 = Krakowiec beds; 15 = sulfur-bearing limestones; 16 = algal-vermetid reefs limestones; 17 = Biocalcarenity of the Chmielnik Formation; 18 = Sloboda Conglomerate; 19 = Dobrotiv Formation; 20 = Stebnyk Formation; 21 = Balych Formation, 22 = Bohorodchany Formation; 23 = Kalush (Tyras) Formation; 24 = Kosiv Formation; 25 = Dashava Formation; 26 = Radych beds; 27 = Zhuriv Formation; 28 = Tyras Formation; 29 = Ratyn limestones; 30 = Pistyn conglomerates; 31 = Nahoryany (Oncophora) beds; 32 = Berezhany beds; 33 = Baraniv beds; 34 = Mykolaiv beds; 35 = Naraiv beds; 36 = Rostoche and Kaiserwald beds; 37 = Kryvchytsi beds; 38 = Ternopil beds; 39 = Buhliv beds; 40 = Volyn beds.

clastic deposits was at least 40 km (25 mi) wide. East of the Zawoja–Sucha area, this foredeep extended toward the Sambir basin (Ukraine), whereas its prolongation beneath the Carpathians is probably manifested by the magnetotelluric depression and a gravimetric low (Oszczypko, 1998).

In the Sambir unit (UCFB), the lower Miocene deposits are represented by coarse-grained clastic deposits. Their relation to the platform or flysch basement is not clear. This fining- and thinning-upward sequence begins with the Sloboda conglomerates (Table 2; Figure 12C). These clast- to matrix-supported conglomerates are composed of pebbles and boulders of the Proterozoic metaargilites, mainly green phyllithes and Jurassic limestones, derived from the platform basement of the Carpathian Foredeep. The roughly estimated thickness of these conglomerates is as much as 2000 m (6600 ft). The transition of the Sloboda conglomerates to Dobrotiv Formation (suite) is exposed at the Oslava River outflow to Prut River (Figure 13A). In this place, the dark, fine-grained conglomerates pass upward into coarse- to fine-grained, cross-laminated sandstones.

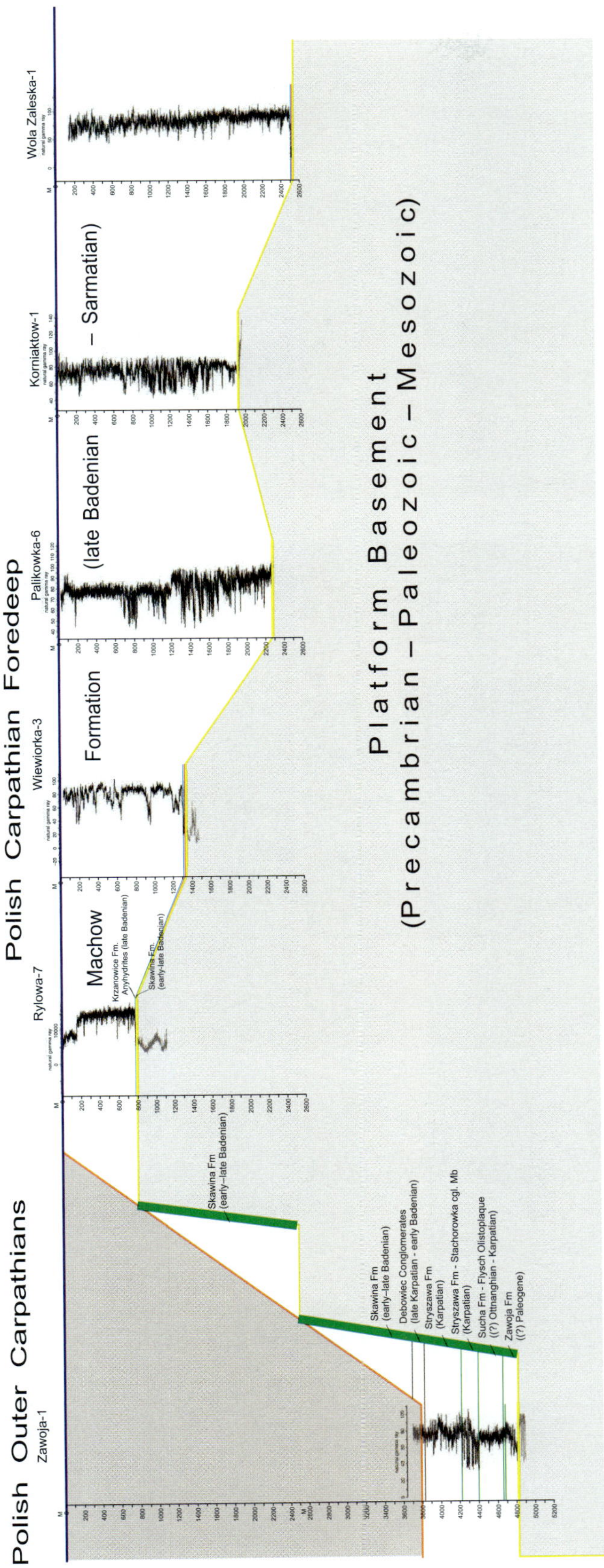

Figure 11. Gamma-ray log cross section of the Polish Carpathian Foredeep basin (compiled by Krzywiec and Oszczypko).

The sandstones are followed by a few meters-thick interval, represented by repetition of thin-bedded sandstones and mudstones. The higher part of the section belongs to the channelized thick- and very thick-bedded, amalgamated, medium- to coarse-grained sandstones, with thin intercalations of green and reddish mudstones. The basal surface of sandstone beds reveals a flute-cast and a load-cast structure. The Dobrotiv Formation, 600–800 m (2000–2600 ft) thick, is known from vertebrate (birds, artiodactyla, and felines) footprints as well as rain prints (Andreyeva-Grigorovich et al., 1997). The upper part of the Dobrotiv Formation is represented by thick-bedded parallel and cross-laminated sandstones that are exposed in the Prut River at the Lanchyn. The total thickness of the Dobrotiv Formation is as much as 700 m (2300 ft). In the Lanchyn section, the transition from the Dobrotiv Formation to Stebnyk Formation (suite) is manifested by the appearance of the red (rose) and dark and gray calcareous mudstones with thin intercalations of fine sandstones. The lower part of the Stebnyk Formation (Figure 13B) contains intercalation of thick-bedded, Dobrotiv type of sandstones. The flute casts and cross-lamination reveal the paleotransport direction from the northwest (290–300°). In the middle part of the Lanchyn section, the Stebnyk variegated deposits include a layer of salt breccia (at least 10 m [33 ft] thick), which probably is derived from the Vorotyshcha Salt Formation.

Middle Miocene (Langhian and Serravalian)

The Badenian strata rest directly on the platform basement with the exception of the southwest part of the inner foredeep, where they cover the lower Miocene deposits (Table 2; Figure 11). Commonly, the lower Badenian (Ney, 1968; Ney et al., 1974) begins with a thin layer of conglomerates; however, in the western part of the foredeep, the Debowiec conglomerates attain a thickness as much as 100 m (330 ft). The conglomerates pass upward into dark, clayey-sandy sediments (Skawina Formation). The thickness of the lower Badenian deposits is variable, reaching as much as 1000 m (3300 ft) in the western inner foredeep, whereas in the remaining parts of the foredeep, it does not often exceed 30–40 m (100–130 ft) (Ney et al.,1974). The sedimentation of the Skawina Formation began in the inner foredeep with *Praeorbulina glomerosa* zone (N8), whereas in the outer foredeep, the sedimentation began with the *Orbulina suturalis* (N9 or N10) zone (Garecka et al., 1996; see also Oszczypko, 1997). According to the calcareous nannoplankton studies, this formation belongs to the NN5 zone and, in its uppermost part, to the

A

B

C

Figure 12. (A) Fine conglomerates of the Stachorowka Member of the Stryszawa Formation (lower Miocene), Zawoja 1 borehole, depths 4301 – 4309 m (14,110–14,137 ft). (B) Thin-bedded sandstones and variegated mudstones of the Stryszawa Formation (lower Miocene), Zawoja 1 borehole, depths 4137 – 4144 m (13,573–13,596 ft). (C) Uppermost part of the Sloboda conglomerate (lower Miocene) of the Sambir unit of the UCFB, Oslava River at Dobrotiv.

A

B

Figure 13. (A) Basal part of deltaic sequence of the Dobrotiv Formation (lower Miocene) of the Sambir unit of the UCFB, Oslava River at Dobrotiv. (B) Lower–middle Miocene variegated shales with sandstone intercalations of the Stebnyk Formation, left bank of Prut River at Lanchyn.

NN6 zone (Andreyeva-Grigorovich et al., 1997, 2003). The radiometric age of a tuffite from the uppermost part of the Skawina Formation in the Wieliczka salt mine (WT-1, Bukowski and Szaran, 1997) has been determined as 12.5 ± 0.9 m.y. The upper Badenian evaporitic horizon either overlies these deposits or rests directly on the platform basement. This horizon consists of rock salts, claystones, anhydrites, gypsum, and marls. Between Wieliczka and Tarnow, the thickness of salts attains 70–110 m (230–360 ft) (Garlicki, 1968; Bukowski and Szaran, 1997) and decreases toward the east to a few dozen meters, whereas the thickness of gypsum and anhydrites commonly ranges between 10 and 30 m (33 and 100 ft). In the UCFB (Sambir Unit)

A

B

Figure 14. (A) The late Badenian Kalush Salt Formation of the Sambir unit of the UCFB, Kalush open salt mine, general view toward the south. (B) The late Badenian Kalush Salt Formation of the Sambir unit of the UCFB. View on northwest wall of the Kalush open salt mine.

the evaporitic horizon is represented by the Kalush Formation (Figure 14). The evaporitic horizon passes upward into the upper Badenian–Sarmatian (NN6–7 to NN8–9 zone, see Gazdzicka, 1994; Andreyeva-Grigorovich et al., 2003) sandy-silty deposits with a thick sandstone complex at the base. Their thickness ranges from a few hundred meters in the Tarnow area to as much as 3000 m (1000 ft) near Przemysl. In the Rzeszow area, these deposits rest directly on the platform basement. In the UCFB, the thickness of the upper Badenian–Sarmatian deposits reaches 4000 m (13,200 ft) (see Andreyeva-Grigorovich et al., 1997). In the Krakow–Bochnia region at the top of evaporitic horizon, there occur silty-sandy deposits (Chodenice beds) with a few intercalations of tuffites. The radiometric age of these tuffites is around 12 Ma (Van Couvering et al., 1981).

In the northern, marginal part of the PCFB, the early Sarmatian littoral carbonate and clastic deposits are well preserved (Figure 15A).

The development of the folded Miocene units (Figures 2, 4) in the PCFB was strongly controlled by both the slope of the Carpathian overthrust surface and the depth of the platform basement (Oszczypko and Tomas, 1985). The Sambir unit has been recognized at the front of the Skole nappe southeast of Przemysl, as well as beneath this nappe (Ney, 1968). The Sambir unit (as much as 2200 m [7200 ft] thick, Ney, 1968) is composed of both the early and middle Miocene up to the Sarmatian (Andreyeva-Grigorovich et al., 1997; Garecka and Olszewska, 1997). In the Ukrainian Carpathians (Figures 6–9), the folded early Miocene molasses are known from the Boryslav–Pokuttya as well

A
B

Figure 15. (A) Early Sarmatian littoral cross-laminated calcareous sands from the Stawiany–Samostrzalow area. Marginal part of the PCFB, northeast from Pinczow. (B) Early Sarmatian thin- to medium-bedded fine-grained sands and siltstones of the Sambir unit of the UCFB. Right bank of the Vyrva River at Bonevychy near Dobromil.

as Sambir units (Andreyeva-Grigorovich et al., 1997). These deposits belong to the Stebnyk and Balich formations. The Stebnyk Formation, as much as 3000 m (10,000 ft) thick, is composed of variegated clays with intercalations of sandstones and gypsum lenses. The formation contains the Ottnangian, probably recycled foraminifera (see Stryszawa Formation) and, probably, Karpatian calcareous nannoplankton (Andreyeva-Grigorovich et al., 1997). The Stebnyk Formation is overlain by the Balych formation, as much as 600 m thick, and it is represented by gray and green-gray limy clays with intercalations of sands and sandstones. In the lower part of the formation, the intercalations of pink clays and argillites are observed. The calcareous nannoplankton studies determine the age of the formation as the NN4 zone (Ottnangian–lower Badenian), but in some localities, the early Badenian species *P. glomerosa* and *Orbulina universa* have been discovered (see Andreyeva-Grigorovich et al., 1997; Garecka and Olszewska, 1997). The Balych formation is overlain by the Badenian strata. The upper Badenian deposits of the Sambir unit are composed of potash and rock salts, claystones, and anhydrites of the Kalush formation (Table 2). The uppermost part of the upper Badenian–Sarmatian sequence is represented by the sandy-silty deltaic deposits (Figure 15B) with the flysch-derived Radych conglomerates at the top. The Stebnyk Formation of the Sambir unit could be compared with the Stryszawa Formation, whereas Balych formation is probably an equivalent of the Bielsko and partly Skawina formations (Oszczypko, 1997).

Between Przemysl and Krakow, along the Carpathian frontal thrust, a narrow zone (as much as 10 km [6 mi]) of folded Badenian and Sarmatian deposits (Zglobice unit, see Kotlarczyk, 1985) occurs. In our opinion, this unit can be regarded as a prolongation of the Sambir unit (Figure 2). The Badenian and Sarmatian strata are also preserved as erosional outliers in the Polish Outer Carpathians. The southernmost occurrence of the upper Badenian–Sarmatian marine sediments

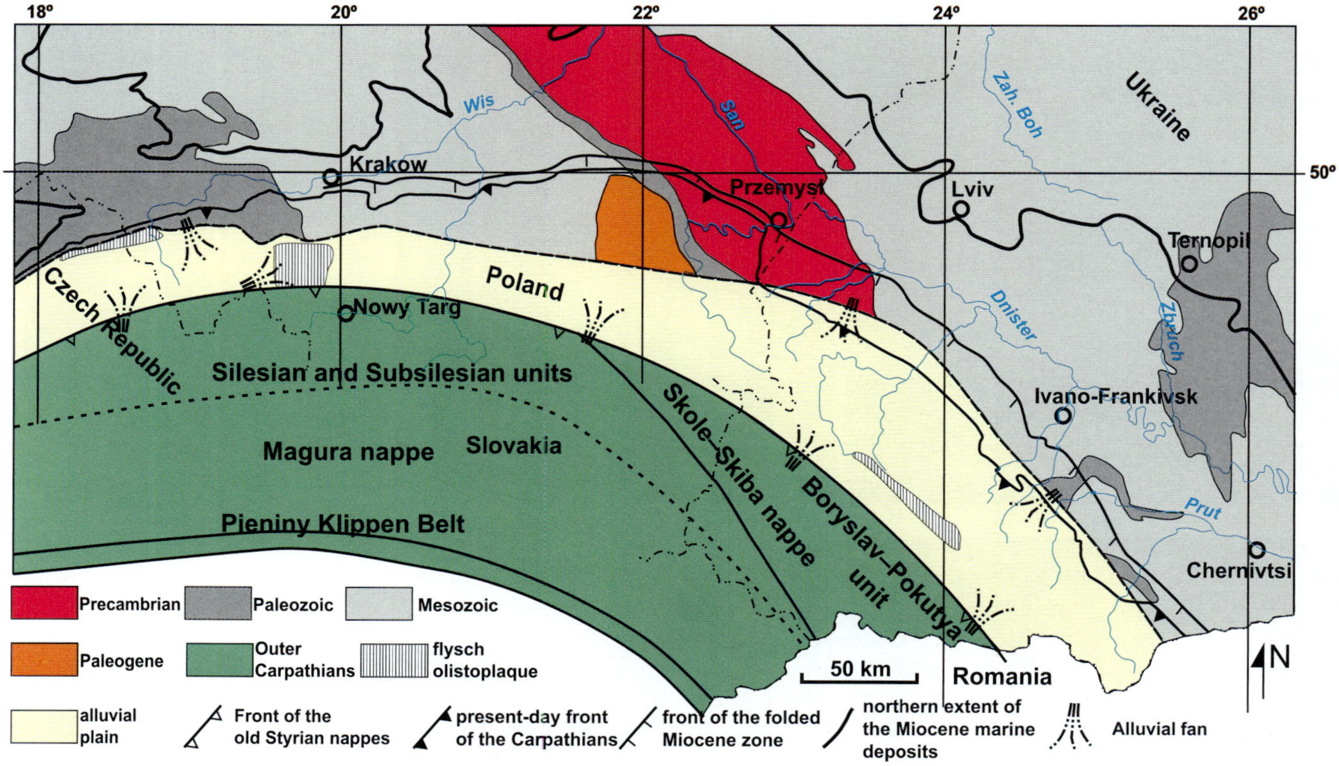

Figure 16. Late Burdigalian (Karpatian) palinspastic paleogeography of the Polish and Ukrainian Carpathian Foredeep (compiled by Oszczypko and Popadyuk).

is known from the Nowy Sacz basin (Oszczypko et al., 1992). In the Cieszyn–Wadowice area, the lower and upper Badenian(?) deposits are incorporated into the Subsilesian and Silesian units. The Zawoja 1 borehole reached the parautochthonous lower Badenian(?) deposits beneath the Magura nappe (Figure 3) (see also Moryc, 1989; Baran et al., 1997).

MAIN STAGES OF THE CARPATHIAN FOREDEEP DEVELOPMENT

Early Miocene (Late Burdigalian)

During the late Burdigalian, tectonic movements in the front of the Western Outer Carpathians overrode the platform, and the thin-skinned Subsilesian nappe (northeast Moravia and south Silesia) reached approximately the present-day position of the Carpathians (Figure 16) (see also Oszczypko, 1998; Oszczypko et al., 2000). All buried fragments of the Subsilesian and Silesian units (Sucha and Zamarski formations) recognized by drillings are erosional remnants of the Old Styrian nappes, which were at least 25 km (16 mi) wide (Figure 16). The load of these nappes caused the development of the Karpatian flexural depression (Inner Carpathian Foredeep) 25–50 km (16–31 mi) wide. During this time, the axis of subsidence was situated about 20–50 km (12–31 mi) south from the present-day position of the Carpathian thrust and extended toward the east (Boryslav–Pokuttya and Sambir units in the Ukraine; see Oszczypko, 1997, 1998; Zytko, 1965). This basin developed both on the top of the advancing Carpathian front as well as on the platform basement. The relatively narrow brackish- or freshwater, moat-like basin was filled with coarse clastic sediments, derived both from the emerged front of the Outer Carpathians as well as its platform (Figure 16). In the Sucha–Cieszyn area, the Old Styrian nappe are overlapped by the Stryszawa Formation. The basal part of the formation is represented by conglomerates as much as 650 m (2100 ft) thick, which pass upward into variegated, conglomerate-sandy-mudstone strata. The lower part of the formation shows features of the fan-delta deposits, whereas the upper part was probably deposited as an alluvial fan. In the Ukrainian segment, the basal part of the inner foredeep deposits begins with the Sloboda conglomerates, which overlapped the Vorotyshcha Salt Formation. These conglomerates are composed mainly of the clasts of the Upper Jurassic limestones derived from the platform basement (Andreyeva-Grigorovich et al., 1997). These matrix-supported conglomerates (Figure 12C) were deposited by the cohesive gravitational flows in the fan-delta environments. The conglomerates are followed by the Dobrotiv Formation, represented by thick-bedded, sandstone-dominated sequences. These well-organized sequences show features of deltaic deposits (Figure 13A). The uppermost part of the lower Miocene molasse sequence in the UCFB belongs to the Stebnyk Formation (Figure 13B). This formation is dominated by variegated (rose, green, and gray) marly claystones, with intercalations of the coarse-grained, medium- to thick-bedded, mainly cross-laminated sandstones. In Przemysl in few boreholes, the Eocene–Oligocene flysch deposits were pierced at the base of the Sambir–Stebnyk unit (see Zielinski, 1963; Wdowiarz, 1976; Ksiazkiewicz, 1977). These flysch blocks, like the Sucha and Zamarski formations of the western PCFB, can be interpreted as being of olistholithe origin (see Oszczypko, 1997, 1998; Oszczypko and Oszczypko-Clowes, 2003). In the Romanian Foredeep (Micu, 1982), the lower Miocene sequences are represented by terrestrial red beds (Magiresti and Hirja beds). The lower Miocene deposits of the inner foredeep in the Czech Republic, Poland, and Ukraine formed a clastic wedge along the Carpathians, comparable with the lower freshwater molasse of the Alpine foreland.

In the north Moravian–upper Silesian area, an intra-Karpatian uplift and an erosion of the Cieszyn–Slavkov ridge occurred after deposition of the Stryszawa Formation (Elias and Palensky, 1997; Palensky et al., 1995; Oszczypko and Lucinska-Anczkiewicz, 2000). It explains the erosive character of the Zamarski Formation in the Cieszyn area. This period of erosion could be correlated with the discordance below the terminal Karpatian strata in southern Moravia (Jiricek, 1995).

Early Badenian (Langhian and Early Serravalian)

The new period of the late Karpatian–early Badenian subsidence was characterized by the development of southwest–northeast- and northwest–southeast-trending grabens at the front of the old Styrian thrust (Oszczypko and Lucinska-Anczkiewicz, 2000). The subsidence affected both the inner as well as outer foredeep that began to developed. This was accompanied by the early Badenian regional transgression (Oszczypko, 1998), which flooded both the foredeep and marginal part of the Carpathians. Simultaneously, the seashore line shifted northward from 30–40 to 100 km (19–25 to 60 mi) in the western and eastern part of the basin, respectively. The paleobathymetry of the early Badenian basin varied between upper bathyal depths in the axis of the basin to neritic (Figure 17) (see also Rogl, 1999) and littoral in the northern and southern (Carpathian) marginal part of basin. The northern shore of the early Badenian sea was occupied by littoral sands and lithothamnium limestones of the Pinczow Formation as well as Mykolaiv sands and Naraiv limestones. These limestones are composed mostly of red algae with bryozoans and other organisms. On the relatively wide (50–100 km; 30–60 mi) northern shelf (outer foredeep), glauconitic sandstones, muddy shales, and sandstones of the Zhuriv and Baranow beds were deposited. The deepest part of the basin (inner foredeep), which reached midbathyal depth, was located about 20 km (12 mi) south of the present-day position of the Carpathian front (Sucha Beskidzka–Nowy Sacz–Krosno and Drohobych). This part of the basin was dominated by a deposition of dark-gray calcareous mudstones with sporadic intercalations of fine sandstones (Skawina, Balych, and Bohorodchany formations). These deposits sometimes revealed basinal turbidite characteristics. The southern slope and its shelf with small embayments were dominated by calcareous mudstones. The littoral zone of the Carpathians was composed of biohermal, organodetritic limestones and quartz sandstones with rodoliths (Golonka, 1982; Rajchel, 1989). More locally, the zone was composed of lagoonal and paludal deposits with lignite intercalations (Grudna Dolna). In the Carpathian offshore area, lower Badenian deposits onlapped deformed and eroded flysch deposits of the Skole and Silesian and Subsilesian nappes with angular unconformity. The Serravalian (ca. 15 Ma) gradual shallowing of bathymetry in the Carpathian foreland basin coincided with a gradual global fall of relative sea level (Oszczypko, 1997, 1998). This was accompanied by the development of the large, marine fan delta at the front of the ancient Carpathians (Figure 17). These types of flysch-derived coarse-clastic deposits were recognized by Doktor (1983) in the area of Gdow, southeast from Krakow. The same deposits were also recognized west of Krakow in the locality of Bacharowice (see Doktor, 1983). The shallowing of the early Badenian sea caused a partial isolation of the basin and initiated the Badenian (ca. 13.0–12.5 Ma) salinity crisis in the shallow basin [inner-middle neritic(?) depths].

Late Badenian Salinity Crisis

Introduction

The Badenian evaporite deposits in the Polish part of the Carpathian Foredeep basin, commonly referred to as chemical deposits, are included in two formations: the Krzyżanowice Formation and Wieliczka Formation (Alexandrowicz et al., 1982). In addition, they

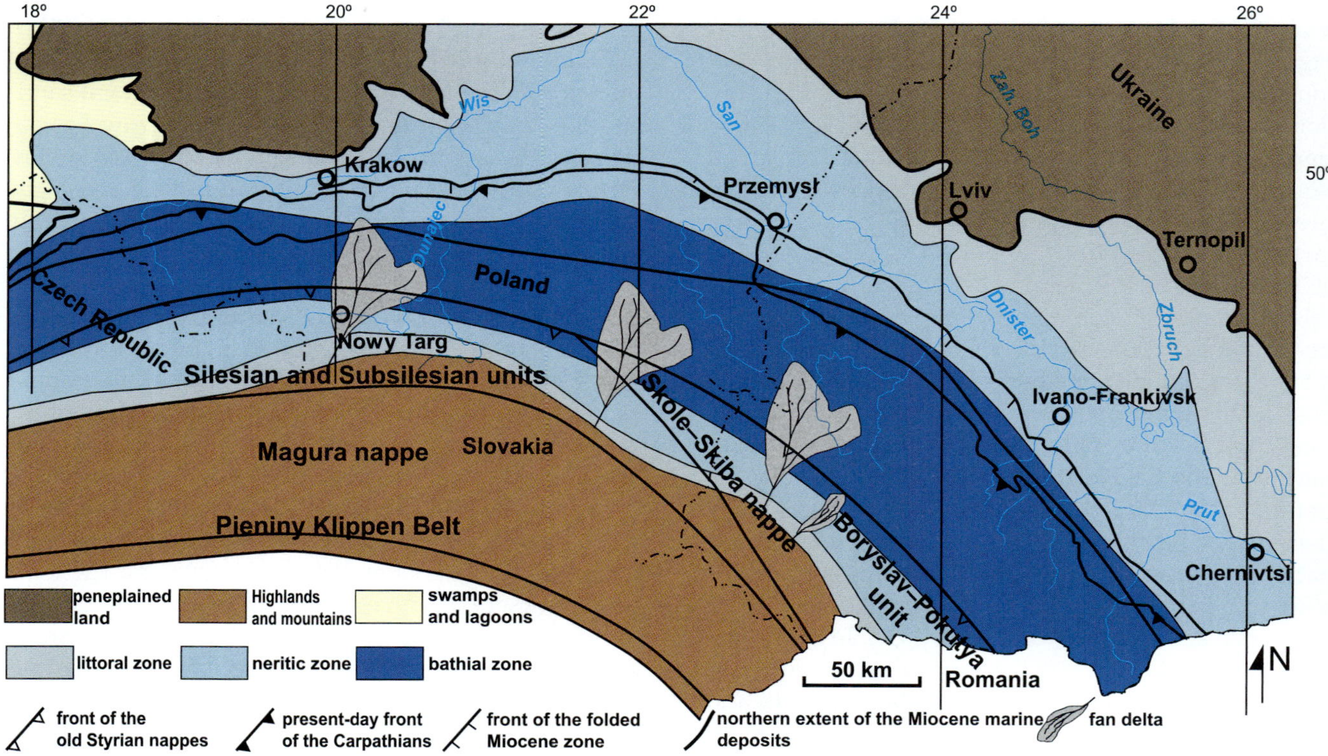

Figure 17. Langhian (early Badenian) palinspastic paleogeography of the Polish and Ukrainian Carpathian Foredeep (compiled by Oszczypko and Popadyuk).

occur in the foreland, in the upper Silesia in Poland, and in the Opava region in the Czech Republic (Peryt et al., 1997). In Ukraine, the evaporites and related deposits are included into the Tyras Suite (Formation) (Petryczenko et al., 1994) (Figure 18).

The Krzyzanowice Formation comprises sulfate deposits: gypsum and anhydrite, as well as the associated siliciclastic deposits (mainly claystones) and barren and sulfur-bearing limestones that are commonly considered as those that originated from the metasomatosis of sulfates (Figure 18). It occurs in almost the entire area of the Carpathian Foredeep except in small areas near Zory (upper Silesia) and in the region of the Carpathian overthrust between Krakow and Tarnow, where the Wieliczka Formation occurs. In addition, the Krzyzanowice Formation is missing north of Rzeszow, in the so-called Rzeszow Island, from where it was subsequently eroded (Ney et al., 1974), although another interpretation relates the lack of evaporites in that area to the great depth of water in the sedimentary basin (Karnkowski and Ozimkowski, 2001). The thickness of the formation is several tens of meters in the marginal part of the basin and commonly 10–20 m (33–66 ft) in the central part.

The Wieliczka Formation (Garlicki, 1994a) consists of rock salt with intercalations of claystones and minor sulfates (Figure 19). The stratotype of the formation occurs in the Wieliczka salt mine, and the total thickness is 30–100 m (100–660 ft). Garlicki (1979, 1994a) recognized five cyclothems that are present throughout the Polish part of the Carpathian Foredeep except for the upper Silesia and Wieliczka, where the lowermost cyclothem is lacking (Garlicki, 1994b). Marker beds are present in some cyclothems that facilitate the stratigraphic correlation of distant sections. The Wieliczka Formation occurs in the upper Silesia in the triangle area between Rybnik, Zory, and Orzesze, then in the form of narrow band along the Carpathian overthrust from Krakow to Brzesko (Garlicki, 1979, 1994b), and further to the east, it occurs under the overthrust Carpathian nappes (Poltowicz, 1993).

In Poland and the Czech Republic, evaporites are underlain by the lower Badenian Skawina Formation (marly mudstones, siltstones, and claystones) and are overlain by siliciclastics of the Machow Formation (Chodenice beds).

The Tyras suite is overlying the lower Badenian Bohorodchany suite or Baraniv beds in the Bilche–Volytsa zone of the Ukrainian part of Carpathian Foredeep, and it consists of sulfates with claystone intercalations. In local depressions (see Panow and Plotnikow, 1996, their table 1), sulfates and rock salt with claystone

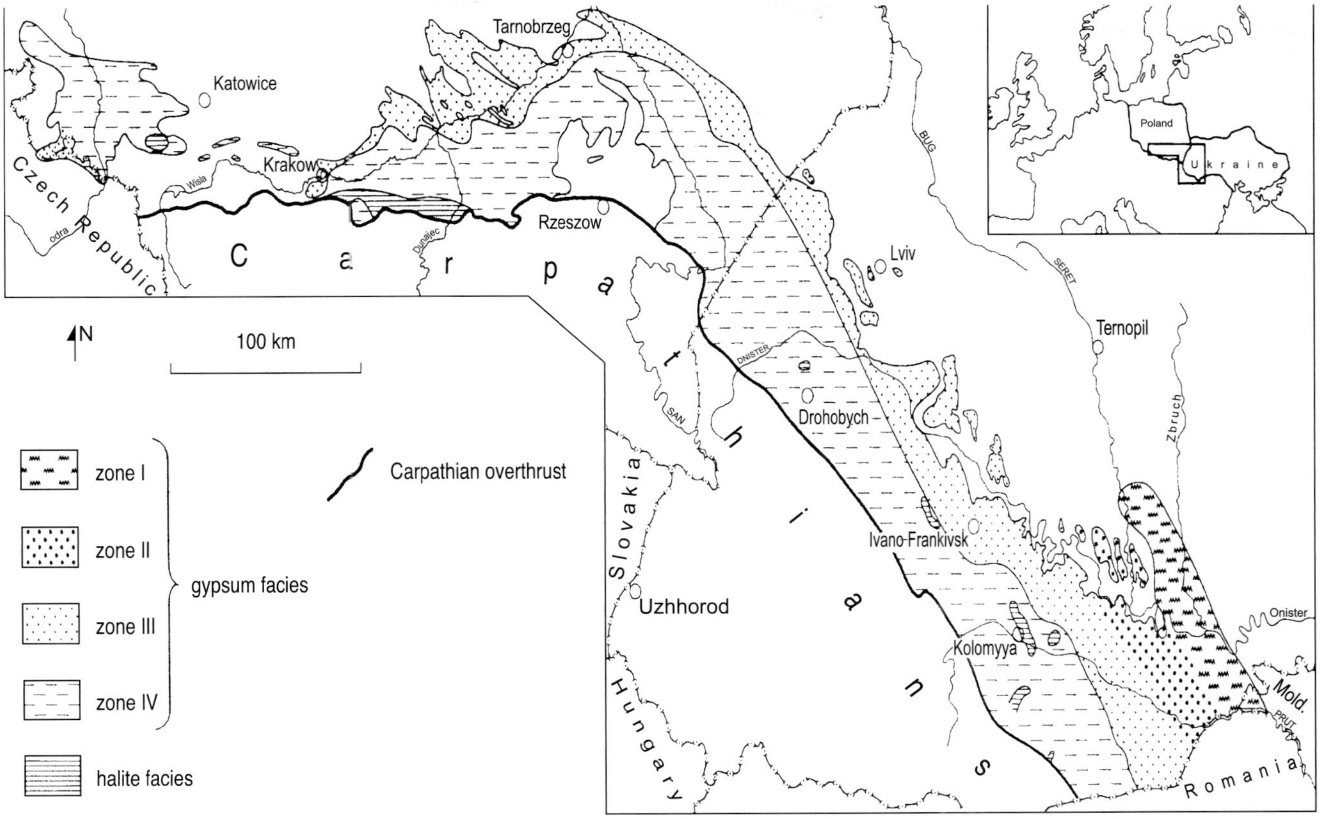

Figure 18. Distribution of Badenian evaporite facies in the Carpathian Foredeep basin.

intercalations of considerably bigger thickness occur (as much as 275 m [902 ft] in the Korshiv-Ispas 22 borehole). In the part of the Badenian basin located on the East European platform, the suite consists of gypsum and, in places, sulfur-bearing limestones. These deposits overlie Devonian, Cretaceous, or lower Badenian Opole suite and are overlain, in turn, by the Ratyn Limestone (a few tens of centimeters to a few meters thick) that is related to the next marine transgression (Peryt and Peryt, 1994). The present pattern of gypsum thickness is complex, mostly because of erosion prior to the deposition of the Ratyn Limestone, but commonly, the thickness of gypsum is around 20 m (66 ft), except for local grabens where the gypsum sections are more complete and reach 60 m (220 ft) in thickness. The Ratyn Limestone is overlain by the Kosiv Suite (Formation), which is a few tens of meters thick in the marginal part of the Carpathian Foredeep and increases in thickness abruptly (reaching as much as 1200 m [4000 ft]) toward the central part of the Carpathian Foredeep.

The lateral counterpart of the Tyras suite of the Bilche–Volytsa zone are, in the Sambir zone, Kalush beds consisting of interbedded salt claystones, potash salts (exploited in the Kalush mine), gypsum, and anhydrite, as well as sandstones 200–400 m (660–1300 ft) thick (Petryczenko et al., 1994). However, many authors consider that the Kalush deposit belongs to the Stebnyk suite (Ottnangian) (e.g., Klimov, 1968).

Age of Evaporites

As in many evaporite basins, the age and correlation of evaporites occurring in different parts of the Badenian basin and, hence, the reconstruction of basin history during deposition of evaporites are subject to controversies.

Based on the calcareous nannoplankton study, the evaporites were included in the NN6 zone (Dudziak and Laptas, 1991), with a reservation that the lowermost part of the complex may represent the NN5 zone (Dudziak and Luczkowska, 1991), or in the NN7 zone (Gazdzicka, 1994), and the NN6–NN7 zone (Peryt and Peryt, 1994; Prisyazhniuk et al., 1997). Peryt (1997) proved that the Badenian sulfates in the Gliwice region represent the lower part of the NN6 zone, and subsequently, she also found that the evaporites in the Bochnia salt mine and in several boreholes in southern Poland contain calcareous nannoplankton assemblage (and especially *Cyclicargolithus floridanus* and *Discoaster*

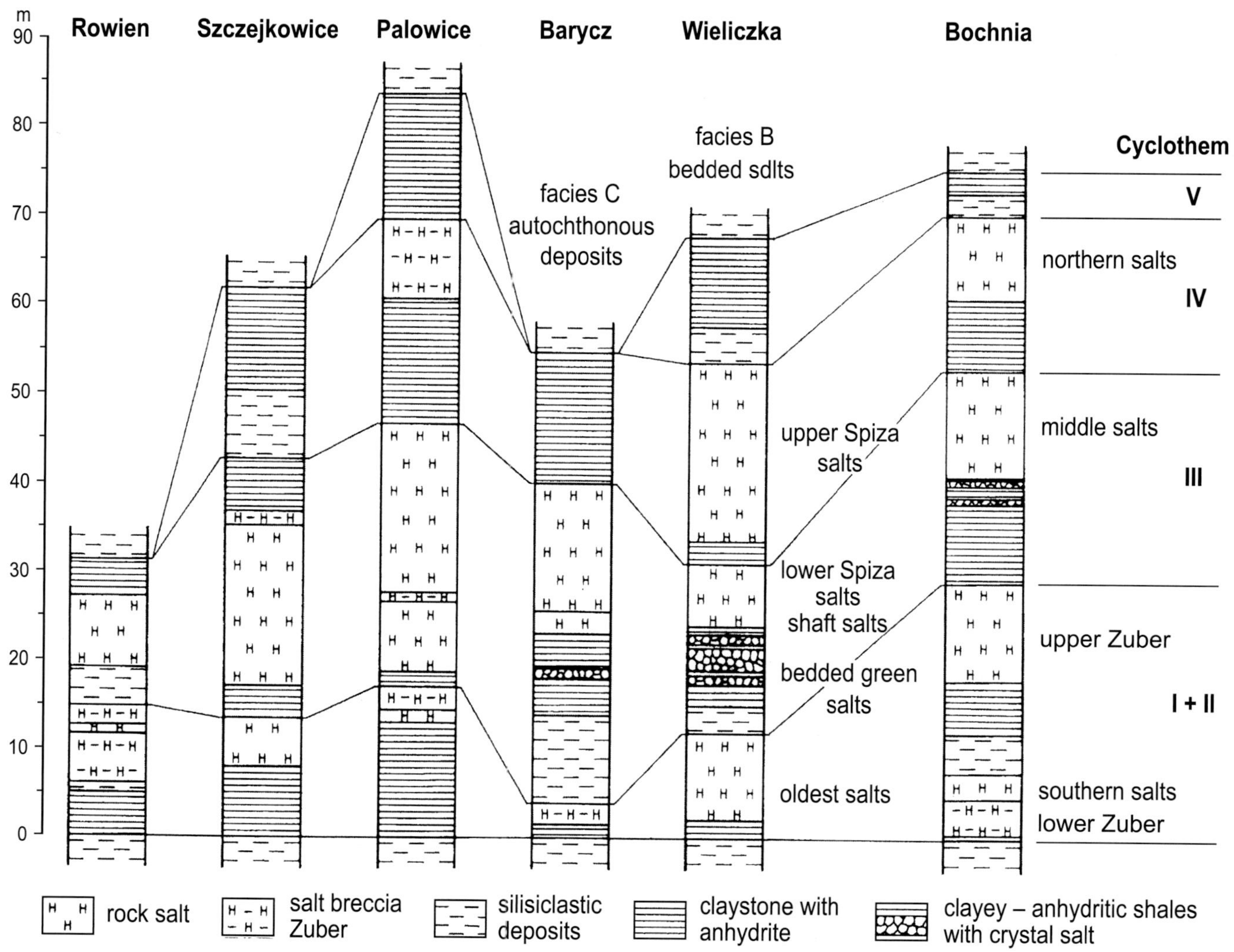

Figure 19. Correlation of the Badenian salt sequences in Poland (modified after Garlicki, 1979, 1994b; Tobola, 2000).

exilis; Peryt et al., 1998b) characteristic for the NN6 zone (Peryt, 1999).

Andreyeva-Grigorovich et al. (1997) recorded the nannoplankton association characteristic for the NN5 zone in clay intercalations in evaporites of the Tyras suite. Recently, Andreyeva-Grigorovich et al. (2003) found that the evaporites in the Bochnia and Wieliczka deposits belong to the NN6 zone. They also assume that the evaporites in Kalush belong to the NN6 zone, which is based on the occurrence of the boundary between NN5 and NN6 zones in the Kalush 340 borehole, 32 m (105 ft) below the evaporites, and the zone NN6–7 in clays above evaporites in the same borehole (Andreyeva-Grigorovich et al., 2003). However, the structure of the Kalush deposit is very complicated (Korin, 1994), and the borehole section may not provide a reliable stratigraphic section. Considering a complex geologic structure, the questions about the number of evaporite formations, their thicknesses, stratigraphy, and tectonics are still being discussed. According to recent views (Dzhinoridze et al., 1980; Korin, 1994), only two evaporite formations in the Carpathian Foredeep exist: Vorotyshcha of Eggenburgian age and Tyras of Badenian age. The Tyras age of the Kalush potash deposit is in contrast to what is known about the relatively low concentration of brines during Badenian halite precipitation (Garlicki and Wiewiorka, 1981; Galamay, 1997) and the K/Ar age of the langbeinite (13.5–15.9 m.y.; Wojtowicz et al., 2002) that is considerably older when compared to the radiometric ages of amphibole occurring in the tuff of Wieliczka salt (12.5 m.y.; Bukowski, 1999) and biotite occurring in the tuff of the Chodenice beds near Bochnia (Wieser et al., 2000). Accordingly, geochemical arguments and radiometric data suggest that the third evaporite formation is also present in the Carpathian Foredeep: it would be older than the Tyras Formation and younger than the Vorotyshcha Formation.

Facies Distribution

Badenian evaporites in the Polish and Ukrainian parts of the Carpathian Foredeep show a regular spatial facies pattern. Primary gypsum, which forms a wide (10–50 km [6–31 mi] in Poland and as much as 100 km [60 mi] in Ukraine) marginal sulfate platform, occurs in the most marginal part of the Carpathian Foredeep (Figure 18). It passes basinward into the sulfate basin, 20–60 km (12–37 mi) wide in Poland and Ukraine. In the southern part of sedimentary basin, in the narrow axial part of the basin, halite deposits (sparsely exceeding 100 m in thickness) occur in local salt basins (Figure 18). Close to the halite facies field, halite and anhydrite intercalations of variable thickness and number are also present (Poltowicz, 1993).

As far as the marginal sulfate platform is concerned, the greatest facies variation was recorded in the Ukrainian part of the basin (Peryt, 1996, 2001). Three main types of gypsum section exist (Figure 20). The first one consists entirely of stromatolitic gypsum (Figure 21A) and is characteristic of the area (>15 km [>9 mi] wide) bordering the nearshore sulfate facies limits. The stromatolitic gypsum exhibits a considerable variety of forms, although planar or slightly crenulated stromatolites prevail. The second type of gypsum section is located further toward the basin center. This type is characterized by the occurrence of stromatolitic gypsum in the lower part of the section and sabre gypsum (Figure 21B) (occasionally with a clastic gypsum unit above the sabre gypsum) in the upper part. The width of the zone where the second type of gypsum section occurs is more than 40 km (25 mi). The third type occurs in further basinward locations and is composed of giant gypsum intergrowths (or secondary nodular gypsum pseudomorphs of that facies) in the lower part (Figure 22), overlain by stromatolitic gypsum, sabre gypsum (Figure 21C), and clastic gypsum units (Figure 20). In Poland and the Czech Republic, only the third type of gypsum section development was recorded (Peryt et al., 1998a) (Figure 20). Very characteristic for that zone is a possibility of correlation of particular gypsum units. Kasprzyk (1993) recognized 18 units (units a–r) in the complete gypsum section south of the Holy Cross Mountains (Poland), and many of them can be traced throughout the northern Carpathian Foredeep in Poland, although differences in the succession, thickness, and component lithofacies of these gypsum units exist (Figure 20) (see Kasprzyk, 1999; Peryt and Kasprzyk, 1992). South of the Holy Cross Mountains, Kasprzyk (1993, 1999) recorded the occurrence of the nearshore facies association (comprising units b–e and j–l) that is characterized by common microbial and nodular structures and a large content of terrigenous material. In the deeper subsurface, gypsum is partly or completely replaced by anhydrite and associated secondary gypsum, but it is possible to distinguish anhydrite lithofacies that are equivalent to parent gypsum lithofacies (Kubica, 1992).

In general, the Badenian sulfate section in the marginal sulfate platform can be divided into two parts, although in many cases, especially in west Ukraine, the upper part is absent, having been deposited only locally in tectonically controlled troughs (Peryt, 2001). The gypsum of this upper, clastic part of the section consists of detritus derived from erosion of the lower part of gypsum section and possibly of primary precipitates.

In some parts of the marginal sulfate platform, postsulfate (gypsum-ghost) limestones with prevalent gypsum lithofacies occur. Locally, these limestones contain native sulfur (Kubica, 1997; Sul and Poberezski, 1995).

The anhydrite sections in the central part of the Carpathian Foredeep (called here the sulfate basin) consist mainly of interbedded laminated anhydrite and anhydrite breccia. These facies display features characteristic of redeposited sediments such as distortions caused by slumping, erosional boundaries, and graded-bedded laminae (Peryt, 2000). Lamination in laminated anhydrite is sparsely regular and commonly accompanied by disruptions. No major trend in the development of profiles from that area is present.

The best recognized facies changes of Badenian salts come from the historical Wieliczka salt mine, where two units are distinguished (Gawel, 1962). The lower unit (stratiform deposit-bedded part) is a complex of strongly folded, deformed, and overthrusted salt beds interbedded with anhydrite and anhydritic clays. The upper unit (boulder deposit) is developed as coarse breccia composed mainly of salt clay with blocks (as much as >150 m [500 ft] long) of coarse-grained salt. In addition, three sequences of evaporates distinguished in the Wieliczka area are regarded as individual facies, deposited in various parts of the primary Wieliczka subbasin (Figure 19).

The lithostratigraphical correlation in a regional scale, based on sedimentary cyclothems and marker beds originally distinguished in the bedded part of the Wieliczka salt mine, made it possible to state that salts of the first cyclothem were deposited only in the area between Bochnia and Tarnow (Garlicki, 1979, 1994b). In cycle 2, the salt sedimentation occurred in an area from upper Silesia to Tarnow, and at that time, the oldest salts of Wieliczka were deposited. The salts of the cyclothem 3 have the widest extent, from the upper Silesia to the area east of Tarnow, and in the Wieliczka mine, the cyclothem includes green-bedded salt, shaft

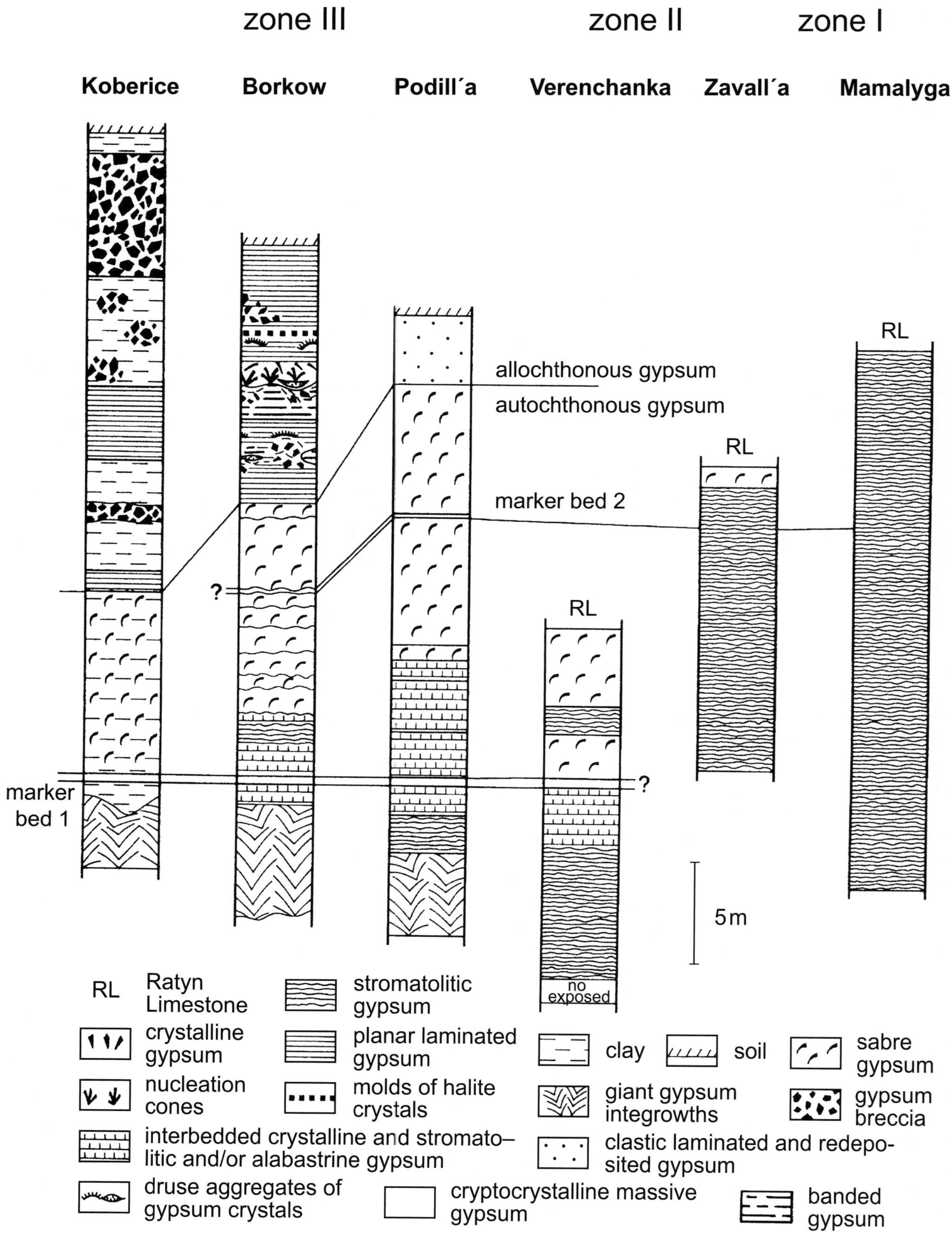

Figure 20. Correlation of Badenian sulfate sections in the Carpathian Foredeep basin (Czech Republic, Poland, and Ukraine).

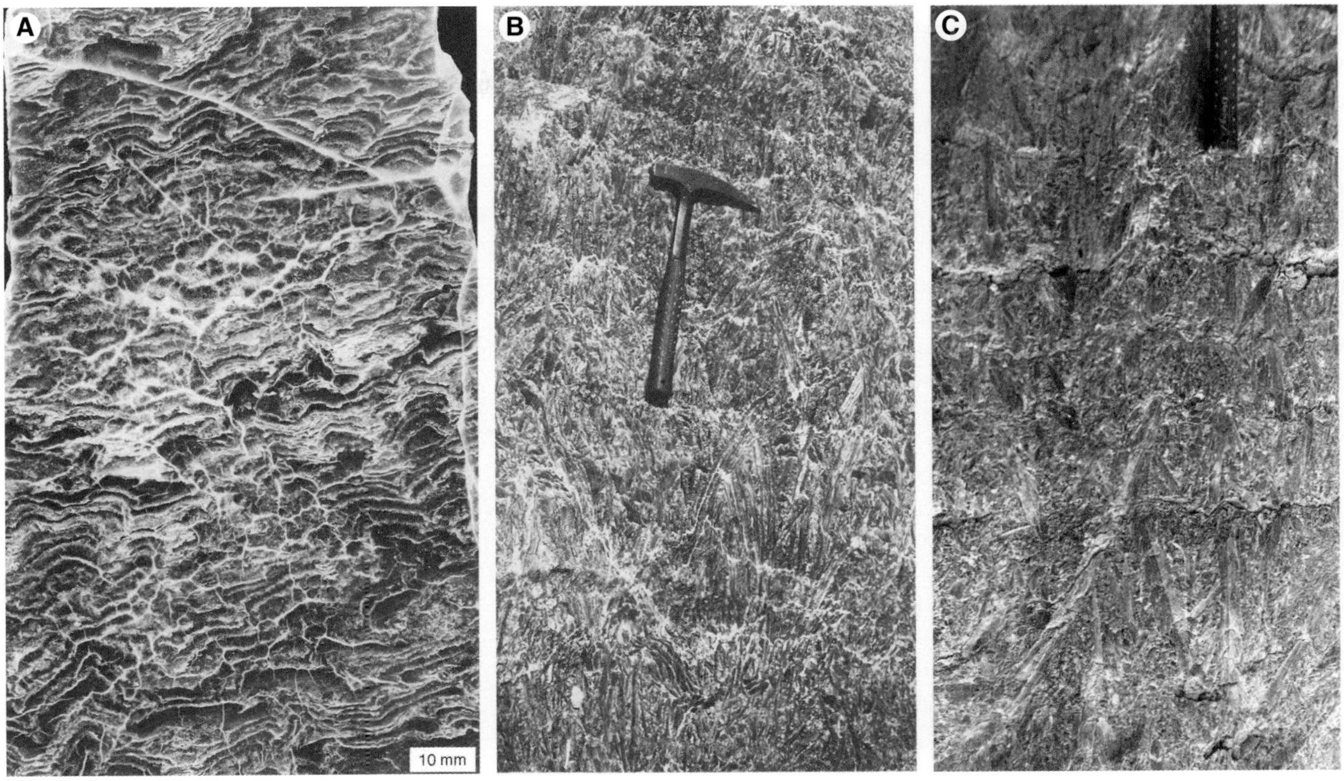

Figure 21. (A) Stromatolitic gypsum, Mamalyga; (B) sabre gypsum, Verenchanka; (C) sabre gypsum, Piski.

salt, and lower spiza salt. The cyclothem 4 in Wieliczka includes the upper spiza salt. The deposits of the youngest cyclothem 5 are known only from some boreholes (Figure 19) (Garlicki, 1979, 1994b).

The deposits of the boulder and stratiform units have been identified as facies varieties (Garlicki, 1979). The boulder unit deposits precipitated in a more shallow evaporite basin situated closer to the Carpathian border (Bukowski, 1994), and the stratiform unit deposits accumulated in the deeper and more distal part of the subbasin. Both facies exhibit sedimentary structures indicating that most of the halite layers have been formed from redeposition by gravity mass movements (e.g., Slaczka and Kolasa, 1997).

Sedimentary Environments

Autochthonous gypsum facies (crystalline gypsum and stromatolitic and massive alabastrine gypsum) were deposited in two main environments (Peryt, 1996). One variety of crystalline gypsum (giant gypsum intergrowths) precipitated from highly concentrated brines in moderate water depths at the initial stage of gypsum precipitation. Other varieties of autochthonous gypsum facies represent very shallow subaqueous gypsum deposited in a vast brine pan characterized by a facies mosaic that reflected an interplay of concentrated brines from the central part of the evaporite basin, diluted brines resulting from the influx of continental meteoric waters, and local evaporation of a restricted, shallow-water body. Accordingly, the salinity oscillated (Peryt, 1996; Rosell et al., 1998). A facies continuum, microbial gypsum-grasslike selenite-skeletal selenite-sabre gypsum, indicates increasing salinity of the brine with time. The sabre gypsum unit contains an intercalation of gypsarenitic, mostly laminated gypsum (marker bed 2 in Figure 20), which is correlated with the limestone intercalations recorded in some outcrops (such as Mamalyga Figure 20). The inflow waters were normal marine, as indicated by the presence of marine fossils in the limestones, so in such cases, gypsum precipitation resumed once the entire brine again became supersaturated with gypsum. The nearshore facies association distinguished by Kasprzyk (1999) originated in the sabkha environment. The origin of gypsum-ghost limestones and sulfur is commonly attributed to metasomatic alteration of sulfates (see Kubica, 1997, for a review), although Gasiewicz (2000) stresses the differences between gypsum-ghost limestones and selenitic gypsum facies with regard to their sedimentological features and major element compositions. Gasiewicz

(2000) supposes that both the lithological units developed in coeval depositional domains.

The gypsum of the upper, clastic part of the section was deposited in a basin with a bottom topography of at least a few tens of meters. In contrast, during the deposition of the lower part of the gypsum section, the basin relief was minor, and the differences in water depth between facies zones I and III probably did not exceed 10 m (33 ft). During the formation of the allochthonous gypsum (encompassing clastic gypsum and gypsum breccias), an influx of seawater is indicated by a fauna of mollusks, foraminifera (mostly pelagic globigerinids), ostracods, and pteropods, recorded from clays intercalated in the gypsum.

Basin-marginal Badenian evaporites formed in a standing body of water, as well as in desiccated environments subject to floods. The lateral persistence of thin beds over large areas with only minor changes in thickness and facies indicates that they formed on broad, very low-relief areas that were affected by rapid transgressions that led to major changes in brine chemistry. Babel (1999) suggested two major salinity rises during gypsum deposition and correlated them with two cyclothems recognized by Kubica (1992). A different scenario results from the detailed analyses of the strontium content in gypsum (Rosell et al., 1998; Kasprzyk, 1999).

The anhydrite sections in the central part of the Carpathian Foredeep display features characteristic of redeposited sediments. A great deal of redeposition combined with the in-situ brecciation suggests that the triggering mechanism for redeposition could be earthquakes. They generated mass flows and eventually turbidites, so that earlier deposited grains could be redeposited. The redeposition processes were also a very important factor during the deposition of halite and associated siliciclastics (Kolasa and Slaczka, 1985; Bukowski, 1997; Slaczka and Kolasa, 1997).

The studies on fluid inclusions in halite (Galamay et al., 1997; Kovalevich, 1997; Kovalevich and Petrichenko, 1997) showed that the brines belonged to Na–K–Mg–Cl–SO_4 type. The ratios of major ions in the Badenian brine are close to those in modern seawater evaporated to the initial or middle stages of halite precipitation. Significant fluctuations in composition and concentration of brines in the vertical profile of salt deposit are not observed. The increased content of K in brines might be caused by a partial dissolution of earlier precipitated potash salts of the older evaporite formations. The reason of the decreased content of

Figure 22. Giant gypsum intergrowth facies and the marker bed 1 (arrowed) in Koberice (A), Borkow (B), and Oleshiv (C).

SO_4 ion was probably the inflow of large amounts of surface water from the land. A very high contribution of continental water was also indicated by Bukowski et al. (2001) based on the study of stable isotope composition of fluid inclusions in halite from Wieliczka. Low values of bromine content (Garlicki and Wiewiorka, 1981; Garlicki et al., 1991; Galamay, 1997) indicate that halite is partly recycled in origin (cf. Garcia Veigas et al., 1997). Cendon et al. (1999) deduced from the study of solute concentrations and the different evaporation scenarios modeled that during halite precipitation, recycling of previously formed evaporites occurred with an important input of water of continental origin. In addition, the $\delta^{37}Cl$ values imply a persistent nonmarine chloride source, and in the case of the shaft salt (cyclothem 3) from Wieliczka, the introduction of nonmarine chloride was abrupt (Eastoe and Peryt, 1999).

Halite and associated deposits in the central part of the Badenian evaporite basin show that the same facies successions and marker beds can be traced across and between individual basins (Garlicki, 1994b). It is commonly assumed that the depth of those basins was a few hundred meters deep (Garlicki, 1979), and such a conclusion is supported by geochemical data (Gawel, 1965; Kovalevich, 1997). Marine incursions were related to the extrabasinal factors, and, based on rapid facies changes, it seems that the Badenian evaporite basin was located in a depression in which the brine top level was located below the contemporaneous sea level. Accordingly, during sea level rise, new marine water could enter this depression and bring with it a temporary pulse of marine fauna.

Correlation

Although the correlation of different Badenian evaporite facies is difficult (if possible at all), no such problems are present in the case of particular evaporite facies zones.

The correlation between facies zones of the marginal sulfate platform was based on the occurrence of characteristic marker beds, as well as the observations made in areas transitional between the facies zones (Peryt, 2001). However, this is possible only if the major part of the section is composed of primary gypsum (or anhydrite). When only secondary (nodular) gypsum occurs, it is not possible to subdivide the sulfate succession.

A characteristic marker bed (commonly 20–40 cm [8–16 in.] thick) occurring a few tens of centimeters to a few meters above the giant gypsum intergrowths unit in zone III is of particular importance for correlation. This marker (bed 1 in Figure 20) is developed across much of the Carpathian Foredeep basin (Peryt et al., 1998a). The marker bed is commonly underlain by a unit consisting of interbedded crystalline and stromatolitic and/or alabastrine gypsum, which is, in turn, underlain by stromatolitic gypsum. In the Koberice section, the marker bed is underlain by clay unit (Peryt et al., 1997).

Marker bed 1 consists of cryptocrystalline massive gypsum with a distinctive finely crenulated lamination and has a domelike to cauliflowerlike top (Figure 22). It was possible to distinguish the same marker bed in zone II where it overlies a unit consisting of interbedded crystalline and stromatolitic and/or alabastrine. The marker was not distinguished in zone I perhaps because the entire gypsum section of this zone is younger than the marker bed in zones II and III (Peryt, 2001).

Above marker bed 1, selenitic and stromatolitic facies occur in most parts of the marginal sulfate platform (Figure 20); their upper part is developed as sabre gypsum. Continual lateral facies changes between stromatolitic and selenitic gypsum occur, which can be observed in particular outcrops (Peryt, 1996), and thus, the distance between the marker bed and the sabre gypsum unit fluctuates. Within zone III, it is possible to distinguish a thin (commonly 20 cm [8 in.]) unit of gypsarenitic (marker bed 2 in Figure 20), mostly laminated gypsum in the sabre gypsum unit. This unit is also sparsely recorded in zone II, and in zone I, at approximately the same position in the gypsum section, a distinct surface or a limestone intercalation occurs as recorded in west Ukraine (Peryt, 2001). Finally, another important correlative boundary is the autochthonous-clastic gypsum contact. This facies boundary seems to reflect a major change in the basin history (see Peryt, 1996) and, as such, is considered to be essentially isochronous.

The marker beds seem to reflect events that may be related to sudden and widespread changes in water chemistry, which, in turn, imply major changes in basin hydrology (Peryt, 1996, 2001). Correlation between the facies zones based on characteristic marker beds strongly suggests that the deposition of giant gypsum intergrowth facies and stromatolitic gypsum facies was coeval. Limestone intercalations in the form of lenticular bodies (commonly centimeters thick) in the upper part of the gypsum sections have been recorded in some sections in zones I and II. The microfacies (mainly the presence of marine fauna) of these limestones indicate a marine provenance.

In zone III, it is commonly possible to correlate the particular units and even beds (Kubica, 1992; Kasprzyk, 1993, 1999; Babel, 1999). Within the giant gypsum intergrowth facies, Babel (1999) distinguished coeval growth zones of crystals. In the case of sections made of anhydrite and secondary gypsum, the geophysical logs are an important tool in helping in correlation. Even in the case of gypsum-ghost limestones, it is possible to relate the limestones with the particular primary gypsum lithofacies (Kubica, 1992). It should be mentioned that in the Ukrainian part of the Carpathian basin, many authors relate gypsum-ghost limestones with the Ratyn Limestone that is of clearly different age and origin when compared to the gypsum-ghost limestones (Peryt and Peryt, 1994).

Physical stratigraphic relationships between sulfates occurring in the marginal sulfate platform and sulfates from sulfate basin are uncertain. Kasprzyk and Orti (1998) correlate this upper (clastic, allochthonous) gypsum member of the marginal sulfate platform with the entire sulfate section in the sulfate basin area. The correlation of particular halite cyclothems, units, and marker beds was done for the Polish part of the Carpathian Foredeep basin (Figure 19) (Garlicki, 1979, 1994b). However, the relation of the areas of halite and sulfate deposition remains enigmatic. In some boreholes in the inner zone of the Ukrainian part of the Carpathian Foredeep (Wyszynski, 1939; Petrichenko et al., 1994; Panow and Plotnikow, 1996), sulfates occur in the lower and upper parts of the evaporite sequence, and halite occurs in its middle part. Petrichenko et al. (1997) suggested that gypsum deposits in the marginal parts of the basin correspond only to the upper part of the evaporite sequence in those boreholes. The interpretation assumes that the beginning of precipitation caused by evaporation of seawater occurred in a narrow zone that was adjacent to the rising Carpathian Mountains. The precipitation of terrigenous sulfate and sulfate, mostly gypsum, deposits occurred, and subsequently, it was changed by sodium chloride precipitation. The halite stage was followed by calcium sulfate stage and terrigenous sedimentation, but those sulfates were deposited only in the outer part of the Carpathian Foredeep basin because of migration of evaporite basin, pushed by the migrating Carpathian nappes. In the Polish part of Carpathian Foredeep, similar relations are observed in some borehole sections; however, in others, the evaporite succession is far more complex (Poltowicz, 1993, their figure 7). If resedimented evaporites from the upper part of the sulfate sequence in the marginal sulfate platform correspond to the entire sequence from the sulfate basin, and the latter (considering the omnipresence of the redeposition phenomena in the halite sequence) correspond to the entire halite sequence, the autochthonous gypsum deposits from the marginal sulfate platform would be represented in deeper water settings only by the organic-rich shales underlying the redeposited evaporites.

The Origin of the Salinity Crisis

The estimated duration of the Badenian evaporite deposition of the Central Paratethys was 20,000–35,000 yr (Garlicki, 1979; Niemczyk, 1995; Petrichenko et al., 1997). The scenario of events leading to the deposition of widespread evaporites, being a synchronous event, in the Badenian is not established in detail. It is assumed that the interrupted communication of the Paratethys with the ocean was a consequence of eustatic sea level fall, possibly related to climatic cooling (Gonera et al., 2000). Such a conclusion resulted from the isotopic studies of Badenian foraminifera occurring below and above evaporites that indicated the lowest temperatures in a part of the Badenian section where evaporites occur (Gonera et al., 2000). Quantitative reconstructions of the middle to late Miocene climate evolution in the southern Carpathian Foredeep basin in northwest Bulgaria (Ivanov et al., 2002) indicated the existence of a warm, subtropical climate during the Badenian, with mean annual temperature between 16 and 18°C and mean annual precipitation between 1100 and 1300 mm (43 and 51 in.); however, small climate fluctuations existed as well. The reconstruction based on paleobotanic study of the Ciscarpathian study showed that immediately and after evaporite formation, a warm temperate climate prevailed (Demarq, 1987; Syabryay and Teslenko, 2000). However, a tectonic closure of the connection with the Tethys could contribute to the origin of the salinity crisis as well.

In some places in the marginal sulfate platform area, a thin sandy or marly coquina layer (Ervilia bed) occurs below the gypsum sequence. This layer, containing fauna almost exclusively composed of two mollusk species (*Modiola hoernesi* Reuss and *Ervilia pusilla* Philippi), is attributed to restricted water circulation and density stratification that followed sea level fall and an increase in salinity because those species are particularly adapted to an increased salinity and low oxygenation (Kowalewski, 1966). The density and brine stratification, once established, continued during the gypsum precipitation in the marginal sulfate platform zone (cf. Peryt et al., 1997; Babel, 1999; Kasprzyk, 1999) and during the deposition of halite-bearing series (Garlicki, 1979).

The study of brine inclusions in sedimentary halite indicated that the parent basin water did not differ significantly from evaporated marine water (Petrichenko et al., 1974; Kovalevich and Petrichenko, 1997), although as mentioned before (Garcia Veigas et al., 1997; Cendon et al., 1999), during halite precipitation, recycling of previously formed evaporites with an important input of water of continental origin occurred. The halite precipitation in different parts of the basin was alternated with precipitation of calcium sulfate and terrigenous material. The data on chemical composition of brine inclusions in gypsum deposits indicate that the Badenian water, during gypsum precipitation, essentially differed from modern seawater and showed some similarity to some continental and continental-marine basins, such as the Aral Sea (Petrichenko et al., 1997, their table 3). Total mineralization of the Badenian water during gypsum precipitation was 3–10 times lower than is characteristic for modern, calcium sulfate-saturated seawater, and the water salinity varied strongly throughout the entire basin (Petrichenko et al., 1997, their table 1). The waters were enriched in calcium sulfate (as much as 6 g/L; Petrichenko et al., 1997).

Sedimentological and geochemical data indicate the cannibalization of evaporites throughout most of the evaporite deposition time. Primary shallow-water evaporites from the margin of the evaporite basin that was adjacent to the Carpathian orogen have gone and were deposited in deep-water settings. The orogen was also controlling the sedimentary history in the basin foreland as indicated by the shoreward migration of the sabre gypsum facies in the Ukrainian part of the Carpathian Foredeep basin during the course of Badenian gypsum deposition. A general transgressive sequence of evaporites, as found in the marginal sulfate platform, results from the migration of facies zones that was induced by the nappe movement. Another result of the influence of the Carpathian orogen for the evaporite deposition was the supply of continental water as well as brines from the orogen. It should be mentioned that Liszkowski (1989) proposed that the main source of brine in Badenian evaporite basins was pore waters expelled from the flysch deposits during folding in the Carpathians. The importance of such a source of brines was most probably more minor than was envisaged by Liszkowski (1989), but the $\delta^{37}Cl$ values of halite in Wieliczka clearly indicate the subsurface water component (Eastoe and Peryt, 1999).

The cannibalization at the end of gypsum deposition in the marginal sulfate platform was accompanied by the block tectonics that resulted in the creation of a bathymetric difference at least a few tens of meters as indicated by the occurrence of Ratyn Limestone on stratigraphically different parts of the gypsum section. It was accompanied by a change in the hydrology of the Central Paratethys that was tectonically driven and possibly related to the block tectonic phase manifested in the marginal part of the Carpathian Foredeep basin. The change in hydrology implied the dilution of brines by inflowing marine water, and this terminated the Badenian salinity crisis.

Late Badenian and Sarmatian

Interpretation of dense coverage of seismic lines, coupled with interpretation of cores and well logs, provided information on the late Badenian–Sarmatian (postevaporitic) depositional pattern of the PCFB (Krzywiec, 1997, 2001). In the central part of this basin, the Miocene (foredeep) sediments can be divided into three major lithological intervals belonging to the Skawina (Baranow), Krzyzanowice (Wieliczka), and Machow formations (Alexandrowicz et al., 1982; Piwocki et al., 1996). Lower Badenian shales are relatively thin (10–40 m [33–130 ft] thick on average; Jawor, 1970), and commonly too thin to be distinguished on seismic data. Overlying evaporites are also thin, in the order of 10–50 m (33–160 ft) (Jawor, 1970; Ney et al., 1974), but because of their increased seismic velocity, they form an excellent seismic marker. Lower Badenian siliciclastics and upper Badenian evaporites are present across the entire central part of the PCFB, and in most cases, both these intervals can be correlated with one strong seismic reflector. Upper Badenian evaporites are covered by thick (as much as 1.5 km [0.9 mi] in the central part and as much as 2.5 km [1.6 mi] in the eastern part) upper Badenian siliciclastics built of sandstones and shales. In the axial part of the eastern PCFB, these siliciclastics are made of the so-called Krakowiec shales belonging to the Machow Formation (Ney et al., 1974; Piwocki et al., 1996). Lower and upper Badenian sediments, similar to the central part of the PCFB, can be correlated with a single, strong reflector generated by both the top of the basement and the relatively thin evaporitic horizon.

Within the central part of the Carpathian Foredeep, a regional angular unconformity between upper Badenian evaporites and upper Badenian siliciclastics (Figure 23) is the most striking seismic feature that could be observed on all the seismic lines from this part of the basin. To present a large-scale depositional pattern of the study area, a composite seismic line located along the axis of one of the major paleovalleys (so-called Szczurowa Paleovalley; Poltowicz, 1964, 1995; see also below) was constructed (Figure 23). All the upper Badenian reflectors are strongly inclined toward the north; only in the vicinity of the Carpathian front did the upper Badenian reflectors change their inclination and become nearly horizontal (Figure 23). No similar change of inclination is observed in the Mesozoic basement. Mesozoic seismic reflectors gently dip toward the south. This gentle dip is related to the long-wavelength flexure of the foreland plate beneath the Carpathians. Lack of short-wavelength variations (in order of 1–2 km [0.6–1.2 mi]) of inclination of the Mesozoic reflectors, similar to variations observed in

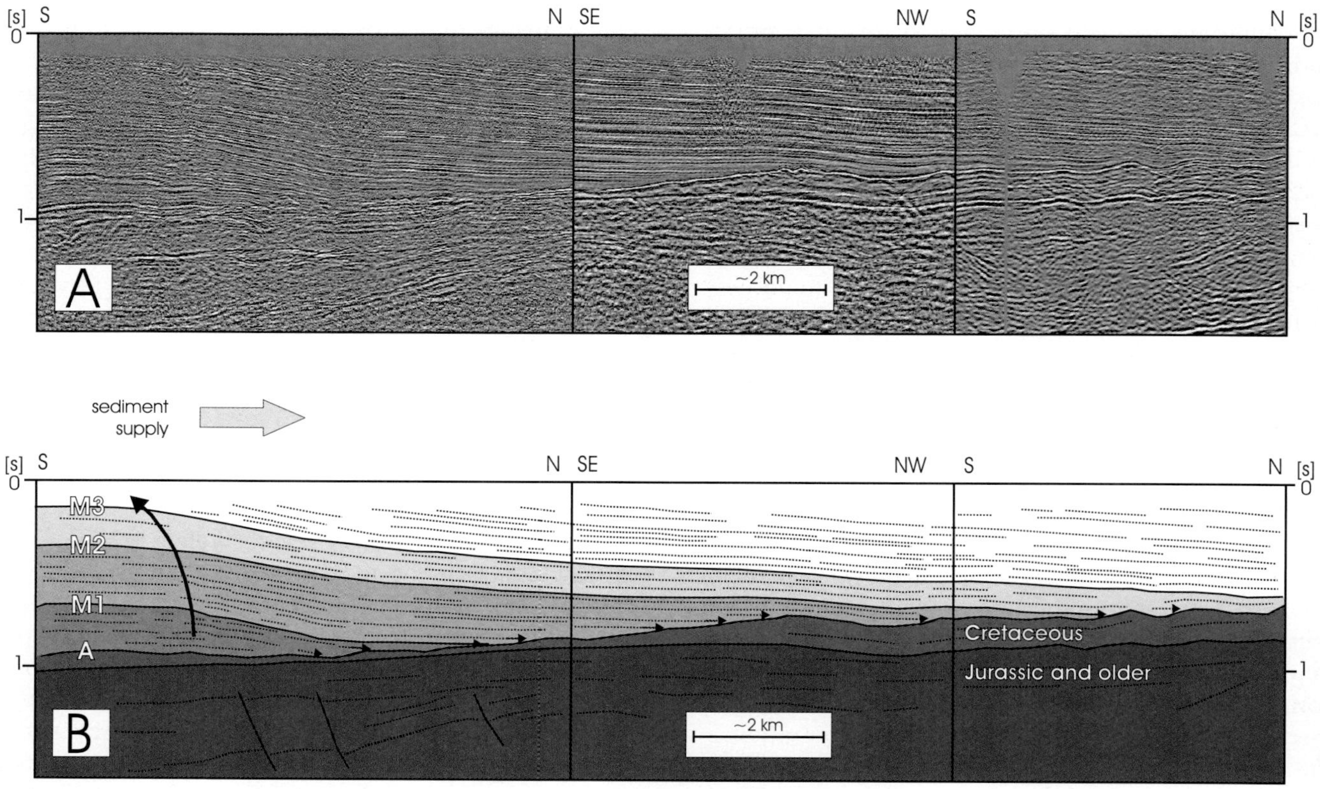

Figure 23. Regional seismic profile located along the axis of the Szczurowa paleovalley cut in the Mesozoic basement documenting late Badenian–early Sarmatian(?) (postevaporitic) sediment progradation from the Carpathians toward their foreland (modified after Krzywiec, 2001). Arrow indicates migration of the late Badenian–early Sarmatian(?) offlap break. Vertical scale is two-way traveltime (TWT). See Figure 10 for location.

the upper Badenian section, proves that the observed geometry of the postevaporitic siliciclastic sedimentary section is not related to any local tectonic movements like, e.g., short-wavelength bending of the foreland plate. Observed unconformity between upper Badenian evaporites (seismic horizon [A]) and upper Badenian siliciclastics is marked by numerous downlap seismic contacts. Particular seismic packages pinch out toward the north. This pattern can be observed on all of the seismic lines from the study area and was interpreted as a large-scale prograding clinoform, with the main direction of sediment progradation to the north. A set of horizontal reflectors was interpreted as the partly preserved top set part of prograding clinoform related to the shelf area of the presumed deltaic system entering the marine environment of the foredeep basin. At the southernmost part of the composite seismic line, migration of the offlap break toward the south can be observed, with all the top set reflectors preserved (Figure 23). This suggests that the upper Badenian foredeep sediments developed because of aggradation and retrogradation of the entire progradational system toward the south (compare Posamentier et al., 1988). Such a depositional scenario would imply relative sea level rise and/or high subsidence rate. Increased tectonic subsidence might have been related to a late Badenian increased rate of overthrusting movements in the Carpathian accretionary wedge and/or increased slab-pull processes in the Carpathian subduction zone.

The same architecture of the supraevaporite sequence (upper Badenian Chodenice and Grabowiec beds, 300–1000 m [1000–3300 ft] thick) has been recognized in the Krakow–Brzesko area (central part of PCFB) on the basis of borehole wire logs interpretation (Porebski and Oszczypko, 1999). These deposits make up the north-sloping clinoform. The clinoform reflects the deposition on the deltaic shelf, slope, and basin plain, which was supplied from the southern (Carpathian) active margin. These seismic-scale deltaic clinoforms grew from the south and developed a shelf-to-basin floor relief of more than 300 m (1000 ft) (Porebski et al., 2003).

The seismic interpretation of the late Badenian foredeep depositional pattern is compatible with sedimentological studies of numerous cores available from the central PCFB. Sandstone bodies present in the upper

Badenian section were interpreted as deltaic distributary channels separated by prodelta shales (Hollendonner 1995). Heavy-mineral assemblages from the upper Badenian core samples from wells located in the vicinity of the present-day Carpathian front are very similar to the assemblages described from the Carpathian nappes (Krysowska, 1966; Hollendonner, 1995). Toward the present-day Carpathian front, significant increase of redeposited flysch material was identified (Kirchner and Moryc, 1966). Correlation of well logs from numerous exploration boreholes showed a significant increase of sandstone content toward the Carpathians (Jucha, 1974, 1985). A significant amount of redeposited Carpathian foraminifera (Gradzinski, 1957; Otfinowski, 1981; Doktor, 1983; Krzywiec et al., 1995) and calcareous nannoplankton assemblages (mainly middle to upper Eocene and upper Oligocene; also upper Cretaceous, Paleocene, and lower Miocene; Hollendonner, 1995; Krzywiec et al., 1995) was identified in the upper Badenian and Sarmatian foredeep deposits of the central PCFB. All these results point to the eroded Carpathian nappes as a source area for the foredeep sediments. Large-scale seismic progradational clinoform described in front of the present-day Carpathians in the central PCFB formed a distal part of the entire deltaic system that supplied siliciclastic material to this part of the basin. Its more proximal parts have most probably covered Carpathian nappes and formed a piggyback basin carried on top of the thrust sheets (Oszczypko and Slaczka, 1985, 1989; compare Ori et al., 1986). Because of postorogenic uplift of the Carpathians, sediments deposited in this basin were mostly eroded, but some remnants are still preserved in the vicinity of Nowy Sacz, directly to the south from the study area (Cieszkowski et al., 1988; Oszczypko et al., 1991, 1992).

The sedimentological studies of upper Badenian coarse clastics on the southeast periphery of Krakow made the reconstruction of the paleogeographical patterns after intralate Badenian compressional event possible (Porebski and Oszczypko, 1999). In this area, the Bogucice sands make up an east-northeast–west-southwest-trending, lenticular body 175–310 m (574–1017 ft) thick. The sands belong to the basal part of the upper Badenian Grabowiec beds. The sand body is based in the southwest by an angular unconformity that passes to the northeast (basinward) into a prominent erosional disconformity, which is still replaced further basinward by a correlative conformity. The wide spectrum of depositional processes recognized in exposures, gamma-ray-log shapes, and prominent basal relief of the Bogucice sands points to shelf incision by the southwest–northeast-flowing Carpathian River, which initiated filling of the valley [canyon(?)] by mass-flow deposits. These deposits pass upward into the mouth bar and shoreface complex of a shelf-margin delta and increased shelf reworking of the mouth bar complex in unconfined flow condition after the relative sea level rose above the offlap break. In the Rzeszow area, the wave-dominated sandy fan delta developed simultaneously with the deposition of Bogucice sands (Doktor, 1983). These deposits, as much as 330 m (1082 ft) thick, rest with angular unconformities on evaporites and marly mudstones with sandstone intercalations. The same channelized deltaic deposits have also been recognized in the area of the Husow–Albigowa–Krasne area (Oszczypko and Slaczka, 1989). In the eastern, marginal part of the PCFB, the erosional disconformity has been described by Dziadzio (2000) at the top of the upper Badenian deposits.

Unlike in the central part of the PCFB, no significant sediment progradation from the Carpathians toward their foreland was detected on seismic profiles from the eastern PCFB. Instead, a gentle onlapping of the upper Badenian evaporites by the upper Badenian–Sarmatian fine-grained siliciclastics (Krakowiec shales) can be seen on all seismic lines oriented southeast–northwest. This feature was interpreted as an effect of relative sea level rise and gradual, northwest-directed marine ingression after upper Badenian salinity crisis that eventually led to the establishment of open-marine connections with eastern Paratethys. Relative sea level rise was most probably caused by enhanced tectonic subsidence related to Carpathian thrusting movements and/or intensification of slab-pull processes in the Carpathian subduction zone.

TECTONIC EVOLUTION

Introduction

The Carpathian Foredeep basin could be regarded as a typical foredeep basin, i.e., sedimentary basin that developed as a result of the flexure of the lithospheric plate (cf. Allen and Allen, 1990; Miall, 1995). Typical features of such sedimentary basins include the displacement of the zone of maximum subsidence toward the foreland of the migrating thrust belt, uplift, migration, and erosion of the flexural forebulge and consecutive foreland-directed onlapping of the foreland plate by foredeep sediments. Because of foreland-directed propagation of the thrust belt, older foredeep sediments become progressively involved in thrust-related folding, and synorogenic sediments, such as conglomerates or fan deltas developed in front of the active thrust-related growing folds, form during particular stages of deformation in the orogenic belt. Within the foredeep sedimentary infill, local and regional unconformities develop because of tectonic movements in the thrust belt, various related subsidence rates, and forebulge migration. Normal faults that developed in the foreland plate are related to its flexural extension. During the development of foredeep basins, older foreland plate structures are commonly reactivated because of compressional stresses transferred from the collision zone. The present-day structure of the basement is a result of several deformation stages connected with the Mesozoic extension and later basin inversion. The evolution of the Carpathian Foredeep was controlled by synsedimentary fault tectonics. The majority of the northwest–southeast- and west-northwest–east-southeast-trending faults were reactivated during the Late Cretaceous–Paleogene inversion. These faults are very well documented in the western (upper Silesian) and easternmost part of PCFB (Figure 2). In the northeast of Rzeszow and Przemysl, the faults of this system bounded the Wielkie Oczy graben and Ryszkowa–Wola high. In the southeast prolongation of the Wielkie Oczy graben, the thickness of the middle Miocene (mainly Sarmatian) deposits reached 2500 m (8200 ft). In the Krukenychi depression in Ukraine, the thickness of these deposits is as much as 6 km (4 mi). Toward the southeast, the Krukenychi depression dips beneath the Sambir unit. The younger, northeast–southwest-trending fault generation in PCFB is observed in the Krakow and Rzeszow areas (see also Moryc, 1992).

During the overthrusting of the Outer Carpathians, the platform basement collapsed at the front of the Carpathians. As a result, the southern margin of the platform dropped considerably. For example, 50 km (31 mi) south from the present margin of the Carpathians, the depth of the platform basement oscillates from about 3–6 km (1.8–4 mi) in the west to about 8–10 km (5–6 mi) in the east, and as much as 20 km (12 mi) in the middle part of Ukrainian Carpathians (Figure 3, 7). During the course of overthrusting, the platform basement became locally involved in the process.

Reactivated Foreland Structures

Introduction

During the Miocene development of the Carpathians, tectonic structures of the foreland (lower) plate have been significantly reactivated, although the degree of reactivation and influence of reactivated foreland plate structures on the development of the foredeep basin was considerably different in its various parts (Krzywiec, 2001). This reactivation was mainly caused by flexural extension, i.e., extension of the upper continental crust during its flexure beneath the thrust belt (Bradley and Kidd, 1991). Additionally, especially during the latest stages of Carpathian collision, foreland structures were reactivated because of compressional stresses generated in the collision zone.

Structure of the Foreland Plate beneath the Carpathians

The relationship between the Outer Carpathian accretionary wedge and the Miocene molasse deposits has been very well recognized in the PCFB by the deep boreholes as far as 30–40 km (18–25 mi) from the present-day Carpathian frontal thrust (see Wdowiarz, 1976; Oszczypko and Tomas, 1985; Zytko et al., 1989). The surface of the overthrust of the Outer Carpathian nappes is of regular shape and rather gently inclined. Beneath the Outer Carpathians, faults that are recognized in the platform basement as a factor do not cut the surface of overthrust and are older than the Carpathian overthrust (Oszczypko and Tomas, 1985). It is very well known that beneath the western part of the Polish Outer Carpathians (Figure 3), the system of northeast–southwest-trending normal faults progressively lowered the platform basement from a depth of 2 to 10 km (1.2 to 6 mi). All these faults are blind and do not cut the Carpathian sole thrust. West of Tarnow, the northwest–southeast-trending normal faults dominate beneath the Carpathians and folded Miocene units. This also occurred and was also very well documented in the Przemysl gas area and in the western part of UCFB. In these areas, the progressive increase of the amplitude of the fault increased toward the east by as much as 4 km (2.5 mi) (Figures 5–7).

In areas not penetrated by deep boreholes, the structure of the Carpathian basement can be roughly interpreted on the basis of magnetotelluric soundings, which reveal a high-resistivity horizon, being correlated with the top of the crystalline basement. The Carpathian basement lies flat in the northernmost part of the magnetotelluric profiles, whereas in its central part, a significant low in the basement was detected. The lowering of the basement ranges from a few kilometers in the western part of the PCFB to as much as 15–17 km (9–11 mi) south of Krosno (Figures 2, 3, 5). The magnetotelluric depression coincides with the gravity minimum. Its origin was interpreted as being related to the Mesozoic extension of the southern margin of the North European platform and the formation of the passive margin. Therefore, it can be postulated that in this area, relatively thick Mesozoic (Upper Jurassic to Cretaceous) and Paleogene sediments could be present above the crystalline basement. These may be covered by older (Burdigalian) inner foredeep deposits. Along the Rzeszow–Bardejov geotraverse (see Oszczypko et al., 1998) to the south, from the described basement depression, a significant uplift of the crystalline basement can be observed on magnetotelluric data. This was interpreted as a result of the thick-skinned, basement-involved early Miocene thrusting and inversion of the Mesozoic normal faults.

The interpretation of the subthrust structure of the Ukrainian Carpathians is based mainly on geophysical data (Figures 7, 9). This interpretation revealed significant differences between the deep structures of the Ukrainian and Polish Carpathians. The inner part of UCFB is bordered from the north by a steep-dipping edge, formed by a system of normal faults that lowered the platform basement down to 6–7 km (3.6–4.2 mi) below sea level on the eastern geotraverse and 20 km (12 mi) in the western part. According to these estimations, the width of the inner foredeep, where the deposition of the Sambir sequence occurred, is around 30 km (19 mi). With the exception of the northern edge of the inner foredeep, the sole thrust surface of the Ukrainian Carpathians, like in Poland, is more or less flat. The amplitude of the overthrust of the Ukrainian Carpathians is at least 50 km (31 mi). This suggests that the northern margin of the residual flysch basin (Boryslav–Pokuttya sequence) was located at least 50 km (31 mi) south from the present-day position of the front of the Boryslav–Pokuttya unit.

Structure of the Foreland Plate in Front of the Carpathians

Foreland plate underlying PCFB is characterized by a complicated system of deformations related to various stages of its geological evolution. Within its topmost part, a regional trend of northwest–southeast deformations related to the development and inversion of the most southern part of the mid-Polish Trough is observed, and the top of the foreland plate is characterized by prominent morphological lows. These lows have already been mapped with first seismic reflection profiles acquired by the oil industry in this area (Karpala and Lapinkiewicz, 1962). They were interpreted either as purely erosional features (ancient paleovalleys) (Poltowicz, 1964; Jawor, 1970; Karnkowski, 1974) or as tectonic negative features (grabens) related to the Miocene extension of the Carpathian foreland plate (Oszczypko and Slaczka, 1985). Very similar features, filled by Paleogene (foreland-derived) and Neogene (Carpathian-derived) sediments, were described in the more western part of the Carpathian Foredeep in the Czech part of this basin (Picha, 1979). They were interpreted as ancient, tectonically controlled, submarine canyons of the Tethyan margin. Large paleovalleys present in the central PCFB (e.g., Szczurowa paleovalley; see also Krzywiec, 1997, 2001) belong to the same system of ancient canyons of the ancient Tethyan margin. They were most probably incised also in the Paleogene after inversion and erosion of this part of the mid-Polish Trough because in this period, the central part of the future PCFB was a place of enhanced tectonic activity and erosion (Dzulynski, 1953; Bogacz, 1967; Rutkowski, 1989; Felisiak, 1992). No Paleogene sediments were found in the paleovalleys from the study area. However, at their southern extension, remnants of Paleogene sediments have been recently identified, as indicated by core studies from wells located in the vicinity of Rzeszow (Moryc, 1995). Erosion followed older faults zones, and only subordinarily did they develop as purely erosional features.

In the central part of the Carpathian Foredeep, basin-extensional deformations related to subduction and flexure of the foreland plate beneath the Carpathian orogenic belt were rather minor. Normal faults, commonly of listric geometry, observed in the pre-Miocene basement have been slightly reactivated after deposition of upper Badenian evaporites (Figure 24). This normal faulting affected only the lowest part of postevaporite siliciclastic succession; hence, it can be assumed that normal faulting in this area related to flexural extension was mainly restricted to immediate post-Badenian evaporite event. Their upper Badenian activity and related earthquakes could have caused a redeposition of evaporites— a well-documented feature in the PCFB (Peryt and Kasprzyk, 1992; Peryt and Jasionowski, 1994; Peryt, 2000). An observed small degree of reactivation could be partly explained by the character of older foreland faults and its relationship to the regional mode of flexural extension. Normal faults of the foreland plate in the central part of the PCFB (Krakow–Tarnow–Debica area) dip generally toward the northeast, which is related to the development of the southwest part of the mid-Polish Trough; however, flexure of the foreland beneath the Carpathian orogenic belt preferred the development of normal faults dipping toward the south. Therefore, interplay of such opposite directions allows only for a slight reactivation of foreland structures.

Very different modes and degrees of tectonic deformations could be observed in the foreland plate in the eastern part of the Polish Carpathian Foredeep as well as in its Ukrainian part where, however, the top of the foreland plate demonstrates the morphological lows similar to that identified westward in Polish and Czech segments. These features, commonly of zigzag running and filled by foreland-derived Neogene sediments as much as 2.5 km (1.6 mi) thick, were interpreted as paleocanyons incised deeply into the Mesozoic before the Miocene (Utrobin, 1960; Vyshniakov et al., 2000; Andreychuk et al., 2001).

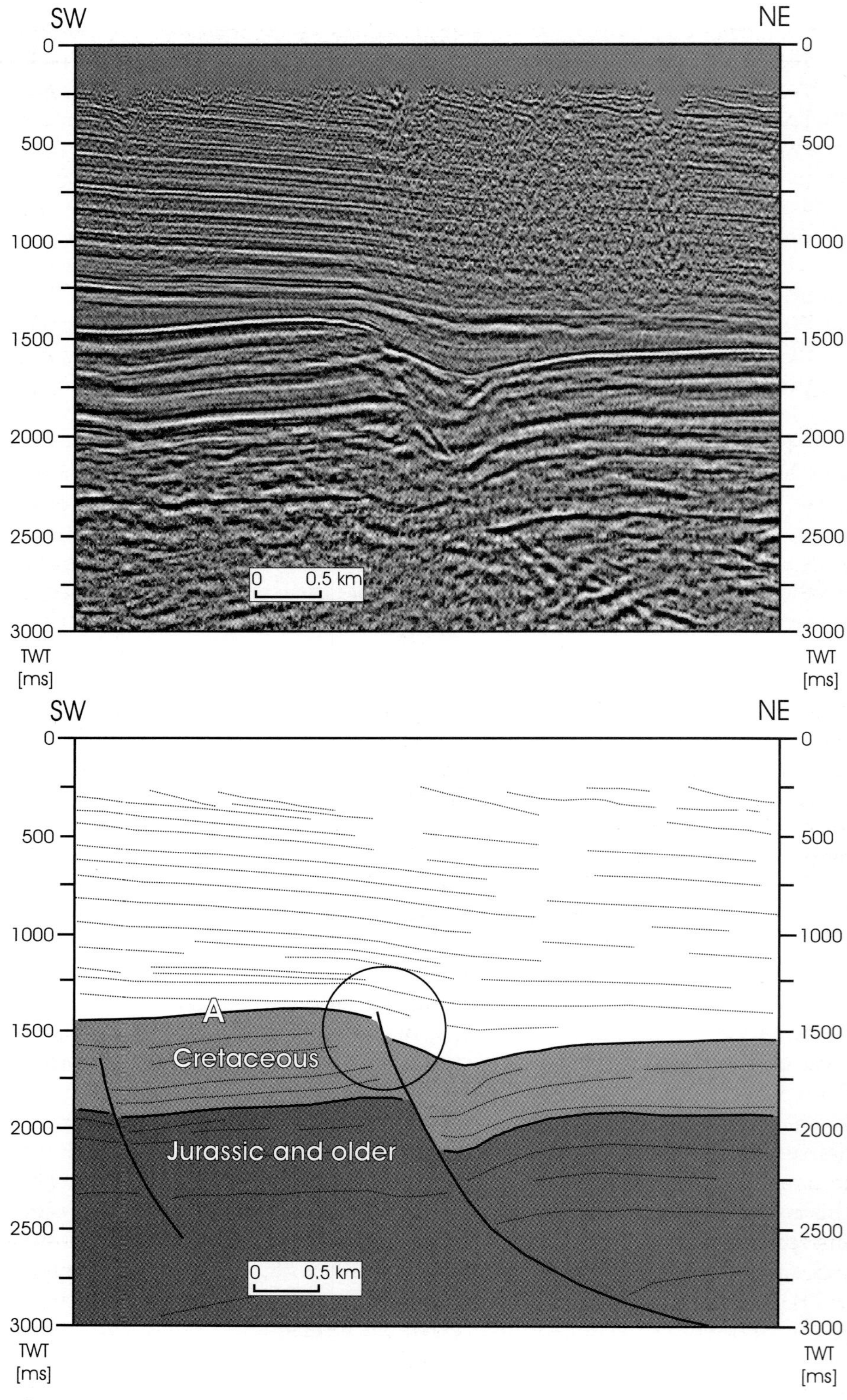

Figure 24. Seismic line documenting minor reactivation of basement normal fault after deposition of the evaporates in central (vicinity of Krakow) part of the Polish Carpathian Foredeep basin. Vertical scale is two-way traveltime (TWT). A = upper Badenian evaporates. See Figure 10 for location.

The main tectonic deformations identified in this area include large normal faults and antithetic reverse faults present in the hanging-wall block (Figure 25). Both hanging-wall and footwall blocks are relatively uniformly covered by evaporites, so the onset of faulting activity could be dated as postevaporitic (post-late Badenian). The identified faults have considerable displacement, and total stratigraphic throw for evaporitic

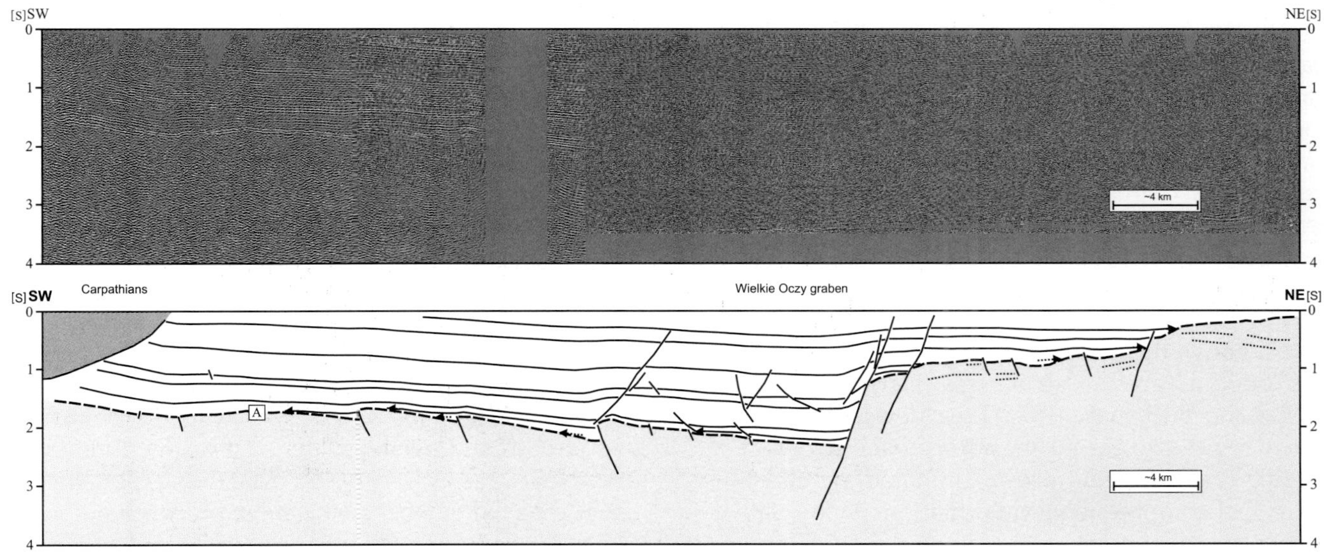

Figure 25. Regional (Przemysl–Lubaczow) seismic profile from the eastern Polish Carpathian Foredeep basin (modified from Krzywiec, 1999). Vertical scale is two-way traveltime. A = upper Badenian evaporates. Observed large-scale normal faults were interpreted as Mesozoic faults reactivated during Miocene flexural extension of the foreland plate. See text for more explanations. See Figure 10 for location.

horizon is in the order of 2–2.5 km (1.2–1.6 mi). Recently, the zone of maximum tectonic subsidence related to these faults (the so-called Wielkie Oczy graben; Figure 25) is located some 30 km (18 mi) toward the foreland from the present-day thrust. During their formation, the Carpathian front was located considerably further to the south, about 50 km (31 mi) from its present-day position as indicated by palinspastic reconstructions (Oszczypko and Slaczka, 1989); therefore additional mechanism apart from simple flexural extension (cf. Bradley and Kidd, 1991) must have been invoked to explain the development of these large normal faults. It was proposed that brittle extensional deformations were concentrated in this area because of the combined effect of flexure of the foreland Carpathian plate and the related reactivation of older tectonic discontinuities. The area of Miocene normal faulting belonged during the Mesozoic to the southeastern extension of the mid-Polish Trough. It has been suggested that in the Jurassic, extension affected the southeastern part of the mid-Polish Trough (Kutek, 1994; Hakenberg and Swidrowska, 1997; cf. Krzywiec, 2002). This inherited fault pattern was reactivated during the Miocene flexure of the foreland plate and the recently formed system of large normal and reverse faults.

All the basement normal faults propagate into the Miocene foredeep section. Significant thickness variations of the Miocene section were described and attributed to Miocene syndepositional activity of major normal faults. However, most probably, a considerable component of these thickness variations is also caused by differential compaction. In this area, conditions were favorable to initiate such a process: the presence of the rigid and fully compacted lower Paleozoic basement of the foredeep basin and a very thick package of soft sediments (the so-called Krakowiec shales) being deposited above the up- and downthrown sides of the normal faults. Studies of differential compaction above normal faults (Skuce, 1996) suggest that the discussed process could largely influence thickness variations across the normal faults without their tectonic activity. Differential compaction could also produce significant normal drag above the hanging-wall feature that can be clearly seen in proximity of the normal faults developed in the PCFB.

The thickness variations of Miocene packages described across the major basement-involved normal faults can be attributed both to their syndepositional activity and differential compaction. However, several compaction-related faults solely caused by the latter process have also been identified. These faults developed above the basement uplifts (e.g., the so-called Ryszkowa Wola block) produced because of inversion of antithetic faults in the hanging wall (Figure 25; see also below). Such normal faults and related small-scale anticlines have recently become a target for hydrocarbon exploration because they are related to important gas accumulations (Borys, 1996).

Other tectonic features identified on many seismic profiles located in the eastern part of the PCFB include slightly inverted antithetic basement faults and associated folds developed in the Miocene sediments

(Figure 25). They are characterized by gently inclined, nearly planar backlimb dips and short, hooked forelimbs, i.e., geometry considered typical for rotated blocks (McClay and Buchanan, 1992; Coward, 1996). These basement blocks were most probably rotated shortly after deposition of evaporites because they were unconformably onlapped by younger siliciclastics. Thinning of the sedimentary packages, at least at the lowermost part of the postevaporitic succession, above their crests and slightly divergent onlap pattern above their backlimbs indicate that they have developed as growth structures (Cartwright, 1989; McClay and Buchanan, 1992). Their development could also be related to the formation of backthrusts, i.e., hanging-wall vergent thrusts, initiated by the buttressing of compressional deformations by a relatively steep surface of major normal faults (McClay and Buchanan, 1992). Development of reverse faults (backthrusts) in the hanging wall of the compressed normal fault has also been successfully modeled in sandbox experiments (Koopman et al., 1987).

During the inversion of normal faults, the so-called inversion anticlines located above inverted normal faults are formed (Hayward and Graham, 1989). Such anticlines, when observed on seismic data, are commonly regarded as indicators of the structural positive inversion of the normal fault. Gentle anticlines can be observed above some of the normal faults in the eastern PCFB (Figure 25). This suggests that inversion of the entire normal fault system produced by flexural extension was accommodated not only by basement hanging-wall backthrusts described above but also by minor movements along major normal faults.

Miocene tectonic activity in this area also included strike-slip movements because deformations present in the pre-Miocene basement include basement pop-up structures (e.g., Ryszkowa Wola high; Figure 26) that are indicative of strike-slip movements. Above the uplifted basement block, a complicated zone of normal and reverse faults developed in the Miocene foredeep sedimentary section. Their development was attributed both to strike-slip basement movements and to differential compaction. It can be postulated that oblique compression and reactivation of the older, Mesozoic faults in the foreland plate resulted in dextral strike-slip movements along these faults. Miocene (late Badenian–Sarmatian) tectonic activity in various parts of the PCFB has already been postulated on the basis of detailed field studies, and a strike-slip component of these deformations has been suggested (Jaroszewski, 1977; Jarosinski, 1992; Dobrowolski, 1995), and this observation is compatible with postulated strike-slip tectonic activity in the foreland plate of the Eastern Carpathians.

To explain observed large-scale tectonic deformations in the basement and the sedimentary infill of the eastern PCFB, it was proposed that shortly after deposition of the middle Badenian evaporites, normal faults developed at a considerable distance (at least 50–80 km [31–50 mi]) from the Carpathian thrust front. Their development was attributed to the reactivation of the inherited Mesozoic [Jurassic(?)] tectonic zone resulting from the combined effect of nappe emplacement and slab pull. This zone was intrinsically weakened by Mesozoic extension, the mid-Polish Trough development, and its subsequent inversion; hence, it focused Miocene extension related to the flexure of the Carpathian foreland plate. At the same time, compressional stresses transferred from the Carpathian collision zone were buttressed against preexisting normal faults. Interplay of flexural extension and compressional deformations resulted in basement block rotation and/or formation of small-scale reverse faults and backthrusts that developed as growth structures. Later, because of significantly increased sediment loading, most of the deformation was taken up by normal faults. Secondary deformations include numerous compaction-related normal faults. Some form of later inversion of the main normal faults could have occurred because gentle anticlines are observed above these large normal faults. Because of the oblique orientation of regional compressional stresses related to the Carpathian collision and regional tectonic grain of the foreland plate, a small reactivation of the foreland structures included strike-slip movements that resulted in the formation of basement pop-up structures and related fault system in the Miocene foredeep siliciclastics (Krzywiec, 2001).

Folded Miocene Zone

Along the frontal Carpathian thrust, a zone of folded Miocene deposits developed. The width of this zone increases toward the east. It is very visible on the so-called "Przemysl sigmoide" along the Polish–Ukrainian boundary (Figure 2). In this area, the width of the folded Miocene zone drastically increased from a few hundred meters in Przemysl meridian to more than 20 km (12 mi) on the Ukrainian side. In the Eastern Carpathians (Ukraine and Romania), the broad zone of folded Miocene is divided into two tectonic units. The more southern unit is known as the Boryslav–Pokuttya unit (marginal fold unit in Romania), where folds of flysch beds are enveloped by lower Miocene strata. This unit, very well known from its hydrocarbon potential, is regarded as the most marginal part of the Outer Carpathians. The more external part of

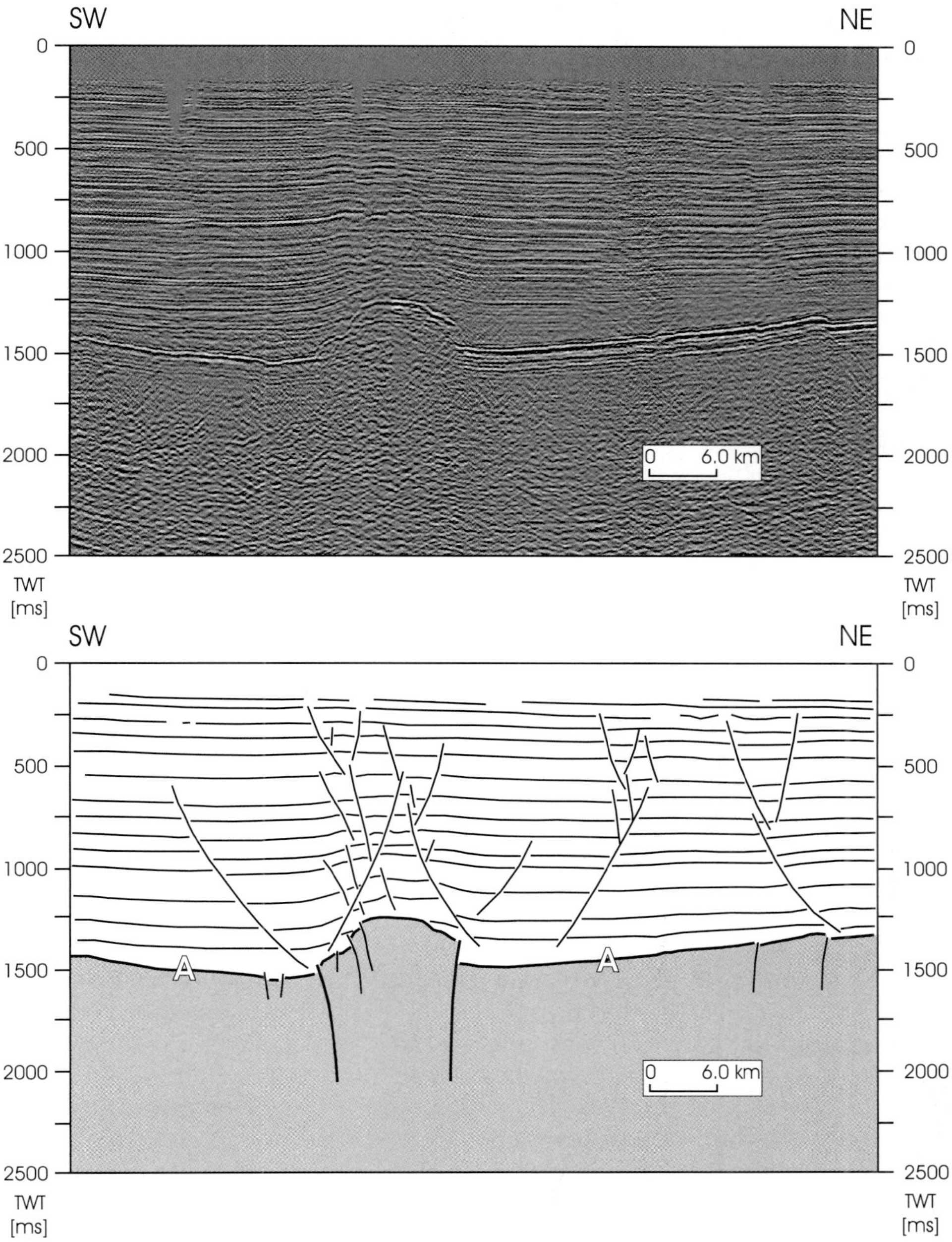

Figure 26. Seismic line showing basement pop-up structure (the so-called Ryszkowa Wola horst) developed because of Miocene strike-slip movements in the eastern Polish Carpathian Foredeep. Vertical scale is two-way traveltime (TWT). A = upper Badenian evaporates. See Figure 10 for location.

the folded Miocene zone belongs to the Sambir unit (sub-Carpathian unit in Romania and Stebnyk unit in Poland). This unit is composed of the lower to middle Miocene [Ottnangian(?)–Karpatian to Sarmatian], folded deposits, which overthrust onto the Sarmatian deposits of the outer foredeep. The Sambir unit is more traditionally regarded as a large-scale synform with an overturned southern limb, located at the front of Boryslav–Pokuttya unit (Figures 6, 7). In the middle part of UCFB, the Kalush salt mine cross section is very well recognized (see Korin, 1994; Andreyeva-Grigorovich et al., 2003). This cross section reveals at least three thrust sheets, which built up as a stack of these thrust sheets as much as 700 m (2300 ft) thick and flatly overthrust on the Dashava Formation (Sarmatian) of the outer UCFB (Bilche–Volytsa zone). In the UCFB, the amplitude of overthrust of the Sambir unit onto the Sarmatian of the outer foredeep is at least 20 km (12 mi).

In Poland, the Sambir–Stebnyk unit occurs at the surface east of Przemysl (see Ney, 1968; Ksiazkiewicz, 1977). The unit is composed of a few imbricated folds with the lower Miocene–lower Badenian deposits (Stebnyk and Balych formations) in the broad anticlinal core and the upper Badenian and Sarmatian deposits in the narrow synclines. In the Przemysl area, the extent of the Sambir–Stebnyk overthrust onto the Sarmatian deposits of the outer foredeep is at least 15 km (9 mi) (Figure 5). In a few boreholes in the

Przemysl–Jaksmanice gas field, the Eocene–Oligocene flysch deposits were pierced at the base of the Sambir–Stebnyk unit (see Zielinski, 1963; Ksiazkiewicz, 1977). In the same position, the Oligocene beds were found in borehole Cisowa IG-1 (Figure 5) (see also Wdowiarz, 1976), 15 km (9 mi) south of the Carpathian front. These flysch blocks, like the Sucha and Zamarski formations in the western PCFB, can be interpreted as of olistholithe origin (Figure 3) (see also Oszczypko, 1997, 1998; Oszczypko and Oszczypko-Clowes, 2003). West of the Przemysl, the Sambir–Stebnyk unit disappears beneath the Skole nappe, although the blocks of Stebnyk Formation were found in a few boreholes in the marginal part of Outer Carpathians between Przemysl and Ropczyce (see Ksiazkiewicz, 1977).

At the front of the Polish Outer Carpathians, a relatively narrow belt of deformed, upper Badenian–Sarmatian foredeep deposits exists (Ksiazkiewicz, 1977) and is known as the Zglobice unit (Kotlarczyk, 1985; cf. Krzywiec, 2001).This unit consists of several fault and fold structures of a maximum width of as much as 10 km (6 mi) (Figure 4) (see also Kirchner and Poltowicz, 1974). The Bochnia salt mine (40 km [24 mi] southeast of Krakow) is located at the front of the Carpathians. In this area, two steep folds are edged between the Carpathian front and the autochthonous deposits in the outer PCFB (see Ksiazkiewicz, 1977, and references therein). The cores of these folds are made up of the Carpathian flysch deposits and are covered by the Skawina Formation (lower Badenian), salt deposits of the Wieliczka Formation, and Chodenice beds of upper Badenian age. A few kilometers west of Bochnia, salt deposits and Chodenice beds form several imbricated folds overthrust onto the upper Badenian deposits of the outer foredeep. West of the Raba River, the front of Carpathians retreats almost 10 km (6 mi) southward (Figure 2), forming the so-called "Gdow embayment" (tectonic semiwindow; see Ksiazkiewicz, 1977) occupied by slightly tilted deposits of the Skawina (lower Badenian), Krzyzanowice (gypsum), and Chodenice beds of upper Badenian age. During the late Badenian (before deposition of the Bogucice Member of the Grabowiec Formation), the deposits of the Gdow embayment, together with their Mesozoic basement, were involved in syndepositional, compressional movements (see Porebski and Oszczypko, 1999, and references therein). This structure was interpreted by Jones (1997) as the triangle zone. Situated on the western tip of the Gdow embayment and at the front of the Carpathians is the Wieliczka salt mine with folded deposits of the Skawina Formation (lower Badenian) and the salt deposits of the Wieliczka and Chodenice beds (upper Badenian). West of Krakow occurs a very narrow zone of folded Miocene deposits (see Ksiazkiewicz, 1977). In the marginal western part of the Carpathians (Andrychow–Kety area), several thrust sheets of the Subsilesian unit, covered by lower and upper Badenian deposits, are known. These thrust sheets are thrust onto the Badenian deposits in the outer PCFB (Figure 3) (see also Ksiazkiewicz, 1977). More recently, these flysch slices were described by Wojcik and Jugowiec (1998) as flysch olistostromes, inside the upper Sarmatian [Pannonian(?)] deposits, and overthrusted onto the outer foredeep.

The frontal, most northern part of the Zglobice unit is formed by fault-propagation folds that evolved at the tip of propagating thrust faults detached in the mid-Badenian evaporites (Figure 27). They are characterized by minor backthrusting and slight imbrication that resulted in the formation of a small blind imbricated fan system. The front of Zglobice unit is developed as a growth structure, i.e., a structure that developed during sedimentation. This is evidenced by the thinning of synkinematic foredeep packages from the limb toward the crest of the growth fold (Figure 27). In front of the described fault-propagation folds belonging to the Zglobice unit, small prograding clinoforms were identified on numerous seismic lines (Figure 27). They were interpreted as a seismic expression of fan deltas that developed because of thrusting, folding, and erosion of the foredeep sediments. Similar fan deltas were also described in several outcrops located in the vicinity of the Carpathian front (Doktor, 1983). The continuity of a majority of reflectors above the crest of the fold, only locally developed progressive unconformities or fan deltas (sensu Ori et al., 1986), and thickness changes of particular sedimentary packages above the crest of this fold all suggest that although accommodation space was locally significantly reduced above the crests because of growth of this structure, no major erosion occurred. The overall geometry of the frontal part of Zglobice unit suggests that during the late Badenian–Sarmatian, the buried thrust front of the Carpathians evolved, with the rate of thrusting movements equal to or lower than the sea level rise. This has resulted in the preservation of accommodation space above the crest of this structure. Such interpretation is also compatible with considerations on offlap-break migration and an aggradational-retrogradational pattern of the foredeep deposits development, implying a relatively high sea level during the deposition of the upper Badenian strata.

The lateral extent of the frontal Zglobice thrust was, to a large degree, controlled by the morphology of the Mesozoic basement, especially by incised paleovalleys cut deep into the foreland plate. Relatively steeper northern slopes of these paleovalleys formed rigid buttresses during thrust movements, concentrated compressional stresses, and focused development of frontal

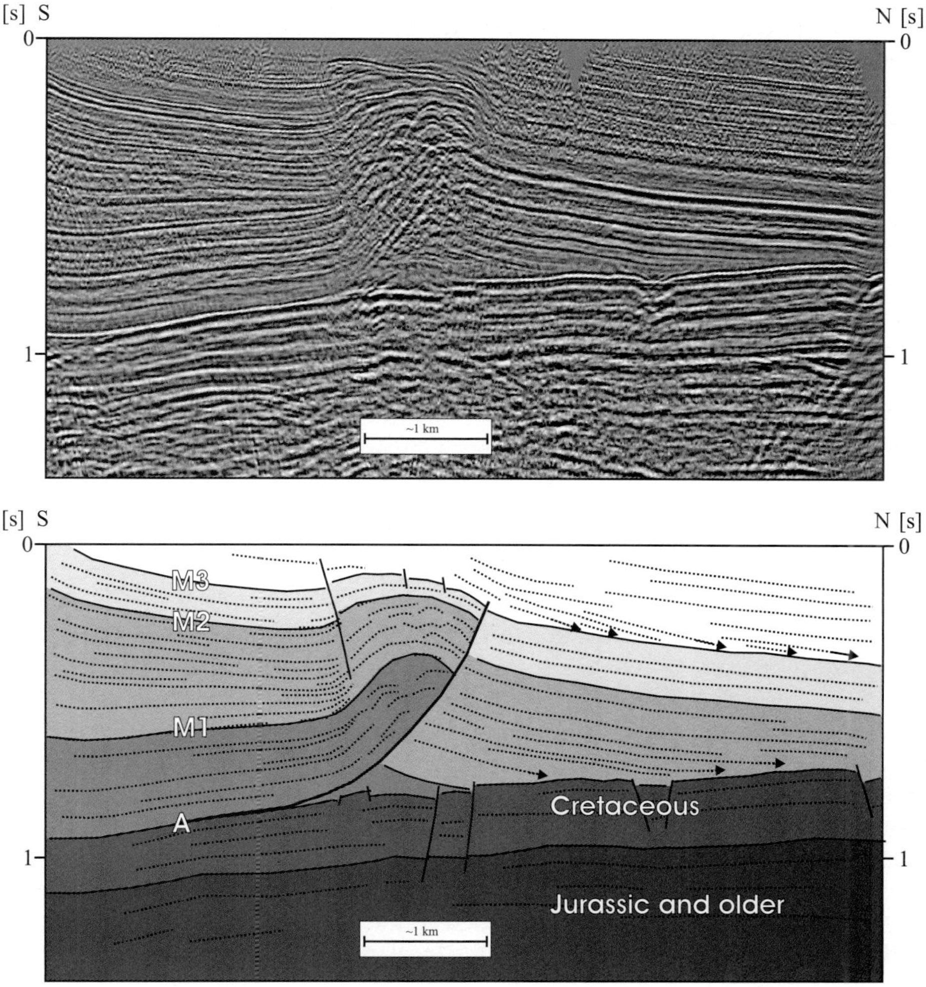

Figure 27. Seismic example of the Carpathian orogenic front developed in the Miocene Zglobice unit. Note the significant thickness reductions of the upper Badenian deposits above the crest of the fault-propagation fold pointing to syntectonic sedimentation. Vertical scale is two-way traveltime. A = upper Badenian evaporates. See Figure 10 for location.

foredeep thrust and fold structures. Within the axial parts of paleovalleys or between them, no buttressing effect occurred; thus, foredeep folds did not develop.

According to recent biostratigraphical data, the youngest deposits of the Sambir–Stebnyk and Zglobice units are of the same age (Andreyeva-Grigorovich, et al., 2003). These units in Poland and Ukraine were overthrust onto the autochthonous (Badenian and Sarmatian) deposits of the outer foredeep after the deposition of the Dashava Formation (Krakowiec beds) and Radych conglomerates of the Sarmatian age. This suggests that the age of these overthrust movements is late Sarmatian or even post-Sarmatian. This provides the possibility of correlating the Sambir–Stebnyk and Zglobice units, although the stratigraphic composition of these units reveals a different development. The Sambir sequence revealed a strong connection with the inner foredeep development, whereas the Zglobice sequence is typical for the southern part of the outer foredeep development.

EVOLUTION OF THE CARPATHIAN FOREDEEP BASIN

Since the fundamental paper of Price (1973), the development of the foredeep basins has been regarded as a result of the lithosphere flexure caused by the loading effect of the growing thrust belt (Beaumont, 1981). The Carpathian Foredeep basin was formed in front of the advancing Carpathians; hence, geodynamic processes in this orogenic belt were a prominent factor in shaping this sedimentary basin. However, the arcuate shape of the Carpathians and their foredeep basin was primarily caused by the significant curvature of the S margin of the foreland plate. The rigid behavior of the Bohemian Massif led to a significant bending of the Alpine–Carpathian transition zone and the opening of the pull-apart Vienna basin (Royden, 1985, 1988a). Further to the east, in the central PCFB (vicinity of Krakow), inherited foreland structures were also an important factor in shaping the PCFB (Ney, 1968). In this region, erosional remnants of the marine Miocene foredeep deposits can be found further to the north from the present-day distal foredeep margin, and this clearly indicates that the ancient basin margin was located further to the north. Despite this fact, in the central PCFB (Figure 1), significant narrowing of the PCFB is observed, which is connected to the northwest–southeast-located foreland structures. Further to the east, the basin becomes wider, and another zone of prominent shape changes, a transition from generally west–east orientation to a clearly northwest–southeast orientation, lies in the vicinity of the Polish–Ukrainian border (Figure 1). In this region, the Carpathians and their foredeep basin are superimposed on the northwest–southeast-trending Tornquist–Teisseyre zone, i.e., the major crustal boundary between the East European craton and Paleozoic platform, along which the Mesozoic mid-Polish Trough also developed. During the Miocene evolution of the PCFB, this crustal boundary significantly influenced the overall shape of this basin.

Main Basin-forming Mechanisms

Foredeep basins form because of the flexure of the foreland lithospheric plate beneath the orogenic belt; hence, the load related to the weight of the thrust belt could be regarded as one of the driving mechanisms responsible for the creation of the accommodation space in the basin (Allen and Allen, 1990). Additional loads could be attributed to the weight of the water possibly present in the foredeep basin, as well as the sediments already deposited in it. Another class of loads, the so-called subsurface loads, acts at depth on the end of subducted lithospheric plate (Royden and Karner, 1984). Subsurface loads could be attributed to the negative buoyancy of the subducted oceanic or thinned continental crust (Royden and Burchfiel, 1989). In many cases, the estimated weight of the thrust and foredeep sediments (i.e., surface or topographic loads) is not sufficient to explain the observed flexure of the foreland (lower) lithospheric plate (Moretti and Royden, 1988; Royden 1988b). Several modeling studies of continental collision zones suggested that in such cases, subsurface loads should be invoked to explain the observed crustal and lithospheric configuration. Flexural modeling studies completed for the Polish (Krzywiec and Jochym, 1996, 1997) and Ukrainian (Royden and Burchfiel, 1989) Carpathians suggest that in the case of this orogenic belt, deep processes and associated subsurface loads were most important for the observed present-day flexure of the foreland lithospheric plate. In the case of the Ukrainian segment of the Carpathian orogenic belt, some other opinions have also been postulated invoking lower crustal processes (contraction of gabbro and formation of dense garnet granulites during nappe emplacement) responsible for the observed crustal structure of the Carpathian region and development of the foredeep basin (Artyushkov et al., 1996).

In Figure 28, the results of flexural modeling along two regional profiles from the western and eastern parts of the PCFB are presented (from Krzywiec, 1997, after Krzywiec and Jochym, 1997). This modeling has been completed using the assumption that the continental collision zone and related subduction of lower lithospheric plate can be adequately approximated by the flexure of a thin, elastic plate of uniform thickness (lithosphere) floating on the fluid half-space of zero viscosity (asthenosphere). In both cases, significant subsurface loads were necessary to maintain the observed flexure of the foreland lithosphere, whereas topographic loads provided only a small fraction of forces necessary to maintain this flexure. Similar results were obtained for other profiles from the Polish Carpathians (Krzywiec and Jochym, 1996, 1997), as well as from the Ukrainian Carpathians (Royden and Karner, 1984; Royden and Burchfiel, 1989). Results of flexural modeling also showed that along the Carpathians, elastic properties of the lower (foreland) lithospheric plate vary significantly. Effective elastic thickness was estimated to be in the range of 8–16 km (5–10 mi) in the western Polish Carpathians and 20–25 km (12–16 mi) in their eastern part. Such variations can be attributed to the fact that in the western Polish Carpathians, the foreland plate belongs to the Western European platform (younger age of consolidation and lower flexural rigidity), whereas in the eastern segment of the thrust belt, the foreland plate belongs to the Tornquist–Teisseyre zone and the Eastern European craton (older age of consolidation and significantly higher flexural rigidity). Effective elastic thickness estimates that show significant lateral variations of the lithosphere strength along the strike of the PCF derived from flexural models (Krzywiec and Jochym, 1997; Zoetemeijer et al., 1999) are compatible with estimates derived from the rheological studies of the Carpathian foreland plate based on failure criteria and temperature models (Lankreijer et al., 1999).

The dominance of the subsurface loads for the Carpathian collision zone in Poland and, at least partly, in Ukraine, combined with other Neogene features of the Carpatho-Pannonian realm like the significant back-arc extension and basin development generally synchronous with main episodes of compression (Royden, 1988b; Horvath, 1993; Kovac et al., 1993; Meulenkamp et al., 1996), the rather small topographic elevations and small degree of erosion in the Carpathians, and the lack of metamorphism and significant influence of deeper water sediments in the foredeep basin

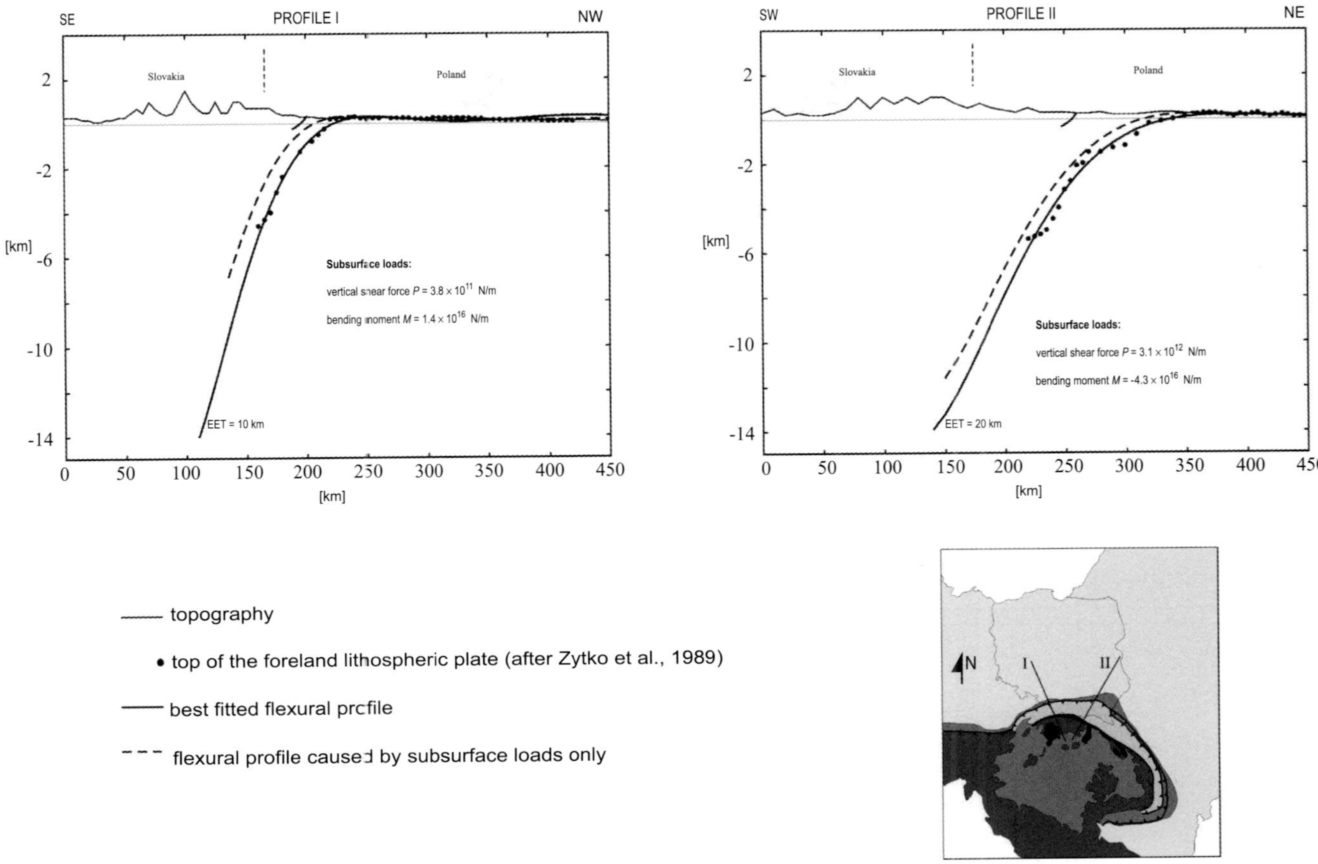

Figure 28. Flexural profiles across the Polish Carpathians (modified after Krzywiec, 1997).

(Oszczypko and Slaczka, 1985; Dziadzio, 2000) all suggest that the Carpathian Neogene collision zone developed in relation to the retreating subduction zone (Royden, 1993a). The slab-pull mechanism, possibly caused by negative buoyancy of the subducted oceanic or thinned continental crust, can be postulated as the main subduction-driving mechanism for the Neogene evolution of the Polish Carpathians and related subsidence in the Carpathian Foredeep basin (cf. Royden, 1993b).

Subsidence History of the Carpathian Foredeep Basin

During the initial stage of the inner foredeep (Ottnangian–Karpatian) development, the rate of total subsidence (1000–1400 m/m.y.; 3300–4600 ft/m.y.) in the Czech and Polish segments was fully compensated by the rate of deposition (Vass and Cech, 1983; Meulenkamp et al., 1996; Oszczypko, 1997, 1998). This resulted in terrestrial and shallow-marine sedimentation. Toward the east (Sambir basin), the rate of the total subsidence increased to at least 2000 m/m.y. (6600 ft/m.y.). The new period of the late Karpatian–early Badenian subsidence was characterized by the development of southwest–northeast- and northwest–southeast-trending grabens at the front of the old Styrian thrust (Figure 16). The early Badenian axis of subsidence was shifted at least 30–40 km (18–25 mi) (rate of migration = 15–20 mm/yr [0.6–0.8 in./yr]) north from the early Miocene position (Figure 16). At that time, in the axial part of the basin, the rate of sedimentation reached 250–500 m/m.y. (820–1640 ft/m.y.) in the Moravia segment (Vass and Cech, 1983; Meulenkamp et al., 1996) to 200 m/m.y. (660 ft/m.y.) in the Polish part (Oszczypko, 1997, 1998). Simultaneously, in the area of the northern stable shelf of PCFB, the subsidence was extremely low (and the rate of sedimentation oscillated from a few dozen to 50 m/m.y. [160 ft/m.y.]) (Figure 29). This period of very low subsidence and accumulation is marked in all of the Central Paratethys foreland basins (see Meulenkamp et al., 1996).

Since the early Badenian (Langhian) marine transgression, the rate of subsidence was higher than the rate of deposition. This resulted in marine sedimentary conditions during the Badenian and Sarmatian. The paleobathymetry of the early Badenian basin varied from the upper bathyal depths in the axis of the basin to neritic (Figure 29) (see also Kovac et al., 1993) and littoral

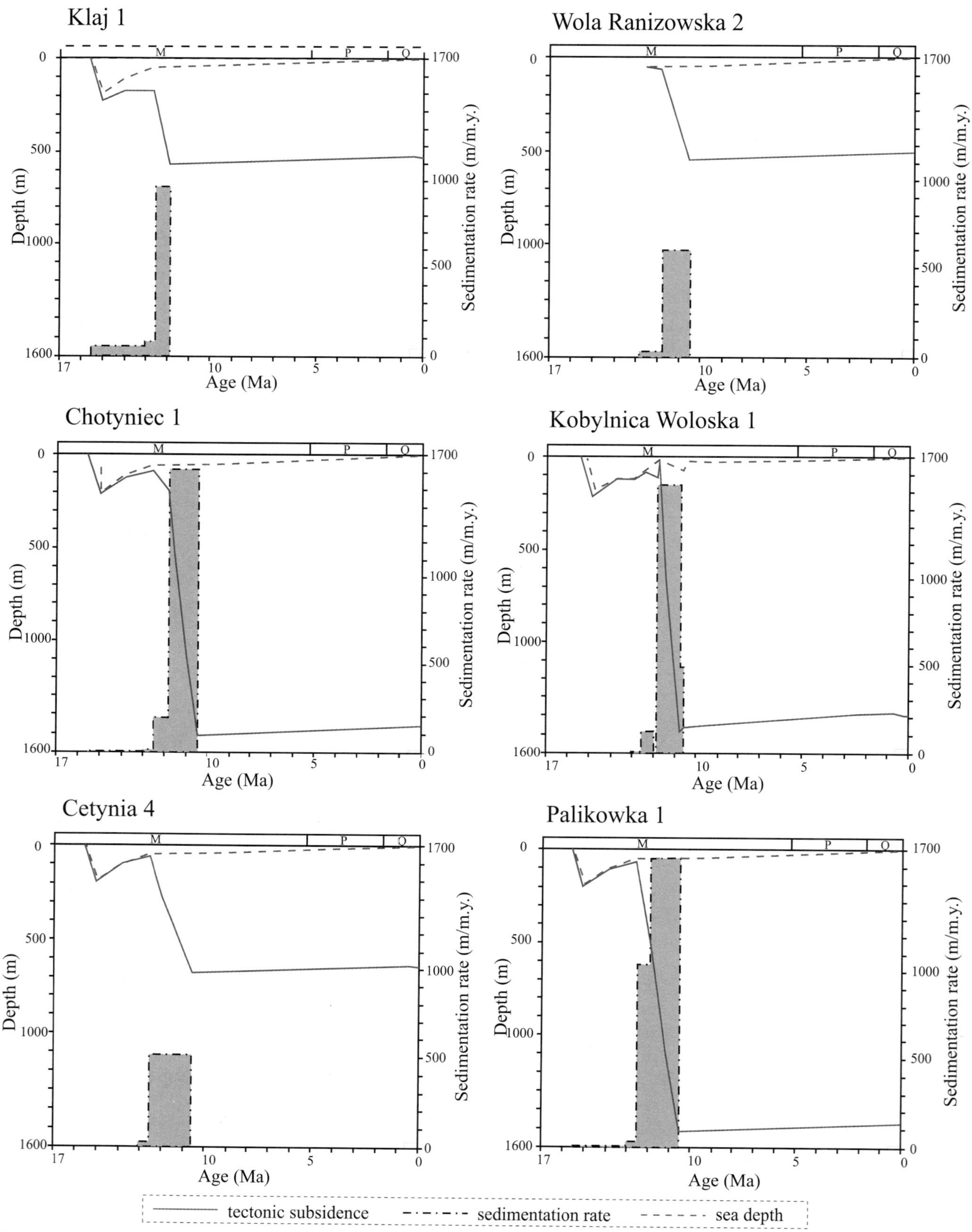

Figure 29. Backstriped burial diagrams of the selected boreholes from the outer part of Polish Carpathian Foredeep (modified after Oszczypko, 1997). See Figure 10 for location.

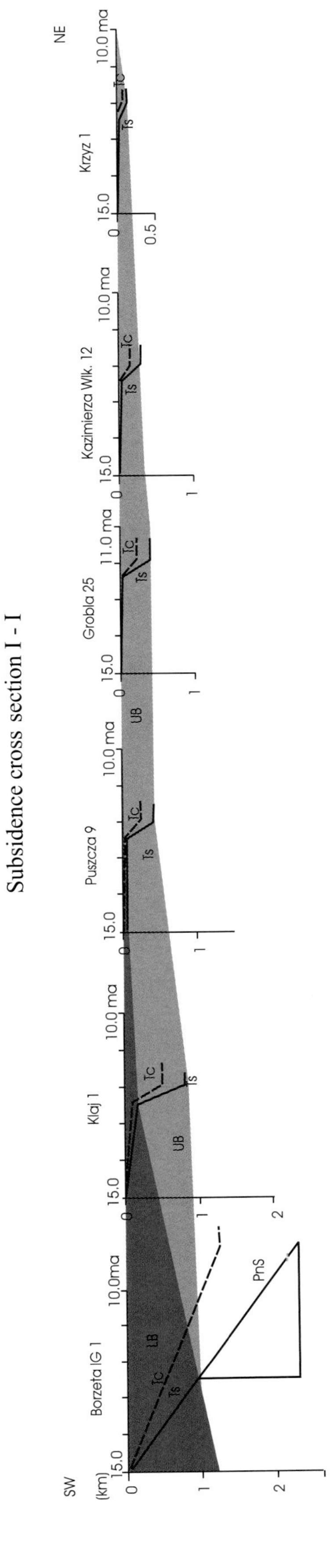

Figure 30. Subsidence cross section I-I (modified after Oszczypko, 1997). Abbreviations: PnS = postnappe subsidence; Ts = total subsidence; Tc = tectonic subsidence; LB = lower Badenian; UB = upper Badenian; S = Sarmatian. See Figure 10 for location.

in the northern and southern (Carpathian) marginal part of the basin.

Since the Serravalian (ca. 15 Ma), the gradual shallowing of bathymetry in the Carpathian foreland basin, coinciding with gradual global fall of relative sea level, was observed. This caused partial isolation of the basin and initiated the Badenian (ca. 13.0–12.5 Ma) salinity crisis in the shallow basin [inner-middle neritic(?) depths, Figure 29]. For that period, diagrams of tectonic subsidence reveal a tendency of progressive uplifting of the sea floor, whereas the sedimentation rate ranges from a few dozen meters per million years in the area of sulfate facies to as much as about 50 m/m.y. (160 ft/m.y.) (Figure 29) in the area of chloride facies. It is a very rough approximation because episodes of chemical accumulation could have been very short-lasting (25–35 k.y.; see Garlicki, 1968, Petrichenko et al., 1997) and, respectively, the rates of subsidence and sedimentation could have been one order higher, as much as 500 m/m.y. (1640 ft/m.y.) After evaporite deposition (ca. 12.5 Ma), when the basin was the shallowest (Figure 29) (see also Kasprzyk, 1993; Gonera, 1994; Czepiec, 1996), tectonic uplift of the foreland reached its maximum. This resulted locally in a post-evaporite erosion of at least 50–100 m (160–330 ft) of evaporitic and subevaporitic deposits (Rzeszow paleoridge; see also Komorowska-Blaszczynska, 1965). The subareal and/or areal erosion of the anhydrites before deposition of the Ratyn Limestone or before deposition of the Kosiv formation has also been reported from the Ukrainian part of the foredeep (see Peryt and Peryt, 1994; Panow and Plotnikow, 1996; Andreyeva-Grigorovich et al., 1997). The middle Serravalian relative fall of the sea level could have caused a basinward shift of clastic sedimentation (Krzywiec, 1997). A new period of intense subsidence was initiated during the late Badenian (Figures 29, 30). The depth of the late Badenian sea oscillated between the outer neritic and inner neritic depths (Gonera, 1994; Czepiec, 1996). The subsidence occurred both in the inner and outer foredeep (Figure 2). In the eastern segment of PCFB, the axis of subsidence shifted 15 km (9 mi) (rate of migration = 3.75 mm/yr [0.15 in./yr]) north from the early Badenian position and was approximately in line with the present-day position of the Carpathians (Figure 2). The highest rate of subsidence, as much as 2000 m/m.y. (6600 ft/m.y.), was estimated in the southeast part of PCFB in the boreholes Przemysl 136 and Drohobyczka 1, situated 10–15 km (6–9 mi) south of the Carpathian front. Toward the northwest, in the Rzeszow area (boreholes Husow 39, Husow 25, and Palikowka 1; Figures 29, 31), the rate of total subsidence reached about 1200–1300 m/m.y. (3900–4300 ft/m.y.), whereas the rate of sedimentation

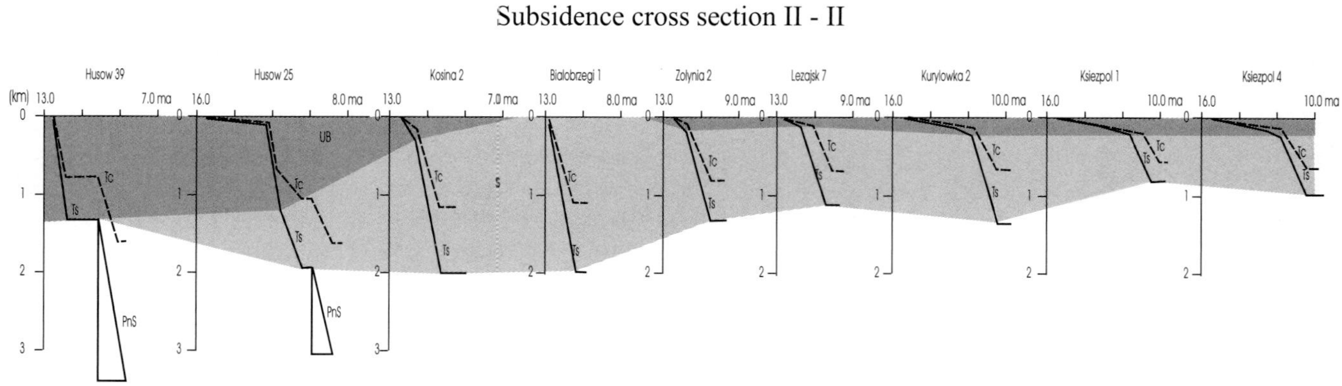

Figure 31. Subsidence cross section II-II (modified after Oszczypko, 1997). Abbreviations: PnS = postnappe subsidence; Ts = total subsidence; Tc = tectonic subsidence; LB = lower Badenian; UB = upper Badenian; S = Sarmatian. See Figure 10 for location.

oscillated around 1000 m/m.y. (3300 m/m.y.) (Figure 29). The slightly lower rates of subsidence and sedimentation are observed in the Tarnow and Bochnia areas. In the area effected by postevaporitic erosion, for example, Rzeszow Island, the late Badenian deposits transgressively overlapped the Precambrian and Paleozoic basement. Toward the northeast margin of the PCFB, the rate of the late Badenian subsidence and sedimentation decreased to 100–200 m/m.y. (330–660 ft/m.y.). The late Badenian subsidence continuously passed into Sarmatian subsidence, whereas the locations of depocenters were distinctly changed (Figures 29–31). In comparison to their former position, the Sarmatian depocenters were shifted 40–50 km (25–31 mi) (rate of migration = 20–25 mm/yr [0.8–1 in./yr]) toward the northeast, and the axis of subsidence rotated clockwise by as much as 20° (Figure 2) (see also Oszczypko and Zytko, 1987). The zone of the Sarmatian maximum subsidence was connected with the so-called Wielkie Oczy Graben. The total subsidence in this zone varied from 1500 m (5000 ft) in its northeast part to as much as 3000 m (10,000 ft) in its southeast part (Figures 29, 31), whereas the maximum values of tectonic subsidence reached 1500 m (5000 ft). Toward the northern margin of PCFB, the total subsidence decreased to a few hundred meters. The high rate of the Sarmatian subsidence (Figure 32) was compensated by the high rate of sedimentation, which reached 1700–2400 m/m.y. (5600–7900 ft/m.y.). During the Sarmatian, the depth of the basin oscillated between outer neritic and littoral depths (Czepiec, 1996). Toward the southeast, in Ukraine, where the

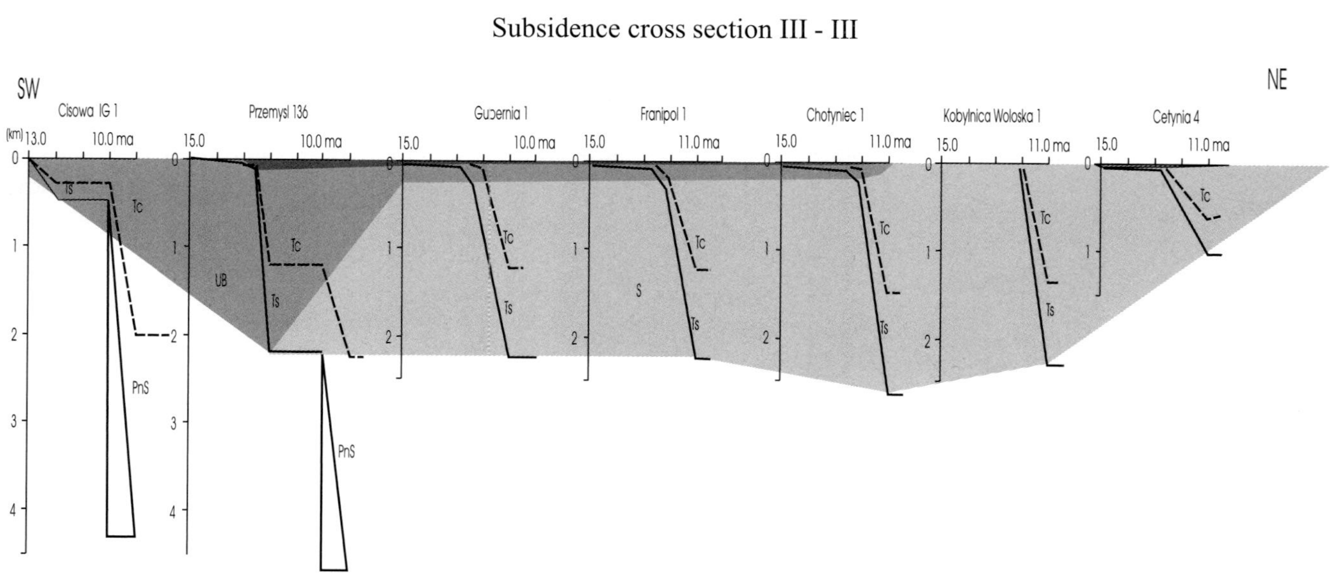

Figure 32. Subsidence cross section III-III (modified after Oszczypko, 1997). Abbreviations: PnS = postnappe subsidence; Ts = total subsidence; Tc = tectonic subsidence; LB = lower Badenian; UB = upper Badenian; S = Sarmatian. See Figure 10 for location.

Sarmatian deposits reached a thickness of as much as 4000 m (13,100 ft) (see Andreyeva-Grigorovich et al., 1997), subsidence and rate of deposition were as much as 4440 m/m.y. (14,570 ft/m.y.) (Meulenkamp et al., 1996).

According to Bubniak et al. (2001), in the outer part of Ukrainian Foredeep, the rate of tectonic subsidence progressively increased during the late Badenian and Sarmatian from less than 10 m/m.y. (33 ft/m.y.) during the deposition of the Tyras Formation to 700 m/m.y. (2300 ft/m.y.) during the deposition of the Kosiv formation to as much as 1800 m/m.y. (5900 ft/m.y.) during the deposition of the Dashava Formation. The rates of deposition for the above span of time were 50, 1200, and 3500 m/m.y. (160, 3900, and 11,500 ft/m.y.) for the Tyras, Kosiv, and Dashava formations, respectively. Both sedimentation and subsidence rates in UCFB increased from northeast to southwest.

The numerous curves obtained during the analyses of the Miocene tectonic subsidence of the Polish and Ukrainian foredeep (Oszczypko, 1997, 1998; Bubniak et al., 2001) revealed their convex shape typical for increasing subsidence in time. This type of tectonic subsidence diagrams, as well as increasing in time rates of deposition, is commonly regarded as a characteristic for flexural basins developed in a compressional regime.

In the Polish Carpathians, the multistage overthrusting has been documented by the occurrence of the flysch-derived olistostromes in the autochthonous and parautochthonous Miocene strata of the different ages (Oszczypko and Slaczka, 1985; Oszczypko and Tomas, 1985; Oszczypko, 1997, 1998) and by the relation of the upper Badenian autochthonous deposits to blind faults (Krzywiec, 1997). These overthrusting occurred during the following periods: (1) before the deposition of the Stryszawa Formation (after Ottnangian and before Karpatian); (2) during the late Badenian; and (3) after Sarmatian.

As a rule, with the exception of the Sarmatian, the area of maximum subsidence was located at the front of the Carpathians. The subsidence occurred not only on the foreland plate but also on the marginal part of the Carpathian accretionary wedge (see also Oszczypko and Slaczka, 1989). During the Karpatian–Sarmatian (7 m.y.), the axis of maximum subsidence in the PCFB moved about 85 km (53 mi) toward the north and northeast. The mean rate of migration of the axes of subsidence could be estimated as 12 mm/yr (12 km/m.y.; 7.5 mi/m.y.). This rate is similar to the mean rate of the Carpathian overthrusting (12 mm/yr [0.5 in./yr]; see Oszczypko, 1997, 1998) and lower than that of the pinch-out migration (13.8 mm/yr; 0.5 in./yr). During the late Badenian–Sarmatian, this resulted in the widening of the eastern part of PCFB. A similar event was described by Homewood et al. (1986) from the Swiss Molasse basin. The main episodes of intense subsidence in the PCFB correspond to the periods of progressive emplacement of the Western Carpathians onto the foreland plate (Oszczypko, 1997), whereas the middle Badenian event of low subsidence or uplifting is related to the period of dormant Carpathian frontal thrust (see also Meulenkamp et al., 1996). A distinct relationship between the periods of the Carpathian overthrusting and the subsidence of the PCFB suggests that the significant driving force of subsidence was the emplacement of the tectonic nappe load related to subduction rollback. Each forward emplacement of the Carpathian front was related to a new period of subsidence. During the early–middle Miocene, the loading effect of the thickening Carpathian wedge on the foreland plate and/or the slab-pull effect increased and caused a progressive increase of the total subsidence in the PCFB.

ACKNOWLEDGMENTS

The preparation of this chapter was supported by the Adam Krzyzanowski's Grant from the Jagiellonian University to Nestor Oszczypko and by the Polish Committee for Scientific Research (Komitet Badan Naukowych), the Polish Geological Institute Project No. 6.20.9071.00.0 to Piotr Krzywiec. Thanks are also due to Victor Antonyshyn for his kind assistance in the preparation of figures illustrating the Ukrainian segment geology. The authors express their gratitude to A. Garlicki and M. Kovac who have critically read the manuscript and offered constructive remarks.

REFERENCES CITED

Alexandrowicz, S. W., 1965, Development of the Carpathian Foredeep in the light of stratigraphic studies of Miocene sediments in southern Poland (in Polish with English summary): Geofizyka i Geologia Naftowa, v. 7–9, p. 243–247.

Alexandrowicz, S. W., A. Garlicki, and J. Rutkowski, 1982, Podstawowe jednostki litostratygraficzne miocenu zapadliska przedkarpackiego: Kwartalnik Geologiczny, v. 26, p. 470–471.

Allen, P. A., and J. R. Allen, 1990, Basin analysis: Oxford, Blackwell Scientific Publishing, 451 p.

Andreychuk, M., Y. Krupski, and V. Kotyk, 2001, The pre-Miocene paleorelief of the outer zone of the Precarpathians foredeep, *in* XII International Scientific Technical Conference "New Methods and Technologies in Petroleum Geology, Drilling and Reservoir Engineering," Krakow, 21–22 czerwca, 2001: Krakow, Akademia Gorniczo-Hutnicza, v. I, p. 23–27.

Andreyeva-Grigorovich, A. S., et al., 1997, Regional stratigraphic scheme of Neogene formations of the Central Paratethys in the Ukraine: Geologica Carpathica, v. 48, p. 123–136.

Andreyeva-Grigorovich, A. S., N. Oszczypko, A. Slaczka, N. A. Savitskaya, and N. A. Trofimovich, 2003, Correlation of the late Badenian salts of the Wieliczka, Bochnia and Kalush areas (Polish and Ukrainian Carpathian Foredeep): Annales Societatis Geologorum Poloniae, v. 73, p. 67–89.

Artyushkov, E. V., M. A. Baer, and N.-A. Morner, 1996, The East Carpathians: Indications of phase transitions, lithospheric failure and decoupled evolution of thrust belt and its foreland: Tectonophysics, v. 262, p. 101–132.

Babel, M., 1999, History of sedimentation of the Nida gypsum deposits (middle Miocene, Carpathian Foredeep, southern Poland): Geological Quarterly, v. 43, p. 429–447.

Baran, U., E. Jawor, and W. Jawor, 1997, Geological and hydrocarbon exploration results in the western part of the Polish Carpathians (in Polish): Przeglad Geologiczny, v. 45, no. 1, p. 66–75.

Beaumont, F., 1981, Foreland basins: Geophysical Journal of the Royal Astronomical Society, v. 65, p. 291–329.

Belka, Z., P. Valverde-Vaquero, H. Arendt, W. Dorr, W. Framke, and K. Wemmer, 2000, Accretion of first Gondwana derrived terranes at the margin of Baltica— New data and open questions: Joint Meeting of Europrobe (Trans European Suture Zone) and Paleozoic Accretion of Central Europe Projects, Zakopane-Holy Cross Mountains, Poland, September 16–23, 2000, p. 8–9.

Berggren, W. A., D. V. Kent, C. C. Swisher, and M. P. Aubry, 1995, A revised Cenozoic geochronology and chronostratigraphy: SEPM Special Publication 54, p. 129–212.

Bogacz, K., 1967, Budowa geologiczna północnego obrzezenia rowu krzeszowickiego: Prace Geologiczne Komisji Nauk Geologicznych Polska Akademia Nauk, O/Krakow, v. 41, p. 1–89.

Borys, Z., 1996, Aktualne problemy poszukiwań weglowodorów we wschodniej części przedgórza Karpat (in Polish): Przeglad Geologiczny, v. 44, no. 10, p. 1019–1023.

Bradley, D. C., and W. S. F. Kidd, 1991, Flexural extension of the upper continental crust in collisional foredeeps: Geological Society of America Bulletin, v. 103, p. 1416–1438.

Bubniak, A., P. Poprawa, and I. Bubniak, 2001, Tectonic evolution of the Carpathian Foredeep in Ukrainian segment: Implications from subsidence analysis (abs.), *in* A. Adam, L. Szarka, and J. Szendroi, eds., Pancardi 2001, Sopron, Hungary, Abstracts Book: Budapest, Hungarian Geological Institute, p. CP-3.

Bukowski, K., 1994, Srodowisko sedymentacji i geneza brylowej części zloża w Wieliczce: Przeglad Geologiczny, v. 42, p. 754–758.

Bukowski, K., 1997, Sedimentation of clastic strata associated with Miocene salts in Wieliczka (southern Poland): Slovak Geological Magazine, v. 3, p. 157–163.

Bukowski, K., 1999, Porownanie badeńskiej serii solonośnej z Wieliczki i Bochni w świetle nowych danych: Prace Panstwowego Instytutu Geologicznego, v. 168, p. 43–56.

Bukowski, K., and J. Szaran, 1997, Oxygen and sulfur isotopes in anhydrites of the salt complex of Wieliczka and Bochnia (southern Poland) (in Polish): Przeglad Geologiczny, v. 45, no. 8, p. 816–818.

Bukowski, K., M. Dulinski, and K. Rozanski, 2001, Origin of the Badenian salts from Wieliczka as indicated by stable isotope composition of fluid inclusions: Biuletyn Panstwowego Instytutu Geologicznego, v. 396, p. 27–29.

Bula, W., and D. Jura, 1983, Lithostratigraphy of the Miocene deposits of the Carpathian Foredeep in the Cieszyn area (in Polish): Geologia, v. 9, p. 5–27.

Bula, Z., 2000, Dolny Paleozoik Górnego Ślaska i zachodniej Malopolski: Prace Państwowego Instytutu Geologicznego, v. 171, p. 1–69.

Bula, Z., 2001, Litologia i stratygrafia klastycznych utworów dolnego permu i karbonu z otworu Tarnawa 1: Prace Panstwowego Instytutu Geologicznego, v. 174, p. 61–66.

Bula, Z., M. Jachowicz, and J. Zaba, 1999, Principal characteristic of the upper Silesian block and Malopolska Block border zone (southern Poland): Geological Magazine, v. 134, p. 669–677.

Cartwright, J. A., 1989, The kinematics of inversion in the Danish Central Graben, *in* M. A. Cooper and G. D. Williams, eds., Inversion tectonics: Geological Society (London) Special Publication Classics, 1989, p. 153–175.

Cendon, D. I., C. Ayora, J. J. Pueyo, C. Taberner, and T. M. Peryt, 1999, Geochemical evidences of recycling in Badenian evaporates: Biuletyn Panstwowego Instytutu Geologicznego, v. 387, p. 18–20.

Cieszkowski, M., M. Gonera, N. Oszczypko, J. Slezak, and W. Zuchiewicz, 1988, Lithostratigraphy and age of upper Miocene deposits at Iwkowa, Polish West Carpathians: Bulletin of the Polish Academy of Sciences. Earth Sciences, v. 36, no. 3–4, p. 309–329.

Coward, M. P., 1996, Balancing sections through inverted basins, *in* P. G. Buchanan and D. A. Nieuwland, eds., Modern developments in structural interpretation, validation and modelling: Geological Society (London) Special Publication 99, p. 51–77.

Czepiec, I., 1996, Paleoecology and bathymetry of marine Miocene deposits of the Carpathian Foredeep (in Polish): Proceedings of the Seminar "Hydrocarbon Potential and Origin of the Natural Gases Accumulated in Miocene Sequence of the Polish and Ukraine Part of the Carpathian Foredeep," *in* M. Kotarba, ed., Abstract book: Krakow, Polish Oil and Gas Company, unnumbered.

Dadlez, R., M. Narkiewicz, R. A. Stephenson, M. T. M. Visser, and J.-D. van Wees, 1995, Tectonic evolution of the mid-Polish Trough: Modelling implications and significance for Central European geology: Tectonophysics, v. 252, p. 179–195.

Demarq, G., 1987, Paleothermic evolution during the Neogene in the Mediterranean through the marine megafauna: Annales Institutum Geologorum Publice Hungariae, v. 70, p. 371–375.

Dobrowolski, R., 1995, Drobne struktury tektoniczne w skalach górnokredowych wschodniej czesci Wyzyny Lubelskiej a dyslokacje podloza platformy wschodnioeuropejskiej w kenozoiku: Annales Societatis Geologorum Poloniae, v. 65, p. 79–91.

Doktor, M., 1983, Sedymentacja osadów zwirowych w miocenie na przedpolu Karpat (Sedimentation of Miocene gravel deposits in the Carpathian Foredeep): Studia Geologica Polonica, v. 78, no. 2, p. 1–107.

Dudziak, J., and A. Laptas, 1991, Stratigraphic position of Miocene carbonate-siliciclastic deposits near Chmielnik (Świetokrzyskie Mountains area, central Poland) based on calcareous nannofossils: Bulletin of the Polish Academy of Sciences. Earth Sciences, v. 39, p. 55–66.

Dudziak, J., and E. Luczkowska, 1991, Biostratigraphic correlation of foraminiferal and calcareous nannoplankton zones, early–middle Badenian (Miocene), southern Poland: Bulletin of the Polish Academy of Sciences, Earth Sciences, v. 39, p. 199–214.

Dulub, V. G., N. M. Zhabina, M. E. Ohorodnyk, and S. E. Smirnov, 2000, Stratigraphic scheme of Jurassic deposits of Precarpathians (in Ukrainian): Ukrainian Stratigraphical Committee approved poster.

Dzhinoridze, N. M., S. D. Gemp, A. F. Gorbov, and V. I. Rayevskiy, 1980, Zakonomernosti razmeshcheniya i kriterii poiskov kaliynykh soley SSSR: Tbilisi, Metsniereba, p. 1–375.

Dziadzio, P., 2000, Depositional sequences in Badenian and Sarmatian deposits in the SE parts of the Carpathian Foredeep (SE Poland): Przeglad Geologiczny, v. 48, no. 12, 1124–1138.

Dzulynski, S., 1953, Tektonika poludniowej cześci Wyzyny Krakowskiej: Acta Geologica Polonica, v. 3, no. 3, p. 1–440.

Eastoe, C. J., and T. Peryt, 1999, Stable chlorine isotope evidence for non-marine chloride in Badenian evaporites, Carpathian mountain region: Terra Nova, v. 11, p. 118–123.

Elias, M., and P. Palensky, 1997, The model for development of the Miocene foredeep in the Ostrava area: Zpravy o geologickym vyzkumech v roce 1997: Praha, Cesky geologicky ustav, p. 65–66.

Felisiak, I., 1992, Osady krasowe oligocenu i wczesnego miocenu oraz ich znaczenie dla poznania rozwoju tektoniki i rzeźby okolic Krakowa: Annales Societatis Geologorum Poloniae, v. 62, no. 2, p. 173–207.

Finger, F., P. Hanzl, C. Pin, A. Von Qyadt, and H. Sreyer, 2000, The Brunovistulicum: Avalonian Precambrian sequence at the eastern end of the Central European Variscides?, *in* W. Franke, W. Haak, V. Oncken, and D. Tanner, eds., Orogenic processes: Quantification and modelling in the Variscan belt: Geological Society (London) Special Publication 179, p. 103–112.

Galamay, A. R., 1997, Pochodzenie soli badeńskich regionu karpackiego: Przegląd Geologiczny, v. 45, p. 1012–1017.

Galamay, A. R., K. Bukowski, and J. Przybylo, 1997, Chemical composition and origin of brines in the Badenian evaporite basin of the Carpathian Foredeep: Fluid inclusion data from Wieliczka (Poland): Slovak Geological Magazine, v. 3, p. 165–171.

Garcia Veigas, J., L. Rossel, and A. Garlicki, 1997, Petrology and geochemistry (fluid inclusions) of Miocene halite rock salts (Badenian, Poland): Slovak Geological Magazine, v. 3, p. 181–186.

Garecka, M., and B. Olszewska, 1997, Stratigraphy of the Stebnik unit in Poland (in Polish): Przeglad Geologiczny, v. 45, no. 8, p. 793–798.

Garecka, M., P. Marciniec, B. Olszewska, and A. Wojcik, 1996, New biostratigraphic data and attempt to correlation of the Miocene deposits in the basement of the Western Carpathians (in Polish): Przeglad Geologiczny, v. 44, no. 5, p. 495–501.

Garlicki, A., 1968, Autochthonous salt series in the Miocene of the Carpathian Foredeep, between Skawina and Tarnow (in Polish with English summary): Biuletyn Instytutu Geologicznego, v. 215, p. 5–77.

Garlicki, A., 1979, Sedymentacja soli mioceńskich w Polsce: Prace Geologiczne, v. 119, p. 1–66.

Garlicki, A., 1994a, Formalne jednostki litostratygraficzne miocenu-formacja z Wieliczki (fm): Przeglad Geologiczny, v. 42, p. 26–28.

Garlicki, A., 1994b, Porównanie osadów solnych Górnego Śląska i okolic Wieliczki: Przeglad Geologiczny, v. 42, p. 752–753.

Garlicki, A., and J. Wiewiorka, 1981, The distribution of bromine in some halite rock salts of the Wieliczka salt deposit (Poland): Rocznik Polskiego Towarzystwa Geologicznego, v. 51, p. 353–359.

Garlicki, A., A. Szybist, and A. Kasprzyk, 1991, Badania pierwiastków śladowych w złożach soli i surowcow chemicznych: Przeglad Geologiczny, v. 39, p. 520–527.

Gasiewicz, A., 2000, Comparative study of major element geochemistry of gypsum-ghost limestones and selenite lithofacies from the Miocene of northern Carpathian Foredeep: Implication to the model of massive replacement of solid sulphates by calcium carbonates: Chemical Geology, v. 164, p. 183–218.

Gawel, A., 1962, Budowa geologiczna zloza solnego Wieliczki: Prace Instytutu Geologicznego, v. 30, no. 3, p. 305–331.

Gawel, A., 1965, Dekrepitacja soli z Wieliczki i Barycza: Sprawozdania z Posiedzeń Komisji Nauk Polska Akademia Nauk, Oddzial Krakow, v. 482, p. 485.

Gazdzicka, E., 1994, Nannoplankton stratigraphy of the Miocene deposits in Tarnobrzeg area (northeastern part of the Carpathian Foredeep): Geological Quarterly, v. 38, p. 553–570.

Gedl, P., 1997, Palynological study of an olistholite from the so-called Sucha Formation, Zawoja-1 borehole (Flysch Carpathians, Poland): Age and palaeoenvironment: Annales Societatis Geologorum Poloniae, v. 67, nos. 2–3, p. 203–216.

Gedl, P., 2000, Newly found marine Oligocene deposits in the Carpathian foreland and its paleogeographical consequences: Slovak Geological Magazine, v. 6, nos. 2–3, p. 155–157.

Glushko, V. V., and S. S. Kruglov, eds., 1977, The evaluation of ultradeep plays in Ukrainian Carpathians (in Russian): Kiev, Naukova Dumka, 176 p.

Glushko, V. V., and S. S. Kruglov, eds., 1986, Tectonic map

of the Ukrainian Carpathians: Kiev, Ministry of Geology of the UkrSSR, scale 1:200,000, p. 485–487.

Golonka, J., 1972, Osady doggeru w południowo-zachodniej części niecki nidzianskiej: Zeszyty Naukowe Akademia Gorniczo-Hutnicza, Geologia, v. 312, p. 65–86.

Golonka, J., 1978, Upper Jurassic microfacies in the Carpathians foreland (English summary): Biuletyn Instytutu Geologicznego, v. 310, p. 1–3.

Golonka, J., 1982, Algae and biosedimentation of Miocene limestones in the vicinity of Rzeszow (English summary): Biuletyn Instytutu Geologicznego, v. 332, p. 5–46.

Golonka, J., 2000, Cambrian–Neogene plate tectonic maps: Krakow, Wydawnictwa Uniwersytetu Jagiellonskiego, p. 1–125.

Golonka, J., L. Gahagan, M. Krobicki, F. Marko, N. Oszczypko, and A. Slaczka, 2006, Plate-tectonic evolution and paleogeography of the circum-Carpathian region, *in* J. Golonka and F. J. Picha, eds., The Carpathians and their foreland: Geology and hydrocarbon resources: AAPG Memoir 84, p. 11–46.

Gonera, M., 1994, Paleoecology of marine middle Miocene (Badenian) in the Polish Carpathians (Central Paratethys) foraminifera record: Bulletin of the Polish Academy of Sciences. Earth Sciences, v. 42, p. 107–125.

Gonera, M., T. M. Peryt, and T. Durakiewicz, 2000, Biostratigraphical and palaeoenvironmental implications of isotopic studies (^{18}O, ^{13}C) of middle Miocene (Badenian) foraminifers in the Central Paratethys: Terra Nova, v. 12, p. 231–238.

Gradzinski, R., 1957, Uwagi o sedymentacji miocenu w okolicy Proszowic: Rocznik Polskiego Towarzystwa Geologicznego, v. 26, no. 1, p. 3–23.

Gradzinski, R., 1982, Explanatory notes to lithotectonic Molasse profiles of the Upper Silesian Basin (Upper Carboniferous–Lower Permian) Carpathian Foredeep and in the Polish part of the Western Carpathians (comment to annex 20), *in* H. Lutzner and G. Schwab, eds., Tectonic regime of Molasse Epochs: Veroffentlichungen des Zentralinstituts fur Physik der Erde, v. 66, p. 225–235.

Gradzinski, R., and S. Porebski, eds., 1995, Carboniferous upper Silesian coal basin: Case studies in sedimentology and basin evolution: Studia Geologica Polonica, v. 108, p. 1–219.

Hakenberg, M., and J. Swidrowska, 1997, Propagation of the south-eastern segment of the Polish trough connected with bounding fault zones (from the Permian to the Jurassic): Compte Rendu Sommaire Seances de la Societe Geologique de France Academie des Sciences, Paris, v. 324, serie IIa, p. 793–803.

Hakenberg, M., and J. Swidrowska, 1998, Evolution of the Holy Cross segment of the mid-Polish Trough during the Cretaceous: Geological Quarterly, v. 42, no. 3, p. 239–262.

Hayward, A. B., and R. H. Graham, 1989, Some geometrical characteristic of inversion, *in* M. A. Cooper and G. D. Williams, eds., Inversion tectonics: Geological Society (London) Special Publication Classics, 1989, p. 17–39.

Heflik, W., and K. Konior, 1974, Obecny stan rozpoznania podłoża krystalicznego w obszarze Cieszyn-Rzeszotary: Biuletyn Instytutu Geologicznego, v. 273, p. 195–221.

Hollendonner, M., 1995, Sedimentologie und Petrologie des autochtonen Miozans der polnischen Karpatenvortiefe in Raum Bochnia, Sudpolen: Ph.D. thesis, Clausthal University, Clausthal, 100 p.

Homewood, P., P. A. Allen, and G. D. Williams, 1986, Dynamics of the Swiss Molasse basin, *in* P. A. Allen and P. Homewood, eds., Foreland basin: International Association of Sedimentologists Special Publication 8, p. 199–217.

Horvath, F., 1993, Towards the mechanical model for the formation of the Pannonian Basin: Tectonophysics, v. 226, p. 333–357.

Ivanov, D., A. R. Ashraf, V. Mosbrugger, and E. Palamarev, 2002, Palynological evidence for Miocene climate change in the Forecarpathian basin (Central Paratethys, NW Bulgaria): Palaeogeography, Palaeoclimatology, Palaeoecology, v. 178, p. 19–37.

Izotova, T. S., and I. V. Popadyuk, 1996, Oil and gas accumulations in the Late Jurassic reefal complex of the West Ukrainian Carpathian Foredeep, *in* P. A. Ziegler and F. Horvath, eds., Peri-Tethys Memoir 2: Structure and prospects of Alpine basins and forelands: Memoires du Museum National D'Histoire Naturelle, v. 170, p. 375–390.

Izotova, T. S., I. V. Popadyuk, and S. D. Riabczun, 1991, Sedimentology and exploration promises of Jurassic siliciclastics; Bilcze-Wolycia zone of the Carpathian Foredeep, *in* V. G. Sviridenko, ed., Geological and geophysical criteria for new prospects evaluation (in Russian): Ukrainian State Geological Research Institute Published Papers, L'viv, p. 100–110.

Jarosinski, M., 1992, Tektonika ilastych skal nadkladu zloza siarki w Machowie k. Tarnobrzegu w swietle analizy mezostrukturalnej: Kwartalnik Geologiczny, v. 36, no. 1, p. 121–150.

Jaroszewski, W., 1977, Sedimentary record of tectonic activity in the middle Roztocze region (in Polish with English summary): Przeglad Geologiczny, v. 25, no. 8–9, p. 418–427.

Jawor, E., 1970, The structure of the deep substratum in the region East of Cracow (in Polish with English summary): Acta Geologica Polonica, v. 20, no. 4, p. 710–769.

Jiricek, R., 1995, Stratigraphy and geology of the lower Miocene sediments of the Carpathian Foredeep in south Moravia and adjacent part of lower Austria, *in* B. Hamrsmid, ed., New results in Tertiary of West Carpathians II: Hodonin, Knihovnicka Zemni Plyn Nafta, v. 16, p. 37–65.

Jones, P., 1997, The Carpathians of the southern Poland: Thrust tectonic or wedge tectonics (abs.): 1997 AAPG International Conference and Exhibition, Austria, Vienna, Abstract book, p. A27–A28.

Jucha S., 1974, Ułozenie przestrzenne i rozwój facjalny miocenu autochtonicznego oraz jego stosunek do nasuniecia karpacko-stebnickiego: Geologia (Zeszyty Naukowe Akademii Gorniczo-Hutniczej), v. 22, p. 9–20.

Jucha S., 1985, New features of structure of the Carpathian

Foredeep and basement of the Carpathian Mountains: Przeglad Geologiczny, v. 33, no. 6, p. 333–344.

Jurkova, A., 1971, Die Entwicklung der badener Vortiefe im Raum der Mährischen Pforte und Gebiet von Ostrava (German summary): Geologicke Prace, v. 57, p. 155–160.

Jurkova, A., 1979, Confrontation of geological structure of Neoid and Variscan structural levels in the Moravian–Silesian Beskides and their foothils, *in* M. Mahel, ed., Tectonic profiles through the Western Carpathians: Bratislava, Geologicky Ustav D. Stura, p. 31–36.

Jurkova, A., V. Molcikova, P. Ctyroky, and J. Policky, 1983, New Eggenburgian finds in NE Moravia (English summary): Geologicke Prace, v. 79, p. 153–168.

Kalvoda, J., 2001, Upper Devonian–Lower Carboniferous foraminiferal paleobiogeography and Perigondwana tarranes at the Baltica–Gondwana interface: Geologica Carpathica, v. 52, p. 205–215.

Karnkowski, P., 1974, Zapadlisko przedkarpackie-Część wschodnia (na wschod od Krakowa), *in* W. Pozaryski, ed., Budowa Geologiczna Polski, IV (1) Tektonika-Niz Polski, Wydawnictwa Geologiczne (in Polish): p. 402–416.

Karnkowski, P., 1999, Oil and gas deposits in Poland: Krakow, Geosynoptics Society "Geos," p. 1–380.

Karnkowski, P. H., and W. Ozimkowski, 2001, Ewolucja strukturalna podłoża mioceńskiego basenu przedkarpackiego (obszar pomiedzy Krakowem a Przemyslem): Przeglad Geologiczny, v. 49, p. 431–436.

Karpala, A., and A. Lapinkiewicz, 1962, Kilka uwag o interpretacji materiałów sejsmicznych w czesci srodkowej i zachodniej przedgórza Karpat (in Polish): Geofizyka Poszukiwawcza i Wiertnicza, v. 8–10, no. 68–70, p. 3–16.

Kasprzyk, A., 1993, Lithofacies and sedimentation of the Badenian (middle Miocene) gypsum in the northern part of the Carpathian Foredeep, southern Poland: Annales Societatis Geologorum Poloniae, v. 63, p. 33–84.

Kasprzyk, A., 1999, Sedimentary evolution of Badenian (middle Miocene) gypsum deposits in the northern Carpathian Foredeep: Geological Quarterly, v. 43, p. 449–465.

Kasprzyk, A., and F. Orti, 1998, Paleogeographic and burial controls on anhydrite genesis: The Badenian basin in the Carpathian Foredeep (southern Poland, Western Ukraine): Sedimentology, v. 45, p. 889–907.

Kiersnowski, H., 1991, Litostratygrafia pemu północnowshodniego obrzezenia Górnośląskiego Zaglebia Weglowego: Przeglad Geologiczny, v. 4, p. 198–203.

Kirchner, Z., and W. Moryc, 1966, Zarys budowy geologicznej okolic Tarnowa-Ladnej (in Polish): Geofizyka i Geologia Naftowa, v. 7–8, no. 115–116, p. 155–166.

Kirchner, Z., and S. Poltowicz, 1974, Geological structure of the area between Brzesko and Wojnicz (southern Poland) (in Polish with English summary): Annales de la Societe Geologique de Pologne, v. 44, no. 2–3, p. 293–320.

Klimov, M. A., 1968, Prykarpats'kyy kalienosnyy baseyn i perspektyvy rozshyrennia syrovynnoi bazy vyrobnytstva kaliynykh dobryv v URSR: Geologicnij Zurnal, v. 28, no. 2, p. 56–60.

Kolasa, K., and A. Slaczka, 1985, Sedimentary salt megabreccias exposed in the Wieliczka mine, Fore-Carpathian depression: Acta Geologica Polonica, v. 35, p. 221–230.

Komorowska-Blaszczynska, M., 1965, The anhydrite-less island in the profiles of the Rzeszow Foreland: Bulletin of the Polish Academy of Sciences Series. Science, Geology, Geography, v. 13, p. 273–280.

Koopman, A., A. Speksnijder, and W. T. Hortsfield, 1987, Sandbox model of inversion tectonics: Tectonophysics, v. 137, p. 379–388.

Korin, S. S., 1994, Budowa geologiczna mioceńskich formacji solonośnych ukraińskiego Przedkarpacia: Przeglad Geologiczny, v. 42, p. 744–747.

Koszarski, A., L. Koszarski, J. Slezak, and M. Iwaniec, 1995, Calcareous nannoplankton from the terminal deposits of the Silesian nappe, Polish Flysch Carpathians: Stratigraphic implications, *in* J. A. Flores and F. J. Sierro, eds., 5th International Nannoplankton Association Conference in Salamanca Proceedings, p. 115–123.

Kotas, A., 1973, Wystepowanie utworów kambru w podlozu GZW: Przeglad Geologiczny, v. 1, p. 37–38.

Kotas, A., 1995, The Carboniferous system in Poland: Prace Panstwowego Instytutu Geologicznego, v. 148, p. 1–215.

Kotlarczyk, J., 1985, An outline of the stratigraphy of marginal tectonic units of the Carpathian orogene in the Rzeszow–Przemysl area, *in* J. Kotlarczyk, ed., Geotraverse Krakow–Baranow–Rzeszow–Przemysl–Komancza–Dukla: Guide to Excursion 4: XIII Congress Carpathian–Balkan Geological Association, Krakow, Poland, p. 21–32.

Kovac, M., I. Cicha, I. Krystek, A. Slaczka, Z. Stranik, N. Oszczypko, and D. Vass, 1989, Palinspastic maps of the Western Carpathian Neogene: Prague, Geological Survey, scale 1: 1,000,000, 5 sheets.

Kovac, M., A. Nagymarosy, J. Sotak, and K. Sutovska, 1993, Late Tertiary paleogeographic evolution of the Western Carpathians: Tectonophysics, v. 226, p. 401–415.

Kovac, M., A. Nagymarosy, N. Oszczypko, A. Slaczka, L. Csontos, M. Marunteanu, L. Matenco, and E. Marton, 1998, Palinspastic reconstruction of the Carpathian–Pannonian region during the Miocene, *in* M. Rakus, ed., Geodynamic development of the Western Carpathians: Bratislava, Slovak Geological Survey, p. 189–217.

Kovalevich, V. M., 1997, Fluid inclusions in Badenian (Miocene) halite of Bochnia (in Polish): Przeglad Geologiczny, v. 45, p. 822–825.

Kovalevich, V. M., and O. I. Petrichenko, 1997, Chemical composition of brines in Miocene evaporite basins of Carpathian region: Slovak Geological Magazine, v. 3, p. 173–180.

Kowalewski, K., 1966, Trzeciorzed Polski Południowej, Tom I, Stratygrafia trzeciorzędu poludniowego obrzeżenia Gór Świętokrzyskich i Roztocza, Część IV, Miocen odcinków poludniowo-sandomierskiego, tarnobrzeskiego i chmielowsko-baranowskiego i ich stosunek do obszarów sasiednich: Biuletyn Państwowego Instytutu Geologicznego, v. 189a, p. 1–143.

Krhovsky, J., M. Bubik, B. Hamrsmid, and M. Stastny, 1995, Lower Miocene of the Pouzdrany unit, the West Carpathian flysch belt, southern Moravia, *in* B. Hamrsmid, ed., New results in Tertiary of West Carpathians II: Hodonin, Zemni Plyn Nafta, p. 73–83.

Krysowska, M., 1966, Mineraly ciężkie w utworach mioceńkich obszaru śląsko-krakowskigo (in Polish): Prace Geologiczne Komisji Nauk Geologicznych Polskiej Akademii Nauk, Oddzial Krakow, v. 36, p. 1–71.

Krzywiec, P., 1997, Large-scale tectono-sedimentary middle Miocene history of the central and eastern Polish Carpathian Foredeep basin— Results of seismic data interpretation: Przeglad Geologiczny, v. 10, no. 2, p. 1039–1053.

Krzywiec, P., 1999, Miocene tectonic evolution of the Eastern Carpathian Foredeep basin (Przemysl-Lubaczow) in light of seismic data interpetation (English summary): Prace Panstwowego Instytutu Geologicznego, v. 168, p. 249-275.

Krzywiec, P., 2001, Contrasting tectonic and sedimentary history of the central and eastern parts of the Polish Carpathian Foredeep basin— Results of seismic data interpretation, *in* S. Cloetingh, M. Nemcok, F. Neubauer, F. Horvath, and P. Seifert, eds., The hydrocarbon potential of the Carpathian–Pannonian region: Marine and Petroleum Geology, v. 18, no. 1, p. 13–38.

Krzywiec, P., 2002, Mid-Polish Trough inversion-seismic examples, main mechanisms and its relationship to the Alpine–Carpathian collision, *in* G. Bertotti, K. Schulmann, and S. Cloetingh, eds., Continental collision and the tectonosedimentary evolution of forelands: European Geophysical Society Special Publication 1, p. 151–165.

Krzywiec, P., and P. Jochym, 1996, Characteristics of the subduction zone in the eastern Polish Carpathians— Preliminary results of flexural modelling: Publications of the Institute of Geophysics of the Polish Academy of Sciences, v. M-18, no. 273, p. 155–161.

Krzywiec, P., and P. Jochym, 1997, Characteristic of the Miocene subduction zone of the Polish Carpathians: Results of flexural modelling (in Polish with English summary): Przeglad Geologiczny, v. 45, no. 8, p. 785–792.

Krzywiec, P., E. Luczkowska, A. Sadowska, and J. Slezak, 1995, Depositional model of the central part of the Carpathian Foredeep basin, south Poland— Results of integrated bio- and seismostratigraphic study, *in* XV Congress Carpatho-Balcan Geological Association, Athens: Geological Society of Greece Special Publication 4, no. 3, p. 1113–1118.

Ksiazkiewicz, M., 1977, The tectonics of the Carpathians, *in* W. Pozaryski, ed., Geology of Poland, IV: Tectonics, Warsaw, Publishing House Wydawnictwa Geologiczne, p. 476–620.

Kubica, B., 1992, Rozwój litofacjalny osadów chemicznych badenu w północnej części zapadliska przedkarpackiego: Prace Panstwowego Instytutu Geologicznego, v. 133, p. 1–64.

Kubica, B., 1997, Relation of sulfur-forming processes to lithofacies and structural features of Badenian chemical sediments in the Carpathian Foredeep (Poland): Slovak Geological Magazine, v. 3, p. 121–130.

Kudrin, L. N., 1964, Paleogene deposits of the southwestern margin of Russian Craton and the external part of Precarpathian deep (in Russian): Bulletin of the Academy of Sciences of the USSR, Geological Division, v. 12, p. 72–80.

Kutek, J., 1994, Jurassic tectonic events in south-eastern cratonic Poland: Acta Geologica Polonica, v. 44, no. 3–4, p. 167–221.

Kutek, J., and J. Glazek, 1972, The Holy Cross area, central Poland, in the Alpine cycle: Acta Geologica Polonica, v. 22, no. 4, p. 603–653.

Lankreijer, A., M. Bielik, S. Cloetingh, and D. Majcin, 1999, Rheology predictions across the western Carpathians, Bohemian Massif, and the Pannonian Basin: Implications for tectonic scenarios: Tectonics, v. 18, no. 6, p. 1139–1153.

Laskowicz, I., 1997, Miocene sediments under the Carpathian overthrust in the Sucha Beskidzka area (Carpathian Foredeep) (English summary): Biuletyn Panstwowego Instytutu Geologicznego, v. 376, p. 61–74.

Liszkowski, J., 1989, Orogeniczno-descensyjny model genetyczny mioceńskich formacji salinarnych regionu karpackiego wschodnich obszarów Centralnej Paratetydy: Prace Naukowe Uniwersytetu Slaskiego w Katowicach, v. 1019, p. 1–96.

Maksym, A., A. Baszkiewicz, Z. Gregosiewicz, B. Liszka, and P. Zdanowski, 2001, Facies and reservoir potential of uppermost Jurassic and Lower Cretaceous deposits in the Brzezowka–Zagorzyce region versus geological structure of the southern Carpathian Foredeep (English summary): Przeglad Geologiczny, v. 49, no. 5, p. 401–407.

Marunteanu, M., 1999, Litho- and biostratigraphy (calcareous nannoplankton) of the Miocene deposits from the outer Moldavides: Geologica Carpathica, v. 50, no. 4, p. 313–324.

Marunteanu, M., M. Crihan, and C. Chira, 1999, Badenian zonation— The Carpathian area, Romania: Acta Palaeontologica Romaniae, v. 2, p. 261–267.

McClay, K. R., and P. G. Buchanan, 1992, Thrust faults in inverted extensional basins, *in* K. R. McClay, ed., Thrust Tectonics: London, Chapman & Hall, p. 93–104.

Meulenkamp, J. E., M. Kovac, and I. Cicha, 1996, On late Oligocene to Pliocene depocenter migration and the evolution of the Carpathian–Pannonian system: Tectonophysics, v. 266, p. 301–317.

Miall, A. D., 1995, Collision-related foreland basins, *in* C. A. Busby and R. Ingersoll, eds., Tectonics of sedimentary basins: Chichester, Blackwell Scientific Publishing, p. 393–424.

Micu, M., 1982, Explanatory notes to lithotectonic profiles of Miocene Molasses from central Moldavia (Eastern Carpathians, Romania), *in* H. Lutzner and G. Schwab, eds., Tectonic regime of Molasse epochs: Veroffentlichungen des Zentralinstituts fur Physik der Erde, v. 66, p. 117–136.

Moretti, I., and L. Royden, 1988, Deflection, gravity anomalies and tectonics of doubly subducted continental lithosphere: Adriatic and Ionian seas: Tectonics, v. 7, p. 875–893.

Moryc, W., 1971, Geological structure of the region of Lubaczow

(in Polish with English summary): Annales de la Societe Geologique de Pologne, v. 31, no. 1, p. 47–83.
Moryc, W., 1989, Tectonics of the Carpathians and their foreland in the light of geophysical and geological investigations (in Polish), *in* Referaty sesji Krakofsw 30 III 1989: Krakow, Komisja Tektoniki Komisja Nauk Geologicznych Polska Akademia Nauk, p. 170–195.
Moryc, W., 1992, Budowa geologiczna podloza miocenu w rejonie Sędziszów Mlp.-Rzeszów i ich perspektywiczność (in Polish): Nafta Gaz, v. 9–10, p. 205–223.
Moryc, W., 1995, Terrestrial Palaeogene formations of in the area of Carpathian Foredeep (in Polish with English summary): Nafta Gaz, v. 61, no. 5, p. 181–195.
Moryc, W., and W. Heflik, 1998, Metamorphic rocks in the basement of the Carpathians between Bielsko-Biala and Krakow: Geological Quarterly, v. 42, no. 1, p. 1–14.
Moryc, W., and H. Senkowiczowa, 1968, O wieku pstrych utworów z Liplasu: Kwartalnik Geologiczny, v. 12, no. 3, p. 537–545.
Moryc, W., and W. Wisniowska, 1965, Neocomian deposits at Basznia near Lubaczów (southeast Poland) (in Polish with English summary): Rocznik Polskiego Towarzystwa Geologicznego, v. 35, no. 1, p. 55–77.
Morycowa, E., and W. Moryc, 1976, Rozwój utworów jurajskich na przedgórzu Karpat w rejonie Dabrowa Tarnowska-Szczucin: Rocznik Polskiego Towarzystw Geologicznego, v. 46, p. 231–288.
Ney, R., 1968, The role of the "Cracow Bolt" in the geological history of the Carpathian Foredeep and in the distribution of oil and gas deposits (in Polish with English summary): Prace Geologiczne PAN, v. 45, p. 1–82.
Ney, R., W. Burzewski, T. Bachleda, W. Gorecki, K. Jakobczak, and K. Slupczynski, 1974, Outline of paleogeography and evolution of lithology and facies of Miocene layers in the Carpathian Foredeep (in Polish with English summary): Prace Geologiczne Komisji Nauk Geologicznych Polskiej Akademii Nauk, Oddzial w Krakowie, v. 82, p. 1–65.
Niemczyk, J., 1995, Profil litostratygraficzny serii gipsowej w Krzyżanowicach koło Pińczowa jako podstawa do rozważań sedymentologicznych: Geologia, v. 21, p. 183–196.
Olszewska, B., 2001, Stratygrafia malmu i neokomu podłoża Karpat fliszowych i zapadliska w swietle nowych danych mikropaleontologicznych: Przeglad Geologiczny, v. 49, no. 5, p. 451.
Ori, G. G., M. Roveri, and F. Vannoni, 1986, Plio-Pleistocene sedimentation in the Apenninic–Adriatic foredeep (central Adriatic Sea, Italy), *in* P. A. Allen, P. Homewood, and G. D. Williams, eds., Foreland basins: International Association of Sedimentologists Special Publication 8, p. 183–198.
Oszczypko, N., 1982, Explanatory notes to lithotectonic Molasse profiles of the Carpathian Foredeep and in the Polish part of the Western Carpathians (comment to annex 6–8), *in* H. Lutzner and G. Schwab, eds., Tectonic regime of Molasse epochs: Veroffentlichungen des Zentralinstituts fur Physik der Erde, v. 66, p. 95–115.
Oszczypko, N., 1996, The Miocene dynamics of the Carpathian Foredeep in Poland (English summary): Przeglad Geologiczny, v. 44, no. 10, p. 1007–1018.
Oszczypko, N., 1997, The early–middle Miocene Carpathian peripheral foreland basin (Western Carpathians, Poland): Przeglad Geologiczny, v. 45, no. 10/2, p. 1054–1063.
Oszczypko, N., 1998, The Western Carpathian Foredeep—Development of the foreland basin in front of the accretionary wedge and its burial history (Poland): Geologica Carpathica, v. 49, no. 6, p. 1–18.
Oszczypko, N., and A. Lucinska-Anczkiewicz, 2000, Early stages of the Polish Carpathian Foredeep development: Slovak Geological Magazine, v. 6, no. 2–3, p. 136–138.
Oszczypko, N., and M. Oszczypko-Clowes, 2003, The Aquitanian arine deposits in the basement of Polish Western Carpathians and its paleogeographical and paleotectonic implications: Acta Geologica Polonica, v. 53, no. 2, p. 101–122.
Oszczypko, N., and A. Slaczka, 1985, An attempt to palinspastic reconstruction of Neogene basins in the Carpathian Foredeep: Annales Societatis Geologorum Poloniae, v. 55, no. 1–2, p. 55–76.
Oszczypko, N., and A. Slaczka, 1989, The evolution of the Miocene basin in the Polish Outer Carpathians and their foreland: Geologica Carpathica, v. 40, p. 23–36.
Oszczypko, N., and A. Tomas, 1985, Tectonic evolution of marginal part of the Polish Flysch Carpathians in the middle Miocene: Kwartalnik Geologiczny, v. 29, no. 1, p. 109–128.
Oszczypko, N., and K. Zytko, 1987, Main stages in the evolution of the Polish Carpathians during the late Palaeogene and Neogene times, *in* Y. G. Leonov and V. E. Khain, eds., Global correlation of tectonic movements: Chichester, John Wiley & Sons, p. 187–198.
Oszczypko, N., R. Zajac, I. Garlicka, E. Mencik, J. Dvorak, and O. Matejovska, 1989, Geological map of the substratum of the Tertiary of the Western Outer Carpathians and their foreland, *in* D. Poprawa and J. Nemcok, coordinators, Geological atlas of the Western Outer Carpathians and their foreland: Warsaw, Panstwowy Instytut Geologiczny, scale 1:500,000, 7 sheets.
Oszczypko N., L. Stuchlik, and A. Wojcik, 1991, Stratigraphy of fresh-water Miocene deposits of the Nowy Sacz basin, Polish Western Carpathians: Bulletin of the Polish Academy of Sciences, Earth Sciences, v. 39, no. 4, p. 433–445.
Oszczypko, N., B. Olszewska, J. Slezak, and J. Strzepka, 1992, Miocene marine and brackish deposits of the Nowy Sacz basin (Polish Western Carpathians)— New lithostratigraphic and biostratigraphic standards: Bulletin of the Polish Academy of Sciences. Earth Sciences, v. 40, p. 83–96.
Oszczypko, N., P. Krzywiec, M. Lemberger, M. Stefaniuk, K. Pietsch, and H. Trygar, 1998, Integrated geological-geophysical interpretation of the Rzeszow-Smilno profile (Western Carpathians) (abs.): Carpatho-Balkan Geological Association, XVI Congress, Vienna, 1998, Abstracts, p. 446.
Oszczypko, N., A. Lucinska-Anczkiewicz, P. Gedl, and E. Malata, 2000, Paleogene autochthonous deposits at the

basement of the Polish Outer Carpathians and their paleogeographical implications: Slovak Geological Magazine, v. 6, no. 2–3, p. 143–145.

Otfinowski, M., 1981, Zespoly otwornic i sedymentacja piaskow bogucickich: Biuletyn Instytutu Geologicznego, v. 22, p. 53–88.

Palensky, P., J. Sikula, and E. Novotna, 1995, Paleogeography and lithology of autochthonous Miocene at north Moravia (English Summary), *in* B. Hamrsmid, ed., New results in Tertiary of West Carpathians II: Hodonin, Knhovnicka Zemni Plyn Nafta, v. 16, p. 119–128.

Panow, G. M., and A. M. Plotnikow, 1996, Badenian evaporites of the Ukrainian part of the Carpathian Foredeep (in Polish): Lithofacies and thickness: Przeglad Geologiczny, v. 44, no. 10, p. 1024–1028.

Pasternak, S. I., Y. M. Senkovski, and V. I. Havrylyszyn, 1987, Volyn-Podolia during Cretaceous (in Ukrainian): Kiev, Naukova Dumka, 260 p.

Peryt, D., 1997, Calcareous nannoplankton stratigraphy of the middle Miocene in the Gliwice area (upper Silesia, Poland): Bulletin of the Polish Academy of Sciences, Earth Sciences, v. 45, p. 119–131.

Peryt, D., 1999, Calcareous nannoplankton assemblages of the Badenian evaporites in the Carpathian Foredeep: Biuletyn Panstwowego Instytutu Geologicznego, v. 387, p. 158–161.

Peryt, T. M., 1996, Sedimentology of Badenian (middle Miocene) gypsum in eastern Galicia, Podolia and Bukovina (west Ukraine): Sedimentology, v. 43, p. 571–588.

Peryt, T. M., 2000, Resedimentation of basin centre sulphate deposits: Middle Miocene Badenian of Carpathian Foredeep, southern Poland: Sedimentary Geology, v. 134, p. 331–342.

Peryt, T. M., 2001, Gypsum facies transitions in basin-marginal evaporites: middle Miocene (Badenian) of West Ukraine: Sedimentology, v. 48, p. 1103–1119.

Peryt, T. M., and M. Jasionowski, 1994, In-situ formed and redeposited gypsum breccias in the middle Miocene Badenian of southern Poland: Sedimentary Geology, v. 94, p. 153–163.

Peryt, T. M., and A. Kasprzyk, 1992, Earthquake-induced resedimentation in the Badenian (middle Miocene) gypsum of southern Poland: Sedimentology, v. 39, p. 235–249.

Peryt, T. M., and D. Peryt, 1994, Badenian (middle Miocene) Ratyn Limestone in Western Ukraine and northern Moldavia: Microfacies, calcareous nannoplankton and isotope geochemistry: Bulletin of the Polish Academy of Sciences. Earth Sciences, v. 42, p. 127–136.

Peryt, T. M., S. Karoli, D. Peryt, O. I. Petrichenko, P. Gedl, W. Narkiewicz, J. Durkovicova, and Z. Dobieszynska, 1997, Westernmost occurrence of the middle Miocene Badenian gypsum in Central Paratethys (Koberice, Moravia, Czech Republic): Slovak Geological Magazine, v. 3, p. 105–120.

Peryt, T. M., M. Jasionowski, S. Karoli, O. I. Petrichenko, A. V. Poberegski, and I. I. Turchinov, 1998a, Correlation and sedimentary history of the Badenian gypsum in the Carpathian Foredeep (Ukraine, Poland, and Czech Republic): Przeglad Geologiczny, v. 46, p. 729–732.

Peryt, T. M., D. Peryt, J. Szaran, S. Halas, and M. Jasionowski, 1998b, Middle Miocene Badenian anhydrite horizon in the Ryszkowa Wola 7 borehole (southeast Poland) (in Polish with English summary): Biuletyn Panstwowego Instytutu Geologicznego, v. 379, p. 61–78.

Petrichenko, O. I., V. M. Kovalevich, and W. N. Chaly, 1994, Geokhimicheskaya obstanovka soleobrazovania v tortonskom evaporitovom bassejne severo-zapadnogo Predkarpatya: Geologia i Geohimia Gorucih Iskopaemyh, v. 41, p. 74–79.

Petrichenko, O. I., T. M. Peryt, and A. V. Poberegsky, 1997, Peculiarities of gypsum sedimentation in the middle Miocene Badenian evaporite basin of Carpathian Foredeep: Slovak Geological Magazine, v. 3, p. 91–104.

Petryczenko, O. I., G. M. Panow, T. M. Peryt, B. I. Srebrodolski, A. W. Poberezski, and W. M. Kowalewicz, 1994, Zarys geologii mioceńskich formacji ewaporatowych ukraińskiej części zapadliska przedkarpackiego: Przeglad Geologiczny, v. 42, p. 734–737.

Picha, F., 1979, Ancient submarine canyons of Tethyan continental margins, Czechoslovakia: AAPG Bulletin, v. 63, no. 1, p. 67–86.

Picha, F., 1996, Exploring for hydrocarbons under thrust belts— A challenging new frontier in the Carpathians and elsewhere: AAPG Bulletin, v. 80, no. 10, p. 1547–1564.

Picha, F., and Z. Stranik, 1999, Late Cretaceous to early Miocene deposits of the Carpathian foreland basin in southern Moravia: International Journal of Earth Sciences, v. 88, p. 475–495.

Piwocki, M., B. Olszewska, and G. Czapowski, 1996, Korelacja litostratygraficzna neogenu Polski z krajami sasiednimi (in Polish), *in* L. Malinowska and M. Piwocki, eds., Budowa Geologiczna Polski, III, Atlas Skamienialosci Przewodnich i Charakterystycznych, 3a: Kenozoik-Trzeciorzed-Neogen, p. 517–529.

Poltowicz, S., 1964, Przedtortońska dolina rzeczna między Rabą i Dunajcem (in Polish): Geofizyka i Geologia Naftowa, v. 6–7, no. 90–91, p. 176–181.

Poltowicz, S., 1993, Palinspastic paleogeography reconstruction of Badenian saline sedimentary basin in Poland (in Polish with English summary): Kwartalnik AGH, Geologia, v. 19, no. 44, p. 203–233.

Poltowicz, S., 1995, Miocene deposits in the basement of the Polish Western Carpathians (in Polish with English summary): Kwartalnik AGH, Geologia, v. 21, no. 2, p. 117–168.

Popadyuk, I. V., and O. V. Bondarenko, 2001, Karst reservoir in the Upper Jurassic–Neocomian carbonate complex of the Precarpathians: Visnyk of the Lviv University, Series Geology, v. 15, p. 104–108.

Poprawa, P., M. Jarosinski, A. Pepel, H. Kiersnowski, and E. Jawor, 2001, Tectonic evolution of the Liplas–Tarnawa area— Analysis of subsidence, mesostructures, seismic and gravimetric data (English summary): Prace Panstwowego Instytutu Geologicznego, v. 174, p. 143–160.

Porebski, S., and N. Oszczypko, 1999, Lithofacies and origin of the Bogucice sands (upper Badenian, Carpathian Foredeep, Poland) (English summary): Prace Panstwowego Instytutu Geologicznego, v. 168, p. 57–82.

Porebski, S., K. Pietch, R. Hodiak, and R. J. Steel, 2003, Origin and sequential development of Badenian–Sarmatian clinoforms in the Carpathian Foreland basin (southeast Poland): Geologica Carpathica, v. 54, no. 2, p. 119–136.

Posamentier, H. W., M. T. Jervey, and P. R. Vail, 1988, Eustatic control on clastic deposition I— Conceptual Framework, *in* C. K. Wilgus, B. S. Hastings, C. Kendall, H. W. Posamentier, C. A Ross, and J. C. Van Wagoner, eds., Sea-level changes: An integrated approach: SEPM Special Publication 42, p. 109–124.

Pozaryski, W., and W. Brochwicz-Lewinski, 1978, On the Polish trough: Geologie en Mijnouw: Journal of the Royal Geological and Mining Society of the Netherlands, v. 57, no. 4, p. 545–557.

Pozaryski, W., and K. Zytko, 1980, On the mid-Polish aulacogen and the Carpathian geosyncline: Bulletin of the Academy of Polonish Sciences, Series: Science Terre, v. 28, no. 4, p. 303–316.

Price, R. J., 1973, Large scale gravitational flow of supracrustal rocks, southern Canadian Rockies, *in* K. A. De Jong and R. Scholten, eds., Gravity and Tectonics: New York, Wiley, p. 491–502.

Prisyazhniuk, V. A., S. A. Lul'eva, S. V. Syabryaj, O. P. Olshtynskaya, and L. V. Stupina, 1997, Novi dani z paleontologii ta stratygrafii miotsenovykh vikladiv Ustechkivskogo grabena Volyno-Podillya: Geologichnyi Zhurnal, v. 1–2, p. 57–64.

Rajchel, J., 1989, Budowa geologiczna doliny Sanu w rejonie Dynów-Dubiecko: Biuletyn Instytutu Geologicznego, v. 361, p. 11–53.

Rogl, F., 1996, Stratigraphic correlation of the Paratethys Oligocene and Miocene: Mitteilungen der Gesellschaft der Geologie- und Bergbaustudenten in Osterreich, v. 41, p. 65–74.

Rogl, F., 1999, Mediterranean and Paratethys, Facts and hypotheses of an Oligocene to Miocene paleogeography (short overview): Geologica Carpathica, v. 50, no. 4, p. 339–349.

Rosell, L., F. Orti, A. Kasprzyk, E. Playa, and T. M. Peryt, 1998, Strontium geochemistry of Miocene primary gypsum: Messinian of southeast Spain and Sicily and Badenian of Poland: Journal of Sedimentary Research, v. 68, p. 63–79.

Royden, L. H., 1985, The Vienna basin: A thin-skinned pull-apart basin: SEPM Special Publication, 37, p. 319–338.

Royden, L. H., 1988a, Late Cenozoic tectonics of the Pannonian Basin system, *in* L. H. Royden and F. Horvath, eds., Pannonian Basin: AAPG Memoir 45, p. 27–48.

Royden, L. H., 1988b, Flexural behaviour of the continental lithosphere in Italy: Constraints imposed by gravity and deflection data: Journal of Geophysical Research, v. 93, p. 7747–7766.

Royden, L. H., 1993a, Evolution of retreating subduction boundaries formed during continental collision: Tectonics, v. 12, no. 3, p. 629–638.

Royden, L. H., 1993b, The tectonic expression of slab-pull at continental convergent boundaries: Tectonics, v. 12, no. 2, p. 303–325.

Royden, L. H., and B. C. Burchfiel, 1989, Are systematic variations in thrust belt style related to plate boundary processes? (The Western Alps versus the Carpathians): Tectonics, v. 8, no. 1, p. 51–61.

Royden, L. H., and G. D. Karner, 1984, Flexure of lithosphere beneath Apennine and Carpathian Foredeep basins: Evidence for insufficient topographic load: AAPG Bulletin, v. 68, no. 6, p. 704–712.

Rutkowski, J., 1989, Budowa geologiczna regionu Krakowa (in Polish): Przeglad Geologiczny, v. 37, no. 6, p. 302–308.

Rylko, W., and A. Tomas, 1995, Morphology of the consolidated basement of the Polish Carpathians in the light of magnetotelluric data: Kwartalnik Geologiczny, v. 39, p. 1–16.

Skuce, A. G., 1996, Forward modelling of compaction above normal faults: An example from the Sirte basin, Libya, *in* P. G. Buchanan and D. A. Nieuwland, eds., Modern developments in structural interpretation, validation and modelling: Geological Society (London) Special Publication, v. 99, p. 135–146.

Slaczka, A., 1976, New data on the structure of the basement of the Carpathians south of the Wadowice: Rocznik Polskiego Towarzystwa Geologicznego, v. 46, no. 3, p. 337–350.

Slaczka, A., 1977, The Miocene deposits in the Sucha IG-1 borehole (in Polish): Kwartalnik Geologiczny, v. 21, p. 405–406.

Slaczka, A., and K. Kolasa, 1997, Resedimented salt in the Northern Carpathians Foredeep (Wieliczka, Poland): Slovak Geological Magazine, v. 3, p. 135–155.

Slaczka, A., and N. Oszczypko, 1987, Olistostrome and overthrusting in the Polish Carpathians: Annales Instititum Geologorum Publice Hungariae, v. 70, p. 282–292.

Strzepka, J., 1981, The lower Miocene microfauna from the Sucha IG 1 borehole, Poland (in Polish with English summary): Biuletyn Instytutu Geologicznego, v. 331, p. 117–122.

Sul, M. F., and A. W. Poberezski, 1995, O warunkach powstawania złóz siarki na ukraińskim Przedkarpaciu: Przeglad Geologiczny, v. 43, p. 1035–1040.

Syabryay, S. V., and Y. V. Teslenko, 2000, Paleobotanicheskiye materially k paleoklimaticheskim rekonstruktsiam vremeni soleobrazovania v badenii Zakarpatya: Geolohichny y Zhurnal, v. 1, p. 15–23.

Tari, G., F. Horvath, and L. Csontos, 1997, Palinspastic reconstruction of the Carpathian/Pannonian system (abs.): AAPG Annual Meeting Program, p. A57.

Tobola, T., 2000, Badania koncentracji potasu i magnezu w solach kamiennych złóza Bochni: Przeglad Geologiczny, v. 48, p. 1163–1168.

Tomek, C., and J. Hall, 1993, Subducted continental margin imaged in the Carpathians of Czechoslovakia: Geology, v. 21, no. 2, p. 535–538.

Utrobin, V. N., 1960, The morphology and origin of the pre-Tertiary surface in the outer part of Precarpathian deep and SW margin of Russian craton (in Russian): USSR Academy of Science Reports, v. 130, no. 2, p. 396–399.

Van Couvering, I. A., M. P. Aubry, Q. A. Berggren, J. P. Bujak, C. W. Naesen, and T. Wieser, 1981, Terminal Eocene event and the Polish connections: Palaeogeography, Palaeoclimatology, Palaeoecology, v. 36, p. 321–362.

Vass, D., and F. Cech, 1983, Sedimentation rates in Molasse Basins of the Western Carpathians: Geologica Carpathica, v. 34, no. 4, p. 411–422.

Vyshniakov, I. B., V. O. Vashchenko, H. A. Havrylko, and O. V. Honyk, 2000, The nature of the Kolomyia–Zabolotiv channel-like structure (in Ukrainian): Geology and Geochemistry of Combustive Minerals, v. 1, p. 3–12.

Wdowiarz, S., 1976, On the relation of the Carpathians to the Carpathian Foredeep in Poland (in Polish with English summary): Przeglad Geologiczny, v. 6, p. 350–357.

Wieser, T., K. Bukowski, and A. Wojtowicz, 2000, Korelacja mineralogiczna i wiek radiometryczny tufitu z warstw chodenickich z okolic Bochni: V Ogolnopolska Sesja, Datowanie Mineralow i Skal, Krakow, February 10–12, 2000.

Wojcik, A., and M. Jugowiec, 1998, The youngest members of the folded Miocene in the Andrychow region (Southern Poland): Przeglad Geologiczny, v. 8, no, 2, p. 763–770.

Wojtowicz, A., S. P. Hryniv, and P. M. Bilonizhka, 2002, Radiometryczne datowanie wybranych minerałow potasowych z mioceńskich formacji solonośnych wschodniego Przedkarpacia: Przeglad Geologiczny, v. 50, p. 625–629.

Wyszynski, O. W., 1939, Przedgórze okolic Kosowa: Przemysł Naftowy, v. 14, p. 7–13.

Zaba, J, 1999, Ewolucja strukturalna utworów dolnopaleozoicznych w strefie granicznej bloków gornoslaskiego i malopolskiego: Prace Panstwowego Instytutu Geologicznego, v. 166, p. 1–162.

Zajac, R., 1981, Korelacja utworów dewonu i karbonu dolnego podłoża środkowej części zapadliska przedkarpackiego: Biuletyn Instytutu Geologicznego, v. 331, p. 39–56.

Zdanowski, P., A. Baszkiewicz, and Z. Gregosiewicz, 2001, Facies analysis of the uppermost Jurassic and the Lower Cretaceous deposits in the Zagorzyce region (southern Poland) (English summary): Przeglad Geologiczny, v. 49, no. 2, p. 161–178.

Ziegler, P., 1990, Geological atlas of western and Central Europe. Shell International Petroleum Company and Geological Society Publishing House, 239 p.

Zielinski, J. J., 1963, The discovery of buried flysch fold in the Region of Przemysl: Rocznik Polskiego Towarzystwa Geologicznego, v. 33, no. 1, p. 387–397.

Znosko, J., 1983, Tektonika środkowo-południowej Polski pozakarpackiej: Kwartalnik Geologiczny, v. 27, no. 3, p. 457–470.

Zoetemeijer, R., C. Tomek, and S. Cloetingh, 1999, Flexural expression of European continental lithosphere under the Western Outer Carpathians: Tectonics, v. 18, no. 5, p. 843–861.

Zytko, K., 1965, Sur le rapport de la formation du petrole et l'orogenese des Karpates: Reports of the VII Congress Carpathian-Balkan Geological Association, Sofia, Part IV, p. 75–81.

Zytko, K., 1997, Electrical conductivity anomaly of the Northern Carpathians and the deep structure of the orogen: Annales Societatis Geologorum Poloniae, v. 67, no. 1, p. 25–44.

Zytko, K., W. Rylko, N. Oszczypko, R. Zajac, I. Garlicka, J. Nemcok, M. Elias, E. Mencik, J. Dvorak, and Z. Stranik, 1989, Geological map of the Western Outer Carpathians and their foreland without Quaternary, *in* D. Poprawa and J. Nemcok, coordinators, Geological atlas of the Western Outer Carpathians and their foreland: Warszawa, Panstwowy Instytut Geologiczny, scale 1:500,000, 7 sheets.

Myśliwiec, Michal, Zenon Borys, Beata Bosak, Bogusław Liszka, Kazimierz Madej, Andrzej Maksym, Krystyna Oleszkiewicz, Małgorzata Pietrusiak, Bożena Plezia, Grzegorz Staryszak, Grażyna Świętnicka, Czesława Zielińska, Krystyna Zychowicz, Piotr Gliniak, Radosław Florek, Jarosław Zacharski, Andrzej Urbaniec, Adam Górka, Piotr Karnkowski, and Paweł H. Karnkowski, 2006, Hydrocarbon resources of the Polish Carpathian Foredeep: Reservoirs, traps, and selected hydrocarbon fields, *in* J. Golonka and F. J. Picha, eds., The Carpathians and their foreland: Geology and hydrocarbon resources: AAPG Memoir 84, p. 319–320.

10

Hydrocarbon Resources of the Polish Carpathian Foredeep: Reservoirs, Traps, and Selected Hydrocarbon Fields

Michał Myśliwiec, Zenon Borys, Beata Bosak, Bogusław Liszka, Kazimierz Madej, Andrzej Maksym, Krystyna Oleszkiewicz, Małgorzata Pietrusiak, Bożena Plezia, Grzegorz Staryszak, Grażyna Świętnicka, Czesława Zielińska, Krystyna Zychowicz
Polish Oil and Gas Company, Exploration Department, South Regional Division, Jaslo, Republic of Poland

Piotr Gliniak, Radosław Florek, Jarosław Zacharski, Andrzej Urbaniec, Adam Górka
Polish Oil and Gas Company, Exploration Department, South Regional Division, Krakow, Republic of Poland

Piotr Karnkowski, Paweł H. Karnkowski
Polish Oil and Gas Company, Exploration Department, Warsaw, Republic of Poland

ABSTRACT

The intention of this chapter is to present a short description of the reservoir rocks, recognized types of the hydrocarbons traps, and a few chosen oil and gas fields in the basement and in the Miocene cover of the Carpathian Foredeep.

Most of the oil produced from the Carpathian Foredeep basin has come from Oxfordian carbonates and Cenomanian sandstones, two of the most oil-productive reservoir rocks. The Devonian, Carboniferous, and Cretaceous carbonates are the secondary basement reservoirs. However, from the commercial point of view, the overlying Miocene clastic deposits cover has the prominent status. It consists of excellent source and reservoir rocks that have produced large amounts of gas. The primary reservoirs are sandstones of different depositional elements of submarine fans, sandstones of deltaic environments (large mouth bar, distributary channels, and others), the shallow-marine clastic deposits of estuaries, and sandy barriers. Sporadically, the gas accumulations are located in the secondary porous anhydrites.

The most common basement oil and gas trap is the combination structural-stratigraphic type with varying systems of sealing. Traps are related mainly to the sub-Miocene and less to the sub-Cretaceous unconformity. The pinching-out stratigraphic traps are known from the Cenomanian sandstones. The stratigraphic traps in the carbonate buildups (reefs) have still-undiscovered potential.

DOI:10.1306/985613M843073

The traps for gas accumulations in the Miocene (Badenian and Sarmatian) deposits of the Carpathian Foredeep are related, first of all, to the paleomorphology of the pre-Miocene basement formed by the erosion supported by the faulting processes. This type of trap is classified as the compactional trap. The second, very productive structural traps were recognized beneath the Carpathian sole thrust, at the front of the Carpathians. The fault-related trapping mechanism is also known from several gas fields. The pinching-out trap types caused by the horizontal and the vertical facies changes are sparse and occur most commonly in the topmost part of the Miocene succession because of the more favorable facies and less compactional deformation of the strata.

During last 50 yr of exploration, more than 120 gas and oil fields were discovered in the Carpathian Foredeep. The bulk of the produced hydrocarbons (97%, mainly gas) were contained in the Miocene deposits, with a further 3% in the Miocene basement. The Miocene contains only gas fields, whereas both oil and gas were found in the basement rocks.

The Miocene gas fields are typically multihorizontal and saturated by gas with a very high methane content, normally from 95 to 99%. About 90 billion m^3 (3.2 tcf) of this kind of gas has already been produced. The Przemysl gas fields group is the largest in Poland with gas initially in place (GIIP) of nearly 71 billion m^3 (2.5 tcf), and the cumulative field production, as of December 31, 2002, amounted to 55 billion m^3 (1.9 tcf).

The Grobla and Plawowice are the biggest oil fields accumulated in the Oxfordian carbonates and Cenomanian sandstones of the Carpathian Foredeep basement. A few important gas fields, like Tarnow and Lubaczow, were also founded in the Oxfordian carbonates.

According to the results of the latest deep wells, the basement of the Carpathian Foredeep is still highly prospective for hydrocarbons, especially the Devonian and Carboniferous carbonates in the central part of the foredeep and the Oxfordian buildups in the more western part. The deep wells like Hermanowa proved to be a very high source potential of the lower Paleozoic rocks, which allows for the prediction of new significant oil and gas discoveries in the nearest future.

The improvement of methods, particularly the direct hydrocarbon indicators method, opened a new stage of exploration for gas accumulations in the Miocene deposits. In only the last 8 yr, nearly 20 new gas fields were discovered on the basis of such interpretation results.

11

Kotarba, M. J., and Y. V. Koltun, 2006, The origin and habitat of hydrocarbons of the Polish and Ukrainian parts of the Carpathian Province, *in* J. Golonka and F. J. Picha, eds., The Carpathians and their foreland: Geology and hydrocarbon resources: AAPG Memoir 84, p. 321–368.

The Origin and Habitat of Hydrocarbons of the Polish and Ukrainian Parts of the Carpathian Province

M. J. Kotarba

Faculty of Geology, Geophysics, and Environmental Protection, AGH–University of Science and Technology, Krakow, Republic of Poland

Y. V. Koltun

Institute of Geology and Geochemistry of Fossil Fuels, Ukrainian Academy of Sciences, L'viv, Ukraine

ABSTRACT

Multiple petroleum systems occurring in different tectonic elements of the Polish and Ukrainian parts of the Flysch Carpathians, their foredeep, and the Paleozoic–Mesozoic basement have been identified and characterized using various geochemical methods. The Carpathian flysch sequence, containing the main oil fields in the area, contains two main potential source rock intervals: the Early Cretaceous and Oligocene. The Oligocene Menilite shales, which are widespread in the Polish and Ukrainian parts of the Carpathian flysch belt, contain high-quality source rocks, with mainly type II kerogen and high petroleum potential. A suitable tectonic position in several units of the flysch belt, especially in its frontal nappes of the Ukrainian part, provided the maturation level of this series, corresponding to different parts of the oil window. Modeling of hydrocarbon generation and expulsion in Menilite rocks shows that these processes are related mainly to Miocene overthrusting. Geochemical studies of oils from different fields occurring in the flysch sequence show that they belong to the same family. The presence of oleanane probably indicates that they have been generated from Menilite rocks. Lower Cretaceous organic-rich rocks, which contain mainly type II and III kerogen with good petroleum potential and a maturation level corresponding to the oil window, should be considered as another important potential source rock sequence in both the Polish and Ukrainian Carpathian flysch belt. The autochthonous Miocene nonfolded molasse sequence contains most of the gas fields in the Polish and Ukrainian Carpathian Foredeep. Geochemical studies of Badenian and Sarmatian strata show that they contain sufficient amounts of immature terrestrial organic matter, which generated methane-rich microbial gases. Molecular and isotopic compositions of natural gases from Miocene reservoirs confirm that they were produced during microbial carbon dioxide reduction. Molecular and stable isotope compositions of gases from Paleozoic–Mesozoic reservoirs of the Polish part of the basement show that they were generated by both microbial and thermogenic processes. Gas from the Lachowice field in Devonian rocks is a

DOI:10.1306/985614M843074

typical nonassociated thermogenic gas, generated from type III kerogen with a high degree of maturation. This gas probably migrated from Carboniferous source rocks. The oils accumulated in Upper Cretaceous and Upper Jurassic reservoirs in the Polish part and, southeast of the Ukrainian part of the basement, contain oleanane and other specific biomarkers and belong to the same geochemical family. They were probably generated from the Oligocene flysch rocks and Middle Jurassic beds.

INTRODUCTION

The Polish and Ukrainian Carpathians represent the central segment of the Carpathian arc that extends between Slovakia and Romania (Figure 1). The frontal part of the flysch belt of the Polish and Ukrainian Carpathians constitutes one of the oldest petroleum-producing regions in the world. The exploitation of oil started in 1854 and that of natural gas in 1921.

Because of the high drilling density and detailed mapping, geology and petroleum occurrence of the area have been extensively studied and well described in numerous publications (e.g., Swiderski, 1952; Dolenko, 1962; Glushko, 1968; Danys et al., 1974; Ksiazkiewicz, 1977; Slaczka, 1980, 1996; Burchfiel and Royden, 1982; Kotlarczyk, 1985; Kruglov et al., 1985; Wdowiarz, 1985; Vialov et al., 1988; Kusmierek, 1990, 1992; Slaczka and Kaminski, 1998). However, only a few studies have been published on the organic-rich formations of both the Ukrainian (Grinberg, 1957; Shevchenko et al., 1984; Gabinet, 1985) and Polish areas (Hempel, 1955; Marzec et al., 1971; Karpinski, 1978; Gucwa and Wieser, 1980). More detailed studies of the Carpathian source rocks and origin of petroleum have been published recently (Kotarba, 1987, 1992, 1998a, b, 1999; Kotarba et al., 1987, 1998a, b; Koltun, 1992; Kotarba and Jawor, 1993; ten Haven et al., 1993; Lafargue et al., 1994; Popescu, 1995; Bessereau et al., 1996; Kruge et al., 1996; Koester et al., 1998a, b; Koltun et al., 1998; Matyasik et al., 2000, 2001).

The main petroleum fields are concentrated in the frontal part of the Carpathian overthrust, the adjacent part of the foreland, and the Transcarpathian depression (Figures 1, 2). They occur in four different tectonic settings and represent different petroleum systems: (1) the Cretaceous–Paleogene flysch sequence, which contains most of the oil fields; (2) the Mesozoic–Paleozoic autochthonous sequence of the southwest margin of the East European platform, which is partly covered by the Carpathian allochthon, with several oil and gas fields; (3) the gas-bearing Miocene Molasse sequence of the Carpathian Foredeep; and (4) the Miocene Molasse of the Transcarpathian depression, containing several gas fields.

The purpose of this chapter is to evaluate the source rocks in the flysch sequence, Mesozoic–Paleozoic autochthonous basement, and Miocene Molasse strata, along with their relationship to petroleum occurrences in the Polish and Ukrainian parts of the Carpathian petroleum province. The location of the studied areas, wells sampled for cores and outcrops for source rock evaluation, and sampled wells for oils and gases are presented in Figures 1 and 2.

The frontal part of the Flysch belt, especially its deep horizons, has the potential for new discoveries. Petroleum potential of the internal part of the flysch belt is still unexplored, and this area is an important exploration target. Thus, the understanding of the processes of hydrocarbon generation in the entire flysch belt will help to evaluate the full potential of the region.

The Mesozoic autochthonous sequence under the Carpathian overthrust represents a new promising exploration target, where the hydrocarbon generation studies can determine the possible extent of occurrence of petroleum prospect. The Carpathian Foredeep, which also has a long history of gas production, also seems to be prospective.

THE SOURCE ROCKS

The aim of this section is to present a geochemical characterization and genetic identification of organic-rich petroleum source rocks (quantity, genetic type, and maturity) and their hydrocarbon potential. They occur in the lithostratigraphic series of the Polish and Ukrainian part of the Flysch orogenic belt, the autochthonous Miocene Molasse sequences of the Carpathian Foredeep and Transcarpathian depression, and the Mesozoic–Paleozoic autochthonous sequence of the southwest margin of the East European platform.

The Flysch Carpathians

The Flysch Carpathians is the outer part of a young Alpine orogenic belt of nappe-thrust structure composed of numerous complexes of generally calcareous sandstones and shales (mudstones and claystones), ranging in age from Late Cretaceous to Oligocene–early Miocene. The claystones and/or mudstones form Lower Cretaceous and Oligocene–lower Miocene strata of flysch profiles of Zglobice, Stebnik (Sambir), Boryslav–Pokuttia, Skole (Skiba), Silesian (Krosno), Subsilesian, Dukla, Grybow, Magura, and Chornogora units of the Polish and Ukrainian Carpathians. The equivalent units to the Skiba and Krosno units in Ukraine are termed the Skole and Silesian units in Poland. It is worth noting that Skole and Krosno are names of two towns in Ukraine and in Poland, respectively.

The Polish Area

The flysch sequence in the Polish Carpathians ranges in age from Lower Cretaceous to Oligocene–lower Miocene. During this time, as much as 4000 m (13,000 ft) of flysch sediments accumulated in the generally southward-deepening basins. The Polish Flysch Carpathians includes the following structural units, from northeast to southwest: Zglobice, Stebnik, Skole (Skiba), Subsilesian, Silesian (Krosno), Dukla, Grybow, and Magura (Figure 1). The Zglobice and Stebnik (called Sambir in Ukraine) units, for tectonic reasons, belong to the Carpathian overthrust, but their profiles are composed of lower Miocene strata of the Inner Carpathian basin (Ney et al., 1974; Oszczypko, 1997). In the Polish Carpathian Province, in the folded Miocene strata of these structural units, hydrocarbon occurrences have not been observed; hence, they were of no interest for petroleum explorations. The Grybow unit lies between the Silesian and Magura nappes (Figure 1), and its strata occur on the surface mainly in tectonic windows in the Magura unit. It is similar to the Dukla unit as far as stratigraphic position and structure are concerned. Nonetheless, as opposed to the Dukla unit, the lithology of the Grybow unit is mainly made of claystones and marls with small quantities of sandstones. In the lithological profile of the Magura unit, sandstone complexes dominate. Shales are generally scarce there, suggesting no potential source rocks in this area. Generalized lithostratigraphic columns and petroleum reservoirs in selected structural units of the Polish Flysch Carpathians are presented in Figure 3.

Previous geochemical and mineralogical studies revealed that petroleum source rocks occur in the following Carpathian Flysch lithostratigraphic members: Spas shales (Lower Cretaceous), Verovice shales (Lower Cretaceous), Lgota beds (Lower and Upper Cretaceous), Inoceramian (Paleocene–Upper Cretaceous), Istebna shales (Paleocene), Hieroglyphic beds (Eocene), Menilite shales (Oligocene) and Krosno beds (Oligocene and lower Miocene), and the most significant, the Oligocene Menilite shales (Hempel, 1955; Wdowiarz, 1960; Gucwa and Wieser, 1980; ten Haven et al., 1993; Bessereau et al., 1996; Kruge et al., 1996; Koester et al., 1998a, b; Karnkowski, 1999; Matyasik et al., 2001). Lower Cretaceous Spas shales in the Skole unit and other Lower Cretaceous series in the remaining tectonic units are also considered to be a good source rock (e.g., Kozikowski, 1978; Koltun et al., 1995, 1998; Bessereau et al., 1996).

The evaluation of source rocks from 52 well profiles from the Flysch Carpathians between Krakow and Przemysl was based on the results obtained from 1730 Rock-Eval analyses, as well as a few hundred bitumen extracts, fractions, stable carbon isotope, and biomarker analyses (Koester et al., 1998a; Kotarba, 1998b, 2000; Kotarba et al., 1999; M. Kotarba, 2004, unpublished data) (Figure 1). Elemental (C, H, N, and O) composition of kerogen, organic sulfur content in kerogen, and petrologic analyses (vitrinite reflectance and maceral composition) for the Menilite shales were also tested. In the case of the Menilite shales, analyses were also carried out on the samples (after removing the weathered material) taken from 29 outcrops in the Skole, Silesian, Dukla, and Grybow units (Figure 1).

Results of geochemical analyses of organic matter in individual beds of the Skole, Silesian, Subsilesian, Dukla, and Grybow units are summarized in Tables 1–3. The tables contain basic values of Rock-Eval parameters and indices (total organic carbon [TOC], T_{max}, S_2, production index, hydrogen index, and oxygen index). Additionally, the type of kerogen, maturity, and hydrocarbon potential were also defined on the basis of results of more sophisticated geochemical and petrological methods.

Recent geochemical studies reveal that organic sulfur content in kerogen functions as a catalyst by changing its kinetic parameters and lowering oil-generating threshold, which can begin maturation of organic matter at about 0.35% vitrinite reflectance (Rock-Eval T_{max} about 428–430°C) (e.g., Bazhenova and Arefiev, 1990; Hunt et al., 1991).

The thickness of the Oligocene Menilite shales ranges from 140 m (460 ft) in the Fore-Dukla (also called Pre-Dukla) zone of the Silesian unit to 550 m (1800 ft) in the Skole unit (Kusmierek, 1990). The TOC content ranges from 0.01 to 17.21% (mean 4.64%) in the Skole unit, from 0.18 to 17.25% (mean 4.48%) in Silesian unit, from 0.36 to 5.65% (mean 3.09%) in the Dukla unit, and from 1.76 to 9.03% (mean 4.52%) in the Grybow unit (Tables 1–3) and locally may reach 22% (Karpinski, 1978; ten Haven et al., 1993; Koester et al., 1998a). Total organic carbon content was higher in the upper part of the Menilite shales of the Skole unit than in the lower part. In contrast, in the Silesian and Dukla units, the TOC was higher in the lower part of the profile (Koester et al., 1998a; Wieclaw, 2002).

Algal, oil-prone type II kerogen, as identified by Rock-Eval, dominates (Figure 4), and locally, mixed

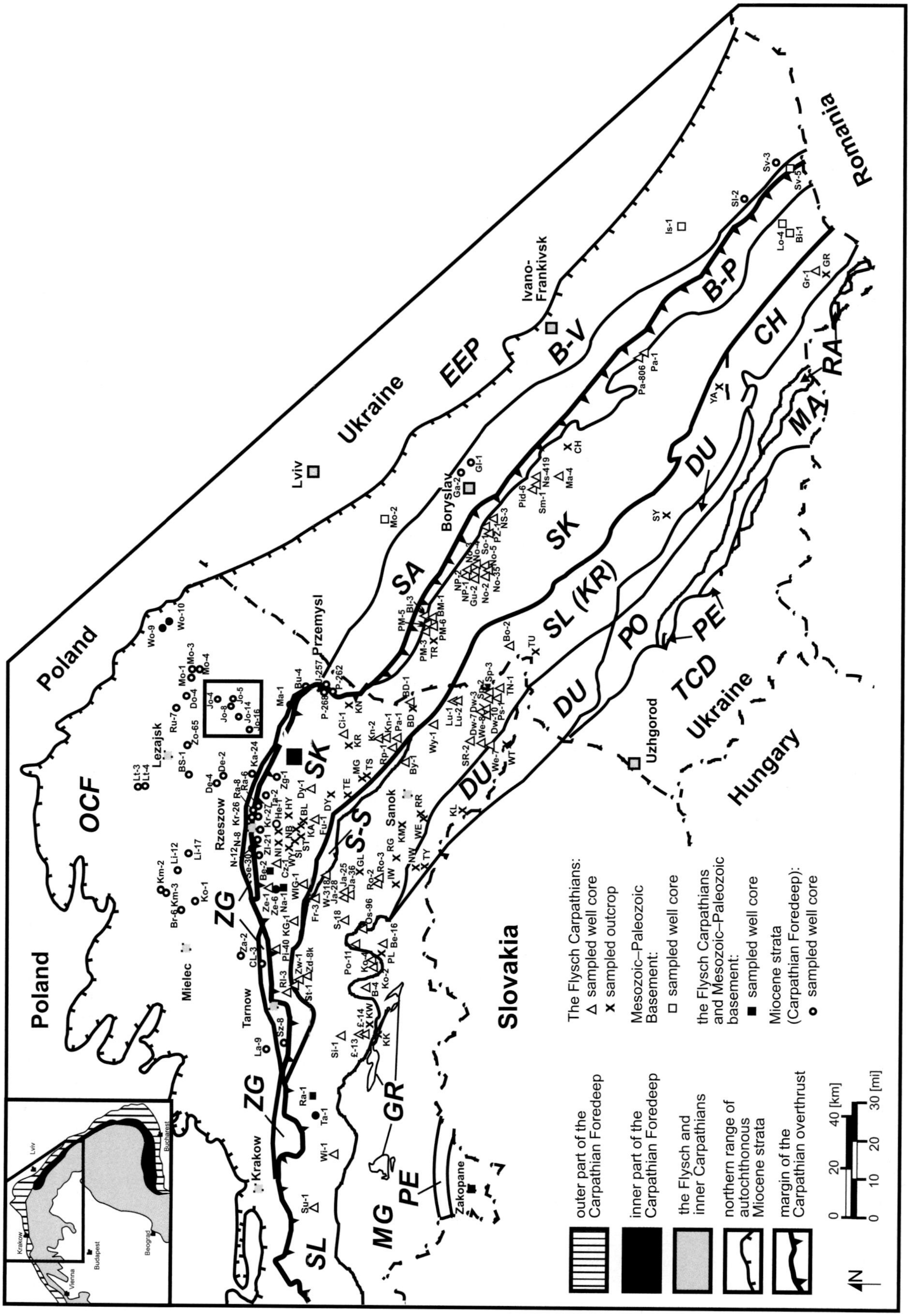
The Flysch Carpathians:
△ sampled well core
x sampled outcrop
Mesozoic–Paleozoic Basement:
□ sampled well core
the Flysch Carpathians and Mesozoic–Paleozoic basement:
■ sampled well core
Miocene strata (Carpathian Foredeep):
○ sampled well core
outer part of the Carpathian Foredeep
inner part of the Carpathian Foredeep
the Flysch and inner Carpathians
northern range of autochthonous Miocene strata
margin of the Carpathian overthrust
0 20 40 [km]
0 10 20 30 [mi]
N
Poland
Ukraine
Slovakia
Hungary
Romania
Krakow
Tarnow
Mielec
Rzeszow
Lezajsk
Przemysl
Sanok
Lviv
Borysla
Ivano-Frankivsk
Uzhgorod
Zakopane
OCF
EEP
ZG
SK
S-S
SA
B-V
B-P
CH
DU
SL (KR)
PO
PE
TCD
MA
RA
SL
MG
GR

types II–III and III–II kerogens occur in the Menilite shales of the Skole, Silesian, Dukla, and Grybow units. Types I and III kerogens are sparse (Figure 4). In the upper part of the Menilite shales of the Skole unit, oil-prone type II kerogen dominates, whereas in the lower part, mixed types II–III and III–II kerogens are abundant. In the Silesian unit, type II kerogen dominates throughout the Menilite shales (Wieclaw, 2002).

Black shales of the Oligocene Menilite shales display a large variability in their bulk and molecular geochemical parameters (Koester et al., 1998b). Biomarker and stable carbon isotope analyses (Koester et al., 1998b) indicate a varying contribution of different algae (particularly dinoflagellates and diatoms) and cyanobacteria. The presence of a highly branched isoprenoid (HBI) C_{25} alkane, C_{35} homohopanes, phytane, and n-alkanes of marine origin (Koester et al., 1998b) indicates intense bacterial sulfate reduction. Oxygen-depleted and even euxinic conditions in the Menilite basin existed (Bessereau et al., 1996; Koester et al., 1998b; Leszczynski, 2001). The Menilite shales contain abundant diatoms (Kotlarczyk, 1961), which could be the source of the C_{25} HBIs (ten Haven et al., 1993) and C_{26} HBI thiophenes (Rospondek et al., 1997). Koester et al. (1998b) also found that hopanoids, depleted in ^{13}C down to $\delta^{13}C$ −57‰, are most probably derived from methanotropic bacteria. This implies that a temporarily enhanced full methane cycle in a marine environment affected the isotopic composition of organisms dwelling in the upper photic zone. The occurrence in the Menilite shales of specific biological markers derived from gymnosperms and angiosperms (ten Haven et al., 1993) indicates the presence of some terrigenous organic matter.

The organic sulfur content, which influenced the kinetics of hydrocarbon generation processes and kerogen classification (e.g., Hunt et al., 1991), in the Menilite shales ranges from 0.5 to 10.1 wt.% (Wieclaw, 2002). Using organic sulfur analyses, it was possible to identify the following types of kerogen: type IID and, locally, type IIB and IIA in the Skole unit, with only type IID in the Silesian, Dukla, and Grybow units (Wieclaw, 2002).

The low maturity in the outer Skole unit is indicated by T_{max} values of 395–444°C (mean 419°C) (Table 1) but increases toward the inner nappes, T_{max} 396–460°C

Figure 1. Sketch map showing the major tectonic units of the Polish and Ukrainian Carpathians and their foreland and location of source rock samples. Tectonic units: EEP = East European platform; OCF (B-V) = outer part of the Carpathian Foredeep (Bilche–Volytsia unit in Ukraine); ZG = Zglobice unit; SA = Sambir (Stebnik) unit; B-P = Boryslav–Pokuttya unit; SK = Skole (Skiba) unit; S-S = Subsilesian unit; SL (KR) = Silesian (Krosno) unit; CH = Chornogora unit; DU = Dukla unit; GR = Grybow unit; MG = Magura unit; PO = Porkulets unit; RA = Rakhiv unit; MA = Marmarosh unit; PE = Pieniny Klippen Belt; TDC = Transcarpathian depression. The Flysch Carpathians sampled wells: (1) Polish part: B-4 = Bystra; BD-1 = Brzegi Dolne; Be-16 = Bednarka; By-1 = Bykowce IG-1; Ci-1 = Cisowa IG-1; Cz-1 = Czudec; Dw-3, Dw-7, and Dw-10 = Dwernik; Dy-1 = Dynow; Fr-3 = Frysztak; Fu-1 = Futoma; He-1 = Hermanowa; Ja-25, Ja-28, and Ja-36 = Jaszczew; KG-1 = Kamienica Gorna; Kn-1 and Kn-2 = Kuzmina; Ko-2 and Ko-4 = Kobylanka; Lu-1 and Lu-2 = Lutowiska; L-13 and L-14 = Lyczana; Os-96 = Osobnica; Pa-1 = Paszowa; Pi-40 = Pilzno; Po-11 = Podlas; Ps-1 = Pszczeliny-1; Rl-3 = Radlna; Ro-2 and Ro-3 = Rowne; Rp-1 = Rozpucie; S-18 = Sobniow; Si-1 = Siekierczyna IG-1; SR-2 = Suche Rzeki; St-1 = Stawiska; Sp-2 and Sp-3 = Stuposiany; Su-1 = Sulkowice; TN-1 = Tarnawa Nizna; W-318 = Weglowka; We-7 and We-8 = Wetlina; Wi-1 = Wisniowa-1; WIG-1 = Wisniowa IG-1; Wy-1 = Wydrne; Ze-1 = Zagorzyce; Zw-1 = Zalasowa-1; Zd-8K = Zawada. (2) Ukrainian part: Bl-3 = Blazhiv; BM-1 = Blazhiv–Monastyrets; Bo-2 = Borynia; Gr-1 = Gryniava; Gu-1 = Guta; Ma-4 = Maxymivka; Ns-419 = Novoselytska; No-2, No-3, No-4, No-5, and No-35 = Novoskhidnytsia; NP-1 and NP-2 = Nyzhni Popeli; NS-3 = Nyzhnia Stynava; Pa-1 and Pa-806 = Pasichna; Pid-6 = Pidborivska; PM-3, PM-5, and PM-6 = Pivdenny Monastyrets; PZ-1 = Pivnichna Zavoda; Sm-1 = Smolianska; So-1 = Sokolovets. The Flysch Carpathians sampled outcrops: (1) Polish part: BD = Brzegi Dolne; BL = Blazowa–Leg; NI = Niechobrz; DY = Dydnia; GL = Glowienka; HY = Hyzne; IW = Iwla; KA = Kakolowka; KK = Karczyska; KL = Komancza–Letnisko; KM = Klimkowka; KN = Kniazyce; KR = Krepak; KW = Kawiory; LZ = Leszczawka; MG = Mrzyglod; NB = Nad Budami; NW = Nowa Wies; PL = Pod Lasem; RG = Rogowska Gorka; RR = Rudawka Rymanowska; SI = Siedliska; ST = Straszydle; TE = Temeszow; TS = Tyrawa Solna; TY = Tylawa; WE = Wernejowka; WT = Wetlina; WY = Wyzne. (2) Ukrainian part: CH = Chechva; GR = Gryniava; SY = Synievir; TR = Tershiv; TU = Turka; YA = Yasynia. Mesozoic–Paleozoic basement sampled wells: (1) Polish part: N-12 = Nosowka; Ta-1 = Tarnawa; Wo-9 and Wo-10 = Wola Obrzycka; Ze-6 = Zagorzyce-6. (2) Ukrainian part: Bi-1 = Biskiv; Is-1 = Ispas; Lo-4 = Lopushna; Mo-2 = Mostivska; Sv-5 = Sviatoslavska. The Flysch Carpathians and Mesozoic–Paleozoic basement sampled wells: Polish part: Be-2 = Bedzienica; Na-1 = Nawsie; Ra-1 = Rajbrot. Autochthonous Miocene strata sampled wells: (1) Polish part: Br-6 = Brazylia; BS-1 = Brzoza Stadnicka; Bu-4 = Buszkowiczki; CL-3 = Czarny Las; De-2 = Debina; De-4 = Debina; Do-4 = Dobra; J-257 = Jaksmanice; Jo-4, Jo-5, Jo-8, Jo-14, and Jo-16 = Jodlowka; Ka-24 = Kanczuga; Km-2 and Km-3 = Komorow; Ko-1 = Kosowy; Kr-26 and Kr-27 = Krasne; La-9 = Lazy; Li-12 and Li-17 = Lipnica; Lt-3 and Lt-4 = Letownia; Ma-1 = Mackowice; Mo-1, Mo-3, and Mo-4 = Molodycz; N-8 = Nosowka; P-262 and P-268 = Przemysl; Ra-6 and Ra-8 = Raczyna; Ru-7 = Rudka; Se-30 = Sedziszow; Sz-8 = Szczepanow; Ta-2 = Tarnawka; Za-2 = Zasow; Zg-1 = Zagorze; Zl-21 = Zalesie; Zo-65 = Zolynia. (2) Ukrainian part: Ga-2 = Gaiska; Gl-1 = Glynkivska; Sl-2 = Slavetska; Sv-3 = Sviatoslavska.

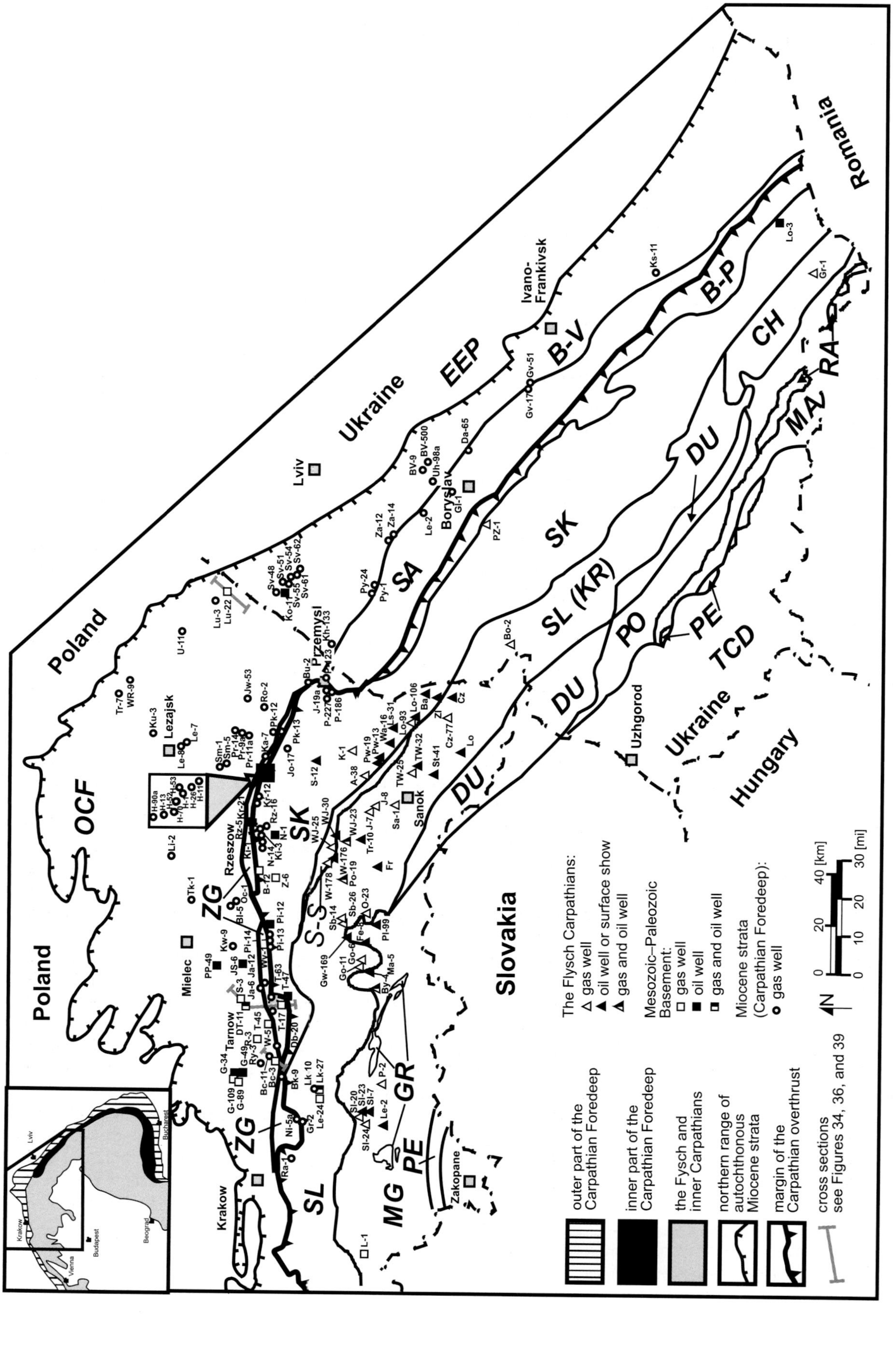
Poland
Ukraine
Slovakia
Hungary
Romania
Krakow
Tarnow
Mielec
Rzeszow
Lezajsk
Przemysl
Sanok
Zakopane
Uzhgorod
Lviv
Boryslav
Ivano-Frankivsk
OCF
EEP
ZG
SK
SL
S-S
SA
B-V
B-P
CH
DU
SL (KR)
PO
PE
TCD
MA
RA
MG
GR
outer part of the Carpathian Foredeep
inner part of the Carpathian Foredeep
the Fysch and inner Carpathians
northern range of autochthonous Miocene strata
margin of the Carpathian overthrust
cross sections see Figures 34, 36, and 39
The Flysch Carpathians:
gas well
oil well or surface show
gas and oil well
Mesozoic–Paleozoic Basement:
gas well
oil well
gas and oil well
Miocene strata (Carpathian Foredeep):
gas well
N
0 20 40 [km]
0 10 20 30 [mi]

(mean 431°C), in Silesian unit (Table 2) and 428–483°C (mean 442°C) and 446–452°C (mean 449°C) in Dukla and Grybow units (Table 3), respectively.

Pyrolysis parameters as a function of depth for the Menilite shales from the Skole and Silesian units are shown in Figures 5 and 6, respectively, and in Figure 7 for Dukla and Grybow units. In the Skole unit (Figure 5), organic matter is immature to about 5000-m (16,400-ft) depth, and in the Silesian unit, organic matter is immature to a depth of about 1200 m (3900 ft) (Figure 6). In the Dukla unit, immature organic matter could be found only in a few samples from outcrops (Figure 7). Organic matter in core samples is generally mature (T_{max} from 435 to 460°C) but is overmature (T_{max} > 460°C) at a depth below 1100 m (3600 ft). Rock samples from outcrops in the Grybow unit are mature. The highest value of hydrogen index (761 mg HC/g TOC) was found in the Silesian unit (Table 2; Figure 6). With the increasing depth, the hydrogen index values drop in all tectonic units (Figures 5–7). The variability of hydrogen index, S_1/TOC and production index with depth, illustrates the variability of hydrocarbon potential. Organic geochemistry studies reveal that Oligocene Menilite beds have a variable (good and/or moderate) hydrocarbon potential in the whole flysch sequence of the Polish Carpathians (both as a function of depth and lateral extent).

Oligocene–lower Miocene Krosno beds with the so-called Transitional beds form a sandstone-shale complex, which is as much as 3000 m (9800 ft) thick in the Skole unit. Total organic carbon content ranges from 0.00 to 7.24% (mean 1.00%) in the Skole unit (Table 1) and 0.05 to 3.65% (mean 0.80%) in the Silesian unit (Table 2). Mainly gas-prone (mixed type III–II kerogen) and sometimes oil-prone (mixed type II–III) organic matter from the Krosno beds is both immature and mature (T_{max} ranges from 397 to 468°C). Deposition of organic matter under oxidized conditions resulted in lean hydrocarbon potential for these strata. Only the bottom part of the profile, mainly in the transition beds, was deposited under reducing conditions, which resulted in moderate source rock potential.

Albian–Aptian Spas shales, as much as 500 m (1600 ft) thick, are predominantly black shales, which, in their lower part, contain sandstone intercalations. Spas beds exclusively occur in the Skole unit. Total organic carbon content ranges from 0.04 to 11.9% (mean 1.20%). Terrestrial and mixed terrestrial-algal (type III and sometimes type III–II) organic matter of Spas beds is mature because of both low- and high-temperature thermogenic processes (T_{max} values range from 420 to 458°C) (Table 1; Figure 8). Results of isotopic and biomarkers analyses confirm those genetic characteristics (Kotarba, 2000). Organic matter of the Spas shales has moderate hydrocarbon potential.

Verovice beds in the Silesian unit contain TOC values that range from 0.09 to 2.49 wt.% (mean 0.61%). Gas-prone (terrestrial) type III kerogen is immature to

Figure 2. Sketch map showing major tectonic units of the Polish and Ukrainian Carpathians and their foreland and location of oil and gas samples. Tectonic units: EEP = East European platform; OCF (B-V) = outer part of the Carpathian Foredeep (Bilche–Volytsia unit in Ukraine); ZG = Zglobice unit; SA = Sambir (Stebnik) unit; B-P = Boryslav–Pokuttya unit; SK = Skole (Skiba) unit; S-S = Subsilesian unit; SL (KR) = Silesian (Krosno) unit; CH = Chornogora unit; DU = Dukla unit; GR = Grybow unit; MG = Magura; PO = Porkulets unit; RA = Rakhiv unit; MA = Marmarosh unit; PE = Pieniny Klippen Belt; TDC = Transcarpathian depression. The Flysch Carpathians sampled wells: (1) Polish part: A-38 = Artur; By-4 = Bystra; Cz-77 = Czarne; Fe-8 = Felnerowka; Go-6 and Go-11 = Gorlice; Fr = Franek; Gw-169 = Gwarectwo; J-7 and J-8 = Jurowce; K-1 = Kuzmina; Ls-31 = Leszczowate; Le-2 = Lesniowka; L-106 = Lodyna; Ma-5 = Magdalena; Lo-93 = Lopiennik; O-23 = Osobnica; Pw-13 and Pw-19 = Paszowa; Pl-99 = Pielgrzymka; Po-19 = Potok; P-2 = Przyszowa; Sa-1 = Sanok; S-12 = Slonne; Sl-7, Sl-20, Sl-23, and Sl-24 = Slopnice; Sb-14 and Sb-26 = Sobniow; St-41 = Stegan; TW-25 and TW-32 = Tarnawa Wielopole; Tr-10 = Trzesniow; Wa-16 = Wankowa; W-178 = Weglowka; WJ-23, WJ-25, and WJ-30 = Wola Jasienicka. (2) Ukrainian part: Bo-2 = Borynia; Gr-1 = Gryniava; PZ-1 = Pivnichna Zavoda. The Flysch Carpathians surface oil shows: Polish part: Ba = Bandrow; Cz = Czarna; Lo = Lopienka; Zl = Zlobek. Mesozoic–Paleozoic basement sampled wells: (1) Polish part: B-12 = Brzozowka; DT-11 = Dabrowa Tarnowska; G-34, G-49, G-89, and G-109 = Grobla; JS-6 = Jastrzabka Stara; L-1 = Lachowice; Le-24 = Leszczyna; Lk-27 = Lakta; N-1 = Nosowka; PP-49 = Partynia–Podborze; Pi-12 = Pilzno; Ry-3 = Rylowa; S-3 = Smegorzow; T-17 and T-47 = Tarnow; W-5 = Wierzchoslawice; Z-6 = Zagorzyce. (2) Ukrainian part: Ko-11 = Kohanivka; Lo-3 = Lopushna. Autochthonous Miocene strata sampled wells: (1) Polish part: Bl-5 = Blizna; Bk-9 = Borek; Bc-11 = Brzezowiec; Bu-2 = Buszkowiczki; Db-20 = Dabrowka; Gr-2 = Grabina; Hu-1, Hu-11, Hu-13, Hu-26, Hu-52, Hu-53, Hu-70, and Hu-90a = Husow; J-19a = Jaksmanice; Jw-53 = Jaroslaw; Ja-6 and Ja-12 = Jasniny; Jo-17 = Jodlowka; Ka-7 = Kanczuga; Ki-1 and Ki-3 = Kielanowka; Kw-9 = Korzeniow; Kr-12 and Kr-21 = Krasne; Ku-3 = Kurylowka; Le-7 and Le-8 = Lezajsk; Li-2 = Lipnica; Lk-10 = Lakta; Lu-2 and Lu-22 = Lubaczow; Ni-5 = Nieznanowice; N-14 = Nosowka; Oc-1 = Ocieka; Pi-13 and Pi-14 = Pilzno; Pk-12 and Pk-13 = Pruchnik; P-123, P-186, and P-227 = Przemysl; Pr-9a, Pr-11a, and Pr-14 = Przeworsk; Ra-1 = Raciborsko; Ry-3 = Rysie; Rz-5 and Rz-16 = Rzeszow; Ro-2 = Rozwienica; Sm-1 and Sm-5 = Smolarzyny; T-45 and T-63 = Tarnow; Tr-7 = Tarnogrod; Tk-1 = Trzesnik; U-11 = Uszkowce; WR-9 = Wola Rozaniecka; Wy-1 = Wygoda-1. (2) Ukrainian part: BV-9 and BV-500 = Bilche Volytsia; Da-65 = Dashava; Gl-1 = Glynkivska; Gv-17 and Gv-51 = Grynivka; Kh-133 = Khodnovychi; Ks-11 = Kosiv; Le-2 = Letnia; Py-1 and Py-24 = Pyniany; Sv-48, Sv-51, Sv-54, Sv-55, Sv-61, and Sv-62 = Svydnytsia; Uh-98a = Uhersko; Za-12 and Za-14 = Zaluzhany.

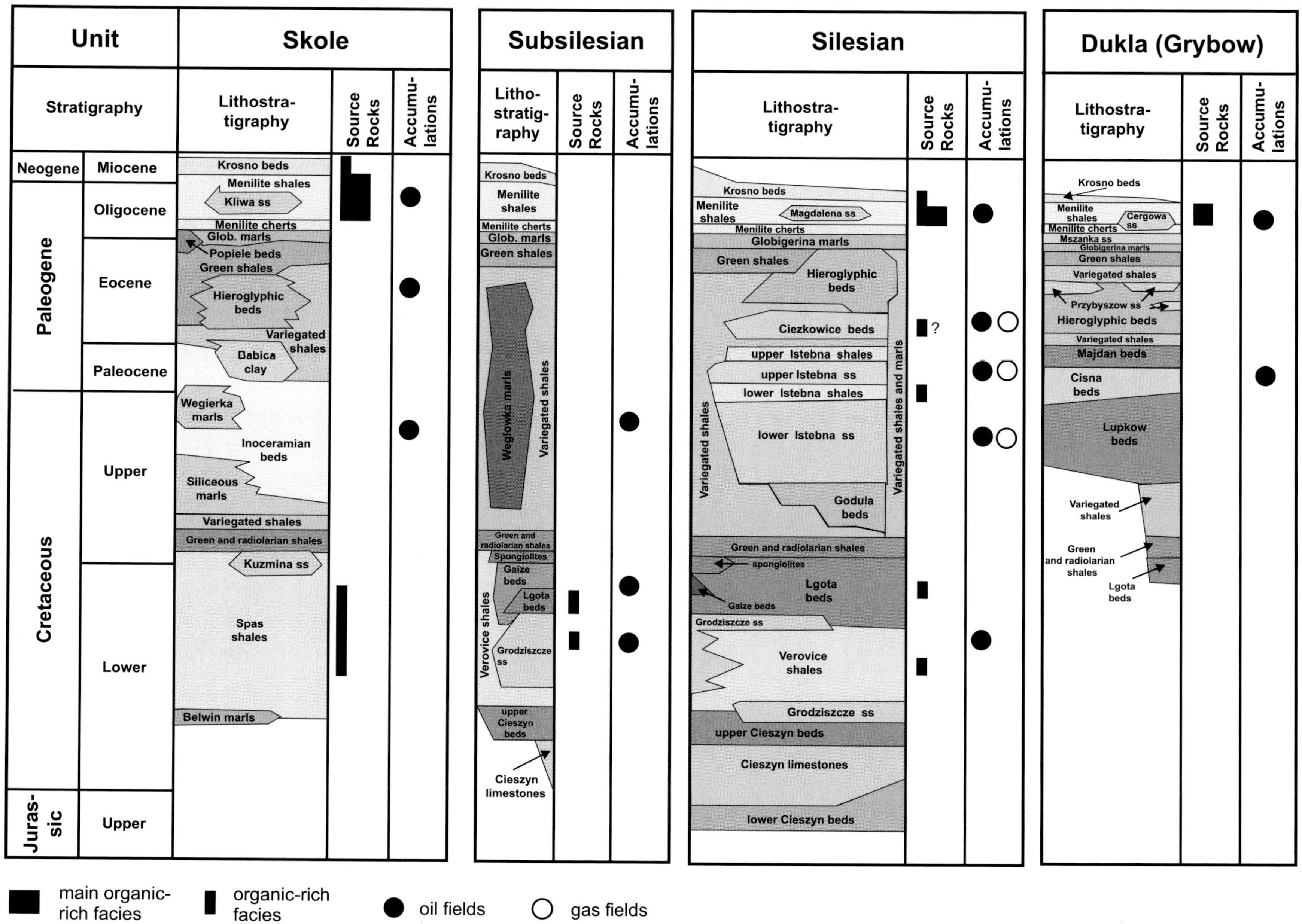

Figure 3. Generalized lithostratigraphic column showing the distribution of petroleum accumulations and the main organic-rich facies in the flysch sequence of the Polish Carpathians. Modified from Slaczka and Kaminski (1998). Glob. = *Globigerina*; ss = sandstones.

Table 1. Geochemical characteristics and hydrocarbon potential of individual formations in the Skole unit. Geochemical parameters and indices are given as numerators of minimum and maximum values, mean values in denominator, and parenthesized (numerator of number of samples from wells, italicized and bold—from outcrops, number of sampled wells in denominator and italicized and bold number of sampled outcrops). The identification of type II kerogen in square brackets A, B, C, or D takes into consideration the organic sulfur content of kerogen after Hunt et al. (1991).

Beds	Krosno	Menilite	Variegated Shales	Hieroglyphic	Inoceramian	Spas
Age	Oligocene-Lower Miocene	Oligocene	Eocene	Paleocene	Upper Cretaceous-Paleocene	Lower Cretaceous
Total organic carbon (TOC) (wt.%)	0.00 – 7.24 / 1.00 (142) / (7)	0.01 – 17.2 / 4.64 (193, ***133***) / (9, ***16***)	0.00 – 0.75 / 0.36 (20) / (3)	0.02 – 1.73 / 0.38 (30) / (6)	0.0 – 3.09 / 0.41 (334) / (12)	0.04 – 11.86 / 1.20 (118) / (5)
Rock Eval T_{max} (°C)	397 – 442 / 425 (130)	395 – 444 / 419 (316)	424 – 445 / 430 (12)	417 – 441 / 432 (23)	416 – 463 / 433 (240)	420 – 458 / 438 (86)
S_2 (mg HC/g TOC)	0.06 – 20.9 / 2.12 (132)	0.17 – 125.8 / 16.0 (316)	0.06 – 0.67 / 0.33 (11)	0.01 – 2.53 / 0.47 (23)	0.01 – 8.44 / 0.30 (249)	0.01 – 22.8 / 1.45 (109)
Production index (PI)	0.00 – 0.50 / 0.15 (132)	0.01 – 0.44 / 0.06 (316)	0.03 – 0.20 / 0.09 (11)	0.0 – 0.73 / 0.17 (22)	0.0 – 0.80 / 0.17 (243)	0.00 – 1.0 / 0.25 (109)
Hydrogen index (HI) (mg HC/g TOC)	28 – 374 / 136 (132)	80 – 769 / 313 (316)	30 – 141 / 77 (11)	8 – 186 / 85 (23)	0 – 273 / 57 (249)	10 – 310 / 97 (109)
Oxygen index (OI) (mg CO_2/g TOC)	0 – 273 / 41 (131)	0.0 – 223 / 39 (315)	3 – 89 / 35 (11)	0 – 200 / 65 (23)	0 – 200 / 43 (249)	0 – 260 / 36 (104)
Type of kerogen	III – II and II – III gas- and/or oil-prone	II and/or II – III [IID, IIB and IIA]	III gas-prone	III gas-prone	III (III – II) gas-prone (oil-prone sporadically)	III (III – II) gas-prone (oil-prone sporadically)
Maturity	immature and low mature	immature and low mature	immature and low mature	immature and low mature	immature and low mature	low and high mature
Hydrocarbon potential	lean or moderate	good or moderate	lean	lean	lean or moderate	moderate

mature (T_{max} from 421 to 459°C) and has moderate to lean hydrocarbon potential.

The Albian Lgota beds have questionable source rocks in the Silesian and Subsilesian units (Bessereau et al., 1996). Gas-prone (terrestrial) type III kerogen ranges from 0.57 to 2.05 wt.% TOC (mean 1.13%) and from 0.51 to 3.65 wt.% (mean 1.66%) in the Silesian and Subsilesian units, respectively. Kerogen in Lgota beds is immature to mature. T_{max} values range from 425 to 469°C and from 417 to 434°C in the Silesian and Subsilesian units, respectively.

The Lower Cretaceous Cieszyn beds are also questionable source rocks in the Subsilesian unit (Table 3). Type III or mixed type III–II organic matter ranges from 0.79 to 1.63 wt.% of TOC (mean 1.12%). Kerogen in the Cieszyn beds has a low thermal maturity (T_{max} values range from 436 to 438°C) (Table 3). They are insignificant petroleum source rocks because of the small thicknesses of the series and considerable tectonic involvement.

Paleocene–Upper Cretaceous Inoceramian beds are lean to moderate source rocks in the Skole unit (Table 1). Considerable variability in the organic carbon content (0.0–3.09 wt.%; mean 0.41%) is observed. Gas-prone (terrestrial kerogen type III) and, infrequently, oil-prone mixed (type III–II) organic matter of Inoceramian beds is immature or mature (T_{max} range from 416 to 463°C).

The Paleocene Istebna beds, Eocene Variegated shales, and Eocene Ciezkowice beds in the Silesian unit have lean and sometimes moderate source rock potential (Table 2). Gas-prone (terrestrial) type III kerogen or mixed type III–II kerogen ranges from 0.03 to 7.07 wt.% (mean 1.22%), from 0.04 to 4.03 wt.% (mean 1.11%), and

Table 2. Geochemical characteristics and hydrocarbon potential of individual formations in the Silesian unit. Geochemical parameters and indices are given as numerators of minimum and maximum values, mean values in denominator, and parenthesized (numerator of number of samples from wells, italicized and bold—from outcrops, number of sampled wells in denominator and italicized and bold number of sampled outcrops). The identification of type II kerogen in square brackets A, B, C, or D takes into consideration the organic sulfur content of kerogen after Hunt et al. (1991).

Beds	Krosno	Menilite	Ciezkowice	Variegated Shales	Istebna	Lgota	Verovice shales
Age	Oligocene-Lower Miocene	Oligocene	Eocene	Eocene	Paleocene	Lower Cretaceous	Lower Cretaceous
Total organic carbon (TOC) (wt.%)	0.05 – 3.65 / 0.80 (132/12)	0.18 – 17.2 / 4.48 (10, ***125*** / 2, ***11***)	0.06 – 10.18 / 1.32 (47/5)	0.04 – 4.03 / 1.11 (32/5)	0.03 – 7.07 / 1.22 (246/12)	0.57 – 2.05 / 1.13 (13/2)	0.0 – 2.49 / 0.61 (33/6)
Rock Eval T_{max} (°C)	421 – 468 / 436 (125)	396 – 460 / 431 (134)	409 – 438 / 428 (39)	414 – 430 / 426 (26)	417 – 465 / 433 (218)	425 – 469 / 458 (12)	421 – 459 / 432 (15)
S_2 (mg HC/g TOC)	0.03 – 9.98 / 1.36 (129)	0.05 – 103.7 / 22.7 (134)	0.03 – 21.58 / 2.70 (39)	0.12 – 6.93 / 1.89 (26)	0.02 – 15.36 / 1.67 (228)	0.24 – 4.15 / 1.07 (13)	0.0 – 0.27 / 0.10 (21)
Production index (PI)	0.01 – 0.45 / 0.17 (129)	0.01 – 0.45 / 0.09 (134)	0.03 – 0.72 / 0.18 (39)	0.03 – 0.43 / 0.09 (26)	0.0 – 0.92 / 0.15 (228)	0.02 – 0.39 / 0.25 (13)	0.11 – 4.26 / 0.90 (21)
Hydrogen index (HI) (mg HC/g TOC)	25 – 628 / 146 (129)	25 – 761 / 452 (134)	25 – 271 / 124 (39)	52 – 226 / 114 (26)	6 – 480 / 94 (227)	39 – 251 / 80 (13)	35 – 179 / 84 (21)
Oxygen index (OI) (mg CO_2/g TOC)	6 – 217 / 41 (128)	1 – 124 / 20 (133)	6 – 233 / 43 (38)	12 – 162 / 30 (26)	0 – 258 / 35 (226)	0 – 22 / 5 (13)	2 – 174 / 29 (21)
Type of kerogen	III – II and II – III gas- and/or oil-prone	II and/or II – III [II D]	III (III – II) gas-prone (oil-prone sporadically)	III (III – II) gas-prone (oil-prone sporadically)	III (III – II) gas-prone (oil-prone sporadically)	III gas-prone	III gas-prone
Maturity	immature and mature	immature and mature	immature and low mature	immature and low mature	immature and mature	immature and mature	immature and mature
Hydrocarbon potential	lean or moderate	good or moderate	lean or moderate	lean or moderate	moderate or lean	moderate or lean	lean or moderate

from 0.06 to 10.18 wt.% (mean 1.32%) in the Istebna beds, variegated shales, and Ciezkowice beds, respectively. Kerogens are immature to mature. T_{max} values range from 417 to 465°C in the Istebna beds, from 414 to 430°C in variegated shales, and from 409 to 438°C in Ciezkowice beds. The best hydrocarbon potential seems to be in the Paleocene Istebna beds. Ciezkowice beds are predominantly sandstones with shale intercalations. Variegated shales commonly exhibit evidence of oxidation.

The Ukrainian Area

Two main organic-rich episodes are recorded in the Cretaceous to Neogene sequences of the flysch belt of Ukraine (Figure 9): (1) the Lower Cretaceous series at the base and (2) the Oligocene–lower Miocene series at the top.

Both the Lower Cretaceous and Oligocene–lower Miocene organic-rich strata were deposited over the large part of the basin, and the area of their occurrence in the present-day sequence of the Ukrainian part of the flysch belt extends from the external to the internal tectonic units (Figure 1). Overthrusting of the flysch units during the late Miocene resulted in stacking of the various nappes and subsequent vertical repetition of stratigraphic units. Because of this stacking, organic-rich strata occur at different depths in most of the flysch belt, including the main oil-producing areas, i.e., the Boryslav–Pokuttya unit and the outer part of the Skiba (Skole) unit. Hence, they constitute the main potential source rock intervals in the region.

The Oligocene–early Miocene Menilite shales have a maximum thickness of 1500 m (4900 ft) (Andreeva-Grigirovich et al., 1986; Vialov et al., 1988). Within the outer tectonic units, it consists of three parts, namely, the lower, middle, and upper Menilite shales (Figure 9). The lower Menilite shales, the principal source rock interval in the Ukrainian Carpathians, have a basin-wide extent. The middle Menilite shales are organic

Table 3. Geochemical characteristics and hydrocarbon potential of individual formations in the Subsilesian, Dukla, and Grybow units. Geochemical parameters and indices are given as numerators of minimum and maximum values, mean values in denominator, and parenthesized (numerator of number of samples from wells, italicized and bold—from outcrops, number of sampled wells in denominator and italicized and bold number of sampled outcrops). The identification of type II kerogen in square brackets A, B, C, or D takes into consideration the organic sulfur content of kerogen after Hunt et al. (1991).

Unit	Subsilesian				Dukla	Grybow
Beds	Weglowka Marls	Lgota	Grodziszcze	Cieszyn	Menilite	Menilite
Age	(Upper Cretaceous)	(Lower Cretaceous)	(Lower Cretaceous)	(Lower Cretaceous)	(Oligocene)	(Oligocene)
Total organic carbon (TOC) (wt.%)	0.0 – 0.53 / 0.16 (16)/(4)	0.51 – 3.65 / 1.66 (43)/(1)	0.19 – 1.04 / 0.71 (4)/(1)	0.79 – 1.63 / 1.12 (5)/(1)	0.36 – 5.65 / 3.09 (4, ***13***)/(4, ***4***)	1.76 – 9.03 / 4.52 (10)/(1)
Rock Eval T_{max} (°C)	423 – 436 / 429 (3)	417 – 434 / 429 (43)	437 – 438 / 437 (4)	436 – 438 / 437 (5)	428 – 483 / 442 (17)	446 – 452 / 449 (10)
S_2 (mg HC/g TOC)	0.05 – 0.22 / 0.37 (3)	0.33 – 5.37 / 1.57 (33)	0.16 – 1.76 / 1.19 (4)	1.42 – 2.22 / 1.81 (5)	0.18 – 38.5 / 12.7 (17)	2.94 – 16.6 / 8.78 (10)
Production index (PI)	0.05 – 0.96 / 0.37 (3)	0.02 – 0.13 / 0.04 (43)	0.06 / 0.06 (4)	0.06 – 0.09 / 0.07 (5)	0.02 – 0.39 / 0.10 (17)	0.04 – 0.17 / 0.12 (10)
Hydrogen index (HI) (mg HC/g TOC)	14 – 68 / 45 (3)	39 – 217 / 95 (43)	84 – 231 / 154 (4)	136 – 230 / 170 (5)	15 – 688 / 390 (17)	96 – 324 / 207 (10)
Oxygen index (OI) (mg CO_2/g TOC)	63 – 132 / 111 (3)	13 – 56 / 26 (43)	17 – 200 / 64 (4)	15 – 29 / 23 (5)	4 – 30 / 18 (17)	0 – 29 / 14 (10)
Type of kerogen	III gas-prone	III gas-prone	III (III – II) gas-prone	III (III – II) oil-prone	II and II – III [IID] oil-prone	II and II – III [IID] oil-prone
Maturity	immature and low mature	immature and low mature	low mature	low mature	immature and mature	immature and mature
Hydrocarbon potential	lean	lean or moderate	lean	lean	good or moderate	good or moderate

poor. The organic-rich upper Menilite shales occur in a localized area.

Total organic carbon content in the black shales of the Menilite shales in the outer tectonic units locally exceeds 20% but normally ranges between 4 and 8%. Total organic carbon values range from less than 1 to more than 20%. Rock-Eval pyrolysis has been used to study core samples from several wells (Figure 1) from the Boryslav–Pokuttya unit (including the main oil-producing areas) and rock samples from outcrops of the Skiba (Skole) unit (Koltun et al., 1998). The results show that type II kerogen (Figures 10, 11) is predominant, with a good to excellent petroleum potential (S_2 reaches locally 97.3 mg HC/g rock and commonly largely exceeds 10 mg HC/g rock).

Figure 12 shows the pyrolysis parameters as a function of depth for the Menilite shales from the Boryslav–Pokuttya unit and the outer part of the Skiba (Skole) unit. These rocks still are immature with respect to oil generation to a depth of 4 km (2.4 mi) and commonly preserve some petroleum potential at depths of about 6 km (3.6 mi). Figure 12 shows that both the values of pyrolysis parameters and the trend of their depth-related evolution do not show lateral variation.

Significant lateral changes of pyrolysis parameters are observed in Menilite shales toward the internal tectonic units of the flysch belt (Koltun et al., 1998). The thickness of these deposits decreases to a few tens of meters (Figure 9). However, TOC content remains rather high. In the Krosno (Silesian) unit (Synievir and

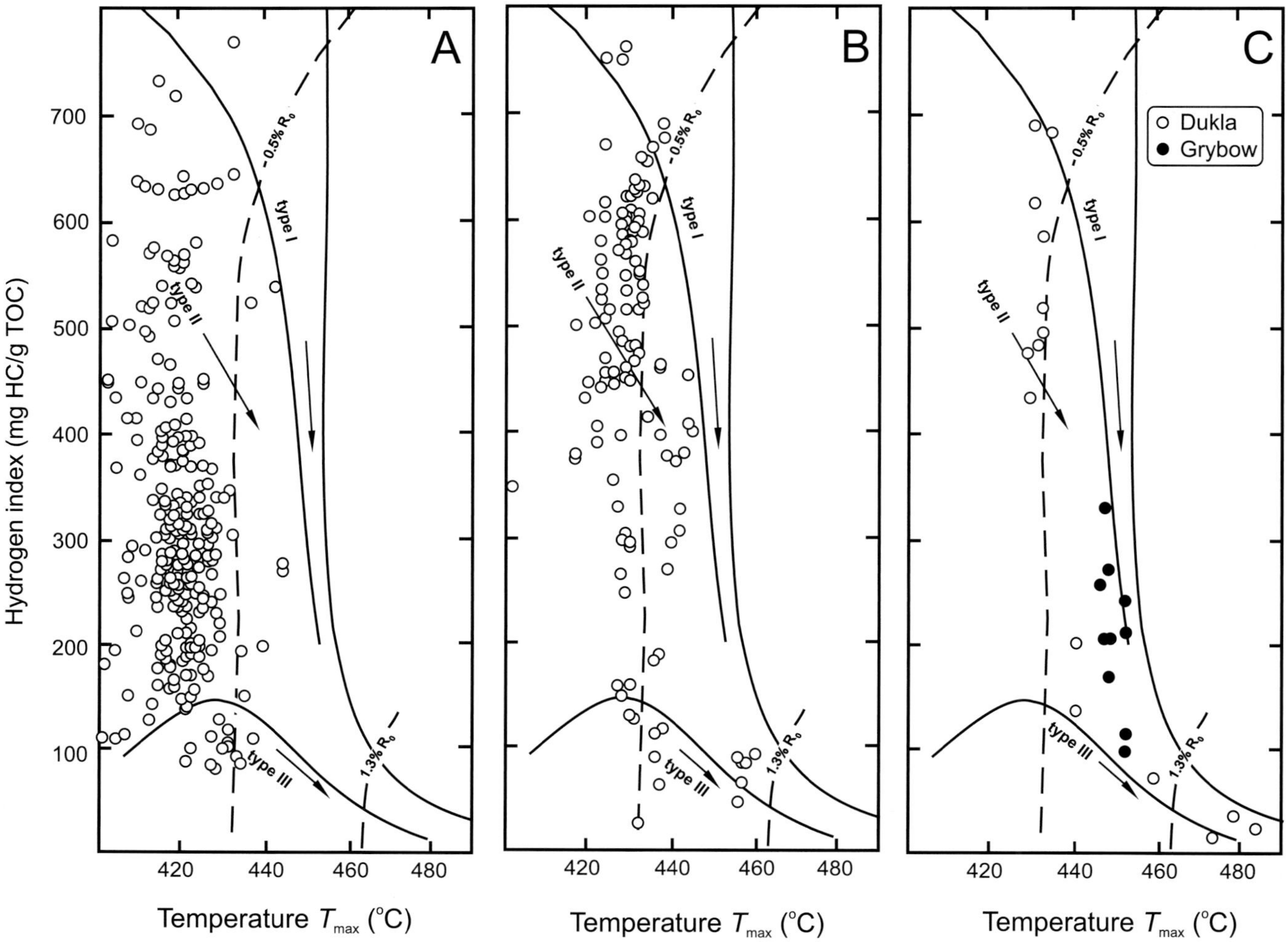

Figure 4. Plot of hydrogen index vs. Rock-Eval T_{max} temperature for the Oligocene Menilite shales in (A) Skole, (B) Silesian, and (C) Dukla and Grybow units (the Polish Flysch Carpathians).

Yasynia outcrops, Gryniava-1 well, Figure 1), TOC values range from 0.47 to 6.39%. They contain a composite type III–II kerogen (Figure 13). The immature rocks from the Yasynia outcrop show a fair to good petroleum potential (S_2 ranges from 2.33 to 12.25 mg HC/g rock). In other sites, the Menilite series are overmature, but it can be assumed that their initial (preoverthrusting) petroleum potential was similar to what is observed in the Yasynia outcrop.

Lower Cretaceous (Barremian to Albian) organic-rich facies are spread over the major part of the Ukrainian Carpathian flysch belt. They comprise two main black shale sequences: the Spas and Shypot shales. The Spas shales occur in the Skiba nappe (Vialov et al., 1988). They crop out at a few sites in the northwest part of the Skiba unit and have been penetrated in some wells in the central part of the same unit. The Shypot shales, occurring in the inner part of the flysch belt (Chornogora, Dukla, and Krosno units), crop out in many sites in this area and were encountered in the Gryniava-1 well.

In its lower part, the Spas shales are as much as 200 m (660 ft) thick, predominantly black shales with minor sandstone and siltstone intercalations. Total organic carbon content in the black shales is commonly greater than 2%, sometimes exceeding 7%. Black shales from the Maxymivka-4 well and coeval outcrops from the Dnister riverbed near Tershiv village (Figure 1), analyzed by Rock-Eval pyrolysis (Koltun et al., 1998), found hydrogen index values commonly less than 200 mg HC/g TOC, indicating type II and III kerogen (Figure 14), and S_2 values as high as 9.38 mg HC/g rock, indicating fair to good petroleum potential. The T_{max} values (from 429 to 431°C) for the outcrop samples indicate an immature level, whereas the Maxymivka-4 well samples from 4374- to 4429-m (14,350- to 14,530-ft) depth show T_{max} values from 445 to 451°C, a maturation level characteristic of the oil window.

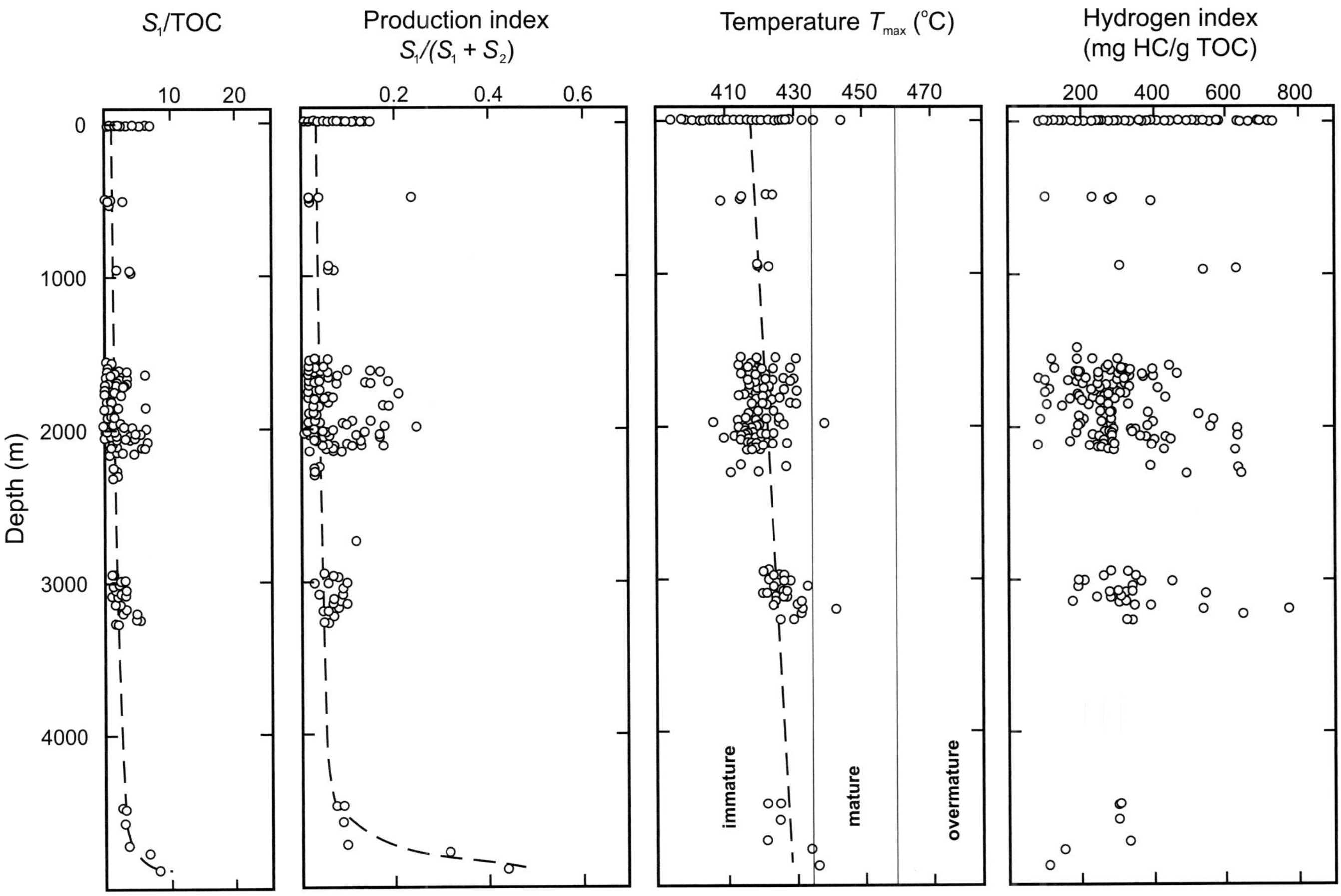

Figure 5. Plot of the Rock-Eval pyrolysis parameters (S_1 normalized to total organic carbon, production index, T_{max}, and hydrogen index) vs. depth for the Oligocene Menilite shales in the Skole unit (the Polish Flysch Carpathians).

The Shypot shales have facies similar to the Spas shales. The lower Shypot beds are made of a 300-m (1000-ft)-thick, predominantly black shale sequence with minor sandstone and siltstone intercalations. The upper Shypot beds are 200 m (6600 ft) thick, with mainly sandstone strata. Black shales of the Shypot shales contain as much as 8% of TOC but average from 2 to 4%. Shypot shales of the Chornogora unit were studied by Rock-Eval pyrolysis in four outcrops as well as in the Gryniava-1 well (Figure 1) (Koltun et al., 1998). Hydrogen index values ranging from 58 to 319 mg HC/g TOC indicate a mixed type II–III kerogen (Figure 14), with fair to good petroleum potential (S_2 ranges from 0.27 to 15.64 mg HC/g rock). T_{max} values of the outcrop samples range from 433 to 441°C, corresponding to the beginning of the oil window. In Gryniava-1 well, at depths from 3496 to 3661 m (11,470 to 12,011 ft), T_{max} reaches 454–458°C, characteristic of the central part of the oil window.

The Lower Cretaceous series of the Carpathian flysch should be considered as a potential source rock.

The Carpathian Foredeep

The Polish Area

The Carpathian Foredeep is divided into two basins: the outer and inner basins (Oszczypko, 1997). The folded Miocene strata of the Stebnik (Sambir) and Zglobice units known from the inner part of the Polish Carpathian Foredeep are thought to be unimportant for petroleum potential (Kotarba, 1999; Kotarba et al., 1998b). The eastern part of the outer basin (the Bilche–Volytsia unit in the Ukrainian part) between Krakow and Przemysl (Figure 1) is filled with Badenian and lower Sarmatian sediments (Figure 15) of the following thicknesses: lower and middle Badenian, from 0 to 300 m (0 to 1000 ft); upper Badenian, from 0 to 1700 m (0 to 5600 ft); and lower Sarmatian, from 0 to 2900 m (0 to 9500 ft) (Ney et al., 1974). The upper Badenian and lower Sarmatian strata developed as clay-sandy, mainly deltaic facies, whereas the lower and middle Badenian encompass shallow-water, psammitic, argillaceous, and

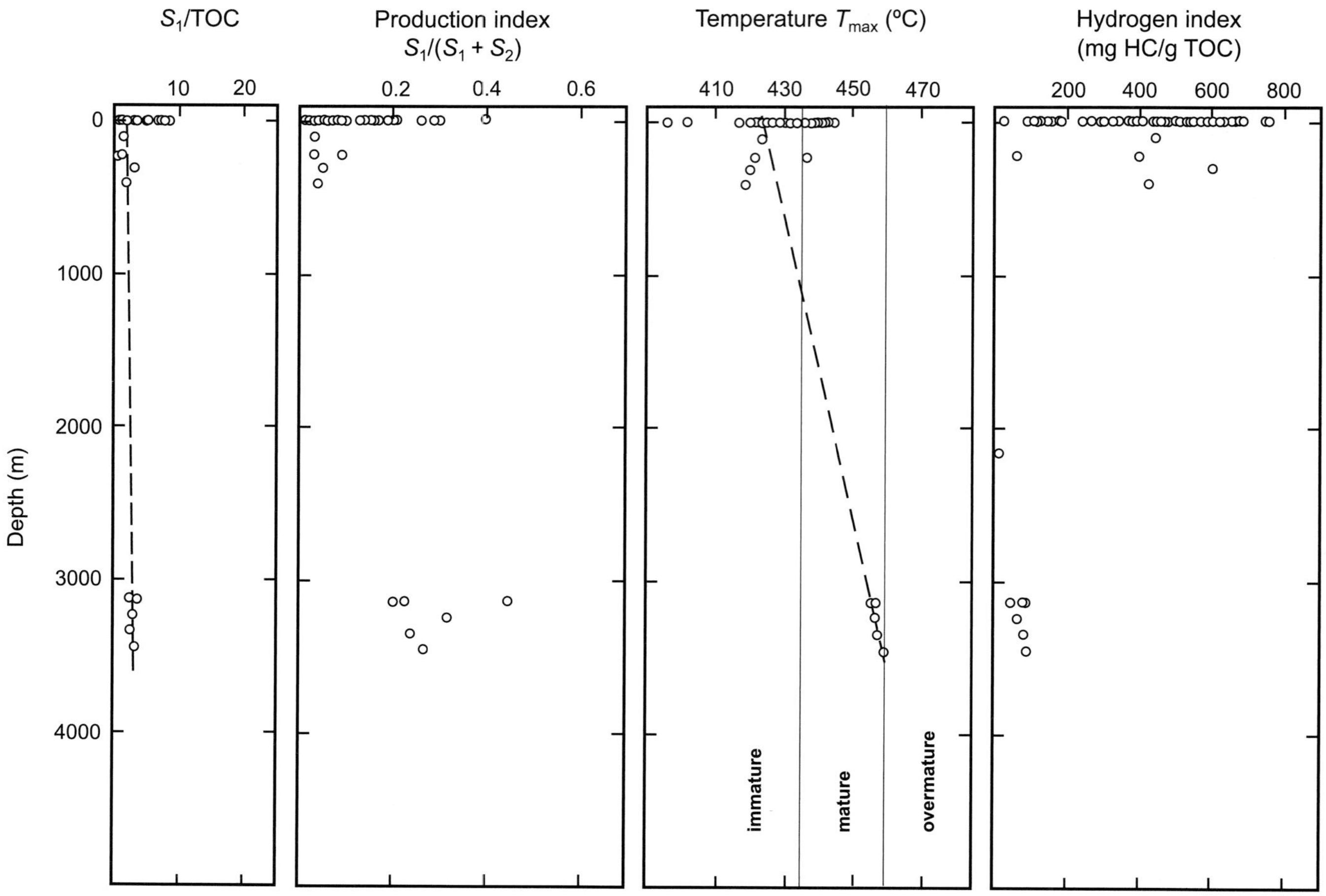

Figure 6. Plot of the Rock-Eval pyrolysis parameters (S_1 normalized to total organic carbon, production index, T_{max}, and hydrogen index) vs. depth for the Oligocene Menilite shales in the Silesian unit (the Polish Flysch Carpathians).

chemical sediments (Figure 15). The autochthonous Miocene sediments of the outer basin of the Carpathian Foredeep have been affected by Alpine orogenic movements and rest almost horizontally on the Precambrian–Paleozoic–Mesozoic basement (Oszczypko, 1997).

Two hundred and thirty-seven claystone and mudstone core samples were collected for organic geochemical analyses (Kotarba et al., 1998b), with 55 samples from the upper Badenian strata and 182 from the lower Sarmatian strata. A total of 41 wells were sampled (Figure 1).

In the upper Badenian strata, TOC content ranges from 0.30 to 3.33 wt.% (mean 0.80 wt.%), and in the lower Sarmatian, TOC content ranges from 0.02 to 3.33 wt.% (mean 0.69 wt.%). T_{max} temperature vs. hydrogen index diagram (Figure 16) shows that it is generally typically gas-prone terrestrial organic matter. Such an origin has been confirmed by maceral composition, elementary composition (H, C, O, and N) and biomarker distribution (Kotarba et al., 1998b). The occurrence of algal organic matter is sparse.

Rock-Eval T_{max} temperature for Miocene samples ranges from 415 to 435°C (Kotarba et al., 1998b). Only in a single sample from the upper Badenian strata (Jo-4 well, depth 3336 m [10,944 ft]) was a T_{max} of 438°C observed (Kotarba et al., 1998b). A plot of depth vs. TOC, T_{max}, total extract/total organic carbon (TE/TOC), and hydrogen index (Figure 17) indicates that organic matter is immature, and down to depths of about 3200 m (10,499 ft), deposition conditions of terrestrial organic matter were generally homogenous in the entire thickness of Miocene strata. The strata generated the microbial methane almost exclusively. Vitrinite reflectance (R_o) values remain low, ranging from 0.29 to 0.55%, at depths of 903–3412 m (2960–11,194 ft), indicating the immaturity of organic matter (Kotarba, 1999).

The rocks with TOC content greater than 0.5% can be considered as source rocks for microbial methane generation (Dickey and Hunt, 1972; Rice, 1992). According to Clayton (1992), the rocks, containing as much as 0.2% of TOC, can saturate the pore waters and release free gas during microbial methanogenesis.

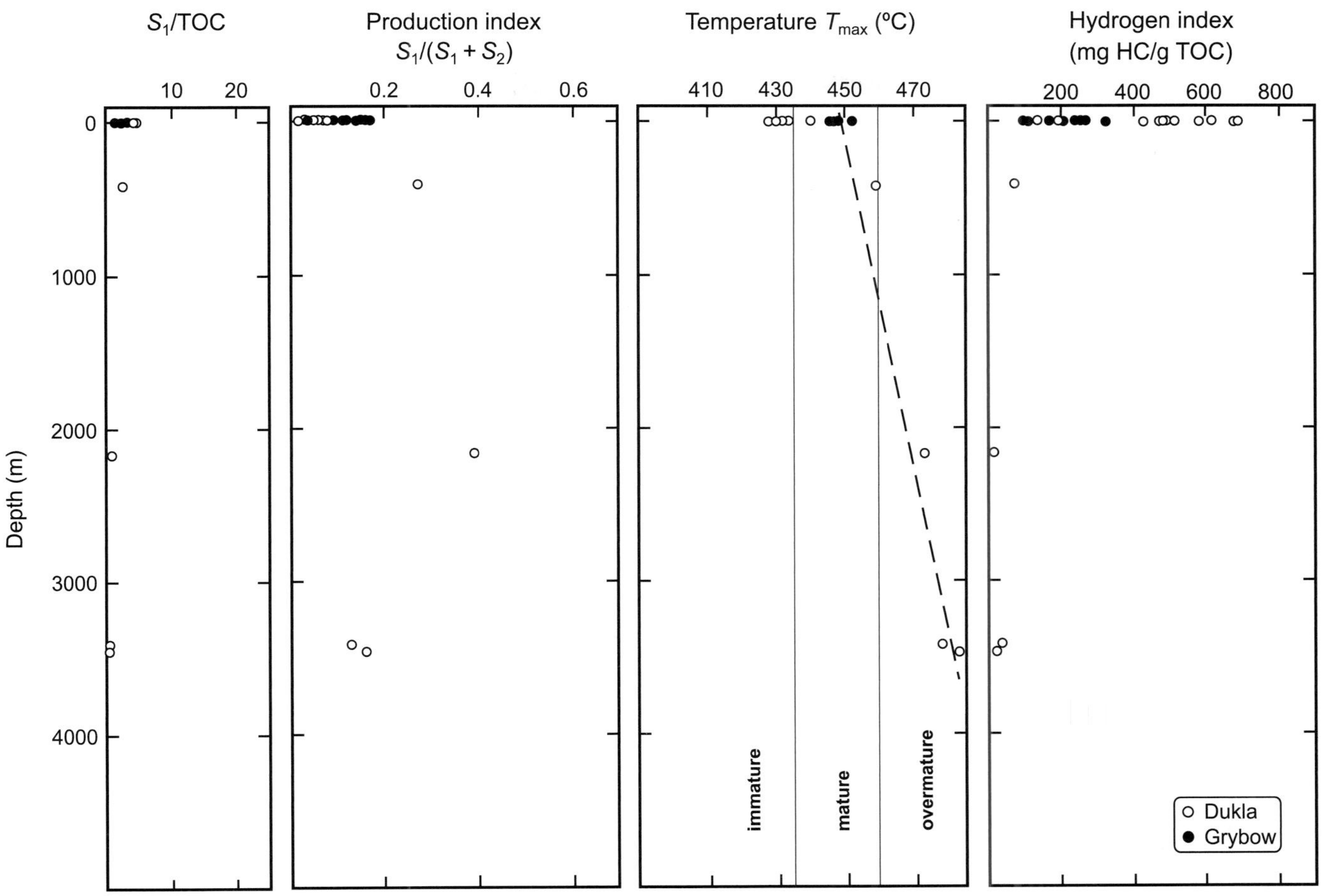

Figure 7. Plot of the Rock-Eval pyrolysis parameters (S_1 normalized to total organic carbon, production index, T_{max}, and hydrogen index) vs. depth for the Oligocene Menilite shales in the Dukla and Grybow units (the Polish Flysch Carpathians).

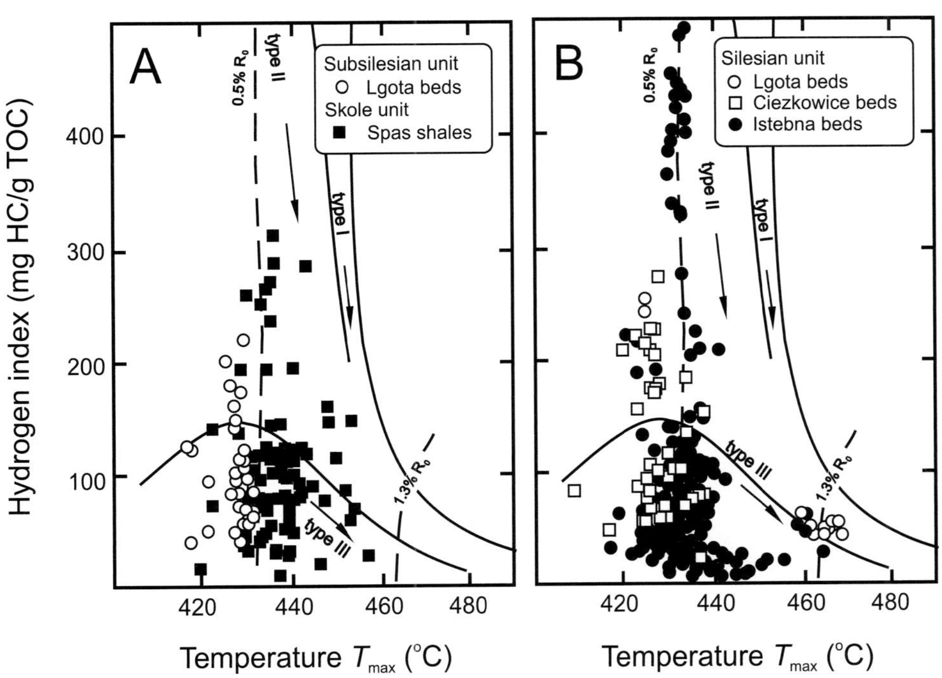

Figure 8. Plot of hydrogen index vs. Rock-Eval T_{max} temperature for (A) Lgota beds in Subsilesian unit and Spas shales in Skole unit and (B) Lgota, Ciezkowice, and Istebna beds in the Silesian unit (the Polish Flysch Carpathians).

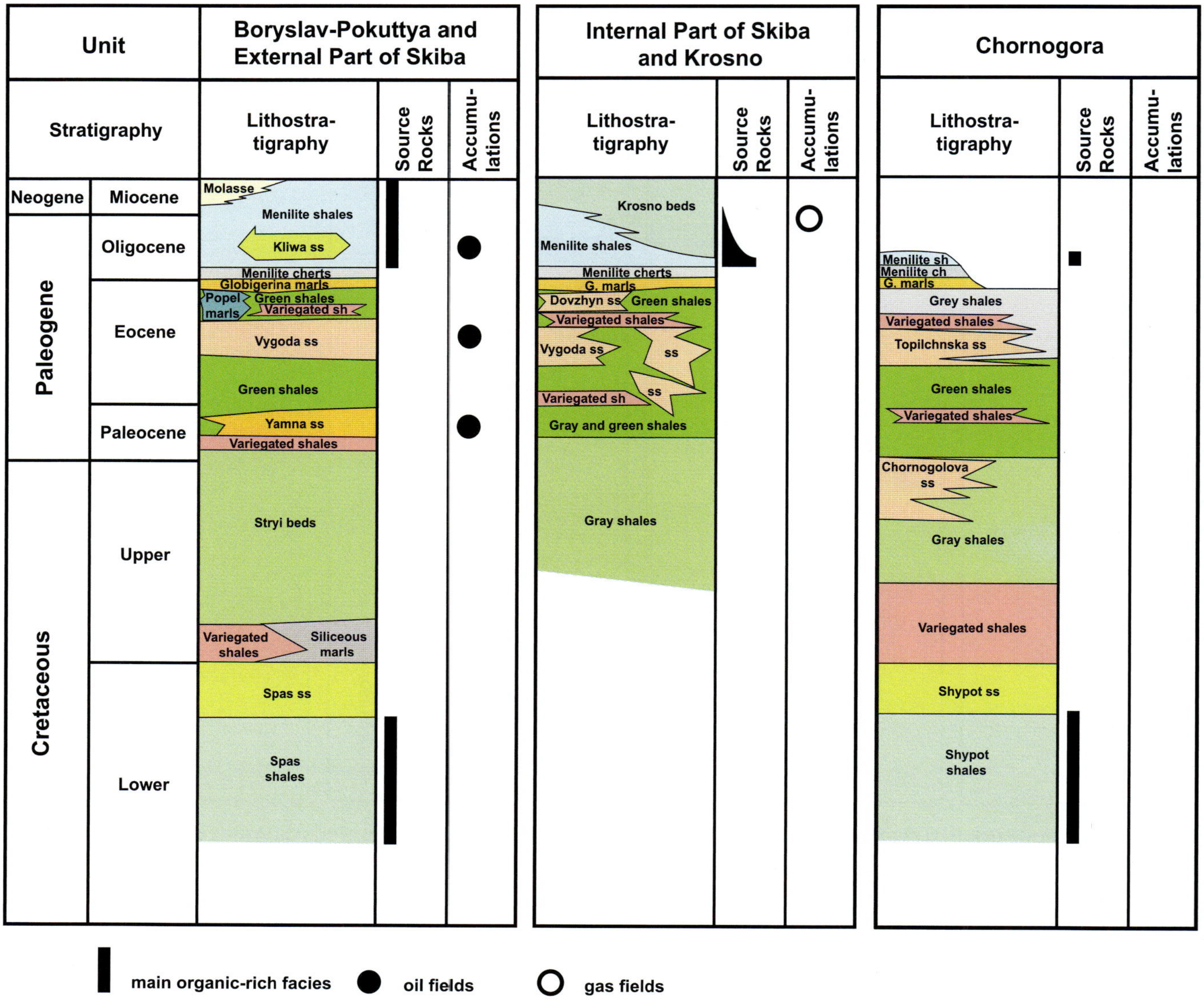

Figure 9. Generalized lithostratigraphic column showing distribution of petroleum accumulations and the main organic-rich facies in the flysch sequence of the Ukrainian Carpathians. Lithostratigraphy after Vialov et al. (1988). ss = sandstones; sh = shales; ch = cherts; G = Globigerina.

The Ukrainian Area

Middle Miocene rocks from four wells (Glynkivska-1, Gaiska-2, Sviatoslavska-3, and Slavetska-2), located in the central and extreme southeastern parts of the Ukrainian segment of the Carpathian Foredeep (Figure 1), were studied by Rock-Eval pyrolysis.

Total organic carbon content varies but shows no significant differences between the central and southeastern part of the foredeep and is also similar to values in the Polish part of the foredeep. Lower Sarmatian rocks (10 samples) contain from 0.49 to 1.44 wt.% (mean 0.99%) of TOC, upper Badenian (11 samples) from 0.12 to 0.86 wt.% (mean 0.63%), and Karpatian (3 samples) from 0.52 to 0.8 wt.% (mean 0.66%). Petroleum potential (S_2) of the samples is from 0.29 to 2.11 mg HC/g rock, being, in most cases, less than 1 mg/g rock. We believe that these rocks, similar to the Polish part of the Carpathian Foredeep, can be considered as potential source rocks for microbial methanogenesis.

T_{max} values for the Miocene rocks range between 409 and 431°C, indicating that the whole sequence is immature. Plots of hydrogen index vs. oxygen index (Figure 18) and hydrogen index vs. T_{max} (Figure 19) for the samples show that they all contain type III terrestrial kerogen. Type III organic matter is the most widespread source of microbial gas (Rice, 1992). The hydrogen index vs. T_{max} diagram (Figure 19) also shows

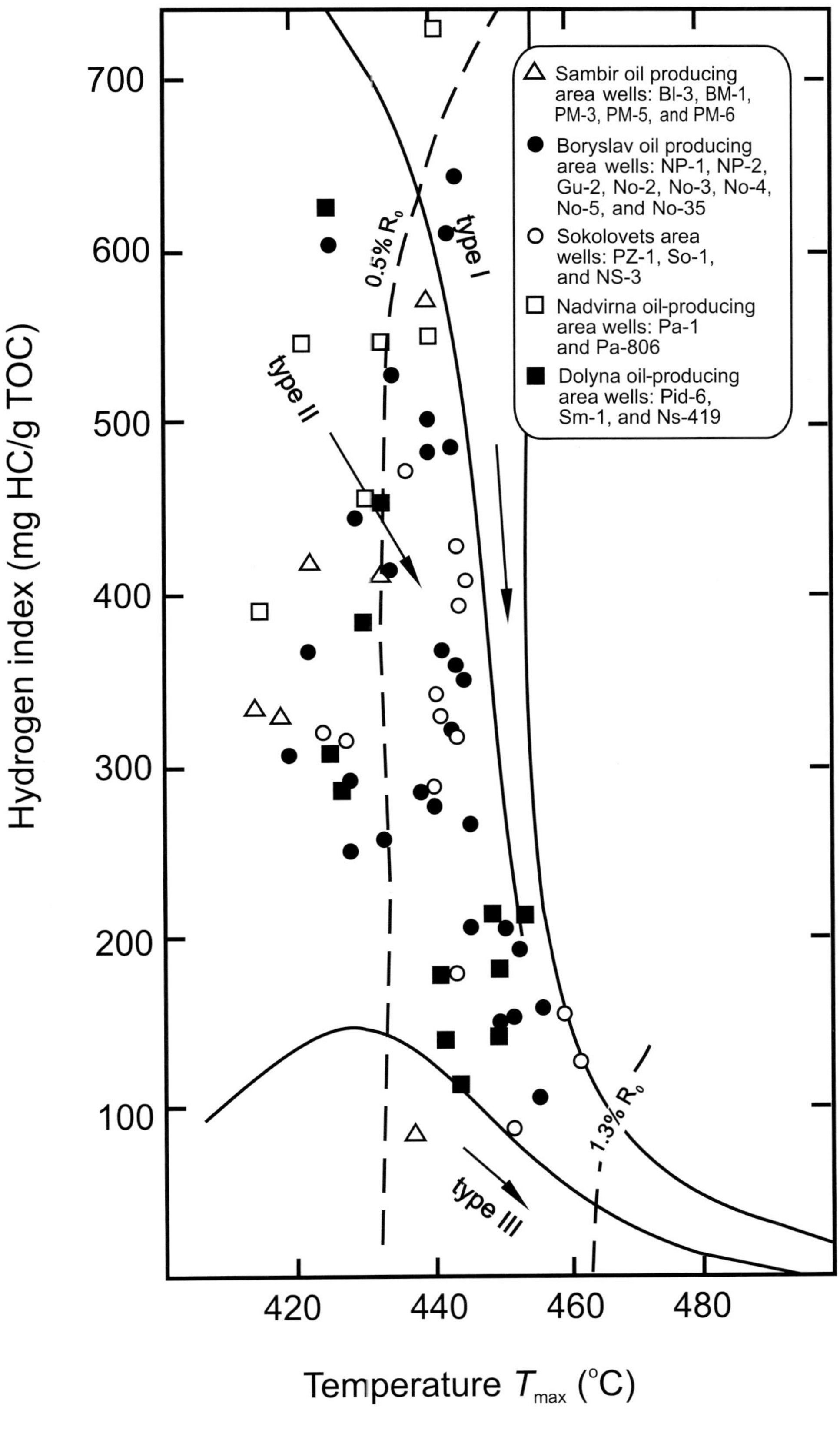

Figure 10. Plot of hydrogen index vs. Rock-Eval T_{max} temperature for the Oligocene Menilite shales in the Boryslav–Pokuttya unit (the Ukrainian Flysch Carpathians).

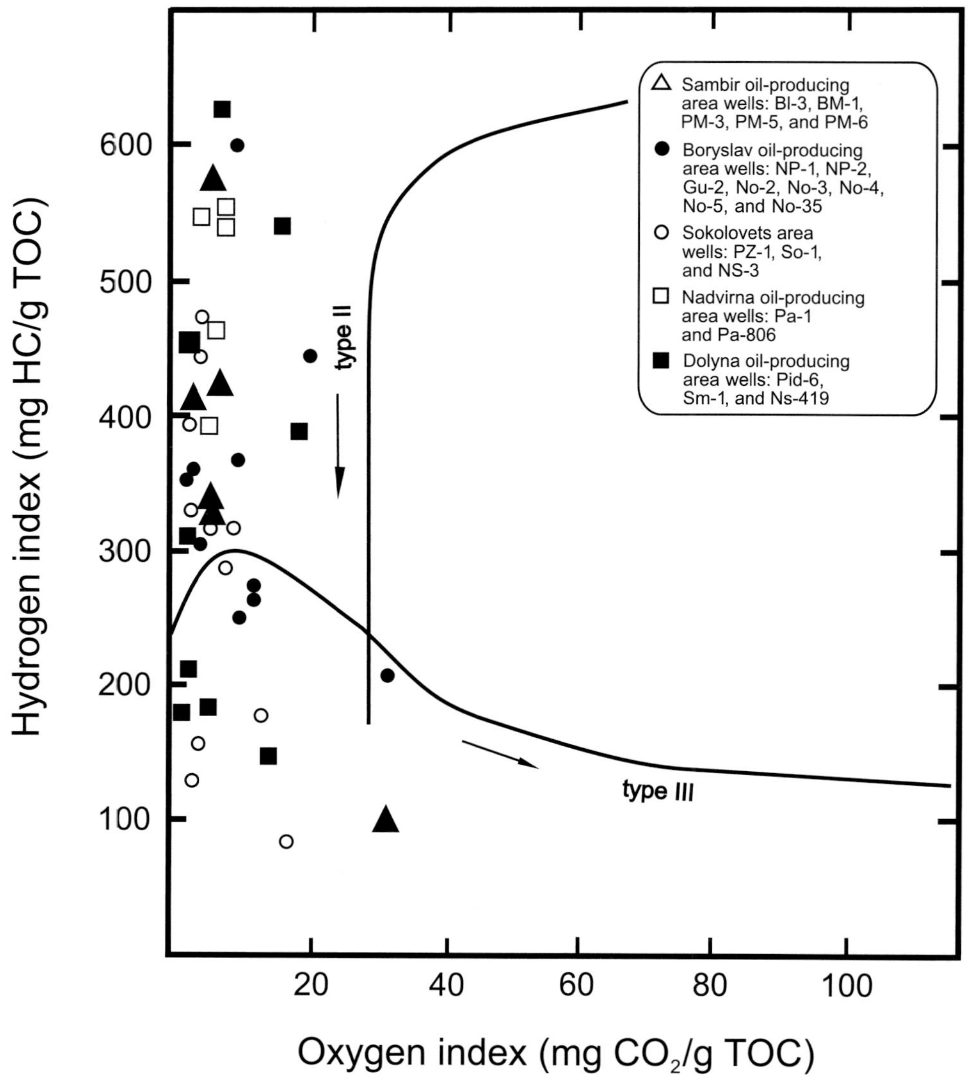

Figure 11. Plot of hydrogen vs. oxygen index for the Oligocene Menilite shales in the Boryslav–Pokuttya unit (the Ukrainian Flysch Carpathians).

that the thermal cracking of kerogen did not occur in the sequence under study.

Figure 20 represents the depth-related evolution of the pyrolysis parameters, T_{max} and production index, which are also thermal maturity indicators. Both parameters show that no significant increase of thermal maturity has occurred at depths of 1700 m (5600 ft). This confirms the immature stage of kerogen thermal evolution of the Miocene rocks.

The Mesozoic–Paleozoic Basement

The Polish Area

The geochemical and petrological studies (Kotarba et al., 2001; Matyasik et al., 2001; M. Kotarba, 2004, unpublished data) reveal that potential source rocks may occur in Ordovician, Silurian, and Middle Jurassic strata of the Paleozoic–Mesozoic basement (Figure 15). Moreover, the Lower Carboniferous and Cretaceous strata may locally contain slightly higher TOC (as much as 2.02% TOC) (Matyasik et al., 2001).

Ordovician strata, as much as 60 m (200 ft) thick, consist of dark-gray limestones, gray sandstones, and shales and conglomerates. Results of Rock-Eval pyrolysis on Ordovician shale intercalations from Hermanowa-1, Nosowka-12, Wola Obszanska-9, Wola Obszanska-10, and Zawada-8K wells (Figure 1) (depth range 956–4838 m [6136–15,873 ft]) indicate that TOC values range from 0.13 to 1.62 wt.% (mean 0.52 wt.%; 10 samples). T_{max} values for these samples range from 436 to 445°C, indicating the initial stage of oil generation (low-temperature thermogenic processes) (Figures 21, 22). Kerogen is type II (Figures 21, 22), with an average hydrocarbon potential (Figure 23).

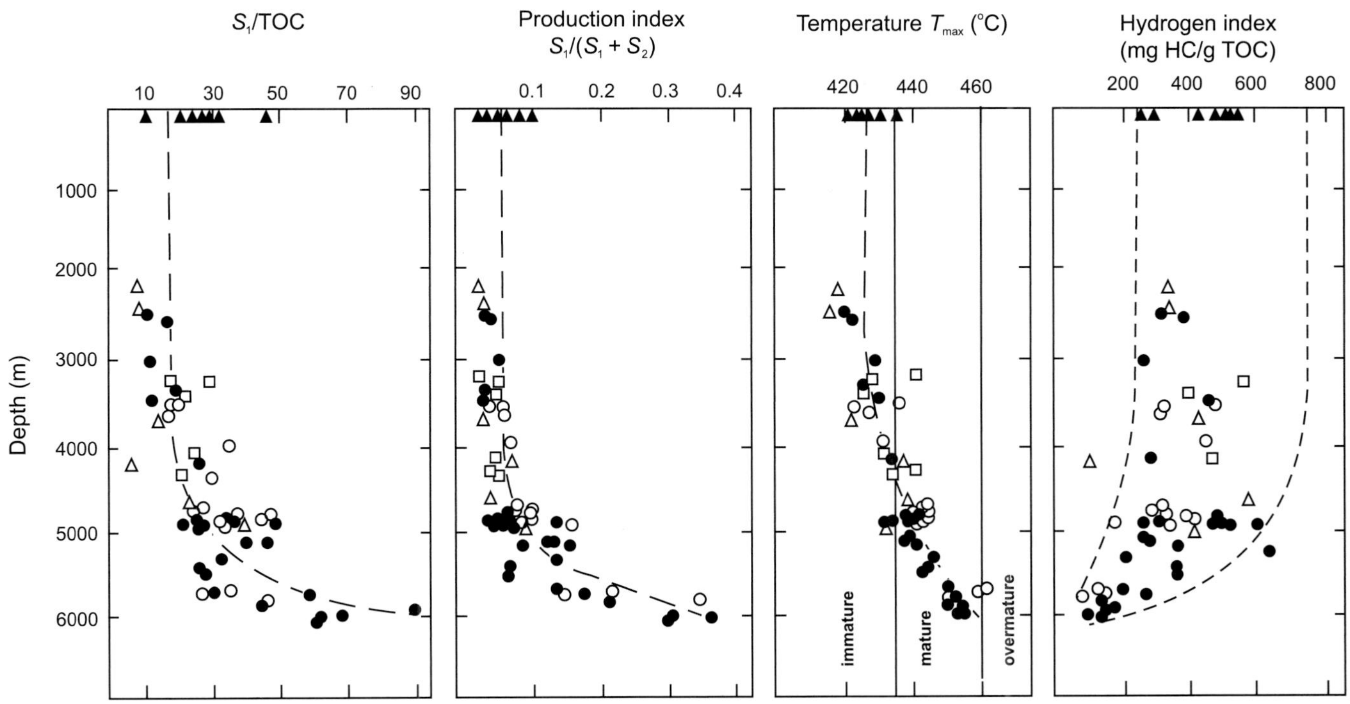

Figure 12. Plot of the Rock-Eval pyrolysis parameters (S_1 normalized to organic carbon, production index, T_{max} temperature, and hydrogen index) vs. depth for the Oligocene Menilite shales in the Boryslav–Pokuttya unit (the Ukrainian Flysch Carpathians).

Silurian rocks, represented by as much as 500 m (1600 ft) of mainly dark-colored graywackes, conglomerates, sandstones, and limestones are widespread in the Carpathian foreland. Rock-Eval pyrolysis results on samples from the shale intercalations from Silurian strata of Hermanowa-1, Pilzno-40, and Zagorzyce-1 wells (Figure 1), at depths ranging from 3552 to 4707 m (11,653 to 15,442 ft), indicate that TOC values range

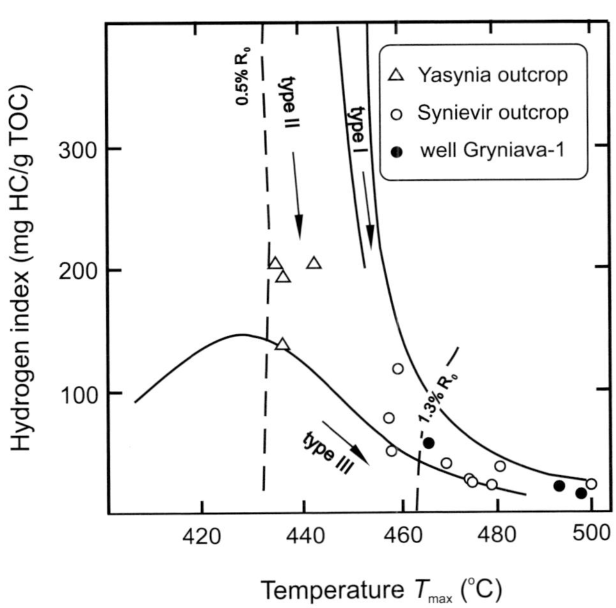

Figure 13. Plot of hydrogen index vs. Rock-Eval T_{max} temperature for the organic-rich Oligocene Menilite shales of the Krosno unit (the Ukrainian Flysch Carpathians).

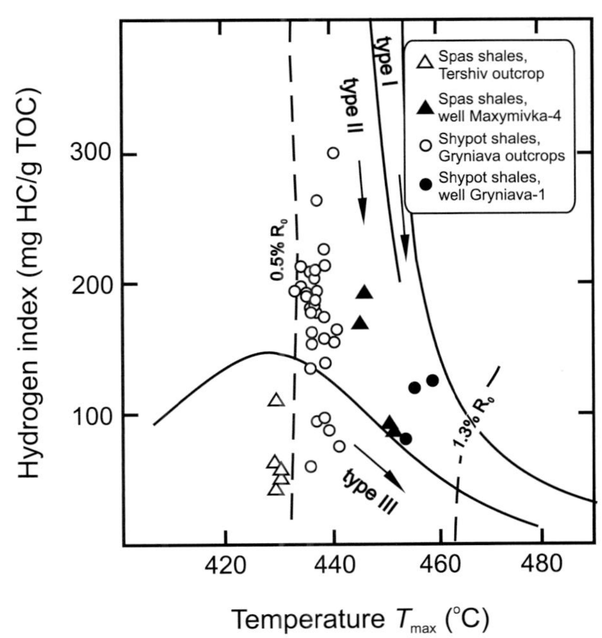

Figure 14. Plot of hydrogen index vs. Rock-Eval T_{max} temperature for the Lower Cretaceous organic-rich rocks, Spas shales in the Skiba nappe and Shypot shales in the Charnogora nappe (the Ukrainian Flysch Carpathians).

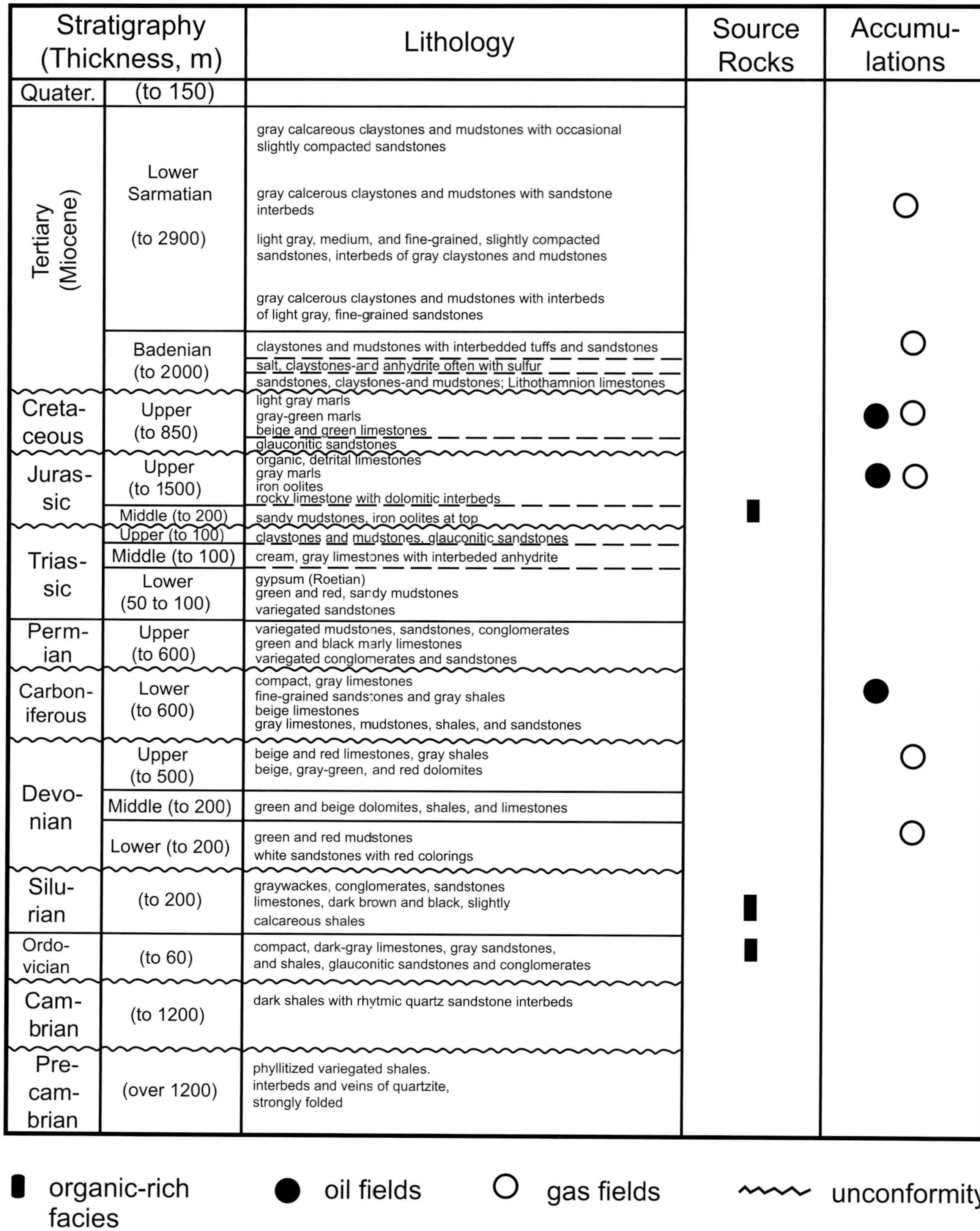

Figure 15. Generalized lithostratigraphic column showing distribution of oil reservoirs and the main organic-rich facies in the sequence of the autochthonous Miocene rocks and Paleozoic–Mesozoic basement (the Polish Carpathian Foredeep). Modified from Karnkowski (1999) (Quater. = Quaternary).

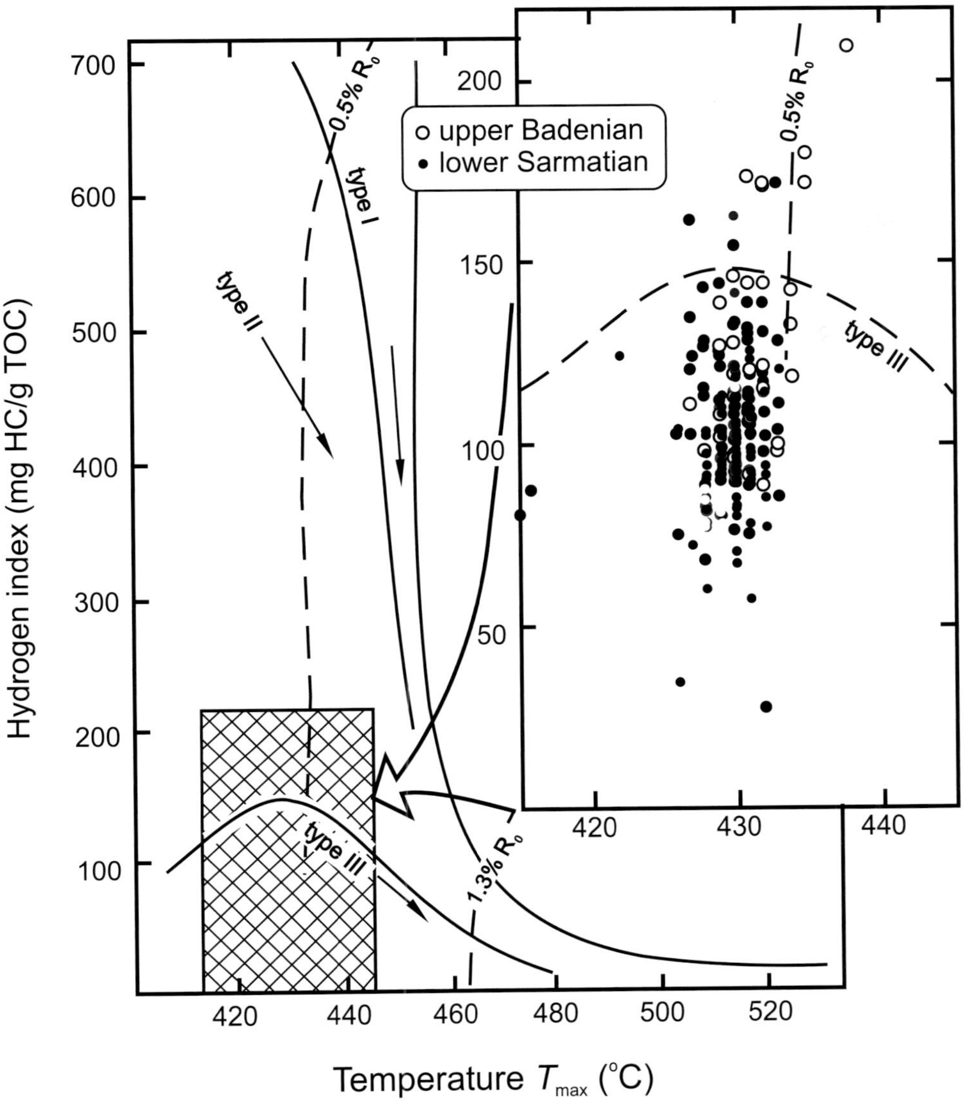

Figure 16. Plot of hydrogen index vs. Rock-Eval T_{max} temperature for the autochthonous Miocene rocks of the Polish Carpathian Foredeep.

from 1.00 to 2.36 wt.% (mean 1.64 wt.%, 7 samples) (Figures 22, 23). T_{max} values of 438–442°C also indicate that they are in the initial stage of oil generation (low-temperature thermogenic processes) (Figures 21, 22). This type II kerogen (Figures 21, 22) has average hydrocarbon-generating potential (Figure 23).

The most significant organic-rich sequence in the Mesozoic section of the foreland platform is the Middle Jurassic (Bathonian–Aalenian) strata, represented by as much as 200 m (660 ft) of sandy mudstones and claystones with iron oolite intercalations in the upper part (Karnkowski, 1999). The mudstone and claystone (Middle Jurassic) strata of Bedzienica-2, Nawsie-1, Tarnawa-1, Zagorzyce-6, and Zalasowa-1 wells (Figure 1), at depths ranging from 2974 to 4601 m (9757 to 15,095 ft) (Figure 23), contain varying amounts of organic matter (0.35–15.7 wt.% of TOC, mean 3.94 wt.%, 32 samples) of algal character (oil-prone type II kerogen) in the Tarnawa-1 well and terrestrial character (gas-prone type III kerogen) in the remaining well profiles (Figures 21, 22). The type II kerogen from the Tarnawa-1 well is immature (T_{max} values from 407 to 430°C), whereas the type III kerogen in the Bedzienica-2, Nawsie-1, Zagorzyce-6, and Zalasowa-1 wells is in the initial stage of low-temperature thermogenic processes (oil window). The Middle Jurassic strata in the Tarnawa-1 well contain immature, oil-prone type II kerogen, but it seems quite probable that organic matter in these strata is present at greater depths. In the southern part, under the Carpathian overthrust, organic matter is probably at a higher stage of maturation, and larger quantities of hydrocarbons were generated.

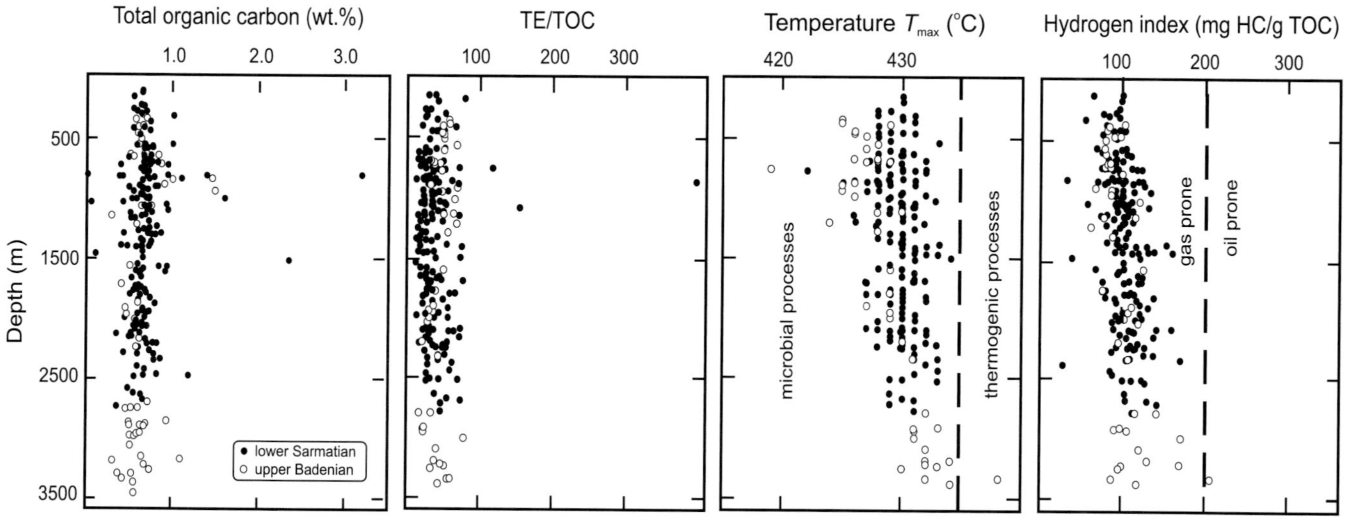

Figure 17. Plot of the Rock-Eval pyrolysis parameters (total organic carbon, total extract/total organic carbon, T_{max} temperature, and hydrogen index) vs. depth for the autochthonous Miocene rocks of the Polish Carpathian Foredeep.

The Ukrainian Area

The sequence of the platform margin, which is partly covered by the allochthonous Carpathian flysch belt, contains organic-rich strata in several different superimposed lithostratigraphic assemblages (Figure 24).

Silurian rocks, represented by more than 500 m (1600 ft) of mainly dark-colored argillites, are widespread in the Carpathian foreland. They were studied by Rock-Eval pyrolysis in the southeast part of the Carpathian foredeep (Ispas-1 well) at depths ranging from 1400 to 1592 m (4593 to 5223 ft) (Koltun et al., 1998). Total organic carbon content in these rocks ranges from 0.24 to 0.68%. The lack of an S_2 peak indicates that these rocks are presently overmature and can no longer account for additional liquid hydrocarbon generation.

The most significant organic-rich sequence in the Mesozoic strata of the foreland platform is the Middle Jurassic Kokhanivka Formation, represented by as much as 500 m (1600 ft) of black shales with siltstone and sandstone intercalations (Dulub et al., 1986). These shales are widespread in a limited area in the

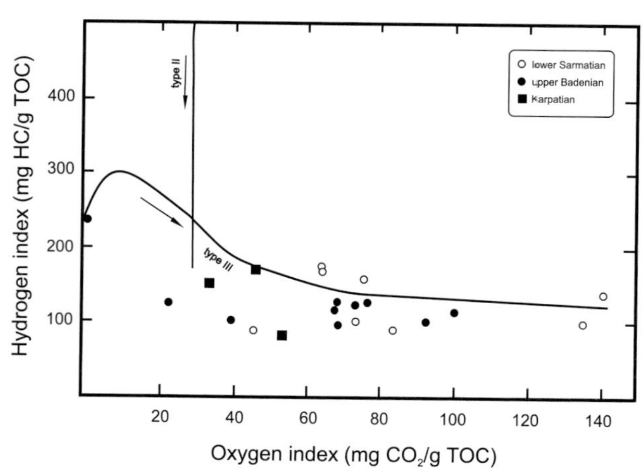

Figure 18. Plot of hydrogen index vs. oxygen index for the autochthonous Miocene rocks of the Bilche–Volytsia unit (the Ukrainian Carpathian Foredeep).

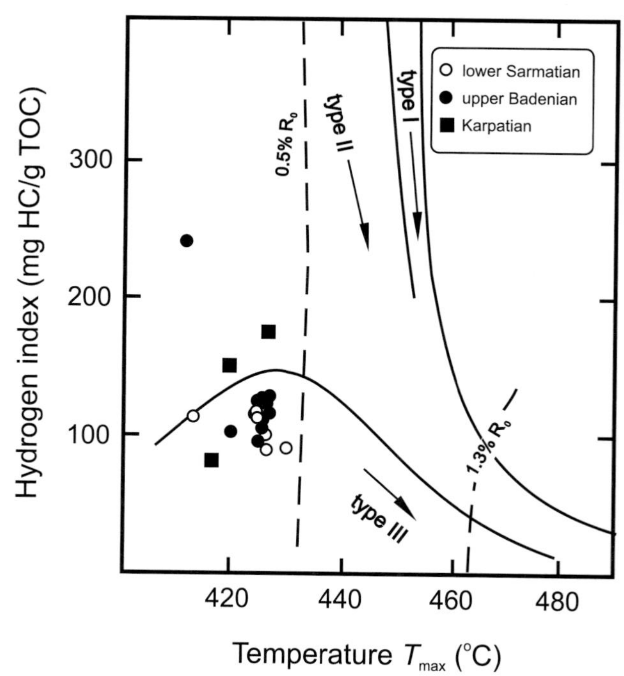

Figure 19. Plot of hydrogen index vs. Rock-Eval T_{max} temperature for the autochthonous Miocene rocks of the Bilche–Volytsia unit (the Ukrainian Carpathian Foredeep).

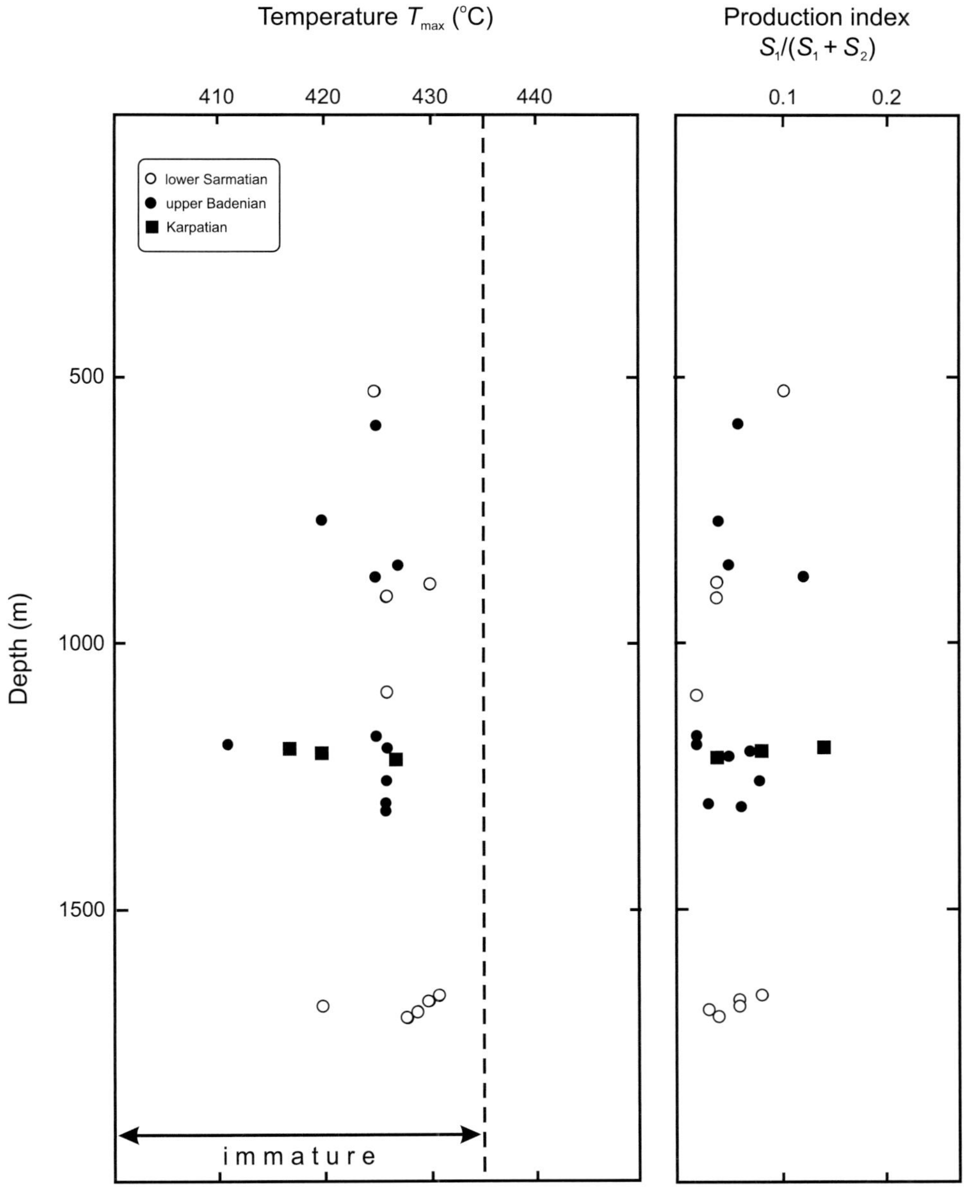

Figure 20. Plot of Rock-Eval pyrolysis parameters T_{max} temperature and production index for the autochthonous Miocene rocks of the Bilche–Volytsia unit (the Ukrainian Carpathian Foredeep).

northwest part of the Carpathian Foredeep and probably extend beneath the Carpathian allochthon. Total organic carbon contents exceed 12%. Pyrolysis results on samples from the Mostivska-2 well, at depths ranging from 2360 to 2549 m (7742 to 8362 ft) (Koltun et al., 1998), indicate a type II kerogen with commonly excellent petroleum potential (S_2 values as much as 31.62 mg HC/g rock). T_{max} values ranging from 434 to 438°C indicate the initial stage of oil generation.

The Cretaceous platform sequence is widespread in the Carpathian foreland, but it is commonly organic poor. Some local enrichment with dispersed organic matter is observed in the Lower Cretaceous series, where TOC values reach 0.87%. However, pyrolysis studies on samples from the Sviatoslavska-5 well with a depth of 1613 m (5291 ft) indicate that these rocks are still immature (T_{max} = 423°C) and contain a type III kerogen with a poor hydrocarbon potential (S_2 = 0.76 mg HC/g rock and HI = 87 mg HC/g TOC).

ORIGIN OF PETROLEUM

The aim of this section is to explain the origin of oils and natural gases and the genetic correlation between petroleum and potential source rocks based on geochemical data and results of time-temperature index (TTI) and Genex™ modeling of hydrocarbon generation and expulsion processes (Waples, 1980; Burnham and Braun, 1990) and the geological setting.

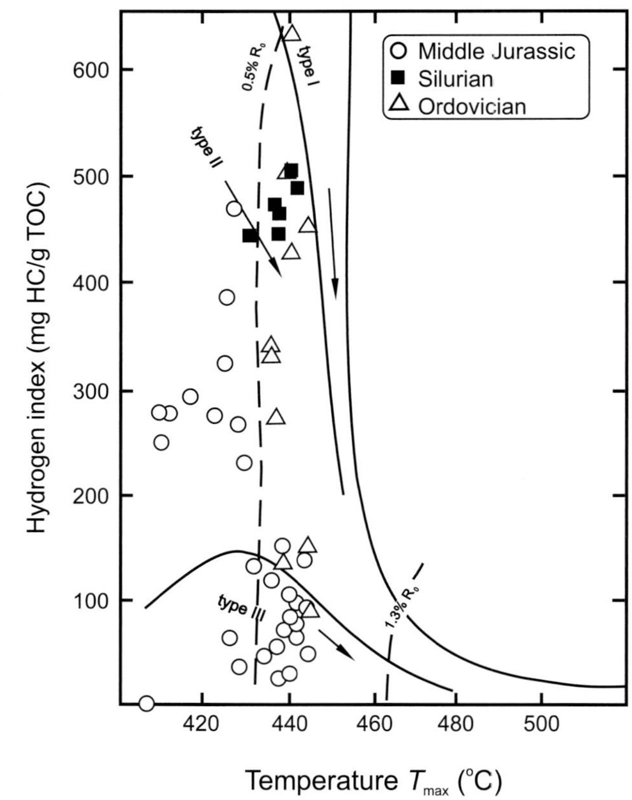

Figure 21. Plot of hydrogen index vs. Rock-Eval T_{max} for the Mesozoic–Paleozoic basement of the Polish Carpathian Foredeep.

The Flysch Carpathians

The Polish Area

Sixty-seven oil and sixteen gas deposits have been discovered in the flysch strata of the Polish part of the Carpathians, most of them in sandstone reservoirs in the Silesian and Skole units (46 and 14, respectively) (Karnkowski, 1999). The sandstone reservoirs occur in several different lithostratigraphic beds from the Lower Cretaceous to Oligocene (Figure 3).

Origin of Oil. The Carpathian oils have low sulfur (S content ranges from <0.1 to 0.56 wt.%). Their specific gravity ranges from 0.750 to 0.943 g/cm^3. Geochemical studies (especially biomarkers and isotope analyses) of 26 oils, accumulated in both Oligocene Kliwa sandstones and in Paleocene and Cretaceous flysch strata in the Skole, Silesian, Subsilesian, and Grybow units (ten Haven et al., 1993), indicate that they were generated from type II kerogen, and all belong to one oil family characterized by similar stable carbon composition and the presence of 28, 30-dinor-hopane, oleanane, and a HBI C_{25} alkane. Within this family, three subfamilies were recognized. A total of 10 of the 26 oils have been biodegraded. A frequent occurrence of secondary degradation processes of the oils, such as biodegradation and water washing in the Flysch Carpathians, was observed by Gondek and Pomykala (1982), Pomykala (1982), and ten Haven et al. (1993). The presence of oleanane and HBI C_{25} alkane suggests that they were mainly derived from the Oligocene Menilite shales (ten Haven et al., 1993; Koester et al., 1998b). According to Matyasik et al. (2000), the presence of higher plant biomarkers, e.g., oleanane from 19 oils occurring in the Potok fold area of the Silesian unit, may support the concept that the Upper Cretaceous lower Istebna beds were the potential source rocks. Because of the lean and sometimes moderate hydrocarbon potential, the Istebna beds cannot be excluded as potential source rocks for oils in the Silesian unit. This problem can be solved after further geochemical effort on oil-oil and oil-source rock correlations.

Origin of Natural Gases. Gases accumulated both in Oligocene Krosno and Kliwa sandstones and in Paleocene and Cretaceous flysch strata in Skole (Lopiennik deposit and two influxes in Kuzmina-1 well) and Silesian (Bystra, Tarnawa–Wielopole, Czarna, Lodyna, Sanok–Zablotce, Jurowce, Strachocina, Osobnica, and Gorlice), Subsilesian (Wola Jasieniecka deposit), and Grybow (Slopnice deposit) units vary both in their molecular and stable isotope compositions. Geochemical indices and stable isotope ratios vary in the following ranges:

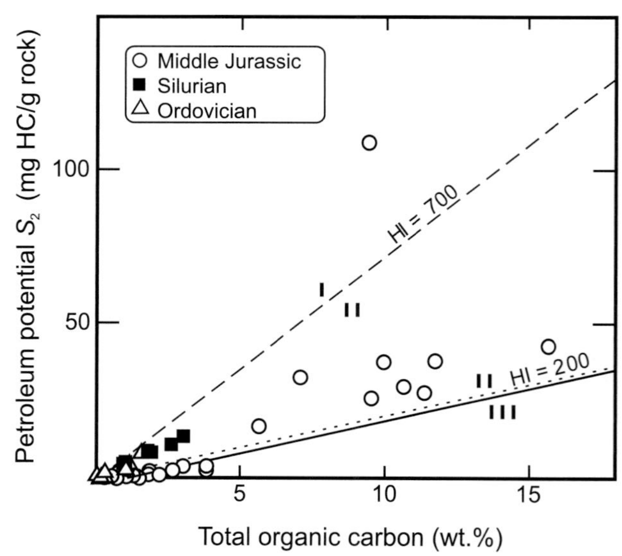

Figure 22. Petroleum potential vs. total organic carbon for the Mesozoic–Paleozoic basement of the Polish Carpathian Foredeep. Boundaries between types I and II kerogen and between types II and III kerogen after Langford and Blanc-Valleron (1990).

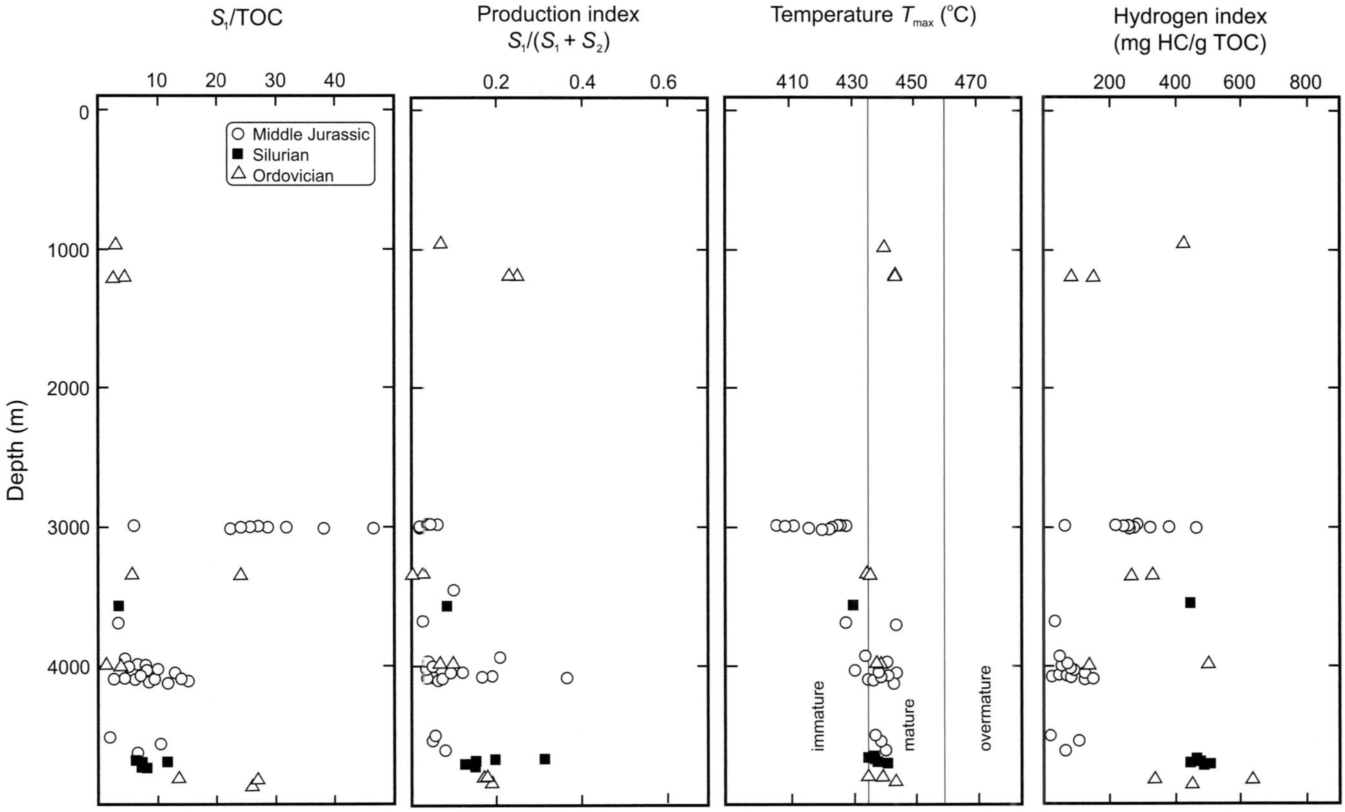

Figure 23. Plot of the pyrolysis parameters (S_1 normalized to total organic carbon, production index, T_{max} temperature, and hydrogen index) vs. depth for the Mesozoic–Paleozoic basement of the Polish Carpathian Foredeep.

hydrocarbon (C_{HC}) index [$C_{HC} = CH_4/(C_2H_6 + C_3H_8)$] from 5 to 159; $\delta^{13}C(CH_4)$ from −65.8 to −34.2‰; $\delta D(CH_4)$ from −205 to −151‰; $\delta^{13}C(C_2H_6)$ from −47.2 to −26.2‰; and $\delta^{13}C(C_3H_8)$ from −29.0 to −24.2‰ (Kotarba, 1987, 1992, 1998b). Practically all analyzed gases are probably low-temperature, oil-associated thermogenic gases generated from mainly type II kerogen and scarce type III kerogen. Gas from the Gorlice deposit is the only exception. In this accumulation, reservoir rocks are Upper Cretaceous Istebna beds (Go-6 well) and Oligocene middle Krosno beds (Go-11 well) (Figure 2). Isotopic and molecular compositions indicate (Figures 25A–29A) that these reservoirs contain microbial methane very similar to gases from autochthonous Miocene strata of the Carpathian Foredeep (Kotarba, 1992). However, migration of Miocene microbial gases does not seem possible because the Gorlice deposit occurs 35 km (21 mi) south of the edge of the Flysch overthrust. Instead, a local zone of immature Oligocene sediments must have existed there, where microbial methane was generated. The complicated tectonic history of this area (folding and thrusting) resulted in only insignificant amounts of gas accumulating and/or migrating to the overlying, even older, overthrusted Istebna beds (Kotarba, 1992).

Isotopic studies show (Figures 25A–29A) that natural gas from Spas beds (Lower Cretaceous) in the Skole unit in the Kuzmina-1 well was generated from the oil-prone type II kerogen. Because mainly gas-prone type III kerogen occurs in the Upper Cretaceous Spas beds, it seems most likely that this gas migrated from the Silesian unit and filled traps in Spas beds (Kotarba, 1998b).

Modeling of Hydrocarbon Generation and Expulsion Processes. The TTI (Waples, 1980) and Genex modelings were performed to describe a time frame for the generation process and the expulsion of hydrocarbons. The results from the organic sulfur content in kerogen and kinetic studies from the Optkin® method (Koester et al., 1998a) were also used. Modeling was prepared for profiles of the following wells—Skole unit: Brzegi Dolne IG-1, Dynow-1, Kuzmina-1, Kuzmina-2, Nawsie-1, Paszowa-1, and Wisniowa-1; Silesian unit: Dwernik-3, Wetlina-1, Bykowce IG-1, and Rowne-1 (Kotarba et al., 1998a; 1999) (Figure 1).

The process of hydrocarbon generation in the folded areas has the individual character of thermal-time conditions of both sedimentation and orogenic stages. Hence, the quantitative evaluation of the process

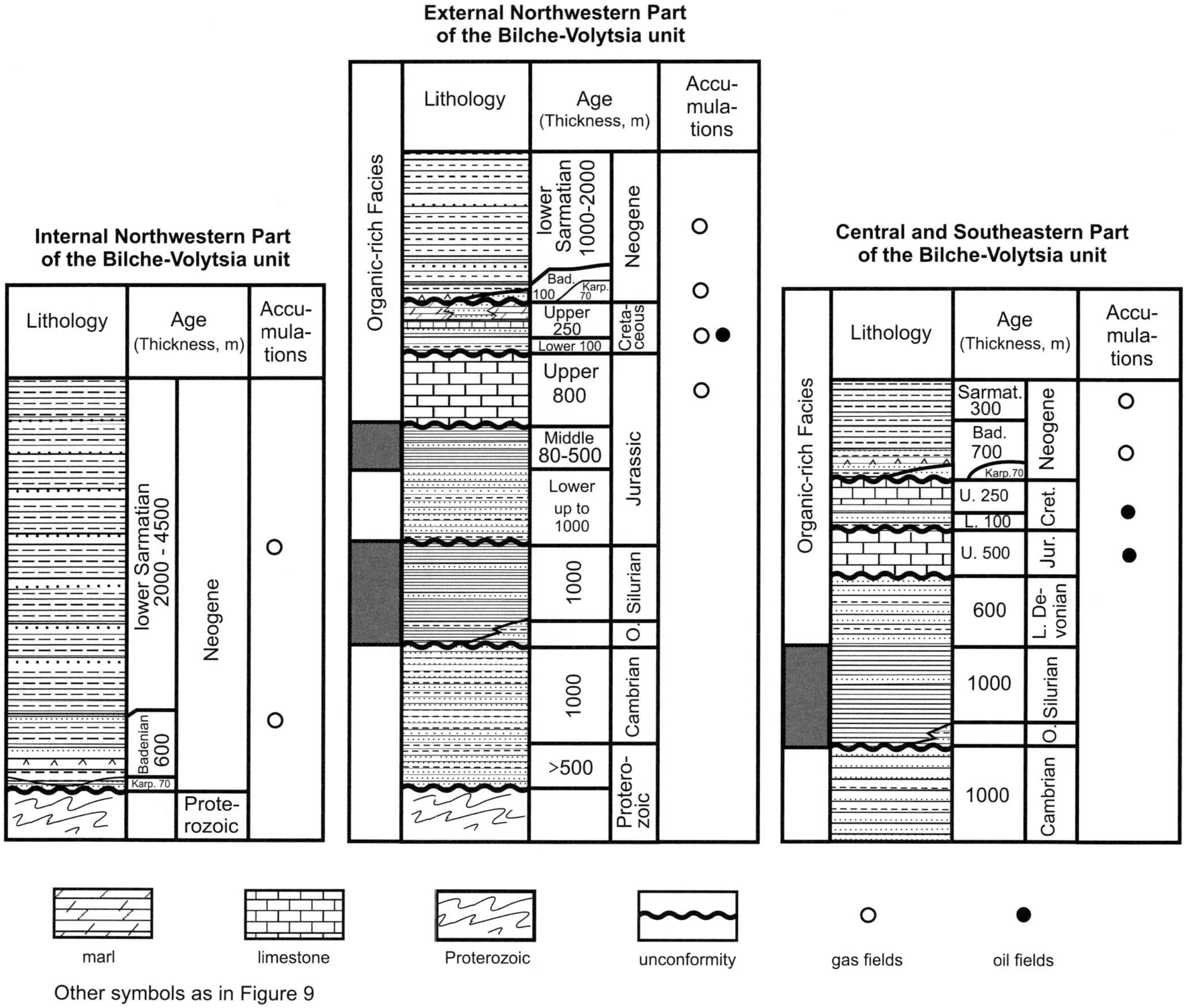

Figure 24. Generalized stratigraphic columns showing distribution of the oil and gas reservoirs and the main organic-rich facies in the autochthonous Miocene strata and the Mesozoic–Paleozoic basement of the Ukrainian Carpathian Foredeep. (A) Internal northwestern part of the Bilche–Volytsia unit; (B) external northwestern part of the Bilche–Volytsia unit; (C) central and southeastern part of the Bilche–Volytsia unit (Sarmat. = Sarmatian; Bad. = Badenian; Karp. = Karpatian; Cret. = Cretaceous; Jur. = Jurassic; O. = Ordovician; L. = Lower).

of hydrocarbon generation in selected locations in the Carpathians was reduced to the reconstruction of time of duration of sedimentation, orogenic, and degradation (postinversion) stages.

The time when sedimentation ended in the individual flysch basins is related to the evolution of folding activity. Precise chronostratigraphic time evaluations are difficult because of erosion of the youngest sediments in the Krosno series. The following times can be defined: the end of the Oligocene (24 Ma) in the south Fore-Dukla zone of the Silesian basin, the earliest Miocene (21.5 Ma) in the north (external) zone of the Silesian basin, and the early Miocene (19.5 Ma) in the external zone of the Skole basin.

The original thickness of the youngest flysch Krosno–Menilite complex was basically disturbed by the main phase of folding, overthrusting, and orogenic inversion from 19 to 5.5 Ma. The flysch profiles in the individual tectonic units in the Carpathians were reduced in thickness to a varying degree because of postorogenic erosion, lasting from about 5.5 Ma until the present. The most complete sections of the Krosno–Menilite complex, preserved in synclines, are a basis for reconstruction of the original thickness of individual

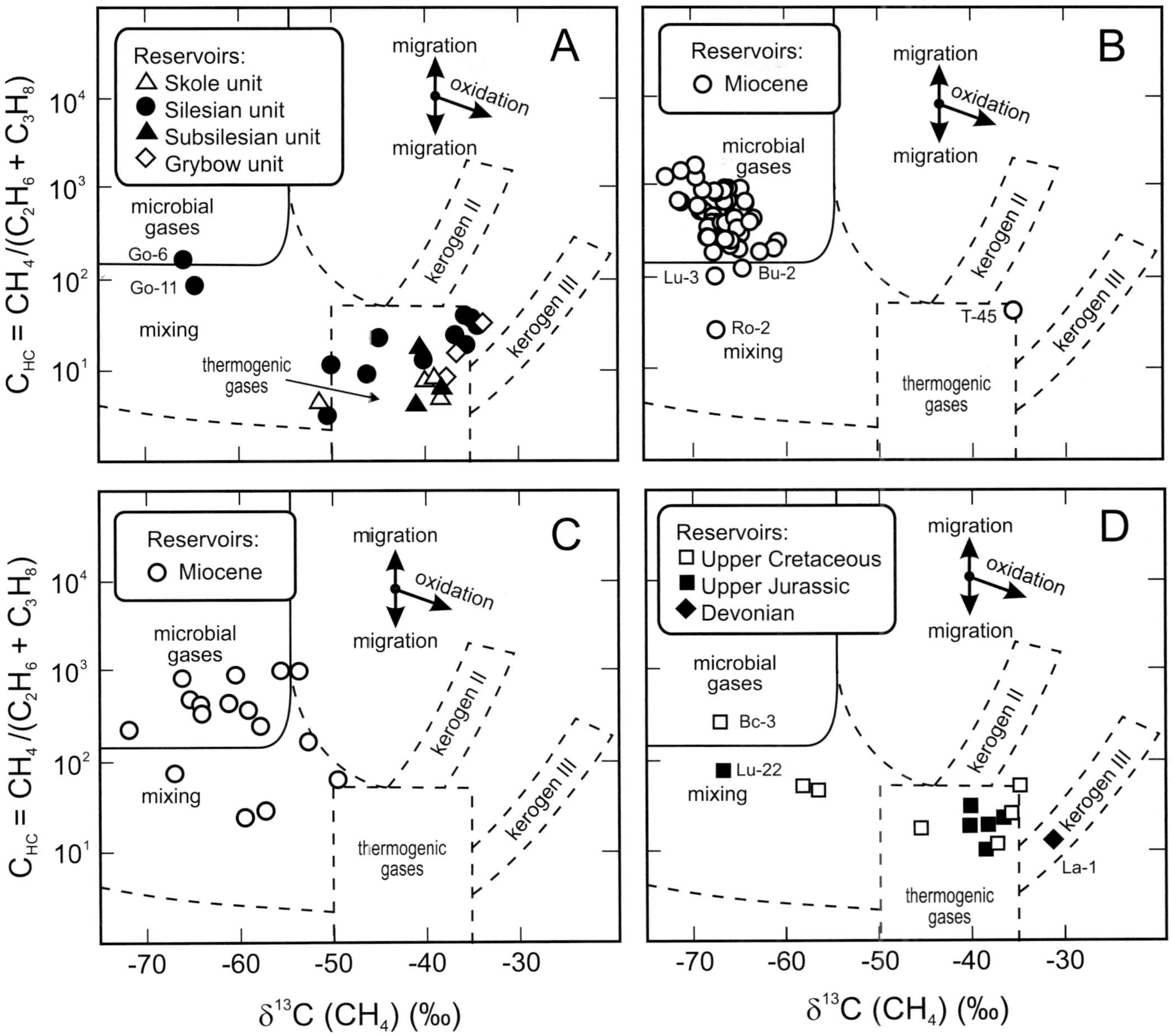

Figure 25. Genetic characterization of natural gases from (A) the Polish Flysch Carpathians; (B) the autochthonous Miocene rocks of the Polish Carpathian Foredeep; (C) the autochthonous Miocene rocks of the Ukrainian Carpathian Foredeep (Bilche–Volytsia unit); and (D) the Mesozoic–Paleozoic basement of the Polish Carpathian Foredeep using $\delta^{13}C(CH_4)$ vs. hydrocarbon index (C_{HC}). Compositional fields from Whiticar (1990).

lithofacies of this series. Dating of the main folding and magnitude of the original uplift is related to the evolution of the Carpathian tectogene that began in the early Styrian phase, about 19.5 Ma (Kotlarczyk, 1988), and finished in the middle Sarmatian (Wdowiarz, 1985; Oszczypko, 1997) at about 5.5 Ma.

Genex modeling of hydrocarbon generation and expulsion processes for source rocks in the Menilite, Inoceramian, and Spas shales of the Skole unit indicates that the process might have started at the sedimentary stage of the Carpathian formation. However, its main development was probably associated with the orogenic stage. During the orogenic phase, structural rebuilding of the Skole depositional basin occurred along with an increase of the thermal maturity of organic matter, thus initiating the hydrocarbon generation process. Its intensity changed with burial depth and reached its maximum in the southeast part of the study area. For example, in the Brzegi Dolne IG-1 well in the southeast part of the basin, at the end of the sedimentation stage at 3000 m (9800 ft) of depth, source rock horizons in the Inoceramian beds and Menilite

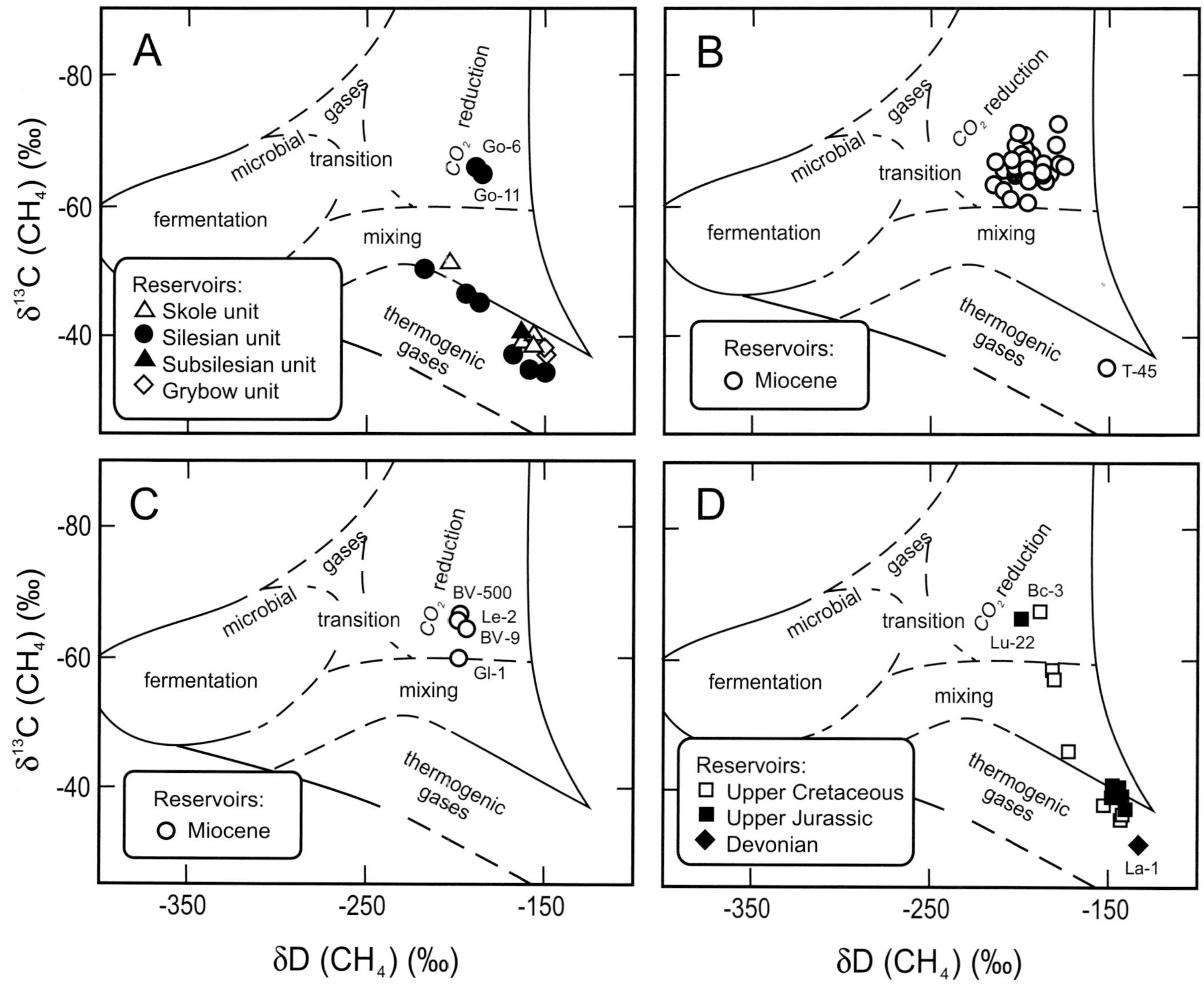

Figure 26. Genetic characterization of natural gases from (A) the Polish Flysch Carpathians; (B) the autochthonous Miocene rocks of the Polish Carpathian Foredeep; (C) the autochthonous Miocene rocks of the Ukrainian Carpathian Foredeep (Bilche–Volytsia unit); and (D) the Mesozoic–Paleozoic basement of the Polish Carpathian Foredeep using $\delta^{13}C(CH_4)$ vs. $\delta D(CH_4)$. Compositional fields from Whiticar et al. (1986).

shales reached a maturity level corresponding to low-temperature thermogenic processes (oil window) (Figure 30).

At the sedimentation stage of the Skole basin, the Menilite organic matter was not sufficiently thermally mature to expel hydrocarbons. This process, on a small scale, probably happened only in the orogenic stage. The folding movements of the Carpathians caused a considerable change of the structural system. Menilite shales were displaced to different depths. They appeared in anticlinal structures in the near-surface or surface conditions (exposed by erosional processes). In other places, Menilite shales were buried in synclines at considerable depth. Moreover, the thickness of Menilite strata varied, and the folding processes acting on the flysch complex commonly caused an increase in thickness. With knowledge of this geological and thermal system, plus the present position of the Menilite shales, it is possible to reconstruct the processes of hydrocarbon generation. With modeling of hydrocarbon generation and expulsion processes in orogenic areas, the erosion timing is of special importance.

A similar trend can be observed in well profiles in the Silesian unit, where, during the sedimentation stage, source rocks in the Menilite, Verovice, and Lgota beds reached oil generation to the extent that expulsion occurred. This process was completed in the orogenic stage. Optkin kinetic parameters indicate the likely

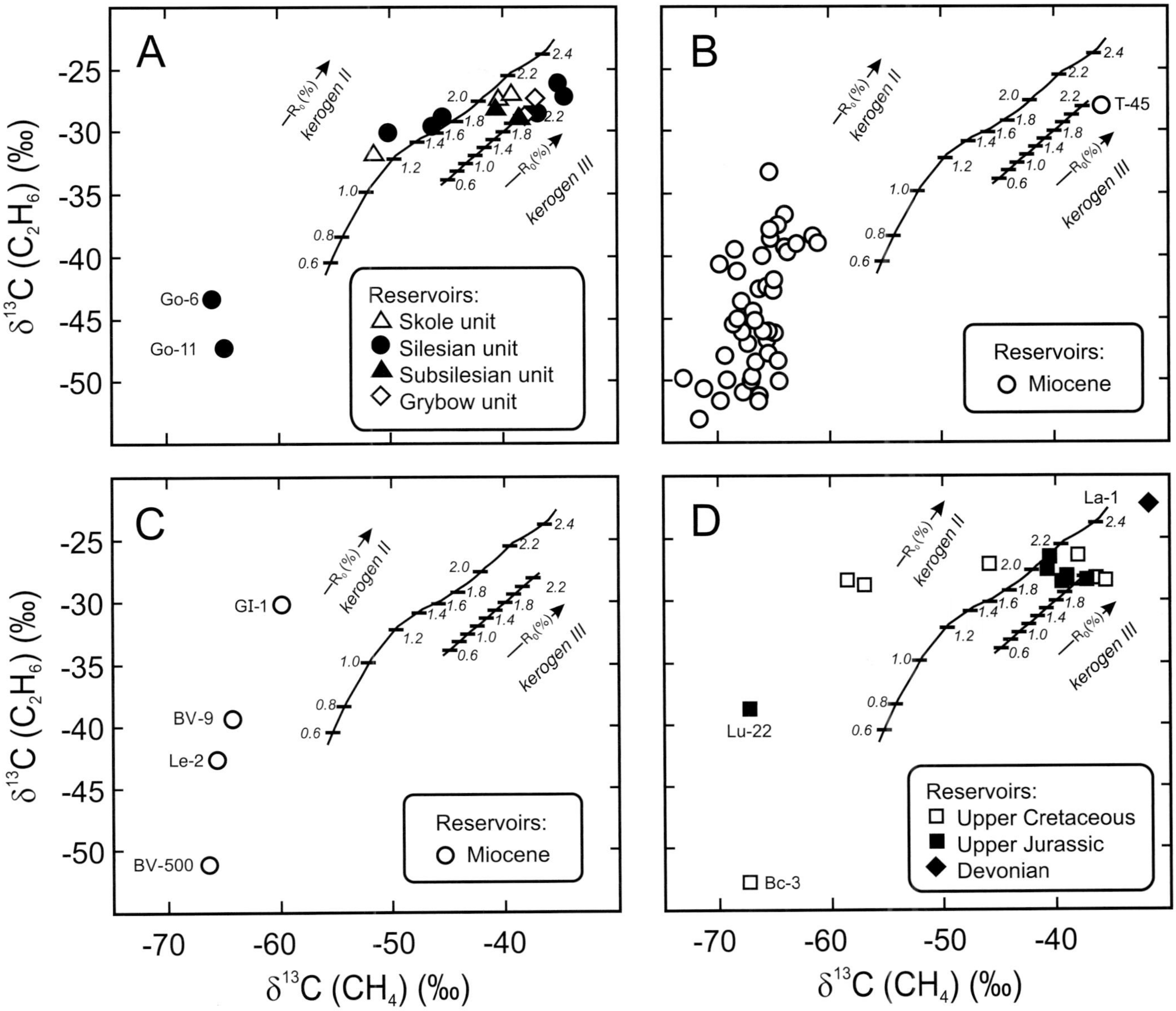

Figure 27. Genetic characterization of natural gases from (A) the Polish Flysch Carpathians; (B) the autochthonous Miocene rocks of the Polish Carpathian Foredeep; (C) the autochthonous Miocene rocks of the Ukrainian Carpathian Foredeep (Bilche–Volytsia unit); and (D) the Mesozoic–Paleozoic basement of the Polish Carpathian Foredeep using $\delta^{13}C(CH_4)$ vs. $\delta^{13}C(C_2H_6)$. Position of the vitrinite reflectance curve for type III kerogen after Berner and Faber (1996).

occurrence of multiphase hydrocarbon generation from kerogen of Menilite beds (Koester et al., 1998a), occurring during both sedimentation and orogeny.

In the Polish Flysch Carpathians, in the Skole, Silesian, and Dukla units, the Oligocene Menilite shales are good or moderate oil-prone source rocks. Genex and Optkin modelings indicate the occurrence of multiphase hydrocarbon generation and expulsion, during both sedimentation and orogenic stages. Hydrocarbons generated during the sedimentary stage migrated into interbedded flysch strata and were either trapped or released and persed into the atmosphere. However, for present accumulations, the most important processes are hydrocarbon generation and expulsion during orogenic and postinversion stages. Expelled hydrocarbons migrated mainly along the zones of crevices and fractures associated with thrusts or thrust slices.

The Ukrainian Area

The burial history of the Carpathian flysch sequence of all depositional basins comprises two main stages: preoverthrusting and postoverthrusting. The pre-emplacement subsidence was controlled only by the sedimentary

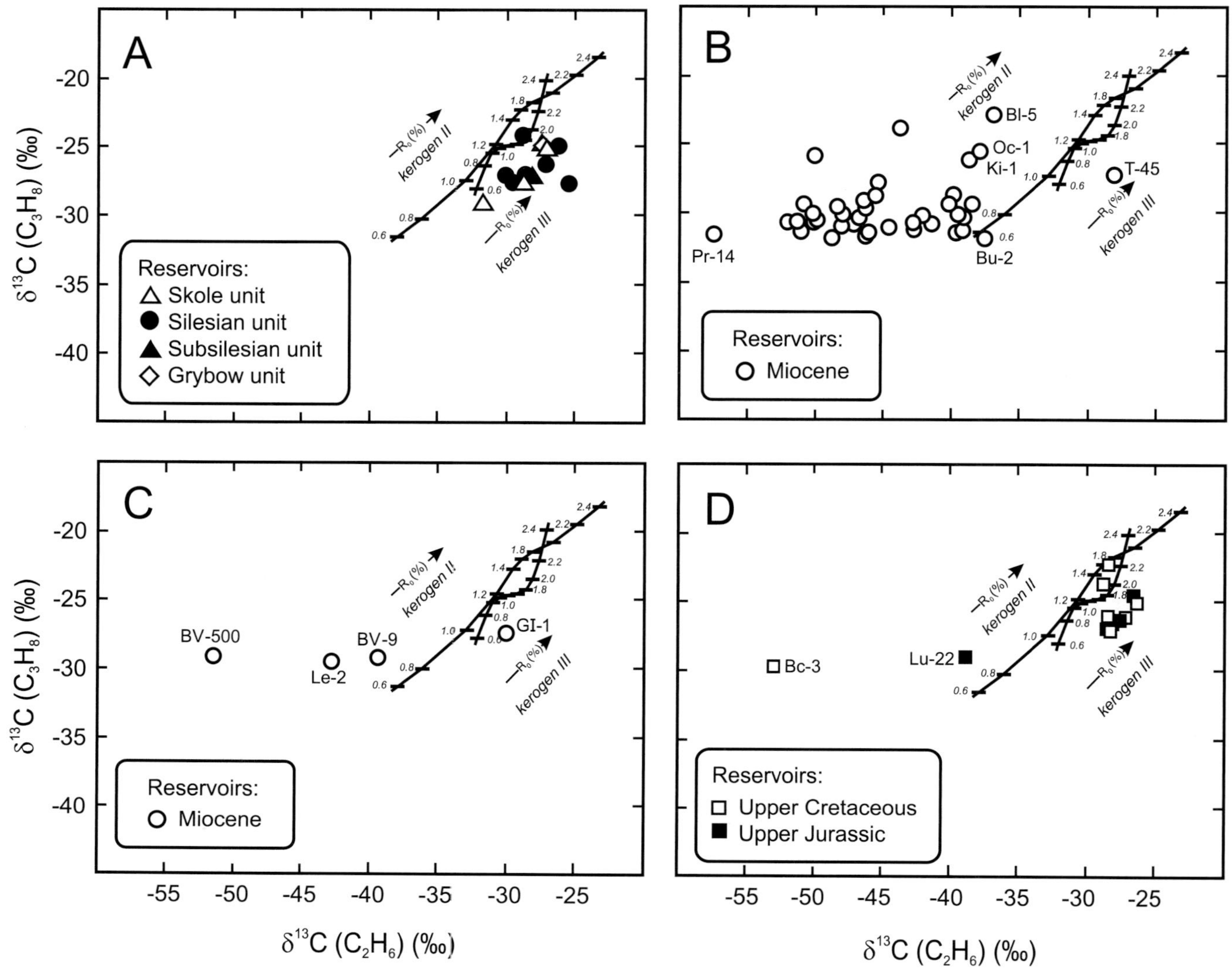

Figure 28. Genetic characterization of natural gases from (A) the Polish Flysch Carpathians; (B) the autochthonous Miocene rocks of the Polish Carpathian Foredeep; (C) the autochthonous Miocene rocks of the Ukrainian Carpathian Foredeep (Bilche–Volytsia unit); and (D) the Mesozoic–Paleozoic basement of the Polish Carpathian Foredeep using $\delta^{13}C(C_2H_6)$ vs. $\delta^{13}C(C_3H_8)$. Position of the vitrinite reflectance curve for type III kerogen after Berner and Faber (1996).

burial and subsidence rates and was relatively uniform throughout the region. In contrast, the overthrust-induced burial history was highly variable among the different nappes and individual folds. This is reflected in the complex interactions between the overlying and the underlying blocks during thrusting and erosion events.

The Carpathian flysch belt consists of several allochthonous tectonic units that have been thrust northeastward. Maturation history of Lower Cretaceous and Oligocene source rocks since the beginning of deformations evolved in connection with the tectonic emplacement of individual nappes. The present-day structure of the Carpathian fold and thrust belt resulted from early Miocene tectonic events, the timing of which is indicated by the age of the unconformable molasse deposits. Thrust emplacement started in the Eggenburgian, after the deposition of the Menilite shales, and lasted until post-early Sarmatian.

Boryslav–Pokuttya Unit. The Boryslav–Pokuttya nappe constitutes the frontal tectonic unit (Figure 1). It is made up of Upper Cretaceous to lower Miocene flysch strata and is unconformably covered by the folded molasse sequence and overthrust by the Skiba nappe from the southwest. The Boryslav–Pokuttya nappe is the main oil-bearing unit in the Ukrainian Carpathians. Major oil fields occur in deeply buried anticlines along the front of the Skiba nappe. The concentration of oil

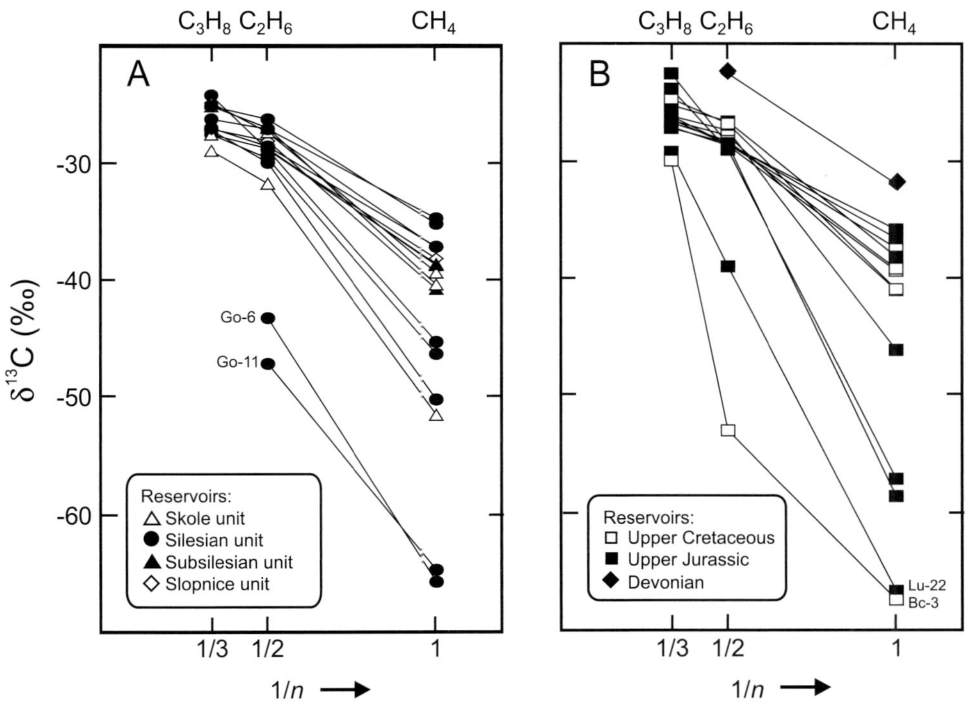

Figure 29. Stable carbon isotope composition in methane, ethane, and propane in natural gases from (A) the Polish Flysch Carpathians and (B) the Mesozoic–Paleozoic basement of the Polish Carpathian Foredeep.

fields in the Boryslav–Pokuttya unit can be explained in two ways:

- The frontal part of the flysch belt is also the main area where the Menilite shales occur. Thus, it contains most of the source rock deposits of the Ukrainian Carpathians.
- The Boryslav–Pokuttya unit represents a complex stack of duplexes commonly involving the Menilite shales. As a result, the latter occur within a great depth interval from the surface to more than 8–10 km (5–6 mi), covering a wide range of thermal maturity.

Hence, the overthrust-induced burial and maturation history of the Menilite shales has resulted in favorable conditions for petroleum generation in the Boryslav–Pokuttya unit.

The preoverthrusting burial history of the Menilite shales (Figure 31) comprises two significant hiatuses, corresponding to the erosion surfaces of the Menilite shales and of the unconformably overlying Miocene molasse sequence (Koltun et al., 1998). The thicknesses of both successions are highly variable, and the reconstruction of their pre-erosional values is difficult. Therefore, the average values of the present-day thicknesses were used here for the reconstruction of the preoverthrusting maturation history of the Menilite shales. In general, burial temperatures during this stage were too low to cause any significant hydrocarbon generation in the Menilite shales before the beginning of the overthrusting (Figure 31).

Thrust emplacement in the frontal part of the Ukrainian Carpathians began at 18 Ma (Oszczypko and Slaczka, 1986) and lasted until the post-early Sarmatian. It consisted of several tectonic episodes of thrust reactivation (Kulchytsky, 1986). Although the main thrust emplacement stopped after the early Sarmatian (Kulchytsky, 1986; Roure et al., 1993), residual fold and thrust movements persisted until at least the Pliocene (Kulczytsky, 1986; Oszczypko and Slaczka, 1986) and perhaps until the present day.

Figure 31 shows the time-subsidence curve for the preoverthrusting stage and several curves for subsequent burial history of the Menilite shales, occurring in depths ranging from the surface to 8 km (5 mi). Very little geological evidence is available to reconstruct the detailed, thrust-induced burial history of the Menilite shales, and therefore, the time-subsidence curve used in this study is only a rough approximation, assuming that the subsidence was uniform and lasted from 18 to 10 Ma. To determine the reliability of this time-subsidence plot, the computed TTI and the measured T_{max} parameter were used as two independent thermal maturity parameters for the Menilite samples, collected at different depths in several wells of the Boryslav–Pokuttya unit (Figure 31). The average present-day geothermal gradient value of 24°C/km was accepted for TTI calculations. Both parameters show a similar trend of depth-related thermal maturation.

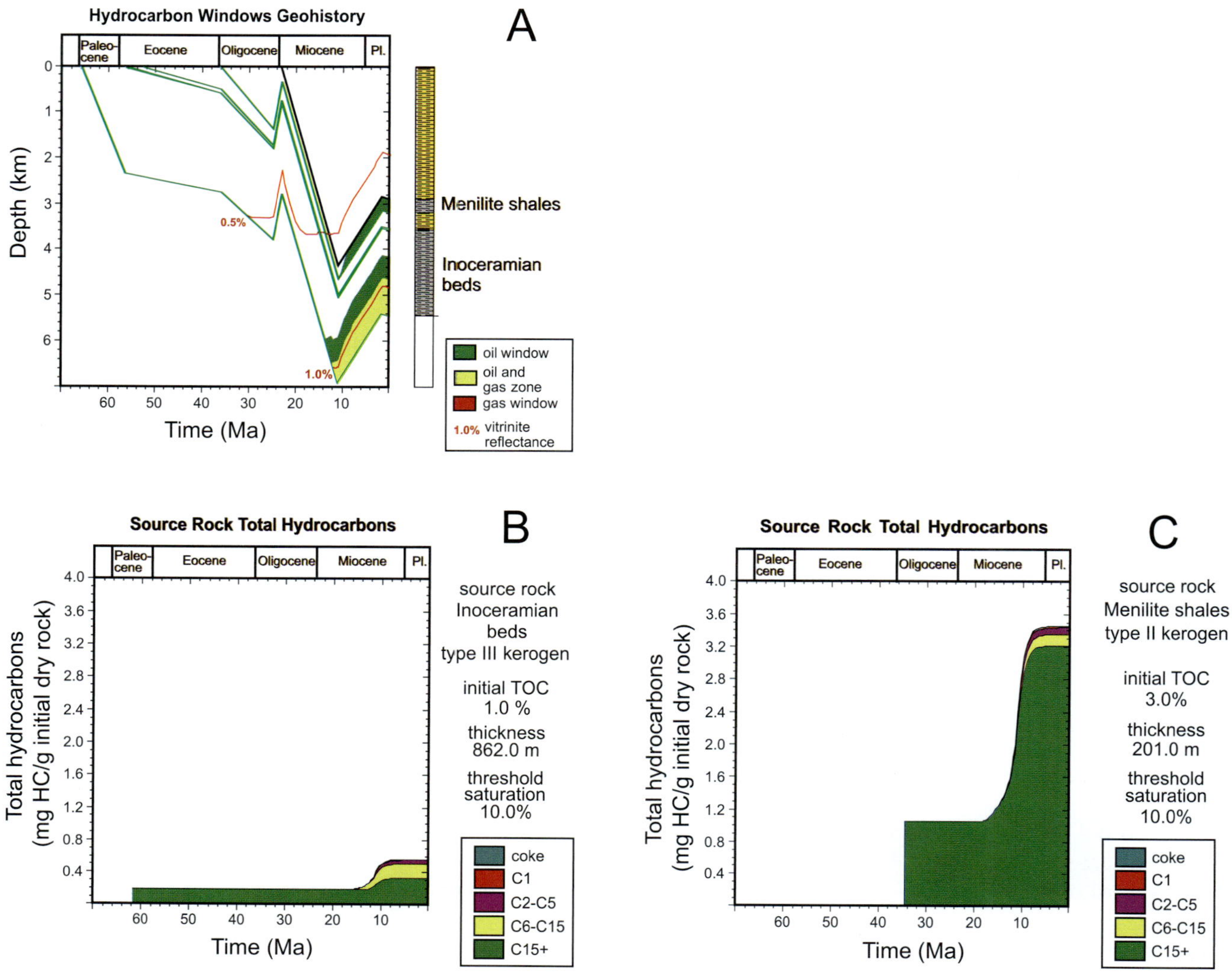

Figure 30. (A) Burial-history curves with thermal maturity windows for Brzegi Dolne IG-1 (BG-1) well and amount of hydrocarbons generated from (B) Menilite shales and (C) Inoceramian beds source rocks in Brzegi Dolne IG-1 well (Skole nappe, the Polish Flysch Carpathians). Modeling using Genex program. For the location of BG-1 well, see Figure 1 (Pl. = Pliocene; TOC = total organic carbon).

According to TTI values, the oil window is located in the depth range of 4.2–5.8 km (2.6–3.6 mi) (Koltun et al., 1998). Peak oil generation occurs at a depth of 5.3 km (3.3 mi). T_{max} values also indicate the beginning of the oil window at depths of 4 km (2.5 mi) and its end at about 6 km (4 mi).

The evolution of the depth of oil-window occurrence through time is also plotted in Figure 31, showing the timing of oil generation by the deeply buried Menilite shales. If we consider that the main thrust episode was finished 10 m.y., then the main traps were already formed by that time and could accumulate hydrocarbons, generated at depths greater than the present-day deadline of the oil window. This means that the present-day occurrence of the petroleum fields is not limited by the depth range of the present-day oil window. According to TTI values, the maximum depth of oil occurrence, with API gravity less than 50°, can be expected at a depth of 6.9 km (4.28 mi) (TTI = 1000), whereas wet gas should still be found at 7.1 km (4.41 mi) (TTI = 1500). Because the Menilite shales in the frontal part of the flysch belt can occur at every depth ranging from the surface to more than 10 km (6 mi) because of a complex fold and thrust structure, this plot can be used as a rough estimation of their maturation level at any depth in the Boryslav–Pokuttya unit.

The above data, involving several wells throughout the Boryslav–Pokuttya unit, show that the time-subsidence plot (Figure 31) generally corresponds to the depth-related evolution of the measured thermal maturity parameter, T_{max}, and sometimes slightly overestimates but never underestimates the maturation

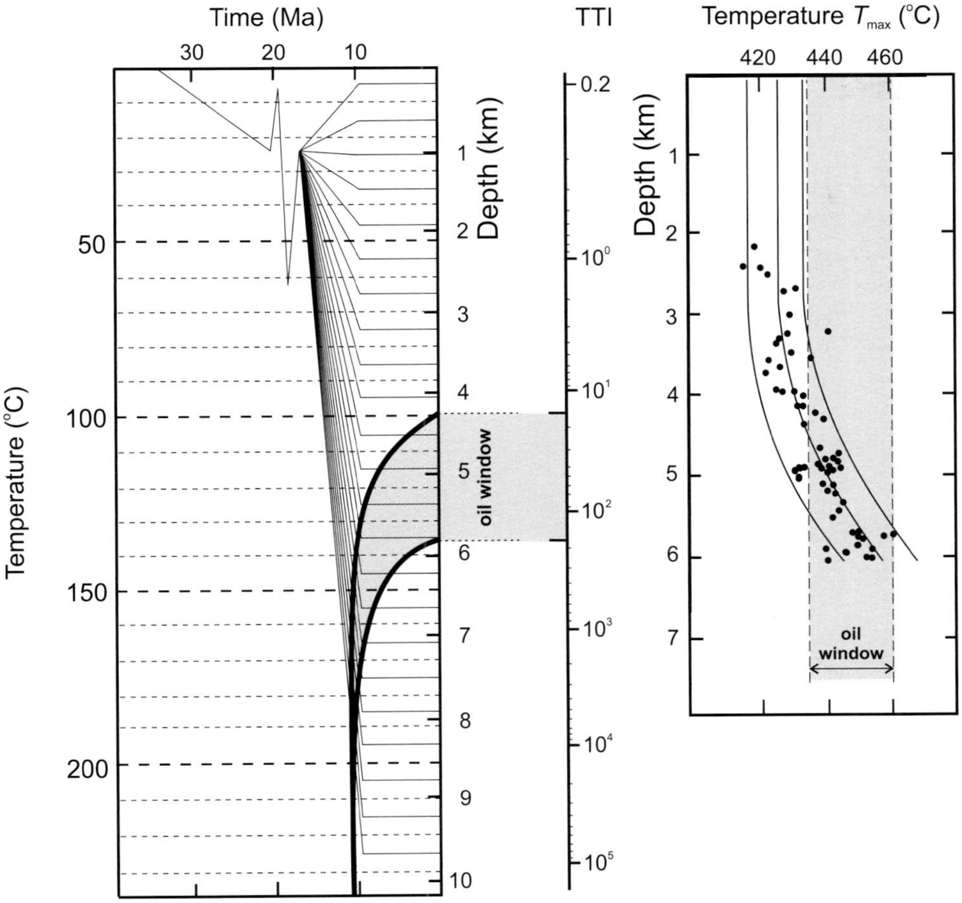

Figure 31. Time-subsidence plot, evolution of the oil window, and correlation of the calculated TTI and measured T_{max} values for the Oligocene Menilite shales of Boryslav–Pokuttya unit (the Ukrainian Flysch Carpathians). Burial curves are shown for the Menilite shales occurring within depth range from the surface to 8 km (5 mi).

level of the Menilite shales in the lower part of the oil window. Two assumptions were made when constructing the plot:

- The heat flow was already as low as today, at least since the Oligocene.
- The present burial of the Menilite shales was reached during the thrust emplacement and remained constant since the post-early Sarmatian.

Hence, although this time-subsidence plot represents the minimum extent of thermal maturation in the Menilite series while subsiding to the present-day depth, it corresponds to the measured (T_{max}) maturation level. If we assumed that the maximum overthrust-related burial of these beds was generally deeper, and then, because of erosion, they were uplifted to the present-day depth, their maturation level would significantly exceed the observed one. This indicates that the present-day occurrence of thrust slices of the Boryslav–Pokuttya unit, which were buried by the overthrust movements, represents their maximum burial depth since the beginning of the emplacement. We assume then that the erosion of the frontal part of the Skiba nappe, which was synchronous with thrust emplacement (Roure et al., 1993), constantly balanced the tectonic burial of the Boryslav–Pokuttya unit. The second conclusion is that the heat flow in the frontal part of the flysch belt has not undergone significant changes since the Oligocene.

In contrast to the Boryslav–Pokuttya unit, a small number of wells have been drilled in the more internal parts of the flysch belt. Only some general assumptions about the maturation history of these rocks can be drawn because of the scarcity of data on thermal maturity of the rocks from these areas.

Skiba (Skole) Unit. Because of erosion, different formations from Early Cretaceous to Oligocene–early Miocene age crop out in the Skiba unit. In places where the older rocks crop out, erosion of as much as several kilometers is evident. Menilite shales, cropping out along the Chechva River (Figure 1), were studied by means of Rock-Eval pyrolysis (Koltun et al., 1998). The rocks are immature. The mean T_{max} value of 426°C shows that their maturation level does not significantly differ from their preoverthrusting level.

Lower Cretaceous organic-rich rocks of the Spas shales, cropping out near the Tershiv village (Figure 1), have T_{max} values ranging from 429 to 431°C (Koltun et al., 1998). Their burial curve is plotted in Figure 32. The TTI value of 6, calculated using the present-day geothermal gradient of 20.3°C/km, corresponds to the maturation level measured by T_{max} parameter, suggesting that the heat flow most likely did not change significantly during that time. Spas shales of the Skiba unit were also studied at depth intervals ranging from 4374 to 4429 m (14,350 to 14,530 ft) in the Maxymivka-4 well. T_{max} values range from 445 to 451°C, indicating their occurrence in the oil window around the peak of oil generation, prior to subsequent erosion and uplift to their present-day depth. This burial curve (Figure 32) shows that 2 km (1.2 mi) of additional tectonic burial was necessary prior to erosion to account for the observed maturity.

Krosno (Silesian) Unit. Several sites in the Krosno unit (Figure 1) show the advanced maturation level of the Oligocene series relative to the more external units (Koltun et al., 1998).

In the central part of the Krosno unit, near Synievir village (Figure 1), the outcropping Oligocene beds are overmature as indicated by T_{max} values ranging from 457 to 500°C.

In the southwest part of the Krosno unit, outcrop Oligocene samples near the town of Yasynia (Figure 1) show a maturation level corresponding to the beginning of the oil window (T_{max} values ranging from 428 to 442°C).

In the Gryniava-1 well, southwest Yasynia (Figure 1), the Oligocene rocks of the Krosno unit occur under the Chornogora nappe at a depth of 4180–4361 m (13,713–14,307 ft). They are presently overmature, as indicated by T_{max} values ranging from 465 to 498°C.

The maturation levels of the Oligocene series in all of these sites indicate that significant erosion must have occurred in the Krosno unit. By the end of overthrusting in the early Sarmatian, the southwest part of this unit was probably still buried beneath the thick frontal part of the adjacent Chornogora and Dukla nappes, prior to subsequent erosion.

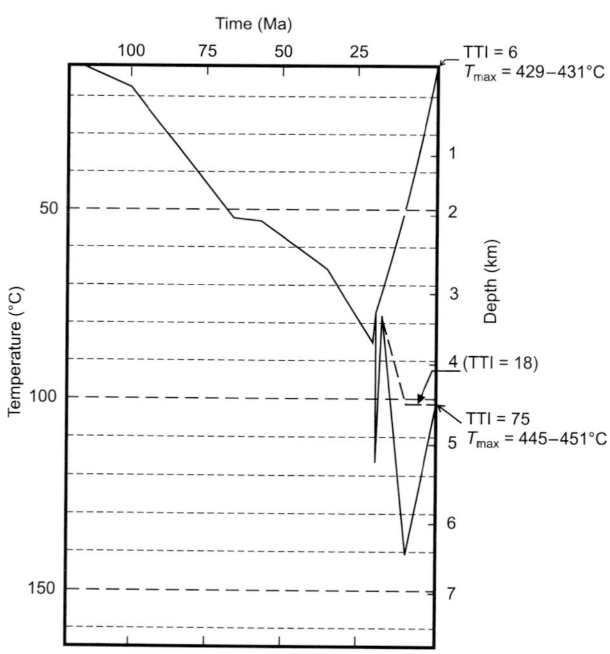

Figure 32. Time-subsidence plot for the Lower Cretaceous Spas beds from Tershiv (TR) outcrop and Maxymivka-4 (Ma-4) well, Skiba unit (the Ukrainian Flysch Carpathians). Dashed line shows the burial curve without the postoverthrusting erosion. For the location of TR outcrop and Ma-4 well, see Figure 1.

Chornogora Unit. The Chornogora unit is thrust southeastward over the Krosno unit. Large outcrops of the Lower Cretaceous organic-rich Shypot series occur throughout the area. Several outcrop samples indicate that their maturation level corresponds to the beginning of the oil window (T_{max} ranges from 433 to 441°C) (Koltun et al., 1998). The Paleogene to Lower Cretaceous flysch sequence of the Chornogora nappe was drilled in the Gryniava-1 well, and the Oligocene beds of the underlying Krosno unit were also penetrated. Shypot shales, from a depth interval of 3496 to 3661 m (11,469 to 12,011 ft), show T_{max} values ranging between 454 and 458°C, corresponding to the central part of the oil window. Figure 33 represents the burial curves of the Shypot shales from both the Gryniava-1 well and adjacent outcrops. If we assume that the heat flow remained constant, then 3 km (1.8 mi) of additional burial is required prior to late-stage erosion to account for the observed maturation level in the Shypot shales from the Gryniava-1 well. Presently outcropping beds have probably been subjected to an earlier tectonic burial to reach their maturation level (Figure 33). Thus, after the end of the overthrusting, all the area must have been subjected to significant uplift and erosion.

The Carpathian Foredeep

The Polish Area

More than 70 gas fields discovered in the autochthonous Miocene strata contain practically only methane (>98 vol.%) accompanied by small amounts of higher gaseous hydrocarbons (Kotarba, 1998a).

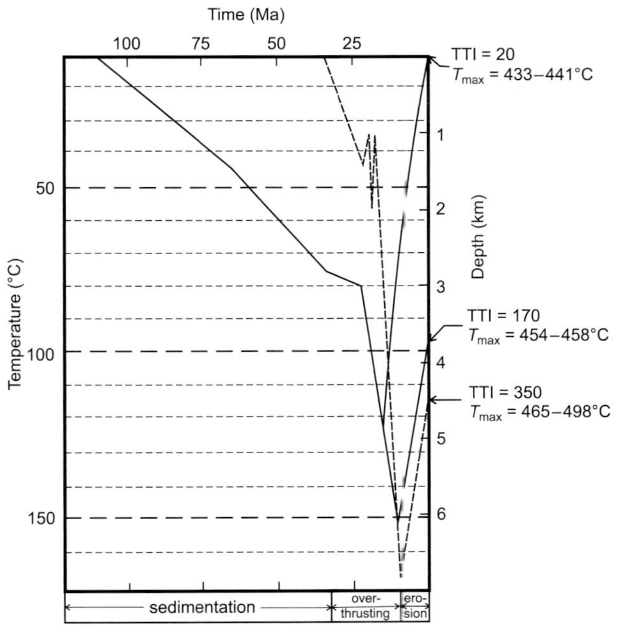

Figure 33. Time-subsidence plot for the Lower Cretaceous Shypot beds from Gryniava (GR) outcrop and Gryniava-1 (Gr-1) well, Chornogora unit (the Ukrainian Flysch Carpathians). Dashed line shows the time-subsidence plot for the Oligocene beds from Krosno unit, discovered by Gryniava-1 well under the Chornogora nappe. For the location of GR outcrop and Gr-1 well, see Figure 1.

The stable carbon isotope composition of methane, ethane, and propane and the stable hydrogen composition of methane allow for the identification of genetic types of natural gases, migration, and mixing processes, as well as evaluation of source rock maturity (e.g., Schoell, 1983; Whiticar et al., 1986; Whiticar, 1990, 1994; Rice, 1992; Berner and Faber, 1996). Gas, generated at different stages of diagenetic evolution of sedimentary rocks, has a distinct molecular and isotopic composition. Microbial gas has several features, which allow one to distinguish it from gas of thermogenic origin. First of all, microbial gas consists mainly of methane. The isotopic composition provides the best evidence for the origin of gas. Methane in an ancient microbial gas is isotopically light ($\delta^{13}C$ values in the range of -90 to -55‰) because it is enriched with light isotope ^{12}C by microorganisms. In contrast, $\delta^{13}C$ values for thermogenic methane are generally heavier than -50‰ (e.g., Schoell, 1983; Whiticar, 1990, 1994). Microbial gases can also be differentiated depending on the environments of their generation (Rice, 1992).

Molecular and stable isotope compositions were determined for 59 gas samples from accumulations in the autochthonous middle Miocene strata of the Carpathian Foredeep (Kotarba et al., 1987; Kotarba, 1992, 1998a). Gases from middle Miocene deposits vary both in their molecular and stable isotope compositions. Geochemical indices and stable isotope ratios vary in the following ranges: hydrocarbon (C_{HC}) index [$C_{HC} = CH_4/(C_2H_6 + C_3H_8)$] from 26 to 1667; $\delta^{13}C(CH_4)$ from -72.6 to -35.7‰, $\delta D(CH_4)$ from -215 to -151‰; $\delta^{13}C(C_2H_6)$ from -57.4 to -28.1‰; and $\delta^{13}C(C_2H_6)$ from -31.8 to -23.8‰. The depth of sampled gas accumulations in Miocene strata ranges from 161 to 2640 m (528 to 8661 ft). Most of the analyzed gases were collected from the lower Sarmatian and upper Badenian strata. One exception is the gas from the Rozwienica field, which comes from the lower (and partly middle) Badenian sequence. In the Lubaczow and Uszkowce deposits, natural gas occurs in the lower Badenian strata and in the upper parts of the Upper Jurassic carbonates. The stable carbon isotope composition of methane, ethane, and propane and the stable hydrogen composition of methane (Figures 25B–28B) indicate that the hydrocarbons were generated by microbial processes and, occasionally, in the initial phase of thermogenic processes. Methane was produced during microbial carbon dioxide reduction (Figure 26B). A very small component of diagenetic and/or low-temperature thermogenic gas was identified in Buszkowiczki (Bu-2 well), Rozwienica (Ro-2 well), and Lubaczow (Lu-3 and Lu-22 wells) fields (Figure 25B). In the Lubaczow field, the Upper Jurassic carbonates also contained gas. Genetically, this gas is a typical microbial methane that migrated to the Upper Jurassic trap from the autochthonous Miocene strata along the fault zone (Figure 34). Minor ethane and propane were generated both in the

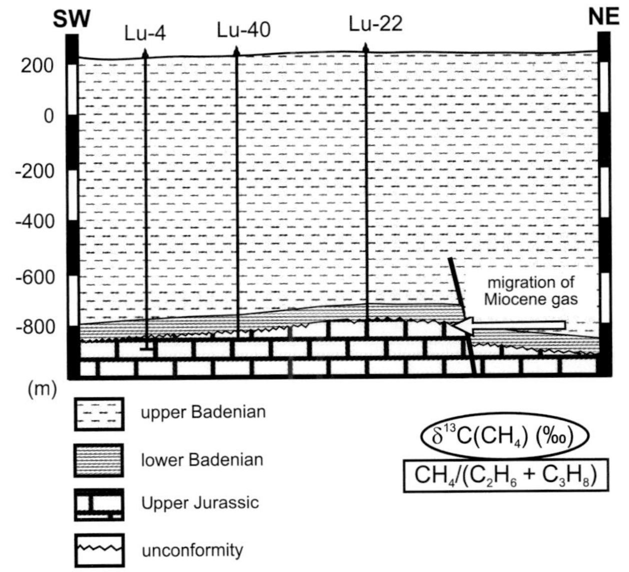

Figure 34. Schematic geological cross section through the Lubaczow field in the Mesozoic–Paleozoic basement of the Polish Carpathian Foredeep. Isotopic data after Kotarba (1998a).

microbial and diagenetic phase and in the initial phase of low-temperature thermogenic processes. The highest content of the thermogenic component was reported from Kielanowka (Ki-1), Blizna (Bl-5), and Ocieka (Oc-1) wells, whereas the largest concentration of microbial (diagenetic) ethane was found in the Przeworsk (Pr-14) well (Figure 26B). Insignificant changes in values of geochemical indices and isotopic ratios with depth (Figure 35) indicate uniform generation conditions of the microbial gases in the whole Miocene succession.

The only exception in the origin of gases accumulated in the autochthonous Miocene strata is gas from the Tarnow-45 well, which is typical of a thermogenic origin (T-45 sample, Figures 25B–28B). This gas was generated from the oil-prone type II kerogen. It migrated with the oil to the autochthonous Miocene strata from the Upper Jurassic carbonates (Figure 36) in which several deposits accumulating oil and gas produced from this genetic type of organic matter were discovered (Kotarba and Jawor, 1993; ten Haven et al., 1993). Unexpectedly, an oil show was encountered in the Tarnow-47 well at the bottom part of the autochthonous Miocene strata (Figure 36). It is surprising for the Carpathian Foredeep.

The TTI hydrocarbon generation modeling in the autochthonous Miocene strata was analyzed in five selected areas (Kotarba et al., 1998b). The Tarnow and Mielec areas represent the part of the Carpathian Foredeep located north of the present edge of the Carpathian Foredeep, and the Rzeszow and Przemysl areas are located in the outer autochthonous Miocene basin along and beneath the Flysch overthrust (Figure 1). The Sanok area was chosen as an example of hypothetical deepest levels, based on the present depth about 11,000 m (36,000 ft), where the relics of autochthonous lower Miocene strata might have existed (Kotarba et al., 1998b). The results of the modeling (Figure 37) revealed that for both the Tarnow and Mielec areas, the TTI values were 0.1 and 1, respectively. In the Rzeszow and Przemysl areas, the maximum TTI values were 21 at a depth of about 3600 m (11,800 ft) and 4 at a depth of about 2600 m (8500 ft), respectively. These values suggest the possibility of low-temperature thermogenic processes in the autochthonous Miocene strata beneath the Carpathian overthrust, i.e., below 2500 m (8200 ft). In the Sanok hypothetical areas, thermogenic gaseous hydrocarbons could have been generated at more than 7500 m (24,600 ft), provided that the autochthonous Miocene strata in that area were rich in organic matter. The result of TTI modeling suggests the generation of microbial methane in almost the full thickness of the outer basin to the north from the edge of the Carpathian overthrust.

Glogoczowski (1980) and Kotarba (1992) suggested that the microbial methane might have accumulated as hydrate beneath the Miocene seafloor. However, the results of new bathymetric analyses (Czepiec and Kotarba, 1998) revealed that the Miocene sea was very shallow, commonly 10–50 m (33 to 160 ft) deep, and, in the Badenian, was only slightly deeper, about 300 m (1000 ft). In addition, the Miocene sea was a warm basin with surface-water temperatures ranging from

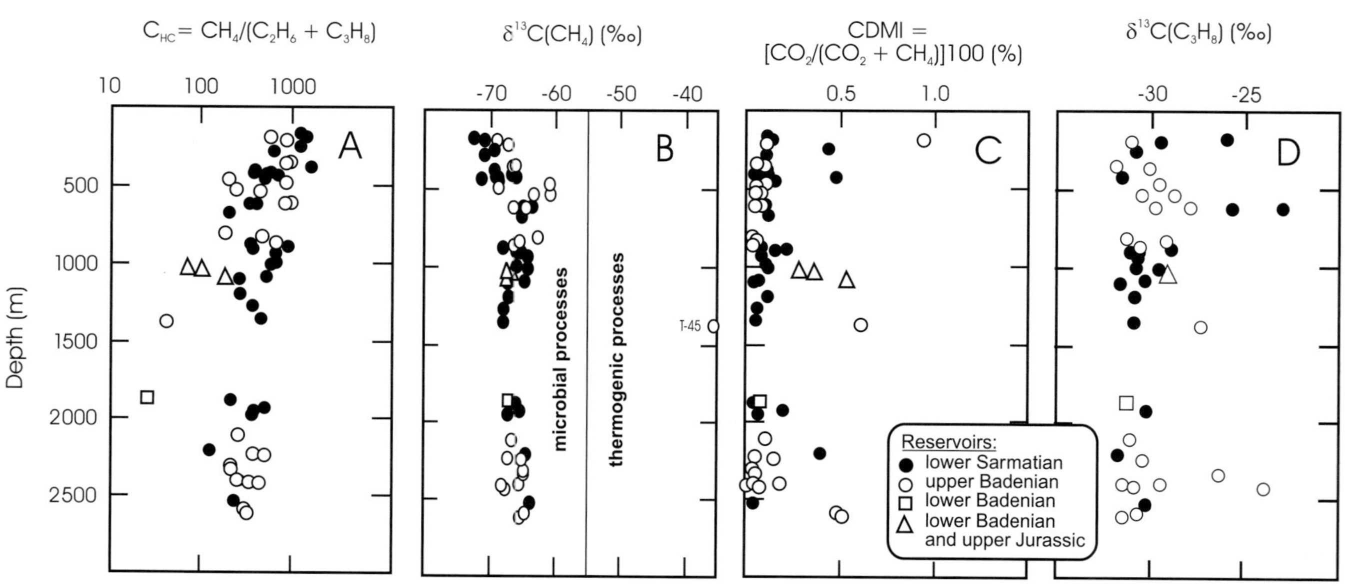

Figure 35. (A) Hydrocarbon index, (B) $\delta^{13}C(CH_4)$, (C) carbon dioxide-methane index, and (D) $\delta^{13}C(C_3H_8)$ vs. depth for natural gases from the autochthonous Miocene rocks of the Polish Carpathian Foredeep.

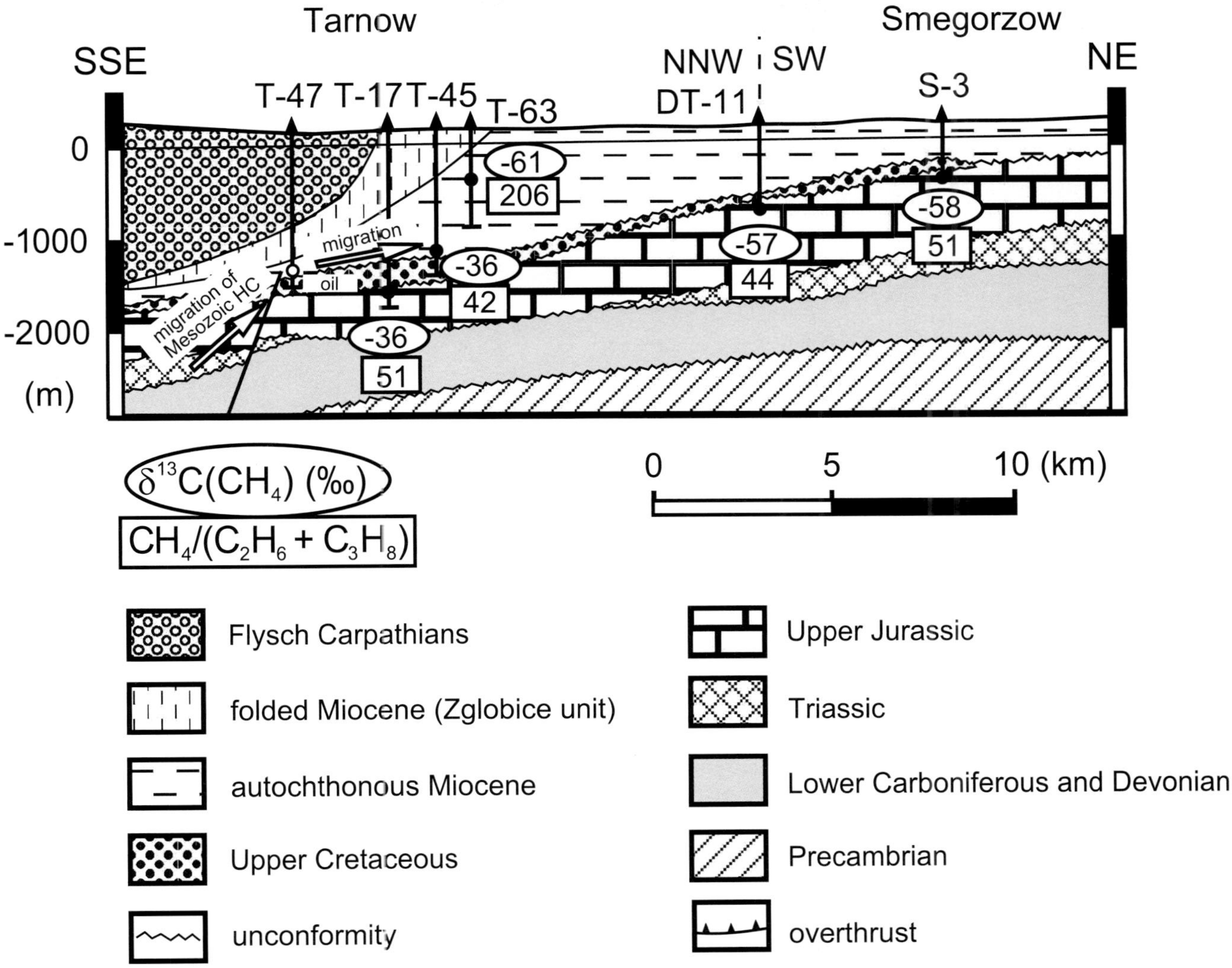

Figure 36. Schematic geological cross section through the Tarnow, Dabrowa Tarnowska, and Smegorzow fields in the Mesozoic–Paleozoic basement of the Polish Carpathian Foredeep. Isotopic data after Kotarba and Jawor (1993). For the location, see Figure 2.

17 to 20°C. Considering these data, the formation of the methane hydrate zone beneath the Miocene seafloor probably did not occur (Kotarba et al., 1998b).

The Ukrainian Area

Results of geochemical investigations of the Miocene rocks discussed above show the possibility of microbial gas generation in the Neogene sequence of the Bilche–Volytsia unit. Geochemical study of gas samples from the same unit was performed to confirm this concept and to indicate the processes that resulted in the formation of this significant gas-producing region.

Molecular and isotope composition studies of gas samples from several Miocene fields of the Bilche–Volytsia unit were carried out (Tables 4, 5). Published isotopic data (Shabo and Mamchur, 1984) are also presented (Table 6).

Tables 4–6 show that for most of the samples, methane content exceeds 98% and hydrocarbon index $CH_4/(C_2H_6 + C_3H_8)$ is much greater than 100, indicating their microbial origin. However, several samples show somewhat higher content of gaseous C_2–C_4 alkanes. These are the Gl-1 sample from Glynkivska field, (Table 4) with methane content of 92.6%, and samples from Zaluzhany and Pyniany fields (Table 6), with hydrocarbon index $CH_4/(C_2H_6 + C_3H_8)$ of 63 and 29, respectively. The Glynkivska field occurs in the Karpatian reservoir at the base of the Miocene sequence and evidently is sourced at least partly by the Middle Jurassic rocks of the basement, which contains an admixture of thermogenic gases. The samples of Zaluzhany and Pyniany fields differ from the other samples

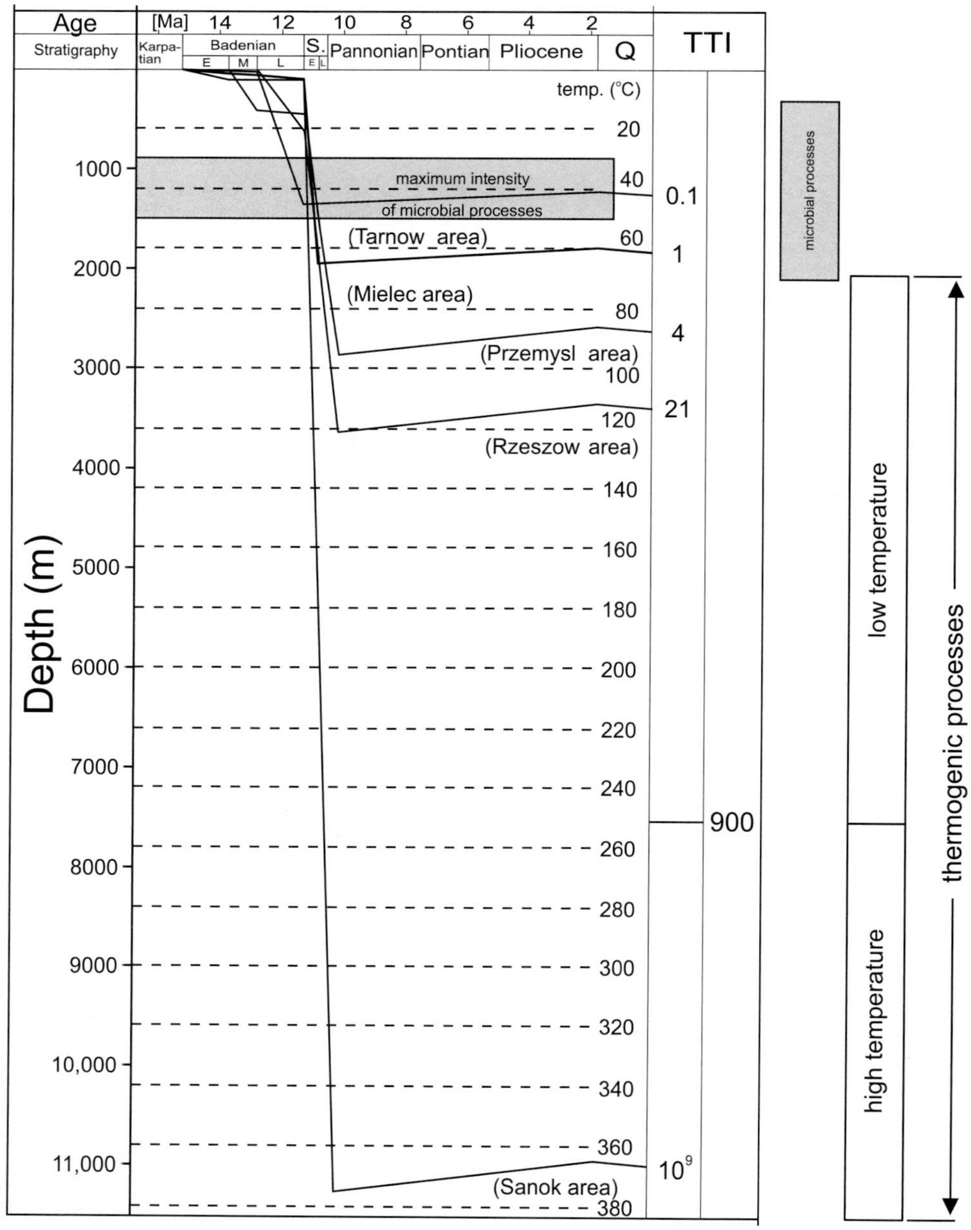

E = early; M = middle; L = late;
S = Sarmatian; Q = Quaternary

Figure 37. Model of gas generation in the autochthonous Miocene rocks of the Polish Carpathian Foredeep. Family of burial-history curves, subsurface temperature, and time-temperature index (TTI) values for the Tarnow, Mielec, Rzeszow, and Sanok areas is after Kotarba et al. (1998a).

also by their isotopic composition and by their depth of occurrence (Table 6) and, hence, may contain thermogenic hydrocarbons generated by the same Miocene rocks.

Figures 25C–28C represent the genetic characterization of natural gases in terms of stable carbon isotope composition of methane, ethane, and propane and stable hydrogen isotope composition of methane. They show that the major part of the gases was generated during microbial processes. However, some samples representing the deeper Miocene fields Pyniany (Py-24) and Zaluzhany (Za-12 and Za-14), as well as the Karpatian reservoir, tested by well Glynkivska-1 (Gl-1), show an admixture of thermogenic gas, as noted in Tables 5 and 6. Figure 38 shows the depth-related evolution of hydrocarbon index and carbon isotopic composition of methane for the Miocene gases. An evident increase of both parameters with depth indicates the contribution of thermogenic gas.

Figure 26C shows the plot of carbon vs. hydrogen isotopic composition of methane in the samples under study. It shows that the Miocene gases were formed

Table 4. Molecular composition of natural gases from gas fields of the Bilche–Volytsia unit (the Ukrainian Carpathian Foredeep).

Well	Age of Reservoir	Reservoir Depth Interval (m)	Molecular Composition (vol.%)											
			CH_4	C_2H_6	C_3H_8	i-C_4H_{10}	n-C_4H_{10}	C_5H_{12}	C_6H_{14}	N_2	CO_2	Ar	He	H_2
Bilche–Volytsia-9 (BV-9)	lower Sarmatian	841–940	98.4	0.18	0.052	0.015	0.006	0.009	–	1.21	0.14	0.008	0.013	–
Bilche–Volytsia-500 (BV-500)	lower Sarmatian	636–650	98.8	0.11	0.010	0.005	0.001	–	–	0.95	0.14	0.008	0.008	–
Letnia-2 (Le-2)	lower Sarmatian	1245–1267	98.6	0.16	0.045	0.020	0.007	0.009	–	1.09	0.08	0.008	0.007	–
Glynkivska-1 (Gl-1)	Karpatian	1192–1222	92.6	2.26	1.65	0.36	0.52	0.33	0.11	2.00	0.08	0.013	0.036	–

as a result of microbial carbon dioxide reduction and not microbial fermentation.

Summarizing the above-mentioned data, it should be emphasized that a major part of the gas from the Miocene beds of the Ukrainian Carpathian Foredeep was generated by the same Miocene strata by microbial processes of carbon dioxide reduction. However, in part, the gases contain an admixture of thermogenic components generated either in the deeper horizons of the same Miocene beds or in older source rocks from the pre-Miocene basement.

To understand the variations of the isotopic and molecular composition of gases from different fields, they were considered with respect to their geological structure and, hence, the peculiarities of gas generation.

1) In fields where the Neogene sedimentary cover overlies the metamorphosed Proterozoic basement, only the Miocene rocks can be the source of hydrocarbons. This group, in turn, can be subdivided into two subgroups:
   a) The Pyniany and Zaluzhany gas fields, where the Neogene sequence exceeds 3-km (1.8-mi) thickness, and therefore, in its lower part, the thermal maturation level is sufficiently high to generate thermogenic hydrocarbons generation. Methane $\delta^{13}C$ values from these fields are from −57.5 to −50.0‰, indicating a thermogenic gas contribution. The content of higher hydrocarbons in these deposits increases with depth, and condensate appears in the deepest gas reservoirs. Pyniany and Zaluzhany fields occur in the Krukenychi depression, where the thickness of the Neogene sedimentary sequence is much greater than in all other parts of the Bilche–Volytsia unit. The Neogene thickness reaches 4 and 4.5 km (2.5 and 2.8 mi), whereas the thickness of the Miocene strata of the other fields, occurring outside this depression, normally does not exceed 1.5 km (0.9 mi) and never 3 km (1.8 mi). Considering the present-day geothermal gradient in the area (around 34°C/km), the thermogenic hydrocarbon generation may occur in the Neogene sequences where thickness exceeds 3 km (1.8 mi). These hydrocarbons can make a minor contribution to the microbial gas in the overlying sequence.
   b) Other gas fields, where the Neogene sequence is commonly much less than 3 km (1.8 mi) in thickness, should contain only an isotopically

Table 5. Isotopic composition of natural gases from gas fields of the Bilche–Volytsia unit (the Ukrainian Carpathian Foredeep).

Well	Mode of Gas Accumulation	Indices			Stable Isotopes (‰)			
		$\frac{CH_4}{C_2H_6 + C_3H_8}$	$\frac{iCH_4}{nCH_4}$	CDMI	$\delta^{13}C(CH_4)$	$\delta D(CH_4)$	$\delta^{13}C(C_2H_6)$	$\delta^{13}C(C_3H_8)$
Bilche–Volytsia-9 (BV-9)	free	424	2.5	0.14	−64.4	−193	−39.5	−29.4
Bilche–Volytsia-500 (BV-500)	free	823	5.0	0.14	−66.4	−196	−51.4	−29.2
Letnia-2 (Le-2)	free	481	3.0	0.08	−65.6	−198	−42.8	−29.6
Glynkivska-1 (Gl-1)	free	24	0.7	0.09	−59.8	−197	−30.1	−27.5

CDMI = $[CO_2/(CO_2 + CH_4)]$ 100 (%).

Table 6. Natural gases from Miocene reservoirs of the Ukrainian Carpathian Foredeep. The isotopic data is according to Shabo and Mamchur (1984).

Field Wells	Depth (m)	Indices		Stable Isotopes (‰)	
		$\frac{CH_4}{C_2H_6 + C_3H_8}$	CDMI	$\delta^{13}C(CH_4)$	$\delta^{13}C(CO_2)$
Uhersko-98a	365–369	244	0.20	−58.0	n.a.
Kosiv-11	389–485	74	0.30	−67.2	n.a.
Svidnytsia-54	350–450	432	0.08	−61.4	n.a.
Svidnytsia-48, Svidnytsia-51, Svidnytsia-61	640–644	904	0.08	−60.7	n.a.
Svidnytsia-55, Svidnytsia-62	660–667	368	0.10	−59.4	n.a.
Dashava-65	754–770	331	0.20	−64.3	n.a.
Khodnovichi-133	719–830	224	0.10	−72.0	−12.5
Grynivka-17, Grynivka-51	1100–1300	993	0.10	−56.0	−18.1
Zaluzhany-12	1400–1600	163	4.90	−53.1	−13.6
Zaluzhany-14	3200–3400	63	0.94	−50.0	n.a.
Pyniany-24	1712–1735	29	0.42	−57.5	−17.1
Pyniany-1	2000–2200	980	0.76	−54.0	n.a.

CDMI = $[CO_2/(CO_2 + CH_4)]$ 100 (%).
n.a. = not analyzed.

light microbial gas that normally exceeds 99% CH_4. Methane $\delta^{13}C$ values in the fields under study are from −72 to −54‰, showing the microbial origin of this gas. Molecular composition of gas in fields of this group also confirms their microbial origin.

2) The fields whose occurrence includes the Miocene reservoirs, as well as older deposits (Upper Cretaceous and Upper Jurassic). They are Pivnichni Medynychi, Malogorozhanske, Bilche–Volytsia, Uherske, Pivdenno Uherske, Letnia, and Rudki fields. These fields occur in the erosional uplifts of the Upper Cretaceous and Upper Jurassic reservoirs, just underlying the Neogene cover. They are sealed by the Miocene clayey or evaporitic sediments and contain gas that migrates into the lowest Neogene (Karpatian) reservoirs. The location of these fields correlates with the band of the organic-rich Middle Jurassic beds occurrence. However, no geochemical investigations have been performed for these fields. The $\delta^{13}C$ values of methane in Badenian and Sarmatian reservoirs of these fields show that gas is mostly of microbial origin, but in some of them, the thermogenic gas contribution is noticeable (Svidnytsia, Uherske, and Glynkivska) (Tables 5, 6). In other fields, the influence of the thermogenic gas in the Badenian–Sarmatian sequence is not noticed (Biltshe–Volytsia and Letnia).

3) Gas fields stretching further to the southeast toward the Romanian border occur in the Miocene succession and are underlain by Cretaceous, Upper Jurassic, and Paleozoic deposits. No significant organic-rich beds are present in the Mesozoic sequence, but most of these deposits show the $\delta^{13}C$ values characteristic for microbial gas (Dashava and Kosiv); however, some thermogenic gas input from the Mesozoic Beds could occur (Grynivka).

The Mesozoic–Paleozoic Basement

The Polish Area

Seventeen oil and gas fields were discovered in the Mesozoic basement. They mainly occur in Upper Jurassic carbonate (e.g., Leszczyna–Lakta, Lubaczow, Partynia–Podborze, and Tarnow) and Upper Cretaceous sandstone (e.g., Grobla–Plawowice and Swarzow) reservoirs (e.g., Karnkowski, 1999). Only the Nosowka oil field and the Lachowice gas field were discovered in Paleozoic strata. The Lachowice gas field occurs in Devonian carbonate reservoirs, and the Nosowka oil field occurs in Lower Carboniferous carbonate reservoirs of the southeastern part of the foreland autochthon, beneath the Carpathian flysch allochthon. In addition, large coal-bed methane resources were found in the southwestern part of the Polish Carpathian Province in the Upper Carboniferous coal-bearing strata of the upper Silesian basin (Kotarba, 2001).

Gases from Paleozoic–Mesozoic deposits vary in both their molecular and stable isotope compositions. Geochemical indices and stable isotope ratios vary in the following ranges: hydrocarbon (C_{HC}) index

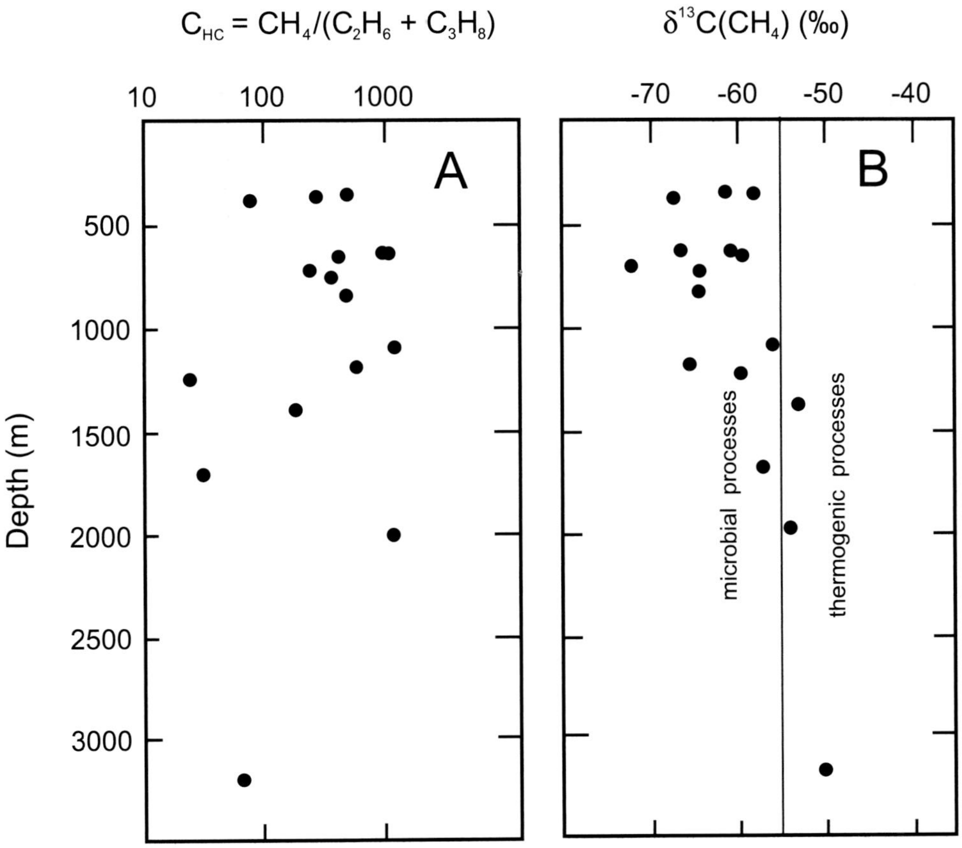

Figure 38. Plots of wet-gas index and stable carbon isotope composition of methane vs. depth for the Miocene gases of the Ukrainian Carpathian Foredeep.

[$C_{HC} = CH_4/(C_2H_6 + C_3H_8)$] from 10 to 51; $\delta^{13}C(CH_4)$ from −58.5 to −31.7‰; $\delta D(CH_4)$ from−182 to −134‰; $\delta^{13}C(C_2H_6)$ from −28.8 to −22.3‰; and $\delta^{13}C(C_2H_6)$ from −27.0 to −22.3‰ (Kotarba and Jawor, 1993; M. Kotarba, 2004, unpublished data).

Natural gases accumulated in the Devonian strata of the Lachowice field (Figure 2, La-2 sample) were generated from type III kerogen, which presumably contained an algal type II kerogen component (Figures 25D–28D). This matter has reached a high degree of maturation (about 2.4% on the vitrinite reflectance scale) (Figure 27D). This natural gas appears to be a typical, nonassociated, thermogenic gas. However, a relatively high content of higher gaseous hydrocarbons [$CH_4/(C_2H_6 + C_3H_8) = 13$] and the presence of condensate presumably reflect the long migration that resulted in the removal of higher hydrocarbons formed from type II organic matter at an earlier maturation stage. The data suggested that the Devonian strata beneath the Carpathian Foredeep should contain both associated and nonassociated gases, which may be accompanied by condensate and even oil (Kotarba and Jawor, 1993).

Gases accumulated in both the Upper Jurassic and Upper Cretaceous suites are generated by microbial and thermogenic processes (Figures 25D–28D). The gases present in the Brzezowka, Leszczyna–Lakta, Groble, Rylowa, Tarnow, Wierzchoslawice, and Zagorzyce fields do not reveal the presence of a microbial methane component, which suggests that the trap did not exist during the microbial and diagenetic stages of organic matter transformation. They are typical thermogenic gases generated from oil-prone type II kerogen sometimes with a minor terrestrial admixture (Figures 25D–28D). The gases were originated during at least two phases: both low-temperature thermogenic (gases associated with oil and condensate) and high-temperature thermogenic (nonassociated gases) (Figure 29B).

A significant microbial methane component (Figures 25D–27D) was found in the Upper Jurassic strata in the Smegorzow (S-3 well) and Dabrowa Tarnowska (DT-11 well) fields. The dominant thermogenic component was generated from the same types of kerogen and processes as mentioned above. These data prove that for the Smegorzow and Dabrowa Tarnowska fields, the traps had already been formed and sealed as early as the initial, microbial stage of the Middle Jurassic organic matter. The traps were successively supplied with hydrocarbons generated from the same source organic matter at successively higher maturation stages or/and migrated from the flysch complex.

Isotopic studies reveal (Figures 25D–28D) that the Upper Cretaceous (Cenomanian) sandstone trap in the Brzozowiec field (Bc-3 sample) was filled with the microbial methane that probably migrated from the autochthonous Miocene strata through the contact on the erosion slope (Figure 39).

Isotopic data from gases accumulated in the Paleozoic–Mesozoic basement complex suggest that the thermogenic generation of hydrocarbons was a multiphase process, and that traps in all deposits were supplied at various times during successive hydrocarbon phases.

Isotopic and biomarker studies of oils from Paleozoic–Mesozoic reservoirs reveal that they are multiple origins (ten Haven et al., 1993). The oils accumulated in Upper Cretaceous and Upper Jurassic reservoirs in the Grobla, Lakta, Tarnow, Partynia–Podborze, Pilzno, Jastrzabka Stara, and Dabrowa Tarnowska fields

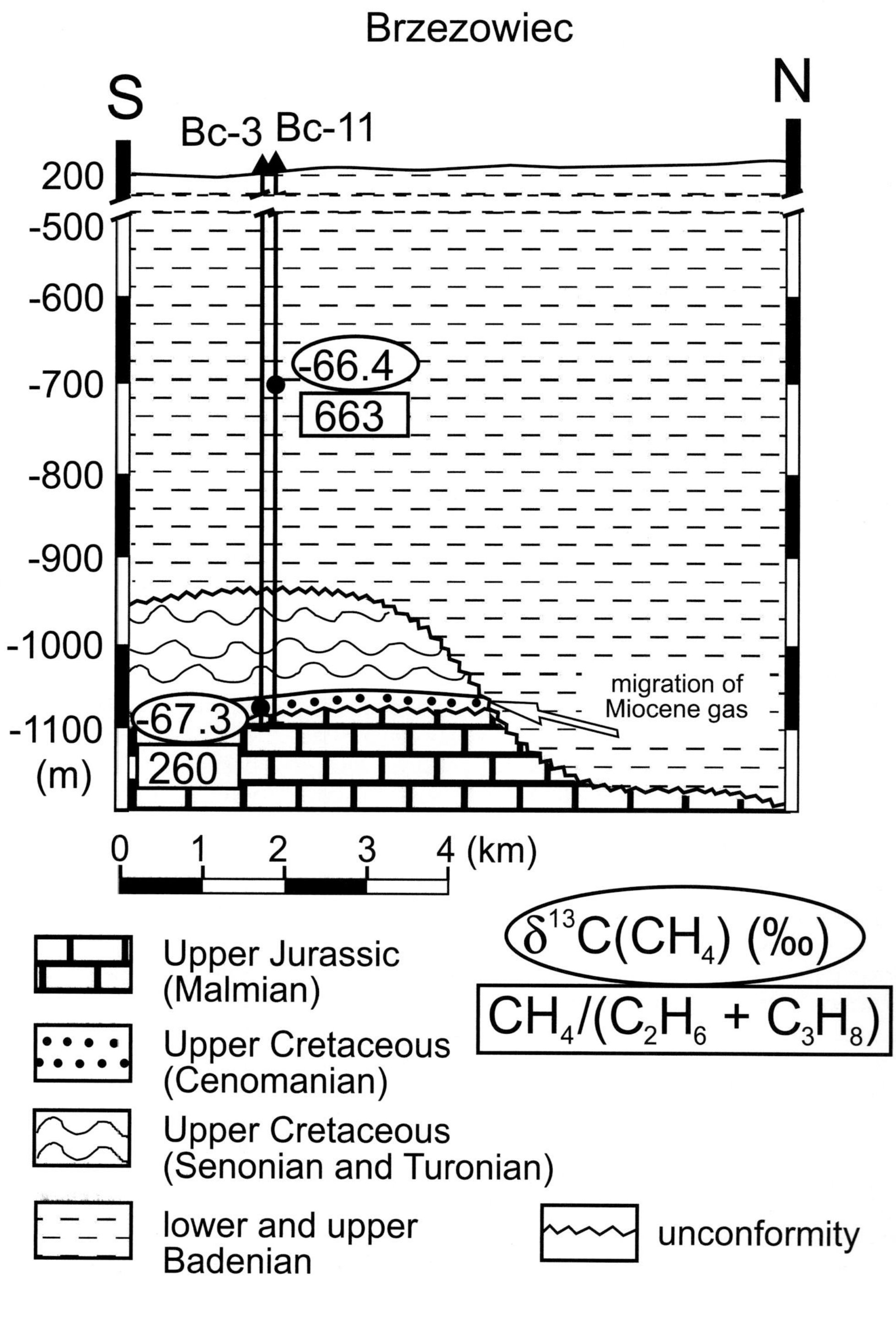

Figure 39. Schematic geological cross section through the Brzezowiec field in the Polish Carpathian Foredeep. Isotopic data after Kotarba (1998a). For the location, see Figure 2.

contain oleanane and other specific biomarkers and have a similar stable isotope composition. They belong to the same geochemical family. Higher hydrocarbon components of these oils most probably were generated from organic matter in the Middle Jurassic mudstones. However, the presence of oleanane indicates another genetic component that generated in Oligocene Menilite shales and migrated from the Flysch Carpathians (ten Haven et al., 1993). However, some systematic variations in the biomarker and isotopic composition of these oils allow for an alternative interpretation. For example, Picha and Peters (1998) explain the differences in composition of oleanane-rich oils accumulated in the Moravian Precambrian–Paleozoic basement by the existence of organic-rich, marginal, autochthonous Paleogene strata below the flysch belt in the Czech Carpathian Province. These rocks, after being buried below the thrust belt, might have produced hydrocarbons that migrated preferentially into reservoirs in the subthrust foreland plate and, only partly, along the faults into the thrust belt. A similar geological situation may have existed below the flysch belt in the Polish Carpathians. However, the oil in Lower Carboniferous carbonates from Nosowka field that accumulated does not contain oleanane, has a different stable carbon composition, and belongs to a different geochemical family of oil (ten Haven et al., 1993). This oil was probably generated from the type II kerogen in the Silurian–Ordovician strata.

The Ukrainian Area

The main organic-rich levels of the autochthonous foreland, i.e., the Silurian and the Middle Jurassic, respectively beneath and above a major unconformity, have quite different maturation history.

Silurian. Thermal maturity of lower Paleozoic rocks may significantly decrease northwest, which is evidenced by much lower maturation level of this sequence in Polish territory. In contrast, S_2 peaks from Rock-Eval pyrolysis no longer exist for the Silurian rocks from the southeastern part of the Carpathian Foredeep (Ispas-1 well) despite the comparatively good TOC content (as much as 0.68%). These data suggest that these rocks are overmature, and their petroleum potential is completely exhausted (Koltun et al,. 1998). The studies of the conodont color alteration index (CAI) of the Paleozoic sequence of the central and southeastern part of the Carpathian foreland (Drygant, 1992) corroborate these data. The value of CAI in Ordovician and Silurian rocks here is 5, corresponding to paleotemperatures far exceeding 200°C. Most likely, this was a result of a deep burial of these beds and may be the results of a high paleoheat flow during the late Paleozoic.

Assuming a type I kerogen for the Silurian, the observed TOC values, ranging from 0.24 to 0.68%, attest for a good source rock potential at the onset of the maturation stage, with the initial TOC content probably exceeding 3% (Espitalie and Bordenave, 1993). However, as their maturation occurred during the pre-Mesozoic, coeval hydrocarbon products could not be preserved in the present-day sequence, because the overmature Paleozoic rocks are separated from the overlying Mesozoic–Cenozoic sequence by a great stratigraphic and thermal unconformity.

Mesozoic. Only three oil fields and seven gas fields were discovered in the Mesozoic sequence of the Ukrainian Carpathian foreland, attesting, however, to the occurrence of potential source rocks in this sedimentary sequence.

Maturation history of the Mesozoic rocks of the autochthon depends on their position relative to the Carpathian allochthon. West of the thrust front, the burial was controlled by the sediment-induced subsidence. This is seen in the Bilche–Volytsia unit, where the rapid deposition of a thick early Miocene molasse sequence induced the maturation of the Mesozoic rocks. The burial depth of the autochthon rapidly increases southwest, where it occurs under the Carpathian allochthon. In this area, tectonic burial was the primary factor for the maturation of the source rocks.

The organic-rich rocks of the Middle Jurassic Kokhanivka beds show T_{max} values from 434 to 438°C in a depth range between 2360 and 2549 m (7742 and 8362 ft) (Mostivska-2 well), which indicate the beginning of the oil window (Koltun et al,. 1998). A thick Cretaceous–Paleogene sequence was required prior to its recent erosion to account for the observed maturity. The 2-km (1.8-mi)-thick Miocene molasse was not sufficient to balance the eroded sediments and increase the maturity of Jurassic rocks, which indeed relates to their pre-Miocene burial.

Considering good source rock quality, Middle Jurassic Kokhanivka beds constitute a potential source rock in the northwestern part of the Carpathian foredeep. The Kokhanivka and Sudova Vyshnia heavy-oil fields, occurring in the Upper Jurassic and Cretaceous reservoirs in the same area, might have been sourced by these rocks at an early maturation stage. Further toward the Carpathian overthrust, with the increase of burial depth, more advanced maturation levels may occur in Middle Jurassic beds.

Cretaceous beds constitute a significant part of the autochthonous sequence. Although no evident source rocks are present, several oil and gas fields have been discovered in the Upper Cretaceous. These are sealed by Miocene molasse strata. Thermal maturity of the Cretaceous rocks from the southeastern part of the Carpathian Foredeep was studied using vitrinite reflectance (Koltun et al., 1998). In the Slavetska-4, Chornoguzy-6, and Sviatoslavska-5 wells at depth intervals from 1432 to 1636 m (4698 to 5367 ft), R_o values range from 0.52 to 0.80%. In the Lopushna-4 and Biskiv-1 wells, where the autochthon is buried deeper beneath the Carpathian allochthon, R_o values for coeval Cretaceous beds range between 0.62 and 1.13% at depth intervals from 4210 to 4728 m (13,812 to 15,511 ft). The difference in maturation rank between the location of these two wells shows that the maturation level of the foreland was strongly influenced by the overthrusting events. Because the depths and, hence, the maturation level of the autochthonous sequence rapidly increase toward the southwest, the area of possible oil occurrences in the underthrust foreland may be restricted within a narrow band paralleling the Carpathian thrust front.

The Lopushna oil field occurs in the Upper Cretaceous and Upper Jurassic reservoirs of the southeastern part of the foreland autochthon, beneath the Carpathian flysch allochthon. Because Middle Jurassic beds are not recognized in this area, no evident source rocks are present in the underlying sequence. According to biomarker studies, which outlined the presence of oleanane, this oil was probably generated from the Oligocene Menilite shales from the Carpathian flysch sequence (ten Haven et al., 1993; Lafargue et al., 1994; Roure and Sassi, 1995; Picha, 1996).

CONCLUSIONS

Lower Cretaceous and Oligocene organic-rich rocks represent the main potential source rock intervals in the flysch sequence of the Polish and Ukrainian Carpathians. Oligocene Menilite shales are the principal source rocks in the area. Biomarker studies suggest that most oils, occurring in flysch reservoirs, have been derived from this series. The frontal part of the flysch belt shows the most favorable conditions for petroleum generation, especially in Ukraine, where a complex overthrust structure distributed thick, organic-rich Menilite shales with high petroleum potential through a wide range of depths, encompassing the entire oil window. The main part of the maturation history and the consequent petroleum generation and expulsion are related to Miocene overthrusting.

The maturation level of the Menilite shales increases from the outer to inner tectonic units of both the Polish and Ukrainian parts of the flysch belt. In the internal part, the postoverthrusting erosion and uplift are more evident here, and the source rocks display higher maturity at less depth, being in places overmature even at the surface. However, even in the internal tectonic units of the flysch belt, the Menilite shales contain potential source rocks.

Lower Cretaceous organic-rich rocks of the Spas shales are widespread in both the Polish and Ukrainian parts of the flysch belt and have significant petroleum potential. These beds, as well as the Shypot shales, are widespread in the internal tectonic units of the Ukrainian Carpathians, where they are considered important potential source rocks in the flysch sequence.

Other stratigraphic units of the flysch sequence, which are enriched with organic matter, are less important as source rocks, even if they may have generated oil and gas.

Gas fields, occurring in the Miocene strata of the Polish and Ukrainian parts of the Carpathian Foredeep, contain gas (mainly methane >98%) that was generated in the Badenian and Sarmatian sequence by the microbial processes of carbon dioxide reduction. In some gases, mixing of thermogenic components is generated either in deeper horizons of the same Miocene beds or in source rocks from the pre-Neogene basement.

Gases accumulated in the Upper Jurassic and Upper Cretaceous reservoirs of the Polish part of the Mesozoic basement were generated by both microbial and thermogenic processes. They either were generated by the pre-Neogene basement rocks or migrated to the pre-Miocene reservoirs from the adjacent Miocene sequence through faults. Gas fields, occurring in similar Mesozoic reservoirs of the Ukrainian area, have not been studied but are assumed to have the same origin.

According to biomarker studies, oils from Paleozoic–Mesozoic reservoirs of the Polish part of the basement, as well as Lopushna field in Mesozoic reservoirs of the Ukrainian part, are assumed to have been generated at least partly by the Oligocene flysch sequence.

Lachowice field gas accumulated in Devonian strata in the Polish part of the basement, has a thermogenic origin, and was generated from type III kerogen of the Carboniferous strata, whereas Nosowka field oil accumulated in lower Carboniferous carbonates and was generated from type II kerogen of the Silurian-Ordovician strata.

Petroleum occurrences of the foreland autochthon of the Ukrainian area are most likely localized in the Mesozoic–Cenozoic cover, because the Paleozoic sequence, at least in the central and southeastern part of the Ukrainian Carpathian foreland, became overmature in the pre-Mesozoic. Middle Jurassic beds, widespread in the northwestern part of the Ukrainian Carpathian foredeep, display a high petroleum potential and are the main potential source rocks in this area. In the Polish part of the foredeep, Middle Jurassic beds also are organic rich, but they are mainly immature.

ACKNOWLEDGMENTS

John Curtis from the Colorado School of Mines, Golden, Colorado, and Mark Pawlewicz from the U.S. Geological Survey, Denver, Colorado, gave very constructive reviews that greatly improved the discussion and the possible consequences of the hypotheses presented in the manuscript.

REFERENCES CITED

Andreeva-Grigorovich, A. S., A. D. Gruzman, L. M. Reifman, and S. E. Smirnov, 1986, Biostratigraphical characteristics of the type sequence of the Menilite shales on Chechva River (Ukrainian Carpathians) (in Russian): Paleontologiceskij Zurnal, v. 23, p. 83–89.

Bazhenova, O. K., and O. A. Arefiev, 1990, Immature oils as the products of early catagenetic transformation of bacterial-algal organic matter: Organic Geochemistry, v. 16, p. 307–311.

Berner, U., and E. Faber, 1996, Empirical carbon isotope/maturity relationships for gases from algal kerogens and terrigenous organic matter, based on dry, open-system pyrolysis: Organic Geochemistry, v. 24, p. 947–955.

Bessereau, G., F. Roure, M. Kotarba, J. Kusmierek, and W. Strzetelski, 1996, Structure and hydrocarbon habitat of the Polish Carpathians, *in* P. A. Ziegler and F. Horvath, eds., Structure and prospects of Alpine basins and forelands: Peri-Tethys Memoirs 2: Memoires du Museum d'Histoire Naturelle, Paris, v. 170, p. 343–373.

Burchfiel, B. C., and L. Royden, 1982, Carpathian foreland fold and thrust belt and its relation to other basins: AAPG Bulletin, v. 66, p. 1179–1195.

Burnham, A. K., and R. L. Braun, 1990, Detailed model of petroleum formation, destruction and expulsion from lacustrine and marine source rocks, *in* B. Durand and F. Behar, eds., Advances in organic geochemistry 1989: Organic Geochemistry, v. 16, p. 27–40.

Clayton, C., 1992, Source volumetrics of biogenic gas generation, *in* R. Vially, ed., Bacterial gas: Paris, Editions Technip, p. 191–204.

Czepiec, I., and M. J. Kotarba, 1998, Paleoecology and organic matter in the late Badenian and early Sarmatian marine basin of the Polish part of the Carpathian Foredeep: Przeglad Geologiczny, v. 46, p. 732–736.

Danys, V. V., J. O. Kulcickij, V. A. Sakin, and O. S. Vialov, 1974, Flysch Carpathians, *in* M. Mahel, ed., Tectonics of the Carpathian–Balkan regions: Bratislava, Geological Institute of Dionyz Stur, p. 210–217.

Dickey, P. A., and J. M. Hunt, 1972, Geochemical and hydrogeological methods of prospecting for stratigraphic traps: AAPG Memoir 16, p. 36–67.

Dolenko, G. N., 1962, Oil and gas geology of the Carpathians (in Russian): Kyiv, Publishing House of the Ukrainian Academy of Science, 368 p.

Drygant, D. M., 1992, Conodont colour as the indicator of the geological processes (Volyn–Podillia) (in Ukrainian with English abstract): Paleontologiceskij Zurnal, v. 23, p. 35–37.

Dulub, V. G., M. I. Burova, V. S. Burov, I. B. Vishniakov, 1986, Explanatory note to the regional stratigraphic scheme of the Jurassic deposits of the Carpathian foredeep and Volyn–Podolian margin of the East European platform: Leningrad, Vsesojuzny Geologichnyi Institut, 58 p.

Espitalie, J., and M. L. Bordenave, 1993, Rock-Eval pyrolysis, *in* M. L. Bordenave, ed., Applied petroleum geochemistry: Paris, Editions Technip, p. 237–278.

Gabinet, M. P., 1985, Post-sedimentary transformation of the Ukrainian Carpathians Flysch (in Russian): Kyiv, Nauka dumka, 148 p.

Glogoczowski, J. J., 1980, Versuch der Aufstellung eines neuen "Hydrat" modells der Erdgasgeneriierung in Miozän der Karpatenvorsenke: Zeitschrift fur Angewandte Geologie, v. 26, p. 156–158.

Glushko, V. V., 1968, Tectonics and oil and gas deposits of the Carpathians and the adjacent basins (in Russian): Moscow, Nedra, 264 p.

Gondek, B., and Z. Pomykala, 1982, The Carpathian oils based on study of n-alkanes and isoprenoids (in Polish with English abstract): Przeglad Geologiczny, v. 30, p. 64–69.

Grinberg, I. V., 1957, Study of the chemical nature and genetic correlations of the organic matter of the Carpathian shales and oils (in Russian): Kyiv, Ukrainian Academy of Science, 230 p.

Gucwa, I., and T. Wieser, 1980, Geochemistry and mineralogy of sedimentary rocks of the Carpathian flysch rich in organic matter (in Polish with English abstract): Mineralogical Transactions Polish Academy of Science, v. 69, 43 p.

Hempel, J., 1955, Source rocks for the Carpathian oils (in Polish with English abstract): Przeglad Geologiczny, v. 3, p. 468–473.

Hunt, J. M., M. D. Lewan, and R. F. Hennet, 1991, Modeling oil generation with time-temperature index graphs based on the Arrhenius equation: AAPG Bulletin, v. 75, no. 4, p. 795–807.

Karnkowski, P., 1999, Oil and gas deposits in Poland: Krakow, The Geosynoptics Society "Geos," 380 p.

Karpinski, T., 1978, Organic matter and bitumens of rocks from the Carpathian region, *in* F. Mitura and F. Kozikowski, eds, Basic criteria of evaluation of perspectives oil and gas occurrence in Poland (in Polish): Prace Instytut Gornictwa Naftowego i Gazownictwa, v. 25, p. 6–12.

Koester, J., M. Kotarba, E. Lafargue, and P. Kosakowski,

1998a, Source rock habitat and hydrocarbon potential of Oligocene Menilite Formation (Flysch Carpathians, southeast Poland): An organic geochemical and isotope approach: Organic Geochemistry, v. 29, no. 1–3, p. 543–558.

Koester, J., M. Rospondek, S. Schouten, M. Kotarba, A. Zubrzycki, and J. S. Sinninghe Damste, 1998b, Biomarker geochemistry of a foredeep basin: The Oligocene Menilite Formation in southeast Poland: Organic Geochemistry, v. 29, no. 1–3, p. 649–669.

Koltun, Y., M. Kotarba, P. Kosakowski, and J. Espitalie, 1995, Hydrocarbon potential of the Menilite and Spas beds in the Polish and Ukrainian parts of the Flysch Carpathians: Extended Abstracts Book, Conference and Exhibition "Modern Exploration and Improved Oil and Gas Recovery Methods," Krakow, Poland, p. 147–150.

Koltun, Y., J. Espitalie, M. Kotarba, F. Roure, N. Ellouz, and P. Kosakowski, 1998, Petroleum generation in the Ukrainian external Carpathians and their adjacent foreland: Journal of Petroleum Geology, v. 21, no. 3, p. 265–268.

Koltun, Y. V., 1992, Organic matter in Oligocene Menilite Formation rocks of the Ukrainian Carpathians: Palaeoenvironment and geochemical evolution: Organic Geochemistry, v. 18, p. 423–430.

Kotarba, M., 1987, Stable carbon isotope composition of natural gases from the central part of the Polish Flysch Carpathians: Bulletin of the Polish Academy of Sciences, Earth Science, v. 35, p. 335–342.

Kotarba, M., 1992, Bacterial gases in the Polish part of the Carpathian foredeep and the Flysch Carpathians: Isotopic and geologic approach, *in* R. Vially, ed., Bacterial gas: Paris, Editions Technip, p. 133–146.

Kotarba, M., 1999, Generation and accumulation of hydrocarbons and geochemical characteristics of organic matter in autochthonous Miocene strata of the Polish Carpathian Foredeep (in Polish with English abstract): Prace Panstwowy Instytut Geologiczny, v. 168, p. 277–296.

Kotarba, M., 2001, Composition and origin of coalbed gases in the upper Silesian and Lublin basins, Poland: Organic Geochemistry, v. 32, p. 163–180.

Kotarba, M., and E. Jawor, 1993, Petroleum generation, migration and accumulation in the Miocene sediments and Paleozoic–Mesozoic basement complex of the Carpathian Foredeep between Cracow and Pilzno (Poland), *in* A. M. Spencer, ed., Generation, accumulation and production of Europe's hydrocarbons, III: Special publication of the European Association of Petroleum Geologists, no. 3, Heidelberg, Springer, p. 295–301.

Kotarba, M., S. Szafran, and J. Espitalie, 1987, A study of organic matter and natural gases of the Miocene sediments in Polish part of the Carpathian Foredeep: Chemical Geology, v. 64, p. 197–207.

Kotarba, M., P. Kosakowski, D. Wieclaw, and A. Kowalski, 1999, Hydrocarbon potential of source rocks from the Skole unit of the Flysch Carpathians (in Polish with English abstract): Proceedings of the Carpathian Scientific Conference "Petroleum Industry and Science Together to 21st Century," Raba Nizna, p. 143–152.

Kotarba, M. J., 1998a, Composition and origin gaseous hydrocarbons in the Miocene strata of the Polish part of the Carpathian Foredeep: Przeglad Geologiczny, v. 46, p. 751–758.

Kotarba, M. J., 1998b, Isotopic and pyrolysis studies of organic matter and natural gases from the Flysch Carpathian strata of the Kuzmina-1 deep well (abs.): AAPG Annual Meeting Program, v. 375, p. 1–4.

Kotarba, M. J., 2000, Source rock habitat and hydrocarbon potential of Oligocene–Lower Cretaceous flysch formations in Skole unit (Flysch Carpathians of southern Poland), *in* L. A. Trindade, A. C. Macedo, and S. M. Barbanti, eds., Proceedings of the 7th Latin American Congress on Organic Geochemistry "New Perspective on Organic Geochemistry for the Third Millenium," Brazil, p. 361–364.

Kotarba, M. J., W. Burzewski, T. Wilczek, K. Slupczynski, P. Kosakowski, and D. Botor, 1998a, Model of gaseous hydrocarbon generation in the Miocene strata of the Polish part of the Carpathian Foredeep: Przeglad Geologiczny, v. 46, p. 737–742.

Kotarba, M. J., T. Wilczek, P. Kosakowski, A. Kowalski, and D. Wieclaw, 1998b, A study of organic matter and habitat of gaseous hydrocarbons in the Miocene strata of the Polish part of the Carpathian Foredeep: Przeglad Geologiczny, v. 46, p. 742–750.

Kotarba, M. J., M. Wagner, D. Wieclaw, P. Kosakowski, and A. Kowalski, 2001, Hydrocarbon potential and geochemical characteristics of organic mater in the Tarnawa 1 borehole section (in Polish with English abstract): Prace Panstwowy Instytut Geologiczny, v. 174, p. 161–172.

Kotlarczyk, J., 1961, Carpathian diatoms and some problems of petroleum geology (in Polish with English abstract): Nafta, v. 17, p. 123–127.

Kotlarczyk, J., 1985, Remarks on the tectonics of the Carpathian orogen between the Wislok and Wiar rivers, *in* J. Kotlarczyk, ed., Guide book to excursion, v. 4, Carpatho-Balcan Geological Association XIII Congress: Krakow, Geological Institute, p. 32–64.

Kotlarczyk, J., 1988, Problems of sedimentology, stratigraphy and tectonics of the Przemysl Carpathians and their nearest foreland, *in* J. Kotlarczyk, ed., The Przemysl Carpathians, Guide book to 57th Meeting of the Polish Geological Society, Przemysl (in Polish), p. 23–62.

Kozikowski, H., 1978, Bitumo-chronological methods for example of the Polish Outer Carpathians (in Polish): Prace Instytutu Gornictwa Naftowego i Gazownictwa, v. 16, 74 p.

Kruge, M. A., M. Mastalerz, A. Solecki, and B. A. Stankiewicz, 1996, Organic geochemistry and petrology of oil source rocks, Carpathian overthrust region, southeastern Poland—Implications for petroleum generation: Organic Geochemistry, v. 24, p. 897–912.

Kruglov, S. S., S. E. Smirnov, S. M. Spitkovskaya, L. E. Filshtynski, and A. V. Hizhniakov, 1985, Geodynamics of the Carpathians (in Russian): Kyiv, Nauka dumka, 136 p.

Ksiazkiewicz, M., 1977, The tectonics of the Carpathians, *in* W. Pozaryski, ed., Geology of Poland, v. 4, Tectonics: Warsaw, Wydawnictwo Geologiczne, p. 476–620.

Kulchytsky, A. Y., 1986, Geology and environments of formation of the evaporite-bearing molasse of pre-Carpathian and Trans-Carpathian Neogene basins (in Russian): Ph.D. thesis, Lviv University, Lviv, 200 p.

Kusmierek, J., 1990, Outline of geodynamics of the Central Carpathian oil basin (in Polish with English abstract): Krakow, Prace Geologiczne Polish Academy of Science, v. 135, 85 p.

Kusmierek, J., 1992, Regularities in the distribution of oil- and gas-bearing zones in the allochthonous structural units of the Polish Eastern Carpathians: Polish Journal of Mineral Research, v. 1, p. 111–119.

Lafargue, E., N. Ellouz, and F. Roure, 1994, Thrust-controlled exploration plays in the Outer Carpathians and their foreland (Poland, Ukraine, and Romania): First Break, v. 12, no. 2, p. 69–79.

Langford, F. F., and M. M. Blanc-Valleron, 1990, Interpreting Rock-Eval pyrolysis data using graphs of pyrolizable hydrocarbons vs. total organic carbon: AAPG Bulletin, v. 74, p. 799–804.

Leszczynski, S., 2001, Carpathian flysch sea in the late Eocene–early Oligocene: Example of bottom-water oxygenation development: Biuletyn Panstwowy Instytut Geologiczny, v. 396, p. 95–96.

Marzec, A., H. Kozikowski, J. Glogoczowski, and W. Kisielow, 1971, Problems of migration of Polish crude oils on the basis of geochemical correlation data and carbon isotope composition: Chemical Geology, v. 8, p. 197–217.

Matyasik, I., A. Steczko, and R. P. Philp, 2000, Biodegradation and migrational fractionation of oils from the Eastern Carpathians, Poland: Organic Geochemistry, v. 31, p. 1509–1523.

Matyasik, I., A. Steczko, M. Mastalerz, and S. C. Brassell, 2001, Petrographic and geochemical characterization of source rock variability in the Carpathian region of Poland: Implication for oil generation: Abstracts of 18th Annual Meeting of the Society for Organic Petrology, Houston, Texas, v. 18, p. 78–83.

Ney, R., W. Burzewski, T. Bachleda, W. Gorecki, K. Jakobczak, and K. Slupczynski, 1974, Outline of paleogeography and evolution of lithology and facies of Miocene layers in the Carpathian Foredeep (in Polish with English abstract): Krakow, Prace Geologiczne Polish Academy of Science, v. 82, 65 p.

Oszczypko, N., 1997, The early–middle Miocene Carpathian period foreland (Western Carpathians, Poland): Przeglad Geologiczny, v. 45, p. 1054–1063.

Oszczypko, N., and A. Slaczka, 1986, An attempt to palinspastic reconstruction of Neogene basins in the Carpathian foredeep: Annales Societatis Geologorum Poloniae, v. 55, p. 55–76.

Picha, F. J., 1996, Exploring for hydrocarbons under thrust belts—A challenging new frontier in the Carpathians and elsewhere: AAPG Bulletin, v. 80, p. 1547–1564.

Picha, F. J., and K. E. Peters, 1998, Biomarkers oil-to-source rock correlation in the Western Carpathians and their foreland: Czech Republic: Petroleum Geoscience, v. 4, p. 289–302.

Pomykala, Z., 1982, Degradation of the Carpathian oils (in Polish with English abstract): Nafta, v. 38, p. 135–139.

Popescu, B. M., 1995, Romania's petroleum systems and their remaining potential: Petroleum Geoscience, v. 1, p. 337–350.

Rice, D. D., 1992, Controls, habitat, and resource potential of ancient bacterial gas, *in* R. Vially, ed., Bacterial gas: Paris, Editions Technip, p. 91–118.

Rospondek, M. J, J. Koester, and J. S. Sinninghe Damste, 1997, Novel C_{26} highly branched isoprenoid thiophenes and alkane from the Menilite Formation, Outer Carpathians, southeast Poland: Organic Geochemistry, v. 26, p. 295–304.

Roure, F., and W. Sassi, 1995, Kinematics of deformation and petroleum system appraisal in Neogene foreland fold-and-thrust belts: Petroleum Geoscience, v. 1, p. 253–269.

Roure, F., E. Roca, and W. Sassi, 1993, The Neogene evolution of the Outer Carpathian flysch units (Poland, Ukraine and Romania): Kinematics of a foreland/fold-and-thrust belt system: Sedimentary Geology, v. 86, p. 177–201.

Schoell, M., 1983, Genetic characterization of natural gases: AAPG Bulletin, v. 67, p. 2225–2238.

Shabo, Z. V., and G. P. Mamchur, 1984, The perspectives of gas bearing of the deep horizons of the Carpathian Foredeep on the base of carbon isotopic composition (in Russian): Geologicheski Zhyrnal, v. 1, p. 50–57.

Shevchenko, E. F., A. P. Shcherbakova, and V. V. Kvak, 1984, Distribution and appraisal of the characteristics of hydrocarbons in the Carpathian petroleum province (in Russian): Geologiya i geokhimia goriuchikh iskopaemykh, v. 63, p. 89–93.

Slaczka, A., 1980, Les Carpathes flyscheuses, *in* R. Osika, ed., La geologie de les gites mineraux en Pologne: Biuletyn Instytut Geologiczny, v. 251, p. 491–497, 606–616.

Slaczka, A., 1996, Oil and gas in Northern Carpathians, *in* G. Wessely and W. Liebl, eds., Oil and gas in Alpidic thrust belts and basins of Central and Eastern Europe: European Association of Geoscientists and Engineers Special Publication 5, p. 187–195.

Slaczka, A., and M. A. Kaminski, 1998, A guidebook to excursions in the Polish Flysch Carpathians: Grzybowski Foundation, Krakow, Special Publication 6, 171 p.

Swiderski, B., 1952, Outline of the tectonics of the Northern Carpathians (in Polish): Prace Instytut Geologiczny, v. 7, p. 1–109.

ten Haven, H. L., E. Lafargue, and M. Kotarba, 1993, Oil/oil and oil/source rock correlations in the Carpathian Foredeep and overthrust, south-east Poland: Organic Geochemistry, v. 20, p. 935–959.

Vialov, O. S., S. P. Gavura, V. V. Danysh, O. D. Lemishko, R. Y. Leshchukh, L. P. Ponomariova, A. M. Romaniv, S. E. Smirnov, N. I. Smolinska, and P. N. Tsarnenko, 1988, Stratotypes of the Cretaceous and Paleogene deposits of the Ukrainian Carpathians (in Russian): Kyiv, Nauka dumka, 204 p.

Waples, D. W., 1980, Time and temperature in petroleum

formation: Application of Lopatin's method to petroleum exploration: AAPG Bulletin, v. 64, p. 916–929.

Wdowiarz, S., 1960, Oil and gas in the geology of the Carpathians (in Polish with English abstract): Biuletyn Instytut Geologiczny (special issue), p. 1–67.

Wdowiarz, S., 1985, On some aspects of geological structure and oil and gas accumulation of the Central Carpathian Synclinorium in Poland: Biuletyn Instytut Geologiczny, v. 350, p. 5–52.

Whiticar, M. J., 1990, A geochemical perspective of natural gas and atmospheric methane: Organic Geochemistry, v. 16, p. 531–547.

Whiticar, M. J., 1994, Correlation of natural gases and their sources, *in* L. B. Magoon and W. G. Dow, eds., The petroleum system—From source to trap: AAPG Memoir 60, p. 261–283.

Whiticar, M. J., E. Faber, and M. Schoell, 1986, Biogenic methane formation in marine and fresh water environment, CO_2 reduction vs. acetate fermentation—Isotopic evidence: Geochimica et Cosmochimica Acta, v. 50, p. 693–709.

Wieclaw, D., 2002, Origin of Oligocene oils from the Polish Flysch Carpathians: Organic sulfur in the kerogen of the Menilite shales and kinetics of hydrocarbon generation process (in Polish): Ph.D. thesis, AGH University of Science and Technology, Krakow, 131 p.

12

Popadyuk, I., M. Vul, G. Ladyzhensky, and P. Shpak, 2006, The petroleum geology of the Ukrainian Carpathian Foredeep, *in* J. Golonka and F. J. Picha, The Carpathians and their foreland: Geology and hydrocarbon resources: AAPG Memoir 84, p. 369.

The Petroleum Geology of the Ukrainian Carpathian Foredeep

I. Popadyuk
Ukrainian State Geological Research Institute, L'viv, Ukraine

M. Vul
Ukrainian State Geological Research Institute, L'viv, Ukraine

G. Ladyzhensky
Institute of Geological Sciences, Kiev, Ukraine

P. Shpak
Institute of Geological Sciences, Kiev, Ukraine

ABSTRACT

The aim of this chapter is to describe the key elements of the petroleum systems responsible for the accumulation of hydrocarbons as well as the most significant discoveries in the Ukrainian Carpathian Foredeep. The petroleum potential of the foredeep is contained in the Jurassic to Miocene sedimentary rocks. A total of 40 gas fields, 4 gas-condensate fields, and 2 oil fields have been discovered in the foredeep to date. Several of the most typical fields have been described in this chapter. These fields produce from the Upper Jurassic to middle Miocene reservoir rocks, sealed by the rocks occurring throughout the Mesozoic–Cenozoic sedimentary column. The traps can be summarized as follows: stratigraphic trapping beneath unconformity; stratigraphic trapping above unconformity; thrust-related trapping; and normal-fault-related trapping. Recently, the basin is at a mature stage of exploration, but new exploration opportunities continue to exist.

DOI:10.1306/985615M843075

13

Popadyuk, I., M. Vul, G. Ladyzhensky, and P. Shpak, 2006, Petroleum geology of the Boryslav–Pokuttya zone, the Ukrainian Carpathians, *in* J. Golonka and F. J. Picha, eds., The Carpathians and their foreland: Geology and hydrocarbon resources: AAPG Memoir 84, p. 371.

Petroleum Geology of the Boryslav–Pokuttya Zone, the Ukrainian Carpathians

I. Popadyuk
Ukrainian State Geological Research Institute, L'viv, Ukraine

M. Vul
Ukrainian State Geological Research Institute, L'viv, Ukraine

G. Ladyzhensky
Institute of Geological Sciences, Kiev, Ukraine

P. Shpak
Institute of Geological Sciences, Kiev, Ukraine

ABSTRACT

The modern industrial hydrocarbon exploration started in the Boryslav–Pokuttya zone in 1904. In 1905, the well Oil City produced approximately 3000 t of oil/day and 900,000 m^3 (31.8 mmcf) of gas per day from the Oligocene Boryslav sandstone, a very significant amount at the beginning of 20th century. After the Second World War, the Dolyna and Bytkiv–Babche oil fields were discovered. The Boryslav–Pokuttya unit in the Ukrainian Carpathians is interpreted as a complex thrust system consisting of the five north-verging folded thrust sheets. The stratigraphic column of this zone includes the Upper Cretaceous deposits, but the Paleogene reservoirs contain almost 95% of the total recoverable hydrocarbons. These deep-marine deposits form the bodies of genetically related mass-flow and turbidity currents. A series of palinspastic maps interpreting the main depositional features was prepared to illustrate the distribution of main turbiditic systems. Traps in all fields are structural or combined structural-stratigraphic, dominated by faulted anticlines bounded by the detachments. Three major fields are described in detail as the classic Eastern Carpathian hydrocarbon accumulations with the long history of exploration and production.

DOI:10.1306/985732M843075

14

Dziadzio, Piotr S., 2006, The geology of the Weglowka oil field, Subsilesian unit, Polish Outer Carpathians, *in* J. Golonka and F. J. Picha, eds., The Carpathians and their foreland: Geology and hydrocarbon resources: AAPG Memoir 84, p. 373.

The Geology of the Weglowka Oil Field, Subsilesian Unit, Polish Outer Carpathians

Piotr S. Dziadzio

Polish Oil and Gas Company, Exploration Department, South Regional Division, Jaslo, Republic of Poland

ABSTRACT

The Weglowka field is one of the biggest oil fields in the Outer Polish Carpathians, situated north of Krosno between Frysztak and Brzozow in the Subsilesian tectonic unit. The field was discovered in 1888 and has produced 998,220 t of oil and 214.52 million m^3 (7.575 bcf) of gas since exploitation began. More than 350 wells have been drilled in the Weglowka oil field; it was completed in the reservoir intervals that range in depth from 100 to 1200 m (330 to 3900 ft). The Weglowka oil field is now in the final phase of exploitation.

Oil is accumulated in several Lower Cretaceous sandstone bodies. The trap is an anticline that is cut by two second-order longitudinal, small thrust faults that subdivide the field between two thrust sheets. Stratigraphic traps probably exist in the Lgota sandstone (the main reservoir), associated with lowstand systems tract sandstones deposited in basin-floor fans. The main seals in the field are the Verovice shales, Lgota Shales, and the younger Godula Shales and Weglowka marls.

DOI:10.1306/985616M843071

15

Dziadzio, Piotr S., Mark A. Enfield, Matthew P. Watkinson, and Szczepan J. Porębski, 2006, The Ciezkowice Sandstone: Examples of basin-floor fan-stacking patterns from the main (upper Paleocene to Eocene) reservoir in the Polish Carpathians, *in* J. Golonka and F. J. Picha, eds., The Carpathians and their foreland: Geology and hydrocarbon resources: AAPG Memoir 84, p. 375 – 376.

The Ciezkowice Sandstone: Examples of Basin-floor Fan-stacking Patterns from the Main (Upper Paleocene to Eocene) Reservoir in the Polish Carpathians

Piotr S. Dziadzio

Polish Oil and Gas Company, Exploration Department, South Regional Division, Jaslo, Republic of Poland

Mark A. Enfield

PDF Ltd., United Kingdom

Matthew P. Watkinson

Department of Geology, University of Plymouth, United Kingdom

Szczepan J. Porębski

Institute of Geological Sciences, Kraków Research Center, Polish Academy of Sciences, Republic of Poland

ABSTRACT

The Ciezkowice Sandstone is a lithostratigraphic term used to define a sand-prone succession of late Paleocene–middle Eocene age and is one of the main producing petroleum reservoirs in the Silesian unit of the central and eastern Polish Outer Carpathians. The excellent reservoir potential of these sandstones and their widespread occurrence make them a key objective for future exploration. In this study, the Ciezkowice Sandstone was analyzed using geophysical logs and nearby exposures in the Silesian unit between the towns of Ciezkowice and Krosno.

The analysis presented in this chapter has identified as much as four laterally persistent, sand-prone units in the Ciezkowice Sandstone. The dominant lithology of these units is coarse-grained, thick-bedded sandstones interbedded with fine-grained turbidites. The thickness of individual sandstone bodies ranges from 20 to more than 50 m (66 to more than 160 ft) within a gross sand-prone interval of as much as 350 m (1100 ft), in which sand/shale ratios can be as much as 9:1. Sandstone beds are mostly massive to normally graded and commonly display shale clasts, horizontal planar stratification, and water escape structures. The thickest beds are commonly amalgamated and locally display very large-scale trough cross-bedding. These sandstones are interpreted as high-density turbidites.

The Ciezkowice Sandstone units are interbedded with pelagic mudstone-dominated tongues, lithostratigraphically termed the "Variegated shales." Sedimentological features of

DOI:10.1306/985617M843071

the sandiest units in the Ciezkowice Sandstone indicate that they were deposited in basin-plain settings during lowstand phases of basin evolution.

The Ciezkowice Sandstone represents a second-order lowstand sequence set (basin-floor fan equivalent). The Variegated shales were deposited during transgressive and highstand systems tracts, when sand supply to the basin was reduced significantly. Each Ciezkowice Sandstone–Variegated shale cycle is interpreted to represent a third-order depositional sequence in a lowstand sequence set of latest Paleocene to middle Eocene age. Several subunits of the Ciezkowice Sandstone provide good examples of basin-floor fan-stacking patterns characteristic for deposition in lowstand basin conditions.

16

Matyasik, Irena, and Piotr S. Dziadzio, 2006, Reconstruction of petroleum systems based on integrated geochemical and geological investigations: Selected examples from the middle Outer Carpathians in Poland, *in* J. Golonka and F. J. Picha, eds., The Carphathians and their foreland: Geology and hydrocarbon resources: AAPG Memoir 84, p. 377–378.

Reconstruction of Petroleum Systems Based on Integrated Geochemical and Geological Investigations: Selected Examples from the Middle Outer Carpathians in Poland

Irena Matyasik
Oil and Gas Institute, Krakow, Republic of Poland

Piotr S. Dziadzio
Polish Oil and Gas Company, Exploration Department, South Regional Division, Jaslo, Republic of Poland

ABSTRACT

A genetic analysis of oils and potential source rocks and their mutual correlation from three areas of hydrocarbon occurrence in three tectonic units (thrust nappes), Subsilesian, Silesian, and Dukla nappes of the Outer Carpathians in Poland, between Gorlice in the west and Krosno to the east, revises existing opinions on the genesis, migration, and accumulation of hydrocarbons in this part of the Carpathians.

One of the important hydrocarbon accumulations occurs in the Subsilesian unit (Weglowka oil field). Detailed geochemical studies show that the oil that accumulated in the Weglowka oil field is the same type as the oil that accumulated in the Potok anticline located about 10 km (6 mi) to the south in the Silesian nappe and derived from the same Oligocene Menilite shales source rock.

The oldest known source-prone interval in the Subsilesian nappe is the Verovice shales (Lower Cretaceous), which are likely to have produced the hydrocarbons found in the nearby Wola Jasienicka field located 8 km (5 mi) to the east from the Weglowka oil field.

The lower Istebna Shales (Upper Cretaceous) that occurred in the Silesian nappe in the area of Potok anticline could be good source rocks for oil. It is also possible that hydrocarbon generation occurred during several stages, and that mixing of oils derived from Menilite beds and Istebna Shales could have occurred in this area.

Similar studies have been made in the southern part of the study area around the oil-rich Gorlice region in the Silesian nappe. This study indicates that the oils accumulated in fields of this region were generated from the Menilite shales of the Dukla nappe, which occurred in tectonic windows in the Magura nappe.

DOI:10.1306/985618M843076

The tectonic relationship of the producing fields in the Weglowka–Potok clearly shows the occurrence of several major syncline around the fields. The deepest are to the south of the Wola Jasienicka and Potok fields, which could represent the generative kitchen for these fields during the different stages of Carpathian thrusting.

In addition, computer-generated simulations of oil generation in the vicinity of Gorlice and Potok areas suggest hydrocarbon generation and expulsion during the middle Miocene to Pliocene (Holocene).

THE GEOLOGY AND HYDROCARBON RESOURCES OF THE EASTERN AND SOUTHERN CARPATHIANS ON THE TERRITORY OF ROMANIA

17

Stefanescu, Mihai, Oprea Dicea, Alexandru Butac, and Daniel Ciulavu, 2006, Hydrocarbon geology of the Romanian Carpathians, their foreland, and the Transylvanian Basin, *in* J. Golonka and F. J. Picha, eds., The Carphathians and their foreland: Geology and hydrocarbon resources: AAPG Memoir 84, p. 381 – 427.

Hydrocarbon Geology of the Romanian Carpathians, Their Foreland, and the Transylvanian Basin

Mihai Stefanescu
Forest Romania Corporation, Bucharest, Romania

Oprea Dicea
Prospectiuni S. A., Bucharest, Romania

Alexandru Butac*
Petrom S. A., Bucharest, Romania

Daniel Ciulavu
Shell Canada, Calgary, Canada

ABSTRACT

The Romanian territory had a very complex geological evolution. This evolution generated a diversified geology with extended platforms limiting, to the east and south, a folded chain where spectacular overthrusts are present. More than this, posttectonic covers (basins) common to different tectonic units are also present. Some sequences of these posttectonic covers are extended from the frontal part of the folded chain over the platforms. Most of the tectonic units, as well as the post-tectonic covers, host hydrocarbon fields. On some of the tectonic units, the oil generation is proved only by the seep presence.

The aim of this chapter is to briefly present the geology of the Romanian onshore, except its western part, namely, the Apuseni Mountains and the Pannonian basin.

Each platform (East European, Scythian, and Moesian), each the folded chains (North Dobrudjea Orogen and groups of nappes from the East and South Carpathians [Transylvanides, Median Dacides, Outer Dacides, Marginal Dacides, Moldavides]), as well as the posttectonic basins (Transylvanian Basin, Getic Depression, Focsani Depression), have been described. For all the above-mentioned units, the standard lithostratigraphy and their internal structure are discussed. More details are given for those units that proved to be important hydrocarbon producers, namely, the Moesian Platform, Getic Depression, Subcarpathian nappe, Tarcau nappe, Diapiric fold zone, and Transylvanian Basin. The petroleum systems (including source rocks, reservoirs, and traps) of most of the units are also analyzed. From almost 1000 oil and gas fields already discovered, a reduced number of fields are quoted, just in order to give an idea on the diversity of traps.

*Deceased.

DOI:10.1306/985619M843077

The history of the geologic evolution of the Romanian onshore related to hydrocarbon genesis and accumulation conclude the chapter.

Finally, it is to remark that, in spite of its character of synthesis, the chapter could be a useful tool for those scientists who are interested in an overview of Romanian geology and the related hydrocarbon fields.

INTRODUCTION

The Romanian territory displays a very complex geology both from the stratigraphic and structural point of view. This geological situation obviously determined the hydrocarbon pool formation and field distribution across the territory.

Because of the existing geological complexity, thousands of pure geological or hydrocarbon-related papers have been published since the beginning of the 19th century.

The geology that will be presented in this chapter is practically based on most of the existing published papers and unpublished geological reports. The limited number of pages dedicated to the present chapter imposes a limit on quotations of preexisting publications only to those that are directly related to its main topics. Nevertheless, we will quote even here in this introductory section some important published maps, papers, and books that marked the development stages in Romanian geology as follows: the printing of the first geological map of Romania (scale 1:175,000; Geological Bureau, 1882); Munteanu Murgoci (1905; nappe presence in Southern Carpathians); Stefanescu et al. (1905; map of the Romanian oil areas); Mrazec (1907, 1927; diapirism theory); Mrazec and Popescu-Voitesti (1914; nappe structure of the Eastern Carpathians and the migration of the depositional axes of the flysch basin); Macovei (1938; a worldwide synthesis of the oil fields); Bancila (1958; the geologic synthesis of the Eastern Carpathians); Vancea (1960; synthesis of the Transylvanian Basin); Dumitrescu and Sandulescu (1970; Romania tectonic map); Patrulius (1969; an overview on the western side of the east Carpathian Bend); Burchfield (1976; a modern approach of Romanian geology); Paraschiv (1979; all Romanian oil and gas fields synthesis); Sandulescu (1984; an analysis of all aspects of the complex Romanian geology); Stefanescu and Working Team (1986; a graphic synthesis materialized by more than 39 geological sections crossing the whole Romanian onshore); and Dicea (1995; structure of the Eastern Carpathians based on seismic data).

For the same reason, we avoided quoting the lithostratigraphic local denominations, especially in the case of the tectonic units that do not have importance from the hydrocarbon point of view. Systematized information on these topics can be found in the following papers: Dumitrescu (1952), Popescu (1952, 1954), Grigoras (1955), Krautner et al. (1981), Marinescu et al. (1981), Sandulescu et al. (1981a, b), Stefanescu and Micu (1987), and Butac et al. (1997).

Here, we mention several important data about the Romanian oil industry evolution beginning with the geologists that partially or completely dedicated their professional interest to this industry: Grigore Cobalcescu, Gregoriu Stefanescu, Henry Coquand, Wiaceslav Teisseyre, Ludovic Mrazec, Gherghe Munteanu Murgoci, David Preda, Dumitru Stefanescu, Iulian Gavat, Ion Bancila, Neculai Grigoras, Mihai Popescu, Grigore Popescu, Florian Olteanu, Justin Gherman, Ion Patrut, Gheorghe Olteanu, and Elefterie Hristescu. The results of their research, published or mentioned in different reports, decisively pushed forward knowledge of the geology of the area of interest for hydrocarbons.

The first written documents mentioning the mazout (black oil) presence can be dated as far back as the 15th century for the northeastern part of the country (Moldavia) and as far back as the 16th century for the southern part of the country (Valachia). Initially, production began by collecting the oil from the excavations dug next to the multiple seeps existing both in the Eastern and Southern Carpathians. Later on, beginning in the 16th century, the production was carried out from pits as deep as 18 m (59 ft) and in square sections (4 × 4 m; 13 × 13 ft). It is interesting to mention that in this primitive stage of black oil production, it was also extracted by washing tar sands. In 1861, the first attempt to mechanically drill a well was made in Bacau County (Mosoare). Among the exploitation methods, it is an unusual one, namely, the production of oil using a mining style, digging vertical pits and/or galleries (drifts). Steaua Romana Company used this method for the first time in the Moinesti area, where two galleries were dug (the longest being 275 m [902 ft]). They have been shut down because of water influx. In 1927, the same oil company dug a third mine in the east Carpathian Bend area (Sarata Monteoru). This mine is producing even today. A visit to this location could be an exciting experience for every specialist from the oil industry because here, one can see the oil coming out everywhere and flowing freely (in the production) in the collecting ditches.

During the last decades of the 19th century, several of more than 23,000 exploration and development wells have been drilled in the country. Among these wells, 400 are deeper than 4000 m (13,000 ft), and 17 are deeper than 6000 m (19,700 ft). The deepest well was drilled in the east Carpathian Bend area (Baicoi) and bottomed at 7025 m (23,047 ft). Before 1948, the wells were drilled by private companies and later by the state companies. A total of 465 oil pools and 430 gas pools have been discovered thus far. Aside from the economic importance of the wells, we must also emphasize the essential part that they play in improving seismic and geologic interpretations.

All the mazout production was mostly used as wax for different mechanisms or as medicine (salve) for both humans and animals, as well as fuel for lamps and torches. A first step toward its real industrial use was marked in 1840, when a distillery for producing lamp oil was built in the Moinesti area (Lucacesti). Later, in 1856, T. Mehedinteanu built a bigger plant in Ploiesti, which was considered as the first Romanian refinery.

In 1854, the production of Bacau County (Moinesti area) was officially recorded. It was estimated that the 100 pits that were dug in the last two centuries produced approximately 77,000 t (Stefanescu et al., 1980). Three years later, in 1857, Romania was the first country with an internationally recorded industrial production (275 t). The highest oil production of 14.7 million t was recorded in 1976, whereas the highest gas production of 36.2 billion m^3 (1.278 tcf) was recorded in 1986. During the last few years, the oil production was constantly kept at a level of about 6 million t, whereas the gas production dropped down to 14.6 billion m^3 (515.5 bcf) in 2000.

After 1990, some modern studies have been carried out either by foreign geologists alone or in cooperation with Romanian scientists affiliated with either academic institutions or oil companies. We mention some of them here because they brought a new fashion of approaching the geology of folded chains and their forelands: Roure et al. (1993), G. B. Gibson (1994, personal communication), Hippolyte and Sandulescu (1996), Linzer (1996), Morley (1996), Proust and Hosu (1996), Dallmmeyer et al. (1998), and Sanders (1998).

Most of the geophysical methods have been used to help understand the deep structure of the territory. The accuracy of the knowledge about the underground structures is based on the very intensive seismic activity studies mostly carried out by the Romanian enterprise Prospectiuni. During the last decade, foreign companies, such as Western Geophysical and Companie Generale de Geophysique, were also active both in the Transylvanian Basin and in the outer part of the Carpathians. About 276,442 km (171,773 mi) of seismic lines onshore and 57,976 km (36,024 mi) offshore have been recorded. Among the different geophysical methods used for borehole investigations, it is worth mentioning that for the first time in the world, electronic logging has been experienced and applied afterward by Schlumberger in fields located at the east Carpathian Bend area.

We briefly presented the above information to emphasize that the present knowledge level of Romanian geology is based on a large number of geological, geophysical, and well data. Despite all these data, further research is still offering a lot of surprises, especially regarding the paleodepositional pattern of the rocks and the structures in which they are involved today, leading to new approaches on the prospectivity of the less explored areas and depths.

GENERAL GEOLOGICAL FRAMEWORK OF THE ROMANIAN ONSHORE

The morphology very well reflects the general structure of the country as follows (Figures 1–3):

- The mountains correspond to the fold and thrust structures (Apuseni Mountains, Carpathians, and Macin Mountains).
- The hilly zone mostly overlies the frontal part of the fold and thrust area.
- The flat areas in front of the mountains superpose on the foreland platforms (Moesian, Scythian, and East European).
- The large Transylvanian Plateau located between the Apuseni Mountains and the eastern and western branches of the Carpathians represents an intermountain basin.

Romania has a very complex geology. Deposits ranking from Precambrian to Quaternary are present in its territory. Most of the Precambrian to lower Paleozoic rocks are metamorphosed. The Paleozoic to Quaternary rocks are sedimentary and have accumulated in many diversified paleoenvironmental conditions. Consequently, a large variety of terrigenous (including flysch and molasse), calcareous, and evaporitic rocks is outcropping or has been penetrated by wells. All these rocks are involved in a very complicated structure generated in successive stages of tectonic movements, among which we will mention here only those that essentially contributed to the present structure, namely, the middle Cretaceous, Subhercynian, Styrian, Moldavian, and Valachian stages. During their evolution, the formations underwent different types of transformation. While on the platform territories, they have mostly been moved vertically or rotated

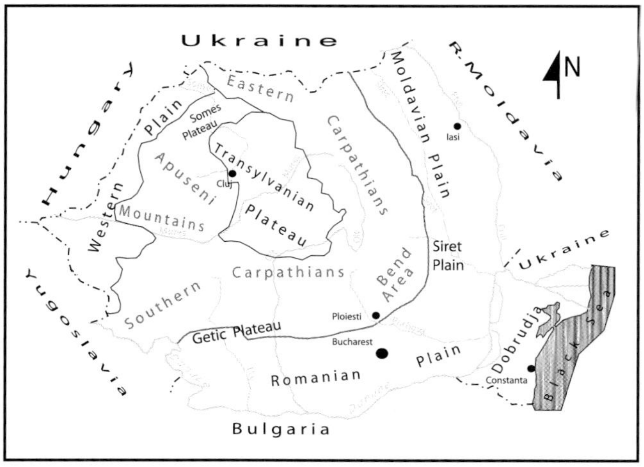

Figure 1. A Romanian map showing the geographic names used in this chapter.

in the mobile zones in which the deposits were strongly compressed, which generated a very complex structure including a large number of nappes.

In this chapter, all the deposits and tectonic units present in the area of interest will be described but not at the same level of detail. Those that have certain importance in relation to the hydrocarbon genesis would be obviously described in greater detail than those that have minor or no relevance in regards to hydrocarbons.

East European Platform

The east European platform is present in the northeast corner of the Romanian territory, where it is described as the Moldavian Platform. The Moldavian Platform is between Prut River to the east, Bistrita fault to the south-southwest, and the Romanian–Ukrainian

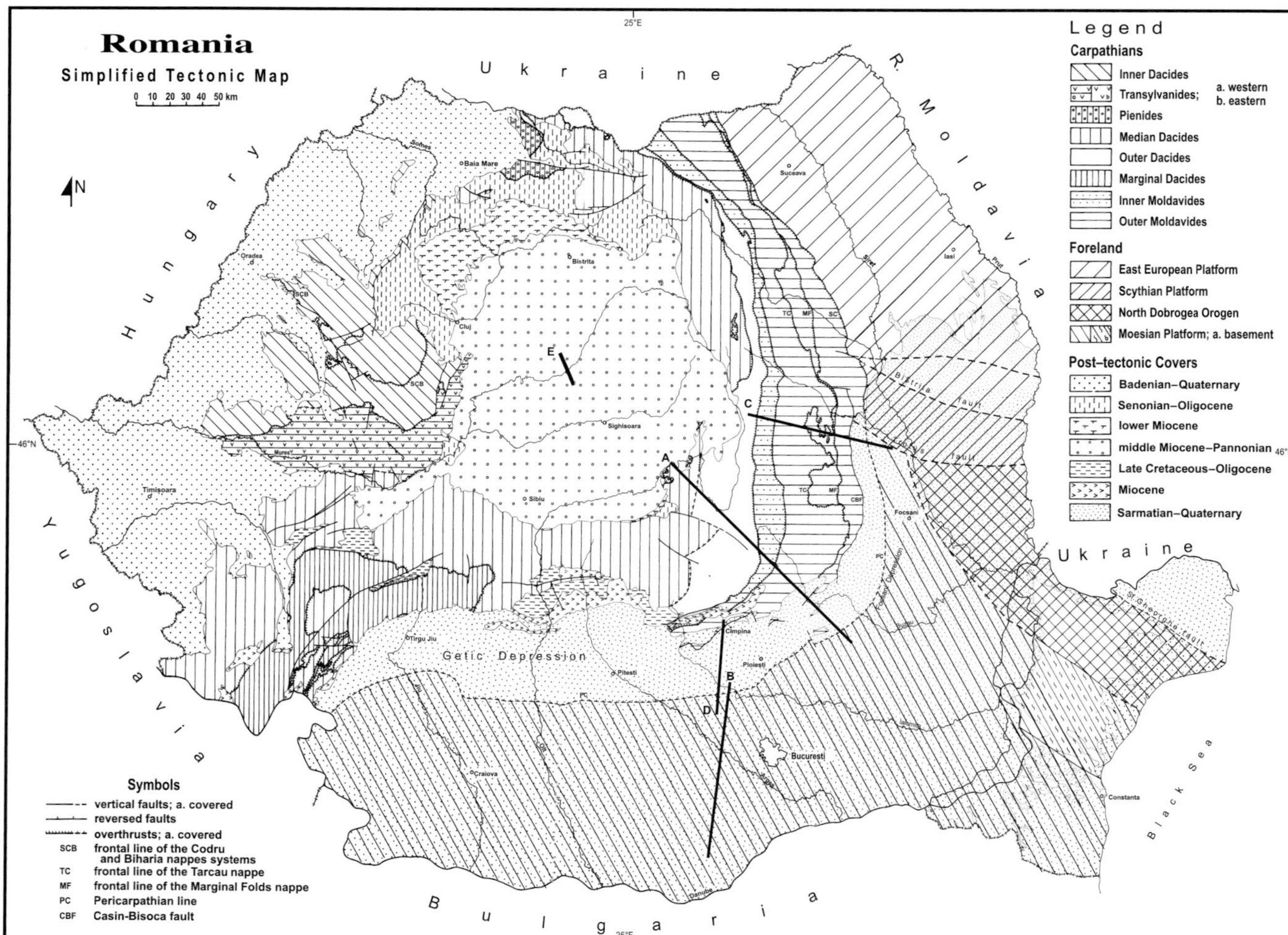

Figure 2. A simplified geologic map of Romania indicating the position of the cross sections.

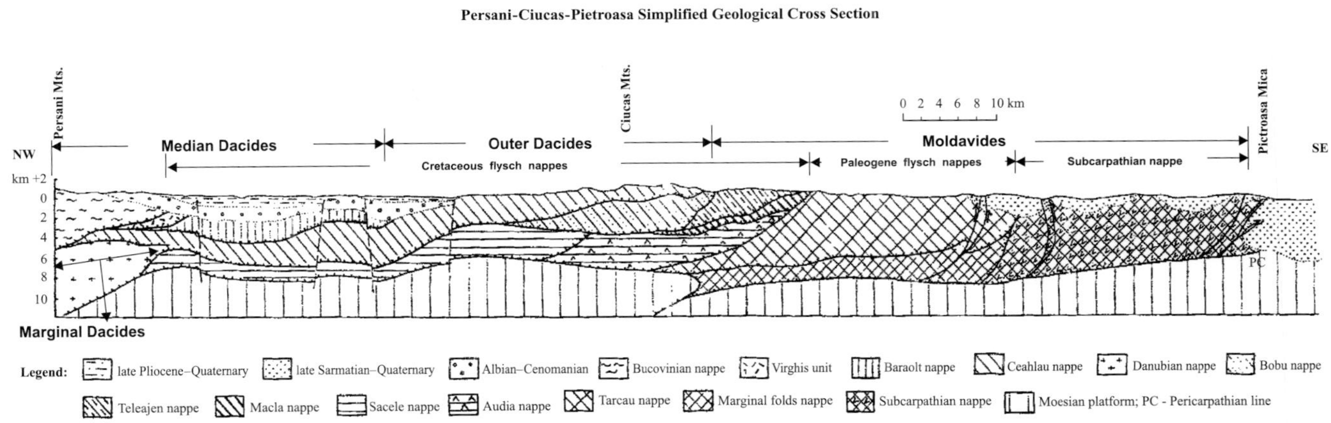

Figure 3. Northwest–southeast cross section showing the main structural elements at the east Carpathian Bend area.

boundary to the north. Westward, it sinks beneath the Carpathian orogen (Figure 2).

The platform basement consists of a mesometamorphic rock association (gneisses and migmatized granites) that is strongly folded. Sedimentary rocks that encompass a large stratigraphic interval, from Vendian to Neogene, unconformably cover the basement (Barbu et al., 1970). The sedimentary sequence starts with blackish-clayey schists and gray or red hard sandstones that have been assigned to the Vendian. The Silurian is not completely developed, being represented only by its middle (Wenlockian) and upper (Ludlovian) part where limestones, quartz sandstones, and hard clayey shales accumulated. The Devonian begins with continental sandstones and clays (Lower Devonian) and continues with reddish or gray sandstones and limestones (Middle Devonian). After a long break, evidence of the sedimentation resumes in the Middle Jurassic, with predominantly calcareous formations (more or less massive limestones, calcareous sandstones, and marly limestones), and continues in the Late Jurassic with red breccia conglomerates, marly limestone clays, and marls with anhydrite nodules. Marly limestones, chalky marls, and glauconitic limestones represent the Lower Cretaceous. Following a short sedimentation break, Cenomanian marly limestones accumulated over almost all of the platform. Turonian and Senonian pelitic deposits have been proven in places. The Eocene sandy limestones and limestones are preserved today only as isolated patches. Only part of the Tertiary is present on the Moldavian Platform, namely, the Badenian–Meotian interval. The Badenian disconformably overlies different older deposits. It consists of reeflike limestones (only in the northeastern corner of the platform), siliceous sandstones, sands, marls, anhydrites, and again marls. The Sarmatian sequence begins with a marly package with sand interlayers followed by a predominantly sandy package. Here and there, tuff interlayers occur. The Moldavian Platform lithostratigraphic column closes up with a Meotian clayey-sandy sequence that is limited only to the southern part of this structural unit but largely develops southward.

It is worth mentioning that the above sequence was mostly restored from the isolated well data, and consequently, it could show some differences along the platform.

During its evolution, the Moldavian Platform underwent alternating vertical oscillations with different amplitudes above and below sea level, a situation that explains both the existing gaps as well as the facies of the accumulated deposits that varies from continental to marine.

The Moldavian Platform shows a general homocline structure that dips from east-northeast to west-southwest toward the Carpathian Foredeep. The homocline is commonly affected by disjunctive faults with different orientation. The frequency of the vertical faults increases in the vicinity of the Scythian platform.

Petroleum system

1) Source rocks. Paleozoic dark shales, Sarmatian shales, and silts are supposed to generate hydrocarbons. Some very dry gas showed up in a few wells that penetrated the Silurian and Devonian shales. The hydrocarbon pools located in Badenian and Sarmatian reservoirs are supposed to be sourced mostly by Sarmatian shales.
2) Reservoir rocks. Badenian and Sarmatian sands and sandstones are the effective reservoirs. They have porosities between 2 and 20% and permeabilities between 0 and 100 md. The Silurian and Devonian sandstones and limestones could be considered as potential reservoirs.

3) Traps. The seismic surveys discovered a number of faults with two main orientations: north-northwest–south-southeast and northeast–southwest. The northeast–southwest faults are younger, displacing the blocks formed by other faults. Lithologic traps are also considered because of the facies variation both in Badenian and Sarmatian.
4) Hydrocarbon fields. Oil, condensate, and gas fields were discovered in the territory of the Moldavian Platform. The gas from the pools located in terrigenous (sands, sandstones, and silts) Badenian and Sarmatian reservoirs (Malini-1, Valea Seaca-2, and Suceava-3; Figure 4) has been generate by unsaturated source shales of the same age. It was supposed that both the condensate (Gura Humorului field-4; Figure 4) and the oil (Cuejdiu field-5; Figure 4) from the present pools were generated far away from their present position by the Oligocene–lower Miocene bituminous shale existing in the Carpathian domain. The liquid hydrocarbon migrated long distances, along faults and/or erosion surfaces, from the initial kitchen up to the present traps (most structural).

Scythian Platform

On the Romanian onshore, the Scythian platform (Figure 2) is located between the East European platform to the north and the north Dobrudja orogen to the south. Its boundaries are tectonic. To the north, it is separated from the Moldavian Platform by Bistrita sinistral strike-slip fault, and to the south, it is separated from north Dobrudja by the north-vergent Sfintu Gheorghe fault and by the Trotus fault. Its western prolongation sinks below the Eastern Carpathian wedge.

Rocks older than the Devonian are unknown in the Romanian territory. The oldest deposits known from wells are the Middle Devonian red sandstones and limestones with clay interlayers. A sequence of dolomites and brown limestones was assigned to the Carboniferous (Dinantian). The Permian–Triassic deposits unconformably cover the older deposits. Two detrital continental piles separated by limestones were accumulated during this interval. The Jurassic locally begins with calcarenites (assigned to Lias), followed by black argillaceous shales interbedded with siliceous sandstones (Dogger), and ends with a thick pile of limestones associated with marls (Malm). Cretaceous consists of more or less bedded limestones in its lower part (Neocomian–Barremian) and sandstones and marls in the upper part of the sequence (Aptian–Cenomanian). The Tertiary unconformably covers older deposits. As in the Moldavian Platform, it begins with its middle part, namely, with Badenian deposits represented by reef-like limestones, siliceous sandstones, anhydrites, marls, and sandstones. The Sarmatian unconformably covers older deposits. It consists of calcareous sandstones, marls, and blackish clays with thin sand interlayers.

The platform underwent more important deformations during the Hercynian and Kimmerian stages. It was slightly deformed during the middle Cretaceous stage and completely uplifted between Late Cretaceous and lower Miocene. Afterward, the area corresponding to the Scythian platform subsided from Badenian to Quaternary.

A network of more or less vertical faults that affected the Neogene deposits also generally characterize the present structure of the Scythian platform. These are faults that were active during the entire life of the platform, separating blocks with different evolution. Sometimes, the blocks have different stratigraphic sequences or different lithologies at the same stratigraphic level.

Petroleum System

1) Source rocks. The Paleozoic source rocks (black and dark gray shales) reached the maturity level all over the platform, whereas the Lower–Middle Jurassic source rocks (dark shales) reached the maturity level only in its western part. Badenian and Sarmatian source rocks (marls, clays, and silts) are not mature, but they generated biogenic gas.
2) Potential reservoirs (detrital and/or calcareous) are present all across the stratigraphic sequence of the Scythian platform, but only Triassic, Jurassic (detrital), Cretaceous (detrital), Badenian (detrital), and Sarmatian (detrital) are saturated with oil and gas (Glavanesti-6, Huruesti-7, Contesti-8, and Bacau-9; Figure 4). In all discovered field so far, the hydrocarbons have been structurally trapped. Theoretically, the Scythian platform has its own petroleum system, represented by the Paleozoic and Jurassic reservoirs and source rocks that reached the oil window and generated at least the hydrocarbon that accumulated in the Contesti field. The other existing fields seem to be supplied by the Tertiary source rocks that belong to a larger petroleum system that covers the entire Carpathian foreland.

North Dobrudja Orogen

A peculiar morphological and structural aspect occurs in the eastern extremity of the Romanian onshore. Here, between the eastward bend of the lower Danube and Black Sea, a hilly area rises from the flat land. It corresponds to a northwest–southeast-trending Alpine thrust-fold chain located between two major platforms

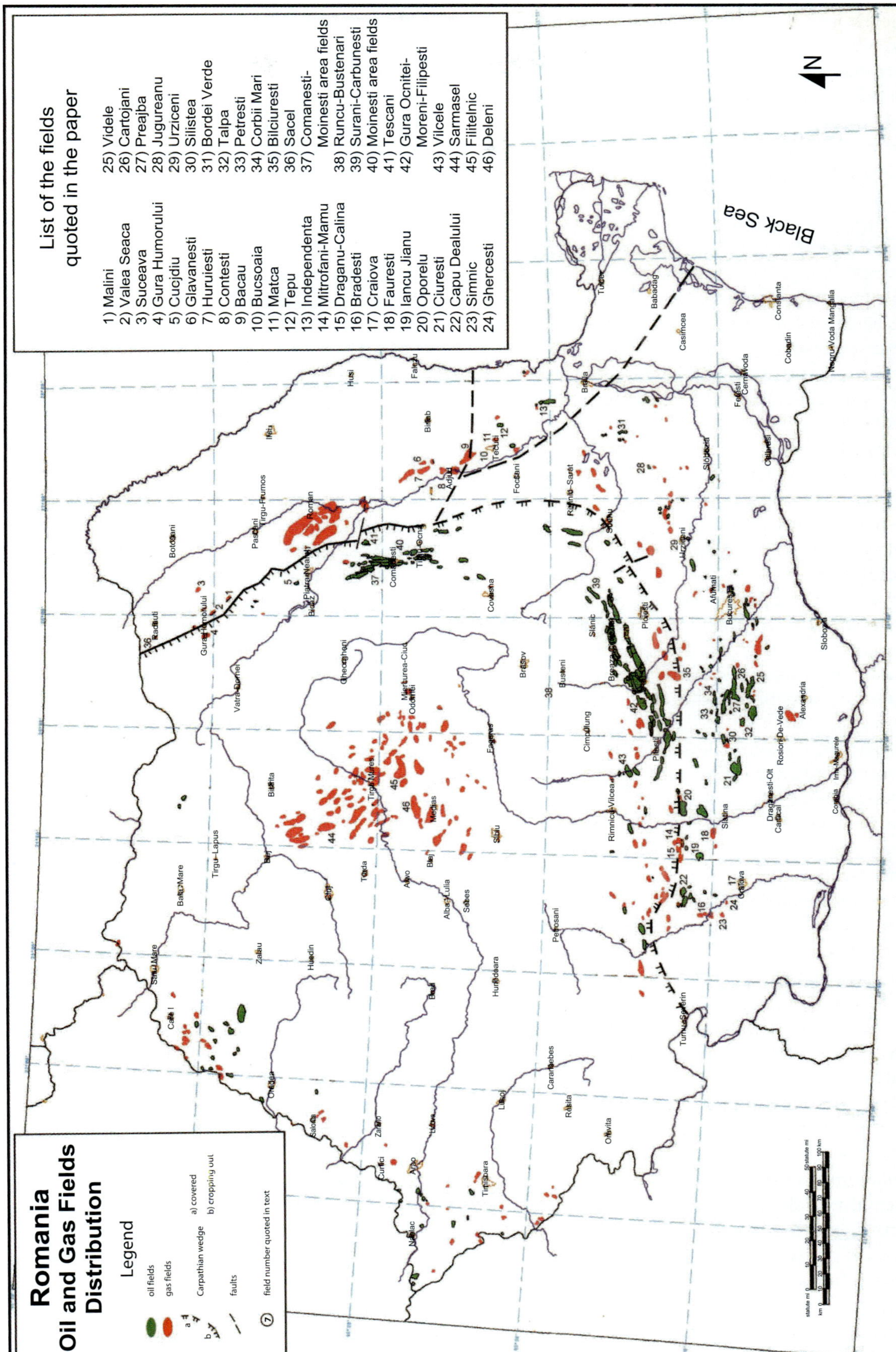

Figure 4. Map of Romanian oil and gas fields.

(intracratonic; Dumitrescu and Sandulescu, 1968): Moesian to the southwest and Scythian to the northeast. The boundaries of these two platforms correspond to two major faults, which are, respectively, the Peceneaga–Camena and Sfintu Gheorghe (Figure 2).

Three distinct imbrications (Mirauta in Sandulescu, 1984) have been distinguished in this area.

1) Macin nappe. This tectonic unit has been assigned the following formations: Late Precambrian terrigenous, amphibolite, calcareous, and basaltic mesometamorphic rocks, Cambrian terrigenous epimetamorphic rocks, Silurian slightly metamorphosed limestones and shales, Carboniferous polymictic conglomerate, sandstones, and shales, sometimes flyschlike, generally affected by cleavage and associated with rhyolitic or basaltic ignimbrites, Permian ignimbrites (rhyolite), Triassic (middle) bedded limestones, and finally, Upper Jurassic marly limestones.
   Granite bodies intrude the Macin nappe deposits. They were emplaced before and after the Lower Carboniferous.
   The nappe inner structure is complicated by a few major reverse faults. Its inner structure was caused by a long tectonic evolution that began in the Hercynian orogeny, was active again during Kimmerian stages, and was finalized during the Austrian stage.
2) Niculitel nappe. Its complete sequence consists of a flysch-type sequence with coarser or finer alternating calcareous turbidites and marls associated with intrusive felsic and mafic rocks, red massive limestones interbedded with basalts and basic pyroclastites, the latter ones practically replacing the limestones toward the middle to outer part of the unit but where they contain limestone blocks or calcareous breccias (Lower Triassic), and finally, a sandy or sandy-shaly flysch (Upper Triassic).
   In the internal side of the Niculitel nappe territory, an important imbrication was distinguished. The inner structure of this unit, including the abovementioned imbrication, was established during the Austrian stage along with the nappe emplacement.
3) Tulcea nappe. Its lithostratigraphic sequence covers a long stratigraphic interval being represented by the following entities (Mirauta 1966; Gradinaru 1984): a rhythmic alternation (flysch type) of metagrauwacke and chlorite schists, sericite phyllites, sericite or graphitous quartzites, black quartzites, locally black limestones, dark slates (lower Paleozoic), flysch-type alternation of siliceous sandstones, siltstones and shales, shaly calcareous limestones, and siliceous shales and jaspers (Devonian).
   Red granites and light-colored porphyries are associated with previous rocks. After a long gap (also marked by an unconformity), the sedimentation resumed in the Triassic when the following deposits accumulated: reddish polymictic conglomerates, sandstones, shales and thin-bedded limestones and marly limestones (Werfenian), dolomites, various limestones of platform type, and marls (Middle–Upper Triassic). The sequence of this nappe is completed by thick sandstones of midfan-type turbidites alternating with gray shales that laterally grade into distal or interfan medium- to thin-bedded turbidites with thicker shale interlayers, calcareous turbidites and dark-colored shales (Lower–Middle Jurassic), gray crinoidal calcarenites, and dark-gray marly limestones, in places with cherts (Upper Jurassic). Scarce rhyolites and basic rocks are associated with the Triassic deposits of this unit.
   The inner structure of the Tulcea nappe shows two different aspects. The pre-Triassic formations are strongly deformed and sometimes slightly metamorphosed, whereas the Triassic–Jurassic deposits are implicated in large anticlines and synclines. This situation is caused by the fact that the pre-Triassic formations were affected by Breton movements, and later on, before the Albian, the Tulcea domain was again deformed less intensely during the Austrian stage. It is worth noting that during the Early Triassic, a slight compression affected the area and generated growing anticlines with Paleozoic deposits in the core.

Post-thrusting Cover

A common sedimentary sequence covers all three nappes (the Macin, Niculitel, and Tulcea) and also extends southward on the central Dobrudja. It begins with calcareous and crinoidal limestones, coarser up to conglomerates toward the northwest, and marls [Vraconian(?)–Cenomanian], followed by thick- to thin-bedded calcareous sandstones, sandy limestones, marly limestones, and marls (Turonian–Coniacian). Chalky limestones assigned to the Santonian (Lower Senonian) crop out on a restricted area in the eastern side of this cover.

North Dobrudja also extends northwestward of the Danube below the Tertiary sequence that covers the platforms. It forms what is described in Romanian geological literature as the north Dobrudja Promontory. Its presence is proved by the data acquired by many wells that penetrated either in different strongly folded, partially metamorphosed Paleozoic rocks (Carboniferous, Devonian, Silurian, and Ordovician–Cambrian)

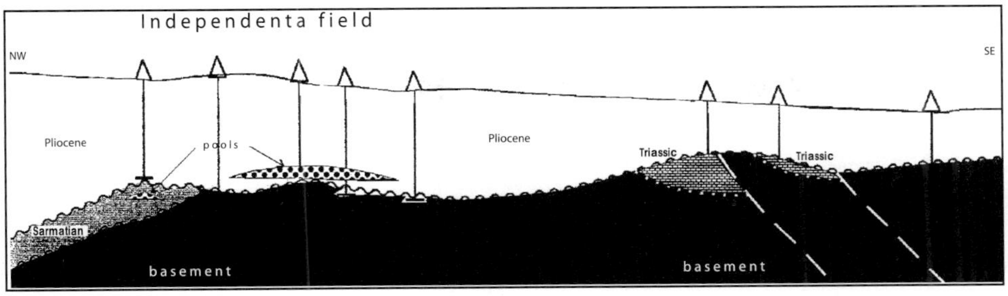

Figure 5. Trap types of the North Dobrudja Promontory fields: Independenta field.

or in meso- and epimetamorphic rocks. Magmatic rocks intrude them. All these rocks are similar to those cropping out again only in Macin nappe. Before ending up in its present position, the north Dobrudja Promontory was, for a long time, uplifted and exposed to erosion until it was covered by Badenian and Sarmatian–Quaternary deposits.

Petroleum System

Oil and gas fields are aligned along the northwest–southeast trend of the north Dobrudja Promontory (Bucsoaia-10, Matca-11, Tepu-12, and Independenta-13; Figure 4). It is considered that the source rocks located in the neighboring depressions (pre-Dobrudja to the northeast and foredeep to the west) have sourced their reservoirs.

Matca, Tepu, and Bucsoaia are predominantly gas fields, their production coming from detrital Badenian and Sarmatian reservoirs. Independenta field (Figure 5) is the most important in this area. The structure is a gentle uplift of the north Dobrudja Promontory with three distinct morphostructural highs. It mostly produces both gas and oil from Pliocene sands. Only locally is the Sarmatian saturated with oil. The pay zones are grouped into 17 intervals, their depths ranging between 450 and 650 m (1476 and 2132 ft). The effective thickness is between 1.5 and 10 m (5 and 33 ft), with an average porosity of 30% and a permeability of 700 md.

Moesian Platform

On the Romanian onshore, the Moesian Platform is limited by the Danube to the south and by the Subcarpathian to the north (Paraschiv, 1965). It plunges northward below the Carpathian thrust wedge (Figure 6). Eastward, the Black Sea shore limits its cropping area, but the platform prolongs into the offshore area. The northeastern limit corresponds to the Peceneaga–Camena fault.

Over a complex metamorphic basement (with slightly metamorphosed to mesometamorphic rocks), a Cambrian–Quaternary sedimentary cover accumulated. It has a variable thickness that sometimes surpasses 10,000 m (33,000 ft). As a result of the differences of both the subsidence and erosion, the sediments accumulated during four distinct sedimentary cycles (Paraschiv, 1982) as follows (Figure 7):

1) The Paleozoic cycle, including the lower detrital (Ialomita) group (Cambrian–Lower Devonian), composed of Cambrian black argillitic deposits with trilobites (Mutiu, 1991), quartzitic sandstones (Mangalia Formation), black, argillaceous shales (Tandarei Formation), and sandstones (Smirna Formation); middle carbonate group or Calarasi Formation (Middle Devonian–Carboniferous), represented by limestones and dolomites with bituminous and evaporitic sequences interlayered; and upper detrital group or Vlasin Formation, of middle–Late Carboniferous age, made up of clay and sandstones with coal interlayers. The cycle reaches 5000-m (16,400-ft) thickness in the south-central part of the platform (Alexandria Depression) but is thinning up on the uplifted areas.
2) The upper Paleozoic (Permian)–lower Mesozoic (Upper Triassic) cycle, subdivided in the Rosiori Formation (Permian–Lower Triassic), composed of quasicontinental deposits with vulcanogen interlayering, basic and acid at the lower and middle parts, respectively; the Alexandria Formation (carbonatic-evaporitic; Middle Triassic), with some acid-effusive volcanic rocks at the upper part; and the Segarcea Formation (Upper Triassic, in places up to upper Lias) with brown-reddish continental, evaporitic, marine, and volcanic deposits.
3) The Middle Jurassic–Upper Cretaceous, predominantly represented by marine sequences with a continental-lacustrine episode, during the Aptian, mainly in the eastern margin of the platform.

The cycle generally begins with Middle Jurassic (Dogger), but toward the south, it starts even with Lower Jurassic (Lias). The upper Lias–Dogger sequence was separated (Paraschiv, 1982) as the Bals Formation. It consists of siliceous sandstones and black shales. The Upper Jurassic–Cretaceous interval is predominantly

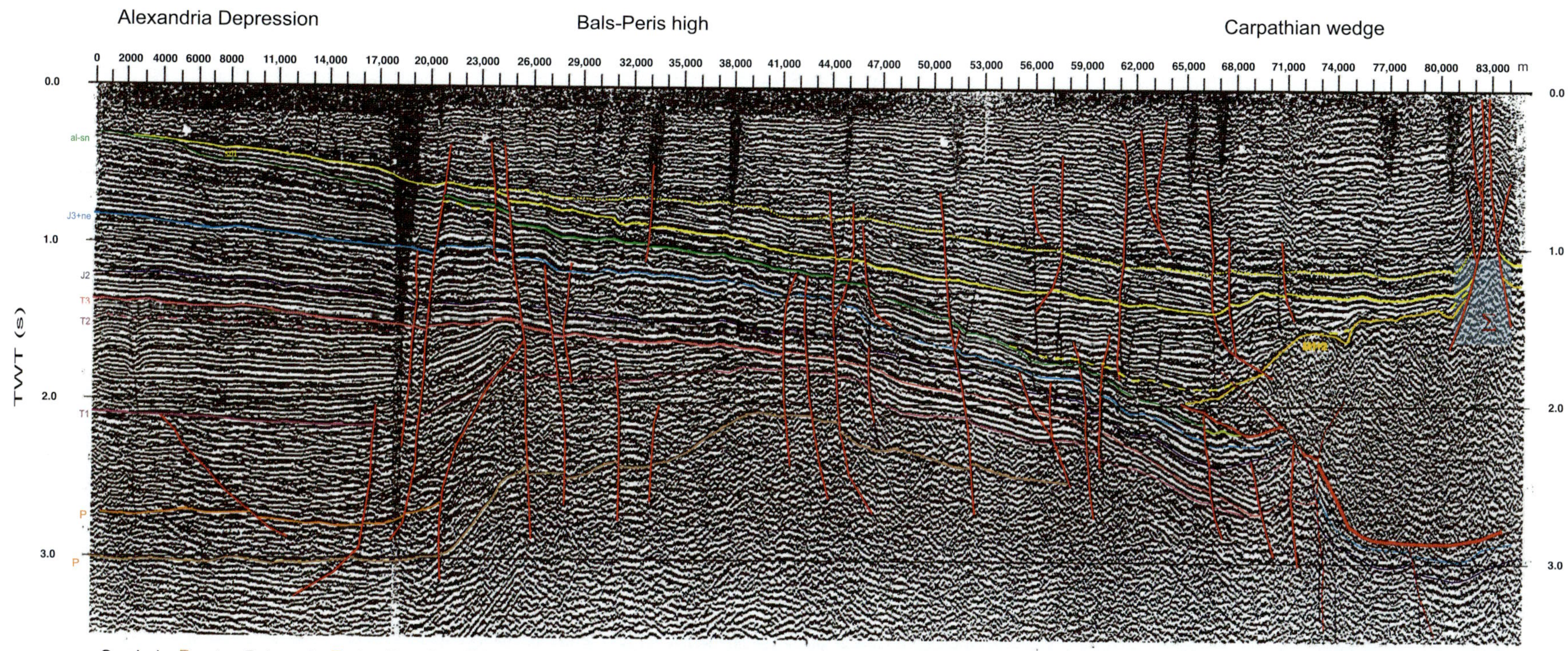

Symbols: Pz = top Paleozoic; P = top Permian; T1 = top Lower Triassic; T2 = top Middle Triassic; T3 = top Upper Triassic; J2 = top Middle Jurassic; J3-ne = top Upper Jurassic – Neocomian; al-sn = top Albian until Senonian; M1+2 = top lower–middle Miocene; sm2 = top Sarmatian; me = top me; po-Q = Pontian–Quaternary; Σ = Salt

Figure 6. Interpreted seismic section across the central-eastern Moesian Platform and its contact with the Getic depression. TWT = two-way traveltime.

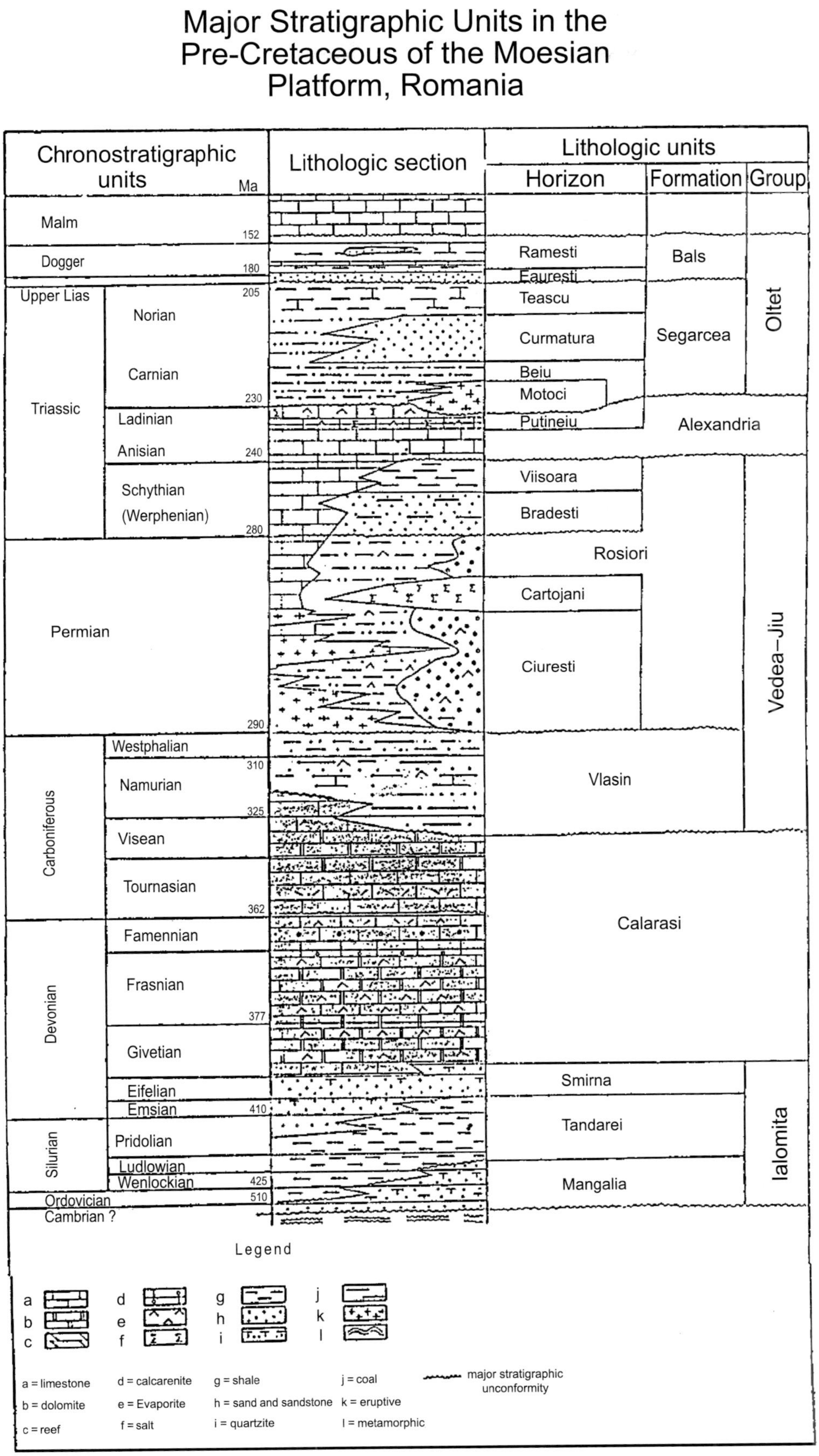

Figure 7. Lithostratigraphic column of the Moesian Platform.

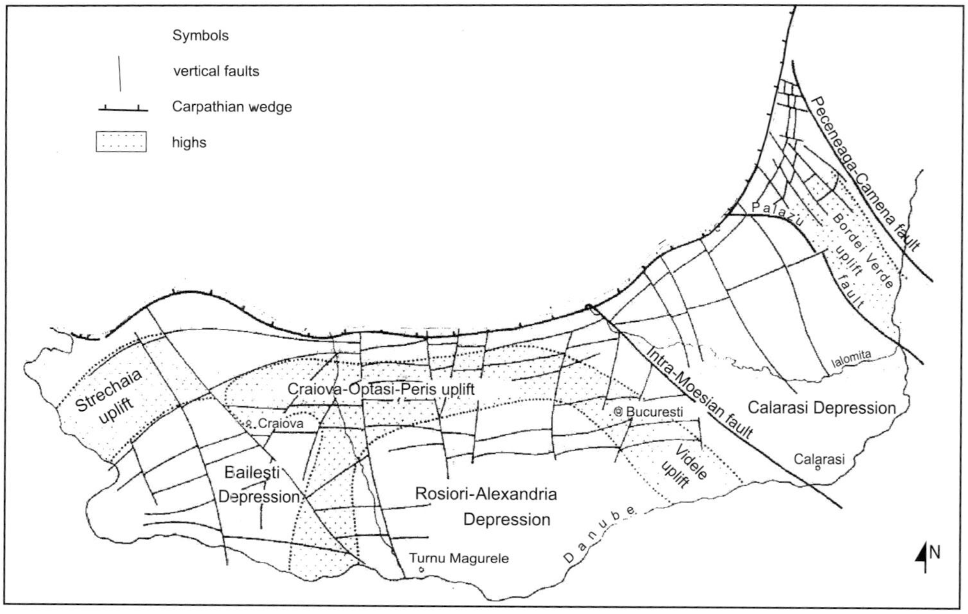

Figure 8. Sketch showing the main fractures, uplifts, and depressions of the Moesian Platform. Modified after Sandulescu and Visarion, 1984.

calcareous (in places with carbonate buildups). Sands, sandstones, marls, and calcarenites do occur in the Albian–Senonian interval.

4) The Neogene cycle. At the end of Senonian, the Moesian Platform uplifted, and Cretaceous formations were partially eroded. In places, the sedimentation resumed for a short interval during the Eocene when a 70-m (230-ft)-thick sequence of green-gray marls, argillaceous limestones, and sandy limestones accumulated on the eastern side of the platform. On the western side of the Moesian Platform, in the prolongation of the Lom Depression, Eocene gray marls, argillaceous limestones, and sandstones (25–250 m [80–820 ft] thick) also deposited. A long period of nondepositional (exposed platform) conditions followed. Beginning with the late Badenian, shallow-marine conditions resumed over the Moesian Platform. The Badenian stage is completely developed only in the northern and northwestern part of the platform, whereas on the east and central areas, only the median section interval is present. Marls, clays, limestones, and sandstones form the lithologic background of the Badenian. Subsequently, the platform was covered by a Sarmatian–Pliocene sequence that is common also both to the Getic depression and to the southwestern part of the Subcarpathian nappe. The Sarmatian is represented by a detrital facies, coarser at the lower section and gradually finer upward, where a predominantly pelitic sequence is present. Locally (Ghimpati area), an oolithic facies is developed. The Meotian is a regressive formation that reaches its maximum thickness of about 1000 m (3300 ft) in the northern part of the platform. Marls, silty marls, sands, and sandstones are present in this interval. The Pontian is mostly pelitic, and thins out southward from hundreds to zero meters. Gravels, sands, sandstones, and silts ranging from hundreds to thousands of meters thick represent the Dacian and Romanian.

A dominant tear structure with different vertical movements (Sandulescu and Visarion, 1978; Paraschiv, 1979; Tari et al., 1997) of the blocks is characteristic to the Moesian Platform (Figure 8). An important part in the platform's tectonic evolution seems to be played by the Intra-Moesian fault that separates two sectors of the platform, each of them with distinct features (Figure 8):

- The western sector with a mesometamorphic (Moesian or Wallachian) basement type and with a more complete Paleozoic–Mesozoic stratigraphic cover. The most remarkable structural feature of this sector is Craiova–Bals–Optasi uplift with its branches, namely, the Strehaia–Vidin uplift to the southwest and Slatina–Ghigheni and Videle–Vetriano uplifts, trending, respectively, to the south and to the southeast. At the same time, several spurs branch out northward, sinking below Jurassic, Cretaceous, or Neogene formations. During the Paleozoic and Triassic, some depressions such as Bailesti and Rosiori developed between these spurs (Figure 8).
- The eastern sector, with a central–south Dobrudja (green schists) basement type. Generally, it has a higher structural position than the western sector,

and its stratigraphic column is characterized by more sedimentary breaks. Pre-Jurassic tectonics is of outcropping Dobrudja type, with northwest–southeast-trending major structural elements. Sinking northwestward from central and south Dobrudja, several basement spurs prolong beneath the Moesian Platform's Paleozoic, Mesozoic, and Neogene formations. Calarasi–Urziceni and Movila Miresii–Ghergheasa depressions developed between the spurs.

West–east (in the western sector) and north–west (in the eastern sector) disjunctive faults cross the platform (Figure 8). The uplifts and the main faults commonly affecting several structural stages are an important factor in the formation and conservation of hydrocarbon fields.

Petroleum Systems

Source rocks. In the sedimentary column of the Moesian Platform (Figure 7), several sequences showing source rock properties have been recognized:

- The Silurian Tandarei Formation reaches about 1000-m (3300-ft) of thickness in the northern part of the platform and mostly (65%) consists of black clayey shales with graptolites. The geochemical analyses revealed values of 1.0–1.4% total organic carbon. In the median part of the platform, the Silurian shales could only reach the top of the oil window, whereas in the northern, sunken part of the platform where they are buried deeper than 3000 m (10,000 ft) (Bradesti-16 and Capu Dealului-22 fields area), the shales got into the oil-phase generation during the upper Miocene.
- The bituminous black limestones of the Calarasi Formation have 0.8–1.25% total organic carbon content derived from type I–II kerogen and reach thicknesses of as much as 600 m (2000 ft). Favorable conditions for hydrocarbon generation were realized in the Alexandria Depression, where the Calarasi Formation could reach the oil window during the Triassic and Jurassic.
- The middle–Upper Carboniferous Vlasin Formation shows an argillaceous facies southwest of Bucharest and a bituminous-carbonate facies in the central part of the platform. The last mentioned facies has total organic carbon values ranging between 0.35 and 9.3%. During the Jurassic and Cretaceous, the oil generation window could be reached in places but only on relatively restricted areas.
- The Middle Triassic Alexandria Formation is dominantly calcareous (limestones and dolomites). It has 0.50–1.40% total organic carbon and a kerogen of type I–II. With an effective thickness of source rocks of about 400 m (1300 ft), it had good potential for oil generation. It is well developed in the Alexandria Depression. During the Miocene, the organic matter could reach maturity levels at depths ranging between 2500 and 4500 m (820 and 1470 ft).
- The Middle Jurassic Bals Formation includes a sequence of bituminous black lutites as much as 400 m (1300 ft) thick, developed in the western third of the platform, where three to four sandstone packages were also interlayered; eastward, only the sandstone facies persists. The black lutites total organic carbon content ranges between 1.2 and 2.0% derived from II to III organic matter type. In the northern sector, the middle Sarmatian Subcarpathian overthrust buried that sequence to maturity depths of more than 3500 m (11,500 ft).
- Badenian–Sarmatian pelitic sequences are continuously developed with total organic carbon values between 1.3 and 2.5% (average 2.0%) on the northern and northeastern areas of the platform; the effective thicknesses range from 200 to 300 m (660 to 1000 ft) in the northwest and as much as 400–800 m (1300–2600 ft) in the northeast. The organic matter is of type I–II in the northwest, becoming more terrestrial toward the east (type II–III; Figure 9). The oil window was entered after middle Sarmatian tectonic events but only at depths greater than 3500 m (11,500 ft); these depths were realized beneath the Getic depression overthrust, in the western and central parts, and in the Carpathian Bend, sunken area, in the east of the platform (Figure 10).

Despite the existence of a wide range of potential source rocks, geochemical biomarkers and source rock-oil correlations performed so far revealed the contribution of only two effective oil facies to the existence of the platform's fields: one represented by the Dogger shales and the other related to Badenian–Sarmatian bituminous pelites. The functions of the other possible source rocks mentioned above need to be solved in the future.

Consequently, the petroleum systems of the Romanian sector of the Moesian Platform could be separated into the following:

- Proven (better studied) as the Jurassic and Neogene systems

The Jurassic petroleum system is present in the western area, and it is related to Middle Jurassic source rock sequences as well as to detrital carrier and reservoir rocks of Lias–Dogger age (Mitrofani–Mamu-14, Draganu–Calina 15, Bradesti-16, Craiova zone-17, Fauresti-18, Iancu Jianu-19, Oporelu-20, and Ciuresti-21; Figure 4).

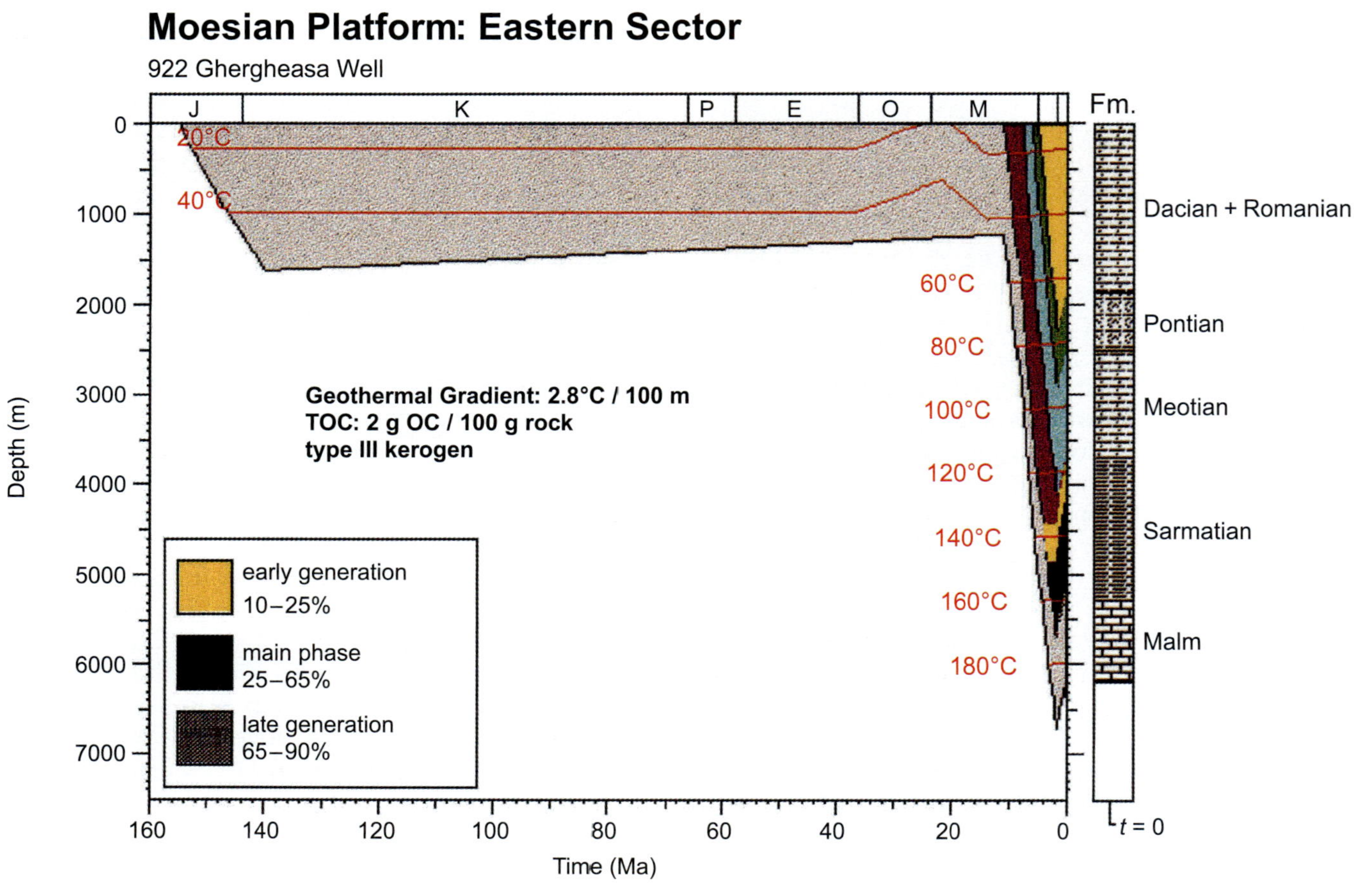

Figure 9. Eastern Moesian Platform burial history.

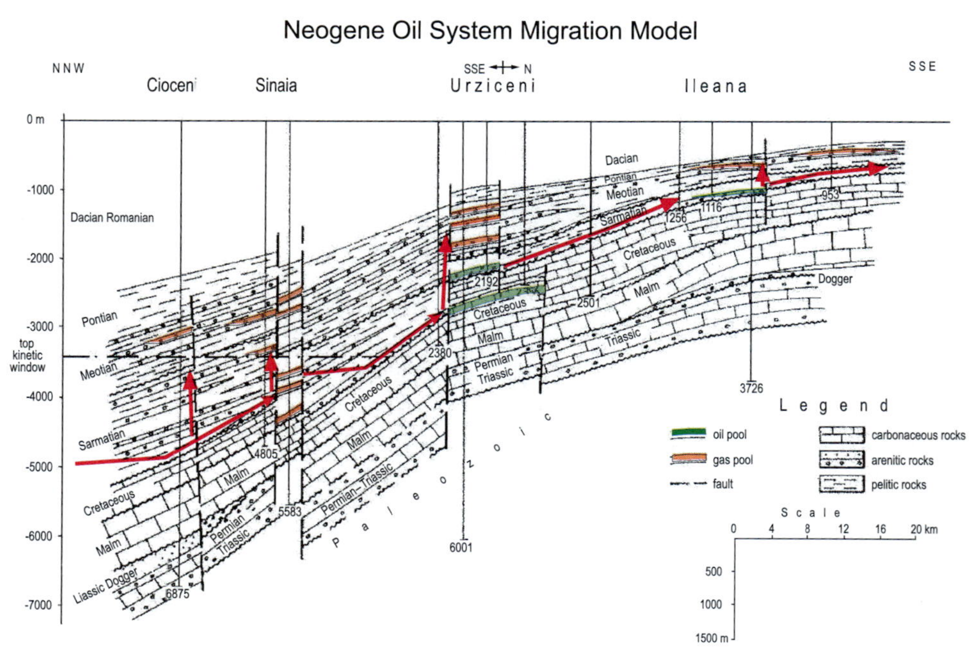

Figure 10. Hydrocarbon migration scheme in the Moesian Platform.

The Upper Jurassic carbonate reservoirs (Ciuresti) are assigned to the same system.

The Neogene petroleum system covers the entire platform, with oil and gas accumulations entrapped into reservoir rocks of ages ranging from Triassic to Neogene.

- Possible (less studied) as the Paleozoic, Triassic, and Cretaceous systems

Paleozoic petroleum systems might occur in the western and southern areas of the platform related to Silurian–Upper Carboniferous organic-rich rocks, but their effectiveness has not been satisfactory proven so far.

Triassic and Cretaceous petroleum systems might be present in the western and central areas, respectively, but further studies are needed to clear their relationships with the other systems of the platform.

As regards the eventual participation of all source rocks to the platform field formation, two scenarios could be considered. In the first scenario, the Paleozoic, Mesozoic (Middle Jurassic), and late Miocene, located on the northern slope of the platform, went into the oil window conditions between the upper Miocene and Pliocene. In the second scenario (which is still theoretical) regarding the southern side of the platform, the Silurian, Devonian, Carboniferous, and, in places, Middle Triassic source rocks could generate hydrocarbons at very different moments from late Paleozoic to Cretaceous, depending on the evolution of the local basins (Bailesti, Alexandria, Lom, and Calarasi).

Reservoir rocks. Commercial hydrocarbon accumulations have been discovered in reservoirs of various age and lithology as follows: in Middle–Upper Devonian to Lower Carboniferous, Middle Triassic, Malm–Neocomian, and Turonian–Senonian carbonates; in Barremian and Albian calcarenites; in lower Sarmatian (upper Miocene) limestones; in sandstones of Permian–Lower Triassic and Dogger; in Albian glauconitic sandstones; and in poorly consolidated sands and sandstones of the Miocene–Pliocene (Paraschiv, 1979, 1982).

Migration. As we have mentioned above, the source rocks accumulated at different stratigraphic intervals in the Silurian to the Sarmatian. In the northern part of the platform (beneath the foredeep zone), all the source rocks (Paleozoic, Mesozoic, and Tertiary) reached maturation stages very late only during the late Neogene (Figure 9). It seems that in the southern area, namely, in the Alexandria Depression, the maturation is Cretaceous.

A lateral migration that could be supposed for both Dogger and stages had been reached much earlier even during the Paleozoic, Triassic, and Neogene source rocks. It was active only beginning with the upper Sarmatian. Later on, a vertical migration began during Valachian tectogenesis coevally with the last folding in the foredeep and the disjunctive tectonics on the Moesian Platform. A second stage of oleogenesis and migration happened during the Pliocene, when hydrocarbons migrated from the foredeep southward along the morphological surface of the Cretaceous formations. The distributions of the oil qualities on the platform and the presence of a large area of weathered oil in the southern part (around the Bucharest town) could be considered as proof of this assumption.

Traps. During the exploration for hydrocarbons, all types of traps were discovered in the Moesian Platform as follows (Popa and Paraschiv, 1974; Paraschiv, 1975):

1) Structural traps: anticlines and hemianticlines (structural nose) were discovered on the northern part of the north Craiova–Bals–Optasi–Peris uplift. Simnic, Ghercesti, Iancu Jianu, Fauresti, Oporelu, and Ciuresti South and North fields (nos. 23, 24, 19, 18, 20, and 21, respectively, in Figure 4) could be included in this category.
2) The most frequent traps are the faulted monoclines located next to the central zone of the platform as in Videle (Figure 4, no. 25; Figure 11A), Cartojani, Preajba, Jugureanu, and Urziceni fields (nos. 26, 27, 28, and 29, respectively, in Figure 4).
3) Stratigraphic traps below and above the unconformity (disconformity) are represented in Bradesti (Figure 11B), Iancu Jianu, Oporelu, Ciuresti, Silistea, Jugureanu, and Bordei Verde (no. 31, Figure 4).
4) Lithologic traps represented by both sand lenses developed in a pelitic sequence (Bilciuresti-35, Ghercesti fields) and reef buildings that grew in the Upper Jurassic–Necomian limestones (Talpa-32, Petresti-33, and Corbii Mari-34 fields; Figure 4).
5) Paleogeomorphologic traps are present as fossilized relief (highs and valleys). The valleys are filled with sands (Preajba, Silistea, etc.).

Fields. More than 100 oil fields and more than 100 gas fields have been discovered in the Moesian Platform (Paraschiv, 1984). Most of the platform fields are located southward of the Carpathian wedge, but also, fields situated on the underthrust (as deep as 5000 m [16,000 ft]) part of the platform are present. They generally show a simple structure with combinations between structural and stratigraphic traps.

Eastern Carpathians

Transylvanides

Nappes (Haghimas and Persani) that are represented today only by some remnant outliers are included in this group. They originated from an extensional oceanic

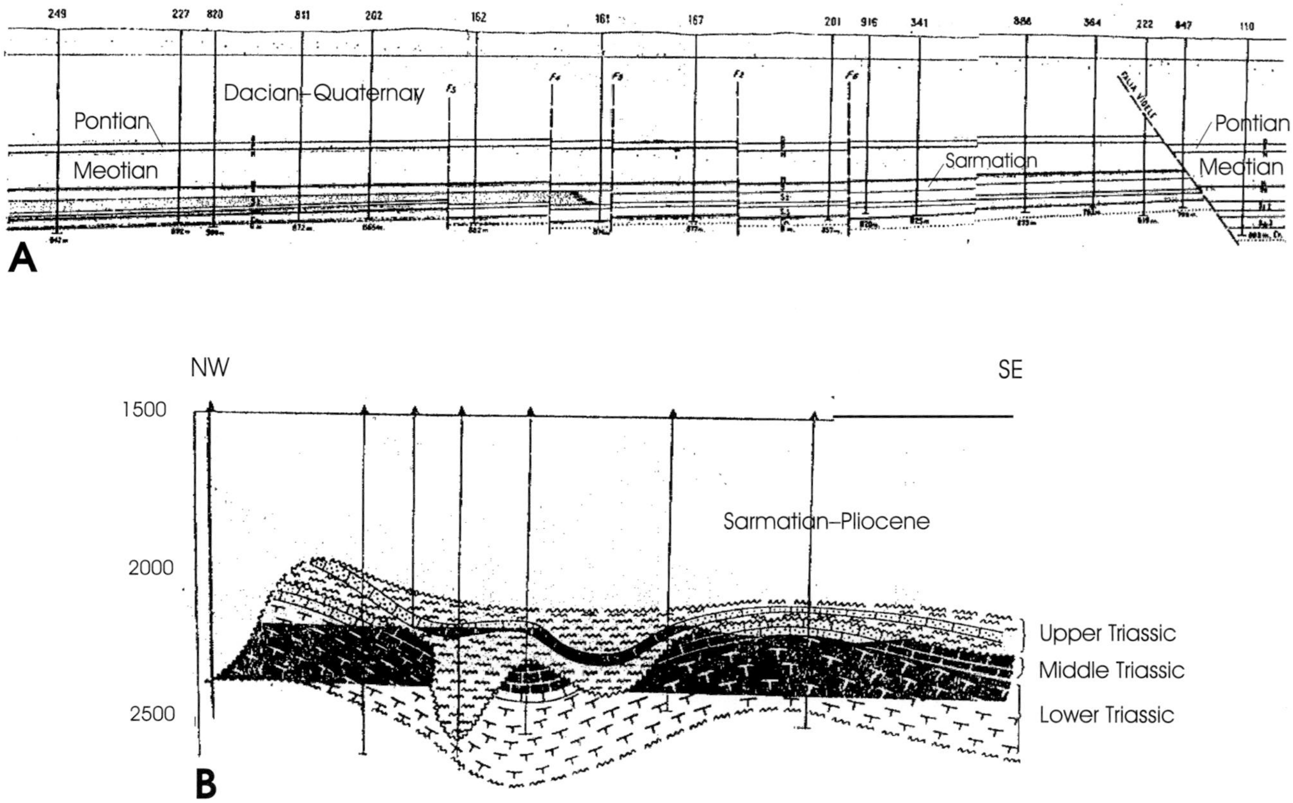

Figure 11. Moesian trap-type examples: (A) faulted monocline, Videle field; (B) stratigraphic, Bradesti field.

crust (Sandulescu and Visarion, 1978), nowadays located below the Transylvanian Depression.

Haghimas nappe. Its lithostratigraphic column, restored from different outliers cropping out in the Central–Eastern Carpathians (Figure 2) (Sandulescu, 1967, 1975), has the following sequence: siliciclastic (Werfenian), calcareous (upper Campilian–Anisian), calcareous with cherts and jaspers (Ladinian), gray-bedded limestones with cherts, white massive limestones (Carnian), red nodular limestones (Norian), limestones and red shales (lower Lias), greenish nodular limestones and sandy limestone interbeds (Kimmeridgian), massive Stramberg-like limestones (Tithonian–Neocomian), and massive organogenous limestones (Urgonian).

Persani nappe. The deposits assigned to this unit are widespread in the Persani Mountains but also appear as isolated elements in the Haghimas and Rarau Mountains. The lithostratigraphic column (Patrulius et al., 1969; Sandulescu, 1975b) includes the following entities: siliceous and calcareous sandstones, gray marls, thin-bedded limestones and shales (Werfenian), bedded, gray, vermicular limestones, in places dolomites, red nodular limestones (Anisian), serpentinite, gabbro basalts, spilite, trachyte and oligophyres, red massive or bedded limestones (Ladinian), red massive limestones and white organogenous limestones (Carnian–Norian), black nodular limestones (Rhaetian), marly limestones, and red shales (lower Lias).

Post-thrusting Covers. Two different levels of deposits cover the Transylvanides. The older one (upper Aptian) is developed only in the east Carpathian Bend area and begins with more or less calcareous conglomerates, associated in places with reef limestones. The molasse conglomerates grade upward into sandy-shaly and shaly flysch facies. The second post-thrusting level unconformably overlies the previous flysch facies. In the east Carpathian Bend area, it begins with polymictic conglomerates (Albian) followed either by calcareous or siliciclastic massive sandstones (upper Vraconian–Cenomanian). The upper Vraconian–Cenomanian sequence is common both for the bend and the central Eastern Carpathians.

The presence of these post-thrusting deposits allows the establishment of the age of the Transylvanides emplacement: older, pre-late Aptian at the east Carpathian Bend area and younger, pre-late Vraconian northward. If one also considers the Transylvanides type of the sedimentary klippen reworked in the Wildflysch Formation, then we can expect that the first horizontal displacement began earlier, most probably since the Barremian.

Median Dacides

As the whole chain, a pile of nappes superimposed with each other makes up the Eastern Carpathians (Figure 2). The analysis is made starting with the higher (more internal) toward the lower (more external) nappes.

Bucovinian Nappe. Both crystalline basement and its sedimentary cover are involved in this nappe. The basement is made up of meso- and epimetamorphic schists. The sedimentary sequence (Patrulius et al. 1966, 1969; Sandulescu 1967, 1975, 1984) is made up of sedimentary breccias solely composed of crystalline schist elements [Upper Carboniferous(?)], siliceous limestones, conglomerates, shales, bedded limestones (Werfenian), white dolomites (Campilian–Anisian), mostly red shales and jaspers (Ladinian), massive limestones (Ladinian), reddish dolomites and limestones (Carnian), bedded limestones with cherts, massive limestones (Carnian–Norian), hematitic, oolitic limestones (Sinemurian–Carixian), calcareous conglomerates (Domerian), limestones with cherts and calcareous sandstones (Dogger), red shales and jaspers (Callovian–Oxfordian), and sandy limestones (Kimmeridgian). After a short gap, three different coeval facies accumulated during the Tithonian–Valanginian interval: inward, marly limestones with calcarenite interlayers that grade upward into flysch facies; in the middle, a sequence of gray, red marls, sometimes red silts, and gray calcarenites; outward, a shaly, sandy, more or less calcareous flysch accumulated. Some of these flysch deposits have been detached from their original basement, because today, they form independent tectonic imbrications (Sadova, Tarnita, Virghis, and Baraolt). During the Hauterivian, polymictic conglomerates are present in the outer part of the Bucovinian domain. These follow a peculiar chaotic sequence that covers the upper Barremian–Albian interval. The Wildflysch Formation consists of polymictic rudites, red green radiolarites overlain by black clays, and siltstones in which polymictic lenticular conglomerates develop. Both Transylvanian and Bucovinian nappes have supplied the conglomerate's elements. The upper part of the Wildflysch Formation follows a calcarenitic and then a flyschlike sequence that is affected by submarine slidings. The Wildflysch is unconformably overlain by Vraconian–Cenomanian conglomerates in the Central–Eastern Carpathians and by Aptian–Cenomanian conglomerates that grade upward into flysch deposits in the east Carpathian Bend area.

Sub-Bucovinian Nappe. Meso- and epimetamorphic schists mainly represent the sub-Bucovinian nappe. They are unconformably overlain (Sandulescu, 1975a) by Verrucano-type conglomerates (Permian), siliceous conglomerate and sandstones, thin-bedded marls and limestones (Werfenian), massive or bedded dolomites (uppermost Campilian–Anisian), jaspers, red radiolaritic siltstones (lowermost Ladinian), limonite sandstones, residual ferruginous deposits, blackish marls (Lias), marly nodular or bedded limestones (Dogger), and finally, calcareous breccias with crystalline schists and dolomite elements (Neocomian).

Infra-Bucovinian Nappe. This encompasses (Sandulescu, 1975a, 1984) all the formations that occur in isolated tectonic windows from below the sub-Bucovinian nappe. This fact makes it very difficult to correlate all the formations assigned to the infra-Bucovinian nappe. It has a very complex structure characterized by internal imbrications and different sedimentary sequences. Mostly epi- and mesozonal schists crop out in the windows. They are unconformably covered by sedimentary sequences that are different from area to area. The sedimentary sequences began either with Permian or directly with Lias deposits and accumulated until the Tithonian or even until the Lower Cretaceous.

Post-thrusting Covers. The Transylvanides postoverthrusting covers are also common for the Median Dacides. Consequently, their emplacement age is the same as it was discussed for the Transylvanides (see above). On the western slope of the northern segment of the Eastern Carpathians, the post-thrusting cover (Transcarpathian Flysch Zone; Sandulescu, 1984) starts (Figure 12) with polymictic conglomerates and massive sandstones (upper Albian–Cenomanian); these are overlain by siltstones and/or marly pelagic deposits (Turonian–Paleocene; Szasz, 1974). After the upper Paleocene–lower Eocene gap, sedimentation resumes with calcareous conglomerates, limestones, sandstones with *Nummulites*, and marls (Dicea et al., 1980) sequence that accumulated during the Eocene. A chaotic formation (wildflysch type) with a blackish marly matrix accumulated at the Eocene–Oligocene boundary. Lenses of sandstones are also locally developed in the Wildflysch. The allochthonous rocks reworked in the matrix are of various ages from Senonian to lower Oligocene. This is followed by black marls and clays (Oligocene). The post-thrusting cover ends with a sandy or shaly sandy flysch sequence (Oligocene–lower Miocene).

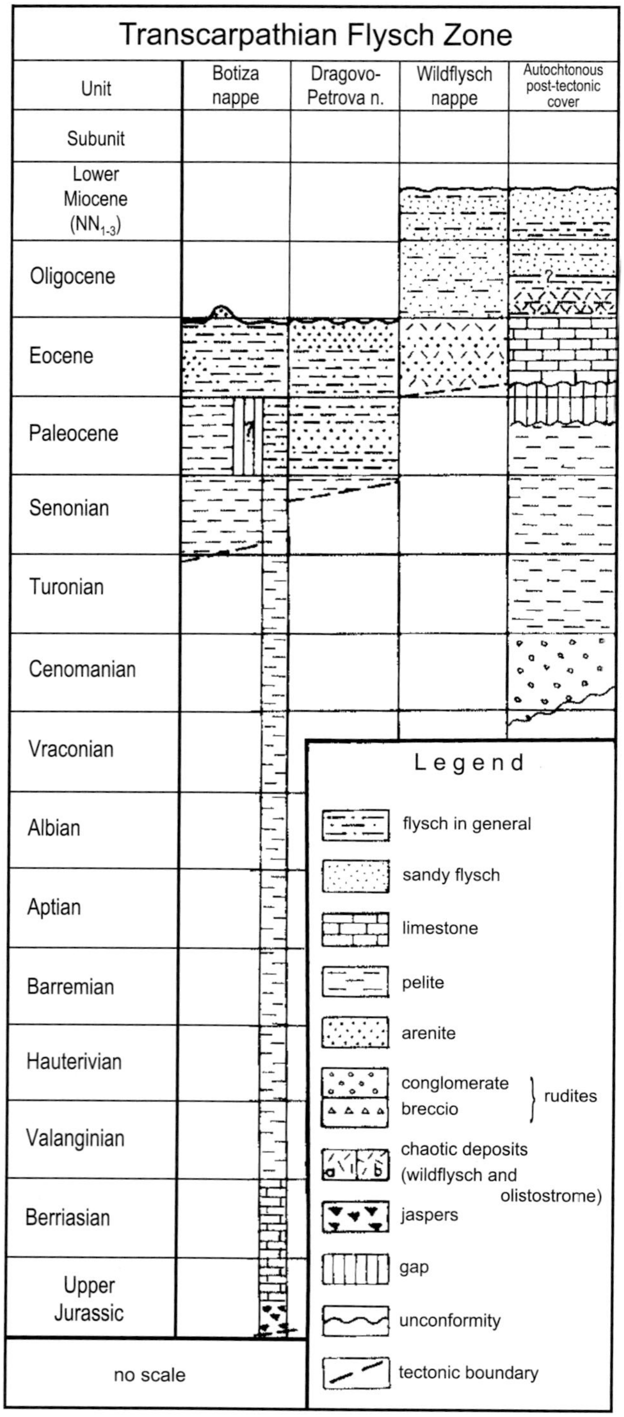

Figure 12. Lithostratigraphic column of the Transcarpathian Flysch Zone.

The autochthonous post-thrusting cover of the Median Dacides was deformed before the middle Miocene and overthrusted from the west by the nappes belonging to the Pienides System (Figure 2). Because these units do not present an immediate interest on the Romanian territory from the hydrocarbon viewpoint, we will not go into further detail about them.

Petroleum System

Source rocks. The Oligocene black marls and clays have been considered as source rocks for the hydrocarbon existing in this area by simply lithologic comparison with the similar rocks from the outer Moldavides. This initial assumption was proven by analyses made on Oligocene pelites that showed total organic carbon content of 6%.
Migration. A short lateral and/or a vertical migration could easily explain the oil fields existing in this area.
Reservoirs. Despite the fact that many of the terrigenous rocks of the post-thrusting cover present reservoir characteristics, only those of the Oligocene practically proved their capability to accumulate hydrocarbons. They have a porosity that ranges between 0.78 and 5.60% and a permeability of 10 md. A secondary (fractures and fissures) permeability could also be considered.
Trap type. Only a structural-type trap has been proven so far.
Oil fields. Only three small, faulted anticlines (e.g., Sacel, Saliste, Figure 4, no. 36) have been proved as oil bearing.

Outer Dacides

Black Flysch Nappe. The Black flysch nappe crops out only in the northern segment of the Eastern Carpathians (Figures 2, 3). The oldest member of the sequence assigned to the Black flysch nappe is represented by an intraplate (Sandulescu et al., 1981a) complex of basalts (massive flows or pillow lava), tuffs, and tuffites (Figure 13). Normally, it stays below the sedimentary rocks, but in places, they interfinger each other. The age of this complex is considered to cover the entire Jurassic, from Lias(?) to Malm. The basalts' complex is covered by bedded limestones, siliceous sandstones, stromatites (Malm), or directly by flysch deposits. The flysch sequence begins with shaly sandy, and it is then followed by sandy to shaly sandy at the upper part of the sequence. The later ones accumulated from the Tithonian until Aptian.

The Black flysch nappe has a very complex inner structure characterized by the presence of four major tectonic imbrications (Bleahu, 1962; Sandulescu, 1975a). The deposits from all four imbrications are affected both by cleavage and high-pressure and low-temperature metamorphism.

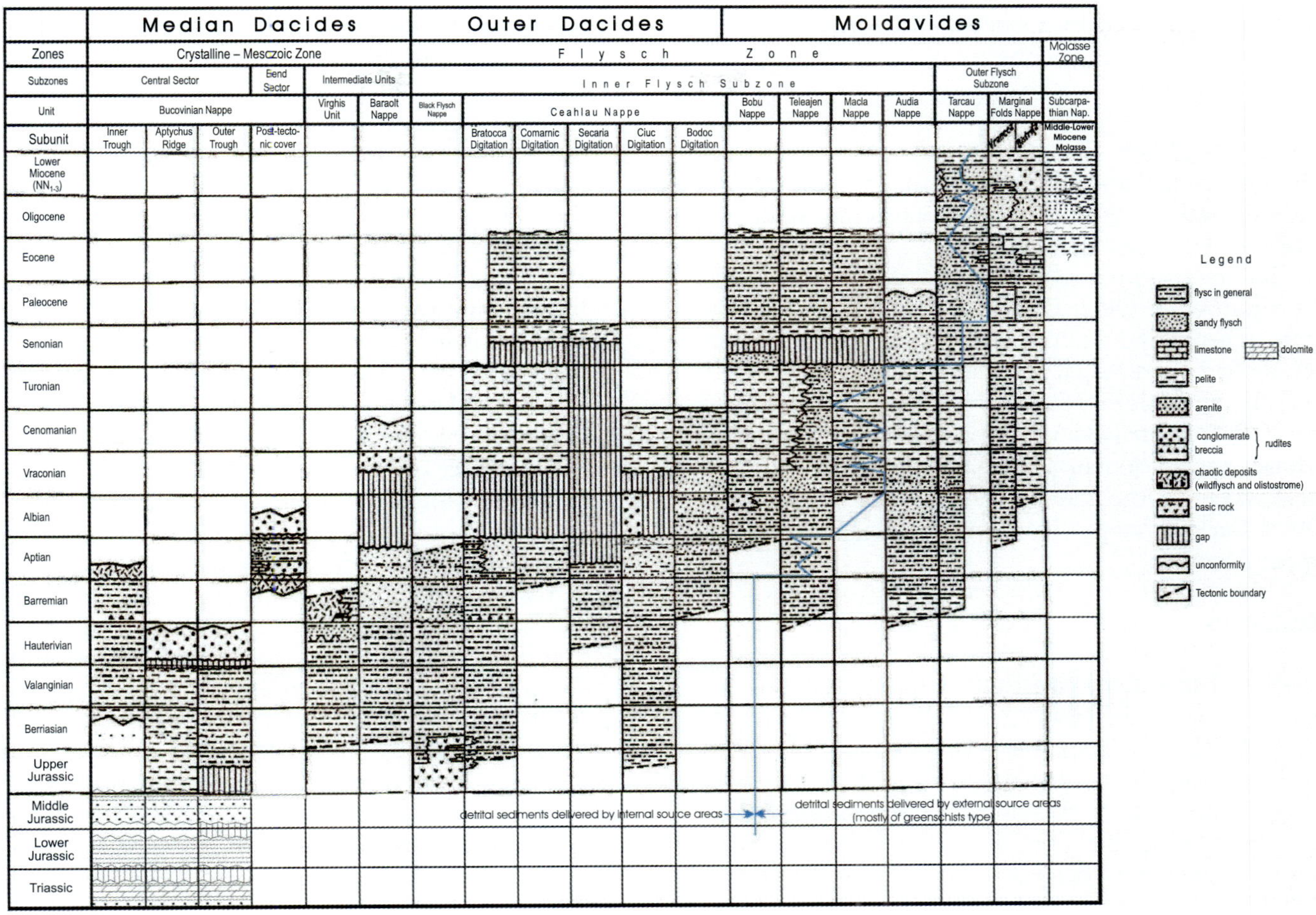

Figure 13. Simplified columns of the Dacides and Moldavides from the Eastern Carpathians.

Ceahlau Nappe. The Ceahlau nappe is one of the most important nappes of the Eastern Carpathians and Outer Dacides because of its length and its complexity. It is essentially made up of flysch deposits. The Tithonian–Neocomian sequence begins with a shaly member (preflysch) that grades upward into a typical sandy-calcareous flysch in which even the limestone interlayers have a turbidity character. In the inner part of this nappe, both the preflysch and the lower part of the sandy-calcareous flysch are associated with red and green siliceous shales, jaspers, and even basalts. During the Barremian–Albian interval, thick flysch also accumulated. Based mostly on the thickness and the frequency of the turbiditic sandstones, it is separated by a few different facies that are characteristic for each existing tectonic imbrication (digitations). In the inner part of the unit, polymictic conglomerates with Urgonian limestone sedimentary klippen intercalate in the terminal part of the Aptian flysch. Flysch facies continued to accumulate during the Albian but only in the external part of the unit. The entire Ceahlau domain was uplifted during the lower interval of the upper Albian. Afterward, the sedimentation resumed during the uppermost Albian, when sediments associated with terrigenous rocks and sedimentary breccias accumulated.

The Ceahlau nappe inner structure is very complicated. Several second-order overthrust sheets (digitations: Bratocea, Comarnic, Secaria, Ciuc, Durau, Ticos, and Bodoc) have been mapped on its territory.

Bobu Nappe. To date, the Bobu nappe has been recognized only in the east Carpathian Bend area, where it largely developed. Its sequence has been divided (Popescu 1958) into the following members: sandy or shaly-sandy flysch (Aptian); shaly-sandy flysch that alternates with sandy flysch; massive sandstones and polymictic conglomerates with a large-scale lenslike development (laterally grading southwestward into sandy and then into shaly-sandy flysch); and a peculiar shaly-micaceous flysch (Albian). The flysch facies grade upward into a shaly pelagic facies with gray, violet, and red marls and clays (upper Vraconian–Turonian). The youngest member (lower Senonian) of the Bobu nappe body is represented by a green sandy or shaly-sandy flysch with lenticular conglomerates and sedimentary breccias.

Post-thrusting Cover. This cover is upper Senonian–early Miocene in age, and it is common both for the Outer Dacides and for the Inner Moldavides, the reason of which is described later below.

The first compressional movements occurred in the inner part of the Outer Dacides during the late Hauterivian, when large sedimentary klippen of crystalline schists had been emplaced. The Dacides emplacement over the inner side of the Ceahlau nappe was finalized before the Albian. Later on, during the upper Albian, the whole Ceahlau domain was again affected by compression, when folding and uplifting occurred. Finally, the Outer Dacides nappes have been emplaced before the upper Senonian.

Moldavides

Teleajen Nappe (Convolute Flysch Nappe). The Teleajen nappe is one of the most constantly developed nappe along the Eastern Carpathians (Figures 2, 3, 13, 14). Its lithostratigraphic sequence (Bancila, 1958; Popescu, 1958; Sandulescu and Sandulescu, 1965; Alexandrescu, 1966; Stefanescu, 1967; Bucur, 1971) is made up of the following members: black, sometimes silty clays (Hauterivian); shaly-sandy flysch with black shale interlayers (Barremian–Aptian); a thick pile of shaly-sandy flysch with gray or green shales; and sandy or even massive sandy flysch and, in places, associated with conglomerates. Sedimentary breccias with exotic blocks occur both in the shaly-sandy flysch and in massive sandstones in the east Carpathian Bend area. Toward the upper part of the Teleajen nappe sequence, either at the Cenomanian or Turonian levels, thin red clays and siltstones intercalate. It is worth mentioning that the massive sandstones laterally grade into sandy-shaly, and further on into shaly-sandy, flysch.

This sequence is involved in a structure with large synclines and anticlines sometimes faulted, or it is affected by reverse faults with different amplitudes. The Teleajen nappe inner structure was acquired during the pre-Senonian shortening stage when the nappe was emplaced.

Petroleum System

Source rocks. The Hauterivian and Barremian black shales as well as the Albian gray shale interlayers are considered to be source rocks. Their cumulative thickness is about 700 m (2200 ft). Some samples collected from the Albian shales showed a total organic carbon ranging between 0.8 and 1.35% and a kerogen of sapropelic type (Stefanescu and Baltes, 1996). R_o values indicate that the Albian–Cenomanian deposits are in the upper part of the oil window.

Potential reservoir rocks. Based only on microscopic analyses, the thick sandstones of the late Albian–Cenomanian sandy flysch (1000 m [3300 ft] thick) could be considered as reservoir candidates.

Trap type. Only structural traps related to anticlines and reverse faults could be considered for this tectonic unit. Unfortunately, most of the structures are deeply eroded. Only the western Teleajen nappe structures located below the Outer Dacides thrust could be tectonically sealed by these units.

Hydrocarbon seeps. The effectiveness of Teleajen nappe source rocks is proved so far only by the existing seeps found in the area of the east Carpathian Bend.

Macla Nappe. The unit consists of an Albian–Turonian shaly flysch (Stefanescu, 1971) with gray, green, purplish, and black shale interlayers. At the highest part of the sequence and only locally, a level of sandy flysch is recognized. It outcrops as a narrow and discontinuous strip between Teleajen and Audia nappes.

The structure of this nappe is characterized as being highly imbricated. It was acquired during the lower Burdigalian, coeval with the nappe emplacement.

Petroleum System

Source rocks. By lithologic comparison with similar rocks from the Teleajen nappe, the gray and black shales of the Macla sequence could be considered as source rocks.

Potential reservoir rocks. The top sequence sandy flysch (maximum 50 m [160 ft] thick) as well as the thicker turbidites (as much as 30 cm [12 in.] thick) intercalated in the shaly flysch are to be considered as reservoirs.

Trap type. As with the highly imbricated structure of the nappe, only the structural trap could be taken into consideration.

Hydrocarbon shows. Despite unfavorable conditions (lack of thick reservoirs and the complicated structure), some of the wells drilled in this unit discovered important gas shows. The wells were located in the southern segment of the east Carpathian Bend area, on the outer thin margin of the unit, because the existing gas was able to migrate from the underthrust unit, namely, from the Tarcau nappe.

Post-thrusting Cover. All the Outer Dacides and the Moldavides described so far are covered by a common sequence that accumulated after the nappe emplacement.

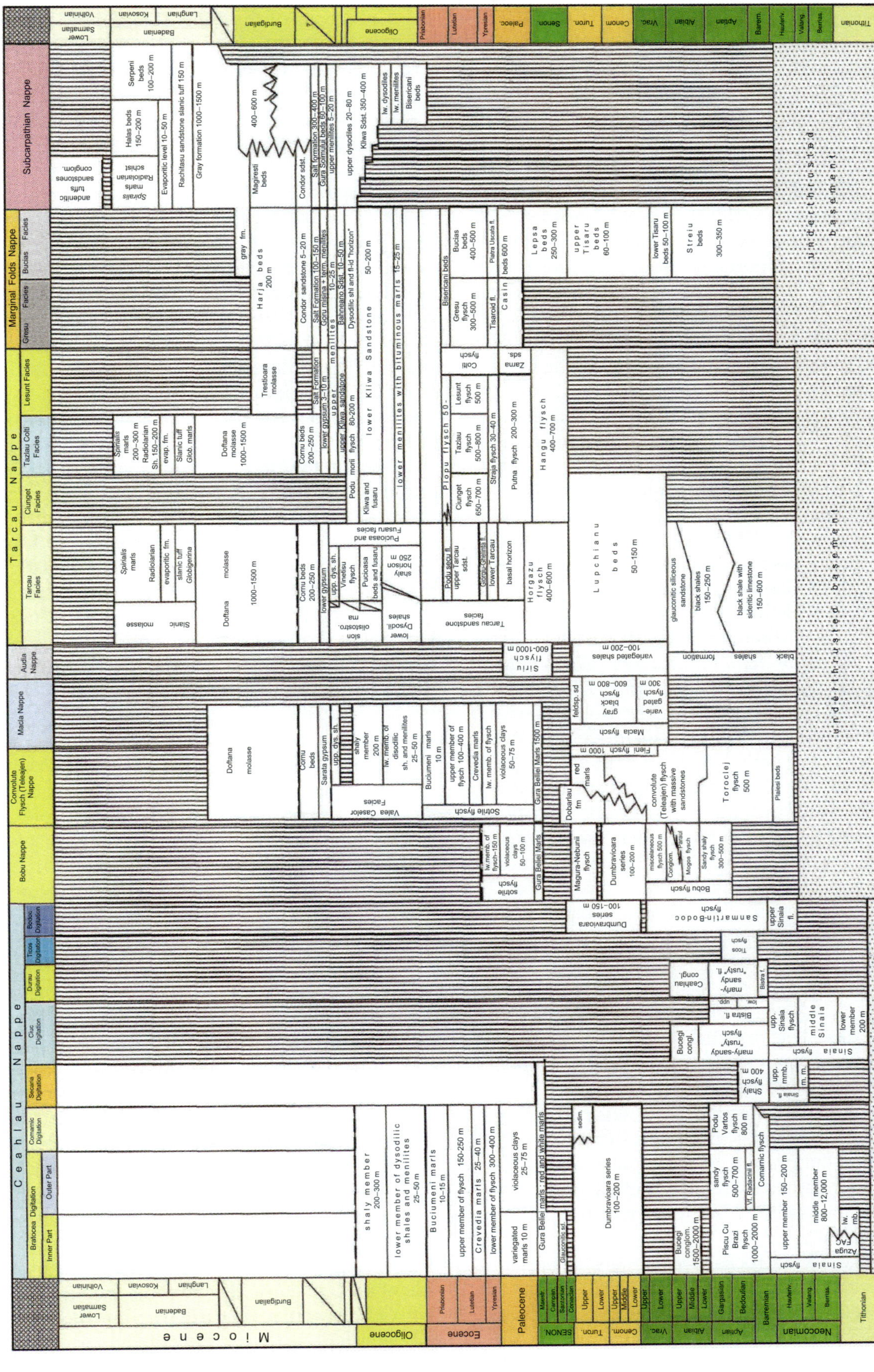

Figure 14. Stratigraphy of the flysch and molasse nappes.

It begins with a package of prevailing red pelagic marls (Senonian–lowermost Paleocene) that grade upward into a predominantly shaly-sandy flysch (Paleocene–Eocene) with two levels (one in the middle and another on top) of white pelagic marls. The youngest (Oligocene–lower Miocene) part of this post-thrusting cover consists of cherts, black shales, and gray and green shales with dacitic tuff interlayers in the upper part. Because of the lack of reservoirs and its complete exposure, the upper Senonian–lower Miocene post-thrusting cover does not have any importance from a hydrocarbon point of view.

Audia Nappe. The lithologic sequence (Filipescu, 1955; Bancila, 1958) of this unit includes the following members: a black shale package with more or less frequent calcareous sandstone interlayers and breccia with granodioritic elements (Valanginian–Albian); glauconitic, siliceous sandy flysch (Albian); gray, green, and red clays with thin dacitic tuff and sedimentary breccia (with granodiorites) interlayers (Vraconian–Turonian); and again, a sandy massive flysch with sandy-shaly flysch interlayers in places associated with red silty marls and thin dacitic tuffs (Senonian–Eocene).

Two different structural aspects are present in the Audia unit. In areas where shales crop out, the structure is characterized by tight imbrications. When either the Albian or the Senonian–Eocene sandy flysch crops out, the structure is represented by large folds affected (or not) by reverse faults. A first stage of deformation was produced in the early Senonian, coeval with the movement from more internal flysch nappes. After that, the main intra-Burdigalian stage of deformation followed, during which the nappe was emplaced. It seems that during its last stage of evolution, the Audia nappe had a gravitational component. Thus, the external part of its red clay series slid down in a chaotic manner, forming what was mapped as the Red Clays nappe.

Petroleum System

Source rocks. The organic content of the black Valanginian–Albian shales (200 m [660 ft] thick) was systematically studied (Baltes et al., 1984). The analyses showed a residual total organic carbon range of 1.05–3.35%. Because of the different burial history and the present tectonic position, the organic matter shows extremely variable values of organic metamorphism ranging between R_o values of 0.55 and 2.02.

Potential reservoir rocks. Despite its important thickness, the Albian siliceous sandy flysch has to be excluded from the potential reservoir because of their very low initial porosity. By contrast, the sandstones of the Senonian–Eocene sandy flysch (800 m [2600 ft] thick) show a good porosity.

Trap type. Only a structural-type trap could be envisaged.

Oil seeps. A few seeps have been discovered related to the black shales' outcrops.

Tarcau Nappe. The Tarcau nappe is the best developed tectonic unit in Eastern Carpathians, both from the areal and the overthrusting extension points of view. It stretches from the northern boundary of Romania (Suceava valley) to the western limit of the Eastern Carpathians (Dambovita Valley), with a maximum width of about 40 km (25 mi), in the bend area, where the overthrust amplitude reaches 35 km (22 mi), as proven by borehole data. North of the Suceava Valley, the Tarcau nappe correlates with the Skibas zone and the central part of the Krosno Depression (Sandulescu, 1984).

Although the lithostratigraphic succession (Figure 13) encompasses a large interval, from Lower Cretaceous to lower Miocene, the Paleogene formations are prevailing, displaying several facies zones and thus conferring on the nappe a polyfacial characteristic (Dumitrescu, 1948, 1952).

The Hauterivian–Turonian deposits are quite similar to those of the Audia nappe, consisting of a predominantly black shaly lower sequence overlain by variegated shales and marly limestones. The Senonian–lower Paleocene is developed in flysch facies, sandy marly in the inner zone of the nappe and more calcareous outward.

In the external zones, the Lower and sometimes Upper Cretaceous formations are absent because of the shallower position of the overthrust sole.

The Paleocene–middle Eocene formations are divided in four facial zones, among which the most internal (Tarcau) is developed on about two-thirds of the nappe area. In that zone, the above-mentioned stratigraphic interval is dominated by proximal sand- and gravel-rich turbidites of Carpathian origin (Tarcau sandstones), forming layers of polymictic, micaceous sandstones, sometimes microconglomeratic, of 1–10-m (3.3–33-ft) thickness, separated by gray and green siltstones, clays, and shales.

Sandy-shaly flysch sequences made up of micaceous sandstones with cross- or convolute lamination and green and red pelites are interbedded with the thick sandstone intervals. In the more external zones, Tarcau sandstones are replaced by distal-type flysch sequences, gradually poorer in massive or coarse arenites.

Irrespective of the earlier facial zones, the upper Eocene consists of a Hieroglyphic bed-type flysch. Outside of the Tarcau sandstone zone, the Eocene ends off with *Globigerina* marls and a thin (as much as 40 m [130 ft]) but persistent quartzitic Kliwa-like sandstone.

The Oligocene–lowermost Miocene is developed into two main lithofacies. In the inner part of the nappe, corresponding to the Tarcau facies area, a thick sandy flysch (the Fusaru Sandstone) followed by a convolute sandy-marly flysch similar to the middle and upper Krosno formations dominates the whole sequence. For both types of flysch, the clastic material was supplied by Western Carpathian sources. Lithologically, the Fusaru is similar to the Tarcau sandstones, but besides their paleontological content, the lutites are interbedded with arenitic layers or intervals and are black or gray, differentiating the two series. Southwest and westward of the bend area, the Fusaru Sandstones and the overlying convolute flysch are gradually replaced by gray and blackish mudstones and shales, interbeddings with uncommon and discontinuous sandstone intercalations (Pucioasa facies), which extend farther more into the Getic depression.

In the external part of the nappe, the second main facies of the Oligocene–lower Miocene, which is characterized by the abundance of oligomictic quartzitic (Kliwa) sandstones and whose clastics are of foreland origin, developed. The sandstones group together into two intervals (lower and upper Kliwa Sandstone) separated by a convoluted flysch sequence.

Kliwa Sandstones are composed of quartz arenites having a bimodal distribution, with fine-grained, less rounded clasts and coarse-grained, better rounded clasts, suggesting an eolian origin; the binder is mixed, made up of silica and clay minerals, sometimes associated with iron oxides and hydroxides. In general, the upper Kliwa Sandstones are less cemented than the lower sandstones.

Both in the Fusaru and Kliwa facies areas, shaly, bituminous deposits are persistent, either as compact sequences at the lower part of the Oligocene (lower Dysodiles) or overlying the convolute flysch in the inner zone and upper Kliwa Sandstones in the external zone (upper Dysodiles). Similar types of rocks are interbedded with thick sandstone layers as well as pelitic components of the convoluted flysch.

The dysodilic sequences underlying and overlying the Kliwa Sandstones are accompanied by two levels of bituminous silicolites (lower and upper menilites) derived from diatom frustulas diagenesis. In the lower sequence, bituminous marls are also present.

An intermediate transition zone between the two facies, where Fusaru and Kliwa Sandstones alternate, could be distinguished.

On the whole, the Oligocene–lowermost Miocene formations were deposited under restrictively anoxic conditions, with condensed intervals, rich in organic matter (bituminous marls, lower and upper menilites, and dysodilic shales), formed during highstand stages, sharply alternating with sandy turbidites (Kliwa and Fusaru sandstones) or sandy-marly flysch sequences, generated during lowstand stages. The series ended with regressive evaporitic deposits (gypsum, salt breccia, and salt).

In the innermost zone of the Tarcau nappe, a singular chaotic formation, within which Upper Cretaceous and Eocene deposits of internal origin are included as olistolites, substitutes the uppermost part of the Fusaru Sandstones and the overlying convoluted flysch. Its origin was related with the rising of westward-neighboring nappes (Audia and Macla) as a result of the upper Oligocene (Savian) tectonic movements.

In the Burdigalian, during the Old Styrian tectogenesis, the Tarcau nappe largely emerged, and the first stage of folding and detachment occurred. As a result, the sedimentary basin receded to the southwest, where molassic and evaporitic formations as well as cinerites and breccias of upper Burdigalian and Badenian age were deposited.

The intra-Badenian (New Styrian) tectonic events led to the main overthrust, when the Tarcau nappe covered and even overpassed the whole Marginal Folds nappe and, locally, the inner part of the Subcarpathian nappe. Later on, in the middle Sarmatian, a new tectogenesis (Moldavian) displaced the Tarcau nappe, together with the Marginal Folds and Subcarpathian nappes, over the foreland units. During the upper Sarmatian, the uppermost Miocene and Pliocene of the molasse formations of the inner flank of the foredeep extended up to the outer margin of the Tarcau nappe, unconformably overlapping its frontal line, westward of the bend area. Northward, the frontal line outlines several tectonic half-windows (Vrancea, Oituz, Bistrita, and Putna) and windows (Dumesnic and Mitocul lui Balan), where Marginal Folds nappe formations outcrop.

A post-tectonic cover made up of upper Sarmatian, upper Miocene, and Pliocene deposits also developed on several restricted areas.

The present-day structural configuration was finished during the lower Pleistocene movements (Wallachian), when the youngest deposits were also involved in folding and faulting processes (Dicea, 1995).

The structural style of the nappe is related to the main lithologic features of the Paleogene–lowermost Miocene formations. In the areas where the Tarcau and Fusaru massive sandstones are prevailing, large vertical or tilled folds, generally faulted, were building up. As a whole, the Tarcau–Fusaru realm was separated as a second-order tectonic unit (digitation), slightly overthrusting the more external facial zones.

Outside the Tarcau digitation, where the sandy flysch ration decreases, narrower and imbricated folds are common. In those areas, three more digitations were

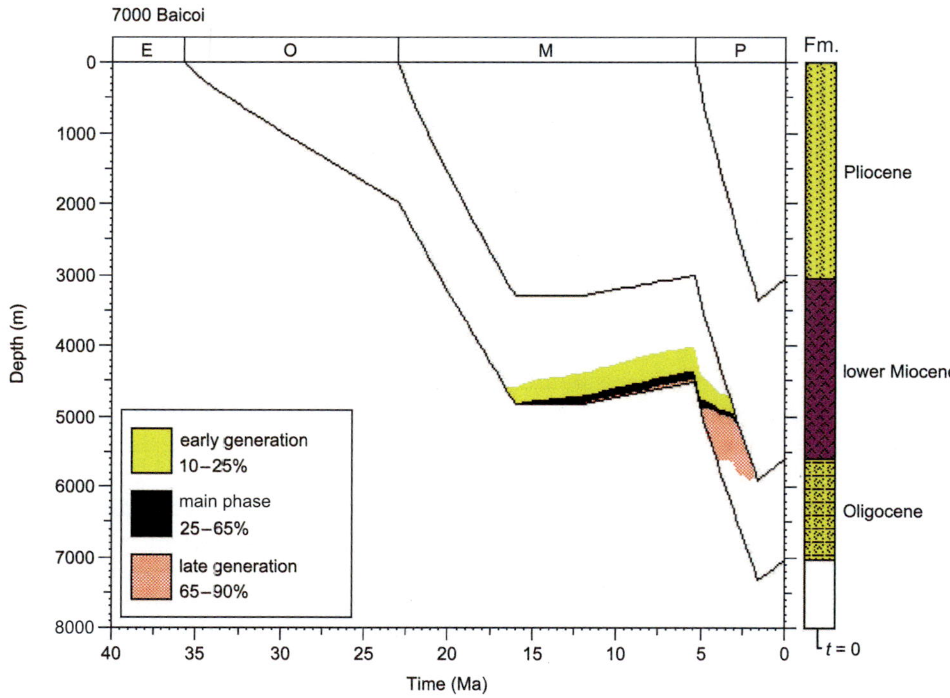

Figure 15. Burial history in the area (Baicoi) where the deepest well was drilled in Romania.

distinguished. In some areas of the outermost zone of the nappe, along the overthrusting frontal line, a tectonic pile of two to as much as four superposed folds or groups of folds involving Eocene and Oligocene–lowermost Miocene deposits is present on a vertical interval exceeding 3000 m (10,000 ft) and sometimes 5000 m (16,000 ft).

Petroleum System

Source rocks. The most important source rocks are associated with the bituminous sequences of the Oligocene–lowermost Miocene, i.e., the lower and upper dysodilic and menilitic intervals as well as the argillaceous shales and marls interlaid with the massive sandstones or forming the pelitic constituents of the convoluted flysch.

The average aggregate thickness of source rock sequences is about 600 m (1900 ft). The organic matter is of type II, partly type II–III. The total organic carbon contents alter from 0.7 to more than 10%, the highest values pertaining to bituminous shales and marls of the external Kliwa facies. As the existing data are not sufficient to characterize the entire area covered by the Oligocene deposits, an average total organic carbon value of 2% was normally assumed for maturation modeling and quantitative approaches.

The Oligocene–lowermost Miocene source rocks show different maturity levels, depending on the burial history, organic matter type, and geothermal gradients. Under the conditions of a generally low geothermal gradient (2.5°C/100 ms), the oil window could be reached only at depths greater than 4300–4500 m (14,100–14,700 ft) or even greater in the diapiric fold zone (Figure 15).

Because of the early uplifting of the nappe, favorable burial depths were realized on restricted areas where the Paleogene deposits were tectonically piled up during the Moldavian (intra-Burdigalian, Badenian, and middle Sarmatian) tectogeneses or were overthrusted by more internal nappes. Those areas are located in the outermost part of the nappe, mainly in the Moinesti region (Dofteana) and between Buzau and Prahova valleys, as well as beneath the Audia or Macla nappes overthrusting.

Southwest of Buzau Valley, favorable maturation conditions could also be realized in two regional syncline areas (Drajna and Slanic) where deposition and subsidence continued after the Old Styrian folding and uplifting of the nappe until the Badenian.

According to thermal maturation modelings, the oil generation began 17 Ma; depending on the burial depths, the main phase of generation was completed or has continued to the present (Figure 16).

Reservoirs. All the producing reservoirs are of clastic origin (sandstones, sands, marly sandstones, and microconglomerates), covering a large stratigraphic interval from Eocene (Tarcau sandstones) to upper Miocene. Among them, the most prolific proved to be Kliwa Sandstones (upper and lower) and the

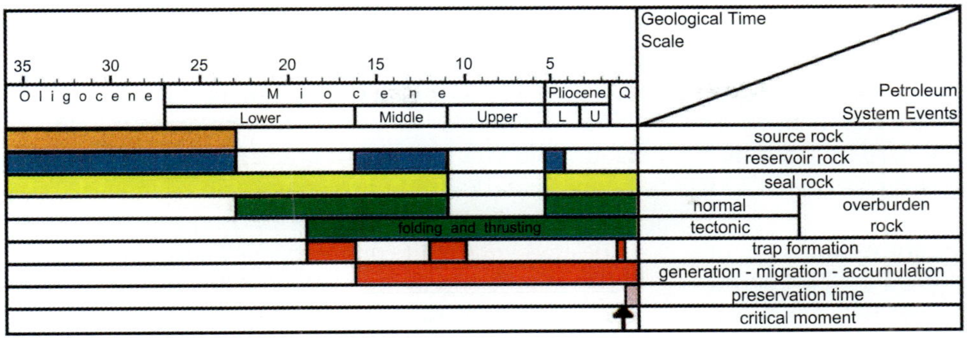

Figure 16. Outer Tarcau nappe and inner flank of the foredeep: Petroleum system events chart.

Meotian sandstones. To a lesser extent, Tarcau, Burdigalian, and post-tectonic upper Sarmatian sandstones are also effective reservoirs.

Seal. After the intra-Burdigalian tectogenesis, the reservoir sequences of the Paleogene–lowermost Miocene were largely exposed to erosion, except the outer- and innermost parts of the nappe, where tectonic conditions of protection were realized locally. Later on, similar conditions were created by the upper Miocene–Pliocene post-tectonic cover, developed as small basins or overlapping the outer margin of the nappe. Within the traps, individual reservoirs are protected by shaly or flysch impervious sequences.

Traps. Structural traps are prevailing in all the productive areas. The Paleogene reservoir rocks are involved in longitudinally and transversely faulted anticlines, parallel to the regional structural trend. As mentioned before, on the outer margin of the Tarcau nappe, two or more anticlines or groups of anticlines are sometimes overthrusting each other, separated by low dipping (10–20°) faults. The upper Miocene (Sarmatian and Meotian) reservoirs are only gently folded but commonly faulted.

The angular unconformity at the base of the upper Miocene screens structural-stratigraphic traps, both for older and younger reservoirs beneath it.

Migration and entrapment. The main migration paths are represented by deep faults connecting the maturation areas with the shallower traps.

At the basin scale, the continuity of the Oligocene–lowermost Miocene source rocks was broken early (in the Burdigalian), and the maturation–expulsion processes in Tarcau and Marginal Fold nappes evolved separately afterward. Consequently, the hydrocarbon fields of Tarcau nappe are parts of two petroleum systems, each one depending on different source rock pods.

Southwest of the Buzau Valley, the hydrocarbons were supplied by the Oligocene–lowermost Miocene source rocks that reached the maturity stage in the deepest positions in the nappe. Northward, these rocks are, in general, immature because of the shallower burial depths, but the fault system allowed the migration from source rocks located deeper in the underthrusted Marginal Folds nappe directly or by secondary migration of hydrocarbons already accumulated in the fields of that unit.

Secondary or even tertiary migration could be also assumed in the case of the youngest reservoirs (Meotian) overlying Oligocene producing reservoirs, sourced by Tarcau nappe mature source rocks.

Hydrocarbon fields. To date, 19 oil fields and 2 gas fields were discovered. Eight oil fields, located in the central zone of the nappe (Moinesti, Figure 4, no. 37), belong to the petroleum system related to the overthrusted Marginal Folds nappe source rocks. The host rocks are Eocene, Oligocene, and Sarmatian (post-tectonic) sandstones. Most of the remaining fields (eight) are located on the outer margin of the Tarcau nappe, southwest of Buzau Valley (as, for instance, the most important fields Runcu Bustenari-38 and Surani–Carbunesti-39; Figure 4), and only two are on more internal alignments, connected by important faults with generating source rocks buried beneath Miocene deposits in regional syncline areas. On the same alignments, several noncommercial oil pools and relatively numerous seeps are witness to the lack of adequate preservation conditions because of the long erosion time of the seal rocks. All these fields and shows were assigned to the second petroleum system, related to the Tarcau nappe source rocks. The initial oil reserve per field was 2 to as much as 13 million t with daily flow rates of 3–30 t/well.

Marginal Folds Nappe. With respect to the previous nappe, the Marginal Folds nappe shows a smaller outcropping development. In outcrops, it can be followed from the Putna Valley basin northward up to the border of the country and continues further in the Pocutia folds zone. It is worth mentioning that despite its limited development at the surface, mostly in half-windows (Putna, Bistrita, Oituz, and Vrancea), the Marginal Folds nappe has an important development westward beneath the Tarcau nappe, where its presence was proven by numerous wells as far as 35–40 km (21–25 mi) from its present wedge.

Its stratigraphic sequence, initially established in the Vrancea, Oituz (Dumitrescu, 1948, 1952), Bistrita (Bancila 1958), and Putna (Joja, 1954) half-windows, starts (Figure 14) with Lower Cretaceous deposits (Barremian–lower Albian). They consist of dark, starved shaly flysch with lenses of siderites, being similar to the deposits of the same age developed in Tarcau and Audia nappes. It follows a sequence of black radiolarites, green and blackish argillites, locally hard glauconitic sandstones, and red, green, or variegated radiolarites interbedded with red marls and, locally, conglomerates with greenschists pebbles (upper Albian–Turonian). Toward the northern segment of the Eastern Carpathians (Bistrita half-window), this sequence is replaced by an essentially pelitic one consisting of red, green, and dark clays with interbeds of spongolites and radiolarites. Both the above-described sequences are normally covered by thin-bedded, whitish, marly limestones, sometimes with cherts, whitish detrital limestones, and gray calcareous-graded sandstones alternating with gray marls bearing the rest of Inoceramus (Senonian).

Concurrent with the Cretaceous–Paleogene boundary, a diversification of facies occurred. Thus, in the Vrancea half-window, two distinct facies have been mapped, whereas the third facies crops out in the Bistrita half-window. The internal (western) facies separated in the Putna half-window was divided into the following members: bituminous sandy limestones and marly limestones (Paleocene); siliceous sandstones and red clays (upper Paleocene–lower Ypresian); rhythmic alternation of gray calcareous sandstones, green marls, and organogenous conglomerates with greenschists also interbedded at different levels (Ypresian–Lutetian); red and green marls and clays with uncommon intercalations of thin green sandstones (Lutetian); alternation of green clays and calcareous sandstones, followed by a monotonous package of dark gray and green clays and siltstones (Lutetian); and white or greenish siliceous sandstones and white marls with *Globigerina* (Priabonian). The external (eastern) facies (Bucias) in the Vrancea half-window consists of the following members: well-bedded gray limestones and/or conglomerates, gray platy marls alternating with calcareous sandstones (turbidites), and dark-gray bituminous clays with intercalations of sandy limestones (Paleocene); gray sandy-shaly flysch that sporadically has the top marked by reddish clayey marls and conglomerates with greenschist elements (upper Paleocene–lower Ypresian); green or greenish white hard marls with subordinate interbeds of calcareous turbiditic sandstones and thin layers of red marls that are grouped at the middle and upper part of the member (Ypresian–Lutetian); dark gray marls, clays, and siltstones (Priabonian); and white or greenish marls, rich in *Globigerina* associated with white or greenish siliceous sandstones of Kliwa type (Priabonian).

Within the Bistrita half-window, the equivalent facies of the above-described deposits consists of the following members: gray calcareous sandstones rhythmically alternating with marly limestones, gray marls, and clays with sporadic microconglomerate interlayers (Paleocene); shaly greenish marls alternating with thin green quartzitic sandstones and sparse microconglomerates with greenschists interlayers (Paleocene–lower Ypresian); green or gray-greenish sandstones and siltstones with subordinate interlayers of green clays and marls, and locally toward the upper part, white metric quartzitic sandstones develop (Ypresian); white cherty limestone (Lutetian); red and green clays alternating with thin greenish sandstones (Lutetian); a monotonous sequence of dark-gray clays, marls, and siltstones (Priabonian); and white or greenish marls, rich in *Globigerina* associated with white or greenish siliceous sandstones of Kliwa type (Priabonian).

Beginning with the *Globigerina* marls, the facies of the deposits became more or less uniform all over the Marginal Folds nappe, essentially being represented by the same black bituminous lithologic background already described for the external part of the Tarcau nappe. Within the bituminous Oligocene–lower Miocene sequence, the following members have been separated, from bottom to top: lower menilites (black, hard siliceous rocks) with hard bituminous marls (white when weathered); lower dysodilic shales (black, paperlike bituminous shales); the Kliwa Sandstone (white, more or less massive siliceous sandstones); upper dysodilic shales, gray flysch with tiny interlayers of white dacitic tuffs; Bahneanu Sandstone (massive sandstone of Kliwa type and conglomerates); upper menilite and dysodiles; Goru–Misina Formation (gray calcareous sandstones, marls, and sandy marls); and terminal menilite or bituminous shales that close the bituminous sequence. In places, and especially north of the Bistrita Valley, the bituminous facies is invaded more or less by conglomerates with greenschists pebbles. The above-described sequence is covered by an early Miocene succession represented by the Salt Formation (dark-gray marls and clays, salt bodies, and important volumes of sedimentary breccias) and Hirja beds (a rhythmical molasse with red marls and clays).

The Marginal Folds nappe shows a very complex inner structure (Figure 17). Within this unit and mostly in its external part, overturned and recumbent folds, in places overlapping each other, have been detected either by mapping or by wells. Second-order thrusting (digitations) sheets with horizontal displacement larger than 15 km (9 mi) have been mapped in the Vrancea half-window. More than this, slices of the internal part of the Marginal Folds nappe have been detached from their initial position by the Tarcau nappe and were pushed in front of it because they crop out today as tectonic outliers.

This complicated structure was generated during successive stages of deformation undergone by this unit. During the lower Miocene (Burdigalian), a first significant stage of deformation occurred when the deposits were folded and coevally uplifted (Old Styrian). Then, the formed relief was eroded before the Tarcau nappe thrust over the Marginal Folds nappe, and those together thrust over the inner side of the Subcarpathian nappe. This event occurred during the lower part of the middle Miocene (Young Styrian). During the early Sarmatian (Moldavian phase), the Marginal Folds nappe thrust over the Subcarpathian nappe and, along with it, then thrust over the foreland.

Petroleum System

Source rocks. Similar to the Tarcau nappe, the most important (if not the only) effective source are the bituminous shaly, marly, and silicolithic sequences of the Oligocene–lower Miocene interval. The average cumulative thickness is about 500 m (1600 ft), lower than in the Tarcau nappe. The geochemical properties (type of organic matter and total organic carbon content) are also similar as well as the regional geothermic gradient. Consequently, the maturity conditions could be realized at depths greater than 4000 m (13,000 ft). Tectonic evolution, i.e., Tarcau nappe overthrusting associated with the thrust sheets building up with the Marginal Folds nappe itself, allowed the existence of several pods of effective rocks. The most important was identified in the area between the Bistrita and Oituz–Slanic half-windows, where the Oligocene source rocks and noncommercial oil flows from the associated Kliwa Sandstones have been recorded as deep as 5500 and 5350 m (18,000 and 17,500 ft), respectively. In general, generating conditions were realized at a larger scale by comparison with the Tarcau nappe. The organic matter maturation stage started about 16 m.y., during the Young Styrian tectonic event, when the Marginal

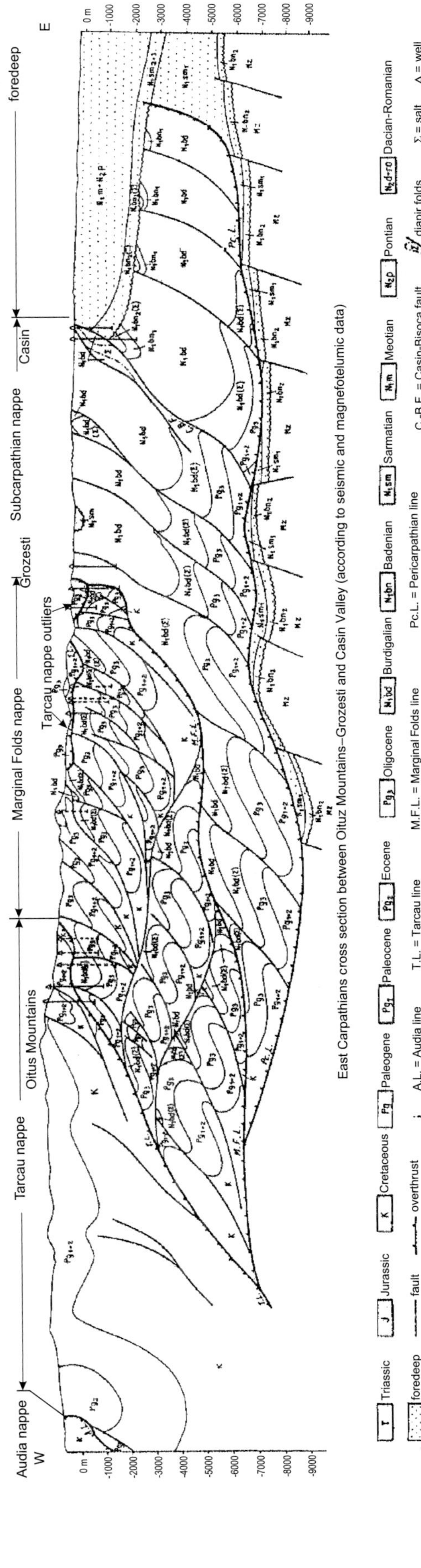

Figure 17. Geological section across the Paleogene flysch including the Marginal Folds nappe.

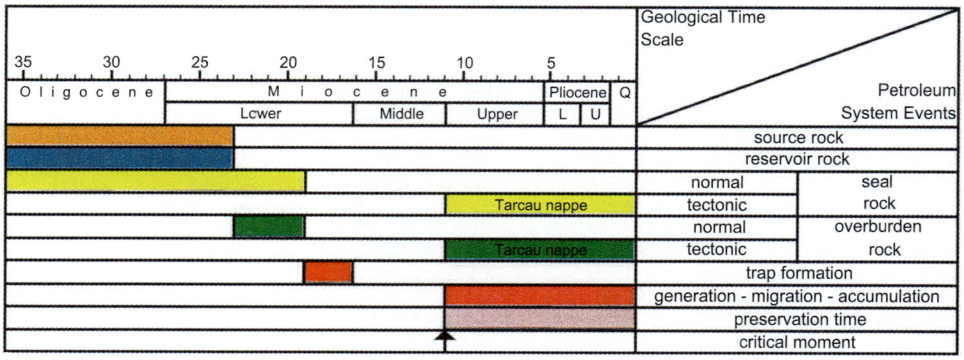

Figure 18. Marginal Folds nappe: Petroleum system events chart.

Folds sunk under the thick pile of deposits belonging to the Tarcau nappe. According to a still insufficient amount of geochemical data, the Lower Cretaceous black shales and the middle–upper Eocene pelites have some source rock properties, but their effective contribution to the generated hydrocarbons is still questionable.

Reservoirs. The main hydrocarbon host rocks are the Kliwa Sandstones. At a lesser extent, sandstones and conglomerates associated with the Goru–Misina Formation and, in a few cases, middle Eocene limestones, are also producing.

Seals. The main, regional, preservation conditions are provided by the Tarcau nappe overthrusting. The largest area with such conditions is located between Slanic–Oituz and Bistrita half-windows. Nevertheless, several fields are nowadays protected only by the lower Miocene normal cover, after the local erosion of the Tarcau nappe during the Pleistocene tectonic uplifting. The normal common seals of the Kliwa Sandstone reservoirs are the upper dysodiles. Within the Kliwa Sandstone complex, a few massive clayey intervals separate the pools, some of them having separate oil-water contact. The younger reservoirs developed in the Goru–Misina sequence are normally sealed by the lower Miocene salt or tectonically by the Tarcau nappe.

Traps. All the traps are of structural type. The most common are longitudinally and transversally faulted anticlines that form regional alignments parallel to the general structural trend of the nappe. Very commonly, the outer, reversed flanks are laminated along longitudinal thrust faults as the inner limbs thrust each other, forming an imbricated structure of scale folds.

Migration. The migration mechanism (Figure 18) was dominated by upward circulation of the hydrocarbons, along the regional, deep faults connecting the profound kitchen areas with the above traps, some of them located even in the higher Tarcau nappe. This is why some of the fields are very shallow (less than 1000 m [3300 ft]). Long distance, longitudinal migration along the carrier rocks (the same as the reservoirs) was prevented by the grid of numerous transverse tight faults that acted as hydrodynamic screens.

Hydrocarbon fields. The Marginal Folds nappe proved to be more prolific than the Tarcau nappe. About 30 oil fields and 2 gas fields have been discovered so far, most of them in the area between the Bistrita and Slanic–Oituz tectonic half-windows, protected by the Tarcau nappe tectonic cover (Comanesti–Moinesti area, Figure 4, no. 40). In the Marginal Folds outcropping territory, only four fields were preserved, but the oil seeps and the impregnated Kliwa Sandstones suggest that a greater number of fields existed before the erosion of the Tarcau nappe. Proven oil reserves range between 1 and 35 million t per field, and the wells' initial oil flow rates varied from 10 to more than 100 t/day.

Subcarpathian Nappe. The Subcarpathian nappe is the most external nappe of the Carpathian orogen. It crops out between the northern border of the country up to the west of the Buzau Valley. From here, it extends westward, beneath the Sarmatian–Pliocene deposits, up to the Getic depression.

In the outcropping area, it is mostly made up of lower Miocene molasse to which, subsequently, Oligocene bituminous rocks are associated.

The oldest deposits (Priabonian) belonging today to the Subcarpathian nappe are known only from wells and are represented by a monotonous sequence of dark-gray clays, marls, and siltstones. They are normally covered by bituminous facies that can be divided into the following members: lower menilites (black, hard siliceous rocks); lower dysodiles (black bituminous clayey shales) (Oligocene); Kliwa Sandstone (Oligocene); upper dysodiles (Oligocene–lower Miocene); and upper menilites (lower Miocene). The upper menilites practically end up as the bituminous facies. Then, sandstones

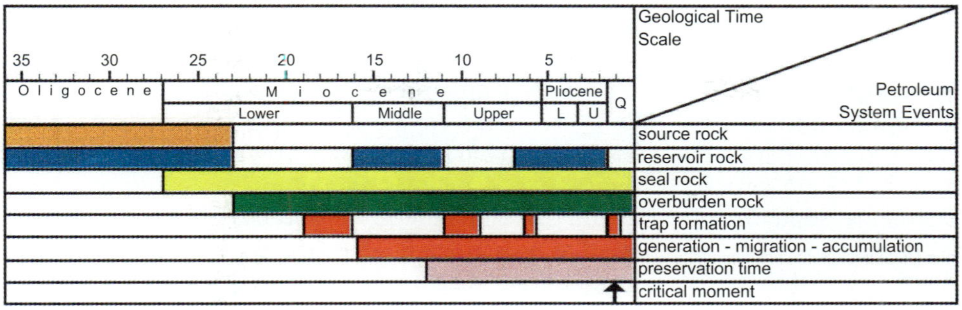

Figure 19. Subcarpathian nappe and inner flank of the foredeep: Petroleum system events chart.

and/or conglomerates with greenschists and Salt Formation (lower Miocene) follow, which, aside from the salt bodies, are represented by dark shale anhydrite and thin sandstones as well as by large masses of sedimentary breccia (salt breccia). The normal sequence continues with a thick pile (more than 2000 m [6600 ft]) of deposits in molasse facies (lower–middle Miocene) that begins with sandstones in the inner part or conglomerates in the outer part, followed by a rhythmic alternation of sandstones, siltstones, and pelites. The rocks' color is predominantly gray, but at certain levels, especially toward the lower part of the sequence, red sandstones and pelites occur. Interlayers of more or less constant anhydrites, tuffs, and calcareous shales are also present. It is followed by white greenish tuffs associated with calcareous-siliceous sandstones and with white *Globigerina*-bearing marls, covered by evaporites (anhydrites and salt), marls, and quartzitic sandstones associated with marls (Badenian). The Subcarpathian nappe sequence ends up with polymictic conglomerates and sandstones (lower Sarmatian) preserved only in deeper and narrow synclines.

Southward of the Trotus Valley, the outer part of the Subcarpathian nappe is covered by the upper Sarmatian–Pliocene deposits belonging to the foredeep. Southwest of the Buzau Valley basin, the foredeep completely covered the Subcarpathian nappe and extended inward over the outer part of the Tarcau nappe. This part of the foredeep is characterized today by the presence of folds with salt core, the reason for which is known in Romanian geological literature as the diapiric folds zone. It is important for two reasons: here, diapiric folds have been first described (Mrazec 1907, 1927) and, for its numerous oil and gas fields, related to the diapirs' presence.

Within the Subcarpathian nappe, three distinct major imbrications (digitations) have been separated (Sandulescu et al., 1977), namely, Magiresti–Perchiu, Pietricica, and Valea Mare. Each of these digitations has its own inner structure. The Magiresti–Perchiu digitation generally shows an imbricated structure conditioned by a detachment layer represented by the lower Miocene salt. The Pietricica digitation is characterized by faulted anticlines. Finally, the Valea Mare, the most external digitation, is squeezed in front of the Moldavides, which is the reason why it has a complex and discontinuous structure.

The inner structure of this nappe was generated in two different tectonic stages, namely, during the middle Miocene and intra-Sarmatian (Moldavian phase).

Petroleum System

Source rocks. Although from the source rock point of view, a single petroleum system might be acknowledged on the Subcarpathian nappe territories, the other elements (reservoir rocks, sealing, and traps), as well as the main processes (migration and entrapment; Figure 19), and the most important feature, the amount of generated reserves, are outstandingly different between the northern and central areas, lacking a post-Moldavian cover, and the Carpathian Bend area, where a thick pile of the inner foredeep upper Sarmatian–Pliocene formations covered the nappe. For these reasons, two distinct subdivisions of the petroleum systems were separated: the first, involving only the Paleogene–lower Miocene (pre-Moldavian) formations, in the areas without younger formations, being specific only to the Subcarpathian nappe; the second, also including the upper Miocene–Pliocene post-tectonic cover of the Bend area, conventionally assigned to the inner flank of the foredeep. The subdivision related to the Subcarpathian nappe could only be characterized as follows: The Oligocene bituminous sequences, with similar geochemical parameters to those in two more internal nappes and matured under a low geothermal gradient (2.5°C/100 ms), were responsible for the oil and gas generated in the area. The oil window depths (more than 4000 m [13,000 ft]) could be reached in rather restricted areas, mainly between the Bistrita and Trotus valley basins.

Reservoirs. The Oligocene Kliwa Sandstones and post-saliferous Burdigalian red sandstones are the host rocks for the hydrocarbon pools discovered so far.

Seals. Sealing conditions were provided by the upper dysodiles and marly interlayers for the Kliwa and Burdigalian reservoirs, respectively.

Traps. Only structural traps, namely, normal or reverse faulted anticlines, were formed. Parallel alignments of relatively narrow imbricate structures develop. Diapir phenomena are sometimes present along the longitudinal faults.

Migration. The oil window depths were realized in the westernmost sunken and partially underthrust part of the nappe but only in the area corresponding to the Bistrita and Trotus valley basins. Northward, the nappe lies in a shallower position, and the Oligocene is missing in most parts. Under these circumstances, major faults were the hydrocarbon pathways to the younger reservoirs (Burdigalian), but shorter distance migration along the reservoirs was needed in the case of Kliwa Sandstone traps.

Fields. Only two relatively small fields have been discovered so far: Tescani (Figure 4; no. 41) and Campeni; the former with both Kliwa (gas and condensate at 2700 m [8800 ft]) and Burdigalian reservoirs (oil) producing and the latter with oil in Burdigalian. The younger reservoirs are located at much shallower depths (less than 1000 m [3300 ft]). The scarcity of the fields discovered in the Subcarpathian nappe could be explained by the small number of exploratory wells drilled in the most favorable areas because of the great depths of the Oligocene (3500–5000 m; 11,400–16,400 ft) and of the limited capacity of the seismic technologies in deciphering the very complex tectonic setting of the nappe. Even under these circumstances, oil was tested, and oil-impregnated cores were recently extracted at 3670 m (12,040 ft) from the Oligocene in two different wells.

Eastern Carpathian Foredeep

From the paleotectonic point of view, the Eastern Carpathian Foredeep represents a post-tectonic cover common both to the Carpathian orogen (Outer Moldavides) and its foreland. It extends all along the Eastern Carpathians (Figure 2), being narrower and incomplete to the north of the Trotus Valley (Sandulescu, 1984) and getting its maximum extension in front of the east Carpathian Bend, where it forms a deep elongated depression, known as the Focsani Depression. In this area, the foredeep continuously subsided from the Sarmatian until the Pleistocene, a time interval when more than 10,000 m (32,800 ft) of sediments accumulated. The sequence is homogenous, consisting mostly of gray sandstones, siltstones, marls, and clays. Some differences in lithology are given by sand and sandstone thickness and frequency. Thus, they are more frequent in upper Sarmatian–lower Meotian as well as in Dacian. Toward the upper part of the sequence, pebbles are getting more and more frequent, prevailing in the Pleistocene. It is worth mentioning that both in the Pontian and Dacian, coal interlayers (even thicker than 1 m [3.3 ft]) are present. Toward the inner part of the foredeep (diapiric folds zone), a more diversified lithology was recorded (Figure 20). Among other rocks worth mentioning are those that have a certain genetic significance, namely, the middle Sarmatian coquina neritic limestones and locally developed anhydrites.

The foredeep has an asymmetrical inner structure, initially made evident by field data and lately well proven by seismic lines. Its outer (eastern and southeastern) limb is very simple, gently dipping from east to west toward the basin axis. Westward of the basin axis, the dips are gradually increasing because, in the vicinity of the present inner border of the Focsani Depression, the Casin–Bisoca fault, they are vertical or even upside down. Here, the internal limb mostly overlies the outer part of the Subcarpathian nappe. Westward of Buzau Valley, it gradually extends inward, covering the inner part of the Subcarpathian nappe up to the inner part of the Tarcau nappe. In this segment, the deposits that belong to the post-tectonic cover (foredeep) have a more complicated inner structure (Figure 20). More or less elongated synclines are separated by growing anticlines. Most of the anticlines are generated by the salt. The salt shows different stages of the diapirism evolution, sometimes up to the complete piercing of the normal sedimentary cover.

Within the paleogeographic area corresponding to the foredeep, the subsidence was the most active tectonic event. Sedimentation began in the Sarmatian after the Carpathian nappes were emplaced over the foreland. In most parts of the foredeep, sedimentation continued until the Pliocene except for its western part, where some uplifting is recorded, being marked by unconformities at the Meotian and Pontian base. The last folding that determined the structure of the internal limb of the foredeep is intra-Pleistocene (Wallachian phase).

Petroleum System

In the Bend area, upper Sarmatian–Pliocene deposits extensively overlapped the Subcarpathian nappe on an area of about 130 km (81 mi) long and 30–40 km (18–25 mi) wide. Within this area (also called the

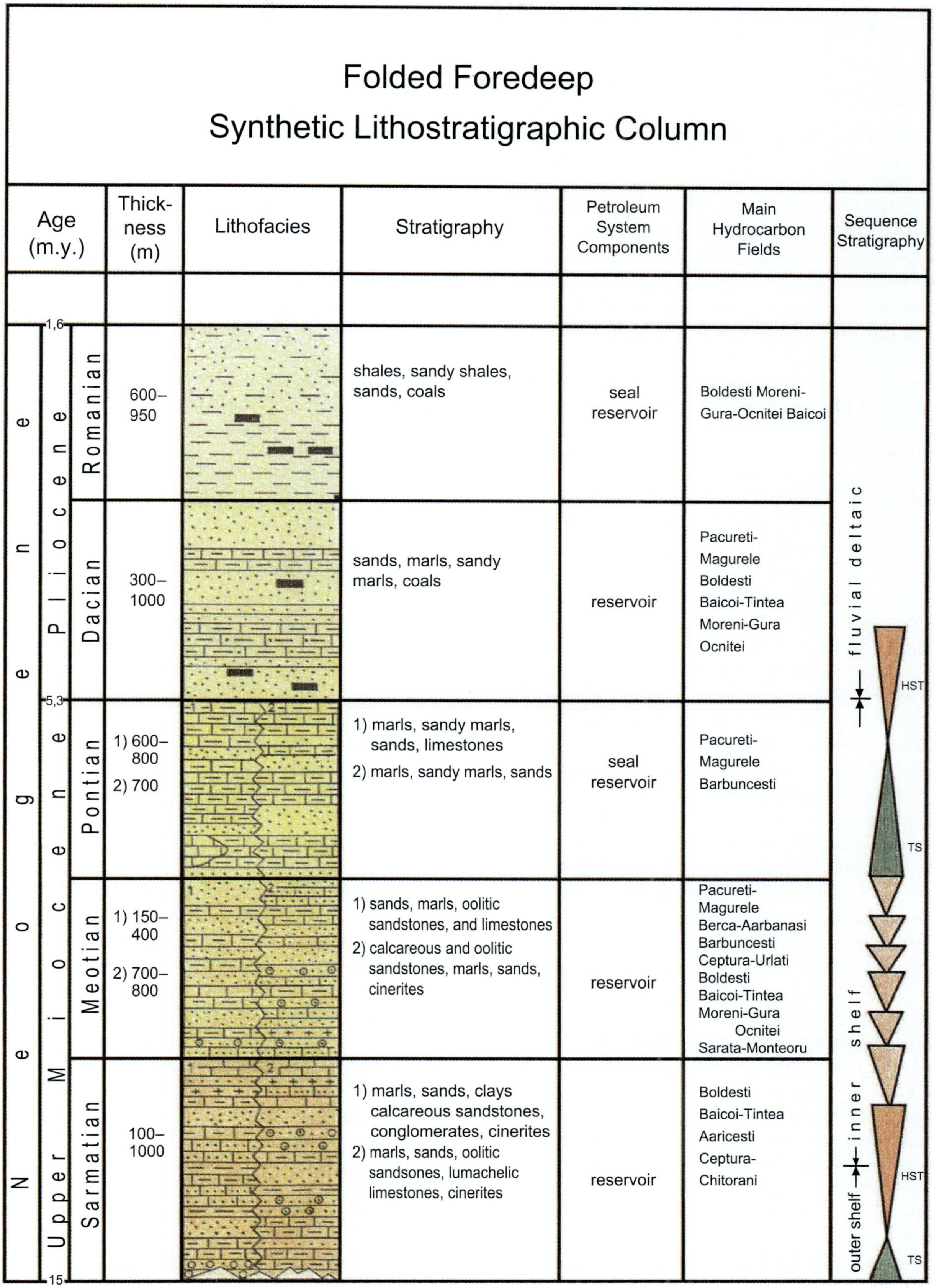

Figure 20. Lithostratigraphic column of the inner (folded) flank of the foredeep. HST = highstand tract; TS = transgressive system tract.

Miocene–Pliocene or diapiric folds zone), these deposits were an important factor as components of the petroleum system related to the Subcarpathian nappe source rocks as previously mentioned.

Source rocks. Again, the Oligocene provided, almost exclusively, the effective source rocks. Unlike the central and northern parts of the Subcarpathian nappe, the Oligocene is continuously developed and deeply buried in more extended areas than in those lacking inner foredeep cover. Consequently, the volume of high organic matter bituminous rocks that entered the oil window (4000–6000 m [8200–19,700 ft]; Figure 15) was considerably larger. The best generating conditions were realized in the area between Prahova (in the east) and Dambovita (in the west) valleys, corresponding to the highest concentration of oil reserves and to the greatest number of wells that crossed the Oligocene. In the deepest well ever drilled in Romania (Baicoi), the Oligocene was encountered at 5600 m (18,300 ft) and opened down to a total depth of 7025 m (23,047 ft); from the 5600–6200-m (18,300–20,300-ft) interval, oil and gas shows in formation water were tested, proving that the oil window is in the deep position because of the low geothermal gradient (2.5°C/100 ms).

East and northeastward of the Prahova Valley, the Oligocene was opened in several deep exploratory wells, and gas was tested in one of these wells at 4850-m (15,900-ft) depth.

Reservoirs. Effective reservoirs of clastic origin (sandstones, marly sandstones, and sparse microconglomerates) covering a large stratigraphic interval (Oligocene, Burdigalian, Badenian, upper Miocene, and Pliocene) are parts of the petroleum system. Among them, the most prolific belong to the Meotian (upper Miocene) and Dacian (Pliocene).

Seals. Numerous pelitic interlaying provided sealing conditions for sandstone bodies. At a larger scale, two to four sandstone packages were separated in each reservoir-bearing formation.

Traps. Structural traps, of which four diapiric anticline alignments offered the best trap conditions, prevail. The diapiric processes were related to the most recent, Wallachian (Pleistocene) tectonic events, when the lower Miocene salt pierced the younger overlying formations, even up to surfaces (on the two northern alignments). Apart from the diapiric traps, faulted anticlines with or without salt cores are also developed, as well as stratigraphic traps above or below the Miocene unconformities.

Migration and entrapment. The petroleum system underwent a very complex evolution because of the many stages of reservoir deposition and trap formation. The first stage started after the beginning of the generation-expulsion-migration phases and lasted up to the Wallachian tectonic events, when the diapiric processes occurred. During this time interval, the first entrapment stage in the Oligocene, Burdigalian, and Sarmatian reservoirs must have occurred (Figure 19). Later on, the diapiric ascent of the salt offered excellent migration pathways to the oils newly generated by the Oligocene source rocks to the upper Miocene–Pliocene reservoirs, but it also led to the redistribution of some volumes of the oil previously entrapped in older reservoirs. Therefore, secondary and tertiary migration occurred when the diapiric rising of the salt accompanied by deep, regional faulting allowed the redistribution of hydrocarbons.

Hydrocarbon fields. The Subcarpathian nappe-folded foredeep (Miocene–Pliocene zone) area was, by far, the most prolific in Romania. To date, about 35% of the proven oil resources of Romania (initial oil in places) are related to the 40 fields discovered in that area. The biggest oil field in the Carpathian units, Gura Ocnitei–Moreni–Filipesti (Figure 4, no. 42), with more than 100 million t recoverable reserves, is also located in the area. On each field, the number of producing formations ranges from one to five. Initial flow rates were 10–100 t/day, except Moreni, with 350–700 t/day, during the fourth decade of the last century. The number of gas fields is small (12), with modest and moderate amounts of reserves.

Southern Carpathians

The Southern Carpathians is a typical Alpine chain with a generally east–west trend, except that their western segment bends first to the southwest and then to the south before reaching the Danube. This chain has a very complex structure characterized by large overthrustings that later on were affected by strike-slip faults. On their territory, the Median, Outer, and Marginal Dacides crop out. Sandulescu (1975b, 1976) did the structural correlation between the Eastern Carpathian and Southern Carpathian Dacides (Figure 1).

Median, Outer, and Marginal Dacides

Median Dacides. Two major nappes (Getic and Supragetic) have been separated (Munteanu Murgoci, 1905; Streckeisen, 1934) in Southern Carpathians. Later research proved that the higher unit has a more complex structure than was initially considered. We will not go into the detailed examination of these units because, for many reasons (mostly made up by basement metamorphic rocks, deeply eroded structures, etc.), they are of very low practical significance from a hydrocarbon point of view. The stratigraphic and structural data are systematized in Cioflica et al. (1981), Krautner et al. (1981), and Nastaseanu et al. (1981) (Figure 2).

Supragetic Nappe— The Supragetic nappe mostly consists of Precambrian–middle Paleozoic(?) rocks in meso- or epimetamorphic facies. They are unconformably overlain by a sedimentary sequence preserved in limited areas. Practically, different and discontinuous sequences are present in each of these areas. A synthetic sequence for the west part of Supragetic nappe includes the following members: Lower Carboniferous molasse detrital rocks; Middle Jurassic–Lower Cretaceous calcareous rocks; Albian–Cenomanian calcarenites, sandstones, and marls; and Turonian–Senonian marls and sandstones, as well as Senonian red marls in a very restricted area. The eastern segment of the Supragetic nappe sedimentary sequence includes Middle Triassic thin-bedded siliceous sandstones and dolomites and Barremian–Aptian organogenous limestones. It is worth mentioning that the Late Triassic, Lower Jurassic, and Tithonian–Neocomian are completely missing from these units.

Within the Supragetic nappe body, several imbrications have been described mostly involving only metamorphic rocks and having a more or less constant development. This complicated structure is mainly caused by the middle Cretaceous stage when the nappe and the imbrications have been emplaced. Later on, during the Laramian stage, some of the thrusting plains have been reactivated.

Getic Nappe— The Getic nappe represents the most important nappe of the Southern Carpathians. On its territory, mostly Precambrian meso- and epimetamorphic rocks crop out. The sedimentary cover is unevenly developed both in time and space. In its western side, an Upper Carboniferous–Lower Permian sandy-conglomerate molasse with coal interlayers crops out. It is followed by sandstones, shales and coal-bearing beds, bituminous shales, gray marls (Lias), marls, sandy marls (Middle Jurassic), limestones with cherts, bedded marls, nodular limestones (Upper Jurassic) and white limestones with cherts, marls, alternating marls and limestones, and massive reef limestones (Tithonian–Aptian). After a short gap, Albian–Cenomanian conglomerates and sandstone were deposited. In the central segment of the Getic nappe, the sedimentary sequence began as predominantly calcareous (more or less in-reef) facies that accumulated during the Upper Jurassic and Lower Cretaceous. Vraconian–Cenomanian polymictic conglomerates and sandstones, sometimes marls, unconformably overlie these limestones. The eastern segment of the Getic nappe is characterized by a more complete sequence. It begins with predominantly red-colored, coarse to fine detrital rocks (Upper Permian–Lower Triassic), bedded marly limestones and shales, black bituminous limestones, and massive reef limestones (Campilian–Ladinian). Then, hard sandstones, trachytic and porphyric tuff interlayers, coal-bearing beds, calcareous and/or micaceous sandstones, shales (Lias–Dogger), red oolithic limestones, gray marls, limestones, jaspers, and different types of bedded or massive limestones (Malm) unconformably follow. After a sedimentary gap, a sequence (Hauterivian–lower Aptian) of marls and limestones, and sometimes small reef limestones, accumulated. Upper Aptian–Albian polymictic conglomerates and Vraconian–Cenomanian massive sandstones unconformably cover the older formations. The sequence ends with bedded gray marls and siltstones (upper Cenomanian–Turonian), as well as red marls (Senonian). Because of the tectonic imbrications and the initial paleogeographic distribution, this synthetic column does not entirely crop out in one and the same place.

The first important horizontal displacement of the Getic nappe occurred during the latest Hauterivian, when the unit delivered large sedimentary klippen into the Ceahlau nappe. The nappe was entirely emplaced over the Outer Dacides before the Albian (middle Cretaceous stage).

Outer Dacides

Severin Nappe— The Severin nappe was separated (Codarcea, 1940) between the Outer and Marginal Dacides. Its sequence, similar to that of the Ceahlau nappe, is mostly made up of a Tithonian–Neocomian calcareous flysch. Variegated shales, siliceous rock, and tuffs underlie it. In turn, these rocks are underlain by ophiolite rock as peridotites, dolerites, gabbros, and basalts. On top of the Severin nappe sequence, Barremian–Aptian marls alternate with calcareous sandstones. The nappe was emplaced before the upper Senonian. During this stage of deformation, a high structural relief generated. Later on, the relief was as deeply eroded as the Severin and Danubian units that cropped out and delivered pebbles into the Senonian–Eocene conglomerates accumulated on the Getic nappe westward of the Olt Valley.

Marginal Dacides

Danubian Unit— The Danubian unit has a very complex inner structure with more or less developed imbrications that involve both basement and sedimentary rocks. Precambrian metaclastic and metabasite rocks, along with Hercynian epimetamorphic rocks, represent the basement. It was intruded in many places by Hercynian granitoid massifs. The older sedimentary deposits that overlie the basement are Upper Carboniferous oolithic conglomerates, sandstones, shales, and coal seams. A Permian sequence with sandstones, black shales, red conglomerates, sandstones, and shales covered by a pyroclastics complex follows. After a gap corresponding to the entire Triassic, the sedimentation resumes with Lias quartzite conglomerates, oolithic limestones, dark shales, coal interlayers, and calcareous sandstones. Concurrent with the Dogger, the sedimentation extends over larger areas, the same being more diversified. Here, where the sedimentation was more complete (Svinita Zone), arkose sandstones, spathic limestones, oolitic limestones or marls, and red nodular limestones represent the Dogger. The Upper Jurassic (Malm) mostly consists of a nodular calcareous sequence with jasper interlayers at the base. Because the Upper Jurassic–Lower Cretaceous is predominantly limy, this is represented by bedded limestones, marls, and reef limestones. After a new gap corresponding to the Aptian–Albian interval, Cenomanian thin-bedded marly limestones accumulated. They are overlain by a thick Wildflysch Formation (Turonian–Senonian) that includes sedimentary klippen delivered by Getic and Severin nappes as well as by the Danubian nappe itself. During the same stratigraphic interval in the westernmost part of the unit, massive polymictic conglomerates and sandstones accumulated (Figure 2).

The inner structure of the Danubian domain is very complex with reverse faults, thrust plains, and some outliers or overthrust slices. The ages assigned to these tectonic accidents are either Hercynian reworked in Alpine or only Alpine. Debates on the ages still exist, but part of them should be coeval with the Danubian underthrust below the Severin and Getic nappes, which means their age is intra-Senonian.

The southern limit of the Danubian unit does not crop out, it being covered by the northern limb of the Getic depression. It was interpreted in many ways with respect to the arguments considered. Nevertheless, the Getic depression, along with its pre-Senonian formations (including Getic and Danubian nappes), thrust over the Moesian Platform. The southern edge of this thrust is very well documented by a lot of recorded seismic lines. A deep well drilled in its eastern side proved that the thrust amplitude is as large as 20 km (12 mi). However, below −5000 m (−16,000 ft), even the seismic data do not offer clear information about the thrust attitude and amplitude toward the north. If we consider the general trend shown by the Southern Carpathians in their western segment (south-southeast–north-northeast) and the Moesian Platform development (east–west), it is clear that their boundary should be tectonic. This is why a thrust or overthrust is to be considered between the Danubian unit and the Moesian Platform. The age of this tectonic contact could be considered initially as being middle Cretaceous and reworked either during the intra-Senonian or pre-Eocene stages.

The structural pile formed by the Danubian unit, Severin, and Getic nappes was affected by the Sarmatian strike-slip fault (Berza and Draganescu, 1988; Neubauer et al., 1994; Matenco, 1997).

Getic Depression

Southward of the Southern Carpathian high mountains, a hilly zone develops. The hills' elevation gradually decreases toward the Getic plateau and then toward Romanian land (Figure 2). No sharp morphologic boundary exists between these entities. In geological context, the Getic depression is delineated as follows: to the north by the southern slope of the Dacides, namely, by pre-Senonian sedimentary deposits and crystalline rocks involved in the nappes pile belonging to Median, Outer, and Marginal Dacides; to the south by the vertical projection of the folded Carpathians wedge; to the east by the Dimbovitza Valley; and to the west by the Danube. It is worth mentioning that the east boundary is conventional as long as part of the deposits cropping out in the bend area are prolonging into the Getic depression. The prolongation of the east Carpathian Bend deposits toward the Getic depression could be envisaged as follows. The Senonian–lower Miocene sequence from the northern part of the Getic depression represents the western paleogeographic prolongation of the post-thrusting cover of the Cretaceous Flysch nappes. The Cretaceous–lower Miocene Tarcau nappe sequence is prolonging toward the central and southern part of the depression (Stefanescu and Polonic, 1996). The lower Miocene–lower Sarmatian deposits are prolongations of the post-thrusting covers common both to Median–Outer Dacides and to Moldavides. Unfortunately, for objective reasons, the detailed manner in which the Eastern Carpathian units are prolonging westward into the Getic depression is not yet completely solved.

Thus, in the area, delimited Senonian to Quaternary deposits crop out. The lithostratigraphy of these deposits (Popescu-Voitesti, 1910; Popescu, 1959a, b;

Szasz, 1975; Motas, 1983; Stefanescu et al., 1986) has been used to interpret the data taken from both the wells and seismic lines. The morphologic depression successively filled up with a sequence of post-thrusting covers the younger ones extending eastward into the bend area.

The sequence assigned to the Getic depression (Figures 21, 22) generally begins with a thick pile of marine polymictic conglomerates and more or less sorted and massive sandstones. The conglomerates consist of a gray sandy matrix that contains crystalline and sedimentary elements supplied by the Dacides. The conglomerates both laterally and vertically grade into sandstones and then into sandstones alternating with gray marls. On the northern paleogeographic margin of the Senonian basin, the conglomerates are completely replaced by a thin (15 m; 49 ft) sequence of gray and red silty marls with calcareous sandstones and calcarenite interlayers. The Senonian sequence ends with a thick pile of marls alternating with medium- and thin-bedded sandstones.

Eocene conglomerates unconformably cover the Senonian. They are polymictic with elements delivered by the Dacides. It is worth mentioning that along with elements coming from the Getic nappes are also reworked elements supplied by both the Danubian and Severin nappes. The presence of the element coming from the last two quoted units proves that, at least partially, the large tectonic window drawn by the Getic nappe was already open before the accumulation of the conglomerates. Westward of the Olt Valley, the conglomerates have a sandy matrix, whereas eastward, the matrix is getting more and more calcareous. A thin package of red marls and thin sandstones separates the above conglomerates from the predominantly pelitic sequence (as much as 800 m [2600 ft] thick) that closes up the Eocene.

The Oligocene–lower Miocene sedimentary cycle begins again with a package of polymictic conglomerates that have a lenticular development because laterally, they can completely disappear. Then a package of blackish, bituminous shales with thin, laminated limestone interlayers follows. This lithologic assemblage, no thicker than a few tens of meters, is a clear reminder of the lower dysodilic member constantly developed along the Eastern Carpathians. The package grades upward into a predominantly gray, dark gray shaly sequence with thin gray sandstone interlayers. Locally, large-scale lenses of conglomerates occur. Westward of the Olt Valley, on top of this sequence, a sedimentary breccia outcrops with blocks from older rocks. It follows a thick pile of massive polymictic sandstones that both laterally and vertically grade into a gray, shaly formation. It is normally covered by a thick

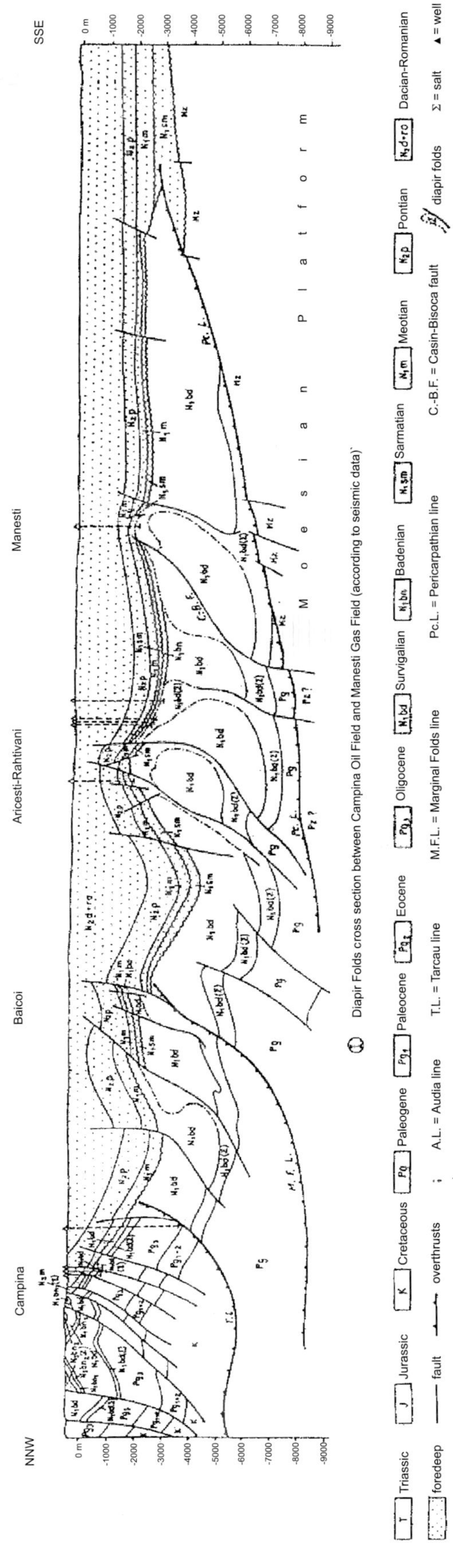

Figure 21. Geological section across the diapiric folds zone.

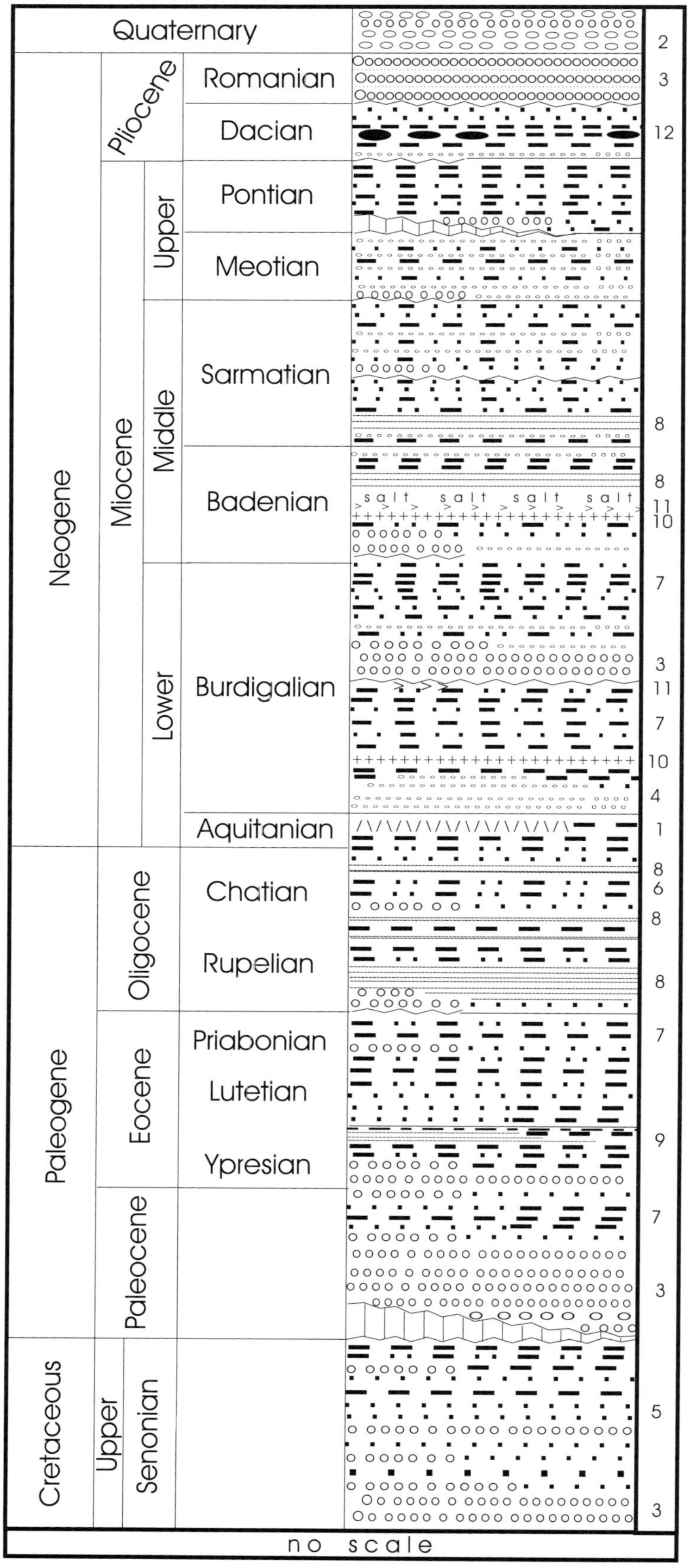

Figure 22. Getic depression lithostratigraphic column.

(over 1000 m [3300 ft]), monotonous sequence consisting of gray, medium- to thin-bedded, sometimes calcareous sandstones and gray marls. Within the lower part of the sequence, white, dacitic tuff interlayers are almost constantly present. On top of this sequence, on very restricted areas, an anhydrite package has been discovered.

Following the rule existing on the northern slope of the Getic depression, the lower–middle Miocene cycle begins again with an extended pile of polymictic conglomerates and massive sandstones. They grade upward into gray and sometimes red shales and sandstones with white tuff interlayers.

The lower part of the middle Miocene (Badenian) begins with a new pile of unconsolidated, polymictic conglomerates and sandstones, which unconformably covers older deposits and is conformably covered by the following entities: white tuffs associated with *Globigerina*-bearing marls, evaporites (salt and anhydrites), black dysodilelike shales with Radiolaria, and gray, gray whitish marls with *Spirialis*. All these packages, except for the evaporites, have locally more or less developed sandstone interlayers.

The upper part of the middle Miocene (Sarmatian) can be divided in two separate sequences. The lower one is represented by sandstones, locally conglomerates, followed or laterally replaced by finer deposits (mostly marls) containing marine bivalves. Southward and westward, these deposits are replaced by dark gray to black pelites with thin distal sandstones and laminar limestones. The upper sequence consists of unconsolidated conglomerates, coarse sandstones, marls, and thin tuff interlayers. As for the previous sequence, it contains marine mollusks. The upper sequence unconformably overlies older formations.

The upper Miocene begins with Meotian deposits that unconformably overlie the older ones. It generally consists of gray silts, marls, and more or less massive sandstones.

The uppermost Miocene (Pontian) marks an important ingression to the north because its deposits are lying even on the Oligocene. It has a lithologic background consisting of marls, silts, and clays and also has sandstone interlayers.

The Pliocene is represented by both stages, namely, by the Dacian and Romanian. The Dacian consists of thick- to thin-bedded sandstones, marls, clays, coaly clays, coals, and locally pebbles. Beside the green, gray marls and sandstones, the Romanian is characterized by the presence of variegated pelites.

Finally, the whole sequence is capped by flat-lying pebbles, sandstones, and loesslike deposits that are assigned to the Pleistocene.

In the outcropping area, the Getic depression shows a relatively simple structure characterized by a southward-dipping homocline that, westward of the Olt Valley basin, is affected by gentle folds and is sometimes affected by vertical faults. The data acquired by wells and seismic acquisition proved that southward, the structure is more complicated than would be expected. Based on these data, eight anticlinal lineaments have been detected. They are developed in echelons (Paraschiv, 1979) trending east–west, parallel to both the Southern Carpathians and the Carpathian wedge. Before reaching the present shape and structure, the Getic depression underwent a complex evolution. Modern structural studies have been performed during the 1990s either northward of the basin margin (Ratschbacher et al., 1993) or into the basin itself (Dicea, 1996; Matenco, 1997; Rabagia and Matenco, 1999). Using all the available data (surface, wells, seismic, and kinematic), the following sequence of the main stages of the basin evolution could be restored: during the Senonian, the Getic nappes along with the Severin nappe thrust the Danubian; a long extensional and subsidence period follows, which continues until the end of the Oligocene or even until the early Miocene; in the tectonic units located northward of the basin, a 50° clockwise rotation is documented toward the end of the Oligocene, a rotation that is coeval with the Cretaceous Flysch nappe emplacement in the bend area (Old Styrian phase) and with slight folding in the Getic depression; these events generated a structural morphology that later on was deeply eroded (Figure 23) because in some instances, even the upper Miocene deposits directly overlie the Oligocene ones; during the early Miocene and the lower part of the middle Miocene, a new extension stage occurred, an extension that was not only limited to the Getic depression but was also active in the bend area (Stefanescu et al., 2000); according to the data acquired by Rabagia and Matenco (1999), some dextral shearing transpressional structures generated during the Moldavian phase before the main stage of deformation (middle Sarmatian) when the depression thrust over a long distance across the foreland; the Getic basin-foreland contact shows a complex structure characterized by the presence of a few imbrications developed on the hanging wall of the master thrust plane and which were generated out of sequence (Stefanescu and Dicea, 1995); a new extensional period follows, generating a slow subsidence until the Romanian when, during the Wallachian compression phase, small-scale folding and thrusting occurred.

Petroleum System

Source rocks. The potential source rocks are developed at three different intervals during the Oligocene–Miocene interval. The Oligocene black

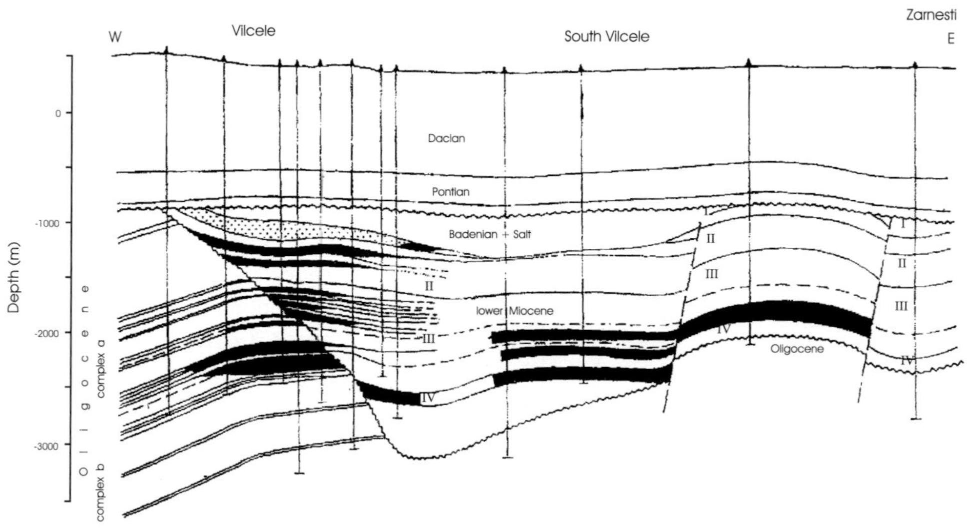

Figure 23. Geological section along the Vilcele field showing intralower Miocene paleorelief.

dysodilic shales have been compactly accumulated at the lower part of the sequence. They are present only in the internal part of the depression, where they generated oil and gas at depths situated between 4000 and 7000 m (13,000 and 23,000 ft) (Figure 24), but they are present in the higher part of the sequence. The middle Miocene (Badenian) black shales that developed above the evaporite level are also considered as one possible source formation. An important source sequence is located at the lower part of the Sarmatian, it being represented by dark-gray to black shales that are similar to Oligocene source rocks. The other Sarmatian–Pontian shales could also be considered as potential source rocks. In this last group of potential source, the Lower Cretaceous black shales discovered by a well in the eastern extremity of the Getic depression could be considered.

Reservoir rocks. All of the sequence assigned to the Getic depression has interlayers of polymictic clastics. Some of them are thicker bedded (from decimeters to meters) as in the Oligocene or lower Miocene, others being much thinner bedded (centimeters to as much as tens of centimeters) as in

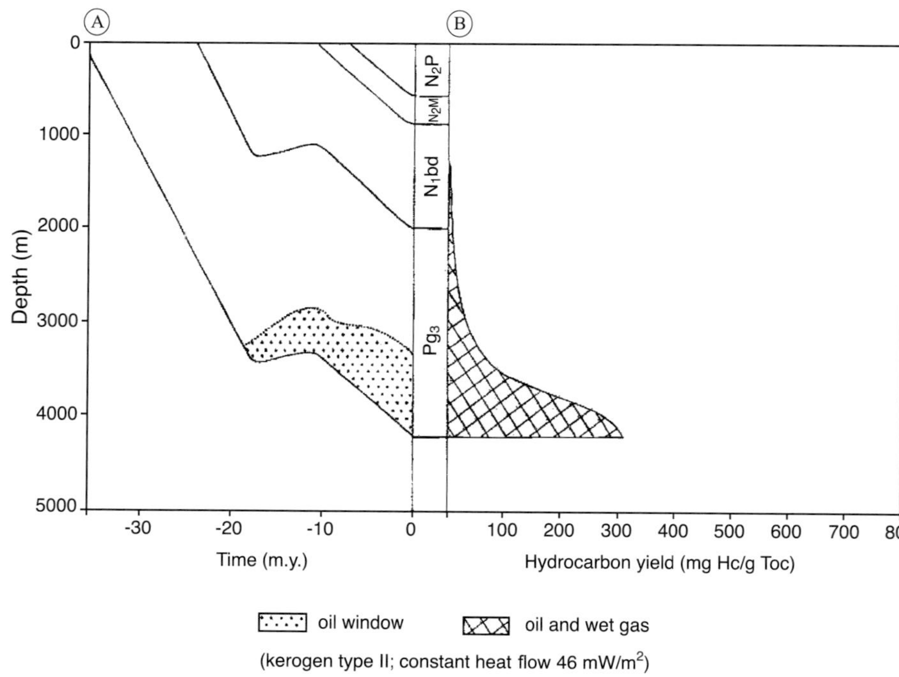

Figure 24. Getic depression: (A) burial history and (B) hydrocarbon kinetics and yield diagram for Oligocene source rocks. Pg_3 = Oligocene; N_1bd = Burdigalian; N_2M = Meotian; N_2P = Pontian.

Sarmatian, Meotian, and Pontian. No matter what the thickness is, they proved their capability to store hydrocarbons in the numerous fields discovered in the depression. The porosities vary as follows: between 24 and 26% for the Oligocene, 7 and 27% for the lower Miocene, 20 and 33% for the Sarmatian, and 16 and 35% for the Meotian. The porosities also have a large variation as follows: around 500 md in Oligocene, 0–4700 md in lower Miocene, 0–700 md in Sarmatian, and 5–2450 md in Meotian.

Trap types. Because of its complex evolution in the Getic depression, almost all classical types of traps from stratigraphic to structural were formed. Within the stratigraphic traps, the reservoirs pinch-out at different levels and the unconformities at the Oligocene–lower Miocene, lower Miocene–Sarmatian, and Meotian–Pontian boundaries are worth being mentioned. It is also worth mentioning that sometimes, the pay intervals are located both on top and beneath the unconformities (Figure 23). Among the structural traps that have been detected are the simple anticlines, faulted anticlines (both longitudinally and transversally), and faulted homocline. In most of the fields, the structural traps are combined with stratigraphic ones.

Fields. Both oil and gas (about 40) and gas (more than 75) fields have been discovered in the Getic depression. It is worth mentioning that the oil fields are generally located in the inner (northern) part of the depression, whereas the gas fields are situated in the outer part of the depression, where the Oligocene source rocks are missing.

Transylvanian Basin

The Transylvanian Basin (Figures 1, 2) is a major sedimentary basin surrounded by the Eastern Carpathians, Southern Carpathians, and the Apuseni Mountains. It has a roughly circular shape with an area of around 20,000 km^2 (7700 mi^2).

The complex tectonic evolution of the Transylvanian Basin has been presented in several papers since the beginning of the 20th century. The theory supposes a tectonic trigger mechanism for the salt diapirism that developed in this basin (Mrazec, 1907). The economical importance (Transylvanian Basin is the largest gas province in Eastern Europe) led to an intensive exploration program, but only few papers about this basin were published before the 1970s. Ciupagea et al. (1970) published the first book with data acquired by the industry about the Transylvanian Basin. After this book, several papers, using geophysical data, presented the deep structure of the basin (Sandulescu and Visarion, 1978; Stefanescu and Working Team, 1986). Papers published about the regional geology of the Carpathians mentioned the Transylvanian Basin as a basin with a poorly understood evolution (Royden, 1988). These few publications was the result of a small amount of data released by the industry. After 1994, new data were released by the industry, and structural work has been undertaken on the basin rim. Using seismic lines, de Broucker et al. (1998) summarized the pre-Badenian tectonic evolution of the northern sector of the Transylvanian Basin. Structural work performed on the basin rim helped put more constraints on the basin evolution (Ciulavu and Bertotti, 1994; Ciulavu et al., 2000). Fission-track data combined with structural data highlighted the similar tectonic evolution of the Transylvanian Basin and of the Carpathians (Sanders, 1998). Using both seismic and structural data acquired in the field, Ciulavu et al. (2000) presented a new image about the tectonic evolution of the Transylvanian Basin, integrated in the tectonic evolution of the Carpathians. These papers reveal a very complex tectonic evolution of the Transylvanian Basin, which is summarized below.

The Transylvanian Basin basement (Figure 25), as found in the deep wells drilled in the basin, is made up of a nappe pile consisting of metamorphic or ophiolitic rocks and a discontinuous Mesozoic cover, similar to some sequences cropping out in the surrounding mountains. In the western sector of the depression, the metamorphic basement is of the Apuseni Mountains type, whereas in the eastern sector, it is of the Eastern Carpathians type. Between these two blocks, the Mures ophiolitic zone developed. The basement is unconformably overlain (Figure 25) by a Triassic–Lower Cretaceous sedimentary cover consisting of continental deposits (Lower Triassic), followed by marine carbonate rocks (Middle–Upper Triassic, Upper Jurassic, Lower–middle Cretaceous) alternating with shaly formations (Middle and Upper Triassic, Lower Cretaceous). The oldest faults, as imaged on the seismic lines, affect the Lower Cretaceous deposits and are east- and southeast-verging thrust faults. They have been detected only in the western sector of the Transylvanian Basin, where they are related to the closure of the Mures zone.

The postmiddle Cretaceous sedimentary fill (5–8 km [3–4.8 mi] thick) of the Transylvanian Basin covers the entire Late Cretaceous–Quaternary interval (Figure 25). Late Cretaceous, northeast-trending extensional basins have been developed. Senonian shallow-water sediments are present in the western sector of the basin, whereas Cenomanian (probably) deep-water sediments are present in the eastern sector.

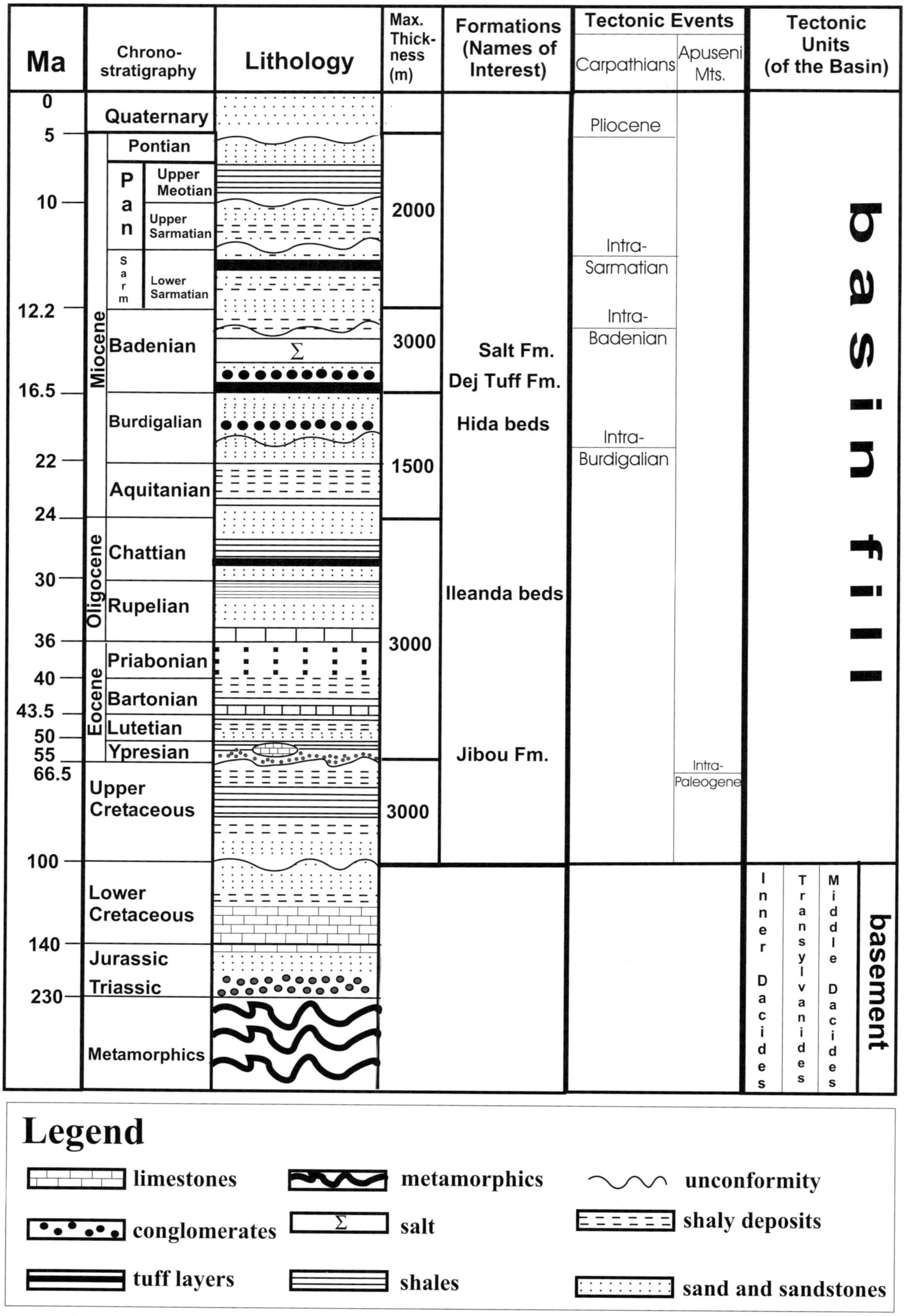

Figure 25. Transylvanian Basin simplified lithostratigraphic column.

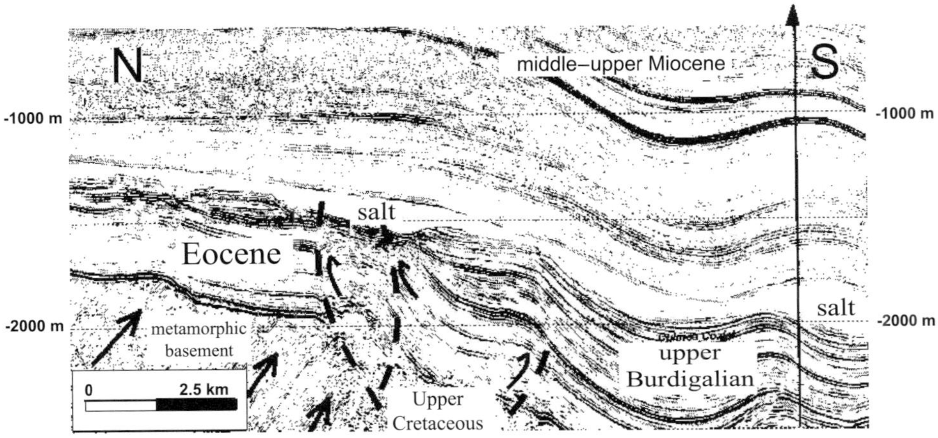

Figure 26. Seismic line across the central slope of the Transylvanian Basin. Arrows represent the tectonic movements of the different blocks.

The Late Cretaceous normal faults have been compressionally inverted in the northern sector of the basin. The continuous tectonic inversion resulted in thrust emplacement with metamorphic basement and pre-Oligocene series. These thrusts have the same north-northeast orientation as the previously described structures.

An important change in tectonics and sedimentation occurred during the Oligocene. The depocenter of the basin was shifted toward a roughly east–west orientation. The Oligocene starts with sandstones, limy sandstones, and continues with marls and limestones. The overlying Ileanda beds consist of bituminous shale and limy deposits. The Oligocene ends with sandstones and sands showing lateral variation to marls.

The east–west-trending orientation of the basin was also maintained during the early Miocene, but the tectonic setting was changed. A southward-thinning clastic wedge developed during the late Burdigalian in the northern part of the basin. This clastic wedge acted as a foreland basin for the Pienides.

The Badenian represents the beginning of a new sedimentary cycle, covers the entire area of the basin, and is similar in character to the Badenian from the Carpathians. It is the time when the Salt Formation has been deposited. After the salt deposition, shortening again occurred. This shortening led to the onset of basin border uplift and, in the basin, represents the trigger mechanism for the salt diapirism. An intra-Badenian tectonically induced unconformity is documented in the northern sector of the basin. Since the Badenian, the basin started to acquire its present shape.

The Sarmatian deposits are made of sands and marls. Several tuff layers are present both in the Badenian and Sarmatian deposits.

The Pannonian and Pliocene deposits onlap onto the older sedimentary deposits and consist of marls and sands.

Tectonically, since the Badenian, the Transylvanian Basin has acted as a rigid block (Figure 26). Some of the inherited faults have been reactivated, mainly as strike-slip faults and also as reverse faults. Structural data indicate a compressive-transpressive tectonic regime since the Badenian.

Petroleum Systems

Source rocks. The Transylvanian area was largely explored for its gas reserves. The sediments that accumulated are naturally separated in two parts: the presalt sequence (the Transylvanian Depression) and the postsalt sequence (the Transylvanian Basin). Despite the fact that all the discovered fields so far are located in the postsalt sequence, hydrocarbon (heavy-oil) indications (some small production during the 19th century) far away westward from the border of the basin are located in the Eocene continental sandstones. We mentioned this fact to prove the presence of two petroleum systems in the Transylvanian area. The potential source rocks that could be considered for the pre-Badenian petroleum system are the following: the Senonian dark gray shales, the black bituminous Eocene limestones (Jibou limestone), and eventually, the lower Miocene shales. The only rocks that have the character of source formation for certain are the Oligocene black bituminous shales (Ileanda shales).

Middle Miocene–Pannonian shales represent the gas source rocks (with type III kerogen) for the postsalt petroleum system. Special mention is given to the Badenian black, Radiolaria-bearing shales that show a typical source rock facies. The analyses performed on samples both from outcrops and cores revealed an immature stage of the organic matter except for the Upper Cretaceous core samples (that are buried deeper than the other samples) that seem to be situated at the immature-mature stage.

Traditionally, it was accepted that the gas from the Transylvanian fields is biogenic. Recent study performed by foreign companies proved that thermogenic gas is also present. The presence of condensate in some fields also proves that at least some of the postsalt source rocks were buried deep enough to cross the biogenic-thermogenic boundary of the oil window.

Reservoir rocks are practically spread all over the sedimentary sequence as follows: fractured Mesozoic carbonate formations, Upper Cretaceous sandstones, Eocene sandstones and limestones, and Oligocene to Pannonian sandstones and silts. Badenian and Sarmatian reservoirs have average porosities of 20% and permeabilities ranging between 0.1 and 1000 md (Vancea, 1960). This reminds us that the reservoir physical properties deteriorate both with depth and from the east toward west and northwest.

Despite the fact that many source-to-oil correlation studies have been performed so far, it is assumed that the oil from the seeps was generated by one of the presalt source rocks. This assumption is also based on the fact that only these source rocks have been buried deep enough in the central part of the basin to reach the oil window.

Seals. The thicker intervals of pelitic rocks function as seals. Sometimes, their effectiveness is not perfect because several gas seeps have been recorded on the surface of the basin.

Traps. They are mostly structural: domes, brachyanticlines, and anticlines. Most of them are more or less transversely and/or longitudinally faulted. The types and amplitudes of the structures are conditioned by the intensity of the salt movement. Thus, most of the structures located in central Transylvania are of dome or brachyanticline type being generated by salt cryptodiapirs that just pop up the overlying deposits without piercing them. By contrast, on both the east and west limbs of the basin, two lineaments are present along which the upward salt movement was more intense, piercing partially or even completely the overlying sequence or even reaching the surface. For further exploration, the stratigraphic and/or subtle traps also have to be considered.

Migration. For most of the fields, a vertical or lateral short-distance migration is accepted. As regards the oil seeps (some of them located in continental deposits), it is assumed that a lateral migration on long distance occurred.

Fields. Regardless of the different hypotheses, the effectiveness of the petroleum system in the Transylvanian Basin is proven by the large number of fields (more than 100) discovered thus far. The first economic discovery, the Sarmasel field (Figure 4, no. 44), was done in 1909. With respect to their lithostratigraphic sequence peculiarities, the structural type and amplitudes of the field are geographically grouped as follows: the central group, the northern group, the western group, the southern, and the eastern group. The most prolific fields are located in the central group. From these fields, we will discuss only Filitelnic (Figure 4, no. 45) and Delenii (Figure 4, no. 46; Figure 27). We will discuss only these two fields because Filitelnic is the biggest field discovered in the Transylvanian Basin, whereas Delenii is the dome on which the deepest well in the Transylvanian Basin was drilled (5200 m; 17,000 ft), a well that reached the ophiolitic basement of the central zone.

The gas is entrapped in Badenian to Pliocene reservoirs that are grouped in several complexes (as much as 25) forming pay zones vertically stacked one on top of the other (Figure 27). Generally, the gas is of very good quality, containing more than 99% methane and only small quantities of ethane and nitrogen.

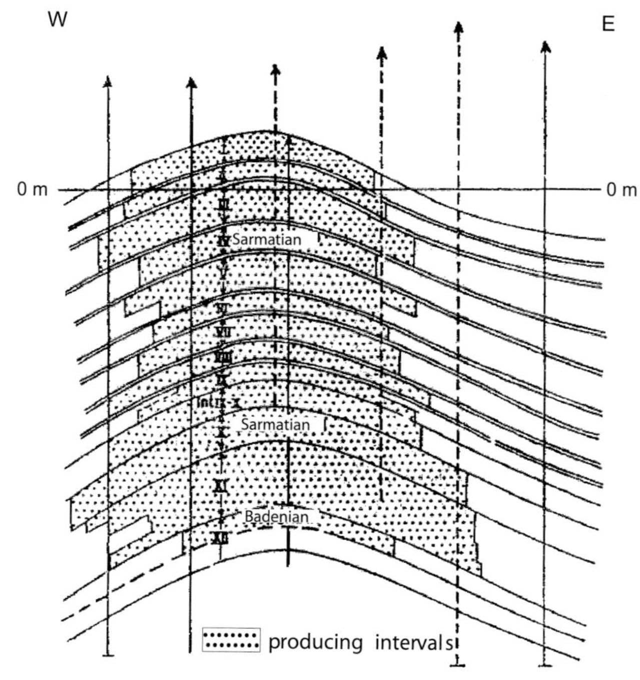

Figure 27. Example of stocked reservoirs in the Transylvanian Basin.

THE HISTORY OF THE GEOLOGIC EVOLUTION AND HYDROCARBON GENESIS AND ACCUMULATION

As with several territories all over the world, the Romanian onshore has long and complicated geological and human histories. Researchers who try to clarify these two histories are facing the same issues, beginning with the gathering of data and finishing with their interpretation. As the data become older, they present a higher degree of uncertainty, and consequently, their interpretation becomes more difficult. Their accuracy increases as they are gathered closer to modern times. Despite the fact that the newest data are the most reliable, we cannot fully integrate them in a unitary model because some of them are either incomplete or, for political reasons, are delivered in a disturbed manner. Some years ago, a revolution occurred in Romania. Despite the fact that we witnessed this event, we are still not able to exactly restore, in a coherent manner, the entire picture of what was happening. Similarly, we could see and record some of the present-day geologic events evolving around us, but it is not easy to correctly apply what we are seeing back to the geological records. We said all of this in order to emphasize the difficulties and the different degrees of certainty of the history of the hydrocarbon genesis that will be explained further on.

The present Romanian territory showed its hydrocarbon vocation a long time before its present architecture could be accomplished. During the Precambrian, westward and southwestward, far away from the country's present geographic delineation, somewhere on the paleo-European domain, thick piles of rich organic matter rocks (potential source) accumulated. Later on, before the Silurian, these rocks, along with their host formations, were metamorphosed (sometimes by multiple phases) as they reached the graphite stage. Despite the fact that they were useless for hydrocarbon genesis even since that time, they were tectonically imported through clockwise rotation and horizontal displacement and incorporated in the Carpathian structure (both east and south) during the middle Cretaceous and Late Cretaceous tectogeneses.

The Paleozoic subsidence generated basins with favorable conditions for accumulation of sediments, along with an important organic matter supply. Thus, on the Moesian Platform, bituminous shales accumulated during the Silurian and Devonian–Carboniferous intervals in a marine environment. It is not clear what function they had in the present oil and gas fields either because, in some parts, they were overcooked, or in the uplifted parts of the platform, they have not been buried enough to generate hydrocarbons. Both on the Danubian and Getic domains, the organic-rich sedimentation resumed with Variscan molasse (following Asturian deformation stage). A limnic (lacustrine) environment was set up during the Late Carboniferous, when coal (sometimes fat coal) and carbonaceous shale sequences accumulated. In places, the black, bituminous coal almost reached a semianthracite stage of transformation. It is difficult to say if these rocks generated economic quantities of hydrocarbons, but it is obvious that because of their current structural position, we could not present any data on them, at least for the time being.

Triassic rocks mostly show a Germanic facies, and consequently, they could not be connected with the oil generation. An exception should still be mentioned. We are referring to a thin Middle Triassic sequence of blackish thin-bedded limestone that is obviously bituminous. It accumulated on the infra-Bucovinian domain that was located far westward from its present position in the Eastern Carpathians. They underwent strong transformation during the overthrusting, because they are very mature today.

With the Jurassic, effective source rocks began to accumulate, some of them contributing to the hydrocarbon genesis. During Lias, in paralic conditions, along with coarse terrigenous rocks, both coal deposits and bituminous shale accumulated on the external side of the Getic domain. The bituminous shales have a bitumen content that can reach 20% (Borcos et al., 1984). It is assumed that these rocks generated hydrocarbons in the past, but no oil or gas fields have been discovered in relation to these source rocks. The same paralic environmental conditions were also present on the Danubian domain. Here, in terrigenous sequences (Gresten facies), both coal and coaly shales accumulated. Because of the strong tectonic deformations that affected these rocks, the coal was overcooked, reaching the anthracite stage. The Gresten facies with brown coal and black, bituminouslike shales also occurred on the external nappe (infra-Bucovinian) of the Median Dacides from the Eastern Carpathians. Because of its limited extension and the complicated structure in which it is involved, the importance of the Gresten facies for economic hydrocarbon generation is negligible.

As a result of the old Kimmerian tectonic movements, the Romanian part of the Moesian Platform was uplifted above sea level and was submitted to erosion almost during the entire Lias. The sedimentation resumed (during the uppermost Lias) first on the central part of the platform and then extended step by step toward the east and west. Siliciclastic sequence followed by black bituminous shales accumulated during the Middle Jurassic (Dogger). This reservoir-source rocks couple proved to be effective in the formation of some fields in the Moesian Platform.

The Upper Jurassic shows two distinct facies. A shallow-marine facies, predominantly represented by bedded or massive limestones, in places with reefs, accumulated on the platforms located on both sides of the Tethyan oceanic crust. A second facies mostly consisting of terrigenous rocks accumulated in two parallel rifting, extensional troughs. One corresponds to the Tethys rifting zone (that began to open since the Middle Triassic), and the second developed on the Outer Dacides paleogeographic territory.

We will focus only on the second subsiding zone, because its evolution was crucial in forming some of the important oil fields on the Romanian onshore. Thus, on its external side (corresponding today to Moldavides), more or less important volumes of source rocks accumulated since the Early Cretaceous until middle Miocene. The source rocks accumulated either in the normal sequence of the basin (Teleajen, Macla, Audia, Tarcau, and Marginal Folds nappes) or in sequences belonging to the post-thrusting covers accumulated after important shortening events (post-Laramian). Two intervals with source rocks accumulated in the normal sequences are worth mentioning, namely, the Lower Cretaceous black shales and the Oligocene–lower Miocene dysodilic shales. They overlap only in the external part of the Moldavides, namely, in the Tarcau and Marginal Folds nappes. Although it seems that the Lower Cretaceous black shales have been buried enough as to reach (at least partially) the oil window even during the subsiding stage, the Oligocene–lower Miocene source rocks certainly got into the oil window after the nappes emplacement. Similar Oligocene rocks accumulated in the Transylvanian Depression, but they did not reach the oil window.

Once generated, the hydrocarbons migrated vertically, laterally (on large), or both to the trapped reservoirs (Dicea et al., 1991). In the Romanian fields, a large variety of reservoirs have been discovered as follows: weathered basement (crystalline schists), conglomerates, sandstones, sands, siltstones, calcareous sandstones, and limestones. The detrital material forming the reservoirs was delivered by two main source areas (Figure 11): one represented by the Carpathians and their paleorelief forms and an external source hidden today but was very active during the Lower Cretaceous to middle Miocene. The detrital reservoirs turned out to be the most prolific ones.

The present structure of the Carpathians is caused by a few successive shortening stages during the middle Cretaceous, Late Cretaceous (intra-Senonian), and early and middle Miocene. During the last two mentioned stages, the Moldavides nappes thrust each other and, altogether, thrust over the foreland. During the thrusting processes, the Oligocene–lower Miocene source rocks were buried deep enough to generate predominantly oil. The oil migrated along the carrier rocks (the same as the Oligocene–lower Miocene reservoirs) and/or the fracture systems and accumulated into the already existing traps, even if they were located (commonly) in the underlying sheet (Marginal Folds nappe) or in the thrust sheet (very rear; Tarcau nappe).

After each important compression stage, more or less extended subsiding basins successively occurred (Transcarpathian Flysch, Transylvanian Depression, Getic depression, and Dacic Basin). The most important, from the hydrocarbon genesis point of view, are the Transylvanian Basin and the Transylvanian foredeep. In the Transylvanian Basin, the Badenian–Sarmatian source rocks only partially reached the oil window top because mostly dry gas with some condensate was generated.

A depression formed outside the Carpathians as a result of the thrusting of these nappes over the foreland. The depression corresponded to the external part of the Carpathians and the external part of the platforms in front of them. This depression, known as the foredeep, was fielded up with molasse-type deposits that included both source and reservoir rocks. The reservoirs of the fields existing in the foredeep have been filled up either with the hydrocarbon generated by its own source rocks or by hydrocarbon migrated from kitchens located in the flysch zone area. The structural traps were formed during the tectonic deformations that followed the overthrusting events. Here, the function of the salt in generating hydrocarbon traps should be emphasized. The salt accumulated during both the early Miocene and middle Miocene. Both salt intervals were successively involved in alternating extensional and compressional stages that deformed it from the original flat position to the present-day diapirs. During their post-Sarmatian evolution, the salt pushed up its cover, forming the diapir flanks excellent for structural traps. In cases where the salt popped up to the surface (exaggerated diapirs), the traps are laterally sealed by the salt itself.

The youngest deformations (Wallachian stage) that could be documented are recorded in the outer part of the Carpathian Bend area. They are Pleistocene in age, and they practically placed the finishing touches to the hydrocarbon traps. The stratigraphic traps exist in most of the deposits, but mostly in those of Miocene–Pliocene age. Obviously, most of them have been affected by the tectonic deformation, including the Wallachian stage.

Here ends the history of the hydrocarbon generation and accumulation along with the geological history of the Romanian territory, but here also begins a new fascinating adventure in which we are deeply involved: the discovery and use of what was generated during this long geologic history.

ACKNOWLEDGMENTS

This chapter succeeded to see the printing light thanks to the support and help of many people. We express our gratitude to all of them. We especially thank Jan Golonka and Frank Picha for their technical advice and support.

REFERENCES CITED

Alexandrescu, G., 1966, Stratigrafia "Flisului curbicortical" din partea de nord a Moldovei (Carpatii Orientali): Studii si Cercetari de Geologie, Geofizica si Geografie, v. 3, p. 275–535.

Baltes N., E. Antonescu, D. Grigorescu, G. Alexandrescu, and M. Micu, 1984, The Black Shales Formation of the East Carpathians, lithostratigraphy and oil potential: Bucharest, Annale of the Institute of Geology and Geophysics, v. 59, p. 79–88.

Bancila, I., 1958, Geologia Carpatilor Orientali: Bucharest, Editura Tehnica, 368 p.

Barbu, C., N. Alimehmed, and C. Paraschiv, 1970, Le Paleozoique de l'Avant Pays Des Carpates Orientales de la Romanie: Budapest, Carpatho-Balkan Geological Association, Xth Congress, v. II, p. 242–250.

Berza, T., and A. Draganescu, 1988, The Cerna-Jiu fault system (South Carpathians, Romania), a major Tertiary transcurrent lineament: Dari de Seama ale Institutului de Geologie si Geofizica, Romania, v. 72–73, p. 43–75.

Bleahu, M., 1962, Cercetari geologice in bazinul superior al vaii Ruscovei (Muntii Maramuresului): Dari de seama ale Comitetului Geologic, v. 45, p. 298–308.

Borcos M., G. Udubasa, M. Sandulescu, H. Krautner, S. Nastaseanu, C. Bitoianu, and S. Radan, 1984, Harta substantelor minerale utile — Nota explictiva: Bucuresti, Institute of Geology and Geophysics, 237 p.

Bucur, I., 1971, Noi date faunistice si structurale in flisul cretacic si paleogen de la nord de depresiunea Bretcu-Tirgu Secuiesc: Dari de seama ale Institutului Geologic, v. 57, no. 5, p. 27–34.

Burchfield, B. C., 1976, Geology of Romania: Geological Society of America Special Paper 158, 82 p.

Butac, A., et al., 1997, East Carpathian Bend area, Romania: Oligocene–Pliocene source rocks, reservoirs and hydrocarbon fields (abs.): AAPG International Conference Vienna, Field Trip, v. 10, p. 43.

Cioflica, G., H. Savu, I. Nicolae, M. Lupu, and S. Vlad, 1981, Alpine ophiolitic complexes in South Carpathians and South Apuseni Mountains: Carpatho-Balcan Geological Association, XII Congress, Institute of Geology and Geophysics, v. 18, 80 p.

Ciulavu, D., and G. Bertotti, 1994, The Transylvanian Basin and its Upper Cretaceous substratum: Alcapa II: Romanian Journal of Tectonics and Regional Geology, v. 75, no. 2, p. 59–65.

Ciulavu D., C. Dinu, A. Szakacs, and D. Dordea, 2000, Neogene kinematics of the Transylvanian Basin (Romania): AAPG Bulletin, v. 84, p. 1589–1615.

Ciupagea, D., M. Pauca, and T. Ichim, 1970, Geologia Depresiunii Transilvaniei: Bucharest, Editura Academiei Romane, p. 255.

Codarcea, A., 1940, Vues nouvelles sur la tectonique du Banat meridional et du plateau du Mehedinti: Anuarul Institutului Geologic al Romaniei, v. 20, p. 111–121.

Dallmmeyer, R. D., F. Neubauer, H. G. Krautner, H. Fritz, and A. V. Bojar, 1998, Variscan and Alpine tectonic processes in the Eastern Carpathians orogen: Evidence from $^{40}Ar/^{39}Ar$ mineral ages and structural analysis (abs.): Carpathian–Balkan Geological Association, XVI Congress, Abstract volume, p. 21.

De Broucker, G., A. Mellin, and P. Diundam, 1998, Tectonostratigraphic evolution of the Transylvanian Basin, presalt sequence, Romania: Bucharest Geoscience Forum, v. 1, p. 36–69.

Demaison, G., and B. J. Hiuzinger, 1991, Genetic classification of petroleum system: AAPG Bulletin, v. 75, no. 10, p. 1626–1643.

Dicea, O., 1995, The structure and the hydrocarbon geology of the Romania East Carpathians border from seismic data: Petroleum Geosciences, v. 1, p. 135–143.

Dicea, O., 1996, Tectonic setting and hydrocarbon habitat of the Romania external Carpathians: Peri-Tethys Memoir, v. 2, no. 170, p. 403–425.

Dicea, O., P. Dutescu, F. Antonescu, G. Mitrea, R. Botez, I. Donos, V. Lungu, and I. Morosanu, 1980, Contributii la cunoasterea stratigrafiei zonei transcarpatice din Maramures: Dari de seama ale Institutului de Geologie si Geofizica, v. 45, no. 5, p. 21–85.

Dicea, O., N. Ionescu, and D. C. Morariu, 1991, Cadrul geologic de formare a ascumularilor de petrol si gaze in principalele bazine sedimentare din Romania (abs.): AAPG Annual Meeting Program, p. 46.

Dumitrescu I., 1948, La Nappe du Gres de Tarcau, la zone marginale et la zone neogene, entre Casin et Putna: Comptes Rendus deo Seances — Institut Geologique de Roumanie, v. 29, p. 84–105.

Dumitrescu, I., 1952, Etude Geologique de la region entre l'Oituz et la Coza: Annuaire du Comite Geologique, v. 14, p. 195–270.

Dumitrescu, I., and M. Sandulescu, 1968, Problemes structuraux fondamantaux des Carpates Roumaines et de leur anant-pays: Anuarul Comitetului de stat al geologiei, v. 36, p. 195–218.

Dumitrescu, I., and M. Sandulescu, 1970, Harta tectonica a Romaniei: Bucharest, Geological Institute of Romania, 1 sheet.

Filipescu, M., 1955, Contributii la orizontarea Cretacicului din flisul Carpatilor Orientali: Revista Universitatii si a Politehnicii, v. 8, p. 293–315.

Geological Bureau, 1882, Geological Map of Romania, scale 1:175,000, 3 sheets.

Gradinaru, E., 1984, Jurassic rocks of north Dobrudja. A depositional-tectonic approach: Revue Roumaine de Geologie, Geophysique et Geographie, v. 28, p. 61–72.

Grigoras, N., 1955, Studiul comparativ al faciesurilor dintre Putna si Buzau: Bucuresti, Anuarul Comitetului Geologic, v. 28, p. 99–219.

Hippolyte, J.-C., and M. Sandulescu, 1996, Paleostress characterization of the "Wallachian phase" in its type area (Southern Carpathians, Romania): Tectonophysics, v. 263, p. 235–248.

Joja, T., 1954, Structura geologica a Flisului marginal de pe Putnisoara si din cursul inferior al Putnei: Dari de seama ale Comitetului Geologic, v. 38, p. 183–202, Bucuresti.

Krautner, H., S. Nastaseanu, T. Berza, I. Stanoiu, and V. Iancu, 1981, Metamorphosed Paleozoic in the South Carpathians and its relations with the pre-Paleozoic basement: Carpatho-Balkan Geological Association, XII Congress, Institute of Geology and Geophysics, v. 16, p. 116.

Linzer, H. G., 1996, Kinematics of retreating subduction along the Carpathian arc, Romania: Geology, v. 24, p. 167–170.

Macovei, G., 1938, Les gisement de Petrole: Paris, Masson et Co., 502 p.

Marinescu, F., C. Ghenea, and I. Papaianopol, 1981, Stratigraphy of the Neogene and Pleistocene boundary: Bucharest, Institute of Geology and Geophysics, 111 p.

Matenco, L. C., 1997, Tectonic evolution of the outer Romanian Carpathians: Amsterdam, Free University, 160 p.

Mirauta, O., 1966, Devonian and Triassic in the Mahmudia Hills (northern Dobrudja): Dari de Seama ale Institutului Geologic, v. 52, no. 2, p. 115–134.

Morley, C. K., 1996, Models for relative motions of crustal blocks within the Carpathian region, based on restoration of the Outer Carpathian thrust sheets: Tectonics, v. 15, no. 4, p. 885–904.

Motas, C., 1983, Nouvelles donnes sur les rapports structuraux entre les Carpathes Meridionales et la Depression getique: Carpatho-Balkan Geological Association, XII Congress, v. 60, p. 141–146.

Mrazec, L., 1907, Despre cute cu simbure de strapungere: Buletinul Societatii de Stiinte, Romania, v. 16, p. 6–8.

Mrazec, L., 1927, Les plis diapir et le diapirism en general: Comptes Rendus des Seances-Institut Geologique de Roumanie, v. 6, p. 226–270.

Mrazec, L., and I. Popescu-Voitesti, 1914, Contribution a la connaissance des nappes du Flysch Carpatique en Roumanie: Annuaire de l'Institut Geologique de Roumanie, v. 5, p. 495–527.

Munteanu Murgoci, G., 1905, Sur l'existence d'une grande nappe de recouvrement dans les Carpates Meridionales: Comptes Rendus de l'Academie des Sciences, v. 7, no. 31, p. 50–52.

Mutiu, R., 1991, On the presence of the Cambrian in the Moesian Platform (Romania): Revue Roumaine de Geologie. Tom, v. 35, p. 75–85.

Nastaseanu, S., I. Bercia, V. Iancu, S. Vlad, and I. Hartopanu, 1981, The structure of the South Carpathians (Mehedinti-Banat area): Carpatho-Balkan Association, XII Congress, Geological Institute of Romania, v. 22, 100 p.

Neubauer, F., H. Fritz, E. Wallbrecher, and A. V. Bojar, 1994, Paleostress patterns in the South Carpathians: Strike-slip and extensional by oroclinal bending, *in* The geological evolution of the Alpine–Carpathian–Pannonian system: Alcapa II, Romanian Journal of Tectonics and Regional Geology, v. 75, p. 39–40.

Paraschiv, D., 1965, Din evolutia paleogeomorfologica a Cimpiei Romane: Anuarul Stiintific al Univsitatii Iasi, v. 11, no. 2b, p. 139–145.

Paraschiv, D., 1975, Tipuri de capcane cu petrol si gaze in Romania: Revista Mine, Petrol si Gaze, v. 26, no. 5, p. 270–279.

Paraschiv, D., 1979, Romanian oil and gas fields: Bucharest, Institute of Geology and Geophysics, 383 p.

Paraschiv, D., 1982, The principal lithostratigraphic units in the pre-Jurassic of the Moesian Platform: Revista Mine, Petrol si Gaze, v. 33, p. 134–136.

Paraschiv, D., 1984, The evolution of the hydrocarbon field distribution in the Moesian Platform: Annuaire de l'Institut Geologique de Roumanie, v. 64, p. 205–213.

Patrulius, D., 1969, Geologia masivului Bucegi si a culoarului Diombovicioarei: Bucharest, Editura Academiei, 324 p.

Patrulius, D., E. Popa-Dimian, and E. Dimitriu-Popescu, 1966, Seriile mezozoice autohtone si pinza de decolare transilvana in imprejurimile Comanei (Muntii Persani): Anuarul Comitetului Geologic, v. 35, p. 397–444.

Patrulius, D., E. Popa, and I. Popescu, 1969, Structura pinzei bucovinice in partea meridionala a masivului cristalin moldav (Carpatii Orientali): Anuarul Institutului Geologic, v. 37, p. 71–117.

Popa, Gh., and D. Paraschiv, 1974, Tipuri de capcane combinate si subtile, campul petrolifer Bradet: Revista Mine, Petrol si Gaze, v. 12, p. 578–582.

Popescu, Gr., 1952, Zona flisului Paleogen intre Valea Buzaului si Valea Varbilaului: Dari de Seama ale Institutului Geologic, v. 36, p. 43–125.

Popescu, Gr., 1954, Cercetari Geologice in regiunea Govora-Rimnicu Vilcea–Olanesti (Depresiunea Getica): Dari de seama ale Comitetului Geologic, v. 38, p. 118–136.

Popescu, Gr., 1958, Contributii la stratigrafia flisului cretacic dintre V. Prahovei si V. Buzaului cu privire speciala asupra vaii Teleajenului: Studii si cercetari de geologie, v. 3, no. 3–4, p. 159–199.

Popescu, Gr., 1959a, Recherches geologiques dans la region de Ciocadia–Piticu–Baia De Fier (Depression getique): Comptes Rendus du Comite Geologique, v. 37–39, p. 179–182.

Popescu, Gr., 1959b, Recherches geologiques dans la region Govora-Rimnicu Vilcea-Olanesti (Depression getique): Comptes Rendus du Comite Geologique, v. 37–39, p. 117–123.

Popescu-Voitesti, I., 1910, Contribution a l'etude stratigraphique du Nummulitique de la Depression Getique (Roumanie occidentale): Annuaire de l'Institut Geologique de Roumanie, v. 3, no. 2, p. 98.

Proust, J. N., and A. Hosu, 1996, Sequence stratigraphy and

Paleogene tectonic evolution of the Transylvanian Basin: Sedimentary Geology, v. 105, p. 117–139.

Rabagia, T., and L. Matenco, 1999, Tertiary tectonic and sedimentological evolution of the South Carpathians foredeep: Tectonic versus eustatic control: Marine and Petroleum Geology, v. 16, p. 719–740.

Ratschbacher, L., H. G. Linzer, F. Moser, R. O. Strusievicz, H. Bedelean, N. Har, and P. A. Mogos, 1993, Cretaceous to Miocene and wrenching along the central South Carpathians due to a corner effect during collision and orocline formation: Tectonics, v. 12, p. 855–873.

Roure, F., E. Roca, and W. Sassi, 1993, The Neogene evolution of the Outer Carpathian flysch units (Poland, Ukraine, and Romania): Kinematics of a foreland/fold-and-thrust belt system: Sedimentary Geology, v. 86, p. 165–185.

Royden, L., 1988, Late Cenozoic tectonics of the Pannonian Basin system: AAPG Memoir 45, p. 27–48.

Sanders, C. A. E., 1998, Tectonics and erosion, competitive forces in a compressive orogen. A fission-track track study on the Romanian Carpathians: Ph.D. thesis, Vrije Universiteit, Amsterdam, 204 p.

Sandulescu, M., 1967, La nappe de Haghimas–une nouvelle nappe de decollement dans les Carpates Orientales: Association Carpatho-Balcanique, VIII Congress, v. 1, p. 179–185.

Sandulescu, M., 1975a, Studiul geologic al partii centrale si nordice a sinclinalului Haghimas: Anuarul Institutului de Geologie si Geofizica, v. 45, 200 p.

Sandulescu, M., 1975b, Essai de synthese structurale des Carpathes: Bulletin de la Societe Geologique de France, v. 17, p. 258–299.

Sandulescu, M., 1976, La correlation structurale de troncon oriental avec celui meridional des Carpates Roumaines: Dari de seama ale Institutului de Geologie si Geofizica, v. 62, no. 5, p. 173–189.

Sandulescu, M., 1984, Geotectonica Romaniei: Editura Tehnica, 336 p.

Sandulescu, M., and J. Sandulescu, 1965, Les nappes interne de la zone du flysch dans la partie central des Carpathes Orientales: Association Geologique Carpatho-Balcanique, VIII Congress, p. 351–356.

Sandulescu, M., and M. Visarion, 1984, La structure des plateformes situees dans l'avant pays et au-dessous des nappe du flysch des Carpathes Orientales: Bucharest, Studii tehnice si economice, Institutului Geologic al Romaniei, p. 62–68.

Sandulescu, M., and M. Visarion, 1978, Consideration sur la structure tectonique du sou-bassement de la depression de Transylvanie: Dari de seama ale Institutului de Geologie si Geofizica, v. 64, p. 153–173.

Sandulescu, M., M. Micu, and B. Popescu, 1977, La structure et la paleogeographie des formations miocenes des Subcarpathes moldaves: Carpatho-Balkan Geological Association, XI Congress, v. Tectonics, p. 184–197.

Sandulescu, M., M. Stefanescu, A. Butac, I. Patrut, and P. Zaharescu, 1981a, Genetical and structural relations between flysch and molasse (the East Carpathians model): Romania, Institute of Geology and Geophysics, 95 p.

Sandulescu, M., H. G. Krautner, I. Balintoni, D. Russo-Sandulescu, and M. Micu, 1981b, The structure of the East Carpathians (Moldavia–Maramures area): Bucharest, Institute of Geology and Geophysics, 92 p.

Stefanescu, I., M. Popa, E. Ionescu Balea, M. Stefanescu, and N. Socolis, 1980, Petroleum republican museum of Ploiesti: Bucharest, Ministry of Petroleum and Geology, 99 p.

Stefanescu, M., 1967, Les nappes interne du Flysch dans l'extremite nordique des Carpathes Orientales: Association Geologique Carpatho-Balkanique, VIII Congress, v. 1, p. 187–192.

Stefanescu, M., 1971, Structura geologica a regiunii dintre valea Talea si valea Ialomitei: Dari de seama ale Institutului Geologic, v. 57, no. 5, p. 191–219.

Stefanescu, M., and N. Baltes, 1996, Do hydrocarbon prospects still exist in the East-Carpathians Cretaceous flysch nappes?: Memoires du Museum National d'Histoire Naturelle, v. 170, p. 427–438.

Stefanescu, M., and O. Dicea, 1995, Multiple thrust events at the Carpathian orogen/foreland contact in Romania (abs.): AAPG Bulletin, v. 79, no. 8, p. 1250.

Stefanescu, M., and M. Micu, 1987, Flysch deposits in the East Carpathians: Romania Academy of Science, special issue on flysch deposits from the Hartz, the Thuringian and Voghtlandian Slate Mountains, the Carpathians, the Balkans and the Caucasus, p. 65–98.

Stefanescu, M., and P. Polonic, 1996, Deep structure in Romania and the related hydrocarbon prospects: First Break, v. 2, no. 3, p. 99–105.

Stefanescu, M., and Working Team, 1986, Elaborarea profilelor geologice, scara 1:200,000 pentru completarea imaginii geologice a teritoriului Romaniei: Romania, Institutul de Geologie si Geofizica.

Stefanescu, M., L. Szasz, E. Bratu, M. Stefanescu, A. Rusu, and A.-M. Piliuta, 1986, Geology of the Titesti depression: Dari de seama ale Institutului de Geologie si Geofizica, v. 70–71, no. 4, p. 133–146.

Stefanescu, M., O. Dicea, and G. Tari, 2000, Influence of extension and compression on salt diapirism in its type area, East Carpathian Bend area, Romania: Geological Society (London) Special Publication 174, p. 131–147.

Stefanescu, S., G. Munteanu Murgoci, L. Mrazec, and V. Tacit, 1905, Map of Romanian oil areas: Oil Official Gazette.

Streckeisen, A., 1934, Sur la tectonique des Carpathes meeridionales: Annuaire de l'Institut Geologique, v. 16, p. 327–417.

Szasz, L., 1974, Pozitia stratigrafica a "gresiilor si conglomeratelor de Prislop" in estul bazinului Borsa si unele consideratii asupra Neocretacicului din Maramures si muntii Birgaului: Dari de seama ale Institutului Geologic, v. 60, no. 5, p. 143–164.

Szasz, L., 1975, Biostratigrafia si paleontologia Cretacicului superior din Bazinul Brezoi: Dari de seama ale Institutului Geologic, v. 62, p. 189–220.

Tari, G., O. Dicea, J. Faulkerson, G. Georghiev, S. Popov, M. Stefanescu, and G. Weir, 1997, Cimmerian and Alpine stratigraphy and structural evolution of the Moesian Platform (Romania/Bulgaria): AAPG Memoir 68, p. 63–90.

Vancea, A., 1960, Neogenul din bazinul Transilvaniei: Bucharest, Editura Academiei Romane, 262 p.

THE GEOLOGY AND HYDROCARBON RESOURCES OF THE INNER CARPATHIANS AND THE PANNONIAN BASIN IN THE TERRITORY OF SLOVAKIA AND HUNGARY

18

Janočko, J., M. Pereszlényi, D. Vass, V. Bezák, S. Jacko Jr., S. Jacko, M. Kohút, M. Polák, and J. Mello, 2006, Geology and hydrocarbon resources of the Inner Western Carpathians, Slovakia, and Poland, *in* J. Golonka and F. J. Picha, eds., The Carpathian and their foreland: Geology and hydrocarbon resources: AAPG Memoir 84, p. 429–463.

Geology and Hydrocarbon Resources of the Inner Western Carpathians, Slovakia, and Poland

J. Janočko
Department of Geology and Mineralogy, Technical University, Košice, Slovak Republic

M. Pereszlényi
Eurogeologic a.s., Bratislava, Slovak Republic

D. Vass
Faculty of Forestry, Technical University, Zvolen, Slovak Republic

V. Bezák, M. Kohút, M. Polák, J. Mello
Geological Survey of Slovak Republic, Slovak Republic

S. Jacko Jr.
Department of Geology and Mineralogy, Technical University, Košice, Slovak Republic

S. Jacko
Department of Geology and Mineralogy, Technical University, Košice, Slovak Republic

ABSTRACT

The Western Carpathians are a part of the extensive Alpine–Carpathian mountain system composed of the Western and Eastern Alps passing eastward into the Carpathians and Dinarides. The Western Carpathians represent the northernmost part of the Alpine orogen adjacent at its foredeep to the North European and Russian platforms. They are divided into two belts: the Outer Western Carpathians, consisting mostly of Neoalpine nappes, and the Inner Western Carpathians, with essentially a Paleoalpine structure overlain by Tertiary postnappe deposits. The Hercynian basement of the Inner Western Carpathians is covered by late Paleozoic and Mesozoic rock sequences that either are autochthonous or form allochthonous nappes. The tectogenesis of the Tertiary postnappe basins is mainly related to the convergence of the Carpathian–Pannonian block and North European lithospheric plate, the tectonic escape of lithosphere fragments from the Alpine realm, as well as the rise of the Pannonian asthenolite.

The Paleozoic units of the Inner Western Carpathians have their hydrocarbon potential practically exhausted. The oldest Paleozoic units of the Tatricum, Veporicum, and Gemericum are altered by different grades of metamorphism. The Mesozoic units are the most prospective

DOI:10.1306/985620M843078

in the western part of the region, where total possible resources of natural gas are estimated to be about 50 × 10^9 m^3 (1.76 × 10^{12} ft^3). The highest potential for hydrocarbon exploration has Tertiary basins represented by Inner Carpathian Paleogene basins and Neogene basins, particularly Vienna, Danube, and the East Slovakian basins. Although knowledge on the Neogene basins is relatively good (existence of three-dimensional seismic data and many boreholes), the area of the Inner Paleogene basin is still at the early stage of prospection.

INTRODUCTION

Eastern continuation of the Alpine mountain system is represented by the Carpathian Mountains. The Western Carpathians, stretching from the Eastern Alps in Austria to the easternmost part of Slovakia where they pass into the Eastern Carpathians, form the northernmost west–east-trending orocline of European Alpides (e.g., Plasienka et al., 1997) (Figure 1). The typical structural pattern of the Western Carpathians and their arcuate shape resulted from complex Mesozoic–Tertiary orogenic processes related to the convergence and collision between the stable North European platform and the Carpathian–Pannonian plate. The Western Carpathians are divided into Outer and Inner Western Carpathians separated by the Pieniny Klippen Belt (Figure 1). The Outer Western Carpathians consist of Neoalpine nappes, whereas the structure of the Inner Western Carpathians is essentially Paleoalpine and was been formed before the Paleogene (e.g., Biely et al., 1996).

The Outer Western Carpathians form the northernmost zone of Carpathians. They predominantly consist of deep-water, Jurassic to Oligocene deposits (flysch deposits) forming external fold-thrust belts. A minor part of the Outer Western Carpathians is composed of shallow-marine deposits belonging to the late Tertiary Carpathian Foredeep. Based on lithostratigraphy, several precursor basins were divided in the thrust belts from south to north represented by the Magura, Dukla, Silesian, Subsilesian, and Skole basins. The sedimentary fill of these basins has been arranged into a series of north- to northeast-vergent stacks of nappes overlapping the foredeep deposits. The Outer Western Carpathians are separated from the Inner Western Carpathians by the Pieniny Klippen Belt, which is traditionally interpreted as an expression of a suture after the closure of the Penninic-related Vahicum Ocean. The belt is composed of Jurassic to Paleogene deposits, which are strongly deformed in a typical klippen style.

The Inner Western Carpathians consist of three basic crustal-scale units represented from north to south by the Tatricum, Veporicum, and Gemericum (e.g., Mahel, 1986; Plasienka, 1997) (Figure 2). Each unit is formed by early Paleozoic crystalline basement and its late Paleozoic and Mesozoic sedimentary cover overlain by a system of Mesozoic nappes. This is partly covered by postnappe Late Cretaceous and Tertiary sedimentary and volcanic formations. The Paleogene deposits are represented by the Inner Carpathian Paleogene basin-fill deposits attached to the southern margin of Pieniny Klippen Belt and by the deposits of the Hungarian (Buda) Paleogene basin. Neogene deposits mainly occur in the four largest Slovakian Neogene basins: Vienna, Danube, South Slovakian, and East Slovakian basins. Minor Neogene deposits also occur in the Inner Carpathian intramountain depressions.

The main aim of the chapter is to describe the structure of Inner Western Carpathians and to show their hydrocarbon potential in relation to their structure. The geology of the Inner Western Carpathians is described chronologically from Paleozoic units through the Mesozoic to the Tertiary units.

GEOLOGY OF INNER WESTERN CARPATHIANS

Paleozoic Units of Inner Western Carpathians

The Hercynian basement in the Western Carpathians is part of three main Paleoalpine crustal units of the Inner Western Carpathians— the Tatricum, Veporicum, and Gemericum (Figure 2). These units, composed of crystalline basement and late Paleozoic–Mesozoic cover, were overthrusted above each other in the Cretaceous. The Tatricum and Veporicum units had a common development during the Paleozoic when they formed Tatric–Veporic domain. In contrast, the Hercynian units of Gemericum originated in a separate domain. The basement rocks of these two domains differ in age, premetamorphic composition, metamorphic T-P conditions, and evolution history, as well as in volume, age, and composition of granitic plutonism. From this point of view, the Tatric–Veporic basement belongs to the internal and the Gemeric basement to the external geodynamic realms of evolved European Hercynides (Bezak et al., 1997).

The internal structure of the Inner Western Carpathians crystalline basement is complicated because it involves several lithotectonic units. The superposition of the crystalline complexes essentially dates back to as early as the Hercynian tectonic processes, although, especially in the Veporicum, it was considerably rejuvenated or destroyed by Alpine disturbances.

Structurally, the basement of Tatric–Veporic zones consists of four south-vergent lithotectonic (nappe) units (Bezak et al., 1997) situated from the top to the bottom in the following order (Figure 3): (1) the low-grade metamorphites in the upper position, (2) the upper unit, (3) the middle unit, and (4) the lower unit.

1) The low-grade metamorphites in the upper position. The low-grade metamorphic, probably early Paleozoic rocks, with common greenschist facies metamorphism, only sporadically occur in the Tatric and Veporic basements. They are commonly associated with higher grade metamorphic units caused by tectonic deformation. The early Paleozoic age of the unit, based on palynological data, was confirmed at several localities (e.g., Klinec et al., 1975; Planderova, 1986). The rocks may have been napped during the late Hercynian stage (Bezak et al., 2002).
2) The upper unit. As follows from the reconstruction of tectonic relations, the high-grade upper unit commonly lies on medium-grade metamorphics. Structural research confirmed a ductile deformation with the southeast direction of thrusts (e.g., Fritz et al., 1992).

The unit consists of paragneisses, orthogneisses, and amphibolites. Metamorphism of the unit is high grade with abundant migmatitization phenomena. In the unit, relics of higher grade metamorphism (granulite facies) preceding the amphibolite facies metamorphism (e.g., Hovorka and Meres, 1989; Vozarova, 1993) as well as banded amphibolites (also referred to as the leptino-amphibolite complex) (e.g., Spisiak and Pitonak 1992; Janak et al., 1996; Hovorka et al., 1997) containing high-pressure relics, probably retrograded eclogites (Janak et al., 1997), were described.

The age of the older metamorphic event is not clear, and preliminary ages cluster around 390–420 Ma (Burchart, 1968; Cambel et al., 1990). The younger, medium-pressure metamorphism was replaced by contact effect of granitoids, which intruded 360–340 Ma. Cooling and exhumation of the whole basement, including the granitoids, occurred at 330–310 Ma (muscovite and biotite Ar/Ar data of Dallmayer et al., 1993; Maluski et al., 1993).

3) The middle unit. Lithologically, mica schist and gneisses are concerned here, along with amphibolites and orthogneisses, minor metaquartzites, graphitic gneisses, and metamorphosed Fe ores. The orthogneisses occur mainly in the hybridic complex of the Southern Veporicum.

The metamorphism of this unit was in amphibolite facies having two observable stages: a prograde phase at a higher amphibolite facies and a retrograde phase at a lower amphibolite facies (e.g., Siman et al., 1996). The end of the second phase coincides with granite intrusions.

An important feature of this unit, especially in the Veporicum, is the late Hercynian and Alpine retrograde reworking. This resulted in the origin of phyllonites in shear zones. Hercynian structures are commonly fully overridden by the Alpine structural plane.

Ar/Ar amphibole data indicate the end of tectonometamorphic processes about 350–340 Ma (Kral et al., 1996) when production of S-type granitoids culminated. Exhumation and cooling before some 300 Ma are documented by mica K/Ar and Rb/Sr ages (Cambel et al., 1990).

4) The lower unit. The unit, representing tectonically the lowest subautochthonous unit, comprises sediments of early Paleozoic age deposited on continental crust probably belonging to the southern Cadomian basement. The metamorphism of sediments consisting of claystones containing basic volcanic rocks, carbonates, and graywacke did not exceed the greenschist facies condition. The unit was structurally and metamorphically reworked during the Alpine phase of the Inner Western Carpathians' evolution (e.g., Kovacik, 1991, 1996; Kral et al., 1996; Janak et al., 2001).

In the Gemericum, three main early Paleozoic units have been distinguished (e.g., Bajanik et al., 1984; Vozarova, 1993): the Gelnica, Rakovec, and Klatov groups.

The Gelnica Group represents a huge (approximately 4500–8000 m [14,700–26,200 ft] thick) volcanic-sedimentary succession related to the active continental margin (Vozarova, 1993). In the vertical profile of the unit, three local formations have been distinguished (e.g., Bajanik et al., 1983). The basal part of each of them generally consists of flysch mesorhythmic sandstone and claystone sediments alternating upward by mainly acidic volcanites (prevailingly volcanoclastics). On the top of the formations, black phyllites intercalated by metalydites and reef-provenance metacarbonate lenses prevail. Late Cambrian to Early Ordovician age of the formation is based on palynological data (e.g., Snopkova, 1964; Corna, 1972). Regional metamorphism of the Gelnica Group did not exceed the lower part of the greenschist facies (Sassi and Vozarova, 1987; Faryad, 1991; Vozarova, 1993). The Gelnica

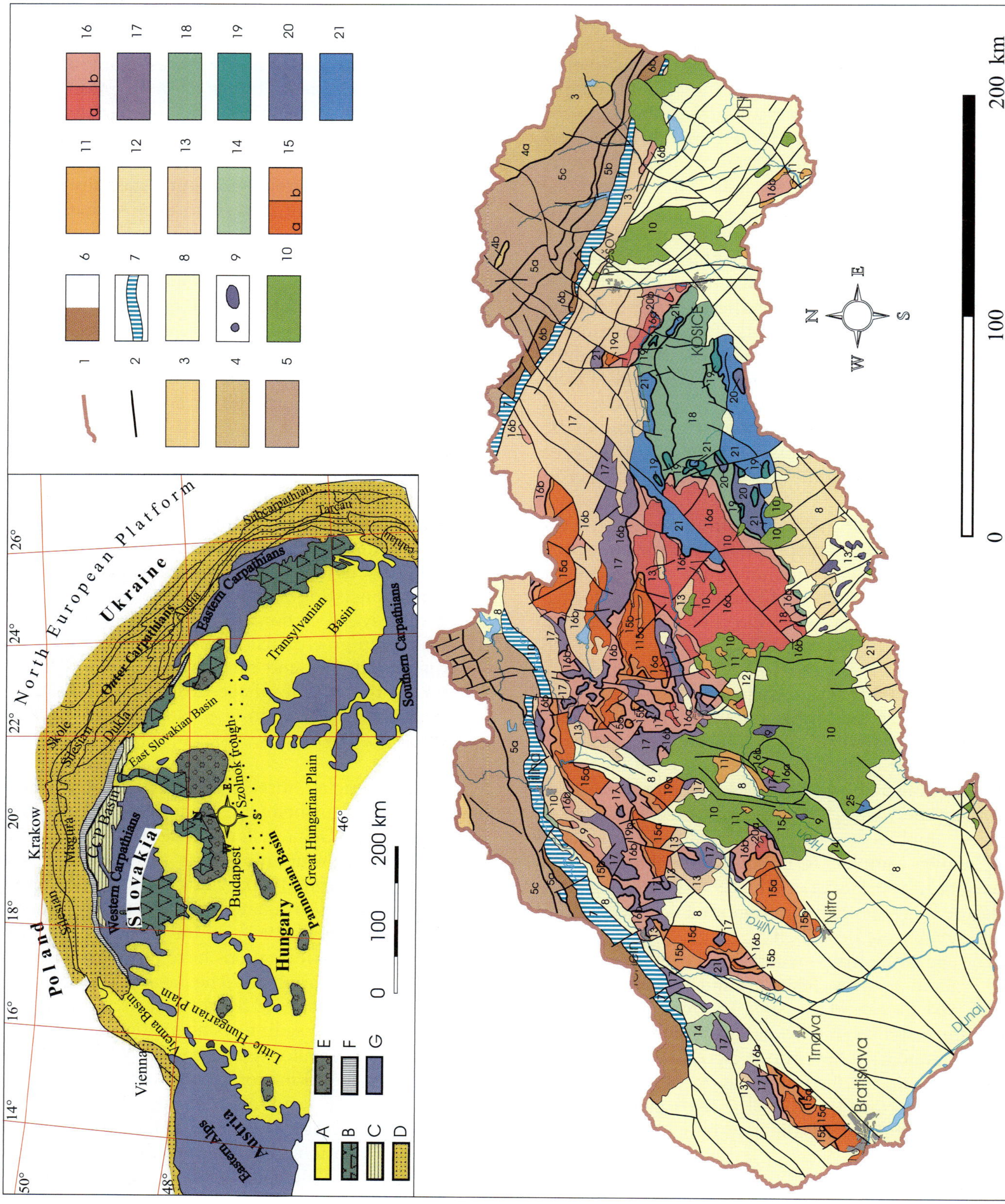
Poland
Ukraine
Slovakia
Hungary
Austria
North European Platform
Krakow
Vienna
Budapest
Western Carpathians
Outer Carpathians
Eastern Carpathians
Southern Carpathians
Eastern Alps
Vienna Basin
Little Hungarian Plain
Great Hungarian Plain
Pannonian Basin
Transylvanian Basin
East Slovakian Basin
Szolnok trough
Skole
Silesian
Dukla
Magura
Audia
Tarcău
Subcarpathian
0 100 200 km
A
B
C
D
E
F
G
Bratislava
Trnava
Nitra
Košice
Prešov
Dunaj
Váh
Nitra
Hron
0 100 200 km

Group of the Gemeric basement domain is exclusively penetrated by specialized (Ss type) leucogranites (e.g., Broska in Petrik et al., 2001) of the Permian–Triassic age (Poller et al., 2000), forming small bodies accompanied by tiny exocontact zones.

The Rakovec Group, overlying the northern flank of Gelnica Group, probably represents a remnant of back-arc basin oceanic crust (Ivan, 1994) filled mainly by tholeiitic metabasalts and underlain subordinately by metasandstone-phyllite horizon. The Middle and Late Devonian age of the 2500–3000-m (8200–10,000-ft)-thick unit, which is bare of fossils, is only based on its stratigraphic position.

The Klatov Group forms nappe klippes on either the Rakovec or Gelnica units in the northern part of the Gemeric basement zone. Composed mainly of amphibolites and gneisses with subordinate serpentinized peridotite bodies and sparse calc-silicate layers, it is the only unit of the Gemeric basement zone metamorphosed under amphibolite facies P-T conditions (e.g., Spisiak et al., 1985). The rock succession of the Klatov unit probably represents a metamorphosed lower continental crust (Hovorka et al., 1997) or island arc crust with some relics of the oceanic crust (Vozarova, 1993; Ivan, 1994), respectively. The clasts of the unit occur in the Westphalian conglomerates (Vozarova, 1993).

Different interpretation concerning evolution, lithostratigraphy, and structure of the Gelnica and Rakovec units was presented by Grecula (1982). He assumes a common early Paleozoic evolution of these units in a rift basin where the early Paleozoic Volovec volcano-sedimentary group originated. The base of the Volovec Group is generally formed by black phyllite formations intercalated by metalydites and metacarbonate lenses. This is generally overlain by the formation of green phyllites, tuff interlayers, and flysch sandstones. Spatially different fault activities in the basin have caused temporarily synchronous development of either acidic or basic volcanism in the basin, which terminated the basin fill. The basin-fill succession is structurally arranged into seven pre-Carboniferous fold nappes thrusted northward. Alpine destruction of the nappes is significant in shear and transpression zones of northwest–southeast and southwest–northeast directions (Grecula, 1997).

Late Paleozoic successions are represented in the northern part of the Gemericum by the Early Carboniferous Ochtina Group (Tournaisian–Visean) containing mainly deep-water facies and basic volcanics, as well as subsidiary acid volcanoclastics; volcanic-sedimentary formations of the Dobsina Group (Visean–Stephanian) formed in a foreland basin and, finally, the Krompachy Group (Early and Late Permian) consisting of terrestrial and volcanogenic formations (Vozarova and Vozar, 1988). In the southern part of the Gemericum, the late Paleozoic rocks are represented by Permian continental rift volcanic-sedimentary succession (Gocaltovo Group) overlying the early Paleozoic rocks by angular unconformity.

The important parts of Hercynian units are granitic rocks (Figure 2). The voluminous felsic Hercynian magmatism resulted in four main granite types in the Western Carpathian basement as a consequence of progressed tectonic evolution. Identified Devonian to older S-type granites (later transformed in orthogneisses), Early Carboniferous (Visean) S-type peraluminous granites, Late Carboniferous (Westphalian–Stephanian) calc-alkaline I-type granites, and Permian postorogenic subalkaline A/S-type granites in the Western Carpathian Belt are present (e.g., Petrik and Kohut, 1997).

In the Gemericum, only small apophyses emerge on the surface. A huge, composite granodiorite-tonalite intrusion, affected by the Alpine tectonics, dominates the Veporicum. In the Tatricum, these rocks together with metamorphites of the upper units form backbones of the so-called Core Mountains.

Mesozoic Units of the Inner Western Carpathians

Mesozoic units of the Inner Western Carpathians (Figure 1) are represented by sedimentary covers of Tatricum, Veporicum, and Gemericum domains, a system of nappes

Figure 1. Position of Western Carpathians in the Carpathian mountain system. Note the European platform in the north. The schematic geologic map of the Slovakia is also shown. A = Neogene sedimentary rocks; B = Neogene volcanics; C = Inner Carpathian Paleogene basin; D = Outer Flysch belt; E = buried Neogene volcanic rocks; F = Pieniny Klippen Belt; and G = inner units of Carpathians. 1 = State border; 2 = faults; 3 = Silesian nappe of the Outer Carpathian Flysch belt (OCF belt); 4 = Foremagura units of the OCF belt: 4a = Dukla unit; 4b = Grybow unit; 5 = external Magura nappes of the OCF belt: 5a = Bystrica unit; 5b = Krynica unit; 5c = Raca unit; 6 = internal Magura nappes of the OCF belt: 6a = Biele Karpaty unit; 6b = equivalents; 7 = Pieniny Klippen Belt; 8 = Neogene and Quaternary rocks; 9 = Alkali basalts; 10 = andesitic volcanic rocks (Neogene); 11 = rhyolitic volcanic rocks (Neogene); 12 = Eocene to early Miocene sediments of the Hungarian (Buda) basin; 13 = sediments of the Inner Carpathian Paleogene basin; 14 = sediments of the Gosau, Myjava, and Hricov groups; 15 = Tatricum: 15a = crystalline basement; 15b = sedimentary cover; 16 = Veporicum, Zemplinicum: 16a = crystalline basement; 16b = sedimentary cover and Krizna nappe; 17 = Hronicum; 18 = Gemericum; 19 = Meliaticum; 20 = Turnaicum; 21 = Silicicum.

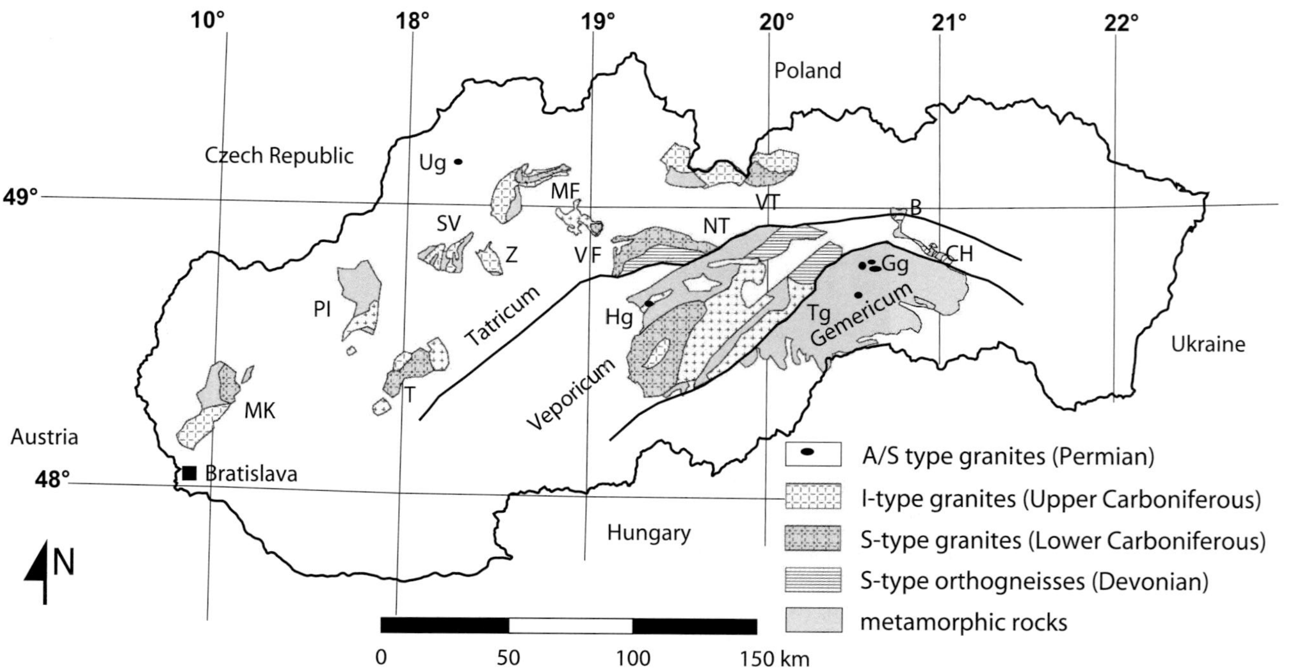

Figure 2. Location of the Western Carpathian core mountains and position of the Tatricum, Veporicum, and Gemericum units. Main locations of the Carpathian granites are also shown. OWC = Outer Western Carpathians; CWC = Central Western Carpathians; MK = Male Karpaty Mountains; T = Tribec Mountains; PI = Povazsky Inovec Mountains; SV = Strazovske vrchy Mountains; Z = Ziar Mountains; MF = Mala Fatra Mountains; VF = Velka Fatra Mountains; NT = Nizke Tatry Mountains; TM = Tatra Mountains; B = Branisko Mountains; CH = Cierna Hora Mountains; Ug = Upohlav granite; Hg = Hroncok granite; Tg = Turcok granite; Gg = Gemer granites.

(Fatricum and Hronicum), which partly overlie the sedimentary cover, and units of Meliaticum, Turnaicum, and Silicum. The structure of the Pieniny Klippen Belt separating the Outer and Inner Western Carpathians is also outlined here.

Pieniny Klippen Belt

The morphology and intensive tectonic deformation of the Pieniny Klippen Belt, which separates the Outer and Inner Western Carpathians, affected its unique position in the Carpathians. The belt (as much as 15 km [9 mi] wide and more than 600 km [370 mi] long) emerges from beneath the Vienna basin Neogene fill near Podbranc and extends as a narrow belt, arched to the north, between the Outer and Inner Carpathians as far as Romania (Figure 1). It consists of Jurassic and Cretaceous deposits prevailingly composed of limestone and tectonically incorporated Paleogene, mostly deep-water deposits. Typically, the Jurassic and Early Cretaceous carbonate sediments are strongly deformed into slabs or cliffs, which are chaotically arranged in the Late Cretaceous and Paleogene matrix.

The klippen belt comprises two distinct Jurassic–Cretaceous successions: the Czorstyn and Kysuca (Pieniny) successions, which were designated by Mahel (1986) as Oravicum units. The Czorstyn unit represents a pelagic swell environment with Jurassic crinoidal and nodular sandstones overlain by variegated marlstone facies (couches rouges) and distal flysch sediments (Albian to Maastrichtian). A typical hiatus in the Early Cretaceous occurred. The Kysuca (Pieniny) sequence is characterized by pelagic sedimentary succession capped by couches rouges and flysch complexes with frequent conglomerates.

In the Vah Valley and Orava region, the klippen belt structure also includes the Klape unit composed mostly of Cretaceous flysch formations. The Senonian part also has couches rouges layers. In the Vah Valley, the klippen belt comprises the Manin unit, which is mainly characterized by shallow-water limestone of Early Jurassic (Hierlatz Limestone) or Early Cretaceous age (Urgonian limestone) and by Cenomanian and Senonian flysch suggesting its Inner Carpathian origin.

The complicated structure of the Pieniny Klippen Belt hinders the unambiguous interpretation of its origin.

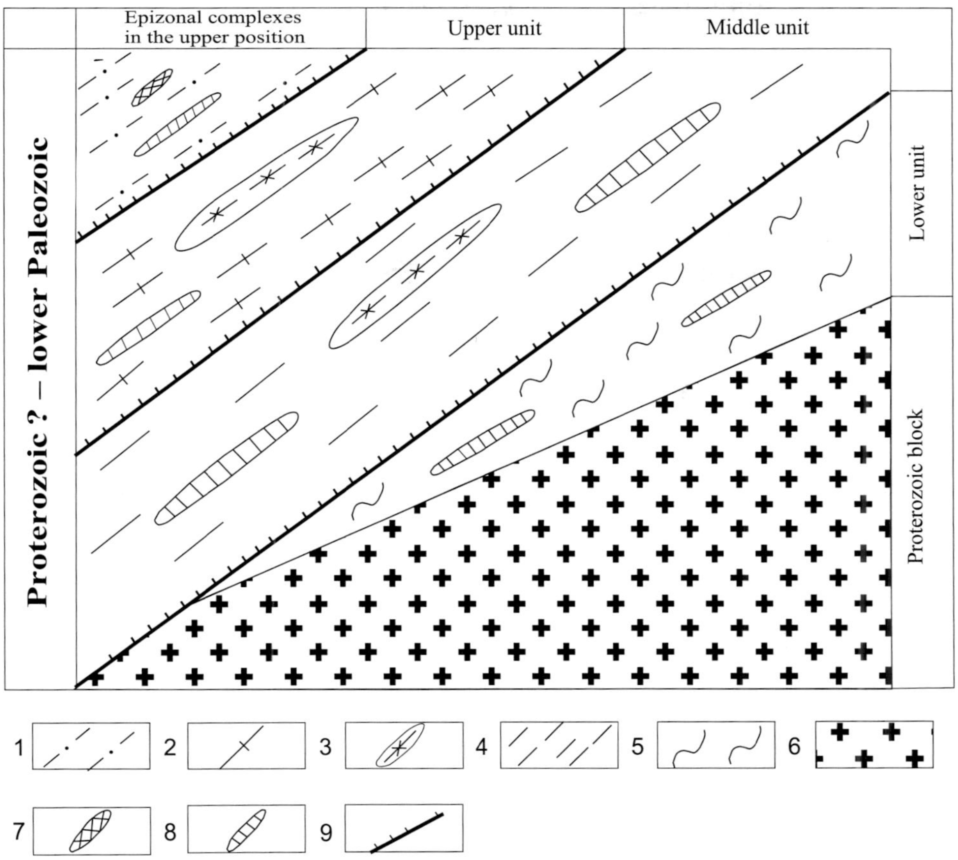

Figure 3. Hercynian lithotectonic units in the Alpine Tatric and Veporic units. 1 = Phylites and schists; 2 = paragneisses and migmatites; 3 = orthogneisses; 4 = mica schist, paragneisses, and products of their diaphtoresis; 5 = phyllites and mica schists; 6 = gneisses, amphibolites, and migmatites; 7 = metacarbonates; 8 = metabasics; 9 = overthrusts.

Palinspastic reconstruction reveals that the outermost section of the sedimentary basin was a Czorsztyn elevation, and zones situated farther to the south from the external basin edge comprised bathyal Pieniny–Kysuca trough. At least since the Albian, the Klape basin, which is linked by some authors to an ultrapieniny cordillera (e.g., Andrusov, 1968), points out that in the Cretaceous, the cordillera supplied exotic detritus, inclusive of Jurassic and Early Cretaceous acid, mafic, and ultramafic igneous rocks and glaucophanites.

The character of the Czorsztyn sequence suggests that it was deposited in a basin underlain by the continental crust. Some authors (Birkenmajer, 1977, 1986) assume that the Pieniny–Kysuce trough may have rested on a crust of oceanic character, but a similar sequence (Veporic Zliechov sequence) originated on a thinned but undoubtedly continental crust.

The position of the Pieniny Klippen Belt between the North European platform and the Carpathian–Pannonian block determined the extreme tectonic deformation of its sedimentary units. The present-day morphostructure of the belt is a result of a complex tectonic deformation with a substantial horizontal component of tectonic movement.

Tatricum

The Tatricum Mesozoic sedimentary cover consists of Siprun, Cervena Magura, and Vysoke Tatry groups, which differ in some details; however, the general sedimentary succession is similar in all three groups.

Triassic sedimentation began in all regions with deposition of the Lower Triassic fluvial and shallow-marine deposits of the Luzna Formation. It is composed of medium- to fine-grained quartzose sandstones and sandstones with conglomerate layers in the basal parts and prevailing pelitic sediments in the upper part. The thickness of the Luzna Formation ranges from 30 to 100 m (100 to 330 ft).

Platform carbonate sedimentation started in the Middle Triassic. It is represented by Anisian, as much as 150 m (500 ft) thick, dark-gray and black Gutenstein Limestone deposited in the deeper part of the basin. The Ramsau Dolomites (as much as 200 m [660 ft] thick) of Anisian to Ladinian age are gray to dark gray and mostly fine crystalline. They occasionally contain Dasycladaceae.

In the Upper Triassic (Norian), the Carpathian Keuper, as a continental facies, points to a regressive period. In the Tatric cover units, it mostly consists of variegated quartzose sandstones. The total thickness of the deposits does not exceed 120 m (400 ft). The uppermost Triassic (Rhaetian) deposits occur only scarcely.

The Jurassic sedimentary cycle in the Tatric cover sequences commences with transgressive deposits. The base is commonly composed of dark-gray, sandy, and sandy-crinoidal limestone, which vertically passes into crinoidal and fine-grained, locally organogenic limestone. They normally contain nodules and layers of dark and black cherts (Trlenska Formation) (Bujnovsky et al., 1979). The age of this transgressive sequence is Hettangian–Sinemurian and is from 50 to 150 m (150 to 500 ft) thick.

During the Lotharingian, the basin deepened, and the euxinic conditions resulted in the deposition of as much as 1000 m (3300 ft) of dark-gray to black shale with marly limestone intercalations that are defined as the Allgau Member (Fleckenmergel). These conditions persisted until the Toarcian. This stratigraphic interval is represented by the Hierlatz Formation in the Cervena Magura sequence. In the Vysoke Tatry sequence, it is represented by the Pisana Sandstone (Sinemurian–Domerian) and crinoidal limestone containing a large amount of detritic material. Subsidence of the basin continued and reached its maximum in Dogger, represented by only 20-m (66-ft)-thick silicified radiolarian limestone and radiolarite in the Siprun facies area. In contrast, the cephalopod limestone was deposited in the Cervena Magura sequence, and in the Tatric cover sequence of the Vysoke Tatry Mountains, gray nodular limestones of Bathonian to Callovian age are found. The overlying lithostratigraphic unit is represented by thin-bedded Aptychi–Sacocoma limestones, which only occasionally develop nodules. The age of the basal aptychi limestones is determined by microfauna as Kimmeridgian. They do not exceed 25 m (80 ft) and pass upward into gray marly Calpionella Limestones of Tithonian age. In the Vysoke Tatry sequence, limburgites and their tuffs are also found in the Tithonian.

The overlying Lucivna Formation (Berriasian–lower Aptian) contains gray marly limestone with nodules of dark-gray to black silicites and attains a maximum thickness of 150 m (500 ft). In the Vysoke Tatry Mountains, the Barremian–Aptian interval is composed of Urgonian limestone with a maximum thickness of 60 m (200 ft). The Aptian of the Siprun and Cervena Magura sequences is as much as 50 m (160 ft) thick and is represented by dark-gray to black, marly to cherty limestone with intercalations of marly shale. This sequence of variable limestone and marly shale grades into the overlying Poruba Formation. It is characterized by an increasing supply of clastic material and the appearance of sandy components in the limestones. It is transitional to the characteristic flysch-type sedimentation, with an alternation of sandstone and claystone. The age of Poruba Formation is Albian to lower Turonian (Vysoke Tatry Mountains), and the thickness of the sequence does not exceed 150 m (500 ft).

Veporicum

The Mesozoic succession of the Veporicum domain is subdivided into the northern Vel'ky Bok Group and the central and southern Federata Group. The Mesozoic succession of the Vel'ky Bok Group is characterized by Early Triassic to Early Cretaceous deposits, whereas the deposits of the Federata Group are only represented by Triassic deposits. The Veporicum also involves a system of superficial nappes overlying Tatricum (Figure 4), which were designated as Fatricum by Andrusov et al. (1973).

The Northern Veporicum Mesozoic Sedimentary Cover: Vel'ky Bok Group (Nizke Tatry, Branisko, and Cierna Hora Mountains). The Mesozoic Northern Veporicum sedimentary cover, represented by the Vel'ky Bok Group, occurs in the Nizke Tatry, Branisko, and Cierna Hora Mountains. The rocks, displaying dynamic metamorphism, are in a para-autochthonous position on the crystalline complexes.

The basal complexes are formed by the Lower Triassic Luzna Formation consisting of variegated quartzose sandstone and shale. The Middle Triassic has carbonate development. The Anisian Gutenstein Limestone attains a maximum thickness of 50 m (300 ft). The main rock type is the Anisian to Ladinian Ramsau Dolomite, containing intercalations of dark shale in the upper parts. Generally, the dolomites are strongly recrystallized. The lower Carnian [Norian(?)] is represented by the Carpathian Keuper, consisting of a complex of variegated shales, dolomites, and quartzites. Claystones are distinctly dynamically metamorphosed as shown by the linear distribution of fabric components. The Rhaetian sediments (Kossen Member) were only deposited in the Cierna Hora Mountains.

Distinct differentiation of the sedimentation, reflected in the variety of Jurassic lithofacies, began in the early Liassic. Lower Liassic sedimentation in the Branisko Mountains is characterized by cordillera facies (black crinoidal cherty limestones). In the lower part of the Vel'ky Bok Group, black marly shale with intercalations of crinoidal limestone occurs.

The Upper Liassic is represented by the Allgau Member. Already in the Dogger, variegated radiolarian

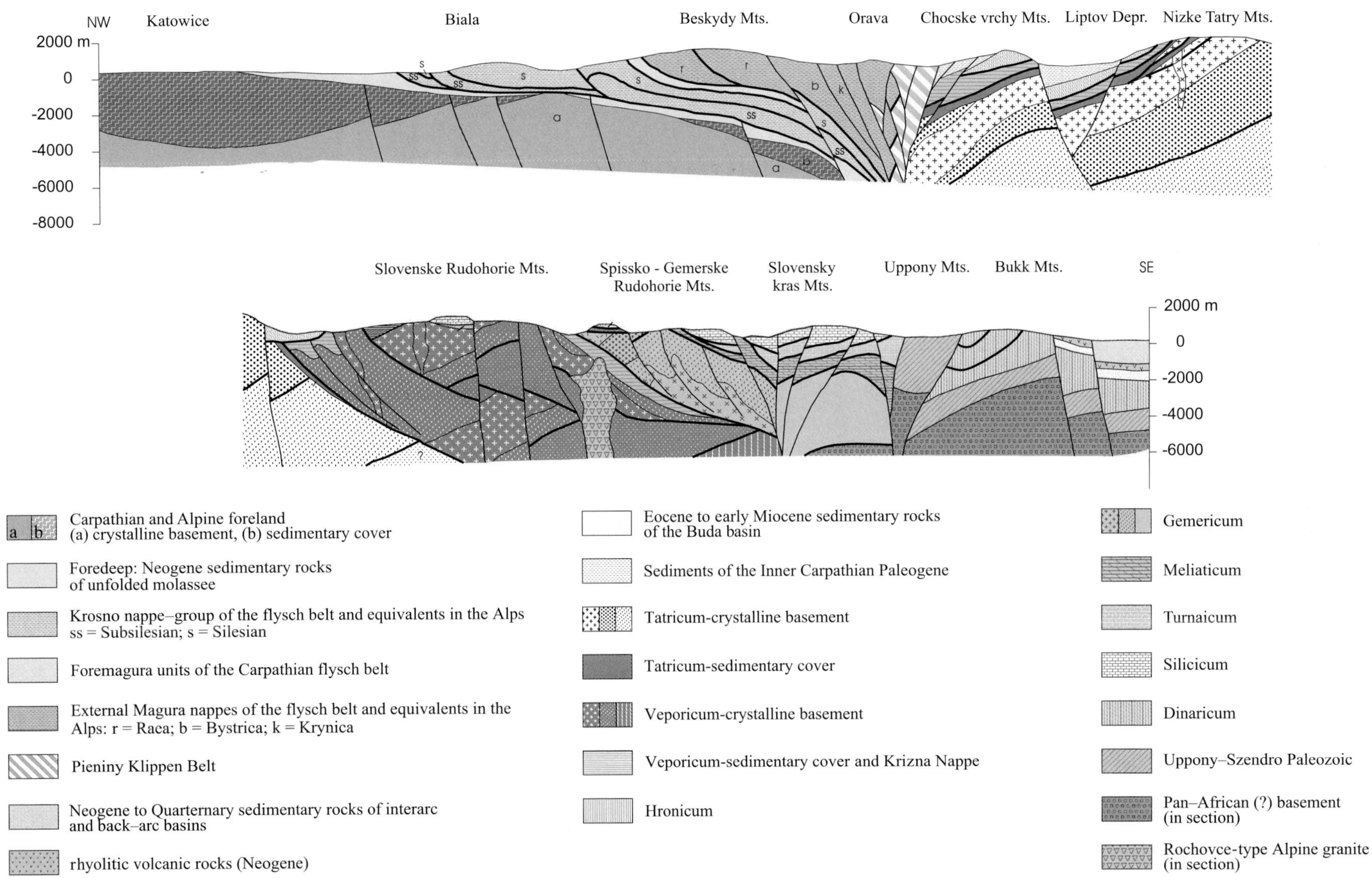

Figure 4. Schematic south–north section of the Western Carpathians (vertical scale exaggerated) (modified after Lexa et al., 2000).

limestones and radiolarites are known, which are overlain by light gray, pinkish, thin-bedded, marly to shaly limestones. Light-gray, weakly marly Calpionella Limestone, occurring in the Vel'ky Bok Group, corresponds to the Tithonian. The overlying sequence of gray, marly limestone, marl, and shale is of Berriasian–lower Aptian age. It is intensively folded and distinctly phyllitized.

Fatricum. The Fatricum, defined by Andrusov et al. (1973), is a system of detached nappes overlying the Tatricum (Figure 4). The sedimentation of the Fatricum commenced with the Early Triassic Luzna Formation, which has a maximum thickness of 20 m (66 ft). It is poorly preserved because of tectonic shearing, as the main detachment plane of the nappes lies above the Luzna quartzites in the Werfenian shales. Lithologically, it is very similar to the Luzna Formation in the Tatricum unit.

The Middle Triassic is characterized by shallow-water, carbonate platform sediments, the lower part being the Anisian Gutenstein Limestones.

The massive Ramsau Dolomite of Anisian to Ladinian age attains a maximum thickness of 200 m (660 ft). Julian is characterized by typical sediments of the Lunz Member with a maximum thickness of 20 m (66 ft). The upper part of the Carnian consists of 50–80 m (160–260 ft) of dolomite. The Norian is represented by the Carpathian Keuper Formation (as much as 100 m [330 ft] thick), which is composed of lagoonal and shallow-water deposits (variegated shales, primary dolomites, and occasional sandstone beds). The Rhaetian is represented by Kossen Member–Fatra Formation of dark organodetritic and commonly organogene (coral) limestones with intercalations of dark marly shales. The thickness of the Fatra Formation does not exceed 80 m (260 ft). The predominance of shale and dolomite in the Carpathian Keuper and Kossen Member in the Fatricum (Krizna nappe) is the main difference compared to the Tatric cover units, where Carpathian Keuper is typical by abundant sandstones.

The Jurassic sedimentary succession may be divided into deep-water (Zliechov) and ridge-type (Vysoka) developments. The sedimentary succession of widespread Zliechov facies starts with as much as 80 m (260 ft) of transgressive Kopienec Formation (Hettangian to Sinemurian). The sediments consist of sandstones, dark crinoidal limestones, shales, organodetritic limestones, and occasionally spongolites. The depositional environment was well aerated and strongly agitated with significant input of clastic material. The upper part of the Liassic (Lotharingian–Domerian) in the Zliechov facies contains 100 m (330 ft) of euxinic Allgau Member (Fleckenmergel), which consists of alternating dark, marly, spotty limestones and shales. The uppermost Liassic (Toarcian) is commonly represented by 15–20 m (50–66 ft) of locally deposited red nodular limestones (Adnet Member). This is overlain by Aptychi–Saccocoma limestones as much as 30 m (100 ft) thick that gradually developed from the radiolarites, showing variegated, gray, commonly nodular limestones of Kimmeridgian age. The Mesozoic sedimentary cycle of the Zliechov facies terminates with the deposition of the 200-m (660-ft)-thick Poruba Formation, represented by flysch sediments of Albian–Cenomanian age.

The sedimentary succession of limited Vysoka facies consists of crinoidal limestones and red nodular limestones of upper Liassic age. Some 30-m (100-ft)-thick red nodular (Czorsztyn) limestones of Kimmeridgian–Tithonian age are typical. The upper Berriasian–Barremian (Neocomian) is represented by marly limestones, shales, and marls of about 300-m (1000-ft) thickness. The Padla Voda Formation (Aptian), which is as much as 50 m (160 ft) thick, consists of massive black, organodetritic shallow-water limestones and marly limestones with siliceous cherts.

The Southern Veporicum Mesozoic Sedimentary Cover: Federata Group. Although many Mesozoic facies of the Northern and Southern Veporicum sedimentary cover are similar, the occurrence of the late Paleozoic Stephanian and Permian rocks and the absence of the Carpathian Keuper facies in the southern zone, as well as the intensity of the Alpine tectonometamorphic reworking (Meres and Hovorka, 1991; Plasienka, 1993), substantiate the division of the Veporicum sedimentary cover in its northern and southern part. The Mesozoic cover of the Southern Veporicum is represented by the Federata Group. It is only represented by the Triassic sediments. Low-grade metamorphic overprint is characteristic. The Early Triassic sediments are similar to the north Veporic and Tatric units: quartzose sandstones, sandstones, and shales (Luzna and Werfen formations). The Middle and Late Triassic rocks are represented by dolomites (rauwacks), dark and light-colored crystalline limestones (Anisian), clayey and siliceous cherty limestones (Ladinian), and clayey-sandy and calcareous shales with lenses of dark cherty limestones (Carnian), capped by dolomites equivalent to the Hauptdolomite (Norian). The thickness of the Federata Group is 200–450 m (660–1400 ft).

The Mesozoic rocks of the Zemplinicum, assigned to the Veporicum, are mainly represented by Lower Triassic quartzites and shales as well as by Middle Triassic epiplatform carbonate successions (Boucek and Pribyl, 1959; Grecula and Egyud, 1977, 1982; Vozarova and Vozar, 1988).

Hronicum

The Hronicum represents a system of imbricated nappes commonly only preserved as nappe remains in individual Inner Western Carpathian mountains (Havrila in Kovac and Havrila, 1998) (Figures 1, 4). The nappes are characterized by a unified development of Permian–Carboniferous sedimentary-volcanic formations, differentiated Oberostalpine type of Triassic, and only locally preserved Jurassic and Early Cretaceous formations. Based on facial analysis, two basins (Dobra Voda and Biely Vah basins) and two carbonate platforms (the Cierny Vah platform and the carbonate platform of higher nappes) were divided (Kovac and Havrila, 1998).

The lithologic character of the late Paleozoic sediments and volcanic rocks (Nizna Boca and Maluzina formations) suggests their formation in a rift basin developed on continental crust (Vozarova and Vozar, 1988, 1996). This interpretation is also supported by findings of tectonic fragments of mylonitized granitoids in the basal part of the lower partial nappe of the Hronicum (Andrusov, 1938; Vozarova and Vozar, 1979).

The Mesozoic sedimentary rocks are commonly represented by the clastic Sunava Formation consisting of Early Triassic quartzose sandstones, sandstones, and shales. The Middle Triassic commences by platform Gutenstein Dolomites and early Anisian limestones. The basic architectural element is composed of Ramsau Dolomite of late Anisian–early Ladinian age. Typical deep-water deposits are represented by late Anisian and Ladinian Reifling and Scheyeralm limestones. In some parts, pale Wetterstein limestones represent this time horizon. The lower part of the Carnian mostly consists of the clastic Lunz Member. The upper Carnian and Norian is made up of thick and massive Hauptdolomite, which represents, in most of the Core Mountains, the uppermost lithostratigraphic unit of the Hronicum.

The upper stratigraphic units are only developed in the Strazov Mountains and in the Nizke Tatry Mountains. The Jurassic sedimentary cycle commences by Hettangian and Sinemurian gray crinoidal limestones with spongoliths. This is overlain by Domerian–Toarcian pink crinoidal Hierlatz Limestones. The late Bathonian and Callovian are represented by a marker horizon of radiolarite limestones. The uppermost part is composed of red nodular limestones overlain by Tithonian gray Calpionella Limestones passing into gray marly limestones and marlstones of the Berriasian and Hauterivian. The youngest formation is represented by Barremian and early Albian organodetritic limestones.

Gemericum

The Mesozoic sedimentary cover of the Gemericum is poorly preserved. It consists of lagoonal, sebkha, and shallow-marine Scythian deposits (Kosa and Janocko, 1999) and presumably Middle Triassic carbonate platform deposits that occur in the western part of the Gemericum beneath Borka nappe.

Meliaticum

The Meliaticum is defined as a tectonic unit derived from an oceanic or paraoceanic domain closed in the Late Jurassic. The unit crops out in several tectonic windows from beneath the Silicic nappe and in a discontinuous system of slices bordering the southern margin of the Gemeric Paleozoic rocks (Figure 1). Lithologically, the Meliaticum is defined as a succession of several hundred meters of thick Jurassic (Bathonian–Callovian) dark clayey shales alternating with thin beds of dark and greenish-red radiolarites and spotted marlstones with sandstones (Mello et al., 1996; Mock et al., 1998). The succession contains huge olistostromes composed of Triassic rocks, mostly limestones, radiolarites, and volcanics. The whole sedimentary succession is strongly tectonically deformed.

According to Mello et al. (1996) and Mock et al. (1998), the Meliaticum contains two subunits. The first one is represented by the Meliata Group sandstones and represents relics of sediments and volcanics formed in a basin with an oceanic crust. The second unit, which an immediately overlies the Gemeric Paleozoic and partly also Mesozoic rocks, represents a remnant of the subduction-accretionary complex between Gemericum and Meliaticum designated as the Borka nappe (Lesko and Varga, 1980; Mello et al., 1996, 2000). It comprises late Paleozoic–Mesozoic metamorphic successions occurring along the northern margin of the Slovensky Raj Mountains between the Jasov and Sirk villages and in the Nizna Slana Depression. The characteristic feature of these rocks is Alpine, medium- to high-pressure metamorphosis at low thermal gradient (Mazzoli et al., 1992).

Turnaicum

The Turnaicum, represented by the Turna and Slovenska Skala nappes, occurs above the Meliaticum and beneath the Silicicum in the area of Slovak Karst. It contains rocks with a stratigraphic range from the middle Carboniferous up to the Late Triassic or even up to the Jurassic (Mello et al., 1996). A characteristic feature of the Turnaicum is a metamorphic reworking of rocks in anchizone up to greenschist conditions. The Turnaicum is also known from the Hungarian territory, where it participates in the structure of the Ruda Banya Mountains.

The oldest parts of the Turnaicum are made up of a middle Carboniferous flysch unit (Turiec Formation) and a continental, conglomeratic-sandstone Permian unit (Brusno Formation). The Early Triassic is represented by the Werfenian unit: variegated sandstones and shales in the lower part and gray shales, marls, and limestones in the upper part, resembling the sedimentary development in the Silicicum. In contrast to the Silicicum, the facies of initial Anisian carbonate platform were not rebuilt after the collapse in the Pelsonian, resulting in the development of slope and deep-water facies in the Middle and Late Triassic. These facies are represented by Reifling and Potschen gray cherty limestones. Alkaline volcanic products are known from the Carnian.

Silicicum

The Silicicum is the uppermost and innermost unit of the Inner Western Carpathians. It is thrusted over the Turnaicum and Meliaticum as well as over the Gemericum, Hronicum, and Veporicum in the north. The northernmost part of the Silicicum is represented by Drienok nappe located south of the Banska Bystrica. The Silicicum is mainly represented by Scythian shallow-marine deposits and Triassic platform carbonates of the outer shelf environment. The youngest known sediments are Oxfordian radiolarites and olistostromes.

Late Cretaceous Postnappe Formations

The deposits of the Late Cretaceous postnappe formations have only minor occurrences in the Inner Carpathians (Figure 1). A thick succession of Cretaceous sediments occurs in the Brezovske Karpaty Mountains, where they comprise the Brezova Group. These rocks are correlated by some authors to the Gossau Cretaceous sediments of the Eastern Alps (e.g., Wagreich, 1995). The sedimentary succession consists of Campanian flysch containing red marlstones, conglomerates, and orbitoid limestones.

Santonian and Campanian marine limestones and marlstones as much as 100 m (330 ft) thick intercalated with sandstones, and sparse conglomerates occur in the surroundings of the Stratena, Sumiac, and Cierna Luka villages and Miglinc Valley. These sediments overlie the substratum commonly represented by Silicicum.

Tertiary Units of the Inner Western Carpathians

Tertiary units of the Inner Western Carpathians are represented by deposits of the Inner Western Carpathian Paleogene basin and deposits of Neogene basins and intramountain depressions. Deposits attached to the southern margin of the Pieniny Klippen Belt and deposits of the Hungarian (Buda) Paleogene basin related to the Pannonian domain also have minor occurrences.

Paleogene Deposits Attached to the Southern Margin of the Pieniny Klippen Belt

The southern margin of the Pieniny Klippen Belt is discontinuously rimmed by Paleocene to Eocene deposits defined as the Myjava–Hricov–Zilina facies (Andrusov, 1965) or sediments of the Povazie–Hanusovce belt (Samuel, 1972). The variable thickness (with a maximum of some 1200 m [4000 ft]) of prevailingly deep-water sediments and their facial features (mostly gravity-flow deposits with large olistostromes) suggest intensive subsidence typical for marginal troughs of mobile mountain belts (e.g., Marschalko and Samuel, 1993). The subsidence was probably related to strike-slip movements along the Pieniny Klippen Belt (e.g., Ratschbacher et al., 1993). Although the basin origin is still under discussion, one of the explanations is that during the Paleocene and Eocene, the area of the future Pieniny Klippen Belt marking the suture zone between the Carpathian–Pannonian block and the North European platform was rimmed by several small strike-slip basins characterized by fast subsidence and uplift (Janocko, 2002). The continual process of oblique convergence of the North European platform and the Carpathian–Pannonian block resulted in the destruction of older strike-slip basins, their incorporation into their own (Mesozoic) klippen belt, and the opening of new basins southward.

Inner Carpathian Paleogene Basin

The Inner Carpathian Paleogene basin lies in the northern part of the Slovakian Inner Western Carpathians (Figure 5). To the west and south, it neighbors the pre-Paleogene, Mesozoic, and Paleozoic formations of the Inner Carpathians. In the southeast, it passes into the East Slovakian basin. In the north, the Pieniny Klippen Belt separates it from the Outer Carpathian Flysch zone. The basin evolution was affected by the convergence between the Carpathian–Pannonian plate and the North European platform (Figure 1), the pull of the subducted slab, and the sediment load. The elongated,

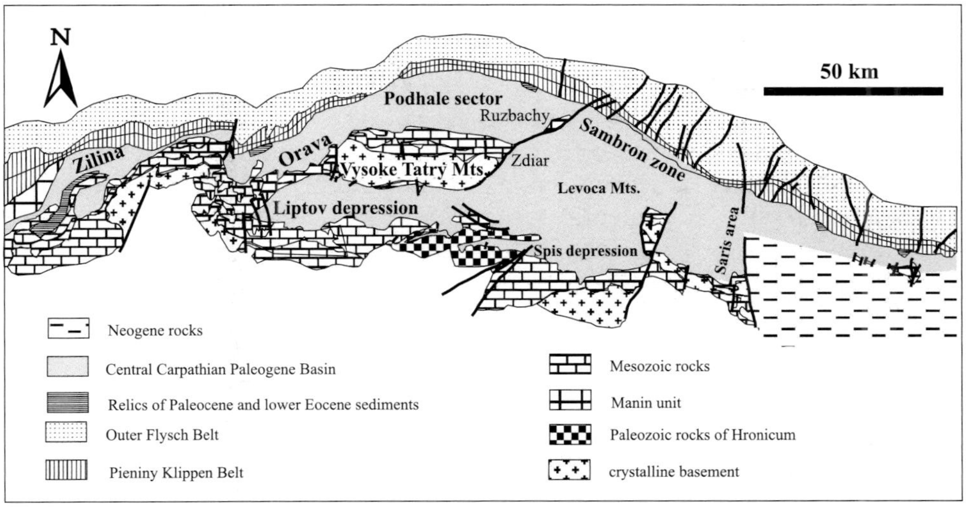

Figure 5. Geologic position of the Inner Carpathian Paleogene basin.

crescent-shaped basin is about 200 km (120 mi) long and about 50 km (30 mi) wide. The northernmost part of the basin extends to Poland, where it is called the Podhale basin (e.g., Westwalewicz-Mogilska, 1986; Wieczorek, 1989; Kepinska, 1998) (Figure 5). The maximum thickness of the basin fill with a stratigraphic range of Eocene–early Miocene is as much as 3500 m (11,500 ft), and its local preservation depends on the postdepositional tectonic history. The clay mineral analyses and vitrinite reflectance data suggest a removal of 1.5–3 km (0.9–1.8 mi) of deposits after deposition because of basin uplift in the Neogene and Quaternary (Francu and Muller, 1983; Kotulova et al., 1998; J. Milicka, 1999, personal communication).

Conventionally, the opening of the Inner Carpathian Paleogene basin is assumed to have occurred in the middle Eocene (e.g., Gross et al., 1980) resulting in a southward-directed sea transgression. The oldest, middle Eocene deposits occur in the northernmost part of the basin, rimming the Pieniny Klippen Belt. However, continuation of the convergence mechanism between the Carpathian–Pannonian block and the North European platform, already acting in the Paleocene, suggests a relation between the Paleogene rocks attached to the southern margin of the Pieniny Klippen Belt and the Inner Carpathian Paleogene basin. We assume that the prevailingly strike-slip movements during Paleocene and early Eocene, active along the Pieniny Klippen Belt (e.g., Ratschbacher et al., 1993; Nemcok and Nemcok, 1994), changed into an extensional regime since the middle Eocene. During this regime, the Inner Carpathian Paleogene basin widened southward, and the sea gradually flooded older terrestrial deposits.

The deposits of the Inner Carpathian Paleogene basin are formally assigned to the Subtatric Group in the Slovakia, which consists of four formations and several lower lithostratigraphic units (Gross et al., 1984; Filo and Siranova, 1996, 1998). According to previous research (Gross et al., 1984), the formations are superimposed on each other. However, a detailed analysis of the basin fill showed rather lateral relationships between the formation deposits (e.g., Janocko et al., 1998; Janocko and Jacko, 1999; Janocko, 2002) comprising individual facies tracts. Most of the sediments are represented by turbidites that originated in a deep-water environment. The shallow-water and terrestrial deposits form the base of the sedimentary basin fill, and they crop out at the basin margin.

The lowermost Borove Formation (Nummulite Eocene in the Polish, Podhale part of the basin; Golab, 1959; Olszewska and Wieczorek, 1998) consists of basal terrestrial and shallow-marine deposits, which are as much as 200 m (660 ft) thick. The age of the formation, based on large foraminifera analyses, varies from middle Eocene to early Oligocene. Genetically, the deposits are represented by terrestrial slope deposits, fluvial and alluvial-fan deposits, and fan-deltaic and shallow-marine deposits.

The terrestrial slope deposits consist of weathered rocks derived from the immediate basement. The sediments are massive or crudely bedded, and they were commonly reworked when flooded by a transgressing sea. Occasionally, fluvial deposits filling paleovalleys occur. Such deposits, mostly deposited by ephemeral streams, were defined as the Hornad Member by Filo and Siranova in 1998. Their thickness is as much as 150 m (300 ft), and they laterally pass into shallow-marine or fan-deltaic deposits. They consist of rocks derived from both local bedrocks and from far provenances. Rugged, high-energy relief along the coastline of the basin is indicated by the development of coarse-grained, fan-deltaic deposits. Well-exposed deposits show alternation of deltaic and shoreface facies, suggesting shift of deltaic lobes or oscillation of relative sea level. A considerable part of the Borove Formation is represented by shallow-marine deposits. These deposits, originally defined as the Chrast Member and interpreted as deltaic deposits by Filo and Siranova (1998), are characterized by large-scale planar and trough cross-stratified sandstones (Janocko, 2002). According to some boreholes, the uppermost part of the Borove Formation is formed by medium- and fine-grained sandstones and siltstones defined by Filo and Siranova (1996) as the Tomasovce Member (Figure 6). The sediments, as much as 150 m (300 ft) thick, are massive and occasionally fill large and shallow scours. They are interpreted as distal parts of deltas entering the basin (e.g., Janocko, 2002).

The Huty Formation (late Eocene–Oligocene, Zakopane Formation in the Polish part of the basin; Golab, 1959; Olszewska and Wieczorek, 1998) consists of dark shales with interbedded thin sandstone and conglomerate beds. The formation overlies the basal Borove Formation with sharp and transitional boundary and is deposited below the wave base on slope and basin plain. The thickness of deposits is as much as 3000 m (10,000 ft) (Figure 7).

Dark, black, and blackish brown shales have massive appearances, and in thin sections, they occasionally show faint parallel lamination. The carbonate content is low, having a mean value of 7%. The mean total organic carbon (TOC) content is 0.7%, with a maximum of 1.5% (J. Milicka, 1999, personal communication). The shales alternate with thin sandstone beds and occasional conglomerate beds. The medium- and fine-grained lithic sandstone is massive, sparsely parallel, and ripple cross-laminated, showing Bouma's $T_{b,c,d}$ and $T_{c,d}$ divisions (facies F9 of Mutti, 1992). The sharply based beds are 5–30 cm (2–12 in.) thick. The conglomerate beds are sharply bounded, and their thickness varies from several centimeters to a few meters. The conglomerates are internally disorganized, mostly showing massive structure and only occasionally horizontal- or inclined-bedded. The roundness, grain size, and composition of clasts strongly vary in individual conglomerate beds.

The shales of the Huty Formation also contain conglomerate bodies believed to be the fill of submarine canyons, and this is defined as the Pucov Member (Gross et al., 1984) (Figure 6). The best outcrops of such bodies occur in the surroundings of the Pucov and Zdiar villages (e.g., Marschalko and Radomski, 1970; Gross et al., 1984; Janocko and Jacko, 1999; Janocko et al., 2000). Coarse-grained conglomerate and sandstone deposits, as much as 200 m (660 ft) thick and as much as 2 km (1.2 mi) wide, are incised into the underlying shales and sometimes even into deposits of the Borove Formation and pre-Tertiary basement. The clast size of conglomerates is variable and does not show any trend along the profile. The base of thick beds is sharp and/or scoured, and erosive bases are less frequent. The conglomerates are prevailingly normally graded; inverse grading and massive structure are also locally developed. The conglomerate beds are separated by pebbly sandstones and medium- to coarse-grained sandstones. The sandstones are horizontally and ripple cross-laminated, and some beds are normally graded (facies F4, F5, and F6 of Mutti, 1992). Water escape structures are common.

The Huty Formation also includes the deposits of the Sambron Member defined by Chmelik (1959) and commonly correlated with the Szaflary Member in the Polish part of the basin. They are characterized by coarse-grained, mostly gravity-flow deposits forming thick intervals in surrounding shales (e.g., T. Durkovic, E. Lukacik, O. Samuel, R. Halouzka, M. Kovacik,

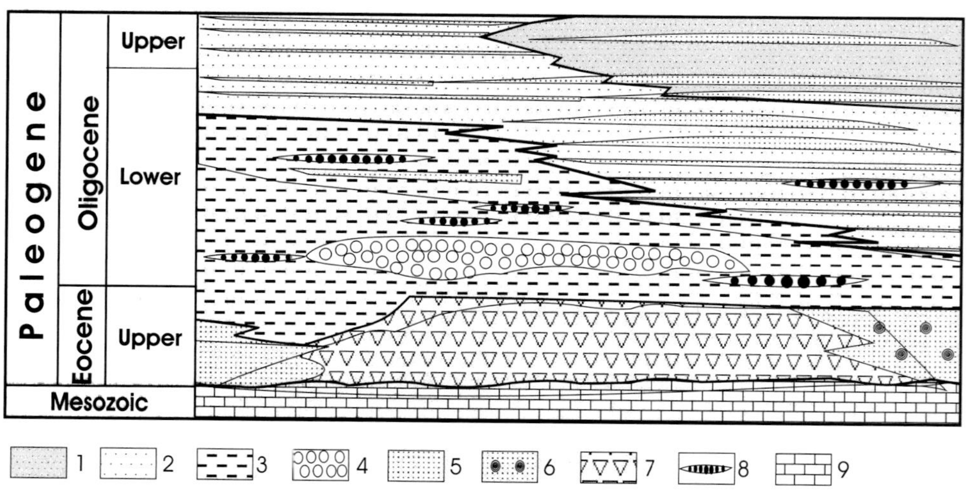

Figure 6. Lithostratigraphic units of the Spisska Magura region in the Inner Carpathian Paleogene basin. 1 = thick sandstone beds alternating with mudstone layers (Kezmarok Member); 2 = alternating sandstone and mudstone beds (Zuberec Formation); 3 = mudstone with sandstone and conglomerate layers (Huty Formation); 4 = coarse-grained fill of submarine canyon; 5 = sandstone (Tomasovce Member); 6 = nummulitic sandstone; 7 = breccias and conglomerate (Borove Formation); 8 = conglomerate; 9 = Mesozoic rocks.

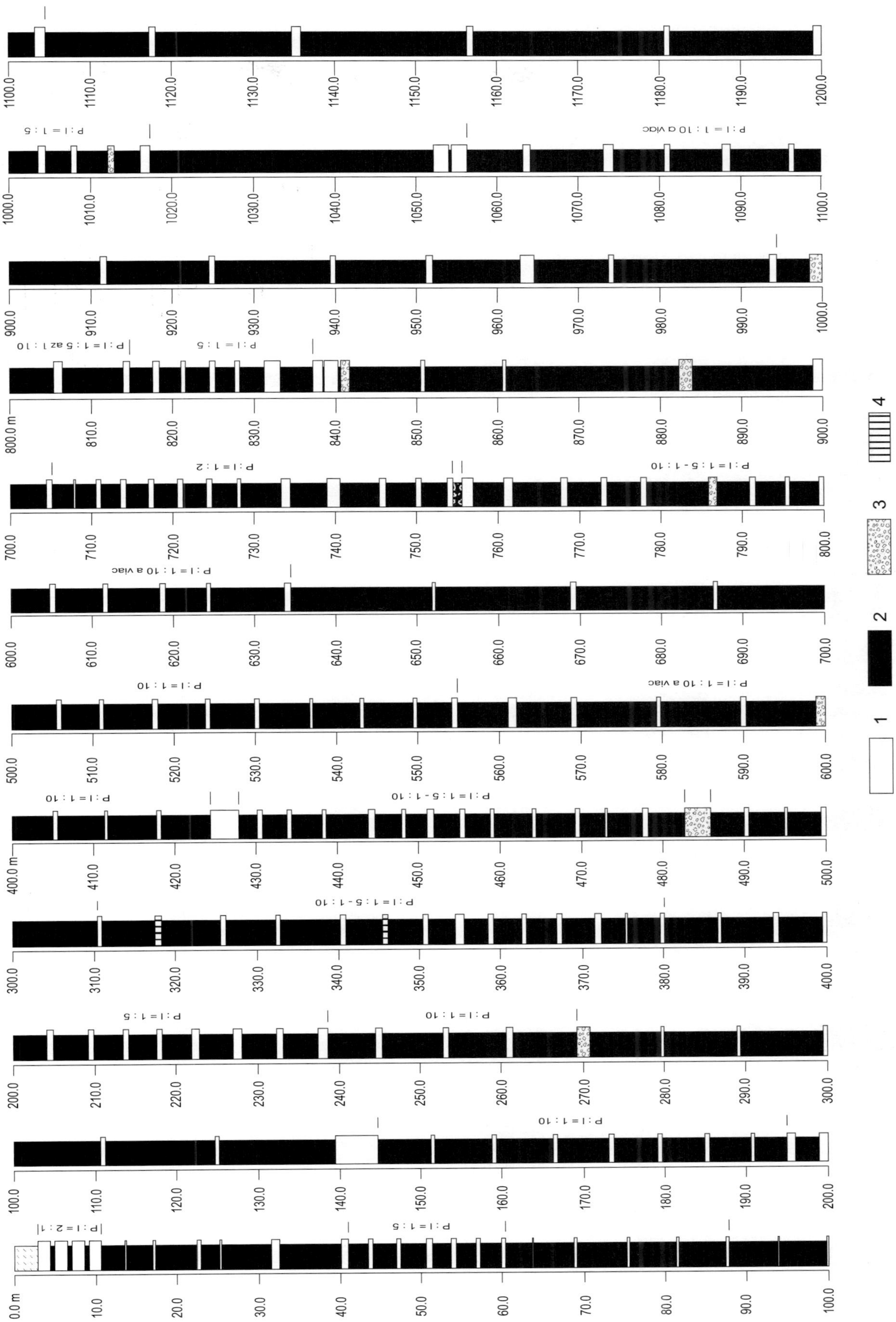

Figure 7. Thick, fine-grained deposits in the northeastern part of the Inner Carpathian Paleogene basin. Borehole VR-7. 1 = sandstone; 2 = mudstone; 3 = conglomerate; 4 = concretions.

and J. Jetel, 1995, personal communication). The sediments, which occur in the Sambron–Kamenica zone attached to the Pieniny Klippen Belt (Figure 5), are strongly deformed by postdepositional tectonics. The thickness of these deposits is more than 3500 m (11,500 ft) (e.g., borehole Nafta Gbely L-3, Lipany, Figure 8). The thickness of the deposits decreases southward of the Pieniny Klippen Belt.

The coarse-grained part of the member consists of clast-supported, minor matrix-supported conglomerates. They are massive and inversely graded, and normal grading occurs scarcely. The size of angular, subangular, and subrounded clasts strongly varies from bed to bed; the maximum clast size observed is about 120 cm (47 in.). The conglomerates comprise as much as 4 m (13 ft) thick and normally amalgamated beds. The

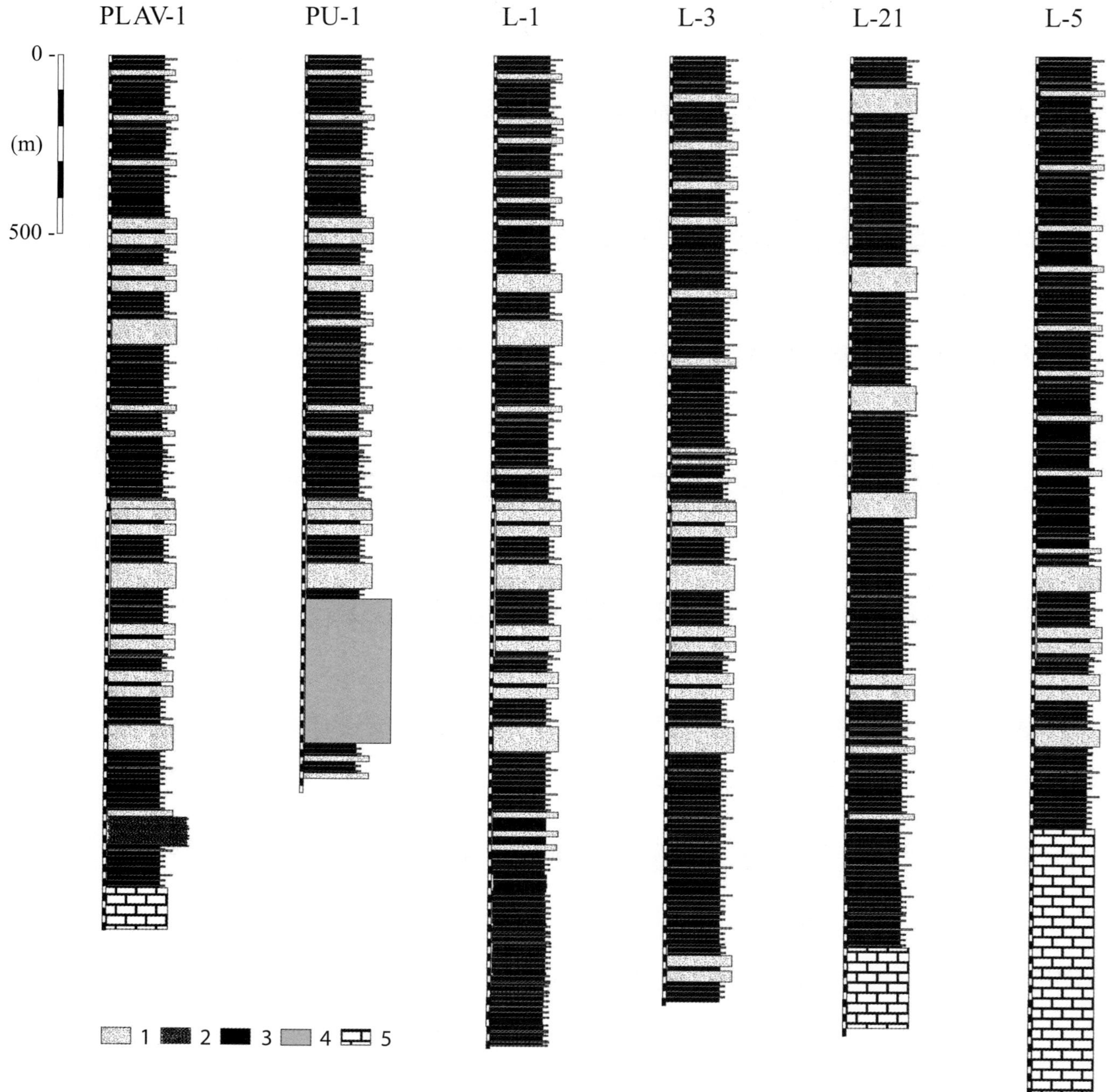

Figure 8. Deep boreholes drilled in the northeastern part of the Inner Carpathian Paleogene Basin showing typical lithological succession in the basin fill. 1 = sandstone; 2 = conglomerate; 3 = mudstone; 4 = slump; 5 = Mesozoic basement.

base of beds is sharp, erosive, and sometimes loaded. The paleoflow direction is polymodal. The sandstones and pebbly sandstones only have minor occurrence.

The Zuberec Formation (upper Eocene–Oligocene, nanoplanktonic zones NP 18–NP 24, Chocholowo Member in the Polish sector of the basin; Golab, 1959; Olszewska and Wieczorek, 1998) prevailingly consists of alternating thin sandstone and shale beds (commonly called rhythmical flysch). The prevailingly medium-grained sandstone normally shows Bouma's $T_{a,b,c}$ divisions, suggesting their deposition by turbidity currents. The shale is massive and parallel laminated. The thin sandstone and shale beds occasionally pinch out laterally. Soft-sediment deformation and starving ripples are common. These features show deposition in levee and interdistributary channel areas. The formation contains the Kezmarok Member (Gross, 1998), which is typically thick (1–3-m; 3.3–10-ft) sandstone beds (probably channel fills, e.g., Janocko et al., 1998) alternating with rhythmical flysch (zebra facies-levee; e.g., Janocko et al., 1998). The thickness of the Zuberec Formation is as much as 300 m (1000 ft). The position of the Zuberec Formation in the basin fill is variable; locally, it laterally interfingers with the deposits of the Huty Formation, or it overlies this formation. It also may laterally pass into the Biely Potok Formation.

The Biely Potok Formation (Ostrysz Member in the Polish part of the basin; Golab, 1959; Olszewska and Wieczorek, 1998) forms the uppermost part of the basin fill. The biostratigraphical analyses indicate the late Oligocene–early Miocene age of the formation, whose maximum thickness is estimated to be about 700 m (2300 ft) (e.g., Gross et al., 1999). The deposits mostly consist of sandstones and minor conglomerates and shales. The thickness of thick sandstone beds ranges from 20 cm to 1 m (8 in. to 3.3 ft). The beds are commonly amalgamated and erosively and sharply based. The sandstones are medium and coarse grained, massive or normally graded. Normally, they represent Bouma's T_a division, and $T_{a,b}$ divisions occur only scarcely. They are classified as medium-grained, subithic arenites in the sense of Pettijohn et al. (1972). The thick sandstone beds alternate with thin sandstone and shale beds commonly marked as zebra facies (Nelson and Nilsen, 1984). The sandstones comprising thin beds are fine to medium grained and show Bouma's $T_{a,e}$, $T_{b,e}$, and $T_{b,c,e}$ divisions. Starving ripples are common. Sharply based sand beds are 2–20 cm (0.8–8 in.) thick. Their geometry is very irregular, and they commonly pinch out laterally. Soft-sediment deformation is very common. The shales comprise sharply based beds of the same thickness like sandstone beds. Shales are massive, and locally, ripple cross-laminations occur. The thickness of these thin sandstone-shale complexes ranges from 20 to 50 cm (8 to 20 in.).

Hungarian (Buda) Paleogene Basin

The Hungarian (Buda) Paleogene basin originated on segments of the elastic lithosphere subjected to tectonic escape from the original position in the Alpine and Dinarides domains (e.g., Kazmer and Kovacs 1985; Csontos et al., 1992; Haas et al., 1995). The basin, which is faulted and deformed by young brachysynclinal updomings, is filled by the Eocene, Oligocene, and Eggerian sediments as much as 1500 m (5000 ft) thick. The Eocene formation, consisting of claystone, marlstone, limestone conglomerate, and coal seams, is only known on the Slovak territory from the surroundings of Sturovo and from Turna depression. The Oligocene and Eggerian Ciz and Lucenec formations are made up mostly of calcareous siltstone and claystone and occur in all partial depression of Southern Slovakia.

Inner Western Carpathians Neogene Basins

The formation and development of the Western Carpathian and Pannonian basins (Figure 1) during the Neogene were controlled by four geodynamic processes acting during the Cenozoic in Central Europe:

- Alpine collision of the Afro-Arabic lithospheric plate and/or its Apulian promontory with the Bohemian Massif
- Carpathian south-directed subduction of the North European lithospheric plate and/or the lithospheric basement of the Outer Carpathian flysch troughs, below the Carpathian–Pannonian lithospheric plate
- tectonic escape of elastic lithosphere fragments from the Alpine collisional region to the Pannonian region
- rise of the asthenosphere (Pannonian asthenolith) and consequent collapse of the Pannonian lithosphere

The basins of the Western Carpathians may be divided into the following classes:

- basins of the foredeep that formed as a consequence of the North European lithospheric plate flexure induced by the weight of accretionary prism and/or flysch nappes and by subsurface load (slab-pull mechanism) (these basins belong to the Outer Carpathians and are not discussed in the following text)

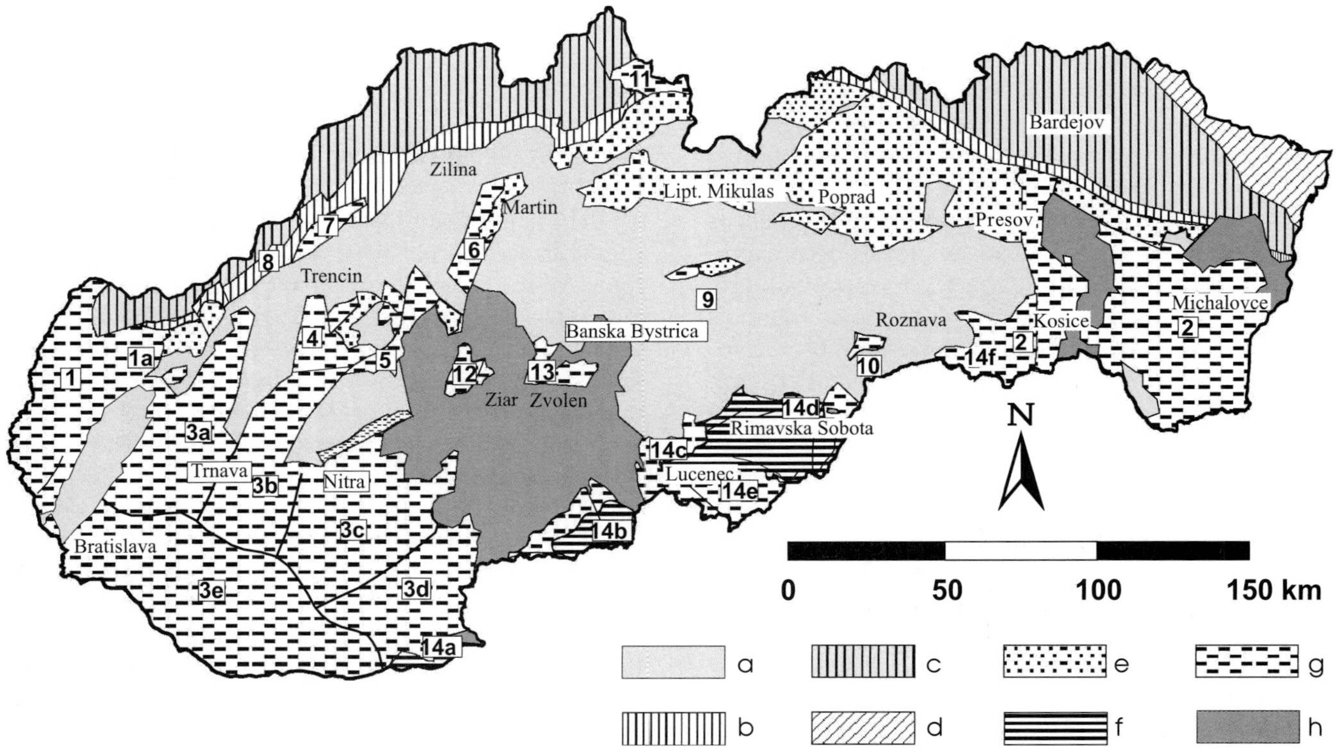

Figure 9. Location of individual geologic units in the Slovakian part of the Western Carpathians. The partial Neogene depressions are also shown. 1 = Vienna basin: 1a = Senica sector; 2 = East Slovakian basin; 3 = Danube basin: 3a = Blatne depression; 3b = Risnovce depression; 3c = Komjatice depression; 3d = Zeliezovce depression; 3e = Gabcikovo depression; 4 = Banovska kotlina depression; 5 = Horna Nitra depression; 6 = Turcianska kotlina depression; 7 = Trencianska kotlina depression; 8 = Vad'ovce depression; 9 = Horny Hron depression; 10 = Roznava depression; 11= Oravska kotlina depression; 12 = Ziarska kotlina depression; 13 = Zvolensko–Slatinska kotlina depression; 14a = Sturovo Paleogene; 14b = Ipel'ska kotlina depression; 14c = Lucenecka kotlina depression; 14d = Rimavska kotlina depression; 14e = Cerova vrchovina upland; 14f = Moldavska kotlina depression.

- piggyback basins on the flysch accretionary prism that originated from the crustal extension at the trailing edge of the prism as a consequence of tension generated by the prism thrust process
- wrench fault furrows at an active convergent margin of the overriding Carpathian–Pannonian lithospheric plate
- transtensional fore-arc and intra-arc basins that originated synchronously with the formation and growth of the Miocene volcanic arc and opened as a result of oblique convergence of the North European lithospheric plate and Carpathian–Pannonian lithospheric plate by the mechanism of strike-slip pull-apart
- back-arc basins that originated from the mechanics of back-arc extension and of thermal lithospheric stretching as a consequence of heat flow from the rising Pannonian asthenolith

Piggyback Basins on Accretionary Prism. The early Miocene Western Carpathians piggyback basin is represented by the western part of the Vienna basin, i.e., the part of the basin that lies on the Outer Carpathian nappes. Similar small basins could also evolve along the northern boundary of the Inner Western Carpathians.

Wrench Fault Furrow Basins at Active Convergent Margins of Overriding Lithospheric Plate. The oblique convergence of the North European lithospheric plate and the Carpathian–Pannonian lithospheric plate resulted in shear strain at the leading edge of the overriding plate. Under such conditions, the early Miocene pull-apart basins opened along the inner side of the klippen belt, which show characteristics of typical wrench fault furrows. In the Western Carpathians, several wrench fault furrows may be distinguished:

1) In the Western Slovakia shear zone, at the Alpine–Carpathian boundary, relics of wrench furrows are represented by the Senica and Kovalov depressions (northeast part of the Vienna basin) and the small Dobra Voda and Vad'ovce partial depressions (Figure 9). The depressions were filled by early Miocene clastic deposits of the Luzice, Laksary,

and Zavod formations (northeast Vienna basin) and the Causa and Planinka formations (Dobra Voda and Vad'ovce depressions).

2) The regime of oblique convergence and transpression at the inner side of the klippen belt and/or at the leading edge of the overriding Carpathian–Pannonian lithospheric plate opened the small depressions of Beckov, Trencin, and Ilava (Figure 9). The sedimentary fill of these depressions is incomplete and composed by clastic marine deposits of the Causa Formation (Eggenburgian) and the Laksary Formation (Karpatian).
3) The wrench faulting influenced the area of the northern part of the present-day East Slovakian basin in the early Miocene. During the Eggenburgian, several hundred meters thick marine clastic deposits of the Presov and Celovce Formations originated.

Formation of the Carpathian–Pannonian Volcanic Arc– The Carpathian–Pannonian volcanic arc began to rise in the Karpatian–Early Badenian. The arc may be generally subdivided in a western, older segment with areally distributed volcanics and an eastern, younger segment parallel with the subduction zone of the Eastern Carpathians (Figure 9). The western segment consists of magmas generated by the subduction of the lithospheric basement of the Outer Flysch belt and magmas generated by a partial melting of the Pannonian lithosphere by rising Pannonian asthenolith. The thermal effect of the asthenolith is at least partly responsible for the areal and random distribution of the volcanics.

The eastern segment resulted from the subduction of the lithospheric basement of the Outer Flysch belt. The volcanoes show spatial and time relationship to the subduction zone (Figure 9) (Vass et al., 1988; Lexa and Konecny, 1998).

Transtensional (Pull-apart) Fore-arc Basins. This type of basin is represented by the Vienna basin sandstones and the East Slovakian Basin.

Vienna Basin– The present-day Vienna basin lies mostly on the East Alpine and Western Carpathian pile of nappes built up by Mesozoic rocks. However, the western part of the basin is floored by Outer Carpathian Flysch belt nappes and by Mesozoic rocks of the Klippen Belt. The basin is faulted, and the maximum thickness of the sedimentary basin fill is about 6000 m (20,000 ft). In the basin fill, the sediments of the early Miocene Vienna piggyback and the Senica–Kovalov wrench furrow are included.

During the Karpatian, the basin was filled by clastic material of extensive alluvial fans and/or deltas, which entered the basin from the south (Aderklaa Conglomerate) as well as from the north (Jablonica Conglomerate; Kovac, 1986).

The significant strike-slip opening of the Vienna basin during the Badenian formed a steep relief at the basin margins, which was the source of the coarse clastic material that formed huge cones and aprons quickly wedging out into the basin (Devinska Nova Ves Member: thick coarse conglomerates and breccias). Basinward, the basal and marginal coarse clastics as well as delta accumulations (as Matzen Sand) are replaced by pelites that are the main lithotypes of the Lanzhot, Jakubov, and Studienka Formations. The tuff layers and algal reef limestones are scarce.

During the late Miocene, the activity of faults faded, and the process of basin opening and subsidence slowed down. During the Sarmatian, basinal hyposaline pelitic deposits of the Holic Formation were replaced by a prograding delta (Skalica Formation), and Pannonian Bzenec and Pontian Cary Formations originated under lacustrine conditions (basinal pelites and sandy deltas with lenses of coal/lignite). The Dacian Gbely Formation was deposited in a shallow lake, whereas the Romanian Brodske Formation originated mostly in a shallow lake and/or river environment.

East Slovakian (Transcarpathian) Basin– The East Slovakian basin is the real mismatch in the basement. The Neogene sediments and volcanics of the basin overlie various basement units. From the north, the Mesozoic formations of the Tatro–Veporicum and of the Humenne Mesozoic rocks are partly covered by the Inner Paleogene deposits. The former unit is separated by a steep strike-slip fault from the slightly metamorphosed Mesozoic and Paleogene rocks of the Inacovce–Kritchevo unit considered to be either the deepest tectonic unit of the Western Carpathians (the Penninicum) (e.g., Sotak et al., 1993) or the equivalent of the Szolnok Flysch (Vozar et al., 1993), the Silicicum and/or Turnaicum, Meliaticum, and Gemericum floor the western part of the basin. The maximum thickness of the sedimentary basin fill is about 8000–9000 m (26,200–29,500 ft).

The East Slovakian basin was opened after the disappearance of the early Miocene transpressional regime at the beginning of Karpatian. The basin has an autonomous position in the Transcarpathian basin stretching from Eastern Slovakia to Ukraine (Vass et al., 1988, 2000).

The opening of the East Slovakian basin compensated coeval space reduction resulting from the thrusting of the Outer Flysch belt frontal nappes between Tarnow and Rzeszow. The main factor during the opening was the pull of the subducted plate in combination with thermal extension induced by the rising asthenosphere in the hinterland.

The thermal extension resulted in the attenuation of the crust to some 27–30 km (16–18 mi) below the East Slovakian basin. Stretching of the crust also affected the upper crustal pre-Neogene rocks. Some of the thrust planes of the Inacovce–Kritchevo unit, as a consequence of crust stretching, began to behave as listric normal faults. These extensional faults of northwest direction opened the East Slovakian basin in the Karpatian.

The initial subsidence in the early Karpatian was slow as suggested by only a 250–400-m (800–1300-ft)-thick early Karpatian clastic Teriakovce Formation (Vass and Cvercko, 1985). The subsidence even diminished in the middle Karpatian, and the basin was affected by a salinity crisis. The sea retreated from the larger part of the basin, leaving in its central depression mud and salt flats and/or lagoons where the evaporates of the Solna Bana Formation originated.

In the late Karpatian, the basin subsidence started in transtensional conditions, changing the direction of maximum compression from northwest–southeast to east-northeast–west-southwest. During this time, the shallow-water Kladzany Formation, prevailingly consisting of silt and clay, was formed (Vass et al., 1996). The transtensional regime also governed explosive volcanism with acid magma of crustal origin (Fintice Tuff).

During the early Badenian, the basin was filled by Nizny Hrabovec Formation, comprising siltstone and claystone interbedded by fine- and medium-grained sandstone. At the end of the early Badenian, the crustal, acid (rhyodacite), highly explosive volcanic activity with wide areal extent appeared.

Fading subsidence during the middle Badenian resulted in hypersaline conditions and a second salinity crisis during the basin evolution as represented by the Zbudza (evaporate) Formation. The stress relaxation terminated the older stage of pull-apart basin evolution.

From the late Badenian to the middle Sarmatian, transtension governed extremely rapid basin subsidence and a revival of the pull-apart opening mechanism of the basin. The basin depocenters shifted to the central part (late Badenian) and later to the southeast part of the basin (early Sarmatian), where the measured rate of subsidence in the basin culminated with a sedimentation rate of some 390 cm/1000 yr [154 in./1000 yr] (= 267 cm/1000 yr [105 in./1000 yr] uncorrected for compaction).

Sinistral northwest and west-northwest strike-slip faults opened space in the late Badenian for the large Klcovo delta progradation into the basin (Janocko, 1990; Rericha, 1992), whereas in the southeast, prodelta and outer shelf pelites of the Lastomir Formation were deposited.

In the early and middle Sarmatian, the Kosice delta prograded into the basin (Janocko in Kovac et al., 1994). The progradation was hindered by the newly formed barrier of the Presov–Tokaj Mountains volcanics. In the basin center, the pelites with acid and andesite tuffs of Stretava Formation were deposited. Because of the isolation of the Central Paratethys from the open sea, the Stretava Formation was deposited in the hyposaline sea.

The coarse-clastic character of deltaic sediments at the western margin of the basin (Kosice Gravel) suggests an intense uplift of the source area at the western edge of the basin. Otherwise, similar uplift probably also occurred during the sedimentation of the basal deposits of the Klcovo delta, which are prevalently composed of gravel (Varhanovce Gravel).

The development of the basin in the late Badenian was accompanied by increasing volcanic activity that culminated in the middle Sarmatian. Predominantly andesitic volcanism accompanied by acid differentiates formed the stratovolcanoes of the Presov–Tokaj Mountains, Vihorlat Mountains, and Popriecny Mountains and the buried stratovolcanoes Malcice and Besa–Cicarovce. Volcanic centers were also situated on and near the Zemplin–Beregovo Horst. The volcanoes typically originated at the intersection of longitudinal and transversal faults. They belong to the areal andesite type as well as to the volcanic arc type (basalt andesites and andesites). Three arcs of volcanics in the basin (the oldest one along the southern margin, the middle one in the basin center, and the youngest one along the northern margin of the basin) imply the intra-arc position of the basin at that time (Vass et al., 1988). Toward the end of middle Sarmatian, the pull-apart basin mechanics definitively faded out.

In a new paleostress field of the late Sarmatian, the rate of subsidence significantly dropped to 60.3 cm/1000 yr (24 in./1000 yr) (Kral et al., 1990). The hyposaline marine environment was progressively replaced from the west to the east by the lacustrine environment. This is reflected by the middle and late Sarmatian, lacustrine Kochanovce Formation, consisting of pelites with rhyolite and andesite tuff layers (locally altered into bentonite) and lenses and seams of coal and/or lignite that originated in swamps and bogs. The late Sarmatian sediments, mostly sandy deposits of the Ptruksa Formation, were restricted to the eastern part of the basin. They originated in a hyposaline environment. The marginal Zavadka Member, containing coal-lignite seams, originated in swamp and bog environments.

The normal faults had restricted activity, and basin subsidence slowed down. The pelitic Secovce Formation with coal-lignite seams, andesite, and rhyolite tuffs (Albinov and Hazin tuffs) onlapped Sarmatian lithostratigraphic units.

To the end of the Miocene, in the Pontian, the basin underwent the last stress shock with east-northeast–west-southwest compression. Lateral compression of the basin caused a slight folding of its sedimentary fill. Minute fold structures are found in the Secovce Formation (Pannonian) and also in the Teriakovce Formation (Karpatian; Kovac et al., 1994). We suppose that both fold disturbances were induced by the same late Miocene compressional event. The Pontian deposits (kaolinitic pelites and gravels) are grouped in the Senne Formation (Vass and Cvercko, 1985).

In the Pliocene, residual lacustrine sedimentation persisted only in the southeast basin sector. The deposition in the basin terminated by the Pliocene Cecehov Formation.

Inner Depressions. Inner depressions of the Western Carpathians are represented by small fore-arc and/or intra-arc basins with a complex tectonic history. The older part of the basin fill is marine, and the younger one is continental. Some of the inner depressions are situated along the continuation of the Western Slovakia shear zone. After the first common stages of their development, they became separated. They are mainly filled by early Miocene marine sediments (Trencianska kotlina, Ilavska kotlina, and Beckovska kotlina depressions; Figure 9). Other depressions (the Banovska kotlina, Hornonitrianska kotlina, and Turcianska kotlina depressions) passed through a similar early Miocene evolution. Later, during the middle Miocene, the development of these depressions was identical to the evolution of Ziarska kotlina and Zvolensko–Slatinska kotlina depressions, where the early Miocene marine deposits are missing and continental middle–late Miocene deposits are highly influenced by contemporaneous andesite and rhyolite volcanism of the Central Slovakia. These depressions have intra-arc character. They were formed in a horst and graben system under conditions of thermal extension.

Back-arc Basins. The back-arc area south of the North European lithospheric plate and the Carpathian–Pannonian lithospheric plate boundary is associated with several consecutive basins. At the beginning of the Eggenburgian, the Filakovo–Petervasara Basin opened. The Filakovo–Petervasara Basin had a configuration of a bay opened to the northeast and north-northwest (Baldi, 1986; Sztano, 1994; Halasova et al., 1996). At the bay margins, shallow-marine sandstone and siltstone were deposited in a tidal-dominated coastal zone (Filakovo–Petervasara Formation). Other lithotypes of the littoral zone are represented by the Budafok and Lipovany sandstones.

Late Eggenburgian termination of deposition in the Filakovo–Petervasara Basin was caused by the regional uplift of the whole Pannonian region, including the southern part of Western Carpathians, where continental sedimentation and/or erosion on the extensive territory occurred (Vass 1995). The regional uplift was the first manifestation of the Pannonian asthenolith rise. The rising asthenosphere heated the crust, giving way to the crustal magmas formation and to the following areal rhyolite-rhyodacite volcanism extending over the whole Pannonian region (Lexa et al., 1993). All these events fall in the period of late Eggenburgian–early Ottnangian (20–18 m.y.).

The rifting related to a lithosphere rise caused by the Pannonian asthenolith opened the Novohrad–Nograd Basin in the late Ottnangian (Vass, 1995), when the Salgotarjan Formation was deposited. The formation is composed of coal seams bearing Potor Member and lacustrine Plachtince claystone. The Ottnangian rifting continued during the Karpatian, giving way to frontal marine transgression and deposition of the Modry Kamen Formation. The transgressive succession commences with the lagoonal Medokys Member, followed by the littoral clastics of the Krtis Member, and culminates with the basinal calcareous siltstone and claystone of the Secianky Member. The evolution of the Novohrad–Nograd Basin terminated because of regional uplift.

The area of Southern Slovakia, particularly the northern part of the Lucenska and Rimavska kotlina depressions, underwent a weak subsidence compensated by fluvial deposition of the kaolinite-bearing Poltar formation accompanied by basalt volcanism of the Podrecany Formation in the Pontian. The basalt volcanism continued to the Pliocene and Pleistocene, when the Cerova Formation originated.

Thermal Extensional Back-arc Basins. Thermal extensional back-arc basins related to the Outer Carpathian subduction mostly occurred in the Pannonian realm on the Hungarian territory. However, the northern part of the Danube basin partially stretches into the Western Carpathian area north of the Danube River. The Northern Danube basin (Figure 9) was opened by asymmetrical stretching and/or lithosphere thinning in the middle Miocene. The basin northern embayments are faulted, and the central Gabcikovo depression has a large brachysynclinal structure only slightly disturbed by faults. The Northern Danube basin is filled by 5000–7000-km (3100–4300-ft)-thick, middle to late Neogene sediments and contemporary volcanics. The synrift deposits originated in marine and hyposaline environments, and the postrift deposits are hyposaline and lacustrine.

In the area of the Gabcikovo depression, i.e., in the central part of the Northern Danube basin, the thickness of the crust is 27.5 km (17 mi) or less, and the thickness of lithosphere is around 80 km (50 mi). The synrift middle Miocene sediments (Badenian and Sarmatian) are hundreds of meters thick; however, their thickness in partial depressions, namely, in the Blatna and Risnovce depressions, exceeds 2500 and/or 3000 m (8200 and/or 10,000 ft). Synrift sediments are disturbed by a dense network of synsedimentary faults with great throws (hundreds of meters).

However, less common sediments of the postrift thermal stage (Pannonian to Pliocene), which are sparsely deformed by faults, have considerable thickness (more than 5000 m [16,000 ft]).

HYDROCARBON POTENTIAL OF THE INNER WEST CARPATHIANS

Paleozoic and Mesozoic units of the Inner Western Carpathians, formed in Hercynian and primarily Paleoalpine phase of orogenic development, i.e., prior to the Paleogene, have their hydrocarbon potential practically exhausted. Old Paleozoic units of the Tatricum, Veporicum, and Gemericum are affected by a different grade of metamorphism. In addition, the young Paleozoic and Mesozoic sediments of the Tatricum, Veporicum, Gemericum, Fatricum, Hronicum, Meliaticum, Turnaicum, and Silicicum are affected by anchimetamorphism and cannot be considered as source rocks. Organic-geochemical analytical data (content of TOC, vitrinite reflectance, maximum temperature of pyrolysis effect, content of free and fixed hydrocarbons, etc.) obtained from surface and deep borehole samples from the basement of Tertiary basins suggest that the hydrocarbon potential of rocks was exhausted already before the Late Cretaceous, maybe already in the original sedimentation basins. The partial exception is the pre-Tertiary basement of the Vienna basin, comprising a part of the Northern Calcareous Alps.

From an organic-geochemical viewpoint, the rocks are occasionally sufficiently rich in organic carbon (some samples of Carboniferous and Permian) corresponding to a phase of black coal, anthracite to graphite. Within the framework of free or fixed hydrocarbon analysis or gaining of the extract, these samples mostly do not provide any analytical response. However, microphotometric measurements indicate an anisotropic organic matter that confirms a high grade of alteration from the viewpoint of the potential production of hydrocarbons.

Some rocks of the Inner Western Carpathian Paleozoic and Mesozoic units, mostly carbonates with secondary porosity acquired by tectonic strain during folding and thrusting or by weathering in a suitable structural position beneath the Tertiary basin fill, can present reservoir rocks.

Considering the complex geological development of the Pieniny Klippen Belt and the low level of investigation, the problem of the origin, migration, and accumulation of hydrocarbons in this unit can be treated only tentatively. Jurassic and Lower Cretaceous sediments have organic matter in a high stage of alteration and do not have any hydrocarbon potential.

The depth distribution of TOC content in the Nafta Gbely, Han-1, and Hanusovce wells, located in the eastern sector of Pieniny Klippen Belt, indicates potential source rocks in some samples of the Upper Cretaceous and Paleogene rocks in the interval of 1300–4000 m (4300–13,000 ft) (content of TOC in some samples is 0.5–0.8 vol.%). The quality of these rocks expressed on the basis of Rock-Eval pyrolysis parameters in a given interval shows a generally acceptable tendency (content of fixed hydrocarbons in some samples is 0.5–0.8 vol.% mg HC/g of rock; hydrogen index is 100–150 mg HC/g TOC; and vitrinite reflectance is 0.6–1%).

However, the Pieniny Klippen Belt is too narrow and too complex a zone for effective hydrocarbon prospection at the moment. This should be executed in connection with adjacent areas.

Upper Cretaceous postnappe formations are only preserved in small remnants in the Inner Western Carpathians, and from the viewpoint of hydrocarbon potential, except for the pre-Tertiary basement of the Vienna basin, they are practically insignificant.

The Inner Carpathian Paleogene basin represents one of the most prospective areas in the Inner Western Carpathians besides the classic areas of today's hydrocarbon exploration in the Inner Western Carpathians, i.e., Neogene basins and the Flysch belt (Outer Western Carpathians). From the viewpoint of total carbon content (Table 1), the predominant part of the analyzed surface and well samples of the Paleogene sedimentary fill meets the criteria for potential source rocks (content of TOC is above 0.5 vol.%), and some of them can be evaluated as very good (content of TOC is above 2%). Kerogen type, upon which directly depends the hydrocarbon potential, is predominantly terrestrial (type III) and locally mixed terrestrial marine (type III–II).

The content and depth distribution of free hydrocarbons or vitrinite reflectance (Table 1) indicates the

Table 1. Basic organic-geochemical data of the Inner Carpathian Paleogene basin sedimentary rocks.

Stratigraphy	TOC (%)	S_1 (mg/g)	S_2 (mg/g)	HI (mg/g)	T_{max} (°C)	R_o (%)
Paleogene rocks	0.0*X* – 2.50	0.0*X* – 0.46	0.0*X* – 1.14	13–200	434–464	0.44–1.27
Mesozoic rocks	0.0*X* – 0.20	0.0*X*	0.0*X* – 0.24	9.89	–	1.60–4.10

TOC = total content of organic carbon (mass %).
S_1 = content of free hydrocarbons [HC = hydrocarbons] per gram of rock (mg HC/g rock).
S_2 = content of hydrocarbons fixed in kerogen (mg HC/g rock).
HI = hydrogen index; HI = $[S_2/\text{TOC}] \times 100$ in mg HC/g TOC.
T_{max} = temperature of pyrolytic effect maximum (°C).
R_o = vitrinite reflectance (%).

reaching of an oil hydrocarbon generation stage. The content of fixed hydrocarbons in kerogen is not exhausted, i.e., in higher depths of burial, it can generate hydrocarbons. Mean hydrocarbon potential varies around 1 kg of hydrocarbons per tonne of rock; however, considering the present grade of alteration, it is possible to assume that a part of it was already realized and potentially could be tied in traps. This is evidenced by oil occurrences in some deep wells, even if no commercial significance exists.

The study of n-alkanes, steranes, and triterpanes distribution suggests positive genetic correlation among oils gained from the Paleogene reservoir rocks (sandstones and breccias) and between this oil and hydrocarbons gained by extraction from Paleogene source rocks. However, a distinctive difference in hydrocarbon composition exists between extracts from underlying Mesozoic rocks and the Paleogene extracts.

Organic-geochemical analyses and temperature alteration indicators suggest that the postsedimentary alteration of rocks in the area is not identical. The most intensive alteration underwent organic matter dispersion in sedimentary rocks adjacent to the Pieniny Klippen Belt. According to burial-history modeling and the development of hydrocarbon generation windows, the alteration of the organic matter was reached 2000 m (6600 ft) deeper compared to the present position of these rocks. The alteration occurred during the Oligocene and early Miocene. At present, the dispersion organic matter occurs in the relics stage of maturity in the prevailing part of the territory.

Despite many negative phenomena (low quality of Paleogene reservoir rocks, Mesozoic basement practically without any source rocks, considerable uplift, subsequent erosion of the area, etc.), it seems that the area has a hydrocarbon potential, primarily in areas of the Paleogene depocenters (in depths of some 2500–4000 m [8200–13,000 ft], where, indeed, more favorable conditions for the preserving of hydrocarbon accumulations were formed).

Inner Carpathian basins, primarily the Vienna, Danube, and East Slovakian basins, have the highest hydrocarbon potential of the Inner Western Carpathians units. Good source rocks of hydrocarbons mainly occur in the lower and middle Miocene sediments and partly also in the underlying Inner Carpathian Paleogene sediments (southern part of the Vienna basin, the Danube, and the East Slovakian basins) or in the Upper Cretaceous–Paleogene sediments of the Flysch belt (Outer Western Carpathian units) underlying the Neogene sediments in the northern part of the Vienna basin.

Reservoir and sealing rocks (sandstones and claystones) alternate from Eggenburgian to early Pannonian. Relatively good reservoir rocks are also represented by underlying Mesozoic carbonates with secondary porosity. The thickness of the Neogene sedimentary fill in the center of the basins is more than 5000 m (16,400 ft).

A complex tectonic and sedimentary evolution of the Neogene basins determined the formation of numerous tectonic and stratigraphic hydrocarbon traps. Basic organic-geochemical parameters of the most significant Neogene basins of the Inner Western Carpathians are given in Tables 2–4.

The presence of nonhydrocarbon gases like carbon dioxide and nitrogen (Figures 10, 11) is a negative feature for the prospection and exploration of hydrocarbon reservoirs in the Danube and East Slovakian basins. Those gases occur together with methane in a variable ratio (in the order from 0 to above 90%). Carbon dioxide origin can be deduced in the Danube basin from the buried Badenian and in the East Slovakian basin from the Badenian–lower Sarmatian stratovolcanoes. Volcanic admixture is commonly of deteriorating reservoir rock quality in both basins.

Oil and gas production in Slovakia has been concentrated in classic oil- and gas-bearing areas represented by Neogene sedimentary basins (Figures 10–12). Considering that oil and gas production in these areas had run for more than 50 yr (in the Vienna basin, even 90 yr), most of the large and relatively easily accessible reservoirs with good properties of reservoir rocks are practically depleted, and a lot of them

Table 2. Basic organic-geochemical data of the Vienna basin sedimentary rocks.

Stratigraphy	TOC (%)	S_1 (mg/g)	S_2 (mg/g)	HI (mg/g)	T_{max} (°C)	R_o (%)
Neogene rocks	0.30–2.50	0.0*X* – 0.85	0.0*X* – 4.50	11–225	420–445	–
Paleogene and Upper Cretaceous rocks	0.40–1.15	0.0*X* – 1.18	0.0*X* – 1.18	32–103	427–447	–
Mesozoic rocks	0.0*X* – 0.89	0.0*X*	0.0*X* – 0.78	36–211	435–446	–

TOC = total content of organic carbon (mass %).
S_1 = content of free hydrocarbons [HC = hydrocarbons] per gram of rock (mg HC/g rock).
S_2 = content of hydrocarbons fixed in kerogen (mg HC/g rock).
HI = hydrogen index: HI = [S_2/TOC] × 100 in mg HC/g TOC.
T_{max} = temperature of pyrolytic effect maximum (°C).
R_o = vitrinite reflectance (%).

have been converted into underground gas storage. The balance of oil- and natural gas-proved reserves in individual Neogene basins is given in Table 5.

During the entire exploitation history, a total of 25.5 × 10^9 m^3 (9 × 10^{11} ft^3) of natural gas and approximately 3.4 × 10^6 t of oil was produced from the Neogene basins. Production culminated in the 1960s and 1970s, when, in Slovakia, approximately 1 × 10^9 m^3 (3.5 × 10^{10} ft^3) of natural gas and 80 × 10^3 t of oil were produced every year. In the last few years, the production varied around 0.18 × 10^9 m^3 (6.3 × 10^9 ft^3) of natural gas and 45 × 10^3 t of oil.

According to Table 5, it is clear that in classic, easily accessible, and unproblematic recoverable reservoirs in regions of main Neogene basins, the proven remaining reserves of gas and oil is only 3.5 × 10^9 m^3 (1.2 × 10^{11} ft^3) and 0.35 × 10^6 t, respectively.

Prospects of Reconnaissance of Oil and Gas Reservoirs in the Inner Western Carpathians

The future of exploration and production of hydrocarbons in the Inner Western Carpathians, as clearly shown by the present situation, consists of small, complex, high-pressure, high-temperature, and tight reservoirs. Many of them have been drilled in the past, but economic and technical conditions did not enable (with some exceptions) their effective involvement in production and work was more or less suspended. However, an enormous development of exploration methods, primarily three-dimensional seismic, multitarget-deviated, and horizontal boreholes, new methods of hydraulic fracturing, and the increased price of natural gas jump-started an effective exploration and production of reservoirs in question.

Geological regions of the Inner Western Carpathians with prospective areas and the given types of reservoirs of natural gas in Slovakia are given in Figure 13.

Small Reservoirs in Sedimentary Fill of Neogene and Paleogene Basins

Small reservoirs of oil and natural gas in the sedimentary fill of Neogene and Paleogene basins with a natural gas volume between 20 × 10^6 and 100 × 10^6 m^3 (7.06 × 10^8 and 3.5 × 10^9 ft^3) occur at depths of 500–3000 m (1600–10,000 ft) (Figure 13; reservoir type 1). The exploration methods used in the past did not enable a successful exploration of these reservoirs. However,

Table 3. Basic organic-geochemical data of the Danube basin sedimentary rocks.

Stratigraphy	TOC (%)	S_1 (mg/g)	S_2 (mg/g)	HI (mg/g)	T_{max} (°C)	R_o (%)
Neogene rocks	0.3*X* – 1.94	0.0*X* – 0.55	0.0*X* – 4.80	20–409	425–446	0.30–0.82
Paleogene and Upper Cretaceous rocks	0.0*X* – 1.45	0.0*X* – 0.23	0.12–5.86	62–404	427–444	0.33–0.80
Mesozoic rocks	0.0*X* – 1.40	0.0*X*	0.0*X* – 1.78	10–110	440–530	0.78–1.80

TOC = total content of organic carbon (mass %).
S_1 = content of free hydrocarbons [HC = hydrocarbons] per gram of rock (mg HC/g rock).
S_2 = content of hydrocarbons fixed in kerogen (mg HC/g rock).
HI = hydrogen index: HI = [S_2/TOC] × 100 in mg HC/g TOC.
T_{max} = temperature of pyrolytic effect maximum (°C).
R_o = vitrinite reflectance (%).

Table 4. Basic organic-geochemical data of the East Slovak basin sedimentary rocks.

Stratigraphy	TOC (%)	S_1 (mg/g)	S_2 (mg/g)	HI (mg/g)	T_{max} (°C)	R_o (%)
Neogene rocks	0.0X – 1.8	0.0X – 0.54	0.0X – 2.03	10–353	417–538	0.40–2.86
Paleogene rocks	0.0X – 4.34	0.0X – 0.20	0.0X – 4.34	14–434	427–440	0.57–0.69
Mesozoic rocks	0.0X – 5.18	0.0X	0.0X	–	–	2.35–3.46
Paleozoic rocks	0.0X – 3.20	0.0X	0.0X	–	–	3.36–4.52

TOC = total content of organic carbon (mass %).
S_1 = content of free hydrocarbons [HC = hydrocarbons] per gram of rock (mg HC/g rock).
S_2 = content of hydrocarbons fixed in kerogen (mg HC/g rock).
HI = hydrogen index: HI = $[S_2/\text{TOC}] \times 100$ in mg HC/g TOC.
T_{max} = temperature of pyrolytic effect maximum (°C).
R_o = vitrinite reflectance (%).

an extensive involvement of three-dimensional seismic data, inversion seismic data, amplitude-vs.-offset analyses, etc., enabled the definition, with relative accuracy, of small reservoir contours and establishment of the most appropriate well location.

Exploration and production of small reservoirs can be effective primarily in areas of existing older oil and gas fields with the existing network of pipelines and gathering centers. Such areas in the Western Carpathians are, e.g., the Vienna and East Slovakian basins, where small gas reservoirs are tied to deltaic sand bodies in the Pannonian and late Badenian (Figure 14A, B) or to tectonic traps in facially constant sandy horizons of the late Badenian and Sarmatian (Figure 14C). Clinoform deltaic sand bodies have generally very good reservoir rock properties (a porosity of about 28% and a permeability as much as 1×10^{-12} m^2 [approximately 1 d]). In some cases, several clinoform bodies can be found more or less above each other and can be verified by a deviated well, which distinctively influences the economic affectivity (Figure 14A). Small reservoirs of oil in the northern part of the Vienna basin are tied to the fractured surface of the flysch underlier and basal sands of the Eggenburgian or lenticular sands of the Ottnangian, Badenian, and Sarmatian. Intraformational sandstones

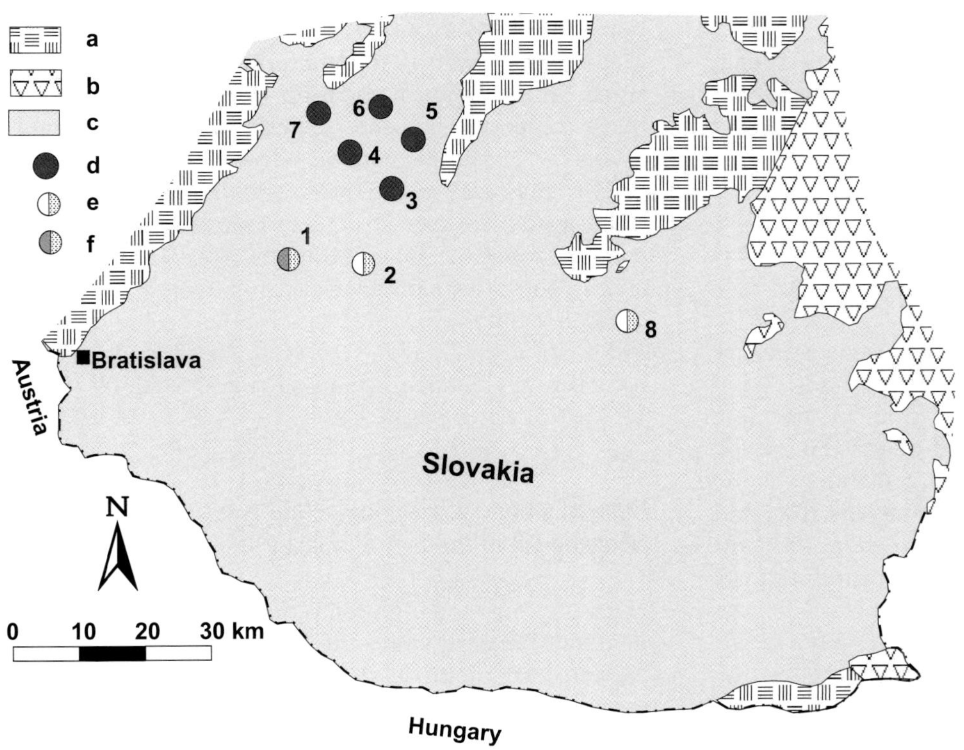

Figure 10. Gas deposits in the Danube basin. a = Undivided Paleozoic, Mesozoic, and Paleogene units of the Inner Western Carpathians; b = Neogene volcanics; c = Neogene sediments; d = mixed gas deposits (nitrogen and natural gas); e = mixed gas deposits (carbon dioxide, natural gas, and nitrogen, in various ratios); f = natural gas deposits. Gas fields: 1 = Cifer; 2 = Sered'; 3 = Trakovice; 4 = Spacince; 5 = Madunice; 6 = Nizna; 7 = Krupa; 8 = Ivanka.

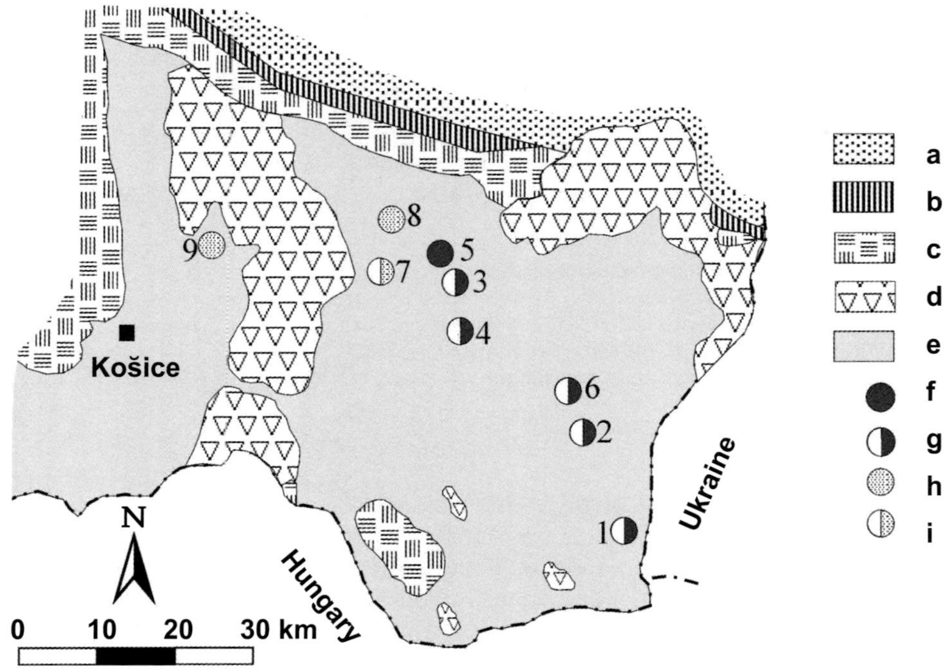

Figure 11. Gas deposits in the East Slovak basin. a = Undivided Mesozoic and Paleogene units of the Outer Western Carpathians (Flysch belt); b = undivided Mesozoic and Paleogene units of the Pieniny Klippen Belt; c = undivided Paleozoic, Mesozoic, and Paleogene units of the Inner Western Carpathians; d = Neogene volcanics; e = Neogene sediments; f = natural gas deposits; g = natural gas and condensate deposits; h = carbon dioxide deposits; i = mixed gas deposits (carbon dioxide, natural gas, and nitrogen, in various ratios). Gas fields: 1 = Ptruksa; 2 = Stretava; 3 = Trhoviste–Pozdisovce; 4 = Banovce; 5 = Rakovec; 6 = Senne; 7 = Visnov–Kravany; 8 = Klcovo Dlhe; 9 = Kecerovske Peklany.

and shales are prospective in the Inner Carpathian Paleogene basin.

In new, little-explored areas, such as the Danube basin and the Paleogene basins (Figure 13), where the existence of small reservoirs of natural gas has been proven, a distinctive advance in the exploration and production can be only brought by the use of three-dimensional seismic survey data.

Possible reserves in regions with small reservoirs in the Slovak territory are estimated to be 10×10^9 m^3 (3.5×10^{11} ft^3) and about 1×10^6 t for natural gas and oil, respectively.

Tight Reservoirs in the Neogene Sedimentary Fill of the East Slovakian Basin

Hydrocarbon exploration in the southwestern part of the East Slovakian basin showed a more than 1000-m (3300-ft)-thick gas-bearing pay zone in the late Badenian and early Sarmatian sediments (Figure 13; reservoir type 2). The characteristic features of the reservoir rocks are low porosity (as much as 16%) and permeability in the order of 1×10^{-15}–1×10^{-14} m^2 (in the order of millidarcys to tenths of millidarcys), which is caused by a tuffaceous admixture. The production of horizons varies in the range of 2–25 × 10^3 m^3 (70–882 × 10^3 ft^3) of natural gas per 24 hr. Recent use of three-dimensional seismic survey data revealed that area has a complex tectonic structure.

Preliminary studies of well cores reveal that significant factors in reservoir rock properties of horizons include the distribution of porosity, swelling, and alteration of clay minerals. The analyses showed that 5% of the total porosity is formed by macropores, 25% by mesopores, and 70% by micropores (i.e., in 16% of the total porosity, 0.8% is formed by macropores, 4% by mesopores, and 11.2% by micropores). The low grade of clay minerals alteration (alteration of smectite into illite) or the high level of expandability of smectite and volcanic material admixture in reservoir rocks causes swelling of rocks.

The gas-bearing horizons are tied to a complex fault system (flower structure; Figure 15a). The gas-bearing zone is 1500–3000 m (5000–10,000 ft) deep, and is comprised of several multilayered reservoirs with an estimated natural gas reserves in the range of 1–5 × 10^9 m^3 (3.5–17.6 × 10^{10} ft^3). Possible reserves of the entire region are estimated to be 60×10^9 m^3 (2.1×10^{12} ft^3) of natural gas.

High-pressure and High-temperature Reservoirs in the Neogene Fill of the East Slovakian Basin

High-pressure and high-temperature reservoirs in sediments of the early and middle Badenian in the northeastern part of the East Slovakian basin (Figure 13; reservoir type 3) represent an unsolved problem at present. These reservoirs are represented by a row of

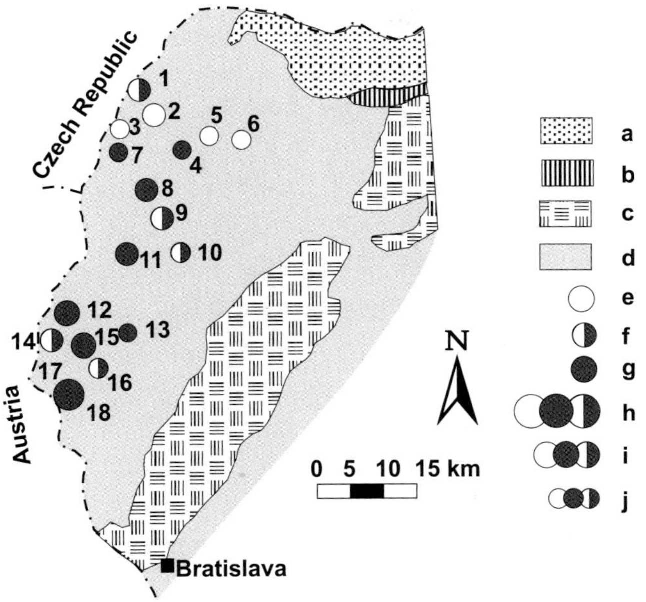

Figure 12. Hydrocarbon deposits in the Vienna basin. a = Undivided Mesozoic and Paleogene units of the Outer Western Carpathians (Flysch belt); b = undivided Mesozoic, and Paleogene units of the Pieniny Klippen Belt; c = undivided Paleozoic, Mesozoic, and Paleogene units of the Inner Western Carpathians; d = Neogene sediments; e = oil deposits; f = oil and natural gas deposits; g = natural gas deposits; h = reserves over 5×10^6 t of oil or 5×10^9 m^3 of natural gas; i = reserves as much as 5×10^6 t of oil or 5×10^9 m^3 of natural gas; j = reserves as much as 0.5×10^6 t of oil or 0.5×10^9 m^3 of natural gas. Oil and gas fields: 1 = Cunin; 2 = Gbely; 3 = Brodske; 4 = Smolinske; 5 = Petrova Ves; 6 = Stefanov; 7 = Kuty; 8 = Borsky Jur; 9 = Zavod; 10 = Studienka; 11 = Zavod (Mesozoic); 12 = Suchohrad–Gajary; 13 = Malacky; 14 = Gajary (Badenina); 15 = Jakubov; 16 = Lab; 17 = Dubrava; 18 = Vysoka.

lens-shaped tight reservoirs of gas in a depth of 2600–4000 m (8500–13,000 ft), with production of about 10×10^3 m^3 (3.5×10^5 ft^3) of natural gas per 24 hr, which are seldom interbedded by gas horizons with high-quality reservoir rock properties. Reservoir pressures are as much as 90% higher than hydrostatic ones, and reservoir temperatures are varying in the range of 140–210°C.

These zones cannot be drilled without special technological equipment (geoservice, rotary preventers, special mud, etc.). Because no three-dimensional seismic investigation was surveyed in the area of interest, the geological structure of the prospects is not entirely clear. Reserves of individual reservoirs are estimated to be in the range of $1–2 \times 10^9$ m^3 ($3.5–7 \times 10^{10}$ ft^3) of natural gas. Possible reserves of the entire region are estimated to be 1×10^{10} m^3 (3.5×10^{11} ft^3) of natural gas. A progress in solving this problem can be reached only by three-dimensional seismic survey use.

High-pressure Reservoirs in the Neogene Fill of the Vienna Basin

High-pressure reservoirs in sediments of the early and middle Badenian and the Karpatian in the central part of the Vienna basin (Figure 13; reservoir type 4) represent a problem very similar to natural-gas reservoirs listed in the previous section and different only by the fact that the reservoir temperature at a depth of 2500–4000 m (8200–13,000 ft) varies in range from 85 to 150°C. This is related to the low-temperature gradient of the Vienna basin.

The reserves volume of individual reservoirs is estimated to be in the range of $0.5–1 \times 10^9$ m^3 ($1.7–3.5 \times 10^{10}$ ft^3) of natural gas, and possible reserves of the entire region are estimated to be 1×10^{10} m^3 (3.5×10^{11} ft^3) of natural gas.

Reservoirs of Mixed Gas in the Neogene Sedimentary Fill of the Danube Basin

Most of the central part of the Danube basin is not yet sufficiently explored (Figure 13; reservoir type 5). Sparse geothermal wells and regional seismic sections only provide an approximate image of the deep geological structure and possible hydrocarbon reservoirs. Gas reservoirs ascertained on the northern slopes of the

Table 5. Balance of the oil and natural gas proved reserves in main Neogene basins in Slovakia.

Basin	Oil in Place	Recoverable Reserves	Produced Reserves	Remaining Reserves
Oil ($\times 10^6$ t)				
Vienna basin	12.2	3.75	3.4	0.35
Natural gas (10^9 m^3)				
Vienna basin	31.4	23.6	22.0	1.6
East Slovak basin	8.1	5.8	4.2	1.6
Danube basin	0.8	0.6	0.3	0.3

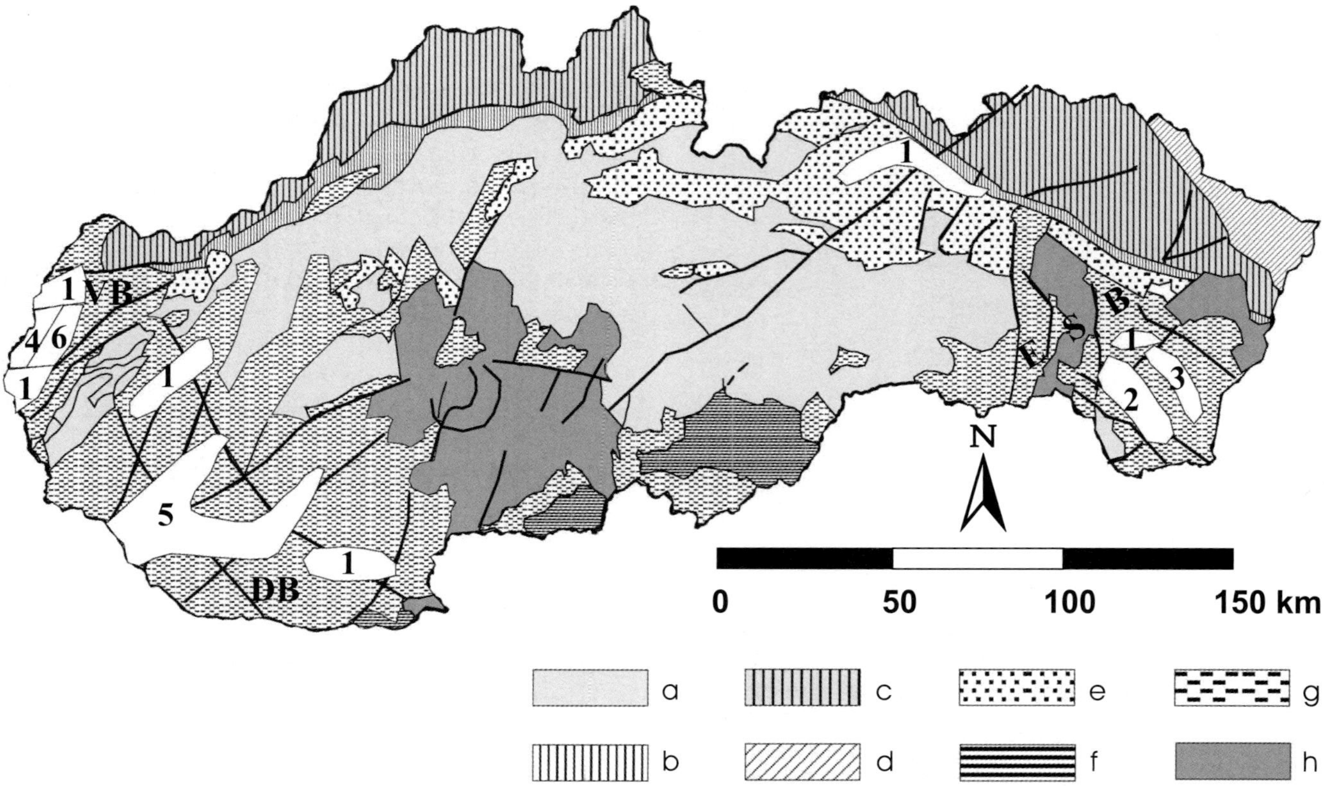

Figure 13. Sketch geological map of the Slovak Republic with hydrocarbon prospective areas in the Inner Western Carpathians. a = Neogene basins; b = Neovolcanics; c = Inner Carpathian Paleogene basins; d = Cretaceous and Paleogene of the Flysch belt; e = Mesozoic and Paleogene of the klippen belt; f = Paleozoic and Mesozoic of the Inner Western Carpathians; types of reservoirs: 1 = small (500–3000 m; 1600–10,000 ft); 2 = tight (1500–3000 m; 4800–10,000 ft); 3 = high pressure and high temperature (2500–4000 m; 8200–13,100 ft); 4 = high pressure (2500–4000 m; 8200–13,100 ft); 5 = mixed gas methane +CO_2 (1000–3000 m; 3300–10,000 ft); 6 = sour gas methane +H_2S (3200–7000 m; 10,500–23,000 ft) (3000–7000 m; 10,000–23,000 ft); DB = Danube basin; VB = Vienna basin, ESB = East Slovak basin.

Central Depression are characterized by the presence of nitrogen and carbon dioxide, commonly in the amount prevailing above the methane. High content of carbon dioxide is related to the buried stratovolcanoes of Badenian age. The origin of the nitrogen is not yet wholly clear, but according to prevailing opinion, it originated from a biogenic activity of anaerobic bacteria causing degeneration of hydrocarbons.

However, the vertical distribution of carbon dioxide suggests its main relation to the Badenian sandy horizons, its minor relation to the Sarmatian, and its sparse relation to the Pannonian horizons. This phenomenon is again related to Badenian volcanism. According to the modeling of sediment burial history and generating windows of hydrocarbon development, the generation of the gas started in the central part of the Danube basin after sedimentation in the Pannonian (approximately before 7 m.y.), i.e., after the termination of volcanism and postvolcanic activity. This suggests that the gas trap is probably represented by the sandy horizons of the early Pannonian deltaic system (Figure 15b), which is spread in the entire central part of the basin in a depth from 1000 to 3000 m (3300 to 10,000 ft). To understand the geological structure of this system and to select particular prospective objects will only be possible after a larger volume of seismic survey data becomes available.

The volume of individual reservoir reserves is estimated to be in the range of $0.5–4 \times 10^9$ m^3 ($1.75–14.1 \times 10^{10}$ ft^3) of natural gas, and possible reserves of the entire region are estimated to be 5×10^{10} m^3 (1.7×10^{12} ft^3) of natural gas.

Sour Gas Reservoirs in the Mesozoic Carbonates of the Vienna Basin Basement

By their areal extent and the level of appraisal by wells, Mesozoic carbonates represent the most explored unit occurring beneath the Neogene sediments in the

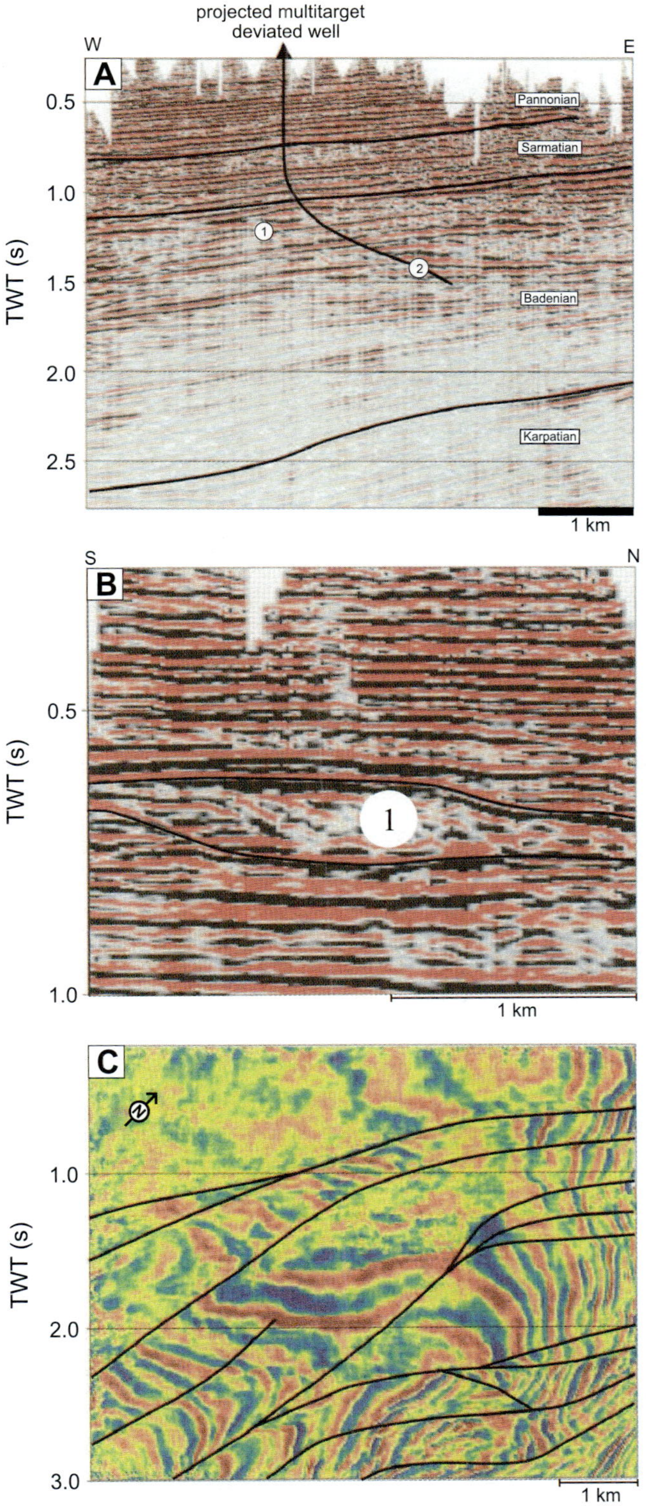

Figure 14. (A) Vienna basin, 3-D seismic Levare, crossline 223. 1 = Deltaic sand body in upper Badenian; 2 = deltaic sand body in middle Badenian. (B) Vienna basin, 3-D seismic Levare, inline 100 (RAP), 1 = deltaic sand body in lower Pannonian. (C) East Slovak basin, 3-D seismic Visoov, time slice 1200 ms, tectonic traps in facially constant horizons of upper Badenian and Sarmatian. TWT = two-way time.

Slovak part of the Vienna basin (Figure 13; reservoir type 6). In 1978, the natural-gas reservoir, called the Zavod–Mesozoic in the Upper Triassic dolomites, was discovered at a depth of approximately 4000 m (13,000 ft). The gas is composed of 95% methane and higher hydrocarbons. A negative feature is the occurrence of hydrogen sulfide (mean content is 120–150 mg L^{-1}). Recoverable reserves of natural gas and oil are 4.2×10^9 m^3 (1.48×10^{11} ft^3) and 140×10^3 t, respectively. The dolomite, which represents the reservoir rock, has two types of porosity: primary matrix porosity, which originated during the sedimentation and subsequent lithification of rocks, and secondary porosity, represented by the complex fracture system. Total porosity and permeability vary to as much as 4.5% and 2.5×10^{-13} m^2 (to 250 md), respectively.

A large spatial variability of petrophysical properties of reservoir rock caused considerable problems in the production phase. Most of the wells were relatively quickly drowned by water after starting production, and the hydrodynamic setting of the reservoir is not entirely clear so far. By the end of 2002, 1.3×10^9 m^3 (4.5×10^{10} ft^3) of natural gas was produced, i.e., 32% of reserves. At present, the reservoir is producing only through three wells, with a total capacity of 2.9×10^6 m^3 (1.02×10^8 ft^3) of natural gas per month.

Based on complex spatial variability of petrophysical properties of reservoir rock, technical exploration characteristics, and production results, it is possible to judge that this reservoir belongs to a special type of natural-gas reservoirs. In such reservoirs with a pressure water regime (active aquifer), the primary production terminates practically without pressure decline. Economic determinants of the termination of production are given by the production decrease and high increase of water production under a constantly high continuous reservoir pressure. As a result of spatial variability of petrophysical properties, the underlying water very rapidly passes through zones with good porosity and permeability into high structural levels of the gaseous part of the reservoir. However, simultaneously in the inhomogeneous reservoir rocks in marginal zones of the reservoir, considerable reserves of isolated natural gas (60–80% of gas in place) remain.

An increase of natural-gas recoverability under such conditions, according to worldwide experience, will be possible using the enhanced recovery method of aquifer depressurization. A very intensive water withdrawal from drowned wells results in a pressure decrease of aquifers and the release of gas from isolated parts of reservoir. The depressurization, realized under such conditions in many reservoirs of the

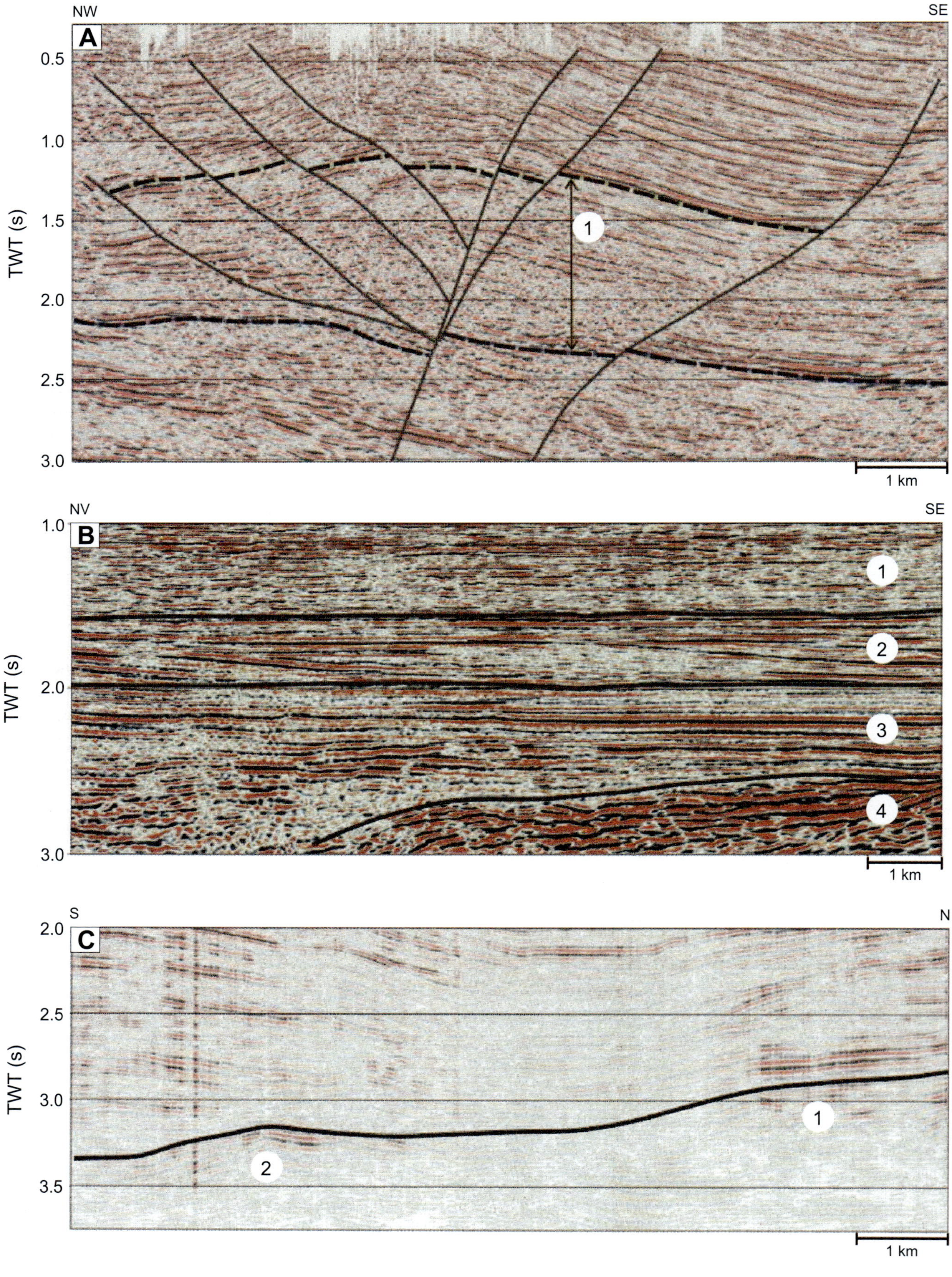
NW
SE
A
0.5
1.0
1.5
2.0
2.5
3.0
TWT (s)
1
1 km
NV
SE
B
1.0
2.0
3.0
TWT (s)
1
2
3
4
1 km
S
N
C
2.0
2.5
3.0
3.5
TWT (s)
1
2
1 km

United States and Russia, resulted in a production increase normally of 10–20%.

The remaining recoverable reserves of the Zavod–Mesozoic reservoir is 2.9 × 10^9 m^3 (1.02 × 10^{11} ft^3) of natural gas. By an increase of recoverability coefficient to 52% in the form of an intensive water withdrawal, it would be possible to recover approximately another 0.9 × 10^9 m^3 (3.1 × 10^{10} ft^3) of natural gas.

The use of a three-dimensional seismic survey data in the Slovak part of the Vienna basin identified a few objects similar to the Zavod–Mesozoic reservoir in the basement of the Vienna basin (Figure 15c). Prospective objects occurred in a depth of 3000–7000 m (10,000–23,000 ft). Reserves volume of individual reservoirs could be expected to be in the range of 0.5–5 × 10^9 m^3 (1.7–17.6 × 10^{10} ft^3) of natural gas, and the total possible resources in the region are estimated to be 5 × 10^{10} m^3 (1.7 × 10^{12} ft^3) of natural gas.

CONCLUSION

The Western Carpathians, forming the northernmost part of the European Alpides, are subdivided into the Outer and Inner Western Carpathians. Although the geologic structure of the Outer Western Carpathians resulted from Neoalpine processes, the Inner Western Carpathians have essentially a Paleoalpine structure remodeled by Neoalpine processes. The Inner Western Carpathians consist of three basic crustal-scale units represented from north to south by the Tatricum, Veporicum, and Gemericum (e.g., Mahel, 1986; Plasienka et al., 1997) (Figure 2). Each unit is formed by an early Paleozoic crystalline basement and its late Paleozoic and Mesozoic sedimentary cover, overlain by a system of Mesozoic (and locally also late Paleozoic) nappes. This is partly covered by postnappe Late Cretaceous and Tertiary sedimentary and volcanic formations. Their tectogenesis is related to complex processes of the convergence between the Carpathian–Pannonian block and the European platform, the tectonic escape of lithospheric fragments from the Alpine domain, and the rise of the Pannonian asthenolith.

The Paleogene deposits are represented by the Inner Carpathian Paleogene basin fill, deposits attached to the southern margin of the Pieniny Klippen Belt, and deposits of the Hungarian (Buda) Paleogene basin. Their thickness is more than 3500 m (11,500 ft).

The Neogene deposits mainly occur in the four largest Slovakian Neogene basins: the Vienna, Danube, South Slovakian, and East Slovakian basins. Minor Neogene deposits also occur in the Inner Carpathian intramountain depressions.

The hydrocarbon potential of the Inner Western Carpathian Paleozoic and Mesozoic units is practically exhausted because of the complex Paleoalpine and Neoalpine tectonometamorphic history. From this point of view, the most prospective regions of the Inner Western Carpathians are mostly concentrated in the Tertiary basins (the Inner Carpathian Paleogene basin and the Neogene basins), the Vienna, the Danube, the South Slovakian, and the East Slovakian basins.

The history of hydrocarbon (gas) exploration in the Neogene basins is relatively long and resulted in good knowledge on the geologic structure. Recently, after the initial use of the structural traps, attention is focused on the prospection of stratigraphic traps. Although the number of survey techniques is relatively high, a precise play analysis still needs development for prospection. Future prospection in the basins should come from better technology available today. It should focus on the identification of small, complex, high-temperature, and tight stratigraphic reservoirs, which have been known from Vienna and East Slovakian basins; however, their exploration has not been possible.

Hydrocarbon prospection in the Inner Carpathian Paleogene basin is just beginning. Preliminary data suggest a high potential for the basin; however, a better estimation needs the application of more high-resolution prospection methods.

ACKNOWLEDGMENTS

The senior author is grateful to scientific grant agency VEGA Grant No. 1/1128/04 financed by the Ministry of Education of Slovak Republic and the project "Tectogenesis of Western Carpathian Sedimentary Basins" financed by the Ministry of Environment of Slovak Republic for supporting this work.

Figure 15. (A) East Slovak basin, 3-D seismic Trebisov, crossline 327, 1 = pay zone (lower Sarmatian–upper Badenian deltaic sequence). (B) Danube basin, seismic section MXS2 (southeast part): 1 = middle and upper Pannonian; 2 = lower Pannonian deltaic bodies; 3 = Sarmatian and upper Badenian; 4 = middle and lower Badenian volcanics. (C) Vienna basin, 3-D seismic Levare, inline 71: 1 = fractured Mesozoic carbonate rocks of Neogene fill underlier (known reservoir Zavod–Mesozoic); 2 = fractured Mesozoic carbonate rocks of Neogene fill underlier (indicated by 3-D seismic survey). TWT = two-way time.

REFERENCES CITED

Andrusov, D., 1938, Geologic investigation of the inner Klippen Belt in the Western Carpathians: Part III—Tectonics (in Czech): Rozpravy statniho geologickeho Ustavu Ceskoslovenske republiky (Praha), v. 9, p. 1–135.

Andrusov, D., 1965, Geology of Czechoslovak Carpathians (in Slovak): Bratislava, Slovenska Akademia Vied, 392 p.

Andrusov, D., 1968, Grundriss der Tektonik der nordlichen Karpaten: Bratislava, Slovenska Akademia Vied, 177 p.

Andrusov, D., J. Bystricky, and O. Fusan, 1973, Outline of the structure of the West Carpathians—Guidebook for Geology Excursion X Congress of Carpathian-Balkan Association: Bratislava, Geological Survey of the Slovak Republic, p. 1–44.

Bajanik, S., V. Hanzel, J. Ivanicka, J. Mello, J. Pristas, P. Reichwalder, L. Snopko, J. Vozar, and A. Vozarova, 1983, Explanation to the geologic map of the Slovenske Rudohorie Mts., eastern part 1:50,000 (in Slovak): Bratislava, Geological Survey of the Slovak Republic, 223 p.

Bajanik, S., J. Ivanicka, J. Mello, J. Pristas, P. Reichwalder, J. Vozar, and A. Vozarova, 1984, Geological map of the Slovenske Rudohorie Mountains: Bratislava, Dionyz Stur Institute of Geology, scale: 1:50,000, 1 sheet.

Baldi, T., 1986, Mid-Tertiary stratigraphy and paleogeographic evolution of Hungary: Budapest, Akademia Kiado, 201 p.

Bezak, V., S. Jacko, M. Janak, P. Ledru, I. Petrik, and A. Vozarova, 1997, Main Hercynian lithotectonic units of the Western Carpathians, *in* P. Grecula, D. Hovorka, and M. Putis, eds., Geological evolution of the Western Carpathians: Bratislava, Mineralia Slovaca Corporation Publishing, p. 261–268.

Bezak, V., J. Michalek, and J. Kral, 2002, Are Klinisko phyllites in Nizke Tatry Mountains Hercynian nappe? (in Slovak with English summary): Mineralia Slovaca, v. 34, p. 203–206.

Biely, A., 1996, Geologic map of the Slovak Republic: Bratislava, Geological Survey of Slovak Republic, scale 1:500,000, 1 sheet.

Birkenmajer, K., 1977, Jurassic and Cretaceous lithostratigraphic units of the Pieniny Klippen Belt, Carpathians, Poland: Studia Geologica Polonica, v. 45, p. 7–159.

Birkenmajer, K., 1986, Stages of structural evolution of the Pieniny Klippen Belt, Carpathians: Studia Geologica Polonica, v. 88, no. 4, p. 7–32.

Boucek, B., and A. Pribyl, 1959, About geologic structure of the Zemplin Mountains in the Eastern Slovakia (in Czech): Geologicke Prace, Zosit, v. 52, p. 185–222.

Bujnovsky, A., M. Kochanova, and M. Pevny, 1979, Trlenska Formation—A new lithostratigraphic unit of Liassic Siprun Group (in Slovak with English summary): Geologicke Prace, Spravy, Bratislava, v. 73, p. 19–59.

Burchart, J., 1968, Rubidium-strontium isochron ages of the crystalline core of the Tatra Mountains, Poland: American Journal of Science, v. 266, p. 895–907.

Cambel, B., J. Kral, and J. Burchart, 1990, Isotope geochronology of the Western Carpathians crystalline basement (in Slovak with English summary): Bratislava, Veda Publishing, 183 p.

Chmelik, F., 1959, Report on geologic investigations of the Central Carpathian Paleogene in the area of the Nizke and Vysoke Tatry Mountains, Spisska Magura and Levocske pohorie Mountains, Reports on geologic investigations in 1956 (In Czech): Manuscript, archive Czech Geological Survey, Praha.

Corna, O., 1972, O nachodke organičeskich ostatkov v betliarskych slojach (Spišsko-gemerskije rudnyje Gory), Slovakija, nižnyj paleozoj: Geologicky Zbornik Slovenskej Akademie Vied, Geologica Carpathica, v. 23, no. 2, p. 379–382.

Csontos, L., A. Nagymarosy, F. Horvath, and M. Kovac, 1992, Tertiary evolution of the Intra-Carpathian area: A model: Tectonophysics, v. 208, p. 221–241.

Dallmeyer, R. D., H. Neubauer, H. Fritz, and M. Putis, 1993, Variscan vs. Alpine tectonothermal evolution within the Eastern Alps and Western Carpathians, Austria-Slovakia: PAEWCR conference, September 1993, Stara Lesna, Slovakia, Geol. Carpathica, v. 44, p. 255–256.

Faryad, S. W., 1991, Metamorphism of mafic rocks in the Gemericum. Mineralia Slovakia v. 23, no. 2, p. 109–122.

Filo, I., and Z. Siranova, 1996,Tomasovce Member—A new lithostratigraphic unit of the Subtatric Group (in Slovak): Geologicke Prace, Spravy, v. 102, p. 41–50.

Filo, I., and Z. Siranova, 1998, Hornad a Chrast Gravels—New regional lithostratigraphic units of the Subtatric Group (in Slovak): Geologicke Prace, Spravy, v. 103, p. 35–51.

Francu, J., and P. Muller, 1983, Organic matter maturity in Perri-Klippen Flysch of the Inner Carpathian Mountains (East Slovakia): Geologica Carpathica, v. 36, p. 483–494.

Fritz, H., F. Neubauer, M. Janak, and M. Putis, 1992, Variscan mid-crustal thrusting in the Carpathians: Part II. Kinematics and fabric evolution of the Western Tatra basement. Terra abstracts: Terra Nova, v. 4, suppl. no. 2, p. 24.

Golab, J., 1959, Outline of geologic characteristics of Western Podhale flysch: Bulletin of the Institute of Geology, v. 149, p. 225–237.

Grecula, P., 1982, The Gemericum—A segment of riftogene basin of the Paleotethys (in Slovak, with English summary): Bratislava, Mineralia Slovaca Corporation Publishing, v. 2, p. 263.

Grecula, P., 1997, Deformation related to the Alpine transpression in the Western Carpathians, *in* A. K. Sinha, F. P. Sassi, and D. Papanikolaou, eds., Geodynamic domains in the Alpine–Himalayan Tethys: New Delhi, Oxford & IBH Publ., Co., p. 339–346.

Grecula, P., and K. Egyud, 1977, Position of the Zemplin Inselberg in the tectonic frame of the Carpathians: Mineralia Slovaca (Bratislava), v. 9, p. 449–462.

Gross, P., 1998, Kezmarok Member—A new member of the Subtatric Group Paleogene (in Slovak): Geologicke Prace, Spravy, v. 103, p. 27–33.

Gross, P., et al., 1980, Geology of the Liptov depression (in Slovak): Bratislava, Dionyz Stur Institute of Geology Publishing, 242 p.

Gross, P., E. Kohler, and O. Samuel, 1984, New lithostratigraphic division of the Inner Carpathian Paleogene (in Slovak): Geologicke Prace, Spravy, v. 81, p. 103–117.

Gross, P., et al., 1999, Explanation to the geologic map of the Poprad and Hornad depressions, Levoca mountains, Spis-Saris Intermountain, Bachuren and Saris mountains (in Slovak): Bratislava, Geological Survey of the Slovak Republic, scale 1:50,000, 239 p.

Haas, J., S. Kovacs, L. Krystyn, and R. Lein, 1995, Significance of Late Permian–Triassic facies zones in terrane reconstructions in the Alpine–North Pannonian domain: Tectonophysics, v. 242, p. 19–40.

Halasova, E., N. Hudackova, K. Holcova, D. Vass, M. Elecko, and M. Pereszlery, 1996, Sea ways connecting the Fil'akovo/Petervasara Basin with the Eggenburgian/Burdigalian open sea: Slovakian Geological Magazine, v. 2, no. 96, p. 123–136.

Hovorka, D., and S. Meres, 1989, Relics of high-grade metamorphic rock in the Tatro–Veporide crystalline basement of the Western Carpathians: Mineralia Slovaca, v. 21, p. 193–201.

Hovorka, D., P. Ivan, and S. Meres, 1997, Leptyno-amphibolite complex of the Western Carpathians: Its definition, extent and genetical problems, *in* P. Grecula, D. Hovorka, and M. Putis, eds., Geological evolution of the Western Carpathians: Bratislava, Mineralia Slovaca Corporation Publishing, p. 269–280.

Ivan, P., 1994, Early Paleozoic of the Gemeric unit (Inner Western Carpathians): Geodynamic setting as inferred from metabasalt geochemistry data: Mitteilungen der Osterreichischen Geologischen Gesellschaft (Wien), v. 86, p. 23–31.

Janak, M., P. J. O'Brien, V. Hurai, and C. Reutel, 1996, Metamorphic evolution and fluid composition of garnet-clinopyroxene amphibolites from the Tatra Mountains, Western Carpathians: Lithos, v. 39, p. 57–79.

Janak, M., D. Hovorka, V. Hurai, B. Luptak, S. Meres, P. Pitonak, and J. Spisiak, 1997, High-pressure relics in the metabasites of the Western Carpathians pre-Alpine basement *in* P. Grecula, D. Hovorka, and M. Putis, eds., Geological evolution of the Western Carpathians: Bratislava, Mineralia Slovaca Corporation Publishing, p. 301–308.

Janak, M., D. Plasienka, M. Frey, M. Cosca, T. Schidt, B. Lutpak, and S. Meres, 2001, Cretaceous evolution of a metamorphic core complex, the Veporic unit, Western Carpathians (Slovakia): P-T conditions and in situ $^{40}Ar/^{39}Ar$ UV laser probe dating of metapelites: Journal of Metamorphic Geology, v. 19, p. 197–216.

Janocko, J., 1990, Sedimentary environment of late Badenian coarse-detritic deposits in northern part of Kosicka kotlina depression (In Slovak with English summary): Mineralia Slovaca, v. 22, p. 539–546.

Janocko, J., 2002, Central Carpathian Paleogene basin—Evolution and depositional systems: Mineralia Slovaca, v. 354, p. 159–180.

Janocko, J., and S. Jacko Jr., 1999, Marginal and deep sea deposits of Central Carpathian Paleogene basin, Spisska Magura region, Slovakia: Implication for basin history: Slovak Geological Magazine, v. 4, p. 281–292.

Janocko, J., B. Hamrsmid, S. Jacko Jr., and Z. Siranova, 1998, Suprafan and channel-and-levee deposits at Tichy Potok section, Levoca Mountains: Central Carpathian Paleogene basin, Slovakia: Slovak Geological Magazine, v. 4, no. 1, p. 3–15.

Janocko, J., et al., 2000, Explanation to geologic map of the Spisska Magura region 1:50,000 (in Slovak with English summary): Bratislava, Geological Survey of Slovak Republic, 174 p.

Kazmer, M., and S. Kovacs, 1985, Permian–Paleogene paleogeography along the Eastern part of the Insubric–Periadriatic Lineament system: Evidence for continental escape of the Bakony–Drauzug unit: Acta Geologica Academiae Scientiarum Hungaricae (Budapest), v. 281, no. 2, p. 71–84.

Kepinska, B., 1998, Model geologiczno-geotermalny niecki podhalanskiej (in Polish): Krakow, Studia, Rozprawy, Monografie, Polish Academy of Sciences, v. 48, 111 p.

Klinec, A., E. Planderova, and O. Miko, 1975, Early Paleozoic age of the Hron complex of Veporicum (in Slovak): Geologicke Prace, Spravy, v. 63, p. 95–104.

Kosa, E., and J. Janocko, 1999, Storm-dominated mixed siliciclastic-carbonate "Szin" ramp (Glac unit of the Silicicum superunit, Inner Western Carpathians): Implications for Lower Triassic eustacy: Slovak Geologic Magazine, v. 5, no. 3, p. 201–212.

Kotulova, J., A. Biron, and J. Sotak, 1998, Organic and illite-smectite diagenesis of the Central Carpathian Paleogene basin: Implications for thermal history (abs.): XVI Congress of Carpatho-Balcan Geological Association Abstracts, p. 293.

Kovac, M., 1986, Lower Miocene sedimentation in the area of Jablonica depression—A model bound to oblique slip mobile zone: Geologicky Zbornik Geologica Carpathica, v. 37, no. 1, p. 3–15.

Kovac, M., J. Kral, M. Marton, D. Plasienka, and P. Uher, 1994, Alpine uplift history of the Central Western Carpathians: Geochronological, paleomagnetic, sedimentary and structural data: Geologica Carpathica, v. 45, no. 2, p. 83–96.

Kovac, P., and M. Havrila, 1998, Inner structure of Hronicum: Slovak Geological Magazine, v. 4, no. 4, p. 275–280.

Kovacik, M., 1991, Micaschists with high content of Al and Fe in southeast Veporicum—Protholite and regional metamorphosis (in Slovak with English summary): Mineralia Slovaca, Bratislava, v. 23, p. 23–32.

Kovacik, M., 1996, Alpine metamorphism of Southern Veporicum—An evolution model and the problem of correlation with the Eastern Alps: Slovak Geological Magazine (Bratislava), v. 96, no. 3–4, p. 215–220.

Kral, J., M. Pereszlenyi, and D. Vass, 1990, Relationship of sediment settling velocity to genesis of East Slovakian basin and to hydrocarbon deposits: 50 years of education of geology and paleontology in Slovakia (in Slovak with English summary): Sedimentation problems of Western Carpathians. Conferences, Symposia, Seminars: Bratislava, Dionyz Stur Institute of Geology, p. 107–114.

Kral, J., W. Frank, and V. Bezak, 1996, $^{40}Ar/^{39}Ar$ spectrums from amphibolite of the amphibolite rocks from the Veporicum (in Slovak): Mineralia Slovaca, v. 28, p. 501–513.

Lesko, B., and I. Varga, 1980, Alpine elements in the West Carpathian structure and significance: Mineralia Slovaca (Bratislava), v. 12, p. 97–130.

Lexa, J., and V. Konecny, 1998, Geodynamic aspects of the Neogene to Quaternary volcanism, *in* M. Rakus, ed., Geodynamic development of the Western Carpathians: Bratislava, Geological Survey of Slovak Republic, p. 219–240.

Lexa, J., J. Konecny, M. Kaliciak, and V. Hojstricova, 1993, Space-time distribution of volcanics in the Carpatho-Pannonian region (in Slovak), *in* M. Rakus and J. Vozar, eds., Geodynamic model and deep structure of the Western Carpathians: Bratislava, Geological Survey of the Slovak Republic, p. 57–70.

Lexa, J., et al., 2000, Geological map of Western Carpathians and adjacent areas: Ministry of Environment of the Slovak Republic and Geological Survey of Slovak Republic, scale 1:500,000, 1 sheet.

Mahel, M., 1986, Geologic structure of the Czechoslovak Carpathians: Part 1—Paleoalpine units (in Slovak): Veda, Bratislava, 503 p.

Maluski, H., P. Rajlich, and P. Matte, 1993, ^{40}Ar-^{39}Ar dating of the Inner Carpathians Variscan basement and Alpine mylonitic overprinting: Tectonophysics, v. 223, p. 313–337.

Marschalko, E., and A. Radomski, 1970, Preliminary results of investigations of current directions in the Flysch basin of the Central Carpathians: Annales Societatis Geologorum Poloniae, Krakow, v. XXX, no. 3, p. 259–272.

Marschalko, R., and O. Samuel, 1993, Sedimentology of the eastern part of the Sulov conglomerates (in Slovak): Zapadne Karpaty, Serie Geologia, v. 17, p. 7–38.

Mazzoli, C., R. Sassi, and A. Vozarova, 1992, The pressure character of the Alpine metamorphism in the Central and Inner Western Carpathians (Czecho-Slovakia), *in* J. Vozar, ed., Special volume to the problems of the Paleozoic geodynamic domains, International Geoscience Program (IGCP) Project No. 276: Bratislava, Geological Survey of Slovakia, p. 109–117.

Mello, J., M. Elecko, J. Pristas, P. Reichwalder, L. Snopko, and A. Vozarova, 1996, Geologic map of the Slovensky Kras Mountains: Bratislava, Geological Survey of Slovak Republic, scale 1:50,000, 1 sheet.

Mello, J., 2000, Explanations to the geologic map of Galmus and Slovensky Raj Mountains: Bratislava, Geological Survey of Slovak Republic, 250 p.

Meres, S., and D. Hovorka, 1991, Geochemistry and metamorphic evolution of the Kohut crystalline complex mica schists (Western Carpathians): Acta Geologica et Geographica Universitatis Comenianae, Geologica, v. 47, p. 15–66.

Mock, R., M. Sykora, R. Aubrecht, L. Ozvoldova, B. Kronome, P. Reichwalder, and J. Jablonsky, 1998, Petrology and stratigraphy of the Meliaticum near the Meliata and Jaklovce villages, Slovakia: Slovak Geological Magazine, v. 4, no. 4, p. 223–260.

Mutti, E., 1992, Turbidite sandstones: Special Publication: Milan, Agip, 275 p.

Nelson, C. H., and T. H. Nilsen, 1984, Modern and ancient deep-sea fan sedimentation: SEPM Short Course 14, 404 p.

Nemcok, M., and J. Nemcok, 1994, Late Cretaceous deformation of the Pieniny Klippen Belt, West Carpathians: Tectonophysics, v. 239, p. 81–109.

Olszewska, B. W., and J. Wieczorek, 1998, The Paleogene of the Podhale basin (Polish Inner Carpathians)—Micropaleontological perspective: Przeglad Geologiczny, v. 46, no. 8-2, p. 721–728.

Petrik, I., and M. Kohut, 1997, The evolution of granitoid magmatism during the Hercynian orogen in the Western Carpathians, *in* P. Grecula, D. Hovorka, and M. Putis eds., Geological evolution of the Western Carpathians: Bratislava, Mineralia Slovaca Corporation Publishing, p. 235–252.

Petrik, I., M. Kohut, I. Broska, P. Uher, L. Hrasko, M. Janak, D. Plasienka, and V. Bezak, 2001, Granitic plutonism of the Western Carpathians: Bratislava, Veda Publishing, 116 p.

Pettijohn, F. J., P. E. Potter, and R. Siever, 1972, Sand and sandstone: New York, Springer-Verlag, 618 p.

Planderova, E., 1986, New knowledge on metasediments age from the Dumbier Mountains crystalline rocks (Jasene area) (in Slovak): Mineralia Slovaca, v. 18, p. 237–251.

Plasienka, D., 1993, Structural pattern and partitioning of deformation in the Veporic Foederata cover unit (Central Western Carpathians), *in* M. Rakus and J. Vozar, eds., Geodynamic model and deep structure of the Western Carpathians: Conferencie, Sympozia, Seminare: Bratislava, Geological Survey of Slovak Republic Publishing, p. 269–277.

Plasienka, D., 1997, The story of a nappe: The Krizna unit of the Western Carpathians, *in* D. Plasienka, et al., eds., Alpine evolution of the Western Carpathians and related areas (international conference): Abstract and introductory articles to the excursion: Bratislava, Geological Survey of the Slovak Republic, p. 37–38.

Plasienka, D., M. Havrila, J. Michalik, M. Putis, and D. Rehakova, 1997, Nappe structure of the Central Carpathians, *in* D. Plasienka, et al., eds., Alpine evolution of the Western Carpathians and related areas (international conference): Abstract and Introductory articles to the excursion: Bratislava, Geological Survey of Slovak Republic, p. 139–161.

Poller, U., I. Broska, F. Finger, P. Uher, M. a Janak, 2000, Permian age of Gemeric granites constrained by single zircon and EMPA monazite dating: Mineralia Slovaca, v. 32, no. 3, p. 189–190.

Ratschbacher, L., W. Frisch, H. G. Linzer, B. Sperner, M. Meschede, K. Decker, M. Nemcok, J. Nemcok, and R. Grygar, 1993, The Pieniny Klippen Belt in the Western Carpathians or northeastern Slovakia: Structural evidence for transpression: Tectonophysics, v. 226, p. 471–483.

Rericha, M., 1992, Late Badenian delta in the East Slovakian Neogene Basin: Mineralia Slovaca, v. 24, p. 63–68.

Samuel, O., 1972, Several remarks to the lithofacial and stratigraphic division of the Klippen Belt Paleogene (in Slovak): Geologicke Prace, Spravy, v. 59, p. 285–298.

Sassi, F. P., and A. Vozarova, 1987, The pressure character of the Hercynian metamorphism in the Gemericum (West Carpathians, Czechoslovakia): Rendiconti della Societa Italiana di Mineralogia e Petrologia, v. 42, p. 73–81.

Siman, P., V. Johan, P. Ledru, V. Bezak, and J. Madaras, 1996, Deformation and P-T conditions estimated in "layered migmatites" from southern part of Veporicum crystalline basement (Western Carpathians; Slovakia): Slovak Geological Magazine, v. 3–4, p. 209–213.

Snopkova, P., 1964, Report on palynologic research of the early Paleozoic samples of the Spissko-Gemerske Rudohorie Mountains, (in Slovak): Spravy o geologickych vyskumoch v roku 1963, Geological Survey of the Slovak Republic, p. 63–64.

Sotak, J., A. Biron, and J. Spisiak, 1993, The Penninic "pull apart" dome in the pre-Neogene basement of the Transcarpathian depression (Eastern Slovakia): Geologica Carpathica, v. 44, p. 11–16.

Spisiak, J., and P. Pitonak, 1992, Banded amphibolic rocks—Pre-Variscan basement of the Western Carpathians? Alcapa Meeting Graz: Terra Nova Abstracts, v. 147, suppl. 2, p. 63.

Spisiak, J., D. Hovorka, and P. Ivan, 1985, Klátovská skupina-reprezentant metamorfitov amfibolitovej fácie vnútorných Západných Karpát: Geologicke Práce, Spravy, v. 82, p. 205–220.

Sztano, O., 1994, The tide-influenced Petervasara sandstone, early Miocene, Northern Hungary: Sedimentology, paleogeography and basin development: Geologica Ultraiectina, v. 120, 153 p.

Vass, D., 1995, The origin and disappearance of Hungarian Paleogene basins and short-term lower Miocene basins in Northern Hungary and Southern Slovakia: Slovak Geological Magazine, v. 2, p. 81–95.

Vass, D., and J. Cvercko, 1985, Lithostratigraphic units of the East Slovakian lowland Neogene (in Slovak): Geologicke Prace, Spravy, v. 82, p. 111–126.

Vass, D., M. Kovac, V. Konecny, and J. Lexa, 1988, Molasse basins and volcanic activity in Western Carpathian Neogene— Its evolution and geodynamic character: Geologicky Zbornik, Geologica Carpathica, v. 39, no. 5, p. 539–561.

Vass, D., M. Elecko, and J. Vozar, eds., 1996, The pre-Tertiary sequences of the Lucenec basin and the Cerova Hills—Explanation notes of the geologic and tectonic maps (in Slovak): Geologicke Prace, Spravy, v. 101, p. 41–42.

Vass, D., M. Elecko, J. Janocko, S. Karoli, M. Pereszlenyi, J. Slavik, and M. Kaliciak, 2000, Paleogeography of the East Slovakian basin: Slovak Geological Magazine, v. 6, no. 4, p. 377–407.

Vozar, J., C. Tomek, and A. Vozarova, 1993, Reinterpretation of pre-Neogene basement of East Slovakian basin (in Slovak): Mineralia Slovaca, v. 25, no. 6, Geovestnik, p. 1–2.

Vozarova, A., 1993, Pressure-temperature conditions of metamorphism in the Northern part of the Branisko crystalline complex: Geologicky Zbornik, Geologica Carpathica, v. 44, p. 219–232.

Vozarova, A., and J. Vozar, 1979, Permian of the West Carpathians: Guidebook for geological excursion: Bratislava, Dionyz Stur Institute of Geology, p. 1–79.

Vozarova, A., and J. Vozar, 1988, Late Paleozoic in the West Carpathians: Bratislava, Dionyz Stur Institute of Geology, 314 p.

Vozarova, A., and J. Vozar, 1996, Terranes of the West Carpathian–North Pannonian domain: Slovak Geological Magazine (Bratislava), v. 1, no. 96, p. 65–85.

Wagreich, M., 1995, Subduction tectonic erosion and Late Cretaceous subsidence along the northern Austroalpine margin (Eastern Alps, Austria): Tectonophysics (Amsterdam), v. 242, p. 63–78.

Westwalewicz-Mogilska, E., 1986, New interpretation of Podhale flysch sediment genesis (in Polish): Przeglad Geologiczcny, v. 12, p. 690–698.

Wieczorek, J., 1989, Model Hecho for Podhale flysch? (in Polish): Przeglad Geologiczny, v. 37, no. 9, p. 419–423.

19

Tari, Gábor C., and Frank Horváth, 2006, Alpine evolution and hydrocarbon geology of the Pannonian Basin: An overview, *in* J. Golonka and F. J. Picha, eds., The Carphathians and their foreland: Geology and hydrocarbon resources: AAPG Memoir 84, p. 465–478.

Alpine Evolution and Hydrocarbon Geology of the Pannonian Basin: An Overview

Gábor C. Tari

Vanco Energy Company, Houston, Texas, U.S.A.

Frank Horváth

Geophysical Department, Eötvös University, Budapest, Hungary

ABSTRACT

The well-understood Neogene Pannonian Basin was superimposed on an earlier, mostly Cretaceous compressional realm. Based on the interpretation of subsurface data sets, such as academic and industry reflection seismic and well data, the Eoalpine units of the Eastern Alps can and should be correlated beneath the northwest Pannonian Basin with corresponding tectonic units of the Western Carpathians. Similarly, the Eoalpine folded belt beneath the southeast Pannonian Basin has a structural style very similar to that of the internal part of the Eastern and Southern Carpathians. The level of understanding of these folded belts subcropping at the base of the Pannonian Basin system is poor compared to those of the existing models of the classical folded belt of the Carpathians. This mismatch is primarily caused by the subsurface nature of the problem and the pronounced overprint by Neogene extensional tectonics.

In the Pannonian Basin, several well-established and a few emerging exploration plays can be outlined. The key elements of the most important plays include Neogene compactional anticlines; structures related to neotectonic inversion and tectonic reactivation; stratigraphic traps in the deepest part of Neogene subbasins; Paleogene sandstones in structural traps; Senonian–Triassic carbonates in complex structural traps; and Mesozoic reservoirs in the Eoalpine (Cretaceous) and Mesoalpine (Paleogene) folded belts. The future of hydrocarbon exploration in the Pannonian Basin is largely the function of better geophysical imaging and advanced geological understanding of what lies beneath the mature Neogene extensional basin.

INTRODUCTION

The Pannonian Basin occupies the central part of the European Alpine belt. At the western margin of the Pannonian Basin, the thrust-fold belts of the Western, Eastern, and Southern Alps branch to the Dinarides to the southeast and to the Western Carpathians to the northeast. Although the Dinarides continue to the southeast with only a minor change in strike into the Hellenides, the Western, Eastern, and Southern

DOI:10.1306/985733M843141

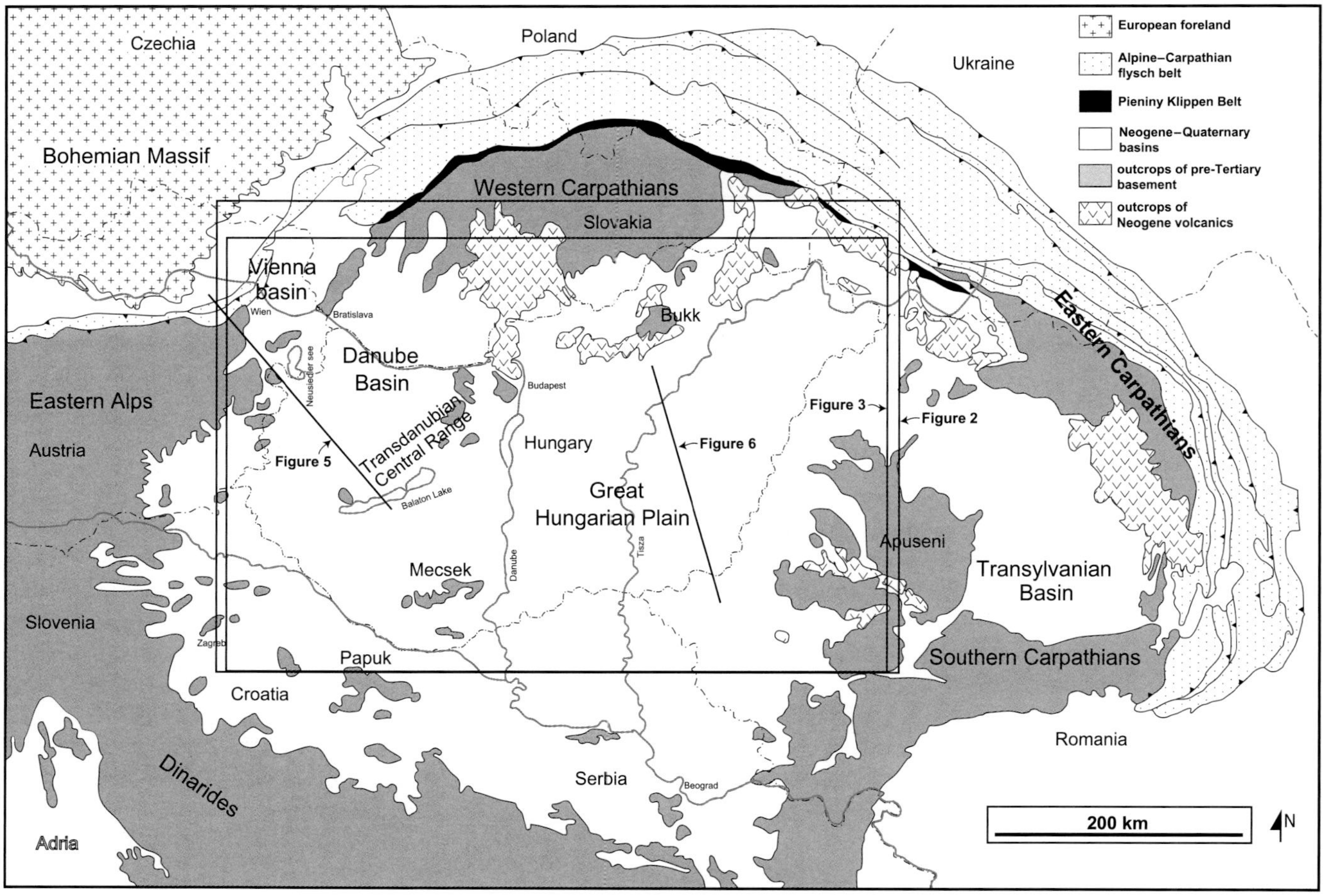

Figure 1. Simplified geology of the Carpathian–Pannonian system. The locations of Figures 2 and 3 are shown as insets.

Carpathians form an almost complete loop before continuing into the Balkans (Figure 1). Thus, the Alpine chain is about 300, 1000, and 400 km (190, 620, and 250 mi) wide in the Alps, Dinarides–Pannonian Basin–Carpathians, and in the Hellenides–Balkans, respectively. The pronounced widening in the Pannonian sector is mainly the result of the Neogene extensional basin formation in this central area. The surrounding Eastern Alps, Carpathians, and Dinarides project below the Pannonian Basin, which is superimposed on these Alpine thrust-fold belts (Bally and Snelson, 1980). However, the existence of Cretaceous Alpine nappes in the pre-Tertiary basement of the Pannonian Basin has been debated since the beginning of the century (for a historical perspective, see Tari, 1994).

The present-day assembly of the superimposed basin complex and the underlying structures of very different style and age in the Pannonian Basin–Alps–Carpathians–Dinarides system suggest a very complicated origin. Individual tectonostratigraphic units can be clearly outlined for distinct Alpine orogenic regimes, but any subdivision is valid only for very limited periods (see below). The major periods of a geodynamic scenario for the Alps–Carpathians–Dinarides–Pannonian Basin system are

1) Late Permian–Early Cretaceous: Two periods of continental breakup occurred during this time. Although the Triassic rifting aborted in most of the area, the Jurassic extension led to the formation of the Tethys Ocean.
2) middle Cretaceous–Paleocene: This was a time of major overthrusting in the Alpine–Carpathian system, when many of the oceanic troughs and passive margins disappeared. This interval is traditionally subdivided into several periods of compressional events.
3) Eocene–early Miocene: This interval represents the second major period of compression and collision in the Alpine thrust-fold belt, following a period of relative tectonic quiescence. This period shares many similarities with the former one.
4) middle Miocene–Holocene: Although continuing convergence between Europe and Africa still forms thrust-fold belts in certain areas, in the Pannonian Basin, this interval is largely characterized by the

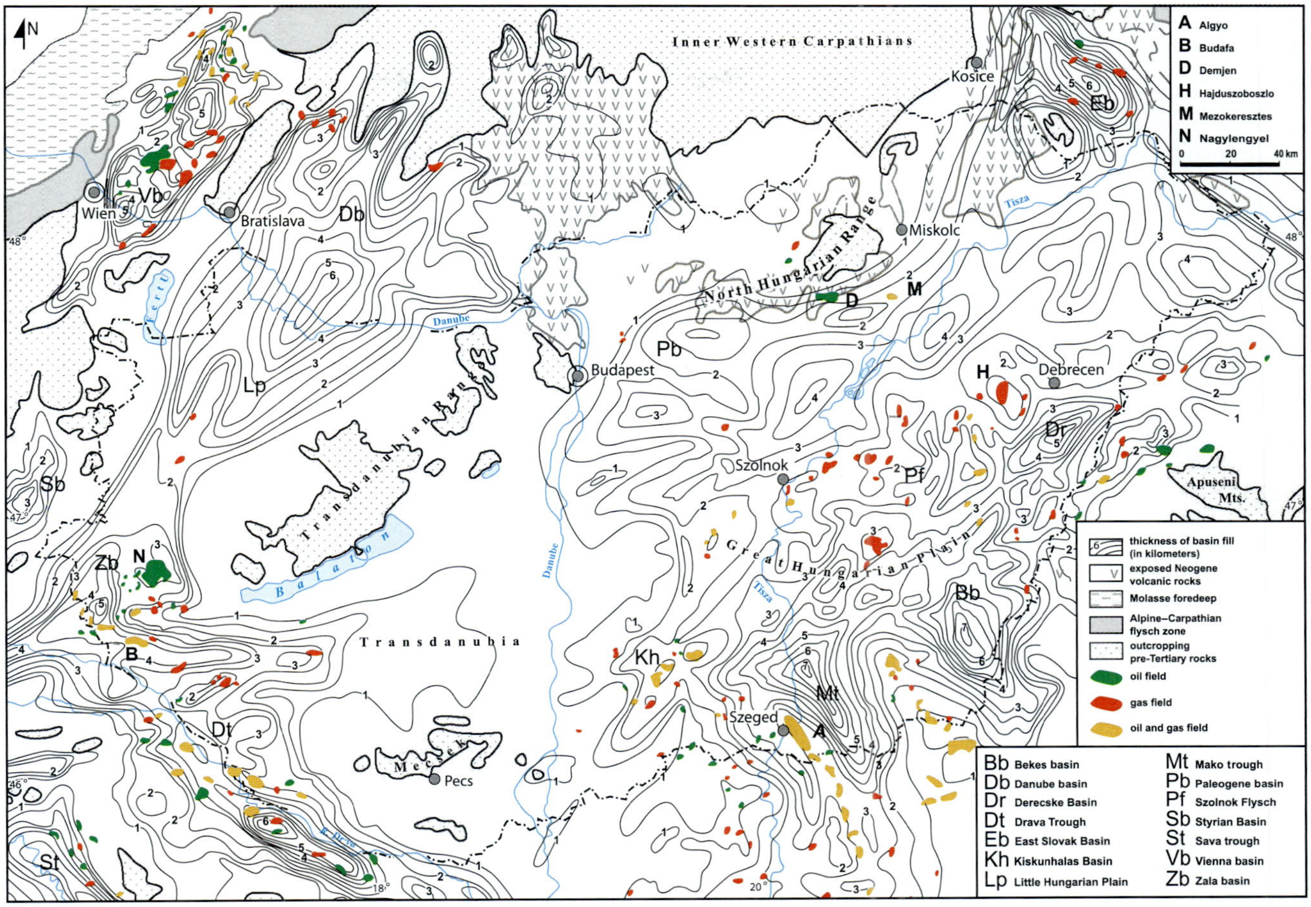

Figure 2. Hydrocarbon fields in the Pannonian Basin (modified after Dank, 1988).

opening of superimposed extensional basins. The marked change from the previous stage was caused by a large-scale continental escape period in the early Miocene, which disintegrated the former Alpine thrust-fold belt.

These broad periods are close to those defined by Trumpy (1980) in the classical Swiss sector of the Alps. Although slight differences in the timing exist, we will refer to his terms for the major Alpine stages outlined above. These are the early Alpine, the Eoalpine (or Paleoalpine), the Mesoalpine, and the Neoalpine stages, respectively.

The structure of the mountain belts surrounding the Pannonian Basin was formed during the above Alpine stages. All these belts have a dominant vergence outward from the intra-Carpathian basins (Figure 1). Within the thrust-fold belts, an internal Eoalpine zone of Mesozoic nappes is distinguished from an outer Tertiary belt of Eoalpine and Neoalpine nappes. The internal zone is somewhat discontinuous, but the outer flysch belt is uninterrupted at the perimeter of the Alps and the Carpathians.

In the intra-Carpathian area, the main Alpine stages outline three major tectonostratigraphic levels. The upper, Neoalpine level comprises the Neogene Pannonian Basin *sensu stricto*; for a detailed description of the stratigraphy, see Vakarcs et al. (1994, 1998). The middle, Mesoalpine level consists of Paleogene basins (e.g., Tari et al., 1993). The underlying basement level is a complicated assembly of Mesozoic successions formed during the early and Eoalpine stages (e.g., Tari, 1995a).

The last major monograph on the Pannonian Basin (Royden and Horvath, 1988) provided a modern understanding of the Neoalpine evolution of the region, also in relation to the surrounding Carpathians. However, the pre-Neogene evolution of the substrata of the Pannonian Basin has not been discussed in a systematic manner with the exception of just a few studies (e.g., Tari, 1994, 1995b). Moreover, the present-day understanding of the hydrocarbons found to date (Figure 2) requires a more in-depth look into the geology and geophysics of the intra-Carpathian area (e.g., Horvath, 1993; Horvath and Tari, 1999).

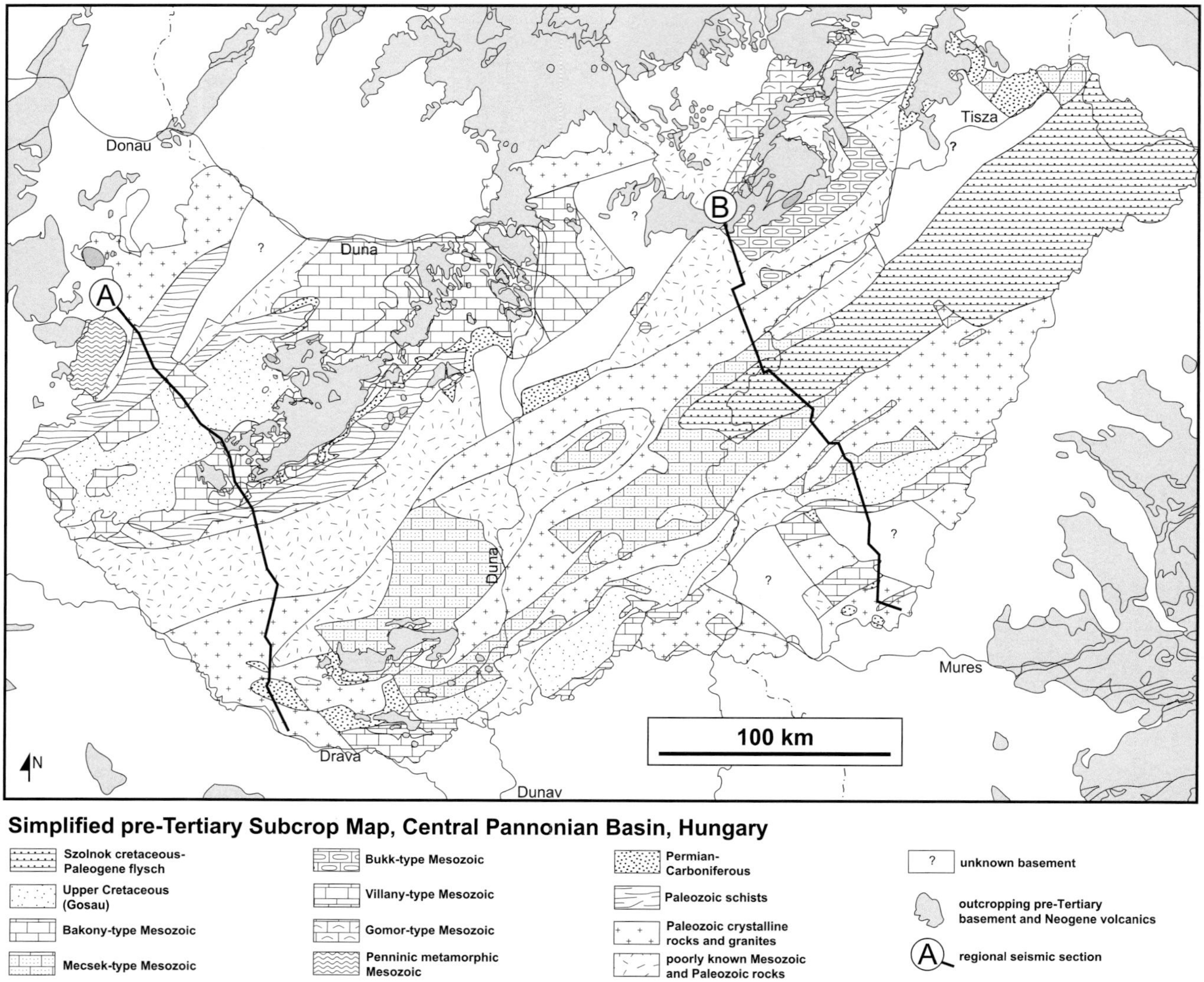

Figure 3. Index map of regional seismic transects shown in Figure 4 superposed on the simplified pre-Tertiary geologic map of the Hungarian part of the Pannonian Basin.

The purpose of this chapter is to emphasize the presence of Eoalpine and Mesoalpine folded belts and nappe systems beneath the Pannonian Basin correlatable with the classical units of the surrounding Alps and Carpathians, which is the main topic of this volume. As most of the pre-Tertiary basement subcrops the Neogene basin fill, our understanding relies dominantly on subsurface data sets, such as reflection seismic, potential field, and well data.

REGIONAL SEISMIC REFLECTION SECTIONS ACROSS THE HUNGARIAN PART OF THE PANNONIAN BASIN

To illustrate the problems outlined above, regional composite seismic transects (Figure 3) were constructed across the central Hungarian part of the Pannonian Basin (see Tari, 1994, for details). These sections are of different vintages, representing different data acquisitions, processing, and displays. The data, however, come mostly from the oil industry, primarily from the Geophysical Exploration Company of Hungary. Supplemental data were obtained from an academic organization, the Eotvos Lorand Geophysical Institute in Budapest.

For brevity, only 2 of the 18 regional transects of Tari (1994) are reproduced here (Figure 4). The first composite profile is located in Western Hungary, trending generally northwest–north-northwest. The line is regarded mainly as a dip line with respect to Eoalpine and Neoalpine structures. Similarly, the other regional transect in southeastern Hungary is also a dip line, but mostly with respect to the Eoalpine edifice.

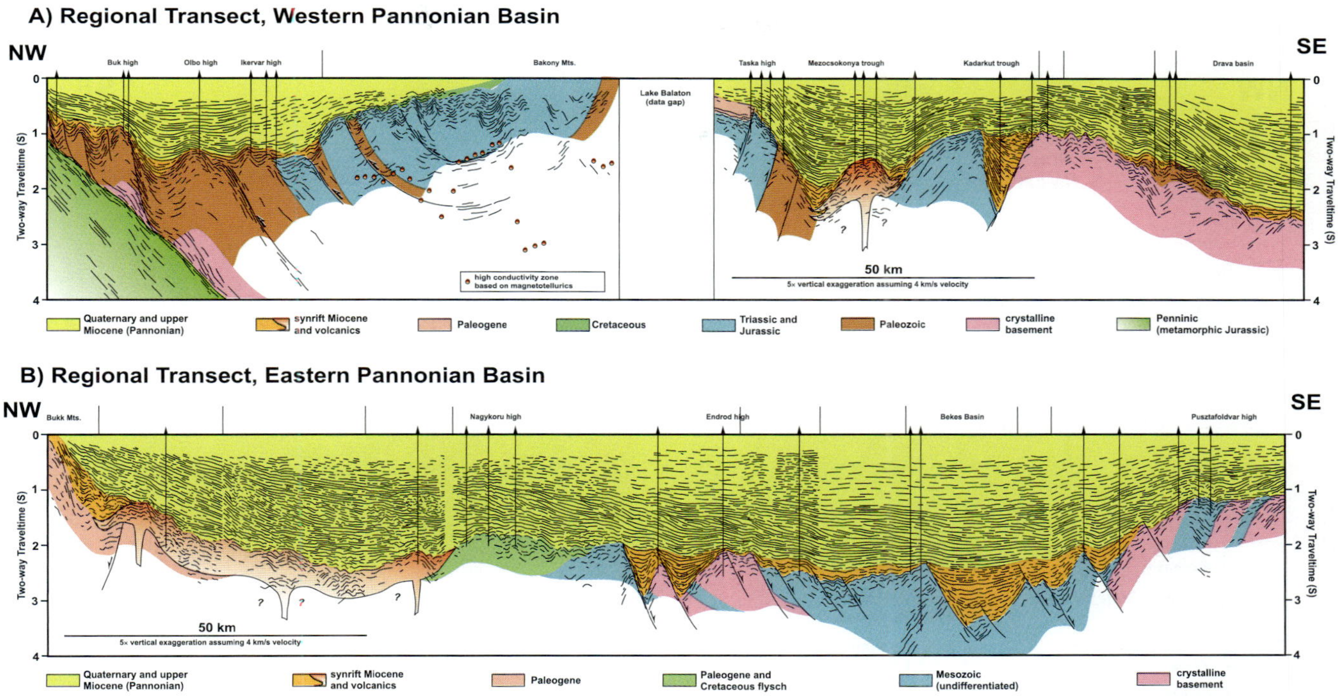

Figure 4. Line drawing interpretation of regional seismic transects (A) in the western Pannonian Basin and (B) in the eastern Pannonian Basin adapted from Tari (1994). For location, see Figure 3.

To give a unified interpretation of these different data sets, the line drawing interpretations of the seismic data were chosen. The detailed interpretation was also aided by the projection of more than 200 wells (listed in Tari, 1994) into the sections.

WESTERN PANNONIAN BASIN

Section A (Figure 4A) covers the Danube and the Drava subbasins of the Pannonian Basin system and the Bakony Mountains, where the pre-Tertiary basement outcrops (Figure 3). As to the subbasins, the postrift sediments tend to be very thick, as much as 5 km (3 mi), whereas the relatively thin synrift succession is just locally present in fault-bounded grabens. Major southeast-dipping low-angle normal faults control the relief of the basement. These detachment faults typically ignore the preexisting Alpine structural grain because the extensional reactivation of Eoalpine overthrust contacts could be demonstrated only in a few cases (Tari, 1996a). The most important of the detachment faults is the Rechnitz fault, which exposed the lowermost Penninic tectonic unit in the Alpine nappe system in a metamorphic core complex (Tari et al., 1992; Tari, 1996a).

On the southeastern flank of the Danube Basin, repetitions of Paleozoic and Mesozoic units were mapped by Tari (1994), suggesting thrust imbrications. Unfortunately, the supposed overthrusts were never penetrated by drilling. However, in the Mesozoic outcrops of the Bakony Mountains and the Balaton Highland, a few southeast-verging thrusts and associated anticlines (e.g., Loczy, 1913; Budai, 1991) can be mapped. A much debated question is whether the thrusts interpreted on seismic data from the northwest flank of the Bakony Mountains and the Liter, Veszprem, and other smaller thrusts in the Balaton Highland should be seen as the manifestation of a large Eoalpine nappe structure. Several authors following Horvath et al. (1987) and Rumpler and Horvath (1988) accepted the large-scale allochthoneity of the Bakony Mountains as a working hypothesis. This hypothesis has been later confirmed by systematic subsurface mapping (Tari, 1994, 1995b, 1996b).

To understand the underlying Eoalpine structure, the very significant footprint of the Neogene tectonics should be removed. The idealized cross section (Figure 5) across the Alpine–Pannonian junction area (Figure 2) based on a regional transect by Tari (1996b) is a qualitative attempt to emphasize the Eoalpine structural picture. Note that the datum level of this transect is the erosional surface at the base of the combined Senonian–Tertiary basin fill.

The most significant element in the regional transect is the structural connection of the Northern Calcareous Alps with the nappe system of the Bakony

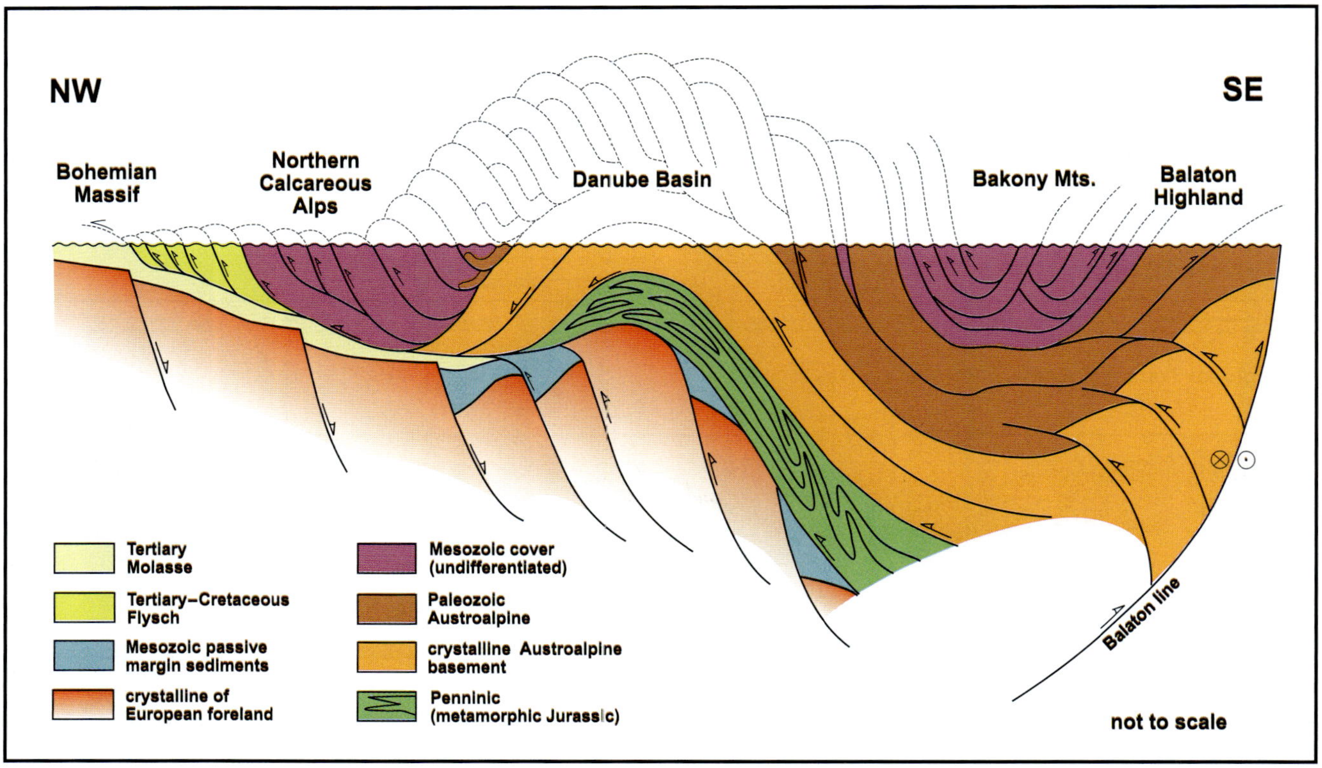

Figure 5. Idealized regional transect through the northwest Pannonian Basin in the junction area of the Eastern Alps and the Western Carpathians; for location, see Figure 1. Note that the reference level is the erosional base of the Tertiary, which, for the most part, coincides with the Neogene basin fill of the Pannonian Basin. For a detailed explanation, see text.

Mountains, both of them in an upper Austroalpine position, i.e., on top of the Alpine nappe system. This once continuous nappe unit disintegrated because of large-scale erosion during the late Oligocene, when basement-involved thrusting occurred at the edge of the European platform. This deep-seated thrusting is responsible for the formation of a large anticlinorium forming a high plateau (i.e., Augenstein Plateau of Tollmann, 1987) following the southwest–northeast trend of the present-day Styrian and Danube subbasins (Figure 1). The very important corollary of this interpretation is that the Eoalpine units of the Eastern Alps can and should be correlated beneath the northwest Pannonian Basin with corresponding tectonic units in the Western Carpathians.

EASTERN PANNONIAN BASIN

The pattern depicted in the pre-Tertiary basement on section B in the eastern Pannonian Basin (Figure 4B) is somewhat similar to that of the Danube Basin. According to the abundant well control, slivers of crystalline basement alternate with Mesozoic cover units with a general northeast–east-northeast trend (Figure 3). The corresponding map view of the pre-Tertiary basement has been interpreted by many as the result of pervasive Neogene strike-slip faulting (e.g., Balla, 1987). However, based on the analysis of the Mesozoic facies, a generally accepted structural correlation exists (e.g., Grow et al., 1994) with the Eoalpine nappe system exposed in the Apuseni Mountains (e.g., Bleahu, 1976) as part of the Inner Carpathians (Figure 1). It is our interpretation that the fundamental reason for the map view of Mesozoic and crystalline basement units is that it is the pre-Tertiary subcrop expression of Eo- and Mesoalpine basement-involved folded belts.

Based on the interpretation of a large number of industry seismic data and well-documented well penetrations of imbricated sequences in the pre-Tertiary basement (e.g., Pap, 1990), an idealized cross section was constructed to emphasize the Eoalpine structure of the eastern Pannonian Basin (Figure 6). Similarly to the section shown in Figure 5, the datum level here is also the base of the Senonian–Tertiary basin fill.

The substratum of the Pannonian here again shows an Eoalpine thrust-fold belt with a pronounced involvement of the crystalline basement. Because the facies of the Mesozoic cover sequence is consistent with the northern (i.e., European) passive margin of the Tethys

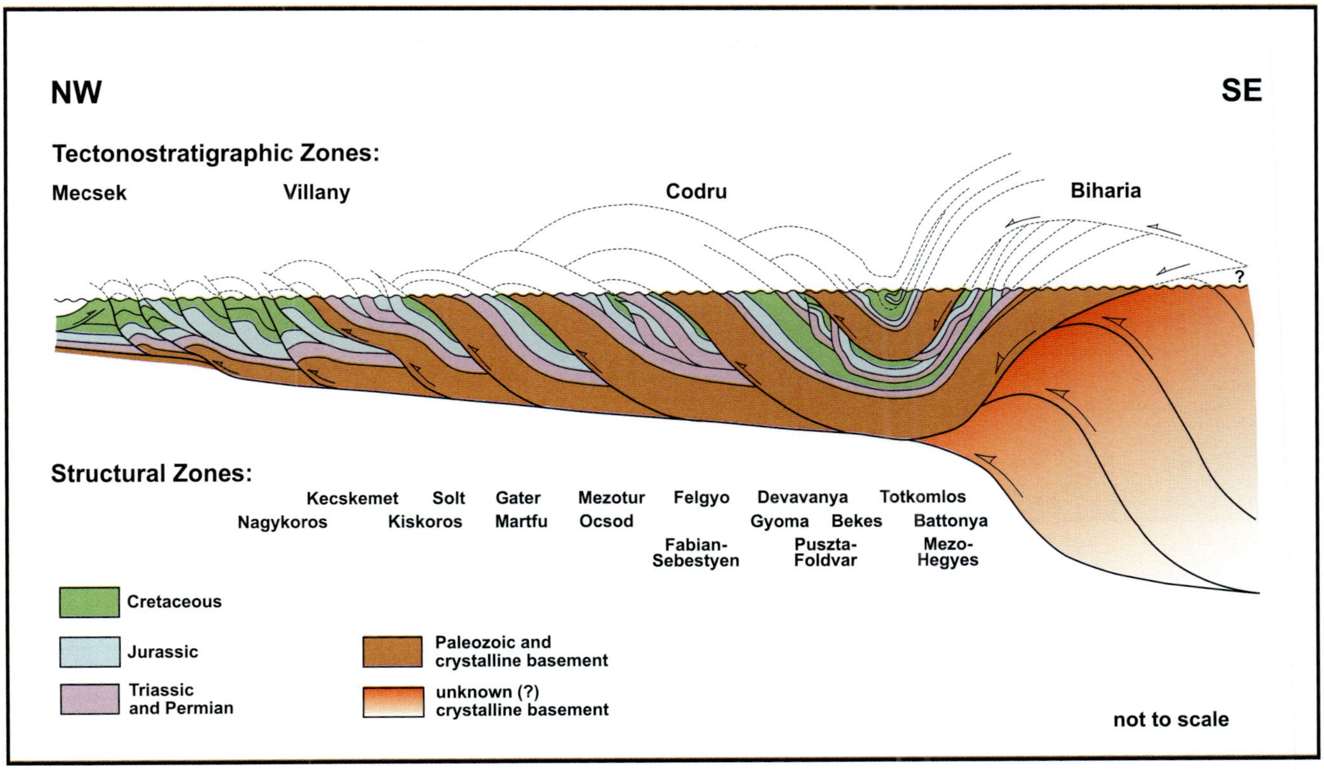

Figure 6. Idealized regional transect through the southeast Pannonian Basin; for location, see Figure 1. Note that the reference level is the erosional base of the Tertiary, which, for the most part, coincides with the Neogene basin fill of the Pannonian Basin. For a detailed explanation, see text.

(e.g., Kovacs et al., 1989), this thrust-fold belt represents the north-vergent foreland deformation during the middle Cretaceous. The major antiformal stack in the southern part of the section is interpreted to be the result of a Late Cretaceous age of compression involving a deeper, possibly middle crustal decollement level. This major detachment is responsible for the pronounced subregional northerly dip of both Mesozoic and crystalline units in southeastern Hungary (Figure 4B). In addition, the fragments of a Senonian basin (Fulop et al., 1987) on top of the Eoalpine folded belt are interpreted in terms of a flexural basin related to this compressional episode.

Comparing the overall Eoalpine structure of the western and eastern Pannonian Basin (Figure 4), differences exist in terms of the involvement of the pre-Alpine basement. The transect between the Eastern Alps and the Western Carpathians shows the typical succession of upper, middle, and lower Austroalpine nappe system. The upper Austroalpine (Gemeric) is dominated by unmetamorphosed cover units with little or no basement involvement; the middle Austroalpine (Veporic) is composed mostly of basement; and the lower Austroalpine (Tatric) is characterized by basement nappes with thin metamorphosed cover sequences. The transect in the southeast Pannonian Basin has a structural style very similar to that of the internal sector of the Eastern and Southern Carpathians, i.e., all the nappe units appear to have crystalline basement and Mesozoic cover units.

Note the important difference as to the origin of the units involved. In the western Pannonian Basin, the Austroalpine units are of African affinity that originated from the southern margin of the Tethys. Therefore, because of the collision between Africa and Europe, elements from both sides of the Tethys and the Penninic ocean in between are present in the Eoalpine structure. In contrast, the Eoalpine structural edifice in the eastern Pannonian Basin is made up of entirely European elements.

BRIEF PETROLEUM GEOLOGY OF THE PANNONIAN BASIN

The hydrocarbon fields of the Pannonian Basin system are shown in Figure 2. The petroleum geology of the Hungarian part of the basin has been discussed in several papers during the last two decades (e.g., Dank, 1988; Kokai and Pogacsas, 1991; Szalay and Koncz, 1991; Kokai, 1994; Teleki et al., 1994). To understand

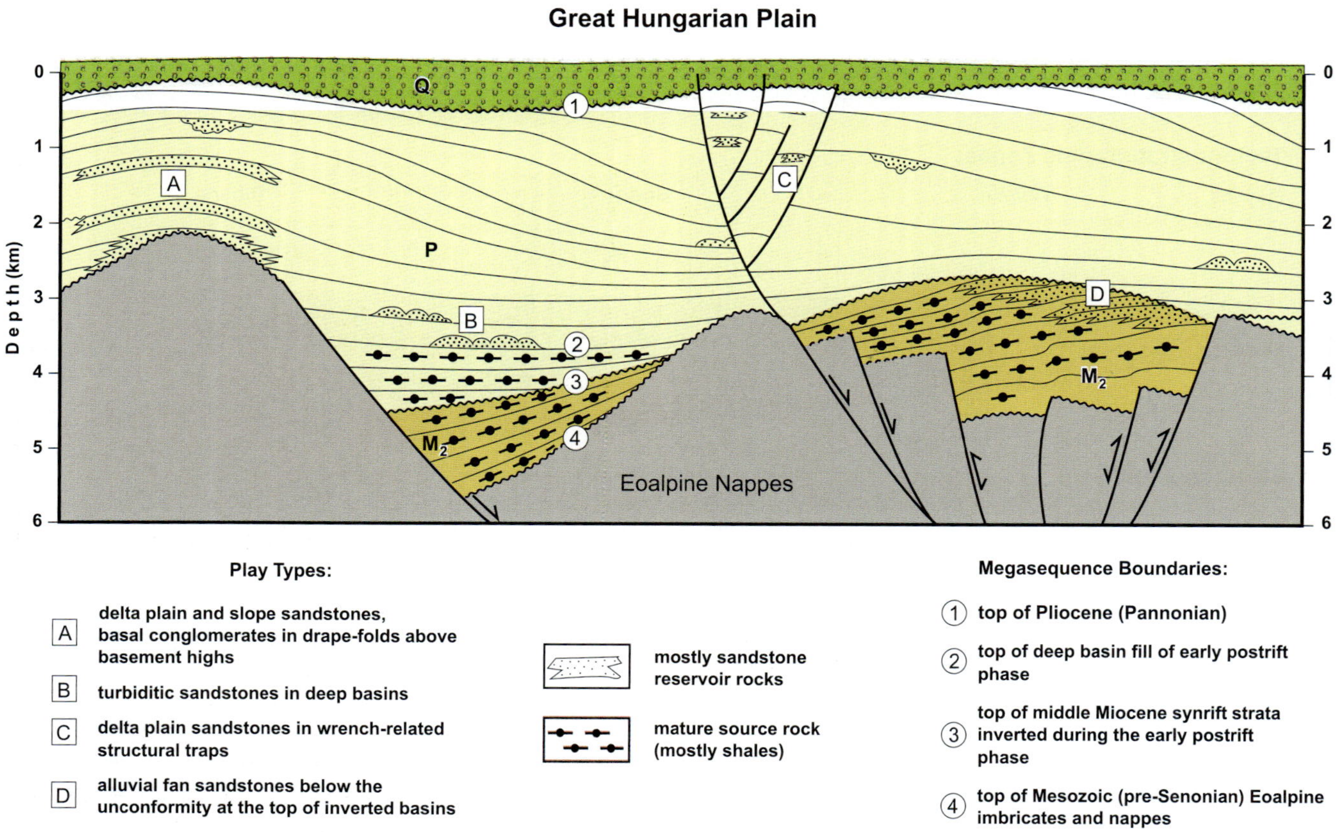

Figure 7. Idealized cross section showing the petroleum systems of the Neogene basin fill at the Great Hungarian Plain (modified after Horvath and Tari, 1999).

the origin of hydrocarbon fields discovered to date, Horvath and Tari (1999) attempted to define petroleum systems in the Pannonian Basin. These systems are reviewed below in a top-to-bottom sequence. The upper system is associated with the Neogene tectonostratigraphic unit; the middle system is related to the Paleogene tectonostratigraphic unit; and the lower system comprises the Senonian and the Mesozoic–Paleozoic tectonostratigraphic units.

The petroleum systems to be defined can be seen in Figures 7–9. It is to be noted that Figures 8 and 9 are not devoted to the illustration of a single petroleum system. Instead, they represent idealized cross sections, showing characteristic structural features and other important elements of petroleum systems along the sections.

Neogene Tectonostratigraphic Unit

Most of the known hydrocarbon fields in the Pannonian Basin are located in the Neogene basin fill. Figure 7 presents an idealized profile in the Great Hungarian Plain showing the distribution of mature source rocks and potential reservoirs. Vitrinite reflectance (R_o) data and maturation history modeling show that, in general, the oil-generation window ($0.6 < R_o < 1.3\%$) is located in the depth range of 2.4–4.3 km (1.5–2.7 mi) (Horvath et al., 1988; Szalay and Koncz, 1991). Middle Miocene to early Pliocene (Badenian, Sarmatian, and locally Pannonian) strata are situated in this range, and their pelagic marls are characterized by total organic carbon content of 1–2%, but locally as high as 5% (Clayton and Koncz, 1994a; Clayton et al., 1994). Rock-Eval pyrolysis studies indicate that they contain type III and also type II–III kerogens; hence, the generation of both oil and gas has occurred (Szalay and Koncz, 1991). Migration pathways were provided by unconformities at the bottom and top of the synrift sequence. Structural closures form traps in different structural settings, most typically above basement highs, because of differential compaction.

The largest fields in compactional anticlines can be found in the southeastern part of the Great Hungarian Plain, above basement highs at the western and eastern flank of the Mako trough. The Algyo field (see Magyar et al., 2006) has been the largest hydrocarbon discovery in Hungary since 1965, with initial recoverable reserves of about 500 million bbl of oil and 3.5 tcf of gas (Kokai and Pogacsas, 1991). The field produces from the

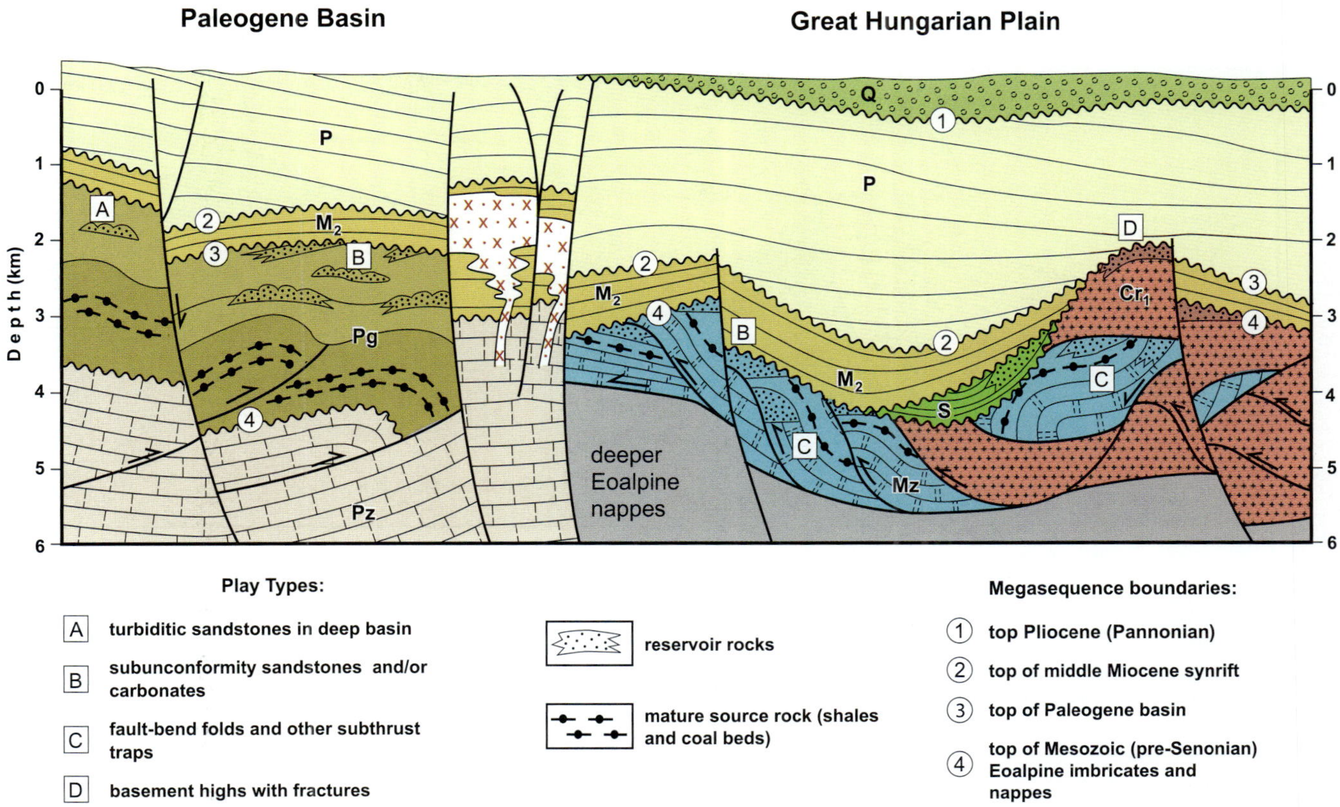

Figure 8. Idealized cross section showing the petroleum system of the Paleogene basin fill in North Hungary and of the Mesozoic–Paleozoic substrata below the Great Hungarian Plain (modified after Horvath and Tari, 1999).

2500- to 1600-m (8200- to 5200-ft) depth interval, and the individual oil, gas, and mixed pools can be found in the following stratigraphic units: fractured basement and overlying basal conglomerate, prodelta turbidites, delta-slope sandstone bodies, and delta-plain channel-fill and point-bar deposits. The most favorable reservoirs are the delta-plain sandstones, with porosities as much as 29% and permeabilities as much as 235 md. Top seals of these reservoirs are pelitic beds associated with the prodelta, delta-slope, and plain environments (Szalay and Koncz, 1993; Mattick et al., 1994).

It should emphasized that the anticline play has been, for many decades, an obvious concept in hydrocarbon exploration in the Pannonian Basin, and every sizable seismically detected structure has been tested by drilling. This applies, however, only to the apex of the structures, and significant downflank satellite fields can be present around known fields. It is also possible that hydrocarbon pools can be found apart from compactional anticlines in stratigraphic traps. Above the deep basins isolated, deep-water sandstone units (i.e., basin-floor fans, turbidites, and slumps; Phillips et al., 1994) occur, and they can trap hydrocarbons migrating upward from deeper source rocks (play B in Figure 7).

Another important type of positive structures in the Neogene strata of the Pannonian Basin was generated by inversion. Examples for early inversion of the basin fill are given by the Kiskunhalas subbasin (Horvath, 1995) and the Paleogene flysch trough (Lorincz and Szabo, 1993). In the Kiskunhalas area (Figure 2), the middle Miocene (Karpatian to Sarmatian) synrift strata were affected by inversion, and the resulting positive forms have been draped by younger sedimentary beds (play D in Figure 7). Sizable fields have been found here in the middle Miocene basal sequence (Karpatian conglomerates and Badenian sandstones). Further to the northeast in the same subbasin, oil accumulations occur in fractured crystalline rocks and Mesozoic carbonates.

A set of fields has been found in positive structures above the Szolnok Paleogene flysch belt (Figures 2, 3), with the dominance of gas and only occasional occurrence of hydrocarbon liquids (Pogacsas et al., 1994). Gas content is highly variable both vertically and laterally. Hydrocarbons alternate and mix with carbon dioxide, nitrogen, and inert gases. Carbon isotope (Szalay and Koncz, 1991) and rare gas (Ballentine et al., 1991) data suggest a deeper crustal origin for the carbon dioxide and nitrogen and thermogenic and biogenic origin

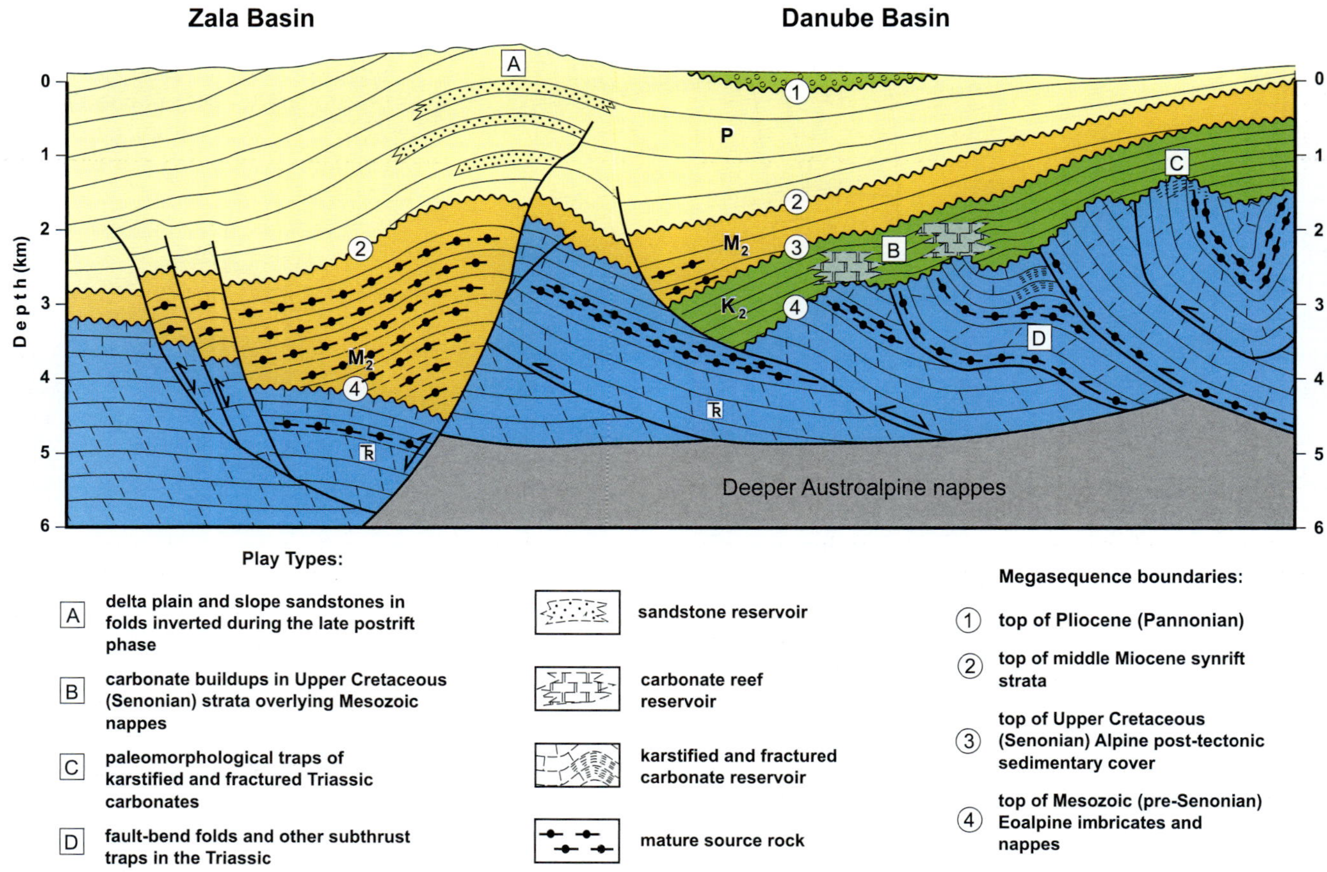

Figure 9. Idealized cross section showing the petroleum system of the Neogene basin fill in the Zala Basin and of the Mesozoic substrata below the Little Hungarian Plain (modified after Horvath and Tari, 1999).

for the hydrocarbon gases. Under favorable conditions, a large amount of gas has been accumulated, like in the Hajduszoboszlo field, which contained 1.1 tcf recoverable gas reserves. This field with stacked Neogene pay horizons is located above and most probably fed by Paleogene source rocks. The field therefore represents an example of possible communication between petroleum systems associated with different tectonostratigraphic units.

The late Pliocene to Quaternary evolution of the Pannonian Basin is characterized by inversion (e.g., Tari, 1994). This includes inversion of the former normal faults and uplift in the western and eastern flanks of the Pannonian Basin, accelerated subsidence, and renewal of strike-slip faulting in its central part (Horvath and Cloetingh, 1996). This leads to two types of structural features that provide excellent targets for hydrocarbon exploration.

Most characteristically, in the Zala Basin (Figure 2), graben inversion occurred by reactivation of the former main normal fault as a reverse fault (play A in Figure 9). Large, four-way closures in Neogene strata offer structural traps for hydrocarbons migrating from the deep Miocene source rocks toward the apex of the structure. In the Budafa field (Figure 2), most of the oil pools were found between 800- and 1300-m (2600- and 4300-ft) depth in Pannonian (Pliocene) prodelta turbidites and delta-plain mouth bar sandstones. Initial in-place reserves amounted to 105 million bbl of oil and 0.17 tcf of gas (Kokai and Pogacsas, 1991).

Another type of structural trap is associated with young wrench fault zones (Pogacsas et al., 1994). These faults commonly produce positive flower structures and en echelon folds in Pannonian to Quaternary strata (play C in Figure 7). Again, turbidites and delta-plain sandstone reservoirs can trap hydrocarbons of deeper origin (Szalay and Koncz, 1993). Within the wrench fault zones, the exploration risk is low, but only small fields can be expected.

Paleogene Tectonostratigraphic Unit

The diverse set of Paleogene basin fragments beneath the Pannonian Basin (Tari et al., 1993) represents another petroleum system. The lower Oligocene Tard Clay is a good source rock in the Paleogene basin of northeastern Hungary with a typical total organic carbon content of 2–4% and type II–III kerogens (Milota et al., 1995). If it reached a depth of more than 2500 m (200 ft), maturity became adequate to start hydrocarbon generation. The Paleogene Basin is interpreted to be a retroarc flexural basin with thrust imbricates and transpressional faults (Tari et al., 1993).

The play concept in Figure 8 depicts a typical structural setting and shows trapping geometries that are known in the area. In the Demjen and Mezokeresztes fields (Figure 2), oil and some gas have been trapped in Oligocene turbiditic sandstones. Unfortunately, the Oligocene turbidites are only fair reservoir rocks because of their low permeability. The average permeability is about 2 md, and even the best values do not exceed 100 md. In addition, the originally small sandstone bodies are typically dismembered and compartmentalized by faults. This can be illustrated by the fact that in the Demjen–East region, 461 wells were drilled in an area of 7 km^2 (2.7 mi^2) to produce only 7 million bbl of oil.

Better reservoir conditions may be present in the deeper part of the Paleogene Basin toward Budapest. If it is not the case, then hydrocarbon prospects in this tectonostratigraphic unit are very limited.

Senonian and Mesozoic to Paleozoic Tectonostratigraphic Unit

These two tectonostratigraphic units generally constitute an interconnected system; these are therefore discussed together below. Because of major differences in Mesozoic stratigraphy, however, we will review separately the petroleum systems in the North Pannonian and South Pannonian terranes.

The petroleum system of the North Pannonian terrane (Figure 9, right side) has been constructed on the basis of data from the Nagylengyel field (Figure 2) and new interpretation of the deep structure below the Little Hungarian Plain (Tari, 1994, 1995b). The field is situated at the central part of the Zala Basin in the subsided flank of the Transdanubian Central Range. This is the largest known oil field of the Pannonian Basin in a Mesozoic reservoir sequence and the second largest of all fields in Hungary. Original in-place oil reserves amounted to 300 million bbl, and 40–60% of it has been recovered (Kokai and Pogacsas, 1991). As a matter of fact, the total production in the period of 1951–1976 provided 120 million bbl (Korossy, 1988).

The main reservoir is a Senonian reefal limestone, and the Upper Triassic Hauptdolomit is also commonly strongly fractured and saturated by oil. The column height of individual pools ranges between 7 and 90 m (23–300 ft) in the Cretaceous and between 20 and 160 m (66–520 ft) in the Triassic reservoirs. The carbonate reservoirs are highly heterogeneous, small intragranular voids are associated with fracture porosity and commonly karstic holes, and even caves were penetrated. This is the explanation for very high initial flow rates (as much as 4500 bbl a day) in several individual wells. The source rock of this petroleum system is the Upper Triassic Kossen Marl overlying the Hauptdolomit and interfingering with the Dachstein Limestone. The Kossen Marl is a good type II source rock with remarkable thickness (as much as 600 m [2000 ft]) and areal distribution. Total organic carbon content ranges between 3 and 6%, with extreme values as high as 19% (Clayton and Koncz, 1994b).

Seismic coverage of the area is quite poor, and only a couple of interpretable seismic sections are available for the Nagylengyel field. Dank (1988) suggested that pools were associated with tilted blocks separated by normal faults of early Miocene age. Kokai and Pogacsas (1991) claim that the recent structure of the Nagylengyel field was generated by synrift extensional block faulting of a large paleohigh. Our model (Figure 9, right side) suggests that the Nagylengyel structure formed in a complex manner, starting with the formation of nappes during the Cretaceous Austrian orogenic phase. The best detachment horizons are the shale layers in the Triassic series, like the Kossen Marl (Tari, 1995b). This resulted in the Cretaceous imbrication of Mesozoic sequences with the repetition of the source rock intervals. The Eoalpine thrust units and nappes were then dissected by listric normal faults during the middle Miocene. Some of these listric faults clearly sole into the Eoalpine detachment surfaces (Tari, 1996b).

It is tempting to speculate that the known and largely exhausted traps in the Nagylengyel region represent only the upper floor of pools, and further reserves can be present in deeper levels below and/or around the known field. Furthermore, the play concept may work elsewhere at the flank of the Transdanubian Central Range, where the Kossen Marl may be present (Mattick et al., 1996).

The petroleum system characterizing the South Pannonian terrane relies on two relatively recent findings. On the one hand, new field observations and reinterpretation of well logs have shown that Toarcian black shales in the Mecsek unit of the South Pannonian terrane are excellent type II source rocks with total organic carbon content of about 4–8%. On the other hand, recent drill-hole data and seismic interpretation (Pap et al., 1992; Grow et al., 1994; Tari et al., 1999) confirmed earlier suggestions (Horvath and Rumpler, 1984) about the presence of Mesozoic and metamorphic crystalline nappes in the substratum of the Great Hungarian Plain (Figures 4B, 6).

The novelty of these interpretations implies that the petroleum system in Figure 8 is rather speculative. However, promising findings are present. These include more than 20 smaller fields that have been found so far in fractured metamorphic basement rocks and the overlying basal conglomerates. These basement rocks have been drilled always at places where they constitute basement highs (play D in Figure 8). The traditional explanation for the formation of these pools, i.e., lateral migration of hydrocarbons out of mature Neogene source rocks in the nearby deep basins, may not be always the case.

Kokai and Pogacsas (1991) showed two examples where the crystalline basement high represents only a relatively thin allochthonous klippen above Mesozoic rocks. During the pre-Senonian orogeny, northwest-vergent nappes formed, and the individual thrust sheets and nappes are composed of Triassic to Lower Cretaceous cover sequences with the crystalline basement involved (Figure 6).

It seems like that the Quaternary tectonic processes have had an impact on the fluid pressure in the fractured metamorphic reservoirs. If one plots the fields on a map showing areas of Quaternary uplift and subsidence, it turns out that most of the geopressured pools are situated in areas of uplift. This suggests that overpressure is mainly controlled by recent uplift and good seal, which hamper fast decay of the excess pressure (Horvath, 1995).

In summary, the implication of this new play concept is to consider traps in the complex Eoalpine imbricates and nappes where hydrocarbon generation out of the Toarcian shales could have occurred as late as the Neogene as a response to synrift subsidence (see analogy with the Vienna basin, Arzmuller et al., 2006). To date, the acquisition and processing of both industry and academic seismic data did not specifically address the optimum imaging of the Eoalpine structure beneath the Pannonian Basin. The obviously high exploration risk associated with this pre-Tertiary petroleum system can only be reduced using sophisticated seismic technology (e.g., prestack depth migration of three-dimensional volumes) and modern subsurface interpretation techniques.

CONCLUSIONS

The Eoalpine thrust-fold belts of the Eastern Alps and the Carpathians do not stop at the edge of the Neogene Pannonian Basin but do project beneath it. The level of understanding of these folded belts subcropping at the base of the Tertiary is poor compared to the existing models of the classical folded belt of the Carpathians (see chapters in this volume). This mismatch is primarily caused by the subsurface nature of the problem and, second, by the dramatic structural overprint by the Neogene extensional tectonics in the Pannonian Basin. Modern geophysical methods, reflection seismic data acquisition in particular, can and will refine the emerging structural picture briefly outlined in this chapter. Hydrocarbon exploration has already reached its maturity in the Pannonian Basin, and significant additional reserves can only be expected from modern and systematic exploration efforts beneath the Tertiary basin system.

ACKNOWLEDGMENTS

This chapter is partly based on the Ph.D. thesis of one of us (G. T.) supervised by Albert Bally at Rice University, Houston, Texas, which is gratefully acknowledged. We also appreciate helpful discussions on the Alpine geology of the Pannonian Basin with Andras Galacz and Attila Voros. Jan Golonka and Frank Picha are thanked for their editorial patience.

REFERENCES CITED

Arzmuller, G., S. Buchta, E. Ralbovsky, and G. Wessely, 2006, The Vienna Basin, *in* J. Golonka and F. J. Picha, eds., The Carpathians and their foreland: Geology and hydrocarbon resources: AAPG Memoir 84, p. 191–204.

Balla, Z., 1987, The tectonics of Hungary: A kinematic point of view: Rendiconti della Societa Geologica Italiana, v. 9, p. 97–102.

Ballentine, C. J., R. K. O'Nions, E. R. Oxburgh, F. Horvath, and J. Deak, 1991, Rare gas constraints on hydrocarbon accumulation crustal degassing and groundwater flow in the Pannonian Basin: Earth and Planetary Science Letters, v. 105, p. 229–246.

Bally, A. W., and S. Snelson, 1980, Realms of subsidence, *in* A. D. Miall, ed., Facts and principles of world petroleum occurrence: Canadian Society of Petroleum Geologists Memoir 6, p. 9–94.

Bleahu, M., 1976, Structural position of the Apuseni Mountains in the Alpine system: Revue Roumaine de Geologie, Geophysique et Geographie: Serie de Geologie, v. 20, p. 7–19.

Budai, T., 1991, Contributions to the knowledge of the structural geology of Felsoors, Balaton Upland (in Hungarian with English summary): Annual Report of the Hungarian Geological Survey for the Year 1989, p. 17–33.

Clayton, J. L., and I. Koncz, 1994a, Totkomlos–Szolnok petroleum system of southeastern Hungary, *in* L. B. Magoon and W. G. Dow, eds., The petroleum system— From source to trap: AAPG Memoir 60, p. 587–598.

Clayton, J. L., and I. Koncz, 1994b, Petroleum geology of the Zala Basin, Hungary: AAPG Bulletin, v. 78, p. 1–22.

Clayton, J. L., I. Koncz, J. D. King, and E. Tatar, 1994, Organic geochemistry of crude oils and source rocks, Bekes basin, *in* P. G. Teleki, R. E. Mattick, and J. Kokai, eds., Basin analysis in petroleum exploration: Dordrecht, The Netherlands, Kluwer Academic Publishing, p. 161–185.

Dank, V., 1988, Petroleum geology of the Pannonian Basin, Hungary, an overview, *in* L. Royden and F. Horvath, eds., The Pannonian Basin a study in basin evolution: AAPG Memoir 45, p. 319–331.

Fulop, J., K. Brezsnyanszky, and J. Haas, 1987, The new map of basin basement of Hungary: Acta Geologica Hungarica, v. 30, p. 3–20.

Grow, J. A., R. E. Mattick, A. Berczi-Makk, Cs. Pero, D. Hajdu, Gy. Pogacsas, P. Varnai, and E. Varga, 1994, Structure of the Bekes basin inferred from seismic reflection, well and gravity data, *in* P. G. Teleki, R. E. Mattick, and J. Kokai, eds., Basin analysis in petroleum exploration, a case study from the Bekes basin, Hungary: Dordrecht, The Netherlands, Kluwer Academic Publishing, p. 1–38.

Horvath, F., 1993, Towards a mechanical model for the formation of the Pannonian Basin: Tectonophysics, v. 226, p. 333–357.

Horvath, F., 1995, Phases of compression during the evolution of the Pannonian Basin and its bearing on hydrocarbon exploration: Marine and Petroleum Geology, v. 12, p. 837–844.

Horvath, F., and S. Cloetingh, 1996, Stress-induced late-stage subsidence anomalies in the Pannonian Basin: Tectonophysics, v. 266, p. 287–300.

Horvath, F., and J. Rumpler, 1984, The Pannonian basement, extension and subsidence of an Alpine orogene: Acta Geologica Hungarica, v. 27, p. 229–235.

Horvath, F., and G. Tari, 1999, IBS Pannonian Basin project: A review of the main results and their bearings on hydrocarbon exploration, *in* B. Durand, L. Jolivet, F. Horvath, and M. Seranne, eds., The Mediterranean basins: Tertiary extension within the Alpine orogene: Geological Society (London) Special Publication 156, p. 195–213.

Horvath, F., A. Adam, and W. D. Stanley, 1987, New geophysical data: Evidence for the allochthony of the Transdanubian Central Range: Rendiconti della Societa Geologica Italiana, v. 9, p. 123–130.

Horvath, F., P. Dovenyi, A. Szalay, and L. H. Royden, 1988, Subsidence, thermal and maturation history of the Great Hungarian Plain, *in* L. Royden and F. Horvath, eds., The Pannonian Basin a study in basin evolution: AAPG Memoir 45, p. 355–375.

Kokai, J., 1994, Exploration history and future possibilities in Hungary, *in* B. M. Popescu, ed., Hydrocarbons of Eastern Central Europe: Berlin, Springer Verlag, p. 147–173.

Kokai, J., and Gy. Pogacsas, 1991, Hydrocarbon plays in Mesozoic nappes, Tertiary wrench basins and interior sags in the Pannonian Basin: First Break, v. 9, p. 315–334.

Korossy, L., 1988, Hydrocarbon geology of the Zala Basin in Hungary (in Hungarian with English summary): General Geological Review, v. 23, p. 3–163

Kovacs, S., G. Csaszar, A. Galacz, J. Haas, E. Nagy, and A. Voros, 1989, The Tisza superunit was originally part of the North Tethyan (European) margin: Memoires de la Societe Geologique de France, v. 154, no. II, p. 81–100.

Loczy, L., 1913, Die geologische Formationen der Balatongegend und ihre regionale Tektonik: Resultaten Wissenschaftlichen Erforschung Balatonsees I/1, Holzel, Vienna, 500 p.

Lorincz, K., and P. Szabo, 1993, Seismic analysis of multiphase tectonism in the central part of the Pannonian Basin in Hungary, *in* A. M. Spencer, ed., Generation, accumulation and production of Europe's hydrocarbons: European Association of Petroleum Geologists Special Publication 3, p. 311–323.

Magyar, I., A. Fogarasi, G. Vakarcs, L. Buko, and G. C. Tari, 2006, The largest hydrocarbon field discovered to date in Hungary: Algyo, *in* J. Golonka and F. J. Picha, eds., The Carpathians and their foreland: Geology and hydrocarbon resources: AAPG Memoir 84, p. 619–632.

Mattick, R. E., J. Rumpler, A. Ujfalusy, B. Szanyi, and I. Nagy, 1994, Sequence stratigraphy of the Bekes basin, *in* P. G. Teleki, R. E. Mattick, and J. Kokai, eds., Basin analysis in petroleum exploration: Dordrecht, The Netherlands, Kluwer Academic Publishing, p. 39–65.

Mattick, R. E., P. G. Teleki, L. Phillips, J. L. Clayton, G. David, Gy. Pogacsas, B. Bardocz, and E. Simon, 1996, Structure, stratigraphy, and petroleum geology of the Little Plain basin, northwestern Hungary: AAPG Bulletin, v. 80, p. 1780–1800.

Milota, K., A. Kovacs, and Z. Galicz, 1995, Petroleum potential of the North Hungarian Oligocene sediments: Petroleum Geoscience, v. 1, p. 81–87.

Pap, S., 1990, Thrust-faulted sequences in the Trans-Tisza region (in Hungarian): Hungarian Geological Survey Special Publication, 36 p.

Pap, S., V. Soreg, and I. Pap Hasznos, 1992, Exploration of the Devavanya south basement structure for hydrocarbons: A case history: Geophysical Transactions, v. 37, p. 211–228.

Phillips, R. L., I. Revesz, and I. Berczi, 1994, Lower Pannonian deltaic-lacustrine processes and sedimentation Bekes basin, *in* P. G. Teleki, R. E. Mattick, and J. Kokai, eds., Basin analysis in petroleum exploration: Dordrecht, The Netherlands, Kluwer Academic Publishing, p. 67–82.

Pogacsas, Gy., R. E. Mattick, G. Tari, and P. Varnai, 1994, Structural control on hydrocarbon accumulation in the Pannonian Basin, Hungary, *in* P. G. Teleki, R. E. Mattick, and J. Kokay, eds., Basin analysis for oil and gas exploration: A case history from Hungary: Amsterdam, Kluward, p. 221–235.

Royden, L., and F. Horvath, eds., 1988, The Pannonian Basin—A study in basin evolution: AAPG Memoir 45, 394 p.

Rumpler, J., and F. Horvath, 1988, Some representative seismic reflection lines from the Pannonian Basin and their structural interpretation: AAPG Memoir 45, p. 153–169.

Szalay, A., and I. Koncz, 1991, Genetic relations of hydrocarbons in the Hungarian part of the Pannonian Basin, *in* A. M. Spencer, ed., Generation, accumulation and production of Europe's hydrocarbons: European Association of Petroleum Geologists Special Publication 3, p. 311–323.

Szalay, A., and I. Koncz, 1993, Migration and accumulation of oil and natural gas generated from Neogene source rocks in the Hungarian part of the Pannonian Basin, *in* A. M. Spencer, ed., Generation, accumulation and production of Europe's hydrocarbons III: European Association of Petroleum Geologists Special Publication 3, p. 303–309.

Tari, G., 1994, Alpine tectonics of the Pannonian Basin: Ph.D. thesis, Rice University, Houston, Texas, 501 p.

Tari, G., 1995a, Phanerozoic stratigraphy of the Hungarian part of the NW Pannonian Basin, *in* F. Horvath, G. Tari, and Cs. Bokor, eds., Extensional collapse of the Alpine orogene and hydrocarbon prospects in the basement and basement fill of the Western Pannonian Basin: AAPG International Conference and Exhibition, Nice, France, Guidebook to Fieldtrip No. 6, Hungary, p. 21–46.

Tari, G., 1995b, Eoalpine (Cretaceous) tectonics in the Alpine–Pannonian transition zone, *in* F. Horvath, G. Tari, and Cs. Bokor, eds., Extensional collapse of an Alpine orogene and hydrocarbon prospects in the basement and basin fill of the Western Pannonian Basin: AAPG International Conference and Exhibition, Nice, France, Guidebook to Fieldtrip No. 6., Hungary, p. 133–155.

Tari, G., 1996a, Extreme crustal extension in the Raba River extensional corridor (Austria/Hungary): Mitteilungen der Gesellschaft der Geologie- und Bergbaustudenten in Osterreich, v. 41, p. 1–17.

Tari, G., 1996b, Neoalpine tectonics of the Danube basin (northwest Pannonian Basin, Hungary), *in* P. Ziegler and F. Horvath, eds., Peri-Tethys Memoir 2: Structure and prospects of Alpine basins and forelands: Memoires du Museum National d'Histoire Naturelle, v. 170, p. 439–454.

Tari, G., F. Horvath, and J. Rumpler, 1992, Styles of extension in the Pannonian Basin: Tectonophysics, v. 208, p. 203–219.

Tari, G., T. Baldi, and M. Baldi-Beke, 1993, Paleogene retroarc flexural basin beneath the Neogene Pannonian Basin: A geodynamic model: Tectonophysics, v. 226, p. 433–455.

Tari, G., P. Dovenyi, I. Dunkl, F. Horvath, L. Lenkey, M. Stefanescu, P. Szafian, and T. Toth, 1999, Lithospheric structure of the Pannonian Basin derived from seismic, gravity and geothermal data, *in* B. Durand, L. Jolivet, F. Horvath, and M. Seranne, eds., The Mediterranean basins: Tertiary extension within the Alpine orogene: Geological Society (London) Special Publication, v. 156, p. 215–250.

Teleki, P. G., R. E. Mattick, and J. Kokai, eds., 1994, Basin analysis in petroleum exploration: Dordrecht, Kluwer Academic Publishing, 300 p.

Tollmann, A., 1987, The Alpidic evolution of the Eastern Alps, *in* H. W. Flugel and P. Faupl, eds., Geodynamics of the Eastern Alps: Wien, Deuticke, p. 361–378.

Trumpy, R., 1980, Geology of Switzerland, a guide book: Part A. An outline of the geology of Switzerland: Basel, Wepf, 104 p.

Vakarcs, G., P. R. Vail, G. Tari, Gy. Pogacsas, R. E. Mattick, and A. Szabo, 1994, Third-order Miocene–Pliocene depositional sequences in the prograding delta complex of the Pannonian Basin: Tectonophysics, v. 240, p. 81–106.

Vakarcs, G., J. Hardenbol, V. S. Abreau, P. R. Vail, P. Varnai, and G. Tari, 1998, Oligocene–middle Miocene depositional sequences of the Central Paratethys and their correlation with regional stages, *in* J. Hardenbol, J. Thierry, M. B. Farley, T. Jacquin, P. C. De Graciansky, and P. R. Vail, eds., Sequence stratigraphy of European basins: SEPM Special Publication 60, p. 209–231.

20

Magyar, Imre, Attila Fogarasi, Gábor Vakarcs, László Bukó, and Gábor C. Tari, 2006, The largest hydrocarbon field discovered to date in Hungary: Algyo, *in* J. Golonka and F. J. Picha, eds., The Carpathians and their foreland: geology and hydrocarbon resources: AAPG Memoir 84, p. 479.

The Largest Hydrocarbon Field Discovered to Date in Hungary: Algyo

Imre Magyar, Attila Fogarasi, Gábor Vakarcs, László Bukó
MOL Hungarian Oil and Gas Company, Budapest, Hungary
Gábor C. Tari
Vanco Energy Company, Houston, Texas, U.S.A.

ABSTRACT

The Algyo field is currently one of the largest oil- and gas-producing fields in the Pannonian Basin, Central Europe. Located in the southern part of Hungary, it was discovered in 1965, and it has produced 31 million t of oil and 70 billion m^3 (2.47 tcf) of gas. Today, it provides more than half of Hungary's oil production and also holds the largest reserves. The areal extent of the field is about 80 km^2 (31 mi^2), and it comprises several dozens of sandstone reservoir beds with 5–30 m (16–100 ft) of gross thickness each. The reservoir sands as well as the probable source rocks were deposited in Lake Pannon and in adjacent deltaic and fluvial environments during the late Miocene and earliest Pliocene. Lake Pannon was a huge, long-lived, brackish lake, similar to the modern Caspian Sea. The hydrocarbon traps formed in a compaction anticline above a northwest–southeast-trending metamorphic core complex. A recently shot three-dimensional seismic survey was a milestone for further development and opened new ways for exploration effort, including the use of direct hydrocarbon indicators, to find additional satellite fields in the vicinity of this significant field. Complex interpretation of all available data and using sequence stratigraphy and its chronostratigraphical framework have been tremendously improving our understanding of the mechanisms controlling the reservoir characteristics, facies development, and the oil- and gas-trapping mechanism.

DOI:10.1306/985734M843142

CRUSTAL AND LITHOSPHERIC STRUCTURE OF THE CARPATHIAN–PANNONIAN REGION: A GEOPHYSICAL PERSPECTIVE

21

Pospíšil, Lubomil, and Antal Ádám, 2006, Review of the crust–lithosphere research in the Carpathians, *in* J. Golonka and F. J. Picha, eds., The Carpathians and their foreland: Geology and hydrocarbon resources: AAPG Memoir 84, p. 481–495.

Review of the Crust–Lithosphere Research in the Carpathians

Lubomil Pospíšil
Geoinform Consultants, Brno, Czech Republic

Antal Ádám
Geodetic and Geophysical Research Institute of the Hungarian Academy of Sciences (MTA GGKI), Sopron, Hungary

ABSTRACT

This chapter summarizes the basic information on the development of views and fundamental geophysical data concerning the lithosphere of the Carpathians and the Pannonian Basin. Mostly, those results and interpretations published in the past are mentioned and still keep their validity. It is obvious that the period of syntheses of geoscience information concerning crustal–lithosphere research in the Carpathian region still has not reached its apex. The main reason for it is the absence and limited availability of data from particular countries because of reasons of legislation (Romania and Ukraine) and commercial problems (Poland, Hungary, and currently, the Czech and Slovak Republics).

INTRODUCTION

If we are to evaluate the history of the development of views in the research of the lithosphere in the Carpathians and the Pannonian Basin, it should be stated that from the very beginning, the search for the upper part of the lithosphere, i.e., the crust, has dominated in the Central and Eastern Europe since the 1920s.

In this period, numerous local geophysical measurements of potential fields, seismic and seismological registrations, etc., were carried out; nevertheless, these geophysical data have not affected greatly the opinions of leading geologists.

After World War II, the largest amount of geophysical work was done. New regional gravity and magnetometric maps were compiled, and new seismological observations and seismic refraction measurements, heat flow measurements, geological surveys, and paleomagnetic investigations were realized.

Research of the Carpathian lithosphere was coordinated by the following international cooperations of Central and Eastern European countries such as the Carpathian–Balkan Geological Associations, the European Seismological Commission, and the Commission of the Academies of Socialist Countries for Planetary Geophysics (KAPG).

In 1963, the Deep Seismic Sounding (DSS) program was started, whose results served as a basis and reference concerning the velocity distribution in the upper part of the lithosphere, the crust, for a long period of time.

In 1981, in the framework of the KAPG program, the "Geophysical and Geodynamic Model of Lithosphere of the Central and Eastern Europe" international

DOI:10.1306/985621M843079

project started. The following topics were solved in the framework of several subprojects:

- methodology of the complex interpretation of the geological-geophysical survey
- structure of the Central and Eastern European lithosphere
- geodynamic models of Central and Eastern Europe

At the beginning of the 1980s, the deep seismic reflection profiling, which was introduced in the United States and the United Kingdom with the Consortium for Continental Reflection Profiling (COCORP) and The British Institutions Reflection Profiling Syndicate (BIRPS) projects, has brought the new view on the deepest horizons of the crust (Tomek, 1988; Tomek et al., 1989; Posgay et al., 1995).

Experiments in Hungary (Posgay et al., 1995) brought the first detailed seismic information concerning the asthenosphere.

Parallel to these projects, a number of results from other geosciences were collected in the framework of the International Union of Geological Sciences–United Nations Educational, Scientific, and Cultural Organization international program of correlations in geology (e.g., geochemistry, ophiolites, development of the Precambrian lithosphere, etc.).

An approach to the research of sedimentary basins and the search for hydrocarbons was fully based on the geosynclinal concept in the 1950s to the 1980s, which was strongly affected by the so-called ''Soviet school.'' For example, up to the 1990s, the Moho discontinuity was considered a dominant boundary whose form and course was considered as a mirror of the geological development of the region.

During this period, the clash of two geotectonic concepts, based on the processes of the differentiation of mantle masses (Belousov, 1975) and the influence of gravitational force (Van Bemmelen, 1960, 1966) in the process of tectogenesis, has come to a close. During this period, discussions were held concerning the existence of nappes in the Carpathians; in the meantime, there were no doubts concerning their existence in the Alps.

It was in the 1970s when the plate tectonics concept started to be applied in geology in the countries of the Central and Eastern Europe.

The first Central European proponents of the new theory were those who suggested and accepted the mantle diapir theory, but the application of plate tectonics (based on meridional mountain chains surrounding the Pacific) for equatorial ranges was not an easy job (Kober, 1921; Staub, 1928; Szadeczky-Kardoss, 1968, 1969, 1971, 1973; Stegena, 1972; Stegena et al., 1975; Channel and Horvath, 1976; Balogh, 1993).

Views of geologists were mostly determined by the papers of Laubsher (1971) and Dietrich and Franz (1976). These papers stated that the sedimentary rocks of the Southern Alps, Eastern Alps, and Western Carpathians were deposited on the southern margin of the Tethys, on the North African shelf, during the middle Cretaceous. These overthrusted the Penninic zone, i.e., the oceanic part of the Tethys, which, together with Bundner Schiefer and ophiolites, were subducted at that time. The shelf of the European continent is represented by the outermost belts of the Alps and the Helvetic and Ultrahelvetic zones. Their Permian and Mesozoic facies are close to the German facies.

The principal change in the approach to solving the development of basins in the Carpathians and in the Pannonian Basin occurred in 1979–1982. Papers of Wein (1978) and Jiricek (1979) evaluating the course of sedimentation and the movement of the Carpathians in space and time in the region of the foredeep and the Outer Carpathians and the Pannonian Basin, respectively, brought the pillars and significant impulse for compiling the complex geodynamic model of the development of the Carpathians.

Wein (1978) was the first who attempted to correlate the plate-tectonic theory with geological facts. He was influenced by the ideas of Geczy (1972, 1973) and Hadzi et al. (1974). He tried to reconstruct the Permian–Triassic position of the Carpathian–Dinaride tectonic-facies units compared to the northern, middle, and southern belts of the Tethys. According to his idea, the central part of the Tethys is represented by the Penninic zone, Fruska gora, Transylvanian Ore Mountains, and the Inner Dinarides. The Permian–Mesozoic succession of southern Transdanubia, together with the Bihar autochthonous and line-overlying nappes, were deposited north of the central Tethyan region, i.e., on the European shelf, whereas sequences of the Transdanubian Central Range, Gemer, and Igal–Bukk–Szendro were deposited south of it, i.e., on the southern African shelf. This pattern has been altered by transcurrent faults, produced by the Jurassic to Early Cretaceous opening of the Tethys (Periadriatic lineament, Balaton–Darno line, Zagreb–Kulcs–Hernad line). The lateral displacement along these lines attained a magnitude of 700–800 km (430–500 mi) from the middle Cretaceous to the Miocene. The northern displacement of the Adriatic indenter and the westward motion of the Moesian microplate helped in changing the original position of crustal fragments: northern units reached a southern position and vice versa.

Thus, the basement of the Pannonian Basin consists of two tectonic units, separated by the Zagreb–Kulcs–Zemplen lineament. The African formations of the northwestern segment overthrusted the Penninic unit of the

Tethys Ocean and the northern shelf units as well during the middle Cretaceous closure of the Tethys.

Later, this model was more precisely analyzed by many geologists from the Carpathian region and was very much influenced by geophysical approaches to the modeling of the structure and processes in the lithosphere (e.g., Vass, 1979; Pospisil, 1980, 1984; Vass and Pospisil, 1984; Csontos et al., 1992).

The period when the process of the synthesis of data started was dated by the workshop entitled the "Evolution of Extensional Basins within Regions of Compression, with Emphasis on the Intra-Carpathians," held in June 20–26, 1982, in Veszprem (Hungary). Here, the fundamentals that were based entirely on the plate tectonics concept were laid down via a new approach to the solution of the setting of the Carpathians and the Pannonian Basin.

The papers read throughout the world (Lexa and Konecny, 1974; Stegena et al., 1975; Ksiazkiewicz, 1977; Royden et al., 1982, 1983a, b; Royden and Horvath, 1988) provided the basics of the extensional Tertiary model of the Pannonian Basin with thermal collapse in the upper Miocene. Although these models were simplified at the beginning, during the time, and after reevaluating a tremendous amount of data, these models were improved (Laubscher, 1971; Sandulescu, 1975, 1988; Lesko and Varga, 1980; Horvath and Royden, 1981; Balla, 1984, 1987a, b; Horvath, 1984, 1986, 1993; Horvath and Rumpler, 1984; Dercourt et al., 1985; Oszczypko and Slaczka, 1985, 1989; Royden and Horvath, 1988; Kovac et al., 1989; Csontos et al., 1992, and others), strongly affecting and unifying some kind of dispersion of opinions of particular traditional national approaches (Roman, 1970; Szadecky-Kardos, 1973; Birkenmajer, 1974; Mahel, 1974; Sandulescu, 1975; Stegena et al., 1975; Ksiazkiewicz, 1976, 1977; Mahel and Reichwalder, 1979; Vass, 1979; Bleahu, 1985, and others).

The start of geophysical research of the lithosphere of the Carpathians (the Pannonian region), with respect to the vast financial support from state budgets, was dated from the acquisition of the first information relating to the thickness of the lithosphere of the Pannonian Basin (magnetotelluric measurements; Adam, 1965, 1974; seismological data; Bisztricsany, 1974). Extensive research of the lithosphere has been realized in the framework of international projects and has recently been summarized in the syntheses of researches on national levels in joint publications and monographs (e.g., Burianov et al., 1987; Tschekunov, 1988a, b; Bucha and Blizkovsky, 1994, and many others).

Direct relation and influence of the lithosphere for the formation of the Neogene basins based on complete geophysical and geological data were first applied and described by Stegena (1972), Stegena et al. (1975), Pospisil (1984), and Vass and Pospisil (1984). In these papers, even the first model of the lithosphere across the Western Carpathians was presented, considering not only geological but also all geophysical data available (Figure 1). Since the mid-1990s, researches about the lithosphere both in the platform regions of the Bohemian Massif and the Eastern European platform and in the regions of the Carpathians and the Pannonian Basin have been conducted extensively.

Research on the lithosphere of the Carpathians was concluded by the preparation and realization of the lithosphere dynamics program EUROPROBE, the Pannonian Basin, Carpathian arc, Dinaride (PANCARDI), and The Central European Lithospheric Experiment Based on Refraction (CELEBRATION) (Guterch et al., 2000) projects that brought evidence on the composition, thicknesses, and processes conducted mainly in the period of the late Tertiary.

Results of the tomographical processing of refraction data (Spakman et al., 1993) brought some clarification even into the asthenosphere, namely, by the information on postsubduction slabs that were broken off, displaying different velocities.

In the following sections, basic information on anomalies and boundaries observed in the whole scope of the lithosphere and their interpreted manifestations are respectively presented in a manner in which these phenomena were detected by particular methods and authors. Several examples of original interpretations and anomalies are accompanied by maps or sections.

SEISMIC AND SEISMOLOGICAL DATA

The data related to the lithosphere were taken from the following sources:

- data related to refraction seismics, detecting the upper part of the lithosphere–crust (Beranek et al., 1972; Szenas, 1972; Zatopek, 1979; Tschekunov, 1989; Posgay et al., 1990)
- data referred to teleseismic P-wave daily times (Babuska et al., 1987; Spakman et al., 1993)
- data referred to reflection seismics that offer information not only on the thickness of the lithosphere but also on the structural elements
- data offering the explanation for complete basin analyses
- earthquakes registration and monitoring

The following main velocity anomalies were described and investigated in the Carpathian crustal–lithosphere. Between the European platform and the

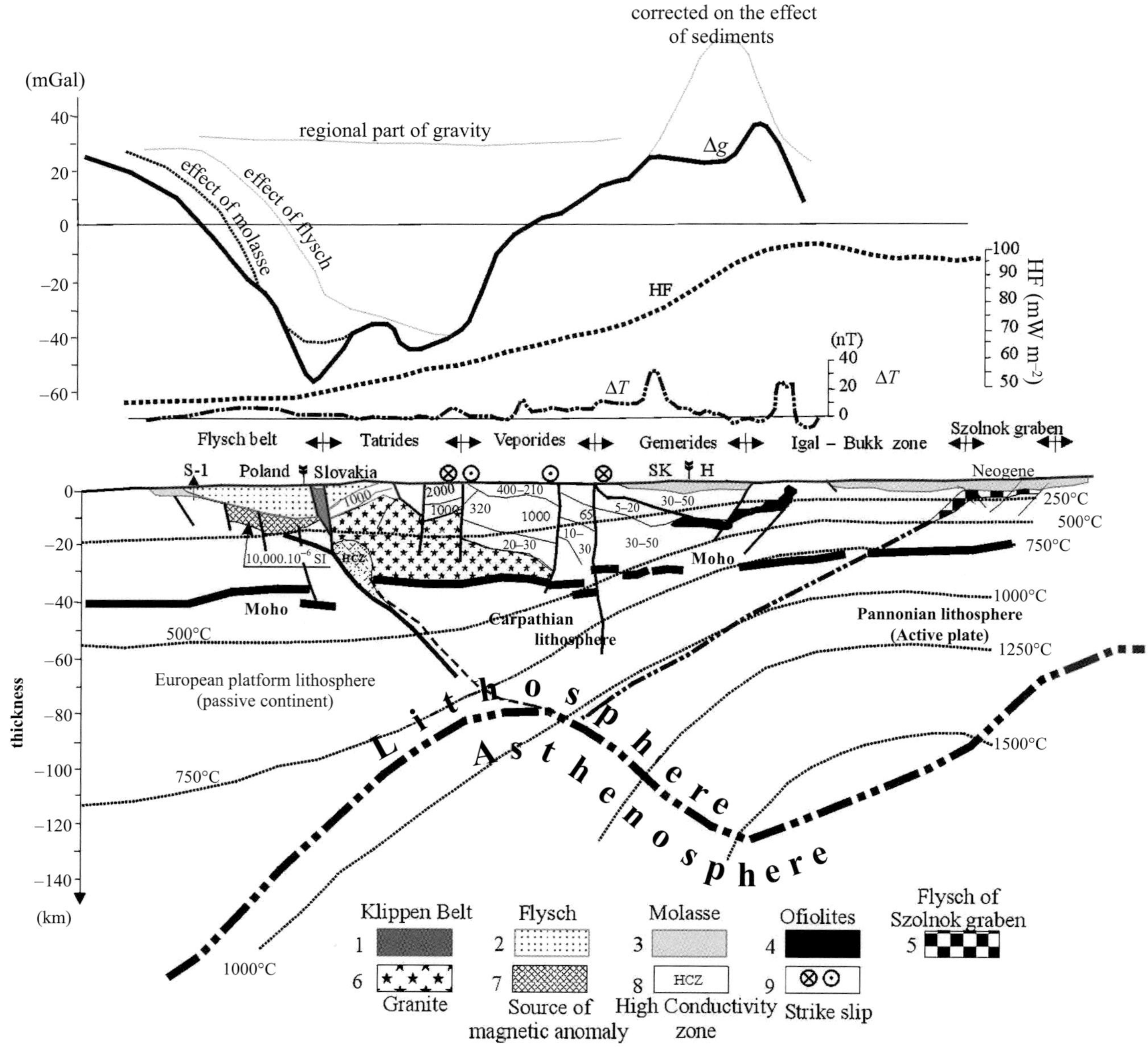

Figure 1. Lithospheric model through the Western Carpathians (modified after Pospisil, 1980). On the section, the calculated 2-D temperature distribution and model of magnetic sources in the left side of section (with parameter of susceptibility) are presented. Numbers in models denote resistivity values (in ohm-meters). HCL = high conductivity zone. Location shown on Figure 2.

Western Carpathians, a barrier exists, which causes energy attenuation of seismic waves of earthquakes originating in the Eastern Alpine area (Zatopek, 1979). A low velocity zone is present in the mantle underneath the Pannonian Basin at the depth of 70 km (43 mi), shown by P- and S-wave arrivals (Bisztricsany, 1974).

In the Western Carpathians, two vertically separated Moho discontinuities exist. Their vertical separation increases progressively from northwest to southeast (Szenas, 1972; Zatopek, 1979; international seismic profiles Deep Seismic Sounding [DSS] nos. III and V).

The lithosphere–asthenosphere boundary rises to a depth of 80 km (50 mi) in the Western Carpathians and 60 km (37 mi) in the Pannonian Basin, as indicated by P-wave traveltime residuals (Pajdusak et al., 1988) (Figure 2). The lower crust shows a higher amplitude subhorizontal reflectivity in the majority of deep reflection seismic profiles made in the Western Carpathians (Subotin, 1975; Ibrmajer and Suk, 1989; Tschekunov, 1989; Bucha and Blizkovsky, 1994).

The mantle lithosphere earthquakes that, according to many geophysicists, indicate postsubduction

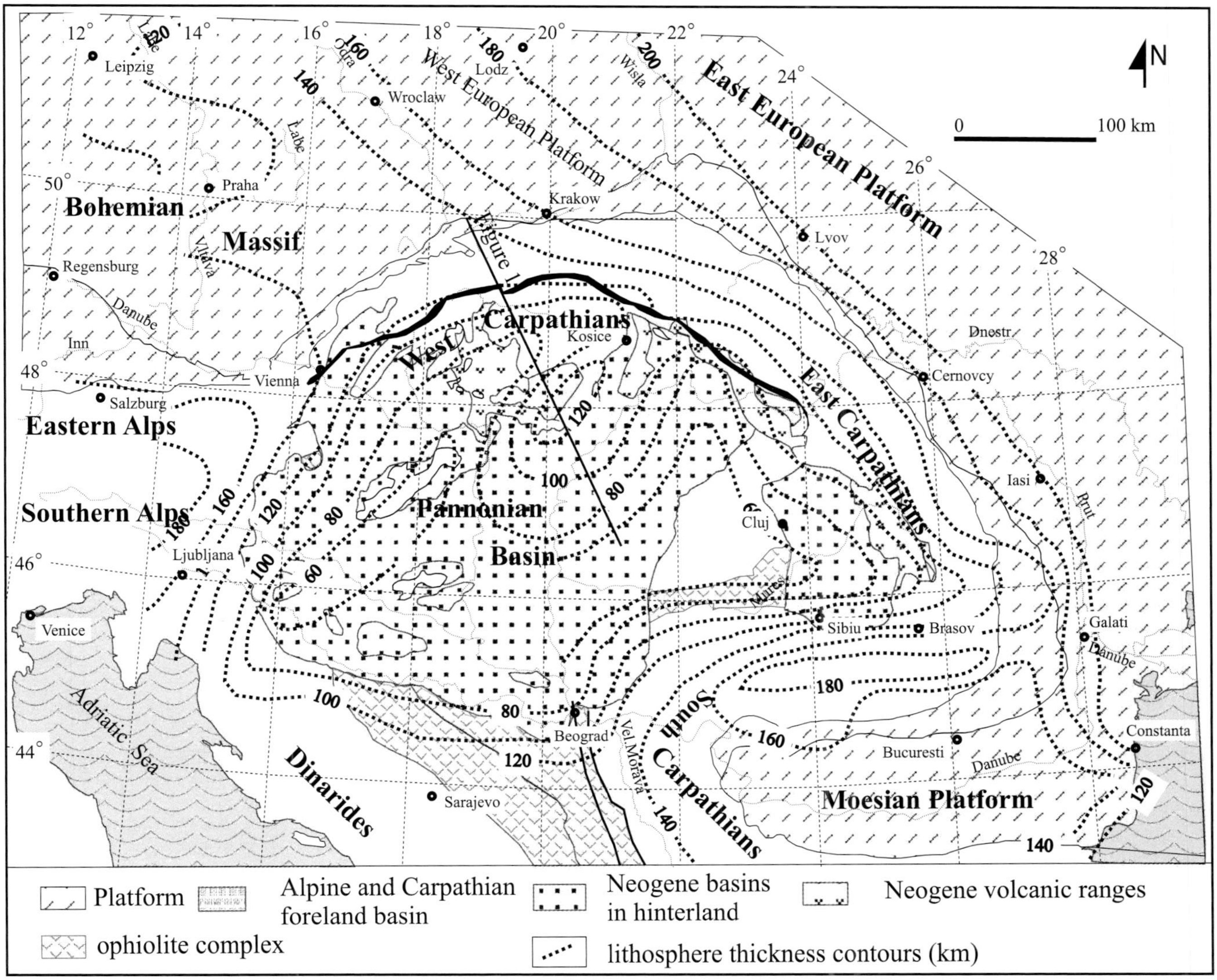

Figure 2. Lithosphere thickness contour map of the Carpathian–Pannonian region (modified after Babuska et al., 1987; Pajdusak et al., 1988). Data were derived by the study of P seismic wave traveltime residuals.

processes are present in the Vrancea region (the Carpathian bend to the north of Bucharest; Enescu et al., 1982). The depth of the seismofocal zone is from 70 to 220 km (43 to 136 mi). Moment magnitudes of earthquakes of the last century range to as much as 7.6 (e.g., Radu, 1979; Enescu, 1987). Focal mechanisms indicate polydirectional horizontal compression and subvertical extension (e.g., Roman, 1970; Iosif and Iosif, 1979; Monfret et al., 1990), which is interpreted as a slab descending under its own weight (Roman, 1970; Tschekunov, 1989) (Figure 3). The distribution of hypocenters outlines the subvertical slab, which dips northwest and trends east-northeast. A seismic gap between 40 and 70 km (25 and 43 mi) exists, between the lithospheric mantle earthquakes and the crustal ones, which is interpreted by Radu (1974), Fuchs et al. (1979), and Constantinescu and Enescu (1984) as a low-viscosity, low-velocity, seismically inactive zone where the subducting plate detaches. Shallow earthquakes as much as 40 km (25 mi) deep have thrust focal mechanisms that have a west–east-oriented compression that plunges at about 49° and drives displacements along subhorizontal planes (e.g., Radu, 1974; Fuchs et al., 1979; Polonic, 1986).

GRAVITY DATA

Regional gravity data show several distinct anomalies.

1) A distinct density minimum occurs along the Carpathian arc, which outlines the flexure zone of

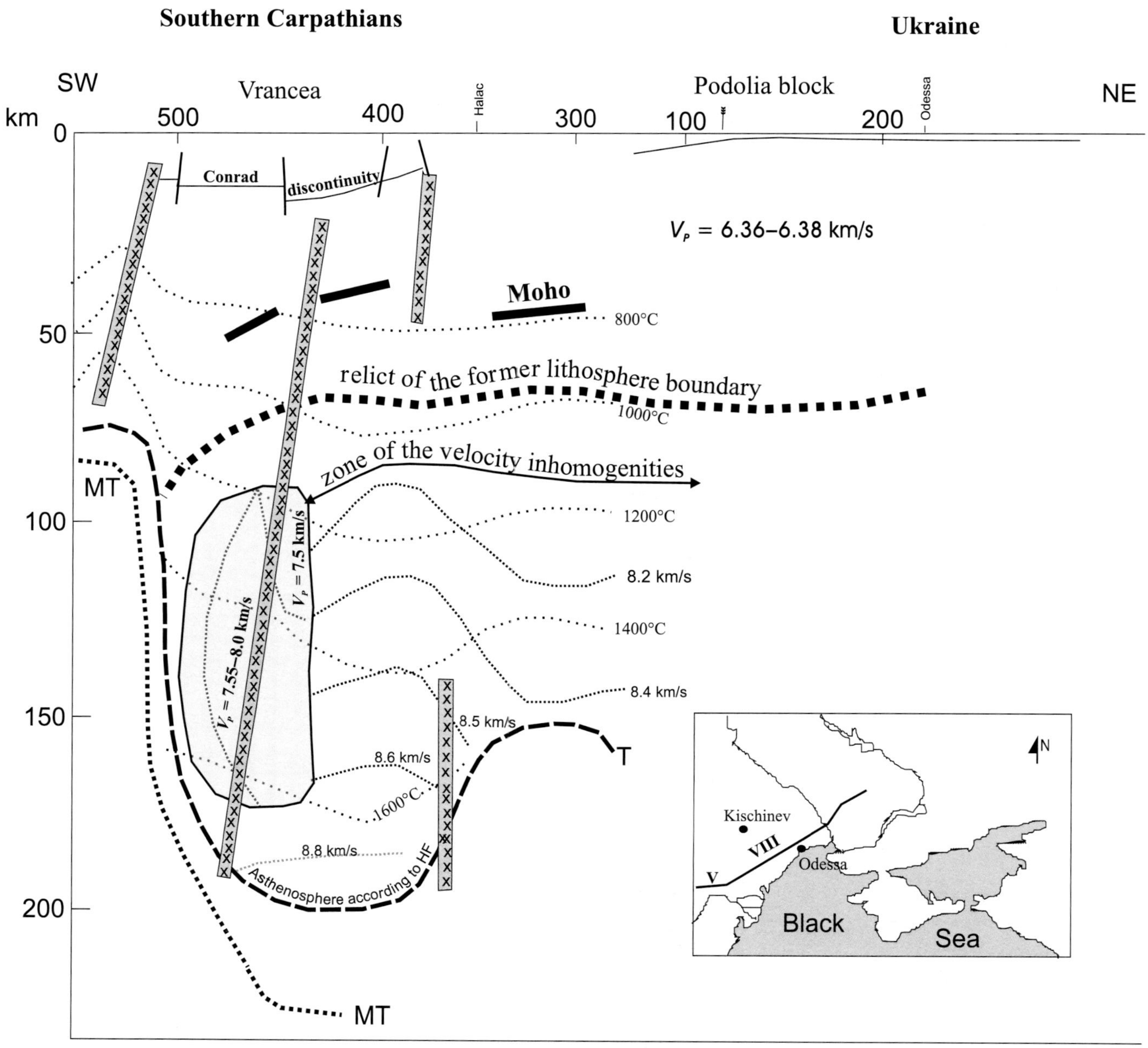

Figure 3. Complex model of the Carpathians–East European platform lithosphere constructed by Tschekunov (1989).

the platform and which is caused by sediments of the Molasse zone and Flysch belt (e.g., Tomek, 1988; Mocanu and Radulescu, 1994);

2) A problematic part of the central gravity low in the central part of the Inner Western Carpathians is probably caused by Cretaceous granitoids (Pospisil and Filo, 1980);
3) The Pannonian Basin shows positive anomalies even after correction for the gravity effect of the Tertiary basin fill (Tschekunov and Boljubach, 1979; Burianov et al., 1980; J. Sefara, 1980, personal communication; Mesko, 1983; Tari et al., 1999), which indicates an elevated asthenosphere.

CONDUCTIVITY BOUNDARY

A high-conductivity zone, determined by magnetotelluric data, is interpreted as the lithosphere–asthenosphere boundary at 50–60 km (31–37 mi) in the Pannonian Basin region (Adam, 1965, 1976) and at 100–150 km (62–93 mi) in the Western Carpathians (Pecova et al., 1979; R. Kucharski, 1980, personal communication). It reaches a depth of 200 km (124 mi) in the Bohemian Massif. A second high-conductivity zone is present at a depth of 100 km (62 mi) in certain parts of the eastern Bohemian Massif (Pecova et al., 1979). Its correlation with a low-velocity zone detected by seismological

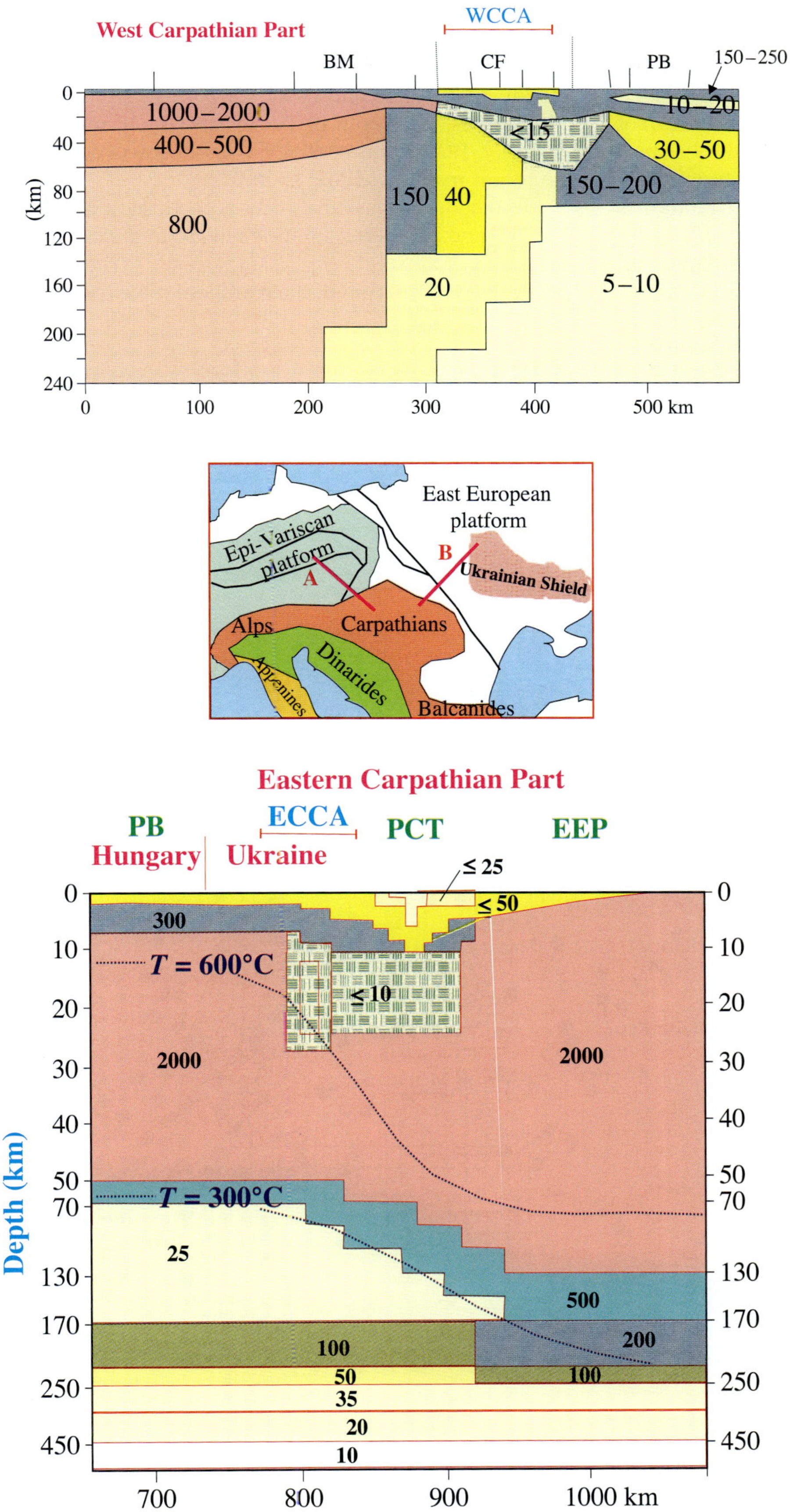

Figure 4. Two magnetotelluric sections through the Carpathians. Numbers in models denote resistivity values (in ohm-meters). The location of the Western Carpathian (WCCA) (Cerv et al., 1987) and Eastern Carpathian (ECCA) (Zhdanov et al., 1986) conductivity anomalies are shown above each model.

analysis was shown by Pecova et al. (1979). This high-conductivity zone disappears at the contact of the Bohemian Massif with the Western Carpathians (Figure 4). Another high-conductivity zone exists at the Pannonian Basin and the Western Carpathians contact at a depth of 40–50 km (25–31 mi) (C. Varga and E. Lada, 1989, personal communication). One of the most characteristic shallow conductors in northwest Transdanubia indicates fracture tectonics caused by low-viscosity graphite and fluid (Adam, 2001). In the lower crust of the Pannonian Basin, at a depth of about 15–17 km (9–10.5 mi), a conductive layer with very small conductance can be supposed because of fluid lost by dehydration (Adam, 1988).

The systematic investigation of the areal distribution of geomagnetic field variations, with several minutes to 2–3-hr periods (Jankowski et al., 1979; Praus et al., 1981) in the Western Carpathians and 20-min periods in the Eastern Carpathians (Rokityansky et al., 1976; Shapiro and Kuznetsova, 1981), revealed two important conductivity zones with zero variation of the geomagnetic field Z-component (Figure 5). The conductivity zone follows the Carpathian arc. The conductor lies at a depth of 16–26 km (10–16 mi) (Rokityansky, 1976; Praus et al., 1981) or shallower according to recent magnetotelluric soundings.

The conductivity anomaly does not correlate with a gravity low and is generally shifted inward to the Inner Carpathians. Adam and Pospisil (1984) relate to the contour of the passive margin of the continent plate (platform).

HEAT FLOW

The highest heat-flow values of the Carpathian–Pannonian region, 100 mW m^{-2}, are located in a northeast-trending broad zone in the Pannonian Bazsin (Cermak and

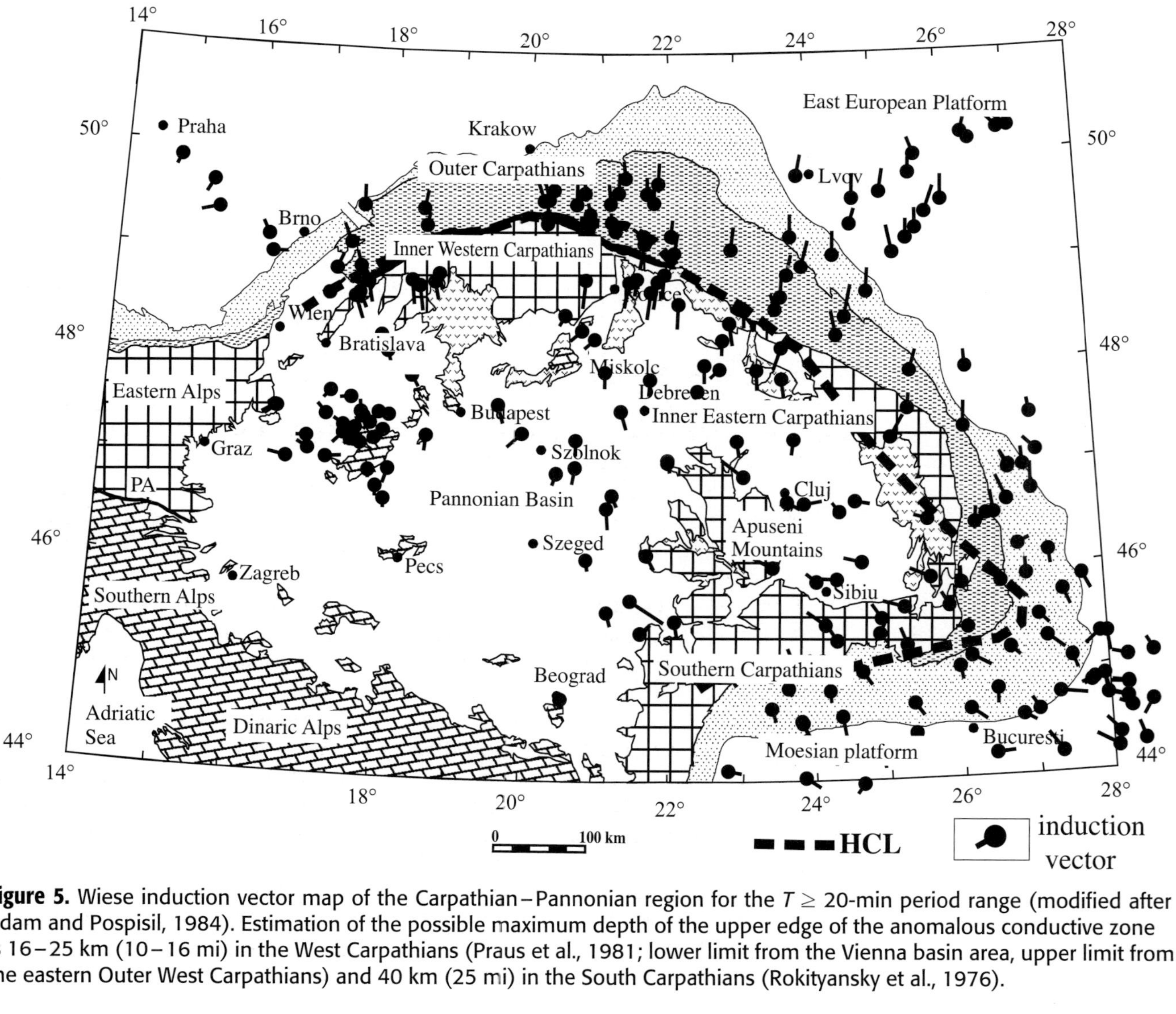

Figure 5. Wiese induction vector map of the Carpathian–Pannonian region for the $T \geq 20$-min period range (modified after Adam and Pospisil, 1984). Estimation of the possible maximum depth of the upper edge of the anomalous conductive zone is 16–25 km (10–16 mi) in the West Carpathians (Praus et al., 1981; lower limit from the Vienna basin area, upper limit from the eastern Outer West Carpathians) and 40 km (25 mi) in the South Carpathians (Rokityansky et al., 1976).

Hurtig, 1977; Horvath, 1979; see subsection "Terrestrial heat-flow density map of the Carpathian–Pannonian region," Pospisil et al., 2006). Their source was interpreted by Stegena et al. (1975) to be in the upper mantle. The whole Pannonian region is characterized by relatively high values (e.g., Horvath, 1979; Dovenyi et al., 1983; Demetrescu and Andreescu, 1994; Mocanu and Radulescu, 1994). Lower values were determined in the Outer Carpathians, and the lowest ones were in the Transylvanian Basin (e.g., Cermak and Hurtig, 1977; Mocanu and Radulescu, 1994). A regionally restricted heat-flow maximum comparable with Pannonian values is present to the north of Bucharest. Another thermal regime of the Eastern Carpathians and Transylvanian Basin was attributed to their different geological histories with processes involving the whole lithosphere (Demetrescu and Andreescu, 1994).

Generally, it is accepted that a zone of partial melting occurs at a boundary where a temperature is near the 1300°C isotherm. Calculated temperatures at a depth of 20 km (12 mi) and Moho discontinuity are 300–350 and 400–500°C for the Western Carpathian accretionary wedge and 400–650 and 500–800°C for the western part of the Pannonian Basin (Cermak, 1986; Lizon, 1986). The zone of melting can be identified just below the Pannonian Basin in the shallow asthenosphere.

RECENT VERTICAL CRUSTAL MOVEMENTS

Remeasurements of national geodetic polygon networks provide a data set on present uplift rates. These crustal vertical movement data (Joo et al., 1981) show a pronounced subsidence in the Danube basin and the Brasov area and uplift in the northern Western Carpathians, Outer Eastern Carpathians, and northwestern Southern Carpathians. The uplift in the Outer Eastern Carpathians is the most prominent Carpathian feature (see subsection "The recent vertical movements in the Carpathian region," Pospisil et al., 2006). It affects the whole Outer Eastern Carpathians and also the narrow adjacent zone of the European Platform.

MAGNETIC DATA

No unified magnetic map for the Carpathian–Pannonian region is available. Maps constructed in different countries show the anomaly of the vertical component of the magnetic field strength (ΔZ), but they do not match because they use different normal fields (Posgay, 1963; Krutichovskaja et al., 1971; Airinei et al., 1983; Sefara et al., 1987). Therefore, we made a map that concentrates on the separation of platform and orogenic sources from Tertiary volcanic sources (Figure 6).

Two types of magnetic anomalies exist in the Western Carpathians (Figure 6). The first type shows the high-intensity, abrupt changes in polarity, concentric structure, large spatial distribution, and volcanic source. The second type has positive anomalies of amplitude 50–300 nT and covers a smaller area. Its source is magnetic rocks of the pre-Tertiary basement. Anomalies of the second type form a zone in front of the Carpathian orogen and a complex pattern inside the orogen (Krutichovskaja et al., 1971). The northern zone extends from the Eastern Alps through the Vienna basin, Outer Flysch belt, to the Eastern Carpathians in Ukraine. Its western and northern parts coincide with the minimum-gravity zone (e.g., Krus and Sutora, 1986). The whole zone follows the zero Wiese vector zone from the platform side, which indicates the contact between the crust of the orogen and its foreland.

Orogenic positive magnetic anomalies comprise anomalies localized along large Raba, Muran, Hurbanovo, Balaton, and Darno faults, which juxtapose crustal blocks with different magnetic character. Based on the analysis of gravity and magnetic anomalies, Nemesi and Stomfai (1992) suggest that these anomalies are caused by the upwarping mantle material under one of the deep troughs of the Pannonian Basin. Based on two-dimensional (2-D) (Pospisil, 1980) and three-dimensional (3-D) gravity modeling (Bielik, 1988; Szafian, 1999), mafic intrusions under all of the deep troughs of the Pannonian Basin were suggested to be a part of the Moho updoming.

PALEOMAGNETIC DATA

An enormous amount of paleomagnetic data has been accumulated for the Carpathian–Pannonian realm. Many syntheses have been presented (Marton and Marton, 1978, 1989; Krs et al., 1986; Balla, 1987a; and others).

The main result of these observations is that two different tectonic units can be recognized in the Carpatho-Balkan region (Marton and Marton, 1978).

The first comprises the Dinarides, the Transdanubian Central Mountains, the Bukk Mountains, and the Inner Western Carpathians, commonly denoted as ALCAPA (earlier North Pannonian plate; Balla, 1987a, b). The second unit comprises the Mecsek and Villany Mountains, the Papuk Mountains, and the Apuseni Mountains and is denoted as Tisza–Dacia.

The Tisza–Dacia unit of the Carpathians was found to have rotated about 100° (clockwise), and the ALCAPA unit rotated about 35° (anticlockwise).

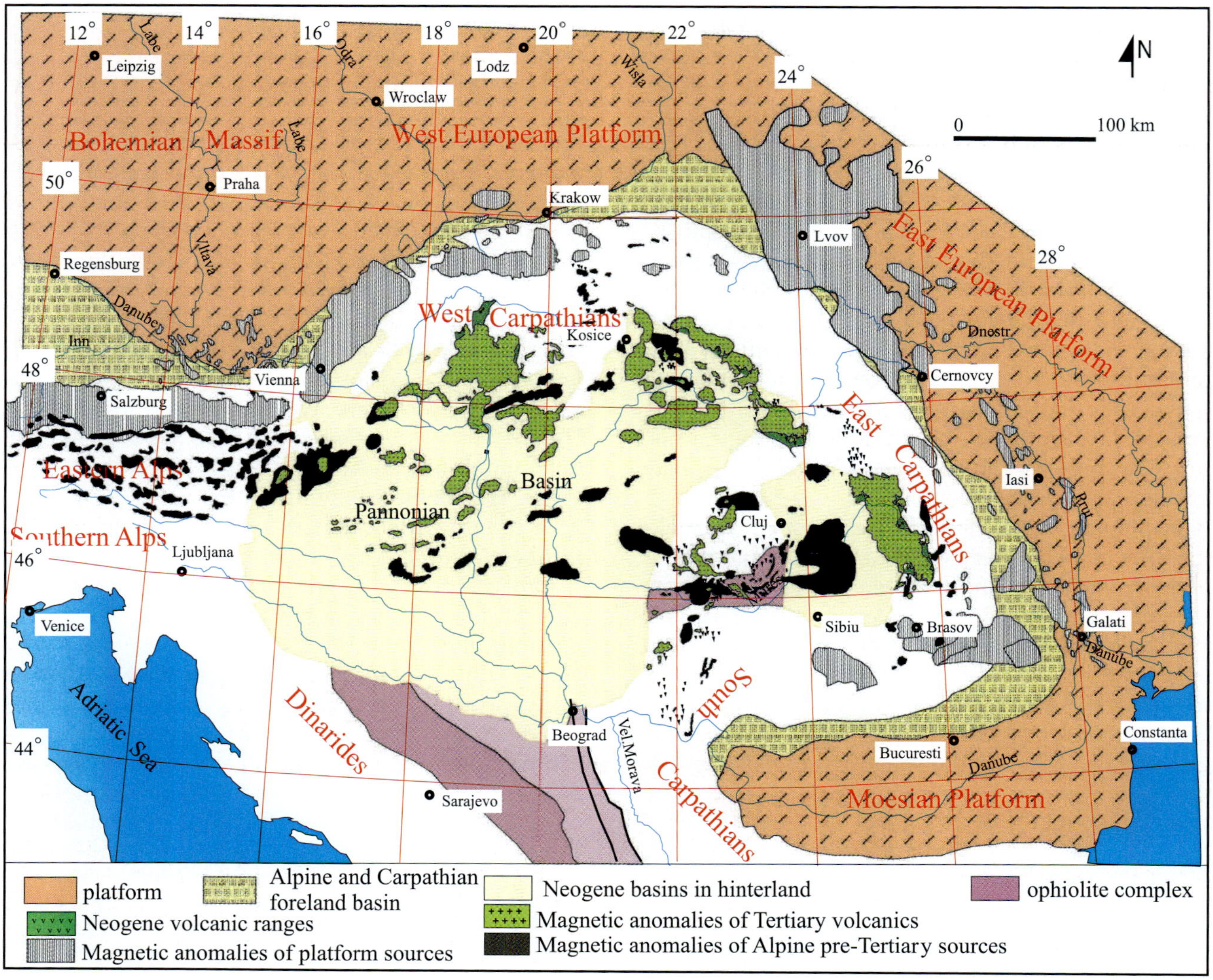

Figure 6. Magnetic anomaly source map of the Carpathian–Pannonian region. Sources are located in the European platform, located in the pre-Tertiary basement of the Carpathians and Pannonian Basin, and are buried volcanic sources.

LITHOSPHERE-ASTHENOSPHERE BOUNDARY

The base of the lithosphere corresponds roughly to a zone of the decrease of the P-wave velocity, an increase of electrical conductivity, and a change in the S-wave velocity, which is being put on the isotherm of the peridotite melting (Blizkovsky, 1986). The thinnest lithosphere is present in the Pannonian area, forming two minimums at 60 km (36 mi) separated by an elongated area of normally thick lithosphere (100–120 km; 62–75 mi) located between Budapest and Kosice (Babuska et al., 1987; Pajdusak et al., 1988) (Figure 3). This separation of the two shallow areas does not appear in the map of the conducting asthenosphere determined by magnetotellurics (Adam and Wesztergom, 2001). A lithospheric root is developed only in two areas of the Eastern Alpine–Carpathian orogen: in the Eastern Alps and Southern Carpathians. The lithospheric root in the Eastern Alps has a maximum thickness of 220 km (136 mi) (Babuska et al., 1984). Gravity modeling indicates an ongoing protrusion of the lithosphere into the asthenosphere (Lillie et al., 1994). The lithospheric root present along the northern margin of the Moesian platform has a maximum thickness of 180 km (111 mi). No lithospheric root developed underneath the Carpathian arc between these two areas.

CONCLUSION

A short review of the research of the Carpathian–Pannonian lithosphere manifests the main periods of the development and change of the character of research. International cooperation of Central and Eastern

European countries enabled a unification of results from the paleomagnetic, magnetotelluric, thermal, recent vertical crustal movements, and seismic and seismological researches.

During this period, the gravity and magnetic data were considered secret. For this reason, the complex analyses in the framework of the 2-D modeling suffered from some inaccuracies.

The period since the 1990s and political changes in the region brought a chance for a unified evaluation (reinterpretation) of all geophysical information that, with their density in space and time, provide an excellent chance for compiling 3-D models of the crust–lithosphere system. As an example of such activity, the prepared processes of all refraction profiles of the western part of the Pannonian Basin (E. Hegedus, Eotvos Lorand Geophysical Institute, 2003, personal communication) or a compilation of a unified magnetic map of the Carpathians is missing thus far.

REFERENCES CITED

Adam, A., 1965, Einige Hypothesen über den Aufbau des oberen Erdmantels in Ungarn: Gerlands Beiträge zur Geophysik, v. 74, p. 20–40.

Adam, A., 1974, Geoelectric deep structure and its geothermal interpretation in the Hungarian basin (in Hungarian): Magyar Tudomanyos Akademia, X. Osztaly Kozlemenyei, v. 7, nos. 3–4, p. 265–287.

Adam, A., 1976, Results of deep electromagnetic investigations, *in* A. Adam, ed., Geoelectric and geothermal Studies: Commission of the Academies of Socialist Countries for Planetary Geophysics Geophysical Monograph: Budapest, Akademiai Kiado, p. 547–560.

Adam, A., 1988, Are there two types of Conductivity Anomaly (CA) caused by fluid in the crust?: Physics of the Earth and Planetary Interiors, v. 45, p. 209–215.

Adam, A., 2001, Relation of the graphite and fluid bearing conducting dikes to the tectonics and seismicity (review on the Transdanubian crustal conductivity anomaly): Earth, Planets and Space, v. 53, p. 903–918.

Adam, A., and L. Pospisil, 1984, Crustal Conductivity Anomalies in the Carpathian Region: Acta Geodaetica, Geophysica et Montanistica Academiae Scientiarum Hungaricae, v. 19, no. 1–2, p. 19–34.

Adam, A., and V. Wesztergom, 2001, An attempt to map the depth of the electrical asthenosphere by deep magnetotelluric measurements in the Pannonian Basin (Hungary): Acta Geologica Hungarica, v. 44, p. 167–192.

Airinei, S., S. Stonescu, G. Velescu, D. Romanescu, M. Visarion, S. Radan, M. Roth, L. Besutin, and G. Besutin, 1983, Delta Z magnetic anomaly map for the territory of Romania (in French): Anuarul Institutului de Geologie si Geofizica, v. 63, p. 7–13.

Babuska, V., J. Plomerova, and J. Sileny, 1984, Spatial variations of P-residuals and deep structure of the European lithosphere: Geophysical Journal of the Royal Astronomical Society, v. 79, p. 363–383.

Babuska, V., J. Plomerova, and J. Sileny, 1987, Structural model of the subcrustal lithosphere in Central Europe, *in* K. Fuchs and C. Froidevaux, eds., Composition, structure and dynamics of the lithosphere–asthenosphere system, American Geophysical Union, Geodynamics, v. 16, p. 239–251.

Balla, Z., 1984, The Carpathian loop and the Pannonian Basin: A kinematic analysis: Geophysical Transactions, v. 30, p. 313–353.

Balla, Z., 1987a, Tertiary paleomagnetic data for Carpatho-Pannonian region in the light of Miocene rotation kinematics: Tectonophysics, v. 139, p. 67–98.

Balla, Z., 1987b, The tectonics of Hungary: A kinematic point of view: Rendiconti della Societa Geologica Italiana, v. 9, p. 97–102.

Balogh, K., 1993, Brief history of Hungarian geology: Budapest, Hungarian Geological Society, 95 p.

Belousov, V. V., 1975, The bases of the geotectonics (in Russian): Moscow, Nedra, 260 p.

Beranek, B., A. Dudek, M. Feifar, A. Hrdlicka, M. Suk, M. Zounkova, and J. Weiss, 1972, The crustal structure of Central and Southeastern Europe based on the results of explosion seismology: Geophysical Transactions, Special Edition, p. 87–98.

Bielik, M., 1988, Analysis of the stripped gravity map of the Pannonian Basin: Geologicki Zbornik, Geologica Carpathica, v. 39, p. 99–108.

Birkenmajer, K., 1974, Carpathians, *in* A. M. Spencer, ed., Mesozoic–Cenozoic Orogenic belts (in Russian): Published by the Scottish Academic Press Edinburgh for the Geological Society (London), p. 163–199.

Bisztricsany, E., 1974, The depth of the LVL in Europe and some adjacent regions: Geophysical Transactions, v. 22, p. 61–68.

Bleahu, M., 1985, The magmatic arc of the East Carpathians: Discussion and a proposal for interpretation: Annales Societatis Geologorum Poloniae, Rocznik Polskiego Towarzystwa Geologicznego, v. 55, nos. 1–2, p. 23–31.

Blizkovsky, M., ed., 1986, Geophysical model of lithosphere: Final report: The State Found of Technical Development, period 1982–1985 (in Czech): Praha, Geofyzika Brno-former Czechoslovak Academy of Science, 423 p.

Bucha, V., and M. Blizkovsky, eds., 1994, Crustal structure of the Bohemian Massif and the West Carpathians: Springer-Verlag and Academia, Berlin, 355 p.

Burianov, V. B., V. V. Gordienko, S. N. Kulik, and I. V. Logvinov, 1980, Complex geophysical model of the Western Carpathians, Czech Massif and Saxon–Thuring zone lithosphere (along the 6th international DSS profile) (in Russian): Geofyzicky Zbornik, v. 2, no. 3, p. 3–13.

Burianov, V. B., B. B. Gordienko, O. V. Zavgorodniaa, S. N. Kulik, I. M. Lovginov, and V. N. Schuman, 1987, Geophysical model of the Europe tectonosphere (in Russian): Kiev, Naukova Dumka, 184 p.

Cermak, V., 1986, Geothermic model of the Earth crust on

the territory of the Central and Eastern Europe, *in* M. Blizkovsky, ed., 1980: Geophysical model of lithosphere: Final report: The State Found of Technical Development, period 1982–1985 (in Czech): Praha, Geofyzika Brno-former Czechoslovak Academy of Science, 423 p.

Cermak, V., and F. Hurtig, 1977, Preliminary heat flow map of Europe: Postsdam, Praha, scale 1: 5,000,000, 2 sheets.

Cerv, P., J. Pek, and O. Praus, 1987, Numerical modelling of geoelectrical structures in Czechoslovakia: Physics of the Earth and Planetary Interiors, v. 45, p. 170–178.

Channel, J. F. T., and F. Horvath, 1976, The African (Adriatic) promotory as a paleogeographical premise for Alpine orogeny and plate movements in the Carpathian–Balcan region: Tectonophysics, v. 35, no. 1–3, p. 71–102.

Constantinescu, L., and D. Enescu, 1984, A tentative approach to possibly explaining the occurrence of the Vrancea earthquakes: Revue Roumaine de Geologie, Geophysique et Geographie. Serie de Geophysique, Bucarest, v. 28, p. 19–32.

Csontos, L., A. Nagymarosy, F. Horvath, and M. Kovac, 1992, Tertiary evolution of the Intra-Carpathian area: A model, *in* P. A. Ziegler, ed., Geodynamics of rifting: I. Case history studies on rifts: Europe and Asia: Tectonophysics, v. 208, p. 221–241.

Demetrescu, C., and M. Andreescu, 1994, On the thermal regime of some tectonic units in a continental collision environment in Romania: Tectonophysics, v. 230, p. 265–276.

Dercourt, J., et al., 1985, Presentation de 9 cartes paleogeographiques au 1/20.000.000 s'etandant de l'Atlantique au Pamir pour la peflode du Lias l 'Actuel: Bulletin de la Societe Geologique de France, 8 Serie, v. 1, no. 15, p. 637–652.

Dietrich, V., and U. Franz, 1976, Alpidische Gebirgsbildung in den Ostaalpen: ein plattentektonisches Modell: Geologische Rundschau, v. 65, p. 361–374.

Dovenyi, P., F. Horvath, P. Liebe, J. Galfi, and I. es Erki, 1983, Geothermal conditions of Hungary: Geophysical Transactions, v. 29, no. 1, p. 3–114.

Enescu, D., 1987, Contributions to the knowledge on the lithosphere structure in Romania on the basis of seismic data: Studii si Cercetari de Geologie, Geofizica, Geografie. Seria Geofizica, v. 25, p. 20–27.

Enescu, D., I. Cornea, F. L. Radulescu, V. Taileanu, and A. Pompilian, 1982, Seismological data on the deep structure of the Vrancea region: Revue Roumaine de Geologie, Geophysique et Geographie. Serie de Geophysique, v. 26, p. 29–36.

Fuchs, K., et al., 1979, The Romanian earthquake of March 4, 1977: II. Aftershocks and migration of seismic activity: Tectonophysics, v. 53, p. 225–247.

Geczy, B., 1972, The formation of the Jurassic fauna provinces and the Mediterranean plate tectonics (in Hungarian): Magyar Tudomanyos Akademia Ipargazdasagtani Kutatocsoport sz. Kozlemeny, v. 5, p. 297–312.

Geczy, B., 1973, Plate tectonics and palaeogeography in the East Mediterranean region (in Hungarian): Magyar Tudomanyos Akademia Ipargazdasagtani Kutatocsoport sz. Kozlemeny, v. 6, p. 219–226.

Guterch, A., M. Grad, G. R. Keller, K. Posgay, J. Vozar, A. Spicak, E. Brueckl, Z. Hajnal, H. Thybo, and S. Oguy, 2000, Celebration 2000: Huge seismic experiment in Central Europe: Geologica Carpathica, v. 51, p. 413–414.

Hadzi, E., N. Pantic, V. Aleksic, and M. Kolenic, 1974, Alpidi Jugoistvene Europe u svetlotectonik ploca, *in* E. Hadzi, ed., Metalogenie; geotektonike koncepciej Jugoslavie: Beograd, Rudarsko-geoloski fak., p. 21–38.

Horvath, F., 1979, Physical quantities used in geothermal researches (in Hungarian): Magyar Geofizika, v. 20, no. 1, p. 31–38.

Horvath, F., 1984, Neotectonics of the Pannonian Basin and the surrounding mountain belts: Alps, Carpathians and Dinarides: Annals of Geophysics, v. 2, p. 147–154.

Horvath, F., 1986, Geophysical model of the formation of the Pannonian Basin (in Hungarian): Ph.D. Thesis, Eotvos University, Budapest, 148 p.

Horvath, F., 1993, Towards a mechanical model for the formation of the Pannonian Basin: Tectonophysics, v. 226, p. 333–357.

Horvath, F., and L. Royden, 1981, Mechanism for the formation of the Intra-Carpathian basin: A review: Earth Evolution Sciences, v. 3–4, p. 307–316.

Horvath, F., and J. Rumpler, 1984, The Pannonian basement: Extension and subsidence of an Alpine orogene: Acta Geologica Hungarica, v. 27, p. 229–235.

Ibrmajer, J., and M. Suk, 1989, Geophysical image of CSSR (in Czech): former Czechoslovak Academy of Science, Prague, 354 p.

Iosif, T., and S. Iosif, 1979, Vrancea earthquake of March 4, 1977—A multiple seismic event and its seismotectonic implications: Revue Roumaine de Geologie, Geophysique et Geographie, Serie de Geophysique, v. 23, p. 3–17.

Jankowski, J., V. Petr, J. Pecova, and O. Praus, 1979, Induction vector estimates in the Polish–Czechoslovak part of the Carpathians: Studia Geophysica et Geodaetica, v. 23, p. 89–93.

Jiricek, R., 1979, Tectonic development of the Carpathian arc during the Oligocene and Neogene, *in* M. Mahel, ed., Tectonic profiles through the West Carpathians (in Czech): Bratislava, Geologicky ustav Dionyze Stura, p. 205–215.

Joo, I., E. Csati, P. Jovanovic, M. Popescu, V. I. Somov, H. Thurm, J. Thury, I. N. Totomatov, J. Vanko, and T. Wyrzykowski, 1981, Recent vertical crustal movements of the Carpatho-Balkan region: Tectonophysics, v. 71, p. 41–52.

Kober, L., 1921, Der Bau der Erde: Berlin, Borntraeger, 324 p.

Kovac, M., I. Cicha, I. Krystek, A. Slazska, Z. Stranik, N. Oszczypko, and D. Vass, 1989, Palinspastic maps of the Western Carpathian Neogene: Prague, Geological Survey, scale 1: 1,000,000, 31 p.

Krs, M., P. Muska, O. Orlicky, and P. Pagac, 1986, Paleomagnetism and paleomagnetic interpretations in the post-Carbon formations of the Czechoslovak West Carpathians, *in* M. Blizkovsky, ed., Geophysical model of lithosphere: Project of The State Found of Technical Development, period 1982–1985, Final Report: Brno/Prague (in Czech): Prague, Geofyzika Brno/former Czechoslovak Academy of Science, p. 397–404.

Krus, S., and A. Sutora, 1986, Geophysical-geological atlas of the Alpine–Carpathian mountain system: Brno, Geofyzika Brno, 12 sheets, 25 p.

Krutichovskaja, Z. A., O. M. Rusakov, I. K. Paskevic, and K. Posgay, 1971, Magnetic field of the Carpathian region and several problems of its geologic interpretations (in Russian): Acta Geologica Hungarica, v. 15, p. 377–382.

Ksiazkewicz, M., 1976, La profondeur du bassin du flysch carpathique: Bulletin de la Societe Geologique de France, v. XVIII, no. 1, p. 191–196.

Ksiazkiewicz, M., 1977, The tectonics of the Carpathians—Geology of Poland, Tectonics: Warsaw, Publishing House Wydawnictwa Geologiczne, Geological Institute, v. 4, p. 476–608.

Laubscher, H. P., 1971, Das Alpen–Dinariden–Problem und die Palinspastik der sudlischen Tethys: Geologische Rundschau, v. 60, p. 813–833.

Lesko, B., and I. Varga, 1980, Alpine elements in the West Carpathian structure and their significance: Mineralia Slovaca, v. 12, no. 2, p. 97–130.

Lexa, J., and V. Konecny, 1974, The Carpathian volcanic arc: A discussion: Acta Geologica Hungarica, v. 18, p. 279–293.

Lillie, R. J., M. Bielik, V. Babuska, and J. Plomerova, 1994, Gravity modelling of the lithosphere in the Eastern–Western Carpathian–Pannonian Basin region: Tectonophysics, v. 231, p. 215–235.

Lizon, I., 1986, Geothermic model and geodynamics of the West Carpathians (in Slovak), *in* M. Blizkovsky, ed., Geophysical model of lithosphere. Project of The State Found of Technical Development, period 1982–1985, Final Report, Brno/Prague: Prague, Geofyzika Brno/former Czechoslovak Academy of Science, p. 292–298.

Mahel, M., 1974, Tectonics of the Carpathian–Balkan regions: Explanations to the tectonic map of the Carpathian–Balkan regions and their forelands: Bratislava, Geologicky ustav Dionyze Stura, 453 p.

Mahel, M., and P. Reichwalder, eds., 1979, Czechoslovak geology and global tectonics: Proceedings of the Conference, Smolenice, May 19–21, 1976, Bratislava, VEDA, Publishing House of the Slovak Academy of Sciences, 285 p.

Marton, E., and P. Marton, 1978, On the difference between the paleomagnetic poles from the Transdanubian Central Mountains and the Villany Mountains, respectively (in Hungarian): Magyar Geofizika, v. 19, p. 129–136.

Marton, E., and P. Marton, 1989, A compilation of paleomagnetic results from Hungary: Geophysical Transactions, v. 35, p. 117–133.

Mesko, A., 1983, Regional Bouguer gravity maps of Hungary: Acta Geodaetica, Geophysica et Montanistica Hungarica, v. 18, p. 187–200.

Mocanu, I. V., and F. Radulescu, 1994, Geophysical features of the Romanian territory, *in* T. Berza, ed., ALCAPA II Field Guidebook: Geological Institute of Romania, p. 17–36.

Monfret, T., A. Deschamps, and B. Romanowicz, 1990, The Romanian earthquake of August, 30 1986: A study based on GEOSCOPE very long-period and broadband data: Pure and Applied Geophysics, v. 133, no. 2, p. 367–379.

Nemesi, L., and R. Stomfai, 1992, Some supplements to the exploration of the basement of the Bekes basin (in Hungarian): Magyar Geofizika, v. 33, p. 70–79.

Oszczypko, N., and A. Slaczka, 1985, An attempt to palinspastic reconstruction of Neogene basins in the Carpathian Foredeep: Annales Societatis Geologorum Poloniae, v. 55, p. 5–75.

Oszczypko, N., and A. Slaczka, 1989, The evolution of the Miocene basins in the Polish Outer Carpathians and their foreland: Geologica Carpathica, v. 40, p. 23–36.

Pajdusak, P., J. Plomerova, and V. Babuska, 1988, Lithosphere deep structure of the West Carpathians determined on the basis of P-Residue (in Czech): Proceedings of the Conference: "Deep structure investigation of Czechoslovakia," May 2–5, 1988, Smolenice, p. 151–158.

Pecova, J., V. Petr, and O. Praus, 1979, Internal electrical conductivity distribution on Czechoslovak territory, Geodynamical investigations in Czechoslovakia: Bratislava, VEDA, Slovak Academy of Science, p. 119–127.

Polonic, G., 1986, On the seismotectonic relations in the Moldavian Platform and adjacent units: Revue Roumaine de Geologie, Geophysique et Geographie, Serie de Geophysique, v. 30, p. 11–17.

Posgay, K., 1963, A comprehensive map of the magnetic masses in Hungary and its interpretation: Acta Technica, v. 43, no. 3–4, p. 271–288.

Posgay, K., I. Albu, E. Hegedus, and Z. Timar, 1990, Deep seismic investigations along the Pannonian Geotraverse: Acta Geodaetica, Geophysica et Montanistica Hungarica, v. 25, p. 267–277.

Posgay, K., T. Bodoky, E. Hegedus, S. Kovacsvolgyi, L. Lenkey, P. Szafian, E. Takacs, Z. Timar, and G. Varga, 1995, Asthenospheric structure beneath a Neogene basin in southeast Hungary, *in* S. Cloetingh, B. D'Argenio, R. Catalano, F. Horvath, and W. Sassi, eds., Interplay of extension and compression in basin formation: Tectonophysics, v. 252, p. 467–484.

Pospisil, L., 1980, Interpretation of the gravity field in the East Slovakian Neogene (in Slovak): Mineralia Slovaca, v. 12, no. 5, p. 421–440.

Pospisil, L., 1984, Some elements of the lithosphere structure of the West Carpathians (in Czech): Proceedings of the Workshop "Research of the Deep Geological Structures of Czechoslovakia," Loucna n/D., February 13–16, 1984, Brno, Geofyzika n.p., p. 19–31.

Pospisil, L., and M. Filo, 1980, The West Carpathian central gravity minimum and its interpretation: Mineralia Slovaca, v. 12, p. 149–164.

Pospisil, L., et al., 2006, Crustal and lithospheric structure of the Carpathian–Pannonian region—A geophysical perspective: Regional geophysical data on the Carpathian–Pannonian lithosphere, *in* J. Golonka and F. J. Picha, eds.,

The Carpathians and their foreland: Geology and hydrocarbon resources: AAPG Memoir 84, p. 651–697.

Praus, O., J. Pecova, V. Petr, K. Pec, M. Hvozdara, V. Cerv, J. Pek, and M. Lastovickova, 1981, Electronic induction and electrical conductivity in the Earth's body, *in* A. Zatopek, ed., Geophysical syntheses in Czechoslovakia: Bratislava, VEDA, Slovak Academy of Science, p. 297–316.

Radu, C., 1974, Contribution a l'e'tude de la se'ismicite' de la Roumanie et comparaison avec la se'ismicite' du basin mediterrane'en et en particulier avec la se'ismicite' de Sud-Estde la France: Ph.D. thesis, University of Strasbourg, France, 404 p.

Radu, C., 1979, Catalogul cutremurelor puternice produse pe teritoriul Romanei: Partea I. Inainte de 1901; Partea II, 1901–1979, *in* I. Cornea and C. Radu, eds., Certetori Seismologice asupra Cutremurului din 4 Martie 1977: Bucharest, Romanian Academy, p. 723–752.

Rokityansky, I. I., V. K. Amirov, S. N. Kulik, I. M. Logrinov, and V. N. Shuman, 1976, The electric conductivity anomaly in the Carpathians, *in* A. Adam, ed., Geoelectric and geothermal studies: Commission of the Academies of Socialist Countries for Planetary Geophysics Geophysical Monograph, p. 604–612.

Roman, C., 1970, Seismicity in Romania—Evidence for the sinking lithosphere: Nature, v. 228, p. 1176–1178.

Royden, L. H., and F. Horvath, eds., 1988, The Pannonian Basin: A study in basin evolution: AAPG Memoir 45, 394 p.

Royden, L. H., F. Horvath, and B. C. Burchfiel, 1982, Transform faulting, extension and subduction in the Carpathian–Pannonian region: Geological Society of America Bulletin, v. 73, p. 717–725.

Royden, L. F., F. Horvath, A. Nagymarossy, and L. Stegena, 1983a, Evolution of the Pannonian Basin system: 2. Subsidence and thermal history: Tectonics, v. 2, no. 1, p. 91–137.

Royden, L. H., F. Horvath, and J. Rumpler, 1983b, Evolution of the Pannonian Basin system: 1: Tectonics, v. 2, p. 63–90.

Sandulescu, M., 1975, Essai de synthese structurale des Carpathes: Bulletin de la Societe Geologique de France 7, v. XVII, no. 3, p. 299–358.

Sandulescu, M., 1988, Cenozoic tectonic history of the Carpathians, *in* L. H. Royden and F. Horvath, eds., The Pannonian Basin: A study in basin evolution: AAPG Memoir 45, p. 17–25.

Sefara, J., M. Kurkin, and, J. Plancar, 1980, Research of the Conrad discontinuity in the West Carpathians (in Slovak): Manuscript, Brno, Geofyzika n.p, 287 p.

Sefara, J., et al., 1987, Structural-tectonic map of the Inner West Carpathians for the purposes of the resource prognoses—Geophysical interpretation: Bratislava, Slovak Ministry of Environmental Affairs, 287 p.

Shapiro, V. A., and V. G. Kuznetsova, 1981, Some results in the investigation of geomagnetic field variation in the Urals and Carpathians: Geophysical Journal, v. 66, p. 71–82.

Spakman, W., S. Van der Lee, and R. Van der Hilst, 1993, Travel-time tomography of the European–Mediterranean mantle down to 1400 km: Physics of the Earth and Planetary Interiors, v. 79, p. 3–7.

Staub, R., 1928, Der Bewegungsmechanismus der Erde: Berlin, Borntraeger, 270 p.

Stegena, L., 1972, Plate tectonics, Tethys and the Hungarian Basin (in Hungarian): Foldtani Kozlony, v. 102, nos. 3–4, p. 280–300.

Stegena, L., B. Geczy, and F. Horvath, 1975, Late Cenozoic evolution of the Pannonian Basin: Tectonophysics, v. 26, p. 71–91.

Subotin, S. I., 1975, The Earth's physics problems in the Ukraine (in Russian): Kiev, Naukova Dumka, 173 p.

Szadeczky-Kardoss, E., 1968, The structure and evolution of the Earth (in Hungarian): Budapest, Akademiai Kiado, 340 p.

Szadeczky-Kardoss, E., 1969, The interaction between the crust and mantle in the surroundings of the Carpathian basins (in Hungarian): Magyar Tudomanyos Akademia Ipargazdasagtani Kutatocsoport sz. Kozlemeny, v. 2, no. 3, p. 197–206.

Szadeczky-Kardoss, E., 1971, Der Karpato-Dinarische Raum vom Gesichtspunte der neuen Globaltektonik aus betrachtet (in Hungarian): Magyar Tudomanyos Akademia Ipargazdasagtani Kutatocsoport sz. Közlemeny, v. 4, p. 64–81.

Szadeczky-Kardoss, E., 1973, Jungalpine Subductionslinien im KBD Gebiet, Section III, Tectonics, Proceedings of the Xth Congress Carpathian-Balkan Geological Association: Bratislava, Geologicky ustav Dionyze Stura, p. 268–276.

Szafian, P., 1999, Gravity and tectonics—A case study in the Pannonian Basin and the surrounding mountain belt: Ph.D. Thesis, Vrije University, Amsterdam, 154 p.

Szenas G., ed., 1972, The crustal structure of Central and Southeastern Europe based on the results of explosive seismology: Geophysical Transactions, 172 p.

Tari, G, P. Dovenyi, I. Dunkl, F. Horvath, L. Lenkey, M. Stefanescu, P. Szafian, and T. Toth, 1999, Lithospheric structure of the Pannonian Basin derived from seismic, gravity and geothermal data, *in* B. Durand, L. Jolivet, F. Horvath, and M. Seranne, eds., The Mediterranean basins: Tertiary extension within the Alpine orogen: Geological Society (London) Special Publication 156, p. 215–250.

Tomek, C., 1988, Geophysical investigation of the Alpine–Carpathian arc, *in* M. Rakus, J. Dercourt, and A. E. M. Nairn, Evolution of the northern margin of Tethys: Memoires de la Societe Geologique de France, Paris, Nouvelle Serie No. 154, v. I, p. 167–199.

Tomek, C., I. Ibrmajer, T. Korab, A. Biely, L. Dvorakova, J. Lexa, and A. Zboril, 1989, Crustal structures of the West Carpathians on deep reflection seismic line 2T: Mineralia Slovaca, v. 1, no. 21, p. 3–26.

Tschekunov, A. V., ed., 1988a, Lithosphere of Central and Eastern Europe—Geotraverse I, II, V (in Russian): Kiev, Naukova Dumka, 168 p.

Tschekunov, A. V., ed., 1988b, Lithosphere of the Central

and East Europe: Geotraverses IV, VI and VIII (in Russian): Kiev, Naukova Dumka, 172 p.

Tschekunov, A. V., 1989, Geodynamic model of geotraverse Dnepr-Donieck paleorift-Ukraine Shield-Southern Carpathians (in Russian): Geotectonika, v. 6, p. 3-11.

Tschekunov, A. V., and K. A. Boljubach, 1979, Geologiceskije aspekty komplexnoj interpretacii seijsmiceskich i gravimetriceskich dannych v zapadnoj casti Ukrainy, *in* Gravitacionnaja model kory i verchnej mantii Zemli: Kiev, Naukova Dumka, 289 p.

Van Bemmelen, R. W., 1960, New views on East Alpine orogenesis: *in* T. Sorgenfrei, ed., 21st International Geological Congress, Copenhagen, v. 18, p. 99-116.

Van Bemmelen, R. W., 1966, On mega-undulations; a new model for the Earth's evolution: Tectonophysics, v. 3, p. 83-127.

Vass, D., 1979, Genesis of inner-molasse basins in West Carpathians in light of leading function of mantle in Earth's crust development: Proceedings of the Conference: Czechoslovak geology and global tectonics: Bratislava, VEDA, Publishing House of the Slovak Academy of Sciences, p. 183-197.

Vass, D., and L. Pospisil, 1984, Influence of the structure of the lithosphere on the formation and development of intramontane and back molasses basins of the Carpathian Mountains: Geophysical Transactions, v. 30, no. 4, p. 355-371.

Wein, G. Y., 1978, Outlines of the development of the Carpathian Basin (in Hungarian): Altalanos Foldtani Szemle, v. 11, 534 p.

Zatopek, A., ed., 1979, On geodynamical aspects of geophysical synthesis in Central Europe: Final report: Geodynamic investigations in Czechoslovakia, Bratislava, VEDA, p. 91-104.

Zhdanov, M. S., N. G. Golubev, I. M. Varentsov, L. M. Abramova, V. S. Shneer, M. N. Berdichevsky, O. N. Zhdanova, V. V. Gordienko, S. N. Kulik, and A. I. Bilinsky, 1986, 2-D model fitting of a geomagnetic anomaly in the Soviet Carpathians: Annales Geophysicae, v. 4, no. B3, p. 335-342.

Pospíšil, L., A. Ádám, J. Bimka, P. Bodlak, T. Bodoky, P. Dövényi, H. Granser, E. Hegedüs, I. Joó, A. Kendzera, L. Lenkey, M. Nemčok, K. Posgay, B. Pylypyshyn, J. Sedlák, W. D. Stanley, G. Starodub, V. Szalaiová, B. Šály, A. Šutora, G. Várga, and T. Zsiros, 2006, Crustal and lithospheric structure of the Carpathian–Pannonian region – A geophysical perspective: Regional geophysical data on the Carpathian–Pannonian lithosphere, *in* J. Golonka and F. J. Picha, eds., The Carpathians and their foreland: Geology and hydrocarbon resources: AAPG Memoir 84, p. 497–499.

22

Crustal and Lithospheric Structure of the Carpathian–Pannonian Region— A Geophysical Perspective: Regional Geophysical Data on the Carpathian–Pannonian Lithosphere

L. Pospíšil
Geoinform Consultants, Brno, Czech Republic

A. Ádám
Geodetic and Geophysical Research Institute, Sopron, Hungary

J. Bimka
Moravske naftove doly (Moravian Oil Company), a.s., Hodonín, Czech Republic

P. Bodlak
Western Ukrainian Geophysical Expedition, Zapadnaja Ukrainskaja Geofyziceskaja Razveaõnaja Expedicija, Lviv, Ukraine

T. Bodoky
Eötvös Loránd Geophysical Institute of Hungary, Budapest, Hungary

P. Dövényi
Academic Research Group of Environmental Physics, Department of Geophysics, Eötvös University, Budapest, Hungary

H. Granser
OMV Aktiengesellschaft Wien, Vienna, Austria

E. Hegedüs
Eötvös Loránd Geophysical Institute of Hungary, Budapest, Hungary

I. Joó
College of Geoinformatics, University of West Hungary, Székesfehérvár, Hungary

A. Kendzera
National Academy of Sciences, Institute of Geophysics, Kiev, Ukraine

L. Lenkey
Eötvös Loránd University, Budapest, Hungary

M. Nemčok
Energy & Geoscience Institute, Salt Lake City, Utah, U.S.A.

K. Posgay
Eötvös Loránd Geophysical Institute of Hungary, Budapest, Hungary

B. Pylypyshyn
Ukrainian State Geological Research Institute, Lviv, Ukraine

J. Sedlák
Geofyzika, a.s., Brno, Czech Republic

DOI:10.1306/985622M843079

W. D. Stanley

U.S. Geological Survey, U.S.A.

G. Starodub

National Academy of Sciences, Institute of Geophysics, Lviv, Ukraine

V. Szalaiová

Geocomplex, a.s., Bratislava, Slovakia

B. Šály

Nafta, a.s., Gbely, Slovakia

A. Šutora

Geoinform Consultants, Brno, Czech Republic

G. Várga

Eötvös Loránd Geophysical Institute of Hungary, Budapest, Hungary

T. Zsíros

Seismological Observatory, Geodetic and Geophysical Research Institute, Hungarian Academy of Sciences, Budapest, Hungary

ABSTRACT

Research of the lithosphere of the Western Carpathians, based on the analysis and interpretation of the complex geological and geophysical data, has brought several remarkable results that enable us to understand more precisely the development of the sedimentary basins. First, a tight relationship between gravity anomalies and sources originated from Tertiary sedimentary basins that creates substantial parts of the gravity effect. The main part of the gravity effects originating from both the Moho discontinuity and the boundary between lithosphere and asthenosphere is mutually compensated for in the region of the Carpathians, and therefore, it can be traced only with difficulties.

In this chapter, basic information regarding the study of the gravity field of the Western Carpathians are presented, and these both describe the most significant anomalies and the features of the gravity field of basins and support the significance of gravity in the process of verification of interpretation of seismic, magnetotelluric, and geological models in the area of contact of the Inner and Outer Carpathians and the models of the lithosphere.

The fundamental results of deep seismic sounding in the Carpathian–Pannonian region constitute the next part of the presented data, especially the observation of the Moho reflection by deep seismic reflection surveys (in 1955), the deepening of the crust–mantle boundary under the surrounding mountain range inferred from international refraction and wide-angle reflection surveys (1964–1975), the detection of seismic reflection arrivals of several kilometers originating from the upper mantle and in quantifying the depth of the lithosphere–asthenosphere boundary by determining the velocity-vs.-depth function of the P waves (1975–1981), the determination of a domal uplift of the lithosphere–asthenosphere boundary, and the existence of shear zone systems cutting through a significant part of the lithosphere (1990–2000).

The seismicity of the Carpathian Basin is summarized based on a comprehensive earthquake catalog with more than 20,000 events. The recurrence curve of earthquakes predicts an earthquake of magnitude *M* 5 every 1 yr, an earthquake of magnitude *M* 6 every 10 yr, and an earthquake of magnitude *M* 7 every 100 yr in the Carpathian Basin on average. Most of the shallow-depth events originate from the 5–15-km (3–10-mi) depth domain, whereas in the Vrancea region, roughly three groups can be outlined with depth centers of about 15, 80, and 130 km (10, 50, and 80 mi), respectively. The average yearly energy release is 6.1×10^{13} J in the whole basin, and the contribution of the Vrancea zone to this amount is 4.8×10^{13} J/yr, whereas 1.3×10^{13} J/yr is the product of the remaining large part of the studied area.

The synthetic seismograms for the International Geotraverse II of the Eastern Carpathian lithosphere structure were calculated. Geology, seismic, heat, gravity, and electromagnetic data were considered. Vertical and horizontal variations of velocity, density, and energy dissipation were approximated by means of the finite-element method. Synthetic seismograms were compared with experimental ones registered by West Ukrainian Uszhorod, Kosiv, and Miszhirja seismic stations. Effects of change of the lithosphere structure characteristics in the Eastern Carpathian cross section were interpreted on the seismic patterns of waves spreading through the explored cross section.

The most important part of the chapter compiles the data on the distribution of the electrical conductivity anomalies, which have been determined by magnetotellurics in the Earth's crust and upper mantle in the Carpathian–Pannonian Basin. Among them, the Carpathian conductivity anomalies, the Transdanubian upper crustal conductor, the middle to lower crustal conducting layer, and the updoming asthenosphere are discussed in detail.

The interpreted resistivity model along the reflection profile 2T, which crossed the whole part of Western Carpathians in 1987–1988, has brought several surprising results that, in combination with reflection seismic, gravity, and other geophysical data, enable us to construct a more precise model of the Western Carpathian lithosphere.

The complex set of basic information supplies heat flow, recent vertical movement, and paleomagnetic data.

A new heat-flow map is presented for the Pannonian Basin, the Carpathians, and the surrounding region. A brief review summarizes the geodynamic and geothermal background of heat-flow pattern.

The investigation on recent vertical crustal movements in the Carpathian region offers very important information. As a first presentation, a new color map of vertical movements is enclosed, with 1-mm/yr isoline interval (see below section on The Recent Vertical Movements in the Carpathian Region). In the investigation, repeated precise geodetic measurements and oceanographic data have been used. The characteristic data of the movement tendencies, including the reliabilities of measurements and adjustment, have also been presented.

The most important paleomagnetic data are compiled and briefly discussed in this chapter. All these data suggest that the ALCAPA terrane rotated consistently counterclockwise but by different amounts, which indicates internal deformation of early to middle Miocene age. The Tisza unit underwent opposite rotation with respect to the northern unit. In conclusion, the rotation of the ALCAPA and Tisza–Dacia units with different sense occurred at a similar time between 19 and 15 Ma.

23

Bielik, Miroslav, and Antal Ádám, 2006, Structure of the lithosphere in the Carpathian–Pannonian region, *in* J. Golonka and F. J. Picha, eds., The Carpathians and their foreland: Geology and hydrocarbon resources: AAPG Memoir 84, p. 501.

Structure of the Lithosphere in the Carpathian–Pannonian Region

Miroslav Bielik
Geophysical Institute, Slovak Academy of Science, Bratislava, Slovakia

Antal Ádám
Geodetic and Geophysical Research Institute of the Hungarian Academy of Sciences, Sopron, Hungary

ABSTRACT

This chapter summarizes the present data on the change of the thickness of the lithosphere in the Carpathian–Pannonian region obtained by seismics and seismology, and magnetotelluric and gravity studies, emphasizing the updoming of the asthenosphere below the Pannonian Basin in accordance with its high heat flow. Important conclusions have been drawn from the stripped gravity map concerning the deep crustal structures, including the intrusion in the narrow subbasins. Special attention is paid to the narrow rift zone of the Bekes subbasin with high-density mafic intrusion in its crust and the elevation of Moho and asthenosphere during the rifting as it was stated by reflection seismics and interpretation of magnetotelluric sounding measurements and supported by stripped gravity map and gravity modeling.

DOI:10.1306/985623M843080

24

Stefaniuk, Michał, 2006, Some results of a new magnetotelluric survey in the area of the Polish Outer Carpathians, *in* J. Golonka and F. J. Picha, eds., The Carpathians and their foreland: Geology and hydrocarbon resources: AAPG Memoir 84, p. 503–512.

Some Results of a New Magnetotelluric Survey in the Area of the Polish Outer Carpathians

Michał Stefaniuk

Faculty of Geology, Geophysics and Environmental Protection, Department of General and Mathematical Geology, AGH University of Science and Technology, Krakow, Republic of Poland

ABSTRACT

In 1998, the Geophysical Exploration Company and the State Geological Institute of Poland started broad-scale magnetotelluric (MT) investigation in the Polish Carpathians. The measurements were made with the use of the MT-1 system designed and produced by Electromagnetic Instruments, Inc., over a frequency range of 300–0.0005 Hz, with remote magnetic reference applied. The results of data interpretation provided new elements in the recognition of the structure of the flysch orogen and its basement and confirmed earlier interpretation of the roof of the high-resistivity horizon.

A considerable resistivity differentiation was observed in the Mesozoic, Paleozoic, and Precambrian formations in the northern part of the cross sections, where the high-resistivity platform-type basement occurs beneath the autochthonous Miocene sediments. The zone of low-resistivity rocks occurs in the Outer Western Carpathians. The great differences in resistivity distribution are observed in the deep basement along the general axis of the flysch orogen in the southern part of the study area. The results along two MT sections and from three detail areas are presented. The general features of the Carpathian basement interpreted from MT sounding data are different for the eastern and the western part of the study area. Low-resistivity complex and overthrust zones are characteristic of the eastern part of the area. The geological identification of the low-resistivity layer is still not clear.

INTRODUCTION

The structure and the dynamics of the basement are of great interest to geological investigations of the Carpathian orogen. That problem is likely to determine the resolution of a wide range of matters and doubts related with the structure, geodynamics, and development of the Carpathian overthrust. Unfortunately, the recognition of the basement structure is not a simple matter. The complex geological structure of the orogen and the limited borehole data on the flysch cover, particularly related to its inner zone, make the interpretation of the basement difficult. Under these circumstances, surface geophysical survey is the basic factor in investigating the Carpathian basement. The flysch orogen, made of a strongly folded series of shale and sandstone layers

DOI:10.1306/985624M843081

with diverse physical properties and cut by faults and overthrusts, generates a very complicated wavefield pattern. As a result, seismic data interpretation is difficult. Therefore, other geophysical methods are employed. Of these, the magnetotelluric (MT) and gravity methods provide the most interesting data for recognizing the Carpathian basement (Stefaniuk, 2001).

Magnetotelluric investigations in the Polish Carpathians have been conducted for 20 yr. Most measurements were made by the Geophysical Exploration Company, Warsaw, for the Geological Institute and Polish Oil and Gas Company (Swiecicka-Pawliszyn, 1980, 1984; Molek and Klimkowski, 1991). Several measurements were conducted by the Institute of Geophysics of the Polish Academy of Sciences, Warsaw (Jankowski et al., 1991), and the Institute of Geophysics of the University of Mining and Metallurgy, Krakow (Miecznik et al., 1995, 1996).

Results of the MT data interpretation were presented by many authors (Swiecicka-Pawliszyn and Pawliszyn, 1978, 1983; Woznicki, 1985; Stefaniuk and Kusmierek, 1986; Jankowski et al., 1991; Rylko and Tomas, 1995; Stefaniuk, 1995, 2001; Zytko, 1997). The main objective of MT data interpretation was to recognize the topography of the high-resistivity horizon related with the top of Mesozoic, Paleozoic, or Precambrian basement of the Flysch Carpathians (Swiecicka-Pawliszyn and Pawliszyn, 1978; Stefaniuk, 1995). A low-resistivity layer resting on a high-resistivity horizon in the southeastern part of the study area was also investigated.

In 1998, the Geophysical Exploration Company started the broad-scale MT investigations in the Polish Carpathians using the state-of-the-art MT system MT-1 designed and produced by Electromagnetic Instruments Inc., Richmond, California, United States. Results of the new data interpretation provided new elements in the recognition of the structure of the flysch orogen and the basement and confirmed earlier interpretations of the top of the high-resistivity horizon. In this chapter, some results of that new MT survey made in the Polish Carpathians are presented.

METHODOLOGY OF MAGNETOTELLURIC SURVEY

A major project of MT surveys has been conducted in the Polish Carpathians since 1998. A network of 25 MT profiles was designed (Figure 1). Nineteen profiles are oriented transversely to the major axis of the flysch orogen, whereas others are parallel to it. The goal of the project was aimed at recognition of the structure of the subflysch basement and the deep-seated parts of the flysch cover (Stefaniuk et al., 1998a; Czerwinski and Stefaniuk, 2001). Measurement lines were about 70–90 km (43–56 mi) long with sounding sites some 3 km (1.8 mi) apart. Measurements were made over a frequency range of 300–0.0005 Hz using the MT-1 system. The remote magnetic referencing was applied to reduce the influence of electromagnetic noise. Field data were processed using the MTR15 computer program incorporated in the MT-1 system. The results of data processing were amplitude and phase MT sounding curves, impedance polar diagrams, and skew. They were then subjected to qualitative and quantitative interpretation. The first stage of interpretation gave apparent resistivity and phase pseudosections (Stefaniuk et al., 1998a; Stefaniuk and Klitynski, 1999). The types of curves, polar diagrams, and skew were analyzed in qualitative interpretation. One-dimensional (1-D) quantitative interpretation was made using Bostick, Occam, and least square (LSQ) methods. Based on 1-D Bostick inversion, pseudo-two-dimensional (pseudo-2-D) resistivity cross sections were computed using the kriging method. Lateral correlation of the results of 1-D LSQ inversion was applied to obtain a 2-D resistivity section (Figures 2, 3). Additionally, geological information on the flysch orogen, gravity data, seismic refraction data, and results of 2-D inversion with the rapid relaxation inversion (RRI) method were also applied (Stefaniuk et al., 1998a, 2002a, b). Based on the interpretation made along MT lines, maps of selected resistivity boundaries were constructed. For the eastern region, the top of the Precambrian and the Mesozoic–Paleozoic basement is reflected in the map (Figure 4). Maps of the tops of Mesozoic–Paleozoic and Precambrian basement were made for the western region (Figures 5, 6). To construct those maps, borehole data and reflection seismic data were used because of irregular MT profiles distribution.

RESULTS OF MAGNETOTELLURIC DATA INTERPRETATION

A magnetotelluric survey was made along 13 transverse profiles and 4 linking lines in 2 areas: west between Zywiec and Nowy Sacz, and east of Jaslo (Figure 1). Additionally, a semidetailed survey was made along several profiles located between Jaslo and Rzeszow. Results of the MT data interpretation provided new factors in recognition of the structure of the flysch orogen and its basement and confirmed earlier interpretations of the top of the high-resistivity horizon (Stefaniuk et al., 1998a, b; Stefaniuk and Klitynski, 1999, 2000).

The MT imaging of the geological medium for the eastern part of the study area is considerably different than it is for the western part. A resistivity differentiation was observed in the Precambrian formations in

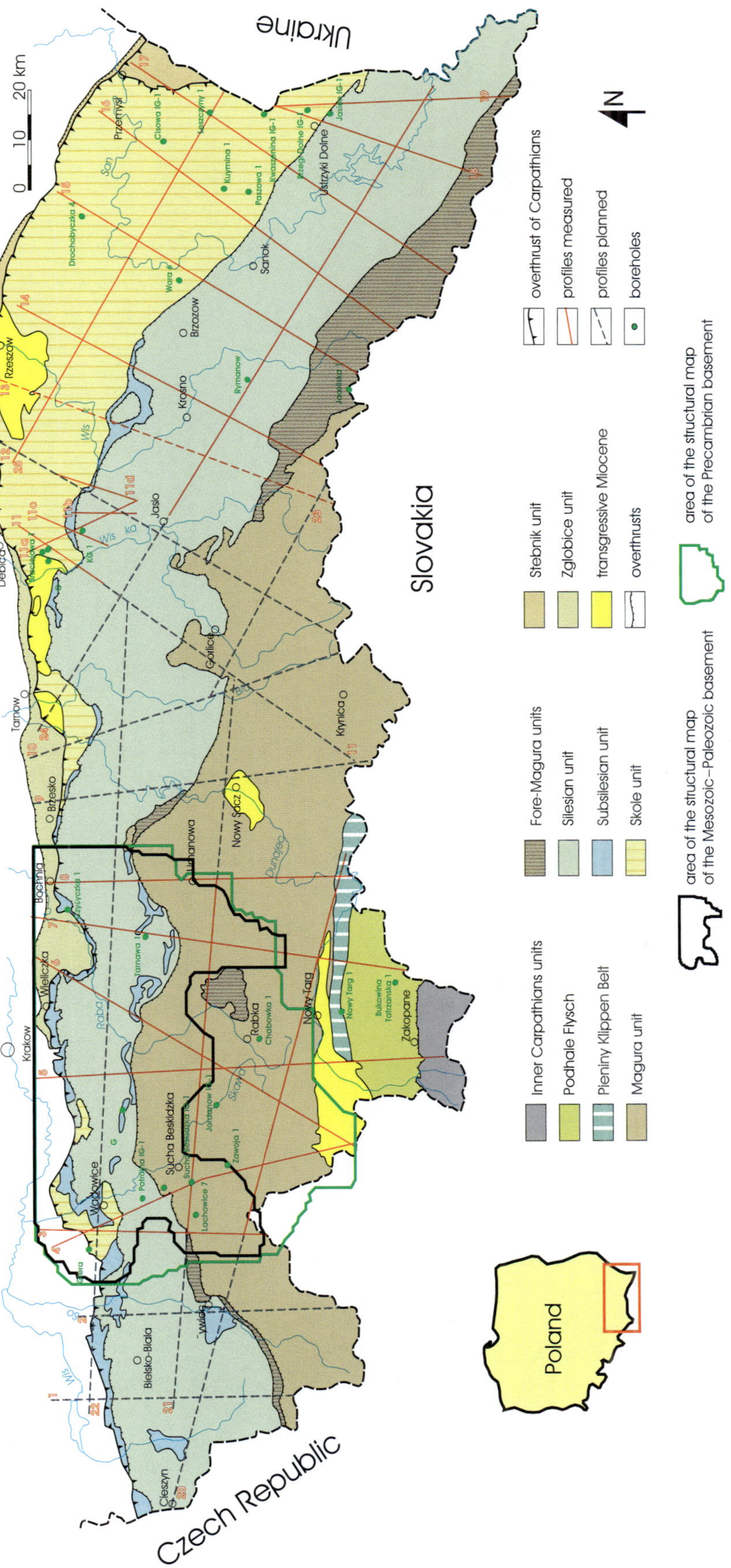

Figure 1. Study area.

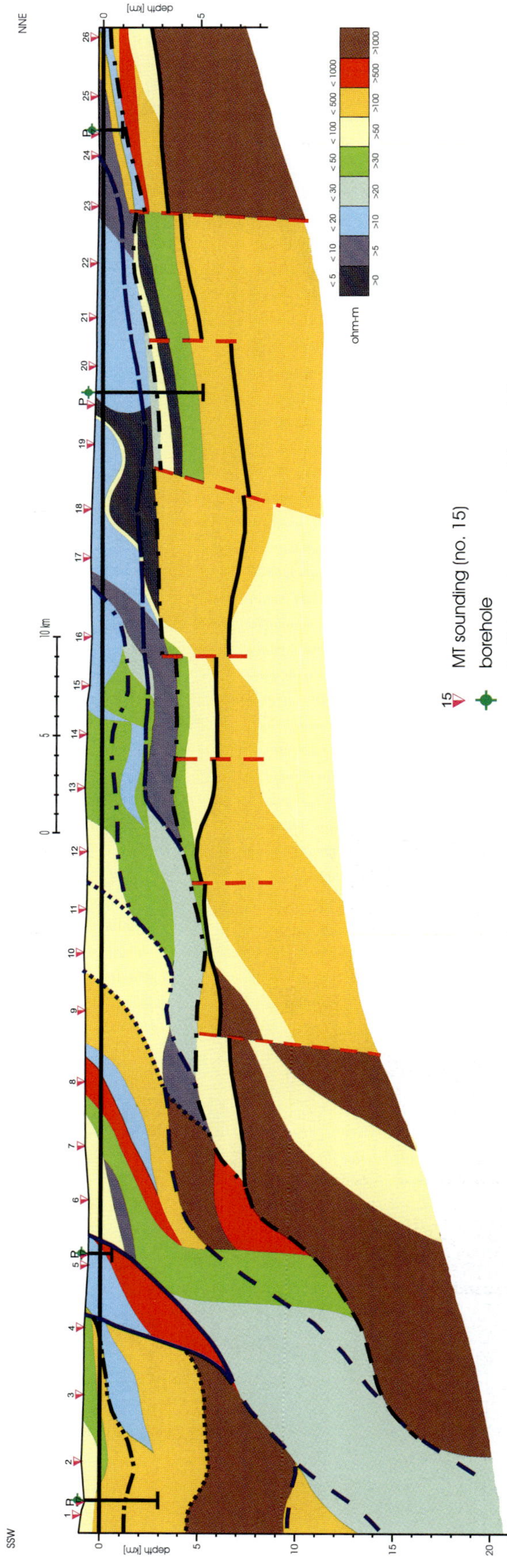

Figure 2. Cross section Bukowina Tatrzanska–Niepolomice, western part of the Polish Carpathians (after Stefaniuk et al., 2002b).

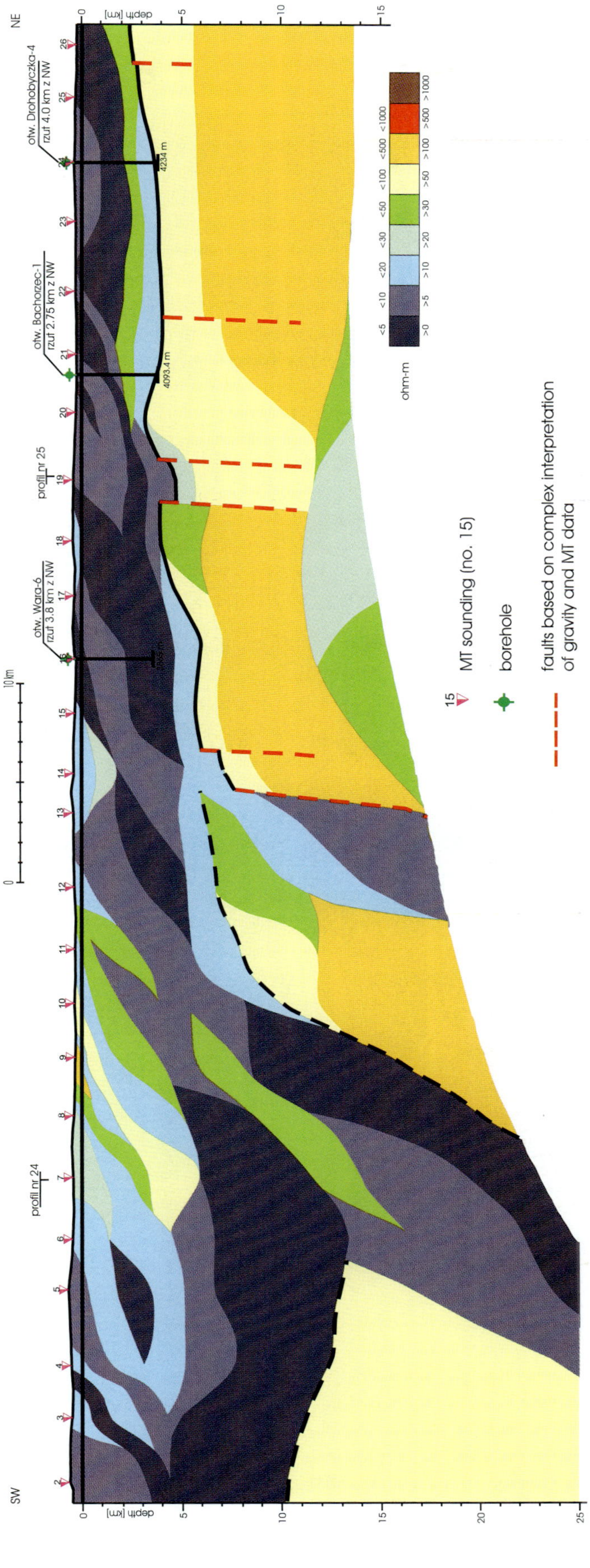

Figure 3. Cross section Wola Wyzna – Rzeplin, eastern part of the Polish Carpathians (after Stefaniuk et al., 2002b).

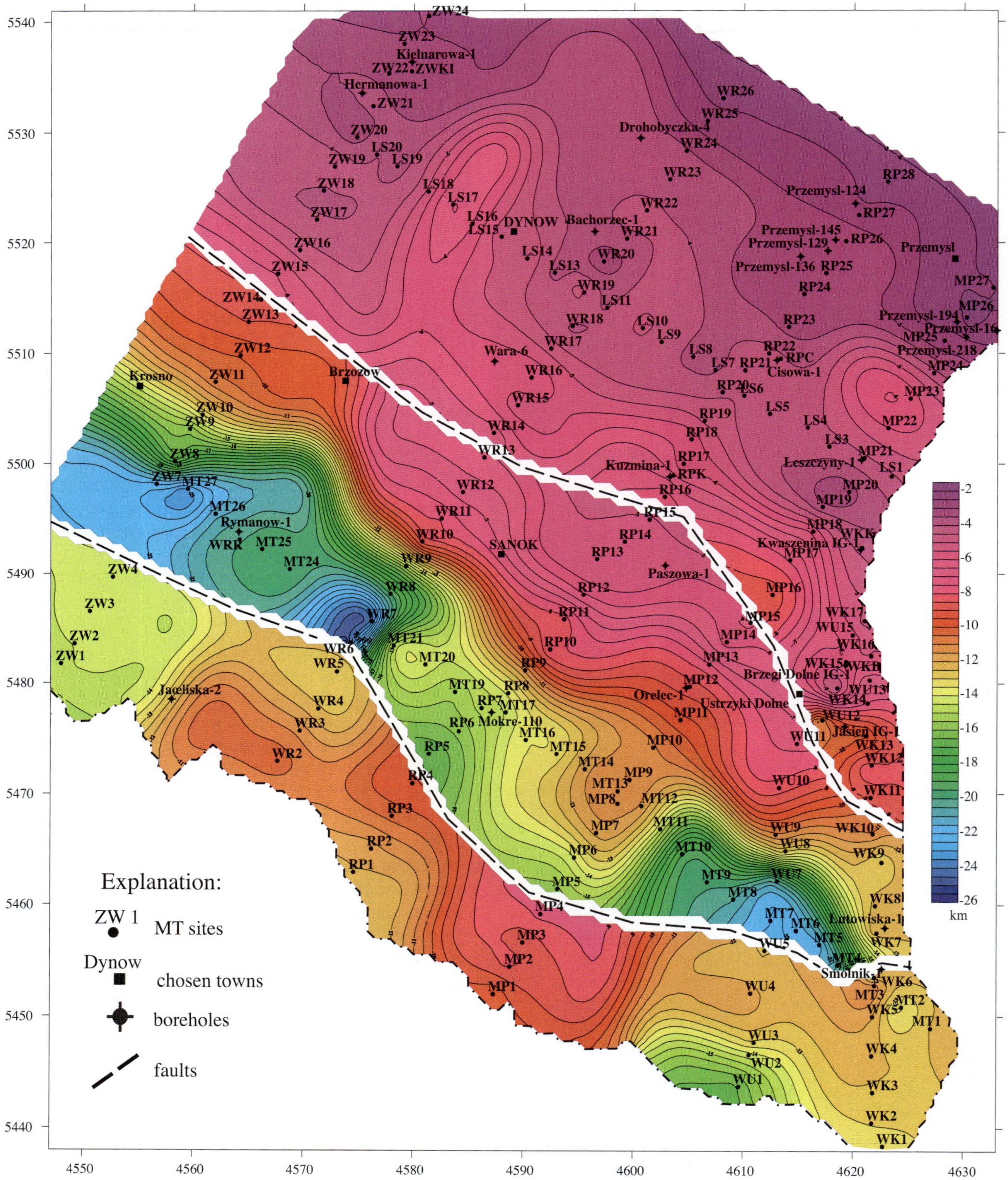

Figure 4. Structural map of high-resistivity basement, eastern part of the Polish Carpathians (after Stefaniuk et al., 2002a).

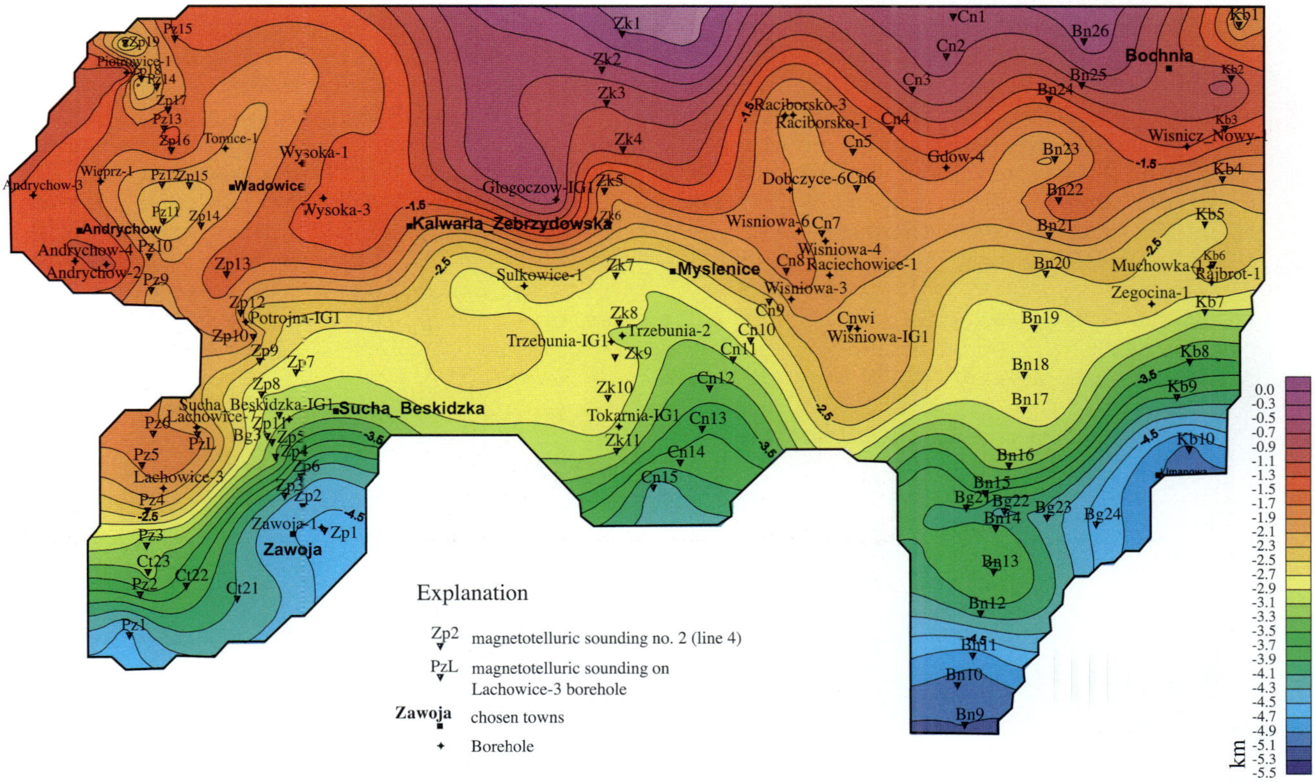

Figure 5. Structural map of the roof of Mesozoic–Paleozoic basement, western part of the Polish Carpathians (after Czerwinski et al., 2002).

the northeast part of the area, where the high-resistivity platform-type basement rests beneath the autochthonous Miocene sediments (Stefaniuk and Klitynski, 1999, 2000). In the roof of that rock complex, some elevations and broad depressions are observed in the high-resistivity horizon. The depressions are filled with relatively low-resistivity deposits and have probably tectonic origin. The zone of low-resistivity rocks occurs to the southwest of well Kuzmina-1. It is likely connected with a major Alpine compression structure (Figures 3, 4). In the southwest part of the area, the high-resistivity horizon is buried deep under a thick low-resistivity complex (Figure 3). In the southern margin of the sections, close to the state border, a thick high-resistivity block is overthrust onto those low-resistivity deposits. Its top is interpreted to rest at a depth of 8–10 km (4.8–6 mi). The arrangement of those high- and low-resistivity complexes suggests that an overthrust zone is present where low-resistivity deposits are impressed under a high-resistivity block overthrust from the south. It is probably an element of the deep-seated basement, uplifted in front of the hypothetical subduction zone, which is generally related with the Pieniny Klippen Belt. Investigations made by Slovak geologists and geophysicists (Bielik, 1999) and a team from the Institute of Geophysics Polish Academy of Sciences (Jankowski et al., 1991) proved this.

Great differences in resistivity distribution are observed in the deep basement along the general axis of the flysch orogen in the southern part of the area. Strong undulations of the roof of the high-resistivity horizon, probably related with deep-seated tectonic zones, are observed (Figure 4). Low-resistivity complexes dominate in the resistivity cross sections.

Resistivities of the flysch orogen differ over a narrow range in the eastern part of the study area. They tend to increase toward the south and east. In the southern part of the sections, thick, low-resistivity complexes, whose geology is not clearly identified, underlie the flysch. The position of the surface of the Carpathian overthrust is also not clear (Stefaniuk and Klitynski 1999, 2000; Czerwinski and Stefaniuk, 2001; Stefaniuk et al., 2002b).

A much different situation is in the western part of the study area. Some general features of resistivity distribution are observed in that region (Czerwinski et al., 2000, 2002; Czerwinski and Stefaniuk, 2001). Geoelectric cross sections may be divided into two parts (Figure 2). The uppermost part, with relatively low resistivity, is related with the flysch cover and Miocene sediments

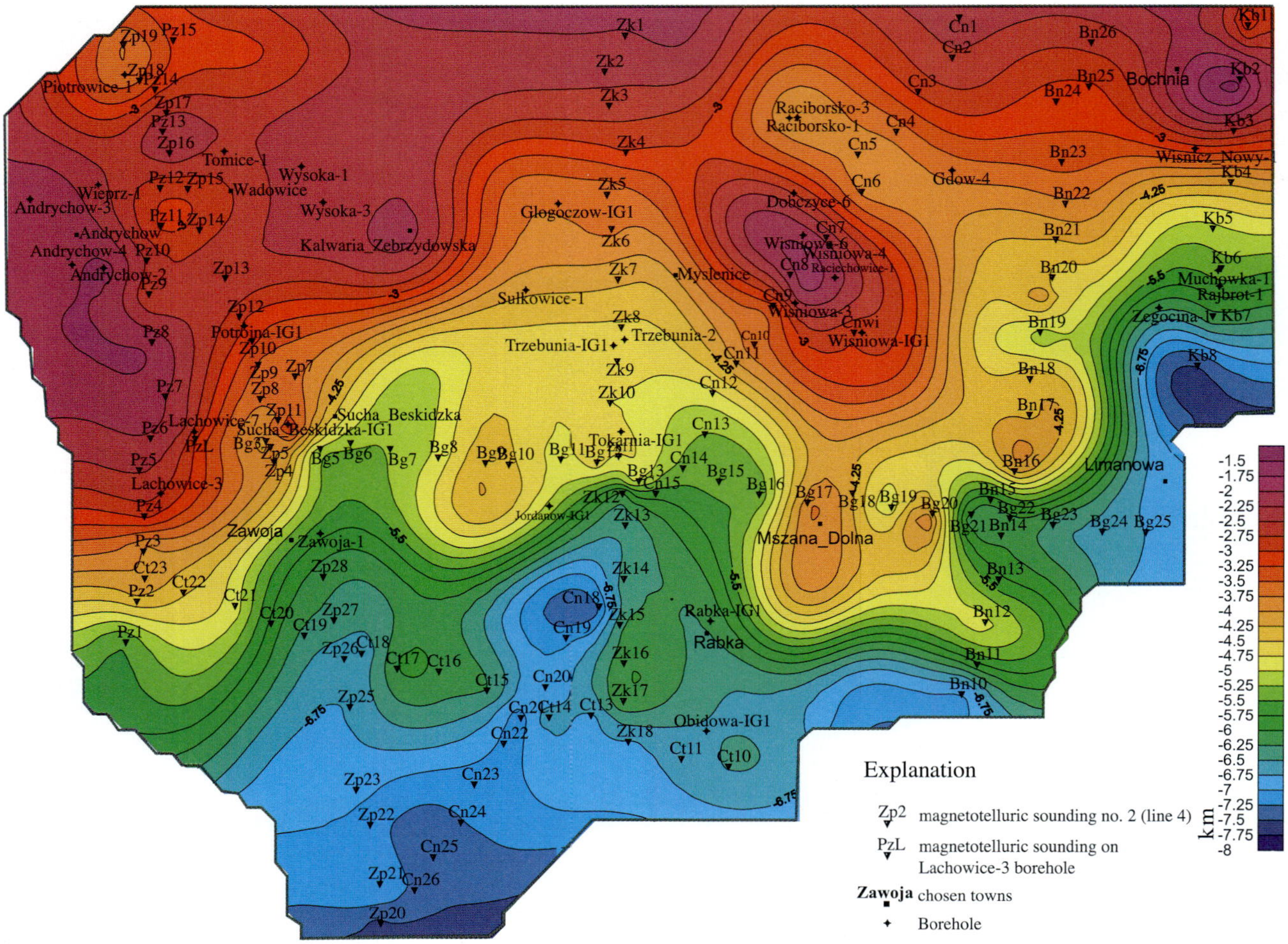

Figure 6. Structural map of the roof of Precambrian basement, western part of the Polish Carpathians (after Czerwinski et al., 2002).

of the flysch basement. Resistivity distribution is much varied in that part of the cross section. Low-resistivity rock complexes dominate in the Silesian and Subsilesian units, whereas a high-resistivity formation occurs in the Magura unit. The lower part of the cross section, with relatively high resistivity, is related to the sub-Miocene and/or subflysch substratum. Results of MT data interpretation allowed the tectonics of the Carpathian basement to be recognized there. In the northern part of the area, the continuation of the structures of the European platform is observed (Figures 5, 6). In the central and southern parts of the area, the major structural directions in the basement generally agree with those in the flysch cover.

Two measurement lines go across the Pieniny Klippen Belt and reach the northern border of the crystalline massif of the Tatra Mountains. A zone of relatively low-resistivity complex associated with the Pieniny Klippen Belt dips abruptly to the south beneath the Tatra Massif (Figure 2). This confirms an earlier hypothesis on the allochthonous character of that massif.

SOME REMARKS ON THE GEOLOGICAL INTERPRETATION OF MAGNETOTELLURIC DATA

Of great importance to MT data interpretation is the geological identification of geoelectric horizons. The main high-resistivity horizon corresponds to the top of the Mesozoic, Paleozoic, or Precambrian basement (Stefaniuk, 1995, 2001). This holds for the outermost zone and has been proven by deep borehole data and refraction seismics.

Still, the geological identification of the low-resistivity rock complex is a key problem. Different interpretations and hypotheses exist on the origin of that complex. Thus, it is thought that the complex is associated with deep oceanic sediments saturated with mineral water (Jankowski et al., 1985; Woznicki, 1985), partial melting of crystalline rocks (Lebiediev and Chitarov, 1964), or graphitized shales or graphites (Zytko, 1997).

The analysis of resistivities of flysch complexes and basement for some 20 boreholes located near the orogen margin shows that resistivity of the Miocene clayey sediments falls in a range of 1–5 ohm-m, thus being the same as that of the low-resistivity layer (Miecznik et al., 1995, 1996). Stanley (1989) states that black shales similar to those Miocene sediments generate conductivity anomalies in the Western Carpathians. It may be concluded that a depression in the southern part of the study area is a foredeep filled with molasse-type sediments and covered by the Carpathian overthrust. It is possible that the conductivity anomaly is generated by a thick complex of deep-marine sediments as well. Geophysical data are insufficient to confirm this; however, the hypothesis may be verified by analyzing the tectonic development of the Carpathian orogen (Stefaniuk, 2001).

CONCLUSIONS

The MT method may be effectively applied to recognize the structure of the Carpathian orogen. Its role in studying the basement structure should be especially emphasized. The results of the MT data interpretation are much more general than good seismic data; however, seismic data are noninterpretable in many cases. However, the MT method can support seismic data interpretation.

General features of the Carpathian basement interpreted from MT sounding data are different for the eastern and western parts of the study area. Low-resistivity complex and overthrust zones are characteristic of the eastern part of the area. Geological identification of the low-resistivity layer is still not clear. Of great interest is a high-resistivity block occurring near the Polish–Slovak border, likely overthrust onto a low-resistivity complex to the north. Based on MT data from profiles located in the Polish territory alone, it is not possible to recognize the genesis and the structure of that block. Hence, the MT profiles should be extended to the southwest beyond the Pieniny Klippen Belt. Generally, the length of the regional profiles is too short to recognize the structure of the deep-seated basement. Therefore, the profiles should be extended south to Slovakia and north to the North European platform and East European platform. A more detailed MT survey is needed to resolve problems of recognizing the flysch structure.

ACKNOWLEDGMENTS

Results of the "Project of Magnetotelluric Survey in Carpathians" financed by the Ministry of Environment through the National Fund of Environmental Protection and Water Resources as well as the Polish Oil and Gas Company were used in this chapter. The author thanks the directors of those institutions for their consent to use the data. The author's experience from the statutory research of the Department of General and Mathematical Geology, No. 11.11.140.159, was of great help.

REFERENCES CITED

Bielik, M., 1999, Geophysical features of the Slovak Western Carpathians: A review: Geological Quarterly, v. 43, p. 251–262.

Czerwinski, T., and M. Stefaniuk, 2001, Recognition of geological structure of the Carpathians as a result of magnetotelluric investigations: Slovak Geological Magazine, v. 7, p. 139–144.

Czerwinski, T., W. Klitynski, and M. Stefaniuk, 2000, Some results of magnetotelluric survey in Polish Western Carpathians: Vijesti Hrvatskoga Geoloskog Drustva, v. 37, no. 3, p. 31.

Czerwinski, T., R. Florek, and W. Klitynski, 2002, Geophysical and geological interpretation of regional magnetotelluric survey in the western Polish Carpathians: Naukovyj Visnik Ivano-Frankivskogo Nacionalnogo Technicnogo Universitetu Nafti i Gazu, v. 3, no. 4, p. 97–103.

Jankowski, J., Z. Tarlowski, O. Praus, J. Pecova, and V. Petr, 1985, The results of deep electromagnetic sounding in the West Carpathians: Geophysical Journal of the Royal Astronomical Society, v. 80, p. 561–574.

Jankowski, J., J. Pawliszyn, W. Jozwiak, and T. Ernst, 1991, Synthesis of electric conductivity surveys performed on the Polish part of the Carpathians with geomagnetic and magnetotelluric sounding methods: Publication of the Institute of Geophysics, Polish Academy of Science, v. A-19, p. 183–214.

Lebiediev, J. B., and N. J. Chitarov, 1964, Načalo plavlenia granita i elektroprovodnosti ego rasplava v zavisimosti od vysokogo davlenia parov vody: Geochimia, v. 3, p. 195–201.

Miecznik, J., M. Stefaniuk, and W. Klitynski, 1995, The methodology of magnetotelluric investigation in heterogenous media: Bulletin of the Polish Academy of Sciences: Earth Sciences, v. 43, p. 99–112.

Miecznik, J., M. Stefaniuk, and W. Klitynski, 1996, Badania magnetotelluryczne w Karpatach fliszowych: Kwartalnik AGH University of Science and Technology, Geologia, v. 22, p. 39–48.

Molek, M., and W. Klimkowski, 1991, Dokumentacja badań magnetotellurycznych, temat: Badania wgłębnej budowy geologicznej Karpat: Karpaty, lata 1988–1990, Przedsiębiorstwo Badań Geofizycznych Warszawa: Archiwum Oddziału Karpackiego Polski Instytut Geologiczny Kraków.

Rylko, W., and A. Tomas, 1995, Morphology of the consolidated basement of the Polish Carpathians in the light of magnetotelluric data: Geological Quarterly, v. 39, p. 1–16.

Stanley, W. D., 1989, Comparison of geoelectrical/tectonic models for suture zones in the western U.S.A. and Eastern Europe: Are black shales a possible source of high conductivities?: Physics of the Earth and Planetary Interiors, v. 53, p. 228–238.

Stefaniuk, M., 1995, Selected problems of the basement tectonics of the Polish Carpathians in the light of magnetotelluric sounding interpretation: XV Congress Carpatho-Balkan Geological Association, Athens, Special Publications of the Geological Society of Greece, p. 1155–1160.

Stefaniuk, M., 2001, Główne elementy strukturalne podłoża wschodniej części Karpat polskich w świetle badań magnetotellurycznych: Kwartalnik AGH University of Science and Technology, Geologia, v. 1, p. 32.

Stefaniuk, M., and W. Klitynski, 1999, Interpretation of magnetotelluric data along the Radoszyce–Przemysl line—eastern part of the Polish Carpathians: Romanian Journal of Tectonics and Regional Geology, v. 77, suppl. 1, p. 32.

Stefaniuk, M., and W. Klitynski, 2000, Selected results of magnetotelluric data interpretation in eastern part of the Polish Carpathians: Vijesti Hrvatskoga Geoloskog Drustva, v. 37, no. 3, p. 121.

Stefaniuk, M., and J. Kusmierek, 1986, Interpretation of the basement roof of the eastern part of Polish Carpathians in the light of magnetotelluric survey and geological premises: Proceedings of 31st International Geophysical Symposium, Gdansk, p. 221–232.

Stefaniuk, M., T. Czerwinski, A. Wajda, and T. Mrzyglod, 1998a, First results of high-frequency magnetotelluric investigations in Poland: Book of Abstracts, the 14th Workshop on Electromagnetic Induction in the Earth, Sinaia, p. 172–173.

Stefaniuk M., T. Czerwinski, A. Wajda, and T. Mrzyglod, 1998b, Perspektywy i problemy wykorzystania badań magnetotellurycznych do rozpoznawania utworów fliszowych na przykładzie przekroju Zawoja-Potrójna: Materiały XIX Konferencji Terenowej Sekcji Tektonicznej PTG-Magura '98, p. 42–43.

Stefaniuk, M., W. Klitynski, and M. Wojdyla, 2002a, New results of regional magnetotelluric survey in the area of eastern part of the Polish Carpathians: Naukovyj Visnik Ivano-Frankivskogo Nacionalnogo Technicnogo Universitetu Nafti i Gazu, v. 3, no. 4, p. 92–97.

Stefaniuk, M., W. Klitynski, M. Wojdyla, A. Wojcicki, T. Adamczak, A. Pepel, and W. Kleszcz, 2002b, Dookumentacja badań magnetotellurycznych w Karpatach cz. VI: Prace objęte Aneksem nr 1, 191 p.

Swiecicka-Pawliszyn, J., 1980, Dokumentacja badań geoelektrycznych, temat: Karpaty, Profile regionalne F i V rok 1975 i lata 1978–79: Archiwum PBG Warszawa, 28 p.

Swiecicka-Pawliszyn, J., 1984, Dokumentacja badań magnetotellurycznych, temat: Karpaty cz. wschodnia, rok 1975 i lata 1978–79, 1981–83: Archiwum PBG Warszawa, 37 p.

Swiecicka-Pawliszyn, J., and J. Pawliszyn, 1978, Zastosowanie badań magnetotellurycznych do rozpoznawania złożonych struktur geologicznych: Biuletyn Przedsiębiorstwo Badań Geofizycznych Warszawa, v. 2, p. 16–25.

Swiecicka-Pawliszyn, J., and J. Pawliszyn, 1983, Magnitelluričeskoe modelirovanie karpackoj anomalii elektroprovodnosti na territorii Polszi, 18 Meždunarodnyj Geofizičeskij Simpozium: Trudy II, p. 848–859.

Woznicki, J., 1985, Low-resistivity element in the Carpathians: Kwartalnik Geologiczny, v. 29, p. 153–166.

Zytko, K., 1997, Electrical conductivity anomaly of the Northern Carpathians and the deep structure of the orogen: Annales Societatis Geologorum Poloniae, v. 67, p. 25–43.

25

Hrušecký, Igor, Dušan Plašienka, and Lubomil Pospíšil, 2006, Identification of the North European platform below the eastern part of the Western Carpathian Flysch belt, *in* J. Golonka and F. J. Picha, The Carpathians and their foreland: Geology and hydrocarbon resources: AAPG Memoir 84, p. 513–514.

Identification of the North European Platform below the Eastern Part of the Western Carpathian Flysch Belt

Igor Hrušecký
Geological Survey of Slovak Republic, Bratislava, Slovakia

Dušan Plašienka
Faculty of Natural Sciences, Comenius University, Bratislava, Slovakia

Lubomil Pospíšil
Geoinform Consultants, Brno, Czech Republic

ABSTRACT

For the identification of the North European platform below the Western Carpathian Flysch belt and for the definition of the lithospheric and crustal relationships in the collisional zone between the Western Carpathian block and the North European platform, two migrated reflection seismic profiles with deep registration were chosen in the eastern sector of the Western Carpathians. Both of them were newly reprocessed (1999–2000), deep and composite seismic profiles, which define some new features in the geological structure in the eastern part of the Western Carpathians.

From a general point of view, both seismic profiles are dominated by a fanwise (flower-like) structure. However, we define this structure as fanwise, because in many features, it looks like a classic flower structure. The described fanwise structure has a great impact on the final form of the collisional zone. The impact of the fanwise structure is also clearly visible on both seismic sections in the area of the Inner Carpathian Paleogene basin. The Inner Carpathian Paleogene basin base dips moderately toward the Pieniny Klippen Belt in the northeast, and according to our interpretation of seismic lines, the Inner Carpathian Paleogene basin sedimentary package is divided into three or four structural levels (the oldest level is built up by basal Paleogene sediments, and the overlying folded Sambron Member beds consist of two structural levels and three structural levels [our opinion] of sediments; see mainly seismic section AB below). What is very interesting is that the uppermost level of the Sambron Member beds displays clear extensional features.

DOI:10.1306/985625M843082

The Pieniny Klippen Belt is a major tectonic unit at the surface, where it separates the Outer and Inner Western Carpathians. However, both seismic lines reveal that the Pieniny Klippen Belt has no continuous depth prolongation as a tectonostratigraphic unit. The seismic image allows an interpretation of an older subhorizontal structure of laterally wedging-out and interfingering rock units that originated by collisional processes in a wider zone associated with the Pieniny Klippen Belt. Younger and steep brittle structures that accompany the Pieniny Klippen Belt and its surroundings may be traced to great depths where they merge into a subvertical fault zone. This fanwise fault zone closely resembles a transpressional flower structure.

From the point of view of hydrocarbon prospection, the most important feature of both seismic lines is the anticline structure of the North European platform below the thrust stack of the Flysch belt. This strong reflection package probably indicates the presence of carbonates in the North European platform cover (see seismic profile AB below). In our opinion, this structure is the most attractive for hydrocarbon exploration in the eastern part of the Slovakian Flysch Belt.

26

Hrušecký, Igor, 2006, Bivergent lithospheric extension in the Slovak part of the Danube Basin: Example of the influence of the lithospheric processes on the style of development of the basin, *in* J. Golonka and F. J. Picha, eds., The Carpathians and their foreland: Geology and hydrocarbon resources: AAPG Memoir 84, p. 515.

Bivergent Lithospheric Extension in the Slovak Part of the Danube Basin: Example of the Influence of the Lithospheric Processes on the Style of Development of the Basin

Igor Hrušecký
Geological Survey of Slovak Republic, Bratislava, Slovakia

ABSTRACT

An evolution of the Slovak part of the Danube Basin has started with transtension-extension spreading of the crust [Paleogene(?) to Ottnangian(?), Karpatian, and lower Badenian]. This protostage of the evolution has been an important factor in the further evolution of the basin, but the dominant element, from the viewpoint of the evolution of the Slovak part of the Danube Basin and also its broader surroundings, appears to be the Ulany ridge. This ridge connected (as an interbasinal ridge) the evolution of the Slovak part of the Danube Basin and of the Vienna basin into one dynamic context until the end of the crustal extension phase (lowest Pannonian). In that way, the Ulany ridge controlled the bivergent extension (toward northwest–west and southeast–east) from the north-northeast–south-southwest axis of the ridge uplift. The recent structural inexpressiveness of this ridge is caused by its tectonic unroofing in the crustal extension phase, erosion, and later (upper lower Pannonian to Holocene) thermal collapse in the basin.

DOI:10.1306/985626M843082

27

Nemčok, Michal, Gyorgy Pogácsás, and Lubomil Pospíšil, 2006, Activity timing of the main tectonic systems in the Carpathian–Pannonian Region in relation to the rollback destruction of the lithosphere, *in* J. Golonka and F. J. Picha, eds., The Carpathians and their foreland: Geology and hydrocarbon resources: AAPG Memoir 84, p. 517–540.

Activity Timing of the Main Tectonic Systems in the Carpathian–Pannonian Region in Relation to the Rollback Destruction of the Lithosphere

Michal Nemčok
Energy and Geoscience Institute, University of Utah, Salt Lake City, Utah, U.S.A.

Gyorgy Pogácsás
Hungarian Oil and Gas Company, Budapest, Hungary

Lubomil Pospíšil
Geoinform Consultants, Brno, Czech Republic

ABSTRACT

The Oligocene–Sarmatian paleostress data and the data on the timing of the main Carpathian–Pannonian fault systems show that tectonic events can be characterized in eight periods. Oligocene was characterized by strike-slip-controlled stretching of the eastern Alpine domain, which continued during the Egerian. Egerian was the onset time for the first strike-slip faults inside the ALCAPA (Eastern Alps–Carpathians–Pannonian Basin) microplate. The eastward Alpine movements slowed down in Ottnangian, whereas the ALCAPA strike-slip system became denser. Karpatian was the time of northeastward movements of ALCAPA along a couple of bounding strike-slip fault systems, which were most effective during early Badenian. The northwestern strike-slip boundary of ALCAPA progressively became inactive in the eastward direction during middle Badenian. The ALCAPA started to experience the eastward expansion of the normal faulting from its western end during the same time. The eastward motion of the ALCAPA microplate further slowed down during the late Badenian, and a separate motion of the Tisza–Dacia microplate initiated. ALCAPA stopped in Sarmatian and accommodated a separate eastward motion of Tisza–Dacia by west–east extension.

The data on fault timing and kinematics indicate that the subduction rollback in the remnant Carpathian Flysch Basin was the most important driving mechanism of the Carpathian–Panonian development, which controlled the northeasterly distance traveled by the ALCAPA microplate and its internal deformation. The terminal collision, oceanic slab break-off, Eastern Alpine lateral extrusion, and the topography and rheology of the orogenic foreland were less important driving mechanisms.

DOI:10.1306/985627M843083

INTRODUCTION

The last 6 yr of research resulted in a generally accepted semiquantitative understanding of the driving mechanisms, boundary conditions, and Cenozoic palinspastic model of the Carpathian–Pannonian region (Figure 1) and their very recent numeric validations. Results also contain numerous palinspastic models of the region, which differ in details inside of the Pannonian region for the Neogene but vary considerably for the Paleogene. Models differ considerably in understanding adjacent regions during the Neogene. Palinspastic models are either based on the restoration of the shortened frontal parts of the system (e.g., Ellouz and Roca, 1994; Morley, 1996) (Figure 2) or on the restoration of the intra-Carpathian area affected by strike-slip and normal faulting (e.g., Csontos, 1995; Tari et al., 1995; Fodor et al., 1999).

The large-scale driving mechanism of the Carpathian–Pannonian development was the general Africa–Europe convergence (e.g., Le Pichon et al., 1988), with a separate northwestward and northward movement of the Apulian microplate (e.g., Laubscher, 1971; Bada, 1999). The important driving mechanisms were a subduction rollback in the remnant Carpathian Flysch Basin area (e.g., Royden et al., 1982; Royden, 1993), a gentle collision at the end of the subduction (e.g., Lillie et al., 1994; Zweigel, 1997; Sperner et al., 2002), and an oceanic slab break-off at the end of the subduction (e.g., Spakman, 1990; von Blanckenburg and Davies, 1995; Sperner, 1996; Nemcok et al., 1998a). However, another driving mechanism was the lateral extrusion in the Eastern Alps, which affected the western part of the ALCAPA (e.g., Neubauer and Genser, 1990; Ratschbacher et al., 1991a, b; Bada, 1999). Boundary conditions of the orogen were the high strength and topographies of the Bohemian Massif and Moesian Platform (e.g., Ratschbacher et al., 1991b; Bada, 1999). The significance of various driving mechanisms was not specifically studied.

Paleogene and early early Miocene crustal deformation was characterized by overall contraction (e.g., Nemcok et al., 1998b; Fodor et al., 1999). The dextral Alpine–Carpathian transpressional arc underwent both thrusting and strike-slip faulting. Basins were forming because of flexural lithosphere loading (e.g., Tari et al., 1993) and strike-slip faulting (e.g., Kovac et al., 1989; Marko et al., 1990, 1991). Contractional deformation slightly changed its local directions and character after the onset of the eastward movements of ALCAPA microplate during the late Oligocene (Ratschbacher et al., 1991a; Tari et al., 1995; Bada, 1999). The function of various faults in the overall deformation remains unknown.

Late early–middle Miocene crustal deformation was characterized by the ALCAPA and Tisza–Dacia microplates progressively advancing into the region of the subducting remnant Carpathian Flysch Basin and rotating counterclockwise and clockwise, respectively (e.g., Kovac et al., 1994; Csontos, 1995; Tari et al., 1995; Fodor et al., 1998). The microplates collided with each other during this time under the overall late Oligocene–Ottnangian dextral and Karpatian–Badenian sinistral displacement along roughly east-northeast-trending faults (e.g., Csontos et al., 1991; Csontos and Nagymarosy, 1998). Each of them experienced a complex deformation because of hinterland extension spreading toward the accretionary wedge in front of the advancing orogen since the Eggenburgian (e.g., Pecskay et al., 1995; Tari and Horvath, 1995; Nemcok et al., 1998a; Fodor et al., 1999; Tari et al., 1999). Each microplate can be characterized by frontal shortening, wrench zones along sides, and complex normal fault systems inside (e.g., Tari et al., 1995; Bada, 1999), but the kinematics of major faults accommodating the described deformation remains to be studied.

Late Miocene crustal deformation was characterized by the end of the subduction in front of the ALCAPA and the ongoing subduction in front of the Tisza–Dacia microplates (e.g., Csontos, 1995; Tari et al., 1995; Fodor et al., 1999). Late Badenian–early Sarmatian was the onset time for the regional northwest–southeast to west–east extension, with the exception of the accretionary wedge in front of the Tisza–Dacia, which was experiencing shortening (e.g., Bergerat, 1989; Nemcok et al., 1989; Huismans et al., 1997; Linzer et al., 1998; Zweigel et al., 1998). This period was also characterized by the first Pannonian Basin inversion event (e.g., Tari et al., 1993; Horvath and Cloetingh, 1996; Bada, 1999; Horvath and Tari, 1999).

The Pliocene–Quaternary was characterized by a regional basin inversion in the Pannonian region (Tari et al., 1993; Horvath and Cloetingh, 1996) under the stress field more or less similar to the modern one.

The aim of this chapter is to describe the data necessary to study problems indicated earlier as open and discuss their interpretation. Specific goals include

1) activity timing of the main fault systems
2) characterization of paleostresses active during different periods
3) relation of the described fault kinematics and paleostresses to the driving mechanisms of the Cenozoic Carpathian–Pannonian development

Mentioned data sets are complemented by a discussion of data on the timing and tectonic setting of Neogene magmatism and facial correlation data.

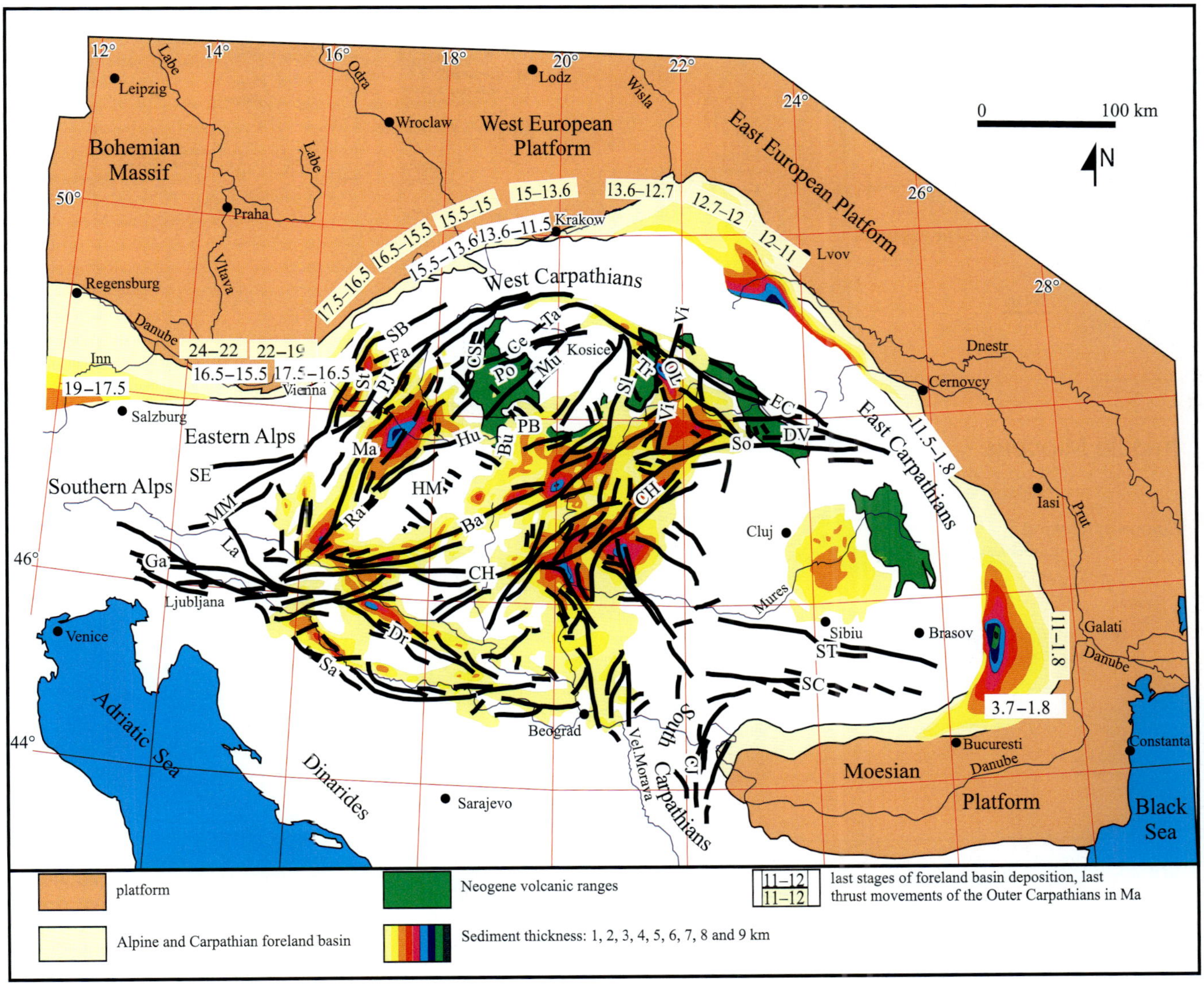

Figure 1. Tertiary sediment thickness map of the Carpathian–Pannonian region with indication of the last period of foreland molassic deposition, thrusting of the Outer Carpathians (modified after Jiricek, 1979; Krus and Sutora, 1986; Meulenkamp et al., 1996), and maximum subsidence period in the intra-Carpathian basins (modified after Krus and Sutora, 1986; Csontos et al., 1992; Meulenkamp et al., 1996). Contoured basins include the Pannonian Basin system, the Vienna basin, the Danube Basin, the Transcarpathian basin, the Sava Basin, the Drava Basin, the Transylvanian Basin, and the molasse foreland basins. The sediment thickness distribution is based on industrial borehole, reflection seismic, and gravity data. Map includes a fault pattern (modified after Mahel, 1974; Kocak et al., 1981; Royden, 1985, 1988; Krus and Sutora, 1986; Ratschbacher et al., 1991a, 1993b; Csontos et al., 1992; Kovac and Hok, 1993; Vass et al., 1993; Linzer, 1996; Nemcok et al., 1998b). SE = Salzachtal–Ennstal–Mariazell–Puchberg fault system; MM = Mur–Murz fault zone; Ga = Gailtal fault; La = Lavanttal fault zone; Sa = Sava fault zone; Dr = Drava fault zone, SC = Southern Carpathian fault; CJ = Cerna–Jiu fault; Ra = Raba fault zone; CH = Central Hungarian fault zone; Ba = Balaton fault zone; PB = faults of both Matra Mountains and north of the Pannonian Basin; Bu = Buda fault zone; HM = fault zones of the Central Hungarian Mountains; Da = Darno fault zone; ST = South Transylvanian fault; DV = Dragos–Voda faults; So = Somes fault; EC = fault contact of External and Internal Flysch in the Eastern Carpathians; Ma = Male Karpaty marginal faults; PJ = Pottendorf–Jablonica fault zone; St = Steinberg fault, Fa = Farske fault zone; SB = Schrattenberg–Bulhary fault zone; Hu = Hurbanovo fault; CS = Central Slovakian fault system; Ce = Certovica fault zone; Po = Pohorela fault; Mu = Muran fault; Ta = Tatry fault; Sl = Slanske Vrchy fault system; Tr = Trebisov fault system; OL = Ondava–Laborec fault system; Vi = Vihorlat fault zone.

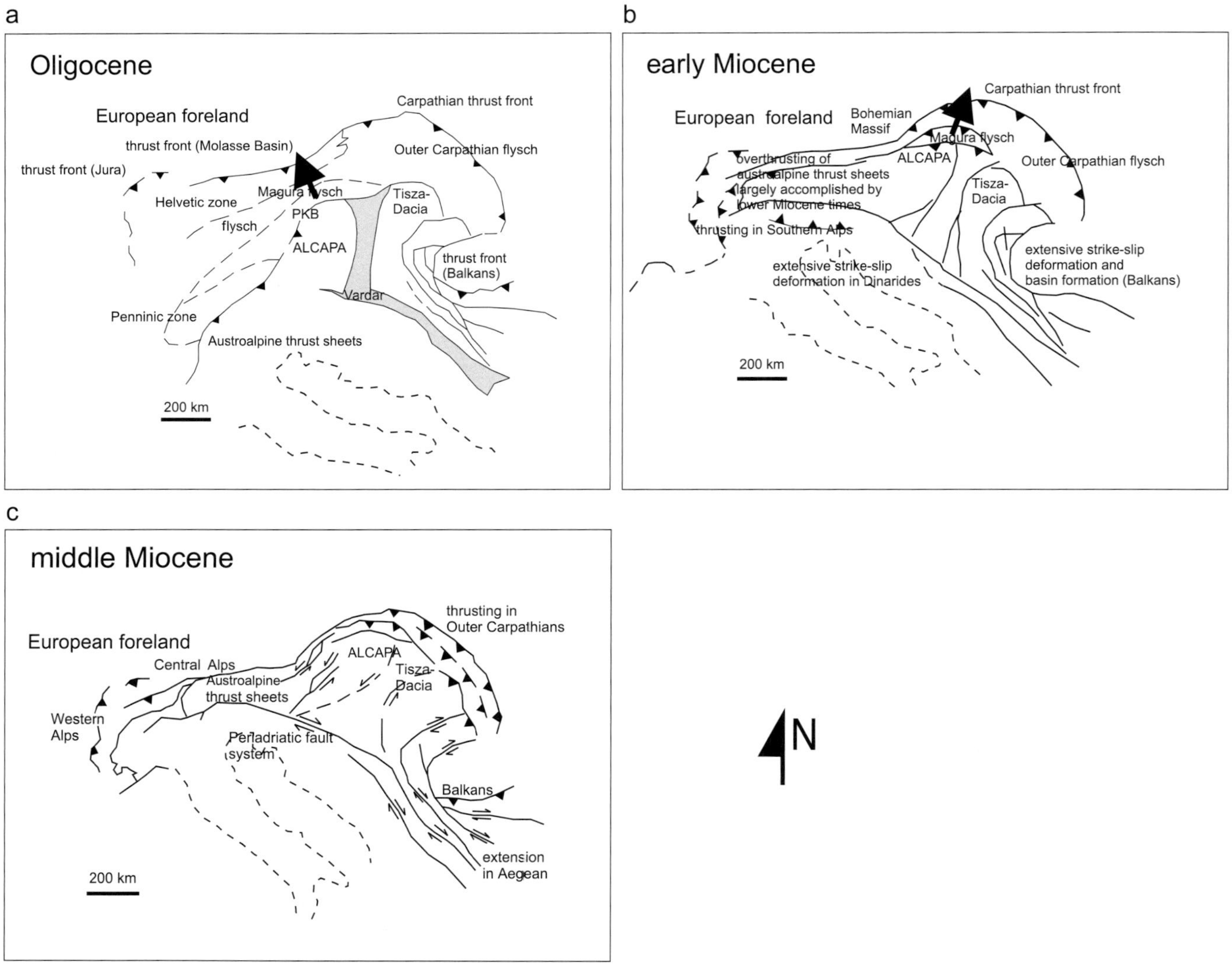

Figure 2. (a) Oligocene, (b) early Miocene, and (c) middle Miocene regional geological setting in the Alpine–Carpathian–Pannonian area (modified after Morley, 1996). (a) Thick arrow indicates convergence in the Alpine area. (b) Thick arrow indicates convergence in the Carpathian area. Note that the advance of the Magura unit in front of the ALCAPA to its new position requires its oblique advance. It is in accordance with predominant sinistral strike-slip faulting along its western part and predominant thrusting in its eastern part (e.g., Nemcok et al., 1998c), larger paleomagnetically indicated counterclockwise mass rotations in its western part and smaller counterclockwise mass rotations in its eastern part (e.g., Korab et al., 1981; Krs et al., 1986, 1991, 1993; Tunyi and Kovac, 1991). PKB = Pieniny Klippen Belt.

METHODS

The relationship of major fault systems to various stratigraphic units, focusing on rocks deformed vs. undeformed by each of them, has been gathered from geologic maps and available published studies. Resultant fault activity timing and associated kinematic data have been grouped into subsets for eight periods, including the Oligocene, Egerian, Ottnangian, Karpatian, early Badenian, middle Badenian, late Badenian, and Sarmatian. Chosen periods document all significant events of the Cenozoic Carpathian–Pannonian development. Faults of each period are located on the same large-scale geologic map. The data from the adjacent Alpine region were also used. To narrow down activity intervals of the Slanske Vrchy fault system and the Vihorlat fault zone, we have made new interpretations of reflection seismics from that area (Nemcok, 2001).

The database of paleostress results was gathered from the literature independently of recent syntheses (e.g., Bada, 1999; Fodor et al., 1999) and represented by simple arrow patterns on the same geologic map as the fault timing. Data have been grouped into subsets for eight periods, including Eocene–early Oligocene, late

Oligocene–early Miocene, early Badenian, middle Badenian, late Badenian–Sarmatian, early Pannonian, late Pannonian, and present. Selected periods do not fully overlap with periods chosen for the fault activity documentation. The reason is that new fault systems have been sometimes initiated under similar stress regimes, or some faults remained active under changed stress regime, but we wanted to document all significant stress changes.

DATA ON FAULT TIMING AND KINEMATICS

Alps

The Salzachtal–Ennstal–Mariazell–Puchberg fault system (Figures 1, 3a, b) was active since the late Oligocene, as indicated by related small basins along the Ennstal fault (their deposition is not well dated covering the late Oligocene to middle Miocene; Tollmann, 1985; Steininger et al., 1988; Linzer et al., 1997) or by the Gesause fault of the Ennstal fault system, which offsets Cretaceous thrusts (Ratschbacher et al., 1991a). Other time constraints for this faulting are provided by the following basins. The Karpatian Wagrain basin close to the northeastern tip of the Tauern window formed at a sinistral overstep and was subsequently inverted by dextral transpression along the Salzachtal–Ennstal–Mariazell–Puchberg fault system (Wang and Neubauer, 1998). The Salzachtal–Ennstal–Mariazell–Puchberg fault system is postdated by the Vienna basin deposition from the Ottnangian to the end of the Pannonian, related to the Hieflau basin with just the Tertiary age of the fill (Tollmann, 1985; Linzer et al., 1997) and Ennstal basins with deposition during the late Oligocene–middle Miocene (Ratschbacher et al., 1991a). After the middle Miocene, the compressional structures overprinted the extrusion-related northwest–southeast-striking dextral and northeast–southwest-striking sinistral strike-slip faults in the surrounding area as sinistral and dextral, respectively (Decker et al., 1993).

The timing of the Mur–Murz sinistral transtensional zone and the Lavanttal dextral transtensional zone (Figures 1, 3d–g) is provided by the Mur–Murz-related Fohnsdorf Basin with the Karpatian–Badenian (17.5–13.6-Ma) fill, the Trofaiach and Aflenz basins with the lower–middle Badenian (16.5–15-Ma) fill (Nievoll, 1985), and the Lavanttal basin with the Karpatian–Pannonian (17.5–6.2-Ma) deposition with the major subsidence in the Badenian (16.5–13.6 Ma). Grabens also opened in the broader vicinity of both faults in the extruding Alpine wedge: the Styrian Basin with several subbasins, being part of the Pannonian Basin system, with the Ottnangian–Pannonian (19–6.2-Ma) fill and the major faulting and subsidence during the Karpatian–Sarmatian (17.5–11.5 Ma) (Kroll et al., 1988; Ebner and Sachsenhofer, 1995).

The Periadriatic fault system (Figures 1, 3a) transects the Cretaceous cover thrust belt of the Drauzug (Ratschbacher et al., 1991a). The Oligocene displacement is dated by the ductile synmagmatic sinistral transpressive deformation of Periadriatic plutons intruded into the wrench zone under a north–south compression (Nemes, 1996). The exhumation of tonalites occurred during this event. The sinistral transpressive shearing was followed by the Miocene–Pliocene top-to-the-northwest nappe stacking to the north of the eastern part of the Periadriatic fault system and dextral west-northwest-trending strike-slip faulting (Polinski and Eisbacher, 1992). A north–south contraction caused brittle transpressional structures, i.e., north-vergent thrusts, south-vergent backthrusts, northwest-striking dextral faults (R faults), and northeast-trending sinistral faults (R′ faults) along the Periadriatic fault system. The final overthrust of the Karawanken Mountains to the north during the Pliocene–Quaternary, which form a positive flower structure, also includes a flexure of the northern foreland and a formation of a narrow Sarmatian to Quaternary (13.6–1.5-Ma) foreland Klagenfurt basin (Nemes et al., 1997). After the Miocene, the Periadriatic fault system acted as a sinistral strike-slip fault.

Western Carpathians

The sinistral displacement of the Male Karpaty marginal faults (Figures 1, 3e–g) is dated by the lower–middle Badenian onset of the Danube basin pull-apart (e.g., Royden et al., 1982) with very fast middle–upper Badenian subsidence and coarse marginal facies along the western side (Kovac et al., 1993). They became normal faults during the upper Miocene (Figure 3h) (Kovac et al., 1993).

The Pottendorf–Jablonica fault zone offsets Pannonian (11.5–6.2-Ma) sands of the Vienna basin, probably as the normal fault. Its Eggenburgian–Badenian (2213.6-Ma) transpressional and later transtensional sinistral activity (Figure 3c–e) is documented by structural and sedimentary data in the Vienna basin area (Marko et al., 1990; Nemcok et al., 1998b, and references therein). The onset of the Vienna basin pull-apart is dated as Karpatian (e.g., Royden, 1985; Lankreijer et al., 1995). The middle Badenian debris aprons outline the faults of the southeastern margin of the Vienna basin (Vass et al., 1988). The Sarmatian–Pannonian deposition in the northeastern parts of the Vienna basin was controlled by normal faulting, reactivating

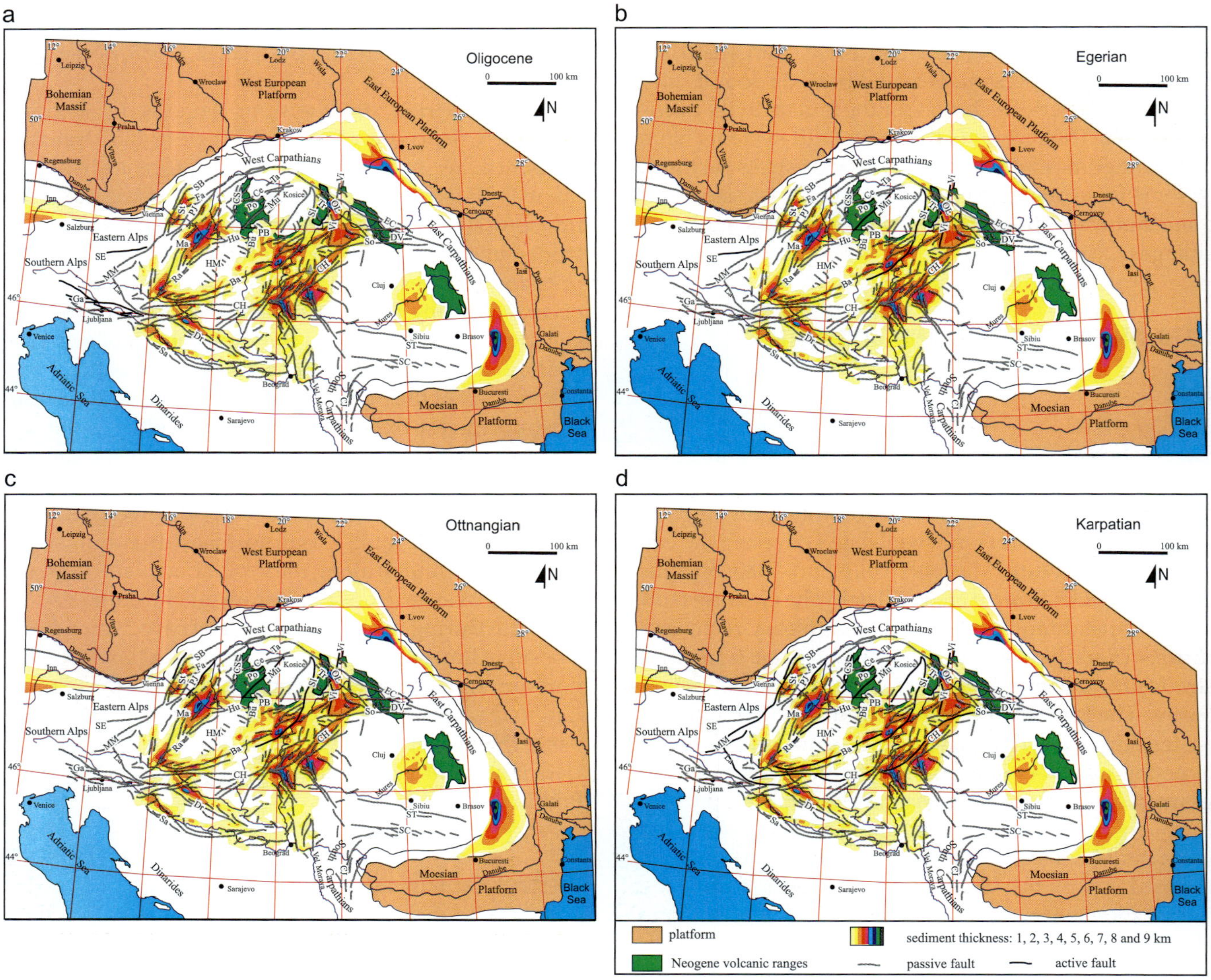

Figure 3. Fault kinematics and timing in the ALCAPA for (a) Oligocene, (b) Egerian, (c) Ottnangian, (d) Karpatian, (e) early Badenian, (f) middle Badenian, (g) late Badenian, and (h) Sarmatian. Thick black solid lines indicate main strike-slip faults, and thick black dashed lines indicate main normal faults active during their respective period. See Figure 1 for fault names and other explanations.

earlier sinistral strike-slip faults (Figure 3h) (Nemcok et al., 1989). Kiscelian (36–26.5-Ma) sediments along this fault underwent northwest–southeast shortening by folding, constrained by the above-mentioned data to have the Oligocene–Ottnangian age. The Eggenburgian northwest–southeast shortening in the Inner Carpathians to the east of the Vienna basin is indicated by the coarse talus cone deposition bounded by northeast–southwest-striking thrusts (Kovac et al., 1989, 1991). Eggenburgian sediments are the youngest rocks deformed by the northwest–southeast thrusting on the southeast margin of the Vienna basin (Nemcok et al., 1989).

The Central Slovakian fault system (Figures 1, 3d–f) was active as the sinistral and dextral strike-slip fault systems during the late Eocene–middle Miocene and middle–late Miocene, respectively, as indicated by fault-slip analyses (Kovac and Hok, 1993).

The Pohorela fault is sealed by the lower–middle Badenian (16.5–15-Ma) volcanics (Figures 1, 3b–d).

The Muran fault is sealed by the lower Badenian (16.5–15.5-Ma) volcanics and sinistrally offsets the middle Eocene facies of the Central Carpathian Paleogene basin (Figures 1, 3b–d).

The Slanske Vrchy fault system offsets the Eggenburgian–Karpatian (22–16.5-Ma) sediments of the small

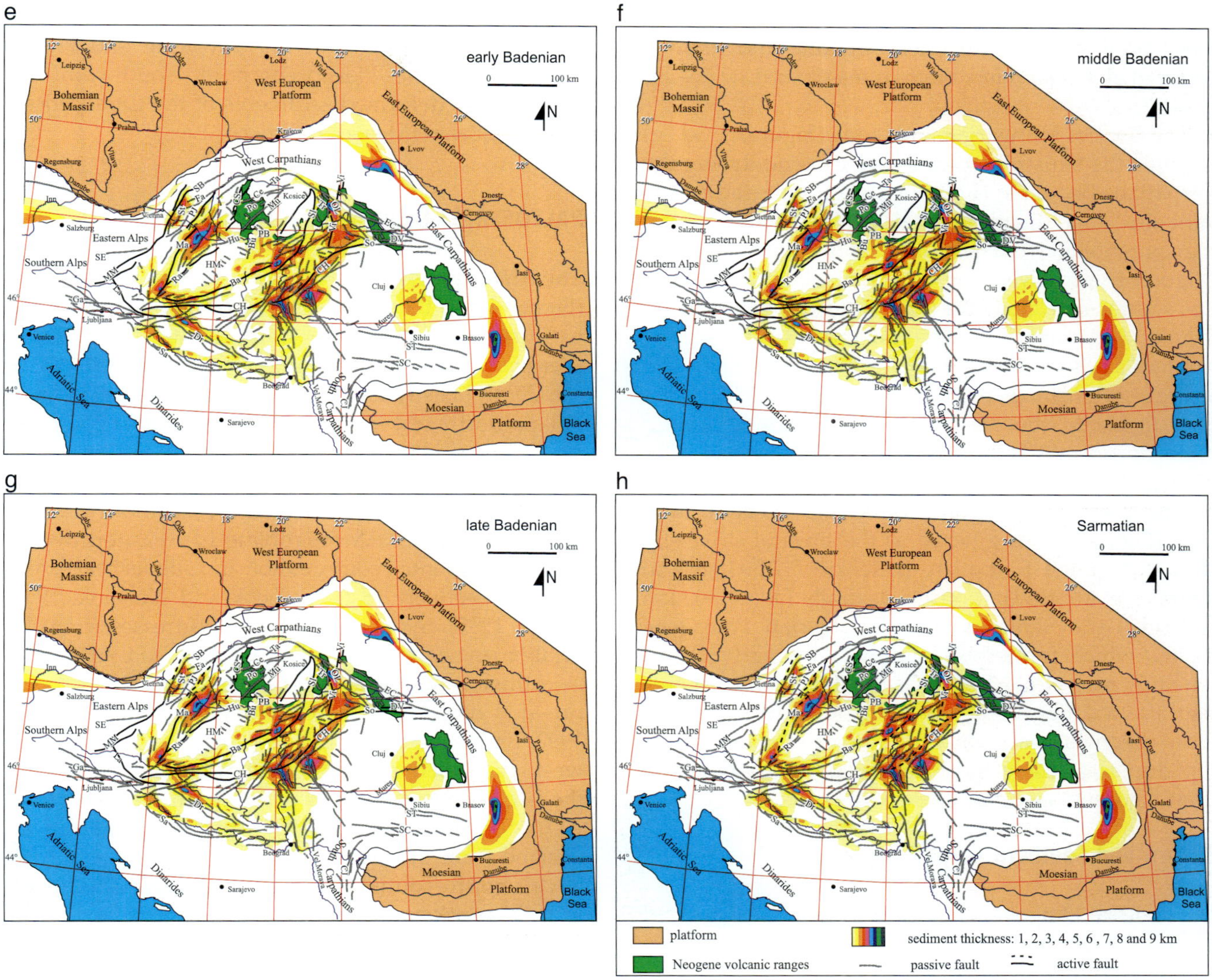

Figure 3. (cont.).

piggyback basin northeast of Kosice (Figures 1, 3b–g). These sediments were deposited during the northeast–southwest thrusting. The northeastward-verging fault-propagation folds are present in the Karpatian (17.5–16.5-Ma) sediments in the northern part of the Transcarpathian basin.

The Vihorlat fault zone offsets the Cretaceous–Oligocene flysch sediments (Nemcok and Nemcok, 1994) and the middle Badenian tuff horizon and is sealed by the upper Sarmatian volcanics (Figures 1, 3b–g).

Pannonian Region

Based on microtectonic data, the Darno fault zone (Figures 1, 3d–g) acted as the sinistral strike-slip fault during the Karpatian–Badenian (17.5–13.6 Ma) (Csontos et al., 1991). Zelenka et al. (1983) and Balla (1987) document the Paleogene dextral and late Miocene sinistral displacement, respectively.

Based on microtectonic data, the low-angle Raba fault zone (Figures 1, 3d–g) acted as the sinistral strike-slip fault during the Karpatian–Badenian (17.5–13.6 Ma) (Csontos et al., 1991). However, its oblique-slip character was revealed by reflection seismic studies that indicate a significant dip-slip component (Horvath et al., 1987; Tari, 1991). Detailed observations from seismics indicate a reactivation of deeper parts of the earlier thrust planes (e.g., Horvath, 1993; Szafian et al., 1999). Some observations concluded only normal faulting during the middle Miocene (e.g., Mattick et al., 1996).

The Buda fault zone (Figure 1) activity is indicated by a northeast-trending late Eocene–early Oligocene antiforms with synsedimentary growth. Tari et al. (1993) relate them to the first-order thrusting and folding. Fodor et al. (1992) relate them to the dextral strike-slip faulting.

The Central Hungarian fault zone (Figures 1, 3d–g) acted as the sinistral strike-slip fault zone during the Karpatian–Badenian (17.5–13.6 Ma), indicated by the fault-slip data (Csontos et al., 1991) as the dextral transpressional strike-slip fault zone during the late Oligocene–early Miocene as described earlier. Reflection seismic data along the zone indicate north-dipping thrusts and normal faults and crosscutting strike-slip faults (Csontos and Nagymarosy, 1998). They also indicate extensional regime during the middle Miocene, local late Miocene inversion, and late Miocene–Pliocene normal faulting and sinistral transtension (Figure 3h).

Based on microtectonic data, the Balaton fault zone (Figures 1, 3d–g) acted as the sinistral strike-slip fault during the Karpatian–Badenian (17.5–13.6 Ma) (Csontos et al., 1991). It offsets dextrally (about 300 km [180 mi]) the Slovenian and Hungarian Paleogene basins, which formed one basin in the early Oligocene (Csontos et al., 1992). The dextral displacements also affected Eggenburgian–Ottnangian (22–17.5-Ma) rocks (Csontos et al., 1992). Sediments of exotic strike-slip duplexes inside the Balaton fault zone have the late Oligocene age (Baldi-Beke, 1984).

Eastern and Southern Carpathians

The sinistral Dragos–Voda faults (Figures 1, 3g) acted during the late Badenian and/or later as indicated by kinematic data and pre- or syntectonic late Badenian sediments (Zweigel, 1997; Linzer et al., 1998).

The dextral southern Carpathian fault (Figure 1) was active during the late Miocene as indicated by the inversion of the Bozovici basin with the upper Badenian (14-Ma) fill (Linzer, 1996; Linzer et al., 1998).

The dextral Cerna–Jiu fault (Figure 1) was active during the Eocene, as indicated by the opening of the Petrosani Basin. The subparallel dextral Baia de Arama fault opened the Baia de Arama graben in the Egerian (24 Ma) Moser and Frisch, 1996). A north–south extension controlling the continuing dextral displacements along the fault system subparallel to the Cerna–Jiu fault opened the Sichevita, Bozovici, and Caransebes–Mehadia Basins in the late Badenian (14 Ma) (Moser and Frisch, 1996). The subsidence of these basins ceased in the Pannonian (8.5 Ma) (Moser and Frisch, 1996). To the south of the Cerna–Jiu fault, the dextral Arama–Severin–Timok fault system opened the Orsova pull-apart basin at the Karpatian–Badenian boundary (Moser and Frisch, 1996). The Bistra–Lotru–Cozia dextral strike-slip fault opened the Sarmatian (13.6–11.5-Ma) Hateg pull-apart basin (Moser and Frisch, 1996). A Pliocene–Pleistocene north–south compression slightly deformed sediments of the Caransebes–Mehadia Basin (Moser and Frisch, 1996).

PALEOSTRESS DATA

The determined late Eocene (40-Ma) stress field (Figure 4a) in the European platform to the north and west of the Alps has a north–south compression (Bergerat, 1987). Stress configurations from the northern Central Eastern Alps and units to the north of them have a northwest–southeast (north–south) compression and both horizontal and vertical extensions (e.g., Decker et al., 1993; Linzer et al., 1995, 1997; Peresson and Decker, 1997a). Stress tensors of the northwestern side of the ALCAPA microplate have a northwest–southeast-oriented compression and subvertical extension, respectively (e.g., Nemcok and Nemcok, 1994; Fodor, 1995). Syndepositional paleostresses from the Priabonian sediments from the central ALCAPA have a northwest–southeast and northeast–southwest compression and extension, respectively (e.g., Fodor et al., 1992, 1994; Kovac and Hok, 1993; Bada, 1999). Roughly west–east-oriented compression was determined from the late Eocene sediments in the southern parts of the ALCAPA (Fodor et al., 1999). Stress configurations derived from the northeastern parts of the ALCAPA have a north–south to northeast–southwest compression (Nemcok and Nemcok, 1994). Determined stress tensors from the Eastern and Southern Carpathians and Transylvanian Basin are scarce, and different authors have contradictory results. Several reasons exist, including the oldest part of the stratigraphic section, which requires the largest amount of filtering from younger data, an incomplete stratigraphic section, a lack of robust data sets, and an inappropriate use of paleostress computer programs designed only for monophase data in simple structural settings (e.g., Angelier and Goguel, 1979, and related program of Angelier et al., 1982, or Sperner et al., 1993). Results include a northeast–southwest (Moser and Frisch, 1996; Matenco et al., 1997; Linzer et al., 1998; own data) or east–west compression in the Southern Carpathians, an east-northeast–northeast-directed compression in the northern Transylvanian Basin (Gyorfi et al., 1999), or a north–south extension in the Transylvanian Basin (Huismans et al., 1997).

The determined Oligocene (35–30-Ma) stress field (Figure 4a, b) in the European platform to the north and west of the Alps has a west–east extension (Bergerat, 1987). Stress configurations along the northern Central

Alpine boundary and to the north of it are characterized by a north–south compression and vertical, later horizontal, extension (e.g., Decker et al., 1993; Linzer et al., 1995, 1997; Peresson and Decker, 1997a). The change from northwest–southeast to north–south compression was coeval with the onset of the dextral strike-slip faulting along the Periadriatic fault system (Peresson and Decker, 1997a). Stress tensors of the northwestern side of the ALCAPA have a northwest–southeast and northeast–southwest compression and extension, respectively (e.g., Marko et al., 1990, 1991; Kovac and Hok, 1993; Nemcok and Nemcok, 1994; Fodor, 1995). A roughly west–east compression was inferred from the dextral offset of the Oligocene basins along the southeastern side of the ALCAPA (Csontos et al., 1992). Northwest–southeast- and northeast–southwest-oriented compression and extension were calculated (Fodor et al., 1992, 1994; Kovac and Hok, 1993; Bada, 1999) from the central parts of the ALCAPA. Stress configurations from the northeastern parts of the ALCAPA have a north–south to northeast–southwest compression (Nemcok and Nemcok, 1994). Stress tensors from the Southern Carpathians have a northeast–southwest compression (Moser and Frisch, 1996; Linzer et al., 1998; own data), a northwest–southeast first-order extension (Matenco et al., 1997), or an east–west-oriented compression (Ratschbacher et al., 1993b). Stress tensors from the northern Transylvanian Basin have north-northwest–south-southeast compression (Gyorfi et al., 1999) or northwest–southeast compression (Huismans et al., 1997).

A reconstructed early Miocene (22–20-Ma) stress field (Figure 4b) in the European platform to the north and west of the Alps has a northeast–southwest compression (Bergerat, 1987). Stress configurations along the northern Eastern Alps and in units to the north of them are characterized by a north–south to northeast–southwest compression (e.g., Decker et al., 1993; Kurz et al., 1993; Nemes et al., 1995) that changed to a west–east extension (Genser and Neubauer, 1989) and north–south extension (Neubauer et al., 1999) in the central Eastern Alps. North–south compression was calculated from the dextral transpressive eastern Periadriatic fault system (Nemes, 1996; Fodor et al., 1998). Stress tensors of the northern and central parts of the ALCAPA have a northwest–southeast and northeast–southwest compression and extension, respectively (e.g., Kovac et al., 1989; Nemcok et al., 1989; Fodor et al., 1990, 1992, 1994; Kovac and Hok, 1993; Vass et al., 1993; Sperner, 1996). Inferred and calculated stresses of the southeastern and northeastern side of the ALCAPA have a west–east and north–south to northeast–southwest compression, respectively (Csontos et al., 1992; Nemcok and Nemcok, 1994; Kovac et al., 1995). Stresses were inferred from the dextral offset of the lower Miocene basins by Csontos et al. (1992). In the southern and central ALCAPA, the sinistral transpressive activity of faults driven by northwest–north-northwest-directed compression finishes by the Ottnangian (Bada, 1999). Stress tensors from the Southern Carpathians have a northeast–southwest to north-northeast–south-southwest compression (Moser and Frisch, 1996; Linzer et al., 1998; own data), a northwest–southeast first-order extension (Matenco et al., 1997), or an east–west-oriented compression (Ratschbacher et al., 1993b). Stress tensors from the northern Transylvanian Basin either have a north-northwest–south-southeast compression (Gyorfi et al., 1999) or a north-northeast–south-southwest compression (Huismans et al., 1997).

Eggenburgian–Ottnangian is the time of rifting initiation inside the Pannonian Basin (Pecskay et al., 1995; Horvath and Tari, 1999). The Pannonian Basin region is characterized by a northeast–southwest to east-northeast–west-southwest extension since the late Ottnangian (e.g., Fodor, 1995). Pull-apart basins opened in the intra-Carpathian region during the Karpatian–middle Badenian (17.5–15 Ma) (Figure 4c, d), controlled by a north–south and west–east compression and extension, respectively (e.g., Bergerat, 1989; Nemcok et al., 1989; Csontos et al., 1991). Karpatian is the timing of the Rechnitz metamorphic core complex uplift because of a major northeast–southwest and a minor northwest–southeast extension dated by fission tracks (Dunkl, 1992; Dunkl and Demeny, 1997), which drove low-angle normal faults (Tari et al., 1999). The Ottnangian–Karpatian (19–16.5-Ma) stress, with a northwest–southeast compression in the area of the Transdanubian Central Range, changed to the Badenian–Sarmatian (16.5–11.5-Ma) stress with a north–south compression and the Pannonian (11.5–6.2-Ma) stress with a northeast–southwest compression (Csontos et al., 1991; Tari, 1991). North–south compression changed to northeast–southwest compression in various Western Carpathian blocks during the middle–late Badenian (15.5–13.6 Ma) (Nemcok et al., 1989; Fodor et al., 1990; Nemcok and Lexa, 1990). This stress rotation finished in the central ALCAPA during the Sarmatian (13.6–11.5 Ma) (Bada, 1999). Ottnangian–lower Badenian sinistral transtension, driven by a northwest–southeast compression and a northeast–southwest extension, was reported from central ALCAPA (Marton and Fodor, 1995). Karpatian–Badenian transtension, driven by a north–south compression and a west–east extension, was present in the southern ALCAPA (Benkovics, 1997). A northwest–southeast contraction was determined for flysch units along the western ALCAPA, changing directions from northwest to north and to the northeast in time during the Karpatian–Badenian (e.g., Nemcok, 1993; Fodor et al., 1995). Extension in the ALCAPA was

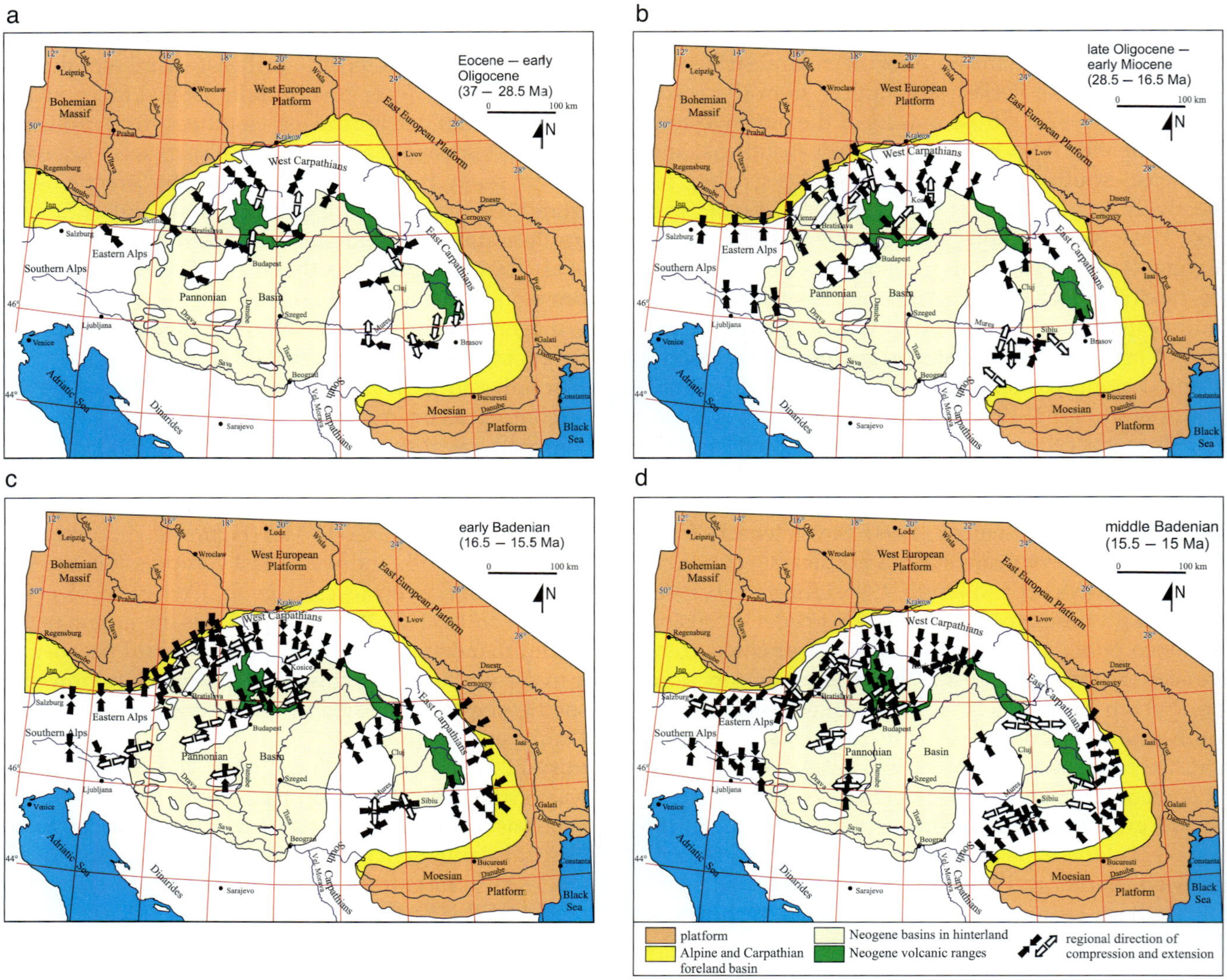

Figure 4. Paleostress directions in the Carpathian–Pannonian region for (a) Eocene–early Oligocene, (b) late Oligocene–early Miocene, (c) early Badenian, (d) middle Badenian, (e) late Badenian–Sarmatian, (f) early Pannonian, (g) late Pannonian, and (h) present. Directions in (a–g) represent average values of stress states calculated in the Alps (Decker and Jarnik, 1992; Decker et al., 1993, 1994; Kurz et al., 1993; Linzer et al., 1995, 1997; Nemes et al., 1995; Nemes, 1996; Peresson and Decker, 1997a), in the broader Pannonian Basin region (Bergerat et al., 1983; Bergerat, 1989; Csontos et al., 1991; Tari, 1991; Csontos and Bergerat, 1992; Fodor et al., 1992, 1994; Marton and Fodor, 1995; Benkovics, 1997; Bada, 1999; Gyorfi et al., 1999), in the Western Carpathians (Kovac et al., 1989, 1990, 1993, 1995; Nemcok et al., 1989, 1998b, 2000a; Fodor et al., 1990, 1995; Marko et al., 1990, 1991; Nemcok and Lexa, 1990; Nemcok and Nemcok, 1990, 1994; Nemcok, 1991, 1993; Kovac and Hok, 1993; Ratschbacher et al., 1993a; Vass et al., 1993; Fodor, 1995; Sperner, 1996; Sperner et al., 2002), in the Eastern and Southern Carpathians (Ratschbacher et al., 1993b; Hippolyte and Sandulescu, 1996; Linzer, 1996; Morley, 1996; Moser and Frisch, 1996; Matenco et al., 1997; Zweigel, 1997; Linzer et al., 1998), and in the Transylvanian Basin (Huismans et al., 1997) by stress inversion techniques or compiled in synthetic papers (Bada, 1999; Fodor et al., 1999). Regional direction of compression is taken from stress states calculated from thrust or strike-slip regimes; regional direction of extension is taken from stress states calculated from normal fault regimes. Principal horizontal stress direction map for present (h) is modified from the World Stress Map Project (1997). Data came from earthquake focal mechanism, overcoring, and borehole breakout studies. Boundary conditions modeled by finite-element approach (Bada, 1999) as fitting the data are shown, indicating fixed or slightly deforming Bohemian Massif, active convergence in the Vrancea region, fixed Moesian Platform, active deformation at the Dinaric front, and rotating Adriatic (Apulian) microplate.

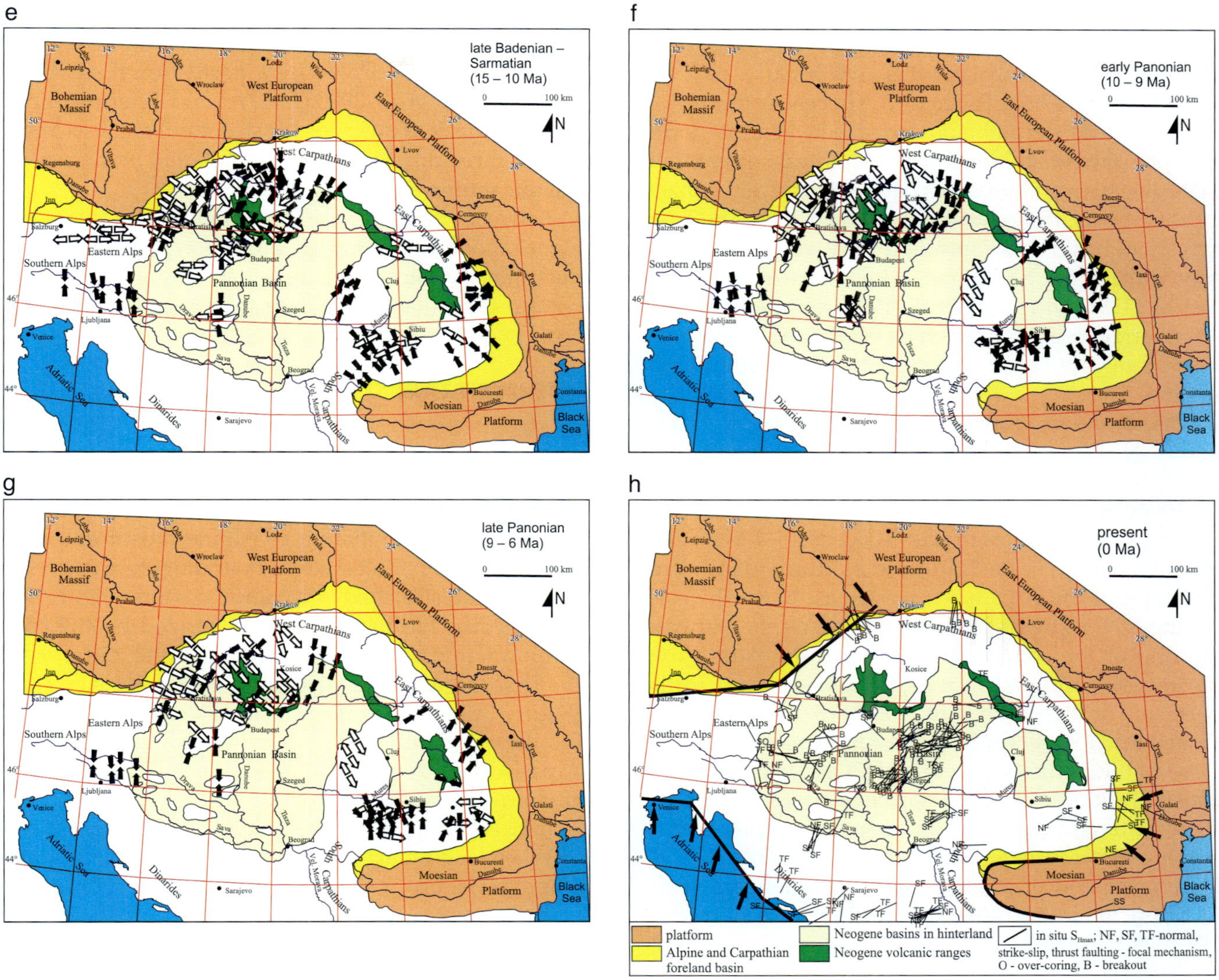

Figure 4. (cont.).

calculated as advancing from its center toward its margins during the Karpatian–middle Badenian (Nemcok et al., 1998b).

Northern Eastern Carpathians were characterized by a west to west-northwest compression, and southern Eastern Carpathians were characterized by a northwest to north-northwest compression during the early Miocene–Sarmatian (e.g., Zweigel, 1997).

The Badenian–Pannonian period (Figure 4c–g) is the time of the intra-Carpathian basin and stress rearrangement, when a very prominent west–east to west-northwest–east-southeast regional extension started during the end of the Badenian–Sarmatian (e.g., Bergerat, 1989; Fodor et al., 1990; Nemcok and Lexa, 1990; Csontos et al., 1991; Bada, 1999), having various timing at various places. A comparison of ratios of principal stress magnitudes of the progressively younger stresses from the middle Slovakian volcanic field indicates a progressively stronger extension (Nemcok et al., 2000a). The middle Miocene stress along the northern Eastern Alps has a north–south to northeast–southwest compression and a subhorizontal and subvertical extension, followed by an east–west extension (e.g., Linzer et al., 1997; Peresson and Decker, 1997a). A west–east extension continued in the central Eastern Alps. Karpatian–early Badenian sinistral strike-slip faults in the Vienna basin area were reactivated as normal faults (e.g., Nemcok et al., 1989; Nemcok, 1993; Fodor, 1995).

The middle Miocene stress field determined in the Southern Carpathians has a northwest–southeast compression (e.g., Ratschbacher et al., 1993b; Girbacea, 1997; Matenco et al., 1997). The middle Miocene compression in the Eastern Carpathians varied between the northeast–southwest and northwest–southeast orientation in the northern and southern parts, respectively (Linzer, 1996; Girbacea, 1997; Zweigel, 1997; own data). The Transylvanian Basin was controlled by a west–east extension during the Sarmatian (Huismans et al., 1997). Nemes (1996) and Fodor et al. (1998) reported repeated dextral transpression driven by north–south compression along the eastern part of the Periadriatic fault system.

The boundary between the ALCAPA and Tisza–Dacia was controlled by a northeast–southwest compression and a northwest–southeast extension since the late Badenian (Gyorfi et al., 1999).

A reconstructed late to post-Miocene (7–4-Ma) stress field (Figure 4g) in the European platform to the north and west of the Alps has a west–east to northwest–southeast compression, characterized by the outward fanning of the compression close to the Alps (from northwest–southeast in the east, to east–west in the west, and north–south in the south) (Bergerat, 1987; Le Pichon et al., 1988). Far from the Alps, compression was close to the northwest–southeast. Late Miocene stress configurations along the northern Eastern Alps are characterized by an east–west compression and a north–south extension (e.g., Peresson and Decker, 1997a). During the end of the Badenian to Pannonian (16.5–11.5 Ma), the intra-Carpathian basins experienced a regionally consistent east–west to east-southeast–west-northwest extension (e.g., Bergerat, 1989; Nemcok, 1993). A north–south to north-northeast–south-southwest compression was determined in the Southern Carpathians (e.g., Ratschbacher et al., 1993b) or a northwest–southeast compression (Girbacea, 1997).

Later during the Pannonian (11.5–6.2 Ma), the Carpathian stress configuration changed to the stress configuration similar to the present one (Figure 4h), as determined in the Vienna basin, southern and northern Hungary, and central Slovakia (Fodor et al., 1990; Csontos et al., 1991; Benkovics, 1997; Nemcok et al., 1998b). The period since the early Pannonian was characterized by short-term intra-Pannonian Basin inversion event and Pliocene–Quaternary inversion (Bada, 1999, and references therein). A large mismatch of timing of the short-term event exists, including the Sarmatian–Pontian (9–5.3 Ma) for the entire region, driven by east–west compression (Peresson and Decker, 1997b), which was proven as physically incorrect (see Bada, 1999), early Pannonian for the Central Hungarian fault zone or Mecsek Mountains, driven by north–south compression (Benkovics, 1997; Csontos and Nagymarosy, 1998), and late Pannonian for the middle Slovakian basins, driven by a north–south compression (Nemcok et al., 2000a). All related determinations indicate a lack of outcrops for structural analysis and a lack of three-dimensional determination in the case of seismic interpretation.

Between the Eastern and Southern Carpathians, an early Pannonian northwest–southeast compression was determined, which was later replaced by a north–south compression (e.g., Hippolyte and Sandulescu, 1996).

INTERPRETATION

Figure 3 shows displacements along the largest Carpathian–Pannonian faults under controlling stresses shown in Figure 4. The Oligocene fault activity can be described as the initiation of Alpine rails, formed by major strike-slip fault systems bounding the eastern Alpine domain stretched from north to south (Figure 3a). The Egerian faulting was characterized by the continuing activity of the Alpine rails and the initiation of the first larger strike-slip faults inside the ALCAPA (Figure 3b). The Ottnangian tectonics indicated that the eastward movements of the Alpine part of the Alps–ALCAPA system slowed down (Figure 3c). The amount of intra-ALCAPA strike-slip faults became larger as they accommodated the ALCAPA stretching.

When fault-timing data are compared to facial correlation data (Figure 5), the upper Egerian–Eggenburgian–Ottnangian (24–18-Ma) period can be characterized by the presence of northwest–southeast and northeast–southwest basins belonging to the early synrift stage. Basins in the orogenic hinterland behind the Carpathian thrust belt were controlled by strike-slip and normal faults, which were connected to the thrust belt. Basins of the Pannonian Basin system underwent fluvial deposition. They were more or less perpendicular to marine basins with flysch and molasse deposition, located in the area of the subducting remnant Carpathian Flysch Basin. Rollback and hinge retreat contributed to their fast subduction.

This period in the Pannonian Basin system was characterized by eruptions of acidic calc-alkaline volcanics of crustal origin (Figure 6), followed by mantle-derived areal andesitic magmatism (Figure 7), which overlapped with the area of extension and elevated asthenosphere.

The Karpatian fault activity can be characterized by the initiation of the ALCAPA northwestern rail formed by the system of sinistral strike-slip faults along its northwestern boundary accommodating its northeastward movements (Figure 3d). The southeastern rail, a

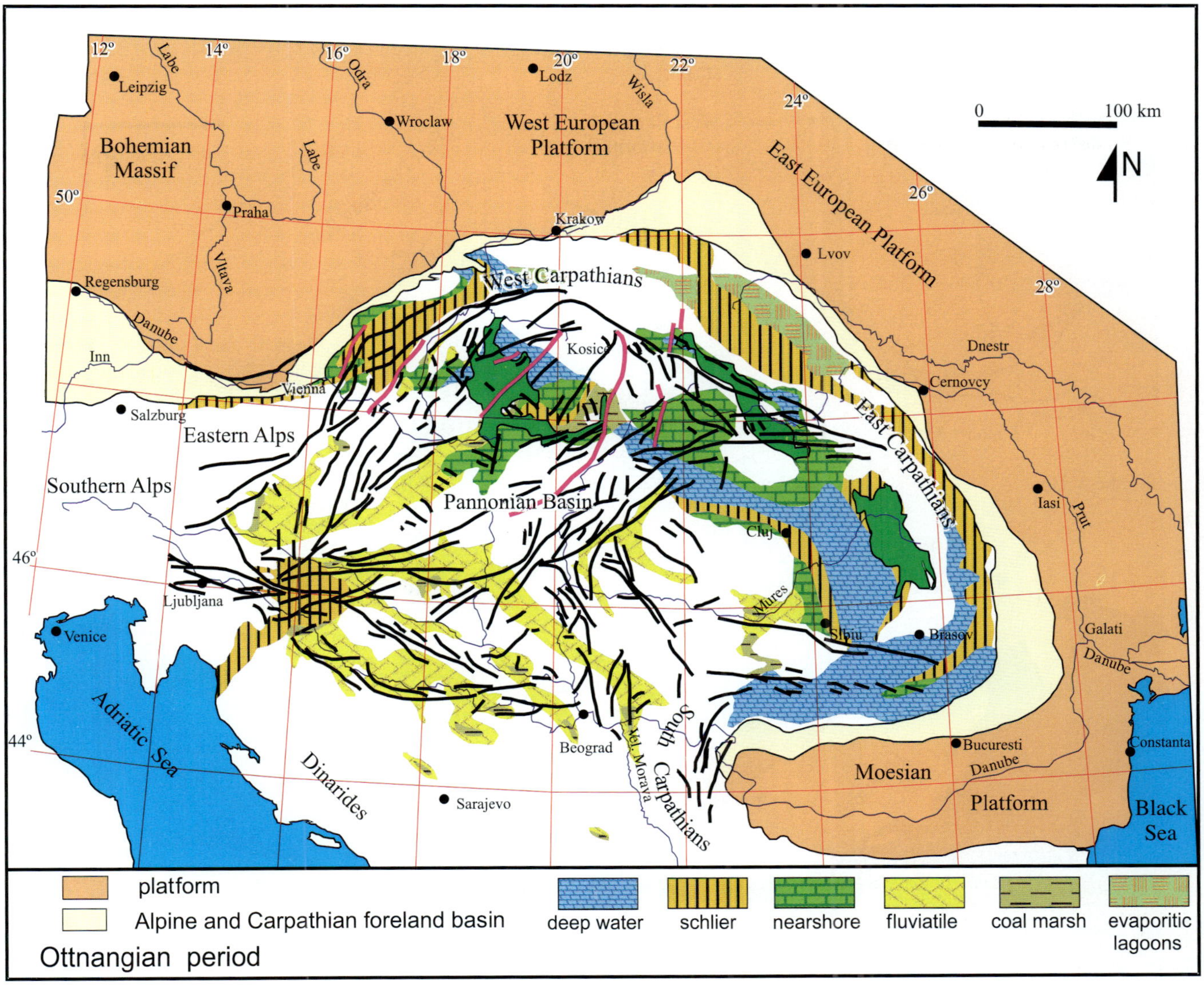

Figure 5. Ottnangian facies map of the Carpathian–Pannonian region (modified from Hamor et al., 2001) combined with fault pattern and fault activity maps from Figures 1 and 3. Northwest–southeast- and northeast–southwest-oriented basins were formed by early synrift stage. Hinterland basins right behind the thrust belt were controlled by strike-slip and normal faults, experiencing marine conditions. Hinterland basins further away from the thrust belt, controlled by the same fault types, underwent fluvial deposition. Hinterland zones of fluvial sediments were extended along main block-bounding strike-slip faults of this period. These sediment zones were more or less perpendicular to marine basins located in the area of subducting remnant Carpathian Flysch Basin, thrust belt, and its immediate hinterland. Subduction rollback and hinge retreat were the main engine of controlling tectonic events. ALCAPA was moving eastward and northeastward over the remnant Carpathian Flysch Basin, undergoing counterclockwise rotation. It was bounded by sinistral and dextral strike-slip systems along its northern side and along the contact with the Southern Alps and the Tisza–Dacia, respectively. Main active strike-slip faults of the ALCAPA are in purple. Tisza–Dacia advanced northeastward at slower rate.

system of dextral strike-slip faults, was already well functioning. The early Badenian faulting is described as the northwestern ALCAPA rail being in its most effective stage and the southeastern rail functioning well (Figure 3e). The middle Badenian tectonic setting indicates that the northwestern ALCAPA rail progressively stopped functioning from west to east (Figure 3f). The ALCAPA started to experience normal faulting in areas where strike-slip faults stopped.

Facial data indicate that the Karpatian–middle Badenian (18–14-Ma) period belonged to a major synrift stage. Northeast–southwest-oriented basins were connected to the Mediterranean Sea. This period was characterized by evaporitic lagoons behind the emerging

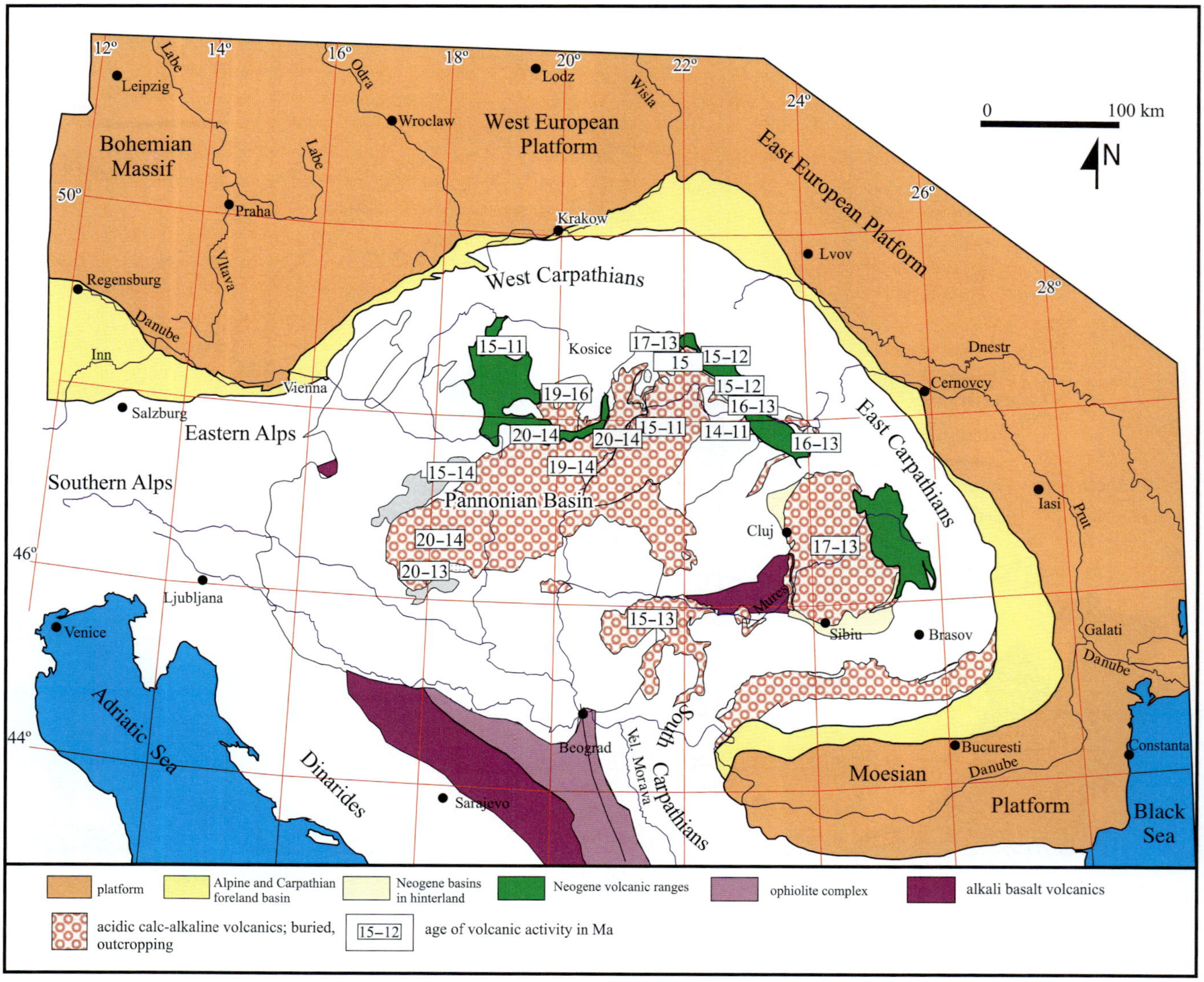

Figure 6. Acidic calc-alkaline volcanics distribution map of the Carpathian–Pannonian region (modified after Pecskay et al., 1995). Rock ages in Pecskay et al. (1995) are derived from the K-Ar dating and relationships with biostratigraphically dated formations of surrounding basins.

Carpathian thrust belt (Figure 8). The foreland basin experienced synorogenic subsidence. Pull-apart basins in the hinterland experienced widening and deepening of their initial narrow fluvial stages. New rifts and pull-apart basins were also opened in the hinterland. The overall paleogeography was dominated by wide marine facies zones separated by eroded areas and areas with nondeposition. Termination of the synrift stage in the late middle Badenian (14 Ma) is indicated by regional unconformity.

This period was characterized by a mantle-derived areal andesitic magmatism in the Pannonian Basin system, which overlapped with the region of extension and elevated asthenosphere (Figure 7). This period was also characterized by volcanic eruptions now recorded as the linear zone of the basaltic andesite to andesite located along the contact of the hinterland, which was related to the oceanic slab break-off (Figure 7).

The late Badenian fault activity can be described as the further eastward cessation of the northwestern ALCAPA rail (Figure 3g). The very western and central parts of ALCAPA were affected by normal faulting. A separate motion of the Tisza–Dacia initiated. The Sarmatian tectonic situation documents a complete stop of the northeastward motion of ALCAPA (Figure 3h). The internal west–east extension of the ALCAPA accommodated the eastward motion of the Tisza–Dacia.

The late Miocene period was dominated by the post-rift thermal subsidence in the Pannonian Basin system

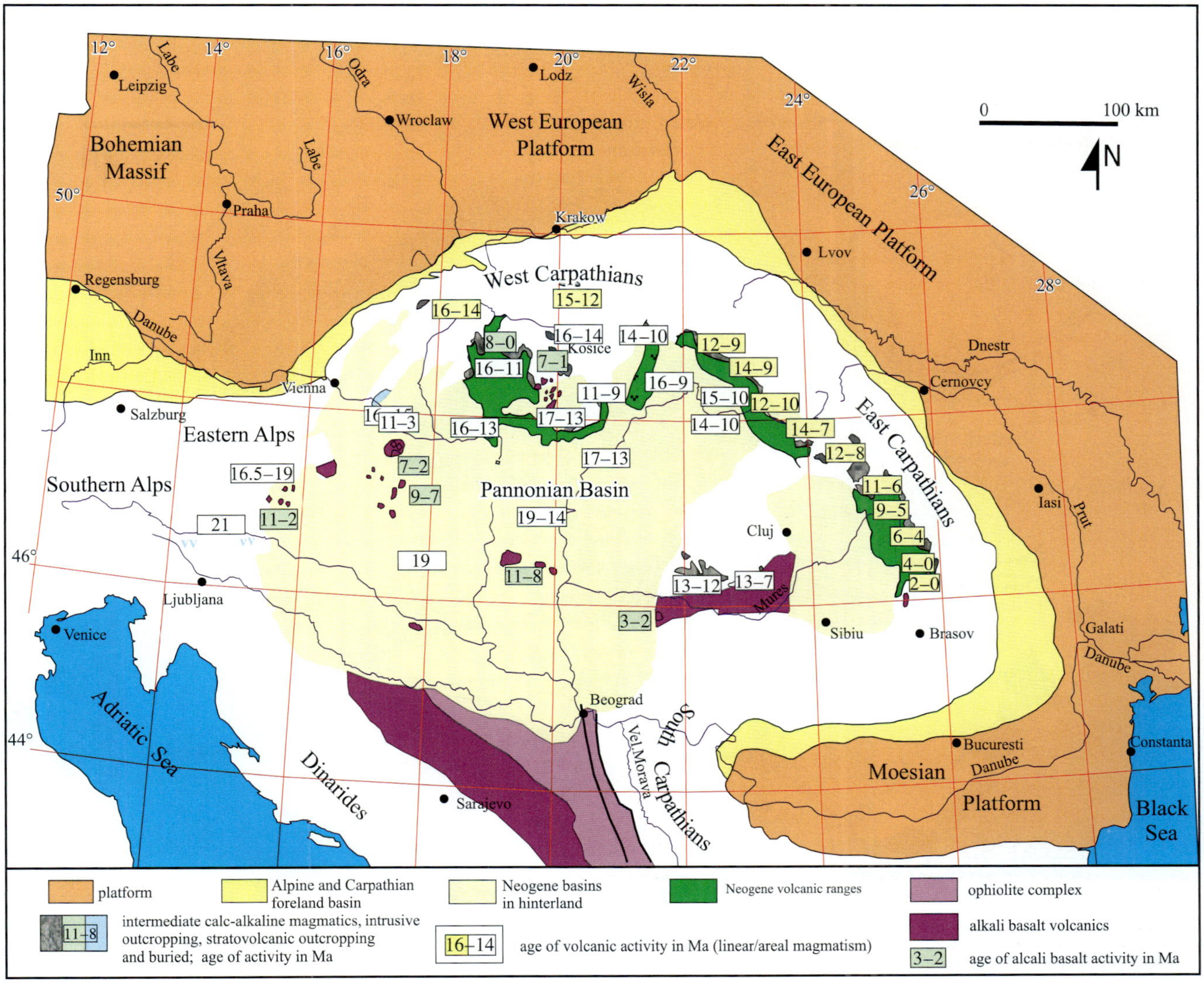

Figure 7. Intermediate calc-alkaline magmatics and alkali basalt volcanics distribution map of the Carpathian–Pannonian region (modified after Pecskay et al., 1995). Rock ages in Pecskay et al. (1995) are derived from the K-Ar dating and relationships with biostratigraphically dated formations of surrounding basins.

and the accelerated erosion of the Carpathian thrust belt. The emerging thrust belt provided sediments for the foreland basin and for the basins in hinterland (Figure 9). Rivers transported a large amount of erosional debris. Nevertheless, the late Badenian and Sarmatian subsidence rate exceeded the rate of the sediment supply. Reef limestones were deposited at margins of uplifted areas. The rate of the sediment supply exceeded the subsidence in the late Pannonian. Southward-prograding deltas of large rivers controlled the deposition of the Pannonian Basin system. Well-developed third-order sequences consisting of lowstand-prograding wedges, backstepping transgressive system tracks, and highstand prograding system tracks characterize the thick postrift sedimentary succession.

The late Miocene was also the period when a mantle-derived areal andesitic magmatism in the Pannonian Basin system ended by 10–9 Ma (Figure 7), which was the time of the structural inversion event. The linear zone of the basaltic andesite to andesite located along the contact of the hinterland, related to the oceanic slab break-off, continued its eastward propagation (Figure 7). Eruptions of alkali basalts postdated the areal calc-alkaline volcanism, indicating extension in the hinterland (Figure 7). Their magmas may be related to a widespread upwelling of the asthenospheric mantle beneath Europe (Hoernle et al., 1995).

The Pliocene–Quaternary paleogeography was characterized by structural inversion, uplift, and erosion in Transdanubia and Danube–Tisza areas. Deposition

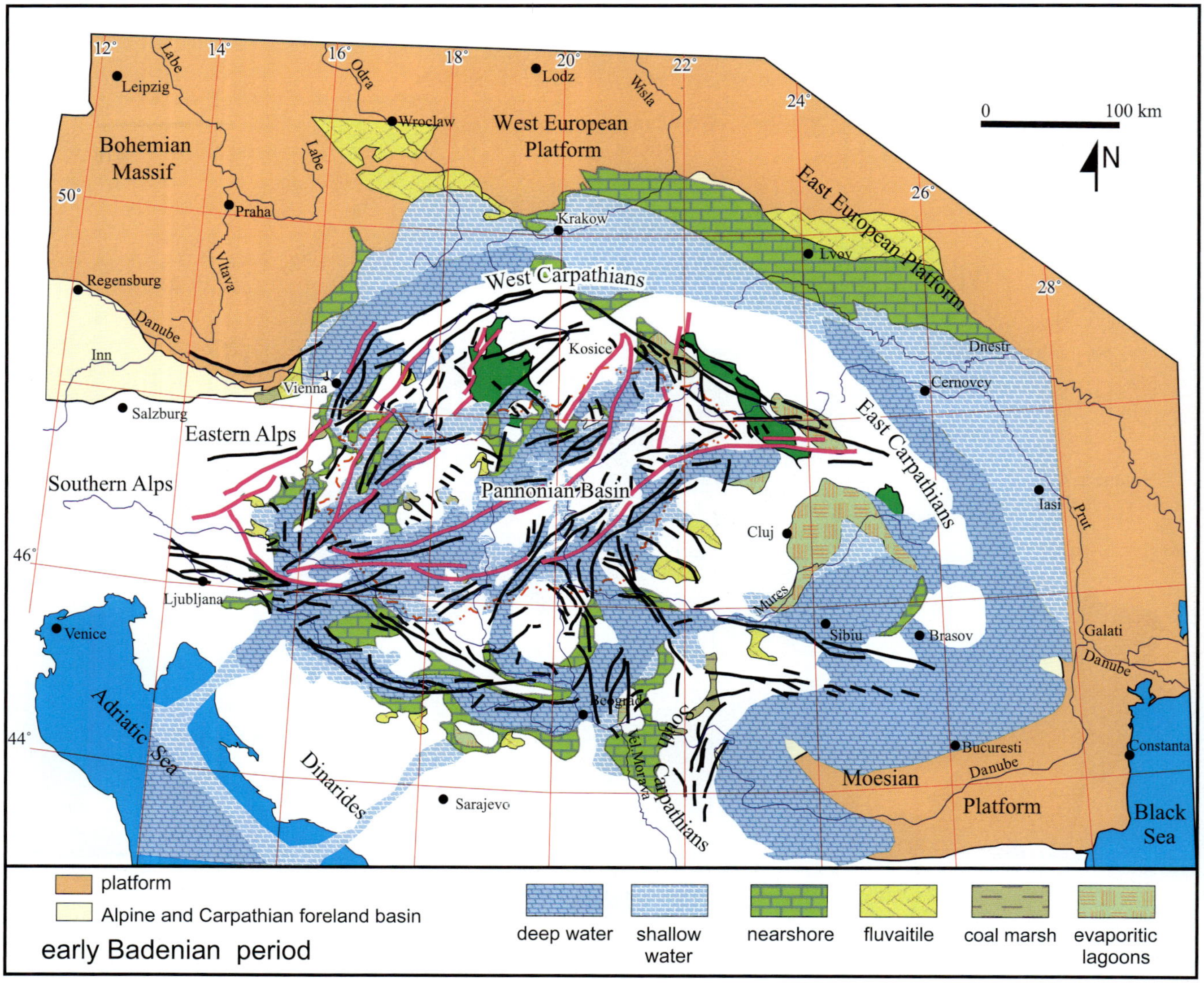

Figure 8. Early Badenian facies map of the Carpathian–Pannonian region (modified from Hamor et al., 2001) combined with fault pattern and fault activity maps from Figures 1 and 3. Northeast–southwest-oriented marine basins in broader orogenic hinterland were formed by major synrift stage. They were controlled by strike-slip and normal faults, many of them having pull-apart mechanisms. They either experienced widening and deepening of their early rift fluvial stages or they were newly formed. Basins were separated by eroding areas and areas of nondeposition. Eastern Carpathian part of the thrust belt was emerging, and adjacent thrust belt and foreland basin parts experienced shallow- or deep-marine conditions. Evaporitic lagoons formed behind the emerging part of the thrust belt. Subduction rollback and hinge retreat were the main engine of controlling tectonic events. Northeastward moving ALCAPA started colliding with the West European platform from west to east along its front. It was bounded and internally deformed by extensive system of strike-slip faults. Main active strike-slip faults of the ALCAPA are in purple. Tisza–Dacia continued in its slower northeastward movement.

continued in the Danube, Drava, and Tisza basins, where progradational delta filling terminated during the late Pliocene, followed by continental, fluvial, and ephemeral lake deposition. Pleistocene paleogeography was characterized by thick alluvial fans in mountain areas. Thick loess sediments were deposited during glacial intervals. Alkali basalt eruptions continued during this period (Figure 7).

DISCUSSION

It is difficult to differentiate the east–west stretching of the Eastern Alps from the eastward lateral extrusion. Structural data indicate both mechanisms. East–west stretching of the Eastern Alps can be driven by their north–south shortening coeval with the northwestward imbrication of the Western Alps and southwestward

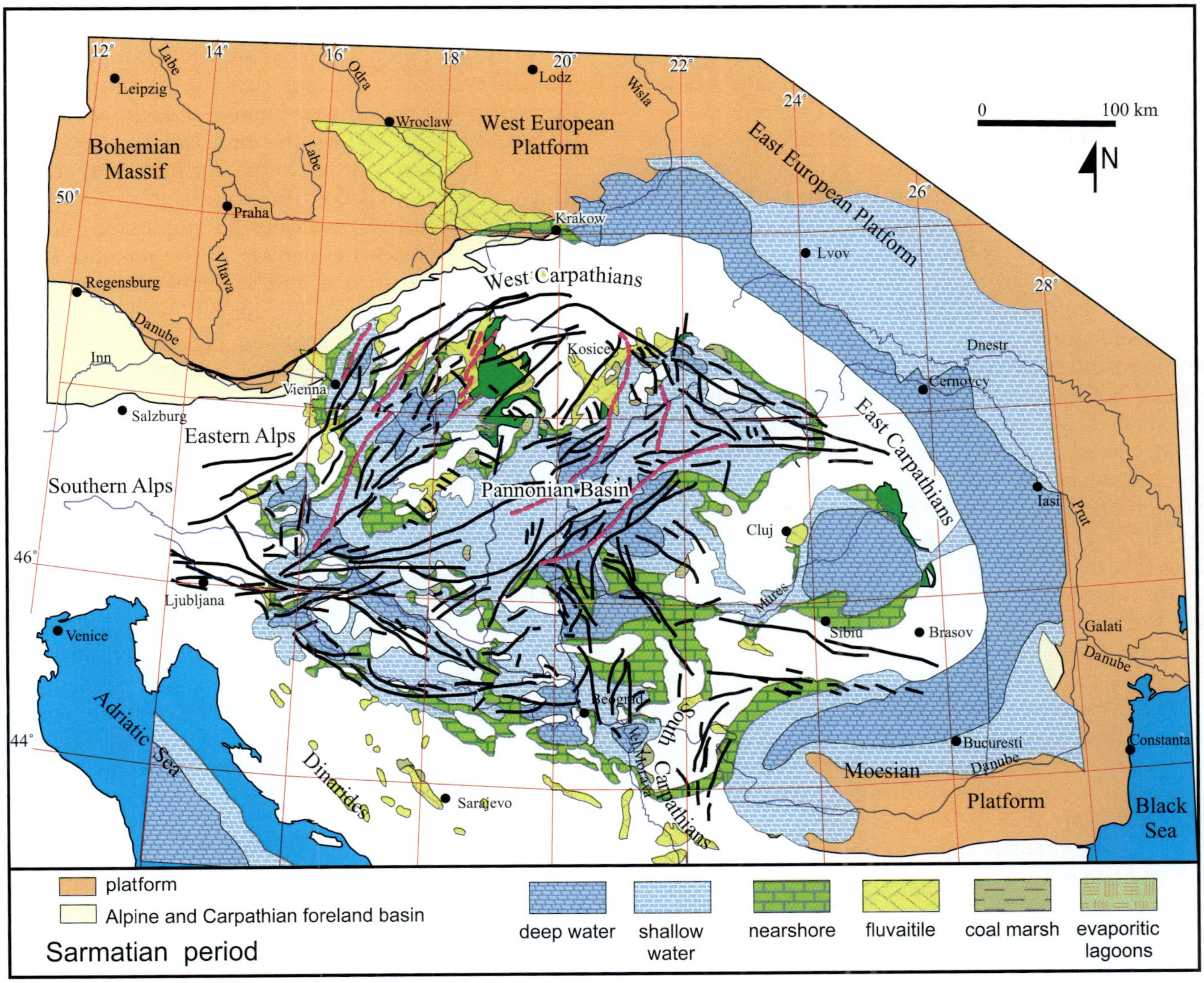

Figure 9. Sarmatian facies map of the Carpathian–Pannonian region (modified from Hamor et al., 2001) combined with fault pattern and fault activity maps from Figures 1 and 3. Northeast–southwest-oriented marine basins in orogenic hinterland were experiencing subsidence controlled by normal faults, which was followed by postrift thermal subsidence. Basins experienced widening, separated by small areas of nondeposition and erosion. Most of the thrust belt was emergent, and its erosion accelerated. Thrust belt provenance sourced deposition in both foreland and immediate hinterland. Rivers transported a large amount of erosional debris. Fluvial systems behind the Western Carpathian part of the thrust belt were especially areally extensive. Basin subsidence in hinterland further from thrust belt exceeded deposition. Reef limestones were deposited at margins of these basins. Last parts of the remnant Carpathian Flysch Basin were overridden by the orogen in western half of the Carpathians. The northeastward motion of ALCAPA came to a complete stop, and the eastward motion of Tisza–Dacia became pronounced. ALCAPA reacted by the internal west–east extension. Its main active normal faults are in purple.

subduction of the remnant Flysch basin located to the north and east of Eastern Alps, respectively. This stretching can be coeval with north–south compression in the zone to the north of the central Eastern Alps during the Oligocene to part of the early Miocene, determined by Decker et al. (1993). The lateral extrusion can be indicated by (1) the distinct period of pull-apart basin opening along the faults bounding the extruding wedge during the Ottnangian–Karpatian (19–16.5 Ma) (see data from Ratschbacher et al., 1991a); (2) the Egerian–Eggenburgian transtension and transpression change of the sinistral Inntal fault (to the northwest of the eastern Southern Alps) observed by Ortner and Sachsenhofer (1996); and (3) the large out-of-sequence oblique thrust of the Magura unit over the units made from the remnant Carpathian Flysch Basin fill and their unconformable lower–middle Badenian molasse seal (see Nemcok et al., 2000b). However, the differentiation of these two mechanisms and their timing requires further study.

The Inner Carpathians underwent deformation driven by the earlier Alpine collision and later consumption of the remnant Carpathian Flysch Basin. The ALCAPA microplate, for example, contains the Paleogene fault pattern created in its ancestral Alpine position. Its eastern part rotated counterclockwise when it moved to the remnant Carpathian Flysch Basin space and disintegrated in this free interface when it lost its northern and southern constraints, the Bohemian Massif and Southern Alps. Paleogene fault systems were reactivated by the Ottnangian–middle Badenian (19–15-Ma) stress fields. As indicated by the paleomagnetic (e.g., Korab et al., 1981; Krs et al., 1986, 1991, 1993; Marton and Marton, 1989; Marton, 1990, 1993; Tunyi and Kovac, 1991; Patrascu et al., 1993, 1994) and fault data (Figure 3), the block mosaic of the Inner Carpathians underwent a large-scale strike-slip faulting during the lower Miocene stage of the accretion of the remnant Carpathian Flysch Basin fill. The counterclockwise rotation of the ALCAPA microplate rotated earlier Cretaceous–Eocene thrusts to orientations suitable for their normal or strike-slip fault reactivation by Karpatian–Pannonian stresses. This reactivation was described in numerous studies (e.g., Horvath, 1993; Nemcok et al., 1998c).

The upper crustal data do not precisely indicate lithospheric processes, described in detail by Nemcok et al. (1998a). Limits for this indication are imposed by the interplay of factors influencing the upper crustal behavior: (1) eastward lateral motion of the ALCAPA from the Alpine collision; (2) separate movements of the blocklike ALCAPA and Tisza–Dacia microplates and their interactions during the lithospheric process; and (3) the barrier effect of the Moesian Platform, imposing a constraint for microplates overriding the remnant Carpathian Flysch Basin. Indications mentioned earlier about the crustal blocks of the intra-Carpathian region being detached at the lower crustal level exist, e.g., based on reflection seismic data (Subotin, 1975; Ibrmajer and Suk, 1989; Bucha and Blizkovsky, 1994) and thermal modeling (Lizon, 1986). According to what was suggested in other regions (Bird, 1991) about the tectonics in areas with lower crustal temperatures above 700°C, the lower crustal temperature present in the wider Pannonian region caused the lower crustal lithology to flow fast enough to substantially reduce topography, to as much as several kilometers laterally, in a few million years. From the above-mentioned arguments, it follows that the intra-Carpathian upper crustal data cannot indicate lithospheric processes precisely.

Finite-element modeling of the eastward Alpine lateral extrusion (Bada, 1999) has shown that this driving mechanism is much less important for the Carpathian–Pannonian tectonics than originally thought (e.g., Peresson and Decker, 1997a). The mere fact that ALCAPA was able to move that far to the north in comparison with its Alpine and Balkan orogenic neighbors after the Oligocene (Figure 2a) indicates the importance of the subduction rollback mechanism in the remnant Carpathian Flysch Basin area. This highlights the concept of a passive subduction of the remaining oceanic slab under the remnant Carpathian Flysch Basin by its own weight, a mechanism that was influencing the Carpathian–Pannonian development the most.

CONCLUSIONS

During the Late Cretaceous–early Paleogene, the Alpine–Carpathian orogen, formed by the subduction of the Penninic realm, comprised future ALCAPA and Tisza–Dacia microplates.

During the late Eocene, the future ALCAPA microplate underwent collisional north–south shortening in its ancestral Alpine position, bounded from the Bohemian Massif by its accretionary wedge, the Magura Flysch zone. The westernmost part of the future Tisza–Dacia microplate was also part of the Alpine realm, and the eastern part was juxtaposed with the remnant Carpathian Flysch Basin that subducted southwestward under the Carpathians. The late Eocene was the last time period with north–south collision until the Pliocene–Quaternary, when the convergence vector again became parallel to the Africa–Europe convergence.

During the Oligocene, both future microplates underwent east–west stretching and segmentation by the strike-slip and normal fault pattern and advanced toward the remnant Carpathian Flysch Basin. During the early Miocene, the ALCAPA moved eastward, bounded by sinistral strike-slip system along its northern side and dextral strike-slip fault system along its contact with the Southern Alps and the Tisza–Dacia. The largest movements took part during the Ottnangian–Karpatian (19–16.5 Ma). At a slower rate, the Tisza–Dacia advanced northeastward over the remnant Carpathian Flysch Basin. The ALCAPA blocks rotated counterclockwise during this time.

During the middle–late Badenian (15.5–13.6 Ma), the ALCAPA collided with the European platform, and the eastward movement of the Tisza–Dacia became pronounced. Earlier development of pull-apart basins, related to the extensive strike-slip faulting inside the ALCAPA, changed to the graben opening driven by the westward subduction and the eastward motion of the Tisza–Dacia. The shortening of the outer wedge was accompanied by the west–east extension in the intra-Carpathian region.

During the Sarmatian–Pannonian (13.6–6.2 Ma), the last parts of the remnant Carpathian Flysch Basin were overridden by the Carpathian orogen. The eastward motion of the Eastern Alps ceased. The Tisza–Dacia collided with the European platform during the Pannonian (11.5–6.2 Ma), and the intra-Carpathian stress field changed to the present stress field.

During the Pliocene–Quaternary, the postrift fill of the Pannonian Basin system, related to the regional thermal subsidence, started to undergo an inversion. Convergence vector again became parallel to the Africa–Europe convergence vector.

The size and shape of facies zones of each development period were controlled by tectonics. Eruptions of acidic calc-alkaline volcanics of crustal origin were associated in space and time with major phases of crustal stretching. Mantle-derived areal andesitic magmatism in the Pannonian Basin system overlapped with the region of extension and elevated asthenosphere. The linear zone of the basaltic andesite to andesite located along the contact of the hinterland with thrust belt is related to the oceanic slab break-off. Eruptions of alkali basalts, which are the youngest volcanics in the Carpathian–Pannonian development, were not affected by subduction.

ACKNOWLEDGMENTS

The authors thank M. Horn, F. Picha, and J. Golonka for their review and members of the Europrobe Pancardi group for fruitful discussions. Assistance with drafting was provided by B. Welker. Support of HNSF (OTKA) grants #T035168 and #T047159 are gratefully acknowledged.

REFERENCES CITED

Angelier, J., and J. Goguel, 1979, Sur une methode simple de determination des axes principaux des constraintes pour une population de failles: Comptes Rendus de l'Academie des Sciences Paris, Serie D, v. 288, p. 307–310.

Angelier, J., A. Tarantola, B. Valette, and S. Manoussis, 1982, Inversion of field data in fault tectonics to obtain the regional stress: I. Single phase fault populations, a new method of computing the stress tensor: Geophysical Journal of the Royal Astronomical Society, v. 69, p. 607–621.

Bada, G., 1999, Cenozoic stress field evolution in the Pannonian Basin and surrounding orogens: Inferences from kinematic indicators and finite element modeling: Ph.D. thesis, Vrije University, Amsterdam, 204 p.

Baldi-Beke, M., 1984, The nannoplankton of the Transdanubian Paleogene formations (in Hungarian): Geologica Hungarica, Series Paleontologica, v. 43, p. 1–307.

Balla, Z., 1987, The tectonics of Hungary, a kinematic point of view: Rendiconti della Societa Geologica Italiana, v. 9, p. 97–102.

Benkovics, L., 1997, Etude structurale et geodynamique des Monts Buda, Mecsek et Villany: Ph.D. thesis, University Flandres–Artois, Lille, 231 p.

Bergerat, F., 1987, Stress fields in the European platform at the time of Africa–Eurasia collision: Tectonics, v. 6, p. 99–132.

Bergerat, F., 1989, From pull-apart to the rifting process, the deformation of the Pannonian Basin: Tectonophysics, v. 157, p. 271–280.

Bergerat, F., J. Geysant, and C. Lepvrier, 1983, Etude de la fracturation dans le bassin pannonien, mecanismes et etapes de sa creation: Annales de la Societe geologique du Nord, v. 103, p. 265–272.

Bird, P., 1991, Lateral extrusion of lower crust from under high topography, in the isostatic limit: Journal of Geophysical Research, v. 96, p. 10,275–10,286.

Bucha, V., and M. Blizkovsky, eds., 1994, Crustal structure of the Bohemian Massif and the West Carpathians: Berlin, Springer-Verlag, 284 p.

Csontos, L., 1995, Tertiary tectonic evolution of the intra-Carpathian area, a review, *in* H. Downes and O. Vaselli, eds., Neogene and related magmatism in the Carpatho-Pannonian region: Acta Vulcanologica, v. 7, p. 1–13.

Csontos, L., and F. Bergerat, 1992, Reevaluation of the Neogene brittle tectonics of the Mecsek–Villany area (southwest Hungary): Annales Universitatis Scientiarum Budapestinensis de Rolando Eotvos Nominatae, Sectio Geologica, v. 29, p. 3–12.

Csontos, L., and A. Nagymarosy, 1998, The mid-Hungarian line, a zone of repeated tectonic inversions: Tectonophysics, v. 297, p. 51–71.

Csontos, L., G. Tari, F. Bergerat, and L. Fodor, 1991, Evolution of the stress fields in the Carpatho-Pannonian area during the Neogene: Tectonophysics, v. 199, p. 73–91.

Csontos, L., A. Nagymarosy, F. Horvath, and M. Kovac, 1992, Tertiary evolution of the intracarpathian area, a model: Tectonophysics, v. 208, p. 221–241.

Decker, K., and M. Jarnik, 1992, Structural analysis of the Late Cretaceous Gosau Group of Rigaus, Fahrenberg and Nussensee (Northern Calcareous Alps, Salzbur, Upper Austria), Tertiary deformation during lateral extrusion illustrated: Mitteilungen der Gesellschaft der Geologie- und Bergbaustudenten, v. 38, p. 93–106.

Decker, K., M. Meschede, and U. Ring, 1993, Fault slip analysis along the northern margin of the Eastern Alps (Molasse, Helvetic nappes, North and South Penninic flysch, and the Northern Calcareous Alps): Tectonophysics, v. 223, p. 291–312.

Decker, K., H. Peresson, and P. Faupl, 1994, Die miocane Tektonik der ostlichen Kalkalpen, Kinematik, Palaospannungen und Deformationsaufteilung wahrend der "lateralen Extrusion" der Zentralalpen: Jahrbuch der Geologischen Bundesanstalt, v. 137, p. 5–18.

Dunkl, I., 1992, Final episodes of the cooling history of eastern termination of the Alps, *in* F. Neubauer, ed., The Eastern Central Alps of Austria, ALCAPA 1992 meeting field trip guidebook: Graz, University of Graz, p. 137–139.

Dunkl, I., and A. Demeny, 1997, Exhumation of the Rechnitz window at the border of the Eastern Alps and Pannonian Basin during Neogene extension: Tectonophysics, v. 272, p. 197–211.

Ebner, F., and R. F. Sachsenhofer, 1995, Paleogeography, subsidence and thermal history of the Neogene Styrian Basin (Pannonian Basin system, Austria): Tectonophysics, v. 242, p. 133–150.

Ellouz, N., and E. Roca, 1994, Palinspastic reconstruction of the Carpathians and adjacent areas since the Cretaceous, a quantitative approach, *in* F. Roure, ed., Peri-Tethyan platforms: Paris, Editions Technip, p. 51–78.

Fodor, L., 1995, From transpression to transtension, Oligocene–Miocene structural evolution of the Vienna basin and the East Alpine–Western Carpathians junction: Tectonophysics, v. 242, p. 151–182.

Fodor, L., F. Marko, and M. Nemcok, 1990, Evolution microtectonique et paleochamps de contraintes du Bassin de Vienne: Geodynamica Acta, v. 4, p. 147–158.

Fodor, L., A. Magyari, M. Kazmer, and A. Fogarasi, 1992, Gravity flow dominated sedimentation on the Buda paleoslope (Hungary), record of the Late Eocene continental escape of the Bakony unit: Geologische Rundschau, v. 81, p. 695–716.

Fodor, L., A. Magyari, A. Fogarasi, and K. Palotas, 1994, Tertiary tectonics and late Paleogene sedimentation in the Buda Hills, Hungary: A new interpretation of the Buda line: Foldtani Kozlony, v. 124, p. 129–305.

Fodor, L., J. Francu, O. Krejci, and Z. Stranik, 1995, Paleogeographic and tectonic evolution of the Carpathian Flysch belt of the southern Moravia (Czech Republic), *in* Proceedings of the XV Carpatho-Balkan Geologic Congress: Geological Society of Greece Special Publication 4, p. 31–33.

Fodor, L., B. Jelen, E. Marton, D. Skaberne, J. Car, and M. Vrabec, 1998, Miocene–Pliocene tectonic evolution of the Slovenian Periadriatic fault, implications for Alpine–Carpathian extrusion models: Tectonics, v. 17, p. 690–709.

Fodor, L., L. Csontos, G. Bada, I. Gyorfi, and L. Benkovics, 1999, Tertiary tectonic evolution of the Pannonian Basin system and neighbouring orogens, a new synthesis of paleostress data, *in* B. Durand, L. Jolivet, F. Horvath, and M. Seranne, eds., The Mediterranean basins, Tertiary extension within the Alpine orogen: Geological Society (London) Special Publication 156, p. 295–334.

Genser, H., and F. Neubauer, 1989, Low angle normal faults at the eastern margin of the Tauern window (Eastern Alps): Mitteilungen der Osterreichischen Geologischen Gesellschaft, v. 81, p. 233–243.

Girbacea, R., 1997, The Pliocene to recent tectonic evolution of the Eastern Carpathians (Romania): Ph.D. thesis, University of Tubingen, Tubingen, 136 p.

Gyorfi, I., L. Csontos, and A. Nagymarosy, 1999, Early Tertiary structural evolution of the border zone between the Pannonian and Transylvanian Basins, *in* B. Durand, L. Jolivet, F. Horvath, and M. Seranne, eds., The Mediterranean basins, Tertiary extension within the Alpine orogen: Geological Society (London) Special Publication 156, p. 251–267.

Hamor, G., G. Pogacsas, and A. Jambor, 2001, Paleogeographic/structural evolutionary stages and the related volcanism of the Carpathian–Pannonian region: Acta Geologica Hungarica, v. 44, p. 193–222.

Hippolyte, J. C., and M. Sandulescu, 1996, Paleostress characterization of the "Wallachian phase" in its type area (southeastern Carpathians, Romania): Tectonophysics, v. 263, p. 235–248.

Hoernle, K., Y. Shang, and D. Graham, 1995, Seismic and geochemical evidence for large-scale mantle upwelling beneath the eastern Atlantic and Western and Central Europe: Nature, v. 374, p. 30–34.

Horvath, F., 1993, Towards a mechanical model for the formation of the Pannonian Basin: Tectonophysics, v. 226, p. 333–357.

Horvath, F., and S. A. P. L. Cloetingh, 1996, Stress-induced late-stage subsidence anomalies in the Pannonian Basin: Tectonophysics, v. 266, p. 287–300.

Horvath, F., and G. Tari, 1999, IBS Pannonian Basin project, a review of the main results and their bearings on hydrocarbon exploration, *in* B. Durand, L. Jolivet, F. Horvath, and M. Seranne, eds., The Mediterranean basins, Tertiary extension within the Alpine orogen: Geological Society (London) Special Publication 156, p. 195–213.

Horvath, F., A. Adam, and W. D. Stanley, 1987, New geophysical data, evidence for the allochthony of the Transdanubian Central Range: Rendiconti della Societa Geologica Italiana, v. 9, p. 123–130.

Huismans, R. S., G. Bertotti, D. Ciulavu, C. A. E. Sanders, S. Cloetingh, and C. Dinu, 1997, Structural evolution of the Transylvanian Basin (Romania), a sedimentary basin in the bend zone of the Carpathians: Tectonophysics, v. 272, p. 249–268.

Ibrmajer, J., and M. Suk, 1989, Geophysical image of CSSR (in Czech): Prague, Czechoslovak Academy of Sciences, p. 354.

Jiricek, R., 1979, Tectonic development of the Carpathian arc during the Oligocene and Neogene (in Czech), *in* M. Mahel, ed., Tectonic profiles through the West Carpathians (in Czech): Bratislava, Dionyz Stur Geological Institute, p. 205–215.

Kocak, A., S. Mayer, and M. Psenickova, 1981, Structural scheme of the Vienna basin on the pre-Neogene base (in Czech): Prague, Geofond, 36 p.

Korab, T., M. Krs, M. Krsova, and P. Pagac, 1981, Palaeomagnetic investigations of Albian(?)–Paleocene to lower Oligocene sediments from the Dukla unit, East Slovakian Flysch, Czechoslovakia: Zapadne Karpaty, Seria Geologia, v. 7, p. 127–149.

Kovac, P., and J. Hok, 1993, The Central Slovak fault system— Field evidence of a strike-slip: Geologica Carpathica, v. 44, p. 155–159.

Kovac, M., I. Barath, I. Holicky, F. Marko, and I. Tunyi, 1989, Basin opening in the lower Miocene strike-slip zone in the southwest part of the Western Carpathians: Geologica Carpathica, v. 40, p. 37–62.

Kovac, M., F. Marko, and M. Nemcok, 1990, Neogene history of intramontane basins in the western part of the Carpathians: Rivista Italiana di Paleontologia e Stratigrafia, v. 96, p. 381–404.

Kovac, M., I. Barath, K. Sutovska, and P. Uher, 1991, Lower Miocene events in the sedimentary record of the Dobra Voda depression: Mineralia Slovaca, v. 23, p. 201–213.

Kovac, M., F. Marko, and M. Nemcok, 1993, Neogene structural evolution and basin opening in the Western Carpathians: Geophysical Transactions, v. 37, p. 297–309.

Kovac, M., J. Kral, E. Marton, D. Plasienka, and P. Uher, 1994, Alpine uplift history of the Central Western Carpathians, geochronological, paleomagnetic, sedimentary and structural data: Geologica Carpathica, v. 45, p. 83–96.

Kovac, M., P. Kovac, F. Marko, S. Karoli, and J. Janocko, 1995, The East Slovakian basin— A complex back-arc basin: Tectonophysics, v. 252, p. 453–466.

Kroll, A., H. W. Flugel, W. Seiberl, F. Weber, G. Walach, and D. Zych, 1988, Erlauterungen zu den Karten uber den pratertiaren Untergrund des steirischen Beckens und der sudburgenlandischen Schwelle: Geologische Karte der Republik Österreich, Geologische Bundesanstalt, Vienna, scale 1:200,000.

Krs, M., P. Muska, O. Orlicky, and P. Pagac, 1986, Paleomagnetism and paleomagnetic interpretations in the post-Carbon formations of the Czechoslovak West Carpathians (in Czech), *in* M. Blizkovsky et al., eds., Geophysical model of lithosphere: Project Oborovy Foud Technickeho Rozvoje, period 1982–1985, final report: Brno and Prague, Geofyzika Brno and Czechoslovak Academy of Sciences Prague, p. 397–404.

Krs, M., M. Krsova, R. Chvojka, and M. Potfaj, 1991, Palaeomagnetic investigations of the flysch belt in the Orava region, Magura unit, Czechoslovak Western Carpathians: Geologicke Prace, Spravy, v. 92, p. 125–151.

Krs, M., M. Krsova, P. Pruner, R. Chvojka, and M. Potfaj, 1993, Palaeomagnetic investigations in the Biele Karpaty mountains, flysch belt of the West Carpathians: Geologica Carpathica, v. 45, p. 35–43.

Krus, S., and A. Sutora, 1986, Geophysical-geological atlas of the Alpine–Carpathian mountain system: manuscript, archive Brno, Geofyzika Brno, scale 1:3,000,000, 12 sheets, 25 p.

Kurz, W., F. Neubauer, H. Genser, and H. Horner, 1993, Sequence of Tertiary brittle deformations in the eastern Tauern window (Eastern Alps): Mitteilungen der Osterreichischen Geologischen Gesellschaft, v. 86, p. 153–164.

Lankreijer, A., M. Kovac, S. Cloetingh, P. Pitonak, M. Hloska, and C. Biermann, 1995, Quantitative subsidence analysis and forward modelling of the Vienna and Danube basins, thin-skinned vs. thick-skinned extension: Tectonophysics, v. 252, p. 433–451.

Laubscher, H. P., 1971, Das Alpen-Dinariden-Problem und die Palinspastik der südlischen Tethys: Geologische Rundschau, v. 60, p. 813–833.

Le Pichon, X., F. Bergerat, and M. J. Roulet, 1988, Plate kinematics and tectonics leading to the Alpine belt formation, a new analysis, *in* S. Clark, Jr., B. C. Burchfiel, and J. Suppe, eds., Processes in continental lithospheric deformation: Geological Society of America Special Paper 218, p. 111–131.

Lillie, R. J., M. Bielik, V. Babuska, and J. Plomerova, 1994, Gravity modelling of the lithosphere in the Eastern–Western Carpathian–Pannonian Basin region: Tectonophysics, v. 231, p. 215–235.

Linzer, H. G., 1996, Kinematics of retreating subduction along the Carpathian arc, Romania: Geology, v. 24, p. 167–170.

Linzer, H. G., L. Ratschbacher, and W. Frisch, 1995, Transpressional collision structures in the upper crust, the fold-thrust belt of the Northern Calcareous Alps: Tectonophysics, v. 242, p. 41–61.

Linzer, H. G., F. Moser, F. Nemes, L. Ratschbacher, and B. Sperner, 1997, Build-up and dismembering of a classical fold-thrust belt, from non-cylindrical stacking to lateral extrusion in the Eastern Alps: Tectonophysics, v. 272, p. 97–124.

Linzer, H. G., W. Frisch, P. Zweigel, R. Girbacea, H. P. Hann, and F. Moser, 1998, Kinematic evolution of the Romanian Carpathians: Tectonophysics, v. 297, p. 133–156.

Lizon, I., 1986, Geothermic model and geodynamics of the West Carpathians (in Slovak), *in* M. Blizkovsky et al., eds., Geophysical model of lithosphere, Project Oborovy Foud Technickeho Rozvoje, period 1982–1985, Final report: Brno and Prague, Geofyzika Brno and Czechoslovak Academy of Sciences Prague, p. 292–298.

Mahel, M., 1974, Tectonics of the Carpathian–Balkan regions, explanations to the tectonic map of the Carpathian–Balkan regions and their forelands: Bratislava, Dionyz Stur Geological Institute, p. 453.

Marko, F., M. Kovac, L. Fodor, and K. Sutovska, 1990, Deformations and kinematics of a Miocene shear zone in the northern part of the Little Carpathians (Buková furrow, Hrabník Formation) (in Slovak): Mineralia Slovaca, v. 22, p. 399–410.

Marko, F., L. Fodor, and M. Kovac, 1991, Miocene strike-slip faulting and block rotation in Brezovske Karpaty Mountains: Mineralia Slovaca, v. 23, p. 201–213.

Marton, E., 1990, Paleomagnetic studies on the Miocene volcanic horizons at the southern margin of the Bukk Mountains: Annual Report of Eotvos Lorand Geophysical Institute of Hungary, v. 1988–1989, p. 307–311.

Marton, E., 1993, Paleomagnetism in the Mediterranean from Spain to the Aegean, a review of data relevant to Cenozoic movements, *in* E. Mantovani, A. Morelli, and E. Boschi, eds., Recent evolution of seismicity of the Mediterranean region, North Atlantic Treaty Organization Advanced Study Institute Series C: Dordrecht, The Netherlands, Kluwer Academic Publications, p. 367–402.

Marton, E., and L. Fodor, 1995, Combination of paleomagnetic and stress data— A case study from North Hungary: Tectonophysics, v. 242, p. 99–114.

Marton, E., and P. Marton, 1989, A compilation of paleomagnetic results from Hungary: Geophysical Transactions, v. 35, p. 117–133.

Matenco, L., G. Bertotti, C. Dinu, and S. Cloetingh, 1997, Tertiary tectonic evolution of the external South Carpathians and the adjacent Moesian Platform (Romania): Tectonics, v. 16, p. 896–911.

Mattick, R. E., P. G. Teleki, R. L. Phillips, J. L. Clayton, G. David, G. Pogacsas, B. Bardocz, and E. Simon, 1996, Structure, stratigraphy, and petroleum geology of the Little Plain basin, northwest Hungary: AAPG Bulletin, v. 80, p. 1780–1800.

Meulenkamp, J. E., M. Kovac, and I. Cicha, 1996, On late Oligocene to Pliocene depocenter migrations and the evolution of the Carpathian–Pannonian system: Tectonophysics, v. 266, p. 301–317.

Morley, C. K., 1996, Models for relative motion of crustal blocks within Carpathian region, based on restorations of the Outer Carpathian thrust front: Tectonics, v. 15, p. 885–904.

Moser, F., and W. Frisch, 1996, Tertiary deformation in the Southern Carpathians— Structural analysis of a brittle deformation, TSK 6 Symposium Erweitere Kurzfassungen: Vienna, Facultas–Universitatsverlag, p. 280–282.

Nemcok, J., and M. Nemcok, 1990, Significance of movement vectors in the Vihorlat–Cirocha fault system (in Slovak), *in* M. Sykora, J. Jablonsky, and O. Samuel, eds., Sedimentologicke problemy Zapadnych Karpat: Bratislava, Dionyz Stur Geological Institute, p. 89–106.

Nemcok, M., 1991, Structural deformation analyses in both Flysch and Pieniny Klippen units along the Vlara River Valley (in Slovak): Geologicke Prace, Spravy, v. 93, p. 55–61.

Nemcok, M., 1993, Transition from convergence to escape, field evidence from the West Carpathians: Tectonophysics, v. 217, p. 117–142.

Nemcok, M., 2001, Timing and mechanisms of Tertiary tectonic events of the ALCAPA megablock (abs.): Pancardi 2001 meeting, Sopron, II Abstracts, p. PO-9.

Nemcok, M., and J. Lexa, 1990, Evolution of the basin and range structure around the Ziar mountain range: Geologica Carpathica, v. 41, p. 229–258.

Nemcok, M., and J. Nemcok, 1994, Late Cretaceous deformation of the Pieniny Klippen Belt, West Carpathians: Tectonophysics, v. 239, p. 81–109.

Nemcok, M., F. Marko, M. Kovac, and L. Fodor, 1989, Neogene tectonics and paleostress changes in the Czechoslovakian part of the Vienna basin: Jahrbuch der Geologischen Bundesanstalt, v. 132, p. 443–458.

Nemcok, M., L. Pospisil, J. Lexa, and R. A. Donelick, 1998a, Tertiary subduction and slab break-off model of the Carpathian–Pannonian region: Tectonophysics, v. 295, p. 307–340.

Nemcok, M., J. Hok, P. Kovac, F. Marko, M. P. Coward, J. Madaras, J. J. Houghton, and V. Bezak, 1998b, Tertiary extension development and extension/compression interplay in the West Carpathian mountain belt: Tectonophysics, v. 290, p. 137–167.

Nemcok, M., J. J. Houghton, and M. P. Coward, 1998c, Strain partitioning along the western margin of the Carpathians: Tectonophysics, v. 292, p. 119–143.

Nemcok, M., J. Nemcok, M. Wojtaszek, L. Ludhova, R. A. Klecker, W. J. Sercombe, M. P. Coward, and F. J. Keith Jr., 2000a, Results of 2D balancing along 20° and 21°30′ longitude and pseudo-3D in the Smilno tectonic window, implications for shortening mechanisms of the West Carpathian accretionary wedge: Geologica Carpathica, v. 51, p. 281–300.

Nemcok, M., P. Konecny, and O. Lexa, 2000b, Calculations of tectonic, magmatic and residual stress in the Stiavnica Stratovolcano, Western Carpathians, implications for mineral precipitation paths: Geologica Carpathica, v. 51, p. 19–36.

Nemes, F., 1996, Kinematics of the Periadriatic fault in the Eastern Alps— Evidences from structural analysis, fission track dating and basin modelling: Ph.D. thesis, University of Salzburg, Salzburg, 225 p.

Nemes, F., W. Pavlik, and M. Moser, 1995, Geologie und Tektonik im Salzachtal (Steiermark): Kinematik und Palaospanungen entlang des Enstal-Mariazell Blattverschiebungsystems in den Nordlichen Kalkalpen: Jahrbuch der Geologischen Bundensanstalt, v. 138, p. 349–367.

Nemes, F., F. Neubauer, S. A. P. L. Cloetingh, and J. Genser, 1997, The Klagenfurt basin in the Eastern Alps; an intra-orogenic decoupled flexural basin?: Tectonophysics, v. 282, p. 189–203.

Neubauer, F., and J. Genser, 1990, Architektur und Kinematik der ostlichen Zentralalpen, eine Ubersicht: Mitteilungen des Naturwissenschaftlichen Vereines fur Steiermark, v. 120, p. 203–219.

Neubauer, F., H. Fritz, J. Genser, W. Kurz, F. Nemes, E. Wallbrecher, X. Wang, and E. Willingshofer, 1999, Structural evolution within an extruding block, model and application to the Alpine–Pannonian system, *in* F. K. Lehner and J. L. Urai, eds., Aspects of tectonic faulting: Berlin, Springer, p. 141–153.

Nievoll, J., 1985, Die bruchhafte Tektonik entlang der Trafaiachlinie (Ostliche Zentral alpen, Österreich): Jahrbuch der Geologischen Bundensanstalt, v. 127, p. 643–671.

Ortner, H., and R. F. Sachsenhofer, 1996, Evolution of the lower Inn Valley Tertiary and constraints on the development of the source area, *in* W. Liebl and G. Wessely, eds., Oil and gas in Alpidic thrust belts and basins of Central and Eastern Europe: European Association of Geoscientists and Engineers Special Publication 5, p. 237–247.

Patrascu, S., M. Seclaman, and C. Panaiotu, 1993, Tectonic implications of paleomagnetism in Upper Cretaceous deposits in the Hateg and Rusca Montana basins (South Carpathians, Romania): Cretaceous Research, v. 14, p. 255–264.

Patrascu, S., C. Panaiotu, M. Seclaman, and C. E. Panaiotu,

1994, Timing of rotational motion of Apuseni Mountains (Romania), paleomagnetic data from Tertiary magmatic rocks: Tectonophysics, v. 233, p. 163–176.

Pecskay, Z., et al., 1995, Space and time distribution of Neogene–Quaternary volcanism in the Carpatho-Pannonian region, *in* H. Downes and O. Vaselli, eds., Neogene and related magmatism in the Carpatho-Pannonian region: Acta Vulcanologica, p. 15–28.

Peresson, H., and K. Decker, 1997a, The Tertiary dynamics of the northern Eastern Alps (Austria), changing paleostresses in a collisional plate boundary: Tectonophysics, v. 272, p. 125–157.

Peresson, H., and K. Decker, 1997b, Far-field effects of late Miocene subduction in the Eastern Carpathians, east–west compression and inversion of structures in the Alpine–Carpathian–Pannonian region: Tectonics, v. 16, p. 38–56.

Polinski, R. K., and G. H. Eisbacher, 1992, Deformation partitioning during polyphase oblique convergence in the Karawanken Mountains, southeastern Alps: Journal of Structural Geology, v. 14, p. 1203–1213.

Ratschbacher, L., W. Frisch, H. G. Linzer, and O. Merle, 1991a, Lateral extrusion in the Eastern Alps: Part 2. Structural analysis: Tectonics, v. 10, p. 257–271.

Ratschbacher, L., O. Merle, P. Davy, and P. Cobbold, 1991b, Lateral extrusion in the Eastern Alps: Part 1. Boundary conditions and experiments scaled for gravity: Tectonics, v. 10, p. 245–256.

Ratschbacher, L., W. Frisch, H. G. Linzer, B. Sperner, M. Meschede, K. Decker, M. Nemcok, J. Nemcok, and R. Grygar, 1993a, The Pieniny Klippen Belt in the Western Carpathians of northeastern Slovakia, structural evidence for transpression: Tectonophysics, v. 226, p. 471–483.

Ratschbacher, L., H. G. Linzer, F. Moser, R. O. Strusiewicz, H. Bedelean, N. Har, and P. A. Mogos, 1993b, Cretaceous to Miocene thrusting and wrenching along the central south Carpathians due to corner effect during collision and orocline formation: Tectonics, v. 12, p. 855–873.

Royden, L. H., 1985, The Vienna basin, a thin-skinned pull-apart basin, *in* K. T. Biddle and N. Christie-Blick, eds., Strike-slip deformation, basin formation, and sedimentation: SEPM Special Publication 37, p. 319–338.

Royden, L. H., 1988, Late Cenozoic tectonics of the Pannonian Basin system, *in* L. H. Royden and F. Horvath, eds., The Pannonian Basin: A study in basin evolution: AAPG Memoir 45, p. 27–48.

Royden, L. H., 1993, The tectonic expression of the slab pull at continental convergent boundaries: Tectonics, v. 12, p. 303–325.

Royden, L. H., F. Horvath, and B. C. Burchfiel, 1982, Transform faulting, extension and subduction in the Carpathian–Pannonian region: Geological Society of America Bulletin, v. 73, p. 717–725.

Spakman, W., 1990, Tomographic images of the upper mantle below Central Europe and the Mediterranean: Terra Nova, v. 2, p. 542–553.

Sperner, B., 1996, Computer programs for the kinematic analysis of brittle deformation structures and the Tertiary tectonic evolution of the Western Carpathians (Slovakia): Ph.D. thesis, University of Tuebingen, Tuebingen, 120 p.

Sperner, B., L. Ratschbacher, and R. Ott, 1993, Fault-strike analysis, a Turbo Pascal program package for graphical presentation and reduced stress tensor calculation: Computers and Geosciences, v. 19, p. 1361–1388.

Sperner, B., L. Ratschbacher, and M. Nemcok, 2002, Interplay between subduction rollback and lateral extrusion, tectonics of the Western Carpathians: Tectonics, v. 21, p. 1-4–1-23.

Steininger, F., C. Muller, and F. Rogl, 1988, Correlation of Central Paratethys, Eastern Paratethys, and Mediterranean Neogene stages, *in* L. H. Royden and F. Horvath, eds., The Pannonian Basin: A study in basin evolution: AAPG Memoir 45, p. 79–87.

Subotin, S. I., 1975, The Earth's physics problems in the Ukraine (in Russian): Kiev, Naukova Dumka, 173 p.

Szafian, P., G. Tari, F. Horvath, and S. Cloetingh, 1999, Crustal structure of the Alpine–Pannonian transition zone, a combined seismic and gravity study: International Journal of Earth Sciences, v. 88, p. 98–110.

Tari, G., 1991, Multiple Miocene block rotation in the Bakony Mountains, Transdanubian Central Range, Hungary: Tectonophysics, v. 199, p. 93–108.

Tari, G., and F. Horvath, 1995, Overview of the Alpine evolution of the Pannonian Basin, *in* F. Horvath, G. Tari, and C. Bokor, eds., Hungary, extensional collapse of the Alpine orogene and hydrocarbon prospects in the basement and basin fill of the western Pannonian Basin: AAPG International Conference and Exhibition, Nice, Guidebook to fieldtrip no. 6, Hungary, p. 7–19.

Tari, G., T. Baldi, and M. Baldi-Beke, 1993, Paleogene retroarc flexural basin beneath the Neogene Pannonian Basin, a geodynamic model: Tectonophysics, v. 226, p. 433–455.

Tari, G., F. Horvath, and G. Weir, 1995, Palinspastic reconstruction of the Alpine/Carpathian/Pannonian system, *in* F. Horvath, G. Tari, and C. Bokor, eds., Hungary, extensional collapse of the Alpine orogene and hydrocarbon prospects in the basement and basin fill of the western Pannonian Basin: AAPG International Conference and Exhibition, Nice, Guidebook to fieldtrip no. 6, Hungary, p. 119–131.

Tari, G., P. Dovenyi, I. Dunkl, F. Horvath, L. Lenkey, M. Stefanescu, P. Szafian, and T. Toth, 1999, Lithospheric structure of the Pannonian Basin derived from seismic, gravity and geothermal data, *in* B. Durand, L. Jolivet, F. Horvath, and M. Seranne, eds., The Mediterranean basins, Tertiary extension within the Alpine orogen: Geological Society (London) Special Publication 156, p. 215–250.

Tollmann, A., 1985, Geologie von Österreich-Ausserzentralalpiner Anteil: Vienna, Deuticke, 710 p.

Tunyi, I., and M. Kovac, 1991, Paleomagnetic investigation of the Neogene sediments from the Little Carpathians: Contributions of the Geophysical Institute of the Slovak Academy of Sciences, v. 21, p. 125–146.

Vass, D., A. Nagy, M. Kohut, and I. Kraus, 1988, The Devinska

Nova Ves Member, olistostromes on the southeastern border of the Vienna basin: Mineralia Slovaca, v. 20, p. 109–122.
Vass, D., J. Hok, P. Kovac, and M. Elecko, 1993, The Paleogene and Neogene tectonic events of the Southern Slovakia depressions in the light of the stress-field analyses (in Slovak): Mineralia Slovaca, v. 25, p. 79–92.
von Blanckenburg, F., and J. H. Davies, 1995, Slab breakoff, a model for syncollisional magmatism and tectonics in the Alps: Tectonics, v. 14, p. 120–131.
Wang, X., and F. Neubauer, 1998, Orogen-parallel strike-slip faults bordering metamorphic core complexes, the Salzach-Enns fault zone in the Eastern Alps, Austria: Journal of Structural Geology, v. 20, p. 799–818.
World Stress Map Project, 1997, http://www-wsm.physik.uni-karlsruhe.de (accessed August 18, 2005).
Zelenka, T., C. Baksa, Z. Balla, J. Foldessy, and K. Jaranyi, 1983, The role of Darno line in the basement structure of northeastern Hungary: Geologica Carpathica, v. 34, p. 53–69.
Zweigel, P., 1997, The Tertiary tectonic evolution of the Eastern Carpathians (Romania), orogenic arc formation in response to microplate movements: Ph.D. thesis, University of Tubingen, Tubingen, 158 p.
Zweigel, P., L. Ratschbacher, and W. Frisch, 1998, Kinematics of an arcuate fold-thrust belt; the southern Eastern Carpathians (Romania): Tectonophysics, v. 297, p. 177–208.

Nemčok, M., L. Pospíšil, I. Hrušecký, and T. Zsíros, 2006,
Subduction in the remnant Carpathian Flysch Basin,
in J. Golonka and F. J. Picha, eds., The Carpathians and
their foreland: Geology and hydrocarbon resources:
AAPG Memoir 84, p. 541–559.

28

Subduction in the Remnant Carpathian Flysch Basin

M. Nemčok
Energy and Geoscience Institute, University of Utah, Salt Lake City, Utah, U.S.A.

L. Pospíšil
Geoinform Consultants, Brno, Czech Republic

I. Hrušecký
Geological Survey of Slovak Republic, Bratislava, Slovakia

T. Zsíros
Seismological Observatory, Geodetic and Geophysical Research Institute of the Hungarian Academy of Science, Budapest, Hungary

ABSTRACT

This chapter serves as an overview of facts on the Carpathian–Pannonian Tertiary development. It discusses about 35 geophysical-geological data layers of the database presented in the entire chapter, correlates them briefly, and notes the main results of such a multidisciplinary synthesis. This synthesis indicates that the remnant Carpathian Flysch Basin (rCFB) was affected by subduction since the early Miocene. Areas to the west and east of the rCFB were undergoing continent-continent collisions at the same time. From the late Oligocene to middle Miocene, crustal wedges moving laterally from these collisions became parts of the Carpathian–Pannonian slab overriding the rCFB. This basin underwent passive subduction characterized by passive sinking of the oceanic slab under its body forces. Subduction setting was characterized by a steep dip of the subducting slab, strong hinterland extension, extensive areal magmatism in the hinterland, and asthenospheric upwelling in the hinterland. The oceanic plate underlying the rCFB subducted southwestward and westward. The subduction rollback of the subducting slab under the advancing Carpathian orogen, the oblique closure of the rCFB, and the progressive change of subduction to collision from west to east along the Carpathian arc drove a scissorlike break-off of the subducting oceanic slab along the Carpathians. The break-off was driven by the weight of the subducting oceanic slab and the buoyancy of the attached continental slab. The break-off ran eastward along the whole Carpathian arc, from the early Miocene to its present position in the bend between the Eastern and Southern Carpathians.

DOI:10.1306/985628M843083

INTRODUCTION

The Tertiary to modern history of the Carpathian–Pannonian region (Figure 1) offers an outstanding opportunity to study the interaction of asthenospheric and lithospheric processes and their mutual dependencies during volcanic arc, fore-arc, and back-arc development. The Carpathian evolution was controlled by the closure of Alpine–Carpathian–Balkan–Dinaridic oceanic basins (e.g., Le Pichon et al., 1988), subduction rollback in the area of the remnant Carpathian Flysch Basin (rCFB) (e.g., Royden and Baldi, 1988), asthenospheric upwelling in the Pannonian region (Lexa and Konecny, 1974; Stegena et al., 1975; Royden et al., 1982), and lateral extrusions of the Eastern Alps and the Balkans on the western and eastern sides of rCFB (Ratschbacher et al., 1991a, b, 1993; Morley, 1996).

The Carpathian arc is particularly useful for a present-day snapshot of the freshly closed oceanic basin between the Carpathian–Pannonian and European continental plates. It is the present-day data, mainly the seismicity in the Vrancea zone, the area where the closure was youngest, and modern vertical movement data that provide key constraints on the mentioned driving mechanisms during the Carpathian development.

The main goal of this chapter is to discuss about 35 geophysical-geological information layers on the present-day Carpathian–Pannonian plate setting, subject them to quick multidisciplinary analysis, and synthesize the gained knowledge of the Carpathian–Pannonian lithosphere in discussional form using the overview of current literature. The overview draws from the earlier version of this geological-geophysical database published by Nemcok et al. (1998a).

EARLIER STUDIES OF THE CARPATHIAN–PANNONIAN TECTONICS

Although palinspastic reconstructions of the whole Tethyan belt resulted in a relatively accepted picture for the first-order space-time arrangement of subduction

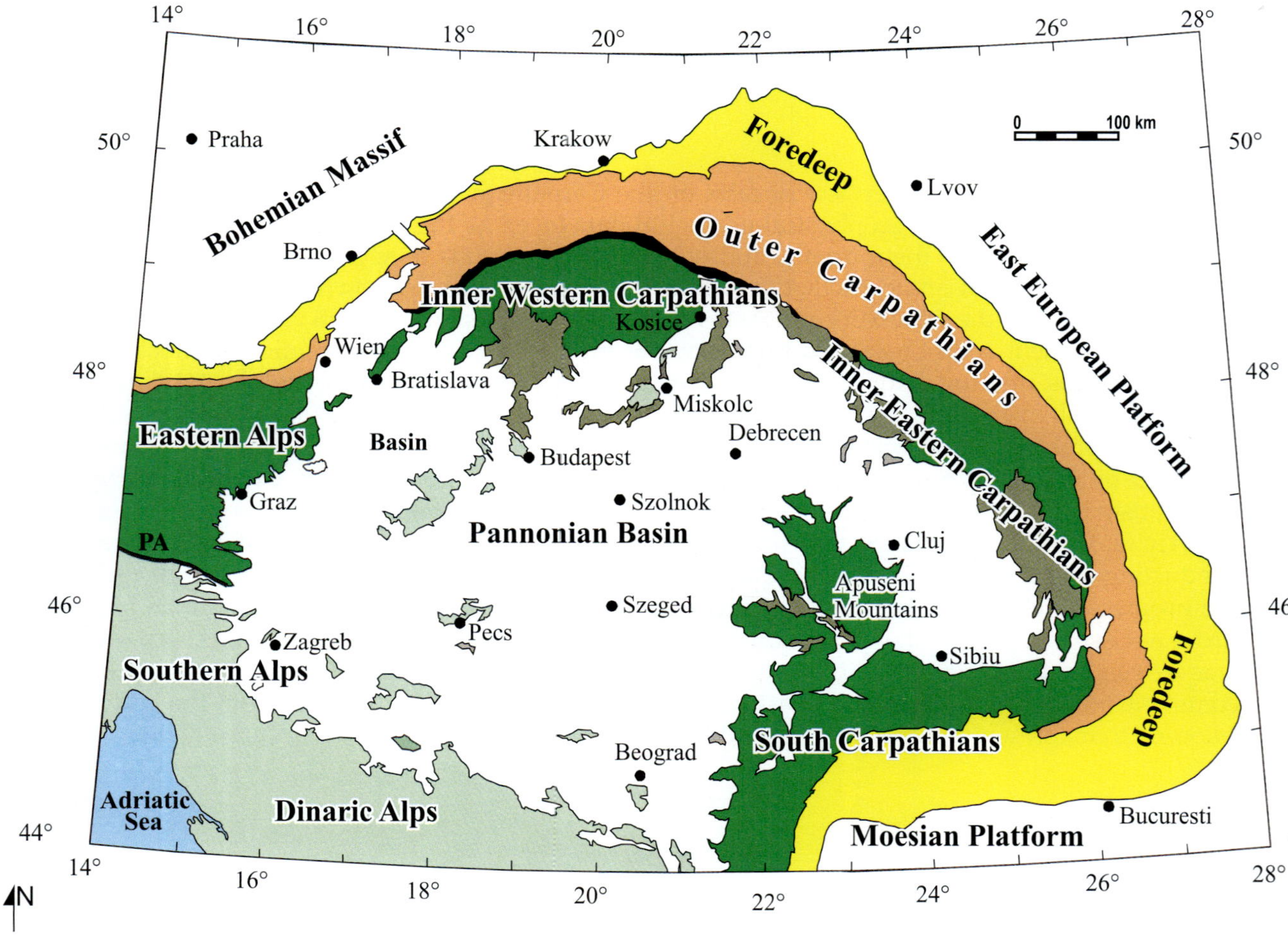

Figure 1. The space of the Carpathian area with surrounding tectonic units.

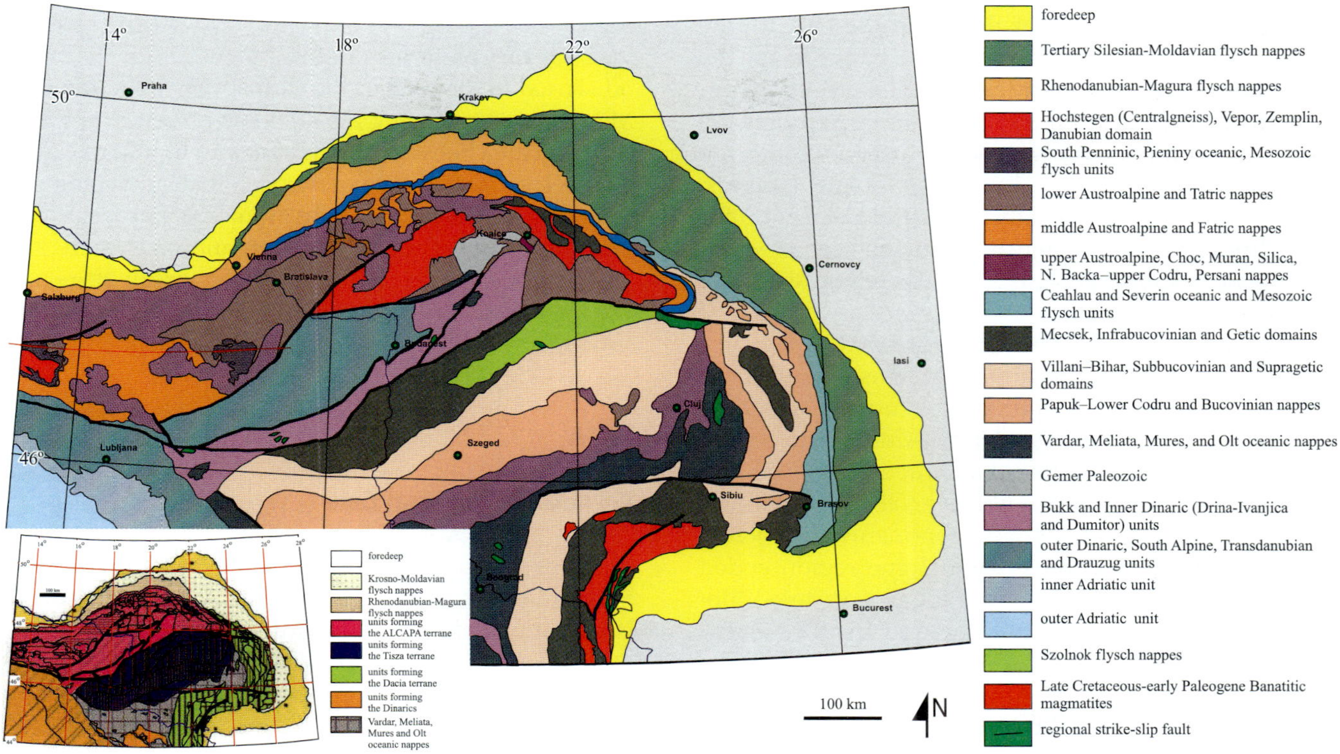

Figure 2. Block pattern map of the Carpathian–Pannonian region based on the correlation of Paleozoic–Mesozoic facies (modified after Csontos et al., 1992).

and collision zones (e.g., Le Pichon et al., 1988), fundamental problems still remain for paleotectonic reconstructions in the Carpathian–Pannonian segment of the Tethys (Figure 1), where the convergent boundaries involved interactions with microplates and irregular margins.

The Carpathian orogen developed during the Tertiary by a northeastward and eastward migration of an accretionary wedge in front of the ALCAPA (Eastern Alps–Carpathians–Pannonian Basin) and Tisza–Dacia microplates (Figure 2). The rCFB, which was floored by oceanic and thinned continental crust and located between the orogen and European platform, was subducted southwestward and westward (e.g., Royden and Baldi, 1988). The late Oligocene–lower Miocene period of the Carpathian evolution was coeval with the eastward material movement from the Eastern Alps toward the Carpathian–Pannonian realm (e.g., Royden et al., 1982; Le Pichon et al., 1988; Ratschbacher et al., 1991a, b). Shortening in the frontal parts of the Carpathian orogen predated and ocurred synchronously with the extension in the oroenic hinterland accompanied by asthenosphere upwelling (e.g., Lexa and Konecny, 1974; Stegena et al., 1975; Royden et al., 1982, 1983a, b; Horvath, 1986; Bergerat, 1989). Subduction, hinterland extension, and asthenospheric upwelling resulted in volcanic activity, particularly pronounced in the Neogene (e.g., Embey-Isztin et al., 1993; Lexa et al., 1993; Pecskay et al., 1995). Subduction was characterized by a northeastward and eastward rollback. It finished by the oceanic slab break-off in its final stages at different times along the Carpathian arc (Decker and Peresson, 1996; Nemcok et al., 1998a).

As mentioned earlier, the attenuated orogenic hinterland is formed by two microplates: the ALCAPA and the Tisza–Dacia (Csontos et al., 1992; Figure 2), defined by characteristic Paleozoic, Mesozoic, and Paleogene facies underneath younger sediments. The detailed discussion of the original facial reconstruction can be found in Csontos et al. (1992) and Haas et al. (1995). The ALCAPA microplate (Balla, 1984) is bounded by the Pieniny Klippen Belt in the north and the Central Hungarian fault zone in the south. The Tisza–Dacia microplate is bounded in the north by the Central Hungarian fault zone and in the south by the Sava–Dealu Mare fault zone (Csontos et al., 1992). These two microplates roughly correspond to two terranes defined by two distinct paleomagnetic declination patterns (e.g., Marton and Marton, 1978; Marton and Marton, 1989).

Mesozoic tectonostratigraphic units of both microplates evolved during the Triassic–Jurassic rifting and

Cretaceous convergence in the Dinaridic and Alpine belts. Each microplate has a facies change from the European to the Mediterranean facies affinity, from north to south in the ALCAPA and the Tisza part of the Tisza–Dacia microplate, and from east to west in the Dacia part of the Tisza–Dacia microplate (Kovacs, 1987). The juxtaposition of the Tisza and Dacia terranes probably occurred during the Late Cretaceous–early Paleogene (Csontos et al., 1992). This chapter will treat them as one microplate.

The main tectonic events that shaped the Carpathian–Pannonian region can be summarized as follows. During the Late Cretaceous–Eocene, the future Carpathian orogen was part of the larger Alpine–Carpathian orogen formed by the southeastward subduction (135–55 Ma) of the Penninic ocean and the ultimate collision (55–40 Ma) of the European and Adriatic continents (Le Pichon et al., 1988). Subduction in the Alpine region ended by the slab break-off of the subducting oceanic lithosphere from the continental one and by the related magmatism (43–24 Ma) (von Blanckenburg and Davies, 1995). While the Eocene Alps was under continental collision, the Eocene rCFB was still subducting.

The Carpathians and the eastern parts of the Alps started to occupy the region of the present Carpathians during the early Miocene (e.g., Marton, 1988, 1993a, b; Tunyi and Kovac, 1991). The oceanic plate underlying the rCFB subducted southwestward and later westward (Royden et al., 1982), whereas the northeastward and eastward migration velocity of the accreting plate boundary decreased through time (Nemcok et al., 1998a, 1999). The slower subduction rate caused the following: the steeper dip of the subducting slab; the Eggenburgian beginning (22–19 Ma) and following duration of extension in the orogenic hinterland accompanied by the asthenosphere upwelling; and the onset of the crustally derived volcanism above the risen asthenosphere and the younger, mantle-derived volcanism, progressively less contaminated by the crustal material, all contemporaneous with a shortening in the Carpathian accretionary wedge (Lexa and Konecny, 1974; Stegena et al., 1975; Horvath, 1986; Poka, 1988; Salters et al., 1988; Lexa et al., 1995; Nemcok et al., 1998a, b).

During the end of the early Miocene, subduction of the rCFB finally led to a collision with the continental margin along the westernmost part of the present Carpathian arc (e.g., Ksiazkiewicz, 1960; Buday, 1965; Jiricek, 1979; Benada, 1986). The subduction rollback of the subducting slab of the rCFB under the advancing Carpathians, the oblique closure of the basin, and the progressive change of the subduction to collision from west to east along the Carpathian arc drove a scissor-like break-off of the subducting oceanic slab along the Carpathians (Nemcok et al., 1998a). The break-off was driven by the weight of the subducting oceanic slab and the buoyancy of the attached continental slab. The break-off started in the west of the Carpathian arc in the early Miocene and ran along the arc to its present position in the bend area between the Eastern and Southern Carpathians. It separated the oceanic slab that subducted passively under its own weight from the European continental margin. The lateral propagation rate of the slab break-off decreased with time. The break-off-related volcanism was synchronous with the final stages of the subduction and the following short collision.

DATA

Regional Geophysical Data

The present-day geophysical data on the Carpathian–Pannonian lithosphere have the following characteristic features shown in Figure 3. Figure 3 contains a lithospheric cross section, which runs from the Bohemian Massif to the East European platform, crossing the Pannonian Basin system, the Apuseni Mountains, and the Transylvanian Basin. This cross section indicates a thermal erosion of platform margins, which are in contact with the orogen formed by the Carpathians in front and the Pannonian Basin system in the hinterland. The thermal erosion is documented by the heat-flow profile (Figure 3). It is also indicated by increased depths of the lithosphere–asthenosphere boundary interpreted at different depth levels from seismological and magnetotelluric data (Pajdusak et al., 1988; Tschekunov, 1989; Adam, 1997), which react to increased temperature differently. Seismological interpretations assume that a depth of the peridotite melting isotherm coincides with the zone of the P-wave velocity decrease or the change in the S-wave velocity, whereas magnetotelluric interpreters put this isotherm at a zone of the electrical conductivity increase (e.g., Blizkovsky et al., 1986).

An interesting comparison emerges when we overlap the heat-flow map (Pospisil et al., 2006), the crustal thickness maps derived from gravity data and gravity data stripped off the sediment effect (Bielik, 1988, 1991; Szafian, 1999; Pospisil et al., 2006), the map of recent vertical movements (Pospisil et al., 2006), the lithospheric thickness map (e.g., Babuska et al., 1987), and the map of the ALCAPA and Tisza–Dacia locations (Figure 2). This comparison indicates that the ALCAPA roughly overlaps with the thermally active part of the Carpathian–Pannonian region and areas of thinnest crust determined from both types of gravity data. The larger part of the Tisza–Dacia overlaps with the area of the cooler heat flow, with the extent of the normal to thicker crust and with the area affected by recent uplift.

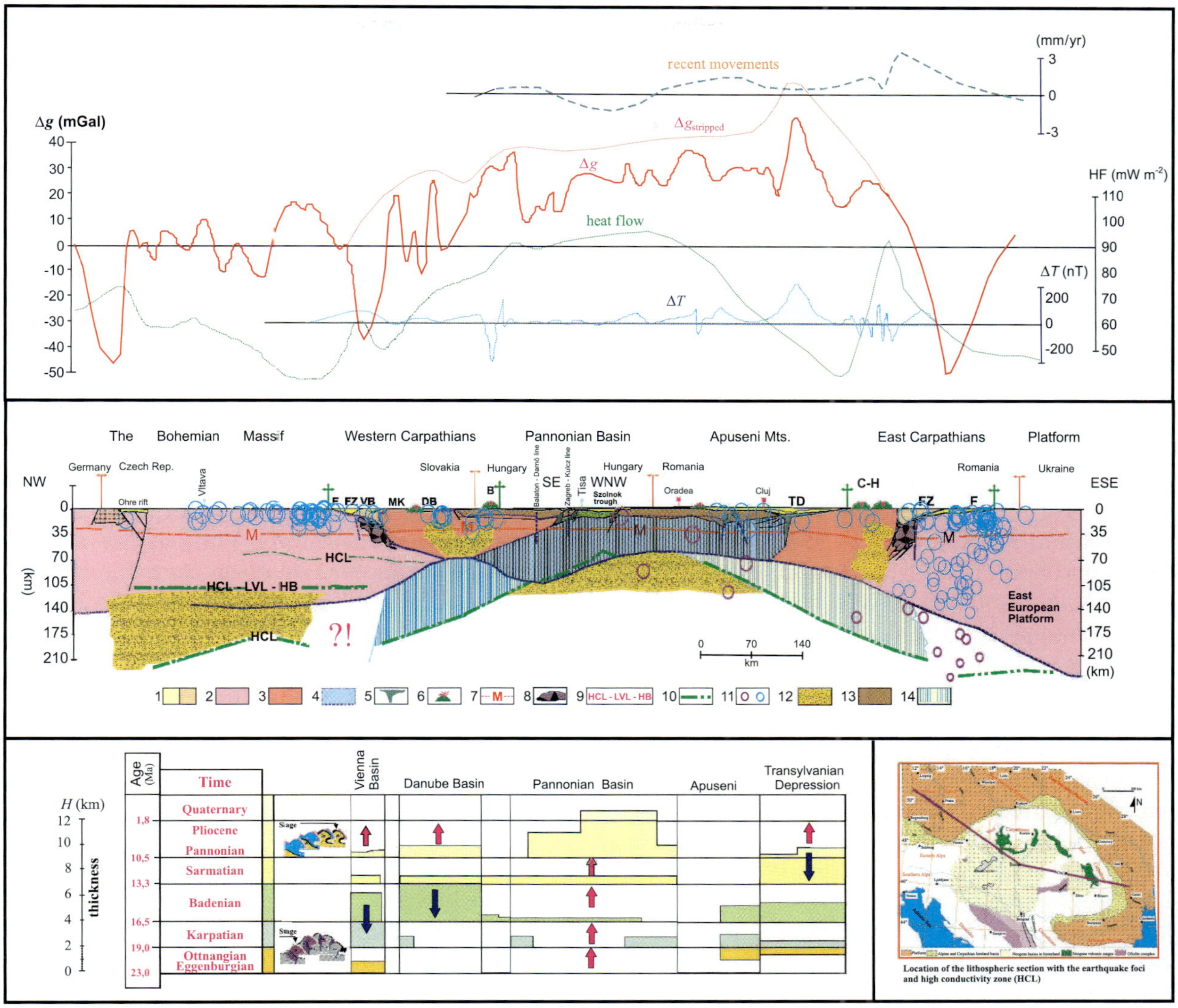

Figure 3. Complete geological-geophysical data converted from the maps to the representative lithospheric profile through the Bohemian Massif, Pannonian Basin system, and East European platform with some earlier interpreted geophysical data, including recent vertical movements (Joo et al., 1981), heat flow (Cermak and Hurtig, 1977), magnetic data (ΔT; Posgay, 1967; Krutichovskaja et al., 1971; Airinei et al., 1983; Gnojek and Janak, 1986), earthquake hypocenters (Pospisil et al., 2006), gravity (Δg; Krus and Sutora, 1986), stripped gravity effect ($\Delta g_{\text{stripped}}$; Pospisil and Vass, 1984), and determined lithospheric boundaries (Adam, 1976; Tschekunov and Boljubach, 1979; Babuska et al., 1987; Pajdusak et al., 1988) compiled by authors. Explanation: 1 = sedimentary rocks: yellow = Neogene Molasse zone; gold = Flysch zone; 2 = platform lithosphere; 3 = transitional lithosphere; 4 = lithosphere of the Pannonian Basin system; 5 = ophiolites; 6 = Neogene–Quaternary volcanics; 7 = Moho; 8 = high-conductivity zone (HCL; Jankowski et al., 1979; Praus et al., 1981); 9 = LVL: low-velocity layer (Bisztricsany, 1974; Pecova et al., 1979); HCL: high-conductivity layer; HB: determined on the basis of the modeling (Cermak, 1986); 10 = lithosphere–asthenosphere boundary determined using magnetotelluric data (Adam, 1976; Tschekunov, 1993); 11 = earthquake hypocenters: violet circles = below seismologically determined L/A boundary; blue circles = in lithosphere (processed on the basis of Zsiros's data in Pospisil et al., 2006); 12 = thermally active part of the lithosphere–asthenosphere system (after Subotin, 1975; Burjanov et al., 1980; Cermak, 1986); 13 = detached blocks combined with strange lithosphere (Drauzug–Bakony zone, Apuseni block); F = foreland basin with Neogene molasse sediments; FZ = flysch zone; VB = Vienna basin; MK = Male Karpaty Mountains; DB = Danube Basin; B = Borzsony Mountains; TD = Transylvanian Depression; C–H = Calimani–Hargita Mountains; 14 = zone where differences between seismological model and the MT model of lithosphere thickness exist. The explanation can be considered in connection with thermal erosion, processes of delamination, or destruction of the slab break-off. The table includes a simplified sedimentary record in space and time with relative thicknesses (modified after Pospisil and Vass, 1984). Blue arrows = subsidence; red arrows = uplift.

Mentioned characteristics document a different character of the ALCAPA and Tisza–Dacia microplates. The thinnest lithosphere is present in the Pannonian area overlapping with the ALCAPA, forming two minimums at 60 km (36 mi) separated by an elongated area of normally thick lithosphere (100–120 km; 60–75 mi) located between Budapest and Kosice (Babuska et al., 1987). A lithospheric root is developed only in two areas: in the Eastern Alps along the northern boundary of the ALCAPA and in the Southern Carpathians along the southern boundary of the Tisza–Dacia (Babuska et al., 1984).

The ALCAPA lithosphere with thinned crust can be further characterized by high seismic velocities (Burjanov et al., 1987; Tschekunov, 1989; Posgay et al., 1998), by high conductivities (Pospisil et al., 2006), and by the occurrence of the extensive and continuous volcanism that lasted from early Miocene to Pliocene–Quaternary (see Pecskay et al., 1995). The ALCAPA crustal thickness reaches values of 22–25 km (13.6–15.5 mi) (Zatopek, 1979; Krus and Sutora, 1986; Szafian, 1999). It is characterized by a large number of extensional basins, belts of magnetic anomalies (Hrusecky et al., 2006), and high values of the gravity field (Sefara et al., 1980; Krus and Sutora, 1986; Szafian, 1999). The whole Pannonian region overlapping with the ALCAPA is characterized by relatively high values of the heat flow (Horvath, 1979; Dovenyi et al., 1983; Demetrescu and Andreescu, 1994; Mocanu and Radulescu, 1994). The Outer Carpathians of the ALCAPA have lower heat-flow values than other ALCAPA regions (Cermak and Hurtig, 1977).

All these properties characterize the ALCAPA microplate, which experienced a lateral extrusion from the ancestral Eastern Alpine realm and traveled as far as to the contact with the West and East European platforms in Poland and Ukraine.

The Tisza–Dacia lithosphere, covering areas of the Apuseni Mountains, the Transylvanian Basin, and the Eastern and Southern Carpathians, can be characterized by the thicker crust (Zatopek, 1979; Krus and Sutora, 1986; Szafian, 1999), the lower values of the gravity field (Sefara et al., 1980; Krus and Sutora, 1986; Szafian, 1999), the lower heat-flow values in comparison with the ALCAPA (Cermak and Hurtig, 1977; Krus and Sutora, 1986), and the high seismicity in its southeast part (World Stress Map Project, 1997). The lowest heat-flow values are recorded in the Transylvanian Basin (Cermak and Hurtig, 1977; Demetrescu and Andreescu, 1994; Mocanu and Radulescu, 1994). A broad zone along the contact with the East European and Moesian platforms has characteristic uplift movements (Pospisil et al., 2006).

All these properties characterize the Tisza–Dacia microplate, which did not experience the level of extension and large-scale faulting comparable to the ALCAPA but only underwent a large clockwise rotation during the later stages of its movements (Marton and Marton, 1989; C. G. Panaiotu, 2001, personal communication).

Data from the Vrancea seismic zone

The Vrancea area, located in the Carpathian bend to the north of Bucharest, is a small part of the Carpathian arc, which is still seismically active (Figures 4, 5). This area experienced strong recent subcrustal earthquakes (1940, Mw = 7.7; 1977, Mw = 7.5; 1986, Mw = 7.2; 1990, Mw = 6.9) situated between 70- and 200-km (43- and 124-mi) depths (e.g., Radu, 1979; Oncescu, 1987) (Figure 3). It is the only Carpathian area where folding and thrusting occurred as late as during the Pleistocene, located in the outermost thrust belt zone between the two northwest–southeast-striking faults that deformed also the Moesian Platform. Recent crustal movements recorded in the Vrancea area are accompanied by the moderate crustal seismic activity. Seismic events have 10–30 (or 40)-km (6–18 [or 25]-mi) depths and magnitudes below 5.5.

Enescu and Enescu (2000) summarized a character of the Vrancea subcrustal seismic zone on the basis of the spatial distribution of hypocenters, the focal mechanism of earthquakes in the investigated zone, the lithosphere structure, the thickness of the lithosphere, and tectonics as follows:

1) The hypocenter zone has the largest horizontal dimension in the northeast–southwest direction. The southeast–northwest dimension is much smaller. Recorded focal mechanisms form a pipelike body that is almost vertical.
2) Data in various depth intervals of this pipelike body have a slightly different character, which allows their division into 60–90-, 90–110-, 110–130-, 130–150-, and 150–180-km (37–56-, 56–68-, 68–80-, 80–93-, and 93–111-mi) intervals. Earthquakes of each depth interval have their own characteristic periodicity and different fault-plane solution, traveltime curves, and stress pattern (Constantinescu and Enescu, 1984). The absence of the reflected wave can prove that the lower boundary of the lithosphere does not mark a velocity contrast in the compressional waves or, more likely, that this contrast is fairly weak (Enescu, 1987).
3) A seismically quiescent zone exists above the subcrustal hypocenter zone between 30 (or 40) and 60 km (18 [or 25] and 37 mi), indicated by a lack of seismic events.
4) Subcrustal focal mechanisms have mostly vertical tension axes and mostly horizontal compression

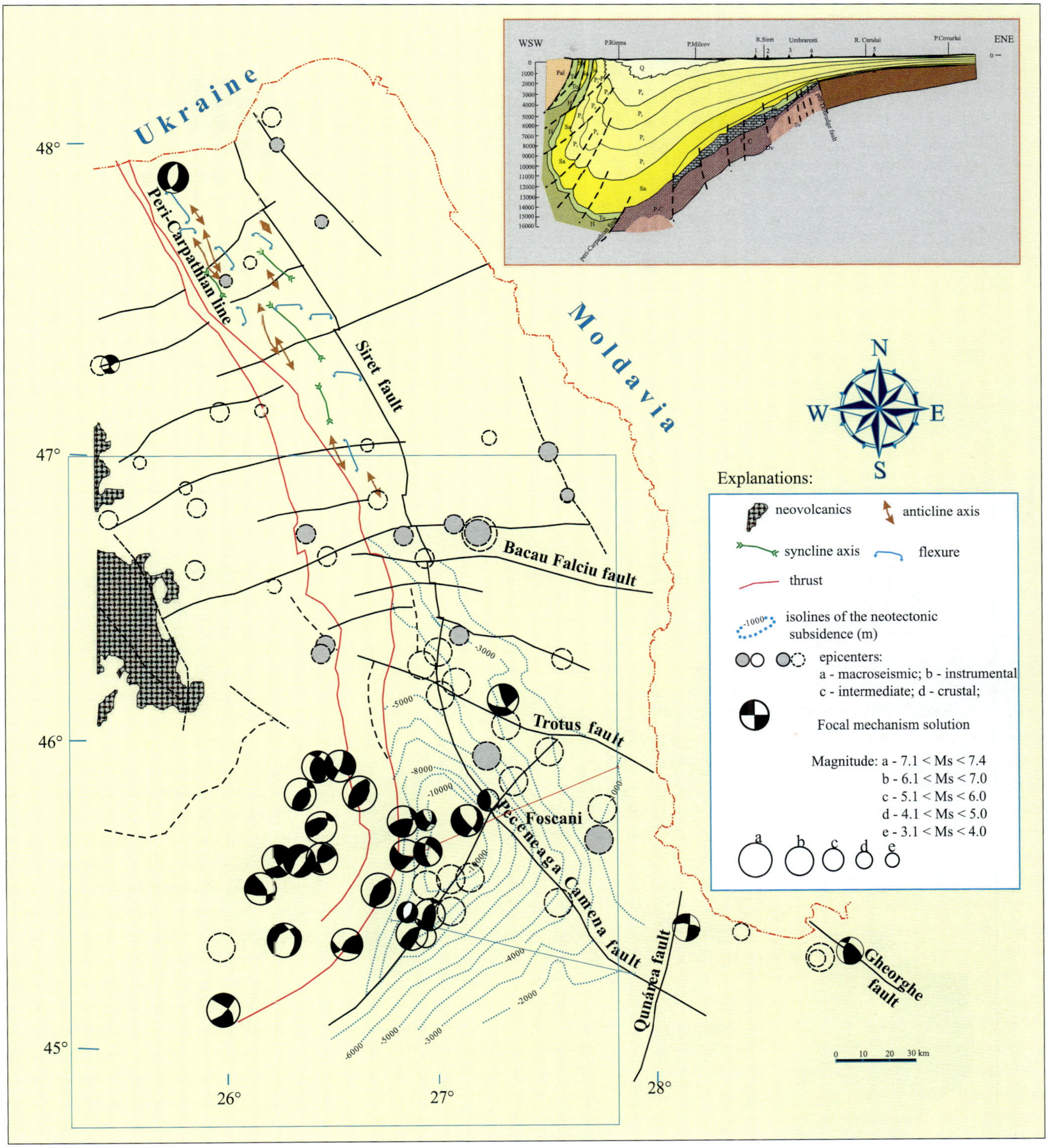

Figure 4. Zone of earthquake hypocenters and focal mechanisms determined in the Vrancea zone with profile of the foreland basin (Polonic, 1986). Blue box indicates the location of Vrancea zone detail in Figure 5. A thin red line in the Focsani region indicates the location of the geological profile (Turculet and Mihail, 1970). Profile crosses the foreland basin where the intensity of the platform flexure can be observed.

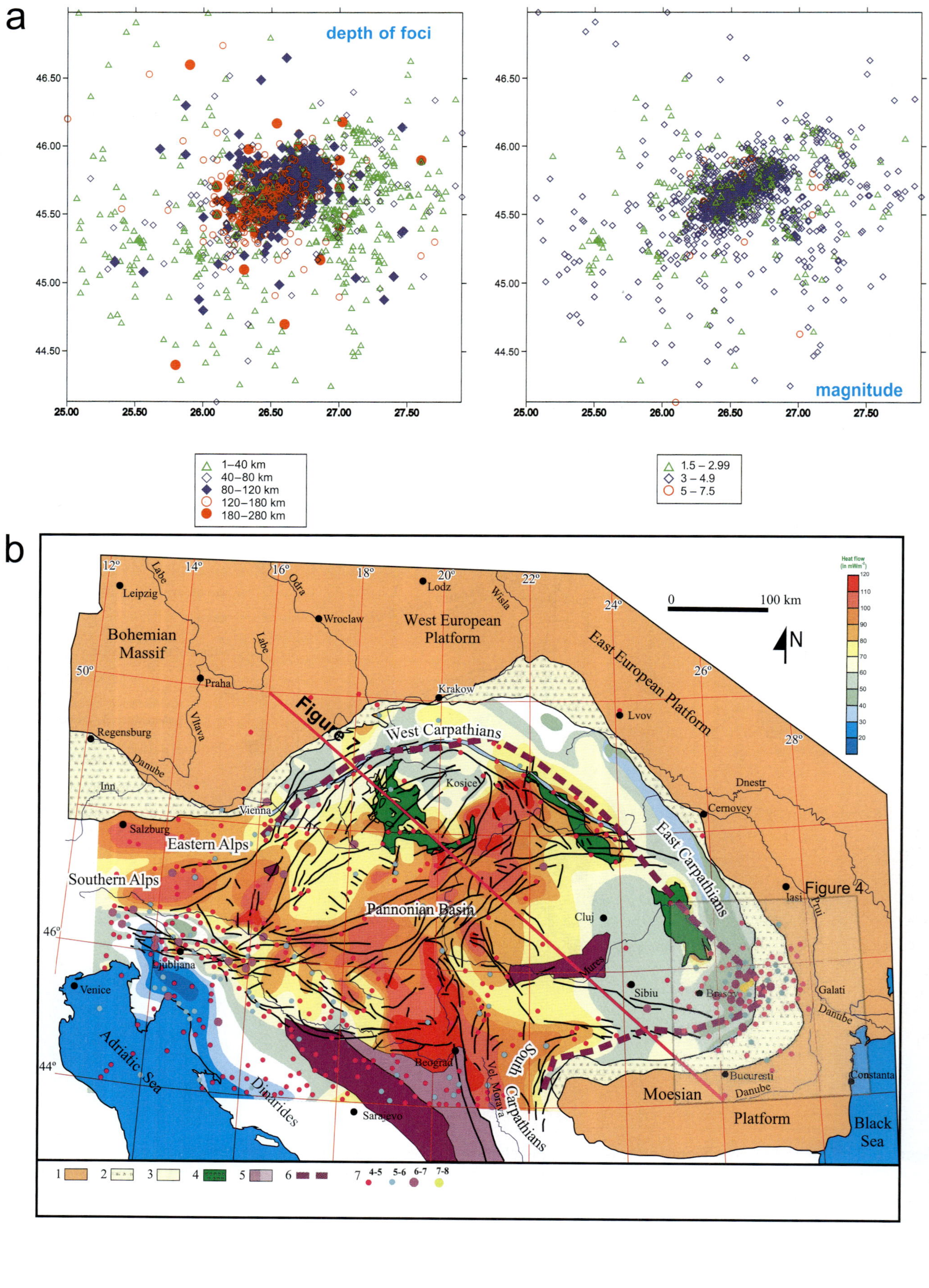
a
depth of foci
magnitude
46.50
46.00
45.50
45.00
44.50
25.00
25.50
26.00
26.50
27.00
27.50
1–40 km
40–80 km
80–120 km
120–180 km
180–280 km
1.5 – 2.99
3 – 4.9
5 – 7.5
b
Heat flow (in mWm⁻²)
120
110
100
90
80
70
60
50
40
30
20
0
100 km
N
Leipzig
Labe
Bohemian Massif
Praha
Wroclaw
Odra
West European Platform
Lodz
Wisla
Krakow
Lvov
East European Platform
Regensburg
Danube
Inn
Vltava
Figure 7
West Carpathians
Kosice
Vienna
Salzburg
Eastern Alps
Southern Alps
Pannonian Basin
East Carpathians
Cernovcy
Dnestr
Iasi
Prut
Figure 4
Cluj
Mures
Sibiu
Brasov
Galati
Danube
Ljubljana
Venice
Adriatic Sea
Dinarides
Beograd
Vel. Morava
South Carpathians
Sarajevo
Bucuresti
Moesian
Platform
Constanta
Black Sea
12°
14°
16°
18°
20°
22°
24°
26°
28°
50°
46°
44°
1
2
3
4
5
6
7
4-5
5-6
6-7
7-8

axes. However, tension axes have a tendency to become horizontal with depth.

5) Two main earthquake categories are present. The first category includes large earthquakes (MGR > 6), some moderate (MGR = 5–6), and the small (MGR < 5) ones. It is characterized by northwest–southeast-trending compression axes. The second category includes only moderate and small earthquakes that have northeast–southwest-trending compression axes. Both categories have subvertical tension axes.
6) These two main categories contain earthquakes with different fault-plane orientations. The first category includes northwest-dipping fault planes. The second category predominantly contains fault planes with southwest and northeast dip directions, but west and east dip directions are also known. Fault planes of both categories contain various dips ranging from shallow to steep.
7) Reverse fault types dominate. Numerous normal faults and some strike-slip faults are also present.
8) The occurrence of the Vrancea subcrustal earthquakes is linked to a deformation of paleosubducted oceanic lithosphere fragments or subcrustal continental lithosphere fragments or both (Enescu and Enescu, 2000). The involved East European plate is known to subduct toward the southwest or, more precisely, toward the west-southwest. Its continental edge, the southwest margin of the East European platform, lies in the broader Vrancea region.

Comparison of Geological and Geophysical Data Sets

The Southern Carpathian crustal root, located along the northern margin of the Moesian Platform, lies above the lithospheric root (see Nemcok et al., 1998a). The lithospheric root has its eastern end in the Brasov bend, but the crustal root continues underneath the Eastern and Western Carpathians, having its northwestern end near Krakow. This end is well imaged by the gravity data corrected for the Moho effect (Bielik, 1991; Pospisil et al., 2006). Lithospheric and crustal roots are not documented underneath the Western Carpathians to the west of Krakow, but both are imaged underneath the Eastern Alps. The area of the thinned crust in the Pannonian Basin region roughly overlaps with the thinner lithosphere, but minimums do not exactly coincide.

Overlapped gravity and crustal thickness data (Nemcok et al., 1998a) indicate that the highest negative gravity anomalies coincide with the Carpathian Foredeep and the crustal root areas. Neogene molasse thickness maximums are located in front of two of the three largest Carpathian negative gravity anomalies. They are also located in front of the ALCAPA and Tisza–Dacia microplates.

The heat-flow maximum overlaps with the western lithospheric minimum (Nemcok et al., 1998a). The eastern lithospheric minimum overlaps with fairly high heat-flow values. The heat-flow maximum overlaps with the crustal thickness minimum. The Romanian heat-flow maximum forms the exception, overlapping with the area of the recent volcanism. Neogene sediment thickness maximums in the hinterland roughly coincide with lithospheric thickness minimums. The Eastern Carpathian and adjacent platform areas of the isostatic rebound overlap with a region having the crustal root present and the lithospheric root missing.

The linear volcanism was active in separate segments (Figure 6). Each of them was relatively short lived. Some of these volcanic segments became nonactive when the activity of the eastern neighbor started; some did not. The linear volcanism and the last thrusting along the Carpathian arc roughly coincide (Figure 6), indicating that each volcanic segment operated during the last stages of the subduction and during the following short-lived collision. The rate of the eastward migration of this linear volcanic activity calculated by Nemcok et al. (1998a) has a decreasing trend from 154 km/Ma (95 mi/Ma) during 16–15 Ma to 10 km/Ma (6 mi/Ma) during 4–0 Ma.

The increased Karpatian eastward shift of the molasse depocenter in front of the ALCAPA correlates with the opening of the Vienna and East Slovakian basins inside the ALCAPA (compare Royden, 1985, 1988; Jiricek and Seifert, 1990; Kovac et al., 1993; Lankreijer et al., 1995; Meulenkamp et al., 1996). The early Badenian extremely fast eastward shift of the molasse depocenter in front of the ALCAPA coincides with pronounced basin opening in the ALCAPA (compare Royden, 1988; Vass et al., 1990; Lankreijer et al., 1995; Meulenkamp et al., 1996). The main molasse depocenter in

Figure 5. (a) Earthquake hypocenters in the Vrancea region sorted according to their depth and magnitude. The region is located in Figure 4. (b) Relationship between map of heat flow (Pospisil et al., 2006) and seismic foci (Pospisil et al., 2006) presents a Transylvanian area as a relatively cool and passive segment of Carpathians. The high-conductivity zone (HCL) separates the Carpathians from the passive margin of the European continent (platform). The red line indicates a location of profile in Figure 7. Explanation: 1 = platform; 2 = Alpine and Carpathian foreland basin; 3 = Neogene basin in hinterland; 4 = Neogene volcanic ranges; 5 = ophiolite complexes; 6 = high-conductivity zone (HCZ), 7 = magnitude foci in Medvedev-Sponheuer-Karnik scale (MSK).

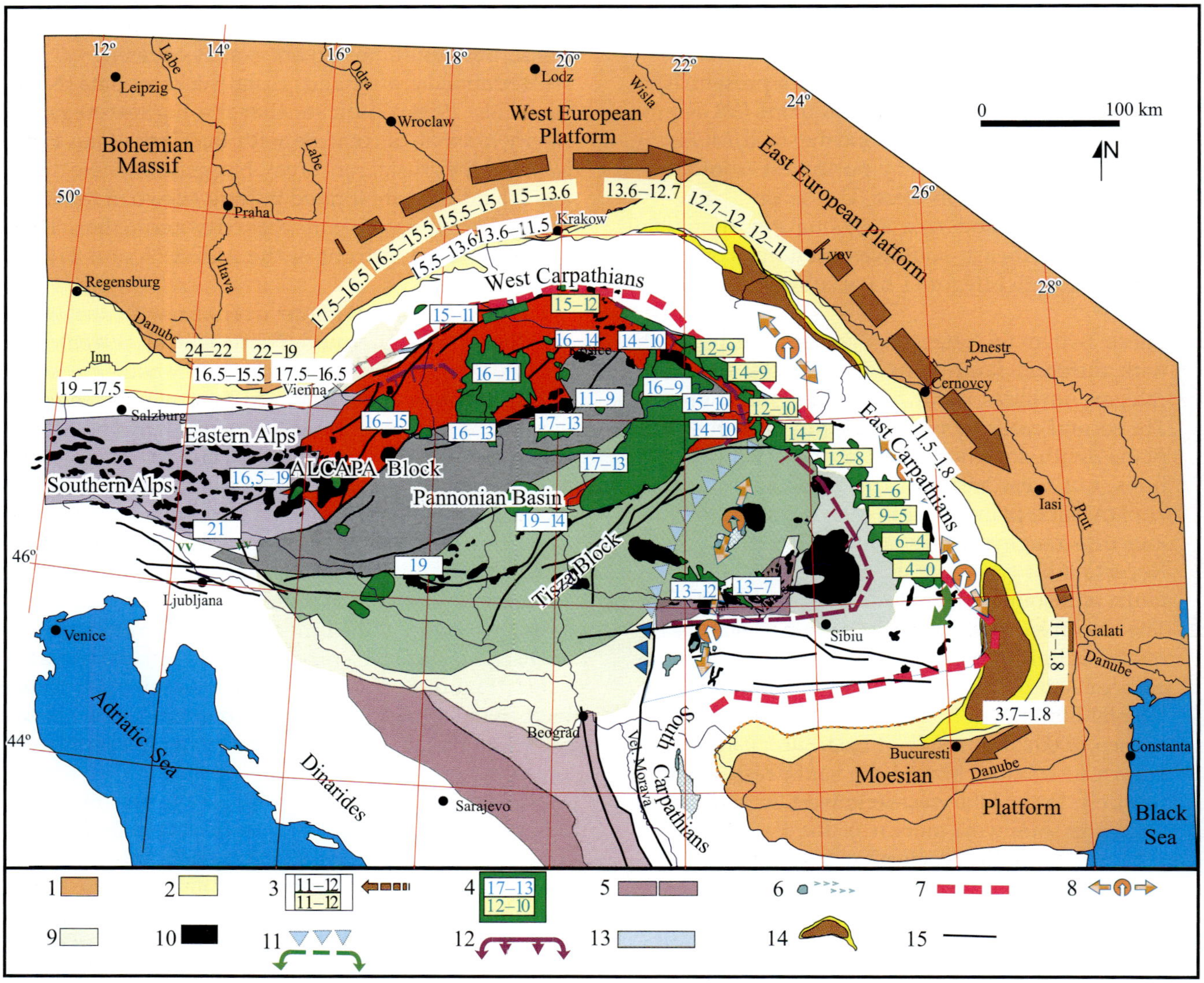

Figure 6. A summary figure with geophysical data, geological data, and speculative boundaries in the Carpathian–Pannonian region. Explanation: 1 = platform; 2 = Alpine and Carpathian foreland basin; 3a = last stage of foreland basin deposition (in Ma); 3b = last thrust movements of the Outer Carpathians (in Ma); 4a = age of intermediate calc-alkaline magmatics in Ma (Pecskay et al., 1995); 4b = age of linear volcanics in Ma (Pecskay et al., 1995); 5 = ophiolite complex; 6 = calc-alkaline magmatics (banatites; 90–60 Ma); 7 = location of the subducted passive continental margin indicated by magnetotellurics (Jankowski et al., 1979; Praus et al., 1981); 8 = platform area affected by isostatic rebound in reaction to slab break-off; 9 = hinterland areas affected by Neogene extension; 10 = magnetic anomalies of Alpine pre-Tertiary sources (Posgay, 1967; Airinei et al., 1983; Gnojek and Janak, 1986), 11 = zone of speculative slab break-off; 12 = marginal zone of thrusted ophiolites in Carpathian basement (lower crust); 13 = remnant of the subducted oceanic lithosphere, including transitional lithosphere and adjacent part of passive continental margin in its northeast and south boundaries; 14 = maximal thickness of sediments; 15 = fault.

front of the ALCAPA underwent a northeastward shift of 1.8 cm/yr (0.7 in./yr) during the early Miocene and 0.6 cm/yr (0.2 in./yr) during the middle Miocene (Nemcok et al., 1998a). The stop of the ALCAPA movement and the existence of the stationary molasse depocenter in front of it correlate with the main basin rearrangement inside the ALCAPA during the middle Badenian–early Sarmatian. The accelerated molasse depocenter shift from being in front of the ALCAPA to being in front of the Tisza–Dacia during the Sarmatian is followed by the overall subsidence in the Pannonian Basin system at the beginning of the Pannonian (compare Royden, 1988; Csontos et al., 1992; Tari et al., 1992; Horvath, 1993; Meulenkamp et al., 1996).

The timing of the acidic areal volcanism and the large-scale normal faulting in the hinterland coincide, having Eggenburgian age (Nemcok et al., 1998a). The acidic volcanism formed by crustal melting above the asthenosphere elevated by extension (Poka, 1988; Salters et al., 1988; Lexa et al., 1993; Pecskay et al., 1995). Areas experiencing progressively larger stretching (Nemcok et al., 2000) roughly correlate with areas experiencing intermediate calc-alkaline volcanism progressively less contaminated by the crust (Lexa et al., 1995). During the Eggenburgian–Badenian, the hinterland extension affected areas progressively closer to the thrust belt, leaving a narrow zone of contraction (Nemcok et al., 1998b). The same spatial progression during the same time is indicated by ages of the areal acidic volcanism, which required both thinned crust and lithosphere for an excess heat (see Pecskay et al., 1995). Spatial and temporal distribution of the areal intermediate calc-alkaline magmatism indicates the same function (Figure 6), spreading northeastward and eastward (see Pecskay et al., 1995).

DISCUSSION

Subduction and Oceanic Slab Break-off Model

The Badenian–Pannonian east–west extension in the hinterland (e.g., Bergerat, 1989; Nemcok, 1993), the Badenian rearrangement of Neogene foreland depocenters (e.g., Meulenkamp et al., 1996), the top asthenosphere morphology (e.g., Babuska et al., 1984, 1987), the distribution of lithospheric and crustal roots, and other data discussed earlier indicate that subduction was southwest and west directed.

The Carpathian subduction was different from the Alpine one. The southeast-vergent Alpine subduction can be characterized by a compression in the hinterland, a lack of the extensive areal volcanism (implied shallower dip of the subducting slab), and followed by a strong collision (e.g., Doglioni, 1992). The southwest- and west-vergent Carpathian subduction can be characterized by strong hinterland extension, extensive areal volcanism (implied subducting slab steepening), and followed by weak collision (e.g., Royden et al., 1982). The Carpathian thrust belt was much less eroded than the Alpine one. It lacks a comparable amount of the shortening and large-scale basement-involved thrusts. The Carpathian Flysch belt was not overthrust by inner thrusts like in the Alpine case.

Although the intra-Carpathian extension in the order of 100–150 km (62–93 mi) was large enough to compensate for the middle Miocene and younger shortening of the thrust belt (Horvath, 1986), the pre-middle Miocene shortening was not fully compensated by hinterland extension. This implies a large early Miocene orogenic advance to the northeast and east by translation (Csontos et al., 1992).

The northeastward and eastward subduction rollback (e.g., Royden et al., 1982) of the rCFB slab subducting under its own weight, the oblique closure of the rCFB, and the progressive change of the subduction to collision eastward along the orogen controlled the location, timing, and mode of the subducting oceanic slab break-off. The eastward-propagating Badenian–Holocene break-off developed by a combination of the opening and scissorlike shearing. Assuming a break-off and volcanism delay (see Davies and von Blanckenburg, 1995), its propagation timing is only very roughly indicated by ages of linear volcanism (Figure 6), overlapping with the Carpathian collision. The segment between Eastern Slovakia and northern Romania is exceptional, comprising probably a mixture of both arc-type and break-off-related volcanisms (Nemcok et al., 1998a).

Linear volcanics are geochemically similar (Nemcok et al., 1998a) to products of evolved (or continental) volcanic arcs, representing mantle-derived magmas from metasomatized asthenosphere wedge and plume, with low to moderate crustal contamination (Lexa et al., 1993; Pecskay et al., 1995). The break-off timing indicates that it followed an earlier steepening of the subducting slab. This is shown by the widespread forelandward progression of the hinterland extension (Nemcok et al., 1998b) that occurred earlier than the oceanic slab break-off.

The slab break-off should be responded to by an isostatic rebound of the buoyant continental neighbor of the detached oceanic slab. This pronounced motion reversal should terminate the subsidence in the foreland basin and cause its uplift. It should also uplift the adjacent thrust belt part and accelerate its erosion. Such an isostatic rebound is indicated by the pronounced present-day uplift of the Eastern Carpathian immediate foreland and adjacent thrust belt front (Pospisil et al., 2006).

A correct separation of areal and linear intermediate calc-alkaline magmatics is critical for the understanding of the slab break-off. The spatial and temporal data from linear volcanics allow a rough determination of the gradual slab break-off propagation and, thus, simplification of preexisting complicated models (e.g., Csontos, 1995; Linzer, 1996). However, the emerging development model complexity in the form of the Vrancea subcrustal earthquakes inside the continental crust and the possibility of the present-day subduction hinge along the Gutii–Apuseni–Vardar line require targeted studies and numerical validation.

The intermediate calc-alkaline areal volcanism can be related to more advanced stages of the subducting slab steepening, extension in hinterland, and mantle upwelling, indicating progressively less contamination by crustal material because of the extension and thinning of the crust (e.g., Lexa et al., 1993, 1995; Pecskay et al., 1995).

Young alkaline basalt occurrences in the hinterland, contemporaneous with and postdating the thermal subsidence in the Pannonian region (e.g., Horvath, 1984, 1993; Horvath and Rumpler, 1984; Nemcok and Lexa, 1990; Csontos et al., 1992; Tari et al., 1992), indicate a magmatism generated only by the extension (Embey-Isztin et al., 1993). The mantle source of the alkali basaltic magmas is considered to be the same as that of calc-alkaline magmas (Downes et al. 1995), i.e., enriched asthenospheric mantle, because basaltic isotopic composition rules out shallow lithospheric mantle and lower crust from being either a major component in the source or a contaminant.

A correlation of the various data sets in our database indicates that the Vrancea area is not an area of active slab break-off as believed earlier (e.g., Fuchs et al., 1979; Constantinescu and Enescu, 1984) because the seismic activity takes part inside the continental lithosphere (Figure 3), far away from the broken-off and sinking oceanic slab (see Spakman, 1990) (Figure 7). Therefore, the scissorlike slab break-off, which ran along the Carpathian arc, stopped somewhere to the north of the Vrancea area, roughly coinciding with the area that has 4–0-Ma ages of linear volcanism (Figure 6). A further problem appears when we could interpret from tomographic images of Spakman (1990) that the sinking broken-off oceanic slab has a north-northeast–south-southwest strike and is located to the west of the imaginary Gutii–Apuseni–Vardar line.

Vrancea Seismic Area

The Vrancea area is the only area in the whole Carpathian–Pannonian region with mantle–lithosphere earthquake data. This allows the insight of plate geometry (Figure 3). However, the earthquake zone has a very small size. Its horizontal cross section is 20 × 60 km (12 × 37 mi), and its depth interval is 70–200 km (43–124 mi) (Figure 3b). Therefore, its help with plate geometry determination is limited. The polydirectional horizontal compression and the subvertical extension in this zone (e.g., Roman, 1970; Iosif and Iosif, 1979; Monfret et al., 1990) (Figure 3) were interpreted as driven by the slab descending under its own weight (Roman, 1970), dipping toward the northwest and having a northeast–southwest to east–west strike. The seismic gap between the lithospheric mantle and crustal earthquakes was understood as a low-viscosity, low-velocity, seismically inactive zone where the subducting plate breaks off (e.g., Radu, 1974; Fuchs et al., 1979; Constantinescu and Enescu, 1984). The fact that location of the seismic gap does not coincide with location of the gap between the East European platform and sinking oceanic slab shown in Figures 3 and 7 was first pointed out by Girbacea and Frisch (1998), who proposed an alternative interpretation: lower crust decoupling. One could try to explain earthquakes as attributed to the behavior of the banded continental margin affected by extension in its upper parts and compression in its lower parts overridden by the Carpathians that are under compression, together with part of the platform immediately under and in front of them. Crustal earthquakes located above the depth of 30 (or 40) km (18 [or 25] mi) have focal mechanisms with west–east compression that plunges as much as 49°, driving thrusts with shallow to subhorizontal dips (e.g., Radu, 1974; Fuchs et al., 1979; Polonic, 1986) (Figure 4). Such a stress distribution is known from thrust belts (e.g., Mandl, 1988). Subcrustal earthquakes would be successfully explained by the sinking slab undergoing subvertical extension (e.g., Roman, 1970) because they were not located inside the East European platform and were a bit further away from its margin and not very steeply dipping (Figures 3, 7). The other discrepancy with the original model is that subcrustal earthquakes with tension axes depart from vertical and approach horizontal as described earlier (see Enescu and Enescu, 2000). The zone of subcrustal earthquakes is located inside the continental slab, which is not vertically dipping, and it is located farther away from the continental margin. A simple behavior of the banded continental margin affected by extension in its upper parts and compression in its lower parts is also not a successful explanation of the subcrustal focal mechanisms, which are clearly in need of further studies. Therefore, instead of trying to explain them, we just focus on the following question: What makes their location different from other locations along the Carpathian arc?

When we now overlap the discussed Vrancea data with their characteristics to regional geophysical data mentioned earlier, we have a following match. The Vrancea area is characterized by several lithospheric anomalies in comparison to the rest of the Carpathian–Pannonian region. One of them is its location near the Tisza–Dacia margin, which has a colder thermal regime than the ALCAPA. The Vrancea area is an important boundary in the distribution of vertical movements. It bounds the uplifting Outer Eastern Carpathians and their foreland, the most prominent Carpathian features on the vertical motions map (Pospisil et al., 2006) from the south. The extent of this uplifting territory stops abruptly in the Vrancea area where the Outer Carpathian wedge subsides, where the adjacent platform parts indicate no vertical movements, and where the linear volcanism has its eastern end. This localized and pronounced subsidence is restricted only to the Brasov bend area and does not extend to the Southern

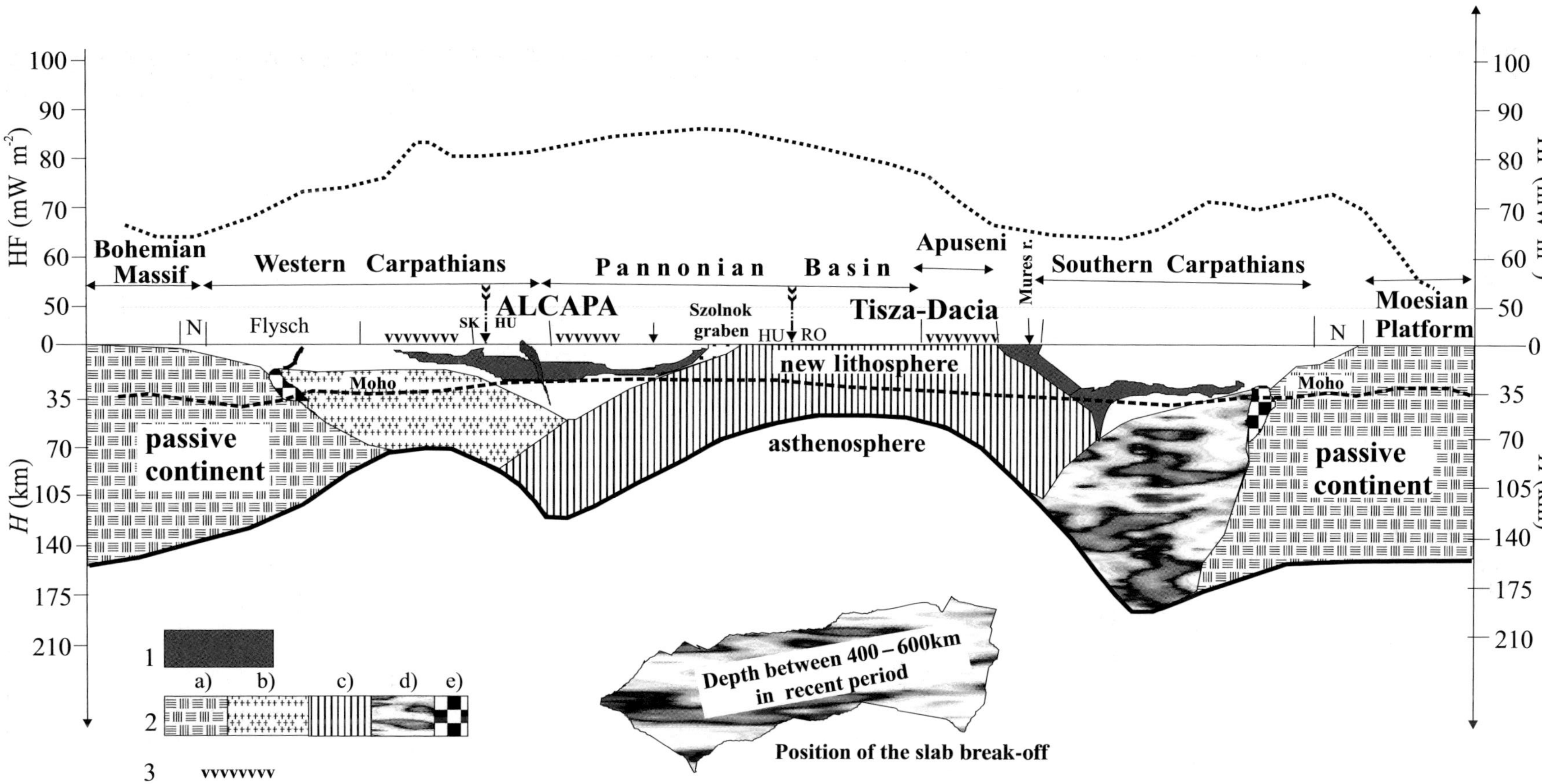

Figure 7. Simplified lithospheric profile through the Bohemian Massif, Pannonian Basin system, and Moesian Platform providing information about lithospheric composition and structure together with heat-flow curve. Green v letters = location of Neogene–Quaternary volcanics; N = sediments of the Neogene foreland basin; 1 = ophiolitic complex; 2 = lithospheric types: (a) continental (platform type, including Bohemian Massif); (b) transitional [underneath rCFB and adjacent to (a)]; (c) new lithosphere of the Carpathian-Pannonian system; (d) broken-off and sunken oceanic crust and remnant of the oceanic crust attached to continental lithosphere; (e) high-conductivity zone interpreted as suture zone; 3 = Neogene volcanic ranges. Remnant of the oceanic lithosphere represents the interpretation of the velocity anomaly from the tomographic experiment of Spakman (1990) from a depth of 400 km (248 mi). HF line = heat-flow curve.

Carpathians. The Vrancea area is located to the east of the Southern Carpathian lithospheric root and at a very sharp corner between the Eastern and Southern Carpathian crustal roots. The Vrancea area lies to the east of the eastern corner of the oceanic slab that subducted underneath the Tisza–Dacia part of the Carpathian orogen. The fact that the triangular region among the eastern half of the Eastern Carpathians, Southern Carpathians, and the imaginary Gutii–Apuseni–Vardar line (see Figure 6) has a geophysical characteristic different from the rest of the Carpathian–Pannonian region would let one speculate about the alternative to the model of Nemcok et al. (1998a), which places a sinking oceanic slab relatively close to the west of the Vrancea region. Let us name it the second alternative. Mainly, a cool nature of this triangular area would let one speculate that this region was not entered by the asthenosphere located underneath the orogen and adjacent to the subducting oceanic slab, which controls a distinctively different, warm character of the ALCAPA region to the west. The subduction-related flexure of this oceanic slab would still require that oceanic slab break-off along its northeast margin propagates in a scissorlike mode as far as the very eastern end of the linear volcanism (Figure 6) to allow this flexure. The flexure hinge Gutii–Apuseni–Vardar placed roughly at the longitude of the western end of the Moesian Platform would not require any break-off along the southern boundary of this triangular remnant of the oceanic slab to allow its flexure. The oceanic slab connection with the Moesian Platform would prevent it from a further eastward subduction rollback. The situation described could explain the zone of the strong vertical isostatic rebound of the East European platform reacting to the scissorlike break-off from the oceanic slab. One could speculate that the oceanic slab pull and buoyancy of the East European platform resulted in the break-off inside the oceanic slab along the Gutii–Apuseni–Vardar line (Figures 6, 8b), which would be indicated by the tomographic results of Spakman (1990) (Figure 7). This situation would also cause an isostatic rebound in the Vrancea segment of the arc to follow a flexed trajectory of the subducting oceanic slab remnant attached to the European platform. This movement would be opposed by the lithostatic conditions in the eastern continuation of the East European platform. These two opposing movements could cause the unique stress regime of subcrustal earthquakes in the Vrancea area, characterized by subhorizontal compression and subvertical extension. We would keep this second alternative only at the speculative level because it requires further elaboration, including a study of its constraints by volcanic data in the Tisza–Dacia region and a numeric simulation, which are both beyond the scope of this chapter.

MAIN CHARACTERISTICS OF THE CARPATHIAN–PANNONIAN DEVELOPMENT

The main evolutionary stages of the subduction-collision process leading to the development of the Carpathian arc (Figure 8) are as follows.

The western (Alpine) part of the Alpine–Carpathian convergence underwent a change from subduction to collision during the Eocene. The end of the subduction was accompanied by a break-off of the subducting oceanic slab and by a break-off-related volcanism of the Eocene–Eggenburgian age (43–19.7 Ma). The break-off propagated eastward. Collision in the Alps was synchronous with the closure of the main Tethyan Ocean in the Alpine–Carpathian region. The rCFB was the only leftover of this ocean, left in front of the overriding plate to the east of the colliding Alps (Figure 8a).

The eastward propagation of the slab detachment continued from the Alps into the area of the rCFB (Figure 6), which was affected by subduction since the early Miocene. From the late Oligocene to middle Miocene, crustal wedges, moving eastward from the Alpine collision, became parts of the Carpathian–Pannonian overriding slab. The rCFB underwent a passive subduction, characterized by a passive sinking of the slab under its body forces, by a steep dip of the subducting slab, by a strong hinterland extension, by an extensive areal magmatism, and by an asthenosphere upwelling in the hinterland. The Carpathian subduction ended during the Pannonian (11.5–6.2 Ma) by weak collision. The subducting slab break-off ran along the whole Carpathian arc, from the early Miocene to its present position in the bend between the Eastern and Southern Carpathians (Figures 6, 8b).

The interpretation of correlated geophysical and geological data on the Carpathian–Pannonian region allows us to define

1) the present-day subduction zone, which is located to the west of the Vrancea region
2) the area above an upwelling asthenosphere located on the orogenic side of the subducting slab, indicated by the extent of the intermediate calc-alkaline areal magmatism
3) the location of the slab break-off, roughly indicated by linear intermediate volcanism
4) the rough location of the passive continental margin, interpreted from magnetic anomalies having different character for the platform and orogen and from the high-conductivity zone that outlines the whole Carpathian arc

The slab break-off-related linear volcanism ends 50 km (30 mi) north of the Vrancea seismoactive area practically

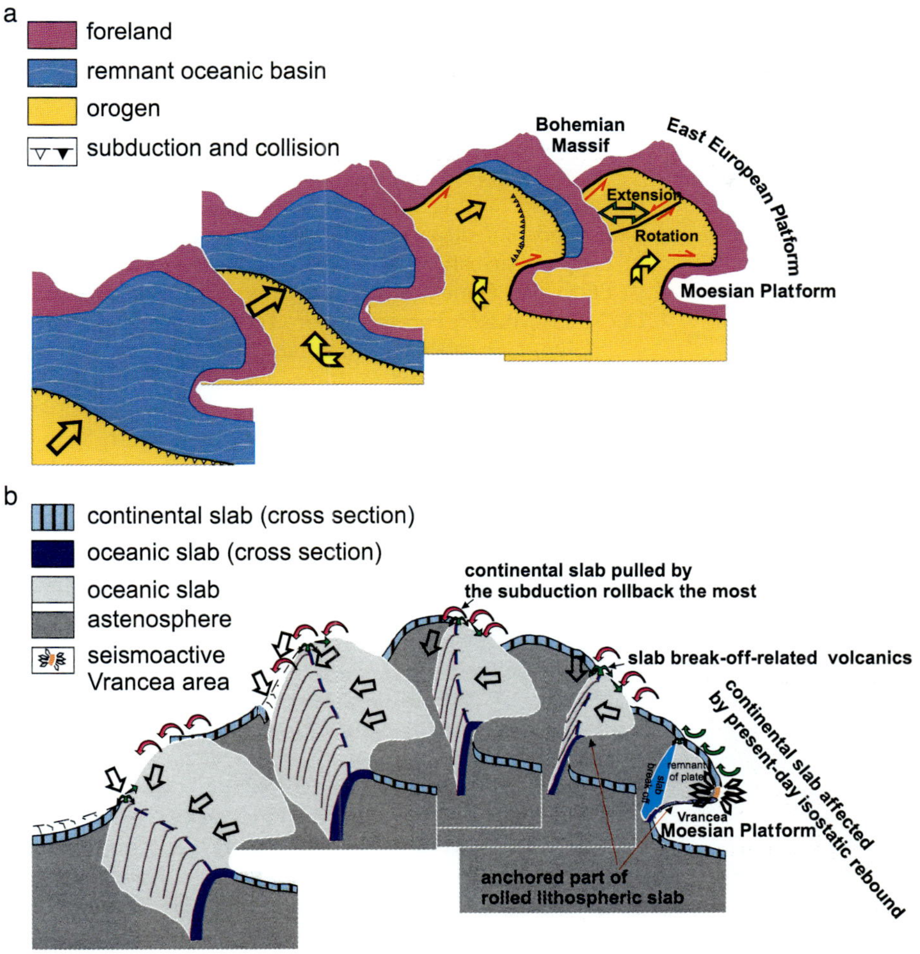

Figure 8. (a) Subduction of the remnant Carpathian Flysch Basin and advance of the Carpathian orogen in map view. (b) Subduction of the oceanic lithosphere attached to the European platform and propagation of the slab break-off. This scheme documents the second of the two alternatives discussed in the text. It is a speculative alternative to the model proposed by Nemcok et al. (1998a), described as the first alternative in the text.

in the Holocene. Some of our data indicate that the subduction hinge zone could be as far to the west from the Vrancea zone as along the Gutii–Apuseni–Vardar line, but this speculation requires further elaboration. Why the slab break-off did not run along the Southern Carpathians remains the open problem. The fact that the subducting oceanic slab underneath the Southern Carpathians is attached to the Moesian Platform in combination with the existence of the slab break-off underneath the Eastern Carpathians seems to be the key to understanding the present-day seismic activity in the Vrancea area.

ACKNOWLEDGMENTS

The chapter was made within the framework of the Europrobe–Pancardi research. The authors thank numerous coworkers from the Europrobe–Pancardi program for fruitful discussions, namely, Jaroslav Lexa, Blanka Sperner, Piotr Krzywiec, Michal Kovac, Cristian Panaiotu, Lothar Ratschbacher, Gyorgy Pogacsas, Cestmir Tomek, Giancarlo Serri, Szabolcs Harangi, Hilary Downes, and Ioan Szeghedi.

REFERENCES CITED

Adam, A., 1976, Results of deep electromagnetic investigations: Geoelectric and geothermal studies: Committee of the Academies of Sciences of Socialist Countries for Planetary Geophysical Investigation Geophysical Monograph, p. 547–560.

Adam, A., 1997, Mantle diapir–mantle plumes in the Pannonian Basin (Investigations on the asthenosphere in Hungary from the sixties till the present), *in* W. Schroder, ed., Physics and geophysics with historical case studies: Germany, Bremen-Roennebeck, p. 71–89.

Airinei, S., S. Stonescu, G. Velescu, D. Romanescu, M. Visarion, S. Radan, M. Roth, L. Besutin, and G. Besutin, 1983, Delta Z magnetic anomaly map for the territory of Romania (in French): Anuarul Institutului de Geologie si Geofizica, v. 63, p. 7–13.

Babuska, V., J. Plomerova, and J. Sileny, 1984, Large-scale oriented structures in the subcrustal lithosphere of Central Europe: Annales Geophysicae, v. 2, no. 6, p. 649–662.

Babuska, V., J. Plomerova, and J. Sileny, 1987, Structural model of the subcrustal lithosphere in Central Europe, *in* K. Fuchs and C. Froidevaux, eds., Composition, structure and dynamics of the lithosphere–asthenosphere system:

American Geophysical Union, Geodynamics Series, v. 16, p. 239–251.

Balla, Z., 1984, The Carpathian loop and the Pannonian Basin: A kinematic analysis: Geophysical Transactions, v. 30, no. 4, p. 313–353.

Benada, S., 1986, New knowledge about the extent of the para-autochthonous sheet of Karpatian sediments in the central portion of the Neogene foredeep in Moravia (in Czech): Zemny plyn a nafta, v. 4, p. 485–492.

Bergerat, F., 1989, From pull apart to the rifting process: The deformation of the Pannonian Basin: Tectonophysics, v. 157, p. 271–280.

Bielik, M., 1988, Analysis of the stripped gravity map of the Pannonian Basin: Geologica Carpathica, v. 39, no. 1, p. 99–108.

Bielik, M., 1991, Density modelling of the Earth's crust in the intra-Carpathian basins, *in* S. Karamata, ed., Geodynamic evolution of the Pannonian Basin: Academic Conference 62, Serbian Academy of Science and the Arts, Beograd: Beograd, Serbian Academy of Science and the Arts, p. 123–132.

Bisztricsany, E., 1974, The depth of the low velocity layer (LVL) in Europe and in some adjacent regions: Geophysical Transactions, v. 22, p. 61–68.

Blizkovsky, M., et al., 1986, Geophysical model of lithosphere: Project Oborovy Fond Technickeho Rozvoje, period 1982–1985, final report (in Czech): Manuscript, Brno, Geofyzika; Prague, Czechoslovak Academy of Sciences, 423 p.

Buday, T., 1965, Die Tektogenese und der Bau der Neogenen Becken der Westkarpaten, *in* T. Buday, I. Cicha, and J. Senes, eds., Miozän der Westkarpaten: Bratislava, Dionyz Stur Geological Institute, p. 169–250.

Burjanov, V. B., V. V. Gordienko, S. N. Kulik, I. V. Logvinov, 1980, Complex geophysical model of the Western Carpathians, Czech Massif and Saxon-Thuring zone lithosphere (along the 6th international deep seismic sounding [DSS] profile): Geofyziceskij zurnal, v. 2, p. 3–13.

Burjanov, V. B., V. V. Gordienko, O. V. Zavgorodnaja, S. N. Kulik, I. V. Logvinov, and V. N. Schuman, 1987, Geophysical model of European tectonosphere: Kiev, Naukova Dumka, 184 p.

Cermak, V., 1986, Geothermic model of the Earth's crust on the territory of the Central and Eastern Europe (in Czech), *in* M. Blizkovsky, ed., Geophysical model of lithosphere. Final Report, Oborovy Fond Technickeho Rozvoje, period 1982–1985: Brno, Geofyzika; Praha, Czechoslovak Academy of Sciences, p. 423.

Cermak, V., and F. Hurtig, 1977, Preliminary heat flow map of Europe: Praha and Potsdam, Geophysical Institute, Czechoslovak Academy of Sciences, Central Earth Physics Institute, Academy of Science of the German Democratic Republic, scale 1:5,000,000, 3 sheets, 58 p.

Constantinescu, L., and D. Enescu, 1984, A tentative approach to possibly explaining the occurrence of the Vrancea earthquakes: Revue Roumaine de Geologie, Geophysique et Geographie, Serie de Geophysique, v. 28, p. 19–32.

Csontos, L., 1995, Tertiary tectonic evolution of the Intra-Carpathian area: A review, *in* H. Downes and O. Vaselli, eds., Neogene and related magmatism in the Carpatho-Pannonian region: Acta Vulcanologica, v. 7, no. 2, p. 1–13.

Csontos, L., A. Nagymarosy, F. Horvath, and M. Kovac, 1992, Tertiary evolution of the intracarpathian area: A model: Tectonophysics, v. 208, p. 221–241.

Davies, J. H., and F. von Blanckenburg, 1995, Slab breakoff: A model of lithosphere detachment and its test in the magmatism and deformation of collisional orogens: Earth and Planetary Science Letters, v. 129, p. 85–102.

Decker, K., and H. Peresson, 1996, Tertiary kinematics in the Alpine–Carpathian–Pannonian system; links between thrusting, transform faulting and crustal extension, *in* G. Wessely and W. Liebl, eds., Oil and gas in Alpidic thrust belts and basins of Central and Eastern Europe: European Association of Petroleum Geoscientists Special Publication 5, p. 69–77.

Demetrescu, C., and M. Andreescu, 1994, On the thermal regime of some tectonic units in a continental collision environment in Romania: Tectonophysics, v. 230, p. 265–276.

Doglioni, C., 1992, Main differences between thrust belts: Terra Nova, v. 4, p. 152–164.

Dovenyi, P., F. Horvath, R. Liebe, J. Galfi, and I. Erki, 1983, Geothermal conditions of Hungary: Geophysical Transactions, v. 29, no. 1, p. 1–89.

Downes, H., G. Panto, T. Poka, D. P. Mattey, and P. B. Greenwood, 1995, Calc-alkaline volcanics of the Inner Carpathian arc, Northern Hungary: New geochemical and oxygen isotopic results, *in* H. Downes and O. Vaselli, eds., Neogene and related magmatism in the Carpatho-Pannonian region: Acta Vulcanologica, v. 7, no. 2, p. 29–41.

Embey-Isztin, A., H. Downes, D. E. James, B. G. J. Upton, G. Dobosi, G. A. Ingram, R. S. Harmon, and H. G. Scharbert, 1993, The petrogenesis of Pliocene alkaline volcanic rocks from the Pannonian Basin, Eastern Central Europe: Journal of Petrology, v. 34, p. 317–343.

Enescu, D., 1987, Contributions to the knowledge on the lithosphere structure in Romania on the basis of seismic data: Studii si Cercetari de Geologie, Geofizica, Geografie, Seria Geofizica, v. 25, p. 20–27.

Enescu, D., and B. Enescu, 2000, Seismotectonic model regarding the genesis and the occurrence of Vrancea (Romania) earthquakes: Proceedings of Society of Exploration Geophysicists/European Association of Geoscientists and Engineers/Romanian Society of Geophysics Bucharest 2000 International Geophysical Conference and Exposition, April 10–14, 2000, Bucharest, Romania, p. 496–497.

Fuchs, K., et al., 1979, The Romanian earthquake of March 4, 1977: II. Aftershocks and migration of seismic activity: Tectonophysics, v. 53, p. 225–247.

Girbacea, R., and W. Frisch, 1998, Slab in the wrong place: Lower lithospheric mantle delamination in the last stage of the Eastern Carpathian subduction retreat: Geology, v. 26, p. 611–614.

Gnojek, I., and F. Janak, 1986, Complex evaluation of the airborne measured geophysical fields of the Inner Carpathians in relation to mineral resources in scale 1:50,000

(Souhrnne zpracovani letecky merenych geofyzikalnich poli vnitrnich Karpat do meritka 1: 50,000) (in Czech): Manuscript, Brno, Geofyzika a.s., p. 317.

Haas, J., S. Kovacs, L. Krystyn, and R. Lein, 1995, Significance of Late Permian–Triassic facies zones in terrane reconstructions in the Alpine–North Pannonian domain: Tectonophysics, v. 242, p. 19–40.

Horvath, P., 1979, Physical quantities used in geothermal researches (in Hungarian): Magyar Geofyzika XX, v. 1, p. 31–38.

Horvath, F., 1984, Neotectonics of the Pannonian Basin and the surrounding mountain belts: Alps, Carpathians and Dinarides: Annales Geophysicae, v. 2, p. 147–154.

Horvath, F., 1986, Geophysical model of the formation of the Pannonian Basin (in Hungarian): Ph.D. thesis, Eotvos Lorand University, Budapest, 148 p.

Horvath, F., 1993, Towards a mechanical model for the formation of the Pannonian Basin: Tectonophysics, v. 226, p. 333–357.

Horvath, F., and J. Rumpler, 1984, The Pannonian basement: extension and subsidence of an Alpine orogene: Acta Geologica Hungarica, v. 27, p. 229–235.

Hrusecky, I., D. Plasienka, and L. Pospisil, 2006, Identification of the North European platform below the eastern part of the Western Carpathian Flysch belt, *in* J. Golonka and F. J. Picha, eds., The Carpathians and their foreland: Geology and hydrocarbon resources: AAPG Memoir 84, p. 717–727.

Iosif, T., and S. Iosif, 1979, Vrancea earthquake of March 4, 1977— A multiple seismic event and its seismotectonic implications: Revue Roumaine de Geologie, Geophysique et Geographie, Serie de Geophysique, v. 23, p. 3–17.

Jankowski, J., V. Petr, J. Pecova, and O. Praus, 1979, Induction vector estimates in the Polish-Czechoslovak part of the Carpathians: Studia Geophysica et Geodetica, v. 23, p. 89–93.

Jiricek, R., 1979, Tectogenetic development of the Carpathian arc during Oligocene and Miocene (in Czech), *in* M. Mahel, ed., Tektonicke profily Zapadnych Karpat: Bratislava, Dionyz Stur Geological Institute, p. 205–215.

Jiricek, R., and P. Seifert, 1990, Paleogeography of the Neogene in the Vienna Basin and the adjacent part of the foredeep, *in* D. Minarikova and H. Lobitzer, eds., Thirty years of geological cooperation between Austria and Czechoslovakia: Prague, Geological Survey, p. 89–105.

Joo, I., E. Csati, P. Jovanovic, M. Popescu, V. I. Somov, H. Thurm, J. Thury, I. N. Totomatov, J. Vanko, and T. Wyrzykowski, 1981, Recent vertical crustal movements of the Carpatho-Balkan region: Tectonophysics, v. 71, p. 41–52.

Kovac, M., A. Nagymarosy, J. Sotak, and K. Sutovska, 1993, Late Tertiary paleogeographic evolution of the Western Carpathians: Tectonophysics, v. 226, p. 401–416.

Kovacs, S., 1987, Plate tectonic position of the basement of the Great Hungarian Plain and its connections (in Hungarian): Hungarian Academy of Science and Communication, Szeged, Special Publication, p. 17–20.

Krus, S., and A. Sutora, eds., 1986, Geophysical-geological atlas of the Alpine–Carpathian mountain system: Manuscript, Archive Brno, Geofyzika, scale 1:3,000,000, 12 sheets, 25 p.

Krutichovskaja, Z. A., O. M. Rusakov, I. K. Paskevic, and K. Posgay, 1971, Magnetic field of the Carpathian region and several problems of its geologic interpretations (in Russian): Acta Geologica Hungarica, v. 15, p. 377–382.

Ksiazkiewicz, M., 1960, Description of the paleogeography of the Polish Flysch Carpathians (in Polish): Prace Institutu Geologicznego, v. 33, p. 209–231.

Lankreijer, A., M. Kovac, S. Cloetingh, P. Pitonak, M. Hloska, and C. Biermann, 1995, Quantitative subsidence analysis and forward modelling of the Vienna and Danube basins: Tectonophysics, v. 252, p. 433–452.

Le Pichon, X., F. Bergerat, and M. J. Roulet, 1988, Plate kinematics and tectonics leading to the Alpine belt formation; a new analysis, *in* S. Clark Jr., B. C. Burchfiel, and J. Suppe, eds., Processes in continental lithospheric deformation: Geological Society of America Special Paper 218, p. 111–131.

Lexa, J., and V. Konecny, 1974, The Carpathian volcanic arc: A discussion: Acta Geologica Academiae Scientiarum Hungaricae, v. 18, p. 279–293.

Lexa, J., V. Konecny, M. Kaliciak, and V. Hojstricova, 1993, Volcanic distribution in the Carpathian–Pannonian region in space and time (in Slovak), *in* M. Rakus and J. Vozar, eds., Geodynamicky model a hlbinna stavba Zapadnych Karpat: Bratislava, Dionyz Stur Geological Institute, p. 57–69.

Lexa, J., V. Konecny, M. Kovac, and M. Nemcok, 1995, Relationship among convergence, back-arc extension and volcanism in the Western Carpathians during Neogene: Europrobe workshops— Pancardi, Stara Lesna 1995, p. 65–66.

Linzer, H. G., 1996, Kinematics of retreating subduction along the Carpathian arc, Romania: Geology, v. 24, no. 2, p. 167–170.

Mandl, G., 1988, Mechanics of tectonic faulting: Models and basic concepts: Amsterdam, Elsevier, 407 p.

Marton, E., 1988, Paleomagnetism: An overview of the central Mediterranean, *in* M. Rakus, J. Dercourt, and A. E. M. Nairn, eds., Evolution of the northern margin of Tethys, v. I: Occassional Publication of Earth Sciences and Resources Institute New Series, v. 3, p. 223–244.

Marton, E., 1993a, Paleomagnetism in the Mediterranean from Spain to the Aegean: A review of data relevant to Cenozoic movements, *in* E. Mantovani, A. Morelli, and E. Boschi, eds., Recent evolution and Seismicity of the Mediterranean Region: North Atlantic Treaty Organization Advanced Study Institute, Series C, v. 402, p. 367–402.

Marton, E., 1993b, The itinerary of the Transdanubian Central Range: An assessment of relevant paleomagnetic observations: Acta Geologica Hungarica, v. 37, no. 1–2, p. 135–151.

Marton, E., and P. Marton, 1978, On the difference between the paleomagnetic poles from the Transdanubian Central Mountains and the Villany Mountains, respectively (in Hungarian): Magyar Geofizika, v. 19, p. 129–136.

Marton, E., and P. Marton, 1989, A compilation of paleomagnetic results from Hungary: Geophysical Transactions, v. 35, p. 117–133.

Meulenkamp, J. E., M. Kovac, and I. Cicha, 1996, On late Oligocene to Pliocene depocenter migrations and the evolution of the Carpathian–Pannonian system: Tectonophysics, v. 266, p. 301–317.

Mocanu, I. V., and F. Radulescu, 1994, Geophysical features of the Romanian territory, *in* T. Berza, ed., ALCAPA II field guidebook: Geological Institute of Romania, p. 17–36.

Monfret, T., A. Deschamps, and B. Romanowicz, 1990, The Romanian earthquake of August 30, 1986: A study based on GEOSCOPE very long-period and broadband data: Pure and Applied Geophysics, v. 133, no. 2, p. 367–379.

Morley, C. K., 1996, Models for relative motion of crustal blocks within Carpathian region, based on restorations of the Outer Carpathian thrust front: Tectonics, v. 15, p. 885–904.

Nemcok, M., 1993, Transition from convergence to escape: Field evidence from the West Carpathians: Tectonophysics, v. 217, p. 117–142.

Nemcok, M., and J. Lexa, 1990, Evolution of the basin and range structure around the Ziar mountain range: Geologica Carpathica, v. 41, p. 229–258.

Nemcok, M., L. Pospisil, J. Lexa, and R. A. Donelick, 1998a, Tertiary subduction and slab break-off model of the Carpathian–Pannonian region: Tectonophysics, v. 295, p. 307–340.

Nemcok, M., J. Hok, P. Kovac, F. Marko, M. P. Coward, J. Madaras, J. J. Houghton, and V. Bezak, 1998b, Tertiary extension development and extension/compression interplay in the West Carpathian mountain belt: Tectonophysics, v. 290, p. 137–167.

Nemcok, M., M. P. Coward, W. J. Sercombe, and R. A. Klecker, 1999, Structure of the West Carpathian accretionary wedge: Insights from cross section construction and sandbox validation: Physics and Chemistry of the Earth, Part A: Solid Earth and Geodesy, v. 24, p. 659–665.

Nemcok, M., P. Konecny, and O. Lexa, 2000, Calculations of tectonic, magmatic and residual stress in the Stiavnica stratovolcano, Western Carpathians, implications for mineral precipitation paths: Geologica Carpathica, v. 51, p. 19–36.

Oncescu, M., 1987, On the stress tensor in Vrancea region: Journal of Geophysics, v. 62, p. 62–65.

Pajdusak, P., J. Plomerova, and V. Babuska, 1988, Deep lithospheric structure of the West Carpathians determined on the basis of P-residua, *in* Proceedings of the Conference "Deep structure investigation of Czechoslovakia," May 2–5, 1988, Smolenice, Bratislava, Veda, p. 151–158.

Pecova J., V. Petr, and O. Praus, 1979, Internal electrical conductivity distribution on Czechoslovak territory, *in* V. Babuska and J. Plancar, eds., Geodynamic investigations in Czechoslovakia: Bratislava, Veda, various scales, 1 sheet.

Pecskay, Z., et al., 1995, Space and time distribution of Neogene–Quaternary volcanism in the Carpatho-Pannonian region, *in* H. Downes and O. Vaselli, eds., Neogene and related magmatism in the Carpatho-Pannonian region: Acta Vulcanologica, v. 7, no. 2, p. 15–28.

Poka, T., 1988, Neogene and Quaternary volcanism of the Carpathian–Pannonian region: Changes in chemical composition and its relationship to basin formation, *in* L. H. Royden and F. Horvath, eds., The Pannonian Basin: A study in basin evolution: AAPG Memoir 45, p. 257–276.

Polonic, G., 1986, On the seismotectonic relations in the Moldavian Platform and adjacent units: Revue Roumaine de Geologie, Geophysique et Geographie, Serie de Geophysique, v. 30, p. 11–17.

Posgay, K., 1967, A magyarorszagi fdldmagneses hatók attekintö vizsgálata: Geofizikai Kozlemenyek (Budapest), v. 16, no. 4, p. 89.

Posgay, K., T. Bodoky, and E. Hegedus, 1998, Investigation of the lithosphere and asthenosphere by seismic method, results and open questions: Magyar Geofizika, v. 39, p. 90–99.

Pospisil, L., and D. Vass, 1984, Influence of the structure of the lithosphere on the formation and development of intramontane and back molasse basins of the Carpathian mountains: Geophysical Transactions, v. 30, no. 4, p. 355–371.

Pospisil, L., et al., 2006, Crustal and lithospheric structure of the Carpathian–Pannonian region— A geophysical perspective: Regional geophysical data on the Carpathian–Pannonian lithosphere, *in* J. Golonka and F. J. Picha, eds., The Carpathians and their foreland: Geology and hydrocarbon resources: AAPG Memoir 84, p. 651–697.

Praus, O., J. Pecova, V. Petr, K. Pec, M. Hvozdara, V. Cerv, J. Pek, and M. Lastovickova, 1981, Electromagnetic induction and electrical conductivity in the Earth's body, *in* A. Zapotek, ed., Geophysical syntheses in Czechoslovakia: Bratislava, Veda, 2 p.

Radu, C., 1974, Contribution a l'etude de la seismicite de la Roumanie et comparaison avec la seismicite du bassin mediterraneen et en particulier avec la seismicite de Sud-Est de la France: Ph.D. thesis, University of Strasbourg, Strasbourg, France, 404 p.

Radu, C., 1979, Catalogul cutremurelor puternice produse pe teritoriul Romanei: Partea I. Inainte de 1901; Partea II. 1901–1979, *in* I. Cornea and C. Radu, eds., Certetori Seismologice asupra Cutremurului din 4 Martie 1977: Bucharest, Center of Earth Physics and Seismology, p. 723–752.

Ratschbacher, L., O. Merle, P. Davy, and P. Cobbold, 1991a, Lateral extrusion in the Eastern Alps: Part 1. Boundary conditions and experiments scaled for gravity: Tectonics, v. 10, no. 2, p. 245–256.

Ratschbacher, L., W. Frisch, and H. G. Linzer, 1991b, Lateral extrusion in the Eastern Alps: Part 2. Structural analysis: Tectonics, v. 10, no. 2, p. 257–271.

Ratschbacher, L., H. G. Linzer, F. Moser, R. O. Strusiewicz, H. Bedelean, N. Har, and P. A. Mogos, 1993, Cretaceous to Miocene thrusting and wrenching along the central south Carpathians due to corner effect during collision and orocline formation: Tectonics, v. 12, p. 855–873.

Roman, C., 1970, Seismicity in Romania— Evidence for the sinking lithosphere: Nature, v. 228, p. 1176–1178.

Royden, L. H., 1985, The Vienna Basin: A thin-skinned pull-apart basin, *in* K. T. Biddle and N. Christie-Blick, eds., Strike-slip deformation, basin formation, and sedimentation: SEPM Special Publication 37, p. 319–338.

Royden, L. H., 1988, Late Cenozoic tectonics of the Pannonian Basin system, *in* L. H. Royden and F. Horvath, eds., The Pannonian Basin: A study in basin evolution: AAPG Memoir 45, p. 27–48.

Royden, L. H., and T. Baldi, 1988, Early Cenozoic tectonics and paleogeography of the Pannonian and surrounding regions, *in* L. H. Royden and F. Horvath, eds., The Pannonian Basin: A study in basin evolution: AAPG Memoir 45, p. 1–16.

Royden, L. H., F. Horvath, and B. C. Burchfiel, 1982, Transform faulting, extension and subduction in the Carpathian–Pannonian region: Geological Society of America Bulletin, v. 93, p. 717–725.

Royden, L. H., F. Horvath, A. Nagymarosy, and L. Stegena, 1983a, Evolution of the Pannonian Basin system: 2. Subsidence and thermal history: Tectonics, v. 2, p. 91–137.

Royden, L. H., F. Horvath, and J. Rumpler, 1983b, Evolution of the Pannonian Basin system: 1. Tectonics: Tectonics, v. 2, p. 63–90.

Salters, J. M., S. R. Hart, and G. Panto, 1988, Origin of late Cenozoic volcanic rocks of the Carpathian arc, Hungary, *in* L. H. Royden and F. Horvath, eds., The Pannonian Basin: A study in basin evolution: AAPG Memoir 45, p. 279–292.

Sefara, J., M. Kurkin, and J. Plancar, 1980, Research of the Conrad discontinuity in the West Carpathians (in Slovak): Manuscript, Brno, Archive Geofyzika, 89 p.

Spakman, W., 1990, Tomographic images of the upper mantle below Central Europe and the Mediterranean: Terra Nova, v. 2, p. 542–553.

Stegena, L., B. Geczy, and F. Horvath, 1975, Late Cenozoic evolution of the Pannonian Basin: Tectonophysics, v. 26, p. 71–91.

Subotin, S. I., 1975, The Earth's physics problems in the Ukraine: Kiev, Naukova Dumka, 173 p.

Szafian, P., 1999, Gravity and tectonics: A case study in the Pannonian Basin and the surrounding mountain belt: Ph.D. thesis, Vrije University, Amsterdam, 154 p.

Tari, G., F. Horvath, and J. Rumpler, 1992, Styles of extension in the Pannonian Basin: Tectonophysics, v. 208, p. 203–219.

Tschekunov, A. V., 1989, Geodynamic model of geotraverse Dnepr–Donieck paleorift–Ukraine Schield–Southern Carpathians (in Russian): Geotectonika, Moscow, v. 6, p. 3–11.

Tschekunov, A. V., ed., 1993, Geophysical studies of the lithosphere (in Russian): Kiev, Naukova Dumka, 156 p.

Tschekunov, A. V., and K. A. Boljubach, 1979, Geologiceskije aspekty komplexnoj interpretacii seijsmiceskich i gravimetriceskich dannych v zapadnoj casti Ukrainy, *in* Gravitacionnaja model kory i verchnej mantii Zemli: Kiev, Naukova Dumka, 289 p.

Tunyi, I., and M. Kovac, 1991, Paleomagnetic investigation of the Neogene sediments from the Little Carpathians: Contributions of the Geophysical Institute of Slovak Academy of Sciences, v. 21, p. 125–146.

Turculet, L., and E. Mihail, 1970, Considérations géologiques sur la Plate-forme Moesique de la partie NE de la Roumanie: Acta Geologica Academiae Scientiarum Hungaricae, v. 14, p. 95–101.

Vass, D., M. Pereszlenyi, M. Kovac, and M. Kral, 1990, Outline of Danube basin geology: Foldtani Kozlony, Buletin of the Hungarian Geological Society, v. 120, p. 193–214.

von Blanckenburg, F., and J. H. Davies, 1995, Slab breakoff: A model for syncollisional magmatism and tectonics in the Alps: Tectonics, v. 14, no. 1, p. 120–131.

World Stress Map Project, 1997, http://www-wsm.physik.uni-karlsruhe.de (accessed August 22, 2005).

Zatopek, A., 1979, On geodynamical aspects of geophysical syntheses in Central Europe, *in* Geodynamic investigations in Czechoslovakia, Final report: Bratislava, Veda-Slovak Academy of Sciences, p. 91–104.

SPECIAL THEMATIC PAPERS

29

Golonka, Jan, Andrzej Ślączka, and Frank J. Picha, 2006, The Western Carpathians and Ouachitas: A comparative study of geodynamic evolution, *in* J. Golonka and F. J. Picha, eds., The Carpathians and their foreland: Geology and hydrocarbon resources: AAPG Memoir 84, p. 561–584.

The Western Carpathians and Ouachitas: A Comparative Study of Geodynamic Evolution

Jan Golonka
AGH University of Science and Technology, Jagiellonian University, Krakow, Republic of Poland
Andrzej Ślączka
Institute of Geological Sciences, Jagiellonian University, Krakow, Republic of Poland
Frank J. Picha
International Petroleum Consultant, Walnut Creek, California, U.S.A.

ABSTRACT

Twelve time-interval maps have been presented that depict the plate-tectonic configuration, paleogeography, and lithofacies for the circum-Carpathian area from the Late Carboniferous through Neogene and for the circum-Ouachita region from the Late Cambrian through Early Permian.

The following geodynamic evolution stages can be distinguished in these two orogens:- stage I: rifting of terranes off the major continent, forming oceanic basins, Triassic–Early Cretaceous in the Carpathian region, Cambrian–Devonian in the Ouachita basin; stage II: formation of subduction zones along the active margin, partial closing of the oceanic basin, development of flysch basin associated with this rifting on the platform (passive margin) with the attenuated crust, Late Cretaceous–Paleocene in the Carpathian region, Early Carboniferous in the Ouachitas; stage III: collision, perhaps terrane-continent, with the accompanying convergence of two large continents, development of accretionary prisms, the Eocene–early Miocene in the Carpathian region, Late Carboniferous in the Ouachitas; stage IV: postcollisional, Miocene–present–future(?) in the Carpathians, Permian–Triassic in the Ouachitas. Both Carpathians and Ouachitas are an accretionary prism formed in response to terrane-continent and continent-continent collision. The paleogeographic approach we have taken shows how these mountain belts were constructed through the orogenic cycle, which reflects complex plate-tectonic processes. The Carpathians and Ouachitas record complete and homologous Wilson cycles.

DOI:10.1306/985629M841463

INTRODUCTION

The aim of this chapter is to compare the plate-tectonic evolution and position of the major crustal elements of the Carpathians and Ouachita Mountains in a global framework and to show the relationships between tectonic processes and sedimentary record during their orogenic cycles. The Carpathian and Ouachita Mountains are geographically distant. The age of the formation of each orogen is also quite different. The Carpathians mountain belt formed during the Mesozoic and Cenozoic, whereas the Ouachitas formed in the Paleozoic; their tectonostratigraphic history, however, displays striking similarities.

Both orogens have been the subject of numerous classic sedimentological studies of the flysch deposits (Ksiazkiewicz, 1954; Dzulynski and Slaczka, 1958; Dzulynski et al., 1959; Cline, 1960, 1970; Dzulynski and Walton, 1965; Morris, 1974, 1989; McBride, 1975; Lowe, 1976, 1989; Moiola and Shanmugan, 1984; Pescatore and Slaczka, 1984; Shanmugan and Moiola, 1995; Picha, 1996). The Ouachita orogenic belt was also a subject of a tectonic synthesis from the point of view of the Wilson cycle (Viele and Thomas, 1989). The history of the whole circum-Carpathian realm is more complex (e.g., see Golonka et al., 2000, 2003; Golonka and Slaczka, 2003); if we concentrate, however, on the Western Carpathians, the orogenic cycle is also clear and evident. Thus, these two orogens could provide a good example to compare the modern and ancient orogens and fulfill the goal of the International Geological Correlation Project (IGCP) 453.

Twelve time-interval maps have been presented that depict the plate-tectonic configuration, paleogeography, and lithofacies for the circum-Carpathian region and adjacent areas from the Late Carboniferous through Neogene and for the circum-Ouachita region from the Late Cambrian through Early Permian. This paleogeographic approach shows how a mountain belt was constructed through the orogenic cycle, which reflects complex plate-tectonic processes. Some unsolved questions and problems could be answered by a comparison between one and another orogen.

TECTONIC SETTING

Western Carpathians

The Carpathians form a great arc of mountains, which stretches more than 1300 km (807 mi) from the Vienna forest to the Iron Gate on the Danube (Figure 1). On the west, the Carpathians are linked with the Eastern Alps and, on the east, pass into the Balkan chain. Traditionally, the Carpathians are subdivided into the Western and Eastern Carpathians (Mahel, 1974). The Western Carpathians consist of an older range known as the Inner or Central Carpathians, and the younger ones are known as the Outer or Flysch Carpathians (Mahel, 1974;

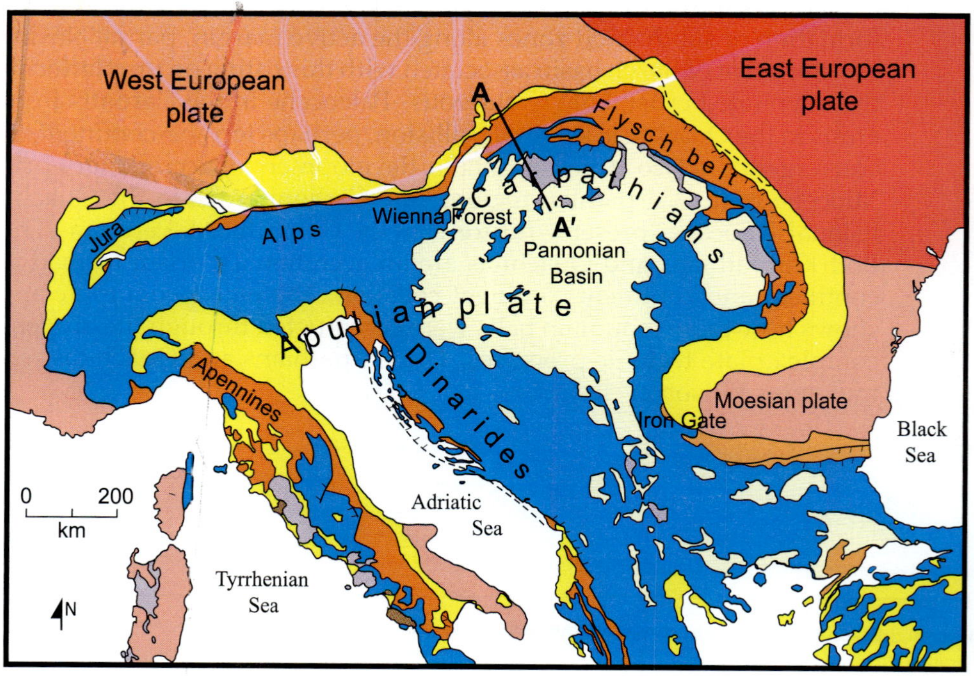

Figure 1. Tectonic sketch map of the Alpine–Carpathian–Pannonian–Dinarides basin system (modified from Picha, 1996). AA′ = localization of cross section (Figure 3).

Ksiazkiewicz, 1977; Slaczka and Kaminski, 1998). At the boundary of these two ranges is the Pieniny Klippen Belt (Figure 2). The Inner Carpathians are regarded as a prolongation of the Northern Calcareous Alps and form part of the Apulia plate that, in a regional sense, is a promontory of the African plate (Picha, 1996). They are divided into the Tatric, Veporic, and Gemeric nappes (Figure 2) and are the prolongation of the Lower, Middle, and Upper Austroalpine nappes, respectively. The Inner Carpathian nappes contact along a Tertiary strike-slip boundary with the Pieniny Klippen Belt (Figures 2, 3). The Pieniny Klippen Belt is composed of several successions (mainly deep- and shallow-water limestones), covering a time span from the Early Jurassic to Paleogene. This strongly tectonized structure is a terrain about 800 km (497 mi) long and 1–20 km (0.6–12 mi) wide, which stretches from Vienna on the west to the Maramures (or Poiana Botizii area, northeast Romania) on the east (Figure 2). The Pieniny Klippen Belt is in the western part of the area thrust over the Outer Carpathian nappes (Figure 3) in Poland, and Eastern Slovakia is separated from the Outer Carpathians by the Miocene subvertical strike-slip fault. The Outer Carpathians are built up of a stack of nappes and thrust sheets changing along the Carpathians, built mainly of continual flysch sequences as much as 6 km (3.6 mi) thick, representing a time span from the uppermost Jurassic up to the lower Miocene. All the Outer Carpathian nappes are overthrust onto the southern part of the North European platform covered by the autochthonous Miocene deposits of the Carpathian Foredeep to a distance of at least 70 km (43 mi). During overthrusting movement, the Northern Carpathian nappes became uprooted from the basement, and only basinal parts were preserved (Figure 3). A narrow zone of folded Miocene deposits developed along the frontal Carpathian thrust, i.e., the Zglobice Wieliczka unit in the Northern Carpathians (ZG; Figure 2) and its equivalent Subcarpathian (or Borislav) or Sambor–Rozniatov unit of the Ukrainian and Romanian parts of the Eastern Carpathians (Figure 2).

Ouachita Fold Belt

The Ouachita Mountains in Oklahoma and Arkansas are a surface expression of the Ouachita belt (Figure 4), which continues in the subsurface to the southwest, surfacing once again in the Marathon region of west Texas near the Mexican border (Arbenz, 1989; Viele and Thomas, 1989). The continuation of the fold belt beneath the surface into Mexico is at the least uncertain and speculative. Structural, lithofacies, and Pb isotopic data (e.g., Dickinson and Lawton, 2001) do not support this continuation. To the east, the Ouachita orogenic belt converges with the southern Appalachian belt in the area of the central Mississippi uplift (beneath a thick Mesozoic cover). Our knowledge of this junction remains speculative (Thomas, 1977, 1989).

Structures in the exposed part of the Ouachita orogenic belt are folds and thrusts vergent toward the North American craton with major decollement thrusts visible especially along some parts of the orogenic front (Figure 5). In most of the Ouachita Mountains, the structures trend east but curve strongly to the southwest at the western end in Oklahoma and slightly southeastward at their eastern end in Arkansas.

Among the structures, several anticlinoria are present such as the Broken Bow, Benton, and Potato Hill (Figure 4) uplifts, where older rocks of the Ouachita facies are exposed (Arbenz, 1968, 1989; Viele and Thomas, 1989).

Generally, all the exposed rocks in the Ouachita Mountains are strongly allochthonous and have been thrust as far as 80 km (50 mi) northward from their former position. The overthrust surface is steep in the northern part of orogen; southward, it is nearly horizontal. The major elevated zone can be distinguished here (e.g., the zone drilled by the Hassie Hunt 1 Carl Neely well in Figure 5). The slope of the southwest surface is strongly dependent on the basement configuration (Lillie et al., 1983; Golonka, 1988).

The orogenic belt faces northwestward and northward toward the craton, where variable thickness of Paleozoic rocks lies on the Precambrian basement. Positive basement areas include the Nashville dome, the Ozark uplift, the Arbuckle and Wichita Mountains (southwestern Oklahoma), and the Llano uplift of central Texas (Figure 4). The negative areas are the Black Warrior, Arkoma, Fort Worth, and Val Verde basins (Viele and Thomas, 1989) (Figure 4).

The Arbuckle and Wichita structures are foreland structures related to the Ouachita orogeny. Generally, two uplifted zones are present: Arbuckle and Wichita Mountains with the exposed Precambrian and Lower Cambrian crystalline rocks, separated by the Ardmore and Anadarko basins (Ham, 1950, 1969). They are characterized by a system of faults trending east and southeast, as well as uplifts and grabens. Structural relief from uplift to basin floor is as much as 9 km (5.4 mi) (Lillie et al., 1983; Golonka, 1988). The faults are high angle, commonly overthrust. These structures continue southward beneath the overthrust, allochthonous Ouachita orogenic belt. Based on seismic and well data, two uplifted zones are separated by a depression zone (Golonka, 1988) (Figure 4). The northern uplift is the continuation of the Arbuckle Mountains. It is bordered on the north by the Anadarko basin. The depression zone

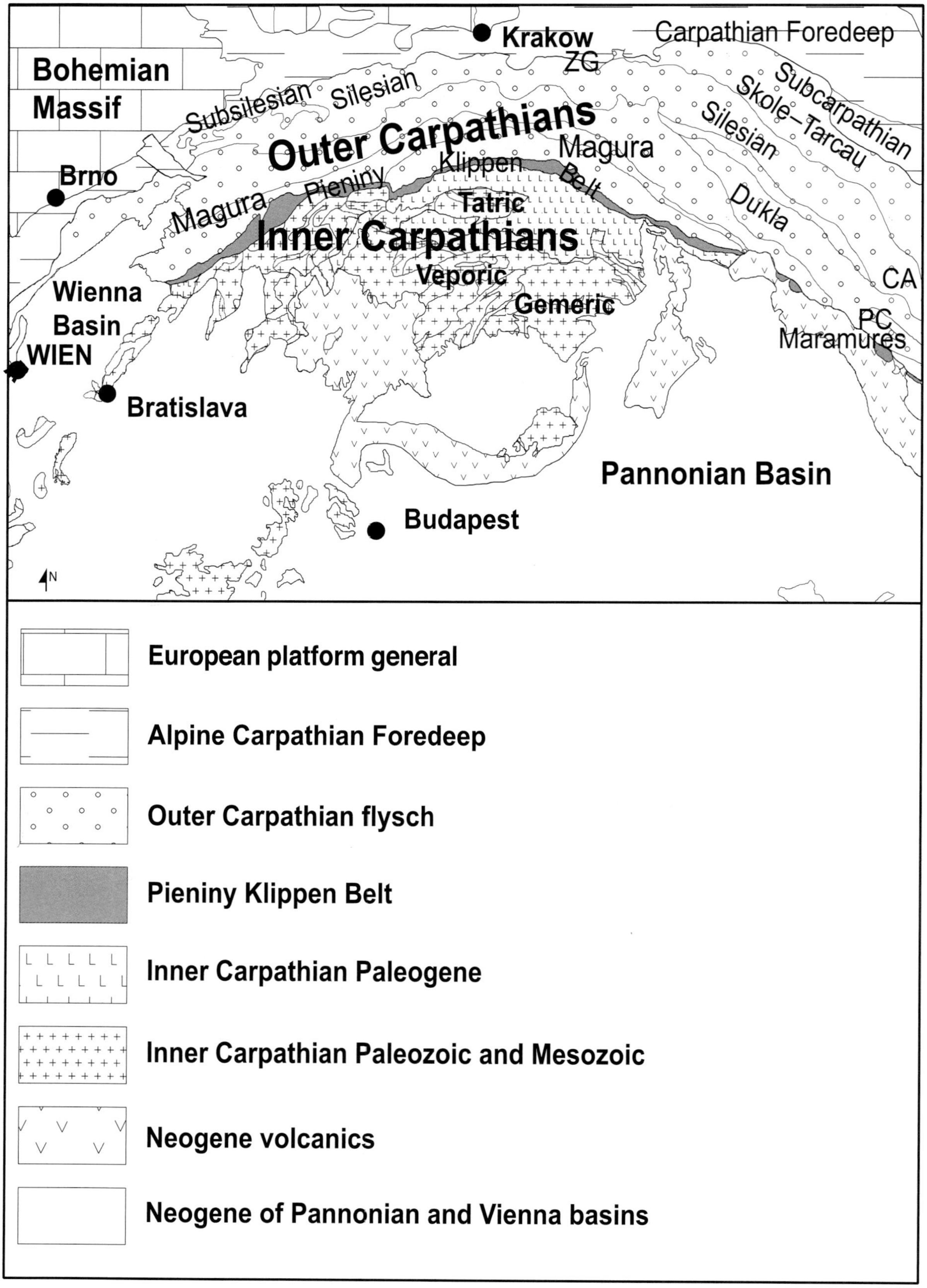

Figure 2. Geological map of Western Carpathians and adjacent areas (modified from Wessely, 1996). ZG = Zglobice–Wieliczka unit; CA = Charnahora–Audia unit; PC = Porkulec–Convolute unit.

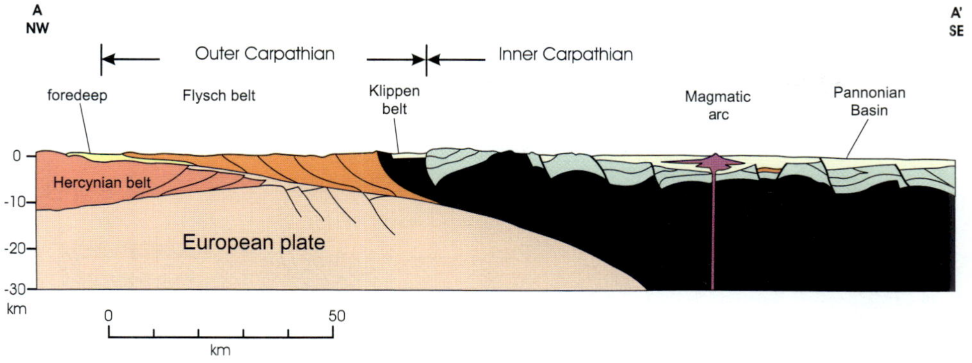

Figure 3. Generalized cross section across the Carpathian–Pannonian region (Picha, 1996).

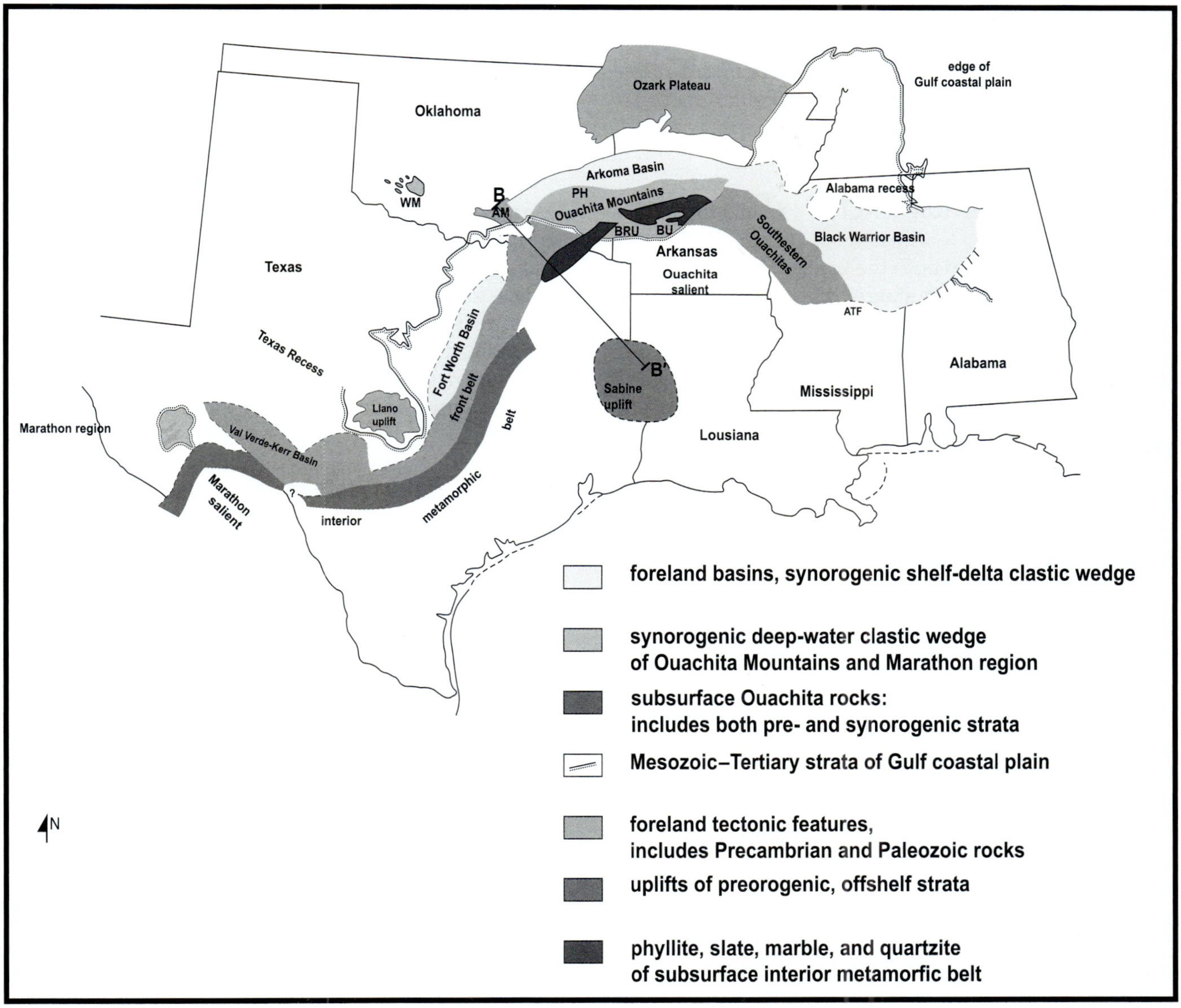

Figure 4. Map of the Ouachita fold belt (modified after Arbenz, 1989; Thomas, 1989; Viele and Thomas, 1989). BB′ = Location of cross section (Figure 5); AM = Arbuckle Mountains; ATF = Appalachians tectonic front; BRU = Broken Bow uplift; BU = Benton uplift; WM = Wichita Mountains; PH = Potato Hill uplift.

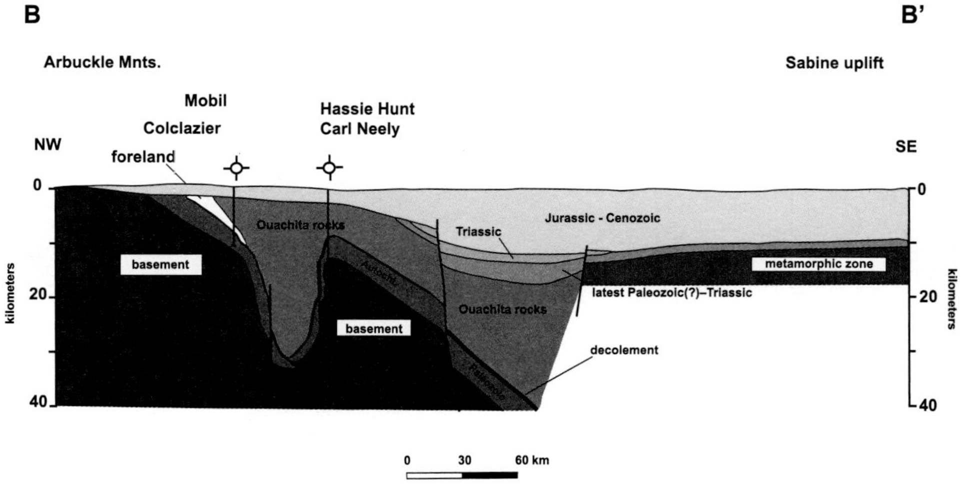

Figure 5. Generalized cross section across Ouachita fold belt between Arbuckle Mountains and Sabine uplift, based partially on seismic reflection line GC36 (courtesy of Mobil Exploration and Producing Technical Center, Dallas, Texas).

is related to the Fort Worth and Ardmore basins. The axis of the uplift trends northeast through Oklahoma and Arkansas beneath the Potato Hill.

GEODYNAMIC EVOLUTION STAGES

During the evolution of both the Carpathians and Ouachita orogens, four main stages can be distinguished that marked the general changes in their developments (Tables 1, 2). The stages normally can be divided into more local substages.

Stage I

This stage is characterized by rifting of terranes off the major continent and formation of oceanic type of basins (Table 2).

The Western Carpathians (Triassic and Jurassic–Early Cretaceous)

The basement of most of the plates that is an important factor in the Mesozoic–Cenozoic evolution of the circum-Carpathian area was formed during the late Paleozoic collisional events (Golonka et al., 2000, 2003). Moesia, the Eastern Alps, the Inner Carpathians, Tisza (Figure 6), and adjacent terranes were sutured to the Eurasian (Laurasian) arm of Pangea (Figures 6, 7), whereas Adria and adjacent terranes were situated near the Gondwana (African) arm. The equatorial position of the Bohemian Massif, adjacent to the Carpathian plates, agrees with the global Pangean model of Golonka (2000, 2002). The Paleotethys Ocean was located south of Eurasia and was established in Southern and Central Europe during the Permian–Triassic. The Meliata–Halstatt Ocean (Figure 8) formed as a result of rifting of the Tisza terrane from Eurasia during the Triassic. The Vardar–Transylvanian Ocean separated the Tisza (Bihor–Apuseni) block from the Moesian–Eastern European platform (Sandulescu, 1988; Sandulescu and Visarion, 2000). The existence of the embayment of Vardar–Transylvanian oceanic zone between the Inner Carpathian and the European platform is possible (Golonka et al., 2000, 2003). Pelagic limestones and Triassic pebbles in the exotic sequences in the Pieniny Klippen Belt (Birkenmajer, 1988; Birkenmajer et al., 1990) and Magura unit (Sotak, 1986) could have originated in this embayment. The embayment position and its relation to the other parts of the Tethys, Vardar Ocean, Meliata–Halstatt Ocean, Dobrogea rift, and Polish–Danish aulacogen (Figure 8) remain quite speculative.

The Triassic shallow-water limestones, marls, and dolomites with numerous reefs (Kiessling et al., 1999) prevailed in the Inner Carpathian platform areas and were underlaid by Upper Permian and Lower Triassic clastic deposits. The sedimentary sequence rests on metamorphic and granitic rocks of the late Paleozoic.

The above-mentioned period of sedimentation was followed by the Jurassic opening of the Ligurian–Penninic–Pieniny Klippen Belt and Magura Ocean (Figures 8, 9). This opening resulted in the rifting of the Alpine–Inner Carpathian terranes off Eurasia. Several basins separated by carbonate platforms developed in the Carpathian region. The local rift with the andesitic volcanics developed also on the southern margin of the North European platform (Slaczka, 1998). Stampfli (2001) recently postulated single Penninic ocean (Pe, Figure 8) separating the Apulia and the Eastern Alps blocks from Eurasia. We proposed a similar model for the Pieniny Klippen Belt (Figure 8) ocean in the Carpathians. The orientation of the Pieniny Klippen Belt ocean was southwest–northeast (see discussion in

Golonka and Krobicki, 2001). This ocean was divided into the northwestern and southeastern basins by the midoceanic Czorsztyn ridge (Figure 9). The deepest part of the southeastern basin is documented by extremely deep-water Jurassic–Early Cretaceous deposits (pelagic limestones and radiolarites) of Zlatna unit (Sikora, 1971; Golonka and Sikora, 1981), later described also as the Ultrapieninic unit or Vahicum (e.g., Plasienka, 1999). The transitional slope sequences between deepest basinal units and ridge units are known as Pieniny, Branisko (Kysuca), Niedzica, and Czertezik successions (Figure 10). The shallowest ridge sequences are known as the Czorsztyn succession. Dark Lower Jurassic shales (Table 1) in this succession are followed by Middle Jurassic–earliest Cretaceous crinoidal and nodular limestones and Cretaceous variegated marls. Sedimentation of pelagic limestones, marls, cherts, cherty limestones, and some synrift turbiditic deposits occurred during this period in the basinal areas, whereas on the uplifted parts, more shallow, calcareous sediments developed. The deepest part of the northwestern basin is represented by extremely deep-water condensed Jurassic–Early Cretaceous deposits (pelagic cherty limestones and radiolarites) of the southern Magura (or Grajcarek or Hulina) unit (Golonka and Sikora, 1981; Birkenmajer, 1986; Golonka et al., 2000, 2003). The paleogeographic extent of the Magura basin remains somewhat enigmatic and speculative. Also speculative is the existence of the oceanic crust below the whole Magura basin. The transitional slope sequence is known from some outcrops located north of the Czorsztyn ridge (such as Zawiasy and Stare Bystre in Poland) (Golonka and Sikora, 1981). Ridge sequences, as well as transitional slope sequences, are also called the Oravicum (e.g., Plasienka, 1999).

During the Late Jurassic (Figure 9), the southern part of the North European platform, north of the Pieniny–Magura realm, started to be rifted, and small basins (e.g., proto-Silesian basin in the Western Carpathians, Figure 9), with black, mainly redeposited marls [Kimmeridgian(?)–Tithonian], were created (Pescatore and Slaczka, 1984). The Western Carpathian Silesian basin probably extended in the Eastern Carpathian Sinaia or Black flysch basin (Sandulescu, 1984; Krautner, 1996; Krautner and Krstic, 2000). The rifting in the Eastern Carpathians was accompanied by the extrusion of diabase–melaphyre volcanics (Lashkevitsch et al., 1995; Golonka et al., 2000). These basins indicate the beginning of the Outer Carpathian realm development. The opening is related to the closing of Pieniny–Magura (North and South Penninic) basin. In the Carpathian region, subduction developed at the end of Jurassic and the beginning of the Cretaceous (Figures 7, 9), along the southern margin of the narrowing basin north of the approaching Inner Carpathians and began to consume the Pieniny Klippen Belt ocean (Birkenmajer, 1986).

The rapid supply of shallow-water calcareous material to the newborn basins could be an effect of the strong tectonoeustatic sea level fluctuations known from that time. Black sediments mark the beginning of a euxinic cycle in the Outer Carpathian basin that lasted until the Albian. The black marls pass gradually upward into calcareous turbidites (Cieszyn limestones–Sinaia beds), which created several submarine fans (Table 1). The occurrence of deep-water microfauna indicates that subsidence of the basins must have been quite rapid. During the early part of the Cretaceous, the calcareous turbidites gave way to black calcareous shales and thin sandstones passing upward into black, commonly siliceous shales. This type of sediment is already known also from the other Outer Carpathian basins. During the Hauterivian, Barremian, and Aptian, several coarse-grained submarine fans developed. The supply of clastic material was probably connected with the Early Cretaceous uplift known from the Bohemian Massif. The rifting of the North European platform lasted during the Early Cretaceous up to the Aptian. The main phase of the hypabyssal intrusion and extrusion alkaline magma (mesocratic teschenites and leucocratic syenites) was connected with the late period of rifting (Lucinska-Anczkiewicz et al., 2000). However, the probability that magmatic eruptions could have started earlier cannot be excluded (Smulikowski, 1980), as in the case of the Eastern Carpathians. Lack of changes in foraminifera assemblages through the Early Cretaceous suggests a lack of pronounced changes of depth of basins that corresponded generally to the *Recurvoides* zone of Haig. It implies a generally continuous tectonic subsidence of the basins during that period. This subsidence was equal to the rate of sedimentation.

In the early Albian in the black shales, widespread turbiditic sedimentation started that can be connected with a compressional period very pronounced in the Eastern Carpathians. In that part of the Carpathians, the compressional movement started during the Aptian and Albian, and the inner part of the Carpathians was folded, nappes formed, and, in front of moving nappes, coarse-grained sediments (Bucegi–Soymul conglomerates) and olistostromes developed (Sandulescu, 1984, 1988; Kruglov, 1989, 2001).

The Ouachita Fold Belt (Cambrian–Devonian)

Two models explain the formation of the Ouachita basin. In one model (e.g., Keller and Cebull, 1973), the ocean with oceanic crust was opened during the early

Table 1. Comparative Stratigraphic Chart.

Ouachitas				Carpathians			
Time		**Ouachita**	**Arbuckle**	**Time**		**Magura –Pieniny**	**Outer subbasins**
Stage IV (290–225 Ma)	Triassic	clastic red beds of Eagle Mills Formation	clastic red beds of Eagle Mills Formation	Stage IV (20–0 Ma)	Pliocene – Quarternary	fluvial and continental strata	fluvial and continental strata
	Permian	fluvial-deltaic and continental strata	Fluvial-deltaic and continental strata		middle Miocene	continental shales and sandstones with marine intercalations	brackish sandstones, shales and limestones; marine, fine-grained, and clastic sediments; limestones, evaporites including salt, locally olistostromes
Stage III (325–290 Ma)	Late Carboniferous (325–290)	poorly sorted sandstones and dark gray shales, sometimes limestones (Atoka Fm.); olistostromes with enormous blocks of limestone, chert, and black shale; black shales and thin-bedded sandstones (Johns Valley); flysch and olistostromes (Jackfork Sandstone)	poorly sorted sandstones and dark gray shales, sometimes limestones (Atoka Fm.); limestone, shale, and sandstone (Wapanucka Limestone)	Stage III (55–20 Ma)	early Miocene	medium- and thin-bedded sandstones (Malcov Fm.)	basal conglomerates passing upward to brackish and marine sandstones, shales and marls, locally salt; olistostromes; sandstones, shales, and gray marls (Krosno beds)
					Oligocene	medium- and thin-bedded sandstones (Malcov Fm.); flysch, marls, proximal turbidites (Magura Fm.)	sandstones, shales, and gray marls (Krosno beds), olistostromes; dark brown bituminous shales and cherts (Menilite shales), locally sandstone submarine fans (Mszanka, Cergowa, Kliwa sandstones), olistostromes
					Eocene	*Globigerina* marls flysch, Variegated shales, proximal turbidites (Magura Fm.)	*Globigerina* marls flysch, Variegated shales, proximal turbidites (Ciezkowice sandstone)
Stage II (365–325 Ma)	Early Carboniferous	flysch with tuffs, shales and cherts (Stanley Shale); black shales, cherts (Woodford Chert)	black shales and sandstones (Springer Fm.) dark-colored Caney Shale; Sycamore Limestone; black shales, cherts (Woodford Chert)	Stage II (100– 55 Ma)	Paleocene	flysch (Ropianka), Variegated shales, marls	thick-bedded, coarse-grained turbidites and fluxoturbidites (Istebna, Ciezkowice), marls (Weglowka), variegated shales
					Late Cretaceous	flysch (Ropianka), calcareous turbidites marls, radiolarites, green and red shales	thick-bedded, coarse-grained turbidites and fluxoturbidites (Godula, Istebna), flysch (Ropianka), marls, (Weglowka, Frydek)

Table 1. (cont.).

Stage I (500–365 Ma)	Devonian	black shales, cherty limestones, cherts (Arkansas Novaculite)	black shales, cherts (Woodford Chert)
	Silurian	turbiditic sandstones, siltstones, and shales (Missouri Mountain Shale, Blaylock Sandstone)	limestones and shales (Hunton)
	Ordovician	turbiditic sandstones and limestones, graptolitic shales, cherts, coarse-grained submarine slides (Blaylock, Polk Creek, Bigfork, Womble, Blakely, Mazarn, Crystal Mountain, Collier)	limestones, dolomites, shales, cherts (Hunton, Sylvan, Viola, Simpson, Arbuckle)
	Cambrian	collier graptolitic shale with lenses and beds of limestones of a distal turbiditic character	Arbuckle Group, Honey Creek limestones, Reagan transgressive sandstone

Stage I (235–100Ma)	E. Cretaceous	black shales, turbiditic sandstones, marls, pelagic cherty limestones, cherts (radiolarites), nodular limestones	black shales, carbonate turbidites (Cieszyn, Sinaia), turbiditic sandstones, coarse-grained submarine fans, olistostomes (Bucegi-Soymul)
	Late Jurassuc	pelagic cherty limestones, cherts (radiolarites), nodular limestones	carbonate turbidites, black shales, and marls
	Middle Jurassic	crinoidal limestones, cherts (radiolarites)	
	Early Jurassic	dark shales, pelagic limestones, turbiditic sandstones and limestones	
	Triassic	pelagic limestones	

Paleozoic. According to Thomas and Astini (1999), the Argentine Precordillera was rifted from the Ouachita embayment of Laurentia during the Cambrian. This model is preferred by the authors (Figures 11–13; note that the North American continent is rotated; see Walker et al., 1995; for example, Austin, Texas, is north of Little Rock, Arkansas). It agrees well with the global Phanerozoic plate-tectonic maps (Golonka, 2000, 2002). In addition, the lithostratigraphic sequences (Tables 1, 2) support the geodynamic evolution of the origin from rift through oceanic passive-margin, accretionary prism related to basin closure to thrusting and inversion. In the other model (e.g., Arbenz, 1989), the Ouachita basin was separated from the main (Iapetus–Rheic) ocean by poorly defined terranes. The present-day Sabine uplift may constitute the remnant of these terranes. Viele and Thomas (1989) stated that Carboniferous volcanics formed a part of the upper southern plate. The fragments of these southern terranes are perhaps included into the Inner Ouachita fold belt in Texas.

According to Thomas and Astini (1999), the distribution of synrift rocks and structures indicates diachronous rifting events along the margin of Laurentia during the Cambrian. The Ouachita rocks are younger than rocks known from the Blue Ridge area in the Appalachians. The time of rifting and formation of the Alabama–Oklahoma transform fault (Thomas, 1991) could be connected with an emplacement of the igneous rocks in the southern Oklahoma. These rocks, as young as 525 ± 25 Ma, are overlain by Upper Cambrian sandstones. For the Late Cambrian beginning of sedimentation in the Ouachita basin (Table 1), we have somewhat arbitrarily assumed time at around 500 Ma according to the Paleozoic timescale by Golonka and Kiessling (2002).

Table 2. Tectonostratigraphic Comparison.

Ouachitas				Carpathians			
Time		Tectonic Events and Elements	Sedimentary Style	Time		Tectonic Events and Elements	Sedimentary Style
Paleozoic–Triassic	Perm. Triassic (290–225 Ma)	thrusting and inversion, final formation of Arkoma foreland basin; formation of postorogenic extensional basins	marine and continental molasse; volcanics; continental red beds	Mesozoic–Cenozoic	middle Miocene–Future(20–? Ma)	thrusting and inversion, final formation of Carpathian Foredeep; formation of Pannonian extensional basin	marine and continental molasse; volcanics
	Late Carboniferous (325–290 Ma)	convergence of South America and North America, Sabine – Eurasia collision, thrusting and inversion with the continuing formation of accretionary prisms; incipient formation of Arkoma foreland basin	flysch, olistostromes; marine and continental molasse; volcanics		Eocene – Early Miocene (55–20 Ma)	convergence of Africa and Eurasi, Alcapa – Eurasia collision, thrusting, and inversion with the continuing formation of accretionary prisms; incipient formation of Carpathian Foredeep	flysch, olistostromes, pelagic shales, carbonates, and biogenic siliceous deposits; marine and continental molasse; volcanics; carbonates on ridges
	Early Carboniferous (365–325 Ma)	the subduction along the active margin of the Ouachita ocean; enigmatic Sabine terrane moving north, partial closing of oceanic basin development of main flysch basin on the platform (passive margin Ma) with the attenuated crust and beginning of formation of accretionary prisms	flysch, olistostromes, pelagic shales, carbonates, and biogenic siliceous deposits; volcanics		Late Cretaceous–Paleocene (100– 55 Ma)	subduction zones along the active margin of the Magura–Pieniny ocean, Inner Carpathian terrane moving north, partial closing of oceanic basin, nappes with northward polarity in the Inner Carpathians, development of main flysch basins on the platform (passive margin) with the attenuated crust and beginning of formation of accretionary prisms	flysch, olistostromes, pelagic shales, carbonates, and biogenic siliceous deposits; volcanics
	Cambrian–Devonian (500–365 Ma)	rifting of Precordilleran terrane off the North America; formation of the Ouachita ocean	passive-margin, deep-water turbidites, pelagic shales, carbonates, and biogenic siliceous deposit, volcanics (?); carbonates on surrounding platform		Triassic–Early Cretaceous (235–100 Ma)	rifting of Eastern Alps, Inner Carpathian terranes: off Eurasia; formation of the Pieniny-Magura, Outer Carpathian oceans and basins	passive-margin, deep-water turbidites, pelagic shales, carbonates, and biogenic siliceous deposit, volcanics; carbonates on surrounding platform

Two different facies were developed in the lower Paleozoic (Table 1): a deep offshelf water passive-margin facies in the Ouachita orogenic belt and a predominantly shallow-water cratonic facies in the Ouachita foreland (Ham, 1950, 1959, 1969; Cline et al., 1959; Cline, 1960, 1970; Flawn et al., 1961; McBride, 1975; Lowe, 1985, 1989; Viele and Thomas, 1989).

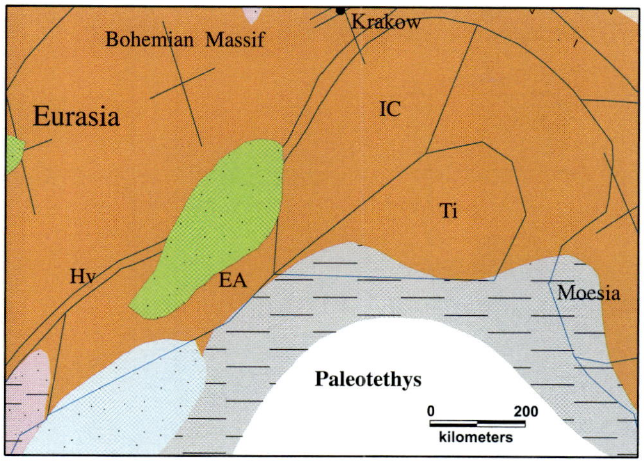

Figure 6. Paleogeography and lithofacies of the circum-Carpathian area during Late Carboniferous (modified from Golonka et al., 2000). EA = Eastern Alps; Hv = Helvetic; IC = Inner Carpathians; Ti = Tisza. The legend to this map is also valid for Figures 8, 9, 12–20.

The oldest rock of the deep-water or so-called Ouachita facies is the Collier Shale of Late Cambrian–Early Ordovician age (Arbenz, 1989; Lowe, 1989). The bulk of the formation is bluish-black, graptolitic shale with lenses and beds of limestones of a distal turbiditic character. The next unit is the Crystal Mountain sandstone, which overlies the Collier Shale. Light-gray, well-sorted, fine- to medium-grained, quartz-cemented, thick-bedded quartz sandstones with minor shales prevail here. The sandstones are proximal turbidites. According to Lowe (1989), they indicate deposition from submarine sediment gravity flows, including both high- and low-density turbidity currents. The Mazarn Shale conformably overlies the Crystal Mountain sandstone. It is chiefly dark-gray, laminated, graptolitic shale with numerous thin interbedded distal turbidites, both quartzose as well as carbonate types. According to Lowe (1989), they were accumulated under low-energy deep-water conditions.

The Blakely Sandstone (Middle Ordovician), which overlies the Mazarn Shale, consists of massive- to thin-bedded, well-sorted, fine-grained quartz sandstones and varicolored silts and shales deposited by sandy high-density turbiditic currents with debris flows and submarine slides (Lowe, 1989). The Womble Shale conformably overlies the Blakely Sandstone. It consists primarily of gray-black graptolitic slaty hemipelagic shale (Lowe, 1989; Gleason et al., 2002). Numerous thin-bedded calcareous distal turbidites occur in the lower part, whereas the upper part contains some turbiditic siltstones, phosphatic breccias, and gray-black, well-sorted, pelletal turbiditic limestones.

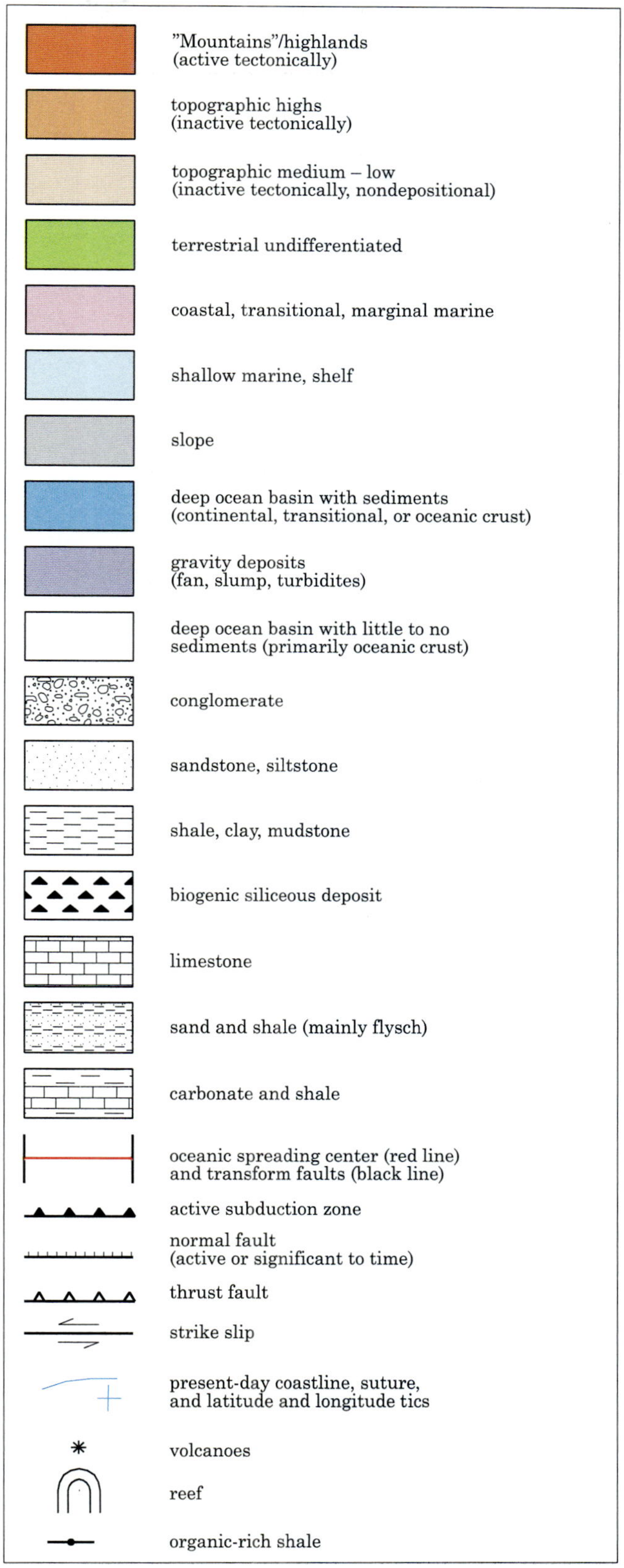

Figure 6. (cont.).

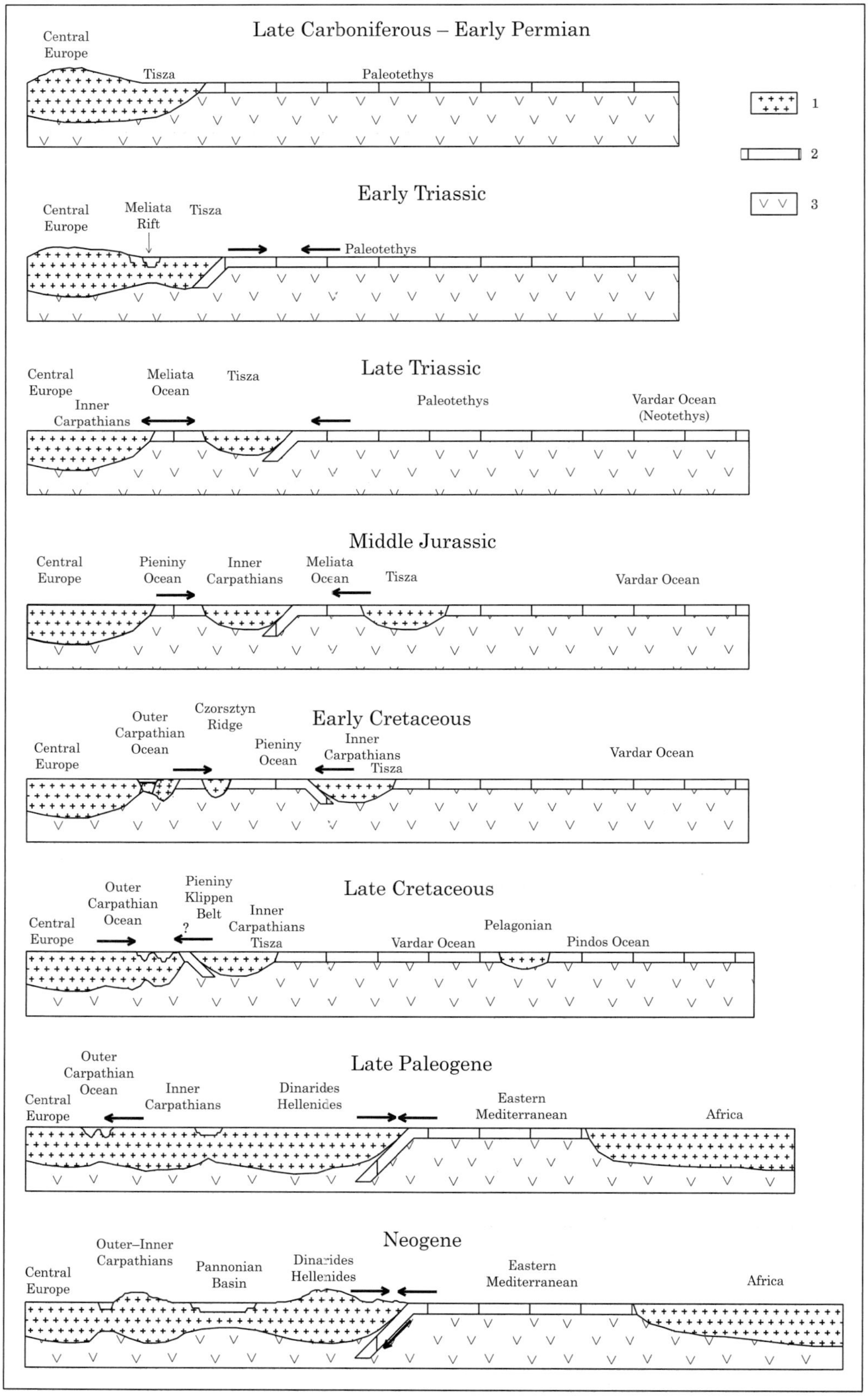

Figure 7. Highly schematic (not to scale) plate-tectonic profiles (modified from Golonka et al., 2000). Central Europe–Carpathians–Greece. 1 = Continental crust (including obducted, allochthonous rocks and sedimentary cover); 2 = oceanic crust (including deposits); 3 = upper mantle.

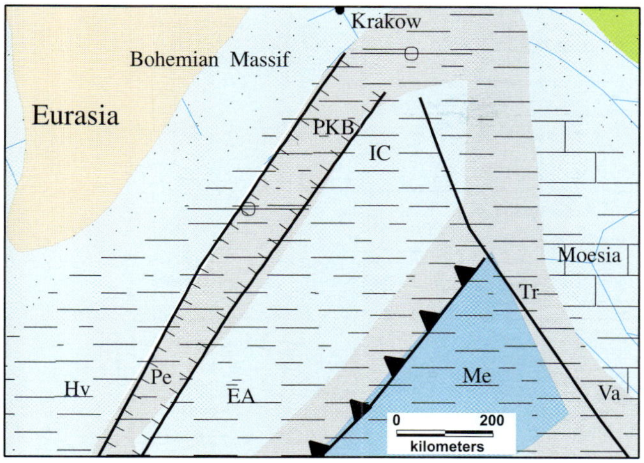

Figure 8. Paleogeography and lithofacies of the circum-Carpathian area during the Early Jurassic. For legend, see Figure 6. Do = Dobrogea; EA = Eastern Alps; Hv = Helvetic; IC = Inner Carpathians; Me = Meliata–Halstatt Ocean; PD = Polish–Danish aulacogen; Pe = Ligurian–Penninic; PKB = Pieniny Klippen Belt basin; Va = Vardar; Ti = Tisza; Tr = Transylvanian.

The Bigfork Chert of Middle Ordovician age conformably overlies the Womble Shale. This resistant unit consists of gray and black, thin-bedded, highly fractured chert with minor interbedded clay shales, siliceous shales, and siliceous limestones. The limestones appear to be turbidites, whereas the other rocks are pelagic in origin.

Conformably overlying the Bigfork Chert is the Polk Creek Shale of Late Ordovician age (Lowe, 1989). This thin shale unit is black, highly fissile, graptolitic, and carbonaceous with a few thin beds of calcareous chert. The Blaylock Sandstone of Silurian and possible partly Late Ordovician age (Lowe, 1989) overlies the Polk Creek Shale conformably or with a stratigraphic break. The formation consists of subequal proportions of olive-gray, thin-bedded, laminated, feldspathic, very fine-grained sandstones and siltstones as well as greenish-gray shales that together comprise a shaly flysch facies.

The Missouri Mountain Shale (Silurian) is, in places, transitional into the underlying Blaylock Sandstone. Varicolored shales (or slates) comprise the bulk of the unit. Near the top are thin interbedded turbidites composed of white, fine- to medium-grained, well-rounded, silica-cemented quartz sandstones.

The next stratigraphic unit is the Arkansas novaculite, which includes rocks of both Devonian and lowermost Mississippi age (Figure 14). This distinctive unit consists chiefly of novaculite, a light-colored, extremely fine-grained, homogenous, highly fractured siliceous rock similar to cherts but characterized by a dominance of quartz instead of chalcedony. The upper and lower members contain laminated beds of rounded and angular quartz grains as well as intraformational breccias of black chert, phosphate, and coarse quartz grains. Thin, graded, laminated siliceous beds and black shales characterize a middle member. According to Lowe (1989), breccias in the novaculite indicate erosion, whereas Viele and Thomas (1989) argue for deep-marine environment of deposition.

The cratonal or Arbuckle (Table 1) facies overlies the Precambrian and Cambrian igneous rocks. The older sedimentary unit is the Timbered Hills Group of Upper Cambrian age. It is composed of the Honey Creek Limestone and the Reagan Sandstone. The Reagan is a basal transgressive sandstone that incorporated much weathered material from underlying igneous rocks during its deposition. The Honey Creek consists chiefly of coarsely to very coarsely crystalline, highly glauconitic fragmental limestone. Thin intercalated middle-grained limestone and sandstone are frequent.

The Arbuckle Group, which overlies the Timbered Hills Group, is composed of more than 2000 m (6600 ft) (in southern Oklahoma) of carbonates, primarily limestones with approximately 15% dolomites. All Arbuckle carbonates are typically similar in lithology,

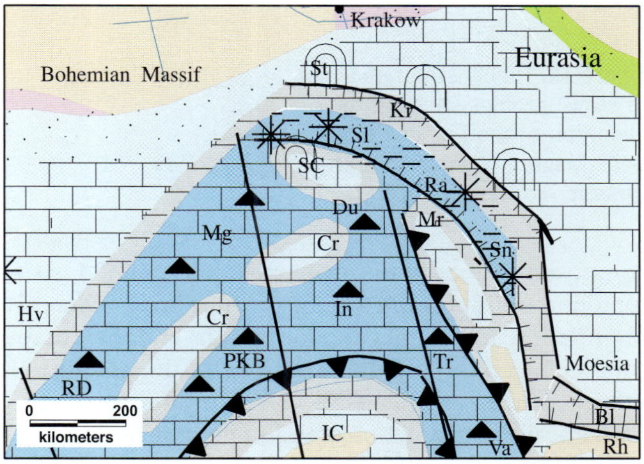

Figure 9. Paleoenvironment and lithofacies of the circum-Carpathian area during the Late Jurassic–Early Cretaceous (modified from Golonka et al., 2000). For legend, see Figure 6. Bl = Balkans; Br = Branisko; Cr = Czorsztyn ridge; Cz = Czertezik; Du = Dukla; Gr = Grajcarek; Hv = Helvetic; IC = Inner Carpathians; In = Inacovce–Kricevo; Kr = Kruhel; Ku = Kurovice; Mg = Magura basin; Mr = Marmarosh; Nd = Niedzica; Pi = Penninic; PKB = Pieniny Klippen Belt basin; Ra = Rakhov; RD = Rhenodanubian basin; Rh = Rhodopes; SC = Silesian ridge (cordillera); Sl = Silesian basin; Sn = Sinaia basin; St = Stramberk olistolith; Va = Vardar; Tr = Transylvanian; Zn = Zlatna.

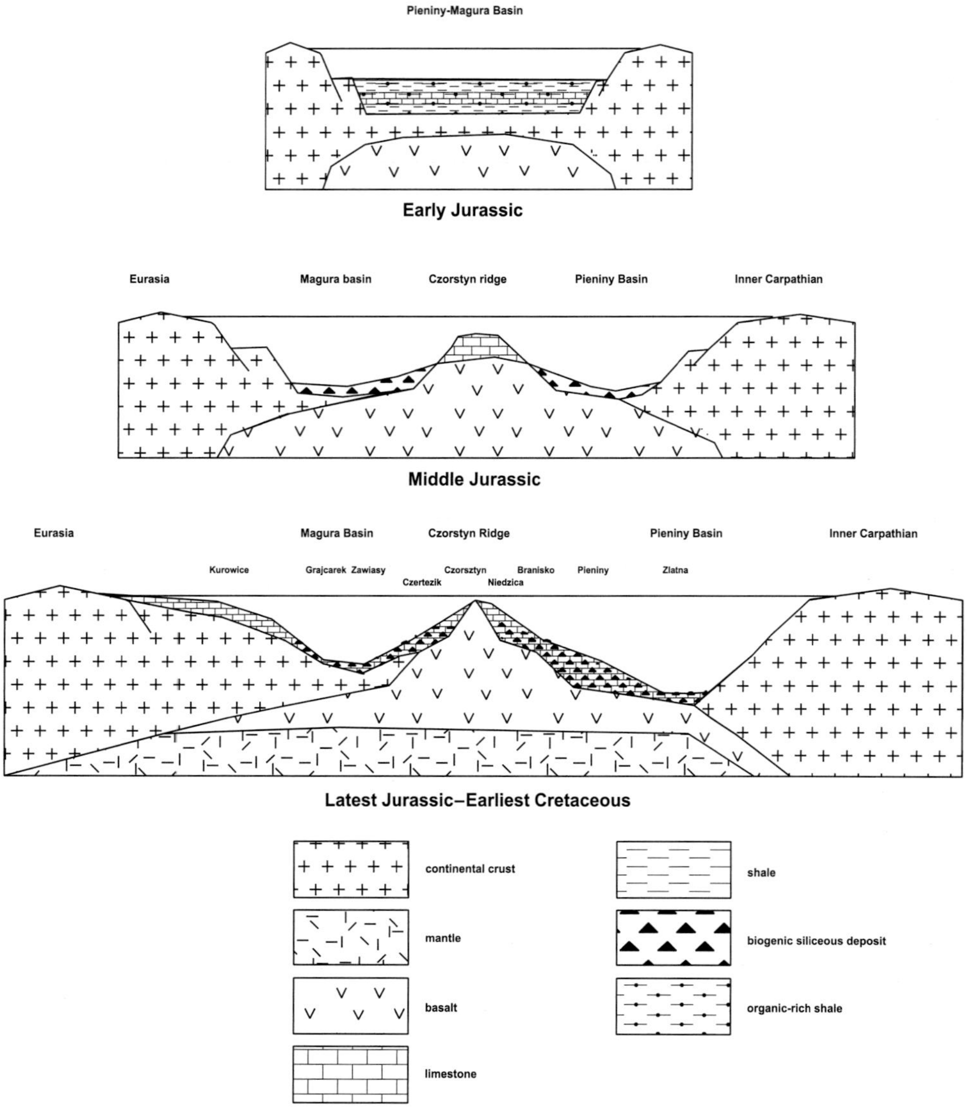

Figure 10. Highly schematic (not to scale) profiles showing the evolution of the Pieniny Klippen Belt–Magura ocean during the Jurassic–earliest Cretaceous.

composed primarily of limestones, fine-grained, stromatolitic, oolitic, or organodetritic limestones. The differentiation of the units into formations and their mapping is based largely on variations in the amount of quartz sand present and also on color and fossil content. Locally, a distinction of dolomite units is also possible. Although several formation names are used in the Ozark uplift–Arkoma basin area (Gasconade, Roubidoux, Jefferson City, Cotter, and Powell formations) in the wells, the Arbuckle Group is undivided. Similar shallow-water carbonate facies of Ordovician age are known in the foreland of Ouachita fold belt in Texas as the Ellenburger Group and in the Appalachians as the Knox Group.

The Simpson Group unconformably overlies the Arbuckle Group. The thickness of the Simpson ranges from 1000 to 2000 ft (3300 to 6600 ft). The basal formation of the Simpson is limestone and dolomitic limestone. The lower part is cream to brown dolomite, very finely sucrose in appearance, and the upper part is limestone and commonly fossiliferous. The Oil Creek Formation consists of gray-green and black shales, white to cream dense- to fine-crystalline limestone, and about 100-ft (30-m)-thick sand (important producing horizon). The following McLish Formation contains more limestone. These limestones are normally ostracodal, commonly contain coarse brown to gray oolites set in a white to creamy matrix, and commonly become sandy and dolomitic. Locally, some maroon, brown, or olive shale can be found. The uppermost Bromide Formation consists mostly of green shales with relatively thin white to cream commonly ostracodal limestones.

The Viola Limestone (Upper Ordovician) conformably overlies the Simpson Group. The lower part is dark brown to almost black and very cherty. Some beds of almost solid brown cherts are present. The upper part

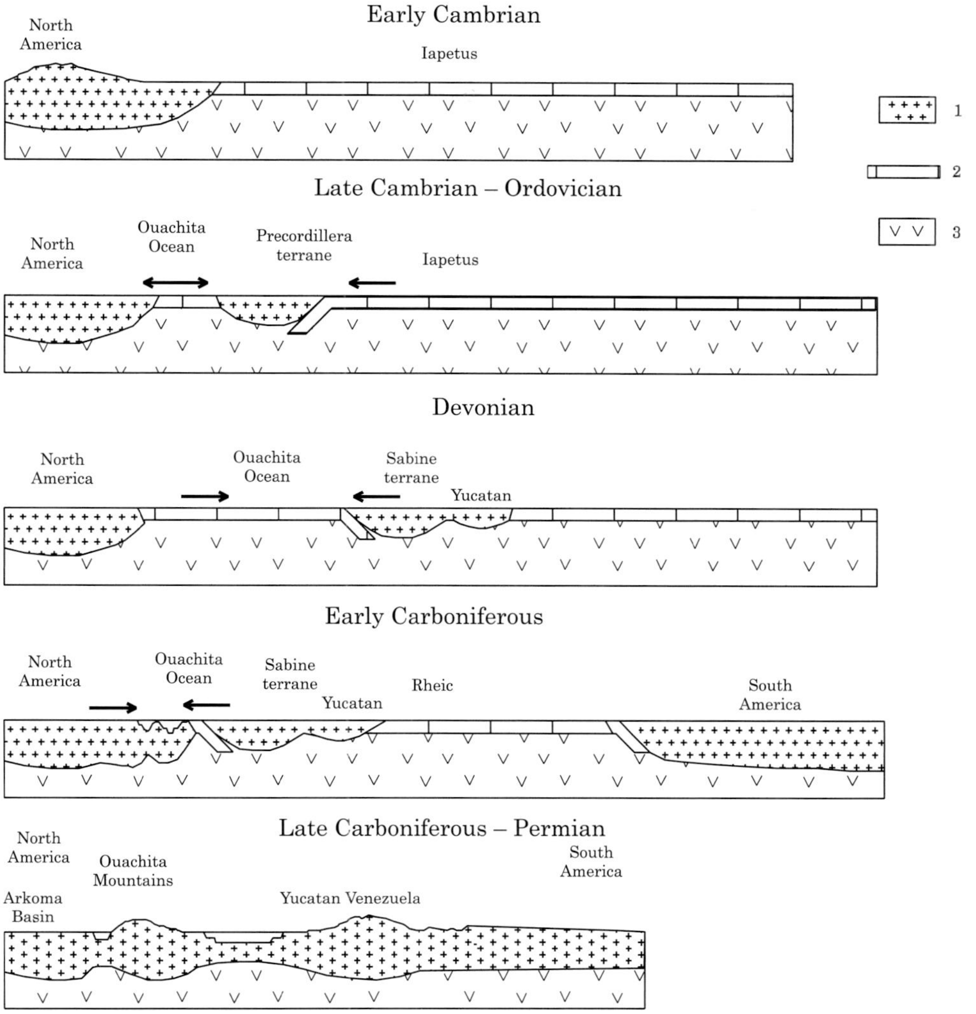

Figure 11. Highly schematic (not to scale) plate-tectonic profiles (modified from Golonka and Slaczka, 2000). North America–Ouachitas–Yucatan–South America. 1 = Continental crust (including obducted, allochthonous rocks and sedimentary cover); 2 = oceanic crust (including deposits); 3 = upper mantle.

is white to pink, coarsely crystalline, and contains a few fossils. The Sylvan Shale represents the uppermost part of Ordovician; it is a light-green and gray, flaky, splintery, soft shale containing graptolites.

The Hunton Formation, which overlies the Sylvan Shale, is of Ordovician–Silurian age. It consists of marls, dense marly thin-bedded limestones, and shales. The Woodford Chert (or Shale) of Late Devonian and early Mississippian(?) age unconformably overlies the Hunton Formation. It consists of alternating beds of black sapropelic papery shale with phosphate nodules and black cherts.

Among the lower Paleozoic rocks of the Arbuckle facies, the Arbuckle Group and Simpson Group merit special consideration. Both of them are the potential petroleum producing horizons, and additionally, some shales in the Simpson Group are potential source rocks. Deposition of graptolitic shales, pelagic limestones, flysch deposits, marls, cherts, and cherty limestones occurred during this period.

Stage II

This stage is characterized by formation of subduction zones along the active margin, partial closing of oceanic basin, development of main flysch basins on the platform (passive margin) with the attenuated crust, and beginning of formation of accretionary prisms (Table 2).

The Western Carpathians (Late Cretaceous–Paleocene)

In the beginning of the second stage, during the Cenomanian and Turonian, compression embraced the Inner Carpathians, and several nappes with northward polarity developed. Subduction consumed the major part of the Pieniny Klippen Belt ocean (Figures 5, 7). Cherty limestones gave way to marls and flysch deposits. With the development of the Inner Carpathian nappes, the fore-arc basin was formed between the uplifted part of the Inner Carpathian terrane (the so-called

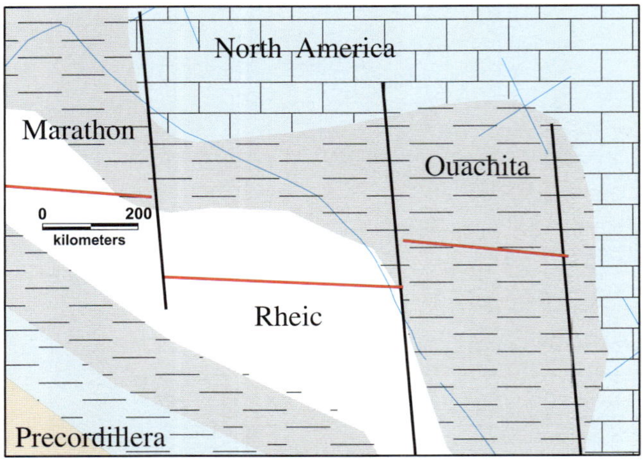

Figure 12. Paleogeography and lithofacies of the circum-Ouachita area during the Late Cambrian (modified from Golonka, 2000, according to Thomas and Astini, 1999). For legend, see Figure 6 [note that North American (Laurentia) continent is rotated (see Walker et al., 1995) and Austin, Texas, projected on the craton is north of Little Rock, Arkansas; spreading and transform faults are from Thomas and Astini, 1999; Thomas et al., 2002].

Andrusov ridge) and subduction zone. The flysch of the Klappe and Zlatne (Figure 15) succession was formed in this area. Behind the ridge, the Manin succession (Figure 15) was deposited in the back-arc basin. As an effect of these movements, the Inner Carpathians and Alps jointed with the Adria plate, and the ALCAPA terrane was created. In the Cenomanian period, subsidence was faster than the sedimentation rate, and uniform, deep-water pelagic sedimentation of radiolarites and

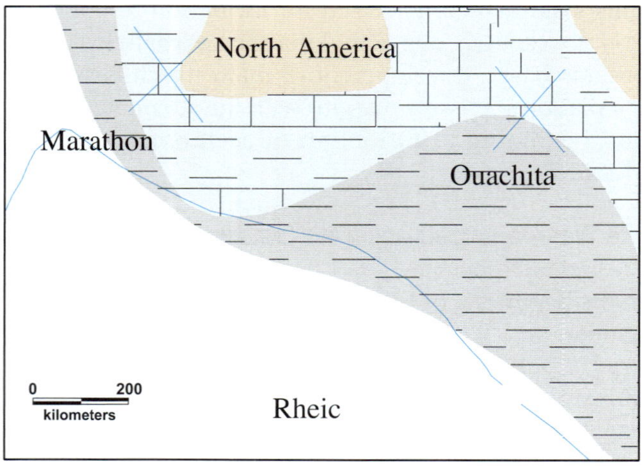

Figure 13. Paleogeography and lithofacies of the circum-Ouachita area during the Early Ordovician (modified from Golonka, 2000). For legend, see Figure 6.

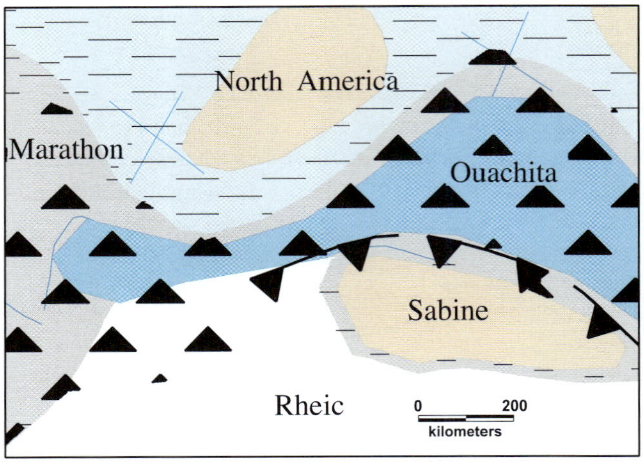

Figure 14. Paleogeography and lithofacies of the circum-Ouachita area during the Late Devonian (modified from Golonka, 2000). For legend, see Figure 6. Barbed line includes everything from the inferred volcanic arch through the trench and foreland basin.

green and red shales embraced a greater part of the Outer Carpathian basins.

In the Outer Carpathians during this stage, several ridges have been uplifted as an effect of the orogenic

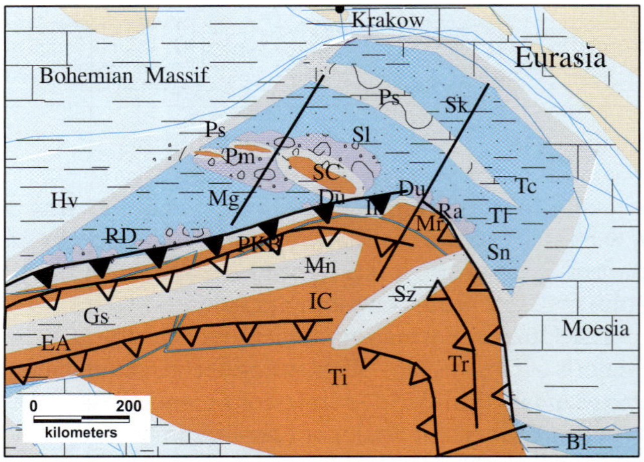

Figure 15. Paleogeography and lithofacies of the circum-Carpathian area during the Late Cretaceous–earliest Paleogene (modified after Golonka et al., 2000). For legend, see Figure 6. Bl = Balkans; Du = Dukla; EA = Eastern Alps; Gs = Gresten; Hv = Helvetic; IC = Inner Carpathians; In = Inacovce–Kricevo; Mg = Magura basin; Mn = Manin; Mr = Marmarosh; PKB = Pieniny Klippen Belt basin; Pm = Fore-Magura; Ps = Subsilesian; Ra = Rakhov; RD = Rhenodanubian basin; Rh = Rhodopes; SC = Silesian ridge (cordillera); Sl = Silesian basin; Sk = Skole; Sn = Sinaia basin; SZ = Szolnok; Tc = Tarcau; Ti = Tisza; Tl = Teleajen; Tr = Transylvanian nappes; Zn = Zlatne.

process. These ridges distinctly separated several subbasins, namely, Magura, Porkulec–Convolute, Dukla–Fore-Magura, Silesian, Charnahora–Audia, and Skole–Tarcau subbasins (Figures 2, 15). More outer subbasins (Skole, Silesian, and Dukla–Fore-Magura) reached diagonally the northern margin of the Outer Carpathians and successively terminated toward the west (Figure 15). From uplifted areas, situated in the Outer Carpathian basin as well as along its northern margin, an enormous amount of the clastic material was transported by various turbidity currents and filled the Outer Carpathian basins (Dzulynski and Slaczka, 1958; Ksiazkiewicz, 1968). Each basin had a specific type of clastic deposits, and sedimentation commenced in different times. It is interesting to note that this sedimentation started earlier in the outer subbasin (Skole–Tarcau subbasin) and migrated diachronously toward the inner subbasins (Bieda et al., 1963; Slaczka and Kaminski, 1998). In the Skole–Tarcau basin (Figure 15), sedimentation started during the Turonian and ended in the Paleocene, and deposits were represented by calcareous turbidites (siliceous marls) and thin- to thick-bedded turbidites (*Inoceramus*–Ropianka beds). In the western part of the area, these turbidites were terminated during the late Turonian–Coniacian by slump deposits. In the Silesian basin, sedimentation started during the late Turonian–early Coniacian and lasted up to the early Eocene and was mainly represented by thick-bedded, coarse-grained turbidites and fluxoturbidites (Godula beds, Istebna beds, and Ciezkowice Sandstone). In the Dukla (Figure 15) (Slaczka, 1971) and Magura (Figure 15), subbasin sedimentation commenced during the Campanian and lasted until the Paleocene (Oszczypko, 1992, 1998), and medium- and thin-bedded, medium-grained turbidites (*Inoceramus*–Ropianka beds) prevailed there. On the underwater ridge that divided Skole and Silesian subbasins (Subsilesian sedimentary area; Figure 15), a sequence of red and green marls (Weglowka marls) of Senonian to Eocene and gray marls (Senonian Frydek marls) was deposited. Small turbiditic fans developed locally in the small basins (Table 1).

In the eastern, Romanian part of the Outer Carpathians, Late Cretaceous–Paleocene sedimentation was not as intensive as in the western part, except for the outermost Tarcau subbasin, where clastic material was derived from the outer margin like in the Skole basin (Figure 15) (Sandulescu, 1984).

Ouachita Fold Belt (Early Carboniferous)

During the Early Carboniferous (Figures 11, 16), the Ouachita basin became a narrowing trough, with the flysch basins receiving vast amount of clastics (Cline et al., 1959; Briggs and Roeder, 1975; Arbenz, 1989). The subduction probably occurred along the southern margin of this narrowing basin north of the approaching Inner Ouachitas (or of the enigmatic Sabine terrane) and began to consume the Ouachita ocean. According to Arbenz (1989), the earliest evidence of the constriction of the Ouachita depositional basin and of the extension in the northern region is from the Lower Carboniferous (Meramecian). The northern margin supplied the clastic material for flysch deposits.

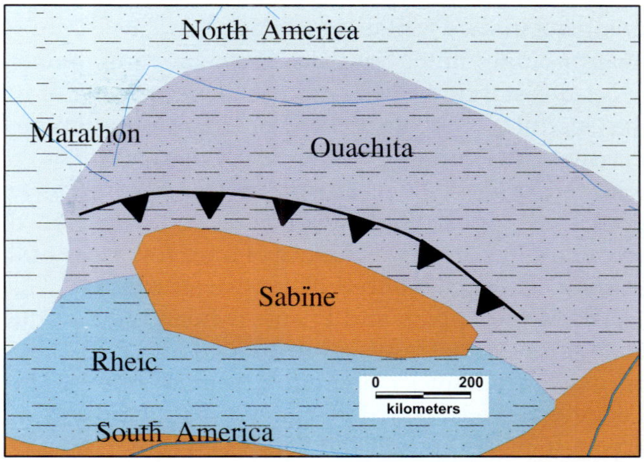

Figure 16. Paleogeography and lithofacies of the circum-Ouachita area during the Early Carboniferous (modified from Golonka et al., 2000). For legend, see Figure 6.

The Mississippian and Pennsylvanian deposits of predominantly flysch character are very thick and are the most widespread rocks in the Ouachita Mountains region (Table 1) as well as in the buried Ouachita fold belt (Cline et al., 1959; Flawn et al., 1961; Morris, 1974, 1989; McBride, 1975; Viele and Thomas, 1989). These deposits overlie the widespread Woodford Chert (Shale) of the Devonian–early Mississippian age described in the previous section. They are divided into the Stanley Shale, Jackfork Sandstone, Johns Valley Shale, and Atoka Formation.

The Stanley is dominantly olive-gray to black shale with subordinate proximal and distal turbidites that are very fine-grained and generally feldspathic sandstones at the base and near the top. Along the southern Ouachitas, sandstones become thicker and more numerous; tuffs and tuffaceous sandstones are prominent at the base; and disturbed bedding, impure cherts, and siliceous shales are sparse. Along the Frontal Ouachitas, a thinner Stanley section has less turbidites at the base and only a minor tuff interval but more intervals of impure chert and siliceous shale.

The Jackfork consists of gray-black shale and rhythmically interbedded whitish-gray, very fine-grained, quartz-rich turbidite packets. Olistostrome-type beds dominate in the Frontal Ouachitas of Arkansas, proximal turbidites in the Southern Ouachitas, and thinner bedded distal turbidites and black siliceous shales in the Ouachitas of Oklahoma. The proximal turbidites are massive, ridge-forming quartz-rich, scour-and-fill units with few or no shale interbeds. Distal turbidites have partial to complete Bouma sequences, better developed sole markings, greater evidence of trace fossils, as well as slope- or shock-induced disruptive structures.

The Mississippian and Pennsylvanian rocks of the cratonal or Arbuckle facies are divided into the Caney, Springer, Wapanucka, and Atoka units (Table 1). The Caney shale consists of dark-colored shale, hard and siliceous in places but soft and flaky elsewhere. The hard, blue Sycamore Limestone can be present at the base of Caney.

The Springer Formation mainly consists of barren black shales with hard ferruginous and calcareous concretions and contains several sandstone members important for oil production. The Wapanucka Limestone consists of limestone, shale, and sandstone. The limestone beds are predominantly dark gray, commonly siliceous, cherty, locally fossiliferous, and oolitic. The shale is gray, clayey, calcareous, and weather-stained. The sandstones and siltstones contain abundant argillaceous material, calcite, and glauconite.

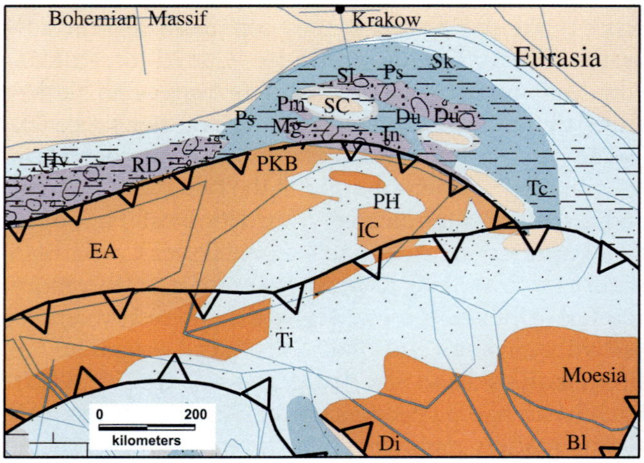

Figure 17. Paleogeography and lithofacies of the circum-Carpathian area during the Oligocene (modified from Golonka et al., 2000). For legend, see Figure 6. Bl = Balkans; Di = Dinarides; Du = Dukla; EA = Eastern Alps; Hv = Helvetic; IC = Inner Carpathians; In = Inacovce–Kricevo; Mg = Magura basin; PH = Podhale flysch (Inner Carpathian flysch); PKB = Pieniny Klippen Belt basin; Pm = Fore-Magura; Ps = Subsilesian; RD = Rhenodanubian basin; Rh = Rhodopes; SC = Silesian ridge (cordillera); Sl = Silesian basin; Sk = Skole; Sn = Sinaia basin; Tc = Tarcau; Ti = Tisza.

Stage III

This stage is characterized by the convergence of two large continents, terrane-continent collision, with the continuing formation of accretionary prisms (Table 2).

Western Carpathians (Eocene–Early Miocene)

In the circum-Carpathian region, the Adria–ALCAPA (Inner Carpathians) terranes continued their northward or northeast movement during the Eocene–early Miocene (Golonka et al., 2000) (Figures 7, 17). Their oblique collision with the North European plate led to the development of the accretionary prism of the Outer Carpathians. During the compressional stage, interbasinal ridges (bulges) were reactivated. Flysch still continued to be deposited in the subbasins. Numerous olistostromes were formed during this time (Slaczka and Oszczypko, 1987).

The process of migration of clastic facies from the inner to the outer part began, connected with the development and migration of accretionary prisms. In the inner part of the Outer Carpathians (Magura subbasin; Figure 17), the migration of clastic facies and development of accretionary prisms started already in the early Eocene and lasted until the Oligocene (Oszczypko, 1998). In this subbasin, varied lithofacies developed. The lateral differences in lithofacies across the basin allowed dividing the Magura subbasin into several sedimentary domains. During the early Eocene, the sedimentation of thick-bedded, coarse-grained turbidites commenced (Magura Formation) in the form of several submarine fans. In more distal parts, medium- and thin-bedded sandstones and shales developed, passing farther toward the north into variegated shales. These lithofacies migrated in time across the Magura subbasin toward the north. Sedimentation in the Magura basin terminated generally by accretionary wedges represented by medium- and thin-bedded sandstones (Malcov Formation) during the Oligocene (Table 1).

Within the more outer parts of the Carpathian realm, from Dukla to Skole subbasins (Figure 17), evidence of migration of depocenter appeared at the Eocene–Oligocene boundary (Slaczka, 1969) as an effect of compressional movements. As a consequence of these movements, the bottom of the basins started to deform, and initial anticlines locally developed; slump and coarse-grained sediments were locally deposited; and volcanic activity increased. Deep-marine connection with the Tethys Sea was closed, and euxinic conditions developed (Slaczka, 1969). In more external basins (Dukla, Silesian, Subsilesian, and Skole; Figure 17), green and gray shales with thin- and medium-bedded sandstones with intercalations of red shales prevailed during the early Eocene. The sedimentation of thick-bedded turbidites continued during the early Eocene in the Silesian subbasin only. Small turbiditic fans developed only locally, except in the eastern part where a huge clastic fan developed (Tarcau sandstones). This type of sedimentation gave way to green shales and yellowish *Globigerina* marls during the late Eocene, marking a period of unification of sedimentation.

The Oligocene sequences commenced with dark-brown bituminous shales and cherts (Menilite shales) with locally developed sandstone submarine fans or a system of fans as much as several kilometers long. The main ones were the Mszanka and Cergowa sandstones in the Dukla subbasin and the Kliwa Sandstones in the Skole–Tarcau subbasin. The upper boundary of the bituminous shales is progressively younger toward the north, and they pass gradually upward into the sequence of micaceous, calcareous sandstones and gray marls (Krosno beds), and they thin upward. The lower part of the sequence is generally represented by a complex of thick-bedded sandstones that pass upward into a series of medium- to thin-bedded sandstones and gray marls that terminated the sedimentation of the whole flysch sequence in the Outer Carpathians, and they can be considered as postflysch deposits. The boundaries between these lithofacies are diachronous across the basins, older in the south and younger in the north. In addition, cessation of deposition is diachronous across the Carpathians because of migration of tectonic activity and formation of trailing imbricate folds and/or accretionary prisms generally from the south to the north. During the sedimentation of the Krosno beds (Table 1), several slump deposits with blocks of shallow-water, Paleogene limestones up to tens of meters long, and smaller blocks of metamorphic and igneous rocks were deposited from intrabasinal cordilleras (Slaczka, 1961, 1969). With the final phase of tectonic movement in front of advancing nappes and/or accretional wedges (prisms), huge (as much as kilometers in size) slumps (olistostromes) with material derived from approaching nappes developed (Slaczka and Oszczypko, 1987).

During the third stage, Africa converged with Eurasia (Table 2). The direct collision of the supercontinents never happened, but their convergence led to the collision of intervening terranes, leading to the formation of the Alpine–Carpathian orogenic system. Through the Miocene, tectonic movements caused a final folding of the basin fill and created several imbricate nappes that generally reflected the original basin margin configurations (Figures 7, 18).

During overthrusting, the outer, marginal part of the advanced nappes was uplifted, whereas in the inner part, sedimentation persisted in the remnant basin.

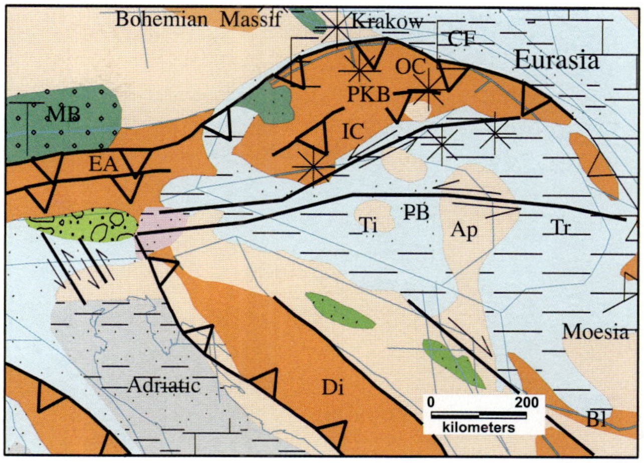

Figure 18. Paleogeography and lithofacies of the circum-Carpathian area during the Miocene (modified from Golonka et al., 2000). For legend, see Figure 6. Ap = Apuseni; Bl = Balkans; CF = Carpathian Foredeep; Di = Dinarides; EA = Eastern Alps; IC = Inner Carpathians; MB = Molasse Basin; OC = Outer Carpathians; PB = Pannonian Basin; PKB = Pieniny Klippen Belt basin; Tr = Transylvanian Basin; Ti = Tisza.

From that uplifted part of the nappes, big olistolites glided down into the adjacent, more distal basins. That uplifted, marginal part of nappes probably had a character of accretionary prisms. The nappes became detached from the basement and were thrust northward in the west and eastward onto the North European platform with its Miocene cover. Overthrusting movements migrated along the Carpathians from the west toward the east. The Outer Carpathian allochthonous rocks, as a result of Miocene tectonic movements, have been overthrust onto the platform for a distance of 50 to more than 100 km (31 to more than 62 mi). In response to the overriding of the platform by the Carpathians, a peripheral foreland basin formed along the moving orogenic front (Oszczypko and Slaczka, 1985, 1989; Oszczypko, 1998); the basin extended along the Carpathian folded zone for at least tens of kilometers. Part of the autochthonous Miocene deposits was detached from the basement and included into the accretionary prism. This Carpathian foreland basin (Figure 18) is an equivalent of the Arkoma basin in the Ouachita region.

In front of the advancing Carpathian nappes, the inner portion of the eastern part as well as the marginal part of the flysch basin (Subcarpathian–Borislav nappe) started to downwarp, and the tectonic depression formed during the early Miocene. Thick molasse deposits filled up this depression. Generally, it started with shallow-water, basal conglomerates passing upward to brackish and marine sandstones, shales, and marls, locally

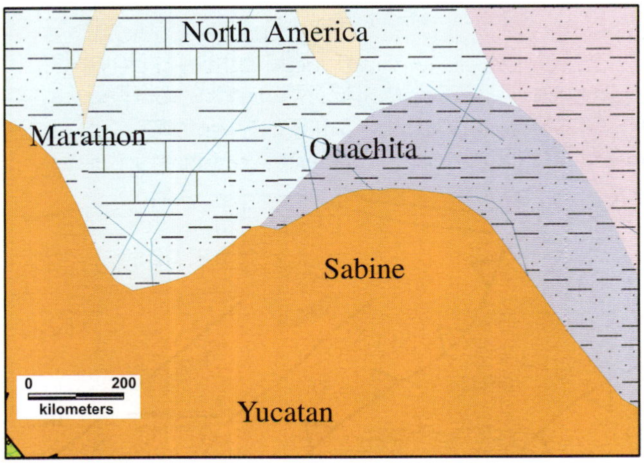

Figure 19. Paleogeography and lithofacies of the circum-Ouachita area during the Late Carboniferous (modified from Golonka and Ford, 2000). For legend, see Figure 6.

with olistostromes containing the material derived from the advancing Carpathians. In the Eastern Carpathians, conditions periodically favorable for precipitation of salt were present.

Ouachita Fold Belt (Late Carboniferous)

In the circum-Ouachita region, the Inner Ouachitas and Sabine terrane continued their northward movement during the Carboniferous (Thomas, 1976; Wickham et al., 1976; Ross, 1979; Thomas and Viele, 1983; Viele and Thomas, 1989). According to Viele and Thomas (1989), the Sabine composite terrane is a collage of tectonic elements, including not only the Sabine uplift but also the Yucatan and Coahuila platforms. In this case, the Yucatan platform would represent the stable crustal element of Pan-African origin. The Sabine uplift contains volcanics related perhaps to the south-dipping subduction. The Sabine terrane collision with the North American plate led to the development of the accretionary wedge of the Ouachita Mountains. During the compressional stage, flysch still continued to be deposited. Olistostromes of the Maumelle chaotic zone formed during this time. The main collisional activity occurred during the Late Carboniferous–Early Permian in Ouachita Mountains (Figures 11, 19, 20). The flysch deposits passed upward into the molasse.

The Late Carboniferous Johns Valley Shale is characterized by a thin distinctive zone of olistostromes with enormous blocks of limestone, chert, and black shale in the Central Ouachitas at Oklahoma. Eastward into Arkansas, the deformational zones were replaced by gray-black shales and rhythmically interbedded, very fine-grained, matrix-rich, thin-bedded sandstones that constitute a distal turbidite facies. Some slump zones also occur in the southern Ouachitas.

The lower part of the Atoka Formation is probably the most classic flysch sequence in the Ouachitas. It consists of a thick sequence of alternating interbeds of dark-gray shale and gray-brown, very fine-grained, laminated, graded subfeldspathic lithic wacke turbidites with common hieroglyphs. Upward, shallow-water sedimentary structures and coal seams indicate basinal filling from the north.

The Atoka group of the foreland is more molasse in character than flysch. The environment of deposition seems to have been a relatively shallow fluvial-deltaic sedimentation ranging to nearshore (swamp, marsh with minor marine influence). Some limestone beds occur especially in the lower member, but the bulk of formation consists of predominantly poorly sorted sandstones and dark-gray shales. The facies boundary between the Atokan rocks of the Ouachita Mountains and those of the foreland of the Arkoma basin and Arbuckle Mountains is sometimes hard to define. Probably because of basinal filling from the north, uniform facies formed during the upper Atokan time.

This major tectonic activity was caused by the collision of the Inner Ouachitas and the Sabine terrane with North America, with the accompanying convergence of Laurasia on the north and Gondwana (composed of Africa and South America, including the Yucatan promontory) on the south. To the north of the Ouachitas and partly beneath them was the North American platform with its autochthonous cover. As a result of Carboniferous–Permian tectonic movements, Ouachita allochthonous rocks have been overthrust onto the platform for a distance of 50 to more than 100 km (31 to more than 62 mi). As a result of the

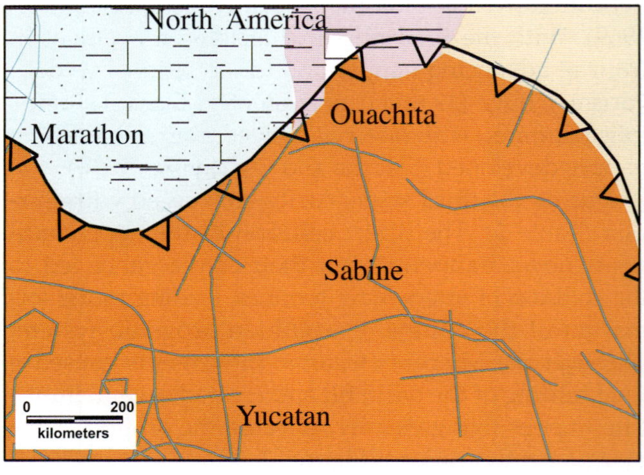

Figure 20. Paleogeography and lithofacies of the circum-Ouachita area during the Early Permian (modified from Golonka and Ford, 2000). For legend, see Figure 6.

Ouachitas overriding the platform, the peripheral foreland Arkoma basin formed along the moving orogenic front. The Ouachita fold belt was a part of the central Pangean mountain range, which extended from west Texas to Poland (Keller and Hatcher, 1999; Golonka, 2000, 2002).

Stage IV

Western Carpathians (Middle–Late Miocene)

During and after the main orogenic phase and the suture of the continents, an initial rifting system was initiated in the early Miocene behind the Carpathian arc in the Pannonian Basin. Extension in the Alpine–Carpathian system continued during the Miocene to Pliocene, forming horst and grabens in the orogen as well as in its foreland.

At the end of the early Miocene, the initial foreland basin became overthrust by the Carpathians, and a new, more external foreland basin developed. This basin and its depocenter migrated outward and eastward, contemporary with the advancing Carpathian nappes. As a result, the Neogene deposits show clear diachrony in the foreland area. In the west, sedimentation terminated already during the middle Miocene and lasted until the Pliocene in the east. Marine, fine-grained, and clastic sedimentation of the Carpathian and foreland provenance prevailed, with a break during the middle Miocene, when a younger salinity crisis appeared. Locally, olistostromes were deposited (Slaczka and Oszczypko, 1987) with material derived as well from the Carpathians as from the inner margin of the molasse deposits. During the middle Miocene, part of the Northern Carpathians also collapsed, and the sea invaded the already eroded Carpathians.

The thrusting of the Carpathian orogen on the platform was diachronous and migrated from the west toward the east, where it terminated during the Pliocene. Part of the autochthonous Miocene deposits were detached from the basement and formed a narrow folded zone in front of the Carpathians. Finally, the Outer Carpathian allochthonous rocks overthrust the southern margin of the Eurasian plate for a distance of 50 to more than 100 km (31 to more than 62 mi). At the end of the Miocene, the vertical, neotectonic movements began and is still occurring today. Further development of these processes could be predicted in the future.

Ouachita Fold Belt (Permian–Triassic)

During and after the main orogenic phase and the suture of the continents, an initial rifting system was initiated in the Permian(?)–Triassic in the Ouachitas (Woods and Addington, 1973). The basin (equivalent of the Pannonian Basin) developed south of the Ouachitas arc (Arbenz, 1989). Subsidence is revealed by Upper Carboniferous and Permian fluvial-deltaic and continental strata beneath southern Arkansas (Lillie et al., 1983). The rift system was strongly developed during the Triassic, and the rifts contain Triassic clastic red beds of Eagle Mills Formation. According to Viele and Thomas (1989), the rifting in the Sabine plate marked the onset of the next Wilson cycle, the Mesozoic opening of the Gulf of Mexico. Transgressive Jurassic, Cretaceous, and Cenozoic deposits of the Gulf Coast Basin partly covered the Ouachita region. The Sabine terrane, as well as the possible latest Paleozoic–Triassic basinal deposits, could be hidden below the thick (locally more than 10 km [6 mi]) Jurassic evaporites and Upper Jurassic, Cretaceous, and Cenozoic marine deposits (Woods and Addington, 1973; Lillie et al., 1983).

CONCLUSIONS

The Ouachita orogenic belt displays features of the complete Paleozoic Wilson cycle, from rifting (Late Cambrian) through the formation of the oceanic type of basin with passive continental margin (Ordovician), the formation of an active margin with subduction zone (Devonian) to the collision of terranes with a continent (Carboniferous), and continent-continent collision with the formation of the fold belt and foreland basin (Late Carboniferous–Permian).

The Western Carpathian orogenic belt shows the features of a Mesozoic–Cenozoic Wilson cycle from rifting (Triassic) through the formation of the oceanic type of basin with continental passive margin (Jurassic), the formation of an active margin with subduction zone (Cretaceous), and the collision of terranes with a continent (Cenozoic). The full continent-continent collision (Eurasia–Africa) remains to be completed in the future. While the Pieniny Klippen Belt and Magura basins were closing, the Silesian and subsequent subbasins were opening. These two processes were closely related. This kind of development is still unknown from the Ouachitas. Some other similarities, however, are possible. The Ouachita basement configuration warrants further investigation of the subject.

Both orogens are characterized by the thin-skinned tectonics. Sedimentation is similar with characteristic deposits like cherty limestones-novaculites, flysch, olistostromes, and molasse in the later stages. Internides-externides relation exists in both orogens: in the Ouachita fold belt, the externides are covered with Mesozoic sediments and are known only from wells. Both orogens have similar foreland features, the Carpathian Foredeep and Arkoma basin in front of Ouachitas.

ACKNOWLEDGMENTS

This work was partly supported by the Grant from the Polish Committee for Scientific Research (Komitet Badan Naukowych grant no. 14 T 12 B 025 28), and AGH University of Science and Technology grant 11.11.140.159. It is also a contribution to the IGCP project 453. We are grateful to Patricia Wood Dickerson and to the anonymous reviewer for their reviews.

REFERENCES CITED

Arbenz, J. K., 1968, Structural geology of the Potato Hills, Ouachita Mountains, Oklahoma, *in* L. M. Cline, ed., A guidebook to the geology of the western Arkansas basin and Ouachita Mountains, southern Oklahoma: Oklahoma City Geological Society Guidebook, Oklahoma City, Oklahoma, p. 109–121.

Arbenz, J. K., 1989, The Ouachita system, *in* A. W. Bally and A. R. Palmer, eds., The Geology of North America—An overview: The Geology of North America, v. A: Boulder, Colorado, Geological Society of America, p. 371–396.

Bieda, F., S. Geroch, L. Koszarski, M. Ksiazkiewica, and K. Zytko, 1963, Stratigraphie des Karpates externes polonaises: Biuletyn Instytutu Geologicznego, v. 182, p. 5–174.

Birkenmajer, K., 1986, Stages of structural evolution of the Pieniny Klippen Belt, Carpathians: Studia Geologica Polonica, v. 88, p. 7–32.

Birkenmajer, K., 1988, Exotic Andrusov ridge: Its role in plate-tectonic evolution of the West Carpathian fold belt: Studia Geologica Polonica, v. 91, p. 7–37.

Birkenmajer, K., H. Kozur, and R. Mock, 1990, Exotic Triassic pelagic limestone pebbles from the Pieniny Klippen Belt of Poland: A further evidence for early Mesozoic rifting in West Carpathians: Annales Societatis Geologorum Poloniae, v. 60, p. 3–44.

Briggs, G., and D. H. Roeder, 1975, Sedimentation and plate tectonics, Ouachita Mountains and Arkoma basin, *in* G. Briggs, E. F. McBride, and M. J. Moiola, eds., Sedimentology of Paleozoic flysch and associated deposits, Ouachita Mountains–Arkoma basin, Oklahoma: Dallas, Texas, Dallas Geological Society, p. 1–22.

Cline, L. M., 1960, Stratigraphy of the late Paleozoic rocks of the Ouachita Mountains, Oklahoma: Oklahoma Geological Survey Bulletin, v. 85, p. 1–158.

Cline, L. M., 1970, Sedimentary features of late Paleozoic flysch, Ouachita Mountains, Oklahoma, *in* J. Lajole, ed., Flysch sedimentology in North America: Geological Association of Canada Special Paper 7, p. 85–101.

Cline, L. M., W. J. Holsweck, and D. E. Ferray, eds., 1959, The geology of the Ouachita Mountains, a symposium: Dallas, Texas, and Ardmore, Oklahoma, Dallas and Ardmore Geological Society, 208 p.

Dickinson, W. R., and T. F. Lawton, 2001, Carboniferous to Cretaceous assembly and fragmentation of Mexico: Geological Society of America Bulletin, v. 113, p. 1142–1160.

Dzulynski, S., and A. Slaczka, 1958, Directional structures and sedimentation of the Krosno beds (Carpathian flysch): Annales Societatis Geologorum Poloniae, v. 28, p. 205–259.

Dzulynski, S., and E. K. Walton, 1965, Sedimentary features of flysch and graywackes: Development in Sedimentology, v. 7, 247 p.

Dzulynski, S., M. Ksiazkiewicz, and P. H. Kueanen, 1959, Turbidites in flysch of the Polish Carpathians: Geological Society of America Bulletin, v. 70, p. 1089–1118.

Flawn, P. T., A. Goldstein, Jr., P. B. King, and C. E. Weaver, 1961, The Ouachita system: Texas University, Bureau of Economic Geology Publication 6120, 601 p.

Gleason, J. D, S. C. Finney, and G. E. Gehrels, 2002, Paleotectonic implications of Mid- to Late Ordovician provenance shift, as recorded in sedimentary strata of the Ouachita and Southern Appalachian Mountains: Journal of Geology, v. 110, p. 291–304.

Golonka, J., 1988, Basement structures beneath thrust Ouachita rocks: AAPG Bulletin, v. 72, p. 100.

Golonka, J., 2000, Cambrian–Neogene plate tectonic maps: Krakow, Wydawnictwa Uniwersytetu Jagiellonskiego, 125 p.

Golonka, J., 2002, Plate-tectonic maps of the Phanerozoic, *in* W. Kiessling, E. Flugel, and J. Golonka, eds., Phanerozoic reef patterns: SEPM Special Publication 72, p. 21–75.

Golonka, J., and D. Ford, 2000, Pangean (Late Carboniferous–Middle Jurassic) paleoenvironment and lithofacies: Palaeogeography, Palaeoclimatology, Palaeoecology, v. 161, p. 1–34.

Golonka, J., and W. Kiessling, 2002, Plate-tectonic maps of the Phanerozoic, *in* W. Kiessling, E. Flugel, and J. Golonka, eds., Phanerozoic reef patterns: SEPM Special Publication 72, p. 1–20.

Golonka, J., and M. Krobicki, 2001, Upwelling regime in the Carpathian Tethys: A Jurassic–Cretaceous palaeogeographic and paleoclimatic perspective: Geological Quarterly, v. 45, p. 15–32.

Golonka, J., and W. Sikora, 1981, Microfacies of the Jurassic and Lower Cretaceous sedimentarily thinned deposits of the Pieniny Klippen Belt in Poland (in Polish with English abstract): Biuletyn Instytutu Geologicznego, v. 31, p. 7–37.

Golonka, J., and A. Slaczka, 2000, Geodynamic evolution stages of two orogens: The Carpathians and Ouachita Mountains: Slovak Geological Magazine, v. 6, p. 139–142.

Golonka J., and A. Slaczka, 2003, Geodynamic evolution stages and climate changes in the Outer Carpathians: Mineralia Slovaca, v. 35, p. 11–14.

Golonka, J., N. Oszczypko, and A. Slaczka, 2000, Late Carboniferous–Neogene geodynamic evolution and paleogeography of the circum-Carpathian region and adjacent areas: Annales Societatis Geologorum Poloniae, v. 70, p. 107–136.

Golonka, J., M. Krobicki, N. Oszczypko, A. Slaczka, and T. Slomka, 2003, Geodynamic evolution and palaeogeography of the Polish Carpathians and adjacent areas during Neo-Cimmerian and preceding events (latest Triassic–earliest Cretaceous), *in* T. McCann and A. Saintot, eds., Tracing tectonic deformation using the sedimentary record: Geological Society (London) Special Publication 208, p. 138–158.

Ham, W. E., 1950, Geology of the Arbuckle mountain region: Oklahoma Geological Survey Guidebook 3, Oklahoma City, Oklahoma, 61 p.

Ham, W. E., 1959, Correlation of pre-Stanley strata in the Arbuckle–Ouachita Mountains region, *in* L. M. Cline, W. J. Holsweck, and D. E. Ferray, eds., The geology of the Ouachita Mountains, a symposium: Dallas, Texas, and Ardmore, Oklahoma, Dallas and Ardmore Geological Society, p. 71–86.

Ham, W. E., 1969, Regional geology of the Arbuckle Mountains, Oklahoma: Oklahoma Geological Survey Guidebook, v. 17, 52 p.

Keller, G. A., and S. E. Cebull, 1973, Plate tectonics and the Ouachita systems in Texas, Oklahoma, and Arkansas: Geological Society of America Bulletin, v. 84, p. 1659–1666.

Keller, G. R., and R. D. Hatcher, 1999, Some comparisons of the structure and evolution of the southern Appalachian–Ouachita orogen and portions of the trans-European suture zone region: Tectonophysics, v. 314, p. 43–68.

Kiessling, W., E. Flugel, and J. Golonka, 1999, Paleoreef maps: A comprehensive database of Phanerozoic reefs with graphic presentations: AAPG Bulletin, v. 83, p. 1552–1587.

Krautner, H. G., 1996, Alpine rifting, subduction and collision in the Romanian Carpathians, *in* G. Amman, K. W. Handler, and H. P. Steyrer, eds., Abstracts of the VI Symposium für Tektonik Struktur- und Kristallingeologie, Salzburg, p. 230–234.

Krautner, H. G., and B. Krsti, 2001, Correlation of Alpine and pre-Alpine structural units within the Southern Carpathians and the Eastern Balkanides, *in* A. Adam, L. Szarka, and J. Szendroi, eds., Pancardi 2001 II. Abstracts: Hungarian Academy of Sciences, Sopron, Hungary, p. CP1.

Kruglov, S. S., 1989, Geodynamics of the Ukrainian Carpathians: Geologica Carpathica, v. 40, p. 101–123.

Kruglov, S. S., 2001, The problems of tectonics and paleogeodynamics of Western Ukraine (a critical survey of the new publications, in Ukrainian): L'viv, Ukrainian Interdisciplinary Committee for Tectonics, 83 p.

Ksiazkiewicz, M., 1954, Graded and laminated bedding in the Carpathian flysch: Annales Societatis Geologorum Poloniae, v. 22, p. 399–449.

Ksiazkiewicz, M., 1968, Evolution structurale des Carpathes polonaises: Memoires de la Societe Geologique de France, v. 42, p. 529–562.

Ksiazkiewicz, M., 1977, Tectonics of the Carpathians, *in* W. Pozaryski, ed., Geology of Poland, Tectonics: Warsaw, Poland, Wydawnictwa Geologiczne, v. 4, p. 476–604.

Lashkevitsch, Z. M., A. P. Medvedev, and Y. Z. Krupskiy, 1995, Tectonomagmatic evolution of Carpathians (in Russian): Kiev, Naukova Dumka, p. 1–113.

Lillie, R. J., K. D. Nelsen, B. de Voogd, J. A. Brewer, J. E. Oliver, L. D. Brown, S. Kaufman, and L. D. Viele, 1983, Crustal structure of Ouachita mountains, Arkansas: A model based on integration of COCORP reflection profiles and regional geophysical data: AAPG Bulletin, v. 67, p. 907–931.

Lowe, D. R., 1976, Grain flow and grain flow deposits: Journal of Sedimentary Petrology, v. 52, p. 279–297.

Lowe, D. R., 1985, Ouachita trough; part of a Cambrian failed rift system: Geology, v. 13, p. 790–793.

Lowe, D. R., 1989, Stratigraphy, sedimentology, and depositional setting of pr-orogenic rocks of the Ouachita Mountains, Arkansas and Oklahoma, *in* R. D. Hatcher, Jr., W. A. Thomas, and G. W. Viele, eds., The Appalachian–Ouachita orogen in the United States: The Geology of North America, v. F-2: Boulder, Colorado, Geological Society of America, p. 575–590.

Lucinska-Anczkiewicz, A., A. Slaczka, I. M. Villa, and R. Anczkiewicz, 2000, $^{39}Ar/^{40}Ar$ dating of the teschenite association rocks from the Polish Outer Carpathians: Mineralogical Society of Poland Special Paper 17, p. 241–242.

Mahel, M., 1974, The Inner Carpathians, *in* M. Mahel, ed., Tectonics of the Carpathian–Balkan regions: Bratislava, Geologicky Ustav Dionyza Stura, p. 91–133.

McBride, E. F., 1975, The Ouachita trough sequence: Marathon region and Ouachita mountains, *in* G. Briggs, E. F. McBride, and R. J. Moiola, eds., Sedimentology of Paleozoic flysch and associated deposits, Ouachita Mountains–Arkoma basin, Oklahoma: Dallas Geological Society Guidebook, Dallas, Texas, p. 23–41.

Moiola, R. J., and G. Shanmugan, 1984, Submarine fan sedimentation, Ouachita Mountains, Arkansas and Oklahoma: Gulf Coast Association of Geological Society Transactions, v. 34, p. 175–182.

Morris, R. C., 1974, Sedimentary and tectonic history of the Ouachita Mountains, *in* W. R. Dickinson, ed., Tectonic and Sedimentation: SEPM Special Publication 22, p. 120–142.

Morris, R. C., 1989, Stratigraphy and sedimentary history of post-Arkansas Novaculite Carboniferous rocks of the Ouachita Mountains, *in* R. D. Hatcher, Jr., W. A. Thomas, and G. W. Viele, eds., The Appalachian–Ouachita orogen in the United States: The Geology of North America, v. F-2: Boulder, Colorado, Geological Society of America, p. 591–602.

Oszczypko, N., 1992, Late Cretaceous through Paleogene evolution of Magura basin: Geologica Carpathica (Bratislava), v. 43, p. 333–338.

Oszczypko, N., 1998, The Western Carpathian Foredeep—Development of the foreland basin in front of the accretionary wedge and its burial history (Poland): Geologica Carpathica, v. 49, p. 415–431.

Oszczypko, N., and A. Slaczka, 1985, An attempt at palinspastic reconstruction of Neogene basin in the Carpathian Foredeep: Annales Societatis Geologorum Poloniae, v. 55, p. 55–75.

Oszczypko, N, and A. Slaczka, 1989, The evolution of the Miocene basins in the Polish Flysch Carpathians and their

foreland: Geologicky Zbornik, Geologica Carpathica, v. 40, p. 23–36.
Pescatore, T., and A. Slaczka, 1984, Evolution models of two flysch basins: The Northern Carpathians and the Southern Apennines: Tectonophysics, v. 106, p. 49–70.
Picha, F. J., 1996, Exploring for hydrocarbons under thrust belts—A challenging new frontier in the Carpathians and elsewhere: AAPG Bulletin, v. 80, p. 1547–1564.
Plasienka, D., 1999, Tektonochronologia a paleotektonicky model jursko-kriedoveho vyvoja centralnych Zapadnych Karpat: Tectonochronology and paleotectonic model of the Jurassic–Cretaceous evolution of the Central Western Carpathians: Bratislava, Veda, 127 p.
Ross, C. A., 1979, Late Paleozoic collision of North and South America: Geology, v. 7, p. 41–44.
Sandulescu, M., 1984, Geotectonica Romaniei (Geotectonics of Romania) (in Romanian): Bucuresti, Editions Technica, 334 p.
Sandulescu, M., 1988, Cenozoic tectonic history of the Carpathians, *in* L. Royden and F. Horvath, eds., The Pannonian Basin: A study in basin evolution, AAPG Memoir 45, p. 17–25.
Sandulescu, M., and M. Visarion, 2000, Crustal structure and evolution of the Carpathian–Western Black Sea area: Romanian Geophysics, v. 7, no. 1, p. 228–231.
Shanmugan, G., and R. J. Moiola, 1995, Reinterpretation of depositional processes in a classic flysch sequences (Pennsylvanian Jackfork Group), Ouachita Mountains, Arkansas and Oklahoma: AAPG Bulletin, v. 79, p. 672–695.
Sikora, W., 1971, Outline of the tectonogenesis of the Pieniny Klippen zone in Poland in the light of the new geological data (in Russian): Annales Societatis Geologorum Poloniae, v. 61, p. 221–238.
Slaczka, A., 1961, Exotic bearing shale from Bukowiec (Polish External Carpathians) (in Polish, English summary): Rocznik Polskiego Towarzystwa Geologicznego, v. 31, p. 129–143.
Slaczka, A., 1969, Final stages of geosynclinal development in the southeast-part of the Polish Carpathians: Acta geologica Academiae Scientiarum Hungaricae, v. 13, p. 331–335.
Slaczka, A., 1971, Geology of the Dukla unit (in Polish with English summary): Prace Instytutu Geologicznego, v. 1, p. 1–63.
Slaczka, A., 1998, Age of the andesitic rocks in the Subsilesian unit (Outer Carpathians) (abs.): Abstracts, XVI Congress, Carpatho-Balkan Geological Association, Vienna, August 30–September 2, 1998, p. 559.
Slaczka, A., and M. A. Kaminski, 1998, Guidebook to excursions in the Polish Flysch Carpathians: Grzybowski Foundation Special Publication 6, 171 p.
Slaczka, A., and N. Oszczypko, 1987, Olistostromes and overthrusting in the Polish Carpathians: Annales Instituti Geologici Publici Hungarici, v. 70, p. 287–292.
Smulikowski, K., 1980, Remarks on Techen magmatic province (in Polish with English summary): Annales Societatis Geologorum Poloniae, v. 51, p. 41–44.
Sotak, J., 1986, Stratigraphy and typology of the Upper Triassic development in outer units in the West Carpathians (reconstructions from redeposited localised in the Silesian Cordillera area): Knihovnicka Zemny Plyn a Nafta, v. 31, p. 1–53.
Stampfli, G. M., ed., 2001, Geology of the western Swiss Alps—A guidebook: Memoires de Geologie, Lausanne, v. 361, p. 1–195.
Thomas, W. A., 1976, Evolution of Ouachita–Appalachian continental margin: Journal of Geology, v. 86, p. 323–342.
Thomas, W. A., 1977, Structural and stratigraphic continuity of the Appalachian and Ouachita Mountains: Arkansas Geological Commission Symposium on the Geology of the Ouachita Mountains, Little Rock, Arkansas, p. 9–24.
Thomas, W. A., 1989, The Appalachian–Ouachita orogen beneath the Gulf coastal plain between the outcrops in the Appalachian and Ouachita Mountains, *in* R. D. Hatcher, Jr., W. A. Thomas, and G. W. Viele, eds., The Appalachian–Ouachita orogen in the United States: The Geology of North America, Boulder, Colorado, Geological Society of America, v. F-2, p. 537–553.
Thomas, W. A., 1991, The Appalachian–Ouachita rifted margin of southeastern North America: Geological Society of America Bulletin, v. 103, p. 415–431.
Thomas, W. A., and R. A. Astini, 1999, Simple-shear conjugate rift margins of the Argentine Precordillera and the Ouachita embayment of Laurentia: Geological Society of America Bulletin, v. 111, p. 1069–1079.
Thomas, W. A., and G. W. Viele, 1983, Tectonic history of the Ouachita Orogen: Geology, v. 11, p. 482–483.
Thomas, W. A., R. A. Astini, and G. Bayona, 2002, Ordovician collision of the Argentine Precordillera with Gondwana, independent of Laurentian taconic orogeny: Tectonophysics, v. 354, p. 131–152.
Viele, G. W., and W. A. Thomas, 1989, Tectonic synthesis of the Ouachita orogenic belt, *in* R. D. Hatcher, Jr., W. A. Thomas, and G. W. Viele, eds., The Appalachian–Ouachita orogen in the United States: The Geology of North America, v. F-2: Boulder, Colorado, Geological Society of America, p. 695–728.
Walker, D. A., J. Golonka, A. Reid, and S. Reid, 1995, The effects of paleolatitude and paleogeography on carbonate sedimentation in the late Paleozoic, *in* A. Y. Huc, ed., Paleogeography, paleoclimate, and source rocks: AAPG Studies in Geology 40, p. 133–155.
Wessely, G., 1996, Geological overview and distribution of oil and gas of Alpidic regions of Central and Eastern Europe, *in* G. Wessely and W. Liebl, eds., Oil and gas in Alpidic thrustbelts and basins of Central and Eastern Europe, European Association of Geoscientists and Engineers Special Publication 5: Geological Society (London) insert map, scale 1:10,000,000, 1 sheet.
Wickham, J., D. R. Roeder, and G. Briggs, 1976, Plate tectonics models for the Ouachita fold belt: Geology, v. 4, p. 173–176.
Woods, R. D., and S. W. Addington, 1973, Pre-Jurassic geologic framework, northern Gulf basin: Gulf Coast Association Geological Societies Transactions, v. 23, p. 92–108.

Sozański, Józef, Stanisław Kuk, Czesław Jaracz, and Piotr S. Dziadzio, 2006, How the modern oil and gas industry was born: Historical remarks, *in* J. Golonka and F. J. Picha, eds., The Carpathians and their foreland: Geology and hydrocarbon resources: AAPG Memoir 84, p. 585.

30

How the Modern Oil and Gas Industry was Born: Historical Remarks[1]

Józef Sozański*

Podchorążych Str., Krosno, Republic of Poland

Stanisław Kuk

Polish Oil and Gas Company, Exploration Department, South Regional Division, Jaslo, Republic of Poland

Czesław Jaracz

The Oil and Gas Institute Krosno Branch, Krosno, Republic of Poland

Piotr S. Dziadzio

Polish Oil and Gas Company, Exploration Department, South Regional Division, Jaslo, Republic of Poland

ABSTRACT

This chapter presents historical remarks and a quick review of knowledge of the occurrence and use of bituminous raw materials in the Carpathians. The most important data on the use of hydrocarbons in the Carpathians in the literature from the 16th century, and on the development of its use, exploration, and exploitation date from the beginning of 19th century. In the laboratories of Lvov and Gorlice, Ignacy Lukasiewicz, the founding father of the Polish oil industry, distilled his first amounts of kerosene and other oil-derivative products. Public lighting of kerosene lamps in a Lvov hospital occurred on July 31, 1853. This date was generally recognized as the beginning of the national oil industry. Also in 1854, the first street kerosene lamp was lit up in Gorlice.

These two events showed how important the research of this raw material is. Several natural oil seeps concentrated on the surface signalled the beginning of discoveries of many shallow oil deposits, starting from Stary Sacz, Gorlice, Krosno, Sanok, Sambor, Drohobych, Nadwornaya, Bogorodchany, Kossovo, and going all the way to the Romanian Carpathian Mountains.

In this chapter, dynamic development of the petroleum industry from the second half of the 19th century up to the beginning of World War II is also presented, concentrating on gradual activities of many oil companies in the Polish part of the Carpathians. At that time, new drilling and exploratory technologies and oil-distillation processes were developed, causing the growth of the oil industry in the Carpathians and in many countries worldwide.

[1]Portions of this text are fragments of the publication *Light from the Earth* (Dziadzio and Letowski, 2003a). Permission to reprint from Wydawnictwo Tekst Sp. z o.o. Bydgoszcz, Poland.
*Deceased.

DOI:10.1306/985630M843071

Index

C

D

E

F

G

H

I

J

K

L

M

N

O

P

Q

R

S

T

U

V